T0406124

Handbook of the Anthropocene

Nathanaël Wallenhorst • Christoph Wulf
Editors

Handbook of the Anthropocene

Humans between Heritage and Future

Volume I

 Springer

Editors
Nathanaël Wallenhorst
Université Catholique de l'Ouest (UCO)
Angers, France

Christoph Wulf
FB Erziehungswissenschaft
Freie Universität Berlin
Berlin, Germany

ISBN 978-3-031-25909-8   ISBN 978-3-031-25910-4   (eBook)
https://doi.org/10.1007/978-3-031-25910-4

This Springer imprint is published by the registered company Springer Nature Switzerland AG
The registered company address is: Gewerbestrasse 11, 6330 Cham, Switzerland

Acknowledgements

We would like to thank all the authors for their contributions. Special thanks go to our editor Christi Lue, who believed in this project from the start and honoured us with her trust, and to Dr. Möller, Deputy Secretary General of the German Commission for UNESCO. We also thank the Université catholique de l'Ouest (Angers, France) for the grant to supervise the translations. Finally, we would like to thank Liz Hamilton for her extremely valuable and thorough help with the translations.

Anthropocene, the Concept of the 21st Century – A General Introduction

What Happened to the Future?

For millennia, the prospect of the future has offered hope. Men and women have gazed, in fascinated anticipation, towards the horizon of the future. 'What new inventions, creations and exploration will humanity achieve in years to come?', 'What wonders await?' and other such questions have engaged our ancestors for centuries.

Human achievements over time have been staggering. We have discovered that the earth is not the centre of the universe. The invention of the steam engine sparked the Industrial Revolution, which changed the face of the world. Quantum physics has totally reshaped the way in which we think. The advent of antibiotics has meant that diseases which would once have been a death sentence are now curable. In the space of only a few short decades since setting foot on the moon, we have built communications networks that keep us connected wherever we are on the planet. Today, thanks to internet search engines, we have practically all human knowledge just a few clicks away. More astonishing still is that we are now able to actually edit the human genome. Dolly the sheep was cloned less than 25 years ago; yet today CRISPR-Cas9 allows us to slice and splice parts of the human genome.

Time has also brought many nasty shocks. The Industrial Revolution, encouraged by Descartes' exhortation that we become 'masters and possessors of nature', was in full swing in the early 20th century, when war broke out on a scale never seen before. Economic globalisation saw that war spread across the face of the planet like a plague. Barely two decades later came a second global conflict, but where WWI had left 18 million dead, manmade technology meant 60 million lost their lives in WWII! However, despite the twists and turns and bumps along the way, the modernist era was characterised by the marriage of technical and social progress. The commitment to build a brighter future elevated vast swathes of humanity out of grinding poverty.

However, three prosperous decades later, when 'grand narratives' (Lyotard, 1979) collapsed and ushered in the era of postmodernism, the timelines began to unravel. Technical and social progress no longer go hand in hand. The idea of linear progress – today is better than yesterday, and tomorrow will be better still – no longer holds true. In addition, the new geological era, the Anthropocene – meaning human activity has now inexorably altered the planet's habitability – threatens the very survival of humanity (Bonneuil & Fressoz, 2013; Bourg & Papaux, 2015; Federau, 2017; Magny, 2019; Wallenhorst, 2019, 2020, 2021; Wallenhorst & Wulf, 2022; Kamper & Wulf, 1989, 1994; Gil & Wulf, 2015; Wulf & Zirfas, 2020; Wulf, 2022a). Our future – whether it holds shock or awe – is fading from view. Greta Thunberg's words to her young peers put the point across in resonant fashion: 'Why should I be studying for a future that soon may be no more?' This is a sobering hypothesis: we may soon have no future at all.

Voices – mainly those of Californian technology giants – occasionally proclaim the opposite: we *do* have a future, and a bright one at that. The future will be *great* or *bigger than ever*, we are told by a chorus of such figures, including: Mark Zuckerberg, Facebook co-founder and CEO; Larry Page and Sergey Brin, founders of Google; Jeff Bezos, Amazon founder and CEO; and Elon Musk, whose groundbreaking businesses include SpaceX, Tesla and Neuralink (the 2016 startup aiming to interface the human brain with digital technology). The future is undoubtedly digital. Ultimately, Musk tells us, this is nothing to fear. His ambition is to establish a human colony on Mars within the foreseeable future, using Artificial Intelligence (AI). Neoliberal capitalism has destroyed our 'shared home', so Musk wants to build us a new one, on Mars, harnessing the power that the coming digital and technoscientific revolution promises. There are no limits – planetary, corporeal or cognitive – that cannot be overcome by human engineering. We shall soon be able to merge with machines, and so massively increase our capabilities. The future will be in the hands of digital technology. Here we have a second hypothesis: our future will be a Promethean digital endeavour. The *hubris* of *Homo oeconomicus*, maximising individual interests, will prevail. The stratospheric rise of the so-called Big 5 (or GAFAM, standing for Google, Amazon, Facebook, Apple and Microsoft) and their Chinese counterparts (BATX – Baidu, Alibaba, Tencent and Xiaomi) is indicative of this trend. These companies are shaping the future for their billions of consumers across the globe.

Clearly, we need a future and must obviously work diligently to bring it about. However, surely that future is not the preserve of those Promethean few who have access to a spacecraft capable of reaching Mars. Might we not have a different future – one which is post-Promethean, post*hubris*, postcapitalist or simply convivial? We may, even now, be seeing such a world begin to emerge. Here is a third hypothesis: the future could be marked by an ethos of coexistence. Of course, in reality, the future is unlikely to be at either extreme of a scale, but rather, somewhere in between. By examining the world today, we can extract clues as to how it will look tomorrow.

The future is not a given – far from it. Perhaps the most pressing questions we face relate to the fate of humanity. *Who will we become? How can we continue to be human in the current context, marked by climate change, ecosystem collapse, but also the augmented humans, profound social transformations and rampant radicalisation?* Our living environment has been inexorably altered, transhumanist theories abound, and researchers are seeking to surpass human limitations through technology. In this world, it is increasingly complex – even, sometimes, impossible – to 'form a society together'. We can no longer take our humanity for granted, but it is crucial to retain it. How do we define ourselves, and what do we want to make of ourselves? This anthropological question is addressed in this handbook, based on three conceptual and epistemic options.

Firstly, it is by becoming more earthly – more connected to the earth – that *Homo sapiens* will become progressively more human. Indeed, the anthropological viewpoint discussed herein is that humanity should immerse itself fully in the natural world. This stands in opposition to the theories, which have been dominant in Western thinking since the Age of Enlightenment (Papaux, 2015), whereby humanity is hierarchically superior to the natural world, and should therefore maintain distance from it.

Secondly, breaking with the essentialism and substantialism currently popular in Western thinking, *Homo sapiens* becomes human as a result of, through and in relationships (with others, with future generations and with non-humans). Humanity is defined not just intrinsically and in static terms, but also on the basis of what it could become and what it makes of itself within its environment. In this context, the modern boundaries between nature and culture break down, as demonstrated by Philippe Descola (2005).

Thirdly, the human race will have a great deal to gain – not least, its own assured survival – from post-Promethean practices.

One thing appears clear: the hypotheses concerning the future and how to prepare for it are shaped by this new context which we must take ownership of and to which we keep returning – the Anthropocene. It concerns what is happening to the climate, changes in the biosphere and the dynamics of our societies. It is the concept of the 21st century, the concept of which it is necessary to have an in-depth understanding in order to understand the transformations that are taking place in the Earth system, to understand the world and allow ourselves to continue the process of becoming human.

Anthropocene: Improvisation of the Term at a Conference in Mexico in 2000

In February 2000 in Mexico, during exchanges at the conference of the IGBP – *International Geosphere-Biosphere Programme* – an interdisciplinary group studying the Earth system, there was discussion of the Holocene, the preceding epoch (lasting from over 11,700 years ago more or less to the current day). Someone stood up and said, 'Stop using the word Holocene. We're not in the Holocene anymore. We're in the…the…the… [searching for the right word] …the Anthropocene!' Later he would say, 'I suddenly thought this was wrong. (…) I just made up the word on the spur of the moment. Everyone was shocked. But it seems to have stuck' (Keats, 2011, p. 19). We can say that the proposal of the term 'Anthropocene' by Crutzen in 2000 was pure improvisation (Zalasiewicz et al., 2017a, p. 56).

Having used the term at the IGBP conference, Paul Crutzen contacted the American biologist, Eugene Stroemer, to suggest collaborating with him on a paper for the IGBP review, because Stroemer had been using this term informally since the 1980s. Here we read how Jacques Grinevald (2007, p.243), the French philosopher who was working in Geneva, reports Stroemer's proposal: 'I started to use the term "Anthropocene" in the 1980s but I never formalised it before Paul [Crutzen] contacted me.' It is clear in this one-page article that one of the characteristics of the invention of the term 'Anthropocene' was that it was put forward before it was accorded a scientific profile and before a precise, exhaustive definition was suggested. Later, in 2002, Crutzen picked up on these elements in a short article in the review *Nature*, where he summarises all the environmental changes brought about by humans since the industrial revolution as evidence that we are entering the 'Anthropocene': a tenfold increase in the world population between 1700 and 2000 along with a similar increase in cattle; depletion of fossil resources and the release of CO_2 into the atmosphere; a large increase in species extinction, by a factor of 1000.

A second feature of the concept of the Anthropocene, besides the fact that it was an improvisation, is that it refers to a geological epoch and is thus part of those time periods that are customarily defined by stratigraphers on the basis of their observations of soils. But one of the first uses of this term was proposed by Paul Crutzen, a Dutch geochemist working in Germany, independently of stratigraphic observations. Originally, the term Anthropocene was not suggested by geologists based on stratigraphic evidence but with reference to changes in processes of the Earth system brought about by human activity. Therefore, in its original form, the concept of the Anthropocene is a systemic concept which has its roots in Earth system science rather than in geology.

The third interesting point is that Crutzen was awarded the Nobel Prize for Chemistry in 1995, which gave a large amount of publicity to this new term in the media and in the scientific world. The fact that a word was coined without being precisely defined (even in 2011 Steffen, Grinevald, Crutzen and McNeil recognised that the term Anthropocene still had an informal flavour) and without the usual respect for academic rules is seen by many as a reason for the success of this term, also in the way it was taken up by a very heterogeneous mixture of disciplines before it became widely used by the stratigraphers. Thus, as Ellis (2017) recognised, we may say that the concept of the Anthropocene was created by Eugene Stroemer, then publicised by Paul Crutzen. Its establishment as a scientific concept, based on stratigraphic observations, has been the fruit of work by many different Earth system scientists, the most notable of whom are Steffen and Zalasiewicz.

To begin with, it was non-geologists (chemists, physicists, biologists) who suggested creating a new geological epoch. This audacity initially took the geologists by surprise, but in 2008 the Geological Society of London instituted an official working group on the Anthropocene with the aim of getting recognition from the International Union of Geological Sciences (IUGS). This working group, composed of brilliant research scientists, published dozens of articles, in the field of Earth system science and in geology, from the establishing of irrefutable truth to stratigraphic foundations, from the end of the Holocene Epoch and from the beginning of the Anthropocene (Zalasiewicz et al., 2011, 2014, 2017a). Thus, the Anthropocene is a new geological epoch, characterised by a change in the habitability conditions of the Earth, impacting in particular human life in society as we know it.

The scientific debates are exciting and are structured within three disciplines which examine three concepts of the Anthropocene: Earth System science, geological stratigraphy and finally the whole field of humanities and social sciences.

The Concept of the Anthropocene Within Earth System Science

The first point is the extent of the systemic upset we have caused to Earth's systems. All publications in this field examine when we passed, or will pass, the boundaries of what the planet can cope with, and the room for manoeuvre that we humans have in which to safely take action. The functioning of the Earth is a complex system, characterised by sudden, irreversible changes when certain bio-geological processes are forced upon it. There are certain boundaries that cannot be passed if we do not wish the Earth system to transgress certain systemic thresholds. There are nine boundaries to have been identified (Rockström et al., 2009; Steffen et al., 2015; Persson, 2022; Erlandsson, 2022): climate change, rates of loss of terrestrial and marine biodiversity, the alteration of biogeochemical cycles such as nitrogen and phosphorous, the impoverishment of the stratospheric ozone layer, acidification of the oceans, the global use of fresh water, land usage, chemical pollution and the discharge of aerosols into the atmosphere. The first two have a special status. Crossing just one of these boundaries is enough to tip the entire Earth system into a way of functioning that is considerably less propitious for life. *Example:* today we find that our principal allies in our economic development over the last two centuries, soil and the plants that grow in it, which act as carbon reservoirs absorbing some of the CO_2 emissions that have been released by human activities, not only capture less CO_2 but may also be releasing more than they actually capture. The Amazonian jungle, for a long time considered to be the 'lungs of the planet', even shows a positive carbon balance (releasing more carbon than it absorbs) in years that are particularly dry (Gatti et al., 2014). Other articles followed, confirming these results (Brienen, 2015; Doughty, 2015; Esquivel-Muelbert, 2019, 2020). We cannot consider ourselves to be outside nature. There are innumerable studies which show how deeply embedded we are in our environment. Barnosky et al. (2011) write about the sixth mass extinction; Rockström et al. (2009) and Steffen et al. (2015) stress the importance of staying within the planet's boundaries. Barnosky et al. (2012) and Steffen et al. (2018) warn that our biosphere is approaching a systemic tipping point, and the planet is in danger of becoming a 'hothouse Earth', incapable of supporting human society. Im et al. (2015, 2017), Mora et al. (2017) and Bador et al. (2017) forecast heatwaves of increasing frequency and intensity in years to come, etc. This branch of the debate also includes publications on what corrective action we can take, by geoengineering, or 'planetary stewardship' (Crutzen, 2006; Steffen et al., 2011b). It is directly connected to normative and prospective political discussion. Often, the question 'What can we do?' is asked. The irreversible corruption of our planet's ability to support life is a matter of concern. Thus, we expect this vein of Earth system science will continue to produce studies grounded in the humanities and social sciences, on how society can be reorganised to exist in the altered biosphere (Eckersley, 2017; Arnsperger & Bourg, 2017; Curnier, 2017; Federau 2017; Lesourt, 2018; Wallenhorst, 2019, 2021, 2022; Wallenhorst & Pierron, 2019; Hétier, 2021; Hétier et Wallenhorst, 2021, 2022a, b, c; Prouteau et al., 2022; Wulf 2022a).

The Concept of the Anthropocene Within the Sciences of Geology and Stratigraphy

The second main branch of debate concerns the date of entry into the Anthropocene. Innumerable articles have been published in this field, although, as they are less 'spectacular' and more technical, they are less well known to the general public. This debate among experts in Earth system science usually draws on the multidisciplinary practice of stratigraphy. While they may indeed be less widely known, these articles are of singular importance. It is these studies (rather than those which focus on humans' indelible and systemic impact on the biosphere) that will perhaps determine the 'official' start of the new geological epoch. That official date, when it is known (and if the International Union of Geological Sciences officially ratifies this new geological epoch) will be included in school curricula and will not be edited to suit a particular political agenda. As pupils study the geological timescale early in secondary school (in different countries), the Anthropocene will become part of the curriculum once officially included in that timeline.

Geologists think of Earth's history in terms of different slices of time, based on shifts in the planet's overall condition, markers of which can be seen in sediment cores. They identify recognisable segments on the basis of climate, sea level and biodiversity. Stratigraphic markers are used to define chronostratigraphic units, giving geologists a common language in which to advance their knowledge of the planet's history. In the debate on how to date the onset of the Anthropocene, Earth system researchers are leading the way, identifying features visible today which we know (or can assume with a high degree of probability) will still be visible in hundreds of thousands – or even millions – of years. We do not yet know precisely what the Anthropocene has in store for the next few hundred or few thousand years; also, the most remarkable anthropic changes made to Earth's system have yet to manifest themselves. However, it is important for Earth system scientists to put a date on the dawn of the new geological epoch, based on sufficiently solid stratigraphic evidence. Most boundaries are defined by identifying a specific point on Earth in a stratotype known to date from a specific geological era. The markers are known as GSSPs: Global Boundary Stratotype Sections and Points. They may be rocks, sediments or glaciers which developed during a given geological period. Stratigraphic commissions aim to locate such points, which are also known as 'golden spikes', because once agreed upon, the stratigraphic section is marked with a plaque, sometimes on the head of a golden spike driven into the rock face.

One of the difficulties in defining a GSSP for the Anthropocene is that the full effects of human activities on the Earth system are not immediately apparent but are spread out across decades or even centuries (Lewis & Maslin, 2015, p. 173). A GSSP places a date on a physical marker that correlates to other, secondary markers. It must be at a particular site on the planet but have proven correlation with a new global context; it must also have 'complete continuous sedimentation with adequate thickness above and below the marker' (Lewis & Maslin, 2015, p. 172).

Indeed, a series of indicators can help date the onset of the Anthropocene, such as deposits from human activities, shifts in the biorecord, geochemical shifts, oceanic shifts (such as changes in ocean geochemistry, ocean biodiversity and sea level) and catastrophic events (be they 'natural' or caused by humans). Isotopes[4] can also contribute to the debate. Changes in isotope levels can yield data about the climate or the chemical composition of the atmosphere. For example, lead (Pb) isotopes clearly indicate the Ancient Greek civilisation, which was marked by innovation and the use of heavy metals. Human activities have altered not only the atmospheric concentrations of different gases but also isotope levels. Whilst the isotope record evidences anthropogenic changes over the past few millennia, the isotopes alone do not clearly demarcate the Anthropocene from the Holocene (Dean et al., 2014, p. 284).

Waters et al. (2014) provide evidence of a stratigraphic basis for the Anthropocene, by identifying a collection of anthropogenic sediments. In their view, there is a clear stratigraphic boundary between the Holocene and the Anthropocene: 'Humans are altering the planet, including long-term global geological processes, at an increasing rate' (p. 137). Stratigraphic indicators of a shift from the Holocene to the Anthropocene have recently been found in sediments from a lake in west Greenland, containing plastics, radionuclides, fly ash, pesticides, reactive nitrogen and metals (Waters et al., 2016, p. 137). They view the anthropic signatures of the new-dawned Anthropocene, distinguishing it from the Holocene, as the combination of 'accelerated technological development, rapid human population growth and increased resource consumption' (Waters et al., 2016, p. 139).

The stratigraphic principles involved in geological methodology are particularly rigorous and based on reason. What is astonishing is how much of human activities the Earth retains. In the scientific literature of the last 20 years, various possible dates have been suggested for the beginning of the Anthropocene: the Stone Age with the controlled use of fire (Doughty, 2013); the emergence of agriculture with the control over ecosystems and the progressive change in the chemical composition of the atmosphere (Ruddiman, 2013; Ruddiman et al., 2014); the encounter between the ancient and the new world which transformed the ecosystems (importation of different species) and reduced the world population (genocide and pandemics) (Lewis & Maslin, 2015); the steam engine and associated capitalism, becoming increasingly hegemonic (Crutzen, 2002; Steffen et al., 2011); the explosion of atomic bombs with radionuclides spreading to all four corners of the earth (Lewis & Maslin, 2015); or when we entered an age of increased consumption with its pre-programmed obsolescence (Steffen et al., 2015). There is no shortage of hypotheses when it comes to dating our entry into this new geological epoch – and the debates surrounding this have been particularly lively (Autin & Holbrook, 2012; Finney & Edwards, 2016; Zalasiewicz et al. 2012, 2017b; Edwards, 2018). Furthermore, each one of these debates contains valuable anthropological insights: human beings are characterised by their control over ecosystems, their imperialism, their unreserved exploitation of resources to the extent of determinedly sawing off the branches which give them life, they are true economic animals seeking to maximise individual interests, they are Promethean, considering all boundaries as obstacles to be overcome, they are capable of associating techno-scientific capabilities with military power, etc.

The Concept of the Anthropocene Stemming from the Humanities and Social Sciences

Since around 2010, a number of criticisms of the concept of the Anthropocene have emerged from the humanities and social sciences. They begin by questioning the *anthropos* of the Anthropocene. Crutzen, in his 2002 article, stipulates that the Anthropocene is the work of 25% and not all of humanity; however, in most of the articles by Earth system researchers, the *anthropos* responsible for our entry into this new geological epoch appears somewhat undifferentiated, far removed from the critical contributions of history and the sciences. The narrative of the Anthropocene produced by Earth system science is constructed within the framework of species and nature. At the Earth Summit in Rio in 1992, the political declaration of the various governments refers to 'common but differentiated responsibilities' (CBDR). It is a question of a weighting that is not completely missing in the narrative of the Anthropocene, but which does not give sufficient consideration to this differentiation.

   In the following quotation from the main authors of the concept of the Anthropocene, Zalasiewicz et al., (2010), we see their analysis of the ultimate cause of our entry into to the Anthropocene and the link between human population growth and the use of fossil fuels in the Industrial Revolution: 'First, how have the actions of humans altered the course of Earth's deep history? The answers boil down to the unprecedented rise in human numbers since the early nineteenth century – from under a billion then to over six billion now, set to be nine billion or more by midcentury. This population growth is intimately linked with massive expansion in the use of fossil fuels, which powered the Industrial Revolution, and allowed the mechanization of agriculture that enabled those additional billions to be fed.' This analysis may seem quite obvious at first glance, but it is worth examining the reasons for the progressive hegemony of fossil fuels that were at the heart of the Industrial Revolution. In an article from 2013, the Swedish historian Andreas Malm shows the way in which the choice of the steam engine rather than hydraulic energy in Great Britain during the 19th century was due to the power of capital over labour. The advantage of the steam engine is that it can be deployed close to towns and workers, and the fact that it does not matter what environment it is in and whether it is close to hydraulic plants. Although in the first few decades, the steam engine's use of fossil fuel resulted in considerably lower energy yields, the capitalist industrialists preferred to use the steam engine because it enabled the energy of the workers to be harnessed to make better use of the workforce. Thus, fossil energy was not created by an undifferentiated *anthropos*.

The humanities and social sciences insist, therefore, on the importance of social and societal forces for the Earth system. Malm and Hornborg remind us that between 1850 and 2000, the "capitalist" (or "northern") countries accounted for 18.8% of the world population and were responsible for 72.7% of CO_2 emissions. Just after 2000, 45% of the poorest people on the planet were responsible for 7% of CO_2 emissions; at the very same time, 7% of the richest people emitted 50%. Furthermore, the British environmentalist David Satterthwaite (who was a member of GIEC) shows in a precise study (2009) that people's responsibility for the emission of greenhouse gases varies by a ratio of 1:1000 according to where they live across the world and the life choices they make. Hence the amount of emissions in tons of carbon per person is 10.94 in Qatar; 4.71 in the USA; 1.57 in France; 0.01 in Chad and Mali. It is this that tells Malm and Hornborg that 'As long as there are human societies on Earth, there will be lifeboats for the rich and privileged' (Malm & Hornborg, 2014, p. 66). They therefore ask this simple question: 'Are these basic facts reconcilable with a view of humankind as the new geological agent?'

Following all these early critical studies concerning the concept of an undifferentiated *anthropos* being the origin of our entering the Anthropocene, Steffen *et al.* gradually introduced elements of differentiation into their analyses, but this remains fairly sketchy and does not go as far as proposing a different concept to the Anthropocene. We can see this in figure 4 (2011b). However, it is possible to see that if the curves of the graph gradually show a differentiated *anthropos*, the accompanying narrative is characterised by a relative lack of differentiation (compared to the contributions of the social sciences).

It is, therefore, not possible to consider the current changes in the Earth system, and the changes that are to come, simply as the product of human nature or of the activity of an undifferentiated or abstract humanity. It is not a question of natural inevitability but a question of social, political, economic or historical causes that can be clearly identified. Earth system science alone is not able to explain the deep-rooted causes for our entry into the Anthropocene. *Anthropos* is not necessarily at the heart of the way the Earth system functions – or in any case would need further in depth analysis.

The changes in the biosphere are evidence of how fragile the political and social world as we know it, and for which we are educating our children, has become. There is little time left to us to ensure that the Earth system can remain within the ranges of climatic variation and the functioning of the biosphere that were those of the Holocene and thus to avoid leaving its trajectory with no possible turning back for millions of years to come. The timescale we are given shows the urgency of the situation. Moreover, the risks currently incurred by the anthropic violation of our ecosystems, our use of the soil, the climate system, the chemical composition of the atmosphere and the oceans or biogeochemical fluxes, all of these are transcendental dangers to the whole of human life in society. The Anthropocene will continue to make us very fearful (something of which we became aware during the Covid 19 pandemic). Finally, lurking in the shadows of the strictly geological threat of the Anthropocene, we can see a threat that is directly political. In fact, in circles that are aware of the real risks of climate change and the collapse of our ecosystems, it is impossible to deny the seductive nature of 'strong' powers, hoping that they will succeed where neo-liberal democracies have failed.

To conclude, it is now up to each disciplinary field to develop its acceptance of the Anthropocene, in other words to take ownership of all the knowledge at their disposal in order to identify the paradigmatic breaks to be constructed within each of the disciplines.

The Future of Humanity Based on the Materiality of the New Geological Context

With human beings having such a central role to play on our planet in the Anthropocene, anthropological research continues to gain in importance. In such research, the human being is rarely seen an entity that is separate from the world, a *homo clausus*. It is far more common to see the human being as entangled with nature, with other creatures and other human beings. There is an increasing awareness of the *dual historicity and culturality* of anthropology. In anthropological research, themes and objects are examined at a specific time in history within a specific cultural frame of reference. There is a focus on both the past and the present, and frequently assumptions and projections of future developments are made. Historical, cultural and philosophical anthropology has offered some important contributions to our understanding of humanity in recent decades (Kamper & Wulf, 1994; Wulf & Kamper, 2002; Wulf, 2013, 2022a; Wallenhorst & Wulf, 2022) that show the importance of anthropological knowledge for our understanding of the Anthropocene. This is also true of the research into the negative developments of the Anthropocene and the global efforts to lessen their impact. Particularly important are the attempts to combine local and regional views with global perspectives. Many problems of the Anthropocene can only be understood and dealt with by taking into consideration the interaction of local and global dimensions. Agenda 2030, with its Sustainable Development Goals, is an example of this.

In view of the demands of the Anthropocene, anthropological research projects use *plural forms of knowledge* that are often interconnected and yet also follow their own developmental logic. They acquire their knowledge from many scientific disciplines for which they use various research methods. Anthropological knowledge is often multi-, inter- and trans-disciplinary. Because of the situation in which our planet currently finds itself, this knowledge is also often inter- or transcultural. Often different, ever-changing points of reference are at the basis of global knowledge which make these research projects more complex. It is a challenge for contemporary anthropological research to make connections between several scientific disciplines and paradigms. For example, investigations by the earth sciences are important now – these include geological, stratigraphic, Earth system science and multi-paradigmatic investigations. New knowledge that is important for our understanding of the Anthropocene is also emerging in almost all sciences.

Anthropology is an *open science*. It has no pre-determined subject area. Generally, its research projects have had to do with the actions and behaviour of human beings in the Anthropocene and contribute to a better understanding of human beings. In principle, this includes all areas of human action and behaviour in relation to the world and other people. Anthropological knowledge is multi-paradigmatic; it includes the natural sciences, and in the humanities and social sciences, it embraces both qualitative and quantitative knowledge. New forms of knowledge arise in anthropology that may also be paradoxical and self-contradictory. If we presuppose a broad understanding of science this can result in new forms of anthropogenic knowledge that change the accepted way of seeing nature, the world and human beings. Examples of this are to be found in climate research, genetics and Artificial Intelligence.

New forms of knowledge are also arising in the cultural and social sciences which are important for the idiographic understanding and the nomothetic explanation of the effects of humans on the planet. Some of these come about in relation to the Sustainable Development Goals and others because of new historical or cultural perspectives. There is also the formation of hybrid forms of knowledge in which dynamics that differ from one culture to another affect each other. This happens to the same extent both in the arts and sciences, and these forms of knowledge create complex processes or "grand narratives" that are contingent on them, and new interpretations, as mentioned above (Lyotard, 1979).

By showing how knowledge relates to human beings and how they act with regard to nature and the world, the anthropological perspective can help to make it easier to understand knowledge in the Anthropocene that is often hard to access and make sense of (Council of Science and Humanities, 2006). Anthropological research projects attempt to dissolve traditional boundaries between the sciences and the humanities through new forms of connecting and combining knowledge. They can also clearly show the importance of anthropological processing of scientific knowledge for bringing the Sustainable Development Goals to fruition.

When faced with the complexity of the Anthropocene, there needs to be *critical reflection* on the limitations of anthropological research. The question is to what extent it is possible, in discourses on the trans-human for example, to see beyond the anthropological perspective (More & More, 2013). This gives rise to doubt. Although the anthropological perspective is often expanded in an appropriate way, the anthropological viewpoint is not, as a matter of course, transcended. Even when these discourses criticise the central role of human beings as the reference point of research, they do refer to them and can transcend neither anthropology nor human beings. For many years now, the goal of anthropological research is no longer abstract human beings, detached from their historical and cultural context. The goal is far more historical-cultural research that has as its focus the specific conditions of human life in the Anthropocene (Kamper & Wulf, 1994; Wulf, 2013, 2022a, b; Wulf et al., 2010, 2011; Wallenhorst & Wulf, 2022).

With this cultural-historical anchorage, the ambition of this *Handbook of the Anthropocene* is to look ahead and to try to identify the main lines of a prospective anthropology. This dynamic joins the structuring elements of the 'new humanism', as defined in 2010 by UNESCO, in a tradition marked by utopian pragmatism, which emphasises humanity's potential creativity, in terms both of devising new projects and goals and of redefining ourselves. These articles attempt to marry new humanism with ecology-based thinking and biogeochemical materiality. Such thinking (which is not seen in new humanism) represents a breakaway from anthropocentrism. This is singularly important when thinking about humanisation – the focus is not on the essence of humanity, but on our relationship with nature (one of dependency), and on accommodation of the human race (or the 'human adventure') as an integral part of the Earth system.

The Anthropocene: A Critical but Potentially Constructive Challenge to Shape the World

This collective work is a step toward prospective anthropological reflection, but it is by no means intended to be exhaustive. Finally, rather than being restricted to a single scientific discipline, this compendium opens doors and encourages debate.

When we refer to the Anthropocene, this often has negative connotations. We think of the great problems of the modern age that have arisen as a consequence of industrialisation, the acceleration of life and the increasing abstraction of modern ways of life. These include, for example, climate change, the acidification of the oceans, the destruction of biodiversity, the development of nuclear energy, the depletion of non-renewable energy and destruction of the environment. Most of this destruction is the result of an increasing number of people wanting a 'comfortable' life, based on exponential consumption, coupled with population growth. Now, however we realise that the ways of life involved in this have led to a high degree of violence towards nature, other people and also individual existence. This has created a situation in which many people see the future of humankind and the planet as being under threat. Given this situation, there is an increase in endeavours that rest on people's creative ability and require intensive efforts to prevent such a negative outcome. If, in the Anthropocene, humans have such a profound influence on everything that concerns the planet, they can surely also use this influence to improve the situation – and, as we have mentioned, the question of temporalities is decisive here. With the work on the 17 Sustainable Development Goals (SDGs), this would seem possible. Indeed, in these articles, education is presented as a tool to help make choices in relation to the medium- and long-term changes we face (Wallenhorst et al., 2018; Wallenhorst & Pierron, 2019; Wulf, 2022a). However, it is by no means certain that these efforts will be successful. And yet it seems reasonable and promising for the nations of the global community to want to work towards achieving these goals in as far-reaching a way as possible. Here it is clear that hopes and utopian dreams have a role to play. These SDGs require fundamental changes in politics, the economy, public awareness and education. It is clear that the education and socialisation of the future generations is important. And for this reason, one of the goals is expressly aimed at education and personal development.

The global community needs to be transformed on the basis of the SDGs. Such a development extends across the following five areas that need to be improved by means of the SDGs. These are *people* (poverty and hunger, dignity, equality, a healthy environment), *planet* (protection of the ecosystems), *peace* (inclusion, peace, justice), *prosperity* (bringing prosperity to all people by means of economic and technical development) and *partnership* (cooperation). These tasks should be fulfilled through the principles of universality, inseparability, inclusion, accountability and partnership (Wallenhorst, 2019; Wulf, 2022a; Wallenhorst & Wulf, 2022, 2023). In every area a programme is being developed, the success of which hinges on correcting the undesirable developments caused by humans. Here educational processes play an important part. They oscillate between perfection and incorrigibility (Kamper & Wulf 1994). Violence, which may be explicit or implicit, plays a central role in these processes. Considerable energy and strength is required to achieve the necessary social transformations and violence is therefore sometimes unavoidable. Paradoxical conflicts that are very difficult to resolve often arise in these processes, and education creates important methods of dealing with these conflicts in a constructive way. There is a need for all-embracing processes of education and learning and the anthropological research that will make them possible. There is still uncertainty about the extent to which these processes are possible in the long term in non-democratic societies. We need to create new images of nature, the world and humanity that are not simply based on utopian ideas but also include elements of human resistance and incorrigibility. The goal is not only to change the individual and collective imaginary but also to change actions and behaviour. There needs to be a critical but constructive change in people's dominant habits, which have at their centre, for example, consumption and utility. There also need to be new approaches to nature and the world (Wulf, 2013, 2022a; Paragrana, 2014). Mimetic processes, i.e. processes of creative imitation of examples of successful practice, are important here. Creative mimetic processes that embrace the sensuality of the body can help to form new habits (Gebauer & Wulf, 1995, 1998). On the one hand, these are the product of embodied practical knowledge, and on the other hand, they also lead to new forms of practice and structures. Rituals and ritualisations play an important role in this (Wulf, 2010; Wulf & Zirfas, 2014). For the SDGs to produce the social transformations for which they strive, they must be performative, i.e. they must lead to the staging and performance of social behaviour, that is geared towards the critical but constructive possibilities inherent in human actions and behaviour in the Anthropocene. These processes require us to embody within us the anthropogenic actions and behaviour to which we aspire (Kraus & Wulf, 2022) and also to realise the importance of what is implicit and tacit (Kraus et al., 2021).

It should be pointed out that the concept of the Anthropocene has been the subject of so many empirical studies, research and scientific publications in almost all disciplinary fields that it has now acquired an autonomy with respect to official geological institutions. The Anthropocene, regardless of its formalisation, can now be understood as a total social fact with a proven systemic scientific basis (Mauss, 1923–24), which is why it is important to explore it in this *Handbook of the Anthropocene*.

This *Handbook* is divided into six sections which constitute six facets of the Anthropocene that it is necessary to understand if we are to consider and prepare a long-term future.

1. The planet: caught between biogeophysical knowledge and the uncertainty of our adventure
2. Moving towards new epistemological paradigms linking the sciences and humanities
3. Human beings: bridging nature and culture
4. Societies: Prometheanism and post-Prometheanism
5. Profound long-term changes: education, apprenticeships and socialisation
6. Peace and violence

The almost 280 articles that make up this *Handbook of the Anthropocene* do not by any means cover the whole spectrum of the Anthropocene and its impact on human societies in a comprehensive way. We would have liked to add many more entries. But the multidisciplinary project had to end somewhere. This end is relative and temporary: as of now this task is continuing in the preparation of an *Encyclopaedia of the Anthropocene* for Springer-Nature. This work will mobilise the authors of this handbook and many others (from all continents and all disciplinary fields) to continue to circumscribe all the novelties and ruptures of the new geological era that is now ours.

References

Arnsperger, C., & Bourg, D. (2017). *Ecologie intégrale, Pour une société permacirculaire*. PUF.

Autin, W. J., & Holbrook, J. M. (2012). Is the Anthropocene an issue of stratigraphy or pop culture? *GSA Today, 22*(7), 60–61.

Bador, M., et al. (2017). Future summer mega-heatwave and record-breaking temperatures in a warmer France climate. *Environmental Research Letters, 12*, 1–2.

Barnosky, A. D., et al. (2011). Has the Earth's sixth mass extinction already arrived? *Nature, 471*, 51–57.

Barnosky, A. D., et al. (2012). Approaching a state shift in Earth's biosphere. *Nature, 486*, 52–58.

Bonneuil, C., & Fressoz, J.-B. (2013). *L'événement anthropocène*. Seuil.

Bourg, D., & Papaux, A. (Eds.). (2015). *Dictionnaire de la pensée écologique*. PUF.

Brienen, R. J. W., et al. (2015). Long-term decline of the Amazon carbon sink. *Nature, 519*, 344–348.

Council of Science and Humanities. (2006). *Recommendations for the development and promotion of the humanities in Germany*. Wissenschaftsrat.

Crutzen, P. J. (2002). Geology of mankind: 'The Anthropocene'. *Nature, 415*, 23.

Crutzen, P. J. (2006). Albedo enhancement by stratospheric sulfur injections: a contribution to resolve a policy dilemma? *Climate change, 77*, 211–219.

Curnier, D. (2017). *Quel rôle pour l'école dans la transition écologique? Esquisse d'une sociologie politique, environnementale et prospective du curriculum prescrit*. Doctoral thesis in Environmental Sciences, University of Lausanne.

Dean, J. R., Leng, M. J., & Mackay, A. W. (2014). Is there an isotopic signature of the Anthropocene? *The Anthropocene Review, 1*, 276–287.

Descola, P. (2005). *Par-delà nature et culture*. Gallimard.

Doughty, C. E. (2013). Preindustrial human impacts on global and regional environment. *Annual Review of Environment and Resources, 38*, 503–527.

Doughty, C. E., et al. (2015). Drought impact on forest carbon dynamics and fluxes in Amazonia. *Nature, 519*, 78–82.

Eckersley, R. (2017). La démocratie à l'ère de l'Anthropocène. *lapenseeecologique.com, 1*(1), 1–19.

Edwards, L. E. (2018). What is the Anthropocene? *Eos*, https://eos.org/opinions/what-is-the-anthropocene

Erlandsson, L. W. et al. (2022). A planetary boundary for green water. *Nature reviews Earth and Environment*, 26 April.

Esquivel-Muelbert, A., et al. (2019). Compositional response of Amazon forests to climate change. *Global Change Biology, 25*, 39–56.

Esquivel-Muelbert, A., et al. (2020). Tree mode of death and mortality risk factors across Amazon forests. *Nature communications, 12*, 1–20.

Federau, A. (2017). *Pour une philosophie de l'Anthropocène*. PUF.

Finney, S., & Edwards, L. E. (2016). The 'Anthropocene' epoch: scientific decision or political statement? *GSA Today, 26*(3–4), 4–10.

Gatti, L. V., et al. (2014). Drought sensitivity of Amazonian carbon balance revealed by atmospheric measurements. *Nature, 506*, 76–80.

Gebauer, G., & Wulf, C. (1995). *Mimesis. Culture, art, society*. University of California Press.

Gebauer, Gunter and Wulf, Christoph (1998). *Spiel, Ritual, Geste. Mimetisches Handeln in der sozialen Welt*. : Rowohlt.

Gil, I. C., & Wulf, C. (Eds.). (2015). *Hazardous future: Disaster, representation and the assessment of risk*. De Gruyter.

Hétier, R., & Wallenhorst, N. (Eds.). (2021). L'éducation au politique en Anthropocène, *Le Télémaque, 58*.

Hétier, R., & Wallenhorst, N. (2022a). The COVID-19 pandemic: a reflection of the human adventure in the Anthropocene. *Paragrana, 30*, 41–52.

Hétier, R., & Wallenhorst, N. (2022b). *Enseigner à l'époque de l'Anthropocène*. Le Bord de l'eau.

Hétier, R., & Wallenhorst, N. (2022c). Promoting embodiment through education in the Anthropocene. In A. Kraus & C. Wulf (dir.) *Learning bodies – Tact, emotion, performance. A european handbook*. Palgrave MacMillan.

Im, E.-S., Pal, J. S., & Eltahir, E. A. B. (2017). Deadly heat waves projected in the densely populated agricultural regions of South Asia. *Science Advances, 3(8)*, 1–7.

Kamper, D., & Wulf, C. (1989). *Looking back at the end of the world*. Massachusetts Institute of Technology.

Kamper, D., & Wulf, C. (Eds.). (1994). *Anthropologie nach dem Tode des Menschen. Vervollkommnung und Unverbesserlichkeit*. Suhrkamp.

Keats, J. (2011). Anthropocene. In J. Keats (Ed.), *Virtual words* (pp. 18–22). University Press of Oxford.

Kraus, A., & Wulf, C. (Eds.). (2022). *Palgrave handbook of embodiment and learning*. London Palgrave Macmillan.

Kraus, A., Budde, J., Hietzge, M., & Wulf, C. (Eds.). (2021). *Handbuch Schweigendes Wissen. Erziehung, Bildung, Sozialisation und Lernen* (2nd ed.). Beltz Juventa.

Lesourt, E. (2018). *Survivre à l'Anthropocène*. PUF.

Lewis, S. L., & Maslin, M. A. (2015). Defining the Anthropocene. *Nature, 519*, 171–180.

Lyotard, J.-F. (1979). *La condition postmoderne*. Les éditions de minuit.

Magny, M. (2019). *Aux racines de l'Anthropocène*. Le Bord de l'eau.

Malm, A. (2013). The origins of fossil capital: From water to steam in the British cotton industry. *Historical Materialism, 21(1)*, 15–68.

Malm, A., & Hornborg, A. (2014). The geology of mankind? A critique of the Anthropocene narrative. *The Anthropocene Review, 1*, 62–69.

Mauss, A. (2012) (or. ed. 1923–1924). *Essai sur le don. Forme et raison de l'échange dans les sociétés archaïques*. PUF.

Mora, C., et al. (2017). Global risk of deadly heat. *Nature climate change, 7*, 501–506.

Nature. (2011). Editorial. The human epoch. Official recognition for the Anthropocene would focus minds on the challenges to come. *Nature, 473*, 254.

Papaux, A. (2015). Homo faber. In D. Bourg & A. Papaux (Eds.), *Dictionnaire de la pensée écologique* (pp. 536–540). PUF.

Paragrana. (2014). Internationale Zeitschrift für Historische Anthropologie. *Art and Gesture, 23(1)*.

Persson, L., et al. (2022). Outside the safe operating space of the planetary boundary for novel entities. *Environmental Science & Technology, 56(3)*, 1510–1521.

Prouteau, F., Hétier, R., & Wallenhorst, N. (2022). *Critique, utopia and resistance: three functions of pedagogy of resonance in the Anthropocene. Vierteljahrsschrift für Wissenschaftliche Pädagogik*.

Renaud, H. (2021). *L'humanité contre l'Anthropocène. Résister aux effondrements*. PUF.

Rockström, J. (2015). Bounding the planetary future: Why we need a great transition. *Great Transition Initiative, 9*, 1–14.

Rockström, J. W., et al. (2009). A safe operating space for humanity. *Nature, 461*, 472–475.

Ruddiman, W. F. (2013). The Anthropocene. *The Annual Review of Earth and Planetary Sciences, 41*, 45–68.

Ruddiman, W. F., et al. (2014). Does pre-industrial warming double the anthropogenic total? *The Anthropocene Review, 1*, 1–7.

Steffen, W., et al. (2011a). The Anthropocene: Conceptual and historical perspectives. *Philosophical Transactions of the Royal Society, 369*, 842–867.

Steffen, W., et al. (2011b). The Anthropocene: From global change to planetary stewardship. *Ambio, 40(7)*, 739–761.

Steffen, W., et al. (2015). The trajectory of the Anthropocene: The great Acceleration. *The Anthropocene Review, 2(1)*, 81–98.

Steffen, W., et al. (2016). Stratigraphic and Earth system approaches to defining the Anthropocene. *Earth's Future, 4*, 1–22.

Anthropocene, the Concept of the 21st Century – A General Introduction   xxiii

Steffen, W., et al. (2018). Trajectories of the Earth system in the Anthropocene. *Proceedings of the National Academy of Sciences, 115*(*33*), 8252–8259.

Visconti, G. (2014). Anthropocene: another academic invention? *Rendiconti Lincei: Science Fisiche e Naturali, 25*(*3*), 381–392.

Wallenhorst, N. (2019). *L'Anthropocène décodé pour les humains*. Le Pommier.

Wallenhorst, N. (2020). *La vérité sur l'Anthropocène*. Le Pommier.

Wallenhorst, N. (2021). *Mutation. L'aventure humaine ne fait que commencer*. Le Pommier.

Wallenhorst, N. (2022). *Qui sauvera la planète ?* Actes Sud.

Wallenhorst, N., & Pierron, J.-P. (Eds.). (2019). *Éduquer en Anthropocène*. Le Bord de l'eau.

Wallenhorst, N., & Wulf, C. (2022). *Humains. Un dictionnaire d'anthropologie prospective*. Vrin.

Wallenhorst, N., Prouteau, F., & Coatanéa, D. (Eds.). (2018). *Éduquer l'homme augmenté*. Le Bord de l'eau.

Wallenhorst, N., Robin, J.-Y., & Boutinet, J.-P. (2019). L'émergence de l'Anthropocène, une révélation étonnante de la condition humaine? In N. Wallenhorst & J.-P. Pierron (Eds.), *Éduquer en Anthropocène* (pp. 23–36). Lormont.

Waters, C. N., et al. (2014). Evidence for a stratigraphic basis for the Anthropocene. In R. Rocha, J. Pais, J. Kullberg, & S. Finney (Eds.), *STRATI 2013* (pp. 989–993). Springer Geology, Springer.

Waters, C. N., et al. (2016). The Anthropocene is functionally and stratigraphically distinct from the Holocene. *Science, 351*, 137–147.

Wulf, C. (2010). *Der Mensch und seine Kultur* (2nd ed.). Anaconda.

Wulf, C. (2013). *Anthropology. A continental perspective*. University of Chicago Press.

Wulf, C. (2022a). *Education as human knowledge in the Anthropocene. An anthropological perspective*. Routledge.

Wulf, C. (2022b). *Human beings and their images. Imagination, mimesis and performativity*. Bloomsbury.

Wulf, C., & Kamper, D. (Eds.). (2002). *Logik und Leidenschaft. Erträge Historischer Anthropologie*. Reimer.

Wulf, C., & Zirfas, J. (Eds.). (2014). *Handbuch Pädagogische Anthropologie*. Springer VS.

Wulf, C., & Zirfas, J. (Eds.). (2020). Den Menschen neu denken. *Paragrana: Internationale Zeitschrift für Historische Anthropologie, 29/2020/1*.

Wulf, C., et al. (2010). *Ritual and identity: The staging and performing of rituals in the lives of young people*. Tufnell Press.

Wulf, C., Suzuki, S., et al. (2011). *Das Glück der Familie: Ethnografische Studien in Deutschland und Japan*. Springer VS.

Zalasiewicz, J., Williams, M., Steffen, W., & Crutzen, P. (2010). The new world of the Anthropocene. *Environmental Science & Technology, 44*, 2228–2231.

Zalasiewicz, J., et al. (2011). The Anthropocene: A new epoch of geological time? *Philosophical Transactions of the Royal Society, 369*, 835–841.

Zalasiewicz, J., et al. (2012). Response to Autin and Holbrook on 'Is the Anthropocene an issue of stratigraphy or pop culture?'. *GSA Today, 22*(7), e21–e22.

Zalasiewicz, J., et al. (2014). When did the Anthropocene begin? A mid-twentieth century boundary level is stratigraphically optimal. *Quaternary international, 30*, 1–8.

Zalasiewicz, J., et al. (2017a). Anthropocene: Its stratigraphic basis. *Nature, 541*, 289–289.

Zalasiewicz, J., et al. (2017b). Making the case for a formal Anthropocene Epoch: an analysis of ongoing critiques. *Newsletters on Stratigraphy, 50*, 205–226.

Contents of Volume I

Part II   The Earth's Surface and Its Elements

Section II   Epistemology, Sciences and Humanities
Nathanaël Wallenhorst and Christoph Wulf

Part V   The Anthropocene as an Interpretative Framework

Part VII   The Paradigmatic Impasses of Technology?

Part VIII   The Acceptance of Limits, Containing and Salutary

Part IX   The Refusal of Limits, Illusory and Destructive

Section III   Human Beings: Bridging Nature and Culture
Nathanaël Wallenhorst and Christoph Wulf

Part X   Humanity as Birth

Part XIV   Humanity as Justice

Contents of Volume II

**Section V   The Profound Changes in the Long Term:
            Education, Learning, and Socialization**
Nathanaël Wallenhorst and Christoph Wulf

Section VI   Violence and Peace
         Nathanaël Wallenhorst and Christoph Wulf

Part XX   The Risk of Violence

Part XXI   The Challenge of International Institutions

About the Editors

Nathanaël Wallenhorst is Professor at the Catholic University of the West (UCO). He is Doctor of Educational Sciences and Doktor der Philosophie (first international co-supervision PhD) and Doctor of Environmental Sciences and Doctor in Political Science (second international co-supervision PhD). He is the author of twenty books on politics, education and anthropology in the Anthropocene. Books (selection): *The Anthropocene Decoded for Humans* (Le Pommier, 2019, *in French*); *Education in the Anthropocene* (ed. with Pierron, Le Bord de l'eau 2019, *in French*); *The Truth About the Anthropocene* (Le Pommier, 2020, *in French*); *Mutation: The Human Adventure Is just Beginning* (Le Pommier, 2021, *in French*); *Who Will Save the Planet?* (Actes Sud, 2022, *in French*); *Vortex: Facing the Anthropocene* (with Testot, Payot, 2023, *in French*); *Political Education in the Anthropocene* (ed. with Hétier, Pierron and Wulf, Springer, 2023, *in English*); and *A Critical Theory for the Anthropocene* (Springer, 2023, *in English*) e-mail:nathanael.wallenhorst@uco.fr.

Christoph Wulf is Professor of Anthropology and Education and a member of the Interdisciplinary Centre for Historical Anthropology, the Collaborative Research Centre (SFB, 1999–2012) 'Cultures of Performance', the Cluster of Excellence (2007–2012), 'Languages of Emotion' and the Graduate School 'InterArts' (2006–2015) at the Freie Universität Berlin. His books have been translated into 20 languages. For his research in anthropology and anthropology of education, he received the title *professor honoris causa* from the University of Bucharest. He is Vice-President of the German Commission for UNESCO. *Major research areas:* historical and cultural anthropology, educational anthropology, imagination, intercultural communication, mimesis, aesthetics, epistemology and Anthropocene. Research stays and invited professorships have included the following locations, among others: Stanford, Tokyo, Kyoto, Beijing, Shanghai, Mysore, Delhi, Paris, Lille, Modena, Amsterdam, Stockholm, Copenhagen, London, Vienna, Rome, Lisbon, Basel, Saint Petersburg, Moscow, Kazan and Sao Paulo e-mail: christoph.wulf@fu-berlin.de.

Contents (Alphabetical Order)

Section I
The Planet: Caught between Biogeophysical Knowledge and the Uncertainty of Our Adventure

Nathanaël Wallenhorst and Christoph Wulf

There is an abundance of scientific studies on the current environmental situation, all of which are very alarming. Let us start by reviewing the key results of the last few years which offer a good interpretation of the current situation and the speed with which it has evolved (Wallenhorst, 2019, 2020a, 2020b, 2021, 2022; Wallenhorst & Wulf, 2022; Kamper & Wulf 1989, 1994; Gil & Wulf, 2015; Wulf, 2022).

2009–2022: *planetary boundaries.* In 2009, the Swedish environmentalist Johan Rockström and his 28 colleagues published "*A Safe Operating Space for Humanity*" (Rockström et al., 2009) in *Nature.* They identify nine planetary boundaries, the crossing of which would render our planet clearly less propitious for human life. The nine boundaries of nine processes of the Earth system which guarantee that humans evolve within a safe operating space. In 2009 three planetary boundaries were crossed: climate change, the erosion of biodiversity and the changing of the biogeochemical cycle of nitrogen.

To put this more simply: the boundaries are estimated within a safe operating space from the threshold (or "systemic tipping point") into dangerous consequences for human life in society. This article presents with great clarity what all Earth system researchers know full well, that the earth functions as a system consisting of numerous sub-systems, which are characterised by reacting in an abrupt, non-linear way where the boundaries are closely intermeshed. Climate and biodiversity have a very singular systemic power in this ensemble – crossing one of these two boundaries is enough to set the whole Earth system into a mode of completely reorganising

N. Wallenhorst (✉)
Catholic University of the West, Angers, France
e-mail: nathanael.wallenhorst@uco.fr

C. Wulf
Freie Universität Berlin, Berlin, Germany
e-mail: christoph.wulf@fu-berlin.de

itself which will complicate the very way that we are able to coexist on Earth. In 2015 the same authors continue their work in *Science* (it is noteworthy that the two most important scientific reviews ratify the existence of planetary boundaries that must not be crossed). A few years on and the situation is now considerably worse. The boundary of biogeochemical cycles has been crossed twice (from now on the boundary of the nitrogen cycle is crossed more than that of phosphorus which was already crossed) and a further boundary has been crossed, that of land-system change. In 2022 we were informed by an international team of 14 researchers, led by Linn Persson, a chemist at the Stockholm Environment Institute, that a fifth planetary boundary had been crossed, which threatens to destabilise the Earth system – the boundary of chemical pollution. In 2022 again, another team of 15 researchers led by Lan Wang-Erlandsson and Johan Rockström showed that a sixth boundary, the boundary of freshwater consumption, has been crossed.

2011. The American biologist Anthony Barnosky and eight of his colleagues pull the alarm cord. We are in the process of causing a new mass extinction of species (2011). Over its entire history the Earth has known five large episodes of extinction of life, which, according to palaeontologists, resulted in the loss of 75% of animal and plant diversity (on Earth and in the oceans) in less than two million years. This article, *"Has The Earth's Sixth Mass Extinction Already Arrived?"* leaves us in no doubt – we are in the middle of a catastrophe, a sixth mass extinction, caused by the only human species and in record time, despite the fact that coexistence with other species is fundamental to our own existence.

2014. Luciana V. Gatti, physicist and chemist in the field of atmosphere, and 16 other researchers published a "carbon balance" of the Amazon in *Nature* (2014). Up until then it was understood that a tree, through photosynthesis, captured CO_2 and replaced it with O_2. The trees, which decarbonised the atmosphere were considered our best friends – they functioned as carbon filters. The tropical forests absorbed about a quarter of the carbon gas released into the atmosphere by human activities. This is what caused the Amazon rainforest to be called the "lungs of the planet". But is that still what it is? Their conclusion was that photosynthesis of the trees in the Amazon lessened in the dry year (2010, in comparison with 2011, a very wet year), with the trees releasing more carbon than they absorbed. Conclusion: the Amazon is approaching a point of no return – and the deforestation that is currently taking place is aggravating the situation still more. Vegetation can change from being our best friend to being our worst enemy depending on climatic conditions which themselves are dependent on the activities of humans.

2017. We learn from the Mexican biologist Gerardo Ceballos and two of his American colleagues that the situation is much more concerning than it first appeared in the analysis of Anthony Barnosky in 2011. Published in the great American review *PNAS*, their article paints a picture of "biological annihilation" unprecedented in the history of the Earth. Their research is focused on the populations, on the occurrence of vertebrates (amphibians, reptiles, birds, mammals), rather than on species (Ceballos et al., 2017). The result? The number of populations threatened with extinction is huge and not commensurate with the number of their species. The researchers concluded that, in terms of numbers and speed, the phenomenon surpasses everything that we have known so far.

2017. A study published in *Environmental Research Letters*, led by the French climatologist Margot Bador et al. (2017). What will the heatwaves and the extreme summer temperatures in France be like in the year 2100, they ask, if global greenhouse gas emissions are not drastically cut? They predict increases of between +6 °C and + 13 °C, depending on region, compared to the heatwave which we experienced in 2002, which they used as a point of comparison. The peaks in temperature could easily go higher than 50°, temperatures which are customary in the desert, and could reach 55 °C in Eastern France in the year 2100.

2018. The US chemist Will Steffen and six other scientists (including Swedish environmentalist Johan Rockström, British oceanographer Colin Summerhayes, American biologist Anthony Barnosky and German physicist Hans Joachim Schellnhuber), stress the fact that the data forecasting the temperature of the Earth are worse than previously envisaged (2018). They study the major risk to be addressed – the possibility that a series of chain reactions, or the domino effect, is pushing the Earth system towards an increase of average temperature that will settle at +5 °C, a threshold that is incompatible with human life in society in most regions of the world. If we are to avoid the planet irrevocably becoming a "Hothouse Earth" for the thousands or tens of thousands of years to come (without the possibility of any reversal) we have two decades left to *radically* change the way we live.

2020. 11,000 scientists from 150 countries all over the world, from the fields of the geosciences and life sciences, co-sign a statement in *Bio Science*, invoking the moral obligation under which they find themselves. The seriousness and the acceleration of the climate crisis are "threatening the destiny of humanity". "We declare clearly and unequivocally that planet Earth is facing a climate emergency. To secure a sustainable future, we must change how we live" (Ripple et al., 2020).

Etc.

To sum up, the countdown has started. In the view of Johan Rockström, we have ten years left before this system crosses different systemic boundaries and the planet becomes irrevocably unfavourable for the human adventure in society. Ten years in which to decarbonise industry, energy, transportation and habitats. Ten years to transform agriculture, which should be storing, and not releasing, CO_2. An indisputable conclusion which necessitates a radical change in how we live (Rockström, 2020). Why? Because humans have become a geological force, having the same impact as other natural forces. Because they pose a threat to their own survival in society.

From now on this threat has a name – Anthropocene. Or the title of the geological epoch in which we live. The consequences for our civilisation are considerable. In fact, the emergence of civilisations was due to the stable and predictable climate of the interglacial period of the Holocene which began 11,700 years ago. If *Homo sapiens*, who had been there for about 350,000 years, hadn't planted seeds and put a few animals in an enclosure during his first 340,000 years, it's not that he didn't think of it. It quite simply wasn't possible. It is the stability of the Holocene that allowed *Homo sapiens* to control the ecosystem in order to build up agricultural surplus, which then allowed some people to do different things, rather than just engage in subsistence farming. This allowed the emergence of our complex societies. Our societies are built on the bedrock of a favourable, stable climate. The question of their collapse is a result of the current changes in this bedrock that enabled them to emerge. The climate is not to be taken lightly. It is directly related to the way in which we coexist, the way in which we all live together as humans – and will be able to continue to do so. The climate is closely connected to biodiversity. If the climate changes it alters the resilience of this solid tissue that is life. However, the erosion of biodiversity reinforces climate change and rising temperatures on Earth.

As mentioned in the general introduction, there is no doubt that the Anthropocene is the concept of the twenty-first century. It is the name of a new geological epoch in which we find ourselves, in view of the impact of humans on the Earth system. The term is part of a tradition that has been taking place over more than 150 years, during the course of which the idea of humanity as a geological force has emerged, as numerous scientific publications have reported for several decades how human activity has brought about changes in the biosphere. Coined informally to begin with in February 2000 at a scientific conference, this term gradually went on to find scientific acceptance and has come to crystallise the magnitude of the human footprint on the Earth as a whole.

The Anthropocene is as simple as it is complex. The simple thing about it is that we have changed the trajectory of the Earth system. Simpler still, from the point of view of the human adventure, is that we might summarise the situation like this: "The situation is serious". From the perspective of the adventure of life as a whole, no panic, living tissue is solid, resilient and creative. Although it severely affected, it will manage to get through the upheavals of the Anthropocene – it has been through others.

The Anthropocene is also complex. As we will discover in the different articles in this *Handbook of the Anthropocene*, it is at the crossroads of many scientific disciplines: geology and stratigraphy, palaeontology, geography, biochemistry and geochemistry, anthropology, social and political sciences ... The Earth and its biosphere have left the various fields where they were up until now (climate, chemical composition of the oceans, extinction of animal and plant species, etc.), and which allowed the possibility of human life in society.

Human activity has changed the habitability conditions of the Earth for all living creatures. Human life in society finds itself jeopardised at this very moment in certain parts of the world, and it will be even more so in the decades to come. The first thing to do is to bring an awareness of the Anthropocene to people's attention. At the present time, where we sometimes find it hard to distinguish between what is true and what is "fake", it is it important to reach agreement on the facts, without which we risk sliding towards a totalitarian regime, as the German philosopher Hannah Arendt clearly demonstrated. It is this in particular that is addressed by the articles in this first section, where scientific knowledge of how the Earth system works is often counter-intuitive. There are very many aspects of the Anthropocene that we must try and understand (with all its new terms and concepts).

References

Bador, M., et al. (2017). Future summer mega-heatwave and record-breaking temperatures in a warmer France climate. *Environmental Research Letters, 12*, 1–12.

Barnosky, A. D., et al. (2011). Has the Earth's sixth mass extinction already arrived? *Nature, 471*, 51–57.

Ceballos, G., Ehrlich, P. R., & Dirzo, R. (2017). Biological annihilation via the ongoing sixth mass extinction signaled by vertebrate population losses and declines. *Proceedings of the National Academy of Sciences, 114*(30), 6089–6096.

Gatti, L. V., et al. (2014). Drought sensitivity of Amazonian carbon balance revealed by atmospheric measurements. *Nature, 506*, 76–80.

Gil, I. C., & Wulf, C. (Eds.). (2015). *Hazardous future: Disaster, representation and the assessment of risk*. De Gruyter.

Kamper, D., & Wulf, C. (1989). *Looking back at the end of the world*. Massachusetts Institute of Technology.

Kamper, D., & Wulf, C. (Eds.). (1994). *Anthropologie nach dem Tode des Menschen. Vervollkommnung und Unverbesserlichkeit*. Suhrkamp.

Ripple, W. J., et al. (2020). World scientists' warning of a climate emergency. *Bioscience, 70*(1), 8–12.

Rockström, J. (2020). 10 years to transform the future of humanity – Or destabilize the planet, *TED Talks*, (online): www.c40knowledgehub.org/s/article/Johan-Rockstrom-10-years-to-transform-the-future-of-humanity-or-destabilize-the-planet?language=en_US

Rockström, J., et al. (2009). A safe operating space for humanity. *Nature, 461*, 472–475.

Steffen, W., et al. (2018). Trajectories of the earth system in the Anthropocene. *Proceedings of the National Academy of Sciences, 115*(33), 8252–8259.

Wallenhorst, N. (2019). *L'Anthropocène décodé pour les humains*. Le Pommier.

Wallenhorst, N. (2020a). *La Vérité sur l'Anthropocène*. Le Pommier.

Wallenhorst, N. (2020b). *Mutation. L'aventure humaine ne fait que commencer*. Le Pommier.

Wallenhorst, N. (2022). *Qui sauvera la planète?* Actes Sud.

Wallenhorst, N., & Wulf, C. (Eds.). (2022). *Humains. Un dictionnaire d'anthropologie prospective*. Vrin.

Wulf, C. (2022). *Education as human knowledge in the Anthropocene. An anthropological perspective*. Routledge.

Nathanaël Wallenhorst is Professor at the Catholic University of the West (UCO). He is Doctor of Educational Sciences and Doktor der Philosophie (first international co-supervision PhD), and Doctor of Environmental Sciences and Doctor in Political Science (second international co-supervision PhD). He is the author of twenty books on politics, education, and anthropology in the Anthropocene. Books (selection): *The Anthropocene decoded for humans* (Le Pommier, 2019, *in French*). *Education in the Anthropocene (*ed. with Pierron, Le Bord de l'eau 2019, *in French*). *The Truth about the Anthropocene* (Le Pommier, 2020, *in French*). *Mutation. The human adventure is just beginning* (Le Pommier, 2021, *in French*). *Who will save the planet?* (Actes Sud, 2022, *in French*). *Vortex. Facing the Anthropocene* (with Testot, Payot, 2023, *in French*). *Political education in the Anthropocene* (ed. with Hétier, Pierron and Wulf, Springer, 2023, *in English*). *A critical theory for the Anthropocene* (Springer, 2023, *in English*).

Christoph Wulf is Professor of Anthropology and Education and a member of the Interdisciplinary Centre for Historical Anthropology, the Collaborative Research Centre (SFB, 1999–2012) "Cultures of Performance," the Cluster of Excellence (2007–2012) "Languages of Emotion," and the Graduate School "InterArts" (2006–2015) at the Freie Universität Berlin. His books have been translated into 20 languages. For his research in anthropology and anthropology of education, he received the title "*professor honoris causa*" from the University of Bucharest. He is Vice-President of the German Commission for UNESCO. *Major research areas:* historical and cultural anthropology, educational anthropology, imagination, intercultural communication, mimesis, aesthetics, epistemology, Anthropocene. Research stays and invited professorships have included the following locations, among others: Stanford, Tokyo, Kyoto, Beijing, Shanghai, Mysore, Delhi, Paris, Lille, Modena, Amsterdam, Stockholm, Copenhagen, London, Vienna, Rome, Lisbon, Basel, Saint Petersburg, Moscow, Kazan, Sao Paulo.

Part I
The Earth as a System

Atmosphere

Lutz Möller

Abstract This article examines the history and the meaning of the term "atmosphere". It describes the structural differentiation of the atmosphere and its character as a relational concept of human beings towards their planetary environment. It argues that the concept of the atmosphere, while being overly present in daily experiences and generally being well comprehensible, is poorly understood by the layperson. It also explains how significant and fragile the atmosphere is. Using the two phenomena of global warming and ozone depletion, it shows the huge impact that human civilization has on the atmosphere and, in turn, the atmosphere's significance for humanity in the Anthropocene.

The Term and Its Conceptual Differentiation

"Atmosphere" is an old and very widely used term. Both is true in contrast to the other three "spheres" that collectively make up today's differentiated geochemical model of planet Earth (geosphere, biosphere, and hydrosphere, compare the articles in this Handbook). "Atmosphere" refers to the layered gas environment around planet Earth, but also to the gas layers of other planets, within the solar system or beyond, that are connected to their planet through gravity.

In addition, the term is used with other meanings. In particular, between human beings or in human-built environment, it can refer to a specific quality of interpersonal relationships or places, their "ambiance", their "mood", their "spirit", or their "aura". Interestingly, relationships or places also can have a "climate" in colloquial language use. There is also a physical quantity, which designates units of pressure, as "standard atmosphere", which refers to the average air pressure at sea level.

L. Möller (✉)
German Commission for UNESCO, Bonn, Germany
e-mail: moeller@unesco.de

© The Author(s), under exclusive license to Springer Nature
Switzerland AG 2023
N. Wallenhorst, C. Wulf (eds.), *Handbook of the Anthropocene*,
https://doi.org/10.1007/978-3-031-25910-4_1

The term dates back to the seventeenth century. However, like all other "spheres", the term has old roots in the concept of "elements" which have been used in cultural traditions from East Asia to the Mediterranean. As has been written in the article on "geosphere" in this Handbook, the Greek philosopher Aristotle used the model of celestial and terrestrial "spheres". Partially independently, partially interconnected and with many important successors, Hugh Doherty (1864), Stephen Pearl Andrews (1871) and Eduard Suess (1875) established the modern version of the geochemical model of the planet.

On planet Earth, as is the case of all other planets, the atmosphere does not have a fixed outer bound. In some definitions, it reaches out into space for some 10,000 kilometres, where its density becomes so low that it is definitely not reasonable anymore to speak of a "gas", since gas particles can travel dozens of kilometres between their mutual collisions and can freely escape into space. This low-density layer of gas called exosphere is without any meteorological phenomena and reaches down all the way down to the thermopause, an abstractly defined boundary at 500–1000 kilometres above sea level. In the thermosphere below the thermopause, the air particles interact much more intensively with the solar radiation; it mostly overlaps with the ionosphere which is important for the propagation of long-distance radio transmission. Aurora Borealis and Aurora Australis form here. Most satellites and the International Space Station orbit in the thermosphere.

There are also other definitions of the outer bound of the atmosphere. An important legal definition is the Kármán line at 100 kilometres above sea level, i.e. in the middle of the thermosphere, about the height where Aurorae are occurring. The Kármán line has been agreed upon by international bodies such as the United Nations. This legal definition is important to delineate international jurisdiction for aeroplanes from jurisdiction for spacecraft above.

Between some 50 and 85 kilometres above sea level, the mesosphere is the layer where the highest clouds can form that are visible to ordinary human beings. Most meteorites burn up here. The stratosphere between some 12 and 55 kilometres above sea level has hardly any meteorological phenomena; it is the place of the ozone layer; weather balloons can travel up here. The troposphere finally is the place of almost all weather phenomena, jet-powered airplanes can fly (only) here, and it is the part of the atmosphere reaching down to the planetary surface. Eighty percent of the 5 trillion megaton of mass of gas is found in the troposphere.

The troposphere is also the place of the most important regular or recurring phenomena in the atmosphere. At the top end of the troposphere, the polar jet stream is a meandering and rather permanent wind phenomenon. The jet stream can be very continuous over long distances and circle the entire planet. The polar jet streams can change quickly in altitude and latitude; they separate the polar wind circulation cells and the other wind circulation cells. The subtropical jet stream is weaker, at lower latitudes and higher altitudes. Another key atmospheric phenomenon is the monsoon, a seasonal changing pattern of wind, which can lead to very high amounts of precipitation; its best known patterns are the South Asian monsoon and the West African monsoon. A third key atmospheric phenomenon are tropical cyclones, strong rotating winds that generate over tropical seas; locally they are called typhoons or hurricanes.

The atmosphere is not a sphere, and all altitudes mentioned above are rough averages. The thickness of the atmosphere varies across the globe; its height varies by a factor of 2 between the poles and the equator. The extension of the atmosphere and its layers changes over time, for example seasonally. The "boundaries" between the layers are not sharply delineated either; the atmosphere, its composition, and the variations of temperature, density and pressure are characterized as much by continuity as by difference. A key difference between the different layers is the prevalence of different gases (compare below). This differing chemical composition can lead to impressive pictures taken from satellites that demonstrate that the layers of the atmosphere are not only a theoretical concept but describe visible approximate realities.

Of course, there is also ambiguity as regards the delineation with all three other "spheres". Dry air on planet Earth consists by some 78% nitrogen, some 21% oxygen, almost 1% argon, 0,04% carbon dioxide plus much smaller amounts of other gases such as Neon, Helium, Methane or Krypton. The different gases dominate different layers of the atmosphere: Nitrogen, Oxygen and Argon dominate within the troposphere, stratosphere and mesosphere. The thermosphere is dominated by atomic oxygen (individual O atoms, not O_2 molecules), the lower exosphere is dominated by helium, and the outer exosphere by atomic hydrogen. The four spheres also depend upon each other, as has been described in the article "hydrosphere" in this Handbook.

Air contains varying amounts of water vapour. At sea level, the global average is 1% – this means that air on planet earth is a mixture of dry air and water vapour. Such atmospheric vapour is generally not counted as part of the hydrosphere, even though it is part of the water cycle of course. Of course, air is also present in every organism (for example in our lungs), dissolved in water bodies and oceans, and also in rocks.

As is the case for all other "spheres", the atmosphere is a relational concept of human beings towards their planetary environment; it is a conceptually organizing principle, not a fully objective independent existence.

The Atmosphere Is a Well-Comprehensible Concept

The atmosphere is a rather graphic concept that is easy to understand. However, "the totality of gases retained by planet Earth" is a concept that is not human-centric and thus needs a conscious act of comprehension. From the perspective of the individual human being, there are several distinct experiences, which all essentially depend on the atmosphere.

Human beings breath, they need air to survive. Without air, human beings die within minutes, for example under water or if strangled. The experience of the necessity of air is as old as mankind. Even if air cannot be seen, it can be felt, for example when breathing out heavily.

Wind is another age-old human experience closely related to the atmosphere. A connection of wind and breathing, for example in Greek philosophy, is likely expressed in the concept of "pneuma" that both Aristoteles and the Stoa used.

The (blue) sky seems to be a quite different human experience. The colour of the sky is obviously a consequence of light diffraction on the gases in the atmosphere. This is everyday knowledge in the twenty-first century – however, the explanation is recent. Only some hundred years ago, Albert Einstein gave a convincing scientific explanation why small molecules such as nitrogen and oxygen could diffract sunlight to form a blue sky. Previous explanation efforts by Euler, Tyndall or Rayleigh effectively missed the key mechanism. The fact that the Earth, if seen from space, is the famous "Blue Marble" is actually a consequence of the reflection of the "blue sky" on the ocean surface.

The concept of the atmosphere connects some very fundamental human experiences with a global perspective. This concept is urgently needed in the Anthropocene.

The Significance and Fragility of the Atmosphere

Obviously, human beings have not really thought too much or even cared too much about the atmosphere, at least until very recently. This having said, there is documented proof that already some 2000 years ago there have been localized issues with air pollution. There are documented complaints about smoke, dirt and smell from glass melting furnaces in Roman antiquity that have led to the relocation of furnaces to the Roman outskirts. Similar instances of complaints about air pollution are documented across the late Middle Ages, from England, Germany and other places. Such complaints have led to the relocation of polluting enterprises or even to bans, such as that of burning sulphurous coal.

However, such localized complaints about air pollution have a vastly different significance from what has happened after the Second World War. It was only then that air pollution became a geographically widespread and widely perceived phenomenon. "The Great Smog" of London in 1952 has been registered far beyond the UK borders. Willy Brandt, who was running for the German chancellorship for the first time in 1961, was requesting that year that the "sky above the Ruhr metropolitan area has to turn blue again". Only one year later, one of the first environmental citizens' initiatives against air pollution was founded in the German city of Essen, and in late 1962, there was the "Smog crisis of the Ruhr area" which increased mortality by 20%.

The occurrence of such widespread Smog events lead to changes in national policies on industrial requirements for filtration, mitigation and containment, as well as to new policies for land use planning in some countries – not all, not even today. Other key events and trends during the "Cold War period" include the 1984 disaster in Bhopal, India, and the forest dieback in the U.S. and Central Europe in the late 1970s and early 1980s due to acid rain caused by air pollution of sulphur dioxide and nitrogen oxide. The introduction of catalytic converters for cars driven by combustion engines was a very effective technological means to address Smog and related air pollution, filtering out carbon monoxide and unburned hydrocarbons. Catalytic converters became standard in the U.S. in the mid 1970s and in Central Europe in the mid 1980s. As from the year 2021, the last remaining country, Algeria, has also banned the sale of leaded petrol.

This period of the 1950s until the 1980s, with the increasingly alarming experience of non-localized air pollution, marks the emergence of a collective awareness of the atmosphere as an "entity".

The continued strengthening of this awareness was massively supported by the detection of two major truly global atmospheric crisis phenomena: ozone depletion and global warming. Both phenomena were detected in the late 1970s and continue to worry us until today.

Global Warming

Global warming is a phenomenon not only of the atmosphere, but also of the hydrosphere and with massive impacts on the biosphere. Its anthropogenic origin, the greenhouse effect due to the emission of gases, mainly carbon dioxide and methane, does not need to be explained here, because it is nowadays standard knowledge in the population. In 2021, the atmosphere at sea level has already warmed by 1.2 °C on average in comparison to pre-industrial levels. However, the oceans have stored more than 90% of the additional energy in the climate system that is a result of the greenhouse gas emissions over the last 50 years.

The greenhouse effect has been understood for a very long time already. Key insights have been provided, amongst others, in 1824 by Jean Baptiste Fourier, in 1862 by John Tyndall, and in 1896 by Svante Arrhenius. Until the middle of the twentieth century, however, there have also been competing views in science and a rather unclear data situation. This changed in the 1950s, in the "International Geophysical Year" 1957/58. That "Science Year", organized by the International Council of Scientific Unions (ICSU) with support from UNESCO, led to massive improvement of Earth System science coordination, in particular across the political divides of the Cold War. The impact of that year with the involvement of thousands of scientists can hardly be overstated; amongst others, the year led to the Antarctic Treaty System, and also facilitated the first satellites in space. The year also had as a result the famous Keeling Curve, the diagram with the measurements of carbon dioxide concentration in the atmosphere on Hawaii ever since this year – named after Charles Keeling who started and coordinated the measurements until his 2005 death.

It was Roger Revelle, who issued the first "political" warning about global warming in 1965 to the U.S. president. However, over the subsequent ten years, average temperatures decreased (likely due to aerosols in the atmosphere) which removed global warming from the political agenda. Global warming returned to the agenda mainly through the impacts of the Sahel droughts in the 1970s and 1980s. A particularly important moment was the first global World Climate Conference, held by the UN special agency mandated for the atmosphere, the World Meteorological Organization (WMO), in February 1979.

This conference resulted in the establishment of the World Climate Research Programme (as a joint effort of WMO, ICSU and UNESCO) and of the Intergovernmental Panel on Climate Change (IPCC, n.d.) several years later, in 1988, as a joint WMO and UNEP effort. Until today, the IPCC is unrivalled in its effectiveness for scientific policy advice. In 1992, the UN Framework Convention on Climate Change was adopted, and in 1997, the Kyoto Protocol followed. Of course, the progress towards mitigating global warming has been enormously insufficient. That is why the 2015 Paris Agreement has been adopted. Efforts to contain global warming at 1.5 °C or at 2 °C are now highly overdue. However, this discussion is widely known in the public and does not need to be repeated here.

What is necessary to emphasize, however, is that the atmosphere (together with the ocean) is the key driver of the global climate system. Whatever happens in the biosphere of the Amazonian rainforest or boreal forests due to global warming is triggered by changes in the atmosphere. There are several atmospheric systems that can become "tipping points" of the Earth System (Schellnhuber, 2009), such as the polar jet stream or the monsoon systems in West Africa or South Asia. If these phenomena are changed too much, they will change permanently, irreversibly, for the worse.

Ozone Depletion

Ozone depletion is the most important example of anthropogenic Earth System change that gives hope to mankind (Morrisette, 1989). In the late 1970s it was discovered that globally ozone concentration in the ozone layer of the stratosphere had declined by some percentage points. Later it was found that in the polar regions, in springtime, there was a hole forming and not disappearing again – that is, more than half of the stratospheric ozone was destroyed in the regions of these "holes" (more than 90% of the atmospheric ozone is found in the stratosphere). The effect was an immediate steep rise in UV (UVB, more specifically) radiation at the surface of planet Earth leading to malignant melanoma and other carcinoma in humans and negative effects in other living organisms as well. The main culprit was quickly and correctly identified, the emission of chemicals such as chlorofluorocarbons (CFCs), which act inter alia as refrigerants, solvents, and foam-blowing agents.

Empirical and theoretical research started in the 1970s. Paul Crutzen, the main populariser of the term "Anthropocene" counts among the pioneers as do Frank Rowland and Mario Molina. Based on their work, in 1978, CFCs were banned for the first time for some uses by some countries. Negotiations for a global treaty were organized and finalized in March 1985, when the "Vienna Convention for the Protection of the Ozone Layer" was adopted and signed. Together with its operative protocol, the Montreal Protocol, adopted in 1987, it was in place "just in time". By coincidence, in early 1985, the discovery of the Antarctic ozone hole was made. The Montreal Protocol led to the necessary reductions in CFC emissions and atmospheric concentrations. The total amount of ozone is now stabilized, and the ozone layer will likely recover.

Fighting the ozone hole has shown that mankind can effectively act on the negative effects it has on the Earth System. As regards global warming, this time for action is long overdue.

The paragraphs in the first paragraphs, presenting knowledge from meteorology and atmospheric science, is actually standard textbook knowledge as presented in: Vallis, G.K..: *Essentials of Atmospheric and Oceanic Dynamics*. Cambridge 2019; or Wallace, J.M. and Hobbs P.V. *Atmospheric Science, an Introductory Survey*, Academic Press 2006; or Carlons, T., Knight, P. Wyckoff, C., *An Observer's Guide to Clouds and Weather*, Washington 2015. The German language-version Wikipedia page on the Air Pollution (accessed on 15 September 2021) has also been a very helpful source for this article: https://de.wikipedia.org/wiki/Luftverschmutzung as has been the English language-version Wikipedia page on the history of Climate Change science: https://en.wikipedia.org/wiki/History_of_climate_change_science. Other key inspiration was the vast literature on the history of the environmental movement, e.g. Radkau, J. *Die Ära der Ökologie*. C.H. Beck, 2011.

References

Andrews, S. P. (1871). *The primary synopsis of Universology and Alwato*. Dion Thomas.

Doherty, H. (1864). *Organic philosophy or Man's true place in nature volume I – Epicosmology*. Trubner & Co.

IPCC. (n.d.). *IPCC History Website*, https://www.ipcc.ch/about/history/. Accessed on 15 Sept 2021.

Morrisette, P. M. (1989). The evolution of policy responses to stratospheric ozone depletion. *Natural Resources Journal, 29*, 793–820.

Schellnhuber, H. J. (2009). Tipping elements in the earth system. *Proceedings of the National Academy of Sciences, 106*(49), 20561–20563.

Suess, E. (1875). *Die Entstehung der Alpen*. Wilhelm Braumüller Universitäts-Verlagsbuchhandlung.

Lutz Möller is Deputy Secretary-General of the German Commission for UNESCO since 2015; he also heads its Division for Education, Science and Culture. He holds a Ph.D. in Theoretical Physics from Munich University (LMU) and has a study background in Physics and Philosophy (Munich, Oxford). At the German Commission for UNESCO, where he has worked since 2004, he guides the policy advice work, is responsible for the overarching cooperation with UNESCO, and with other National Commissions for UNESCO globally, and with the private sector. He has been a member of the German National Committee for the UNESCO Programme "Man and the Biosphere" (MAB) more than 17 years, he has provided decisive input to key policy processes for the MAB Programme at global and national level for this entire period; he has also been a decisive figure in the creation and the implementation of the UNESCO Global Geoparks Programme, globally and nationally. He has also been the representative of the German Commission for UNESCO, for more than ten years respectively, in the relevant bodies for the Intergovernmental Hydrological Programme (IHP) of UNESCO, for UNESCO's Intergovernmental Oceanographic Commission (IOC) and the International Geoscience Programme (IGCP).

Biocapacity and Regeneration

Mathis Wackernagel and David Lin

Abstract Biological regeneration (or biocapacity for short) refers to the amount of biomass ecosystems can replenish within a given time period, typically a year. This regeneration is driven by the solar radiation through photosynthesis. The resulting biomass is the basis of virtually all life on our planet. Regeneration provided by ecosystems is one of humanity's most critical physical inputs, as the foundation of (nearly) all food chains on the planet, including that of the human economy. Biocapacity quantifies the regeneration rate of ecosystems. It does it in a way that enables comparisons of ecosystem productivity and human demand across time and space.

Biological regeneration refers to photosynthetic regeneration in ecosystems. It is the essence of what enables life. Through this process, solar energy powers the conversion of CO_2, water, and other nutrients into biomass. The output of this foundational process provided by our planet's ecosystems, together with water and air, is humanity's most critical physical input. It is far more critical than non-renewable resources such as ores or fossil fuels, which themselves are also limited by regeneration, as explained in the entry on →*ecological footprints*.

Photosynthetic regeneration is the foundation of all life on the planet. This is also true for the entire human economy. There is no single product or service within the economy that does not depend on regeneration, contrary to statements that claim that only half of World GDP depends on biological capital (WEF, 2020). Even mere beautiful thoughts and leisurely daydreams while sitting on a couch require biological resources to maintain the biological metabolism of the thinker.

To measure how much regeneration is required for human activities, we need metrics, including those documenting how much regeneration occurs on Earth. This is what biocapacity is: a quantification of the regeneration rate of ecosystems in a

M. Wackernagel (✉) · D. Lin
Global Footprint Network, Geneva, Switzerland
e-mail: mathis@footprintnetwork.org; david.lin@footprintnetwork.org;
https://www.footprintnetwork.org

© The Author(s), under exclusive license to Springer Nature
Switzerland AG 2023
N. Wallenhorst, C. Wulf (eds.), *Handbook of the Anthropocene*,
https://doi.org/10.1007/978-3-031-25910-4_2

way that can be applied across time and space. The biocapacity metric uses as its reference point ecosystems' inherent potential to regenerate biomass, including soils. Biocapacity depends on many factors such as soil condition, solar radiation, weather conditions, etc.

This potential can then be compared to human demand on biocapacity, also known as people's →*ecological footprint*.

Human activities depend on biological regeneration and are therefore in competition for the regeneration that ecosystems provide. Regeneration is therefore akin to a common "biological currency" that allows analysts to map all human activities against each other. In short, the dynamic is about competition for regeneration. Using this lens of biological regeneration, biocapacity accounting reveals the size of human economies in terms of their requirement for biological regeneration. These requirements include demand for biological products such as food, fibres, or timber; occupation of bioproductive surface areas by dwellings, roads, or other infrastructure; and demand for regeneration to absorb and neutralize waste flows and pollution. The latter encompasses cleaning polluted water, air, or sequestering excess carbon dioxide emissions from the combustion of fossil fuels and disturbed ecosystems.

The size of demand on regeneration can then be compared to the regenerative capacity of the entire planet or any of its regions, i.e., their respective biocapacity. If demand exceeds biocapacity, this indicates →*overshoot*. Global overshoot is one key characteristic of the Anthropocene. Biocapacity changes over time for several reasons. Within the course of a year, seasons have a large effect; during winters in the temperate or frigid zones of the planet, biocapacity is relatively low, and during the summer, biocapacity is relatively high. In subtropical and tropical zones, shifts of productivity during the year are determined by seasonal rain patterns. But there are also longer-term shifts in biocapacity. For instance, changing weather patterns, shifting land-uses, or modified management practices, including water regimes, affect the productivity of the ecosystem.

Comparing demand for biocapacity to availability of biocapacity is relevant for those studying the biological resource dependence of economies. Note that some portion of the planet's biocapacity is also needed to maintain →*biodiversity* (Wilson, 2016). If indeed the biological resources are a critical, limiting factor for the human economy, then mapping the ratio of the ecological footprint to the ecosystem's biocapacity over time becomes foundational for economic considerations, particularly in a time of global →*overshoot*.

To make biocapacity an applicable metric, it is measured in planetary surface area, which in turn is scaled by its relative biological productivity, ideally in terms of the potential →*net primary productivity* of that area. The measurement unit used is "global hectare", which refers to a biologically productive hectare with world-average productivity (if viewed as an area) or the productivity of a world average hectare (if viewed as a flow). In real-world applications, biocapacity estimates are approximations, because there are always data limitations. For instance, the methodology of the National Footprint and Biocapacity Accounts use yield factors and equivalence factors to estimate relative biocapacity of areas. Yield factors describe relative yields among the same area type, while equivalence factors compare productivity across area types. This then allows analysts to scale each hectare to the equivalent value in global hectares. This national accounting approach is described in detail by Wackernagel et al. (2021), Lin et al. (2018), and Borucke et al. (2013). The most common estimates for the biocapacity of countries are provided by those National Footprint and Biocapacity Accounts (Wackernagel and Beyers, 2019). The National Footprint and Biocapacity Accounts are currently produced by York University for the Footprint Data Foundation (York University et al. 2022). These accounts are based on UN statistics. Using those statistics was a deliberate choice to avoid any perception that data selection was arbitrary or biased. At the same time, we recognize that the UN data set may have its own limitations, but this downside is worth the upside of equanimity. UN data are particularly narrow on the biocapacity side, with overemphasis on agricultural yields, while covering little about depletion. As a result, biocapacity estimates are too high. Biocapacity estimates would be more helpful if they could distinguish, what portion of the biocapacity may be fragile due to depletion of key assets like soils and groundwater. But such estimates do not yet exist. Still, in spite of all these limitations, the National Footprint and Biocapacity Accounts show that demand has risen far more rapidly than even the optimistically measured biocapacity of the world.

References

Borucke, M., Moore, D., Cranston, G., Gracey, K., Iha, K., Larson, J., Lazarus, E., Morales, J., Wackernagel, M., & Gall, A. (2013). Accounting for demand and supply of the Biosphere's regenerative capacity: The National Footprint Accounts' underlying methodology and framework. *Ecological Indicators, 24*, 518–533. https://doi.org/10.1016/j.ecolind.2012.08.005

Lin, D., Hanscom, L., Murthy, A., Galli, A., Evans, M., Neill, E., Mancini, M. S., Martindill, J., Medouar, F.-Z., Huang, S., & Wackernagel, M. (2018). Ecological footprint accounting for countries: Updates and results of the National Footprint Accounts, 2012–2018. *Resources, 7*(3), 58. https://www.mdpi.com/2079-9276/7/3/58

Wackernagel, M., & Beyers, B. (2019). *Ecological footprint: Managing the biocapacity budget.* New Society Publishers. https://www.footprintnetwork.org/2019/09/04/18187/

Wackernagel, M., Hanscom, L., Jayasinghe, P., Lin, D., Murthy, A., Neill, E., & Raven, P. (2021). The importance of resource security for poverty eradication. *Nature Sustainability*. https://doi.org/10.1038/s41893-021-00708-4

Wilson, E. O. (2016). *Half-earth: Our planet's fight for life*. Liveright.

World Economic Forum (WEF). (2020). *Nature risk rising: Why the crisis engulfing nature matters for business and the economy* (New Nature Economy series), in collaboration with PwC. https://www3.weforum.org/docs/WEF_New_Nature_Economy_Report_2020.pdf

York University Ecological Footprint Initiative & Global Footprint Network. National Footprint and Biocapacity Accounts, 2022 edition. Produced for the Footprint Data Foundation and distributed by Global Footprint Network. Available online at: https://data.footprintnetwork.org

Dr. Mathis Wackernagel created the *footprint* concept in the early 1990s with Prof. William E. Rees. The carbon footprint has become the most popular variant. In 2003, he founded Global Footprint Network, a sustainability think-tank, making planetary constraints relevant to decision-making. Its largest engagement campaign is its annual Earth Overshoot Day. Mathis's honors include the 2018 World Sustainability Award, the 2015 IAIA Global Environment Award, and the 2012 Blue Planet Prize.

Dr. David Lin leads Global Footprint Network's research team, and contributes to the production, development, and improvement of the National Footprint and Biocapacity Accounts. Prior to joining Global Footprint Network, David earned his Ph.D. and worked as a post-doctoral researcher in the Systems Ecology Laboratory at the University of Texas at El Paso. His research focused on integrating models of ecosystem function with land cover change analysis in Arctic ecosystems. David is a native of California, and holds a BS in ecology, behavior, and evolution from the University of California, Los Angeles.

Biosphere

Lutz Möller

Abstract This article examines the genesis of the term "biosphere", the details of its content as a relational concept of human beings towards their planetary environment, as well as its current scientific and everyday use. It compares the term with similar concepts, in particular "biodiversity" and the advantages and disadvantages of the concepts. It describes when, why and how the term "biosphere" has emerged in the context of UNESCO (for example as regards UNESCO biosphere reserves). The article draws the conclusion that the concept is suitable and relevant in the Anthropocene.

Basics: The Term and Its Conceptual Connotations

The term "biosphere" was defined by the Austrian geologist Eduard Suess in 1875 (Suess, 1875). Some 200 years after the introduction of the term "atmosphere" (compare corresponding article in this Handbook), this was the second term introduced for describing a global functional differentiation of the planet. Today, the terms "hydrosphere" and "geosphere" (compare the two corresponding articles in this Handbook) are used in addition, together constituting the planet as an abstract geochemical model.

In fact, while Eduard Suess used the term "biosphere" to describe a collection of **places** (everywhere on Earth where life is possible), today, the term is typically used to refer to the totality or sum of the **number** and/or **mass** of all living organisms (animals, plants, fungi, and microorganisms).

The individual geographical places of Eduard Suess are called "habitats" or "biotopes" today, depending on the focus: Is the focus on the population of organisms (that is what "habitat" stands for) or is the focus on the **functional interplay** of the organisms, that they are an "biological functional community" (that is what

L. Möller (✉)
German Commission for UNESCO, Bonn, Germany
e-mail: moeller@unesco.de

"biotope" stands for). "Biocoenosis" is a (somewhat outdated) synonym for "biological functional community". If reference should be made to the abiotic factors of the environment of a biotope, then the term "ecosystem" is used.

The term "biosphere" has an additional conceptual ambiguity, which is analogous to the difference between "habitat" and "biotope": In some definitions, it does not only reference the number or the mass of all living organisms, but also the totality of their functional interactions. Both variations of the definition have disadvantages: Organisms and biomass without functional interaction have very little meaning – however functional interaction among organisms is meaningful locally, but not really at the global scale of the term "biosphere". This is why other scientific disciplines use other terms such as "biomass" to refer to the quantitative concept which geochemists call "biosphere".

Obviously, "biosphere" is a concept. It is actually a rather abstract concept. The "biosphere" is not an entity that exists independently of human beings. It is not entirely well defined, as has been described above (places, number, mass, functional relations, etc.). Even if it had a well-defined intension, the concept would not have a well-define intension: There are unclear limits between the biosphere and the geosphere as regards soils; there is an unclear definition of "one organism" – every human being consists of more bacteria then of human cells; habitats, biotopes and ecosystems cannot be well delineated from each other. This means that "biosphere" is a relational concept of human beings towards their planetary environment; it is a conceptually organizing principle, not an objective independent existence.

Even more, a purely quantitative understanding of the term "biosphere" is not too helpful for the urgent *political* discussions held and needed today. Since the 1980ies, the term "biological diversity" or "biodiversity" has thus become the dominant scientific and political term, enshrined for example in the famous 1992 UN Convention on Biological Diversity. The recent decades have demonstrated great progress in our understanding that "biosphere"/"biomass" and "biodiversity" have to go hand in glove for our collective survival.

Two examples: 10 gigatons of biomass of tropical rainforest have much more value for nature and for mankind than 10 gigatons of biomass of a monoculture tree plantation. However, it is equally significant and worrying that both the diversity and the biomass of insect populations in industrialized countries are declining heavily - in German protected areas, according to the famous Krefeld study (Hallmann et al., 2017), the insect biomass declined by 76 percent from 1990 until 2017.

The Term and UNESCO – Part 1

UNESCO focused on the "biosphere" since the 1960ies with UNESCO's landmark 1968 "Biosphere Conference", and ever since with its intergovernmental programme "Man and the Biosphere" (UNESCO, 1971) and the 727 UNESCO Biosphere Reserves (in October 2021) that it has designated. Biosphere Reserves are sites that foster sustainable development in real-live scenario with a focus on biodiversity conservation. They prioritize conservation through sustainable use of natural resources over classical "no touch conservation". Why has UNESCO focused on the term "biosphere"?

The term "biosphere" has advantages over the two similar but less well-defined terms "nature" and "the environment" – and there are no other terms useful for the global scale (better: there have not been other terms, since the term "biodiversity" appeared in the 1980ies). All three terms are always used in the singular. The latter two terms are usually utilized in many different ways; they can refer to an immediate local context or to the global context or to both at the same time. They can be used also refer to human beings and their character. Even beyond an unclear scope, the term "nature" has many different meanings; while many people use it to refer to collectives of living organisms, it actually encompasses much of the hydrosphere and geosphere as well. "Nature" can also refer to the cosmos beyond Earth. The term "environment" is similarly unclear; it can refer to the natural environment (roughly equivalent to "nature") or the biophysical environment. In the scientific discussions of today, "environment" is used often in an anthropocentric way, referring mostly biophysical variables such as water or air pollution and why they are detrimental to human beings.

This means that "biosphere" is much better defined in its reference to living organisms only. In all contexts, the term "biosphere" also has a clearly global connotation, different from "nature" and "environment". Thus, the term fits the usage in the context of a global intergovernmental organization such as UNESCO.

While the term "biodiversity" is dominating the scientific and political discourse today, the term "biosphere" remains important and useful. In particular, the UNESCO programme name "Man and the Biosphere" remains significant since it implies a relation and possibly an ethical responsibility of human beings to all living organisms globally. It will be interesting to observe the future discussion that UNESCO Member States have decided in 2021 to hold about the name of the programme.

The Emergence of the Term

It is not surprising that the term "biosphere" with its clear global connotation was starting to be intensively used in the 1960ies.

In the 1950ies and 1960ies, mankind started to grasp, for the first time, the inextricable interconnectedness of mankind and the biosphere – at the *global* scale, not only at a local scale. That insight was new.

In earlier centuries, human beings in all cultural spaces across the globe had developed diverse, manifold and often contradictory relationships towards the natural environment. To many, nature was holy and/or an image or proof of divine perfection. To others, nature was a threat. To even others, in particular in the European Enlightenment and early industrialization, nature was imperfect and needed human intervention such as breeding and grafting, river straightening and dike construction or later the Green Revolution. To others, such as during the European Romanticism, nature had an aesthetic value, a trend which lead to early nature conservancies. Others looked to nature mostly in terms of the resources such as timber, fish or wild game that humans could extract. In the nineteenth century, the complexity of nature was increasingly understood, which has led to our modern understanding of biotopes and ecosystems.

Almost all relationships of man to nature described in the last paragraph played out at the local level. To the extent that negative impairments of nature were understood at all, these were also negative impairments at the local level and they were addressed locally – for example the introduction of sanitary systems in town or limits to forest logging from the eighteenth century.

The 1950ies and 1960ies saw the emergence of a new, global understanding. In 1948, the first Director-General of UNESCO, Julian Huxley, was instrumental in creating the "International Union for the Protection of Nature" which is now IUCN, the most important normative nature conservation NGO. East Africa was the key region for both Huxley and the German zoologist Bernhard Grzimek in 1959 and 1960 to leverage a global understanding of nature. Grzimek's documentary "Serengeti Shall Not Die" won an Oscar in 1960, the same year that Huxley wrote pieces in "The Observer" after a visit to East Africa. Both described the massive hunt and thus decline of wildlife in East Africa.

Huxley's and Grzimek's engagement in East Africa was a similar trigger for the emergence of global responsibility in a very similar way to UNESCO's Nubia campaign for the cultural heritage of Abu Simbel launched in 1959. Huxley's articles triggered the creation of the World Wide Fund for Nature in 1961 – as the first truly globally oriented nature conservation NGO, reaching out to the public at large.

It is necessary to emphasize that it is clearly no coincidence that the emergence of this new global understanding and responsibility for the biosphere coincided with the end of colonialism. Tanganyika was a UN trust territory under British control since 1947 and became independent in 1961. There clearly was an obviously racist prejudice among many or most Europeans at the time that Africans might not be able to administer well their natural resources that many concerned Europeans loved (and loved to kill in big-game hunting).

The Term and UNESCO – Part 2

Therefore, it is obvious that and why "biosphere" was the right term for the 1960ies from a nature conservation perspective. However, why exactly did it become the title of a UNESCO programme from 1970? Is it true that the only answer is, as has been written above, that the two alternatives "nature" or "the environment" seemed to be less adequate?

Member States of UNESCO such as the Federal Republic of Germany that promoted a new UNESCO programme on human-nature-relationships since 1964 did not really promote a specific terminology. The German Commission for UNESCO hosted two international conferences on the "harmonisation of protection and use of nature" in 1966 on the island of Mainau in Lake Constance and another in April 1968 in Berchtesgaden. One long-standing member of the Commission, Dr. Magda Staudinger, was of particular importance. She focussed on man as a part of the biosphere with a "biological conscience" and a new partnership between man and nature. This approach was politically more important than the terminology of "biosphere".

Lacking full historical analysis, the best assumption for the momentum of the terminology "biosphere" is that in the late 1950ies and 1960ies, UNESCO was successfully conducting international geophysical and geochemical efforts such as the International Geophysical Year 1957/58, the creation of the Intergovernmental Oceanographic Commission in 1960 and the establishment of the International Hydrological Decade in 1965. The terminology "biosphere" fitted perfectly into that setting with much geochemical momentum.

The Significance of the Term "Biosphere" in the Anthropocene

As has been shown above, the term "biosphere" is conceptually very well suited for the age of the Anthropocene: defined in a globalized context, model-driven, and suited for quantitative analysis. At the same time, as 50 years of the experience of the UNESCO Programme "Man and the Biosphere" have shown, the terminology is well understood by the general public and well accepted by practitioners.

It is therefore not surprising that in March 2021, leading scientists have published the white paper "Our future in the Anthropocene biosphere" (Folke et al., 2021); or that the 2015 paper "The Anthropocene biosphere" (Williams et al., 2015) is a very well received paper.

Today, in the Anthropocene, the planetary biosphere, understood as the collection of all living organisms, is under enormous pressure. It has to be noted that in industrialized countries, not only species diversity is dwindling sharply, but also the biomass and head-count of populations of every-day species are dwindling. Mankind is impacting the biosphere in such dramatic dimensions that conservationists need to stop being too picky: yes, the snipe has been disappearing from Central Europe for decades, with a few thousands breeding couples left, which is a tragedy – but do we need to start worrying about pigeons, sparrows and starlings now? As a response to climate change, tree-planting efforts have mushroomed in recent years. But what are they worth? Do they maybe make things worse for the biosphere? What about bioethanol and biodiesel? In most cases, biofuels seem to have a pretty obvious negative sustainability balance sheet.

During the last 50 years, since the broad public uptake of the term "biosphere", its state has deteriorated massively in terms of quality and quantity. It is time to focus on the biosphere much more. The "biosphere" today is a space of suffering.

Therefore, we need creativity and new momentum. We need to start using the mechanisms that we have available in earnest, instead of constantly reinventing the wheel. For example, most experts and many governments agree that we need much more conservation areas globally. The target is to protect at least 30% of the planet by 2030. This discussion is currently held around the "Post-2020 Biodiversity Framework".

However, protecting 30% of the planet may not lead to the other 70% of the planet being even more overused and exploited. Luckily, UNESCO with its UNESCO Biosphere Reserves has exactly the right instrument to accommodate conservation and use of the biological resources of the biosphere. The instrument has been massively improved in terms of quality in recent years. It is time to use this instrument intensively.

It is likely that there are millions or billions of places in the universe which harbour life, maybe even life similar to the one on planet Earth. However, until today, the biosphere of planet Earth is the only place in the universe of which we **know** that it has life, that it is life. It is very likely the only place in the universe with life that can ever be reached by humans. Even more, the life of the Earth's biosphere is rich, astonishing and beautiful; it is life that allows human beings to flourish, that allows dignity and happiness. The most important goal for humanity therefore must always be to maintain and safeguard the biosphere in its wealth and functionality. In the Anthropocene, at least so far, humanity does not score well on this goal. It is maximally urgent to change course.

References

Folke, C., Polasky, S., Rockström, J., et al. (2021). Our future in the Anthropocene biosphere. *Ambio, 50*, 834–869. https://doi.org/10.1007/s13280-021-01544-8

Hallmann, C., Sorg, M., Jongejans, E., Henk, S., Nick, H., Schwan, H., et al. (2017). More than 75 percent decline over 27 years in total flying insect biomass in protected areas. *PLoS One, 12*(10), e0185809. https://doi.org/10.1371/journal.pone.0185809

Suess, E. (1875). *Die Entstehung der Alpen*. Wilhelm Braumüller Universitäts-Verlagsbuchhandlung.

UNESCO. (1971). *International co-ordinating Council of the programme on man and the biosphere* (MAB), first session, UNESCO-MAB report series, no. 1, Paris, France, 61pp.

Williams, M., Zalasiewicz, J., Haff, P. K., Schwägerl, C., Barnosky Anthony, D., & Ellis, E. C. (2015). The Anthropocene biosphere. *The Anthropocene Review, 2*(3), 196–219. https://doi.org/10.1177/2053019615591020

Lutz Möller is Deputy Secretary-General of the German Commission for UNESCO since 2015; he also heads its Division for Education, Science and Culture. He holds a Ph.D. in Theoretical Physics from Munich University (LMU) and has a study background in Physics and Philosophy (Munich, Oxford). At the German Commission for UNESCO, where he has worked since 2004, he guides the policy advice work, is responsible for the overarching cooperation with UNESCO, and with other National Commissions for UNESCO globally, and with the private sector. He has been a member of the German National Committee for the UNESCO Programme "Man and the Biosphere" (MAB) more than 17 years, he has provided decisive input to key policy processes for the MAB Programme at global and national level for this entire period; he has also been a decisive figure in the creation and the implementation of the UNESCO Global Geoparks Programme, globally and nationally. He has also been the representative of the German Commission for UNESCO, for more than 10 years respectively, in the relevant bodies for the Intergovernmental Hydrological Programme (IHP) of UNESCO, for UNESCO's Intergovernmental Oceanographic Commission (IOC) and the International Geoscience Programme (IGCP).

Cosmos

Mark Williams, Tom Stallard, and Jan Zalasiewicz

Abstract Notions of the cosmos are deep-rooted in human consciousness. They are expressed in our earliest monumental constructions as explanations of the world around us and in the practices of many indigenous peoples. Here we examine the physical cosmos from the perspective of life on Earth. We note a central importance for the Earth in the vastness of space, as a planetary oasis for a highly complex and long-lived biosphere, one now being fundamentally altered by humans. We refer to other notions of the cosmos that may enable us to navigate a better path for life in the Anthropocene.

The physical cosmos probably contains many planets lying in the Goldilocks zone around their parent stars, where liquid water can exist at their surface (Kasting et al., 1993). Even beyond this zone many worlds may possess a minimum set of parameters like liquid water (often beneath a carapace of ice), access to the building-blocks of carbon, hydrogen, oxygen, nitrogen, phosphorus and sulphur, a source of energy, transition metals to facilitate biochemical pathways, and the requisite temperature, pH and radiation levels that can support life (Cockell et al., 2021). Thus, life could occur beyond the Earth on other bodies even within our Solar System. But even from the perspective of our own planetary neighbourhood the Earth appears to be something special, nurturing life over billions of years and enabling the development of a complex biosphere that profoundly alters the surface environment and that would eventually evolve a consciousness, one able to contemplate the nature of existence. That consciousness used to place the Earth at the centre of the physical cosmos; then scientific enquiry displaced us to the margins of one galaxy among many; and then further enquiry showed that our planet may not be commonplace.

M. Williams (✉) · J. Zalasiewicz
University of Leicester, Leicester, UK
e-mail: mri@le.ac.uk; Jaz1@le.ac.uk

T. Stallard
Northumbria University, Newcastle upon Tyne, UK
e-mail: tom.stallard@northumbria.ac.uk

© The Author(s), under exclusive license to Springer Nature
Switzerland AG 2023
N. Wallenhorst, C. Wulf (eds.), *Handbook of the Anthropocene*,
https://doi.org/10.1007/978-3-031-25910-4_4

Looked at most simply, the Earth looks quite small and insignificant when compared to the cosmos. The total numbers of galaxies may be somewhere in the range of 200 billion to 2 trillion (Conselice et al., 2016) and each galaxy may contain tens to hundreds of billions of stars. Within our own Milky Way there may be billions of Sun-like stars that burn for several billion years, and some of these may provide stable settings to nurture life on attendant exoplanets in the habitable zone. A recent estimate suggests that many millions of stars will have such planets (Patel, 2020; Bryson et al., 2021), whilst an earlier study suggested a half million planets in our Milky Way could have global biospheres (Franck et al., 2002). Amongst such cosmic numbers, how special is the Earth in the cosmos?

Earth-like planets orbiting Sun-like stars are predicted by science but, so far, they are very difficult to detect, let alone characterise. So, there is little observable astronomical data to address their cosmic abundance or viability as a home for life, and most exoplanets discovered so far could not support Earth-like life. By contrast, we have a vast resource of geological data that demonstrates just how special the Earth is in terms of nurturing and sustaining a biosphere. Life on Earth is very ancient (see Bell et al., 2015; Brasier et al., 2015), and has persisted, despite some periodic setbacks, for over three billion years. Co-evolution between the Earth's atmosphere, lithosphere, hydrosphere, and biosphere – some would call this Gaia – has maintained a habitable zone at the surface and near-surface (Lovelock, 1972; see also Vernadsky, 1998). Other planets in our Solar System may also have originated biospheres early in their development. Venus and Mars may once have been water worlds, but both have long since lost any liquid water on their surfaces. Venus is closer to the Sun and vaporised then lost its water to develop a super-greenhouse climate early on. Mars is much less massive than the Earth, its interior heat cooled quickly, and so it became tectonically dead, or almost so. Both planets have no magnetic field, and so volatiles like water within their atmospheres are blown away by the solar wind (Lundin et al., 2007). Mars's lack of a significant atmosphere prevents it from sustaining a dynamic rock cycle to provide a constant stream of building blocks for a complex biosphere.

Whether Earth-like planets orbiting Sun-like stars in their Goldilocks zone are rare or not, it is unlikely that any other world has evolved quite like ours. The Earth's moon is likely a statistical outlier, but the presence of lunar tides will have provided a uniquely cyclical environmental driver that enhanced the probability of life (Lathe, 2004). Many planets *may* have biospheres, but the temporal pattern of unpredictable forces like massive eruptions of lava, snowball glaciations and asteroid impacts that reset the track of Earth evolution many times, would not be duplicated elsewhere. Indeed, many if not most biospheres might remain in a microbial state, as the Earth's was for billions of years. An intelligence sufficiently self-aware and manipulative to develop a technosphere (Haff, 2014) and thus begin to control its planet's fluxes of matter and energy may be very rare because it took billions of years to develop, as just a single instance, on Earth. Even optimistic estimates suggest only a tiny number of worlds in our galaxy will have sentient intelligence (Westby & Conselice, 2020), and these would lie so far apart that contact between them is unlikely.

If modern science has shown that our world does not lie at the centre of the physical cosmos, this does not diminish the Earth's significance as a cosmic oasis for life in the vast desert of space. None of our neighbouring planets can be made habitable for a complex biosphere like ours: Venus is a hades-like world with a crushing atmosphere while Mars is a cold desert. In any case such an approach poses ethical questions about colonisation. The next world that might be something like ours is possibly thousands of light years away (Westby & Conselice, 2020). Visions of a cosmos where humans have spread to many worlds remain in the realms of science fiction, and if it ever were to happen, how would we secure the integrity of the extra-terrestrial biospheres we encountered, given what we have done to the Earth?

Here on Earth, we have a biosphere that extends through the air, on the land, in the oceans, and beneath the land surface. Its incredible complexity is barely understood. It throws up astonishing surprises, like the degree of interconnectedness of woodland ecosystems with their vital collaborations between plants and fungi. Science has catalogued only a small proportion of the biosphere's millions of species (Mora et al., 2011; Louca et al., 2019). This infinitely complex and self-perpetuating system seems like nothing less than a miraculous component of the cosmos, although it has demonstrably arisen by Darwinian evolution from much simpler beginnings.

How then are we to view our place in the cosmos from the perspective of the physical Anthropocene, one manifesting in immense and damaging impacts to our home planet (Zalasiewicz et al., 2019). Might other, 'non-scientific' interpretations of the cosmos be used to help us navigate a better relationship with the Earth, one that is not based on domination and short-term exploitation? Many human cultures have developed views of the cosmos where the relationships between the human and non-human are much more intertwined, much less predicated on control, and where animals, plants and spirits have agency. Such notions of deep interdependence could help us to develop mutually beneficial relationships with the biosphere (Fernández-Llamazares & Virtanen, 2020), as opposed to viewing it as a resource to be exploited (Sanga, 2020).

The scientist Suzanne Simard and others have shown how the interconnectedness of trees and fungi in forest ecosystems works for the mutual benefit of all. Where healthy trees share their resources through an extended web of fungal mycelia with those trees that are stressed, and where trees help to stabilise the ecosystem and nurture the next generation of plants (Beiler et al., 2010; Gorzelak et al., 2015). Simard and co-workers have used their work to suggest better ways of managing forests. Such interconnected systems of mutual benefit are mirrored in the relationships between many indigenous human cultures and their environments and can lead to demonstrable advantages for land management and biodiversity (Sanga, 2020). Elsewhere in this volume we have argued that city ecosystems should be modelled on mutualism.

What should be our vision of the Earth's place in the cosmos? The Earth is clearly not a small, commonplace terrestrial planet in a run-of-the-mill galaxy, in one corner of the Universe. Our vision, thus, should represent the central importance of the Earth as a rare world within the cosmos, where a complex and long-lived biosphere flourishes. It should be a worldview borrowing from cultures that see an interconnectedness between all things living and non-living, a means to approach the use of the Earth's resources that might enable our longer-term survival. Let us hope that the journey of human consciousness, to rediscover that global worldview, does not run out of time.

References

Beiler, K. J., Durall, D. M., Simard, S. W., Maxwell, S. A., & Kretzer, A. M. (2010). Architecture of the wood-wide web: Rhizopogon spp. genets link multiple Douglas-fir cohorts. *New Phytologist, 185*, 543–553.

Bell, E. A., Boehnke, P., Harrison, T. M., & Mao, W. L. (2015). Potentially biogenic carbon preserved in a 4.1 billion-year-old zircon. *PNAS, 112*, 14518–14521.

Brasier, M. D., Antcliffe, J., Saunders, M., & Wacey, D. (2015). Changing the picture of Earth's earliest fossils (3.5–1.9 Ga) with new approaches and new discoveries. *PNAS, 112*, 4859–4864.

Bryson, S., Kunimoto, M., Kopparapu, R. K., Coughlin, J. L., Borucki, W. J., Koch, D., et al. (2021). The occurrence of rocky havitable-zone planets around Solar-like stars from Kepler data. *The Astronomical Journal, 161*, 36. https://doi.org/10.3847/1538-3881/abc418

Cockell, C. S., Wordsworth, R., Whiteford, N., & Higgins, P. M. (2021). Minimum units of habitability and their abundance in the universe. *Astrobiology, 21*, 481–489. https://doi.org/10.1089/ast.2020.2350

Conselice, C. J., Wilkinson, A., Duncan, K., & Mortlock, A. (2016). The evolution of galaxy number density at z < 8 and its implications. *The Astrophysical Journal, 830*, 83.

Fernández-Llamazares, A., & Virtanen, P. K. (2020). Game masters and Amazonian indigenous views on sustainability. *Current Opinion in Environmental Sustainability, 43*, 21–27.

Franck, S., von Bloh, W., Bounama, C., Steffen, M., Schönberner, D., & Schellnhuber, H.-J. (2002). The evolving sun and its influence on planetary environments. In B. Montesinos, A. Gimenez, & E. F. Guinan (Eds.), *ASP conference proceedings* (Vol. 269, pp. 261–271). Astronomical Society of the Pacific. ISBN: 1-58381-109-5. https://articles.adsabs.harvard.edu//full/2002AS PC..269..261F/0000261.000.html

Gorzelak, M. A., Asay, A. K., Pickles, B. J., & Simard, S. W. (2015). Inter-plant communication through mycorrhizal networks mediates complex adaptive behaviour in plant communities. *AoB PLANTS, 7*, plv050. https://doi.org/10.1093/aobpla/plv050

Haff, P. K. (2014). Technology as a geological phenomenon: Implications for human wellbeing. In C. N. Waters, J. Zalasiewicz, & M. Williams (Eds.), *A stratigraphical basis for the Anthropocene* (Special Publication, 395) (pp. 301–309). The Geological Society.

Kasting, J. F., Whitmire, D. P., & Reynolds, R. T. (1993). Habitable zones around Main sequence stars. *Icarus, 101*, 108–128.

Lathe, R. (2004). Fast tidal cycling and the origin of life. *Icarus, 168*, 18–22.

Louca, S., Mazel, F., Doebeli, M., & Parfrey, L. W. (2019). A census-based estimate of Earth's bacterial and archael diversity. *PLoS Biology, 17*(2), e3000106.

Lovelock, J. (1972). Gaia as seen through the atmosphere. *Atmospheric Environment, 6*, 579–580.

Lundin, R., Lammer, H., & Ribas, I. (2007). Planetary magnetic fields and solar forcing: Implications for atmospheric evolution. *Space Science Reviews, 129*, 245–278.

Mora, C., Tittensor, D. P., Adl, S., Simpson, A. G. B., & Worm, B. (2011). How many species are there on earth and in the ocean? *PLoS Biology, 9*(8), e1001127.

Patel, N. V. (2020). *Half the Milky Way's sun-like stars could be home to earth-like planets*. MIT Technology Review. https://www.technologyreview.com/2020/11/06/1011784/half-milky-way-sun-like-stars-home-earth-like-planets-kepler-gaia-habitable-life/. Accessed Jan 2022.

Sanga, K. K. (2020). Global importance of indigenous and local communities' managed lands: Building a case for stewardship schemes. *Sustainability, 12*(19), 7839.

Vernadsky, V. I. (1998). *The biosphere (complete annotated edition: Forward by Lynn Margulis and colleagues and introduction by Jacques Grinevald)*. Copernicus (Springer-Verlag), 192 pp.

Westby, T., & Conselice, C. J. (2020). The Astrobiological Copernican weak and strong limits for intelligent life. *The Astrophysical Journal, 896*, 58.

Zalasiewicz, J., Waters, C., Williams, M., & Summerhayes, C. (2019). *The Anthropocene as a geological time unit*. Cambridge University Press.

Mark Williams is a palaeobiologist at the University of Leicester. He studies the evolution of life on Earth over hundreds of millions of years and has been a long-time member (and former Secretary) of the Anthropocene Working Group. He has co-authored popular science books with Jan Zalasiewicz on climate change, ocean evolution, and the story of life on Earth, and with Jan and Julia Adeney Thomas he co-authored *The Anthropocene: A Multidisciplinary Approach*. His new book with Jan is called *The cosmic oasis: the remarkable story of Earth's biosphere* (Oxford University Press).

Tom Stallard is a planetary astronomer at Northumbria University. He studies the aurorae and ionospheres of Giant planets using infrared telescopes on both Earth and spacecraft, for which he was awarded the Royal Astronomical Society Chapman medal in 2019. He teaches on a wide range of physics topics, including the formation and evolution of planets, and astrobiology within the galaxy. He is the co-editor of *Saturn in the twenty-first Century*, and a co-author on *Principles Of Heliophysics: a textbook on the universal processes behind planetary habitability*.

Jan Zalasiewicz is a geologist (now retired) at the University of Leicester, and a member (formerly Chair) of the Anthropocene Working Group. He is co-editor of *The Anthropocene as a geological time unit: a guide to the scientific evidence and current debate*, has been involved in other AWG publications, and with Julia Adeney Thomas and Mark Williams has co-written *The Anthropocene: A Multidisciplinary Approach*. His books for a general audience that *inter alia* explore Anthropocene concepts include *The Earth after Us* and (with Mark Williams) *The Goldilocks Planet* and *Ocean Worlds*.

Deep Ecology

Konrad Ott

Abstract The article presents the philosophical essence of the Deep Ecology movement, as expounded by Arne Naess and his followers. The article begins with the general idea of deep ecology. Following the structure of the "apron" model, it moves from ecosophies, to platform principles, to politics. It sums up with an outline of the development of deep ecology after Naess.

The concept of Deep Ecology was coined by the Norwegian philosopher Arne Naess (1912–2009). Naess (1973) opposed "shallow environmentalism" with "deep ecology". While the former type of environmentalism espouses technological reforms and slight adjustments of institutions to cope with the ecological crisis, deep ecology wishes to address the deeper roots of unsustainability in order to cultivate more profound remedies.

Search for deeper roots requires "deep questioning" of the ontological, epistemological, ethical, and political foundations of the project of Western modernity. This project has been based on rationality, liberal market economies, human rights, representative democracy, science driven technology, and the "objectification" of nature. Values are seen as individual preferences, lifestyles are consumptive, cultures are permissive, and morals are human-centered. Nature provides resources (input for production), sceneries for tourism, and sinks for waste. The sharp divide between (human) subjects and (natural) objects has been the predicament of modernity, evident in the philosophical dualism of Descartes (res cogitans versus res extensa), Bacon (science-technology-industry), Kant and Fichte ("I" as supreme category of knowing), and even Marx (human labour). In such a paradigm, subjects will only take the attitude of mastery, exploitation, and control against objects. Some "wise use regulations" based on prudence are the only reasonable restrictions with respect to nature. To Naess, such mindsets misconceive and distort the relations between humans and the more-than-human world. Thus, deep ecology is a quest for post-modern outlooks on nature which might be inspired by pre-modern ones.

K. Ott (✉)
Christian Albrechts University Kiel, Kiel, Germany
e-mail: ott@philsem.uni-kiel.de

In order to understand Naess's conception, his intellectual history is important. In the 1920s, Arne Naess had been trained in logical empiricism of the Vienna Circle. The Vienna Circle conceived logical positivism thusly: concepts should be operationalized, true propositions must refer to empirical facts, while hypotheses must be verifiable. The Vienna Circle was a highlight of epistemology based on hard science. In his academic career, Naess worked on semantics, logic, and epistemology. As an intellectual legacy of the Vienna Circle, Naess adopted scepticism against ethics. To Naess, moral propositions are imperatives being imposed upon people by authorities. To overcome the environmental crisis, ontological shifts look far more attractive than imperatives.

To Naess, deep ecology was rather a movement than a doctrine. Such a movement, however, stands in need of common ground. Together with the American philosopher George Session, Naess conceived the so-called "apron" model of ecosophy. The model distinguishes between four layers. At the supreme or most cosmological layer, different "ecosophies" exist. Any ecosophy is a specific outlook on nature from which emotions, perceptions, attitudes, and commitments might be derived. There is a multitude of ecosophies. At the layer of "ecosophies", one might follow Spinoza, Whitehead, Heidegger, Buddhism, Daoism, Vedic wisdom, indigenous cosmologies, or other comprehensive "green" doctrines. Paul Taylor's biocentric outlook on nature (Taylor, 1986), Hans Jonas' ontological ethics of nature's self-affirmation (Jonas, 1973), and Holmes Rolston's ontology of a "pro-jective" nature (Rolston, 1988), might count as ecosophies. In ecosophies, concepts, attitudes, perceptions, and commitments fuse. There is no sharp divide between facts and values, since the fact-value-divide only holds for modern concepts of nature.

At the layer of ecosophies, Naess conceived his own as "Ecosophy T". "T" represents the name of his hut in the Norwegian mountains: "Tvergastein". Ecosophy T is inspired by Vedic wisdom (the Bhagavad-Gita) and by Spinoza (1967). Naess distinguishes a "narrow ego" from a "widening self" (Naess, 1986a). The former is egotistic and comes close to "homo oeconomicus" wishing to maximize personal utility. To Naess, escaping the narrow cage of egotism requires identification with other beings, be they human or non-humans. The concept of identification denotes an intrinsic relation between beings. Identification is based on experience. Naess (1986a, p. 227) reports a deontic experience with a dying flea as paradigm for identification. Under his microscope, Naess once observed a flea which drowned in a toxic fluid and struggled for survival in vain. Naess suddenly identified with the flea: "I saw myself in the flea". Identification means to see oneself *as* the other. Such identification (in Vedic wisdom: "tat twam asi") is prior to sentiments as compassion. Among humans, identification is the source of solidarity. Identification can and should be widened beyond the human sphere. "Ecosophy T" does not require that natural beings identify with humans. Widening identification is epistemically anthropocentric, but it overcomes anthropocentrism in content.

All beings, one can identify with, become part of the widening Self. Naess divides a narrow and egotistic "ego" from a "Self" which identifies with other beings, either humans or non-human beings. The more the Self widens, the stronger the longing for more widening becomes, in a positive feedback loop. This process is an ongoing and life-long dance of and with "Self-realization". Widening identification has no ontological stop gap. It can encompass animals, plants, and even rivers, forests, mountains, and the ocean. "This large comprehensive Self (with a capital 'S') embraces all the life forms on the planet" (Naess, 1986b, p. 80).

Ecosophy T gave rise to three kinds of criticism. First, widening identification might not respect the otherness of non-human beings but might assimilate them anthropomorphically. Second, the capability of identification might be specific to some spiritual elites and disrespectful against democratic deliberation. Third, identification requires some sort of internal perspective at the side of the beings one identifies with. If the latter obtained, identification might be possible with sentient animals, but it would become dubious in cases of plants and bacteria, and become fanciful and pointless in cases of ecosystems, natural wholes, genes, and non-living objects. Taking the perspective of others presupposes that there "is" some internal perspective. If this is the case, identification cannot be widened without end.

To Naess, widening identification has spiritual meaning. Naess likes to quote Spinoza in this respect: "*Quo magis res singulares intelligimus, eo magis Deum intelligimus*": The more one recognizes individual entities, the more one recognizes god. The fulfilment of widening identification is spiritual bliss from which personal virtues emerge. Imperatives become superfluous as one wishes to care for all beings with which one has identified. Identification nullifies moral distance. The difference between egoism and altruism becomes pointless. Harming others means to harm oneself. In ethical terms, Naess comes close to an ecosophical virtue ethics. Self-Realization terminates in "beautiful actions" (Naess, 1993). According to Kant, beautiful actions are morally obligatory actions being performed out of pleasure and joy.

From a logical point of view, ecosophies are axiomatic stipulations. Naess defines "Self-realization" as supreme axiological axiom of a normative system (Naess, 1989, pp. 84–86). Justification of supreme axioms isn't possible, since one can't derive them, but they can be affirmed by a one-word-sentence as "Good!" Naess offers the analogy of God's very first word "Light!" From this supreme axiom, Naess wishes to derive a comprehensive scheme of norms (rules, commitments) and empirical hypotheses which boil down to political topics. Such a scheme combines an ecosophical top or abstract layer of positing with logical techniques of derivation for practical direction and application.

To Naess, anthropocentrism does not qualify for an ecosophy. From a logical point of view, however, the distinctions between a) "shallow versus deep" and b) "anthropocentrism versus bio/eco/physiocentrism" are independent. If so, four positions in environmental philosophy emerge:

1. Deep physiocentrism
2. Deep anthropocentrism
3. Shallow physiocentrism
4. Shallow anthropocentrism

Naess seems to assume that physiocentrism is generally deeper than anthropocentrism. If high standards of philosophical reasoning are supposed, position 3 remains a possibility since physiocentrism is not as such always philosophically deep. There might be also a deep anthropocentric environmental ethics (Ott, 2016, 2020) that recognizes eudaimonic values, virtues, spiritual attitudes, and is open-minded to the demarcation problem. A similar question asks whether environmental pragmatism might qualify for an ecosophy.

The second layer of the apron-model provides some normative unity. It consists of a list of eight prescriptive statements. The first statement is about the demarcation problem of inherent moral value in nature (see Ott, 2008 for analysis). It commits all deep ecologists to (egalitarian) biocentrism and/or ecocentrism. From an "Ecosophy T" perspective, the inferential rationale is straightforward: If I attribute inherent value to myself, and if I identify myself with X, I must attribute inherent value to X. Such derivation might, however, be less straightforward in other ecosophies. Heideggerians, for example, can't derive specific moral values from the revelation of "Seyn" (Heidegger's artificial word for ontological disclosure).

The second statement on the list is about the intrinsic values of richness and diversity of life. Here, deep ecology touches an ethics of biodiversity. It remains, however, questionable whether diversity as such without any further qualification can be a locus of inherent value. If biodiversity is defined as "the variability among living organisms from all sources" (Convention of Biological Diversity), intrinsic value must be *ex definition* attributed to the disposition called variability. Thus, the second statement opens a vast array of problems which an ethics of biodiversity has to address with respect to genes, populations, species, and ecosystems.

The third statement restricts human entitlements to impair natural richness to the fulfilment of basic needs. This is a rigid version of sufficientarianism. It requires defining what counts as "basic" to a human life. If a human life is essentially a cultural (and political) life, there must be recognition of cultural needs beyond eating, drinking, shelter, and sexual activity. The third statement opens an ongoing debate about what are the bounds of deep ecological desirable lifestyles.

The fourth statement demands a substantial reduction of human population. Such demand was popular in the late 1960s. This commitment to reduce the number of humans requires restriction in proliferation patterns. Such reduction should, as Naess insists, be realized without coercion in the longer run. Persons who live their lives within the horizon of ecosophies, may give less weight to reproduction. In a provocative statement, Naess called for a world population of 100 million people (Naess, 2005, Vol. 10, p. 270) which implies a reduction of 99% in relation to 10 billion people. I take this statement as an open topic since population indeed matters. To Naess, it would be misleading to ignore population growth worldwide and put the blame solely to Western consumption patterns.

The fifth statement sees human impacts in nature as "inappropriate" and states that the situation is constantly worsening. The second part of the statement implicitly denies that environmental reformism can succeed in actual ecological benefit. Without empirical control ("Is the situation really worsening?") the fifth statement directly moves into apocalyptic reasoning. Judgment of our ecological situation should be seen as place-holder for empirical analysis about how environmental states change over time. Such interpretation is in line with Naess's general anti-dogmatism.

The sixth statement broadly demands profound change in political and economic institutions. It points at the third layer of the apron model. The seventh statement distinguishes between standard of living and quality of life. Standard of life as measured by GDP is an insufficient indicator of life quality. Here Naess comes close to proposals for (physical and economic) degrowth.

The final statement commits all persons who agrees to the first seven statements to engage accordingly. It bridges the gap to the next layers. The previous comments on the principles have a common motive: The platform-principles are not firm principles but they rather constitute topics of debate in environmental philosophy.

The third layer of the apron-model is about policy making and political engagement. Here, the apron gives leeway for engagement under different political constellations. In his political writings, Naess argues that there should be, first, a green movement and even a party, while, in the longer run, all political camps (conservatism, liberalism, socialism) should turn "green". Here, Naess opened the field of environmental political philosophy. The fourth layer is about individual choices, actions, and lifestyles. Here, a cultural shift to more "naturalism" serves as the guideline, but Naess refrains from giving moral precepts how exactly one should live. The fourth layer gives room for cultural innovation, new imaginaries, and new ways of living in communities.

Naess dubs the logical relation between those four apron-layers a "loose derivation" if it runs top-down from ecosophy (cosmology) to (practical) lifestyle. The term "loose" indicates that there is no logical stringency but instead plausible inferential connections. If the relation is "bottom-up", Naess refers to this sequence of thought as "deep questioning". One may ponder which beliefs make oneself an environmentalist, protester, campaigner, vegetarian etc. Starting such questioning will end up in the formulation of one's personal ecosophy.

A simplified version of the platform is given by Sessions (1990): (1) Rejection of the man-in-environmental image in favor of the relational total-field image; (2) Biospherical (or ecological) egalitarianism; (3) Principles of diversity and symbiosis; (4) Anticlass posture; (5) Fight against pollution and resource depletion; (6) Complexity, not complication; and (7) Local autonomy and decentralization. The first principle requires clarification. As Naess writes (1989, p. 55): "The term 'relational field' refers to the totality of our interrelated experience". Material things and single organisms are viewed as junctions within ecological relations. This concept is connected to "Ecosophy T": "Identification is a process in which the relations which define the junction expand to comprise more and more" (ibid.) Identification, then, dissolves solid ontological entities into fluid relations. Such relationalism goes beyond widening identification. Individual organisms are seen like waves within a vast sea lasting for a glimpse of time before they are replaced by other waves. Spinoza offers a similar image; all singular entities are manifestations of one single substance which is "Deus sive Natura". Naess's philosophy has deep roots in Spinoza (Ott, 2006).

More concretely, Naess presents a blueprint of how to share planet earth. One-third of the terrestrial sphere should remain in purewilderness – where human beings do not go – or should be restored to wilderness. Another third should be devoted to "free" nature where human interference remains far from dominant but some activities as hiking or skiing remain possible. The final fraction would be utilized for human purposes in a sustainable manner via "bio-culture" (his version of agroecology or permaculture). Such a proposal requires substantial reductions in the human population. The pitch, however, could be realized stepwise by an increasing system of natural reserves worldwide. Reserving roughly 20–30% of the terrestrial space for conservation might be realistic if conservation includes not just wilderness areas, but also restoration projects, organic agriculture, ecological forestry, extensive grazing systems, coastal zones, and even urban greeneries. A system of terrestrial conservation areas should be augmented by a system of marine protected areas, some of them very large. Naess's idea has since been taken up with a much more public reception by E. O. Wilson.

For many years, Naess remained "spiritus rector" of Deep Ecology. Members of the "second generations" of Deep Ecology included George Sessions, Bill Devall, Warwick Fox, Dave Foreman, Gary Snyder, Alan Drengson, and Joanna Macy. The journal *The Trumpeter* publishes many deep ecological articles. Introductory volumes include Drengson and Inoue (1995), Sessions (1995), Katz et al. (2000). At the moment, Deep Ecology is not as prominent as it was some decades ago. Its motives are dispersed throughout the field of environmental ethics, philosophy, and humanities. The dialectical task of Deep Ecology is to reconcile philosophical rigor with spiritualities and worldviews of "deep green" movements.

References

Drengson, A., & Inoue, Y. (Eds.). (1995). *The deep ecology movement*. North Atlantic Books.
Jonas, H. (1973). *Organismus und Freiheit*. Göttingen.
Katz, E., Light, A., & Rothenberg, D. (Eds.). (2000). *Beneath the surface*. MIT Press.
Naess, A. (1973). *The shallow and the deep, long-range ecology movements: A summary*. (Inquiry Vol. 16, Reprinted in "Deep Ecology for the Twenty-First Century", ed. George Sessions. 1995. Boston, pp. 151–155).
Naess, A. (1986a). *Self-realization: An ecological approach to being in the world*. (Lecture Murdock University. Reprinted in Deep Ecology for the Twenty-First Century, ed. George Sessions. 1995. Boston, pp. 225–239).
Naess, A. (1986b). *The deep ecological movement: Some philosophical aspects*. (Reprinted in Deep Ecology for the Twenty-First Century, ed. George Sessions. 1995, pp. 68–84).
Naess, A. (1989). *Ecology, community and life style: Outline of an Ecosophy*. Cambridge University Press.
Naess, A. (1993). Beautiful action. Its function in the ecological crisis. *Environmental Values, 2*(1), 67–71.
Naess, A. (2005). In A. Drengson (Ed.), *Selected works* (Vol. I-X). Springer.
Ott, K. (2006). Zur Bedeutung Spinozas für die Tiefenökologie. In W. van Bunge (Ed.), *Spinoza and Dutch Cartesianism* (= Studia Spinoza Vol. 15 (1999), pp. 153–176). Würzburg.

Ott, K. (2008). A modest proposal of how to proceed in order to solve the problem of inherent moral value in nature. In L. Westra, K. Bosselmann, & R. Westra (Eds.), *Reconciling human existence with ecological integrity* (pp. 39–60). Earthscan.

Ott, K. (2016). On the meaning of eudemonic arguments for a deep anthropocentric environmental ethics. *New German Critique, 43*, 105–126.

Ott, K. (2020). Mapping, arguing, and reflecting environmental values: Toward conceptual synthesis. In O. Lysaker (Ed.), *Between closeness and evil. A Festschrift for Arne Johan Vetlesen* (pp. 263–292). Scandinavian Academic Press.

Rolston, H. (1988). *Environmental ethics*. Temple University Press.

Sessions, G. (1990). Deep ecology and new age. In H. Mesch (Ed.), *Ecoresistance/Ökowiderstand* (pp. 96–105). Hamburg.

Sessions, G. (Ed.). (1995). *Deep ecology for the twenty-first century*. Shambhala.

Spinoza, B. (1967). *Ethica*. Darmstadt.

Taylor, P. (1986). *Respect for nature*. Princeton University Press.

Konrad Ott is full professor of Philosophy and Ethics of the Environment at the Department for Philosophy at Christian Albrechts University Kiel (Germany). He has a background in critical theory. Konrad Ott wishes to reconcile discourse-ethics with environmental ethics via unconstrained reasoning about human-nature-relations. Following Herman Daly, he has developed a theory of "strong" sustainability. Recent research projects are on climate ethics, management of radioactive waste, ocean ethics, Hegel and sustainability, and historical origins of the Anthropocene. Konrad Ott was member of the German environmental advisory board (2000–2008) and is principal investigator of the ROOTS cluster of excellence at Kiel University.

Earth History

Francine M. G. McCarthy

Abstract This article explores the Anthropocene as an interval of geologic time in the context of the 4.6 billion year history of our planet. There is widespread, although not unanimous, agreement that human activities have sufficiently impacted the Earth system to warrant the establishing of a new interval of geologic time.

The geologic timescale evolved to provide an overview of Earth history over the past few centuries, with the major impetus to its erection being production of *The Map that Changed the World* (Winchester, 2001) by William Smith in 1815. Stratigraphers began subdividing geologic time based on changes in the geologic record that can (at least potentially) be correlated globally. In most cases, the criteria that define the bases of the chronostratigraphic units that provide the physical basis for defining intervals of geologic time (eons, eras, periods, epochs) are paleontologic. Recognition of the origination and extinction of species in 'catastrophic events' through the works of Charles Darwin and Georges Cuvier in the 19th C provided the criteria that provide the physical basis for defining most intervals of geologic time (eons, eras, periods, epochs).

Mass extinctions punctuate the history of life through Earth History, typically in response to major environmental change, with five major mass extinction events in the Phanerozoic Eon – six, if you count the ongoing extinction event largely attributed to human activities (Barnosky et al., 2011). All these changes pale in comparison with the single largest environmental change, when the surface of our planet became oxygenated around 2.5 billion years ago. While it eventually allowed complex (eukaryotic) cells to evolve, allowing for sexual reproduction and true multicellularity during the Proterozoic Eon, the Great Oxidation Event (Holland, 2002) was highly toxic to obligate anaerobes. Many anoxic environments still exist – including in the human gut – so anaerobic bacteria continue to thrive, but many species of

F. M. G. McCarthy (✉)
Brock University, St. Catharines, Ontario, Canada
e-mail: fmccarthy@brocku.ca

© The Author(s), under exclusive license to Springer Nature
Switzerland AG 2023
N. Wallenhorst, C. Wulf (eds.), *Handbook of the Anthropocene*,
https://doi.org/10.1007/978-3-031-25910-4_6

prokaryotes undoubtedly became extinct, marking a major stratigraphic boundary in the geologic timescale: that between the Archean and Phanerozoic Eons.

The current version of the geologic time scale places us in the Holocene Epoch, characterised by rapid global warming recorded in an archived core from the Greenland Ice Cap at the NGRIP site. The base of this epoch, and this the end of the Pleistocene Epoch of the Quaternary Period, was placed at 11,700 y before AD2000 +/− 99 y based on multiparameter annual layer counting (Walker et al., 2008). The base of the Quaternary Period, as well as those of the sub-epochs of the Holocene Epoch, were similarly defined on the basis of marked global environmental change rather than on paleontologic criteria. If the International Commission on Stratigraphy (ICS) accepts the proposal of the Anthropocene Working Group (AWG) of the Subcommission on Quaternary Stratigraphy, a new epoch will be added to reflect changes in Earth systems resulting from the rapid growth of the human population, industrialization and globalization. The AWG is currently looking for a Global Boundary Stratotype Section and Point (GSSP) to define the Anthropocene with a base around the mid-twentieth century of the Common Era (AWG, 2020). Many proxies, including microplastics and spheroidal carbonaceous particles resulting from incomplete combustion of fossil fuels, are much more abundant in sediments deposited since the mid-twentieth century, and thus are potentially useful markers of the Great Acceleration (Steffen et al., 2015).

In the context of 4.6 billion years of Earth history, the environmental changes of the last several decades, or even centuries or millennia which some have argued should be viewed as the base of the Anthropocene, may seem relatively insignificant. Earth's climate fluctuated wildly as the sun's luminosity gradually increased, and ice sheets covered much of our 'Snowball Earth' during the late Neoproterozoic Era. Several ice ages punctuated the Paleozoic Era, including during the Late Carboniferous (Pennsylvanian) through Permian, with cyclothems recording frequent glacoacioeustatic fluctuations when most of our planet's continental land mass was united as Pangaea. At other times, including through much of the Mesozoic Era, continental ice volume was negligible and vast epeiric seas covered large part of continents in which many of our petroleum resources originated as algal blooms and accumulated in dysoxic-anoxic environments.

Does anthropogenic impact on Earth systems rise merit recognition as a designated interval of geologic time, and if so, at what level –epoch? period? Colloquially, the loosely defined term 'Anthropocene' is often referred to as an 'era'. While this term might be appropriate in the context of human history, the degree of stress on Earth Systems recorded in geologic strata does not rise to the level of boundaries between eras, marked by large-scale faunal turnover/mass extinction much more intense than that experienced to date in the current 'sixth major mass extinction'. The question of whether the Anthropocene should be treated as a formal chronostratigraphic unit defined by a GSSP was considered at the 2016 meeting of the International Geological Congress in Cape Town, South Africa, and following further investigation and deliberation, the AWG voted in favour of erecting an Anthropocene Epoch within the Quaternary Period. A formal proposal has yet to be presented to the ICS, but the primary isochronous marker for this boundary is likely to be fallout of radionuclides like ^{239}Pu and ^{137}Cs that are abundant in the geologic record between 1952 and 1963, when the *Limited Test Ban Treaty* was signed in Moscow, effectively putting an end to above-ground thermonuclear testing.

Whether the Anthropocene should be defined as a formal interval of geologic time remains an open question, within the geological community and beyond. Time will tell what the ICS decides after deliberating over the proposal of the AWG, but it is difficult to argue that any of the subsystems of the Earth system have not been substantially impacted by human activity.

References

Anthropocene Working Group. (2020). *Newsletter, Volume 10: Report of activities 2020.* Anthropocene-Working-Group-Newsletter-Vol-10-final.pdf (stratigraphy.org)

Barnosky, A. D., Matzke, N., Tomiya, S., et al. (2011). Has the Earth's sixth mass extinction already arrived? *Nature, 471*(7336), 51–57.

Holland, H. D. (2002). Volcanic gases, black smokers, and the great oxidation event. *Geochimica et Cosmochimica Acta, 66*, 3811–3826. https://doi.org/10.1016/S0016-7037(02)00950-X

Steffen, W., Broadgate, W., Deutsch, L., Gaffney, O., & Ludwig, C. (2015). The trajectory of the Anthropocene: The great acceleration. *The Anthropocene Review, 2*, 81–98.

Walker, M., Johnsen, S., Rasmussen, S. O., et al. (2008). The global stratotype section and point (GSSP) for the base of the Holocene series/epoch (Quaternary system/period) in the NGRIP ice core. *Episodes, 31*, 264–267.

Winchester, S. (2001). *The map that changed the world: William smith and the birth of modern geology.* HarperCollins.

Francine M. G. McCarthy is Professor of Earth Sciences, Associate Member of Biological Sciences and Core Member of Environmental Sustainability at Brock University, whose campus lies in the Niagara Escarpment UNESCO World Biosphere Reserve. She is also a Research Associate in Natural History at the Royal Ontario Museum. She and more than 40 researchers have been studying over the past 3 years to assess its potential to define a new interval of geologic time defined by overwhelming anthropogenically-driven changes to Earth systems – the Anthropocene Epoch. She has published more than 50 refereed journal articles on the use of microfossils in marine and freshwater sediments to reconstruct past environmental change. She seeks to bridge the divide between the humanities and the sciences to address our global environmental crisis.

Earth Systems

Francine M. G. McCarthy

Abstract This article examines the role of human agency in altering Earth Systems – the highly complex group of open systems (atmosphere, hydrosphere, geosphere, cryosphere and biosphere) that interact to produce the environment of our planet at any given time.

The *Earthrise* photograph taken by Apollo 8 astronauts in lunar orbit on Christmas Eve, 1968, illustrated our planet – like a Christmas tree ornament – rising above the desolate surface of the far side of the moon. It impacted our collective image of Earth, illustrating the fragility and uniqueness of our blue planet. Its publication in *Life Magazine*, and that of many others that followed as numerous satellites were launched into space, coincided with rapid expansion of the Environmental Movement. Widespread realization that human actions could have profound, yet unintended consequences on our environment arguably began with the publication of *Silent Spring* by Rachel Carson in 1962.

Initially viewed as a subversive movement, recognition that the biosphere (living organisms) is inextricably interconnected with the geosphere (solid rocks and soils), hydrosphere (liquid water), cryosphere (ice) and atmosphere (envelope of gases) that surrounds our planet, became increasingly mainstream. The highly complex interactions between these open systems that are part of the 'Earth System' became obvious, especially in highly populated and industrialized regions where 'dilute and disperse' was no longer a viable option. A striking example of this was the Cuyahoga River that burned for several days in June 1969 (referred to in Randy Newman's 1972 song *Burn On*), although shockingly, this was not the first time that high concentrations of industrial effluent had caused rivers in the Great Lakes basin to catch fire (Hartig, 2018).

F. M. G. McCarthy (✉)
Brock University, St. Catharines, Ontario, Canada
e-mail: fmccarthy@brocku.ca

© The Author(s), under exclusive license to Springer Nature Switzerland AG 2023
N. Wallenhorst, C. Wulf (eds.), *Handbook of the Anthropocene*,
https://doi.org/10.1007/978-3-031-25910-4_7

Most scientists now believe that the Earth System has been dramatically and permanently altered by human agency. For thousands of years, the alteration of landscapes by fixed human settlements with domesticated animals and cultivated plants had noticeable local impacts on the biosphere, geosphere, and hydrosphere, resulting in local extirpation, erosion and depletion of soils, and pollution and eutrophication of aquatic systems (Williams et al., 2022). Significant global-scale anthropogenic impact began with the Industrial Revolution. Large volumes of effluent and emissions, particularly associated with widespread combustion of fossil fuels, produced deadly smog events and acid precipitation that illustrated the interconnectedness between atmospheric emissions and ecosystems and human health, particularly in regions where the geosphere had little capacity to buffer these effects. The exponentially increase in human population and rampant consumerism termed the 'Great Acceleration' (Steffen et al., 2015) has sufficiently altered the Earth System that the Anthropocene Working Group has proposed the formal erection of the Anthropocene Epoch with a base in the mid-twentieth century (Anthropocene Working Group, 2019).

The altered atmospheric composition has far-reaching implications, including ocean acidification stressing all calcareous marine organisms from microscopic pteropods ('sea-butterflies') to reef-building corals (Sutton et al., 2019) and increased rates of weathering of continental rocks (Beaulieu et al., 2012). Lake Erie, famously declared dead (Edmonds, 1965), is becoming eutrophic again despite substantial efforts in abating nutrient influx over the past several decades, as stressors become primarily global rather than local (Watson et al., 2016). Changes in atmospheric composition and climate warming have contributed to blooms of toxic cyanobacteria ('blue-green algae') and deoxygenation of marine and freshwater environments worldwide (Griffith & Gobler, 2020), once again threatening the ecosystem of Lake Erie, with dead zones expanding as decomposition uses up dissolved oxygen.

Concentrations of atmospheric CO_2 measured in 2021 (with little discernible signal of the global economic slowdown due to the COVID-19 outbreak; NOAA Climate.gov) reached levels not experienced in more than four million years (Haywood & Valdes, 2004). The best-known (but not the most easily predictable) consequence of increased fossil fuel combustion is intensification of the Greenhouse Effect. Without greenhouse gases, the mean temperature of our planet would be approximately -18 °C and thus inhospitable to life. A slight increase to the tiny percentage of greenhouse gases in our atmosphere, however, produces substantial global warming, with many associated impacts on Earth systems. The last time atmospheric CO_2 concentrations were as high as they are today, during the Pliocene Epoch, large forests grew in areas of the Arctic that are now tundra and global sea level was approximately 24 metres higher than today (Dumitru et al., 2019). Higher sea levels at times of climate warming are largely due to the impact on the cryosphere, with melting of continental ice, although expansion resulting from warming the huge volume of water in the oceans is also significant.

Although we refer to 'global' warming, the effects are greater at the poles as feedbacks in the climate system produces what has been called polar amplification (Lee, 2014). Additionally, CO_2 emissions from accelerated microbial breakdown of large quantities of organic carbon stored in permafrost are another example of positive feedback and complex interaction between various components of the Earth System (Schuur et al., 2015). The Greenland and West Antarctic Ice Sheets, containing ice equivalent to about 7 m and 3–5 m sea level rise, respectively (Cazenave & Llovel, 2010) have been shown to be highly susceptible to climate warming. Melting and destabilization of the cryosphere resulting from nonlinear threshold response to climate forcing could produce a very rapid sea in level rise, impacting the large percentage of the human population that inhabits coastal regions (Diaz & Keller, 2016). For comparison, global (eustatic) sea level was around 6 m higher than today following the penultimate glacial termination, around 128,000 years ago, when the mean annual temperature of our planet was only a few degrees warmer than today.

Rapid terminations of ice ages since the mid-Pleistocene appear to have resulted from complex synergistic interactions, as collapse of large Northern Hemisphere Ice Sheets disrupted global patterns of ocean and atmospheric circulation, intensifying polar warming (Denton et al., 2010). The feedbacks and interactions between the various subsystems of the Earth System are difficult to predict and model, but accurate predictions are essential to coping with environmental change. It is unlikely to capture the imagination of the world to the degree that the Apollo missions did, but perhaps the formal recognition of the Anthropocene as an epoch on the Geologic Time Scale will help draw attention to the 'wicked problems' facing humanity.

References

Anthropocene Working Group. (2019). *Newsletter, Volume 9: Report of activities 2019*. Anthropocene-Working-Group-Newsletter-Vol-9-final.pdf (stratigraphy.org)

Beaulieu, E., Goddéris, Y., Donnadieu, Y., Labat, D., & Roelandt, C. (2012). High sensitivity of the CO_2 sink by continental weathering to the future climate change. *Nature Climate Change, 2*, 346–349. https://doi.org/10.1038/NCLIMATE1419

Cazenave, A., & Llovel, W. (2010). Contemporary Sea level rise. *Annual Review of Marine Science, 2*, 145–173. https://doi.org/10.1146/annurev-marine-120308-081105

Denton, G. H., Anderson, R. F., Toggweiler, J. R., et al. (2010). The last glacial termination. *Science, 328*, 1652–1656.

Diaz, D., & Keller, K. (2016). A potential disintegration of the West Antarctic Ice Sheet: Implications for economic analyses of climate policy. *American Economic Review, 106*, 607–611.

Dumitru, O. A., Austermann, J., Polyak, V. J., et al. (2019). Constraints on global mean sea level during Pliocene warmth. *Nature, 574*, 233–236. https://doi.org/10.1038/s41586-019-1543-2

Edmonds, A. (1965). *Death of a great lake* |Maclean's|November 1 1965 (macleans.ca)

Griffith, A. W., & Gobler, C. J. (2020). Harmful algal blooms: A climate change co-stressor in marine and freshwater ecosystems. *Harmful Algae, 91*, 101590. https://doi.org/10.1016/j.hal.2019.03.008. Epub 2019 May 21.

Hartig, J. (2018). *From burning river to catalyst for economic revival – The Buffalo River story*. A Great Lakes Moment from John Hartig – Great Lakes Now.

Haywood, A. M., & Valdes, P. J. (2004). Modelling Pliocene warmth: Contribution of atmosphere, oceans, and cryosphere. *Earth and Planetary Science Letters, 218*, 363–377. https://doi.org/10.1016/S0012-821X(03)00685-X

Lee, S. (2014). A theory for polar amplification from a general circulation perspective. *Asia-Pacific Journal of the Atmospheric Sciences., 50*, 31–43.

Schuur, E., McGuire, A. D., Schaedel, C., et al. (2015). Climate change and the permafrost carbon feedback. *Nature, 520*, 171–179.

Steffen, W., Broadgate, W., Deutsch, L., Gaffney, O., & Ludwig, C. (2015). The trajectory of the Anthropocene: The Great acceleration. *The Anthropocene Review, 2*, 81–98.

Sutton, A. J., Feely, R. A., Maenner-Jones, S., et al. (2019). Autonomous seawater pCO_2 and pH time series from 40 surface buoys and the emergence of anthropogenic trends. *Earth System Science Data, 11*, 421–439. https://doi.org/10.5194/essd-2018-114

Watson, S. B., Miller, C., Arhonditsis, G., et al. (2016). The re-eutrophication of Lake Erie: Harmful algal blooms and hypoxia. *Harmful Algae, 56*, 44–66.

Williams, M. et al. (2022). Planetary scale change to the biosphere signalled by global species translocations can be used to identify the Anthropocene. *Palaeontology, 65*(4). https://doi.org/10.1111/pala.12618

Francine M. G. McCarthy is Professor of Earth Sciences, Associate Member of Biological Sciences and Core Member of Environmental Sustainability at Brock University, whose campus lies in the Niagara Escarpment UNESCO World Biosphere Reserve. She is also a Research Associate in Natural History at the Royal Ontario Museum. She and more than 40 researchers have been studying over the past 3 years to assess its potential to define a new interval of geologic time defined by overwhelming anthropogenically-driven changes to Earth systems – the Anthropocene Epoch. She has published more than 50 refereed journal articles and book chapters, most on the use of microfossils in marine and freshwater sediments to reconstruct past environmental change. She seeks to bridge the divide between the humanities and the sciences to address our global environmental crisis.

Environment

Federica Buongiorno and Xenia Chiaramonte

Abstract The term 'environment' is complex and conveys different meanings: the word 'environment' is employed as a synonym for space, territory, place, or ecosystem. A comprehensive definition of environment describes it as the set of conditions in which living takes place: it is the complex system of physical, chemical and biological factors, of living and non-living elements and of the relationships in which all the organisms that inhabit the planet are immersed. While we can envision many types of environment, the term is commonly used in relation to nature, the so-called natural environment. This latter notion encompasses all forms of living and non-living beings; ecology is the science that studies the interactions of organisms with each other and with their physical surroundings.

The environment is a complex system of physical, chemical and biological factors, of living and non-living elements and of relationships in which all the organisms that inhabit the planet are immersed.

The term 'environment' is complex and plurivocal. It conveys different meanings depending on its possible specifications, and it is for this reason that the word 'environment' is often accompanied by an adjective that qualifies its significance. One can think of a natural, social, cultural, urban, or digital environment; in nearly any case, the word 'environment' is employed as a synonym for space, territory, place, or ecosystem. A comprehensive definition of environment describes it as the set of conditions in which living takes place. All these meanings have an element in common: the environment is what surrounds, what is around, the complex set of interconnected factors and conditions that surround an organism, be it vegetable or animal.

F. Buongiorno (✉)
University of Florence, Florence, Italy
e-mail: federica.buongiorno@unifi.it

X. Chiaramonte
ICI Berlin Institute for Cultural Inquiry, Berlin, Germany
e-mail: xenia.chiaramonte@ici-berlin.org

N. Wallenhorst, C. Wulf (eds.), *Handbook of the Anthropocene*,
https://doi.org/10.1007/978-3-031-25910-4_8

While we can envision many types of environment, the term is commonly used in relation to nature, the so-called natural environment. The natural environment encompasses all forms of living and non-living beings. Climate, weather and natural resources, as well as the living species inhabiting the earth, all contribute – in their interrelations – to the composition of the environment. Ecology is precisely the science that studies the interactions of organisms with each other and with their physical surroundings.

Ecology has been conceived as not only the science of the whole (i.e., the environment) but also as an ethics, that is, a way to construct and develop a better relationship between human and non-human entities and their surroundings, a branch of knowledge that aspires to create a fairer and less harmful cohabitation among all the species that inhabit the planet. However, etymologically, ecology also denotes a "science of the house" (οἶκος, oikos, "house"). The problem of establishing an effective environmental ethics lies in this root. In his 1749 magnum opus *De econo-mia naturae*, Isaac Biberg, a pupil of Linnaeus, constructed the domestic metaphor that casts its shadow on today's ecology. Concepts formulated in accordance with this metaphor include that of ecosystem (place or 'household' where certain species live) or the defense of the environment (as a protective modality that encloses the 'household' of the living). Before the theory of evolution, species were thought to be immutable; hence, the relationship between species was explained by the creationist hypothesis, which implies the existence of a common divine creator of all things (Coccia, 2020).

With the German zoologist Ernst Haeckel (1866), we passed from the economy of nature to the new term "ecology". But on closer inspection, the reference to the household – the economy in the mercantile sense – continues to dominate today's ecology, as do numerous related theological conceits. First, order and utility are concepts that issue from this genealogy of the domestic: under these principles, the world would correspond to an ordered set of ecosystems, each assigned to their own "house", that must be protected in order to keep their usefulness intact. The corresponding idea of returning to a pristine condition is full of theological inspiration – but alas "there is no garden and never has been" (Haraway, 1992, 309). This widespread view is misleading for several reasons. First, this perspective implies that human beings are responsible for defending the environment conceived as "nature". And even if the ultimate aim of ecology is to resituate humans as part of the environment – thereby reaffirming them as parts of nature rather as dominant actors – the concept of defending or protecting nature inevitably leads back to the dualism between man and nature and to a human exceptionalism (Vogel, 2002). Second, conceiving the environment as an entity to be defended once again establishes nature in economic and proprietary terms as a resource to be exploited or not exploited, as something to be fenced off for the purposes of containment or defense. This vision, among other things, implies something like the park-form (Martinez-Alier, 2002), the fence of the wilderness (Cronon, 1996), thus refusing to recognize that today, we are faced with an extremely more complex issue: the challenge that Stengers calls 'the intrusion of Gaia' (Stengers, 2015). Third and finally, the metaphor of the house does not work, also because every species migrates by definition.

Moreover, it is the planet that migrates in the first place, as we know from the theory of Continental Drift (Wegener, 1912).

Current research on the environment focuses on overcoming the traditional dualism between a subjective apprehension of nature (as an object of experience) and an objective (scientifically determined) description of it. As observed by Altamirano, "the bifurcation of nature into a physical aspect and a conscious aspect facilitates a picture of the human in confrontation with nature, rather than in an ecological relation, whether humans are conceived as stewards of the earth or human life is conceived as emergent within an environmental system" (Altamirano, 2016, 32). A useful concept increasingly used to frame the non-dualistic relation between subjects and their environment is that of 'milieu', understood as the "relational space in between the organism and its complex environment": so conceived, the milieu represents a 'middle place' that is composed by "the heterogeneous series of organism and environment through the artifices that conduct their mutual eventualities" (Altamirano, 2016, 129). As Canguilhem puts it: "the notion of milieu is becoming a universal and obligatory mode of apprehending the experience and existence of living beings; one could almost say it is now being constituted as a category of contemporary thought. But until now it has been quite difficult to perceive as a synthetic unity the historical stages in the formation of this concept, the various forms of its utilization, and the successive inversions of the relationship in which it is one of the terms-in geography, in biology, in psychology, in technology, in economic and social history." (Canguilhem, 2008, 98).

In other words, the environment as 'milieu' is a space that results from the conjunction of its eventualities and is not given prior to them. Jakob von Uexküll (1921, 1934) has exemplarily argued for the osmosis – a qualitative one – between organisms and the environment: he introduced the notion of *Umwelt* (literally: 'surrounding world', also 'environment') to describe the way organisms subjectively perceive their surroundings through the information provided by senses. Thus, organisms rely on a 'self-in-world' perspective (similar to Bateson's (1972) notion of 'organism-in-environment) that allows them to perceive the environment through species-specific, spatiotemporal frames. Von Uexküll's concept of *Umwelt* reminds of the phenomenological notion of *Umwelt* as it has been presented by Edmund Husserl (1952), according to whom the environment is that wordly-sorrounding experienced by subjects in the first-person perspective: environmental phenomenology has recently emerged as a productive approach in order to study environmental issues from a subjective positioning, that does not give up to a strong scientific commitment, thereby contributing to the overcoming of the strict dualism between subjectivism and objectivism (Seamon, 2006).

The concept of milieu is also pivotal to structuralist theories of the environment. Inquiries into primitive relations between humans and environments, such as those developed by Claude Lévi-Strauss, show the high level of interactions between humans and their surroundings at very early stages: "(…) for the mind not only does not fail to react in the surrounding environment, but it is also conscious that different milieus exist, and that their inhabitants are reacting with them, each group in its own fashion" (Lévi-Strauss, 1985, 154). Dismissing a dualistic interpretation of the environment implies, again, overcoming the traditional ecological discourse about the "defense of nature": as Gregory Bateson (1972) already pointed out, to be stressed is the "interrelation of organism *and* environment as a system with exchange and feedback. Humans are no longer independent agents who act on nature, or impose order as they would like. They are a priori in exchange with their environment" (Andermatt Conley, 1997, 46). This is why, according to Bateson, an organism that destroys its environment destroys itself too.

The potential of the interaction between humans and the environment is fully disclosed by risk-situations: "for a situation to be classified as 'risky', it is crucial that a potential loss is contingent and avoidable – hence the important role that decisions play in an environment of risk" (Lübcken & Mauch, 2018, 8). From a legal point of view, the issues at stake are very complex. Criminal law is based on personal responsibility, whereas ecological disasters qualify as system crime (Ferrajoli, 2019). We have entered the era of catastrophes (Stengers, 2015) but we do not have any legal resources at the moment that can cope with this threatening condition. In fact, disasters cannot be configured as crimes in the legal sense. The victims are entire peoples, all of humanity, but also future generations. The responsibility for disasters can rarely be attributed to an individual. Catastrophic phenomena cannot necessarily be traced back to a single episode; most of the time it is a series of events whose concatenation ends up producing devastating effects. The massive political and economic activity of extraction of resources and consequent pollution carried out by an indeterminate plurality of subjects, along with the agency of non-human actants, coalesce to produce extraordinarily powerful and harmful events. The result of these events cannot be faced with criminal law since they lack all the fundamental principles on which it is based: the principle of personal responsibility, the principle of legality, and the determination of punishable facts.

"Recent research on the history of natural disasters has shown that risk is manufactured by society and it affects different parts of the population in varying degrees" (Andermatt Conley, 1997, 46). This awareness must be expanded today in order to include non-human forms of harm and damage: post-structuralist ecological accounts such as eco-criticism attempt to overcome a biocentric worldview so as to extend the ethics and humans' conception of global community to include both nonhuman forms of life and the physical environment (Oppermann, 2018). From this perspective, "poststructuralism not only exposes the structures which make environmental destruction appear inevitable and which constitute people as the agents of this destruction; it has [also] been called upon to explain why environmentalists are failing in their struggle, and to show them how to be more effective" (Gare, 1995, 92). At the core of poststructuralist theories of the environment lies the idea that human representations must be scrutinized since they produce specifically anthropocentric views (interpretations) of nature: this should lead to gaining a new critical awareness about traditional natural discourses and the way we understand the non-human. Again, it is important to avoid the polarization between a strict constructivist approach and an objectivist one: as suggested by Opperman, "eco-criticism can offer a multiperspectival approach that probes into the problematic relationship of representation and the natural environment" (Oppermann, 2018,

124). In such a 'multiperspectival' account the role played by technological intervention cannot be underestimated: "the electronic revolution produces economic shifts and reorientations that have pronounced (or even devastating) effects on our sense of the environment" (Andermatt Conley, 1997, 74). It is time to "reinstall the human within her concrete environment: technically, in an environment composed of sense-events and effect-events; and machinically, in the milieus that compose that environment as so many eventualities acting transversally across distances of time and space" (Altamirano, 2016, 154). The relation to the environment is conceived not merely in terms of sharing a common world, but becomes a technical problem of intervention upon the environment and its eventualities. This implies putting into question the dichotomy between the natural and the artificial: transhumanism precisely regards human technological enhancement as a necessary condition to improve wellbeing and diminish human suffering. "The concept of human enhancement includes any technique discovered in the course of the human history to increase the chances of living a good and better life" (Garcia-Belaunde, 2009, 2): in this sense, humans have always been more than just natural living beings. Environmental philosophy shall therefore be regarded as a part of philosophy of technology (Vogel, 1999). (Feminist) posthumanism mitigates transhumanist enthusiasm for technological human enhancement and theorizes the 'continuum' of nature and culture as the best strategy to overcome the dualistic account.

References

Altamirano, M. (2016). *Time, technology, and environment. An essay on the philosophy of nature*. Edinburgh University Press.

Andermatt Conley, V. (1997). *Ecopolitics. The environment in poststructuralist thought*. Routledge.

Bateson, G. (1972). *Steps to an ecology of mind: Collected essays in anthropology, psychiatry, evolution, and epistemology*. University of Chicago Press.

Biber, I. (1749). *De economia naturae*. R. and J. Dodsley.

Canguilhem, G. (2008). The living and its milieu. In *Id., The knowledge of life*. Fordham University Press.

Coccia, E. (2020). *Métamorphoses*. Rivages.

Cronon, W. (1996). The trouble with wilderness: Or, getting Back to the wrong nature. *Environmental History, 1*(1), 7–28.

Ferrajoli, L. (2019). I crimini di sistema e il futuro dell'ordine internazionale, *Teoria politica*. *Nuova serie* Annali, 9. http://journals.openedition.org/tp/878 (Last Retrieved July 28th, 2021).

Garcia-Belaunde, V. (2009). *Should we try to overcome nature? A Transhumanistic approach*. www.ssh.org.pe (Last Retrieved July 28th, 2021).

Gare, A. (1995). *Postmodernism and the environmental crisis*. Routledge.

Haeckel, E. (1866). *Generelle Morphologie der Organismen*. G. Reimer.

Haraway, D. (1992). The promise of monsters: A regenerative politics for inappropriate/d others. In L. Grossberg, C. Nelson, & P. Treichler (Eds.), *Cultural Studies* (pp. 295–337). Routledge.

Husserl, E. (1952). *Ideas pertaining to a pure phenomenology and to a phenomenological philosophy – Second book: Studies in the phenomenology of constitution*. Kluwer.

Lévi-Strauss, C. (1985). *The view from Afar*. Basil Blackwell.

Lübcken, U., & Mauch, C. (2018). Uncertain environments: Natural hazards, risk and Insurance in Historical Perspective. *Environment and History, 17*, 1–12.

Martinez-Alier, J. (2002). *The environmentalism of the poor: A study of ecological conflicts and valuation*. Edward Elgar.

Oppermann, S. (2018). Theorizing ecocriticism: Toward a postmodern Ecocritical practice. *Interdisciplinary Studies in Literature and Environment, 13*(2), 103–128.

Seamon, D. (2006). Interconnections, relationships, and environmental wholes: A phenomeno-logical ecology of natural and built worlds. In *Conference: Phenomenology and ecology. Twenty-third annual symposium of the Simon Silverman phenomenology center, 2005*. Simon Silverman Phenomenology Center, Duquesne University.

Stengers, I. (2015). *In catastrophic times: Resisting the coming barbarism*, Open Humanity Press. http://dx.medra.org/10.14619/016, http://openhumanitiespress.org/books/titles/in-catastrophic-times (Last retrieved July 28th, 2021).

von Uexküll, J. (1921). *Umwelt und Innenwelt der Tiere*. 2. Aufl. Springer.

von Uexküll, J. (1934). *Streifzüge durch die Umwelten von Tieren und Menschen: Ein Bilderbuch unsichtbarer Welten*. Springer.

Vogel, S. (1999). For and against nature. *Rethinking Marxism, 11*(4), 102–112.

Vogel, S. (2002). Environmental philosophy after the end of nature. *Environmental Ethics, 24*(1), 23–39.

Wegener, A. (1912). Die Entstehung der Kontinente. *Geologische Rundschau* 3: 276–292 https://doi.org/10.1007/BF02202896 (Last Retrieved July 28th, 2021).

Federica Buongiorno is a post-doc fellow at the ICI Berlin Institute for Cultural Inquiry. She received her PhD in Philosophy from the Sapienza University of Rome in 2013. She has been a post-doc fellow at the TU Dresden (2017–2020), at the IISF of Naples (2017), at the Freie Universität Berlin (2014–2017), and at the IISS of Naples (2012–2013). Her research interests include phenomenology, philosophy of technology, and applied ethics, with a special focus on the epistemological and ethical problems related to the 'digital turn'. She wrote three monographs on phenomenology, including her PhD dissertation (*Logica delle forme sensibili. Sul precategoriale nel primo Husserl* (Rome 2014)).

Xenia Chiaramonte is a jurist and a socio-legal scholar. She is currently a fellow at the ICI Berlin where she develops a project on law and ecology. She has been a visiting scholar at UCBerkeley and a postdoctoral researcher at the University of Bologna and Roma Tre as well as a CAS SEE (Center for Advanced Studies of Southeastern Europe) fellow at the University of Rijeka. She recently published her monograph *Governare il conflitto: La criminalizzazione del movimento No TAV* (Milan, 2019), which analyses the criminalization of one of the most longstanding and high-profile environmental movements in Western Europe.

Geosphere

Lutz Möller

Abstract This article examines the origins and the quality of the term "geosphere". It describes its extension and its character as a relational concept of human beings towards their planetary environment. It argues that unfortunately the geosphere is "invisible" in daily experiences and presents initiatives, including those by UNESCO, to increase the visibility of the geosphere and its constituents. It also presents examples for the considerable impact that human civilization already has on the geosphere and, in turn, the geosphere's significance for humanity in the Anthropocene.

The Term and Its Conceptual Differentiation

"Geosphere" is a terminology with multiple, diverging definitions. However, the differences in definition are not highly significant from a layperson's point of view. For the purposes of this article, only the most important terminological differentiations are highlighted.

The term was first introduced by the US philosopher Stephen Pearl Andrews (Andrews, 1871), almost at the same time that the term "biosphere" was introduced by the Austrian geologist Eduard Suess (Suess, 1875, compare the article "biosphere" in this Handbook). The different content and intent of these two terminological innovations make it likely that both happened independently from each other; however, this is not certain. In turn, it is certain that Stephen Pearl Andrews drew upon the work of the British scientist Hugh Doherty who had introduced in 1864 terms such as "geospheric realm" or "atmospheric realm" (Doherty, 1864).

Even if introduced only in the late nineteenth century (The German language Wikipedia page on the history of the term "geosphere" is an excellent overview), the terminology has an ancient precursor. Many cultural traditions around the globe

L. Möller (✉)
German Commission for UNESCO, Bonn, Germany
e-mail: moeller@unesco.de

© The Author(s), under exclusive license to Springer Nature Switzerland AG 2023
N. Wallenhorst, C. Wulf (eds.), *Handbook of the Anthropocene*,
https://doi.org/10.1007/978-3-031-25910-4_9

55

have described the abiotic nature in terms of a small number of elements. The ancient cultural traditions of China, India, Japan, Persia, Babylonia, Egypt and Greece all had their elements such as wood, fire, metal, earth, water, air, etc. It was the Greek philosopher Aristotle who used to describe the cosmos in the model of "spheres" – celestial and terrestrial spheres. Even if Aristotle did not use the terminology of "geosphere" or "hydrosphere", he seems to have been only one step away from it.

The most typical conceptual content of the term "geosphere" is the one introduced by Stephen Pearl Andrews. Very simply put, geosphere refers to the solid parts of the earth.

More scientifically put, it is the Lithosphere and the parts of the Earth's mantle and core beneath it; they are called Asthenosphere, Mesosphere and Barysphere. This article will not describe in much detail the physical constitution of the Earth's mantle and its core; however, a few highlights need to be emphasized. The mantle has a thickness of about 2900 kilometres, of which the upper 100 kilometres (below the oceans) or 150–200 kilometres are the rigid Lithosphere made up of tectonic plates moving upon the more viscous (almost solid but partially molten) Asthenosphere which again is layered on the Mesosphere or lower mantle. All parts of the mantle have in common is that they are mainly composed of "rock", i.e. of chemical materials such as silicon, oxygen, or aluminium. The Barysphere in turn is the Earth's core (and the size of the moon), consisting of the inner and outer core. Between the mantle and the core, there is a huge difference in terms of chemical composition, since the core is mainly made up of iron. The outer core is composed of fluid iron (and responsible for the magnetic field of the Earth); the inner core is solid. The only observations that mankind can make about the inner structure of the Earth below the Lithosphere is due to seismic waves resulting from earthquakes on the diametrically opposed parts of the planet. Due to its spherically layered structure, the term "geo*sphere*" seems particularly apt.

This most generally used definition of the geosphere, which will also be used in this article, encompasses parts of the Earth which are familiar to human beings, such as rock outcrops, but also other parts about which mankind has very little knowledge and even less collective imagery, such as the Earth's core or the lower mantle. There are some discussions whether soils are part of the "geosphere" and whether very recent sediments are – for the purposes of this article, soils and recent sediments are included.

The concept of the "geosphere" is not a physical quantity. The geosphere is not measured in tons or square meters. It is not a conceptual placeholder for the solid parts of the earth at a particular moment in time; neither is it process-oriented concept that would focus on the constant changes within the Lithosphere, the Earth's mantle and core. Similarly as is the case for all other "spheres", the "geosphere" is a rather abstract concept, it is not an entity that exists independently of human beings – it doesn't have a well-defined intension, it doesn't have a well-defined extension. As is the case for all other "spheres", "geosphere" is a relational concept of human beings towards their planetary environment; it is a conceptually organizing principle, not an objective independent existence.

Among the diverse conceptual differentiations of the term "geosphere", the one introduced by the French anthropologist and philosopher Pierre Teilhard de Chardin (1956), was and continues to be influential as well. It is more encompassing than Pearl Andrews' definition, and contains much of the abiotic factors influencing biological life. Teilhard de Chardin's definition was used in International Geosphere-Biosphere Programme (IGBP), a very influential multi-year global research programme by the then ICSU (International Council of Scientific Unions, today International Science Council).

Currently, a new term is starting to be used: "geodiversity" (Gray, 2004). As the abiotic equivalent of "biodiversity", it describes earth materials and processes from the perspective of their difference in terms of genesis and distribution. While UNESCO has adopted an "International Day of Geodiversity" in 2021, the author of this article sees need to further discuss this concept.

The "Invisibility" of the Geosphere

The biosphere, the hydrosphere and the atmosphere are all very present in the daily lives of human beings. Human beings feel the wind and changes in air temperature, they feel the rain and swim in rivers, they marvel at animals and plants. At first sight, most of the geosphere, however, seems to be absent from our daily lives. Where is it? We need to take only one step further: Hills with sparse vegetation, mountains, steep valleys, deserts, river beds, rock outcrops, beaches – the Earth' surface can be very visible in the landscape, not only in "beautiful" landscapes that human beings attend for vocational purposes. This visible surface of the geosphere is most typically the result of the interaction of the geosphere with the atmosphere (wind erosion), the hydrosphere (water and glacier erosion) and the biosphere (soils and vegetation). The surface of the geosphere is present in our daily lives, it is often attached with positive emotions; when the geosphere's surface enters our conscience, it often does so in an "impressive" way.

Apart from such structural surface phenomena, some human beings will know as well that soils have very much to do with the underlying rock strata and their minerals. Some human beings will be aware about underground mining, they know about earthquakes, geysers and volcanoes. Most people, however, have "underground" experiences in motorway tunnels, metro stations or basements – this is, in culturally prepared spaces that are characterized first of all by the absence of sunlight and tend to be "uneasy". Naked underground rocks or soil profiles are not commonly experienced by human beings. This means that there is very limited emotional connection to the geosphere.

A short side-track: From another perspective, this lack of experience seems to be quite surprising. Stones are highly present in our built environment, they are present in most older business and residential houses, they are present as cobblestones or setts to pave walkways or roads, they are present in many religious buildings. However, the fact that our built environment, at least most of its more traditional parts, is made of stones from the lithosphere, specific stones with specific properties, is one of the facts that most people hardly ever think about. What stones are, what properties they have, where they come from: Most people never think about it. Stones are "culturally invisible" and "emotionally invisible" to a surprising extent.

A particular striking example for this "invisibility" of stones is the following: Cologne Cathedral is a UNESCO World Heritage site. It is a very impressive cathedral, one of the tallest in the world, built mostly from Trachyte, a volcanic stone from the nearby Siebengebirge. The Trachyte was quarried on the Drachenfels, a particularly beautiful hill with a castle ruin on top towering immediately over the river Rhine. Actually the quarry had existed already since the Roman times. In 1824, the Drachenfels was "protected", i.e. the Trachyte quarries had to close; this was one of the first nature conservation interventions in German history. In summary, the stone that Cologne Cathedral is built from, what every visitor *sees* when she or he looks at the Cathedral, is in this case related to a nearby, very specific, very well-known place with a very interesting and important story to tell.

However, in the very long German-language Wikipedia article about the Cathedral with more than 300 sources quoted (accessed on 2 October 2021), there is exactly one short sentence about the material that the Cathedral is built from. There is no mention of the material of the Cathedral in the English version of the Wikipedia article (accessed on 2 October 2021). Consumers in most European countries at least are highly interested into the composition of our food – why are we not interested in what our buildings are made from? The author of this article has not done research on the answer of the question, but it is not farfetched to assume that we take stones *for granted*. In fact, the author is convinced that human beings take the entire geosphere *for granted* and thus do not spend one thought about it.

In recent years, geoscientists and laypeople have started efforts to make stones and the geosphere "culturally and emotionally visible". Instead of relying on the traditional presentation of stones and geological phenomena in Natural History Museums, new initiatives have started.

A very interesting but still small initiative in Germany is "Stones in Cities" ("Steine in der Stadt"). More than hundred enthusiasts work together since 2006 through presenting the stones that the buildings in their respective city or town are built from, for example through weekend walks. This initiative has had predecessors since the 1920s in Germany and there have been and are similar initiatives in countries such as Russia, the U.S., Austria, the UK or Switzerland. The approach of these initiatives is the same in all places, generating interest and visibility for "culturally invisible" elements of our reality. These enthusiasts want people to look at a church and not only see "Romanesque style" but also "sandstone".

Another approach is taken through the identification and designation of geoscientifically important places in the natural landscape. If such places are small-scale and/or focused on one specific geoscientific phenomenon, they are called "geosites" – alternatively but far less frequently, as loanword from German language, the term "geotopes" is used. The term "geotope" is reminiscent of the biological counterpart "biotope".

The following are usually defined as geosites, they can be manmade or they can be natural or a mixture of both: meteor impact craters, rock outcrops, mountains or parts thereof, volcanoes, maars, geysers, ice-age landscape remnants, cliffs, active or historic mines, active or historic quarries, fossil sites, boulders, caves, grottos, river valleys, waterfalls, coastal landscapes, ravines, caves, stone material stockpiles, remnants of pre-industrial agricultural landscapes such as stone walls, remnants of previous land-use like post-mining landscapes, post moor drainage landscapes, etc.

The landscape is full of geosites. In most cases that the term is used today, it refers to geosites with some "significance". A number of countries have established inventories or registers of their "significant geosites". Obviously, "significance" is a property that is not objective but ascribed by human beings in a cultural process. What is "significant" varies by time, place and cultural context. "Significance" can result, amongst others, from (local, national or global) rarity, from scale, from original unaltered quality, from "aesthetic impressiveness", or from economic and/or social usefulness. The identification, description, inventory and comparison of geosites is a comparatively new phenomenon. It is only in the last 20 years that more systematic efforts have been made, for example in Germany by the designation of "National Geotopes" or through comparative studies by IUCN for future World Heritage designation (IUCN, 2005). Therefore it is very interesting to observe these currently ongoing cultural processes of attaching "significance". Within a short time span, names and significance have been attached to places and phenomena that were "culturally invisible" before. The process is definitely not finished yet.

UNESCO Global Geoparks are an additional mechanism to make use of the geosphere. In contrast to typically small-scale geosites, UNESCO Global Geoparks are typically larger-scale areas that bring together a number of significant geosites (DUK, 2020). However, they are understood not only as a regional reference frame or to make geosites visible – they are supposed to be managed with an integrative approach of conservation and sustainable development for the entire region. In essence, UNESCO Global Geoparks are the mechanism to turn geosites into a socio-economic asset for the region. An asset that helps to define the region, that creates identity and social cohesion for the inhabitants, that creates new income and job opportunities.

UNESCO Global Geoparks are the ambitious attempt to take the second step together with the first step: We as human beings should not only be aware of the geosphere, we should make use of it for a more sustainably prosperous future.

Human Impact on the Geosphere and the Geosphere's Significance

As has been described above, the geosphere seems to be the only one of the four "spheres" used in today's standard geochemical model of the Earth that needs to be made culturally and emotionally visible. In the Anthropocene, this is highly important, since the entire concept of the Anthropocene depends upon the conviction that we influence the entire planetary environment. There is no part of the planet that we may take for granted, since we ourselves change the planet through the results of our action.

In contrast to the other three "spheres", the "geosphere" seems to be unchangeable, eternally robust, beyond the influence of human beings. This is not true.

Human beings (start to slowly) prepare themselves against the impact of future meteorites.

Human beings change the surface of the Earth: They build gigantic dams and create huge reservoirs (which usually quickly fill up with sediments). They extract so much water from rivers that huge lakes dry up. They remove not only mountaintops, they remove entire mountains. They build new mountains from garbage. They dig huge holes into the planet, for example in surface coal mining. They build new islands, for military purposes, for agricultural land, for airports or for residential areas.

Human beings have altered the nitrogen and phosphorous flows in ways that endanger life on large parts of the planet.

Human beings change the climate to such an extent that deserts expand quickly, arctic ice and glaciers melt fast and sea-level rises increasingly.

The melting of glaciers in turn can lead to the lifting of landmass.

Human beings extract natural gas from underground caverns to such an extent that regular earthquakes can be the consequence.

Anthropogenic radioactive fallout and plastic are part of today's sediments.

To change the geosphere, human beings need not necessarily drill down to the Earth's core. Actually, the Kola Superdeep Borehole, a scientific drilling project of the Soviet Union only reached little more than 12 km depth. To change the geosphere, human beings need not necessarily influence plate tectonics or the formation of Hotspots in the Mantle, Mantle plumes or other Mantle convection processes. These processes are driven by such enormous amounts of energy that indeed human civilization's impacts represent a late decimal place in comparison.

Human beings need a relationship to the geosphere. While large parts of the geosphere may indeed (luckily) be beyond human reach, there is need to understand that what humans do to the solid parts of the planet is not irrelevant either. Human beings need the geosphere for natural resources such as metal, minerals, stones, fossil fuels; they need the geosphere for storing fossil fuels, intermediary means such as pressurized air or methane to compensate for varying renewable energy availability; they maybe need the geosphere for storing carbon dioxide; they need it to get rid of their garbage (radioactive or not); they desperately need underground water from aquifers; they need healthy soils; they need the underground for transport solutions. In fact, some regions of the world have started "underground spatial planning" already, since the underground might become a scarce resource.

The Anthropocene is here: Human beings heavily influence all parts of the planet, including the geosphere. Unfortunately, as has been described above, most human beings are not even aware of the geosphere. There is urgent need to stop taking the solid parts of the Earth for granted.

The paragraphs in the first section, presenting knowledge from geology and other geosciences, is actually standard textbook knowledge as presented in: Tarbuck, E.J., Lutgens, F.K., Tasa, D.: Earth: An Introduction to Physical Geology. Pearsson College Div., 2004; or Spooner, A.M. Geology for Dummies, 2020; or Marshak, S., Earth: Portrait of a Planet, Norton & Co. 2018. The German language-version Wikipedia page on the history of the term "geosphere" (accessed on 15 September 2021) has also been a very helpful source for this article: https://de.wikipedia.org/wiki/Geschichte_des_Begriffs_Geosph%C3%A4re

References

Andrews, S. P. (1871). *The primary synopsis of Universology and Alwato*. Dion Thomas.

Doherty, H. (1864). *Organic philosophy or man's true place in nature volume I – Epicosmology*. Trubner & Co.

DUK/German Commission for UNESCO. (2020). *Model Regions for sustainable development – UNESCO Global Geoparks: From geological heritage to a sustainable future*. Bonn. https://tinyurl.com/2p83a87p

Gray, M. (2004). *Geodiversity: Valuing and Conserving Abiotic Nature*. Wiley.

IUCN. (2005). *Geological world heritage: A global framework*.

Suess, E. (1875). *Die Entstehung der Alpen*. Wilhelm Braumüller Universitäts-Verlagsbuchhandlung.

Teilhard de Chardin, P. (1956). *La Place de l'Homme dans la Nature*. Albin Michel.

Lutz Möller is Deputy Secretary-General of the German Commission for UNESCO since 2015; he also heads its Division for Education, Science and Culture. He holds a Ph.D. in Theoretical Physics from Munich University (LMU) and has a study background in Physics and Philosophy (Munich, Oxford). At the German Commission for UNESCO, where he has worked since 2004, he guides the policy advice work, is responsible for the overarching cooperation with UNESCO, and with other National Commissions for UNESCO globally, and with the private sector. He has been a member of the German National Committee for the UNESCO Programme "Man and the Biosphere" (MAB) more than 17 years, he has provided decisive input to key policy processes for the MAB Programme at global and national level for this entire period; he has also been a decisive figure in the creation and the implementation of the UNESCO Global Geoparks Programme, globally and nationally. He has also been the representative of the German Commission for UNESCO, for more than ten years respectively, in the relevant bodies for the Intergovernmental Hydrological Programme (IHP) of UNESCO, for UNESCO's Intergovernmental Oceanographic Commission (IOC) and the International Geoscience Programme (IGCP).

Global Change

Zhisheng An

Abstract Global change refers to changes in Earth system – including physical, chemical, and biological processes in the lithosphere (land), atmosphere (air), hydrosphere (water) and biosphere (living things). Humans are part of the Earth system, with significant impacts on natural physical processes and biogeochemical cycles. The concept of the Anthropocene was initially proposed on the basis of global change studies, as scientists became more aware that human activities may rival natural processes. Global change studies that focus on the Anthropocene help us better understand how the Earth system responds to human forcing, and how these responses will shape our future.

Global change is generally defined as planetary-scale changes in the Earth system, with emphasis on the complexity and changes of the earth as a whole. Although past studies on Earth's evolution formed an important conceptual precursor to understanding the Earth system, it was not until the 1980s that Earth System Science (ESS) became recognized as a valuable, deeper view of the planet (NASA advisory council, 1988). ESS aims to understand the structure and function of the earth as a complex, adaptive system. Global change encompasses not only natural changes, including interactions among the lithosphere, atmosphere, hydrosphere and biosphere, but also human impacts manifested by social and economic changes in populations, economies, resources, energy, transportation, and land use, etc.

The evolution and development of life have had profound influence on the Earth system, and studies of global change emphasize the critical role of human forcing on the Earth's environment. The term "Anthropocene" was proposed at the International Geosphere–Biosphere Programme (IGBP) meeting in Cuernavaca, Mexico, in 2000 (Crutzen, 2002; Crutzen & Stoermer, 2000). The Anthropocene was considered a new epoch when the scale of human impacts reached that of many

Z. An (✉)
State Key Laboratory of Loess, Institute of Earth Environment,
Chinese Academy of Sciences, Xi'an, China
e-mail: anzs@loess.llqg.ac.cn

© The Author(s), under exclusive license to Springer Nature
Switzerland AG 2023
N. Wallenhorst, C. Wulf (eds.), *Handbook of the Anthropocene*,
https://doi.org/10.1007/978-3-031-25910-4_10

63

natural impacts on the Earth system. In some aspects, human impacts have exceeded the boundaries of previously natural functions (Rockstrom et al., 2009) and have pushed the Earth system toward planetary thresholds beyond the natural Holocene climate state.

A major consequence and manifestation of global change is global climate change. Although "global change" is a broader term than "climate change", the two terms are used interchangeably in much of the current literature. The concept of global change gained popularity with the discoveries of synchronous global cyclic climate change evidenced by $\delta^{18}O$ in deep sea sediments and an increase of atmospheric carbon dioxide concentration measured at Mauna Loa Observatory, Hawaii, as well as global warming evidence recorded by tree ring studies. With the advance in our understanding of climate change, we have gained a better understanding of climate in the context of systematic and integrated global scale interactions of spheres within the Earth system.

Global climate change is modulated by both external and internal drivers , such as Earth's orbital geometry, ice-sheets and greenhouse gas conditions, at various time scales. Multi-scale interactions and feedbacks increase the complexity in understanding climate change (An et al., 2015). Cyclic changes in the earth's orbit around the Sun, including eccentricity, obliquity and precession, lead to seasonal changes in the sunlight distribution on the earth's surface (Milankovitch, 1941; Berger, 1988). Over the past 4.5 billion years, the Earth's climate has experienced repeated rise and fall of air temperature as the Earth system evolved. These climate changes fluctuated over a wide range of timescales, from intraseasonal to interannual cycles, to interdecadal, multidecadal, centennial, suborbital (millennial), orbital, and tectonic-scale variabilities. The primary drivers of climate variability on different timescales are not independent but involve inherent interactions. The overall climate in the Cenozoic era was characterized by a gradual shift from a greenhouse to an icehouse state (Zachos et al., 2008; Westerhold et al., 2020). For example, during the Early Eocene Climatic Optimum, the global mean air temperature was approximately 15 °C higher than today, and atmospheric CO_2 concentrations were about 2500–4500 parts per million (ppm), far higher than today (Zachos et al., 2008). Even in the Pliocene epoch, atmospheric CO_2 concentrations were about 350–450 ppm and air temperatures were about 2–4 °C higher than today (Martinez-Boti et al., 2015). In the Quaternary period, the climate further cooled, with more predictable glacial-interglacial variations (Lisiecki & Raymo, 2005; An et al., 2014). However, Earth's natural environment does not follow a single state. Global change is nonlinear, with some sustained small changes pushing the system across thresholds that result in abrupt changes. This may be especially true when the changes coincide with extremely cool or warm periods.

About 11,700 years ago, the Earth went into a warm interglacial period, called the Holocene. Human activities such as the development of large-scale agriculture and the Industrial Revolution, when fossil fuel burning emissions increased, have had an impact on Holocene climate. A shift from Quaternary glacial-interglacial variations to a warmer planet during the Anthropocene period is likely if human-induced emissions continue (Steffen et al., 2018). Internal land-air-sea interactions (e.g., thermohaline circulation, sea-ice extent, sea surface temperature) associated with solar output and anthropogenic forcing (e.g., greenhouse gases, aerosols, and land use) are key factors determining the climatic condition during this period.

The most prominent feature of the Anthropocene is global warming. The global temperature has been rising from the mid-twentieth century to the present. As warming ensues, the ocean becomes more acidic, ice sheets begin to melt, glaciers retreat, snow cover decreases, sea level rises, Arctic sea ice declines, and global extreme weather events occur more frequently. The current warming trend is highly likely (greater than 95% probability) to be the result of human activity, especially from burning fossil fuels (e.g., coal, oil, and natural gas). Human activities cause the release of large amounts of greenhouse gases into the atmosphere, including carbon dioxide, methane, nitrous oxide and chlorofluorocarbons (CFCs). Atmospheric CO_2 concentrations increased from 280 ppm before 1950 to 412.5 ppm in 2020 and CH_4 concentrations also increased over this period. These greenhouse gases can trap long-wavelength radiation inside the atmosphere and thus cause a rise in Earth's average temperature. Such an increasing air temperature trend has been occurring at a rate that is unprecedented over the past millennia and even over longer timescales (Allen et al., 2019). The pattern far exceeds previous natural analogues from Quaternary interglacials, and thus the Earth system seems to have reached a "no analogue state." In addition, these changes have influenced carbon, nitrogen and phosphorus cycles on the earth (Rockstrom et al., 2009).

Intense human industrial activities produce materials and pollutants that can be harmful to our environment. The cumulative growth of manufactured aluminum, concrete, plastics, synthetic fiber, and spheroidal carbonaceous particles can now be found in most geological compartments (Waters et al., 2016). Aerosols emitted into the atmosphere are of particular concern. Apart from causing air pollution, they also influence the climate system, although the detailed dynamics remain underexplored. Most aerosol components, such as sulfate, nitrate, ammonium, organics, dust, and sea salt are light-reflecting, and can cause a cooling effect in the atmosphere. However, some other aerosol components, such as black carbon (BC) from both fossil fuels and biomass burning (Han et al., 2018), are light-absorbing, and can lead to a warming effect. Relative to greenhouse gases such as CO_2, the radiation effects of the light-absorbing aerosols are highly uncertain, and vary in different regions and under different meteorological conditions. The uncertainties are further exacerbated when aerosol-cloud interactions are considered. Historically, the emissions of BC increased with the rapid development of industries (McConnell et al., 2007). However, over the past decades, aerosol pollution has decreased in many countries as pollution control technologies advanced. Governments around the world are now facing high pressure to reduce greenhouse gases emissions, and meet the goal of the Paris Agreement to limit a rise in air temperature to 2 °C relative to the preindustrial period. It is expected that many more measures will be implemented to achieve this objective.

Even if all governments were to abide by the Paris Agreement and take immediate actions to reduce greenhouse gas emissions, a simultaneous decrease in air temperature would likely be delayed because of the longevity of greenhouse gases in the environment (Allen et al., 2019). Although a reduction in short-lifetime aerosol pollutants can be expected, their climatic effects are more complex due to the relatively large uncertainties in radiation forcing and interactions with clouds. There has been a recommendation stating that the reduction of light-absorbing BC might be an effective way of mitigating the global warming trend (Bond & Sun, 2005). On the other hand, reduction of light-reflective aerosols may offset such effects. Along with the reduction in greenhouse gases, the co-reduction in aerosols and their climatic effects, especially in interactions with clouds, need to be thoroughly investigated.

In the Anthropocene Epoch, humans have played an important role in the Earth system, causing changes not only in the biosphere but also in the atmosphere, ocean, and land. Although previous paleoclimate studies revealed the functions of external and internal forcings that modulate paleoclimate changes, the course of future climate change still needs to be urgently studied. Human impacts on the Earth system are multiple, complex, and sometimes operate at an unprecedented magnitude, scale, and pace. As an internal factor of humankind that influences the Earth system, it may cause "domino effects" in the system, and in some aspects human impacts have exceeded thresholds seen in nature (Rockstrom et al., 2009). The great challenge we arc facing now hinges on greenhouse gas and aerosol emissions that influence our climate, environment, and human health. To better understand the projection of Earth system change and our shared future destiny, changes in global and regional climate trends and their environmental effects must be evaluated from the perspective of both natural and anthropogenic contributions. Particularly, anthropogenic forcings such as greenhouse gases, aerosols, vegetation, and land use must be fully understood in the context of land-air-sea interactions.

References

Allen, M. R., Dube, O. P., Solecki, W. A., Aragón-Durand, F., Cramer, W., Humphreys, S., Kainuma, M., Kala, J., Mahowald, N., Mulugetta, Y., Perez, R., Wairiu, M., & Zickfeld, K. (2019). Framing and context. In V. Masson-Delmotte, P. Zhai, H.-O. Portner, D. Roberts, J. Skea, P. R. Shukla, A. Pirani, W. Moufouma-Okia, C. Péan, R. Pidcock, S. Connors, J. B. R. Matthews, Y. Chen, X. Zhou, M. I. Gomis, E. Lonnoy, T. Maycock, M. Tignor, & T. Waterfield (Eds.), *Global Warming of 1.5°C. An IPCC Special Report on the Impacts of Global Warming of 1.5 °C above pre-industrial levels and related global greenhouse gas emission pathways, in the context of strengthening the global response to the threat of climate change* (p. 84). Intergovernmental Panel on Climate Change (IPCC).
An, Z., Sun, Y., Zhou, W., Liu, W., Qiang, X., Wang, X., Xian, F., Cheng, P., & Burr, G. S. (2014). Chinese Loess and the East Asian Monsoon. In Z. An (Ed.), *Late Cenozoic climate change in Asia* (pp. 23–143). Springer.
An, Z., Wu, G., Li, J., Sun, Y., Liu, Y., Zhou, W., Cai, Y., Duan, A., Li, L., Mao, J., Cheng, H., Shi, Z., Tan, L., Yan, H., Ao, H., Chang, H., & Feng, J. (2015). Global monsoon dynamics and climate change. *Annual Review of Earth and Planetary Sciences, 43,* 29–77.
Berger, A. (1988). Milankovitch theory and climate. *Reviews of Geophysics, 26,* 624–657.
Bond, T. C., & Sun, H. L. (2005). Can reducing black carbon emissions counteract global warming? *Environmental Science & Technology, 39*(16), 5921–5926.

Crutzen, P. J. (2002). Geology of mankind. *Nature, 415*(6867), 23–23.

Crutzen, P. J., & Stoermer, E. F. (2000). The "Anthropocene". *Global Change Newsletter, 41*, 17.

Han, Y. M., An, Z. S., & Cao, J. J. (2018). *The Anthropocene—A potential stratigraphic definition based on black carbon, char, and soot records*. The Encyclopedia of the Anthropocene. Elsevier.

Lisiecki, L. E., & Raymo, M. E. (2005). A Pliocene-Pleistocene stack of 57 globally distributed benthic delta O-18 records. *Paleoceanography, 20*, PA1003.

Martinez-Boti, M. A., Foster, G. L., Chalk, T. B., Rohling, E. J., Sexton, P. F., Lunt, D. J., Pancost, R. D., Badger, M. P. S., & Schmidt, D. N. (2015). Plio-Pleistocene climate sensitivity evaluated using high-resolution CO_2 records. *Nature, 518*(7537), 49-54.

McConnell, J. R., Edwards, R., Kok, G. L., Flanner, M. G., Zender, C. S., Saltzman, E. S., Banta, J. R., Pasteris, D. R., Carter, M. M., & Kahl, J. D. W. (2007). 20th-century industrial black carbon emissions altered arctic climate forcing. *Science, 317*(5843), 1381–1384.

Milankovitch, M. (1941). Kanon der Erdbestrahlung und seine Anwendung auf das Eiszeitproblem. Belgrade, YU: Royal Serbian Academy, Special Publication, 132. *Section of Mathematical and Natural Sciences, 32*, 1–633.

NASA advisory council. (1988). *Earth system sciences: A closer view*.

Rockstrom, J., Steffen, W., Noone, K., Persson, A., Chapin, F. S., III, Lambin, E. F., Lenton, T. M., Scheffer, M., Folke, C., Schellnhuber, H. J., Nykvist, B., de Wit, C. A., Hughes, T., van der Leeuw, S., Rodhe, H., Sorlin, S., Snyder, P. K., Costanza, R., Svedin, U., Falkenmark, M., Karlberg, L., Corell, R. W., Fabry, V. J., Hansen, J., Walker, B., Liverman, D., Richardson, K., Crutzen, P., & Foley, J. A. (2009). A safe operating space for humanity. *Nature, 461*(7263), 472–475.

Steffen, W., Rockstrom, J., Richardson, K., Lenton, T. M., Folke, C., Liverman, D., Summerhayes, C. P., Barnosky, A. D., Cornell, S. E., Crucifix, M., Donges, J. F., Fetzer, I., Lade, S. J., Scheffer, M., Winkelmann, R., & Schellnhuber, H. J. (2018). Trajectories of the Earth system in the Anthropocene. *Proceedings of the National Academy of Sciences of the United States of America, 115*(33), 8252–8259.

Waters, C. N., Zalasiewicz, J., Summerhayes, C., Barnosky, A. D., Poirier, C., Galuszka, A., Cearreta, A., Edgeworth, M., Ellis, E. C., Ellis, M., Jeandel, C., Leinfelder, R., McNeill, J. R., Richter, D. D., Steffen, W., Syvitski, J., Vidas, D., Wagreich, M., Williams, M., An, Z., Grinevald, J., Odada, E., Oreskes, N., & Wolfe, A. P. (2016). The Anthropocene is functionally and stratigraphically distinct from the Holocene. *Science, 351*(6269), 6269, aad2622.

Westerhold, T., Marwan, N., Drury, A. J., Liebrand, D., Agnini, C., Anagnostou, E., Barnet, J. S. K., Bohaty, S. M., De Vleeschouwer, D., Florindo, F., Frederichs, T., Hodell, D. A., Holbourn, A. E., Kroon, D., Lauretano, V., Littler, K., Lourens, L. J., Lyle, M., Palike, H., Rohl, U., Tian, J., Wilkens, R. H., Wilson, P. A., & Zachos, J. C. (2020). An astronomically dated record of Earth's climate and its predictability over the last 66 million years. *Science, 369*(6509), 1383–1387.

Zachos, J. C., Dickens, G. R., & Zeebe, R. E. (2008). An early Cenozoic perspective on greenhouse warming and carbon-cycle dynamics. *Nature, 451*(7176), 279–283.

Zhisheng An is a world-renowned geoscientist who specializes in monsoon dynamics and global change research. He is internationally known for his work on the Chinese loess and on the dynamics of Asian monsoon system. He was a founder of the Institute of Earth Environment, Chinese Academy of Sciences, in Xi'an, China. He served as the vice chair of the scientific committee of the International Geosphere–Biosphere Programme (IGBP). He is currently a member of the Anthropocene Working Group (AWG), a member of the Chinese Academy of Sciences (CAS), and a foreign associate of the US National Academy of Sciences (NAS). An has authored several books on environment, climate, and Chinese loess, and has published over 470 scientific papers, including 30 in Science, PNAS, Nature and Nature Portfolio.

Hydrosphere

Lutz Möller

Abstract This article examines the content, the history and the implications of the term "hydrosphere". It argues that the term and the concept of the term are relevant and easy to understand, even if the term is not in wide daily use. The article goes on to differentiate between the "ocean hydrosphere" and the "freshwater hydrosphere". It argues how crucially important all parts of the hydrosphere are and that humanity is definitely not attaching sufficient value to them. It also presents efforts, including those by UNESCO, to strengthen the hydrosphere's significance for humanity in the Anthropocene.

The Term and Its Conceptual Differentiation

"Hydrosphere" is a term that is not widely used in the public. It refers to all water on planet Earth.

Together with the other three "spheres" (geosphere, biosphere, and atmosphere, compare the articles in this Handbook), the term is part of today's differentiated geochemical model of the planet.

The term was first introduced by Austrian geologist Eduard Suess (Suess, 1875), who in parallel introduced the much better known term "biosphere". Stephen Pearl Andrews, who had introduced the term "geosphere" 4 years earlier (Andrews, 1871), also had a term for all water on Earth. He called that part of the planet "thallatosphere", a term which did not catch on. Hugh Doherty who had introduced in 1864 terms such as "geospheric realm" or "atmospheric realm" (Doherty, 1864), differentiated, when it comes to water, the "pluvial realm" and the "oceanic realm". The terminology has ancient precursors in cultural traditions around the globe, such as China, India, Japan, Persia, Babylonia, Egypt and Greece. Such ancient cultural traditions all counted "water" among the small number of "elements" that would

L. Möller (✉)
German Commission for UNESCO, Bonn, Germany
e-mail: moeller@unesco.de

N. Wallenhorst, C. Wulf (eds.), *Handbook of the Anthropocene*,
https://doi.org/10.1007/978-3-031-25910-4_11

constitute nature. As has been written in the article on "geosphere" in this Handbook, the Greek philosopher Aristotle used the model of celestial and terrestrial "spheres". Thus, Eduard Suess has to be given credit, but he made an incremental contribution only in a long history of conceptual innovation.

What does the term exactly refer to? As is the case with all three other "spheres", there is some ambiguity. The hydrosphere is the totality of water on the planet, understood either in terms of volume, in terms of mass or as a collection of geographical places. As with all other "spheres", the "hydrosphere" is a rather abstract, not a directly observable concept.

The hydrosphere cannot be sharply differentiated from the geosphere: One reason is that each volume of water contains suspended minerals such as bicarbonate, sodium or calcium that can be of varying grain size according to filtration, and, depending on the water body and its flow velocity, can turn into sediments. Another reason that there is groundwater; tiny pockets of groundwater that do not build aquifers are hard to differentiate from the geosphere.

The hydrosphere cannot be sharply differentiated from the biosphere: One reason is that each volume of water contains bacteria and other microorganisms. Another reason is that each living organism consists largely of water; plant evaporation is a very important contributor to the water cycle. Another reason are wetlands such as swamps, or bogs that accumulate peat through a vegetation process in a wet environment.

The hydrosphere cannot be sharply differentiated from the atmosphere either: Air can be dissolved in water, tiny water droplets such as clouds and fog, but also simply "humid air" are actually counted as part of the atmosphere.

The four "spheres" do not only collectively make up planet Earth – they mutually depend upon each other. For example, the atmosphere depends upon the existence of the magnetic field generated in the outer core of the geosphere; without the magnetic field, the atmosphere would be destroyed by solar wind. Water in turn is protected by the atmosphere from high-energy radiation from space. As is well-known, the oxygen in our atmosphere (the crucial 21 percent) is a result of one of the earliest constituents of the biosphere, the cyanobacteria. Early life, in turn, has likely emerged at special places on the hydrosphere-geosphere nexus, deep in the ocean, on hot hydrothermal vents.

As is the case for all other "spheres", the "hydrosphere" is a relational concept of human beings towards their planetary environment; it is a conceptually organizing principle, not an objective independent existence.

The "Hydrosphere" Is a Well-Comprehensible Concept

Taking into account theses preliminaries, the hydrosphere is a rather graphic and easy to understand concept. Probably a key reason is the importance of the oceans for the planet. The ocean covers some 361 million square kilometres and thus some 71% of the Earth's surface; the ocean contains some 1.3 trillion cubic kilometres and 97% of its water. The rest of the mass of the planet (mainly its geosphere) has 4000 times more mass than the entirety of ocean mass. There is actually just one big ocean covering the planet, also called "the sea". For practical purposes, five main oceans (in the plural) are distinguished: the Atlantic Ocean including parts such as the North and Baltic Sea, the Mediterranean and the Caribbean; the Indian Ocean including parts such as the Red Sea and the Arabian Sea; the Pacific Ocean including parts such as the South China Sea or the Philippine Sea; the Arctic Ocean; and the Southern Ocean.

The ocean has high significance for the majority of the human population on earth. Some 40% of the global population live within 100 kilometres of the ocean coast, a significant additional number of human beings visit the coastline temporarily as workers, travellers or tourists. The vastness of the ocean and the obvious dangers it can present to human life if no care is taken make an impression on every human being. Obviously, the ocean is huge; most humans learn early in life that it is global. Thus, there is a small intellectual and emotional step from the existence of the ocean to the "hydrosphere".

Of course, the ocean is not the entire hydrosphere. There is ice, in particular in glaciers in the Antarctic, in the Arctic and on mountains – actually making up some two thirds of freshwater (non-saltwater). There are springs, lakes, there are rivers, ponds, wetland, estuaries, and floodplains; there are groundwater aquifers. In addition, there are clouds (counted as part of the atmosphere).

Most people will have heard of the water cycle, the process that "starts" with evaporation (plus evapotranspiration and sublimation) – more precisely, every element of the cycle can be considered the starting point of the water cycle. This leads to condensation of clouds in the atmosphere and precipitation. Through surface runoff and streamflow, but also through seepage and soil and rock infiltration into groundwater flow or springs, freshwater surface and subsurface flows are driven.

The water cycle is certainly the best known biogeochemical cycle by far; it can be understood quite intuitively, since most people have made personal encounters with several stages of the cycle such as clouds, rain and rivers. Most people have made the experience of fast-travelling clouds and thus it will not come as a surprise that the water cycle is not a "local phenomenon", but rather a process over thousands of kilometres.

The water cycle indeed as a global system and, thus, the hydrosphere as a global concept are thus not hard to understand. Indeed, the regular weather phenomenon of El Niño (which is the warm phase of the El Niño–Southern Oscillation) is well known even to the general public. It is not well known yet that the El Niño phenomenon has impacts upon the weather almost across the entire planet, not only in Southern America but also in Japan, Southern and Eastern Africa, India and East Russia.

Thus, the concept of the hydrosphere, in spite of the precaution that needs to be taken, is useful because it can be associated rather closely with physical realities. At the same time, it is quite well comprehensible by the general public. Thus, the hydrosphere is a particularly useful concept in the Anthropocene.

The Significance of the "Freshwater Hydrosphere"

A UNESCO report of 1998 summarized the total availability of freshwater: "Every year the turnover of water on Earth involves 577,000 km^3 of water. This is water that evaporates from the oceanic surface (502,800 km^3) and from land (74,200 km^3). The same amount of water falls as atmospheric precipitation, 458,000 km^3 on the ocean and 119,000 km^3 on land. The difference between precipitation and evaporation from the land surface (119,000–74,200 = 44,800 km^3/year) represents the total run-off of the Earth's rivers (42,700 km^3/year) and direct groundwater runoff to the ocean (2100 km^3/year). These are the principal sources of fresh water to support life necessities and man's economic activities." (UNESCO, 1998).

This does not only sound a lot. It is a lot. Per capita of the human population this is more than 100,000 times more than what we need for our household consumption. However, not all precipitation and river streams emerge (e.g. in Siberia or Canada) where we need it. Even more, also forests and all other ecosystems need water, agriculture needs water, industry and energy production needs water.

In many places on Earth with high population figures, there is already "water stress, essentially measured as water use as a function of available supply, affects many parts of the world. Over two billion people live in countries experiencing water stress". "An estimated four billion people live in areas that suffer from severe physical water scarcity for at least one month per year" (Mekonnen & Hoekstra, 2016).

Water scarcity and water stress in many parts of the world is the combined result of population growth, economic development and shifting consumption patterns. "Agriculture currently accounts for 69% of global water withdrawals, which are mainly used for irrigation but also include water used for livestock and aquaculture. This ratio can reach up to 95% in some developing countries" (FAO, 2011). At the same time, "water quality has deteriorated as a result of pollution in nearly all major rivers in Africa, Asia and Latin America. Nutrient loading, which is often associated with pathogen loading, are among the most prevalent sources of pollution" (UNEP, 2016).

Thus, problems with water availability and water quality are mostly directly manmade, they are not (yet) mainly driven by global environmental change processes such as climate change – which are also anthropogenic. The impact of climate change is visible already, however: "Globally, floods and extreme rainfall events have increased by more than 50% over the past decade, occurring at a rate four times greater than in 1980" (EASAC, 2018).

Since "climate change is expected to further increase the frequency and severity of floods and droughts" (IPCC, 2018), human societies are under high pressure to change at least those of their detrimental freshwater consumption patterns that immediately worsen water availability and water quality. Today, human beings use too much water in a much too wasteful way.

What might be the reason for this wasteful overuse? While human beings have a principled awareness of the significance of water and know (at least partially) about the global dimension of the hydrosphere, this principled awareness and knowledge does not translate into action. A likely reason is the following: Water has always been present in cultural terms, and, in many places, has often been present in abundant amounts. Therefore, most human beings still take water for granted, whether they open the faucet, dive into rivers and lakes, or pump water from the ground.

What about "holy water" then? Indeed, water is sometimes considered holy or has liturgic functions, in several religions actually, whether it is Sikhs, Hindus, Buddhists, Muslims or Catholic Christians. But in none of these religions, all water is holy. In some cases, water needs to be blessed to be holy; in other cases it is specific water bodies that are considered holy. Probably there is a creek full of garbage and sewage close to every church or temple with a holy water font.

The key problem with wasteful water misuse and water overuse is the lack of general cultural valuation of water. In 2021, UNESCO has published its annual World Water Development Report with the title "Valuing Water" (UNESCO, 2021). Indeed, this is what is needed – humanity, collectively, needs to attach value to water, to all water, to the hydrosphere. There are several forms of value and we need all of them: cultural value, social value and economic value.

One of the most striking figures from the recent UNESCO report is the following: The typical price for 1 m^3 of freshwater for agricultural purposes is 5 US cents – that is a global average. This price should be compared to the average price in Central Europe for household water which is around 2 to 5 Euro per 1 m^3. While the economic price cannot be and should not be the only driver for change, it definitely contributes. Water, which is so cheap, will never have value; such cheap water will not lead to savings in water.

All freshwater resources are significant, culturally, socially, ecologically and economically, not just "holy water". This is what humanity has to learn quickly.

The Significance of the "Ocean Hydrosphere"

As has been written above, the ocean is generally a space and place, where humanity encounters the planet's grandeur, resulting in awe and humbleness, in inspiration, in joy, in recreational recovery, possibly in dread. The ocean is hardly ever a place that leaves human beings untouched and indifferent.

However, the encounter with the ocean does not seem to lead to much permanent engagement. The ocean in the Anthropocene suffers severely from human activity. Many problems of the ocean are rather invisible and not very vivid for the general public. The ocean is not only polluted by plastic, but also by invisible toxins, heavy metals and nutrients. Habitats are being destroyed, natural resources are being depleted.

The decisive challenge for the ocean, however, is climate change and its impacts. The ocean absorbs 30–40% of the carbon dioxide emitted by humans. When dissolved in water, this substance acts as an acid; this means that the ocean gradually acidifies. At the same time, the ocean absorbs 90% of the heat generated by global warming, which is another reason why the oxygen content decreases. Without the ocean, the world atmospheric temperature would have risen much more already. The ocean provides irreplaceable services for human societies. Meanwhile, the combination of ocean acidification and rising surface temperatures is affecting countless aquatic organisms such as fish, seabirds, mammals, but especially animals with calcareous shells like corals.

Another problem is overfishing. Only for 7% of all fish stocks, theoretically, a small increase of the catch is possible. All other fish stocks are depleted already. Overfishing is fuelling the extinction of species (IPBES, 2019). According to the 2019 IPBES global assessment report, there are 20% fewer species worldwide today than at the beginning of the twentieth century. More than 40% of amphibian species, almost 33% of corals and more than a third of all mammal species in the sea are threatened.

Another key challenge ahead, one of the "tipping points" (Schellnhuber, 2009) of the planetary system as regards climate change, is the potential change in the ocean currents, more specifically thermohaline circulation. While currently still a speculation, more and more scientists are worried about a potential slowing of the Gulf Stream, which transports warm water pole-wards driven by wind, from the equatorial Atlantic Ocean along the Eastern shore of North America and the Western shore of Europe. At present, this Gulf Stream water sinks to the ocean floor East of Greenland due to density and salt content. The resulting deep water currents upwell mostly in the Southern Ocean, but some of that water also upwells in the North Pacific after centuries.

The thermohaline circulation is the most convincing, most important process making the Earth a truly "global system", more comprehensible to human beings at least than (more short-term) Jet Stream atmospheric processes or (more long-term) plate tectonics. The thermohaline circulation is "the hydrosphere at its best". The danger that human civilization might alter this system within the next few decades is breath-taking. There is hardly anything that demonstrates better than this threat to the thermohaline circulation that we are living in the Anthropocene.

Humanity needs to better understand how the global processes in the hydrosphere function – and it needs to understand this knowledge into action. Luckily, there are already frameworks in place for exactly this purpose: knowledge generation and translation into action. It is now upon UNESCO and its Intergovernmental Hydrological Programme (UNESCO-IHP, n.d.), as well as its Intergovernmental Oceanographic Commission (UNESCO-IOC, n.d.) and the UNESCO Member States, to do exactly that.

The hydrosphere is the "Canary in the Coalmine" of the Anthropocene crises. Let us listen to the canary.

References

Andrews, S. P. (1871). *The primary synopsis of Universology and Alwato*. Dion Thomas.

Doherty, H. (1864). *Organic philosophy or Man's true place in nature volume I – Epicosmology*. Trubner & Co.

EASAC. (2018). *Extreme weather events in Europe*. Halle/Saale.

FAO. (2011). *The future of food and agriculture*.

IPBES, 2019. *IPBES global assessment on biodiversity and ecosystem services*.

IPCC, 2018. *Changes in climate extremes and their impacts on the natural physical environment*.

Mekonnen, M. M., & Hoekstra, A. Y. (2016). Four billion people facing severe water scarcity. *Science Advances, 2*, e1500323. https://doi.org/10.1126/sciadv.1500323

Schellnhuber, H. J. (2009). Tipping elements in the earth system. *Proceedings of the National Academy of Sciences, 106*(49), 20561–20563.

Suess, E. (1875). *Die Entstehung der Alpen*. Wilhelm Braumüller Universitäts-Verlagsbuchhandlung.

UNEP, 2016. *A snapshot of the World's water quality: Towards a global assessment*. .

UNESCO. (1998). *World water resources: A new appraisal and assessment for the 21st century*. UNESCO.

UNESCO. (2021). *World water development report: Valuing water*. UNESCO.

UNESCO-IHP. (n.d.). Website of the IHP. https://en.unesco.org/themes/water-security/hydrology

UNESCO-IOC. (n.d.). Website of the IOC. https://ioc.unesco.org/

Lutz Möller is Deputy Secretary-General of the German Commission for UNESCO since 2015; he also heads its Division for Education, Science and Culture. He holds a Ph.D. in Theoretical Physics from Munich University (LMU) and has a study background in Physics and Philosophy (Munich, Oxford). At the German Commission for UNESCO, where he has worked since 2004, he guides the policy advice work, is responsible for the overarching cooperation with UNESCO, and with other National Commissions for UNESCO globally, and with the private sector. He has been a member of the German National Committee for the UNESCO Programme "Man and the Biosphere" (MAB) more than 17 years, he has provided decisive input to key policy processes for the MAB Programme at global and national level for this entire period; he has also been a decisive figure in the creation and the implementation of the UNESCO Global Geoparks Programme, globally and nationally. He has also been the representative of the German Commission for UNESCO, for more than 10 years respectively, in the relevant bodies for the Intergovernmental Hydrological Programme (IHP) of UNESCO, for UNESCO's Intergovernmental Oceanographic Commission (IOC) and the International Geoscience Programme (IGCP).

Life

Catherine Fino

Abstract The contemporary context is characterized by the contrast between liberal biotechnical innovation policies and normative protectionist prescriptions against a backdrop of anthropological insecurity. This article aims to question the philosopher Georges Canguilhem as a precursor of thought on the history of the living that allows for conceptualizing the ontological permanence of a subject at the very place of adaptive processes and displacement of individual anthropological norms, without denying the existence of constraints and limits, to ensure the maintenance of biological or social life. Discernment can take into account the different fields of reality more easily when they are also open to history.

Georges Canguilhem has shifted the definition of health, understood in opposition to disease and as the normal situation to restore, towards a synergy in which disease is perceived as a positive test by which the living one proves its capacity to adapt to new living conditions. According to Canguilhem, in fact, "the normal man is the normative man, the being capable of instituting new norms, even organic ones". Conversely, the rigidity of an organic standard weakens the living being unable to adapt to variations in its internal or external environment: "A unique standard of life is felt privately and not positively" (Canguilhem, 2013, p. 116). For Guillaume Le Blanc, Canguilhem introduces by the fact itself of a history of the living, where "life is by itself creation. The event, before acquiring a historical sense, first takes on a biological meaning" (Le Blanc, 1998, p. 41). This history of life can be described as radical in the sense that the adaptability and creativity of the living, inseparable from its vulnerability, its ability to be altered, is not only what characterizes it, but also what ensures its survival - that is, its very existence, its condition of possibility, at parity with the respect of the limits of viability. According to Pierre Macherey, this inescapable historicity of the living requires an ontological choice from Canguilhem in which "the living cannot be reduced to a material datum, but that it

C. Fino (✉)
Catholic Institute of Paris, Paris, France
e-mail: c.fino@icp.fr

is a possibility in the sense of a power, that is to say, a reality which gives itself from the outset as unfinished because it is confronted intermittently with the risks of disease and permanently with that of death" (Macherey, 2009, p. 137). Jean Gayon emphasizes the relational character that characterizes the ontology of the living, based on the articles in which Canguilhem deals with "cellular theory" (Canguilhem, 1992) and defines individuality not as a limited substance in time and space, but first as "an open set of relationships" (Gayon, 2000, p. 35). The adaptability of the living, at all levels, is inseparable from the resources it receives from its living environment. The relationship between the living themselves is part of this dynamic of competition or mutual support for access to vital resources for everyone. Philosophical and ethical discernment must position itself on the choice to privilege the maximization of Life in its global dimension or the accompaniment of the life course of each individual, of each person in relation to others. The practice of medicine directs Canguilhem towards the latter perspective: his anthropocentrism is first a type of benevolence for each person for whom he is responsible, and not the depreciation of other living beings.

In a biological background characterized by real flexibility or "lability" based on the solicitation of available resources, Canguilhem introduces the subjective freedom that all people have in their singular way of embodying the living in their body, through their practices:

"Each of us sets his standards by choosing his exercise models. The standard of the runner is not that of the sprinter. Each of us changes his standards according to his age and his previous standards "(Canguilhem, 2013, p. 277). Canguilhem nevertheless maintains the objective character of the "valorisation" operated by the body which gives orientations to normative creativity and imposes its limits on science: "It is life itself and not medical judgment that makes normal biological a concept of value and not a concept of statistical reality" (Canguilhem, 2013, p. 107). The bodily and biological framework thus remains the objective regulatory standard underlying the pathways that have had to move away from the norm to survive or that result from a new lifestyle choice. (Fino, 2018, p. 520).

In such a philosophical framework, Céline Lefève can describe a dual function and an ethical posture of medicine, in order to facilitate the process of life in the person who solicits it. Medicine combines a practice and an ethics of "vital duty", defined as "this obligation to respond to vital duties, to the incessant injunctions of the environment and existence" and for the physician, the duty to inform the person, and a practice and an ethics of freedom that favours the restoration of the subject's normativity, of his ability to develop his own standards, and for the physician the duty to accompany and rationalize this rebalancing process (Lefève, 2000, p. 115–119). This dual "pedagogical" function of medicine makes it possible to secure people within the dual responsibility that is now conferred upon them, while maintaining the need for existential insecurity that stimulates the body as the subject to constantly innovate for survival.

From the viewpoint of bioethical discernment, the question remains whether the impact of the contemporary development of biotechnologies is sufficiently taken into account. Canguilhem himself faces the challenge of technical development in his article "Machine and Organism" (1992, p. 121), refusing the "one-way theory" which understands the organism as a machine, to privilege the empirical construction of machines "imitating organic movements", "a technical activity as authentically organic as fruit trees in bloom and, primitively, as little aware as plant life of the rules and laws that guarantee its effectiveness." Canguilhem does not hesitate to carefully evaluate Ernst Kapp's biological philosophy of technology (1877), despite the risk of perpetuating "an original order" (Chamayou, 2007, p. 36). Canguilhem continues the dialogue with Gilbert Simondon (2012, p.200), who describes the "concretization" of complex technical "sets" as the progressive ability to acquire a natural appearance, while not possessing "the faculty of the living to change" to face the future.

Nevertheless, the human person has the capacity and the responsibility to change techniques over the course of history and thereby hope to correct nature. But Dominique Bourg (1996, p.21) notes that a new phase of technology has occurred with genetic engineering which conversely contributes to modifying natural objects by making them closer to machines, or even reducing them to sources of production, while techniques acquire an ability to learn. The erasure of the natural-technical frontier raises the fear of an uncontrollable autonomy of technology capable of undermining the freedom and identity of the human person, or of ruining the ecological balance necessary for life.

Contemporary authors are often limited to denouncing the processes that alienate the individual: forgetfulness of being in favour of the essence of technology for Heidegger (1954); the autonomy of techniques under the primacy of efficiency for Jacques Ellul (1954); the disqualification of language by technology for Gilbert Hottois (1984); capital as a nihilistic fulfilment of metaphysics, for Jean Vioulac (2009). To move from deconstruction to the development of criteria for an ethical use of technology, Canguilhem allows for reaction on two levels. On the one hand, the existential insecurity that characterizes the living must be preserved, as it radically establishes its ability to evolve over the course of history, which makes it possible to denounce both the false promises of technology and its systematic rejection on the grounds of protecting human nature from aggression and change. On the other hand, relational ontology is the necessary springboard to overcome the permanent threat of death, in a perspective that is not the avoidance of inevitable degradation, but rather the responsibility entrusted to each person and generation of assuming the challenge and promise of incompleteness, in order to build up the future of humanity. Thierry Magnin (2017) thus proposes an ethics of limits which does not consist in identifying a minimal base to be preserved from technology, but rather prescribes, in a resolutely historical perspective, to redefine progressively the limits that we wish to impose on techniques in order to preserve or to favour the beneficial relations between the living ones and their environment, and thus to benefit from the resources offered by the technological environment. Here we find, under a new modality, the dual function and ethical posture prescribed to medicine in the perspective of Canguilhem: support innovation that ensures adaptability to the requirements of history, while helping individual subjects to identify at their level and at the societal level the conditions of "substantial" but also "relational" viability that they must respect. It is always a question of taking into account the contingency and the reactivity which characterize living persons in their relation to their environment, which is for them a vocation to responsibility, an ontological rooting in history that forces them to discern progressively the protective measures and innovations they must undertake to take care of the community of the living.

References

Bourg, D. (1996). *L'homme artifice. Le sens de la technique*. Gallimard.

Canguilhem, G. (2013). Essai sur quelques problèmes concernant le normal et le pathologique [original édition 1943] et Nouvelles réflexions concernant le normal et le pathologique [original édition 1963–1966]. In *Le normal et le pathologique*. Presses Universitaires de France.

Canguilhem, G. (1992). La théorie cellulaire. Machine et organisme. Le vivant et son milieu. *La connaissance de la vie* (pp. 43–80; 101–127; 129–154). Vrin. (Original édition 1952).

Chamayou, G. (2007). Introduction. In K. Ernst (Ed.), *Principes d'une philosophie de la technique* (Original édition 1877) (pp. 7–41). Vrin.

Ellul, J. (2018). *La technique ou l'enjeu du siècle*. Economica. (Original édition 1954).

Fino, C. (2018). Penser la normativité sur un arrière-fond substantiel ? La reconnaissance de l'histoire des vivants, fondée sur Thomas d'Aquin et Georges Canguilhem. *Revue théologique de Louvain, 49*, 504–524.

Gayon, J. (2000). Le concept d'individualité dans la philosophie biologique de Georges Canguilhem. In G. Le Blanc (Ed.), *Lectures de Canguilhem. Le normal et le pathologique* (pp. 19–47). ENS Editions.

Heidegger, M. (1954). La question de la technique. In: *Extraits de Essais et conférences* (A. Préaud A., Trans.) (pp. 105–110). Gallimard.

Hottois, G. (2018). *Le signe et la technique. La philosophie à l'épreuve de la technique*. Vrin. (Original édition 1984).

Le Blanc, G. (1998). *Canguilhem et les Normes*. Presses Universitaires de France.

Lefève, C. (2000). La thérapeutique et le sujet dans L'essai sur quelques problèmes concernant le normal et le pathologique de G. Canguilhem. In G. Le Blanc (Ed.), *Lectures de Canguilhem. Le normal et le pathologique* (pp. 105–122). ENS Editions.

Macherey, P. (2009). *De Canguilhem à Foucault, la force des normes*. La fabrique.

Magnin, T. (2017). *Penser l'humain au temps de l'homme augmenté*. Albin Michel.

Simondon, G. (2012). *Du mode d'existence des objets techniques*. Aubier. (Original édition 1958).

Vioulac, J. (2009). *L'époque de la technique. Marx, Heiddeger et l'accomplissement de la métaphysique*. Presses Universitaires de France.

Catherine Fino is a Professor at the *Theologicum* of the Catholic Institute of Paris. She is Doctor of Medicine and Doctor of Theology. Among his publications: "Thinking normativity on a substantial background?" The recognition of the history of the living, based on Thomas Aquinas and Georges Canguilhem", *Revue théologique de Louvain*, 49, 2018, 504–524 (*in French*).

Living

Dorothée Browaeys

Abstract Our era is the era of life. Indeed, the Anthropocene, which now refers to the human footprint on Earth's future, reveals the limits and conditions of life on Earth by revealing how they are threatened. In fact, this confrontation teaches us the absolute necessity of maintaining all ecosystems in a stable balance, that is to say the need to look after all living beings. We will explain here the extent of the landmark change that this priority imposes.

Life is the requirement for our future. It is been on Earth for 3,5 billion years and a driving force behind the biological diversity which humanity is part of: today there is about 8.7 million living species, 6.5 million on land and 2.2 million in fresh or saline waters. It is an integrated system, neither separable nor divisible. No living organism, whether it is a bacterium, a dog or an oak viewed in isolation can account for its vital capacity by itself: self-organization (and development), self-repair (regeneration and tissues renewal), self-regulation (coordination, synchronization) and self-reproduction (self-preservation). This "permanence in change" is due to the openness and responsiveness characteristics of living systems which, through their metabolisms, exchange substances and energies to spread forms and information. We can speak of a capacity of autopoiesis (Varela, 1989) which allows the "production of oneself, permanently and in interaction with the environment".

Still, it remains to explain the nature of the phenomenon of Life. In the western world, two conceptions have been opposed since ancient Greek philosophy. On the one hand, the dualism theory is based on a separation between inert and living matter linked to the spiritual (soul, vital breath, vital impetus …) to the source of vitalism; on the other hand, the monism theory – with reference to Parmenides and more recently to Spinoza – considers life as a manifestation or emergence of matter.

Translation Gaud Luneau

D. Browaeys (✉)
TEK4life, Palaiseau, France
e-mail: d.browaeys@tek4life.eu

Modern scientific thought falls under this materialism theory, in particular after Pasteur's experiments which swept away the idea of a possible spontaneous generation. Researches on the primitive tangible conditions of our planet, or on the possibility of extraterrestrial life but also on the synthetic organisms' fabrication from chemical bricks (Benner & Sismour, 2003) allow us to approach the logics of biological innovation. The DNA molecule, the carrier of heredity, which encodes metabolic recipes, does not aim for perfection, but for robustness and resilience. Life' solutions are contingent – being inscribed in time and space – which Stephen Jay Gould explained as: "Rewind life's tape back to the appearance of modern multicellular animals, then replay the tape and the evolution will repopulate Earth with radically different creatures. The probability of seeing a human-like creature appearing, even a distant one, is effectively zero".

Contingent, life is also a paradox. Made of simple physicochemical bricks, they are nevertheless real geometrical geniuses (René Huygue) – capable of producing the tetrahedron, the octahedron and even the icosahedron – but also thermodynamic geniuses able to maintain order by fighting against entropy. In a world governed by the second principle of thermodynamics, any closed system tends to approach a state of maximum disorder, by creation of entropy. On the other hand, life tends to maintain a highly organized state, because it is not a closed system. Erwin Schrödinger explains that living matter avoids disintegration by feeding on negentropy (negative entropy) which corresponds to the cohesive force for systemic experts. It is therefore by consuming external energy that the living organism maintains its organization, structure, form, and functioning. It is these "forces which oppose death" (M. François Xavier Bichat) that determine the stabilization of stationary states such as homeostasis, a concept coined by Walter Bradford Cannon. Thus, higher animals benefit from vital physiological constants (temperature, blood flow, blood pressure, etc.)

The concept of information, which unfolded with cybernetics, was seen as a physical translation of negentropy (Norbert Wiener, 1948). Dominant, he propelled biology in its genetic and molecular dimension. Particularly since in 1953 James Watson, Francis Crick and Rosalind Franklin discovered the DNA double helix, the safe of the heredity code.

However, the living being seen as simple reservoir of information (Dawkins, 1976) has fizzled out. The integrity of living organisms is not strictly guaranteed by immutable DNA. Counterintuitively, it feeds on variations. For example, in bacteria, genetic amplification mechanisms – genes called mut – protect the population by rapidly acquiring resistance. The fortuitous variation kills some individuals, but acts as a genealogical "life insurance". It is clear that what is essential to "living things" is not to "stay the same", but to adapt. Instability strengthens vitality.

Epigenetics, which varies the expression of genes – through ligands that attach to DNA – is its embodiment. Proposed in 1942 by embryologist Conrad Hal Waddington, this modulation phenomenon shows that genes are not absolute regulators (the program metaphor is being abused): they are expressed according to the environment.

Yet, taking the context into account is stranger to modernity, which has established itself as an overhang rational logic driven by divine ideality. Thus, nature has been considered as a reservoir available, not as a partner with which to compose. Jean-Jacques Rousseau, François Quesnay (and the Physiocrats) or Elisée Reclus tried to bring attention back to the conditions of existence of our societies. Geographer Augustin Berque insists: "the Cartesian cogito, which claims no need for a place to be, is erroneous". He mentions a "deterrestrialisation of being." "The dualism and the objective scientific gaze evacuate the relationship: we speak of the environment as something seen 'out of nowhere', when we need to re-establish the larger reality of our belonging to things". Linguist Hélène Trocmé-Fabre has spent her life describing the "immense gap that separates our European languages from the language of the living beings".

Berque reactivates the science of environments or mesology which considers living beings as subjects. "The animal is not a mechanism but a mechanic. Far from submitting only to their environment (as suggested by adaptation theories), the living beings compose *with* it, even compose *it*, wrote the philosopher Benoit Goetz. This anti-reductionist approach fertilizes "thinkers of the life" such as Georges Canguilhem, Gilles Deleuze, or Giorgio Agamben.

In fact, mesology makes obsolete dualism and its substantialism. With it, reality is neither properly objective nor properly subjective, but traveltive (the fruit of a networking experience). However, it is not relativistic: it does not profess that everything depends on the point of view of the concerned subject. It gives a weight to the signs and not to the signals: because beyond the information being the simple avatar of the protean energy, there is another value of the information which derives its transmission force from the meaning it takes for these receivers. A word, a tiny chemical trace of pheromone can reverse a situation! The living being thus participate in the universe of symbolic and immaterial signs whose laws are not those of mechanics or energy but rather those of learning, exchange, logic or morality which are formal and non-material constraints.

For Georges Canguilhem (Canguilhem, 1952) "Life is experience, that is to say improvisation, utilization of occurrences. It is an attempt in all directions". This observation is the result of three characteristics of living organisms: their interdependence – with a plasticity which makes possible to build a brain – their regeneration – with the permanent renewal of the constituents – and their incompleteness which defines them as beings in the making. "Wherever something lives, there is, somewhere open, a register in which time is being inscribed", wrote Henri Bergson (Bergson, 1907).

Paradoxically, it is certainly in his incompleteness (the dear Neoteny of philosopher Dany Robert Dufour) that man can envision his original potential, stating that The human species is characterized not by its superiority over the rest of creation but by its incomplete form, its 'natural' weakness. For the human baby is a sponge. His brain (and his memory) like his immune system "gorge" themselves with the outside world. This interpersonal skill allows the assimilation of experience and culture.

Thus, human beings are shaped by three processes of "incorporation of the world": the hominization which causes the primitive body to evolve into a human body; the humanization that generates a human environment through the symbol; and finally anthropization which is the effect of technique on the environment, which has become "human environment". We are now evolving in this "eco-techno-symbolic" bath (A.Leroi-Gourhan). All reasoning opposing nature and culture is now out of date.

Thus, the common world takes on the value of a major determinant, disrupted today by the generalized digital language. In the future, it will be necessary to arbitrate between automatons which grow more autonomous and living beings increasingly disturbed by physicochemical jamming. (D. Browaeys, 2018). From now on, the political responsibility is indeed to work so that "the world does not undo" (Camus, 1957). Facing with climatic and biological threats, it is a question of orchestrating a coherent, human and united framework, a condition of our future as Hannah Arendt underlined in 1959: "Politics only exist because of biological necessity by virtue of which all humans need from each other to carry out this arduous task of staying alive".

References

Benner, S. A. H., & Sismour, A. M. (2003). Synthetic biology with artificially expanded genetic information systems. From personalized medicine to extraterrestrial life. *Nucleic Acids Research, Supplement. 3*(3), 125–126.

Bergson, H. (1907). *Creative evolution*. Henry Holt and Company.

Browaeys, D. (2018). *L'Urgence du vivant, vers une nouvelle économie*. François Bourin.

Camus, A. (1957). *Discours de réception du prix Nobel de littérature*. http://www.ac-nice.fr/lettres/valbonne/file/Camus_Discours_de_Suede_1957.pdf

Canguilhem, G. (1952). *Machine and organism*. In J. Crary & S. Kwinter (Eds.), in 1992 *Incorporations*. Zone Books.

Dawkins, R. (1976). *The selfish gene*. Oxfort University Press.

Varela, F. (1989). *Autonomie et connaissance*. Le Seuil.

Wiener, N. (1948). *The human use of human beings*. Houghton Mifflin (US)/Eyre & Spottiswoode (UK).

Dorothée *Browaeys* is a biologist and author of several books who works as a science journalist. She is the director of TEK4life, which works to accelerate the shift towards a society that is readjusted to the living environment. Co-founder of the VivAgora association, she is developing ecological accounting in France.

Nature

Gérald Hess

Abstract This article examines how our entry into the Anthropocene Age impacts on the three main images of nature bequeathed by Western thinking, and in particular on the image of nature that has became dominant in modern thinking.

The Anthropocene defines a new geological period. The concept is introduced to distinguish the geological age of the Holocene, characterized for more than 10,000 years by relative climatic stability, from another one into which we have entered. It indicates that human activities (agriculture, greenhouse gas emissions from industrial activities, urbanization, etc.) now have a terrestrial impact at the geological level. Scientists do not all agree on where to locate the beginning of this age or on the evidence needed to accept its reality (Bonneuil, 2015; Guillaume, 2015; Federau, 2017). The fact remains that the Anthropocene challenges what anthropologist Philippe Descola (2005) calls the 'Great Divide' between man and nature at the foundation of modernity. With the Anthropocene, the opposition between humans and nature seems to become obsolete. Indeed, the modern paradigm that has been in place since the seventeenth century sees humanity freeing itself from natural processes. Its influence grew from the eighteenth century onwards and intensified throughout the nineteenth and twentieth centuries through an ever more advanced instrumentalization of nature. The Anthropocene, on the other hand, *de facto* reintegrates humanity into nature by recognizing its dependence on the Earth's major bio-geophysical processes, such as the carbon, nitrogen and phosphorus cycles, and on certain natural phenomena such as the greenhouse effect, etc. What repercussions does this have on traditional representations of nature developed by Western thought since Antiquity?

G. Hess (✉)
Université de Lausanne, Lausanne, Switzerland
e-mail: gerald.hess@unil.ch

85

Three Representations of Nature

At least three representations of nature can be distinguished that have coexisted in the West since Antiquity (Hess, 2013). Plato's *Timaeus* depicts a demiurge who makes the Cosmos out of matter, based on an ideal. Nature is represented as a product or the result of a transformation whose model is technology. The aim of mastery and domination associated with technology is present in particular among the first Greek engineers responsible for building the temples of Asia Minor in the sixth century BC. From the seventeenth century onwards, it became dominant and found expression in Galileo-Newtonian mechanistic physics: nature as an *artefact* was compared to a machine that we know from the cogs that make it up. It is reduced to a set of physical forces that act causally on each other from the outside. The properties of nature are all physical: movement, spatiality, form, mass, etc.

Another representation of nature is linked to the Aristotelian tradition stemming from *Physics*. It no longer insists on the product, on the constructed object, but on the very process of its production. This does not come from a force outside nature. It is nature itself which, in an immanent way, contains the principle of its movement and stability. Nature appears as the opposite of an artefact, an otherness of which the organism is the model. The Renaissance took up and deepened this idea to see in the cosmos – the macrocosm – a living organism that is more than the sum of its parts. But from the seventeenth century onwards, nature-*poiesis* becomes a marginal representation of modernity. It was relegated to the natural history of species and found expression in Darwin's theory of evolution in the nineteenth century and in scientific ecology in the twentieth century. Both will support the idea of a wilderness.

A third representation of nature comes from the Orphic tradition (Hadot, 2004). It is based on the observation, which is also accepted in the other representations, that nature hides itself and its secrets. However, it is characterized by the idea that the right attitude towards nature is not to wrest its secrets and dominate it, but to initiate oneself into its mystery. It is for art and philosophical speculation to unveil its secrets. This tradition insists on an initiating experience of nature that does not consist in knowing nature from the outside, in an objective third-person perspective, like that of science. This nature is accessed from within, in an experience lived beyond the objective distinction between a knowing subject and an object to be known. Nature is no longer a radical exteriority; it is the place of human existence, both his environment and a mystery. This representation of nature – the nature-*habitat* – finds its expression as early as the seventeenth century in the experience of landscape, garden and pictorial art, in travel writing, in the early botanists and geologists and in the romantic tradition of the philosophers of nature. In sum, if nature-*artefact* comes from a techno-scientific perspective, nature-*poiesis* comes from a genetic perspective and nature-*habitat* from a phenomenological perspective.

The Great Divide of Modernity

In his book *Nature et culture* (2005), Descola observes that the West is the only culture to have conceived of nature by dissociating it from humans. Of the four modes of identifying reality that he considers, the one established by Western thought – naturalism – introduces, with the rise of experimental science in the seventeenth century, a discontinuity between humans and non-humans, while at the same time preserving the physical continuity of the world's entities. While being composed of the same matter, the beings of nature are differentiated from each other according to their interiority. Some of them – humans – are endowed with intentionality, in which their deepest nature lies, while others – the rest of nature – are devoid of it. This pattern of integration that structures experience leads to a clear-cut opposition between humans and human societies and nature (non-human animals, plants, etc.). This opposition is ontological, because what defines the human being is not what he or she has in common with nature, i.e. a body; it is his or her interiority, i.e. his or her spirit, soul, conscience. The ultimate goal of humanity is to emancipate itself from natural laws through culture and civilization. Note that naturalism itself is here an anthropological category that does not prejudge a dualistic or materialist conception of reality.

As soon as man belongs to another order than nature, as long as he differs from it, nature is at his free disposal, transformable at will to satisfy his needs and desires. The representation of a nature-*artefact* is undoubtedly the one that best corresponds to the division consummated by modern thought between man and nature. Nature must be mastered; it must be dominated by making it confess its secrets thanks to the empirical evidence that experimental science can now produce. The domination of nature therefore involves scientific objectification. With the Anthropocene, this representation of a nature-*artefact* that is subject to the goodwill of humans is undermined. The idea that humanity can extract itself from nature is an illusion. In this respect, the Anthropocene emphasizes the meeting of the Earth's long time span – hundreds of millions of years for living beings – and the short time span of human history – some 200,000 years for the species *Homo sapiens* (Chakrabarty, 2009). The impact of human activities on the environment is now such that human history influences the history of the Earth. And, through multiple Earth System feedback loops, it is the Earth itself that produces effects on human history. The Anthropocene recognizes that it is no longer possible to separate humanity from nature and that the fate of humanity depends on the Earth.

Representations of Nature at the Age of Anthropocene

How does the Anthropocene influence the three representations of nature described above? At first glance, one might think that the idea of a nature-artefact is no longer tenable. The interweaving of humans with nature is such that it would seem futile to persist in the modern opposition between humans and non-humans, between culture and nature. However, there is indeed a trajectory that continues along this path. Modern thinking in its relationship with nature continues into *hyper*modernity by seeking to dominate the earth's natural processes like climate (geoengineering) or to extricate itself from the finitude it imposes on living beings (transhumanism). In this way, the traditional separation between human and nature is renewed. But the Anthropocene can also lead to the acceptance of this interweaving of Earth history and human history. In this case, we recognize that the Earth is our 'ecumene' and that at least part of the Earth is constructed. This is the path taken by *post*modernity. But depending on the underlying representation of nature – nature-*habitat* or nature-*poiesis* – it defines two different trajectories. Pushed to the extreme, nature-*habitat* can come to recuperate all forms of exteriority and lead to a radical constructivism. Or, without going that far, it can also lead to welcoming the unassimilable part of nature by relying on nature-*poiesis*.

The Anthropocene invites us to challenge our representations of nature. But it can also be a drive to a hardline attitude. Some people believe that modernity has not yet said its last word. Ecological problems such as climate change or the dramatic decline in biodiversity are merely problems that must be answered by further dominating nature through science and technological innovation. Thus the eco-modernist movement (Asafu-Adjaye et al., 2015) argues that the Anthropocene provides an opportunity to assert domination over nature on a global scale. The techno-scientific progress of the industrial revolution is to be used to improve the living conditions of the entire world population, to stabilize the climate and to protect nature. Nature must be brought under control through optimal agricultural efficiency, energy-saving technologies or the exclusive use of nuclear energy. To combat climate change, it is proposed to use technologies capable of regulating the climate, either by extracting CO_2 from the air and sequestering it in the soil or by acting directly on the material cause of the greenhouse effect – solar radiation (Hamilton, 2013). As for transhumanism, it aims to free humans from the limits of the living being, which always implies, in the end, death. Through progress in the fields of medicine, computer science and robotics, the human species is to become master of its own evolution (Guillebaud, 2011).

Two other trajectories differ from the one described above. They do not belong to hypermodernity but to postmodernity. By implicitly relying on a representation of nature-*habitat*, the first postmodern path explores a return to the Earth or, more precisely, to the Terrestrial (Latour, 2017). With the Anthropocene, the environment is no longer simply the framework for political action; it becomes an actor in its own right. It is no longer an exteriority; it is entirely intertwined with human action, with its history: history becomes 'geohistory'. Whereas modern thinking was about detaching oneself from the Earth to look at it objectively from the outside, it is now a question of attaching oneself to it and leaving behind the illusion of a 'Great Outdoors'. The 'terrestrials' are all those who have understood that they depend on the Earth for their survival and who now oppose the 'humans' who believe they are still living in the Holocene age (Latour, 2015). But the Earth is then reduced to being only a humanity that inhabits the Earth, rather than being also and at the same time, an Earth inhabited by humanity. Nature basically loses any dimension of exteriority. As such, this position would be an expression of 'anaturalism' (Neyrat, 2016).

There is, however, a final trajectory of postmodernity induced by the Anthropocene. With regard to nature, this trajectory consists in conceding a share of exteriority to nature by resorting, again implicitly, to the representation of nature-*poiesis*. One aspect of nature irremediably escapes our control; it is and remains an otherness. According to Virginie Marris (2018), for example, the entry into the Anthropocene does not prevent us from talking about a wild nature. Drawing on the thinking of ecofeminist Val Plumwood (1993), Marris challenges binary thinking that either denies difference or excludes, incorporates, instrumentalizes or homogenizes the other. The idea is to recognize that there is a nature that is external to the human world and autonomous, and that it presents itself to us as an irreducible otherness.

References

Asafu-Adjaye, J., Linus, B., Steward, B., et al. (2015). *An ecomodernist manifesto*. www.ecomodernism.org

Bonneuil, C. (2015). Anthropocène (Point de vue 2). In D. Bourg & A. Papaux (Eds.), *Dictionnaire de la pensée écologique* (pp. 35–40). Puf.

Chakrabarty, D. (2009). The climate of history: Four theses. *Critical Inquiry, 35*(2), 197–222.

Descola, P. (2005). *Par-delà nature et culture*. Gallimard.

Federau, A. (2017). *Pour une philosophie de l'Anthropocène*. Puf.

Guillaume, B. (2015). Anthropocène (Point de vue 1). In D. Bourg & A. Papaux (Eds.), *Dictionnaire de la pensée écologique* (pp. 32–35). Puf.

Guillebaud, J.-C. (2011). *La Vie vivante. Contre les nouveaux pudibonds*. Les Arènes.

Hadot, P. (2004). *Le voile d'Isis. Essai sur l'histoire de l'idée de Nature*. Gallimard.

Hamilton, C. (2013). *Earthmasters, the Dawn of the age of climate engineering*. Yale University Press.

Hess, G. (2013). *Éthiques de la nature*. Puf.

Latour, B. (2015). *Face à Gaïa. Huit conférences sur le nouveau régime climatique*. Les Empêcheurs de penser en rond/La Découverte.

Latour, B. (2017). *Où atterrir? Comment s'orienter en politique*. La Découverte.

Marris, V. (2018). *La part sauvage du monde. Penser la nature dans l'Anthropocène*. Seuil.

Neyrat, F. (2016). *La part inconstructible de la Terre. Critique du géo-constructivisme*. Seuil.

Plumwood, V. (1993). *Feminism and the mastery of nature*. Routledge.

Gérald Hess is senior lecturer at the University of Lausanne. He has a PhD in Philosophy and MA in Law. He published several books, among them: *Éthiques de la nature*, 2013, and is editor (with D. Bourg) of *Science, conscience et environnement. Penser le monde complexe*, 2016, (with C. Pelluchon and J.-P. Pierron) of *Humains, animaux, nature. Quelle éthique pour le monde qui vient?*, 2020. He will publish his next book 2023: *Conscience cosmique. Pour une écologie en première personne*.

Planetary Boundaries

Ulrich Brand, Barbara Muraca, Éric Pineault, Marlyne Sahakian,
Anke Schaffartzik, Andreas Novy, Christoph Streissler, Helmut Haberl,
Viviana Asara, Kristina Dietz, Miriam Lang, Ashish Kothari, Tone Smith,
Clive Spash, Alina Brad, Melanie Pichler, Christina Plank,
Giorgos Velegrakis, Thomas Jahn, Angela Carter, Qingzhi Huan,
Giorgos Kallis, Joan Martínez Alier, Gabriel Riva, Vishwas Satgar,
Emiliano Teran Mantovani, Michelle Williams, Markus Wissen,
and Christoph Görg

Abstract The planetary boundaries concept has profoundly changed the vocabulary and representation of global environmental issues. The article starts by highlighting the strengths and weaknesses of planetary boundaries from a social science perspective. It is argued that the growth imperative of capitalist economies, as well as other particular characteristics detailed below, are the main drivers of the ecological crisis and exacerbated trends already underway. Further, the planetary boundaries framework can support interpretations that do not solely emphasize technocratic operational approaches and costs, but also assume that these alone can be the solution.

This book chapter is an excerpt of the full article Brand et al. (2021), available in open access.
Contact: U Brand, University of Vienna, ulrich.brand@univie.ac.at

U. Brand (✉) · A. Brad
Department of Political Science, University of Vienna, Vienna, Austria
e-mail: ulrich.brand@univie.ac.at

B. Muraca
Department of Philosophy and Environmental Studies Program, University of Oregon,
Eugene, OR, USA

É. Pineault
Institute for Environmental Sciences and Department of Sociology, Université of Québec
à Montréal, Montreal, QC, Canada

M. Sahakian
Department of Sociology, University of Geneva, Geneva, Switzerland

A. Schaffartzik · H. Haberl · M. Pichler · C. Görg
Institute of Social Ecology, University of Natural Resources and Life Sciences,
Vienna, Austria

A. Novy · V. Asara · T. Smith · C. Spash
Institute for Multi-Level Governance and Development, Department of Socio-Economics,
Vienna University of Economics and Business, Vienna, Austria

N. Wallenhorst, C. Wulf (eds.), *Handbook of the Anthropocene*,
https://doi.org/10.1007/978-3-031-25910-4_15

The concept of planetary boundaries was introduced by Johan Rockström and colleagues in 2009 in the wake of the United Nations Climate Change Conference in Copenhagen where countries endeavoured – but ultimately failed – to agree upon a new framework for climate-change mitigation. In contrast to earlier debates on environmental limits, "planetary boundaries" focus less on the exhaustion of natural resources than on the biophysical impacts of resource use and material consumption: the overfertilization of soils, the destruction of ecosystems, and overtaxing the

C. Streissler
Chamber of Labor, Vienna, Austria

K. Dietz
Faculty of Social Sciences, Institute of Political Science, University of Kassel,
Kassel, Germany

M. Lang
Department for Environmental and Sustainability Studies, Universidad Andina Simón
Bolívar, Quito, Ecuador

A. Kothari
Kalpavriksh and Vikalp Sangam, Pune, India

C. Plank
Department of Political Science, University of Vienna, Vienna, Austria

Institute of Social Ecology, University of Natural Resources and Life Sciences,
Vienna, Austria

G. Velegrakis
Department of History and Philosophy of Science, National and Kapodistrian
University of Athens, Athens, Greece

Department of Surveying and Geoinformatics Engineering, University of West Attica,
Athens, Greece

T. Jahn
Institute for Social-Ecological Research, Frankfurt, Germany

A. Carter
Department of Political Science and Balsillie School of International Affairs,
University of Waterloo, Waterloo, ON, Canada

Q. Huan
School of Marxism, Peking University, Beijing, China

G. Kallis · J. M. Alier · E. T. Mantovani
Institute of Environmental Science and Technology, Universitat Autònoma de Barcelona,
Barcelona, Spain

G. Riva
Department of Law, Pontifical Catholic University of Rio de Janeiro, Rio de Janeiro
and Cricare Valley Institute, São Mateus, Brazil

V. Satgar · M. Williams
University of the Witwatersrand, Johannesburg, South Africa

M. Wissen
Department of Business and Economics, Berlin School of Economics and Law,
Berlin, Germany

capacity of sinks to absorb emissions and other effluents produced by human activities.

With the introduction of the planetary boundaries' framework, Rockström et al. (2009a, 472) delineate "the safe operating space for humanity," which lies firmly within the Holocene state. The authors argue, "The evidence so far suggests that, as long as the thresholds are not crossed, humanity has the freedom to pursue long-term social and economic development" (Rockström, et al., 2009a, 475). For each threshold, the authors proposed a quantitative "control variable" (Rockström, et al., 2009a, 472, 473), that is, a universal, robust indicator of system change and for which reliable data exist. A boundary exists then at a distance from a presumed trigger value of the control variable, which may encourage less attention to thresholds that are sufficiently remote and do not require immediate attention (Cohen, 2021). The planetary boundaries framework underscores how non-linear dynamics characterize Earth-system changes and key processes (e.g. global biogeochemical cycles).

Although sometimes difficult to identify exactly – due to incomplete scientific understanding of the complex feedbacks in the Earth system, among other factors – the planetary boundaries concept aims to map the safe operating space based on an appreciation of these thresholds in non-linear system dynamics of the Earth system (Steffen et al., 2015). Boundaries are, as the authors point out, normative judgements for the Earth system in general. Given risks and uncertainties, the authors quantify planetary boundaries by taking a risk-averse and conservative approach (Rockström, et al., 2009a, 473).

Rockström and colleagues are careful to avoid the technocratic hubris of prescribing a level and composition of societal metabolism for humanity. They argue, rightly, that boundaries have to be conceptualized or defined based on the risk tolerance of societies to non-linear and potentially catastrophic change. Rockström et al. (2009b) state that the "predominant paradigm of social and economic development remains largely oblivious to the risk of human-induced environmental disasters at continental to planetary scales" (p. 32). Economic activity is identified as a key driver of anthropogenic environmental change that can push "coupled human-environmental systems" beyond thresholds of known stability and into zones of non-linear and potentially "catastrophic" environmental change (Rockström et al., 2009b), but is also not sufficiently problematized. A recent paper suggests that the boundaries concept should include a consideration for a "just" as well as a "safe" operating space (Rockström et al., 2021), yet it stops short of grappling with the complexities of different forms of justice – not solely distributional, but also procedural – and the political implications of such an approach.

Strengths of the Planetary Boundaries Framework

We identify three main strengths of the framework with regard to its potential contributions to transformative knowledge. First, it has widened the political and academic debate on the ecological crisis beyond climate change, which has dominated much of sustainability discussions since the turn of the century, to a more varied account of ecological and biogeochemical forces induced by societal metabolism, including topics such as biodiversity loss or eutrophication. Planetary boundaries proponents warn that the complexity of, and interlinkages among, different bio-physical subsystems or processes are of utmost importance, and that if tipping points are reached, the resulting changes may be unpredictable and possibly irreversible.

As a second strength, the framework rests on the ontological claim that contemporary human societies have become dependent for their flourishing on the "stable environmental conditions" – i.e. ecological and geological conditions – of the Holocene and that there are identifiable thresholds within which this stability is secured. Framing ecological questions in this way stresses the deep connections between geology, biology, as well as human and environmental history (Chakrabarty, 2020). It has provoked humanists and social scientists to analyze particular socio-historical interconnections between human and nonhuman agents (as in the early colonial plantations) in a critical dialogue with the natural sciences (Haraway & Tsing, 2019).

A third strength lies in the iconic image used to depict planetary boundaries: an infographic with Earth overlaid by concentric orbits representing three spaces as distances from a centre: a safe green zone, a yellow zone of risk, and an outermost red zone of thresholds crossed. The boundaries for the nine key Earth-system processes identified in the framework are presented as dimensions emanating from the centre in a simple and intuitive representation of boundary transgression. The popular success of the planetary boundaries concept can certainly be attributed to the visual power of this illustration that rapidly became standard fare in scientific and educational presentations. Thus, the notion of planetary boundaries went beyond the mere presentation of scientific results to change the frame of popular debates and to inform subsequent research on sustainability issues. However, the diagram is a simplification – while easy to communicate, it suppresses the complexity of different planetary processes as well as their interlinkages.

Weaknesses of the Planetary Boundaries Framework

We also see weaknesses and ambiguities that allow for "business-as-usual" and "pro-status quo" interpretations of the framework. The planetary boundaries concept identified the "predominant paradigm of social and economic development" (Rockström et al., 2009b) as the main driver toward "continental and global" environmental disasters, without explaining which societal, political, and economic conditions lead to unsustainability, and in what way. It is not economic activities in the abstract that lead to ecological crisis but rather economic activities with particular logics and under certain circumstances. More precisely, we argue that the growth imperative of capitalist economies, as well as other particular characteristics detailed below, are the main drivers of the ecological crisis and exacerbated trends already underway. For a more in-depth discussion on the social structures of capital, see Brand et al. (2021). Indeed, even before capitalist growth economies, the enclosures of the natural commons – land, water, biodiversity and creative human labour – as part of transitions from feudalism through to militarized mercantile capitalist conquests and settler colonialism, inscribed global accumulation with a destructive logic for our planetary ecology.

Further, the planetary boundaries framework can support interpretations that do not solely emphasize technocratic operational approaches and costs, but also assume that these alone can be the solution. The technocratic bias embedded in the proposed political solutions that often accompanies planetary boundaries research ranges from including nuclear energy as a replacement for fossil fuels to the deployment of large-scale geoengineering technologies. This technocratic drift is not incidental, but rather is built into the planetary boundaries framework itself, in its view of the Earth from an "astronaut's eye view" that can only be provided by scientists, but which runs a risk of ignoring severe regional or local impacts of global warming triggered long before global thresholds are crossed (Sachs, 1999). From this perspective, Earth is envisioned as a globe that appears – at least in principle – as if it can be managed as a cybernetic system, albeit with the complication of non-linear feedback loops. Technical solutions, however, have been subject to a number of criticisms from social scientists and humanities scholars (Muraca & Neuber, 2018; Pichler et al., 2017).

Political ecologists and social ecological economists have long criticized how the framing of limits as something external that resides in nature and as *given* to humanity "depoliticizes" decisions at stake (Asara et al., 2015; Streissler, 2016; Muraca & Döring, 2018; Lövbrand et al., 2015). The post-political definition of planetary boundaries renders invisible, or at least relativizes, the social conflict embedded in the trajectories that transgress the boundaries, or the distribution of the benefits and impacts that they entail (Kallis, 2019; Brand & Wissen, 2021). Moreover, it threatens to mask economic dynamics such as the increasing competition for scarce resources or what movements have called the "last great dispossession of the commons."

A further limit of the planetary boundaries framework lies in the sociopolitical and socioethical implications of selecting these particular nine boundaries. While Earth-system science presents an important valuation perspective with respect to specific biophysical processes included in the planetary boundaries, it does not discuss the normative and political dimensions involved in selecting these boundaries. For example, in the case of biodiversity loss, "ethics" is mentioned as a dimension of acceptability of species loss, but is mostly intended in terms of traditional conservation biology literature and not further examined. By failing to clarify and critically discuss its normative assumptions, the planetary boundaries concept limits its consideration to a rather narrow spectrum of values and worldviews and neglects perspectives voiced, for example, in environmental justice literature or in feminist and indigenous care ethics (Whyte & Cuomo, 2017) and in other, environmental values literature (O'Neill et al., 2018).

Furthermore, the planetary boundaries concept emphasizes the need to bring the "coupled human Earth System" back into a "safe operating space," which assumes that the Holocene or, at least, the recent past was safe for all people. Given societal structures of power and exploitation, this is definitely not the case. Societal values that address dimensions of climate crisis such as the unequal distribution of risks or other aspects of climate justice may require an adaptation of the variables signalling a "safe operating space." In other words: for which part of the global population and for what purposes is a certain "operating space" safe? What is acceptable for one social group might rely upon unacceptable forms of oppression and exposure to environmental hazards for others. Global "agreement" on the maximum of 1.5 degree of global heating might help sustain living conditions and ecosystem functions in some parts of the world, but puts under severe pressure people living in low lying coastal areas, or those depending on the glacier functions of the Andes.

We agree with the original argument made in the Rockstrom et al. paper in 2009 that boundaries are sociopolitical constructs. While they are informed by science – in other words based on the currently available (necessarily incomplete) understanding of Earth-system dynamics – their definition also requires normative and political assumptions of what are acceptable or "unacceptable" paths for humanity *in general*, to use Rockström et al.'s terms (2009a, 472). Reaching across scales, boundaries also imply a notion of (un)acceptable configurations of limits from the local to the national, regional, and global levels. Yet from a purely global perspective, if those in the global North tried to negotiate the distribution of environmental benefits and burdens within and between societies, given the dominant socioeconomic systems, it would surely result in multiple forms of inequality. This is the case as the very idea of any acceptable or unacceptable distribution path is inescapably tied to unequal gender and class relations, racism, colonialism, and imperialism, to name but a few dimensions of the complexity of social relations across scales.

In the chapter on Societal Boundaries, in this handbook, we bring a critical social science perspective to the Planetary Boundaries framework through the notion of societal boundaries and aim to provide a more nuanced understanding of the social nature of thresholds, one that it has the potential to offer guidelines for a just, social-ecological transformation.

References

Asara, V., Otero, I., Demaria, F., & Corbera, E. (2015). Socially sustainable degrowth as a social–ecological transformation: Repoliticizing sustainability. *Sustainability Science, 10*, 375–384.

Brand, U., & Wissen, M. (2021). *The Imperial mode of living: Everyday life and the ecological crisis of capitalism*. Verso.

Brand, U., Muraca, B., Pineault, E., Sahakian, M. (lead authors), Schaffartzik, A., Novy, A., Streissler, A., Haberl, H., Asara, V., Dietz, K., Lang, M., Kothari, A., Brad, A., Pichler, M., Plank, C., Velegrakis, G., Jahn, T., Carter, A., Qingzhi, H., Kallis, G., Martínez Alier, J., Riva, G., Satgar, V., Spash, C., Teran Mantovani, E., Williams, M., Wissen, M., & Görg, C. (2021). From planetary to societal boundaries: An argument for collectively defined self-limitation. *Sustainability. Science, Practice and Policy, 17*(1), 265–292.

Chakrabarty, D. (2020). The human sciences and climate change. *Science and Culture, 86*(1–2), 46.

Cohen, M. (2021). *Sustainability*. Polity Press.

Haraway, D., & Tsing, A. (2019). Reflection on the plantationocene. *Edge Effects Magazine.* https://edgeeffects.net/wp-content/uploads/2019/06/PlantationoceneReflections_Haraway_Tsing.pdf

Kallis, G. (2019). *Limits: Why Malthus was wrong and why environmentalists should care.* Stanford University Press.

Lövbrand, E., Beck, S., Chilvers, J., Forsyth, T., Hedren, J., Hulme, M., Lidskog, R., & Vasileiadou, E. (2015). Who speaks for the future of earth? How critical social science can extend the conversation on the Anthropocene. *Global Environmental Change, 32,* 211–218.

Muraca, B., & Döring, R. (2018). From (strong) sustainability to degrowth: A philosophical and historical reconstruction. In J. Caradonna (Ed.), *Routledge handbook of the history of sustainability* (pp. 339–361). Routledge.

Muraca, B., & Neuber, F. (2018). Viable and convivial technologies: Considerations on climate engineering from a degrowth perspective. *Journal of Cleaner Production, 197,* 1810–1822.

O'Neill, D., Fanning, A., Lamb, W., & Steinberger, J. (2018). A good life for all within planetary boundaries. *Nature Sustainability, 1,* 88–95.

Pichler, P., Zwickel, T., Chavez, A., Kretschmer, T., Seddon, J., & Weisz, H. (2017). Reducing urban greenhouse gas footprints. *Scientific Reports, 7*(14659), 1.

Rockström, J., Steffen, W., Noone, K., Persson, A., Chapin, F., Lambin, E., Lenton, T., Scheffer, M., Folke, C., Schellnhuber, H., and others. (2009a). A safe operating space for humanity. *Nature, 461,* 472–475.

Rockström, J., et al. (2009b). Planetary boundaries: Exploring the safe operating space for humanity. *Ecology and Society, 14,* 32.

Rockström, J., et al. (2021). Identifying a safe and just corridor for people and the planet. *Earth's Future, 9,* e2020EF001866.

Sachs, W. (1999). *Planet dialectics: Explorations in environment and development.* Zed Books.

Steffen, W., et al. (2015). Planetary boundaries: Guiding human development on a changing planet. *Science, 347,* 6223.

Streissler, C. (2016). Planetarische Grenzen – ein brauchbares Konzept? (Planetary Boundaries – A Useful Concept?). *Wirtschaft und Gesellschaft, 42*(2), 325–338.

Whyte, K., & Cuomo, C. (2017). Ethics of caring in environmental ethics: Indigenous and feminist philosophies. In S. Gardiner & A. Thompson (Eds.), *The oxford handbook of environmental ethics.* Oxford University Press.

Ulrich Brand works as a Professor of International Politics at the University of Vienna. He obtained his doctoral degree at Goethe University Frankfurt and wrote his post-doctoral thesis (second monograph) on the internationalisation of the state at the University of Kassel. He is the author of books and articles on critical international politics, ecological crisis, environmental politics, social-ecological transformations, the imperial mode of living and Latin America. His recent books: *The Imperial Mode of Living. Everyday Life and the Ecological Crisis of Capitalism* (with Markus Wissen, Verso 2021, translated into nine languages) and *Capitalism in Transformation: Movements and Countermovements in the twenty-first Century* (as co-editor, Edward Elgar 2019).

Part II
The Earth's Surface and Its Elements

Air

Andreas Weber

Abstract Air is a physical, gaseous medium surrounding planet earth. It is brought forth and kept in its particular balance by the life-enhancing mutual transformation of all domains of this earth, both the geosphere and the biosphere. Air is the invisible domain of mutual transformation of life into non-life, individuals into other individuals, and solid bodies into invisible potential. Air is thus not only a physical reality, but also the potential of giving life, the breath of "poetic space".

As is characteristic for the description of any phenomenon, structure or relationship in the Anthropocene, what is air can be approached from several different, and often contradictory frameworks. First of all, air is what we breathe, hence it is breath. As such, air is one of the classical elements which to some degree is contained in everything that exists. Air is the "ruach" and "spiritus" of semitic monotheistic religions. Here, it is the stuff the soul is made of, subtle as a faint breeze. Air is the reign of the invisible and therefore the nexus to the realm of forces and spirits which are not from the material dominion. Air is what encompasses and connects everything. This is mirrored in the term "on air" when a broadcasting station is emitting its waves, or in the saying "it's in the air" when something invisible, ungraspable is reaching everyone.

And of course air is what encompasses the components of the planet earth's atmosphere and hence is a conglomerate of several gases and solid particles which we can describe in terms of empirical science. From this standpoint, air is the medium in which the objects and organisms on the surface of the planet are immersed; air is a gas of a certain composition with which those objects interact. As such we can measure that the terrestrial air contains about 78% of the chemical element Nitrogen, 21% of Oxygen, 0.9% of Argon, 0.042% of Carbon Dioxide, plus various trace gases, among them up to 3% of water vapour (Cox, 2000).

A. Weber (✉)
Bard College, Berlin, Germany

© The Author(s), under exclusive license to Springer Nature Switzerland AG 2023
N. Wallenhorst, C. Wulf (eds.), *Handbook of the Anthropocene*,
https://doi.org/10.1007/978-3-031-25910-4_16

It is basically this latter perspective which until today is agreed upon as the "real" description of the term "air". Air is the gaseous mixture which fills the atmosphere of the planet earth. For this reason, one of the most dramatic processes of the Anthropocene, global heating, plays itself out as change of the composition of the air. Air harbours a growing percentage of "greenhouse gases" such as carbon dioxide and methane. Global heating is thus understood as a change of the medium which harbors the terrestrial objects, so that the continuity of these objects, at least some of them, is threatened.

In this view, air is an object outside of ourselves that at best interacts with other objects and ourselves. These interactions are described by the equations of physiology (the citrate cycle, photosynthesis e.g.) and by the computations of the "earth system", the biogeochemical circles. In the former, living bodies are put in a physicochemical relation with the air as a medium which surrounds them. In the latter the liquid and solid parts of the planet (rocks, sediments, oceans, rivers and rainwater) are put into an empirical relation with the air.

The marked change to this which we see in the Anthropocene is that the perspective of understanding air as a huge object on the outside of other objects, including us, and being in interaction with other huge objects, is replaced with the notion that air is not only residual on the outside, but is an active agent in a constant process of mutual transformation with other agents. Air is also inside, and therefore the planetary atmosphere is not a huge background object to human agency, but part of our own doing, of our own self-understanding and of our self-experience. This happens on an empirical-physical plane, so can be shown in terms of biogeochemistry. But it also happens on the plane of experience, so must be shown in terms of an existential understanding of our own existence as part – as inside-outside-transformation – of the interplay of the earth body and its breath.

In this respect, in the radical change of understanding and (self-)experience that is the Anthropocene, the older notions of air as breath, spirit, and element perfusing everything to a certain degree, can be re-validated on an integrated plane, which does not loosen ties to earth system science but rather distributes its radical findings into an integrated vision of the earth (see also Bergthaller & Horn, 2019).

This integrated vision follows from empirical findings as well as it precedes philosophical contemplation, for the human body is already part of this integrated setting, first through its own microbiogeochemical composition, and second also through its experiential particpation in an integrated reality, where the old distinctions between outside (the empirical plane, the plane of material determinancy) and inside (the plane of experience, the plane of organismical agency) can not longer be separated, both on empirical and on experiential grounds (Barad, 2003; Morton, 2017; Latour, 2017; Weber, 2019).

Air is a dimension of life, and as such it covers the profound realities which make up life: It is the space of a profound mutual interaction and reciprocal re-creation, which inevitably yields the domain of individual experience (Weber, 2016a, b). It is inside as well as it is outside. Only from this integrated viewpoint on air we can pay due attention to what is really happening in global heating and climate breakdown and only from this standpoint – not from a technological-instrumental approach to air as an object – we can take a productive part in the dramatic transformation of the earth climate. But let us see step for step how we can get to this insight.

The most important historical move to understand air as an integrator of the manifold existences within a living biogeosphere was provided by the research of James Lovelock in the 1960s. Lovelock was then working for NASA in providing a way how to determine if a planet contained life. It turned out, that a living planet (such as the earth) is surrounded by air: by an atmosphere far from the chemical equilibrium (Lovelock, 1989). If the earth's atmosphere were in equilibrium – its composition composed only by physicochemical reactions – it would contain mostly carbon dioxide, as do the atmospheres of Venus and Mars with more than 90% of carbon dioxide.

Our planet's air, however, is not composed as it should be according to chemical laws alone. Lovelock understood that it was life – the manifold lifeforms in their various ecological compositions – that regulated the composition of gases in the planet's atmosphere. And the regulation was done in such a way that it was maximally friendly for the existence and continuity of life. The high oxygen content of the earth's air is an example for this. It is high – but not that high that fires would immediately consume vast swathes of the biosphere.

Today, the earth sciences have in their majority taken up Lovelock's stance (Lenton & Watson, 2011) – with the exception of his wider and probably most important implications, namely of extending the idea of aliveness to the whole planet. A crucial medium of this aliveness is the air. The planet's air – as atmosphere – covers the earth body like a skin – or a fur – covers a living being.

Lovelock termed his idea about the earth's life-welcoming self-regulation "Gaia-Hypothesis". Together with biologist Lynn Margulis, they showed that the earth can only be understood from the perspective of an integrated whole which self-regulates the conditions needed for a continuity of this self-regulation, and that life is an essential part of this (Lovelock & Margulis, 1974). The air is therefore a medium which comes about through the active self-regulation of the planet as a whole, and in which microbial but also macroscopic life plays an indispensable role. Air is alive, and it grants life-bearing qualities to the remainder of the planet, as it permeates its terrestrial parts.

In the earth system, air forms the link between living beings and the mineral body of the planet. The crucial component of the air here is carbon, in the form of carbon dioxide. In a nutshell, living beings produce carbon through respiration, but they also function as agents which put carbon back in the earth's body, where it is then again continuously weathering away from. Innumerable tiny organisms in the oceans build carbon from the air into their shells and bury it in the sediments, which during geological time become rocks and mountains. Trees dump carbon which they have breathed from the atmosphere into the soil. So air is a non-solid phase of the earth body and of the bodies of living beings, in which a transformation takes place. Through the invisible realm of the air, bodies can shapeshift into other bodies (algae into rock, trees into soil).

This holds also for the mutual transformation between living bodies, which is the core of ecology. In a seemingly very simple relationship of reciprocity, animals (and other heterotrophic organisms) breathe in the air which the plants (and other photo-autotrophic beings) breathe out. Animals use the oxygen the plants liberate through photosynthesis, which is actually a sun-driven solidifcation of air into bodies, as plants use the carbon in the air to build their biomass. Animals, on the other hand, breathe out this carbon in the form of carbon dioxide. This carbon stems from the dissolved parts of the animals' bodies; it is literally their bodies become invisible, but breathable by others.

Air is not simply the medium, where the exchange happens, but the entire domain of this transformation, or, rather, air is the invisible fusion of bodies before they part again into individually distinct shapes. It is life, as potential, and communion. Air is not an object, but an immense subject, made of the intercrossing of all breathing individuals. One could say, air is the other side of embodied biological and geological existence, and as this other side, it is its precondition. We see here that in labelling climate breakdown as a problem with the atmosphere (as an object apart from us) means to grossly misunderstand what the atmosphere truly is: ourselves, in a form that is unseen, but all the more able to permeate all solid bodies. If we do not understand that the air is a commons (Bollier, 2014) shared by all beings who breathe, we will be unable to mitigate climate breakdown. The same holds if we do not understand that the air is not a physical container, but a space of agency and experience.

Air is the invisible domain, in which breath meets breath, in which the flesh breathed out suffuses into one another and becomes one, becomes part of a vast commons of aliveness, the "commonwealth of breath" (Abram, 2010). Air is the invisible substance that distributes life to all its participants. We can feel it, incorporate it, but not see it – unless the tiny particles distributed in it give a rose colour to the evening, unless it scatters the sunlight by the water vapour it contains and becomes the epidermis of the blue planet, the magic "vas", the vessel of alchemic transformation (Harding, 2021). Air is at the threshold between bodies and what is their other side. It is "matter, reduced to an extreme thinness, oh so thin" (Emerson, 1981), on the verge of what is neither visible nor even breathable, and in which all individuality is annihilated into the fecund emptiness, "poetic space" (Weber, 2021).

References

Abram, D. (2010). In the depths of a breathing planet. Gaia and the transformation of experience. In E. Crist & B. Rinker (Eds.), *Gaia in turmoil: Climate change, biodepletion, and earth ethics in an age of crisis*. MIT Press.

Barad, K. (2003). Posthumanist performativity. Toward an understanding of how matter comes to matter. *Signs: Journal of Women in Culture and Society, 28*(3), 801–831.

Bergthaller, H., & Horn, E. (2019). *Anthropozän zur Einführung*. Junius.

Bollier, D. (2014). *Think like a commoner: A short introduction to the life of the commons*. New Society Publishers.

Cox, A. N. (Ed.). (2000). *Allen's astrophysical quantities* (4th ed., pp. 258–259). AIP Press.

Emerson, R. W. (1981). Experience. In: *Emerson's essays*. Harper.

Harding, S. (2021). *Gaia alchemy. The reuniting of science, psyche and soul*. Bear & Company.

Latour, B. (2017). *Facing Gaia. Eight lectures on the new climatic regime*. Polity Press.

Lenton, T., & Watson, A. J. (2011). *Revolutions that made the earth*. Oxford University Press.

Lovelock, J. E. (1989). Geophysiology, the science of Gaia. *Reviews of Geophysics, 17*(11), 215–222.

Lovelock, J. E., & Margulis, L. (1974). Atmospheric homeostasis by and for the biosphere: The gaia hypothesis. *Tellus, 26*(1–2), 2–10.

Morton, T. (2017). *Humankind*. Verso.

Weber, A. (2016a). *The biology of wonder. Aliveness, feeling, and the metamorphosis of science*. New Society Press.

Weber, A. (2016b). *Biopoetics. Towards an existential ecology*. Springer.

Weber, A. (2019). *Enlivenment. Towards a poetics for the Anthropocene*. MIT Press.

Weber, A. (2021). A path to poetic space. *Constructivist Foundations, 16*(2), 192–195.

Andreas Weber is a biologist, philosopher and writer. He teaches at the University of the Arts, Berlin, is Visiting Professor at the UNISG, Pollenzo, Italy, and holds an Adjunct Professorship at the IIT, Guwahati, India. He contributes to major German newspapers and magazines and has published more than a dozen books, most recently *Enlivenment. A Poetics for the Anthropocene*, MIT Press, 2019 and *Sharing Life. The Ecopolitics of Reciprocity*, Boell Foundation, 2020.

Amazon

Pierre-Yves Cadalen

Abstract This article examines the socio-ecological dynamics of power around the Amazon Rainforest. It is necessary to highlight the factors of deforestation, the geophysical effects of forest destruction, as well as the political and social processes which could alter or reorientate the current dynamics, which is why the article redefines the concept of Environmental Commons. The Amazon Rainforest is a perfect case study to think the fundamental changes – as far as political anthropology is concerned – induced by the Anthropocene.

The Amazon Rainforest is the global star of all forests. The ecosystem of this forest is unique. It covers 5.5 million km^2 in South America. Nine nations are sovereign over a part of the Amazon Rainforest: first and foremost, Brazil, and then Peru, Colombia, Bolivia, Ecuador, Venezuela, Surinam, Guyana, France (through French Guiana). France excepted, these countries formed the treaty for Amazonian cooperation in 1978, which aimed primarily at protecting their sovereignty over this symbolic forest (Gerlach, 2003).

Before tackling the deep modification faced by this forest as the Anthropocene has begun, we cannot avoid considering its symbolic importance. The Congo Rainforest, being also huge – three million km^2 –, is not subject to the same international attention and does not appeal to the same emotions, especially in Europe and in the United States. Such an interest is partly due to the mythical appeal of the Amazon rainforest and the narrative about the allegedly "discovery of the New World", a brutal and violent conquest which provoked so many deaths that it eased the global atmosphere (Beau & Larrère, 2018; Grove, 2019; Wallenhorst, 2020).

The mythical position of the Amazon rainforest is deeply entrenched with the beginning of the global capitalist integration. Rubber, bananas and, since the 60's, oil, have been strong incentives for the Amazon rainforest to be integrated to the international dynamics of economic accumulation, which is also true for intensive

P.-Y. Cadalen (✉)
CERI Sciences Po, CRBC UBO and AMURE UBO, Paris, France
e-mail: pierreyves.cadalen@sciencespo.fr

© The Author(s), under exclusive license to Springer Nature
Switzerland AG 2023
N. Wallenhorst, C. Wulf (eds.), *Handbook of the Anthropocene*,
https://doi.org/10.1007/978-3-031-25910-4_17

agriculture (Hecht & Cockburn, 2010; Picq, 2016; Smouts, 2001). Furthermore, the imaginary position of the Amazon rainforest has gone global along the same temporal lines as the "anglocene" (Bonneuil & Fressoz, 2013). The essential position of this forest symbolically is linked to a European imaginary, that of a fascinating as well as frightening forest whose qualities can bring wealth or despair (Harrison, 1992). Captain Aguirre on Herzog's raft is a good approximation of this symbolic dimension. If the Amazon rainforest is that important internationally speaking, it is still linked to this nodal dichotomy. Wealth: raw material. Despair: deforestation and climate change.

The mythological position of the Amazon rainforest made it an international symbol for environmentalists at a global scale. The Amazon Rainforest is often coined as the "planet lung", which is scientifically arguable: though it is indeed a huge carbon sink, its balance – oxygen / carbon dioxide – is slightly positive. In other words, the Amazon rainforest is not the main oxygen producer, the oceans and phytoplankton are far ahead. This element leads us to the changes occurring in the Amazon Rainforest, which are nevertheless quite concerning. The central notion to understand the changes linked to the Anthropocene in the Amazon rainforest is that of tipping-point. Once this point is reached, irreversible changes are to happen, and the ecosystem is engaged in a deep transformation. Many scientists consider the Amazon rainforest has already reached this tipping-point or is about to do so, and is currently becoming a savanna (Alves de Oliveira et al., 2021; Lovejoy & Nobre, 2018; Walker, 2021). This process is known as the savannization of the Amazon.

Reaching this tipping point means the Amazon rainforest could lose its fundamental role of macroregional climate equilibrator, contribute positively to climate change by directly emitting carbon dioxide, which in return could favour forest fires and the emission of the carbon contained in the forest. This phenomenon is mainly linked to the process of deforestation which took place these last decades. Another central stake is biodiversity. There are 400 billion of trees and 16,000 different species in the Amazon rainforest. Livestock farming, road's construction, hydroelectric dams' construction and oil and mines exploitation are the main factors of deforestation; they of course interact one with another, which leads us to consider that the main factor of the Amazon deforestation is its position at the periphery of the global economy, and its progressive integration into capitalist accumulation. Those multiple factors are indeed linked to one coherent process of economic accumulation. The Amazon rainforest is one of many examples of the negative relation between capitalism and the environment (Angus, 2018; Malm, 2017).

The current transformation of the Amazon rainforest implicates a new definition of environmental commons. Indeed, the academic field of research around environmental commons has been structured by an institutionalist perspective that mainly analyzes the institutional resorts of environmental governance on either local or global commons. Both categories are independent in the traditional conception of environmental commons: the atmosphere would be for instance a global common, while a fishery in the Bolivian Chapare would be the classic village common (Ostrom, 1990). The situation of the Amazon rainforest invites to overcome this dichotomy: this mythical forest is both a local and a global common. Indeed, it can suffer from as well as it can alter the dynamics of climate change: the forest is linked to the atmosphere and cannot truly be dissociated from it. It is a fundamental element of global biodiversity. The traditional knowledge of indigenous people who live there is also precious to the entire humankind. All these elements attribute to the Amazon rainforest a global dimension, though it is clear all the stakes of power and power relations have local, national, and international dimensions when the protection or exploitation of the forest are concerned. As a concrete and ecologically determined common, it cannot escape its local dimension and physical dimension. The Amazon rainforest is both a village common and a global common. It points toward a general character of environmental commons.

An environmental common can be defined as a space whose destruction or alteration can negatively affect the reproduction of human conditions of life on Earth (Cadalen, 2020). The Amazon rainforest comes to the Anthropocene as a highly political space and stake. It is often referred as an object of tensions around sovereignty. The above-mentioned treaty for Amazonian cooperation aimed at protecting the sovereign claims of States on the forest. The sovereign rights on the forest are useful for the governments not only to protect them from international interference, but also to oppose the indigenous communities' claims over their "ancestral territories". Apart from Guyana, Surinam and France, the other Amazonian countries have ratified the convention 169 of the International Labor Organization (ILO) which guarantees to the indigenous people a right of free, prior, and informed consultation when a project can affect their collective rights, as, for instance, a project of oil extraction that can seriously disturb the ecosytem. The claims of indigenous communities of Amazonia went global, given that the direct negotiations with the States are in most cases difficult, particularly with neoliberal governments: the indigenous peoples have then become an international actor at the United Nations (Bellier, 2012).

Though the repressions were singularly strong with neoliberal governments, another kind of problems emerged with the Leftist governments, for example in Bolivia or in Ecuador. Indeed, the indigenous blocks were divided about the attitude to adopt with these governments, and the latter were suspicious with the European or US NGOs allied to some indigenous communities for land protection (García Linera, 2012). The memories of US influence in South America are a strong reason why the stakes of sovereignty are so central to the political dynamics of the Amazon rainforest. This situation has produced intense academic debates over the notion of "neo-extractivism", coined by Maristella Svampa as the continuation of developmentalist models by Leftist governments (Poupeau, 2020; Svampa, 2011).

In conclusion, the Amazon rainforest is an environmental common, as the quick and contemporary changes it goes through negatively affect the life of human societies both regionally and globally. The geophysical changes it is experiencing rely on economic and political dynamics. The former point towards the capitalist integration of Amazonia. This process began with the Conquest in the sixteenth century and accelerated considerably during the last century's second part. As for the political dynamics, they implicate all levels of power and can be highly conflictual, as they integrate international power dynamics, and the singularity of the indigenous people claims. The capacity of conflict resolution and the possibility of Amazonian protection are positively linked to the capacity to stop the capitalist integration of the rainforest.

It is indeed what social movements and political changes tried to achieve in Bolivia and Ecuador, when Evo Morales and Rafael Correa became presidents of these countries. Their victories were closely linked to social mobilizations for the nationalizations of resources such as gas, oil, or water, which produced an alliance between neo-indigenist movements, environmentalists' associations, Leftist militancy, peasants and workers' unions. Under their governments, one of the main ambiguities dealt with the definition of the development the engaged political processes were supposed to overcome. The notion of *Sumak Kawsay* in Ecuador, *Suma Qamaña* in Bolivia, were both aiming at the reconfiguration of development models. However, the environmentalists considered the development itself, as a logical relation to nature, had to be overcome, while the socialist tendency considered the capitalist development had to be overcome. The latter could lead before all to overcome the underdevelopment and fight extreme poverty. This tension around Amazon and the notion of development is common to South America as a continent and is one of the key features of contemporary politics under the Anthropocene. In other words, the Amazon rainforest political contradictions must be studied for the lessons they can bring for other places in the world: it is global in this specific meaning. The concrete and political articulation of both perspectives might be one of the best ways to protect environmental commons, as suggested by some ecosocialist works (Bookchin, 2012; Malm, 2017). If the question is still opened, the pressure of time is, here as elsewhere for the Anthropocene, urging for political imagination – so that the Amazon myth does not end with fires.

References

Alves de Oliveira, B. F., Bottino, M. J., Nobre, P., & Nobre, C. A. (2021). Deforestation and climate change are projected to increase heat stress risk in the Brazilian Amazon. *Communications Earth & Environment, 2*, 207. https://doi.org/10.1038/s43247-021-00275-8

Angus, I. (2018). *Face à l'Anthropocène: le capitalisme fossile et la crise du système terrestre*. Écosociété.

Beau, R., & Larrère, C. (Eds.). (2018). *Penser l'anthropocène*. Presses de Sciences Po.

Bellier, I. (2012). Les peuples autochtones aux Nations Unies: Un nouvel acteur dans la fabrique des normes internationales. *Critique internationale, 1*, 61–80.

Bonneuil, C., & Fressoz, J.-B. (2013). *L'événement anthropocène*. Seuil.

Bookchin, M. (2012). *Qu'est-ce que l'écologie sociale ?* Atelier de création libertaire.

Cadalen, P.-Y. (2020). L'Amazonie et le vivant à l'épreuve de l'écopouvoir. *Raisons politiques, 4*, 77–90.

García Linera, A. (2012). *Geopolítica de la Amazonía. Poder hacendal-patrimonial y acumulación capitalista*. Vicepresidencia del Estado plurinacional de Bolivia.

Gerlach, A. (2003). *Indians, oil and politics: A recent history of Ecuador*. SR Books.

Grove, J. V. (2019). *Savage ecology: War and geopolitics at the end of the world*. Duke University Press.

Harrison, R. P. (1992). *Forests: The shadow of civilization*. University of Chicago Press.

Hecht, S., & Cockburn, A. (2010). *The fate of the forest. Developers, destroyers and defenders of the Amazon*. The University of Chicago Press.

Lovejoy, T. E., & Nobre, C. (2018). *Amazon Tipping Point* (Vol. 4, p. 4). Science Advances.

Malm, A. (2017). *L'Anthropocène contre l'histoire: le réchauffement climatique à l'ère du capital*. La Fabrique.

Ostrom, E. (1990). *Governing the commons*. Cambridge University Press.

Picq, M. (2016). Rethinking IR from the Amazon. *Revista Brasileira de Política Internacional, 59*(2), e003. https://doi.org/10.1590/0034-7329201600203

Poupeau, F., (2020). Ce qu'un arbre peut véritablement cacher. Le Monde *Diplomatique*.

Smouts, M.-C. (2001). *Forêt tropicale, jungle internationale : revers de l'écopolitique mondiale*. Presses de Sciences Po.

Svampa, M. (2011). Néo-"développementisme" extractiviste, gouvernements et mouvements sociaux en Amérique latine. *Problèmes d'Amérique latine, 3*(83), 101–127.

Walker, R. T. (2021). Collision course: Development pushes Amazonia toward its tipping point. *Environment: Science and Policy for Sustainable Development, 63*, 15–25. https://doi.org/1 0.1080/00139157.2021.1842711

Wallenhorst, N. (2020). *La vérité sur l'Anthropocène*. Le Pommier.

Pierre-Yves Cadalen is currently post-doc researcher at the CRBC – Université Bretagne Occidentale and associate to the CERI-Sciences Po. He recently obtained his PhD in Political Science and International Relations. His works are mainly related to the relations of power around the Environmental Commons. His former studies, which also included a Bachelor in Philosophy, led him to theorize the new forms of power related to the Anthropocene era. Recent publications: « L'Amazonie et le vivant à l'épreuve de l'écopouvoir », *Raisons politiques*, n° 80, 2020, pp. 77–90; « Le populisme écologique comme stratégie internationale », *Critique internationale*, n° 89, 2020, pp. 165–183; « Republican populism and Marxist Populism: Perspectives from Ecuador and Bolivia », in *Discursive Approaches to Populism Across Disciplines – The Return of Populists and the People*, Palgrave MacMilan, 2020.

Carbon

Pierre Léna

Abstract Carbon is an element that plays an essential role in energy and chemical exchanges inside and on the surface of the planet Earth. Because it can easily combine with other elements or with itself, carbon is the fundamental element for life on Earth. In its combined forms, it is stored in the Earth's surface subsystems, which are the lithosphere, the ocean, the atmosphere and the biosphere. The carbon cycle exchanges carbon atoms between these reservoirs, with characteristic times ranging from short (day or year) to long (millennia or more). Living beings contribute significantly to these exchanges, which result in a global state of a dynamic carbon equilibrium. Human activity since the beginning of the industrial era (1850) has disrupted this balance. Combustions massively inject CO_2 gas into the atmosphere, at the expense of fossil carbon which was stored in a durable way. The consequence is the increase of the Earth's average global temperature and climate change. The carbon balance and carbon neutrality measures have thus become central elements for a climate policy aimed at restoring the balance of the Earth system.

The carbon atom (symbol C) is a chemical element, among the hundred or so stable elements which are the basis of the organization of matter into molecular structures. The most abundant isotope of carbon is ^{12}C, whose nucleus consists of 6 protons and 6 neutrons, surrounded by 6 electrons. Four of these electrons are readily available to form chemical bonds with another C atom, or with other elements. This is reflected in the extreme richness of carbon chemistry, and in particular its fundamental role for the existence of life on planet Earth.

In the Universe, carbon is formed by nuclear reactions at the heart of stars, and its abundance in number of atoms is the third in importance, after those of hydrogen and helium, for about one thousandth of the latter. During the formation of the Earth from an interstellar cloud, which also formed the Sun, the final relative abundance of carbon on Earth decreased considerably. In the Earth's lithosphere, carbon

P. Léna (✉)
Observatoire de Paris, LESIA, Meudon, France
e-mail: pierre.lena@obspm.fr

113

represents only 0.03% by mass, while oxygen and silicon represent 73%; most of the hydrogen, whose atomic mass is too low for gravity to hold it, has been lost in space, except when combined, for example in water (H_2O). The solid forms of carbon are diverse (diamond, graphite, amorphous carbon). The Earth is thus an open system from the energetic point of view, receiving solar energy and losing energy to space, but a closed system from the point of view of its chemical elements, the quantity of which remains essentially constant (except for meteorite impacts) throughout the geological ages (Wikipedia, 2022a; Haynes, 2016).

The main reservoir of terrestrial carbon is the lithosphere, because the superficial carbon in ocean, soil and atmosphere represents only 0.4% of the total: the ocean is the main reservoir (93%), the atmosphere containing 1.4%, the surface biomass (soil and plants) 5.6%.

In the atmosphere, carbon is mainly present in gaseous form, combined with oxygen (carbon dioxide CO_2) or hydrogen (methane CH_4). In the ocean (or hydrosphere), most of the carbon is present in inorganic forms dissolved in water, in dissolved organic forms for 3%, and in living forms (algae, plankton, fish) for a tiny proportion. On continental surfaces, carbon is mainly present in the biomass: soils (75%) in the form of various organic compounds, and vegetation (25%).

In the lithosphere, carbon is combined with oxygen in buried carbonates (mainly calcium carbonate $CaCO_3$), combined with hydrogen in buried hydrocarbons (oil and natural gas), and in more or less pure amorphous form in coals. These combinations originated from the surface biomass (continents, ocean). The buried quantity of hydrocarbons, considered as exploitable, is debated but may be of the order of 10,000 Gt, i.e. 5 times the current surface biomass (DOE, 2022b).

By order of storage importance, lithosphere, ocean, biosphere and atmosphere are the surface reservoirs of terrestrial carbon. To these reservoirs, one must add the very deep reservoir constituted by the Earth's mantle, where the quantity of carbon and the exchanges with the surface are still poorly known.

These atoms and molecules containing carbon are not located for ever in their surface reservoirs. They circulate permanently through natural exchanges between reservoirs and thus go through biogeochemical cycles (e.g. carbon, oxygen and metal cycles). The quantities of a given atomic species present at a given time in each of the reservoirs are therefore fixed by a dynamic equilibrium, characterized by the quantities exchanged and the exchange times between sources (the reservoir gains) and sinks (the reservoir loses). Exchange characteristic times can be extremely different: the lithospheric reservoir evolves only on a geological time scale (millions of years) by surface erosion and sedimentation in the ocean (geological cycle). On the other hand, the exchange (or residence) times between the three surface reservoirs, as well as with the 'fossil resources' part of the lithosphere, exploited by man, can be as short as a year or a century.

Carbon, this essential element for life, has its cycle, the carbon cycle (Wikipedia, 2022b). Inorganic carbon, in the form of carbon dioxide CO_2, is exchanged rapidly (day or more) between the atmosphere and the surface ocean, thus maintaining a permanent balance between these two reservoirs (Fig. 1). Moreover, during a cycle lasting about a thousand years (thermohaline circulation), the surface waters

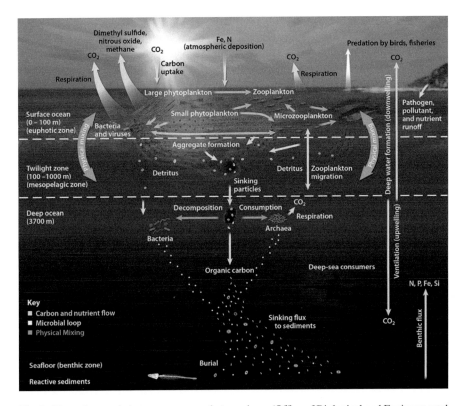

Fig. 1 The carbon cycle between ocean and atmosphere. (Office of Biological and Environmental Research of the U.S. Department of Energy Office of Science (DOE, 2022a))

circulate in surface, dive to the depths and surface, then being able to exchange their carbon with the atmosphere again (Sarmiento & Gruber, 2006). Thus, the deep reservoir, which contains most of the oceanic carbon, can only reach its equilibrium with the atmosphere in several centuries.

Living beings (biosphere) exchange, permanently and rapidly, carbon with the atmosphere (biological carbon cycle). A mass of 60 Gt of carbon is exchanged annually, which is almost one tenth of the total amount of carbon in the atmosphere. The exchange takes place in both directions: fermentation and respiration produce CO_2, while photosynthesis, mainly of chlorophyllous plants, and shell construction fix the carbon from the CO_2 in the biomass (Wikipedia, 2022b).

Finally, human activities contribute to the carbon cycle. This happens mainly through combustions which, in order to produce easily usable energy, remove the fossil carbon which is stored durably in a reservoir, such as coal, gas, oil or wood, and inject it in the form of additional CO_2 into the atmosphere, in increasing quantities with industrialization and population growth (anthropogenic cycle). In 2019, the additional annual injection of carbon into the atmosphere was 12 Gt (or 43 Gt of CO_2), this carbon then entering the general cycle of exchange and storage with the ocean and biomass. 44% of this quantity will remain in the atmosphere for about a 1000 years, enhancing the greenhouse effect, the rest going into the ocean (23%) and in the surface biomass (31%) (IPCC, 2021). This absorption capability of the ocean may decrease in the future, because of the elevation of oceanic temperature on one hand, and the deforestation on the other hand (IPCC, 2019). These natural or anthropogenic cycles, whether biological, oceanic or atmospheric, are mainly composed of chemical transformations and are closely coupled by these exchanges. Therefore, they cannot be treated separately from each other, nor can they ignore extremely variable time scales, ranging from fractions of a year to millennia and beyond. The result is a great complexity of details, whereas the main exchange flows can be described quite simply (Fig. 2).

In the Earth's past, the global balance of the carbon cycle may have changed when external factors altered the conditions governing exchange. For example, intense volcanic activity increased atmospheric CO_2 in the primitive Earth (formation of the second atmosphere), then 3.5 billion years ago, the appearance of

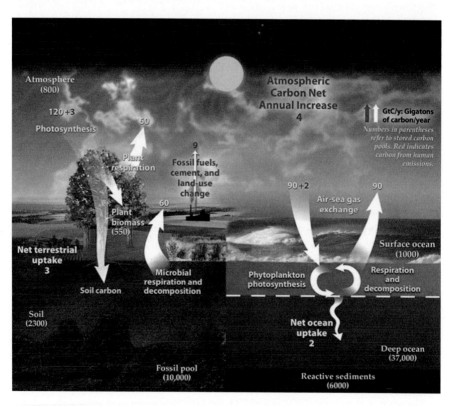

Fig. 2 Fast carbon cycle showing the movement of Carbon between land, atmosphere, and oceans in gigatons per year. Yellow numbers are natural fluxes, red are human contributions, white are stored carbon. The effects of the slow carbon cycle, such as volcanic and tectonic activity are not included. (Adapted from U.S. Department of Energy, Biological and Environmental Research Information System. (DOE, 2022b))

bacteria capable of photosynthesis caused this high concentration to fall (formation of the third atmosphere, comparable to the present one). More recently, the ice ages, caused by the variation of the astronomical parameters setting the Earth's orbital movement, have temporarily lowered the amount of this CO_2. The knowledge, through glacial archives, of the amount of atmospheric CO_2 during the last 10,000 years, allows to describe these variations of the carbon balance.

For the last few decades, interest has focused on the evolution of the quantity of atmospheric CO_2, since the increase of this greenhouse gas leads to a disruption of the Earth's energy balance, a balance between the energy coming from the Sun and absorbed by the Earth and the energy re-radiated to deep space. Resulting mainly from the combustion of carbonaceous fossil resources, this disruption leads to an increase in the Earth's average global temperature, causes climate change and its multiple impacts on human societies (extreme events, rising sea levels, etc.). In 2020, the atmospheric concentration of CO_2 reaches 410 parts per million, a level which has not been reached since 2 billion years (IPCC, 2021). It is therefore invaluable to carry out an annual carbon balance of atmospheric CO_2, which is an essential indicator for assessing all exchanges between reservoirs and the rapid evolution of the carbon stored in each of them.

This assessment of the carbon balance is at the origin of carbon accounting, which measures the exchanges and makes it possible to appreciate the levers, which are likely to mitigate the increase of greenhouse effect and climate change. In particular, this accounting allows financial and social discussions around the carbon tax, proposed as a tool to regulate the injection of CO_2 into the atmosphere. Although CO_2 is the dominant atmospheric gas of the greenhouse effect - water vapour plays an even more important role, but obeys a completely different cycle -, other gases, such as methane (CH_4, therefore carbonaceous) or nitrous oxide (N_2O, therefore non-carbonaceous) also contribute to it. Since 1850, CH_4 injection has increased by 156% and N_2O by 23% (IPCC, 2021). As these gases are also produced by human activity, it is usual to include them in the carbon footprint, as their 'CO_2 equivalent', considering them as equivalent to the amount of CO_2 that would have the same impact on the Earth's energy balance. The carbon footprint of an individual, a group, the production of an object, the activity of a city or a state represent the part of the carbon balance to which this entity contributes annually.

Carbon neutrality is defined as a state in which the global net emissions of CO_2 of human origin into the atmosphere are exactly compensated by the absorption of this gas, either by natural carbon sinks managed by humans (forests, soils and oceans) - the anthropogenic share not always being easy to identify with precision -, or by artificial sinks (so-called negative emission technologies). This neutrality is the objective set by the most ambitious scenarios studied by the IPCC for the year 2050. Carbon neutrality can be global, or it can concern a specific activity, in which case it does not lead to global neutrality but can contribute to it.

For the year 2019, before the impacts of the Covid-19 pandemic on all human activities, the global direct emissions of CO_2 into the atmosphere were 11.8 Gt of carbon (or 43 Gt of CO_2), divided into 85% for the combustion of fossil resources and 15% for changes in land use, which reduce the capacity of absorption of CO_2 by the biomass (photosynthesis). In 2019, the average carbon footprint of each inhabitant of the planet, populated by 7.7 billion people, was therefore of 5.9 tons of CO_2/inhabitant, with a large dispersion between countries and within countries.

The threat posed by climate change has led to investigate whether one of the mitigation measures could be the creation of artificial carbon reservoirs, capable of improving the balance by storing significant quantities of CO_2 over very long periods. A distinction is made between carbon capture and storage, a method that removes CO_2 directly from its source of production (e.g. a cement plant) and recycles it for chemical production purposes, and sequestration, which is capable of storing CO_2 over very long periods (centuries or millennia). Many capture methods, physical or physico-chemical, are being studied (Sorai et al., 2022; Wikipedia, 2022c). Liquefied CO_2 can be sequestrated in geological reservoirs located at a depth of about 1000 m, where the pressure is sufficient to keep it there for geological periods of more than a hundred thousand years. The available storage capacity is estimated at 675–900 Gt of CO_2. Yet, one cannot ignore the fact that this storage, which is affected by uncertainties about its long-term stability, would only cope with the current level of emissions for one or two decades and could dispense with any mitigation effort.

To conclude, the combinations of the element Carbon, built in the heart of stars, have protected the Earth from becoming a frozen body, have allowed life to exist and to develop photosynthesis, have provided the energy for the fire to be domesticated by Homo Sapiens, for the metallurgy to develop, for stainless steel to be invented, for airplane to fly and industry to develop. Today, this element is again at the crossroads of the Anthropocene.

References

Department of Energy (USA) (DOE). (2022a). *Carbon cycle, office of biological and environmental research*. Retrieved from https://commons.wikimedia.org/wiki/File:Oceanic_Food_Web.jpg

Department of Energy (USA) (DOE). (2022b). *Fast carbon cycle between land, atmosphere and ocean*. Retrieved from http://earthobservatory.nasa.gov/Features/CarbonCycle/

Haynes, W. M. (Haynes). (2016). *CRC handbook of chemistry and physics* (Abundance of elements in the Earth's crust and in the sea) (Vol. 97, p. 2402). CRC Press/Taylor and Francis.

IPCC (IPCC). (2019). Special report on the ocean and cryosphere in a changing climate. In H.-O. Pörtner et al. (Eds.), *Summary for policymakers*. Retrieved from https://www.ipcc.ch/srocc/

IPCC (IPCC). (2021). Climate change 2021. The physical science basis. In V. Masson-Delmotte et al. (Eds.), *Contribution of working group I to the sixth assessment report of the intergovernmental panel on climate change summary for policymakers*. Cambridge University Press. Retrieved from https://www.ipcc.ch/report/ar6/wg1/#SPM

Sarmiento, J. L., & Gruber, N. (Sarmiento & Gruber). (2006). *Ocean biogeochemical dynamics*. Princeton University Press.

Sorai, M., et al. (Sorai et al). (2022). CO_2 geological storage. In M. Lackner, B. Sajjadi, & C. Wei-Yin (Eds.), *Handbook of climate change mitigation and adaptation*. Springer Nature Reference, Living Edition 2022.

Wikipedia (Wikipedia). (2022a). *Abundance of the chemical elements*. Retrieved from https://en.wikipedia.org/wiki/Abundance_of_the_chemical_elements

Wikipedia (Wikipedia). (2022b). *Carbon cycle*. Retrieved from https://en.wikipedia.org/wiki/Carbon_cycle

Wikipedia (Wikipedia). (2022c). *Carbon capture and storage*. Retrieved from https://en.wikipedia.org/wiki/Carbon_capture_and_storage

Pierre Léna is an astrophysicist, Emeritus Professor at the Observatoire de Paris and the University of Paris Cité, born 1937. His scientific career dealt with infrared astronomy and the exploration of this domain with instruments on ground-based observatories, aircrafts and satellites. He had a long-lasting interest in the transmission of science to the public, and especially in education, In 1995, he co-founded in France the movement *La main à la pâte* to develop inquiry-based science education in schools, and in 2018 he extended this action to climate issues, with the creation of the Office for Climate education (OCE).

Coral

Joshua Wodak

Abstract Due to its high sensitivity to ocean warming and ocean acidification, coral presents paleoclimatological records that situate the impact of *homo sapiens* as a geomorphic force against the long duration of geology. This article examines coral and its highly particular relevance to the Anthropocene. It offers an overview of coral in terms of its evolutionary biology, the extent of the current existential challenges facing the lifeform, and the emerging field of controversial technoscientific conservation that is proposed for ameliorating these existential challenges. Coral is presented as a discrete means to understand how inadvertent human engineering of the biosphere marks the Anthropocene, such that the future of myriad lifeforms is becoming inextricably entwined with polemical human-engineered evolution that is intended to augment ecosystems eroded by the Anthropocene.

On visiting the Keeling Islands in April 1836, Charles Darwin remarked in his journal that he was "glad we have visited these islands [for] such formations surely rank high amongst the wonderful objects of this world." His profound admiration of coral amounted to "a wonder which does not at first strike the eye of the body, but, after reflection, the eye of reason" (Darwin, 1836, p. 465). Coral appear to be without peer in inducing such wonder – with luminaries such as David Attenborough declaring scuba diving on the Great Barrier Reef to be "the most magical thing" of all his encounters with the more-than-human world (Attenborough, 2014). In expressing their wonder, both naturalists refer not only to the kaleidoscope of colours that coral come in, but also to the myriad forms and sheer scale of "such formations." On comparing their marvel to the "vast dimensions of the Pyramids", Darwin declares "how utterly insignificant are the greatest of these, when compared to these mountains of stone accumulated by the agency of various minute and tender animals" (1836, p. 465). The "various" referred to here actually encompass a holobiont, for coral is literally both animal (the individual polyps), mineral (the calcium

J. Wodak (✉)
Western Sydney University, Sydney, Australia
e-mail: j.wodak@westernsydney.edu.au

© The Author(s), under exclusive license to Springer Nature Switzerland AG 2023
N. Wallenhorst, C. Wulf (eds.), *Handbook of the Anthropocene*,
https://doi.org/10.1007/978-3-031-25910-4_19

carbonate skeleton secreted by the polyps) and vegetable (the zooxanthellae that live within coral tissue). Further, the holobiont is both alive and dead – in that coral reefs consist of an outer layer of living polyps, extending from a substrate of skeletal remains of prior generations of polyps. And, given that individual coral colonies can indeed live for millennia, such formations do indeed grow to the size of "mountains of stone" that Darwin observed.

In Darwin's time, then nascent biological and geological research into coral gave rise to common misunderstandings of how robust the living component of coral is, due to the hardness of their secreted skeleton (McCalman, 2013, p. 15). Nowadays there is a robust appreciation of just how fragile these "tender animals" actually are. For instance, when Attenborough reminisced in 2014 about his first encounter with coral, which took place almost six decades prior, in 1957, he did so with the acute appreciation of the existential challenges facing coral worldwide. Namely, that his answer to the question about "what was the most magical thing you ever saw in your life" comes from the press release for his return to the Great Barrier Reef to make a landmark 3-part BBC documentary series about this largest living formation on earth.

While such existential challenges are faced by untold millions of species at the advent of the Sixth Extinction Event, coral hold particular importance in this regard, as the proverbial 'canary in the coal mine' of what Anthropocene Extinction portends (Kuussaari et al., 2009). Coral hold such a status due to their remarkably high sensitivity to ocean warming and ocean acidification. Whereby, the symbiotic relationship between individual polyps and their zooxanthellae breaks down. Under heat stress, coral eject their zooxanthellae, which then deprives them of their main source of energy and nutrients, leading to imminent death if new, and compatible, zooxanthellae do not re-enter their coral tissue in time. Given that 1/3rd of marine species dependent on coral reefs for their survival, even though reefs constitute 0.1% of ocean surface area, coral indeed lie at the heart of aquatic life the world over (Bowen et al., 2013, p. 360).

Coral also derive their iconic 'canary in the coal mine' status due to how the rising frequency and intensity of global coral bleaching events is showing up just how conservative and inadequate normative climate change modelling by the UNFCCC and IPCC has been. For instance, the unprecedented global coral bleaching event of 2016–17 was followed, for the first time in recorded history, by the first ever back-to-back bleaching event of 2017–18. Then, yet another first, of the global bleaching event of 2019–20 and now another event of 2021–22, which occurred during a Southern Hemisphere La Nina climate pattern, when slightly cooled ocean temperatures lessened the full extent of the bleaching that would have otherwise occurred. According to normative climate modelling, events of such frequency and ferocity were not anticipated to occur until the middle of this century, rendering even Attenborough's (2014) documentary about the Great Barrier Reef as depicting a coral reef that, less than a decade later, quite simply no longer exists.

As both an iconic and totemic organism, coral have thus given impetus toward a substantial body of scientific research into proposed conservation. Such conservation is highly experimental and risky, due both to the sheer scale of the existential plight, as well as the sheer complexity of the holobiont (Wodak, 2017). Two principles means are currently being developed for such conservation: Assisted Evolution and Synthetic Biology. The former is a kind of laboratory-based selective breeding, whereby scientists seek to induce advantageous traits of higher resistance against ocean warming and ocean acidification (Anthony et al., 2017; van Oppen et al., 2015). The latter aims towards the same ends, but through a different means of biologically engineering coral germlines with such elevated resistance (Wodak, 2022). Both proposed means of conservation are still at a relatively nascent stage, and, even if they can be proved to be efficacious in a laboratory, profound moral dilemmas remain as to whether such coral could should be released in the so-called 'wild', as well as debates as to whether such releases could even be technically possible, given the geographic dispersal of coral, as well as the aforementioned unknowns around interactions in the full holobiont, which are not possible to reduce to either computer modelling or laboratory experiments (Redford et al., 2013).

Other conservation proposals take a diverse range of approaches, from bespoke artificial coral reefs, whereby coral are attached to substrates made of metal or plastic, through to the Rigs-to-Reefs program, where decommissioned oil rigs are repurposed into intentional mass-scale artificial reefs, and organisations such as *50 Reefs* which undertook a global survey to identify future refugia of "fifty coral reefs that, together, have the potential of surviving the impacts of climate change and the ability to help repopulate neighbouring reefs over time" (https://www.50reefs.org).

A team of German paleobiologists offer a radically different notion of what may constitute refugia, based on fossil records of coral adaptation to prior periods of abrupt climatic change. In their article "Could 'Ecosystem Atavisms' Help Reefs to Adapt to the Anthropocene?" Reinhold Leinfelder et al. argue that "some modern coral reefs may have the potential to adjust to the Anthropocene by 'atavistically' recombining to lower-complexity reefs with different nutrition strategies." However, they caution that such "readjusted reefs would however be impoverished and would not…aesthetically compensate for the highly structured reefs we still have in some places of the world" (Leinfelder et al., 2012, p. 3). The use of a question mark in the article title further denotes how speculative and conditional their hypothesis is. Wherein, the authors argue that "tipping point–shifts of modern reefs from healthy to sponge/soft-coral/algae to heterotrophic/microbial reefs appear to reflect in-place 'atavistic' switches to an evolutionary less complex state with a reduced number of modules" (2012, p. 6). Although 'ecosystem atavisms' are not a form of human-induced conservation, the hypothesis does constitute inhuman-induced 'conservation', whereby the class Anthozoa (which includes the sponges and soft-coral that are the evolutionary ancestors to existing hard corals) atavistically evolves into ancestral forms. In light of how human-induced coral conservation is drastically insufficient for the rate of global biophysical change currently unfolding, such deep time views offer a caution of humility against the hubris of human agency in the Anthropocene.

Despite their many differences, many of the emerging conservation proposals for coral appear particularly apt for the Anthropocene, in light of how the epoch collapses distinctions between 'nature' and 'culture' or between 'the environment' and 'humans.' For instance, in *The Historical Roots of Our Ecologic Crisis* Lynn White offered one of the earliest articles to situate environmental issues in an adequate historical context. Therein, he argued that

> All forms of life modify their contexts. The most spectacular and benign instance is doubtless the coral polyp. By serving its own ends, it has created a vast undersea world favourable to thousands of other kinds of animals and plants. Ever since man became a numerous species he has affected his environment notably (1967, p. 1203).

Here coral become a kind of synecdoche to explain human-induced environmental modification, whereby Darwin's "mountains of stone" are held up as being analogous to human-built environments.

Such analogies between coral and cities have been instrumental in the development of the Anthropocene hypothesis, with Jan Zalasiewicz, chair of the Anthropocene Working Group for the International Commission on Stratigraphy, arguing that "both skyscrapers and coral reefs are basically large masses of biologically constructed rock, worthy monuments to our respective phyla" (2008, pp. 171–2). True to the credentials for what may constitute strata that bear a distinct anthropogenic imprint, skyscrapers will indeed be reduced to "monuments" of contemporary civilisation so-called. Whether these are "worthy monuments" of *homo sapiens* in the manner of fossilised coral reefs that extend back 480 million years is a topic of considerable debate. Cities, and by extension, skyscrapers, are emblematic of global urbanisation since the European Industrial Revolution. So too are the catastrophic increases in population, resource exploitation, and capital, that are also features of urbanisation. Whereby, these "worthy monuments" are actually emblems of the same industrial processes which unleashed the Anthropocene, and, thus, the existential challenges facing not only coral, but the unfolding Sixth Extinction Event.

Further, analogies between coral and cities also express something of the profound moral dilemmas as to what – if anything – conservation in the Anthropocene should become if it is to be remotely efficacious. Environmental ethicist J. Baird Callicott encapsulates this dilemma when he declared that

> Bluntly put, we are animals ourselves, large omnivorous primates, very precocious to be sure, but just big monkeys, nevertheless. We are therefore a part of nature, not set apart from it. Hence, human works are no less natural than those of termites or elephants. Chicago is no less a phenomenon of nature than is the Great Barrier Reef (1992, p. 17).

Therein, if human-built "mountains of stone" and coral-built "mountains of stone" are both merely a "phenomenon of nature", then the Anthropocene becomes teleological. Wherein, efforts to ameliorate the deleterious effects of the Sixth Extinction Event go against the 'natural' phenomenon that is the Anthropocene. While it may appear at first sight that it is easy to reject such a spurious argument, coral continue to elude both human attempts to understand ecosystem engineering by the more-than-human world, as interest rapidly grows in intentionally bio-engineering organisms such as coral, in an attempt to provide them better means for surviving the Anthropocene itself.

References

A Global Plan to Save Coral Reefs. *50 Reefs – The Ocean Agency* http://www.50reefs.org. Accessed 9 Mar 2022.

Anthony, K., Bay, L., Costanza, R., Firn, J., Gunn, J., Harrison, P., Heyward, A., Lundgren, P., Mead, D., Moore, T., Mumby, P. J., van Oppen, M. J. H., Robertson, J., Runge, M., Suggett, D., Schaffelke, B., Wachenfeld, D., & Walshe, T. (2017). New interventions are needed to save coral reefs. *Nature Ecology & Evolution, 1*, 1420–1422.

Attenborough, D. (2014). *David Attenborough to present new landmark series on the great barrier reef for BBC one.* British Broadcasting Corporation Media Release, 9 September 2014 https://www.bbc.co.uk/mediacentre/latestnews/2014/great-barrier-reef. Accessed 9 July 2021.

Bowen, B., Rocha, L., Toonen, R., & Karl, S. (2013). The origins of tropical marine biodiversity. *Trends in Ecological Evolution, 28*, 359–366.

Callicott, J. B. (1992). La Nature est morte, vive la nature! *The Hastings Center Report, 22*(5), 16–23.

Darwin, C. (1836). *Journal of researches: Coral formations.*

Kuussaari, M., Bommarco, R., Heikkinen, R., Helm, A., Krauss, J., Lindborg, R., Öckinger, E., Pärtel, M., Pino, J., Rodà, F., Stefanescu, C., Teder, T., Zobel, M., & Steffan-Dewenter, I. (2009). Extinction debt: A challenge for biodiversity conservation. *Trends in Ecology & Evolution, 24*, 564–571.

Leinfelder, R., Janina, S., Georg, H., & Struck, U. (2012). Could 'Ecosystem Atavisms' Help Reefs to Adapt to the Anthropocene? In *Proceedings of the 12th International Coral Reef Symposium, Cairns, Australia, 9–13 July 2012* (pp. 1–6).

McCalman, I. (2013). *The reef: A passionate history.* Penguin.

Redford, K., Adams, W., & Mace, G. (2013). Synthetic biology and conservation of nature: Wicked problems and wicked solutions. *PLoS Biology, 11*, e1001530.

van Oppen, M., Oliver, J., Putnam, H., & Gates, R. (2015). Building coral reef resilience through assisted evolution. *Proceedings of the National Academy of Sciences, 112*(8), 2307–2313.

White, L. (1967) The historical roots of our ecologic crisis. *Science 155 (3767)*, 1203-1207.

Wodak, J. (2017). Artificial coral reefs. In R. Emmett, G. Mitman, & M. Armiero (Eds.), *Future remains: A cabinet of curiosities for the Anthropocene* (pp. 99–107). University of Chicago Press.

Wodak, J. (2022). Synthetic biology. In N. Wallenhorst & C. Wulf (Eds.), *Handbook of the Anthropocene.* Springer.

Zalasiewicz. (2008, January). *The earth after us.* Oxford University Press.

Dr. Joshua Wodak works at the intersection of the Environmental Humanities and Science & Technology Studies. His research addresses the socio-cultural dimensions of the climate crisis and the Anthropocene, with a focus on the ethics and efficacy of conservation through technoscience, including Synthetic Biology, Assisted Evolution, and Climate Engineering. He is currently a Senior Research Fellow at the Institute for Culture and Society, Western Sydney University; a Chief Investigator at the Australian Research Council Centre for Excellence in Synthetic Biology; and an Adjunct Senior Lecturer, School of Biological, Earth, and Environmental Sciences, UNSW. He is currently completing a book about what kinds of conservation experimentation we should be considering in response to the unfolding Sixth Extinction Event.

Earthquakes

Gah-Kai Leung

Abstract Earthquakes are among the world's deadliest natural phenomena. On an increasingly crowded Earth, earthquake risk management therefore should be taken seriously as a global policy problem. Thus, this chapter discusses some of the ethical dimensions of earthquakes as a phenomenon of planetary significance in the Anthropocene. In particular, I consider the kinds of harms that may ensue when earthquakes impact human habitation. I distinguish personal, social and environmental harms in turn, before offering some concluding remarks.

Earthquakes are among the world's deadliest natural phenomena. As of 2015, some 2.7 billion people worldwide are at risk from seismic catastrophes (Pesaresi et al., 2017); as many as 3 million people are expected to die in earthquakes by the end of this century (Holzer & Savage, 2013). On an increasingly crowded Earth, earthquake risk management therefore should be taken seriously as a policy problem of planetary significance in the Anthropocene.

As a policy problem, earthquakes raise many ethical issues. For reasons of space, I will not attempt an exhaustive survey, but here I consider one background ethical question: the kinds of harms that occur when an earthquake impacts human habitation. Here, we may distinguish three categories of human-related harms: personal harms, which accrue to individual humans; social harms, which impact on social relationships between humans; and institutional harms, which compromise the ability of institutions to respond to the quake. I will discuss each in turn. Two caveats are in order: (1) though earthquakes may also occur due to manmade sources, naturally-triggered earthquakes from geological faults will be my primary concern here; (2) there will also likely be environmental harms, i.e. those harms inflicted on the non-human world, but I won't consider those here.

G.-K. Leung (✉)
University of Warwick, Coventry, UK
e-mail: Gah-Kai.Leung@warwick.ac.uk

© The Author(s), under exclusive license to Springer Nature
Switzerland AG 2023
N. Wallenhorst, C. Wulf (eds.), *Handbook of the Anthropocene*,
https://doi.org/10.1007/978-3-031-25910-4_20

Personal Harms

Personal harms accrue to individual citizens as a result of earthquake damage. They may be material or psychological. Material harms include those bodily harms such as death, injury and disability. They also include the deprivation of adequate food, water, clothing, supplies and shelter. These undermine citizens' ability to meet their basic human needs, such as subsistence and security (cf. Shue, 1980). Furthermore, citizens also experience a loss of income through destroyed businesses and livelihoods, which further undermine their capacity to achieve a minimally decent standard of living.

Psychological harms include those associated with post-disaster trauma. A survey carried out 4 months after the 2015 Nepal earthquake found that a third of adults had shown symptoms of depression or increased anger and one in ten had experienced suicidal ideation (Kane et al., 2018). Individuals' loss of their sense of place and their consequent rootlessness, as beings with no fixed spatial location in the world, can also bring mental disorientation. Sociological studies of disasters underline the importance of repairing people's sense of place as part of the recovery process (e.g. Carroll et al., 2009; Chamlee-Wright & Storr, 2009; Silver & Grek-Martin, 2015; Winstanley et al., 2015). In this way, the destruction of an earthquake inflicts not only a physical but mental toll on those affected. Such mental costs also extend to those who have to live with the risk of an earthquake, *even if* the risk does not in fact materialize for them (cf. Nakayachi et al., 2015 on the Tohoku earthquake; cf. Jahn, 2015).

Moreover, the material and psychological harms inflicted by earthquakes do not affect all citizens equally, as disaster studies has long recognized. Vulnerability is spatially uneven, because of differences in building design, soil conditions, proximity to the fault, quantities of flammable materials, psychological preparedness and so on, which vary geographically (cf. Daniell et al., 2017). It is socially uneven, to the extent that it is affected by race, gender, ethnicity, class and so on (Peacock et al., 1997; Cutter & Finch, 2008; Bolin & Kurtz, 2018). Certain groups such as children, the elderly and the disabled may be particularly vulnerable (Peek et al., 2018; Ngo, 2001; Stough & Kelman, 2018) and the poorest are often hardest-hit both materially and psychologically (Bolin & Stanford, 1999: 92; Marks, 2018: 349). Such an uneven distribution of vulnerability means that individuals in certain groups will suffer disproportionate harm.

Social Harms

Social harms accrue to interpersonal social relationships. There are at least three main ways in which such harms arise in earthquakes. First, stricken communities can experience temporary isolation from the rest of the country. Bridges may have collapsed, roads may be impassable and telecommunications lines may be out of action. This may seriously impede flows of supplies and information to the disaster area. Supplies flows are important for restoring basic needs such as food, water and shelter. Information flows are vital for a comprehensive assessment of the destruction and for deciding which groups and locations should be prioritized for assistance. This causes a social harm insofar as a chaotic relief effort can impede a community's ability to recover.

Second, there may be intra-community breakdown when earthquakes disrupt the web of social ties that enable citizens to relate to each other on terms of equal concern and respect. Our sociality as human beings is put under severe strain by disasters, which can lead to weeks or months of social isolation, breakdowns in trust, lawlessness, vandalism, looting and a degradation in community spirit. Many survivors move away permanently, never to return (see Morrow-Jones & Morrow-Jones, 1991), which further hampers the ability of the stricken society to restore normality. When intra-community breakdown occurs, the resultant loss of social capital can impede effective disaster recovery (see Aldrich, 2012). Earthquakes may also hinder victims' access to interpersonal contact - for example because communication lines are down or because citizens are physically isolated - and lead to some members of the polity being excluded from community membership and identification. Following the 1923 Kanto earthquake in Japan for instance, false rumours of Korean subterfuge led to mass atrocities against the Korean minority (e.g. Allen, 1996; Aldrich, 2012: 14).

Finally, communities may suffer from ongoing uncertainty regarding the possibility of recovery. Standard models of community resilience portray disasters such as earthquakes as sudden 'ruptures' with well-defined exit paths (Wilson, 2013: 213). This is evident for example in the *Oregon Resilience Plan* for major earthquakes, which depicts high resilience as a period of rapid readjustment and recovery and lower resilience as a slower, lengthier time taken to restore functioning (OSSPAC, 2013: xv). Yet as the 2010–11 Christchurch experience shows, resilience may instead be "an *on-going* process of adjustment," with no clear indication as to whether the community may rapidly improve its standing, recover more slowly or simply wither away (Wilson, 2013: 213–214; author's emphasis).

Institutional Harms

Another kind of harm from earthquakes is *institutional*. Major disasters often overload the capacities of local, regional and national institutions to respond effectively. Earthquakes therefore may be particularly harmful for the ways in which they expose the weaknesses embedded within the institutional system. These institutional failings undermine both the ability of communities to return to normal and citizens' trust in their leaders. If citizens and communities are not reassured that something is being done to help them, they may lose confidence in their institutions. Worse, they may end up taking matters into their own hands: for example, through mass looting, if institutions do not act quickly enough to restore norms of private property. Earthquakes, then, can under certain conditions lead to a crisis of political legitimacy.

Earthquakes damage institutions in three main ways. First, institutional capacity may be overwhelmed in the immediate and long-term aftermath, which compromises the institution's ability to respond successfully to the catastrophe. One way to think about disasters in this respect is as a time-compressed version of the normal process of capital depletion and replacement (Olshansky et al., 2012); thus, institutional damage occurs when this new reality engulfs the institution and devastates its responsive capacities. Furthermore, the institutional setup can itself pose a barrier to effective disaster response. A lack of timely coordination from an institutional standpoint is frequently a key failing in disaster response (Christensen et al., 2016).

Second, institutional functioning can be impeded when disasters allow authorities to circumvent democratic norms, as some have argued (e.g. Honig, 2009). Because emergency powers can give the executive wider latitude than is normally permissible, the fear is then that such powers will be entrenched and eventually normalized; this is known as the *ratchet effect* (Kreuder-Sonnen, 2019: 1). The COVID-19 pandemic may be a vivid illustration of this phenomenon, due to the sheer scale of the powers invoked to control the virus (Kavanagh & Singh 2020: 1007).

Finally, institutions can be harmed when they fail to learn the right lessons either in advance of an impending disaster or in the aftermath of one. There are two reasons for this. The first is *information failure*: for example, the significance of critical factors may not be fully recognized or the authorities may attempt to minimize the risk of a tragedy even in the face of explicit warning signs (Pidgeon & O'Leary, 2000: 19–20; cf. Perrow, 1999; cf. Vaughan, 1986). The second is *depoliticization*: institutional regimes seek to shift blame in an attempt to reassert legitimacy. For instance, in the aftermath of the 2011 Bangkok floods, Thai leaders deflected responsibility by placing blame on external factors allegedly beyond their control, such as climate change. As a result, subsequent flood policy in Thailand focused on blocking and draining floodwater, which simply redistributed the risk rather than reducing the likelihood of future floods (Marks & Elinoff, 2020). The failure to build the right lessons into the institutional culture, then, damages the institution's ability to function properly and so constitutes a harm.

Conclusion

Earthquakes are well-known for their deleterious effects on human societies. Consequently, they merit attention as an understudied policy problem of ethical concern. This chapter overviewed three kinds of harms that earthquakes inflict on human habitation: personal, social and institutional. This analysis will help us better appreciate the moral salience of earthquakes as a global policy imperative in the Anthropocene.

References

Aldrich, D. P. (2012). *Building Resilience: Social Capital in Post-Disaster Recovery*. University of Chicago Press.

Allen, J. M. (1996). The price of identity: The 1923 Kanto earthquake and its aftermath. *Korean Studies, 20*, 64–93.

Bolin, B., & Kurtz, L. (2018). Race, class, ethnicity, and disaster vulnerability. In H. Rodriguez et al. (Eds.), *Handbook of disaster research* (2nd ed.). Springer.

Bolin, R., & Stanford, L. (1999). Constructing vulnerability in the first world: The Northridge earthquake in Southern California, 1994. In A. Oliver-Smith & S. Hoffman (Eds.), *The angry earth: Disaster in anthropological perspective*. Routledge.

Carroll, B., et al. (2009). Flooded homes, broken bonds, the meaning of home, psychological processes and their impact on psychological health in a disaster. *Health & Place, 15*, 540–547.

Chamlee-Wright, E., & Storr, V. H. (2009). "There's no place like New Orleans": Sense of place and community recovery in the Ninth Ward after Hurricane Katrina. *Journal of Urban Affairs, 31*(5), 615–634.

Christensen, T., et al. (2016). Organizing for crisis management: Governance capacity and legitimacy. *Public Administration Review, 76*(6), 887–897.

Cutter, S. L., & Finch, C. (2008). Temporal and spatial changes in social vulnerability to natural hazards. *Proceedings of the National Academy of Sciences, 105*(7), 2301–2306.

Daniell, J. E., et al. (2017). Losses associated with secondary effects in earthquakes. *Frontiers in Built Environment, 3*, 30.

Holzer, T. L., & Savage, J. C. (2013). Global earthquake fatalities and population. *Earthquake Spectra, 29*(1), 155–175.

Honig, B. (2009). *Emergency Politics: Paradox, Law, Democracy*. Princeton University Press.

Jahn, E. (Prod.), & Barrow, B. (Ed.). (2015). *Oregon Field Guide Presents: Unprepared* [television programme]. Oregon Public Broadcasting.

Kane, J. C., et al. (2018). Mental health and psychosocial problems in the aftermath of the Nepal earthquakes: Findings from a representative cluster sample survey. *Epidemiology and Psychiatric Sciences, 27*(3), 301–310.

Kavanagh, M. M., & Singh, R. (2020). Democracy, capacity, and coercion in pandemic response: COVID-19 in comparative perspective. *Journal of Health Politics, Policy, and Law, 45*(6), 997–1012.

Kreuder-Sonnen, C. (2019). *Emergency Powers in International Organizations: Between Normalization and Containment*. Oxford University Press.

Marks, D. (2018). The political ecology of uneven development and vulnerability to disasters. In R. Padawangi (Ed.), *The Routledge handbook of urbanization in Southeast Asia*. Routledge.

Marks, D., & Elinoff, E. (2020). Splintering disaster: Relocating harm and remaking nature after the 2011 floods in Bangkok. *International Development Planning Review, 42*(3), 273–294.

Morrow-Jones, H., & Morrow-Jones, C. R. (1991). Mobility due to natural disaster: Theoretical considerations and preliminary analyses. *Disasters, 15*(2), 126–132.

Nakayachi, K., et al. (2015). Public anxiety after the 2011 Tohoku earthquake: Fluctuations in hazard response after the catastrophe. *Journal of Risk Research, 18*(2), 156–169.

Ngo, E. B. (2001). When disasters and age collide: Reviewing vulnerability of the elderly. *Natural Hazards Review, 2*(2), 80–89.

Olshansky, R. B., et al. (2012). Disaster and recovery: Processes compressed in time. *Natural Hazards Review, 13*(3), 173–178.

Oregon Seismic Safety Policy Advisory Committee (OSSPAC). (2013). *The Oregon Resilience Plan: Reducing risk and improving recovery for the next Cascadia earthquake and tsunami*. OSSPAC.

Peacock, W. G., et al. (Eds.). (1997). *Hurricane Andrew: Ethnicity, gender and the sociology of disasters*. Routledge.

Peek, L., et al. (2018). Children and Disasters. In H. Rodriguez et al. (Eds.), *Handbook of disaster research* (2nd ed.). Springer.

Perrow, C. (1999). *Normal Accidents: Living with High-Risk Technologies, revised edition.* Princeton University Press.

Pesaresi, M., et al. (2017). *Atlas of the human planet 2017: Global exposure to natural hazards.* Publications Office of the European Union.

Pidgeon, N., & O'Leary, M. (2000). Man-made disasters: Why technology and organizations (sometimes) fail. *Safety Science, 34*(1–3), 15–30.

Shue, H. (1980). *Basic Rights: Subsistence, Affluence and U.S. Foreign Policy.* Princeton University Press.

Silver, A., & Grek-Martin, J. (2015). "Now we understand what community really means": Reconceptualizing the role of sense of place in the disaster recovery process. *Journal of Environmental Psychology, 42,* 32–41.

Stough, L. M., & Kelman, I. (2018). People with disabilities and disasters. In H. Rodriguez et al. (Eds.), *Handbook of disaster research* (2nd ed.). Springer.

Vaughan, D. (1986). *The challenger launch decision: Risky technology, culture, and deviance at NASA.* The University of Chicago Press.

Wilson, G. A. (2013). Community resilience, social memory and the post-2010 Christchurch (New Zealand) earthquakes. *Area, 45*(2), 207–215.

Winstanley, A., et al. (2015). Resilience? Contested meanings and experiences in post-disaster Christchurch, New Zealand. *Kōtitui: New Zealand Journal of Social Sciences Online, 10*(2), 126–134.

Gah-Kai Leung is a PhD candidate in the Department of Politics and International Studies at the University of Warwick, UK. His research considers the ethical and political issues in earthquake/tsunami risk management, with an applied case study focusing on the Pacific Northwest USA and Canada. Gah-Kai has general interests in social and political philosophy, science and public policy, applied ethics and disaster risk reduction.

Fire

Christine Eriksen

Abstract This chapter examines how the unfolding, unequal relationship between humans and fire breaks down the perceived dominance of human agency in the Anthropocene. Humans have been a catalyst for change. We have harnessed and, like fire, consumed our world with, among other tools, the help of nuclear fire. However, the ensuing climate change and radioactive pollution have profound social and environmental consequences. From the harnessing of fire, to uncontrollable wildfires, and the curious emergence of 'Involuntary Parks' with nuclear fire, the chapter discusses the seemingly apocalyptic global impacts of human entanglement with fire, and how we can (re)learn to coexist with fire.

When faced with a wildfire – a natural phenomenon also known as forest fire or bushfire, wildland firefighters set up an "anchor point". The anchor point is an advantageous location in a forest, scrub, or grassland from which to start constructing a "fireline". A fireline is a control line that is scraped or dug into the ground, or a wet line where water is used, to create a burn boundary. Fighting a wildfire is known colloquially as being "on the fireline". Both notions hold deep symbolic meaning for wildland firefighters, as they play a crucial role in controlling the spread of wildfire while upholding the safety of firefighters (Desmond, 2007). Notions of anchor points and firelines can also act as poignant metaphors in acknowledging our growing inability to manage fire in the Anthropocene (Eriksen, 2020). Human activity has altered the parameters of Earth systems, making firelines less effective in controlling the known and emerging consequences of climate change, including wildfires. The impacts of human transformation of the planetary atmosphere mean that, even without our direct intervention, life on Earth is feeling the impacts of human activity, such as the use of fossil fuels, through pollution and extreme weather events (Eriksen & Ballard, 2020). In the Anthropocene, environmental management is akin to fighting fires of such escalating intensity that neither

C. Eriksen (✉)
Center for Security Studies, ETH Zürich, Zürich, Switzerland
e-mail: christine.eriksen@sipo.gess.ethz.ch

N. Wallenhorst, C. Wulf (eds.), *Handbook of the Anthropocene*,
https://doi.org/10.1007/978-3-031-25910-4_21

an anchor point can be found, nor a fireline established. Our transformation of Earth systems necessitates new understandings of the relationship between humans and fire.

Like all of the four classical elements, fire is essential to human existence. In some narratives of evolution, the harnessing of fire is what made us human. Fire provides warmth and is a trusted age-old tool for cooking and land management. The relationship between humans and fire, however, has always been profoundly unequal. While fire is a good servant, it is a terrifying master. Under catastrophic fire conditions, people are at the mercy of fire. The only chance humans stand of controlling fire is by removing one of the three essential elements of the fire triangle: fuel, oxygen, and heat. This, however, is easier said than done when faced with a raging wildfire. Climate change exacerbates this inequity due to its cascading effects, with one change resulting in, or interacting with, another change like a series of falling dominoes. For example, a warmer and drier climate results in drier vegetation and soil, water scarcity, along with more high-fire danger days and longer fire seasons in more places than just the regions that traditionally have been considered fire-prone (Bowman et al., 2020). These conditions exponentially increase the risk, frequency, and intensity of uncontrollable wildfires, and the unequal social impacts of wildfire add more inequality to the equation (Simon & Eriksen, 2021).

Climate change is often espoused in public debates as being the main culprit of our fiery climate crisis, but social factors are equally at play. The climate crisis is, in part, a result of the intimate relationship between land use, fire exclusion, tree dieback, and a hotter and drier climate, which acts as a positive feedback loop. Forests are carbon sinks – i.e. they absorb carbon dioxide from the atmosphere. This is why preserving forested habitats is so important to mitigate climate change. However, the suppression of wildfire by land managers in many parts of the world since the mid-nineteenth century, and the stifling of the natural alliance between plants and fire, has resulted in a build-up of fuel that is contributing to the catastrophic megafires we are experiencing today. Moreover, rural abandonment has resulted in both a loss of local environmental knowledge and less fuel management as part of traditional farming practices. Fire exclusion has also made many forests more prone to diseases that naturally occurring fire regimes and colder winters once kept in check. Tree dieback contributes directly to climate change, as tree mortality decreases carbon uptake and increases carbon emissions as the dead trees decay, thereby upsetting forest carbon dynamics. While slow, cool fires are essential for a healthy ecosystem in many forests (e.g., triggering seed dispersal and forest renewal, controlling insects and diseases, and reducing fuel build-up), large-scale hot and intense fires release large amounts of greenhouse gases. And so, the positive feedback loop continues.

Elemental ways of knowing through fire, ultimately also takes us beyond human time – beyond the Anthropocene – in the form of another type of fire created by humans, namely nuclear fire. Nuclear fire has enabled humans to harness and consume like natural fire but, as Anna Tsing (2015, p. 3) argues, grasping the atom was both 'the culmination of human dreams of controlling nature' and 'the beginning of those dreams' undoing'. With scientific advancement, the ultimate catalyst for the creation of the Anthropocene is now considered to be the release of atomic particles into the atmosphere since the mid-twentieth century (Lewis & Maslin, 2015; Zalasiewicz et al., 2017). The consequences of human entanglement with nuclear fire are profound. Due to the varied half-life of different types of radioisotopes, the impact of radiation on life on Earth is both acute and unknown. It is a problem that will not disappear in our lifetime, and possibly not in the lifetime of humanity (see *Nuclear Waste* chapter in this volume). For example, the fire caused by the explosion in the Chornobyl No. 4 Nuclear Reactor in April 1986 contaminated the soil, water, and atmosphere alike with radioactive material at a rate equivalent to twenty times that released during the 1945 atomic bombing of Hiroshima. Thirty-seven years later, wildfires in the Chornobyl Exclusion Zone are still rereleasing radioisotopes originally deposited into local ecosystems during the accident (Eriksen and Turnbull 2022). Like the original Chornobyl nuclear disaster, these wildfires and the drifting, potentially radioactive, smoke are cause for international concern (Eriksen, 2022a, b).

Wildfires reviving the spectre of the Chornobyl nuclear disaster is an extreme – but not a standalone – example of shifting anchor points and control lines. Such disasters – unique to the Anthropocene – have created another phenomenon unique to the Anthropocene; what Bruce Sterling calls 'Involuntary Parks' – previously inhabited areas that, despite being some of the most contaminated areas in the world today, are proving 'fertile' ground for flora and fauna:

> [Involuntary Parks] are not representatives of untouched nature, but of *vengeful* nature, of natural processes reasserting themselves in areas of political and technological collapse. An embarrassment during the 20th century, Involuntary Parks could become a sombre necessity during the twenty-first. (Sterling, 1998–2008)

The radioactive nature of Involuntary Parks, like the Chornobyl Exclusion Zone, makes them too dangerous for people to manage. Instead, nature reasserts itself. Vegetation grows and fuels wildfires, which burn more frequently and intensely with climate change, rereleasing the radioisotopes that created the Involuntary Park in the first place, just as the release of atomic particles by humans into the atmosphere created the Anthropocene (Eriksen & Ballard, 2020). It sounds like science fiction but in the Anthropocene we must learn to grapple with this type of unfathomable positive feedback loops in order to coexist with fire.

What will it take to re-establish anchor points and firelines that we can rely on in the Anthropocene? Scientific knowledge clearly shows that reducing greenhouse gas emissions could effectively stem climate change. If the reduction targets of the Paris Agreement are met, and the international community limit global average surface temperature increases to no more than two degrees Celsius above pre-industrial levels, many of the expected climatic changes will be significantly reduced (UN, 2015). This will directly mitigate the frequency and intensity of future wildfires, although it is too late to return to the more manageable and less smoky landscapes of the early-twentieth century (IPCC, 2022). We also know how to coexist with wildfire – it is not the knowledge that is lacking but rather the will to make it happen (Moritz et al. 2014). For example, we know that how we manage the land and where we live matter. Local and Indigenous environmental knowledge that understands the diverse, interconnected aspects of a landscape are crucial to sustainable land stewardship (Eriksen, 2022a). Knowing the difference between good (cool, slow) and bad (hot, fast) fire regimes is important. Homes built in fire-prone landscapes is another source of fuel unless construction follows fire-resilient building codes. Vegetation – trees in particular, play a vital role in the carbon-climate balance.

Our relationship with nuclear fire is fraught with difficulty. How to manage it in its entirety is a topic beyond the scope of this chapter. However, learning how to manage Involuntary Parks and the leaky materiality of radioisotopes is critical, as radioactive pollution will continue to reshape local ecologies, biologies, and ecosystems in acute and subtle ways for millennia to come (Brown, 2019; Eriksen, 2022b; Eriksen & Ballard, 2020; Eriksen & Turnbull, 2022). In our efforts to (re)learn and (re)appreciate our relationship to fire in the Anthropocene, it is increasingly clear that we are all "on the fireline" in need of an anchor point.

References

Bowman, D. M. J. S., Kolden, C. A., Abatzoglou, J. T., Johnston, F. H., van der Werf, G. R., & Flannigan, M. (2020). Vegetation fires in the anthropocene. *Nature Reviews Earth & Environment 1*(10), 500–515. https://doi.org/10.1038/s43017-020-0085-3.

Brown, K. (2019). *Manual for survival: A Chernobyl guide to the future.* Allen Lane, an imprint of Penguin Books.

Desmond, M. (2007). *On the Fireline: Living and dying with wildland firefighters.* The University of Chicago Press.

Eriksen, C. (2020). *Europe's fiery future: Rethinking wildfire policy* (CSS Policy Perspective, 8/12). https://css.ethz.ch/content/dam/ethz/special-interest/gess/cis/center-for-securities-studies/pdfs/PP8-12_2020-EN.pdf

Eriksen, C., & Ballard, S. (2020). *Alliances in the Anthropocene: Fire, plants, and people.* Palgrave Macmillan Pivot.

Eriksen, C. (2022a). Indigenous lore and the fire knowledge we ignore. 360info. https://doi.org/10.54377/4908-faa1

Eriksen, C. (2022b). Wildfires in the atomic age: Mitigating the risk of radioactive smoke. *Fire 5*(1), 2. https://doi.org/10.3390/fire5010002.

Eriksen, C., & Turnbull, J. (2022). Insure the volume? Sensing air, atmospheres and radiation in the Chornobyl Exclusion Zone. In K. Booth, C. Lucas, and S. French (Eds.), *Climate, society, and elemental insurance: capacities and limitations.* Routledge. https://doi.org/10.4324/9781003157571

IPCC. (2022). Climate Change 2022: Impacts, Adaptation, and Vulnerability. Contribution of Working Group II to the Sixth Assessment Report of the Intergovernmental Panel on Climate Change [H.-O. Pörtner, D.C. Roberts, M. Tignor, E.S. Poloczanska, K. Mintenbeck, A. Alegría, M. Craig, S. Langsdorf, S. Löschke, V. Möller, A. Okem, B. Rama (eds.)]. Cambridge University Press. 3056pp., https://doi.org/10.1017/9781009325844. https://www.ipcc.ch/report/ar6/wg2/.

Moritz, M. A., Batllori, E., Bradstock, R. A., Gill, A. M., Handmer, J., Hessburg, P. F., Leonard, J., McCaffrey, S., Odion, D. C., Schoennagel, T., & Syphard, A. D. (2014). Learning to coexist with wildfire. *Nature, 515*(7525), 58–66. https://doi.org/10.1038/nature13946

Lewis, S. L., & Maslin, M. A. (2015). A transparent framework for defining the Anthropocene epoch. *The Anthropocene Review, 2*(2), 128–146.

Simon, G., & Eriksen, C. (2021). The unequal social consequences of wildfire smoke in California. *Bay Nature*. https://baynature.org/2021/04/20/the-unequal-social-consequences-of-wildfire-smoke-in-california/

Sterling, B. (1998–2008). The world is becoming uninsurable, Part 3 (Viridian Note 23) *The Viridian Project*. http://www.viridiandesign.org/notes/1-25/Note%2000023.txt

Tsing, A. (2015). *The mushroom at the end of the world: On the possibility of life in capitalist ruins*. Princeton University Press.

United Nations. (2015). *The Paris agreement*. UN Framework Convention on Climate Change. https://unfccc.int/process-and-meetings/the-paris-agreement/the-paris-agreement

Zalasiewicz, J., et al. (2017). The working group on the Anthropocene: Summary of evidence and interim recommendations. *Anthropocene, 19*, 55–60.

Dr. Christine Eriksen is a Senior Researcher in the Center for Security Studies at ETH Zürich. She gained international research recognition by bringing human geography, social justice, and environmental hazards into dialogue. With a particular interest in social dimensions of disasters, her widely published work examines social vulnerability and risk adaptation in the context of environmental history, climate change, cultural norms, and political agendas. She is the author of 80+ articles and two books: 'Alliances in the Anthropocene: Fire, Plants, and People' (2020) and 'Gender and Wildfire: Landscapes of Uncertainty' (2014).

Forests

Christian A. Kull

Abstract The Anthropocene draws attention to how humans have increasingly shaped forests in the past, how forest loss and forest planting play a key role in today's climate emergency, and how we think about forests and forest stewardship in the future. In this handbook entry I review the human moulding of forests, both constructive and destructive, since prehistoric times and suggest a conceptualization that explicitly incorporates human elements among the many processes constituting forests.

We all know what forests are. "An extensive tract of land covered with trees and undergrowth…" says the Oxford English Dictionary. But what are forests in the Anthropocene? The Anthropocene concept begs a question not touched by the quotidian use of the term: what is natural, and what is human, in forests? For many, forests have become synonymous with 'nature', for the interplay of land with insects, animals, fungi, plants, and especially trees, creating a canopy over the heads of human visitors, a home for wild creatures, and a provisioner of diverse goods and services. Yet this equivalence with 'nature' obscures the ancient and on-going, persistent and omni-present role of people in shaping forests. It complicates our management of forests in the present-day climate emergency. And it limits conceptions of forest stewardship in an Anthropocene future.

Vegetation communities we would recognize as 'forests' first emerged in the late Devonian and came to dominate the Earth in the Carboniferous era (between 390 and 300 million years ago) before losing ground to glaciers in the Late Palaeozoic. Initially composed of unfamiliar plants similar to giant club mosses and tree ferns, forest species evolved over millennia to include conifers and flowering plants (Willis & McElwain, 2002). Forests came and went with climate variations and continental moves, persisting in certain forest refugia, and developing together with the pathogens, fire, insects, fungi, and animals who shaped them, including

C. A. Kull (✉)
Université de Lausanne, Lausanne, Switzerland
e-mail: christian.kull@unil.ch

© The Author(s), under exclusive license to Springer Nature
Switzerland AG 2023
N. Wallenhorst, C. Wulf (eds.), *Handbook of the Anthropocene*,
https://doi.org/10.1007/978-3-031-25910-4_22

primates and finally humans. At the beginning of the Holocene, some 12,000 years ago, some 44% of the earth's land surface was occupied by forests and woodlands, of which only one quarter was untouched by humans. Today, forests and woodlands occupy 28% of the land surface, including roughly 8% still untouched by humans and 20% with various densities of human residence (Ellis et al., 2021).

Most forests in which people live, wander, observe, cultivate, collect, hunt, exercise, or travel, are shaped in one way or another by the people who came before. Most forests are anthropogenic, some highly so, in ways that are not always visible to the untrained eye or the casual observer from the modern, urbanized world. We collectively shape forests – their extent, the amounts and types of species in them, the shape and health of individual species – through obvious activities like cutting and planting, but also through numerous less direct actions like introducing or removing animals (hunting, grazing livestock), by setting or dousing fires, fertilizing, collecting and spreading seeds, fruit, roots, fungi, insects, and altering the hydrological and climatological regime (extracting water, flooding a reservoir, enriching the carbon plants breath in the atmosphere, and warming, wetting, or drying the climate).

Like the impacts of elephants (pushing over trees, fertilizing forest soils), ants (transporting seeds, structuring soils) or lightning (burning stands), these human impacts extend back to ancient times, but in contrast with other forest shapers, human impacts have accelerated in volume and speed in recent millennia, regularly adding new influences of greater scale (from metal tools to bulldozers). Below follow four examples of how the forests we visit today are shaped by the people that came before us.

1. The forests of the Amazon basin constitute some 40% of the world's tropical forest area and are often seen as a global symbol of primeval, ancient forest. However, archaeological evidence shows that forests were, in impressively widespread regions, shaped by pre-Columbian populations, as shown by the presence of 'dark earths', domesticated forests, and earthworks like raised fields, ditches and embankments. While its precise influence remains debated, the footprint of ancient human presence is clear. In the past century, these forests have become strongly threatened by large-scale conversion to industrial pasture and cropland (Rostain, 2016; WinklerPrins & Levis, 2021).

2. Europe's forests suffered widespread clearing at human hands beginning in prehistoric times. In the past 150 years, forests have regrown in many areas due to abandonment of marginal land uses. Today, signs of human modelling are everywhere in the forest stands of temperature Europe. A casual visitor sees nature; a more observant visitor notices the tree stumps, access roads, signposts, and log piles and understands that they are signs of the ongoing management of the forests that lead to particular species compositions and stand structures. These forests each have unique histories from centuries ago to today: as village commons, feudal estate, crop field, goat pasture, government revenue forest, peri-urban charcoal source, regional natural park, and more. These multitudes of histories, over millennia of use, combined with recent management, fashion today's for-

ests, from the conifers that have replaced hardwoods in the Jura massif, to acacia thickets in Portugal or beech forest parks in Germany (Mather & Fairbairn, 2000; Williams, 2003; Kaplan et al., 2009; Krumm & Vitková, 2016).

3. The humid greenness of south-east Asia hosts diverse human-shaped forests. Dipterocarp natural forests hide histories of logging, trapping, and itinerant slash-and-burn cultivation. While some are now classified as protected areas, deforestation is strong. Many more forests are diverse forms of populated agro-forests, tended to by villagers keen on encouraging diverse harvests of fruits, seeds, poles, and other products (Michon, 2005). Finally, vast acreages of uniform greenery, designated on government maps as 'forestlands', are, when inspected closer, mono-cropped with acacias, rubber, pines, eucalypts, and oil palms (Cochard et al., 2020). It should be noted that such 'economic forests' can be seen around the world, from eucalypts and radiata pines in Chile, New Zealand, and South Africa, to black locust or Douglas-fir in Europe, or Chinese fir and poplar in China, ranging from straight-lined monocrops to more complex sylvicultural arrangements.

4. Around the world, forests reclaim land once cleared by past centuries human uses, retaining signatures of those days. Forests that grew to cover landscapes marred by erosion and a voracious historical appetite for timber and wood during the 1800s now act as scenic amenities (and fire threats) around historical gold mining boom towns in Victoria (Australia) and the Sierra Nevada (US). Old agricultural terraces in Mediterranean Europe now hide under resprouting oak-pine woodlands. In New England, hardwood forests replace the pines that colonized abandoned fields, as pioneer farmers left behind the marginal, stony ground for the deeper soils of the mid-west. In tropical Costa Rica, keen naturalists and government subsidies facilitated a rapid reforestation since the 1990s of pastures previously cleared for beef production. These 'forest transitions' (Mather & Fairbairn, 2000; Kull, 2017) explicitly link forest history with economic and political development.

The Anthropocene concept makes us explicitly recognize that forests have a long, accelerating, sometimes constructive, sometimes destructive relationship with humans. One might debate exactly when human influence on forests pushed the planet into a new Anthropocene era – human impacts on land cover like forests long predate the emergence of industry – but the point is that human influences began a long time ago, have led to the world being dominated by anthropogenic biomes, and have increased exponentially in intensity (Ellis & Ramankutty, 2008; Ruddimann et al., 2015).

Now, as we face daunting human-caused climate change, forests have taken on a new level of significance – irrespective of whether they are natural, or human-shaped, or something of a mix. Cutting these forests and converting them to other land uses takes place on a scale so vast that it annually contributes to 12% to human carbon emissions. This is second only to the much larger contribution of burning fossil fuels sourced in large part from long buried Carboniferous forests (van der Werf et al., 2009). These actions are quite literally contributing to the Anthropocene as a geologic era, causing regional and planetary scale changes visible in the geological record (Butler, 2021).

Planting forests, on top of reversing rapid deforestation, is seen as an important solution to the climate emergency. International agencies and governments have lined up behind numerous initiatives to plan a 'trillion trees' and signed up to a UN Decade on Ecosystem Restoration. So, one could say that the Anthropocene emergency is inspiring even more human shaping of forests. That is, governments and diverse organizations enact policies and undertake projects, funded by donors as well as carbon offset funds, that give forestlands legal protections, that help degraded forests regrow through understory enrichment of important species, and that plant trees anywhere possible. The latter sometimes leads to new forests, often of fast-growing exotics, in easily accessed grassy biomes, leading to criticism of the impacts on biodiversity and livelihoods (Bond et al., 2019).

If the past has been a crescendo of increasingly widespread and heavy-handed human shaping of forests, what does the future hold? Clearly, the human role in shaping where many forests grow, what trees they consist of, and which forests are left alone is likely to continue. If the Anthropocene concept is about recognizing, belatedly, the scale and impact of the human role in the environment in the past and present, then the future goal should be modifying the human role from one of use (and abuse) towards one of stewardship. While specific definitions of stewardship may be debated, the idea signals recognition of the interdependencies of people and forests, a sense of responsibility to the land and earth, and is built and negotiated based on a mix of locally situated traditions, scientific knowledge, and adaptive experience (Gundersen & Mäkinen, 2009; Chapin et al., 2010).

Forests, as we have seen, indeed generally do refer to "tracts of land covered with trees and undergrowth", but this simple definition from the Oxford English Dictionary hides much. This common-sense concept hides debate over what counts as a forest. What is the boundary between a savanna woodland and a forest? Is an exotic tree plantation, planted in rows, devoid of undergrowth, a forest? What makes a forest a forest? Is it simply the trees, or is it something else – its three-dimensional structure, its ecological functions, the forces that shaped it, or a political designation as forestland?

An Anthropocene definition of forest, which should internalize the human contribution (and damage) to forests, needs to go beyond an ecological concept of forests as ecosystems dominated by trees and consisting of biotic and abiotic flows of nutrients, species, and so on. Forests are indeed constituted by trees and the intersection of diverse processes including biogeochemical cycles, plant growth and decay, insect movements and fungal networks. But the human processes of planting, cutting, owning, legislating, harvesting, managing, and, notably of visioning a future also constitute forests. To this end, a forest of the Anthropocene is a tract of land, no doubt with trees, continually reshaped through power-laden and ecologically relevant relationships among various people, trees, understory plants, animals, soils, insects, water flows, and more. Various forest types exist, each differentiated by not only "species composition, structure and function, but also by the actions and actors (human and not) deemed necessary for the forest's persistence, as well as those deemed to threaten it" (Mansfield et al., 2015 p. 287; Kull, 2017).

This definition sees humans not as bad destroyers and good protectors operating 'outside' the forest concept, but instead recognizes humans 'within' its conception of the forest. In this sense, then, forest stewardship, is more than preserving patches of biodiversity, or cultivating woody monuments to capture carbon. Stewardship of Anthropocene forests recognizes forests as complex social-natural assemblages, where the quality of human relationships and human lives is integral to making possible the quality of forests.

References

Bond, W. J., Stevens, N., Midgley, G. F., & Lehmann, C. E. R. (2019). The trouble with trees: Afforestation plans for Africa. *Trends in Ecology & Evolution, 34*(11), 963–965.

Butler, D. R. (2021). The Anthropocene: A special issue. *Annals of the American Association of Geographers, 111*(3), 633–637.

Chapin, F. S., III, Carpenter, S. R., Kofinas, G. P., Folke, C., Abel, N., Clark, W. C., Per Olsson, D., Smith, M. S., Walker, B., Young, O. R., Berkes, F., Biggs, R., Grove, J. M., Naylor, R. L., Pinkerton, E., Steffen, W., & Swanson, F. J. (2010). Ecosystem stewardship: Sustainability strategies for a rapidly changing planet. *Trends in Ecology & Evolution, 25*, 241–249.

Cochard, R., Nguyen, V. H. T., Ngo, D. T., & Kull, C. A. (2020). Vietnam's forest cover changes 2005–2016: Veering from transition to (yet more) transaction? *World Development, 135*, 105051.

Ellis, E. C., & Ramankutty, N. (2008). Putting people in the map: Anthropogenic biomes of the world. *Frontiers in Ecology and Environment, 6*, 439–447.

Ellis, E. C., Gauthier, N., Goldewijk, K. K., Bird, R. B., Boivin, N., Díaz, S., Fuller, D. Q., Gill, J. L., Kaplan, J. O., Kingston, N., Locke, H., McMichael, C. N. H., Ranco, D., Rick, T. C., Rebecca Shaw, M., Stephens, L., Svenning, J.-C., & Watson, J. E. M. (2021). People have shaped most of terrestrial nature for at least 12,000 years. *Proceedings of the National Academy of Sciences, 118*(17), e2023483118.

Gundersen, V., & Mäkinen, K. (2009). Aldo Leopold and stewardship: Lessons for forest planning and management in the Nordic countries? *Norsk Geografisk Tidsskrift, 63*(4), 225–232.

Kaplan, J. O., Krumhardt, K. M., & Zimmermann, N. (2009). The prehistoric and preindustrial deforestation of Europe. *Quaternary Science Reviews, 28*(27–28), 3016–3034.

Krumm, F., & Vítková, L. (Eds.). (2016). *Introduced tree species in European forests: Opportunities and challenges*. European Forest Institute.

Kull, C. A. (2017). Forest transitions: A new conceptual scheme. *Geographica Helvetica, 72*, 465–474.

Mansfield, B., Biermann, C., McSweeney, K., Law, J., Gallemore, C., Horner, L., & Munroe, D. K. (2015). Environmental politics after nature: Conflicting socioecological futures. *Annals of the Association of American Geographers, 105*(2), 284–293.

Mather, A. S., & Fairbairn, J. (2000). From floods to reforestation: The forest transition in Switzerland. *Environment and History, 6*, 399–421.

Michon, G. (2005). *Domesticating forests: How farmers manage forest resources*. CIFOR, ICRAF, and IRD.

Rostain, S. (2016). *Amazonie: Un jardin naturel ou une forêt domestiquée*. Actes Sud/Errance.

Ruddiman, W. F., Ellis, E. C., Kaplan, J. O., & Fuller, D. Q. (2015). Defining the epoch we live in. *Science, 348*, 38–39.

van der Werf, G. R., Morton, D. C., DeFries, R. S., Olivier, J. G. J., Kasibhatla, P. S., Jackson, R. B., Collatz, G. J., & Randerson, J. T. (2009). CO2 emissions from forest loss. *Nature Geoscience, 2*(11), 737–738.

Williams, M. (2003). *Deforesting the earth: From prehistory to global crisis*. The University of Chicago Press.

Willis, K. J., & McElwain, J. C. (2002). *The evolution of plants*. Oxford University Press.

WinklerPrins, A. M. G. A., & Levis, C. (2021). Reframing pre-European Amazonia through an Anthropocene lens. *Annals of the American Association of Geographers, 111*(3), 858–868.

Christian Kull is a geographer with particular interest in the social dimensions of environmental change in developing countries, islands, and highlands. He has investigated the human dimensions of topics like fire, invasive species, afforestation, and conservation in Madagascar, Africa, India, and Vietnam. Educated in the United States, he has held university posts in Canada, Australia, and Fiji. He is now professor in the Institute of Geography and Sustainability at the University of Lausanne, Switzerland. This research was made possible by the Swiss Programme for Research on Global Issues for Development (r4d, grant 194004).

Heat, Heat Wave

Tracey Skillington

Abstract In many ways, rising global temperatures could be said to be produced by the actions of all members of carbon societies. Yet, given what is known regarding the contribution of major carbon emitters to increases in average temperatures, it could be said that the latter are asymmetrically responsible for its detrimental effects, including more frequent and violent wildfires, melting glaciers, rising sea levels, etc. Asymmetry here refers to the specificity of each polluting agent's contribution within a wider configuration of relationships of mutual effect. Considering their scale and intensity, the impacts of rising pollution levels cannot be said to be wholly 'unstructured' or 'accidental'. Rather, they are disproportionately the product of the actions of high emitting agents. In this way, the burning landscapes of the Anthropocene are not wholly 'naturally occurring' phenomena. To describe the type of devastating wildfires witnessed in more recent years as fires that 'start by themselves' is not entirely true in the context of global climate change. Other climate related events, equally, are in part produced by specific political, economic and social arrangements, especially those associated with the historical rise and ongoing expansion of energy capitalism. This chapter notes how lived, embodied experiences of rising temperatures fuel a more heated exchange today on issues of responsibility for what are increasingly transnationally sourced climate harms.

All indications are we are on the cusp of a 'vast planetary burnout' (Marden, 2015, p. 94–5) created almost entirely by human will. In its most recent report, the IPPCC (2021, p. 5) notes with concern how each of the last four decades have been steadily warmer than any decade since 1850, with temperatures during the last decade (2011–2020) also exceeding those of the most recent multi-century warm period approximately 6500 years ago. As a consequence of this warming, the area of late summer Arctic sea ice, for instance, has been smaller this decade than any time in the last 1000 years (ibid., p. 9), while global mean sea levels have risen faster this

T. Skillington (✉)
Department of Sociology & Criminology, University College Cork, Cork, Ireland
e-mail: t.skillington@ucc.ie

145

decade than any period previously in the last 3000 years. All of these changes, it adds, would have been 'extremely unlikely' without 'human influence on the climate system'. But what does this influence consist of? Heat and fire have always been characteristic features of the natural order (e.g. solar and volcanic activity). However, with the discovery and combustion of fossil biomass to fuel the machinery, energy and production of capitalism, to fire and heat was added the dimension of power. Not only would a 'fire-power-capitalism' complex accelerate the need for energy, particularly from the early nineteenth century, as more and more of the earth's hydrocarbons were brought to the surface and set alight, the thermal dimensions of the Anthropocene age began to gain real momentum. In the decades since, temperatures have risen steadily, producing ever new record levels of extreme heat and natural resource destruction.

As the IPCCC (2021) concedes, projections as to future climate harms must be considered also in terms of the impacts rising temperatures are beginning to have on lived embodied experiences of climate change today and how the latter pose a serious challenge to sensory, physical, emotional, mental and social wellbeing. 'Being in this world' (*Dasein*) of climate change is a reciprocal position (Merleau-Ponty, 2002) in which we produce and are produced by interactions with powerful geological forces of change. As members of and participants in the creation of climate changing worlds, our contributions to ecological destruction may flow and shift momentarily but these positions cannot be extracted from more contextualised relations amongst communities, each experiencing the effects of a heating planet to varying degrees of severity.

Commenting on the devastation created by nearly 600 wildfires that raged across Greece, for instance, in August 2021, Greek Prime Minister, Kyriakos Mitsotakis, noted how the climate crisis is 'striking here and now' and in ways that demand an immediate change in modern ways of living (Smith, 2021). Similarly, in Siberia, where wildfires burned through 1.5 m hectares of land, homes and forests in northeast Siberia in the summer months of 2021, scientists and local environmentalists called attention to the links between wildfires, global temperature rises, poor government preparedness and insufficient international cooperation to address these issues (Roth, 2021).

What were once extraordinary events now become regular features of life for those occupying climate vulnerable regions. More frequent wildfires not only increase the volume of pollutants entering the atmosphere (e.g., carbon monoxide, nitrogen oxides, volatile organic compounds, and solid aerosol particles), but in doing so, they also provoke further warming, drier soils and vegetation and, consequently, more fires (data from the Copernicus Atmosphere monitoring Service and NASA's Moderate Resolution Imaging instruments). Particularly vulnerable are the world's peatlands which for thousands of years, have played a key role in cooling the climate and storing the carbon produced by accumulated organic matter. Due to rapid thawing and more intense drying conditions, however, these carbon dense ecosystems are becoming more flammable and prone to wildfires (NASA, 2019). As fires burn across, as well as downwards into the deep layers of peatlands' centuries old carbon-rich matter, more carbon is released into the atmosphere, triggering a feedback loop between warming, more frequent wildfires and fine particle pollution.

Yet rarely in public discourse is the question of responsibility explored from the point of view of the actions of specific wrongdoers. More usually, wildfires are defined as beyond the responsibility of any one group or agent of harm (unless started deliberately). Typically, they are portrayed as fires 'that start by themselves' or as 'non-structured fires' (e.g., Collins English Dictionary, 2021). But what does it mean to define a wildfire as one that 'starts by itself' in the context of global climate change, especially when scientific consensus on the reasons for more frequent wildfires is clear – rising global temperatures? As the product of 'unintended harms', wildfires could be said to be the cumulative outcome of the actions of all members of carbon societies. Yet, given that we know that major emitters, in producing disproportionately more carbon, also contribute overwhelmingly to rising temperatures, it could be said that they are also disproportionately responsible for more intense and frequent wildfires. In this sense, those wildfires that have been linked more directly to climate change (e.g., in Greece, Turkey, Siberia, Italy, Canada, July and August 2021), given their scale and intensity, cannot be said to be wholly 'unstructured' or 'accidental'. Rather, they are the product of the aggregate actions of high emitting agents, a dire effect of pollution acts committed in full knowledge of their long-term impacts. In this sense, the burning landscapes of the Anthropocene are not wholly 'naturally occurring' phenomena but, equally, are produced by particular political, economic and social arrangements, especially those associated with the historical rise and ongoing expansion of energy capitalism.

Furthermore, we may note the societally embedded nature of these arrangements and the fact that they are the product of reciprocal influences deriving from our historical co-presence in one climate changing world. Yet as each agent brings to this shared situation their own particular experiences, symbolic associations and pollution history, relations amongst them are not equal. Some, both historically and today, exert significantly more influence, particularly those associated with the rise and ongoing development of fossil fuel and other extractive economies. Both directly and indirectly, these actors have played a central role in shaping institutional interpretations of freedoms and rights in ways that secure the further development of 'carbon democracies' (Mitchell, 2009). At least initially, criteria for inclusion into the modern democratic project required support for largescale natural resource extraction and the expulsion of peoples and wild nature to make way for mass industrialisation. Only because of the centrality of their labor to the growth of these projects and, in time, the global expansion of capitalist infrastructure, were the rights of workers recognized formally and progressively extended to other social groups. Today, citizens assert these same rights against energy capitalism's continuous exploitation of dwindling reserves of natural resources (the assertion of extraction rights over rare seabed minerals, remote lands and other fragile ecosystems, for instance). Many of the non-reflexive, ahistorical assumptions embedded in carbon democratic reasoning are subject to increasing challenge today. Yet to enable a more complete transformation of its reasoning, current political, economic, social and legal institutional arrangements must also change. There are some moves at present, chiefly legal, to challenge governments' efforts to control emissions levels and keep global temperatures within safe limits (daring on various human rights law and international climate change treaties, particularly the Paris Agreement), and assert communities' rights to self-determine their own social, cultural and economic development. In doing so, these actors demonstrate how it is still possible for Anthropocene societies to initiate change when responding to new justice imperatives and strengthen commitments to carbon reduction targets.

When campaigners insist that a major source of threat to livelihoods and wellbeing today is the unregulated pollution activities of dispersed agents of global climate harms, they face the formidable task of proving how a geo-political separation of peoples, resources and state territories is of increasingly less relevance than what all share in common. A second challenge they face is proving how specific groups of states and corporate actors actively contribute to warming climate conditions (through unsustainable rates of greenhouse gas emissions) and further, how these contributions negatively impact the lives of multiple communities. To meet these challenges, campaigners draw on specific principles in law, including the principle of 'presumptive responsibility' to clarify how in situations where there are a number of wrongdoers who collectively through their pollution endeavours contribute to rising temperatures and there is uncertainty as to which of them is disproportionately responsible, that each wrongdoer is presumptively responsible. Each party in this instance has to show how they, in fact, did not cause the relevant harms, unlike the more usual scenario where it is the injured party who carries the burden of proving harms are traceable to the actions of specific actors.

Although still in the early stages of development, this approach, aligning the question of responsibility with plural polluting agents, is an important step forward, especially as ecological conditions continue to decline, and the life circumstances of communities residing in various climate vulnerable regions becomes ever more conditional on the extraction plans and carbon expenditure patterns of others. Plans to further invest in ecologically damaging resource extraction projects cannot be said to be 'relationally justifiable' (Forst, 2017) in this instance in the sense that the reasonableness of their outcomes has been validated by those most affected. In that, they also cannot be said to be fair or democratic. Similarly, in the case of rising global temperatures and more regular encounters with wildfires and storm surges, the question of responsibility can legitimately be posed, however much their victims and perpetrators are dispersed across regions, populations or generations (Skillington, 2012, 2017, p 116; 2019, p 25–7).

References

Collins English Dictionary. (2021). Wildfires. https://www.collinsdictionary.com/dictionary/english/wildfire. Accessed 26 August 2021.

Forst, R. (2017). *Normativity and power: Analysing social orders of justification*. University of Oxford Press.

Intergovernmental Panel on Climate Change (IPPCC). (2021). Climate change 2021: The physical science basis. https://www.ipcc.ch/report/ar6/wg1/downloads/report/IPCC_AR6_WGI_SPM. pdf. Accessed 25 August 2021.

Marden, M. (2015). *Pyropolitics: When the world is ablaze*. Rowman & Littlefield.

Merleau-Ponty, M. (2002). *Phenomenology of perception*. Routledge.

Mitchell, T. (2009). Carbon democracy. *Economy and Society, 38*(3), 399–432. https://doi.org/10.1080/03085140903020598

NASA. (2019). Boreal forest fires could release deep soil carbon. https://climate.nasa.gov/news/2905/boreal-forest-fires-could-release-deep-soil-carbon/. Accessed 20 July 2021.

Roth, A. (2021, July 20). Everything is on fire: Siberia hit by unprecedented burning. *The Guardian*.

Skillington, T. (2012). Climate change and the human rights challenge: Extending justice beyond the borders of the nation state. *The International Journal of Human Rights, 16*(8), 1196–1212. https://doi.org/10.1080/13642987.2012.728859

Skillington, T. (2017). *Climate justice & human rights*. Palgrave.

Skillington, T. (2019). *Climate change and intergenerational justice*. Routledge.

Smith, H. (2021, August 22). Know your enemy: Greece plans to name heatwaves in the same way as storms. *The Observer*.

Tracey Skillington is Director of the BA (Sociology), Department of Sociology & Criminology, University College Cork, Ireland. Recent monographs include *Climate Justice & Human Rights* (Palgrave), *Climate Change & Intergenerational Justice* (Routledge) and forthcoming, *A Critical Theory of Climate Trauma* (Routledge). Her publications have appeared in many journals over the years, including the *European Journal of Social Theory*, the *British Journal of Sociology*, the *International Journal of Human Rights*, *Distinktion: Journal of Social Theory*, *Sociology*, the *Irish Journal of Sociology* and *Sustainable Development*.

Landscape

Matt Edgeworth

Abstract This essay explores aspects of landscapes relevant to the study of the Anthropocene. Anthropogenic landscapes, it suggests, are more than just palimpsests of marks, traces, and residues of human activity. Whatever happens to land – through the entangled actions of human and non-human forces – is inextricably interconnected with what happens to the whole Earth system.

'Landscape' is a key term in Anthropocene research. In its broadest sense, it denotes any part of Earth's land surface not covered by extensive bodies of ice or water. A landscape may be urban or rural, or a mixture of both. It may be densely occupied, recently settled/abandoned, or completely uninhabited. Forests, fields, rivers, mountains, lakes, buildings, towns, roads, railways, electricity pylons, factories, car parks and fences are all elements of landscapes, as are communities of people that live there, plants that grow and animals that graze. In this essay, when speaking of human impact or influence, domesticated plants and animals are included in the sphere of the 'human' for two main reasons. First, they are under human control, serving as instruments as well as objects of landscape management. Second, their physical form and behaviour is determined by cultural as well as natural selection. Biomass of domesticated livestock, for example, is much greater than that of all humans put together: the visible presence and impact of grazing animals on/in landscapes is correspondingly large. Their part in the human transformation of landscapes should not be underestimated.

One dictionary defines landscape as everything one can see when looking over a piece of land. But a crucial aspect of landscapes is that they are multi-scalar. Depending on whether the observer takes up an embodied and situated standpoint *within* a landscape or a more detached vantage point *without,* looking down from a plane (perhaps through the medium of an aerial photo) or (via computer or television images) from an orbiting satellite, landscapes can be apprehended on a range of

M. Edgeworth (✉)
University of Leicester, Leicester, UK
e-mail: me87@leicester.ac.uk

scales. Google Earth and other GIS computer programs that make use of satellite imagery enable investigators to shift rapidly from local to global and through intermediate scales at the touch of a key. The multi-scalar structure of landscapes is now routinely experienced as part of everyday tasks, such as when driving a car using a GPS satellite navigation device.

Up to now the Anthropocene has been studied mainly in terms of global scale phenomena, yet its distinctive patterns manifest on local scales too. Study of landscape transformation could turn out to be central to a more textured and finely grained multi-scalar understanding of the Anthropocene.

In an important paper, Ellis and Ramankutty (2008) map the global landscape in terms of anthropogenic biomes or 'anthromes'. Anthromes are broadly defined types of landscapes characterized by specific kinds of land use: croplands, rangelands, managed forests, urban and suburban areas, and so on. In the last few hundred years such anthromes have spread to cover well over half of the Earth's ice-free terrestrial surfaces, destroying some habitats, transforming others, and creating new ones, leaving few parts of the world entirely unaffected.

A difficulty in mapping landscapes in the 21st millennium is the rapidly increasing – though highly variable – rates of change. All landscapes change through time. Anthropogenic landscapes – those heavily shaped by human activity – typically undergo transformation at much faster rates than wholly natural ones. Some of the changes are due to the action of natural geomorphological forces (Goudie & Viles, 2010). But it is becoming obvious that humans – together with their 'camp followers' of domesticated plants and animals, entourage of earthmoving machines and sprawling technological infrastructures, along with the layers of cultural debris they leave behind in the ground (Edgeworth, 2018) – are the main agents of transformation.

As Goudie and Viles suggest with their concept of 'anthrogeomorphology', humans are themselves geomorphological agents, and should be recognized as such. Quarries and mines are dug for mineral extraction, creating voids in the Earth's surface. Dumps of landfill waste, spoil from extractive industries and industrial waste are used to fill in hollows and voids and sometimes to make artificial hills, significantly altering topography. Rain forests are logged to provide land for cattle farmers. And so on. Such active forms of landscape modification may greatly speed up natural processes of erosion and sedimentation, as when riverbanks made unstable by deforestation are eroded by water, leading to deposition of substantial 'legacy sediments' (James, 2013) further downstream. Or they can lead to disruptions in natural processes, as when construction of large dams stops river sediment from reaching deltas, which are sinking as a result (Syvitski et al., 2009). Even landscapes not directly shaped by human activities can be radically impacted by knock-on effects.

Despite the manifest evidence of human transformation, a persistent trope – embedded in popular culture, and rooted in landscape art and poetry (Johnson, 2007) – is that of the 'natural landscape'. The form this often takes in Anthropocene discussions is that landscapes were in a near pristine state up to modern times. For instance, a common assumption is that much of the Amazon rainforest was pristine until recent incursions/degradations. But archaeological and other forms of evidence are challenging that view. Large areas of tropical rain forests throughout the world are revealed to have been actively managed and maintained by indigenous peoples for long periods of time (Roberts et al., 2021), and their wide range of biodiversity today is partly a product of those past human-environment interactions.

The reason for the oversight may be that human impact on landscapes tends to be framed in terms of destruction or harm. So even though indigenous peoples may have inhabited and made extensive use of landscapes in sustainable ways for thousands of years, their positive influence on environments does not register on the standard measures of negative impact. Learning from indigenous practices of sustainable living with people as *part* of landscapes rather than *a-part* from them, and associated beliefs about the interconnectedness of things, combined with rationales of custodianship of land, is perhaps key to looking after landscapes in the future.

As the above discussion indicates, the nature/culture dichotomy manifests itself in landscape studies in a particularly strong form. It certainly raises its head whenever an attempt is made to restore landscapes to their supposed natural state, as in river restoration or 're-wilding' projects (Monbiot, 2013). For all that these efforts are valuable and worthwhile, the very measures necessary to artificially recreate *natural* or *wild* environments are generally the most *cultural* interventions in landscape modification it is possible to make.

Given that landscapes are typically mapped or photographed as two-dimensional surfaces, it is good to occasionally ask the question – how deep are landscapes? Common sense tells us they extend at least to the top of tree canopies or roofs of buildings, and down some way into the soil, perhaps corresponding in the vertical dimension roughly to the Critical Zone – the dynamic surface area or skin of the Earth where the geosphere, atmosphere, hydrosphere, and biosphere all intermesh (Richter & Billings, 2015). One thing is for sure: landscapes are not just surfaces: they have depth to them and can be complexly layered. The ground itself is deeply stratified, as soil scientists and archaeologists will testify, with stratigraphic traces of multiple events laid down in an aggregative or cumulative fashion. A single event or phase of human activity leaves its traces and residues, which are then partially erased and/or overlain by the traces and residues of subsequent activities, which in turn get partially erased and/or overlain by later ones.

The landscape thus shaped is often construed as *palimpsest*, like a Roman writing tablet on which layer upon layer of writing may survive, each layer overwriting the previous one. The metaphor of landscape as *text*, formed through multiple acts of *inscription*, is invoked, along with the idea that the landscape can be *read* (Muir, 2000).

Ways in which huge tracts of landscapes are shaped through the accumulation and compounding of smaller effects could provide models for Anthropocene transformations generally. Consider terraced landscapes (Tarolli et al., 2014). At first sight it seems that enormous amounts of work must have been involved in sculpting a landscape into terraces. But such landscapes are rarely formed all at once. In some cases, only one or two terrace walls may be constructed by a community each year, with soil moved partly by enlisting gravity-driven material flows such as hillwash or moving water to carry and deposit the sediment used to create fields. It may take decades for a whole hillside to be terraced, and much longer periods for multiple individual systems of terracing to intermesh into composite systems. Additional terraces are added onto existing ones in modular fashion, extending the system further. Terraced hillsides of rice fields aggregated in this way extend over vast areas of Asia and other parts of the world. "The Anthropocene begins to emerge", as one author puts it, "when we consider human-environmental activity at a local level, compounded by thousands of years, affecting vast areas of interlocking landscapes" (Periman, 2006: 562).

One rice paddy field may have little effect on climate on its own. But tens of thousands of square kilometres of terraced paddy fields generate large amounts of methane which contribute significantly to global warming. Ploughing of soil over large areas has similar effects, releasing carbon from soils faster than plants can replace it (Haddaway et al., 2017). Agriculture over 12,000 years is estimated to have resulted in a net loss of 133 billion tonnes of carbon from soils (Sanderman et al., 2017), with most of that going into the atmosphere as CO_2. Once appreciated that soil (through the activity of plants, soil bacteria, etc) is in gaseous exchange with the atmosphere, it becomes clear that what happens to land is inextricably connected to other parts of the Earth system.

Numerous disciplines take landscapes as the object of study – from earth sciences such as geomorphology, soil science and archaeology to humanities such as landscape history and cultural theory, with geography and landscape ecology somewhere in the middle. The arts, too, have their own interests in landscape aesthetics. Each subject brings its own perspectives, methods, and scales of analysis to bear, with much overlap between fields. That makes landscape studies a common ground for exactly the kind of collaborative inter-disciplinary research widely held to be necessary for holistic investigation of the Anthropocene to take place.

References

Edgeworth, M. (2018). More than just a record: active ecological effects of archaeological strata. In M. A. Torres de Souza & D. M. Costa (Eds.), *Historical archaeology and environment* (pp. 19–40). Springer.

Ellis, E. C., & Ramankutty, N. (2008). Putting people in the map: Anthropogenic biomes of the world. *Frontiers in Ecology and the Environment., 6*(8), 439–447.

Goudie, A., & Viles, H. (2010). *Landscapes and geomorphology: A very short introduction.* Oxford University Press.

Haddaway, N. R., Hedlund, K., Jackson, L. E., et al. (2017). How does tillage intensity affect soil organic carbon? A systematic review. *Environmental Evidence, 6*, 30.

James, L. A. (2013). Legacy sediment: Definitions and processes of episodically produced anthropogenic sediment. *Anthropocene, 2*, 16–26.

Johnson, M. (2007). *Ideas of landscape.* Blackwell.

Monbiot, G. (2013). *Feral: Searching for enchantment on the frontiers of rewilding.* Penguin Books.

Muir, R. (2000). *The new reading the landscape: Fieldwork in landscape history*. University of Exeter Press.

Periman, R. D. (2006). Visualizing the Anthropocene: human land use history and environmental management. In C. Aguirre-Bravo, P. J. Pellicane, D. P. Burns, & S. Draggan (Eds.), *Monitoring science and technology symposium: unifying knowledge for sustainability in the western hemisphere*. U.S. Department of Agriculture.

Richter, D., & Billings, S. A. (2015). 'One physical system': Tansley's ecosystem as Earth's critical zone. *New Phytologist, 206*, 900–912.

Roberts, P., Hamilton, R., & Piperno, D. R. (2021). Tropical forests as key sites of the "Anthropocene". *Proceedings of the National Academy of Sciences of the United States of America, 118*(40), e2109243118. https://doi.org/10.1073/pnas.2109243118

Sanderman, J., Hengl, T., & Fiske, G. J. (2017). Soil carbon debt of human land use. *Proceedings of the National Academy of Sciences of the United States of America, 114*(36), 9575–9580.

Syvitski, J., Kettner, A., Overeem, I., et al. (2009). Sinking deltas due to human activities. *Nature Geoscience, 2*, 681–686.

Tarolli, P., Preti, F., & Romano, N. (2014). Terraced landscapes: From an old best practice to a potential hazard for soil degradation due to land abandonment. *Anthropocene, 6*, 10–25.

Matt Edgeworth is Honorary Visiting Research Fellow in the School of Archaeology and Ancient History at the University of Leicester. He also works in the commercial domain as a field archaeologist. His research interests include urban stratigraphy, archaeology of waste landscapes, and the ecological effects of anthropogenic strata that now cover large parts of Earth's land surfaces. He is author of numerous papers on the significance of archaeological evidence in understanding the Anthropocene.

Ocean

Colin P. Summerhayes

Abstract The ocean plays a key role in the climate system by storing heat and by transporting heat and salt around the world, thus influencing the atmosphere and the cryosphere. Ocean circulation links the two poles via the Atlantic in the Atlantic Meridional Overturning Circulation. 90% of the heat in the climate system due to global warming resides in the ocean, where most of it has been stored since 1950. This constitutes heat 'in the pipeline' that will eventually influence the global climate, raising sea level and melting ice on a timescale of centuries to millennia even after we stop our CO_2 emissions.

The ocean plays a key role in the climate system by storing heat and by transporting heat and salt around the world, thus influencing the atmosphere and the cryosphere. For example, the Gulf Stream transports heat toward northwest Europe, making it much warmer than Labrador. Salt plays a key role in ocean circulation by creating the differences in density that allow one water mass to dive below or rise above another. When warm, salty Gulf Stream water reaches the cooler, fresher Norwegian-Greenland Sea, the cold makes the Gulf Stream water dense enough to sink to depths of around 2500 m. There it forms North Atlantic Deep Water (NADW), which moves south through the Atlantic to the Southern Ocean where it wells up and mixes with the Circumpolar Deep Water (CDW) that circles Antarctica in the Antarctic Circumpolar Current. Wind-driven upwelling brings CDW to the surface where it moves north away from the Antarctic coast under the influence of the Coriolis Force of the Earth's rotation. The ocean circuit is closed in two ways. At the surface warm surface water from the Pacific moves west through the Indian Ocean and across the Atlantic to join the Gulf Stream before sinking in the Norwegian-Greenland Sea. This forms the upper circulation cell. At the deep ocean floor Antarctic Bottom Water (AABW) moves north through the Atlantic before welling up to connect with south moving NADW, forming the lower circulation cell. AABW

C. P. Summerhayes (✉)
Scott Polar Research Institute, Cambridge, UK
e-mail: cps32@cam.ac.uk

157

also moves north through the Pacific and Indian Oceans, eventually welling up to join the west-moving surface circulation and/or the east-moving subsurface CDW. The AABW originates close to the Antarctic coast where the formation of sea ice in a belt 1000 km wide in winter excretes salt into subsurface waters making them dense enough to sink down the continental slope as bottom water. The connections between the Norwegian-Greenland Sea and Antarctica form the Atlantic Meridional Overturning Circulation (AMOC), a unique pole-to-pole climate connection via the ocean. The circulation of a water particle from the Norwegian Sea via the Pacific Ocean and back takes about 1000 years. The AABW ranges in temperature from 1–3 °C with an average of about 2 °C. During times of glacial maxima within the Ice Age, deep ocean temperatures were close to 0 °C.

Solar energy warms the surface waters of the ocean, which are then mixed down into the interior by waves, eddies, storms and density-driven sinking. Owing to these processes, 90% of the heat that has accumulated in the Earth System due to global warming is in the ocean, where it has accumulated mostly since 1950 (when more than 90% of the fossil fuels burned by human agency were burned). The ocean holds this heat because it has enormous heat capacity (the top 3.5 m of the ocean contains as much heat as the entire atmosphere). The capture and retention of this heat makes a primary contribution to the so-called Earth Energy Imbalance, or EEI (the difference between incoming solar radiation and the sum of outgoing radiation), which is what determines the evolution of Earth's climate (Von Schuckmann et al., 2020). Today's EEI is positive because there is less energy going out than coming in, which creates global warming. Given time, the climate will adjust to an EEI of zero as all its components come back into equilibrium. Although much of the ocean's heat lies above the thermocline (<100 m deep on average), heat has penetrated to well over 2000 m since 1950, making water expand and thus raising sea level. The full extent of sea level rise will emerge over coming centuries as the ocean heats to full ocean depth. The ocean has been warming at an increasing rate, increasing the EEI from 0.47 W/m^2 in 1971–2018 to 0.87 W/m^2 in 2010–2018 (Von Schuckmann et al., 2020). It is significant that oceanic heat storage continued unabated through 2000–2012, a period in which the atmospheric temperature rise 'paused'. Evidently, global warming did not stop during the 'pause'; instead the locus of heat storage shifted from the atmosphere into the ocean. The rise in ocean warming was accompanied by an acceleration of sea level rise and surface warming, record temperatures and ice loss in the Arctic, and increased loss of ice from the Greenland and Antarctic Ice Sheets. The warming of the global ocean explains the largely Anthropocene growth in the EEI. In the years since 1950, the global average climate has been warmer than it was in the Medieval Warm Period, and it is now warmer than in the warmest period of the Holocene (Kaufman et al., 2020).

If we stop emitting CO_2 the climate will continue to warm until the EEI falls to zero (i.e. there is warming still 'in the pipeline'). If the heat stored in the ocean (i.e. 'in the pipeline') were released, it would raise temperature by some 0.5 °C. To bring the EEI down to zero we now need to increase radiation to space by 0.87 W/m^2, which would lower atmospheric CO_2 to its 1988 level of 350 ppm (Von Schuckmann et al., 2020).

To achieve climatic equilibrium, the air and the ocean exchange gases across the ocean surface. When we begin reducing CO_2 emissions, this process will keep the CO_2 content of the atmosphere from declining rapidly. Calculations suggest that given the ocean's role in CO_2 exchange, 50% of our CO_2 emissions will still be in the air 300 years from now, decreasing to 17–33% 1000 years from now, 10–15% 10,000 years from now, and 7% in 100,000 year's time (Archer, 2005). CO_2 decline evidently has a very long tail, making the mean lifetime of our fossil fuel CO_2 in the air some 30–35,000 years, vastly longer than the public perception of what is likely (Archer, 2005). This will keep our climate warm for centuries, melting more land ice and raising sea level further.

The ocean has two key roles in melting ice. Firstly, the warming ocean surface provides heat and moisture to be taken by clouds over land, bringing heat to places like western Europe and the Arctic and helping to melt ice there. Secondly, oceanic heat in the subsurface reaches beneath the termini of Greenland's tidewater glaciers, and beneath Antarctica's fringing ice shelves, thinning them from beneath. The thinned glaciers and ice shelves discharge more land ice into the ocean, raising sea level. Moving heat into the polar regions via the ocean helps to melt sea ice, thus reducing albedo (the Arctic's cooling mechanism), and warming exposed dark ocean surfaces. This positive feedback amplifies the global warming signal, doubling average Arctic temperatures compared with those of the globe as a whole. Antarctica suffers less from these feedbacks because its strong surrounding winds and currents act as 'walls' preventing substantial heat input from the north.

Oceans are also important because warm water contains less gas, so a warming ocean absorbs less CO_2, leaving more of our emissions in the air. A warming ocean also enhances the evaporation of water vapour, another greenhouse gas, as well as raising sea level.

During the Anthropocene, the injection of warm air carried by the Gulf Stream, plus the injection of warm Atlantic water in the subsurface, have helped to melt Arctic sea ice, whose area has been reduced by 50% during summer months. The Arctic summer sea ice has also lost volume, most now being a mere 1 m thick instead of 5 m as formerly. Arctic glaciers and ice caps have been melting on land, as has the Greenland Ice Sheet, stratifying the Norwegian Greenland Sea by providing it with a cold freshwater cap. This has diminished the process by which northward moving Gulf Stream water sank to form NADW. The resulting reduction of northward heat transport has cooled the waters of the North Atlantic immediately south of Greenland's southernmost point by about 1 °C. Although this slight cooling should ameliorate the expected effect of global warming on surrounding coasts, a side effect is that heat that would have been transported north will stay in the equatorial Atlantic where it will spawn stronger hurricanes. These changes are still rather new, so their regional effects are not yet very clear.

The ocean experiences natural oscillations whose effects are superimposed on the upward curve of global warming. The Atlantic experiences the Atlantic Multidecadal Oscillation (AMO), whose positive and negative phases last roughly 20–25 years. During the twentieth century the AMO was positive (i.e. warm sea surface temperatures) from 1925 to 1965. This warming also affected Arctic temperatures. For example, Greenland experienced warming and its glaciers receded between 1930 and 1950; its temperature then cooled to a low in 1970 then rose to present values by 2005 in response to the AMO moving into another positive phase from 1990 onward (Kobashi et al., 2011). The AMO is related to changes in the AMOC. When the AMOC is weak, the uptake of heat by the deep ocean decreases and surface air and water temperatures warm more than when the AMOC is strong (Chen & Tung, 2018). The AMOC was weak in the 1930s and has been weak since 1980, explaining surface warming at those times (Chen & Tung, 2018). Oscillations in the AMO and AMOC show connections between ocean, atmosphere and ice.

The AMO's Pacific equivalent is the Pacific Decadal Oscillation (PDO), which was also in a positive phase (warm equator and cold Gulf of Alaska) between 1920 and 1945. However, the PDO and AMO are not always in phase; the PDO began moving into its next positive (warm) phase in 1970, and returned to a negative (cold) phase between 1998 and 2013. The PDO radically changes the coastal ecology of the west coasts of the Americas (Chavez et al., 2003). The cool phase brings more anchovies and salmon, and fewer sardines to California and Peru, while the Gulf of Alaska warms and gets fewer salmon. The warm phase brings more sardines, but fewer salmon to California and Peru, while the cold Gulf of Alaska gets more salmon. The PDO is in now in its negative cool phase, warming the Gulf of Alaska and causing widespread seabird deaths and glacial retreat there.

A minor version of the PDO is the El Niño-La Niña couplet in the equatorial Pacific - a cycle lasting between 4–7 years, which may be larger than usual at times (e.g. the 'Super El Niños over the winters of 1982–83, 1997–98 and 2016–17). El Niños have a wide reach, bringing drought to northeastern Brazil and southern Africa, and wet conditions to northwest Europe. Predicting natural oscillations like these or the PDO is difficult, because their origins are not fully understood. But they can now be predicted about a year ahead of time through careful monitoring of changes in the Pacific Ocean. It is not yet known if there will be more Super El Niños as global warming continues.

Has the Sun been a key driver of ocean/global warming? Modern evaluations of sunspot history show that sunspots peaked at about the same level in the 1780s (within the Little Ice Age), in the 1840s–1860s (at the supposed end of the Little Ice Age), and in 1980–1990, since when sunspots have been in decline) (Clette et al., 2014). Global ocean and atmospheric temperatures in 1980 should therefore have been about the same as they were in those two former periods of peak sunspots. In fact they were very much warmer, and the warming has continued despite sunspot decline from 1990 onwards. During sunspot minima or maxima, total solar irradiance changes by tiny amounts (between 0.1 to 0.3%), which is not enough to explain current ocean/global warming, although it does help to explain fluctuations like the Medieval Warm Period and the Little Ice Age, whose effects were largely confined to the Northern Hemisphere. Projections of sunspot activity suggest there may be a substantial new solar minimum sometime after 2070 (Steinhilber & Beer, 2013), which might give rise to a fall in global temperature of no more than about 0.3 °C.

We can also expect other changes apart from those already mentioned. For example, some models suggest that the melting of Antarctica's ice shelves from the incursion of warm water in the subsurface will provide enough cold fresh water to increase the formation of sea ice in the Southern Ocean, creating regional cooling. Alternative models suggest that global warming will reduce sea ice there by 30% or so. Which of these models will turn out to be closest to reality? Time will tell.

References

Archer, D. (2005). Fate of fossil fuel CO_2 in geologic time. *Journal of Geophysical Research, 110*, C09S05. https://doi.org/10.1029/2004JC002625

Chavez, F. P., Ryan, J., Lluch-Cota, S. E., & Niquen, M. (2003). From anchovies to sardines and back: Multidecadal change in the Pacific Ocean. *Science, 229*, 217–221.

Chen, X., & Tung, K.-K. (2018). Global surface warming enhanced by weak Atlantic overturning circulation. *Nature, 559*, 387–391.

Clette, F., Svalgaard, L., Vaquero, J. M., & Cliver, E. W. (2014). Revisiting the sunspot number. *Space Science Review*. https://doi.org/10.1007/s11214-014-0074-2.

Kaufman, D., McKay, N., Routson, C., et al. (2020). Holocene global mean surface temperature, a multi-method reconstruction approach. *Nature, Scientific Data, 7*, 201. https://doi.org/10.1038/s41597-020-0530-7

Kobashi, T., Kawamura, K., Severinghaus, J. P., et al. (2011). High variability of Greenland surface temperature over the past 4000 years estimated from trapped air in an ice core. *Geophysical Research Letters, 38*, L21501. https://doi.org/10.1029/2011GL049444

Steinhilber, F., & Beer, J. (2013). Prediction of solar activity for the next 500 years. *Journal of Geophysical Research, Space, 118*, 1861–1867. https://doi.org/10.1002/jgra.50210

Von Schuckmann, K., Cheng, L., Palmer, M. D., et al. (2020). Heat stored in the earth system: Where does the energy go? *Earth System Science Data, 12*, 2013–2041.

Colin P. Summerhayes is a marine geochemist researching past global climate change in oceans and icy regions. He is an Emeritus Associate of Cambridge University's Scott Polar Research Institute, and a member of the Anthropocene Working Group. His publications include "*Oceanography: an Illustrated Guide*" (1996), "*Antarctic Climate Change and the Environment*" (2009), "*Earth's Climate Evolution*" (2015), "*The Anthropocene as a Geological Time Unit*" (2019), "*Paleoclimatology: from Snowball Earth to the Anthropocene*" (2020), and "*The Icy Planet: Saving Earth's Refrigerator*" (2023). He has worked for academia, government, UNESCO, and industry, and has been President of the international Society for Underwater Technology, and Vice President of the Geological Society of London.

Permafrost

Martha Jimenez-Castaneda and Rattan Lal

Abstract With a rapidly growing population, the human impact on the environment is undeniable. Despite models and predictions showing the degradation of permafrost, the global consequences of permafrost thawing are little understood by the broader society. At the current rate, Earth's temperature will cause not only an increase in greenhouse emission feedback but will also modify human lifestyles worldwide. Global decarbonization efforts and strategies to offset anthropogenic emissions are needed to minimize the loss of the permafrost and to stabilize the planet temperature at no more than 2 °C above pre-industrial temperatures by 2100.

An Overview of Permafrost

Permafrost is defined as the ground that has been at or below 0 °C for a minimum of two consecutive years to millennia. It consists of a combination of soil, sediments, rocks, organic matter (OM) and varying amounts of ice that bind these elements together. The vertical structure of permafrost consists of an active layer that freezes and thaws seasonally. The permafrost table marks the upper limit of the permafrost, below which there may be unfrozen zones, known as taliks, that can be important water sources in the region. The permafrost base occurs where the ground temperature rises above 0 °C. The depth to the base depends on a balance between the freezing temperatures of the top zones of the permafrost and the intrinsic temperature of the Earth.

The permafrost regions cover about 22.8 million km^2, equivalent to 25% of the exposed land surface in the Northern Hemisphere (Zhang et al., 2008). Recent simulations estimate the permafrost distribution in 0.05 million km^2 in the Southern Hemisphere, and 0.28 million km^2 in Antarctica (Gruber, 2012). Based on the

M. Jimenez-Castaneda (✉) · R. Lal
CFAES Rattan Lal Center for Carbon Management and Sequestration, The Ohio State University, Columbus, OH, USA
e-mail: lal.1@osu.edu

© The Author(s), under exclusive license to Springer Nature Switzerland AG 2023
N. Wallenhorst, C. Wulf (eds.), *Handbook of the Anthropocene*,
https://doi.org/10.1007/978-3-031-25910-4_26

163

percentage of the land that is underlain permafrost, permafrost zones are divided into continuous (90–100%), discontinuous (50–90%), sporadic (10–50%), and isolated (<10%). Within the Northern Hemisphere, 47% of the permafrost zone is classified as continuous, 19% as discontinuous, and 34% can be either sporadic or isolated. According to the geographical location of the permafrost, two basic distributions can be distinguished: latitudinal and altitudinal. The latitudinal, artic, or polar permafrost represents approximately 12.8–17.8% of the Earth surface. It occurs from 26°N in the Himalayas to 84°N in northern Greenland (Zhang et al., 2008) in territories of Siberia, part of Scandinavia, the Tibetan Plateau, Alaska, Canada, Greenland, and in continental shelves below the Arctic Ocean. The altitudinal, alpine or mountain permafrost is found in mountain ranges at middle and low latitudes in Asia and other regions worldwide.

Permafrost in a Warming Climate

Permafrost Carbon

Permafrost covers almost 25% of the terrestrial surfaces in the Northern Hemisphere and accounts for nearly half of all organic carbon (OC) stored within the planet's soil (Tarnocai et al., 2009). As long as the permafrost is stable, its OC content will be stable. However, in a warming climate, frozen OM will thaw, and greenhouse gases (GHG) will be released to the atmosphere, triggering further feedbacks on climate change and modifying other global processes.

Estimations of the northern circumpolar permafrost soil OC (SOC) pools indicate a mean value of 1400–1600 Pg-C (Schädel et al., 2018). Most of the permafrost C (65–70%) is located in the surface (0–3 m depth), with SOC stocks calculated in 1024 Pg C (Tarnocai et al., 2009); and 1035 Pg C (Hugelius et al., 2014; Schuur et al., 2015). Another significant amount of SOC (25–30%) is stored deeper (> 3 m depth). In particular, the Yedoma domain that covers approximately 1.4 million km^2 in Siberia and Alaska, stores about 327–466 Pg C (Strauss et al., 2017). The remaining deep C (< 10%) is contained in Arctic river delta deposits (91 Pg C; Hugelius et al., 2014).

The amount of C totally contained in permafrost is approximately twice the amount present in the atmosphere (829 Pg C) and is approximately 10 times larger than the C contained in both above- and belowground live vegetation biomass and litter in tundra and boreal forest (Schuur & Mack, 2016).

GHG Emissions

As permafrost thaws, microbes that were in perpetual suspension can be reactivated and resume the degradation of available OM. The main products of these microbial processes are carbon dioxide (CO_2) and methane (CH_4), which in turn amplify the anthropogenic warming caused by GHG emissions and exacerbate climate change. Only 5–15% of the permafrost OC could produce a large enough atmospheric GHG flux to increase global temperatures 0.3–0.4 °C by 2100 (Plaza et al., 2019). Keeping the current global warming trajectory (IPCC RCP8.5), >90% of near surface permafrost in the Arctic will be degraded by the end of the century releasing 37–174 Pg C (Schuur et al., 2015).

Despite most of the permafrost C emissions being in the form of CO_2, CH_4 represents significant biogeochemical feedback to the climate system because of its global warming potential (GWP) equivalent to 28–36 Mg CO_2 for a 100-year time horizon. CH_4 is produced in anaerobic microbial processes where methanogens (hydrogenotrophs, aceticlastics or methylotrophs) use an oxidized C source, such as CO_2, as a terminal electron acceptor (Conrad, 2020; Lyu et al., 2018). Reactions can:

1. Involve the oxidation of H_2, formate (-CHO_2) or a few simple alcohols (-OH) and the reduction of CO_2 to produce CH_4

$$H_2 + CO_2 \rightarrow CH_4 + 2H_2O$$

2. Use of acetate to form CH_4 and CO_2

$$CH_3COOH \rightarrow CH_4 + CO_2$$

3. Use of methylated compounds (CH_3-) to form CH_4

$$4CH_3OH \rightarrow CO_2 + 2H_2O + 3CH_4$$

or

$$CH_3OH + H_2 \rightarrow CH_4 + H_2O$$

Besides the potential formation of new CH_4, permafrost contains ancient CH_4 hydrates (clathrates) that were formed during anaerobic decomposition of plant and animal remains in lake bottoms and are stable under high pressures and low temperatures. Permafrost clathrates represent approximately 1% of the global CH_4 hydrates equivalent to 16–124 Pg C (Ruppel, 2015).

In addition, permafrost nitrous oxide (N_2O) emissions are linked to the C cycle and to the ratio of N_2O being consumed in the active layer. The decomposition of nitrous acid (HNO_2) in soil solution can be represented as

$$2HNO_2 \rightarrow NO + NO_2 + H_2O$$

From where

$$2NO_2 + H_2O \rightarrow NO_2^- + NO_3^- + 2H^+$$

These fluxes could add up to the N_2O global budget, but the extent of the permafrost N-pools and their quantification require further investigations. Estimations suggest that permafrost could store approximately 67 Pg N. The release of only 1% of that N stock in the form of N_2O, would be equivalent to 10 times the global annual rate of N_2O emissions from soils under natural vegetation, 67 TgN_2O nitrogen (Ramm et al., 2020).

Impact of Permafrost Change

For thousands of years, the Artic region was inhabited only by indigenous people. Currently the permafrost region houses about five million inhabitants. Here the temperatures have risen much faster than the rest of the world, about 0.6 °C per decade over the last 30 years (IPCC, 2013). Estimations indicate that 9–15% of the top 3 m of permafrost will degrade by 2040, and 47–61%, by 2100; these changes will cause the release of 63 PgC and 232–380 PgC, respectively (Schuur & Abbott, 2011). As permafrost degrades, human settlements in the Artic will become more vulnerable and by 2050 the permafrost population will decrease 61.2% (Ramage et al., 2021).

Processes governing vulnerability of permafrost generally take place in the form of sudden and gradual changes. Immediate responses include process such as thermokarsts and fire events that cause active layer thickening with thaw settlement in supersaturated materials. Intermediate responses occur in a timescale from years to decades, causing disturbance of temperature distribution at depth; and the permanent responses cause the basal melting of permafrost with thaw settlement in supersaturated materials taking several decades to millennia. Thermokarsts cover about 20% of the northern permafrost region with large implications for soil hydrology and the C cycle (Olefeldt et al., 2016), whereas wildfires can promote the renewal of boreal forest and plague control. Under climate change conditions, fire has a dramatic impact on soil conditions, consuming the soil cover, reducing the albedo, increasing the active layer thickness, and causing thawing of near-surface permafrost. In consequence, wildfires can produce a substantial development of taliks, increase deep soil temperatures and contribute to CO_2 emissions.

Warmer temperatures not only affect the permafrost stability but may also trigger other process that can cause the reactivation of ancient, uncharacterized microorganisms and viruses that could exhibit a pathogenic behavior and could affect nutrient cycling. Additionally, climate change can release the vast reservoir of mercury (Hg) that exists in the 0–3 m depth layer of the permafrost (1700 Gg Hg) where most of the SOC occurs. This reservoir is considered to contain twice as much Hg present in the rest of all soils, the atmosphere, and ocean combined (Schuster et al., 2018). Microbial mineralization of permafrost OM can cause a dramatic increase of

Hg in water bodies and therefore a significant accumulation in human and other animal populations through the food chain.

The destructive impact of permafrost loss on infrastructure is not necessarily abrupt and can be caused either by (a) ground subsidence or (b) bearing capacity failure. Ground subsidence is associated with ice melting, accompanied by the consolidation of sediments under progressive thickening of the active layer, whereas the bearing capacity failure involves the reduction of permafrost ability to support buildings and structures, leading to deformations and ultimately structural failures. Some examples of the permafrost thawing impact in modern lifestyles can be seen in Russia, with a large part of its territory in permafrost regions, and several cities already damaged by permafrost changes. Russian oil and gas infrastructure is also vulnerable and represents environmental threats associated with mechanical failures. In Canada, the permafrost loss has caused a housing shortage and millionaire investments in public infrastructure reparation, whereas in China the production and distribution of goods and services is problematic. Indigenous communities also face displacement from their traditional territories in Scandinavia and Alaska and consequently changes in their culture and traditions.

Conclusions

The term Anthropocene confronts human and geological timeframes. Setting aside the controversy over the definition of this epoch, the impact of humans on the global environment is undeniable. Human activities have resulted in the planet warming at a faster rate than at any point in the history of modern civilization over the past century, especially in regions perennially frozen. The constant rise of Earth's temperature will cause not only an increase in GHG emission feedback, but also and in consequence thereof, change of global rain patterns, loss of extensive forest cover, potential expansion of pests and loss of indigenous and industrial settlements in permafrost regions.

In the Policy Implications of Warming Permafrost (Schaefer et al., 2012), the United Nations proposed to set a global target warming of 2 °C above pre-industrial temperatures by 2100. By 2015, the Paris Agreement pledged to limit the global warming, if possible, at 1.5 °C above pre-industrial temperatures. Policy and decision makers of the permafrost regions need to address preventive economic, social, and environmental impacts of permafrost degradation, but global decarbonization efforts and strategies to compensate C emissions are needed. Additionally, it is necessary to develop more accurate inventories of the C and N pools stored in the permafrost region to simulate the GHG emission under different scenarios. Maintaining the current global warming trends, the 1.5 °C limit is likely to be exceeded between 2030 and 2052.

References

Conrad, R. (2020). Importance of hydrogenotrophic, aceticlastic and methylotrophic methano-genesis for methane production in terrestrial, aquatic and other anoxic environments: A mini review. *Pedosphere, 30*, 25–39. https://doi.org/10.1016/S1002-0160(18)60052-9

Gruber, S. (2012). Derivation and analysis of a high-resolution estimate of global permafrost zona-tion. *The Cryosphere, 6*, 221–233. https://doi.org/10.5194/tc-6-221-2012

Hugelius, G., Strauss, J., Zubrzycki, S., Harden, J. W., Schuur, E. A. G., Ping, C.-L., Schirrmeister, L., Grosse, G., Michaelson, G. J., Koven, C. D., O'Donnell, J. A., Elberling, B., Mishra, U., Camill, P., Yu, Z., Palmtag, J., & Kuhry, P. (2014). Estimated stocks of circumpolar perma-frost carbon with quantified uncertainty ranges and identified data gaps. *Biogeosciences, 11*, 6573–6593. https://doi.org/10.5194/bg-11-6573-2014

IPCC. (2013). *Climate change 2013: The physical science basis. Contribution of working group I to the fifth assessment report of the intergovernmental panel on climate change.* Cambridge University Press.

Lyu, Z., Shao, N., Akinyemi, T., & Whitman, W. B. (2018). Methanogenesis. *Current Biology, 28*, R727–R732. https://doi.org/10.1016/j.cub.2018.05.021

Olefeldt, D., Goswami, S., Grosse, G., Hayes, D., Hugelius, G., Kuhry, P., McGuire, A. D., Romanovsky, V. E., Sannel, A. B. K., Schuur, E. A. G., & Turetsky, M. R. (2016). Circumpolar distribution and carbon storage of thermokarst landscapes. *Nature Communications, 7*, 13043. https://doi.org/10.1038/ncomms13043

Plaza, C., Pegoraro, E., Bracho, R., Celis, G., Crummer, K. G., Hutchings, J. A., Hicks Pries, C. E., Mauritz, M., Natali, S. M., Salmon, V. G., Schädel, C., Webb, E. E., & Schuur, E. A. G. (2019). Direct observation of permafrost degradation and rapid soil carbon loss in tundra. *Nature Geoscience, 12*, 627–631. https://doi.org/10.1038/s41561-019-0387-6

Ramage, J., Jungsberg, L., Wang, S., Westermann, S., Lantuit, H., & Heleniak, T. (2021). Population living on permafrost in the Arctic. *Population and Environment, 43*, 22–38. https://doi.org/10.1007/s11111-020-00370-6

Ramm, E., Liu, C., Wang, X., Yue, H., Zhang, W., Pan, Y., Schloter, M., Gschwendtner, S., Mueller, C. W., Hu, B., Rennenberg, H., & Dannenmann, M. (2020). The forgotten nutrient—The role of nitrogen in permafrost soils of northern China. *Advances in Atmospheric Sciences, 37*, 793–799. https://doi.org/10.1007/s00376-020-0027-5

Ruppel, C. (2015). Permafrost-associated gas hydrate: Is it really approximately 1% of the global system? *Journal of Chemical & Engineering Data, 60*, 429–436. https://doi.org/10.1021/je500770m

Schädel, C., Koven, C. D., Lawrence, D. M., Celis, G., Garnello, A. J., Hutchings, J., Mauritz, M., Natali, S. M., Pegoraro, E., Rodenhizer, H., Salmon, V. G., Taylor, M. A., Webb, E. E., Wieder, W. R., & Schuur, E. A. G. (2018). Divergent patterns of experimental and model-derived per-mafrost ecosystem carbon dynamics in response to Arctic warming. *Environmental Research Letters, 13*, 105002. https://doi.org/10.1088/1748-9326/aaae0ff

Schaefer, K., Lantuit, H., Romanovsky, V., Schuur, E., 2012. Policy implications of warming permafrost.

Schuster, P. F., Schaefer, K. M., Aiken, G. R., Antweiler, R. C., Dewild, J. F., Gryziec, J. D., Gusmeroli, A., Hugelius, G., Jafarov, E., Krabbenhoft, D. P., Liu, L., Herman-Mercer, N., Mu, C., Roth, D. A., Schaefer, T., Striegl, R. G., Wickland, K. P., & Zhang, T. (2018). Permafrost stores a globally significant amount of mercury. *Geophysical Research Letters, 45*, 1463–1471. https://doi.org/10.1002/2017GL075571

Schuur, E. A. G., & Abbott, B. (2011). High risk of permafrost thaw. *Nature, 480*, 32–33. https://doi.org/10.1038/480032a

Schuur, E. A. G., & Mack, M. C. (2016). Ecological response to permafrost thaw and conse-quences for local and global ecosystem services. *Annual Review of Ecology, Evolution, and Systematics, 49*, 279–301. https://doi.org/10.1146/annurev-ecolsys-121415-032349

Schuur, E. A. G., McGuire, A. D., Schädel, C., Grosse, G., Harden, J. W., Hayes, D. J., Hugelius, G., Koven, C. D., Kuhry, P., Lawrence, D. M., Natali, S. M., Olefeldt, D., Romanovsky, V. E., Schaefer, K., Turetsky, M. R., Treat, C. C., & Vonk, J. E. (2015). Climate change and the permafrost carbon feedback. *Nature, 520*, 171–179. https://doi.org/10.1038/nature14338

Strauss, J., Schirrmeister, L., Grosse, G., Fortier, D., Hugelius, G., Knoblauch, C., Romanovsky, V., Schädel, C., Schneider von Deimling, T., Schuur, E. A. G., Shmelev, D., Ulrich, M., & Veremeeva, A. (2017). Deep Yedoma permafrost: A synthesis of depositional characteristics and carbon vulnerability. *Earth-Science Review, 172*, 75–86. https://doi.org/10.1016/j.earscirev.2017.07.007

Tarnocai, C., Canadell, J. G., Schuur, E. A. G., Kuhry, P., Mazhitova, G., & Zimov, S. (2009). Soil organic carbon pools in the northern circumpolar permafrost region. *Global Biogeochemical Cycles, 23*. https://doi.org/10.1029/2008GB003327

Zhang, T., Barry, R. G., Knowles, K., Heginbottom, J. A., & Brown, J. (2008). Statistics and characteristics of permafrost and ground-ice distribution in the northern hemisphere. *Polar Geography, 31*, 47–68. https://doi.org/10.1080/10889370802175895

Martha Jimenez-Castaneda is a Research Scientist at CFAES Dr. Rattan Lal Carbon Management and Sequestration Center (C-MASC), The Ohio State University (USA). She is a geochemist interested in the origin and stabilization of soil nutrients at molecular level.

Rattan Lal is the Director of the CFAES Dr. Rattan Lal Carbon Management and Sequestration Center (C-MASC), The Ohio State University (USA). Distinguished University Professor of Soil Science. Adjunct Professor, University of Iceland, Reykjavík, Iceland and the Indian Agricultural Research Institute (IARI), New Delhi, India. He is also Chair in Soil Science and Goodwill Ambassador for Sustainable Development Issues, Inter-American Institute for Cooperation on Agriculture (IICA), San José, Costa Rica.

Plant

Andreas Weber

Abstract A plant is a biological subject whose self-realization transforms mostly invisible inorganic matter and energy (CO_2 and light plus water) into organic matter – flesh – which in turn nourishes other subjects and therefore enables terrestrial ecosystems to exist. Plants are able to perceive, to communicate and have the inner experience of being a self. Plants are the prime metamorphic agents of the biosphere, mediating the potential for life of the anorganic world with fully blossoming aliveness. Plants are expressive of their embodied metamorphosis and therefore lend themselves as "primordial metaphors" to an experience of reality as shared and transformational in most cultures of human history.

Plants are embodied subjects who create their bodies mainly from water, carbon dioxide (a component of the air), and light. The chemical equation for this process, called photosynthesis, is: $6CO_2 + 6H_2O \rightarrow (\text{light}) \rightarrow C_6H_{12}O_6 + 6O_2$. By this transformation plants supply the energetic and material basis for all terrestrial ecosystems. They are the way the sun becomes flesh, thereby cycling a portion of the earth's fixed amount of carbon into living bodies, from which, after having run through diverse transformations into other living beings' bodies, carbon is released into the atmosphere, into soil or sediments again.

Plants are therefore able to transform matter and energy into a conformation which is alive, thus showing and experiencing agency. By this, they are forming the basis for the cascades of agency and incarnation which characterise the biosphere. Plants share those qualities with all other organisms who possess the chemical apparatus necessary for photosynthesis: Cyanobacteria (blue green algae) and photoautotrophic (chlorophyll-bearing) protists (unicellular eukaryotes), who are the main photosynthetic agents in the world's oceans. Together with those, plants enable agency to become manifest from anorganic gradients.

A. Weber (✉)
Bard College, Berlin, Germany

What qualifies as a plant has been changing historically. Here I understand as plants all multicellular beings containing chlorophyll in order to build up their bodies, and variants derived from those who have lost their chlorophyll due to a particularly adapted lifestyle. The chloroyphyll molecules give plants their distinct green colour. Plants build up 80% of the world biomass, about 450 Gigatonnes of carbon (Bar-On et al., 2018).

Historically, plants have arisen through a major symbiotic event about 850 Million years ago, when single cells without cholorphyll incorporated blue-green algae (Sagan [Margulis], 1967; Margulis, 1970; Knauth & Kennedy, 2009; Stadnichuka & Kusnetsova, 2021). The newly integrated cells subsequently became the chloroplasts, the chlorophyll-bearing organelles of plant cells. By this association, the new life form created the precondition for multicellularity and subsequent specialization into various tissues without which the plant bodies we know today could not have been built.

So not only the biochemical self-realization (photosyntesis) of plants, but also their natural history includes several levels of "mutual transformation" (Weber, 2017), where individual poles unite and transmute into emergent new forms. Plants highlight the transformational dimension of biological existence. They are metamorphotic agents. They illustrate that biological individuality is only one level in an intricate interplay of the whole and its parts. In this, plants are no exception from the remainder of life forms, but they highlight the transformational property of life in an important way.

Another crucial transformational dimension of plants is their shared history with the animal reign, particularly through the co-evolution of pollinating insects and blossoms, but also through the relationship between plant-eating insects and secondary plant substances and drugs developed to fend off those insects. These co-evolutionary histories have profoundly formed terrestrial ecosystems, to such a degree that it is apt to speak of a co-evolutionary whole. Plants are implied in the evolution of insects (and other animals, and other groups like fungi), and those groups are implied in the coming about of plant forms.

Another transformational process relates plants and fungi. Fungal mycelia form parts of the mycorrhiza, the plant root–fungal network which proliferates in the soils and enables land plants to grow in the first place. Through the mycorrhizal network plants and fungi organise the re-distribution of nutrients and energy that is the precondition for densely populated terrestrial ecosystems. Fungi also break down plant matter (particularly wood) and thus cycle it back into the domain of living bodies.

The transformational dimension of plants plays out for humans in the fact that many, if not all, plants contain secondary substances of pharmaceutical value and also compounds which are able to induce narcotic states and psychedelic experiences. Most medicine, going back into prehistoric deep time, is plant-based. Nearly all drugs (apart from recent synthetic ones) are plant-derived. Some of the most widely consumed drugs, coffee, tobacco and alcohol, are derived from plants.

Until a few years ago, plants have been widely seen as inert, non-sentient and unable of perception. Indeed they do lack the central nervous system of animals (and as such, humans). Charles Darwin (1880) was one of the first biologists who insisted on the fundamental equivalence of plants and animals in terms of basic functions and experiences of their aliveness (Baluška et al., 2004, 2009). But in spite of Darwin's quickly overarching authority in evolutionary biology, his position concerning plants did not become mainstream. This has changed only recently through the extensive research on plant perception and communication (Wohlleben, 2016).

In occidental philosophy plants were widely viewed as inferior beings. This diverged sharply from animistic concepts, where plant beings played a role in the society of nature equivalent to animals and humans. In Platonic and Aristotelian thinking, plants are to some degree gifted with a soul, but this is of a lower order than that of animals and humans. Early twentieth century Philosophy of nature mostly continued the tradition of placing plants on a lower rung of the ladder of natural history (Plessner, 1928). There were some notable exceptions in romantic thinking, however, particular Gustav Theodor Fechners Fechner (1848) reflections on the "plant soul", in which he ascribed to plants the experiential qualities of animals and humans.

Only in the last ten years or so it has become accepted that plants possess the full spectrum of qualities that living beings have. Plants are subjects with a vulnerable body in which they experience themselves in interaction with what is not themselves and are expressive of these experiences. As such plants have perception (Mancuso & Viola, 2015), can communicate (Trewavas, 2009), learn (Gagliano et al., 2016), actively share resources (Simard et al., 1997), and do even possess motility to some degree (most impressively in the root tips, Baluška et al., 2004). Plants are therefore not fundamentally different from animals, and indeed, from all life forms. To the contrary: In many respects they exemplify qualities shared by all life.

The "neurocentric bias" which had impeded the appreciation of plants as showing the full spectrum of life therefore can today be corrected with a subject-centred stance on plants. This view underscores that all organisms are *ipso facto* self-creating subjects, as they are concerned with establishing their identity through a process termed "Autopoiesis" (Varela et al., 1974). The embodied identity they create is the centre of a self from which this identity is organized, both on embodied and on experiential terms (Weber, 2016a; Weber & Varela, 2002). This identity is built on a comprehensive material exchange of the biological individual with its surroundings.

Goethe Von Goethe (1790) emphasized this metamorphic power in his reflections on the "Urpflanze", the original plant. Different from romantics like Fechner, Goethe did not search for a human-like soul in the plant, but tried to understand the self-creating world as plant-like in its metamorphic plurality. This plurality follows laws in order to make individuality appear and unfold in a temporal succession. This is laid down in visible form in the various parts of a plant, which are all transformations of the underlying growth impulse located in the sprout. The plant's nature is metamorphic, because the cosmos is metamorphic (Coccia, 2018).

Light, one of the dimensions the plants feed upon, thus can be understood as "the other side" of the plant – and the plant exists as the other side of light. Through the plant, light enters a different state with other possibilities of agency. The plant is a manifestation of light, and as all manifestations are, is limited in its potential to a particular way of being. We could also say that plants are "queering" individuality – and ecology (Morton, 2010). They make individuality a collective action, and give collective materiality individual experience. They are threads of an endless tissue in which everything is of flesh and light (Weber, 2021).

From the point of view of enactive cognitive science (Varela et al., 1991), there is no such thing as an individual detachable from the remainder of the world. Each organism is a "meshwork of selfless selves" (Varela, 1991). Each one is agential. Each being has an inward perspective which can be approached through sensing and perceiving, hence is expressive. This is valid for all life, but plants express this enactive power in exemplary ways.

Biological-existential meaning becomes manifest in the exterior form of a plant, as the autopoietic process of creating a self happens in and on matter. Plants therefore exhibit a "gesture of the living" (Weber, 2016b) which can be understood by other living beings. Plants are symbolic of life. They are not so because culture has ascribed meanings to them, but because meanings are generated from biological lifemaking and become transparent for other embodied subjects.

In this respect, plants have brought forth a plenitude of symbolisms in human cultures from the dawn of time. Holy trees – like the world-ash of Norse myths, the Persian Haoma tree, or the Bodhi Fig tree (*Ficus religiosa*), the Buddha's "tree of awakening" – line human cultural history and practice to this day, to the degree that even the Christian Cross is dendromorph (Schama, 1995). Adding to the suggestive power of trees, flowering plants symbolize birth, death, and other stages of transformation.

From the perspective of the plant as metamorphic agent which literally allows the object world to materialize from light and then distributes this solidified energy to other beings, those cultural symbolisms do not seem arbitrary ascriptions, but rather "primordial metaphors" (Weber, 2016b). They are what they profoundly represent. In order to understand them, we need to be alive – of the same flesh that the plant gifts the world by its primordial transformation of potential into presence.

References

Baluška, F., Mancuso, S., Volkmann, D., & Barlow, P. W. (2004). Root apices as plant command centres: The unique 'brain-like' status of the root apex transition zone. *Biologia, 59*(Suppl 13), 9–17.

Baluška, F., Mancuso, S., Volkmann, D., & Barlow, P. W. (2009). The 'root-brain' hypothesis of Charles and Francis Darwin: Revival after more than 125 years. *Plant Signaling & Behavior, 4*(12), 1121–1127.

Bar-On, Y. M., Phillips, R., & Milo, R. (2018). The biomass distribution on earth. *Proceedings of the National Academy of Sciences of the United States of America, 115*(25), 6506–6511.

Coccia, E. (2018). *The life of plants. A metaphysics of mixture*. Wiley.

Darwin, C. (1880). *The power of movements in plants*. John Murray.

Fechner, G. T. (1848 [1908]). *Nanna oder über das Seelenleben der Pflanzen*. Leopold Voß.

Gagliano, M., et al. (2016). Learning by association in plants. *Nature: Scientific Reports, 6*, 38427.

Knauth, L. P., & Kennedy, M. J. (2009). The late Precambrian greening of the earth. *Nature, 460*(7256), 728–732.

Mancuso, S., & Viola, A. (2015). *Brilliant green: The surprising history and science of plant intelligence*. Island Press.

Margulis, L. (1970). *Origin of eukaryotic cells; evidence and research implications for a theory of the origin and evolution of microbial, plant, and animal cells on the Precambrian earth*. Yale University Press.

Morton, T. (2010). Queer ecology. *PMLA, 125*(2), 273–282.

Plessner, H. (1928 [2019]). *Levels of organic life and the human: An introduction to philosophical anthropology*. Fordham University Press.

Sagan [Margulis], L. (1967). On the origin of Mitosing cells. *Journal of Theoretical Biology, 14*, 225–274.

Schama, S. (1995). *Landscape and memory*. Knopf.

Simard, S. W., et al. (1997). Net transfer of carbon between ectomycorrhizal tree species in the field. *Nature, 388*, 579–582.

Stadnichuka, I. N., & Kusnetsova, V. V. (2021). Endosymbiotic origin of chloroplasts in plant cells' evolution. *Russian Journal of Plant Physiology, 68*(1), 1–16.

Trewavas, A. (2009). What is plant behaviour? *Plant, Cell and Environment, 32*, 606–616.

Varela, F. (1991). Organism: A meshwork of selfless selves. In A. I. Tauber (Ed.), *Organism and the origins of self*. Kluwer.

Varela, F. J., Maturana, H. R., & Uribe, R. (1974). Autopoiesis: The organization of living systems, its characterization and a model. *Biosystems, 5*(4), 187–196.

Varela, F. J., Thompson, E., & Rosch, E. (1991). *The embodied mind: Cognitive science and human experience*. The MIT Press.

Von Goethe, J. W. (1790 [2019]). *The metamorphosis of plants*. MIT Press.

Weber, A. (2016a). *The biology of wonder. Aliveness, feeling, and the metamorphosis of science*. New Society Press.

Weber, A. (2016b). *Biopoetics. Toward an existential ecology*. Springer.

Weber, A. (2017). *Matter and desire. An erotic ecology*. Chelsea Green.

Weber, A. (2021). Skincentric ecology. In G. Van Horn, R. W. Kimmerer, & J. Hausdoerffer (Eds.), *Kinship: Belonging in a world of relations, Vol. 4: Persons*. Center for Humans and Nature Press.

Weber, A., & Varela, F. J. (2002). Life after Kant: Natural purposes and the autopoietic foundations of biological individuality. *Phenomenology and the Cognitive Sciences, 1*, 97–125.

Wohlleben, P. (2016). *The hidden life of trees. What they feel, how they communicate – Discoveries from a secret world*. Greystone.

Andreas Weber is a biologist, philosopher and writer. He teaches at the University of the Arts, Berlin, is Visiting Professor at the UNISG, Pollenzo, Italy, and holds an Adjunct Professorship at the IIT, Guwahati, India. He contributes to major German newspapers and magazines and has published more than a dozen books, most recently *Enlivenment. A Poetics for the Anthropocene*, MIT Press, 2019 and *Sharing Life. The Ecopolitics of Reciprocity*, Boell Foundation, 2020.

Sea Level Change

Alejandro Cearreta

Abstract Instrumental and geological records indicate that the rise in the global sea level has been accelerating since the beginning of the twentieth century due to ocean thermal expansion and the melting of the cryosphere. Future sea level increase will depend on the amount of greenhouse gas emissions over the next decades, although geological evidence shows that it will continue to rise for the next millennia due to complex temperature and gas transport mechanisms in the ocean. Furthermore, anthropogenic activities on the coastal area are also exacerbating the effects of sea level rise. Coastal communities and environments are increasingly vulnerable to a rising sea level, and a climate-resilient human society needs to integrate local and regional adaptation measures.

Sea level change has been one of the most important consequences of the orbital climate variability during Earth history, as the sea level gradually decreased during cold glacial intervals and rapidly increased during warm interglacial phases. As human activity has recently become the new climate driver on a planetary scale due to massive emissions of greenhouse gases and deforestation, the Intergovernmental Panel on Climate Change (IPCC) has established a strong link between anthropogenic climate change and global sea level rise.

Sea level is not a constant and planar surface but exhibits spatial and temporal changes at different scales. Geoscientists define sea level as a measure of the position of the sea surface relative to the land, both of which may be moving relative to the centre of the Earth. Sea level can change due to climate processes that include variations in ocean temperature and salinity which cause sea water to expand or contract (steric changes in water volume), changes due to transfer of water between the cryosphere and the ocean (eustatic changes in water mass), and shifts in ocean currents. Additionally, other geophysical factors can affect sea level as, for example,

A. Cearreta (✉)
University of the Basque Country UPV/EHU; Basque Centre for Climate Change BC3,
Bilbao, Spain
e-mail: alejandro.cearreta@ehu.eus

N. Wallenhorst, C. Wulf (eds.), *Handbook of the Anthropocene*,
https://doi.org/10.1007/978-3-031-25910-4_28

177

land subsidence or uplift (tectonic changes) and glacial isostatic adjustments (vertical land motions in response to large-scale changes in surface mass load of land ice and ocean water). Local and regional changes in these climate and geophysical drivers can produce significant deviations above or below the global mean sea level through time (Masson-Delmotte et al., 2013).

The history of systematic sea level observations started in the seventeenth century using tide-gauge measurements and improved in the early 1990s with satellite-based radar altimeter measurements. Tide-gauge data are measures of relative sea level because they are expressed relative to the height of a benchmark on the nearby land. However, satellite altimetry is the only method that provides a measure of absolute height of the ocean surface relative to the centre of the Earth. Two additional observing systems complement these altimetric data: the Argo network, with its series of autonomous floats that monitor temperature and salinity of the ocean since 2000; and the Gravity Recovery and Climate Experiment satellite mission that monitors ice-mass loss from the Greenland and Antarctic ice sheets and mountain glaciers since 2002 (Summerhayes & Cearreta, 2019).

Instrumental records supply high-resolution information, but the temporal and spatial coverage of their measurements is too short to capture pre-anthropogenic trends of sea level. Therefore, geological data become necessary to place instrumental estimates into a longer-term context. The application of a consistent geological methodology to the reconstruction of sea level started in the 1970s. Precise sea level quantitative reconstructions from coastal sediments rely mainly on different groups of microfossils that exhibit vertical zonations within the tidal frame, and preserve in the sedimentary record of coastal wetlands that keep pace with sea level changes (Gehrels & Shennan, 2015).

Both instrumental and geological records show that mean sea level is rising across the Earth. Global mean sea level is currently higher than at any time in the last 115,000 years, after the end of the previous interglacial phase (Church et al., 2013). The current interglacial phase started 11,700 years ago and is characterised by an interval of warming accompanied by ~130 m of sea level rise, at an average rate of about 10 mm/year until about 7000 years ago (Lambeck et al., 2002).

The past 7000 years, when ice volumes stabilized near present-day values, provide the baseline for discussion of anthropogenic contributions, and experienced the transition from relatively low rates of sea level change (<1 mm/year) to modern rates of 3.2 mm/year (Cazenave et al., 2018). Over the period 1900 to 2018, global mean sea level rose by about 19 cm and instrumental records support a continued global sea level acceleration over the twentieth century. The current rate of sea level rise is double than the average rate of rise during the twentieth century, which, in turn, was an order of magnitude larger than the rate of rise over the seven millennia prior to the nineteenth century (Church et al., 2008). Ocean thermal expansion and increasing glacier and ice sheet melting have been the dominant contributors to twentieth century global mean sea level rise in response to greenhouse forcing (Walker et al., 2021).

The magnitude of the sea level rise predicted for the twenty-first century remains a subject of considerable debate. Future variations of sea level have been simulated using different models, and the latest IPCC projections (Oppenheimer et al., 2019) predict an average rise in the sea level at the end of this century of 28–57 cm in a scenario of drastic reduction in greenhouse emissions, and of 55–140 cm if there is growth in emissions. It means that this projected sea level rise ranges from the current rate of 3.2 mm/year up to about four times faster, and implies an upward revision compared to previous IPCC assessments.

The basis for this debate resides on the uncertainties about the dynamical behaviour of the Greenland and *Western* Antarctic ice sheets, which may lead to a non-linear sea level response to climate change. The amount of greenhouse gas emissions in coming decades will be consequential for global mean sea level on a century timescale through a combination of ocean thermal expansion and loss of land ice. If Paris Agreement targets (+1.5 °C and + 2 °C above the pre-industrial level) are exceeded, rapid and unstoppable sea level rise from Antarctica could be triggered around 2060 (DeConto et al., 2021).

Analysis of geological evidence from different warm intervals of the distant past (e.g., the Paleocene–Eocene Thermal Maximum of 55 million years ago or the mid-Pliocene Warm Period of three million years ago) shows that retreat of the large ice sheets and consequent sea level rise will continue for thousands of years, even if human efforts manage to keep global warming at close to +2 °C. Sea level will not follow that global temperature pattern and, by analogy with the current interglacial phase, it will continue to rise for the next 4000 years. Warming of the oceans would take centuries to reverse because the heat and CO_2 transported down to the deep oceans will take hundreds of years to circulate back to the atmosphere (Summerhayes & Cearreta, 2019).

Decelerating sea level rise approximately 7000 years ago allowed coastal communities to become permanently established (Day et al., 2007). Coastal systems are among the most dynamic natural environments on Earth continually evolving at various spatial-temporal scales. However, current sea level rise represents one of the greatest potential threats for coastal communities as a great part of the world's population lives on the coastal areas at densities about three times higher than the global average, and most of the human megacities are also located on the coast. With coastal development continuing at a rapid pace, due to growing human populations and increasing reclamation of environments for urbanization, industrial and agricultural purposes, both human society and coastal environments worldwide are becoming increasingly vulnerable to sea level rise. Vulnerable coastal zones could force significant population migration and socio-economic damage.

Human-induced activities, as the reduced sediment flux to major deltas and estuaries due to sediment entrapment by upstream dams, combined with increasing extraction of groundwater, hydrocarbons and sediments (for aggregates), and construction of heavy buildings, have caused many large deltas to subside at rates much faster (10 mm/year) than current sea level rise (Nicholls et al., 2021). Human effects on subsidence are maximum at some coastal megacities located on deltas and alluvial plains. Reducing city subsidence is feasible as demonstrated in the Netherlands and various Asian cities (e.g., Tokyo or Shanghai), where it involves managing groundwater withdrawal and maintaining high water tables. However, these policies generally reduce rather than stop all subsidence (Mimura, 2021).

Coastal adaptation aims to reduce impacts of sea level rise. Different measures specifically designed to be aware of evolving predictions of sea level rise include defensive engineering structures resilient to periodic flooding (e.g., the Thames Barrier in the UK and the MOSE Project in Italy) or retreat from exposed areas, combined with enhancement of natural defences such as wetlands and vegetated sand dunes (Oppenheimer & Alley, 2016). Adaptation to current and future sea level rise has a close relationship with other development challenges, such as management of water resources or preservation of natural environments, and should therefore be fully integrated in the wider context of the climate-resilient development of human society.

References

Cazenave, A., Meyssignac, B., Ablain, M., et al. (2018). Global Sea-level budget 1993-present. *Earth System Science Data, 10*, 1551–1590.

Church, J. A., White, N. J., Aarup, T., et al. (2008). Understanding global sea levels: Past, present and future. *Sustainability Science, 3*, 9–22.

Church, J. A., Clark, P. U., Cazenave, A., et al. (2013). Sea level change. In T. F. Stocker, D. Qin, G.-K. Plattner, et al. (Eds.), *Climate change 2013: The physical science basis. Contribution of working group I to the fifth assessment report of the intergovernmental panel on climate change* (pp. 1137–1216). Cambridge University Press.

Day, J. W., Gunn, J. D., Folan, W. J., Yáñez-Arancibia, A., & Horton, B. P. (2007). Emergence of complex societies after sea level stabilized. *Eos, Transactions of the American Geophysical Union, 88*, 169–176.

DeConto, R. M., Pollard, D., Alley, R. B., et al. (2021). The Paris climate agreement and future sea-level rise from Antarctica. *Nature, 593*, 83–89.

Gehrels, W. R., & Shennan, I. (2015). Sea level in time and space: Revolutions and inconvenient truths. *Journal of Quaternary Science, 30*, 131–143.

Lambeck, K., Esat, T. M., & Potter, E.-K. (2002). Links between climate and sea levels for the past three million years. *Nature, 419*, 199–206.

Masson-Delmotte, V., Schulz, M., Abe-Ouchi, A., et al. (2013). Information from paleoclimate archives. In T. F. Stocker, D. Qin, G.-K. Plattner, et al. (Eds.), *Climate change 2013: The physical science basis. Contribution of working group I to the fifth assessment report of the intergovernmental panel on climate change* (pp. 383–464). Cambridge University Press.

Mimura, N. (2021). Rising seas and subsiding cities. *Nature Climate Change, 11*, 293–299.

Nicholls, R. J., Lincke, D., Hinkel, J., et al. (2021). A global analysis of subsidence, relative sea-level change and coastal flood exposure. *Nature Climate Change, 338*, 338–342.

Oppenheimer, M., & Alley, R. B. (2016). How high will the seas rise? *Science, 354*, 1375–1377.

Oppenheimer, M., Glavovic, B., Hinkel, J., et al. (2019). Sea level rise and implications for low-Lying Islands, coasts and communities. In H.-O. Pörtner, D. C. Roberts, V. Masson-Delmotte, P. Zhai, et al. (Eds.), *IPCC special report on the ocean and cryosphere in a changing climate.* In press.

Summerhayes, C. P., & Cearreta, A. (2019). Climate change and the Anthropocene. In J. Zalasiewicz, C. N. Waters, M. Williams, & C. P. Summerhayes (Eds.), *The Anthropocene as a geological time unit. A guide to the scientific evidence and current debate* (pp. 200–241). Cambridge University Press.

Walker, J. S., Kopp, R. E., Shaw, T. A., et al. (2021). Common Era Sea-level budgets along the U.S. Atlantic coast. Nature. *Communications, 12,* 1841.

Alejandro Cearreta is Professor of Paleontology at the University of the Basque Country UPV/EHU, Doctorate in Geology (University of Exeter, UK), Head of the Geology Department (UPV/EHU), Coordinator of the Doctorate Program in Quaternary: Environmental Changes and Human Footprint (UPV/EHU), Associate Researcher of the Basque Centre for Climate Change (BC3) and Member of the Anthropocene Working Group (International Commission on Stratigraphy). Author of numerous scientific publications on environmental evolution of the coastal areas during the Holocene and Anthropocene due to sea level change and human impact.

Soil

Daniel D. Richter, Eniko Bihari, and Anna Wade

Abstract The science of pedology has historically studied how soils form in response to natural factors and processes. Today, with soils being so extensively altered by human action, pedology is adapting to the fundamental changes that are ongoing in our objects of study. These scientific adaptations are facilitated by: a quantifiable process model of soil formation, the realization that most soils are polygenetic products of changing environments, and ongoing work to integrate human influence into soil taxonomy. However, the state of long-term soil observatories that provide direct observations of human-influenced soil change is wretchedly inadequate. Because soils evolve in response to high-order interactions, they are changing in diverse, non-linear, and hard-to-predict trajectories that will always require decades to observe before they can be reliably simulated.

In 2002, Rudi Dudal, an expert in pedology felt compelled to ask the World Congress of Soil Science: Are we a soil-forming factor short? The pace and diversity with which human activities were transforming soils drove Dudal to question pedology's object of study: "soil as a natural body." Pedology, Dudal countered, must include human activities as "a fully-fledged factor of soil formation" because human beings were influencing "all natural soils not as deviation but as genetic soil type", i.e., not as disturbing agents but as forming agents.

Others had advanced similar positions before Dudal (Richter, 2020), sometimes with powerful language. Bidwell and Hole (1965) outlined how human activities affected "all natural soil-forming factors." Yaalon and Yaron (1966) were more adamant, asserting that human influence creates a new point of reference for soil formation, a new time zero "from which a new wave of polygenesis has begun." But it was

D. D. Richter (✉) · E. Bihari
Duke University, Durham, NC, USA
e-mail: drichter@duke.edu; enikoebihari@gmail.com

A. Wade
US EPA, Cincinnati, OH, USA
e-mail: awade91@gmail.com

Cline (1961) who wrote that intensifying land management "magnifies man and his activities as factors of soil formation and demands recognition of his work." Cline pointed to a portentous mid-twentieth century development: "If we look at the writing about soil productivity in the 1930's, we see a concept of soil as a stable thing with inherent capacity to produce… More commonly today the question is stated in different terms, 'How much input must be used on a given soil to produce a given volume of product?' This is quite a different matter. It implies that the soil is not a constant thing to be used much in the form in which one finds it, accepting meekly the limitations it may have. Instead, it implies that man remakes the soil to suit his needs…" That soils could be "remade" by management was nothing less than Promethean.

Updating pedology's core concept as a fully integrated human-natural body is an on-going and not easy task but one with a firm foundation. Pedologists from many nations co-authored a book entitled *Global Soil Change* that grappled directly with these issues in the 1980s (Arnold et al., 1990). The USDA Soil Survey Staff (2014) has now worked for decades to modify soil taxa and classification to include major human influences. The cover of the 12th edition of *Keys to Soil Taxonomy* has colour photos of soils composed of dredged and bulldozed materials, of rice and vegetable paddy soils, and of horizons full of bricks and concrete. The World Reference Base for Soil Resources (WRB) has created two new reference soil groups, the Anthrosols and Technisols, to describe soils with substantial human influence (Schad, 2018). Soils textbooks are bolstering discussions of the human influence on soils. Certini and Scalenghe (2011) even argued that the golden spike for the stratigrapher's Anthropocene be placed at the vertical boundary between the natural and human-modified bodies (though see Edgeworth et al., 2015 and Richter, 2020). Entire books have been written about human modified soils and urban soils. The arc of pedology has been bent toward what some call anthropedology.

Given pedology's need to fully integrate human influences, we use a process model adopted from Simonson (1959) that seamlessly combines human and natural influences on soil. Pedogenic processes are organized by four genetic processes – inputs and outputs of materials and energy, and biogeochemical translocations and transformations. These fluxes can be quantified and human influence apportioned. Inputs include organic carbon in plant and animal detritus, atmospheric deposition, and gaseous influxes; biotic immigrations; charcoal additions from fires; sediment depositions from soil creep, colluviation, or alluviation; surface and subsurface water influxes; and mineral weathering release of chemical elements and particles. Outputs include biotic emigrations; erosional losses by wind and water; gaseous effluxes including volatilizations from fires; and surface runoff and groundwater effluxes. Translocations include biotic and physical mixing of materials, dispersion and flocculation of particles; diffusion of gases and solutes; eluviation and illuviation of clays and fine silts; and plant uptake and recycling of nutrients. Transformations include biogeochemical reactions that are reversible and irreversible. Recently, Wade et al. (2021) adopted Simonson's process model to develop a framework for urban-soil pedogenesis to demonstrate how and why concentrations of legacy pollutant lead are decreasing in city soils due to erosion, pedoturbation, and mechanical removals, mixing, and burial over time. Such urban-soil pedogenesis takes us well beyond the formation of soil as a natural body.

Understanding soil bodies in the Anthropocene is also facilitated by late twentieth c. developments known as soil evolution (Johnson & Watson-Stegner, 1987) and soil polygenesis (Targulian & Goryachkin, 2004). Both are fundamental scientific advances and bring new richness to our understanding of soils. Given that soils have lifetimes of up to tens of millions of years, soils evolve in response to processes that vary widely in rate and intensity. Soils evolve polygenetically, even in relatively young Holocene and Late Pleistocene landscapes (Ruhe & Scholtes, 1956; Chadwick et al., 1995; Buol et al., 2011; Richter & Yaalon, 2012). These are major developments, given that soil formation was long thought to be a maturation process toward an equilibrium endpoint often under relatively stable environmental conditions (Marbut, 1935). Today, these ideas seem stiffly linear and deterministic. Soil evolution and polygenesis make soil formation a remarkably nonlinear, unpredictable, and historically contingent process (Phillips et al., 1996).

As soils owe much to processes in the present and the past, a metaphor of palimpsest, ancient Greek for overprint, is entirely applicable. Targulian and Goryachkin (2004) suggested that soils are evolving palimpsests, as opposed to systems that mature chronologically toward endpoints. Speaking metaphorically, most soils are not unlike ancient palimpsests made from animal skins that were written on, erased, and overwritten; skins used and re-used by different peoples writing different and evolving languages. Our pedogenetic task is obvious – to learn to read these various if complex languages.

That soils are evolving palimpsests brings a new dynamism to soil horizons and profiles; to soil features such as organo-mineral complexes, secondary minerals, aggregates, clay skins, complex surface areas, soil pans, and pore and root networks; and to soil properties such as pH, CEC, soil organic matter, and weatherable minerals. As evolving palimpsests, soils are historic bodies as they contain records of past environments with soil features and properties that are constantly forming, persisting, and being erased (Targulian & Goryachkin, 2004; Richter & Yaalon, 2012). In the Anthropocene, human forcings represent an accelerated cultural wave of polygenesis – altering and erasing inherited properties and creating others that are new.

While it may be that pedology is conceptually prepared for soil changes in the Anthropocene, empirically pedology is not well prepared at all. Soil formation and soil time are being radically rescaled by human forcings and there is great need for field studies to support the future of pedology (Richter et al., 2011; Richter & Yaalon, 2012). Considering that about half of the Earth's soils are actively managed for agriculture, pasture, forestry, and for residential, industrial, transportation, and other uses (Ellis & Ramankutty, 2008), we have much to do to better quantify soil changes.

A recent international inventory indicates that several hundred field experiments (iscn.fluxdata.org/network/partner-networks/ltse/) are quantitatively observing human influence and testing critical hypotheses. Known as Long-Term Soil-Ecosystems Experiments, LTSEs, these field studies produce long-term soils data that are invaluable (Jenkinson, 1991; Tilman et al., 1994; Smith et al., 1997; Richter

& Markewitz, 2001; Sachs et al., 2010). LTSEs document soil changes associated with crop-yield gains and declines of intensive cropping systems that provide food for billions of human beings (Chakraborty et al., 2017). LTSE data constrain models of land-atmosphere exchanges of CO_2 and inform climate policy with observational data as well as model simulations (Smith, 2012). LTSEs directly demonstrate: the rates of soil acidification from acidic atmospheric pollutants (Markewitz et al., 1998), long-term nutrient-use efficiencies of intensive management systems (Ladha et al., 2011), and the rates at which legacy soil lead from gasoline and paint is dissipating in urban soils (Mielke et al., 2019).

While the importance of LTSEs in quantifying human influences on soils is amply demonstrated over time scales that range from years (Ward et al., 2021), decades (Mobley et al., 2019), and even centuries (Poulton & Johnston, 2021), these long-term studies require time to mature, and they are highly vulnerable to loss or abandonment. To quote the soil scientist, Ishaku Amapu of Nigeria who recently lost his 60-year LTSE of millet cultivation to university-building construction, "We must make the world's LTSE's work harder!"

Let us conclude with a concrete example of soils undergoing a polygenetic wave of human influence and consider these soils in the light of this essay – the soils of the Mezquital Valley just north of Mexico City. In the Mezquital, each year nearly 100,000 ha of cultivated soils are irrigated with about 2-meters of untreated liquid wastewater from Mexico City. Wastewater irrigation began about 1900 and has greatly expanded with the growth of the megacity. These soils are being transformed in many ways from the valley- to the smallest mineral-grain scale, and have been studied with a flux-process approach that has quantified inputs, outputs, transformations, and translocations (Siebe et al., 2016). Only by periodic soil and water analyses over decades time have depth-dependent soil changes begun to be understood, as the Mezquital soils yearly receive several 100 kgN/ha, and heavy loads of metals, pathogens, and novel chemicals as well. The presence of tens of thousands of farming families make soil and water monitoring absolutely critical for human health and safety. The ability to understand soil formation in soils such as these belongs to pedologists who understand the Mezquital Valley not as deviations but as a wastewater-soil types with a taxonomic home in *Soil Taxonomy* and the World Reference Base.

While it seems in many ways that we are well prepared for a pedology in the Anthropocene, we must better support networks of long-term soil experiments to be able to quantify soil changes as natural, historic, and cultural bodies.

References

Arnold, R. W., Szabolcs, I., & Targulian, V. O. (1990). *Global soil change*. International Institute for Applied Systems Analysis.
Bidwell, O. W., & Hole, F. D. (1965). Man as a factor of soil formation. *Soil Science, 99*, 65–72.
Buol, S. W., Southard, R. J., Graham, R. C., & McDaniel, P. A. (2011). *Soil genesis and classification*. Wiley.

Certini, G., & Scalenghe, R. (2011). Anthropogenic soils are the golden spikes for the Anthropocene. *The Holocene, 21,* 1269–1274.

Chadwick, O. A., Nettleton, W. D., & Staidl, G. J. (1995). Soil polygenesis as a function of Quaternary climate change, northern Great Basin, USA. *Geoderma, 68,* 1–26. https://doi.org/10.1016/0016-7061(95)00025-J

Chakraborty, D., Ladha, J. K., Rana, D. S., Jat, M. L., Gathala, M. K., Yadav, S., Rao, A. N., Ramesha, M. S., & Raman, A. (2017). A global analysis of alternative tillage and crop establishment practices for economically and environmentally efficient rice production. *Scientific Reports, 7,* 1–11. https://doi.org/10.1038/s41598-017-09742-9

Cline, M. G. (1961). The changing model of soil. *Soil Science Society of America Proceedings, 25,* 442–446.

Dudal, R., Nachtergaele, F., & Purnell, M. F. (2002). The human factor of soil formation. In: *17th world congress of soil science*, (CD-ROM, paper 93). .

Edgeworth, M., Richter, D. D., Waters, C., Haff, P., Neal, C., & Price, S. J. (2015). Diachronous beginnings of the Anthropocene: The lower bounding surface of anthropogenic deposits. *Anthropocene Review, 2,* 33–58.

Ellis, E. C., & Ramankutty, N. (2008). Putting people in the map: Anthropogenic biomes of the world. *Frontiers in Ecology and the Environment, 6,* 439–447.

Jenkinson, D. S. (1991). The Rothamsted long-term experiments: Are they still of use? *Agronomy Journal, 83,* 2–10.

Johnson, D. L., & Watson-Stegner, D. (1987). Evolution model of pedogenesis. *Soil Science, 143,* 349–366.

Ladha, J. K., Reddy, C. K., Padre, A. T., & van Kessel, C. (2011). Role of nitrogen fertilization in sustaining organic matter in cultivated soils. *Journal of Environmental Quality, 40,* 1756–1766.

Marbut, C. F. (1935). *Soils of the United States*. USDA.

Markewitz, D., Richter, D. D., Allen, H. L., & Urrego, J. B. (1998). Three decades of observed soil acidification in the Calhoun experimental Forest: Has acid rain made a difference? *Soil Science Society of America Journal, 62,* 1428–1439.

Mielke, H. W., Gonzales, C. R., Powell, E. T., Laidlaw, M. A., Berry, K. J., Mielke, P. W., & Egendorf, S. P. (2019). The concurrent decline of soil lead and children's blood lead in New Orleans. *Proceedings of the National Academy of Sciences, 116,* 22058–22064.

Mobley, M. L., Yang, Y., Yanai, R. D., Nelson, K. A., Bacon, A. R., Heine, P. R., & Richter, D. D. (2019). How to estimate statistically detectable trends in a time series: A study of soil carbon and nutrient concentrations at the Calhoun LTSE. *Soil Science Society of America Journal, 83,* 133–140.

Phillips, J. D., Perry, D., Garbee, A. R., Carey, K., Stein, D., Mordy, M. B., & Sheehy, J. A. (1996). Deterministic uncertainty and complex pedogenesis in some Pleistocene dune soils. *Geoderma, 73,* 147–164.

Poulton, P. R., & Johnston, A. E. (2021). Can long-term experiments help us understand, and manage, the wider landscape—Examples from Rothamsted, England. In *Exploring and optimizing agricultural landscapes* (pp. 233–252). Springer.

Richter, D. D. (2020). Game changer in soil science. The Anthropocene in soil science and pedology. *Journal of Plant Nutrition and Soil Science, 183,* 5–11.

Richter, D. D., & 30 co-authors. (2011). Human-soil relations are changing rapidly: Proposals from SSSA's cross-divisional soil change working group. *Soil Science Society of America Journal, 75,* 2079–2084.

Richter, D. D., & Markewitz, D. (2001). *Understanding Soil Change*. Cambridge University Press.

Richter, D. D., & Yaalon, D. (2012). "The changing model of soil" revisited. *Soil Science Society of America Journal, 76,* 766–778.

Ruhe, R. V., & Scholtes, W. H. (1956). Age and development of soil landscapes in relation to climate and vegetational changes in Iowa. *Soil Science Society of America Proceedings, 20,* 264–273.

Sachs, J., & 24 co-authors. (2010). Monitoring the world's agriculture. *Nature, 466,* 558–560.

Schad, P. (2018). Technosols in the world Reference Base for soil resources—History and definitions. *Soil Science and Plant Nutrition, 64*, 138–144.

Siebe, C., Chapela-Lara, M., Cayetano-Salazar, M., Prado, B., & Siemens, J. (2016). Effects of 100 years of soil-aquifer-treatment of Mexico City's wastewater in the Mezquital Valley. In H. Hettiarachchi & R. Ardakanian (Eds.), *Safe use of wastewater in agriculture: Good practice examples*. UNU-FLORES. Simonson 1959.

Simonson, R. W. (1959). Outline of a generalized theory of soil genesis. *Soil Science Society of America Journal, 23*, 152–156.

Smith, P., & 20 co-authors. (2012). Towards an integrated global framework to assess the impacts of land use and management change on soil carbon: Current capability and future vision. *Global Change Biology, 18*, 2089–2101.

Smith, P., Powlson, D. S., Smith, J. U., Elliott, E. T., & editors. (1997). Evaluation and comparison of soil organic matter models. *Geoderma, 81*, 1–225.

Soil Survey Staff. (2014). *Keys to soil taxonomy* (12th ed.). Natural Resources Conservation Service, USDA.

Targulian, V. O., & Goryachkin, S. V. (2004). Soil memory: Types of records, carriers, hierarchy and diversity. *Revista Mexicana Ciencias Geologia., 21*, 1–8.

Tilman, D., Dodd, M. E., Silvertown, J., Poulton, P. R., Johnston, A. E., & Crawley, M. J. (1994). The park grass experiment: Insights from the most long-term ecological study. In R. A. Leigh & A. E. Johnston (Eds.), *Long-term experiments in agricultural and ecological sciences* (pp. 287–303). CAB International.

Wade, A. M., Richter, D. D., Craft, C. B., Bao, N. Y., Heine, P. R., Osteen, M. C., & Tan, K. G. (2021). Urban-soil pedogenesis drives contrasting legacies of lead from paint and gasoline in city soil. *Environmental Science and Technology, 55*, 7981–7989.

Ward, E. B., Doroski, D. A., Felson, A. J., Hallett, R. A., Oldfield, E. E., Kuebbing, S. E., & Bradford, M. A. (2021). Positive long-term impacts of restoration on soils in an experimental urban forest. *Ecological Applications, 31*, e02336. https://doi.org/10.1002/eap.2336

Yaalon, D. H., & Yaron, B. (1966). Framework for man-made soil changes — An outline of metapedogenesis. *Soil Science, 102*, 272–278.

Daniel D. Richter is Professor of Soils at Duke University. He is author of nearly 200 peer-review papers and the book *Understanding Soil Change* (Cambridge) co-authored with Daniel Markewitz of the University of Georgia. He has taught soil science, pedology, and ecosystem ecology at the University of Michigan and Duke University. He is a member of the Anthropocene Working Group and the Lead Principal Investigator of the Calhoun Critical Zone Observatory in the Southern Piedmont of South Carolina.

Water

Armin Grunwald

Abstract Water as a crucial precondition of life has become a prominent issues in almost all cultures and religions, in particular in regions with water scarcity. The modern attitude to water in the Western world is predominantly economic, regarding water as a resource despite it being a common good. In 2010, the United Nations approved the human right to a water supply and sanitation. In order to counter water pollution by agriculture, industry and households, and severe scarcity of water in large parts of the world, water ethics as well as sustainable and responsible water governance are needed.

Water is, viewed through the glasses of natural science, a chemical substance (H_2O). It is inorganic, tasteless and transparent. Water in its physical representations as ice, fluid and gas forms the Earth's hydrosphere, which is characterized by various cycles and processes. In energetic respect, the water cycle is driven by solar energy and resulting atmospheric processes. The water cycles include precipitation in form of rain, snow or fog, evaporation, transpiration and condensation, clouds as well as ground elements such as rivers, seas and oceans. Water covers about 70% of the surface of the Earth, which gave rise to speak of the Earth as the "Blue Planet", viewed from space. Water is unequally distributed over the World, with the spectrum reaching from extremely dry areas and deserts over arid and temperate zones up to tropical rain forests (Young et al., 1994). Climate change influences this distribution, often accompanied by aridification on the one side, and extreme irrigation events and floods, on the other.

All known forms of life depend on water as a solvent. Water itself neither provides energy nor serves as nutrient. Rather, its property as solvent only makes water vital for all known forms of life. Without water, life is not possible. During usage, water is not consumed but remains the molecule H_2O. Therefore, water is, in principle, a renewable resource. However, water usually is polluted during the usage,

A. Grunwald (✉)
Karlsruhe Institute of Technology, Karlsruhe, Germany
e-mail: armin.grunwald@kit.edu

© The Author(s), under exclusive license to Springer Nature Switzerland AG 2023
N. Wallenhorst, C. Wulf (eds.), *Handbook of the Anthropocene*,
https://doi.org/10.1007/978-3-031-25910-4_30

189

e.g. by wastes and chemicals from households, agriculture and industry, producing wastewater needing sanitation technologies.

Water plays a pivotal role also in the economy at all levels from local to global (Young et al., 1994). Its main user is agriculture as the main source for human nutrition, which consumes about 70% of the freshwater used by humans. Fishery in salt and fresh water also is an essential contribution to global food supply depending on water. Water also provides major transportation routes through oceans, seas, rivers, lakes, and canals, for example for oil and manufactured products. Water in form of ice and steam is used for cooling and heating in industry and homes. Because water is an excellent solvent for both mineral and organic materials, it is widely used in industrial processes as well as for cooking and washing in households. Water is not only a crucial solvent for living organisms for enabling their internal processes but also for human economy. For freshwater supply for the various purposes as well as for sanitation and treating wastewater, complex infrastructures have been established in many countries, in particular in the industrialized ones, operated by private companies or public institutions (Hüttl et al., 2016).

Water as a crucial precondition of life has gained prominent importance in almost all cultures and religions (Grunwald, 2016). In particular, high value is assigned to water in regions with scarcity of water, often accompanied by strict rules on its usage. The three monotheistic religion, Judaism, Christianity, and Islam, have their roots in areas with a harsh desert climate and water scarcity. The Bible and the Quran are full of symbols related with water, pointing to water as source of life, as symbol for purification, or using analogies with water as metaphor for spiritual messages. In Judaism, the Talmud perspective states that while humans may use the world for their needs, they may never irresponsibly damage or destroy the environment. In contemporary Christianity taking care for water resources is seen as part of the duty of stewardship of God's creation building our common home (Pope Francis, 2015, §27–31). In Islam, water is a pivotal issue. It should neither be bought nor sold (Al-Awar et al., 2010, p. 33) but rather belongs to the community. Therefore, no one is allowed to own it unless labour was provided to make it usable, or to distribute it. This labour does not create a right of ownership over water, but rather creates a property on the *value added* to water by labour, which enables pricing and trade only. In Hinduism, water is regarded holy and representing God in a spiritual sense. Therefore, water is believed to provide purifying and cleansing power. In Buddhism, water is a key symbol of life and is, as such, regarded the particular element being able to bring everything together.

In sharp contrast to these cultural roots, the capitalist economic system, which has developed in the Western World over the past two centuries only, regards the value of pieces of the natural environment primarily as *monetary* value. In the assignment of value to water this means: "Water has an economic value in all its competing uses and should be recognized as an economic good" (Young et al., 1994, p. 4). Key to the economic view of water is demarcating it *as a resource*. The US American geographer William McGee expressed in a famous paper (1909) very clearly and early Western thinking about water as an economic resource. In more recent formulation this reads: "Managing water as an economic good is an important way of achieving efficient and equitable use, and of encouraging conservation and protection of water resources" (Young et al., 1994, p. 4). However, this management perspective is only half of the story. Already McGee and his community were open to communitarian ideas and participation, based on the conviction that water belongs to the people. Thus, the attitude of Western modernity to water shows an ambiguity with an inherent tension: on the one hand, the value of water is reduced to that of an economic resource, while on the other also the political and democratic side of water is recognized.

Philosophically, this ambiguity was removed in favour of regarding water as a part of the common goods, motivated by the omnipresent and crucial significance of water for any and in particular human life (Ostrom et al., 2003). While this was a more academic assignment, the United Nations followed this approach at the political level and approved the access to fresh water supply and sanitation as human right. This significant step influenced the United Nations' Sustainable Development Goals postulating: *Ensure access to water and sanitation for all.* Its point of departure is: "Worldwide, one in three people do not have access to safe drinking water, two out of five people do not have a basic hand-washing facility with soap and water, and more than 673 million people still practice open defecation" (UN, 2021).

However, water resources on planet Earth are under increasing pressure and in risk of continuous degradation, according to industrialization and unsustainable economic growth. Water is polluted by wastes and chemicals produced by households, agriculture, and industry. The oceans are burdened e.g. with micro-plastics, wastes from ships and incoming rivers and through acidification caused by the higher carbon-dioxide concentration in the atmosphere. The quality of groundwater and surface freshwater is threatened by pollution from numerous sources, e.g. by chemicals from industry and agriculture such as heavy metals, pesticides, nitrates, and drug residuals from veterinaries and hospitals. Also mining, e.g. for copper and uranium, often leads to water pollution in rivers and groundwater. Groundwater is under pressure also through over-usage in many regions of the world, often to an increasing extent according to climate change. Water scarcity in combination with population growth and industrialization leads to water conflicts, e.g. between Israel and Palestine, Ethiopia and Sudan, and in Central Asia.

The increasing pressure on water resources by ongoing industrialization as well as by climate change and lack of access to freshwater and sanitation for billions of people motivated the call for a new and global water ethics (Brown & Schmidt, 2010; Groenfeldt, 2013). By postulating "A thing is right if it preserves or enhances the ability of the water within the ecosystem to sustain life; and wrong if it decreases that ability" (Armstrong, 2006, p. 13) the awareness of the vulnerability of water systems and the willingness to protecting them shall be enhanced (e.g. Groenfeldt, 2013; UNESCO, 2011). The following set of principles seems to be appropriate to be used as guiding heuristics in water ethics (following Grunwald, 2016):

Human Right to Water and Sanitation: The human right to water and sanitation (UNESCO, 2011) is an implication of the postulate of human dignity not only because water is necessary for survival but also for food production, energy, and human culture. It is closely related to regarding water as a common good and to the value dimension of water, e.g. in Islam.

Sustaining Ecosystem Functions: Ecosystem functions are essential for "keeping nature alive" (Groenfeldt, 2013). Sustaining these functions is an essential element of the imperative of sustainable development and of human long-term responsibility.

Responsible use of Water: Making use of water in a responsible manner is needed for a globally sustainable development. The actual use of water must not endanger the future fulfilment of the right to water, and ecosystem functions must be sustained. These principles may limit the industrial and agricultural use of water in specific cases.

Participatory Water Governance: The distribution of freshwater, the usage of waters for agriculture, industry and energy as well as sanitation and the treatment of wastewaters need care and governance. According to ethical principles of fairness and equity as well to democratic principles of inclusion and deliberation water governance should be participatory, involving people concerned and affected, as well as taking into account future generations.

References

Al-Awar, F., Abdulrazzak, M., & Al-Weshah, R. (2010). Water ethics perspectives in the Arab Region. In P. G. Brown & J. Schmidt (Eds.), *Water ethics. Foundational readings for students and professionals* (pp. 29–38). Island Press.

Armstrong, A. (2006). Ethical issues in water use and sustainability. *Area, 38*, 9–15. https://doi.org/10.1111/j.1475-4762.2006.00657.x

Brown, P. G., & Schmidt, J. (Eds.). (2010). *Water ethics. Foundational readings for students and professionals*. Island Press.

Francis, P. (2015). *Laudato Si. Encyclical letter on care for our common home*. https://www.vatican.va/content/francesco/en/encyclicals/documents/papa-francesco_20150524_enciclica-laudato-si.html. (31.7.2021).

Groenfeldt, D. (2013). Towards a new water ethic. In A. Bhaduri, E. Flinkerbusch, T. Holtermann, & S. Marx (Eds.), *The Bonn declaration on global water security* (pp. 14–15). Bonn.

Grunwald, A. (2016). Water ethics – Orientation for water conflicts as part of inter- and transdisciplinary deliberation. In R. F. Hüttl, O. Bens, C. Bismuth, & S. Hoechstetter (Eds.), *Society – Water – Technology: A critical appraisal of major water engineering projects* (pp. 11–29). Springer.

Hüttl, R. F., Bens, O., Bismuth, C., & Hoechstetter, S. (Eds.). (2016). *Society – Water – Technology: A critical appraisal of major water engineering projects*. Springer.

McGee, W. J. (1909). Water as a resource. *Annals of the American Academy of Political Science, 33*, 37–50.

Ostrom, E., Stern, P. C., & Dietz, T. (2003). Water rights and the commons. In P. Brown & J. Schmidt (Eds.), *Water ethics. Foundational readings for students and professionals* (pp. 147–154). Island Press.

UN – United Nations. (2021). *Sustainable development goals*. Goal No. 6. https://www.un.org/sustainabledevelopment/water-and-sanitation/. (29.7.2021).

UNESCO. (2011). *Water ethics and water resource management*. http://unesdoc.unesco.org/images/0019/001922/192256e.pdf. (1.8.2021).

Young, G. J., Dooge, J. C., & Rodda, J. C. (1994). *Global water resources issues*. Cambridge University Press.

Armin Grunwald is Full Professor of Philosophy and Ethics of Technology at Karlsruhe Institute of Technology (KIT), Germany. He is director of the Institute for Technology Assessment and Systems Analysis (ITAS) at KIT and Director of the Office of Technology Assessment at the German *Bundestag*. His professional backgrounds include theory and practice of technology assessment, ethics of technology and approaches to sustainable development. Armin Grunwald is member of several advisory commissions and committees in various fields of the technological advance, e.g. of the German Ethics Council.

Part III
Life in Abundance

Animal

Thomas Macho

Abstract The question remains controversial as to when the Anthropocene actually began: with the use of fire, with the Neolithic Revolution, or with industrialization and capitalism. But certainly the described ruptures have permanently changed the respective relationships between humans and animals, in the ages of hunting, sacrifice and mass slaughter. At present, we live in a deeply divided world, marked on the one hand by the extinction of species and the cruel techniques of factory farming for increasing meat consumption, and on the other hand by imaginings of an empathic fascination and affection for the animals that populate innumerable households, especially as pets. Can we hope for the dawn of an age of the living after the Anthropocene?

When did the Anthropocene begin? The question is not easy to decide upon (Horn & Bergthaller, 2020, pp. 19–34). Some theories designate the use of fossil energy and the Industrial Revolution as the threshold of the Anthropocene. On the other hand, some theories consider the neolithic period to be the first stage of the Anthropocene, which had already begun around 14,000 years ago in the course of the gradual transition to a sedentary life, i.e., with the domestication of different species, with agriculture and livestock breeding, and the founding of settlements and cities. James C. Scott, in turn, declares the use and mastery of fire – around 400,000 years ago – to be the significant turning point in hominid evolution (Scott, 2017, pp. 37–67). Last but not least, the Anthropocene is described as a sixth mass extinction of numerous animal and plant species (Kolbert, 2014), whereby the exact dating of its beginning seems comparatively secondary given its current acceleration. The *Living Planet Report* for example – published in September 2020 by the environmental foundation World Wide Fund for Nature (WWF) – reported a 70% decline of 14,000 animal populations within 50 years. In May 2019, the U.N. World Biodiversity Council (Intergovernmental Science-Policy Platform on Biodiversity and Ecosystem Services) had published a report with the prognosis that up to one

T. Macho (✉)
International Research Center for Cultural Studies (IFK), Vienna, Austria

197

N. Wallenhorst, C. Wulf (eds.), *Handbook of the Anthropocene*,
https://doi.org/10.1007/978-3-031-25910-4_31

million species were at risk of extinction; around 500,000 species were already classified as "dead species walking" (IPBES, 2019, p. 241).The Anthropocene is the age of a mass species extinction, the reasons for which are largely known: population growth, climate crisis, and the destruction of former animal and plant habitats, such as the rain forests in the Amazon regions. Numerous zoonoses are also attributed to the destruction of these habitats, the transmission of viruses and microorganisms to other animal species and humans; who knows whether the human species itself could now not also be considered a "dead species walking". In any case, the growing threat due to the species extinction, which some organizations consider more dangerous than climate change, is indisputable.

The species extinction can be juxtaposed with the increasing reproduction of animals – cattle, pigs, poultry – for the worldwide meat industry. Recent statistics show that there are ten times as much livestock as there are humans worldwide, in 2020 around 68.8 billion chickens, ducks, geese, and other poultry, 11.8 billion cattle, sheep, and goats, and 1.5 billion pigs. More than half of the agricultural land is used to grow feed for these animals, not to fight famine, instead – in a roundabout way – for the global consumption of meat and dairy products. And this consumption is ever increasing, so that the global meat consumption could increase by another 40 million tons to more than 360 million tons of meat per year by 2029 (Fleischatlas, 2021, p. 7). The consequences of this consumption also include the climate crisis. In *Eating Animals* (2009), Jonathan Safran Foer emphasizes that the agricultural keeping of livestock contributes 40% more to global warming than all of global transport; he considers it a leading cause of the climate crisis (Foer, 2009). Although we know these figures, the annual consumption per capita in the industrialized countries, i.e., in Canada, USA, Europe, Russia, Japan, Israel, South Africa, Australia, and New Zealand, is around 68.6 kg, and around 26.6 kg in all other countries. This shows that poorer countries – despite populations over five times as large – do not even consume (including waste) twice as much meat as the wealthier ones. Although the share of cattle and sheep is decreasing, people are eating more and more pork and poultry, not only in the USA or Europe, but also in China, for example. For the year 2018, the German Federal Statistics Office states an average meat consumption (including wastage) of 78.8 kg per year in Germany, 123.2 kg in the USA, 62.4 kg in China, but only 4.1 kg in India. Meat consumption often does not include meat wastage; in the statistics of meat consumption, meat waste should also be included: in 2015 around 235,000 tons in Germany, 150,000 tons until slaughter (about a quarter of the pigs already die before their slaughter), 32,000 tons during processing, and 53,000 tons in trade. This does not take into account the amount of meat thrown away in restaurants and households (Fleischatlas, 2021, p. 41).

The Anthropocene world is deeply divided. Considerably less meat is consumed in poorer countries than in the wealthier, industrialized countries of the Global North; women eat significantly less meat than men, and a growing number of younger people now have vegetarian or vegan diets. Even more dramatic appears the divide between the industrial practices of cruel factory farming and our collective imagination. While ecological, partly species-appropriate livestock keeping in the European Union affects just 6% of cattle and 1% of pigs (cf. German Federal Statistics Office), sentimental films, novels, children's books, and stuffed animals to cuddle with are produced daily. On social media websites we find countless animal videos which seem to testify to our empathy towards them. The market for pets is growing, especially in wealthy countries: In 2018, a total of 34.4 million pets lived in 45% of all German households: 14.8 million cats, 9.4 million dogs, 5.4 million small animals, 4.8 million birds, ornamental fish in 1.9 million aquariums, and reptiles in one million terrariums. In the USA, there were about 90.5 million families who had a pet at the end of 2021: 69 million dogs, 45.3 million cats, 9.9 million birds, and 3.5 million horses. Meanwhile, the pets – often newly bred to meet the needs of families and children – resemble their inanimate counterparts out of a toy catalogue, with big eyes and cute noses. They are primarily meant to fulfill psychological needs: the longing for companions, siblings, friends. Only rarely though may they be perceived as independent living beings, with needs that are possibly only inadequately adapted to their complex social roles. Before genetic engineering moves on to crafting our ideal dream animals, animals have been relegated to carriers of cultural symptoms and projections that are simply discarded, abandoned, or euthanized when they fail.

A brief glance into the past: The first definition of the Anthropocene, according to which the age of humans began with the use of fire, can be put into relation with a long history of hunting animals; the second definition, which identifies the Neolithic Revolution as the beginning of such a geological era, could seek evidence in the history of agrarian societies, cults, religions, and domesticated animals, as well as of animal sacrifice. The third definition traces the Anthropocene back to the Industrial Revolution and the use of fossil energy. These three definitions correspond to three societally regulated and recognized forms of killing an animal: in hunting, in the context of sacrificial rituals, or in slaughter. Certainly, these three forms were occasionally superimposed upon one another; and yet, they are characteristic of three different ways of life in the human species. Hunting and gathering cultures mostly admired, respected, and hardly ever sacrificed animals; in agrarian societies, on the other hand, hunting mostly remained reserved to a small elite. The technically optimized and mechanized mass slaughter of animals can ultimately be considered as a feature of industrial modernity. Animal sacrifice thus forms the historical middle ground between hunting and slaughter; in this sense, it is not surprising that it shares certain traits with both. Specific rituals of pardon link sacrifice to hunting, and public slaughter is often performed as a part of animal sacrifice. What principally differentiates sacrifice from hunting or slaughter, however, is its exceptionality. Sacrifices are not practiced on everyday occasions; rather, they interrupt the societal routine. Wild animals were hunted; in most cases the animals which were sacrificed were domestic animals. In industrial societies, on the other hand, all animals are slaughtered equally: in terms of the utilization of meat, the difference between wild and domesticated animals practically no longer plays a roll. The domesticated animal of agrarian ways of life have disappeared now for more than

150 years; they have been diversified into farm animals, to which the people have no relationship, and pets, which in turn no longer need to meet any demands of usefulness. Pets (*Heimtiere*) are not domestic animals (*Haustiere*), for the term domestic animal must be taken literally: It denotes a living being that shares a house (lat.: *domos*) with humans, namely not only a territory, but a 'house of language'. Humans obey the same imperatives as their domestic animals; they are subject to the same constraints and working conditions, in agriculture, war, cult, or love. Animal and human sacrifice could therefore often be converted into one another; many myths – the tales of Isaac or Iphigenia – emphasize their mutual possibility to be substituted. We are far removed from such an understanding now.

A brief outlook: What comes after the Anthropocene, whose end seems so ominously close? An uninhabitable planet? Or a new geological era of the living, the encounter, and solidarity with all living beings? In her *Manifeste animaliste* (2017) the French philosopher, environmental and animal ethicist Corine Pelluchon writes: "Our relations to animals test our capacity to experience the common destiny which binds us to other living beings. They also indicate our difficulty to accept alterity. It is a question of war against animals, but also a war against ourselves and amongst ourselves. Therefore, the animal question is and will remain of central importance." (D.S.) (« Nos relations aux animaux mettent à l'épreuve notre capacité à ressentir la communauté de destin qui nous relie aux autres vivants. Elles indiquent également les difficultés que nous avons à accepter l'altérité. Il s'agit d'une guerre contre les animaux, mais c'est aussi une guerre contre nous-mêmes et entre nous. C'est pourquoi la question animale est centrale et le restera […] » p. 10.) And she emphasizes the perception of a vulnerability that we share with all mortal creatures. Pelluchon does not base her argument upon geological eras, but she distinguishes several eras in the history of ideas, namely a first era of religions and theology, a second era of the philosophy of history and political utopias, as well as a third era of nihilism, despair, and of a violence that is practiced above all against the most vulnerable living beings who cannot defend themselves. This violence, practiced in slaughterhouses, experimental laboratories, or wars, arises from a "complete and uninhibited lack of respect for the living" (D.S.) (« un irrespect total et décomplexé du vivant » p. 35.) But a fourth era is dawning, one that Pelluchon characterizes as an "age of the living" (D.S.) (« l'âge du vivant »): "Joining respect of the living and the acceptance of our vulnerability, the new thought voices the aspirations of many people who fear the forms that the obsession of domination is currently taking." (D.S.) (« Associant le respect du vivant et l'acceptation de notre vulnérabilité, elle [la nouvelle pensée] traduit les aspirations de nombreuses personnes qui redoutent les formes que prend actuellement l'obsession de la maîtrise […] » p. 36.) She underlines the connection between justice towards animals and reconciliation with ourselves. (« le lien entre la justice envers les animaux et la réconciliation avec nous-mêmes […] » p. 36.)

Translated by David Saatjian.

References

Fleischatlas. (2021). *Daten und Fakten über Tiere als Nahrungsmittel.* In Heinrich-Böll-Stiftung, Bund für Umwelt und Naturschutz Deutschland and Le Monde diplomatique (Eds.).

Foer, J. S. (2009). *Foer: Eating animals.* Little, Brown and Company.

Horn, E., & Bergthaller, H. (Eds.). (2020). *The anthropocene. Key issues for the humanities.* Routledge.

IPBES. (2019). *Global assessment report of the intergovernmental science-policy platform on biodiversity and ecosystem services.* In E. S. Brondízio, J. Settele, S. Díaz & H. Thu (Eds.). Ngo.

Kolbert, E. (2014). *Kolbert: The sixth extinction. An unnatural history.* Henry Holt.

Pelluchon, C. (2020). *Manifest für die Tiere.* (Michael Bischoff, Trans.). C.H. Beck.

Pelluchon, C. (2021). Das Zeitalter des Lebendigen. *Eine neue Philosophie der Aufklärung.* (Ulrike Bischoff, Trans.). Wissenschaftliche Buchgesellschaft.

Scott, J. C. (2017). *Against the grain. A deep history of the earliest states.* Yale University Press.

Statistisches Bundesamt: Globale Tierhaltung, Fleischproduktion und Fleischkonsum. Ausgabe 2021 [cf https://www.destatis.de/DE/Themen/Laender-Regionen/Internationales/Thema/landwirtschaft-fischerei/tierhaltung-fleischkonsum/_inhalt.html]. Wiesbaden 2021.

Thomas Macho served as professor of cultural history at the Institute for Cultural Studies at Humboldt University in Berlin from 1993 to 2016. Since 2016, he has directed the International Research Center for Cultural Studies (IFK) in Vienna. 2019 he was awarded the Sigmund Freud Prize for scientific prose by the German Academy for Language and Poetry, and in 2020 the Austrian State Prize for Cultural Journalism.

Anthromes

John E. Quinn and Erle C. Ellis

Abstract Anthromes, or anthropogenic biomes, characterize the globally significant ecological patterns shaped by sustained direct human interactions with ecosystems, including agriculture, urbanization, and other land uses. The emergence of anthromes has literally paved the way for the Anthropocene, and now cover more than three quarters of Earth's ice-free land surface, including dense settlements, villages, croplands, rangelands, and cultured lands; wildlands untransformed by agriculture and settlements cover the remaining area.

Human societies and their use of ecosystems have transformed ecological patterns and processes globally for thousands of years (Ellis, 2021). Lands used to sustain hunter gatherers, farmers, pastoralists, foresters, and others have been extensive for millennia, covering nearly three quarters of Earth's land surface for at least 12,000 years (Ellis et al., 2021). Likewise, the cities and towns where most people now live, though less extensive, have an outsized influence on the landscapes around them (McDonald et al., 2016), Consequently, ecosystems, at local to global extents, can no longer be understood without considering how humans have altered them. To address this gap, efforts to deepen scientific understanding of the ecological patterns and processes shaped and sustained by human societies are ongoing at global, regional and local scales, (Ellis, 2015, 2018; Brown & Quinn, 2018; Waters et al., 2016; Stephens et al., 2019).

Anthromes, or anthropogenic biomes, have aided this process by characterizing globally significant ecological patterns created by sustained direct human interactions with ecosystems, including agriculture, urbanization, and other land uses

J. E. Quinn (✉)
Department of Biology, Furman University, Greenville, SC, USA
e-mail: john.quinn@furman.edu

E. C. Ellis
Department of Geography and Environmental Systems,
University of Maryland Baltimore County, Baltimore, MD, USA
e-mail: ece@umbc.edu

N. Wallenhorst, C. Wulf (eds.), *Handbook of the Anthropocene*,
https://doi.org/10.1007/978-3-031-25910-4_32

203

(Ellis & Ramankutty, 2008). Defined as dense settlements, villages, croplands, rangelands, and cultured lands, anthromes have been mapped across Earth's land surface with remaining areas then identified as wildlands without evidence of human populations or intensive land use.

Global changes in anthromes have been mapped based on an increasing wealth of spatial data from remote sensing, government statistics, archaeological evidence, and other sources, including human-altered vegetation cover, built structures, crops, livestock grazing, irrigation, roads and the varying densities of human populations (Ordway et al., 2021). Leveraging these data in 2008, Ellis and Ramankutty made the first map of anthromes using a statistical approach called cluster analysis (Ellis & Ramankutty, 2008). Using this approach, the globally significant patterns of human shaped ecosystems were mapped from data on human populations, land use and vegetation cover. This effort complemented classic maps of natural biomes, from tundra to tropical rainforests, in relation to the globally significant patterns of climate, terrain, and other natural environmental conditions. In 2010, this approach was updated to allow anthromes to be mapped globally over time, from 1700 to 2000 (Ellis et al., 2010). Anthrome maps based on this system of classification have been used widely in teaching, research, and conservation (e.g. Chapin III et al., 2012; Martin et al., 2014; Merritts et al., 2014; National Geographic Society, 2014; Quinn et al., 2014; Miraldo et al., 2016; Gibson & Quinn, 2017; Dinerstein et al., 2017; Smith et al., 2019).

More recently, this system of classification was extended further back in time to map anthromes across the 12,000 years from 10,000 BCE to 2017 CE at 60 time points (Fig. 1, Ellis et al., 2021). Using this system of anthrome classification, six "levels" along a gradient of intensiveness were recognized and mapped globally, as illustrated in Fig. 1. This classification was used recently to show the temporal extent of transformation and the legacy of past changes on nature (Ellis et al., 2021) and to describe variations in birds, mammals, and amphibians between anthromes globally and regionally (Quinn et al., 2021).

Anthromes Along a Gradient of Transformation and Intensification

Urban and other densely settled anthromes are the most highly transformed, sustaining urban population densities well over 1000 persons per square kilometer in landscapes largely converted into cities, suburbs, and other residential and industrial infrastructure. Village anthromes are also highly transformed and densely populated (above 100 persons per square kilometer), but their mostly agricultural landscapes sustain largely rural populations, often in regions inhabited and farmed since ancient times. While these urban and village anthromes covered only about 8% of Earth's ice-free land in 2000 CE, more than 80% of Earth's human populations lived in them. Moreover, ongoing migrations from rural areas into cities are making urban

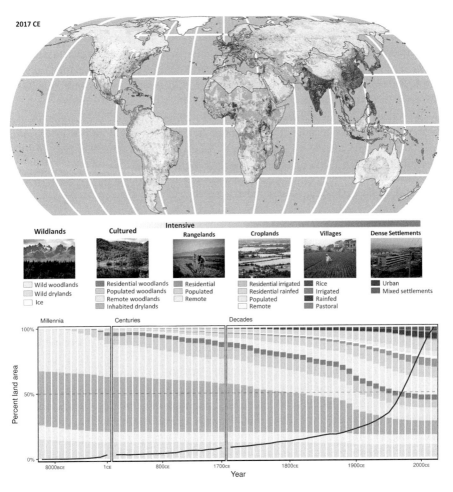

Fig. 1 Global map of anthromes for 2017 CE, and estimated global changes in anthrome areas from 10,000 BCE to 2017 CE, based on Ellis et al., 2021. Red line is global population. Map in Eckert 4 projection (BCE; Before the Common Era, CE; Common Era)

anthromes even denser as human populations continue to concentrate into and expand Earth's relatively small but dynamic area of urban anthromes.

The arable farm fields that produce most food, fibre, and fuel are mostly found in croplands anthromes. In some croplands, crops cover the land almost completely. More commonly though, are croplands formed of a complex mix of crops and settlements together with patches of grass, shrubs, and tree cover managed at lower intensities. Rangeland anthromes sustain grazing livestock and are the most extensive of the intensive anthromes. While these cover more than a quarter of Earth's ice-free land, they also tend to be lower productivity areas with relatively low human population densities and are usually less transformed by human use.

Cultured anthromes are relatively lightly populated and less intensively used landscapes, with a majority of their areas composed of habitats either remaining without direct human use, or modified through uses that maintain native vegetation cover. The spatial extent of these anthromes have declined greatly overtime, mostly through the appropriation, colonization, and intensification of use in lands inhabited and used by prior societies.

Wildlands, without human populations or intensive use of land, have always been rare, covering only about one quarter of Earth's ice-free land over the past 12,000 years, and tend to remain mostly in relatively low productivity regions, like deserts and tundra. As a result, wildlands account for only about 10% of Earth's terrestrial net primary productivity (NPP) even less than one would expect based on their area (Ellis et al., 2010).

Anthromes Are Mosaics of Used and Novel Ecosystems

Scaling down from global patterns, anthromes are best described as heterogeneous multifunctional mosaics combining different land uses and land covers. Today, anthromes retain a greater global area of lightly used habitats, about 37% globally, than the total area remaining in wildlands (around 23%). Moreover, these areas tend to be both ecologically productive and highly biodiverse (Ellis et al., 2012, 2021; Ellis, 2021; Quinn et al., 2021). Remarkably, in 2000, the total area of wild forests remaining on Earth was less than the total area of densely populated village and urban anthromes. More remarkable still, more than 25% of Earth's tree cover is embedded within cropland anthromes, a greater extent than the total area of forests remaining in wildlands (Ellis et al., 2010). This landscape-scale heterogeneity supports biodiversity patterns that vary within and between anthromes with lower species richness in barren and wildlands and higher richness in villages and rangelands (Ellis, 2013; Quinn et al., 2021). Clearly, the intensively used and heterogeneous working landscapes of anthromes demand serious attention in understanding and conserving Earth's ecology (Garibaldi et al., 2021).

Visualizing landscapes across the anthromes (Fig. 1) we see mixtures of native, cultured, and intensively used vegetation that offer diverse benefits to people and other living organisms. Even within cities and intensive farmland, the most densely populated and transformed anthromes, humans rarely convert all land for direct uses leaving small pockets of sometimes unexpected biodiversity (Ellis, 2013, 2019). While transformation and direct use of a given landscape tends to be mostly concentrated in areas where it will be most valuable (e.g., rich grassland soils, floodplains, and wetlands), in adjacent areas people cultivate trees and tend preferred species, create open space, alter fire patterns, and harvest fuelwood, timber, wildlife, plants and other natural resources. Similarly, because of global trade, many ecosystems have increased exposure to and invasion by exotic species. Because of these changes even the least disturbed areas of anthrome landscapes generally have novel biotic communities and ecosystem processes that differ substantially and potentially irreversibly from their prior historical states, broadly fitting the definition of novel ecosystems (Hobbs et al., 2009; Ellis et al., 2010; Ellis, 2015). These novel cultural ecological patterns and processes of anthrome landscapes now represent the terrestrial biosphere in its current, human-altered form, and provide the basis for research, education, and application of ecology and conservation in the Anthropocene (Ellis, 2015).

Ecological Science, Theory, and Education in the Anthropocene

From the basic theoretical view, the anthrome model of global ecology has implications for understanding the fundamentals of ecological science from local landscapes to regional and global scales. Anthromes move ecology away from an outdated view of the world as "natural ecosystems with humans disturbing them" and towards a vision of "human systems entangled with ecosystems" (also known as social-ecological systems, anthroecosystems, or coupled human and natural systems) (Ellis & Ramankutty, 2008; Ellis, 2015). This is a major paradigm shift for most natural scientists concerned with ecology and environment, but it is a shift that will be critical to advance efforts to better understand and help to guide more beneficial human interactions with the biosphere today and in the future.

By incorporating human transformation of terrestrial ecology into a global framework, anthromes offer a more objective view of the contemporary biosphere compared with the traditional biome frameworks taught in classes and emphasized in research (Martin et al., 2012). Nevertheless, anthromes should not be seen as a replacement for the classic biome model, but rather a complimentary model, and this is increasingly being presented and discussed in biology and ecology textbooks (Chapin III et al., 2012; Merritts et al., 2014; Freeman et al., 2016), and in references across a wide range of over 100 different academic disciplines (Fig. 2).

Conservation in the Anthropocene

Human transformation of this planet is causing unprecedented and accelerating changes in Earth's ecology, from climate change to the massive losses of habitat that are causing species extinctions and alarming declines in wildlife populations. While focusing on the negative consequences of human transformation of Earth can help to eliminate or avoid some of these, it is equally, if not more important, to emphasize the unprecedented powers of contemporary human societies to create a better future for both people and nonhuman nature on the only planet we share; the vision of Earth stewardship (Chapin III et al., 2012; DeFries et al., 2012).

With anthromes covering more than three quarters of the terrestrial biosphere, the need to conserve biodiversity and nonhuman habitats in anthrome landscapes is

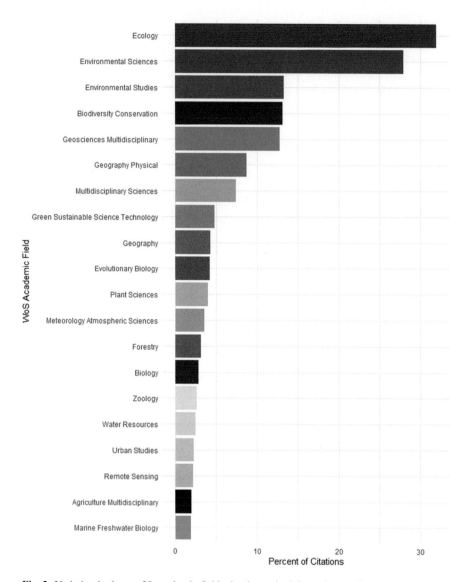

Fig. 2 Variation in the top 20 academic fields that have cited the anthrome literature. (Data from web of science)

increasingly recognized as critical (Martin et al., 2014). Though wildlands are important habitats for biodiversity conservation in large-scale protected areas, anthromes have largely replaced wildlands in Earth's most biodiverse and productive regions (Ellis et al., 2021). Even though many species are poorly adapted to living in close proximity with human societies (and vice versa), recent studies indicate that under appropriate conditions, most native taxa may be sustainable within anthromes, even while increasing anthrome productivity in support of human populations (Ellis et al., 2012, 2021; Quinn et al., 2014, 2017, 2021; Ellis, 2013, 2021).

One key principle of sustainable ecosystem management in the Anthropocene is to develop and sustain pluralistic value structures (Allen et al., 2018; Ellis, 2019). Anthromes can play a key role in this, helping to move conservation discussions beyond the outdated paradigm that human use of ecosystems can only "destroy" them, and to embrace a proactive human role in shaping and sustaining biodiverse and productive ecosystems. There are certainly many examples of human degradation of ecosystems without any long-term benefits to anyone or any being. Nevertheless, human use of land, including traditional hunting and foraging, farming, settlements and other uses are the basis for sustaining human societies, and have also sustained wildlife and habitats for centuries to millennia, long before recent expansion of conservation through protected areas and other strategies.

Earth stewardship in the Anthropocene requires envisioning the future "we" want and strategies to get there (Ellis, 2019). Already, international organizations, agreements and scientific collaborations are focusing on creating a future where people and nature can thrive together over the long term. Making such better futures possible involves negotiations across diverse populations of stakeholders, the "we," who must work together towards these better futures—from farmers to ranchers, urbanists and conservationists, to Indigenous Peoples and governments, nongovernmental organizations and international corporations, and will need to engage multilevel governance solutions that bring people together to negotiate productively across the global supply chains, producers, consumers and conservationists that are now shaping the biosphere (Ellis, 2019).

Do "we" want a sprawling world of human infrastructure that consumes most of Earth's land, or a dense urban world that leaves plenty of shared and wild spaces to sustain the rest of life? The answer will come from all of us, and it will require much more than saving nature in places far away. Ecology and conservation in the Anthropocene will mean governing anthromes to more effectively serve human needs while sharing space to sustain the rest of nature in an increasingly anthropogenic biosphere.

References

Allen, K. E., Quinn, C. E., English, C., & Quinn, J. E. (2018). Relational values in agroecosystem governance. *Current opinion in environmental sustainability., 35*, 108–115.

Brown, M. G., & Quinn, J. E. (2018). Zoning does not improve the availability of ecosystem services in urban watersheds. A case study from upstate South Carolina, USA. *Ecosystem Services, 34*, 254–265.

Chapin, F. S., III, Matson, P. A., & Vitousek, P. M. (2012). *Principles of terrestrial ecosystem ecology* (2nd ed.). Springer.

DeFries, R., Ellis, E., Chapin, F. S., III, Matson, P., Turner, B. L., II, Arun, A., Crutzen, P., Field, C., Gleick, P., Kareiva, P., Lambin, E., Ostrom, E., Sanchez, P., Syvitski, J., & Liverman, D. (2012). Planetary opportunities: A social contract for global change science to contribute to a sustainable future. *Bioscience, 62*, 603–606.

Dinerstein, E., Olson, D., Joshi, A., Vynne, C., Burgess, N. D., Wikramanayake, E., Hahn, N., Palminteri, S., Hedao, P., Noss, R., Hansen, M., Locke, H., Ellis, E. C., Jones, B., Barber, C. V., Hayes, R., Kormos, C., Martin, V., Crist, E., Sechrest, W., Price, L., Baillie, J. E. M., Weeden,

D., Suckling, K., Davis, C., Sizer, N., Moore, R., Thau, D., Birch, T., Potapov, P., Turubanova, S., Tyukavina, A., de Souza, N., Pintea, L., Brito, J. C., Llewellyn, O. A., Miller, A. G., Patzelt, A., Ghazanfar, S. A., Timberlake, J., Klöser, H., Shennan-Farpon, Y., Kindt, R., Barnekow Lillesø, J.-P., van Breugel, P., Graudal, L., Voge, M., Al-Shammari, K. F., & Saleem, M. (2017). An ecoregion-based approach to protecting half the terrestrial realm. *Bioscience, 67*, 534–545.

Ellis, E. C. (2013). Sustaining biodiversity and people in the world's anthropogenic biomes. *Current Opinion in Environmental Sustainability, 5*, 368–372.

Ellis, E. C. (2015). Ecology in an anthropogenic biosphere. *Ecological Monographs, 85*, 287–331.

Ellis, E. C. (2018). *Anthropocene: A very short introduction*. Oxford University Press.

Ellis, E. C. (2019). Sharing the land between nature and people. *Science, 364*, 1226–1228.

Ellis, E. C. (2021). Land use and ecological change: A 12,000-year history. *Annual Review of Environment and Resources, 46*(1), 1–33.

Ellis, E. C., & Ramankutty, N. (2008). Putting people in the map: Anthropogenic biomes of the world. *Frontiers in Ecology and the Environment, 6*, 439–447.

Ellis, E. C., Klein Goldewijk, K., Siebert, S., Lightman, D., & Ramankutty, N. (2010). Anthropogenic transformation of the biomes, 1700 to 2000. *Global Ecology and Biogeography, 19*, 589–606.

Ellis, E. C., Antill, E. C., & Kreft, H. (2012). All is not loss: Plant biodiversity in the Anthropocene. *PLoS One, 7*(1), e30535.

Ellis, E. C., Gauthier, N., Klein Goldewijk, K., Bird, R. B., Boivin, N., Díaz, S., Fuller, D. Q., Gill, J. L., Kaplan, J. O., Kingston, N., Locke, H., McMichael, C. N. H., Ranco, D., Rick, T. C., Shaw, M. R., Stephens, L., Svenning, J.-C., & Watson, J. E. M. (2021). People have shaped most of terrestrial nature for at least 12,000 years. *Proceedings of the National Academy of Sciences, 118*, e2023483118.

Freeman, S., Quillin, K., Allison, L., Black, M., Podgorski, G., Taylor, E., & Carmichael, J. (2016). *Biological Science* (6th ed.). Benjamin-Cummings Publishing Company.

Garibaldi, L. A., Oddi, F. J., Miguez, F. E., Bartomeus, I., Orr, M. C., Jobbágy, E. G., Kremen, C., Schulte, L. A., Hughes, A. C., Bagnato, C., & Abramson, G. (2021). Working landscapes need at least 20% native habitat. *Conservation Letters., 14*(2), e12773.

Gibson, D. M., & Quinn, J. E. (2017). Application of anthromes to frame scenario planning for landscape-scale conservation decision making. *Land, 6*(2), 33.

Hobbs, R. J., Higgs, E., & Harris, J. A. (2009). Novel ecosystems: Implications for conservation and restoration. *Trends in Ecology & Evolution, 24*, 599–605.

Martin, L. J., Blossey, B., & Ellis, E. (2012). Mapping where ecologists work: Biases in the global distribution of terrestrial ecological observations. *Frontiers in Ecology and the Environment., 10*(4), 195–201.

Martin, L. J., Quinn, J. E., Ellis, E. C., Shaw, M. R., Dorning, M. A., Hallett, L. M., Heller, N. E., Hobbs, R. J., Kraft, C. E., Law, E., Michel, N. L., Perring, M. P., Shirey, P. D., & Wiederholt, R. (2014). Biodiversity conservation opportunities across the world's anthromes. *Diversity and Distributions, 20*, 745–755.

McDonald, R. I., Weber, K. F., Padowski, J., Boucher, T., & Shemie, D. (2016). Estimating watershed degradation over the last century and its impact on water-treatment costs for the world's large cities. *Proceedings of the National Academy of Sciences, 113*(32), 9117–9122.

Merritts, D., Menking, K., & DeWet, A. (2014). *Environmental geology: An earth systems approach* (2nd ed.). W. H. Freeman.

Miraldo, A., Li, S., Borregaard, M. K., Flórez-Rodríguez, A., Gopalakrishnan, S., Rizvanovic, M., Wang, Z., Rahbek, C., Marske, K. A., & Nogués-Bravo, D. (2016). An Anthropocene map of genetic diversity. *Science, 353*, 1532–1535.

National Geographic Society. (2014). *National Geographic Atlas of the world* (10th ed.). National Geographic Society.

Ordway, E. M., Elmore, A. J., Kolstoe, S., Quinn, J. E., Swanwick, R., Cattau, M., Taillie, D., Guinn, S. M., Chadwick, K. D., Atkins, J. W., & Blake, R. E. (2021). Leveraging the NEON airborne observation platform for socio-environmental systems research. *Ecosphere., 12*(6), e03640.

Quinn, J. E., Johnson, R. J., & Brandle, J. R. (2014). Identifying opportunities for conservation embedded in cropland anthromes. *Landscape Ecology, 29*(10), 1811–1819.

Quinn, J. E., Awada, T., Trindade, F., Fulginiti, L., & Perrin, R. (2017). Combining habitat loss and agricultural intensification improves our understanding of drivers of change in avian abundance in a north American cropland anthrome. *Ecology and Evolution., 7*(3), 803–814.

Quinn, J. E., Cook, E. K., & Gauthier, N. (2021). Patterns of vertebrate richness across global anthromes: Prioritizing conservation beyond biomes and ecoregions. *Global Ecology and Conservation., 1*(27), e01591.

Sanderson, E. W., Jaiteh, M., Levy, M. A., Redford, K. H., Wannebo, A. V., & Woolmer, G. (2002). The human footprint and the last of the wild. *Bioscience, 52*, 891–904.

Smith, J.A., Powell, L.A., & Brown M.B. (2019). Training wildlife biologists for work in anthromes. In: *Reference module in earth systems and environmental sciences*. Elsevier.

Stephens, L., Fuller, D., Boivin, N., Rick, T., Gauthier, N., Kay, A., Marwick, B., Geralda, C., Armstrong, D., Barton, C. M., Denham, T., Douglass, K., Driver, J., Janz, L., Roberts, P., Rogers, J. D., Thakar, H., Altaweel, M., Johnson, A. L., Sampietro Vattuone, M. M., Aldenderfer, M., Archila, S., Artioli, G., Bale, M. T., Beach, T., Borrell, F., Braje, T., Buckland, P. I., Jiménez Cano, N. G., Capriles, J. M., Diez Castillo, A., Çilingirovglu, Ç., Negus Cleary, M., Conolly, J., Coutros, P. R., Covey, R. A., Cremaschi, M., Crowther, A., Der, L., di Lernia, S., Doershuk, J. F., Doolittle, W. E., Edwards, K. J., Erlandson, J. M., Evans, D., Fairbairn, A., Faulkner, P., Feinman, G., Fernandes, R., Fitzpatrick, S. M., Fyfe, R., Garcea, E., Goldstein, S., Goodman, R. C., Dalpoim Guedes, J., Herrmann, J., Hiscock, P., Hommel, P., Horsburgh, K. A., Hritz, C., Ives, J. W., Junno, A., Kahn, J. G., Kaufman, B., Kearns, C., Kidder, T. R., Lanoë, F., Lawrence, D., Lee, G.-A., Levin, M. J., Lindskoug, H. B., López-Sáez, J. A., Macrae, S., Marchant, R., Marston, J. M., McClure, S., McCoy, M. D., Miller, A. V., Morrison, M., Motuzaite Matuzeviciute, G., Müller, J., Nayak, A., Noerwidi, S., Peres, T. M., Peterson, C. E., Proctor, L., Randall, A. R., Renette, S., Robbins Schug, G., Ryzewski, K., Saini, R., Scheinsohn, V., Schmidt, P., Sebillaud, P., Seitsonen, O., Simpson, I. A., Sołtysiak, A., Speakman, R. J., Spengler, R. N., Steffen, M. L., Storozum, M. J., Strickland, K. M., Thompson, J., Thurston, T. L., Ulm, S., Ustunkaya, M. C., Welker, M. H., West, C., Williams, P. R., Wright, D. K., Wright, N., Zahir, M., Zerboni, A., Beaudoin, E., Munevar Garcia, S., Powell, J., Thornton, A., Kaplan, J. O., Gaillard, M.-J., Klein Goldewijk, K., & Ellis, E. (2019). Archaeological assessment reveals Earth's early transformation through land use. *Science, 365*, 897–902.

Waters, C. N., Zalasiewicz, J., Summerhayes, C., Barnosky, A. D., Poirier, C., Gałuszka, A., Cearreta, A., Edgeworth, M., Ellis, E. C., Ellis, M., Jeandel, C., Leinfelder, R., McNeill, J. R., Richter, D. D., Steffen, W., Syvitski, J., Vidas, D., Wagreich, M., Williams, M., Zhisheng, A., Grinevald, J., Odada, E., Oreskes, N., & Wolfe, A. P. (2016). The Anthropocene is functionally and stratigraphically distinct from the Holocene. *Science, 351*, aad2622.

John E. Quinn is an Associate Professor of Biology at Furman University. At Furman, he leads the CHESS Lab, an interdisciplinary collaboration of faculty and students from across the university. His research investigates landscapes and soundscapes as coupled human natural systems across scales with the objective of providing evidence based solutions for decision-makers. He teaches a diversity of classes at the interface of biology and sustainability.

Erle C. Ellis is a Professor of Geography and Environmental Systems at the University of Maryland, Baltimore County (UMBC) where he directs the Anthroecology Laboratory. His research investigates the ecology of human landscapes at local to global scales towards informing sustainable stewardship of the Anthropocene biosphere. He teaches environmental science at UMBC and has taught ecology at Harvard's Graduate School of Design. His first book, Anthropocene: A Very Short Introduction, was published in 2018.

Biodiversity

Noëlle Zendrera

Abstract Ecology is a science which has become indispensable when thinking about the future of humanity and biodiversity on Earth. In this paper we will recall the links between Diversity, Information, Energy, Matter and the scientific input from Doctor Ramon Margalef (Investigación pesquera 3:99–106, 1956; The American Naturalist 97:357–374, 1963). For a long time science has worked hard to reach an estimate of unknown biodiversity. In this "Anthropocene" era (Crutzen, Nature 415:23, 2002), humans realize that the biosphere and biodiversity are in imminent danger and that they have a serious duty to planet earth. It is evident that the Earth's ecosystem requires our protection, but questions still remain: will mankind be able to do so? Will the transformation of human societies towards sustainable development come about?

Thinking about the future of humanity and the diversity of the biosphere requires indispensable ecological science. The first occurrence of the term Ecology occurred a century and a half ago, stemming from the thought of the zoologist Ernst Haeckel; its etymology refers to the 'study of the house', from the Greek *oikos* (dwelling, home) and *logos* (study, knowledge). Haeckel explained it in 1866 as the "science of an organism's relations with the external world around it". Included in the environmental sciences and earth system sciences, Ecology, rooted in Biology, is interdisciplinary and transdisciplinary. The major concept of Ecosystem (ecological system) is defined as a biological system composed of two intrinsically linked elements: the biocenosis, the living subpart (species sharing a living place), and the biotope, the non-living subpart (habitat). Together all the ecosystems of our planet Earth form the biosphere, the sphere of life. Many researchers have contributed to the emergence of Ecology and Environmental Sciences. Among them, we shall focus on the biologist, limnologist and ecologist Ramon Margalef (1919–2004),

N. Zendrera (✉)
Catholic University of the West (UCO, Université catholique de l'Ouest), Angers, France
e-mail: noelle.zendrera@uco.fr

© The Author(s), under exclusive license to Springer Nature Switzerland AG 2023
N. Wallenhorst, C. Wulf (eds.), *Handbook of the Anthropocene*,
https://doi.org/10.1007/978-3-031-25910-4_33

who curiously, as a relay, was born in the year of the death of Haeckel (1834–1919), the creator of the term Ecology.

From the Chair of Ecology at the University of Barcelona, which he created in 1967, Margalef, who is considered one of the main founders of theoretical ecology and modern ecology, contributed greatly to its development and theorization (Barnaud & Lefeuvre, 1992; Ros, 2005; Prat et al., 2015; Peters & Ruiz, 2019; Pou, 2019; Sherwin & Prat, 2019). Involving systems theory, information theory and other theories relating to ecology in his reasoning, he linked them together to explain ecosystems in terms of energy, entropy, quantity of information (bits) and diversity ('Margalefian ecology', Flos, 2005). Thus, Margalef (1956, 1963) established that diversity in an ecosystem is directly related to the amount of information (bits) it contains: Diversity of an ecosystem = Amount of information contained in the ecosystem.

Diversity concerns both biodiversity within the biocenosis (species and individuals) and diversity within the biotope (habitats, locations of species and individuals). Based on thermodynamics, Margalef also apprehended the general structure of ecosystems in terms of Energy and Matter, by quantifying their 'systemic flows' (energy flows). He expressed the structure of an ecosystem as a whole, composed of two parts; matter, including both living and non-living elements (biocoenosis & biotope = ecosystem), and energy, in this case all interactions and exchanges within the material elements (whether living and/or non-living): "Structure of an ecosystem: energy & matter flows".

The recipient of numerous scientific awards and honours, Margalef warned the academic and media community of the negative impacts of human activities on the biosphere, on biodiversity and on the risks of ecosystem destruction. In 1987 he was awarded an honorary doctorate by the University of Laval (Québec) and he concluded with the following warning: "It would be desirable to stop the accelerated destruction of the pre-human features of the biosphere" (Margalef, 1987). In this, he agrees with the ornithologist Jean Dorst who, as early as the 1960s in his book *Before Nature Dies* (Dorst, 1969), expressed his fears about the consequences of human actions on the environment in an ominous manner. In their words, Dorst, Margalef and many others academics were already anticipating the era of the "Anthropocene", a concept that would later be founded by Nobel Prize winner Paul Crutzen, a chemist and atmospheric scientist (Crutzen, 2002).

For a long time, the count of the diversity of living things remained uncertain. The scientific community estimated the number of living species on Earth very imprecisely between 3 and 100 million, and more precisely between 5 and 12 million (excluding prokaryotes). In 2011 Mora et al. (2011) approached this number more precisely (via statistical and taxonomic simulations) and assume the existence of 8.7 million living eukaryotic species (±1.3 M. standard deviation). This magnitude concerns all species, microscopic and macroscopic, from unicellular algae to mammals, including birds, fishes, insects, spiders, crustaceans, molluscs, flowering plants, mosses, mushrooms; etc. However, of these 8.7 million species, only just over one million (1.2 M.) are currently catalogued. To date, the known biodiversity, the organisms listed, represent only 14% of the species. The unknown biodiversity thus remains the vast majority: 86% of the species probably remain undiscovered to date; they have never been observed or described. It should also be noted that this biodiversity does not include that of prokaryotes (bacteria and archaea), estimated at hundreds or even thousands of billions of species (10^{11}–10^{12}), of which more than 99% remain to be discovered.

In a blunt United Nations (UN) report published in May 2019, the worst fears of Dorst, Margalef, Crutzen and numerous other specialists were confirmed. Like a warning bell sounding for the centenary of the death of Haeckel and the birth of Margalef, the Group of Experts on Biodiversity (IPBES, 2019) provided information on the alarming state of the earth ecosystem and warned in particular of the collapse of biodiversity by accelerating the extinction of species. They described the unprecedented decline that has already occurred in the environment and in nature, of devastated biocenosis and biotopes: 66% of marine ecosystems (oceans, seas) and 75% of terrestrial ecosystems (land) are severely altered, modified. They consider that among animals and plants, one species out of eight will be threatened with extinction in the very short term (decades); this represents one million of the estimated eight million species in danger of extinction. Nearly 40% of amphibians (frogs, salamanders), 33% of coral reefs, 31% of cartilaginous fish (sharks, rays), 27% of crustaceans, 25% of mammals, 14% of birds and 34% of conifers (pines, firs) may no longer exist in a few decades. Among the most affected species, 20,000 are in imminent danger of extinction within a few years. Implicitly, thousands of species among the 7.5 million suspected but as yet unknown species are in danger of disappearing before we humans have even been able to locate them.

Only one species, human beings, *Homo sapiens,* by virtue of their way of functioning, its activities and its disproportionate overexploitation of nature and the terrestrial and marine environment, appear to be responsible for this major decline. The IPBES (2019) calls for a drastic paradigm shift, a major transformation of practices in all fields: "By transformative change, we mean a fundamental, system-wide reorganization across technological, economic and social factors, including paradigms, goals and values". These experts identify the five main factors of this decline: "(1) upheavals in land and seas; (2) direct exploitation of organisms; (3) global warming; (4) pollution; and (5) non-indigenous invasive species". It is undeniable that human society must profoundly change its way of life; this concerns nutrition, fishing, hunting, animal husbandry, agriculture, energy sources, waste management, the use of pesticides, the extraction of materials from ecosystems (wood, mines, oil, gas), and the use of natural resources as a whole.

As Crutzen (2002) explains, in this new geological period, the Anthropocene era, **time is** running out for the entire biosphere, including humanity. Le Le Gall et al. (2017) express in a magnificent formula that it is hoped that "the Anthropocene opens a new era, where humans learn to become Earthlings". Many scholars advocate a political shift and civic education aimed at respecting nature and understanding and recognizing that our own life and survival depends on it: "Earth is our master" (Westbroeck, 2015, p.962, in Wallenhorst, 2018). We humans are both responsible and victims. For the human species, its subjects, its populations, its societies, the consequences are measured in terms of social impacts, psychological impacts, and even in terms of survival (famines, wars, migrations, intoxications, diseases, pandemics, Covid case…?).

At the level of the individual, wellbeing, physical and mental health, are also negatively impacted by this era of disruption, overexploitation and acceleration (Zendrera & Bougeard, 2017). In my opinion it is imperative that we consider the human being as an ecosystem in itself. In this regard, systemic thinking, which is also anthropological, leads us to consider each human individual as an ecosystem, the 'human microecosystem'. The human microecosystem forms a complex ecosystem in itself, and it is an integral part of a much wider macrosystems, human society, living conditions, the environment and the biosphere. And it is in turn made up of multiple microecosystems: on one hand each of its own apparatus (each of its organs, tissues, cells, molecules, ions, functions), and on the other hand the various sets of microorganisms that coexist in our body, the 'human microbiota'. Our microbiota, sometimes unduly called 'bacterial flora', encompass all the microscopic organisms (bacteria, archaea, fungi and others) that sit in our anatomy (intestine, skin, etc.); for mostly in symbiotic association (mutual benefits), where they play an indispensable role in our functioning (David, 2012). In this case, it is estimated, for example, that for each of our human cells 10–100 bacteria cohabit with us. In this perspective, all environmental factors interact with the human microecosystem at all levels and in both directions: from nutrients, to luminosity (natural or artificial light), to temperature (moderate or extreme), to air and water quality, to exposure of toxic products (pollutants, drugs), to pathogenic organisms (bacterial and viral infections), the occurrence of other pathologies, sources of stress (immediacy, productivity, life accidents, deregulated metabolism), burnout at work, modes of transport, rhythms and temporalities, links with nature (animal and plant worlds, landscapes) and relationships between humans.

In particular, for example, cerebral, genetic or immune function can be approached from this angle (Zendrera, 2021). The nervous system, an ecosystem in constant flux, continuously responds, through its plasticity, to the demands of the environment (perception, communication, motricity; learning) by creating and eliminating synapses, neurons, glial cells and networks. The genetic system, our genome (about 20,000 genes) also interacts in a well-known way with the environment, leading to mutations (due to radiation, toxic products), and even epigenetic phenomena (blocking of genes by environmental factors, including psychological factors). In this complexity, neuroscience studies how psychological factors modify biological functioning (the notion of epigenetics) or how the composition of our human microbiota affects our psychobiological functioning. But here begins another story in history...

Thinking about the human species, its past, its present and its future, requires us to be aware of its status: human beings are an integral part of the Earth's ecosystem and it requires our protection and our attention. But questions remain: Will humanity be able to do so? Will the transformation of human societies towards sustainable development come about? Various citizenship actions (like the movement "Fridays for Future" led by the young Swedish citizen Greta Thunberg), many associative dynamics (for example Greenpeace) and some political forces (case of UN) are heading in this direction; however progress remains by now too moderate, too slow.

References

Barnaud, G., & Lefeuvre, J.-C. (1992). L'écologie, avec ou sans l'homme? In M. Jollivet (Ed.), *Sciences de la Nature, Sciences de la Société* (pp. 69–112). CNRS.

Crutzen, P. J. (2002). Geology of Manking. *Nature, 415*(6867), 23.

David, B. (2012). Biodiversité: microbiome et microbiote. 35th Day of GAICRM, 35ᵉ Journée du Groupement d'Allergologie et Immunologie Rhône moyen. https://hal-pasteur.archives-ouvertes.fr/pasteur-01349062/document. Accessed 20 June 2021.

Dorst, J. (1969). *Before nature dies*. (1st French ed., Avant que nature meure, 1965, Neuchâtel: Delachaux et Niestlé). Houghton Mifflin.

Flos, J. (2005). El concepto de información en la ecología margalefiana. *Ecosistemas: Revista Cietifica y Tecnica de Ecologia y Medio Ambiente, 14*(1), 7–17.

IPBES (Intergovernmental Science-Policy Platform on Biodiversity and Ecosystem Services, United Nations, UN). (2019, May 6). Global Assessment Summary for Policymakers. IPBES archive. https://www.ipbes.net/global-assessment. Accessed 29 June 2021.

Le Gall, J., Hamant, O., & Bouron, J.-B.. (2017). Anthropocène. *Géoconfluences, ENS Lyon archive*. http://geoconfluences.ens-lyon.fr/informations-scientifiques/a-la-une/notion-a-la-une/anthropocene. Accessed 29 June 2021.

Margalef, R. (1956). Información y diversidad específica en las comunidades de organismos. *Investigación pesquera, 3*, 99–106.

Margalef, R. (1963). On certain unifying principles in ecology. *The American Naturalist, 97*(897), 357–374.

Margalef, R. (1987). Doctor Honoris Causa Acceptance Speech, Université de Laval (Qc). ICM archives. http://www.icm.csic.es/bio/personal/fpeters/margalef/pdfs/lavalspeech.pdf. Accessed 6 October 2019.

Mora, C., Tittensor, D. P., Adl, S., Simpson, A. G. B., & Worm, B. (2011). How many species are there on earth and in the ocean? *PLoS Biology*. https://doi.org/10.1371/journal.pbio.1001127. Accessed 20 June 2021.

Peters, F., & Ruiz, C.. (2019). 2019, The Ramon Margalef Year. Sibecol archive. http://www.sibe-col.org/news/en/2019/01/29/0003/2019-the-ramon-margalef-year. Accessed 20 June 2021.

Pou, T. (2019, May 10). Un ecòleg universal. *Ara*.

Prat, N., Ros, J.-D., & Peters, F. (2015). *Ramon Margalef, ecólogo de la biosfera. Una biografía científica*. Universitat de Barcelona.

Ros, J.-D. (2005). Ramon Margalef, el científico genial. *Ecosistemas: Revista Cietifica y Tecnica de Ecologia y Medio Ambiente, 14*(1), 52–61.

Sherwin, W. B., & Prat, N. (2019). The introduction of entropy and information methods to ecology by Ramon Margalef. *Entropy, 21*(8), 794 (9p). https://doi.org/10.3390/e21080794

Wallenhorst, N. (2018). Vers un entre nous prométhéen. In N. Wallenhorst, F. Prouteau, & D. Coatanéa (Eds.), *Éduquer l'homme augmenté. Vers Un Avenir Postprométhéen* (pp. 133–150). Le Bord de l'eau.

Westbroeck, P. (2015). Système Terre. In D. Bourg & A. Papaux (Eds.), *Dictionnaire de la pensée écologique* (pp. 957–962). PUF.

Zendrera, N. (2021). Biodiversité et Humanité. In N. Wallenhorst & C. Wulf (to be published) (Eds.), *Dictionnaire d'anthropologie prospective*. Vrin.

Zendrera, N., & Bougeard, A.-S. (2017). Rythme et Apprentissages. Quand la chronobiologie et la psychologie clinique s'interpellent. *Chemins de Formation, 21*, 73–94. (thematic folder: Décélérer pour apprendre?, eds. N. Wallenhorst and J-Y. Robin).

Noëlle Zendrera is Associate Professor at the Catholic University of the West. Biologist and Doctor of Educational Sciences, she studied at the universities of Barcelona (human and animal biology), Sorbonne (Paris VI, oceanology) and Sherbrooke (education to sciences). She works on Psychophysiology, Neuroscience, Statistics and Pedagogy. She's author of a number of publications and claims for transdisciplinary. For N. Zendrera this article is also a personal homage to Dr. Ramon Margalef, a professor who made a strong impact on her. Thanks to R. Jenkins and E. Peace for their help in translating.

Carrying Capacity

Brian D. Fath

Abstract This article examines the concept of carrying capacity as a basic principle of sustainability with applications to ecological and social systems. In the more general sense, carrying capacity refers to the number of individuals of a specific population which can be sustained without degrading the ability of that space to generate and regenerate supporting resources. However, carrying capacity is a dynamic condition that changes over time in response to feedback, growing in times of technological adaptation or innovation and shrinking when degradation harms the ecosystem services. Carrying capacity is an indicator of overshoot when compared to regional or planetary biocapacity. To balance this, one must monitor both the biocapacity generated by ecosystem services and the level of resource use by the population.

Ecological Growth and Development

Life on Earth flourishes because the interconnected and intertwined relations promote self-organization and self-regeneration of simple elements into more complex ones; thus moving the system further from an equilibrium and generating processes that are autocatalytic and self-sustaining in a flow of energy. In short, ecosystems grow and develop, changing in the scale and scope of their active patterns and processes. We can measure these relations in terms of the energy and material exchanges and storages observed.

B. D. Fath (✉)
Towson University, Towson, MD, USA

International Institute for Applied Systems Analysis, Laxenburg, Austria
e-mail: bfath@towson.edu

© The Author(s), under exclusive license to Springer Nature
Switzerland AG 2023
N. Wallenhorst, C. Wulf (eds.), *Handbook of the Anthropocene*,
https://doi.org/10.1007/978-3-031-25910-4_34

At base, the capture and biosynthesis of solar radiation through primary production is the foundation for almost all other life processes.[1] The amount of primary production possible in a specific place depends on the resource availability, namely in terms of temperature, precipitation, and geology. A broad brush distinction is given with the various biomes across the planet (e.g., tropical rain forests, deciduous forests, savannah, temperate grasslands, Chaparral, taiga, tundra, and deserts), where some areas are lush and dense and others sparse and barren. The amount of autotrophic biomass directly controls the consumer biomass in the ecosystem, and thus these factors determine the carrying capacity in the region.

The interplay between growth, development, and stability is evident in the concept of carrying capacity (CC). As stated above, carrying capacity is a measure of the number of individuals that can be supported on a landscape without diminishing the regenerative capacity of that place. One of the impressive characteristics of biological populations without resources constraints is the ability for multiplicative reproduction, which can be modelled using exponential growth – growth at a constant rate – where N is the population and r is the intrinsic rate of growth (Eq. 1). In reality, the CC will eventually provide resource constraints, limiting the growth of the population. Therefore, the population dynamics of the consumers in the region depends on the carrying capacity. Theoretically, one expects that a natural population would grow rapidly when the number of individuals is substantially below the CC, with that growth slowing asymptotically as it approaches the CC. Growth constrained by a carrying capacity can be modelled with a logistic curve (Eq. 2.), where N is the number of individuals, r is the intrinsic rate of growth of the population and K is the carrying capacity. We see that as when N < < K (e.g., K = 1000 deer, and N = 2 deer, and N/K ~ 0), the growth is mostly unaffected and unaware of the

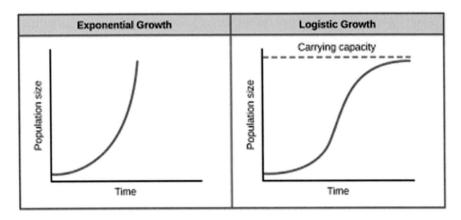

Fig. 1 Exponential and logistic growth curves. The introduction of the carrying capacity restricts growth based on environmental resource availability. (Provided by: OpenStax CNX. Located at: http://cnx.org/content/m44872/latest/Figure_45_03_01.jpg License: CC BY: Attribution)

---

[1] A small amount of chemosynthesis supports ecosystems near deep ocean sulfur vents.

CC. As the population size approaches the CC, this parameter has much more weight in slowing the population growth (e.g., K = 1000 deer, and N = 998 deer, N/K ~ 1), and the overall growth approaches zero (Fig. 1).

$$\frac{dN}{dt} = rN \tag{1}$$

$$\frac{dN}{dt} = rN\left(1 - \frac{N}{K}\right) \tag{2}$$

In reality this smooth, idealized logistic curve is rarely, if ever followed, as the negative feedback controls are too weak, time lags too long, and the built-in momentum of the population demographics is too strong, thus resulting in overshoot of the CC (Fig. 2). At this point, by definition the population is above a sustainable level and therefore, mortality will increase (starvation, additional confrontation, disease, etc.) and/or fertility will decrease (less healthy parents, smaller clutch size, etc.). A feedback between population size and resources tracks accordingly:

When population increases, food supply decreases
When population decreases, food supply increases
When food supply increase, births increase and deaths decrease
When food supply decreases, births decrease and deaths increase.

The carrying capacity becomes a clear "line in the sand" where a bifurcation occurs from a growing population to a shrinking one. The carrying capacity acts as an attractor pulling the population toward it from both directions.

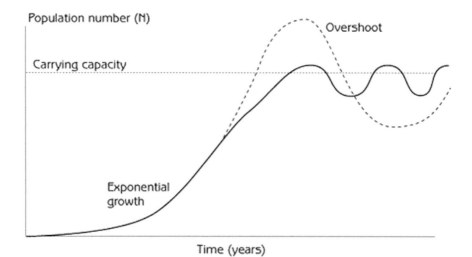

Fig. 2 Population experiencing overshoot, having exceeded its carrying capacity. (From Marten, 2001)

A Changing Carrying Capacity

Having presented the basic structure and behaviour of carrying capacity it is next necessary to consider that it is not a static attractor. The co-evolution of the organisms on the landscape may have a supportive or degrading impact on the environment thus raising or lowering the CC. Species that act to provide additional resources to the space (e.g., plants or decomposers) expand the range of possibilities benefitting the entire ecosystem, thus, over time can raise the carrying capacity. As witnessed through ecological succession, early stages will have lower ability to support populations than at more mature stages.

With human dominated systems, many actions are taken to increase the carrying capacity by supplying inputs that enhance the overall biological productivity (Cottrell, 1955). Clearly, this is the case with agriculture where farmers supply nutrient fertilizers, water, and seeds to maximize production yield at the same time actively interfering with the emergence of any pests or competitors. Conventional agriculture does this using high levels of energy and chemical inputs. This can be seen taken to its grotesque extreme in the case of Confined Animal Feeding Operations (CAFOs) where the animals' connection to the local environment and carrying capacity is severed, with many animals shoulder to shoulder in places devoid of any on site biological productivity and food resources brought directly to the animals from far off distant places. Such a condition is clearly above the local environment's CC with implications on necessity of constant supply chains and waste management.

Natural disasters, disturbances or extended periods over the carrying capacity can result in a degraded environment and lowering of the CC. In other words, if the impact is irreversible, then recovery to the previous level becomes impossible and a new, lower capacity is reached.

Limits to Growth

Although conceptual, abstract, and ephemeral, carrying capacity gives practical guidance into managing resources (Fath, 2020). The earth is a closed system and growth in a finite space is not possible forever (Meadows et al., 1972). The wonderful Blue Marble photo taken of Earth by astronauts in the Apollo 17 mission shows clearly the biophysical resource constraints that all species on the planet contend with. Yet, diverse and complex life flourishes within these limits (Jørgensen et al., 2015). What is the carrying capacity on the global scale? This depends on the species in question. Humans dominate the planetary energy flows utilizing about 1/3 of all terrestrial photosynthesis. Furthermore, industrial scale implementation of Haber-Bosch process and the green revolution allowed human population to increase toward its current level, around 8 billion individuals. Both extensively and intensively, humans are pushing the boundaries of the carrying capacity. Without supplemental inputs, natural processes have been estimated to generate enough surplus energy to support 3 humans per acre (hectare). Even with the advent of technologies, others estimate a carrying capacity of 6–9 billion. Of course, this number is dependent on the resources used per person, namely if the diet is mostly vegetarian or heavily meat based but also pressures for other energy, water, and mineral resources (Cohen, 1995). Therefore the question of "How many" is not answerable without also simply asking "How" are we living? The rapid rise of the human population dependent on non-renewal inputs and the unintended consequences of living near or above the carrying capacity manifests itself in climate change, eutrophication, acid precipitation and other maladies. Therefore, an interesting question arises regarding the sustainability of this elevated population. How will we respond to the pressures of being in overshoot: by innovating new opportunities, by learning to respect limits and living within them, by collapsing back painfully or willingly? Whatever the path ahead, our understanding of the carrying capacity will play an important role.

References

Cohen, J. E. (1995). *How many people can the earth support?* (532 pp). WW Norton & Co Inc.

Cottrell, F. (1955). *Energy and society: The relation between energy, social change, and economic development* (p. xix, 330). Mc-Graw-Hill Book Company.

Fath, B. D. (2020). Limits to growth. In B. D. Fath & S. E. Jørgensen (Eds.), *Handbook of environmental management: Managing human and social systems* (Vol. 6). CRC Press.

Jørgensen, S. E., Fath, B. D., Nielsen, S. N., Pulselli, F., Fiscus, D., & Bastianoni, S. (2015). *Flourishing within limits to growth: Following nature's way* (220 pp). Earthscan Publisher.

Marten, G. G. (2001). *Human ecology: Basic concepts for sustainable development*. Earthscan Publications.

Meadows, D. H., Meadows, D. L., Randers, J., & Behrens, W. W., III. (1972). *The limits to growth; a report for the Club of Rome's project on the predicament of mankind*. Universe Books.

Brian D. Fath is a Professor at Towson University (Maryland, USA) and a Senior Research Scholar at the International Institute for Applied Systems Analysis (Austria) and since 2011 the Scientific Coordinator of IIASA' s Young Scientists Summer Program. He co-authored the books *A New Ecology: Systems Perspective* (2020), *Foundations for Sustainability* (2019), and *Flourishing within Limits to Growth* (2015) and serves as Editor-in-Chief for Frontiers in Sustainable Resource Management. Dr. Fath was twice a Fulbright Distinguished Chair (Parthenope University, Naples, Italy, in 2012 and Masaryk University, Czech Republic, in 2019).

Microbes

Julie Beauté

Abstract Microbes lead to a rethinking of living entanglements from an ecological, anthropological and epistemological perspective. Microcosm ecology highlights microbial metabolisms that reconfigure the traditional concepts of biology and enable us to foresee a more-than-human terraformation with microbes. An anthropology of microbes questions the existence of a microbial turn, from a pathogenic vision to a symbiotic vision of microbial activity. Finally, the epistemology of microorganisms explores connections of scales, decentred and non-dualistic perspectives, and micro-ontologies that attest to the agency of matter.

The notion of microbes generally refers to living entities invisible to the naked eye. With their prodigious ecological variety divided into multiple classifications, microbes include bacteria, viruses, yeasts, protists, amoebas, planaria, archaea, and some algae and fungi. Microbes are omnipresent on Earth—in the environment, on organisms' surfaces or inside all aggregated living beings. They show a remarkable metabolic diversity: they can reconfigure elements, for example by fermentation, sulphuric respiration, photosynthesis, iron fixation or methanization (Margulis & Sagan, 1986). In this way, they combine and regenerate as much as they decompose and destroy matter. Microbiologist Lynn Margulis, with her theory of endosymbiosis, has notably highlighted the importance of microbial symbioses in evolution (Margulis & Sagan, 2000). Microbes have made the planet inhabitable and alive, by modifying the composition of soils, oceans, and the atmosphere. In this perspective, the Gaia hypothesis (Lovelock, 1987) argues that the biosphere's ability to self-regulate depends on the metabolic processes of microorganisms. Because of their crucial role in the biosphere, in living mechanisms and in evolution, microbes appear as a paradigm for the success of life.

J. Beauté (✉)
Aix Marseille Univ, CNRS, EFS, ADES, Marseille, France

ENS Ulm, PSL, Pays Germaniques, Paris, France
e-mail: julie.beaute@univ-amu.fr

N. Wallenhorst, C. Wulf (eds.), *Handbook of the Anthropocene*,
https://doi.org/10.1007/978-3-031-25910-4_35

Microbial activity leads to open up two biological concepts—that of species and that of organism. On the one hand, the microcosm disrupts the traditional species definitions based on interfertility and genetic assortment. Microbes indeed constantly exchange genetic material by horizontal transfer of genes within the same generation, and also with other species and with the environment. They demonstrate that species are not fixed and unchanging categories, but that they do evolve and hybridize. On the other hand, microbes challenge the organism's enclosure (Hird, 2009)—from a triple genetic, immunological and evolutionary perspective. They do not have a genomic identity since the microbiome is made up of heterogeneous elements. The notion of hologenome, that describes the combination of the different partners' genomes, underlines the heterogenomic aspect of living individuals. From an immunological and developmental point of view, many microbes act as stimulators of the development and immune system of other organisms. The organism immunoregulation involves a functionally integrated whole, with endogenous and exogenous elements. Finally, in evolutionary and physiological terms, macroorganisms can be understood as holobionts (Gilbert et al., 2012), i.e., as multi-species and symbiotic complexes including the microbiota. Microorganisms call for considering the organism as an open system and rethinking biological individuality.

The microcosm evidences that mutualistic phenomena, symbiotic assemblages, and multi-species relationships are essential. Since life is deeply relational, the stake is to understand microbial ecologies and the consequent entanglements composing a common world. In this sense, we can only be "with microbes" (Brives et al., 2021). This is also dramatically confirmed by pandemics, which remind of the plurality and ambiguity of our relationships with microbes. The microcosm discourages from glorifying humans as a telluric force, as suggested by the controversial concept of the Anthropocene, and points towards a more-than-human terraformation (Haraway, 2016): in contrast to a systemic and overhanging anthropocene vision of the Earth, the aim is rather to propose a geo-centric multi-species cosmology.

The anthropology of knowledge also emphasizes the ambivalence of microbes, by emphasizing our changing perceptions and conceptions of the microcosm. The discovery—or invention—of microbes in the nineteenth century is linked to the so-called Pasteurian "bacteriological revolution" (Latour, 1984), from which developments in bacteriology have allowed for the optimization of microbial strains and the systematization of taxonomies. However, from food hygiene to human and animal health care organization, microbes have been quickly and mainly presented as dirty, dangerous and contagious pathogens to be controlled. In this pasteurized society, the definition of health and disease has been restricted to specific pathogens and microbes have become the norm to understand and control epidemics. The belief in the possibility of complete pathogen eradication through the invention of biomedical techniques has turned antibiotics into an essential and miraculous tool in health practice (Sariola & Gilbert, 2020). Microbes are thus characterized as enemies of public health—quasi in the military sense—, even more so since their antibiotic resistance has accelerated (Landecker, 2016). A new biosecurity risk rationality is emerging in life management, with specific governance modes and with new surveillance and tracing devices (Brives & Zimmer, 2021).

In the 2000's, metagenomics started developing new methods and technologies in the field of genetic sequencing, making the microbial diversity and ubiquity observable, and better describing microbes' role in organism regulation. These researches challenge the conception of microorganisms as exclusively pathogenic, and propose a new paradigm transforming both life and human sciences. They pave the way for an undetermined biology, which accounts for complex, multiple and even impure relationships to microbes. The conception of microbes as our inevitable partners—to the point that one can speak of Homo microbis (Helmreich, 2018)—inspires new ways of cultivating relationships between humans and microbes: new microbial practices are emerging today in rich, industrialized Western countries, around a craze for fermentation, a fascinated attention to our guts, or probiotic alternative medicines. These new practices involve microbiopolitics, i.e. regulations, governance and moral values—as in the case of raw milk cheese (Paxson, 2008)—as well as do-it-yourself and hacking practices—as in the case of baker's yeast (Sariola, 2021).

Microbes are thus becoming heroes of microbiology, promising future potentialities. Expressions designate this register shift—"post-pasteurian cultures" (Paxson, 2008), "microbial turn" (Paxson & Helmreich, 2014) or "probiotic turn" (Lorimer, 2020)—and stand as transversal proposals to account for the new attention microbes are receiving within the natural sciences, food practices and biomedicine. This turn, however, introduces a dichotomy between a "before"—characterized by a negative and discrete conception of microbes—and an "after"—characterized by a positive and ecological vision—that would benefit from vigilance and critical attention (Brives & Zimmer, 2021). The development of microbial theory has indeed, for example, contributed to minimizing the impact of industrial pollution (Cooper, 2011), research on an allegedly ancestral microbiota tends to renew a racialization of human communities (Benezra, 2020), and works on the beneficial aspects of certain practices, such as vaginal delivery or breastfeeding, lead to social injunctions.

Microbes finally engage epistemological stakes since they require to listen to the invisible lives that shape terrestrial lives and to claim for microscopic worlds an existence for themselves. At one edge of the universe's understanding, microbial epistemology focuses on entities that are problematic because of their size and number. Micro and nowadays nanoscopic worlds, which appeared in laboratory devices using the eye as the main mediator, imbue the world, recalibrate perceptions and reconfigure practices. Microbes' size requires a connection of scales and gives us the opportunity to see life "as networked phenomenon linking the microscopic to the macrocosmic, bacteria to the biosphere, genes to globe" (Helmreich, 2009, 9). Emphasizing the relationality of any knowledge process, microorganisms engage multispecies relationships that open up decentred perspectives, without dichotomizing nature and culture, subject and object, human and other-than-human.

The methodological, conceptual and ontological positions relating to the microcosm require attention to the conditions and logics of knowledge production. In particular, institutions crystallize microgeohistories, i.e., stories of practices, territories and biologies transformations (Brives et al., 2021). Microbial knowledge is thus situated, within regulatory, economic and political structures. In this framework, microbes come to be posited as bio-objects—that is, as technical forms conceived in a metonymic relationship to the living—and as model ecosystems (Paxson & Helmreich, 2014). They appear as both natural and artificial, yet relational and open, and as potential pledges. Parameters for considering microorganisms lead to the conception of a micro-ontology, which goes along with an ethics seriously engaged in the microcosm (Hird, 2009). Micro-ontology emphasizes the need for a transdisciplinary, more-than-human, non-modern epistemology. It involves a large number of actors in troubled relational processes. In these micro-ontological narratives, matter does not appear as a passive and inert product that bends to practices, but as an active agent that constantly tests them. Microbes participate in a dynamic continuation of life, via a process of materialization: they raise the generative powers of matter, thwarting the distinctions between waste and resource, organic and inorganic, animate and inanimate (Coole & Frost, 2010). They witness that matter, in its granularity, its agentic assemblages and its capacity to deflect practices, is always historicity.

References

Benezra, A. (2020). Race in the microbiome. *Science, technology, & human values, 45*, 877–902.

Brives, C., & Zimmer, A. (2021). Écologies et promesses du tournant microbien. *Revue d'anthropologie des connaissances, 15*.

Brives, C., Rest, M., & Sariola, S. (2021). *With microbes*. Mattering Press.

Coole, D., & Frost, S. (Eds.). (2010). *New materialisms: Ontology, agency, and politics*. Duke University Press.

Cooper, M. E. (2011). *Life as surplus: Biotechnology and capitalism in the neoliberal era*. University of Washington Press.

Gilbert, S. F., Sapp, J., & Tauber, A. I. (2012). A symbiotic view of life: We have never been individuals. *The Quarterly Review of Biology, 87*, 325–341.

Haraway, D. J. (2016). *Staying with the trouble: Making kin in the Chthulucene*. Duke University Press.

Helmreich, S. (2009). *Alien Ocean: Anthropological voyages in microbial seas*. University of California Press.

Helmreich, S. (2018). Homo microbis: The human microbiome, figural, literal, political. *Thresholds, 52*–59.

Hird, M. (2009). *The origins of sociable life: Evolution after science studies*. Palgrave Macmillan.

Landecker, H. (2016). Antibiotic resistance and the biology of history. *Body & Society, 22*, 19–52.

Latour, B. (1984). *Pasteur: guerre et paix des microbes*. La Découverte.

Lorimer, J. (2020). *The probiotic planet: Using life to manage life*. Univiversity of Minnesota Press.

Lovelock, J. (1987). *Gaia: A new look at life on earth*. Oxford University Press.

Margulis, L., & Sagan, D. (1986). *Microcosmos: Four billion years of evolution from our microbial ancestors*. Summit Books.

Margulis, L., & Sagan, D. (2000). *What is life?* Press.

Paxson, H. (2008). Post-pasteurian cultures: The microbiopolitics of raw-milk cheese in the United States. *Cultural Anthropology, 23,* 15–47.

Paxson, H., & Helmreich, S. (2014). The perils and promises of microbial abundance: Novel natures and model ecosystems, from artisanal cheese to alien seas. *Social Studies of Science, 44,* 165–193.

Sariola, S. (2021). Fermentation in post-antibiotic worlds: Tuning in to sourdough workshops in Finland. *Current Anthropology, 62,* 388–398.

Sariola, S., & Gilbert, S. F. (2020). Toward a symbiotic perspective on public health: Recognizing the ambivalence of microbes in the Anthropocene. *Microorganisms, 8,* 746.

Julie Beauté is a PhD candidate in philosophy at the ENS Ulm and at Aix-Marseille Université. In her research, she proposes to renew the relation between architecture and ecology. She is particularly interested in the role of material, human and non-human agencies in architecture, and draws on environmental humanities, new materialisms, feminist epistemologies and field philosophy.

New Viruses

Chung-Ming Chang, Riya Mukherjee, and Ramendra Pati Pandey

Abstract The chapter focuses on the interrelation of viruses in this epoch of great acceleration, also known as the Anthropocene. The increased frequency of viral pandemics and epidemics globally is a by-product of the Anthropocene. The severity of epidemics and pandemics have increased, which has magnified the number of cross-species responsible for various infectious diseases that have arisen because of the accelerated frequency of viruses around the globe. As the Anthropocene unfolds, it increases environmental instability, which provokes the microbial population. The frequent evolution and mutations among the viruses are also becoming more prominent. This scenario is rising progressively, and it is a major concern for the scientific world regarding human existence in the future. Globally, implementation of "One Health Approach" should be proposed to understand the current concern of the Anthropocene and emerging new viruses.

C.-M. Chang (✉)
Master & Ph.D. Program in Biotechnology Industry, Chang Gung University,
Taoyuan, Taiwan

Laboratory Animal Centre, Chang Gung University,
Taoyuan, Taiwan
e-mail: cmchang@mail.cgu.edu.tw

R. Mukherjee
Master & Ph.D. Program in Biotechnology Industry, Chang Gung University,
Taoyuan, Taiwan

Graduate Institute of Biomedical Sciences, Chang Gung University,
Taoyuan, Taiwan
e-mail: riya.mukherjee1896@gmail.com

R. P. Pandey
Centre for Drug Design Discovery and Development (C4D), SRM University,
Sonepat, Haryana, India
e-mail: ramendra.pandey@gmail.com

© The Author(s), under exclusive license to Springer Nature
Switzerland AG 2023
N. Wallenhorst, C. Wulf (eds.), *Handbook of the Anthropocene*,
https://doi.org/10.1007/978-3-031-25910-4_36

The Anthropocene era is defined by rising anthropogenic activities that serve as a driver of the threats to biodiversity and, in turn, contribute to the unnerving occurrence and re-emergence of borne viral diseases. Vector-borne diseases, caused by either parasites, bacteria or viruses, account for more than 17% of all infectious diseases, causing more than 700,000 deaths annually, are the main mortality concern for the human population (WHO, 2020). As the globe enters the Anthropocene epoch, we as a species are realising that alleviating all of the negative consequences of an anthropogenic biosphere, such as climate change and large-scale pollution, seems to be beyond our collective capacity. Despite health experts' constant warnings, such as those concerning the dangers of multidrug-resistant bacteria, also the rates of new and emerging viral infectious diseases have been rising in recent decades. The new coronavirus (SARS-CoV-2) epidemic in late 2019 caught the entire world off guard. The worldwide threat of COVID-19 has brought in the aspect of new viruses, which were till now unknown to humankind, but it brings the awareness to the issue of the liberal agenda. The COVID-19 pandemic is a diverse microbial event because it can readily travel between humans and other species, and we may be at fault, too (Tajudeen, 2021).

Viruses are intriguing because of their evolutionary dance with their hosts, the diverse solutions they've devised for replication with a limited repertoire of genes and the unpredictable ways they interact with the immune system, which makes vaccine development difficult and pathogenicity unpredictable. They have such an impact on their surroundings that any biology experiment that ignores the inherent impact of viral infections is certain to fail. There's also the never-ending debate about whether they're truly "alive." Viruses can be found all around the world. The majority of infectious diseases are zoonotic, implying that they can infect humans across species barriers. COVID-19's breakout in 2019 marks the development of a new virus that poses new concerns around the world. In general, we don't know much about virus populations in nature (Emerman, 2011).

There are around 6000 different viral species that infect animals, plants, and microbes. More than one million viral species are thought to occur in mammals, with the number of undiscovered mammalian viruses unknown (Emerman, 2011). The viral inundation from this extensive unknown pool is becoming increasingly common, but we have few technical mitigation strategies to respond to these inundations. Homo sapiens is one of the most powerful species to have ever lived on the planet, and it is currently a major geological and environmental force equal to, if not greater than, natural forces. It has been proposed that the Earth is now in the Anthropocene period. Humans, on the other hand, are bound by the same biological rules as any other entity. As a result, a global perspective is required to comprehend how illness propagation is linked to environmental and socioeconomic issues (Aronsson & Holm, 2020).

Changes in mortality in populations of all creatures, including humans, are determined by new viral/infectious diseases. Bats (Ebola, SARS, and COVID-19), chickens and pigs (avian flu and swine flu), camels (Middle East Respiratory Syndrome, or MERS), chimpanzees (AIDS), mosquitos (Zika), and wild rats (Lassa virus) are among the many viral infections that cause severe disease in humans

(Heyd, 2020). New viruses may pose a threat to health systems and economies, as well as the possibility of pandemics. The COVID-19 pandemic will almost certainly have a wide range of effects on human population functioning and will effect massive ecological ramifications for human-affected ecosystems. Consequently, identification of the elements that contribute to the spread of this virus among human populations is critical. This unique COVID-19 virus is unaffected by whatever immunity people may have had to previous strains, allowing it to spread quickly and infect vast numbers of people in a short amount of time. Typically, urbanisation results in a high density of cars, buildings, people, and factories, which causes pollution. This puts the immune system under even more strain. Therefore, pandemics are more likely to occur in developed countries. (Guynup, 2021; O'Callaghan-Gordo et al., 2020).

Human environments, it is thought, define the habitat for germs or infections. New viral diseases emerge on a regular basis, posing a huge public health risk. The hub for bacteria and virus was found in great quantities in domesticated animals, and as a result, humans came into direct contact with domesticated animals for livestock. This was never a serious worry before, but as the changes progressed, these microbial habitats were discovered to have altered in several cases (Skórka et al., 2020). Transmission from non-human to human species has grown in a variety of ways over time. Massive changes in the environment have resulted in greater rates of diseases that are constantly on the rise in a more forceful way. Viruses are sneaky, and they are shown to have a great capacity for developing medication resistance. They adapt to any situation by modifying their own genetic code, which has proven difficult for researchers trying to combat viruses. Antibiotics are completely worthless in the treatment of viral infections, and their overuse leads to drug resistance in other organisms, weakening patients' immune systems (Kothari et al., 2020).

As humans, we must learn to use the power we have in new ways, and the first step is to recognise our own and our environment's shortcomings. The concept of 'new viruses' refers to the fact that we are in the process of recognising a number of novel species that have the potential to cause serious outbreaks, and that all countries should be prepared to confront them with their health-care systems. The Anthropocene is the epoch in which humans have progressed through continual development. While this has increased the quality of life in terms of progress, it has failed to create a sustainable society. Though we are progressing in civilization, this predicament has caused us to question our own existence from time to time. It is projected that a portion of the virosphere will be shared between humans and non-human creatures in the future, and that this event will continue to grow as the world changes. As a result, a mutation in a different viral species could result in a more lethal multispecies pandemic. To summarize, nonhumans will begin to adapt to the Anthropocene epoch in this scenario, bringing with them several changes as well as a large number of viruses that have the potential to cause havoc in the lives of humans and other domesticated animals. Our presence in the Anthropocene raises unprecedented existential problems, the most pressing of which is the possibility of the human adventures (Skórka et al., 2020).

What can we learn about the Anthropocene from the Covid-19 epidemic or pandemic, and how should we think about the current status of the world? There's a good reason to think of this viral pandemic as one of the most visible expressions of the Anthropocene, or "human period." Can the Coronavirus pandemic prompt us to consider the need for a "geoethic look" at human behaviour in relation to the environment (Heyd, 2020)? It is past time for environmental ethics to take precedence over economic interests, and for an integrated and interdisciplinary approach to solving the immense challenges posed by global change to be adopted. To avoid future pandemics, the One Health strategy is required, which addresses human, animal, and ecosystem health simultaneously, protects mankind and nature, and incorporates disease risk into decision-making. It is past time for environmental ethics to take precedence over economic interests, and for an integrated and interdisciplinary approach to solving the immense challenges posed by global change to be adopted. Policymakers must focus on multilateral cooperation to develop stronger and more equitable public health governance, strengthen health systems for universal health coverage, and adapt viral surveillance to the pressures of a changing world in order to adapt viral pandemic preparedness to the Anthropocene (Guynup, 2021; Keim, 2021).

References

Aronsson, A., & Holm, F. (2020). Multispecies entanglements in the virosphere: Rethinking the Anthropocene in light of the 2019 coronavirus outbreak. *The Anthropocene Review, 9*, 24–36. https://doi.org/10.1177/2053019620979326

Emerman, M. (2011). The little book of viruses. *PLoS Biollogy*. September 6, 2011. https://doi.org/10.1371/journal.pbio.1001139.

Guynup, S. (2021). Address risky human activities now or face new pandemics, scientists warn. *Mongabay*. August 3. https://news.mongabay.com/2021/08/address-risky-human-activities-now-or-face-new-pandemics-scientists-warn/

Heyd, T. (2020). Covid-19 and climate change in the times of the Anthropocene. *The Anthropocene Review, 8*, 21–36. https://doi.org/10.1177/2053019620961799

Keim, B. (2021). Here's how we avoided the worst of zoonotic diseases. *Anthropocene magazine*. https://www.anthropocenemagazine.org/2020/08/a-memo-from-the-year-2050/

Kothari, A. et al. (2020). Coronavirus and the crisis of the Anthropocene. *Eologist informed by nature*. March 27.

O'Callaghan-Gordo, C., Josep, M., & Antó, J. M. (2020). COVID-19: The disease of the anthropocene. *Environmental Research, 187*, 109683–109683. https://doi.org/10.1016/j.envres.2020.109683

Skórka, P., Grzywacz, B., Moroń, D., & Lenda, M. (2020). The macroecology of the COVID-19 pandemic in the Anthropocene. *PLoS One, 15*(7). https://doi.org/10.1371/journal.pone.0236856

Tajudeen, Y. A. (2021). Anthropocene- an era with evil six threats changing the fate of biodiversity: Emerging and re-emerging Aboviruses calls for holistic approach. *Journal of Infectitous Diseases Epidemiology, 7*, 212. https://doi.org/10.23937/2474-3658/1510212

WHO. (2020). *Vector borne diseases*. https://www.who.int/news-room/fact-sheets/detail/vector-borne-diseases

Dr. Chung-Ming Chang is an Assistant Professor of the Master & Ph.D. Program in Biotechnology Industry at Chang Gung University (CGU), Taiwan. He graduated from National Taiwan University with both his bachelor's and master's degrees in veterinary medicine. He also holds a master's degree in veterinary clinical medicine from France's École Nationale Vétérinaire d'Alfort, with a focus in avian pathology. From the French Institut National Agronomique Paris-Grignon, he received his Ph.D. in Animal Genetics. He discovered a single, fully developed endogenous retroviral insertion in chickens that is associated to the recessive white mutation. He worked as a researcher at the Pasteur Institute in Paris and the French National Center for Scientific Research (CNRS), where he was in charge of a research program on influenza viruses in wild birds, before joining the Research Center for Emerging Viral Infections (RCEVI) at Chang Gung University. Chung-Ming Chang is interested in studying human endogenous retroviruses that may be linked to various autoimmune disorders, as well as influenza in both humans and animals with a focus on ecological and epidemiological investigations. Professor Chang has played a key role in developing joint research initiatives between Chang-Gung University and French collaborators. He is the authors of more than 25 articles and one book on virology, biomedicine and pharmacotherapy, One-Health, biosensor, veterinary medicine, genomics, animal genetics and biotechnology. Selected publications: Advanced biosensors for detection of pathogens related to livestock and poultry. 2018. *Veterinary Research*. Nanoparticles as an effective drug delivery system in COVID-19. 2021. *Biomedicine & Pharmacotherapy*.

Miss Riya Mukherjee is a driven and highly motivated research Ph.D. student with a passion for advancing knowledge in Biotechnology. She has graduated from Kalyani University, West Bengal, India, and followed that she received her master's degree in Microbiology from SRM University, Sonepat, Haryana, Delhi-NCR, India. Currently, pursuing a Ph.D. in Biotechnology at the Graduate Institute of Biomedical Sciences, Chang Gung University in Taiwan under the supervision of Dr. Chung-Ming Chang and Dr. Ramendra Pandey. She has already made significant contributions to their area of research through a number of publications and presentations.

Dr. Ramendra Pati Pandey is an Associate Professor in the Department of Biotechnology/Microbiology/Biomedical Engineering at the SRM University, Delhi-NCR, Sonepat. He was a FAPESP Post-Doctoral Fellow (From September 2015 to January 2019), a very prestigious fellowship of Latin America at the Department of Medicine-InCor/HC-FMUSP, University of Sao Paulo, School of Medicine, Brazil.

Viruses

Chung-Ming Chang

Abstract This chapter describes the state of species in the anthropocene. The delicate ecological balance of ecosystems has been disturbed by human activity, allowing viruses to quickly evolve and leap between species. As a result, the Anthropocene has produced the ideal conditions for the spread of viruses. This has caused the creation of novel viruses like COVID-19, which have had a substantial negative impact on the world economy and health. In recent years, viral epidemics have become more frequent. Whether of monetary, medicinal, or biotechnological reasons, viruses are linked to humans or directly human-related hosts. Through raising the likelihood and severity of pandemics, the Anthropocene significantly contributes to the acceleration of the occurrence of viral pathogen-driven cross-species infectious illnesses. Our understanding of the Anthropocene must continue to incorporate "health for all" and "One Health," according to global health.

"The single biggest threat to man's continued dominance on the planet is the virus," said by the Nobel Prize-winning biologist Joshua Lederberg (Pal, 2021). The beginning of virology started at the end of the 19th century with the studies of tobacco mosaic virus, and by the early twentieth century many plant/animal viruses had been discovered (De Pascale & Roger, 2020). Viruses need to replicate themselves with the help of other living cells, they evolve constantly to evade the host's immune system. In animals for example, this kind of evolution allows viruses to move back and forth between animals and humans. There is a vast pool of unknown viruses come from mammals that could potentially infect humans. The viruses that infect humans are the best understood of all, until now, only around 250 species of viruses choose humans as their host (Zimmer, 2021).

C.-M. Chang (✉)
Master & Ph.D. Program in Biotechnology Industry, Chang Gung University,
Taoyuan, Taiwan

Laboratory Animal Center, Chang Gung University,
Taoyuan, Taiwan
e-mail: cmchang@mail.cgu.edu.tw

© The Author(s), under exclusive license to Springer Nature
Switzerland AG 2023
N. Wallenhorst, C. Wulf (eds.), *Handbook of the Anthropocene*,
https://doi.org/10.1007/978-3-031-25910-4_37

Viruses have been known as the most abundant and diverse group of organisms on Earth, forming a virosphere. The *virosphere* describes the parts of the biosphere where multispecies entanglement occurs between viruses and their hosts. Studies show that the known virosphere has expanded based on viruses of human interest, related to many human activities. The analysis of the virosphere network reveals a highly anthropocentric phenomenon, that most viruses are associated with humans or hosts that are directly related to humans by economic, medical or biotechnological interests (Rodrigues et al., 2017).

Zoonosis is the term for infectious diseases caused by pathogens that cross/jump from a non-human animal reservoir to human. Usually, zoonotic events occur sporadically and usually do not trigger epidemics. In most cases, viral epidemics involved an animal host, such as bats (Ebola, SARS and COVID-19), poultry and pigs (avian flu and swine flu), camels (Middle East Respiratory Syndrome, or MERS), chimpanzees (AIDS), mosquito (Zika), and wild rats (Lassa virus) et al. (Ray, 2021.). Since the outbreak of the severe acute respiratory syndrome (SARS), the first pandemic of the twenty-first century, viral epidemics have been appearing with increasing frequency. Studies revealed that up to 75% of new infectious human diseases are zoonotic, and most originate from the tropics (Guynup, 2021). Surveillance and preparedness in preventing the next outbreak of virus disease is the primary task of the virologist.

The term Anthropocene was popularized by Nobel Laureate, Paul Crutzen, to describe the current epoch in which human activities have profoundly altered the planet's geological, hydrological, bio-geochemical and atmospheric cycles (Pal, 2021). The Anthropocene is a matrix in which the development path of humanity and the natural world intersect, with one altering the other (Pal, 2021).

In the normal circumstance, wildlife can be carrier/host of unknown viral pathogens, most of those unknown viral pathogens have few possibility or have not yet crossed to the species boundary to infect human, as the wild animals live and evolve in isolation from other species. Human activities, including wild animal hunting, the destruction of wild animal habitats in forests even deforestation, and construction of anthropogenic landscapes or indirectly through the climate change's side-effects, for example the melting of polar ice caps, forces wildlife to relocate to habitats closer to human community. It can be sure that the Anthropocene plays an important role to accelerate the frequency of cross-species infectious diseases caused by viral pathogens, through increasing the risk and severity of pandemics (Aronsson & Holm, 2020).

'Humans are to blame for the rise in dangerous viral infections', a report published by Smith and Carr in the Conversation Australia and New Zealand Media in 2018 gave this report's title to describe the situation about dangerous viral infections around the world on a regular basis. The report mentioned that although the social media and internet access may be an obvious explanation for their seeming increase. But actually, the number of viruses and the infections they cause are truly increasing. The report pointed out that scientific advances, the way humans live today and virus biology all contribute to the rise of viruses. How humans contribute to the success of dangerous viruses (Smith & Carr, 2018)? Because of the world's population is urbanizing; people live in closer proximity and is conducive to spread of viruses?

Human activity is transforming the earth's natural habitats and ecosystems by intensely altering the patterns and mechanisms of interaction between species. *International travel and the developing global wildlife trade can quickly spread viruses.* Human activity also facilitates the transmission of infectious diseases across species and to humans (O'Callaghan-Gordo & Antó, 2020). Globalization, domestic and international mass transport allow viruses to move between regional populations easily. Many dangerous virus infections are zoonoses, they are diseases transmitted to humans from other animals. Human expanding settlement towards wilderness areas provides more chances for viruses to meet people. Preservation of natural ecosystems and stop the extinction of endangered species will be an efficient method for preventing cross-transmission of viruses from animals to humans. Since we all know that until now, there are no specific drugs and no efficient treatments for most dangerous human viruses. Effective vaccine is the only way to prevent and control viral infection (Smith & Carr, 2018).

The outbreak of the COVID-19 is a model example as an Anthropocene viral disease. It follows a complex sequence involving disruption of the natural, social, economic and governance systems (O'Callaghan-Gordo & Antó, 2020). From the report entitled 'Are we ignoring a black elephant in the Anthropocene? Climate change and global pandemic as the crisis in health and equality', the coronavirus pandemic turned into a serious economic crisis because the political economy *before* the pandemic allowed the virus to deprive economically precarious people of their livelihoods. (Asayama et al., 2021). *"One Health"* is "the collaborative efforts of multiple disciplines working locally, nationally, and globally, to attain optimal health for people, animals and our environment", as defined by the One Health Initiative Task Force. *One Health is based on the* premise that human health, animal health and ecosystem health are inextricably linked. The *One Health approach is needed to prevent future pandemics - simultaneously addressing human, animal and ecosystem health, protecting humanity and nature, and incorporating disease risk into decision-making* (O'Callaghan-Gordo & Antó, 2020).

It is hard to believe how modern societies with so much wealth, power, and knowledge could be so short-sighted and negligent about the threat of diseases emerging in animals and spreading to humans (Keim, 2021). Preventing the next viral pandemic will require global cooperation. It requires bringing together experts from multiple disciplines – medical doctors, public health, epidemiologists, veterinarians, zoologists, agriculture and environment professionals, finance ministers, trade organizations, as well as business leaders, ordinary peoples, and others. The viral outbreak is a sign that by going too far in exploiting the rest of nature, the dominant globalizing culture has undone the planet's capacity to sustain life and livelihoods. The current pandemic is just one aspect of the human-made planetary crisis known as the Anthropocene (Kothari, 2020). What we can learn from the Covid-19 outbreak or pandemic in the Anthropocene, and how we think the present state of the world? There is a reason to think of this viral pandemic as one of the more evident manifestations of 'the human epoch' or Anthropocene (Heyd, 2021).

Can the crisis due to the Coronavirus pandemic make us reflect on the need for a "geoethic look" at human behaviour towards the environment? It is time and necessary to put environmental ethics before economic interests and to adopt an integrated and multidisciplinary perspective to solve the great challenges related to the global change (De Pascale & Roger, 2020). To adapt viral pandemic preparedness to the Anthropocene, policymakers must focus on multilateral cooperation to develop stronger and more equitable public health governance, to strengthen health systems for universal health coverage, and to adapt viral surveillance to the pressures of a changing world (Guynup, 2021).

To achieve viral pandemic preparedness in the Anthropocene, policymakers must take multilateral and multidisciplinary cooperation and to construct the way for stronger and more equitable global health governance, health systems, and scientific research (Carlson et al., 2021). Global health must continue working on building 'health for all' and 'One Health' into our understanding of the Anthropocene.

References

Aronsson, A., & Holm, F. (2020). Multispecies entanglements in the virosphere: Rethinking the Anthropocene in light of the 2019 coronavirus outbreak. *The Anthropocene Review, 9*, 24–36. https://doi.org/10.1177/2053019620979326

Asayama, S., Emori, S., Sugiyama, M., Kasuga, F., & Watanabe, C. (2021). Are we ignoring a black elephant in the Anthropocene? Climate change and global pandemic as the crisis in health and equality. *Sustainability Science., 16*, 695–701. https://doi.org/10.1007/s11625-020-00879-7

Carlson, C. J., Albery, G. F., & Phelan, A. (2021). Preparing international cooperation on pandemic prevention for the Anthropocene. *BMJ Global Health, 6*, e004254. https://doi.org/10.1136/bmjgh-2020-004254

De Pascale, F., & Roger, J. (2020). Coronavirus: An Anthropocene's hybrid? The need for a geoethic perspective for the future of the Earth. *AIMS Geosciences, 6*(1), 131–134. https://doi.org/10.3934/geosci.2020008

Guynup, S. (2021). Address risky human activities now or face new pandemics, scientists warn. *Mongabay*, August 3.

Heyd, T. (2021). Covid-19 and climate change in the times of the Anthropocene. *The Anthropocene Review, 8*, 21–36. https://doi.org/10.1177/2053019620961799

Keim, B. (2021). Here's how we avoided the worst of zoonotic diseases. *Anthropocene magazine*. https://www.anthropocenemagazine.org/2020/08/a-memo-from-the-year-2050/

Kothari, A. (2020). Coronavirus and the crisis of the Anthropocene. *The Ecologist*, March 27.

O'Callaghan-Gordo, C., & Antó, J. M. (2020). COVID-19: The disease of the Anthropocene. *Environmental Research, 187*, 109683. https://doi.org/10.1016/j.envres.2020.109683

Pal, D. (2021). Mankind and the virus. 2021. *The Indian EXPRESS*, November 03.

Ray, S. (2021). The Virality of pandemics: Reassembling the social in the Anthropocene. *Society and Culture in South Asia, 7*(1), 16–31. https://doi.org/10.1177/2393861720975115

Rodrigues, R. A. L., Andrade, A. C. D. S. P., Boratto, P. V. C. M., Trindade, G. S., Kroon, E. G., & Abrahão, J. S. (2017). An anthropocentric view of the virosphere-host relationship. *Frontiers in Microbiology, 8*, 1673. https://doi.org/10.3389/fmicb.2017.01673

Smith, J. R., & Carr, J. (2018). Humans are to blame for the rise in dangerous viral infections. *The conversation*, July 3.

Zimmer, C. (2021). Welcome to the virosphere. *The New York Times*, November 4.

Dr. Chung-Ming Chang is an Assistant Professor of the Master & Ph.D. Program in Biotechnology Industry at Chang Gung University (CGU), Taiwan. He graduated from National Taiwan University with both his bachelor's and master's degrees in veterinary medicine. He also holds a master's degree in veterinary clinical medicine from France's École Nationale Vétérinaire d'Alfort, with a focus in avian pathology. From the French Institut National Agronomique Paris-Grignon, he received his Ph.D. in Animal Genetics. He discovered a single, fully developed endogenous retroviral insertion in chickens that is associated to the recessive white mutation. He worked as a researcher at the Pasteur Institute in Paris and the French National Center for Scientific Research (CNRS), where he was in charge of a research program on influenza viruses in wild birds, before joining the Research Center for Emerging Viral Infections (RCEVI) at Chang Gung University. Chung-Ming Chang is interested in studying human endogenous retroviruses that may be linked to various autoimmune disorders, as well as influenza in both humans and animals with a focus on ecological and epidemiological investigations. Professor Chang has played a key role in developing joint research initiatives between Chang-Gung University and French collaborators. He is now the president of the Association Franco-Taiwanese des échanges académiques, culturels, et économiques (AFTEACE) in Taiwan. He is the authors of more than 25 articles and one book on virology, biomedicine and pharmacotherapy, One-Health, biosensor, veterinary medicine, genomics, animal genetics and biotechnology. Selected publications: Advanced biosensors for detection of pathogens related to livestock and poultry. 2018. *Veterinary Research*. Nanoparticles as an effective drug delivery system in COVID-19. 2021. *Biomedicine & Pharmacotherapy*.

Part IV
Consumable Life?

Energy

Mathis Wackernagel and David Lin

Abstract Energy enables life and any economic process. With few exceptions, the energy used by life on Earth originates from the sun. As energy enables physical work, it is a fundamental concept for anybody analyzing societies or processes, particularly when recognizing that economies are physical subsystems of the biosphere. Focusing on energy flows makes obvious that true economics is a life science.

Energy powers life and any economic process. With few exceptions (such as fusion, fission, Earth's underground heat, and tidal energy) the energy used by life is powered by the Earth's present or past "income" of solar radiation. This energy flux amounts to about 173,000 Terawatt (or TW) electro-magnetic light energy beaming on planet Earth. This flux is far larger than the roughly 18 TW of commercial energy powering the human systems. Currently, humanity eats about 1 TW (or just under 10,000 TW hours per year) of food, with food production occupying about half the planet's $\rightarrow$ *biocapacity*. In other words, $173,000/2 = 87,000$ TW of solar input currently produces about 1 TW of human food.

This last example illustrates the vast cascading effects in energy flows. 87,000 TW of solar energy income (via great amounts of widely available but diffuse sun light) trickle down to 1 TW of concentrated energy (small amounts of highly valuable, concentrated food). This cascading makes energy comparisons in complex systems challenging. Such cascading also illustrates the various levels of energy quality (or low entropy). Low entropy means high level of order or high level of quality, such as concentrated gasoline. High entropy is the opposite, i.e., low quality of energy such as waste heat from a car engine. This "more entropy = less quality" communication challenge led to the use of the term negentropy: more negentropy means more quality.

M. Wackernagel (✉) · D. Lin
Global Footprint Network, Geneva, Switzerland
e-mail: mathis@footprintnetwork.org; david.lin@footprintnetwork.org;
https://www.footprintnetwork.org

© The Author(s), under exclusive license to Springer Nature
Switzerland AG 2023
N. Wallenhorst, C. Wulf (eds.), *Handbook of the Anthropocene*,
https://doi.org/10.1007/978-3-031-25910-4_38

Energy flows can be better understood with the help of the basic laws of thermo-dynamics. The first law stipulates that energy within a system is constant. Hence in an energy cascade as described above, no energy disappears. But there is a large difference between the amount of incoming solar light and the amount of biomass that is being produced using that solar light. This difference is waste energy that eventually turns into low-grade heat. In equilibrium, the amount of waste heat leaving the system is identical to the rate of solar light coming into the system.

The second law of thermodynamic reflects irreversibility or directionality: order erodes spontaneously, meaning entropy never decreases on its own. Here a specific example: when pulling the brakes of a bicycle, you turn the moving energy (the "kinetic energy") of the bicycle into heat in the brake pads. This heat in the brake pads cannot be used again to reaccelerate the bicycle back to its original speed. Regenerative brakes using electric generators and batteries can reuse some of the energy; but also there, the efficiency of reusing this gained electricity is lower than 100%, with some inevitable waste heat being produced. The reason is that the quality of the moving energy is higher than the heat energy. So, the former can be trans-formed into the latter, but very little of the latter can be transformed into the former. This also illustrates that energy flows uni-directionally from higher to lower quality. It is still possible to produce high-quality energy from low-quality energy, but not the same amount. It takes more low-quality energy to produce less high-quality energy, and the difference in energy becomes even lower waste heat.

In a universe governed by the second law of thermodynamics, all isolated systems are expected to approach a state of maximum disorder. Therefore, some wondered whether life contradicts the second law of thermodynamics. Because life produces order, apparently spontaneously. This is sometimes called Schrödinger's "paradox" based on Schrödinger raising the issue in his 1944 book "*What is Life?*" (Schrödinger, 1944). However, life's ability to produce order does not contradict the second law of thermodynamics as life uses energy, primarily solar energy beaming onto the planet, to build that order. In fact, as Eric Schneider and James Kay pointed out, creating order (as life does) enables even higher rates of dissipating high-quality energy (or negentropy) than if there was no life (Schneider & Kay, 1994). In other words, there is no contradiction between the second law of thermodynamics and life. However, our planet is not an isolated system receiving a massive influx of solar energy. The increase of order inside a living organism is more than compensated by an increase in disorder outside this organism, mostly through waste heat.

To be clear, and to be consistent with language in physics, it would have been more appropriate in this article to use the term "power" rather than "energy". Because we are talking here about flows of energy, or how much energy is used per unit of time (i.e. "power"), not the stock or amount of energy. But in the economic or sustainability literature, typically the term "energy" is used, even when "power" is meant. Hence, we are using the term "energy" in this article as well.

As energy is the ability to do work, it is a fundamental concept for anybody analyzing societies or processes. Entire bodies of work have been built on developing ways to evaluate society from the energy lens. Such analysis starts from conventional energy statistics documenting how much coal, gas, oil, and other energies societies consume. There, a distinction is made between primary energy (such as coal, gas, or uranium) and energy carriers produced from these primary inputs (such as electricity or hydrogen). But in reality, the distinction between the two is fluid. For instance: Is solar power a primary energy? It certainly is when it powers a photovoltaic panel, but that solar power would not enter the primary energy statistics of conventional statistical reports, only the produced electricity would.

Others have built entire explanatory systems of ecosystems or societies using energy as their unit of analysis. The most prominent advocates of such thinking may be the Odum family of ecologists and engineers (Odum & Barrett, 2005). H.T. Odum proposed eMergy as a unit of measurement, representing all the energy already embodied in the energy a process consumes, which means going back in the energy cascade all the way to the original sunlight (2007). EMergy therefore would capture all the solar energy required to enable a particular product or activity (Odum, 2007). Combined with the "maximum power principle" proposed by Lotka (1922), Odum and his associates developed a framework to analyze the dynamics of systems such as economies, cities or countries through this energy lens. The maximum power principle is still a hypothesis. It is less established and less universal than the basic two laws of thermodynamics mentioned above. A better name would most likely be "Maximum Entropy Production" (Sciubba, 2011).

Charles Hall developed Odum's approach further. One of his fundamental contributions was the concept of energy return on energy invested (EROI), recognizing that energy used by people comes at an energy cost (Hall & Klitgaard, 2012; Lambert et al., 2014). The less energy people need to invest to harvest a unit of energy, the easier it becomes to operate (and expand) systems. Conversely, when the ratio becomes low, systems become self-limiting. Studies indicate that EROI of 3 and lower self-limit systems (Hall et al., 2009; Lambert et al., 2014).

Energy is also used to track productivity of ecosystems. There, the term "net primary productivity" and its associated concepts are typically used. The net primary productivity of an ecosystem refers to the biomass an ecosystem can accumulate, typically within 1 year. The energy can be measured in terms of the chemical energy of that biomass, the dry weight of the biomass, or the carbon content of the biomass. These three approaches are nearly identical as the energy content of biomass is largely determined by its carbon content, and so is its dry weight. But there are also limitations in this approach, particularly when comparing ecosystem productivity to harvest rates. Harvest rates cannot as easily be expressed in tons removed. For instance, if a tree is harvested, it is not clear which aspects need to be counted: leaves, branches, roots, soil loss? Or just timber that is being carried out of the forest? How much of net primary productivity can be sustainably removed is also unclear; probably significantly less than 100%.

Even when analyzing economies, using classical energy metrics can lead to confusion. For example, some have proposed energy consumption targets. A prominent example is Switzerland's approach of a "2000-Watt society" (Novatlantis, SIA, Swissenergy, 2007). This approach recommends reducing Switzerland's per person primary energy consumption from about 6000 Watt (this was the level of primary energy use in Switzerland when the concept was prosed in 1998) to 2000 Watt. This target gives a clear direction. But the question remains: What Watt? Does the 2000-Watt budget include the sunlight that keep the Earth's temperature above zero Kelvin (the universe's lowest temperature)? Does it count the Watt of solar light that flows through the window and warms the floor of the house? Does it include the sunlight Watt powering the house's own photovoltaic panel? Does it include the sunlight Watt powering the utility size photovoltaic panel? Does it include the ancient sunlight that produced the coal burned in the electric coal plant? These questions demonstrates that even a rigorous physical measurement unit like Watt can become ambiguous when applied to real life situations.

In summary, energy is a foundational concept, and essential for grounding economic theories in physical reality. Sometimes, by using innovative ways of measuring energy in units transcending the traditional Joules (or Watts), such as by tracking carbon weight of biomass, agricultural yields, or →*biocapacity*, comparisons become clearer and more meaningful for such societal analysis. A more robust way of comparing demand on the planet, that even primary school kids understand, may be to put energy flows in their biological context by inquiring: how much do we take from nature compared to how much is being renewed? Or, how much of our planet's regeneration is occupied by a particular Watt of consumption, whether electricity, a piece of firewood, or a particular amount of bioethanol?

References

Hall, C. A. S., & Klitgaard, K. A. (2012). *Energy and the wealth of nations: Understanding the biophysical economy*. Springer Science.

Hall, C. A. S., Balogh, S., & Murphy, D. J. R. (2009). What is the minimum EROI that a sustainable society must have? *Energies, 2*(1), 25–47. https://doi.org/10.3390/en20100025

Lambert, J. G., Hall, C. A. S., Balogh, S., Gupta, A., & Arnold, M. (2014, January). Energy, EROI and quality of life. *Energy Policy, 64*, 153–167. https://doi.org/10.1016/j.enpol.2013.07.001

Lotka, A. J. (1922). Contribution to the energetics of evolution. *Proceedings of the National Academic Sciences, 8*, 147–151. https://www.pnas.org/content/pnas/8/6/147.full.pdf

Novatlantis, SIA, Swissenergy. (2007). *Smarter living: Generating a new understanding for natural resources as the key to sustainable development – the 2000-Watt society*, https://web.archive.org/web/20070927040640/http://www.novatlantis.ch/pdf/leichterleben_eng.pdf

Odum, E., & Barrett, G. W. (2005). *Fundamentals of ecology*, (Cenage learning; 5th ed.).

Odum, H. T. (2007). *Environment, power, and society for the twenty-first century: The hierarchy of energy*. Columbia University Press.

Schneider, E. D., & Kay, J. J. (1994, March–April). Life as a manifestation of the second law of thermodynamics. *Mathematical and Computer Modelling, 19*(6–8), 25–48. https://doi.org/10.1016/0895-7177(94)90188-0

Schrödinger, E. (1944). *What is life?* Cambridge University Press.

Sciubba, E. (2011, April 24). What did Lotka really say? A critical reassessment of the "maximum power principle". *Ecological Modelling, 222*(8), 1347–1353. https://doi.org/10.1016/j.ecolmodel.2011.02.002

Dr. Mathis Wackernagel created the *footprint* concept in the early 1990s with Prof. William E. Rees. The carbon footprint has become the most popular variant. In 2003, he founded Global Footprint Network, a sustainability think-tank, making planetary constraints relevant to decision-making. Its largest engagement campaign is its annual Earth Overshoot Day. Mathis's honors include the 2018 World Sustainability Award, the 2015 IAIA Global Environment Award, and the 2012 Blue Planet Prize.

Dr. David Lin leads Global Footprint Network's research team, and contributes to the production, development, and improvement of the National Footprint and Biocapacity Accounts. Prior to joining Global Footprint Network, David earned his Ph.D. and worked as a post-doctoral researcher in the Systems Ecology Laboratory at the University of Texas at El Paso. His research focused on integrating models of ecosystem function with land cover change analysis in Arctic ecosystems. David is a native of California, and holds a BS in ecology, behaviour, and evolution from the University of California, Los Angeles.

Food

Hannes Bergthaller

Abstract This article outlines the role which changes in food production have played in the transition to the Anthropocene, focusing particularly on plantation agriculture and the impact of fossil fuels on agricultural practices. It argues that food is crucial in order to understand how the relationship between humans and the Earth has been transformed, and that a further transformation of the food system will be necessary in order for human and nonhuman life to flourish in the new geological epoch.

One of the stratigraphic markers separating the Holocene from the new geological epoch of the Anthropocene will be chicken bones. With a standing population of well over 20 billion, the number of domestic chicken today exceeds that of even the most numerous wild birds by at least an order of magnitude; by some estimates, they now account for 70% of the world's total avian biomass. But future archaeologists will recognize Anthropocene chickens not just by the sheer prevalence of their remains: since the middle of the twentieth century, advanced techniques for breeding and stock raising have increased the body mass of individual broiler chickens (i.e., chickens raised for their meat) roughly fivefold, and they reach slaughtering weight within a mere 5–7 weeks. They thus differ markedly from their ancestors with regard both to their anatomy and in the chemical composition of their bones (Bennett et al., 2018).

The story of the modern broiler chicken is a particularly striking example of the central role which food production plays in the ongoing reconfiguration of the biosphere and the larger Earth system. All of the geophysical and most of the socioeconomic indicators included in the iconic set of graphs illustrating the "Great Acceleration" – such as increases in water use and fertilizer consumption, the decline of fisheries, loss of biodiversity and forest cover, rising concentrations of nitrous oxide and methane in the atmosphere – are closely linked to qualitative and

H. Bergthaller (✉)
Department of English, National Taiwan Normal University, Taipei, Taiwan
e-mail: hannes.bergthaller@ntnu.edu.tw

© The Author(s), under exclusive license to Springer Nature Switzerland AG 2023
N. Wallenhorst, C. Wulf (eds.), *Handbook of the Anthropocene*,
https://doi.org/10.1007/978-3-031-25910-4_39

251

quantitative changes in how food is produced, processed, distributed, and consumed (Steffen et al., 2011). These changes are, in turn, deeply entangled with cultural, economic, and political changes, many of which accelerated dramatically around the middle of the twentieth century, although their origins can often be traced much further into the past.

One important strand of these developments is the emergence of plantation-style agriculture in the European overseas colonies. The tremendous profitability of sugar cultivation in the Americas became one of the main drivers of transatlantic trade from the sixteenth century onwards, spurring not only the rapid expansion of the slave economy, but also of European manufacturing and an ever-growing international market for cash crops. Anna Tsing has argued that sugar plantations helped to establish a basic organizational pattern of capitalist modernization: pre-existing social and ecological networks were wiped out and replaced by a radically simplified system whose components were shipped in from abroad. Both labourers and cultivars were regularized in a manner that ensured a high degree of interchangeability and control over the production process. The ease with which this pattern could be replicated, with little regard to socio-ecological context, allowed for seemingly unlimited expansion (Tsing, 2015, p. 37). The "scalability" of plantation agriculture, as Tsing describes these characteristics, is crucial in order to understand the vast homogenization of terrestrial ecosystems which sets in during the sixteenth century (Mann, 2011), and it made the sugar plantation into a model for the emergent industrial factory system (Mintz, 1985, pp. 57–60).

But plantation agriculture also paved the way for industrialization in a more immediate sense. The "ghost acres" of the New World allowed Western European nations to evade the productive constraints of traditional farming on their own territories (Pomeranz, 2000, p. 275). By the late eighteenth century, the most densely populated and technologically advanced areas of the world had hit an ecological bottleneck which limited the possibilities for further urbanization and economic specialization. Britain was able to solve this problem not only through its increased reliance on coal, but just as importantly through trade with the Americas: imports of fibre and fuel (most importantly cotton, hemp, and wood) allowed more land to be used for growing food domestically, including such new cultivars as potatoes and maize, which were much more productive than traditional staple crops such as wheat or rye. Meanwhile, cane sugar gradually turned from a prized luxury item into a major source of carbohydrates for the industrial labour force, accounting for no less than 20% of the daily caloric intake of British workers by 1900 (Pomeranz, 2000, p. 275).

From the very outset, the transition to an energy system based primarily on fossil fuels went hand in hand with equally profound changes in the food system, with both conspiring to lift the age-old "Malthusian curse" (Ladurie, 1971, p. 311) and sustaining a historically unprecedented period of economic and demographic growth. By drastically lowering transportation costs, steam ships and railroads hastened the shift from subsistence to commercial agriculture. Together with new forms of refrigeration, they also enabled long-distance trade in perishable goods such as meat, seafood, and fruit, severing the close linkage between people's diet and the seasons which had prevailed throughout the history of our species. Today, as much as 25% of the world's food is traded on international markets (Clapp, 2020, p. 66).

Perhaps the most consequential confluence of fossil energy and food production, however, resulted from Justus Liebig's discovery of the essential role of fixed nitrogen (e.g. in the form of urea or ammonia, as opposed to molecular nitrogen as it is found in the atmosphere) in plant growth. The usefulness of guano and saltpeter as fertilizers had been recognized well before Liebig published his insights in 1840, and the commercial exploitation of the largest deposits in Peru and Chile was crucial in sustaining high agricultural yields in Europe and North America during the second half of the nineteenth century (Leigh, 2004, pp. 81–86). But it was always clear that these resources were limited, and by the end of the century, a race was on among the world's chemists to devise a formula for creating ammonia in the laboratory. The breakthrough came in 1913, on the eve of World War I, when Fritz Haber and his collaborators developed what came to be known as the "Haber-Bosch process." For the first time, it was now possible to synthesize ammonia at an industrial scale.

Ammonia is also a crucial component in the production of explosives. After WWI, production capacities built up for the war effort were converted to civilian use – a process that repeated itself, at a much larger scale, after WWII. Exporting the model of industrial agriculture around the globe became a key component of US strategy during the Cold War, culminating in the so-called "Green Revolution" of the 1960s and 70s (which also led to the spread to synthetic pesticides, another product of the petrochemical industry; Clapp, 2020, pp. 40–47; Carruth, 2013, pp. 5–6). Over the past century, the annual consumption of artificial nitrogen fertilizers worldwide has risen from 950kT to 118,763kT (Smil, 2004, p. 245; FAO, 2017) – more than a hundredfold. The amount of fixed nitrogen produced by the fertilizer industry is now roughly equivalent to that of total natural nitrogen fixation by terrestrial ecosystems (Smil, 2004, p. 178).

The consequences of this development for the Earth system are hard to overstate. For one thing, it contributes directly to global warming: the most important feedstock for the Haber-Bosch process and related production methods is methane, and the process requires enormous amounts of energy which are usually supplied by fossil fuels. For another, the surfeit of fixed nitrogen has profoundly changed many of the world's ecosystems, leading to the eutrophication of lakes and rivers, the increasing spread of "dead zones" in many coastal areas of the world, and to the release of nitrous oxide, another potent greenhouse gas. While it is conceivable that the world's energy production will someday soon become "carbon neutral," there is no substitute for fixed nitrogen, which represents a hard cap on plant growth. Without artificial nitrogen fertilizers, it would have been impossible to more than triple the average yields for cereal grains since the beginning of the last century; even the most intensive forms of traditional agriculture would not be able to feed much more than half of the current world population (Smil, 2004, p. 204). Despite the fact that the number of people has roughly quadrupled over the past century, the main reason that many millions still go hungry is not an absolute lack of food, but rather its inadequate distribution. In most nations, obesity is now more of a concern than undernutrition (Swinburn et al., 2019). It is not only the anatomy of the broiler chicken that changed in the Anthropocene.

Fossil energy has fundamentally changed the character of agriculture, and thus society's relationship to the biosphere. For much of the Holocene, farming had been the primary source of energy for human societies: it harnessed the capacity of plants to convert solar energy into biomass, feeding the organic bodies, human or animal, which performed most physical labour – and most physical labour was performed on farms so as to satisfy the basic energy needs of society. Today, agriculture has become another branch of industrial production, employing only a miniscule fraction of the labour force, whose purpose is the conversion of biochemical raw materials into food. It has turned from a net producer into a consumer of energy (Sieferle, 2000, p. 15).

In many countries, there is a growing awareness of these issues which finds expression in movements that seek to change how people eat – sometimes for reasons of individual health or animal welfare, but just as often from an understanding of the profound injustices of the modern food system and its destructive effects on the biosphere. Vegan, locavore, or "organic" diets are frequently advocated as ways of opting out of a monstrous system which inundates the world with "Frankenfood," returning us instead to a more "natural" diet. These movements have already had many beneficial effects (such as reversing the long-term trend of rising meat consumption in many Western countries). It is important to remember, however, that they continue to operate within the larger context of a food system that remains dependent on fossil energy. This system has largely untethered dietary habits from local ecological conditions, such that more than at any other time in human history, the question what to eat has become a matter of individual choice – and thus a factor in curating social and political identities, as evidenced by the outsized role that the representation of food has come to play in social media communication (Pancer et al., 2022).

The problem with today's food system is not that it has estranged humans from nature or replaced nature with human artifice, as traditional environmentalist thinking would suggest. Rather the opposite seems to have happened: human activities have become ever more deeply interwoven with biological processes. The vast expansion of human numbers over the past two centuries was made possible not by any belief that humans were somehow separate from the natural world, but on the contrary by the understanding that they are one biological species among others. The story of the modern broiler chicken as recounted above may appear like a particularly ruthless example of human domination over nonhuman nature. However, as Cary Wolfe has argued, the technological interventions which made possible such an "unnatural growth of the natural" (to borrow a phrase from Hannah Arendt, 1998, p. 47) are substantially similar to those which also enabled the proliferation of human bodies (most notably antibiotics, environmental hygiene, scientific management of nutrition; Wolfe, 2012, p. 9). From a biopolitical vantage point, the transition from the Holocene to the Anthropocene is characterized not by the human subjugation of nature, but rather by the penetration of all sorts of biological processes by new forms of power which aim to maximize their productivity (Bergthaller, 2020, p. 164).

For humans and nonhumans to flourish in the Anthropocene, today's food system will have to change profoundly. Intuitions about what is or is not natural which are based on human experience in the Holocene will offer little guidance in this respect. That is not to deny, of course, that the agricultural practices of the past can teach us valuable lessons concerning the long-term viability of human-modified biomes. Yet the regions of the world where traditional subsistence farming is still widely practiced are often the same regions which are already heavily affected by climate change. It will be difficult, if not impossible, to mitigate these effects and avoid hunger on a vast scale without the further spread of scientific farming techniques (including the introduction of new genetically modified organisms) and the global distribution of harvest surpluses. Whatever new shapes the food system will assume as it adapts to the conditions of the Anthropocene, they will amplify the effects of human choices on the Earth, and they will reflect the pressures which the changing Earth system exerts on the structure of our desires. Food, like the Anthropocene itself, is a Gordian knot of nature and culture that will not be cut.

References

Arendt, H. (1998). *The human condition*. Chicago University Press. [Original edition 1958].

Bennett, C. E., et al. (2018). The broiler chicken as a signal of a human reconfigured biosphere. *Royal Society Open Science, 5*(12). https://doi.org/10.1098/rsos.180325

Bergthaller, H. (2020). Ecocriticism, biopolitics, and ecological immunity. *Ecozon@, 11*(2), 162–168. https://doi.org/10.37536/ecozona.2020.11.2.3542

Carruth, A. (2013). *Global appetites: American power and the literature of food*. Cambridge University Press.

Clapp, J. (2020). *Food*. Polity.

Food and Agriculture Organization of the United Nations (FAO). (2017). *World fertilizer trends and outlook to 2020: Summary report*. Rome. https://www.fao.org/3/i6895e/i6895e.pdf

Ladurie, E. L. (1971). *Times of feast, times of famine: A history of climate since the year 1000* (G. May, Trans.). Doubleday. [Original edition 1967]

Leigh, G. J. (2004). *The world's greatest fix: A history of nitrogen and agriculture*. Oxford University Press.

Mann, C. C. (2011). *1493: Uncovering the New World Columbus created*. Vintage.

Mintz, S. (1985). *Sweetness and power: The place of sugar in modern history*. Penguin.

Pancer, E., et al. (2022). Content hungry: How the nutrition of food media influences social media engagement. *Journal of Consumer Psychology, 32*(2), 336–349. https://doi.org/10.1002/jcpy.1246

Pomeranz, K. (2000). *The great divergence China, Europe, and the making of the modern world economy*. Princeton University Press.

Sieferle, R. (2000). Im Einklang mit der Natur. In B.-M. Baumunk & J. Joerges (Eds.), *7 Hügel – Bilder und Zeichen des 21. Jahrhunderts* (pp. 12–16). Henschel.

Smil, V. (2004). *Enriching the earth: Fritz Haber, Carl Bosch, and the transformation of world food production*. MIT Press.

Steffen, W., et al. (2011). The Anthropocene. Conceptual and historical perspectives. *Philosophical Transactions of the Royal Society A, 369*(1938), 842–867.

Swinburn, B. A., et al. (2019). The global syndemic of obesity, undernutrition, and climate change: The Lancet Commission Report. *The Lancet, 393*(10173), 791–846.

Tsing, A. (2015). *The mushroom at the end of the world: On the possibility of life in the capitalist ruins*. Princeton University Press.

Wolfe, C. (2012). *Before the law: Humans and other animals in a biopolitical frame*. Chicago University Press.

Hannes Bergthaller is a professor at the English Department of National Taiwan Normal University in Taipei, Taiwan. His work focuses on the literature and cultural history of modern environmentalism, systems theory and neocybernetics, and environmental philosophy. His publications cover a wide range of issues and authors, including the science fiction of Kim Stanley Robinson, energy in John Updike's *Rabbit* trilogy, and posthumanist perspectives on the Anthropocene. Together with Eva Horn, he co-authored *The Anthropocene: Key Issues for the Humanities* (Routledge, 2020).

Geoengineering

Augustine Pamplany and Bert Gordijn

Abstract Geoengineering is a technological response to anthropogenic climate change. There are two kinds of geoengineering: Solar Radiation Management (SRM) and Carbon Dioxide Removal (CDR). SRM aims at reducing the amount of incoming solar light and CDR at reducing the CO_2 concentration in the atmosphere. Over the past decades, geoengineering has moved from a fringe proposal to a more mainstream contender along with mitigation and adaptation to avert climate change. However, it faces important ethical challenges.

Historical Sketch

The Royal Society defines geoengineering as "the deliberate large-scale intervention in the Earth's climate system, in order to moderate global warming" (Royal Society, 2009: 1). In the early 1970s Russian climatologist, Mikhail Budyko, advanced ideas in Russian publications to modify the global climate focused on enhancing aerosol particle concentrations in the stratosphere (see Budyko, 1977: 239 for references to these earlier publications in Russian). The term 'geoengineering' itself, however, seems to have been first used in relation to the climate by Italian physicist, Cesare Marchetti, in 1977 to refer to a method for getting rid of atmospheric CO_2 by injecting it into sinking thermohaline ocean currents (Marchetti, 1977).

In 1991, the champions of geoengineering received a shot in the arm because of the Mount Pinatubo eruption in the Philippines. It threw up millions of tons of ash serving as an aerosol cloud. Subsequent measurements showed a significant reduction in absorbed solar radiation in the following two years caused by the increase in

A. Pamplany (✉)
Institute of Advanced Interdisciplinary Studies, Aluva, Kerala, India
e-mail: apamplany@gmail.com

B. Gordijn
Institute of Ethics, Dublin City University, Dublin, Ireland
e-mail: bert.gordijn@dcu.ie

albedo (Wielicki et al., 2005; Harries & Futyan, 2006). This was seen as a natural proof for the effectiveness of aerosol geoengineering. In 2005, Russian scientist Yuri Izrael even proposed the idea of burning sulfur in the stratosphere to President Vladimir Putin to contain global warming (Fleming, 2010).

In the same year, Lovelock opined that earth's tipping point had already been crossed (Lovelock, 2006). Further, in 2008, NASA scientist James Hansen wrote that "planet is dangerously near a tipping point" (Hansen, 2008: 6). These warnings gave an element of urgency to the debate on geoengineering.

In 2006, Nobel laureate Paul Crutzen discussed stratospheric albedo enhancement based on sulfur injections as a policy option (Crutzen, 2006), which – given his credibility as a scientist – sparked further serious debate on geoengineering to avert anthropogenic climate change. Several conferences on geoengineering followed at the American Academy of Arts and Sciences (2007), MIT (2009), and at Asilomar in California (2010). In 2009, Royal Society published a special report on geoengineering (Royal Society, 2009). Further, it devoted a special issue of Philosophical Transactions to geoengineering (Latham et al., 2014). In 2011, the Intergovernmental Panel on Climate Change produced an expert meeting report on geoengineering (IPCC, 2011), and in 2014, geoengineering was discussed in its formal report (IPCC, 2014). In 2020, the US Congress approved a grant of $4M to David Fahey at the US National Oceanic and Atmospheric Administration for projects on sulfur dioxide aerosol injection and cloud whitening (Fialka, 2020).

SRM

SRM proposes to confront global warming by enhancing the albedo or reflexivity of the earth by managing radiations. There are surface-based and space-based schemes. Particular surface albedo schemes include brightening urban areas, croplands, and deserts. Albedo could be increased by painting white on rooftops, roads and pavements (Akbari et al., 2009: 275). Covering deserts with a reflective surface is another option, as is the development of modified plants to augment albedo. Cloud albedo could be enhanced by dispersing cloud condensation nuclei of sea salt. Salt crystals would suck up moisture and grow big, thus reflecting more sunlight (Royal Society, 2009: 27–29).

The most discussed geoengineering idea proposes releasing a large quantity of aerosols in the stratosphere after the natural model of the Pinatubo eruption (Crutzen, 2006; Wigley, 2006). More theoretical are proposals to place huge reflective space mirrors in a low earth orbit, which will deflect the incoming radiation, or to create a Saturn-like ring of dust particles with shepherding satellites. Given the difficulties in achieving this, the Royal Society does not consider these proposals to be a "realistic potential contributor" (Royal Society, 2009: 33).

CDR

The Royal Society distinguishes land-based and ocean-based CDR techniques (Royal Society, 2009: 10–18). An example of the first is large-scale air-capture of CO_2 with scrubbers. This can be done by absorption of CO_2 on solids using surfaces derived from commercial resins or by absorption into alkaline solutions (Royal Society, 2009: 15) A second example is 'bioenergy carbon sequestration' where biomass is harvested and used as fuel. The resulting CO_2 is captured and sequestered (Royal Society, 2009: 16). A third approach is 'enhanced weathering'. The common rocks on earth are formed from silicates. Silicates react with CO_2 and form carbonates. Carbon is consumed in this process. However, the rate of this natural reaction is very slow. Enhanced weathering schemes propose to accelerate the rate of these reactions. For example, adding abundant silicate minerals such as olivine to agricultural soil will significantly accelerate the natural weathering process. This method is accepted as chemically capable of reducing the atmospheric concentration of CO_2 (Royal Society, 2009: 12–13).

Ocean based-techniques include iron fertilization where nutrients like iron or nitrogen are dropped into ocean, thus facilitating the growth of the phytoplankton. These then store carbon in their cells in the photosynthesis process, and finally as they die, sequestrate it in the deep sea (Royal Society, 2009: 16–17). Another ocean-based CDR technique involves upwelling and downwelling of ocean currents. Overturning the ocean currents increases the supply of nutrients. This facilitates the growth of phytoplankton, which absorbs more atmospheric CO_2 (Royal Society, 2009: 19).

Ethics of Geoengineering

Understandably, the different geoengineering schemes have been generating a lot of ethical debate with a sharp divide between the proponents and opponents. Dubbed as Plan B, the proponents present geoengineering as a lesser evil and a last resort. As the earth has already crossed several tipping points, geoengineering is seen as an appropriate climate emergency intervention that buys time for mitigation. Its benefits would outweigh the harm. Arguments are also raised from the point of view of feasibility. Given the high reward from every dollar spent, geoengineering easily passes the cost-benefit test. The net benefit is claimed to increase by several trillion dollars if geoengineering is combined with mitigation. As such geoengineering is very attractive from an economic angle. It is argued that a single nation or a group of nations can develop and deploy the SRM schemes, thus avoiding the delay and difficulty in achieving a global consensus. Further, certain variants of geoengineering require only a short-term deployment. Geoengineering research also enhances human innovation and knowledge, creates jobs, and fulfils our obligations to future generations. For all these reasons, for the proponents, research into geoengineering is morally justified (see Pamplany et al., 2020: 3077–3085 for a detailed overview of the pro arguments and further references).

On the other side of aisle, opponents of geoengineering raise several objections. First, as a technology deployed in an open and non-encapsulated system, there will likely be unprecedented levels and scales of harmful side-effects or so the critics argue. This then could especially worsen the plight of the poor and the most vulnerable populations in the world, thus further compounding the injustice meted out to the victims of climate change. Also, geoengineering does not address the root causes of the climate change. Rather it treats the symptom over the cause. Thereby the present generation acquits itself falsely out of its moral responsibilities and passes the buck to the future generations. Moreover, geoengineering runs the risk of a moral hazard in that the idea that there is a technical solution is likely to subvert efforts of mitigation (see Pamplany et al., 2020: 3085–3095 for a detailed overview of the contra arguments and further references).

Geoengineering is an unprecedented technology in its endeavour to change the earth's climate. Indeed, it "would take anthropogenic influence on the earth to a whole new level" (Preston, 2012: 1). An assessment of the geoengineering debate in its present phase suggests that the required scientific evidence and ethical analysis for well-founded policy decisions are still far from complete.

References

Akbari, H., Menon, S., Rosenfeld, A., et al. (2009). Global cooling: Increasing world-wide urban Albedos to offset CO_2. *Climatic Change, 94*, 275–286.

Budyko, M. I. (1977). *Climatic changes*. American Geophysical Union.

Crutzen, P. J. (2006). Albedo enhancement by stratospheric sulfur injections: A contribution to resolve a policy dilemma? An editorial essay. *Climatic Change, 77*, 211–220.

Fialka, J. (2020, January 2020). U.S. geoengineering research gets a lift with $4 million from Congress. *Science.* https://www.sciencemag.org/news/2020/01/us-geoengineering-research-gets-lift-4-million-congress

Fleming, J. R. (2010). *Fixing the sky – The Checkered history of weather and climate control.* Columbia University Press.

Hansen, J. (2008). Tipping point: Perspective of a climatologist. State of the wild 2008–2009. In W. Woods (Ed.), *A global portrait of wildlife, wildlands, and oceans* (pp. 6–15). Wildlife Conservation Society/Island Press.

Harries, J. E., & Futyan, J. M. (2006). On the stability of the earth's radiative energy balance: Response to the Mount Pinatubo eruption. *Geophysical Research Letters, 33*, L23814.

IPCC. (2011). *IPCC expert meeting on geoengineering Lima, Peru.* 20–22 June 2011 meeting report, Edenhofer, O., et al. Eds. https://archive.ipcc.ch/pdf/supporting-material/EM_GeoE_Meeting_Report_final.pdf

IPCC. (2014). *Climate change 2014, mitigation of climate change working group III contribution to the fifth assessment report of the intergovernmental panel on climate change.* Edenhofer, O., et al. Eds. https://www.ipcc.ch/site/assets/uploads/2018/02/ipcc_wg3_ar5_full.pdf

Latham, J., Rasch, P. J., & Launder, B. (2014). Climate engineering: Exploring nuances and consequences of deliberately altering the earth's energy budget. *Philosophical Transactions of the Royal Society – Mathematical Physical and Engineering Sciences, 372*, 2031.

Lovelock, J. (2006). *The revenge of Gaia: Earth's climate in crisis and the fate of humanity.* Penguin.

Marchetti, C. (1977). On geoengineering and the CO_2 problem. *Climatic Change, 1*, 59–68.

Pamplany, A., Gordijn, B., & Brereton, P. (2020). The ethics of geoengineering: A literature review. *Science and Engineering Ethics, 26*(6), 3069–3119.

Preston, C. J. (2012). The extraordinary ethics of solar radiation management. In C. J. Preston (Ed.), *Engineering the climate: The ethics of solar radiation management* (pp. 1–12). Lexington Books. Paperback edition in 2014.

Royal Society. (2009). *Geoengineering the climate: Science, governance and uncertainty.* www.royalsociety.org

Wielicki, B. A., Wong, T., Loeb, N., Minnis, P., Priestley, K., & Kandel, R. (2005). Change in Earth's Albedo measured by satellite. *Science, 308*, 825.

Wigley, T. M. L. (2006). A combined mitigation/geoengineering approach to climate stabilization. *Science, 314*, 452–454.

Augustine Pamplany obtained a PhD in Ethics from Dublin City University with a thesis on the Justice in Geoengineering. He teaches Ethics, Science and Religion, Philosophy of Science, Indian Philosophy, and Scientific Cosmology at various institutions in India. Augustine has published seven books and several peer-reviewed articles covering the same areas.

Bert Gordijn is Full Professor of Ethics and Director of the Institute of Ethics at Dublin City University in Ireland. He has an extensive record of books, edited volumes, peer-reviewed publications and international lectures on a broad range of topics such as Bioethics, Disaster Ethics, Technology Ethics and Research Ethics.

Mass Extinction

Telmo Pievani and Sofia Belardinelli

Abstract The history of life on Earth has been shaken in the last half billion years by five mass extinctions that have killed at least three-quarters of biodiversity in a geologically short time. These five extinctions were due to major ecological upheavals, with endogenous or exogenous drivers (volcanic eruptions, impact of asteroids, etc.). Today, many data show that the current extinction rate is comparable to or even worse than that of the Big Five mass extinctions of the past. The difference is that this time the asteroid is one species: *Homo sapiens*. There are good reasons to include the Sixth Mass Extinction in the definition of the Anthropocene. This drastic reduction in biodiversity will leave an irreversible geological and paleontological mark. It interacts with climate change, with a multiplicative effect. It contributes to the environmental crisis that reduces ecosystem services and threatens human health (i.e., likelihood of spillovers).

Speciation and extinction are two sides of the same coin – evolution. Over 99% of all species that populated the Earth are now extinct. As paleontology and geology have shown, during the last 500 million years of our Planet's history there were five great extinction events, whose magnitude almost eliminated three-quarters of all existing living forms.

However, life has always managed to survive, and even to thrive, after each of these major ecological perturbations, caused by global disruptions like volcanic eruptions, plate tectonic, climate change, oscillations in oxygen concentrations, impacts of asteroids and comets. So, why are we concerned with today's possibility of a Sixth Mass Extinction event? Although it has not yet occurred, we are at the beginning of it. The possibility of this global biodiversity upheaval is a serious

T. Pievani (✉)
Department of Biology, University of Padua, Padua, Italy
e-mail: dietelmo.pievani@unipd.it

S. Belardinelli
Department of Humanities, University of Naples "Federico II", Naples, Italy
e-mail: sofia.belardinelli@unina.it

© The Author(s), under exclusive license to Springer Nature
Switzerland AG 2023
N. Wallenhorst, C. Wulf (eds.), *Handbook of the Anthropocene*,
https://doi.org/10.1007/978-3-031-25910-4_41

threat to human beings and, more broadly, to the current conformation of biodiversity on Earth. The forthcoming extinction event differs from all the previous ones in a crucial feature: a single species, one of the many that comprise Earth's biodiversity – namely, *Homo sapiens* – is the driver.

A mass extinction is a phenomenon that stands out against 'ordinary' extinction events by two main features: rapidity and scale (or magnitude). The background extinction rate – i.e., the average number of species that, under normal circumstances (ecological changes, population reduction, etc.), go extinct every year – is very low. According to De Vos et al. (2015: 459), "there is consistent evidence that background extinction rates are <1 extinction per million species-years (E/MSY) and likely much less than this". Thus, an extinction event can be considered a 'mass' extinction when its overall extinction rate is significantly higher than this background estimate.

A mass extinction also has other unusual features. For example, evolutionary laws follow different patterns during these events. Ecological changes are too rapid and too global for selective processes to be effective as in 'normal times'. The initial trigger might be a geophysical event, both endogenous (such as climate change, continental movements, or volcanic eruptions) or exogenous (an extraterrestrial impact, i.e., the well-known asteroid that likely triggered the Cretaceous-Paleogene extinction 66 Ma, possibly along with other causes (Schulte et al., 2010; Chiarenza et al., 2020)). In such cases, the same extinction risk (probability) is recorded for every taxon, resulting in a general decline in biological diversity (Benton, 1995). As for the timing of these events, contrary to what Raup and John Sepkoski (1984) had proposed, they do not seem to follow a regular pattern of recurrence. Moreover, it is usual for a large adaptive radiation to occur after a mass extinction; then, survivors recover, diversify and occupy the ecological niches that have been left empty.

Therefore, following the definition proposed by Sepkoski (1986: 278), a mass extinction is "any substantial increase in the amount of extinction (i.e., lineage termination) suffered by more than one geographically widespread higher taxon during a relatively short interval of geologic time, resulting in an at least temporary decline in their standing diversity". Whether we consider overall extinction rates or – as proposed by Hull and others (2015) – fluctuation in species abundance ('mass rarity') as a primary proxy, recent observations and data suggest that today an 'accelerated extinction event' is taking place globally.

One of the most common strategies adopted by paleontologists and conservationists to assess whether we are entering a 'Sixth Mass Extinction' is to compare current and past extinction rates. However, some scholars criticize the soundness of this method (Hull et al., 2015). In effect, it might be biased both by the complexity of extrapolating a complete picture of past levels of biodiversity from the fossil record, and by the difficulty of accurately assessing extinction rates while the extinction is happening. Rather, it has been argued that it might be more useful to try to understand the mechanisms that have been at work before, during and after the great extinction events that marked the history of life on Earth, in order to better predict what might happen in the medium- and long-term future: "[…] Paleontologists continue to simply look for patterns in the fossil record, which is incomplete and subject to interpretation. Instead, we should be seeking to study the basic processes that underlie diversity and building models to test those theories. Only by understanding what governs diversity can we settle old arguments about how it has changed over time" (Erwin, 2009: 282). Such rare and extreme events could have multiple causes, not just one, like a 'perfect storm' that includes faster-than-usual climate change, alterations in the composition of the atmosphere, plus an anomalous ecological stress (Brook et al., 2008).

That today's global biodiversity is rapidly declining, both in terms of taxonomic and functional diversity, is indisputable. Recent quantitative studies show how far the crisis has already gone. As summarized in a landmark paper published by Barnosky and colleagues in 2012, biodiversity loss is one of the most severe side effects of human population expansion and activities. This is leading to widespread alteration of ecosystems, sharp reduction in species abundance on the global scale, and weaker ecological networks to support life, all part of a 'highly plausible' global-scale state shift, similar to those that triggered most of the previous great mass extinctions (Kauffman & Erwin, 1995; Barnosky et al., 2012).

As for the claim that we are actually entering the Sixth Mass Extinction, evidence is mounting. Another seminal article published in 2015 shows that, even by assuming the most conservative estimate, the available data strongly suggest that current extinction rates are far higher than normal across nearly all eukaryotic taxa, and have dramatically increased in modern times, mainly from the sixteenth century onwards (Ceballos et al., 2015). "Our results indicate – Ceballos and colleagues conclude – that modern vertebrate extinctions that occurred since 1500 and 1900 AD would have taken several millennia to occur if the background rate had prevailed. The total number of vertebrate species that went extinct in the last century would have taken about 800 to 10,000 years to disappear under the background rate of 2 E/MSY. The particularly high losses in the last several decades accentuate the increasing severity of the modern extinction crisis" (*Id.*, p. 3).

Yet, this is only part of the picture. In fact, the brunt of the extinction crisis could be hiding in understudied taxa, such as small invertebrates and members of other kingdoms of life, whose ecological importance is heavily underestimated. Understanding the scale of the tragedy for those species is difficult, since most of them have not yet been catalogued and described. IUCN *Red List*, which is the most complete assessment of endangered and extinct species, for example, reflects the general lack of knowledge about the number of species that make up the biosphere. Of the estimated 8.7 million existing species (Mora et al., 2011), only 2.12 million have been named and described so far. About half of these are invertebrates; however, the *Red List* is mainly focused on assessing extinction rates in vertebrates, particularly birds and mammals. Thus, both because of this bias towards vertebrates and because of the low number of species assessed (5,6% of known species), it has been recently suggested that IUCN research may not be a good proxy for accurately describing the current biodiversity crisis (Cowie et al., 2022). "Although public perception is that charismatic megafauna are the first victims of anthropogenic extinction and thus attract great concern, the untold thousands of invertebrate species that have gone extinct unknown to humanity vastly outweigh the small number of much better known vertebrate extinctions" (*Id.*, p. 6). Thus, if we consider all the unaccounted-for extinction events occurring every year in the invertebrate realm, we realize that the extinction crisis is far more severe than we previously thought.

One of the reasons why some underestimate (or are even sceptical about) the risk associated with the incipient Sixth Mass Extinction is that we definitely do not know the breadth of our ignorance (Scheffers et al., 2012). This pessimistic scenario also seems to be confirmed by recent estimates (referring to the last 50 years) published in a joint declaration signed by more than 100 scientific societies involved in aquatic sciences (marine biodiversity, fisheries, etc.), which have warned that estimated extinction rates in aquatic environments are substantially rising. As an example, freshwater biodiversity loss amounted to a dramatic 83% from 1974 to present (Wepener, 2020).

One of the properties that allow us to distinguish a background extinction from a mass extinction is the unusual timing of events in the latter: unlike 'common' extinctions, a mass event presents features such as the rapid escalation of the catastrophe and very long recovery times. This is especially true today, since the extinction is likely to occur over a few centuries or millennia, rather than millions of years at the biological scale. What is happening is measurable on the scale of historical, not geological time, and this highlights the worrying gap between anthropic and biological times. As E.O. Wilson repeatedly affirmed in his works, "The previous natural, or Edenic, period of biodiversity, as measured by paleontologists, started with the beginning of the Phanerozoic Era 450 million years ago. It ended fifty to ten millennia ago, with the ascent of Upper Paleolithic and Neolithic peoples, whose improved tools, dense populations, and deadly efficiency in the pursuit of wildlife inaugurated the current extinction spasm" (Wilson, 2003: 103).

Taken together, this growing dataset strongly suggests that biodiversity now faces a critical situation, and that the numerous extinction events that are taking place worldwide are most probably the initial stages of what we might rightfully define the 'Sixth Mass Extinction'. However, as previously stated, this is the first case in which a single species might be causing a global catastrophe and, what is more, is aware of the consequences of its actions.

It might be reasonable, then, to include widespread biodiversity loss, that is now turning into a real mass extinction, among the defining features of the current historical (and perhaps geological) epoch, the Anthropocene (Pievani, 2014). Unless we immediately undertake serious political and technological efforts to try to heal this 'wound', the complex dynamic balance that has sustained life for millions of years will rapidly be disrupted, and we will find ourselves in a radically impoverished planet, to which human life might no longer be able to adapt. The Sixth Extinction will also remain as a mark in the geological record. Already at the stratigraphic levels of the Neolithic transition, paleontologists of the future will observe a substantial reduction in biodiversity, accelerated from the sixteenth century with the globalization of trade and then with the industrial revolution. Moreover, they will notice a further acceleration in the extinction trend after the end of the Second

World War, with the so-called 'Great Acceleration'. It is not an abrupt transition, but it is undoubtedly part of the Anthropocene. If we consider, for example, the entire biomass of mammals living on Earth today, only 4% are wildlife. The rest is us, *Homo sapiens*, plus the weight of all farm animals.

Furthermore, we know that, unfortunately, climate change interacts with and exacerbates all the anthropogenic causes of the Sixth Mass Extinction (deforestation, spread of invasive species, growth of human population, pollution, overexploitation for hunting and fishing) (IPBES, 2019). Some of these causes (i.e., deforestation) are the same for climate change and the Sixth Mass Extinction, and they feed each other. In addition, there is evidence that biodiversity loss and the impoverishment of ecosystems increase the frequency and danger of zoonotic pandemics, threatening human health (Morens & Fauci, 2020). Therefore, including the Sixth Mass Extinction in the definition of the Anthropocene allows us to better frame the latter as a new geological epoch, characterized by a system of interrelated human activities, whose web of relationships produces an indelible geological effect.

Scientists such as Edward O. Wilson, Richard Leakey and Lewin (1995) and Niles Eldredge (2000) already wrote about the Sixth Extinction 25 years ago, with little effect. They were accused of catastrophism, but as Wilson showed in *Half-Earth* (2017), we can calculate the percentage of land and sea surface we should moderately protect to break down the Sixth Extinction (at least 50% of Earth: ambitious, but not impossible).

In 2021, for the first time, politicians have acknowledged the severity of the situation and the urgency to act accordingly. Reversing biodiversity loss has become one of the main objectives of post-2020 global 'green' strategies adopted by COP26, CBD, COP15 and several national Recovery Plans. As scientists have been warning for decades now, it is essential that the Sixth Mass Extinction does not become a reality. We would be the main responsible and among the victims, at the same time.

References

Barnosky, A. D., Hadly, E. A., Bascompte, J., Berlow, E. L., Brown, J. H., Fortelius, M., Getz, W. M., et al. (2012). Approaching a state shift in Earth's biosphere. *Nature, 486*(7401), 52–58. https://doi.org/10.1038/nature11018

Benton, M. J. (1995). Diversification and extinction in the history of life. *Science, 268*(5207), 52–58. https://doi.org/10.1126/science.7701342

Brook, B. W., Sodhi, N. S., & Bradshaw, C. J. A. (2008). Synergies among extinction drivers under global change. *Trends in Ecology & Evolution, 23*(8), 453–460. https://doi.org/10.1016/j.tree.2008.03.011

Ceballos, G., Ehrlich, P. R., Barnosky, A. D., García, A., Pringle, R. M., & Palmer, T. M. (2015). Accelerated Modern Human–Induced Species Losses: Entering the Sixth Mass Extinction. *Science Advances, 1*(5), e1400253. https://doi.org/10.1126/sciadv.1400253

Chiarenza, A. A., Farnsworth, A., Mannion, P. D., Lunt, D. J., Valdes, P. J., Morgan, J. V., & Allison, P. A. (2020). Asteroid impact, not volcanism, caused the end-cretaceous dinosaur extinction. *Proceedings of the National Academy of Sciences, 117*(29), 17084–17093. https://doi.org/10.1073/pnas.2006087117

Cowie, R. H., Bouchet, P., & Fontaine, B. (2022). The sixth mass extinction: Fact, fiction or speculation? *Biological Reviews, 97*, 640–663. https://doi.org/10.1111/brv.12816

Eldredge, N. (2000). *Life in the balance*. Princeton University Press. https://press.princeton.edu/books/paperback/9780691050096/life-in-the-balance

Erwin, D. (2009). A call to the custodians of deep time. *Nature, 462*(7271), 282–283. https://doi.org/10.1038/462282a

Hull, P. M., Darroch, S. A. F., & Erwin, D. H. (2015). Rarity in mass extinctions and the future of ecosystems. *Nature, 528*(7582), 345–351. https://doi.org/10.1038/nature16160

IPBES. (2019). Global assessment report on biodiversity and ecosystem services of the Intergovernmental Science-Policy Platform on Biodiversity and Ecosystem Services. In E. S. Brondizio, J. Settele, S. Díaz, & H. T. Ngo (Eds.), *IPBES secretariat*, Bonn, Germany. https://doi.org/10.5281/zenodo.3831673

Kauffman, E. G., & Erwin, D. (1995). Surviving mass extinctions. *Geotimes, 14*(January), 14–17.

Leakey, R. E., & Lewin, R. (1995). *The Sixth Extinction: Patterns of Life and the Future of Humankind*. Doubleday.

Mora, C., Tittensor, D. P., Adl, S., Simpson, A. G. B., & Worm, B. (2011). How many species are there on earth and in the ocean? *PLoS Biology, 9*(8), e1001127. https://doi.org/10.1371/journal.pbio.1001127

Morens, D. M., & Fauci, A. S. (2020). Emerging pandemic diseases: How we got to COVID-19. *Cell, 182*(5), 1077–1092. https://doi.org/10.1016/j.cell.2020.08.021

Pievani, T. (2014). The sixth mass extinction: Anthropocene and the human impact on biodiversity. *Rendiconti Lincei, 25*(1), 85–93. https://doi.org/10.1007/s12210-013-0258-9

Raup, D. M., & Sepkoski, J.J. (1984). Periodicity of extinctions in the geologic past. *Proceedings of the National Academy of Sciences, 81*(3), 801–805. https://doi.org/10.1073/pnas.81.3.801

Scheffers, B. R., Joppa, L. N., Pimm, S. L., & Laurance, W. F. (2012). What we know and Don't know about Earth's missing biodiversity. *Trends in Ecology & Evolution, 27*(9), 501–510. https://doi.org/10.1016/j.tree.2012.05.008

Schulte, P., Alegret, L., Arenillas, I., Arz, J. A., Barton, P. J., Bown, P. R., Bralower, T. J., et al. (2010). The Chicxulub asteroid impact and mass extinction at the cretaceous-Paleogene boundary. *Science, 327*(5970), 1214–1218. https://doi.org/10.1126/science.1177265

Sepkoski, J. J. (1986). Phanerozoic overview of mass extinction. In D. M. Raup & D. Jablonski (Eds.), *Patterns and processes in the history of life* (Dahlem workshop reports) (pp. 277–295). Springer. https://doi.org/10.1007/978-3-642-70831-2_15

Vos, D., Jurriaan, M., Joppa, L. N., Gittleman, J. L., Stephens, P. R., & Pimm, S. L. (2015). Estimating the normal background rate of species extinction. *Conservation Biology, 29*(2), 452–462. https://doi.org/10.1111/cobi.12380

Wepener, V. (2020). Statement from world aquatic scientific societies on the need to take urgent action against human-caused climate change, based on scientific evidence. *African Journal of Aquatic Science*, September. https://www.tandfonline.com/doi/abs/10.2989/16085914.2020.1824388

Wilson, E. O. (2003). *The Future of life*. Vintage Books.

Wilson, E. O. (2017). *Half-earth: Our planet's fight for life*. Liveright Publishing Corporation.

Telmo Pievani is Full Professor of Philosophy of Biological Sciences at the Department of Biology, University of Padua, since 2015. Past President (2017–2019) of the Italian Society of Evolutionary Biology, he is Fellow of several academic Institutions and scientific societies. He is author of 302 publications, included several books. Fellow of the Scientific Board of science festivals in Italy, since 2014 he is fellow of the International Scientific Council of MUSE in Trento. He is Director of "Pikaia", the Italian website dedicated to evolution, and of the University of Padua web magazine, Il Bo LIVE. He collaborates with RAI radio and TV projects, he is a columnist for Il Corriere della Sera, and the magazines Le Scienze and Micromega.

Sofia Belardinelli is PhD student in Environmental Ethics at the University of Naples 'Federico II' and Research Fellow of the UNEP Seventh Global Environment Outlook. Her lines of research concern philosophy of biology, environmental ethics, and studies on biocultural diversity. During her PhD, she is focusing on investigating the relationship between humans and non-human nature, explored from the perspective of evolutionary biology and environmental ethics. She writes for Il Bo Live, the web magazine of the University of Padova, and for the Italian magazines Micromega and Il Tascabile.

Megafaunal Extinction

Laurent Testot

Abstract The Late Quaternary Megafauna Extinction occurred throughout the 40 millennia before the Pleistocene-Holocene transition (11,700 BP). It resulted in the loss of about 50% of the large-bodied mammals in the world, or 4% of all mammal species (Barnosky, *Proceedings of the National Academy of Sciences of the USA*, 105:543–548, 2008). This extinction event had several consequences, maybe to the point of altering forest cover and climate on a continental scale. Thus it could arguably be presented as a first manifestation of the Anthropocene, providing we could demonstrate that the human factor was instrumental in this extinction. A quick survey of recent studies in this field is, alas, inconclusive. The debate rages to identify the main culprit of the demise of giants, between climatic change and anthropogenic influence. It should be inferred that such a complex event (mass extinction is definitely classified as a complex event) cannot have one cause alone. If climate and human causes, in various combinations, led to the extinction of megafauna, the data as of today are insufficient to declare these extinctions as being the beginning of the Anthropocene.

If we could travel back in time to a 100,000 years BP, choosing to land in North America, we would hold our breath in awe at the sight of the fauna. No humans (*Homo sapiens* will come later, sometime between 35,000 and 23,000 BP) but giants mammals – megafauna (we'll keep with the usual "more than 44 kg – 100 pounds" definition in this article). During this time-travel safari, we would have been able to see three different species of mammoth (including a miniature one), and driving south, mastodons (a kind of elephant with long straight tusks) and two closely related elephant-like *genera*, *Cuvieronius* and *Stegomastodon*. Alongside these pachyderms, many browsers running through the ice-age savannah: three species of wild horses, several kinds of antelope and llamas, one-humped camels 3 meters tall, giant deer and elk weighing 2 tons, with rodents like bear-sized beavers, and beasts as strange as the titan-like armadillos called glyptodons or the giant

L. Testot (✉)
Sciences Humaines, Auxerre, France

sloths (approx. a dozen species known, the biggest of which weighing over 2 tons, some lived on the land, others were able to dig tunnels 4 meters wide in the rock, some living in temperate waters …) Then came the guild of predators, beginning with the star, the 400-kilo sabre toothed tigers. There were also American cheetahs, bears weighing over a ton, lions the size of grizzlies, dire wolves as big as a Saint Bernard (MacPhee, 2019; Testot, 2020; Chansigaud, 2013; also have a look at the Twilight Beasts blog, n.d. for details).

A similar description could apply to any other region in the world, according to the rule that convergent evolution leads to the development of similar animal species in equivalent biological contexts. Hippopotamus swimming in the Thames, lions hunting through present-day France, diprotodon (a marsupial rhino-size creature) foraging in Sydney, etc. This megafauna, prolific 100,000 years ago, vanished abruptly, but depending on the continent considered, at different moments, along different time sequences (Koch & Barnosky, 2006; Barnosky, 2008) and under opposite climate evolution (toward colder or warmer dynamics): 50% of all big mammals – 4% of all the mammals, ninety genera – were wiped out from the surface of the Earth just in the millennia around the Pleistocene-Holocene transition (11,700 BP). Long-lasting in Africa and Eurasia, both peopled by human species for more than 2 million years, this extinction event was stronger in the Americas, with 70% of casualties alleged among the megafauna – in North America, 30 *genera* went extinct; in South America, more than 80 species disappeared between 20,000 and 10,000 BP. The record was set in Australia, where more than 90% of the megafauna seems to have been wiped out around between 45,000 and 20,000 BP – with a *Homo sapiens* arrival recently set before 67,000 BP, which means that extinctions peaked after twenty millennia of coexistence between men and giant beasts.

"We live in a zoologically impoverished world, from which all the hugest, and fiercest, and strangest forms have recently disappeared", wrote Alfred Russel Wallace in 1876. Such a tremendous stab in the tree of life has been detected at the very beginning of sciences like geology and paleontology. As early as the nineteenth century, three hypotheses have been in competition to explain the onslaught: various scientists argued the demise of giants was either climate driven (Charles Lyell) or human driven (John Fleming). The third explanation, variations around Georges Cuvier's idea of a cataclysmic event similar to the biblical Flood, was quickly discarded. It resurfaces periodically under various forms, be it an asteroid impact (for a recent illustration, see Pino et al., 2019) or a sweeping pandemic aiming only at big animals, to be forgotten as soon as the debaters come back to the classic human vs climate arguments.

In the 70's, the American geobiologist Paul S. Martin promoted the overkill hypothesis. His guess was that in the territories recently settled by humans, the big herbivories were hunters' naive: they had no experience of human killers and fell easy prey for our ancestors' cooperative hunting tactics. That explanation was in accordance with the data of his time: it was believed that humans arrived in the Americas around 11.000 BP (Clovis culture), and that coincided with the extinction events. This "blitzkrieg hypothesis" was simple and convenient to explain why big mammals were eradicated from the Americas (or Australia, at a previous time) while some of them survived in the Ancient World – in Africa and Eurasia, they coevoluated with *Homo*'s slow progress in predation.

This hypothesis fired a politically explosive topic, as indigenous people from America and Australia felt blaming their ancestors for overkill was insulting to them. They say in their defence that they had been in custodianship of the land for millennia, with very few (if any) species going extinct … until the Europeans' arrival, with their destructive practices (large-scale agriculture, deforestation, sheep and cow grazing, overhunting – be it bison in North America, or thylacine in Tasmania –, and introduction of a tsunami of invasive species), leading to a quick and irreversible destruction of endemic biodiversity. Interestingly, nobody was tempted to prosecute Europeans' ancestors for the extinction of their mammoths, woolly rhinoceros, cave bears or the like.

Thanks to DNA analysis, we now know that several human species went extinct with the megafauna (Neanderthal, Denisova, *H. Floresiensis*, *H. Luzonensis*) during the last 50,000 years. Thanks to more extensive research in museum collections, systematic ground excavations and interactions with indigenous people (be it rock art paintings or oral traditions), we now know for sure that megafauna survived for a long interval after human arrival (20,000 years for some giant marsupials in Australia). So much for the instant overkill hypothesis. But Paul Martin's move raised new and fundamental questions and called for inquiry.

Could the climate alone have an overkilling effect on megafauna? On a short term approach, the answers sounds like a definitive yes. During the late Quaternary, mean global temperature warmed by 5 °C from peak glacial cold 18,000 BP to peak interglacial warmth 6000 BP, entailing a long-lasting disaster for ecosystems: steppes became forests, many big grazers and browsers should have died from starvation, their extinction spelling death for the bigger meat-eaters. Alas, things aren't so simple. During the last interglacial pulse, between 130,000 and 115,000 years ago, the climate was rather similar to that of the Holocene and even somewhat warmer. These animals had previously survived a dozen or more cold-to-warm and warm-to-cold cycles, and the last deglaciation appears quite ordinary in this aspect. This argument was the strongest trump for the overkilling hypothesis. Something new should have irrupted on the scene, to send so many big animals out of the theatre of life. During the years following the publication of a seminal book (Martin & Klein, 1984) assessing with 40 experts from many fields all hypotheses on the basis of fair discussion, the debate took new colours: both climate and humans should have contributed to the Late Quaternary Megafauna Extinction (LQME). From this moment, many authors described a world were big animals were stressed by the warming climate-driven collapse of the biomes they fed upon, trying to migrate under colder skies, falling easy prey to humans ready to deliver the *coup-de-grâce*.

A thorough attempt to break the mystery was attempted by Lewis J. Bartlett and colleagues (Bartlett et al., 2015). They gathered a large amount of data to establish correlation between megafaunal extinction and climate and/or humans. Starting at 80,000 years ago, they browsed articles for the latest radiocarbon dates of megafauna, reconstructed the climate of the past using all kind of proxies, and hypothesized arrival of humans according to various scenarios where definitive data lacked, to generate 1000 different extinction scenarios. A majority of these simulations show a strong correlation with the arrival of humans and the megafaunal extinction, sometimes upset when the climate changed shortly after the arrival of humans – thus in a minority of *scenarii*, the researchers acknowledged that climate would have played a part too.

Meanwhile, studies underlined that effective overkill happened in some places. Humans' recent arrival on islands, from Madagascar (around 1000 CE or earlier) or Hawaii archipelago (between the first and the fourth centuries CE) to New Zealand (around 1280 CE), was correlated with the extinction of many birds (and, in some cases, especially in Madagascar, of mammalian megafauna). It is suspected that a tenth of all bird species was eradicated these last two millennia as humans settled on islands, European colonialism included – as the dodo could testify (Duncan et al., 2013; see also Flannery, 1994). "Human colonization of remote Pacific islands caused the global extinction of close to 1,000 species of nonpasserine landbirds alone", write Richard P. Duncan and his coauthors, adding that "two-thirds of the populations on these islands went extinct in the period between first human arrival and European contact". But there is one big limit to analogies: islands aren't miniature continents, and their biomes are quite vulnerable to intruders.

The debate burns endlessly. In 2021, Miki Ben-Dor and Ran Barkai advanced a strong argument in favour of the "killing ape hypothesis". They state (Ben-Dor & Barkai, 2021; Ben-Dor et al., 2021) that megafauna consumption was instrumental in our evolution. They underline that the bigger the animal the fatter it is, and that fat is a powerful resource for biological energy acquisition, driving our ancestors towards brain expansion, use of fire and stone tools, collaborative patterns, etc., and ultimately towards megafaunal extinction. Afterwards, human evolved in a new way, downgrading their trophic level from top carnivorous to more energy efficient collecting behaviour – i.e. agriculture.

As of 2021, contradictors continue to come forward. As an example, Cantalapiedra and colleagues (2021) tracked the demise of proboscideans along millions of years – almost all of them disappeared before human intercontinental expansion. According to them, this emblematic case of generic extinction followed a natural pattern of extinctions. They just add, at the end of the article, that human encroachment maybe dealt a final blow to some survivors – but all these elephant-like creatures, who colonized Africa, Eurasia and Americas, couldn't stand long-term (2.6 million years) regular changeover from glacial to warmer climates, which periodically stimulated the rise of forest and the loss of the grasslands necessary for their diet.

Whatever the cause, the main documented impacts of the LQME have been hypothesised as having aggravated aridification and perturbed biological cycles. Chris Johnson (Johnson, 2006, 2009) advanced the hypothesis that the megaherbivores kept down the vegetation by browsing or grazing considerable amounts of biomass. When this balance was lost, thickets grew uncontrollably, and mass-burned repetitively. Biodiversity declined till it reached a new balance. Then another human variable irrupted: the repeated use of partial burning off. This practice encouraged roots and edible plants to grow as well as the proliferation of small game. Evidence of the widespread regional use of this technique has been found as early as

40,000 BCE in Australia, suspected around 125,000 in Southern Africa. According to Johnson, the first *Homo sapiens* and their fires largely altered fauna and flora around the world. Some researchers suspect that these processes altered climates at the continental scale, especially for North America and Australia.

According to Christopher E. Doughty (Doughty et al., 2013), giant herbivores fertilized the Amazonian basin with nitrogen, phosphorous, and other nutrients in their manure. The forest grew from the soil fed with their waste, and the extinction of these species left its mark in the sediments by radically decreasing the spread of nutrients in the soil. For Doughty, the large animals of the Pleistocene acted like "arteries" within ecosystems by dispersing nutrients around the world, and their extinction cut those arteries.

For both reasons (use of firestick farming and disruption of manure fertilisation), if these extinctions were proven to be anthropic, it would champion megafaunal extinction as a firm candidate for a possible Anthropocene mark. Alas, several biais set limits to the enquiry. As an example, the fossil record is far from complete, leaving paleontologists fighting with uncertainties on the precise extinction date for many species. The asynchronous timing of the extinctions signals different trajectories and dynamics of extinction, likely resulting from a combination of climatic, geographic and anthropogenic factors – according to trophic-level analysis, extinctions strike in domino fashion: when some key species are destroyed first, other species can disappear without consequences on the ecosystems – they just get replaced by others.

With the rise of ancient DNA analysis, some answers begin to show. In a world-premier demonstration, Eline D. Lorenzen and her team (Eline D. Lorenzen et al., 2011) combined ancient DNA analysis, species distribution models and human fossil record to elucidate how climate and humans shaped the demographic history of 6 megafauna species: woolly rhinoceros, woolly mammoth, wild horse, reindeer, bison and musk ox. DNA investigation gave information on the stress endured by the species, and its level of interbreeding, before extinction – information correlated with suspected human disruption. The authors estimated that climate has been a major driver of population change over the past 50,000 years, but that each species responded differently to the combined effects of climatic shifts, habitat redistribution and human intrusion. They concluded that climate change alone could explain the extinction of some species, such as Eurasian musk ox and woolly rhinoceros, but that they needed a combination of climatic and anthropogenic effects to explain Eurasian steppe bison and wild horse annihilation. Let's hope some further studies, with a multidisciplinary team backed up with ancient DNA analysis, bring some more light on this subject soon.

Put together, the megafaunal extinction debates chronicle human response to environmental changes. During the late Pleistocene, humans interacted with an environment stricken with major climate oscillation and ecosystem reorganization. Human predation, climate change, and trophic level plant-animal interactions have played a role in the collapse and rebuilding of ecosystems. The most astounding is that our ancestors succeeded in establishing and increasing human populations to a threshold at which they began to play a discernible role as agents of indisputable ecological change: see Neolithics. In many parts of the world, adaptations from hunting and foraging led to agricultural practices, bringing us some millennia later to a world dominated by humans and their domestic species – cows are the last remnant of the powerful aurochs, the last of which died four centuries ago. Though we cannot say for sure if the LQME was the first signal of the Anthropocene, the downfall of big animals carries one powerful message: extinction is on the rise, and we can now be sure that it is both climate + human driven.

References

Barnosky, A. D. (2008). Megafauna biomass tradeoff as a driver of Quaternary and future extinctions. *Proceedings of the National Academy of Sciences of the USA, 105*, 543–548. https://doi.org/10.1073/pnas.0801918105

Bartlett, L. J., et al. (2015). Robustness despite uncertainty: Regional climate data reveal the dominant role of humans in explaining global extinctions of Late Quaternary megafauna. *Ecography, 38*, 1–10. https://doi.org/10.1111/ecog.01566

Ben-Dor, M., & Barkai, R. (2021). Prey size decline as a unifying ecological selecting agent in Pleistocene human evolution. *Quaternary, 4*, 7. https://doi.org/10.3390/quat4010007

Ben-Dor, M., et al. (2021). The evolution of the human trophic level during the Pleistocene. *Yearbook of Physical Anthropology, 175*, 1–30. https://doi.org/10.1002/ajpa.24247

Cantalapiedra, J. L., et al. (2021). The rise and fall of proboscidean ecological diversity. *Nature Ecology and Evolution., 5*, 1266–1272. https://doi.org/10.1038/s41559-021-01498-w

Chansigaud, V. (2013). *L'Homme et la Nature. Une histoire mouvementée*. Delachaux et Niestlé.

Doughty, C. E., et al. (2013). The legacy of the Pleistocene megafauna extinctions on nutrient availability in Amazonia. *Nature Geoscience, 6*, 761–764. http://www.yadvindermalhi.org/uploads/1/8/7/6/18767612/doughtyngeo1895.pdf

Duncan, R. P., et al. (2013). Magnitude and variation of prehistoric bird extinctions in the Pacific. *PNAS, 110*(16), 6436–6441. https://doi.org/10.1073/pnas.1216511110

Flannery, T. (1994). *The future eaters: An ecological history of the Australasian lands and people*. The Text Publishing Company.

Johnson, C. (2006). *Australia's mammal extinction: A 50,000 year history*. Cambridge University Press.

Johnson, C. (2009). Ecological consequences of Late Quaternary extinctions of megafauna (Australia). *Proceedings of Royal Society, 276*, 2509–2519. https://doi.org/10.1098/rspb.2008.1921

Koch, P. L., & Barnosky, A. D. (2006). Late Quaternary extinctions: State of the debate. *Annual Review of Ecology, Evolution, and Systematics, 37*, 215–250. https://doi.org/10.1146/annurev.ecolsys.34.011802.132415

Lorenzen, E. D., et al. (2011). Species-specific responses of late Quaternary megafauna to climate and humans. *Nature, 479*, 359–364. https://doi.org/10.1038/nature10574

MacPhee, R. D. (2019). *End of the megafauna: The fate of the World's hugest, fiercest, and strangest animals*. W. W. Norton & Company.

Martin, P. S., & Klein, R. G. (Eds.). (1984). *Quaternary extinctions. A prehistoric revolution*. The University of Arizona Press.

Pino, M., et al. (2019). Sedimentary record from Patagonia, southern Chile supports cosmic-impact triggering of biomass burning, climate change, and megafaunal extinctions at 12.8 ka. *Scientific Reports, 9*, 4413. https://doi.org/10.1038/s41598-018-38089-y

Testot, L. (2020). *Cataclysms: An environmental history of humanity* (Payot, 2017, translated by Katherine Throssel, Chicago University Press).

Twilight Beasts blog: https://twilightbeasts.wordpress.com

Laurent Testot is a free-lance science journalist who specializes in global history. He is the author of *Cataclysms. An Environmental History of Humanity* (Paris: Payot, 2017, translated by Katherine Throssel, Chicago: Chicago University Press, 2020 – also translated in Italian). Among his other books: *La Nouvelle Histoire du Monde*, Auxerre: Sciences Humaines Editions, 2019 (translated into German); *Homo Canis. Une histoire des chiens et de l'humanité* (Paris: Payot, 2018, translated in Italian). He edited approx. 20 books on history, sciences of religions and geopolitics. His latest book, cowriten with Nathanaël Wallenhorst, is Vortex. Faire face à l'Anthropocène, Paris: Payot, 2023.

Permaculture

Leila Chakroun

Abstract Permaculture offers an ethical, pragmatic, philosophical and technical response to the daunting diagnosis of the Anthropocene. Its comprehension of the Earth draws upon the idea of "working with nature, rather than against it". In that sense, permaculture radically departs from the fixist and homogenizing logics that prevail in agronomic sciences, and in modern territorial planning. By reengaging with the cycle of life and with our own finitude as living beings, permaculture conveys an earthly vision of human that resists the impertinent idea of an Earth becoming the scene solely for *anthropos*.

Permaculture is a call for change. It emerged in reaction to the growing criticism of modern consumerism, petrol-driven society and their socio-agro-environmental consequences. The concept encapsulates its original *raison d'être*, namely, to propose a framework for implementing permanence in agriculture. In reaction to the increasingly recognized limits of industrial agriculture to both healthily feed people and maintain a fertile soil, permaculture encourages diverting large scale mechanized single-crop systems and opting for an "integrated, evolving system of perennial or self-perpetuating plant and animal species" (Mollison & Holmgren, 1978). While addressing global issues from their agroecological aspects, permaculture encompasses more than just a new set of agricultural technics. It was coined in the early seventies by Bill Mollison and David Holmgren and originates from their common desire to propose a rallying concept and a mobilizing project instead of an umpteenth daunting diagnosis of the era – henceforth referred as the Anthropocene. The seminal book *The Limits to Growth* by the Club of Rome had just been released (1972) and alternate routes were dearly needed to prevent the realization of the forecasted consequences of the scenario 'business as usual'. Permaculture filled that void in offering a "both ethical and pragmatic, philosophical and technical" response (Holmgren, 2002, p.xv) – and, I would add, both natural and cultural, analytical and

L. Chakroun (✉)
Institute of Geography and Sustainability, University of Lausanne, Lausanne, Switzerland
e-mail: Leila.chakroun@unil.ch

sensorial. The concept is underpinned by a non-binary vision, which resonates with Augustin Berque's trajectivity of things (Berque, 2019). Permaculture achieves its worldview by revealing multiple layers. It is built on three guiding ethics: Care of the Earth, Care of People and Fair share. It simultaneously designates the method(s) and design principles that will facilitate the realization of its ethics of care. Thus designating a global network of grassroots experimentations happening at different scales and in various fields, connected through virtual social networks and itinerant teachers (Ferguson & Lovell, 2014).

Permaculture conveys a comprehension of the Earth/earth that explicitly draws upon the idea of "working with nature, rather than against it" suggested by Fukuoka Masanobu (1913–2008) the Japanese leader in natural farming (*shizennôhô*) (Mollison, 1988). In Fukuoka's perspective, "working with" signified finding ways of cultivating that do not thwart nature's dynamics. Among permaculture practitioners, this idea has broadened and become an injunction to opt for practices respectful of the existing web of relations and meanings composing social-ecological systems. This implies developing non-exploitative modalities of agriculture work and multidirectional care with the more-than-human. In that sense, permaculture radically departs from the fixist and homogenizing logics that prevail in conventional agronomic sciences, and in the zoning of modern territorial planning (Chakroun, 2020). Instead, it relies on an "epistemology of variation" (Cohen, 2017) based, among other things, on a conception of soil as a living organism mostly determined by the vitality of the edaphon (the soil biota), which in turn determine the soil fecundity (and not its sole fertility; see Verrecchia cited in Chakroun & Linder, 2018). Gestures, techniques, but also work pace (Puig de la Bellacasa, 2015), are therefore expressly adapted to maintain and even boost soil life. For this reason, practitioners avoid tilling and spraying chemical fertilizers and pesticides since they profoundly erode the diversity of the edaphon and destroy the well-structured soil horizons. In this view, permaculture's agricultural strategies and techniques share similarities with other agroecological alternatives: organic agriculture, conservation farming, and agroforestry. To prevent hydrogeochemical erosion and nutrient leaching, soil is kept covered, either by densely interplanting annual plants with perennials, or by mulching when and where necessary. Vegetables requiring rich soil are seeded or planted in permanent raised beds, while cereals are sowed under vegetative cover. Thick, partly edible hedgerows are planted around and within agricultural fields, offering shelter for birds and other small animals and food diversity for humans (fruits, nuts), while acting as a wind erosion barrier, thus enhancing soil moisture. These examples illustrate the intricate interdependency between the land's fecundity and biodiversity. In addition, by consciously caring for the diverse elements of the Earth – the rivers, the woodlands, the fields, wild plants, soil and animals – permaculture creates ecotopian landscapes (Chakroun & Droz, 2020) inspiring future paths towards multispecies commoning and cohabitation (Centemeri, 2018).

Permaculture also refers to a mode of agricultural knowledge that merges abstract reasoning and embodied knowing. Mollison, in his seminal work *A Designer's Manual*, qualifies permaculture as "an experiential system of design" and underlines that to be a good permaculture designer means to "design by natural example, becoming aware, taking notes, sitting a long time in one place, watching the wind behave and the trees respond, thrusting your hand into the soil to feel it for moisture (…) and becoming sensitive to the processes and sights about you" (1988, p.46). This approach to design combines scientific data collected via objectifying methodologies with "data" captured through preconceptual polysensorial (even synesthetic) experiences (Pignier, 2017). The extensive use of patterns is indicative of this profoundly ambivalent mode of relation with the world. Permaculture draws from recurring patterns perceived in nature, using them as a matrix, and implicitly granting nature with a certain purposiveness in the way it evolves and moves. Patterns are a form of what Descola (2005) would call an analogist mode of thinking as they bring heterogenous orders of reality into resonance. For example, the observed analogy between the shapes of trees' branches and roots, and estuaries or blood vessels, is explained by the similar function they fulfil: disseminating energy from a source. This characteristic is mobilized in the design to attain what the Swiss permaculturist Hubert de Kalbermatten calls a "pacification" of the energy flow and force – water in particular – throughout the agroecosystem (Chakroun, forthcoming).

Still, on the contrary to biomimicry and geoengineering instrumentally manipulating nature lore to optimise industrial process or invent the new techno-fix, permaculture emulates nature patterns for the benefit of nature itself, in a counter-spirit of "geo-nurturing" (Taylor, 2014). The design aims at sustaining or recreating the conditions of co-inhabitation between the multiple – human and non-human – entities constitutive of the milieu. Although radically different in theory, the frontiers between geoengineering and geo-nurturing are difficult to delineate in practice. To what extent can a designer re-design an ecosystem? Is it a matter of intention, or is there a threshold that should not be bypassed even if it results in more biodiversity and/or in less CO_2 in the atmosphere? One indicator could be the degree of self-engagement in the design process. In the demiurgic allegations of geoengineering to reconstruct the Earth (Neyrat, 2018), design is understood as the unilateral expression of a human subject over an inert matter. In permaculture, design designates the dialogue initiated with the vibrancy of the Earth matter (Bennett, 2010) through the constant attunement between the designer's body (flesh/animal body) and the "flesh of the world" (Merleau-Ponty, 1964)/the medial body (Berque, 2019). According to Shidara Kiyokazu, one of Japan's permaculture pioneers, we need to feel as one (一体化 / ittaika in Japanese) with the place being designed: "If you design something that is separated from yourself, the design is not good. You should put yourself in the design" (Chakroun, forthcoming). A criteria of a good permaculture design might then be the blurring of frontiers between the designer and the designed, to the point that the design always transforms as much the self as the milieu (Chakroun & Linder, 2018). From this developing interrelation emerges the possibility of re-becoming indigenous of a portion of the Earth, beyond one's ethnical origins or place of birth (Arnsperger, 2019).

The political aspects of transforming identities, cultures and land uses seem to have eluded Mollison and Holmgren. The envisioned change was to be achieved through the peaceful dispersion of permaculture's mindset and the gradual transformation of agricultural landscapes. Field studies show that permaculture provokes contrasting reactions, revaluing farmers work and rural life, especially among young urban dwellers, while crystalizing tensions with the institutions and the farmers of neighbouring plots of lands (Chakroun, 2020, 2022). The permaculture landscapes exhibit an aesthetic that hints at its modes of valuation of the more-than-human world and the gap still existing within today's dominant values. The lushness and untidiness of forest gardens and its soil mulched with the decomposing remains of organic debris clash with the orderly and weeded lines of row crops. Permaculture's care for soil and profound attention to its edaphon is forcing the revamping of landscape aesthetic standards.

At a time when the Anthropocene heralds the end of the world as we knew it, permaculture contributes to reveal the greatest ambivalence of this era: the foreclosing of the importance of decay, decomposition and death in the flourishing of all life forms has resulted in the global ruination of the Earth's habitability (Myers, 2019). By reengaging with the cycle of life and with our own finitude as living beings, permaculture conveys an earthly vision of human that resists the impertinent idea of an Earth becoming the scene solely for *anthropos*. Permaculture's myriad experimentations might well be the prefiguration of what the ecophenomenologist David Abram yearns for: the Humilocene – concept that he coins by playing with the common etymological roots of humanity, humility and humus (Abram et al., 2020), hence encouraging us to acknowledge humbly the earthiness of our existence.

References

Abram, D., Milstein, T., & Castro-Sotomayor, J. (2020). Interbreathing ecocultural identity in the Humilocene. In T. Milstein & J. Castro-Sotomayor (Eds.), *The Routledge handbook of ecocultural identity* (pp. 5–25). Routledge.

Arnsperger, C. (2019). Serons-nous enfin, un jour, indigènes ? Permaculture et éducation des profondeurs. In N. Wallenhorst & J.-P. Pierron (Eds.), *Éduquer en Anthropocène* (pp. 101–112). Le Bord de l'Eau.

Bennett, J. (2010). *Vibrant matter, a political ecology of things.* Duke University Press.

Berque, A. (2019). An enquiry into the ontological and logical foundations of sustainability: Toward a conceptual integration of the interface 'nature/humanity'. *Global Sustainability, 2,* E13. https://doi.org/10.1017/sus.2019.9

Centemeri, L. (2018). Commons and the new environmentalism of everyday life. *Alternative value practices and multispecies commoning in the permaculture movement, Rassegna Italiana di Sociologia, 2,* 289–314. https://doi.org/10.1423/90581

Chakroun, L. (2020). La permaculture au sein des dynamiques territoriales : Leviers pour une mésologisation de l'agriculture suisse. *Développement durable et territoires, 11,* 1. https://doi.org/10.4000/developpementdurable.14784

Chakroun, L. (2022). Entreprendre par et pour le soin Pierre-Alain Indermühle, thérapeute et patient de son organisme agricole. In J. Foyer, A. Choné, & V. Boisvert (Eds.), *Les esprits scientifiques. Savoirs et croyances dans les agricultures alternatives* (pp. 125–147). UGA Editions. https://books.openedition.org/ugaeditions/27117

Chakroun, L. (forthcoming). *S'engager au milieu : Penser l'engagement en Anthropocène à travers la permaculture. Enquêtes en contextes suisse et japonais, Thesis in environmental sciences (dir. Christian Arnsperger and Yoann Moreau).* University of Lausanne.

Chakroun, L., & Droz, L. (2020). Sustainability through landscapes: Natural parks, satoyama, and permaculture in Japan. *Ecosystems and People, 16*(1), 369–383. https://doi.org/10.1080/26395916.2020.1837244

Chakroun, L., & Linder, D. (2018). Le milieu permaculturel comme foyer d'un soi mésologique. In M. Augendre, J.-P. Jean-Pierre Llored, & Y. Nussaume (Eds.), *La mésologie, un nouveau paradigme pour l'Anthropocène ?* (pp. 283–291). Hermann.

Cohen, A. G. (2017). Des lois agronomiques à l'enquête agroécologique. Esquisse d'une épistémologie de la variation dans les agroécosystèmes, Tracés. *Revue de Sciences humaines, 33*, 51–72.

Descola, P. (2005). *Par-delà Nature et Culture.* Gallimard.

Ferguson, R. S., & Lovell, S. T. (2014). Permaculture for agroecology: Design, movement, practice, and worldview. A review. *Agronomy for Sustainable Development, 34*, 251–274. https://doi.org/10.1007/s13593-013-0181-6

Holmgren, D. (2002). *Permaculture. Principles and pathways beyond sustainability.* Holmgren Design Services.

Merleau-Ponty, M. (1964). *Le Visible et l'invisible.* Gallimard.

Mollison, B. (1988). *A Designer's manual.* Tagari Publications.

Mollison, M., & Holmgren, D. (1978). *Permaculture one: A perennial agriculture for human settlements.* Corgi Publishing.

Myers, N. (2019). From Edenic apocalypse to gardens against Eden: Plants and people in and after the Anthropocene. In K. Hetherington (Ed.), *Infrastructure, environment and life in the Anthropocene* (pp. 115–148). Duke University Press.

Neyrat, F. (2018). *The Unconstructable earth: An ecology of separation.* Meaning Systems.

Pignier, N. (2017). *Le design et le vivant: Cultures, agricultures et milieux paysagers.* Connaissances et Savoirs.

Puig de la Bellacasa, M. (2015). Making time for soil: Technoscientific futurity and the pace of care. *Social Studies of Science, 45*(5), 691–716. https://doi.org/10.1177/0306312715599851

Taylor, R. (2014). *Geoengineering, permaculture and transition town.* LabSpace, coll. Natural & Life Sciences. http://doer.col.org/handle/123456789/5493. Accessed 29 July 2021.

Leila Chakroun holds a PhD in environmental humanities from the Institute of Geography and Sustainability of the University of Lausanne. Her research explores, through Augustin Berque's approach of mesology and Christian Arnsperger's notion of existential activism, modalities of self-engaging with the Anthropocene, based on extensive ethnographic fieldwork within the permaculture movement in Switzerland and in Japan. She has published several articles and chapters on permaculture, among which: Sustainability through landscapes: natural parks, satoyama, and permaculture in Japan (In *Ecosystems & People*, with Laÿna Droz, 2020); *Entreprendre par et pour le soin: Pierre-Alain Indermühle, thérapeute et patient de son organisme Agricole* (UGA Editions, 2022).

Renewable Energy

Espen Moe

Abstract Renewable energy can be defined as energy produced by natural resources that are naturally replenished within a time span of a few years (Lund H. Renewable energy systems. Academic, Oxford, 2014). Renewable energy typically includes hydropower ("old renewables"), wind power, solar power, biopower, geothermal power and tidal power ("new renewables") and are among the most important contributors in the battle against global warming. In IEA projections, renewable energy accounts for 32% of the emissions cuts needed for humankind to keep global warming below 1.8 °C (IEA. World Energy Outlook. IEA Publications, Paris, 2019a).

As of 2021, the planet has installed 3146GW of renewable energy, of which 1195GW hydropower, 942GW solar photovoltaic (PV) (+6GW of concentrating solar thermal), 845GW wind power, 143GW biopower, 14.5GW geothermal and 0.5GW tidal power (REN21, 2022). Hydropower is the most mature of these and thus the largest, but consequently also the one with the least growth potential. Thus, while since 2010 global hydropower capacity has increased by 28%, wind power capacity has increased by 325% and solar PV capacity by 2250%. Hydropower's share of electricity generation has long remained at a steady 16–17%, whereas new renewables has increased its share from 3% to 13% in about a decade. Wind power accounts for about 6.5% of electricity generation and solar PV around 3.5%. If we include all renewables, in a decade they have increased their share of electricity generation from 19% to 29%. Yet, these figures hide major differences. Danish wind power in 2020 generated more than 50% of the country's domestic electricity consumption. Including all types of renewables, Germany also broke the 50% barrier in 2020. In contrast, large countries like China, the US, and Japan roughly follow the global average deriving 10–15% of their electricity from wind and solar PV. Some outliers have special geographic conditions, thus, Norway gets 90% of its electricity

E. Moe (✉)
Department of Sociology and Political Science, Norwegian University of Science and Technology (NTNU), Trondheim, Norway
e-mail: espen.moe@ntnu.no

from hydropower, whereas Iceland relies on hydropower for 73% and geothermal for the remaining 27% (Government of Iceland, 2021; REN21, 2022).

Electricity generation only tells the story of the power sector. Final global energy consumption numbers paint a bleaker picture. Here, hydropower accounts for a relatively unchanging 3.9%, with an additional 2.8% from wind, solar, biomass and geothermal. The latter figure has increased by 300% in a little over a decade, up from 0.7% in 2009. Still, since 2009, the share of all renewables in all areas has only increased from 8.7% to 12.6% (REN21, 2022). From a cost competitiveness perspective, it is however worth noting that the costs of wind power have fallen by 70% and utility-scale solar power by 90% since 2009 (Lazard, 2021), making them competitive on price in most countries. One consequence is that almost 70% of all investments in new power capacity now goes to renewable energy compared to 23% for fossil fuels and 8% for nuclear (REN21, 2022).

Is a fully carbon-free, renewable (RE) energy system possible? There is a growing literature on 100% RE, and very few argue that from a technical perspective it is impossible (Hansen et al., 2019). It is however also clear that far more attention has been lavished on the power sector, which accounts for only 17% of final global energy consumption, than on transport, and heating and cooling. The electrification of transport, heating and cooling has so far been a much slower process. Thus, while a 100% RE energy system is possible, a 100% RE power sector is far more achievable.

Creating a 100% RE energy system however comes with imposing challenges. In terms of energy security and geopolitics, a renewable energy world looks very different. Energy security for long meant easy access to affordable petroleum in open world markets. A renewable world is instead a networked world where local fluctuations in wind and sunlight would require major regions to be physically tied together through cross-border, international electricity grids for renewables to deliver reliable energy. However, creating an international grid fundamentally challenges our traditional ways of thinking about energy security and sovereignty (Vakulchuk et al., 2020). Creating such a grid is a slow and arduous process even within the most integrated parts of the world, i.e., the EU. It is far more difficult between less amicable neighbors. An Asian Super Grid, including China, South Korea, Russia, and Japan, has been discussed since 2011, but there are massive political hurdles, and Japan is wary of putting its energy security and energy sovereignty in the hands of China (Midford & Moe, 2021). Desertec was established in 2009 with the purpose of transmitting solar energy from the Sahara to Europe. The Arab Spring however made the challenges of relying on vulnerable grid connections from politically unstable countries very evident (Schmitt, 2018). Since then, the 2022 Russian invasion of Ukraine has made it evident that relying on politically dubious countries for natural gas also brings huge perils. It has made European countries more wary of their energy sovereignty, but also accelerated offshore wind power developments in the North Sea and created an impetus for a European offshore grid net (Hansen & Moe, 2022). A renewable world may easily also depend on rare earth materials. At present China controls the world markets in a host of these metals, and in 2010 put a rare earth embargo on Japan over territorial disputes. It is a common claim that renewable energy installations will trigger geopolitical competition over critical resources. It is however also true that most rare earth materials are expensive rather than geologically scarce, and that technological progress is already enabling manufacturers to significantly reduce the usage of such materials. The literature is divided on whether such materials will become a problem (Overland, 2019).

Among domestic challenges, incumbent or vested interest resistance counts among the obvious. Renewable energy grows in the shadow of some of the biggest and most influential energy giants of all time. Major petroleum companies have a strong interest in perpetuating the existing energy system (Moe, 2015). Mildenberger (2020) argues that the biggest problem for climate policymaking is the dispersion of carbon polluters across political parties. Labor actors and business actors have captured policymaking on the left and the right. No matter who controls government, carbon polluters are part of policy design. Aklin & Urpelainen (2018) suggest that realistically only a major exogenous shock to the system will produce a window of opportunity great enough to allow politicians to unchain themselves from this capture and change renewable energy policy in any dramatic way.

We are still a far cry from a renewable energy *transition*. Yet, renewable energy, primarily wind power and solar PV, stands in a different relationship to the overall energy system than only a decade ago. This brings new challenges. The first phase of renewable energy installations was in many ways simple. Renewable energy grew rapidly from a very low base (with the exception of hydropower) as a *supplement* to the existing energy system, and policy typically consisted of states producing different types of support systems (Feed-in Tariffs, Renewable Portfolio Standards, Certificates, etc.) while leaving the rest to the actors in the market. In the Anthropocene, renewable energy however needs to *replace* rather than supplement the existing energy system. This is more complicated. Infrastructure efforts have typically not been incentivized as renewable energy installations have been, resulting in many countries having severe mismatches between the rapidity of the new energy phased in, and the ability of the existing grid systems to cope with large amounts of fluctuating energy. Thus, this next phase is not just about phasing in more renewable energy, but about upgrading the energy infrastructure to accommodate large amounts of renewable energy – grid lines, energy storage, charging stations for electric vehicles, etc. It will require massive investments and coordination between many different actors, and arguably a more active state (e.g., Midford & Moe, 2021). This next phase represents a structural change in the energy system, with renewable energy serving as disruptive technologies forcing Schumpeterian creative destruction onto the existing energy system.

Structural change tends to trigger interest battles. To reach the 1.5 °C target by 2050, annual wind power installations need to triple and annual solar power installations quintuple (IEA, 2021). If renewable energy is phased in more rapidly it is likely to meet with resistance, as it challenges both existing fossil fuel vested interests and potentially the centralized distribution model of the electric utilities. Upgrading grids and phasing in renewables will also likely meet with resistance from grassroots interests unhappy about massive local infrastructures being erected (Wüstenhagen et al., 2007). Thus, so-called green-on-green conflict is another domestic challenge renewables must overcome in the Anthropocene. Renewable energy has a lower power density than fossil fuels and takes up more space (Smil, 2015). In combination with increased installation rates this has led to conflicts between renewable energy advocates and nature conservationists, especially with respect to wind power, which in many countries have caused serious popular protests (e.g., Moe et al., 2021). Thus, the most significant future obstacles to a renewable energy transition may not be technological or economic, but political, in the form of popular backlashes. As a 100% RE system requires renewable energy far beyond current levels, such backlashes are inevitable if installation processes are not characterized by legitimacy, consensus, and energy justice (e.g., Midford & Moe, 2021; Sovacool, 2014). This puts a major onus on creating licensing systems that take nature conservation interests into account, and on installing renewables in areas where popular backlashes are less of a concern, which often means utilizing offshore areas. While offshore solar PV is in its infancy, offshore wind is growing fast, now accounting for roughly 10% of annual wind power installations (REN21, 2022). Great Britain has almost 50% of its wind power capacity offshore, whereas another frontrunner, Denmark, expects 60% of future capacity to be offshore (Moe, 2015).

The grander plans for offshore wind however links back to geopolitics and international gridlines. WindEurope (2019) for instance has a plan for 450GW of capacity in the North Sea and the Mediterranean, linked together and connected to the Continent through subsea cables. For a renewable energy transition to be realistic most likely such schemes are necessary. This would however require major progress in floating wind power. Of the 845GW of wind power capacity, 57GW is offshore, and only 139MW, i.e., below 0.02% is floating wind (GWEC, 2022). Thus, beyond the coordination problems that often mar international solutions, this is an area where rapid technological progress is still necessary before floating turbine technology is cost effective enough to be realistic.

Finally, renewable energy in the Anthropocene must rely on having a global dimension, as the climate battle may easily be lost in the Global South alone. So far, the overwhelming majority of renewables have been installed in Western countries and China, and to some extent India. The likelihood however is that energy consumption will rise faster in poor countries than in rich countries. The IEA (2019b) for instance estimates that the electricity consumption for Africa overall may triple by 2040. If this consumption growth cannot be met by renewable energy, including the construction of infrastructure for renewables, it will be dramatically harder to limit global warming to 2 °C. This is a daunting challenge for countries that often have weak states and institutions, suggesting that success depends on the recognition that this is also an international energy justice challenge (e.g., Sovacool, 2014).

In conclusion, we see rapid growth in the deployment of renewable energy, first and foremost wind power and solar PV. These technologies are now almost universally competitive on price. In the frontrunner countries, renewable energy has moved from niche player to becoming an integral part of the power sector. Still, with respect to final global energy supply, renewables have not put much of a dent into the existing fossil fuel-based energy system. The challenge is formidable and accentuated by a lack of time. As one of the major components to avoiding catastrophic levels of global warming, renewable energy should be pursued actively. Beyond technology and economics, this puts the onus on politics. Reforming an energy system rests on extremely heavy political processes, of which the political acceleration of the installation of renewable energy is one of the most essential.

References

Aklin, M., & Urpelainen, J. (2018). *Renewables*. MIT Press.

Government of Iceland. (2021). *Energy*. https://www.government.is/topics/business-and-industry/energy/. Accessed 17 Sept 2021.

GWEC. (2022). *Global Offshore Wind Report 2022*. GWEC.

Hansen, K., Breyer, C., & Lund, H. (2019). Status and perspectives on 100% renewable energy systems. *Energy, 175*(15), 471–480.

Hansen, S. T., & Moe, E. (2022). Renewable energy expansion or the preservation of national energy sovereignty? *Political Geography, 99*, 102760, 1–11.

IEA. (2019a). *World Energy Outlook*. IEA Publications.

IEA. (2019b). *Africa Energy Outlook*. https://www.iea.org/reports/africa-energy-outlook-2019. Accessed 17 Sept 2021.

IEA. (2021). *Net Zero by 2050*. https://www.iea.org/reports/net-zero-by-2050?fbclid=IwAR1oCu4NLsoGRYGqfKDKJCUajii9vwPqsNPTjsBKHpPIKZslHLEOvcbtjGQ. Accessed 17 Sept 2021.

Lazard. (2021). *Lazard's levelized cost of energy analysis – version 15.0*. https://www.lazard.com/media/451881/lazards-levelized-cost-of-energy-version-150-vf.pdf. Accessed 19 Feb 2023.

Lund, H. (2014). *Renewable energy systems*. Academic.

Midford, P., & Moe, E. (Eds.). (2021). *New challenges and solutions for renewable energy*. Palgrave Macmillan.

Mildenberger, M. (2020). *Carbon captured*. MIT Press.

Moe, E. (2015). *Renewable energy transformation or Fossil Fuel Backlash*. Palgrave Macmillan.

Moe, E., Hansen, S. T., & Hovland Kjær, E. (2021). Why Norway as a green battery for Europe is still to happen and probably will not. In P. Midford & E. Moe (Eds.), *New challenges and solutions for renewable energy* (pp. 281–318). Palgrave Macmillan.

Overland, I. (2019). The geopolitics of renewable energy. *Energy Research & Social Science, 49*, 36–40.

REN21. (2022). *Renewables 2022: Global status report*. REN21 Secretariat.

Schmitt, T. (2018). (Why) did Desertec fail? *Local Environment, 23*(7), 747–776.

Smil, V. (2015). *Power density*. MIT Press.

Sovacool, B. K. (2014). *Global energy justice*. Cambridge University Press.

Vakulchuk, R., Overland, I., & Scholten, D. (2020). Renewable energy and geopolitics. *Renewable and Sustainable Energy Reviews, 122*, 109547.

WindEurope. (2019). *Our energy, our future*. https://windeurope.org/wp-content/uploads/files/about-wind/reports/WindEurope-Our-Energy-Our-Future.pdf. Accessed 17 Sept 2021.

Wüstenhagen, R., Wolsink, M., & Bürer, M. J. (2007). Social acceptance of renewable energy innovation. *Energy Policy, 35*(5), 2683–2691.

Espen Moe (b. 1972) received his doctorate in Political Science from UCLA in 2004 and is currently Professor of political science at the Norwegian University of Science and Technology (NTNU). He has a particular focus on processes of structural economic change, such as the current renewable energy transition. Moe is the author of *Renewable Energy Transformation or Fossil Fuel Backlash* (2015), and the co-editor of *The Political Economy of Renewable Energy and Energy Security* (2014) and *New Challenges and Solutions for Renewable Energy* (2021).

Resources

Hajo Eickhoff

Abstract The universe is a resource. The source of all beings. From it, the primordial resource, the solar system emerged and the earth, which in turn is an immense resource from which life emerged, giving rise to the diversity of plants and animals as a bio-resource. From the animals, evolution has given rise to humans, who create cultural phenomena as a cultural resource. Their actions are ambivalent. They have created great cultural products, but at the same time damage vital resources such as soils, water and air. Thus, in the Anthropocene, humanity is faced with the decision to continue on this path, or to muster the will to change direction – preserving basic resources by letting go – letting go of excessive consumption of the waste of resources and of the throw-away mentality – so that in a global community effort, a circular ecology and a rebuilding of the world can be achieved.

Resource Species

Humans have invented philosophy and poetry, science and fantastic products, democracy and love, and yet these human beings are frightening in the face of the self-inflicted ecological catastrophe they are hurtling toward.

Resources are potentials. Sources. Possibilities. The origin for everything is the universe – the primordial resource. It brings forth the solar system with the basic resources of sun, soil, water and air, which give rise to the bio-resource of life, which in turn gives rise to cultural resources such as consciousness and education, which give rise to everyday resources such as food, consumer goods and houses, such as money, time and microchips. Thus, there is a step sequence of resources, but basically anything can become a resource.

H. Eickhoff (✉)
Interdisziplinäres Zentrum für Historische Anthropologie, Freie Universität Berlin,
Berlin, Germany

© The Author(s), under exclusive license to Springer Nature
Switzerland AG 2023
N. Wallenhorst, C. Wulf (eds.), *Handbook of the Anthropocene*,
https://doi.org/10.1007/978-3-031-25910-4_45

Some resources are finite, some like soil, water and air can be renewed again and again. Resources like mineral resources are unequally distributed on earth and have eminent geopolitical importance. But resources should not be subjected to arbitrary purposes, but should be treated sparingly for their own sake. Resources belong to no one.

(a) The primordial resource – the universe. The universe is the basis for everything that can come into being. In this primal resource lie the possibilities from which all other resources develop. The processes of the universe as incessant circling, exploding, circulating and drifting, take place under rules like causality, conservation of energy and gravitation. Galaxies and suns are born. Suns can give rise to other resources, which in turn give rise to other resources and forms of being – like the solar system with planet Earth.

(b) The Basic Resources – Sun and Earth. The earth began lifeless and unconscious as a hot stone into which the universe inscribed its laws. Sun as well as soil, water and air are the basic resources for life. The sun gave energy and warmth and drove the development on earth. The cooled soil – the lithosphere – was physically and chemically decomposed by water and air, made permeable by microorganisms and supplied with minerals. The lithosphere gave rise to the first fertile soil: humus – the basic resource for land-dwelling creatures. The atmosphere was formed by volcanic eruptions and the emission of gases from the earth. It consisted first of carbon dioxide and of water vapour, which condensed in the cooling of the earth, fell to the earth as rain and caused the oceans to form. With the invention of photosynthesis, plants converted solar energy into organic matter and energy, releasing the liberated oxygen into the atmosphere. Life needs water because cells are made up of more than 90% water.

(c) The bio-resource – the cell. Life evolved as inorganic molecules that combined to form organic compounds. This chemical evolution produced molecules that could feed and multiply – the first single-celled organisms. These brought other cells and cell elements into themselves and began to combine into multicellular cell systems and evolve into living things such as plants and animals. As their cells were in ceaseless exchange with each other, they formed a brain for coordination. To humans, this organ of coordination has become a special resource, enabling them to live in any place on Earth, to superform it technically, and to penetrate into space. Despite such a brain, they have so far overlooked the fact that forests and green plants are excellent climate protectors.

(d) Culture Resources – Consciousness. Humans are equally beings of nature and culture. Their release from the narrowness of stimulus and reaction and the advantage of being able to pause and have consciousness gave them the possibility to give themselves centre and meaning in rituals, in skills, knowledge and values: visible in architecture, clothing, weapons and musical instruments, invisible in commandments regulating communal life. The capacity of their brain has enabled them to subordinate all bio- and basic resources to human purposes, but what has not been understood is that all other resources are necessary for their existence.

A Resource Becomes Property

(a) The first opening of a resource. The first humans stopped to settle about twelve thousand years ago. A social, architectural, environmental and spiritual process that was followed by a revolutionization of the way of life. They began to produce food, invent tools and build houses and villages. They were the first to dare to dig open the ground. It appeared fertile, as a lot of food grew on a small area. Animal husbandry seemed equally successful, as animals were born into the enclosures and provided energy and meat to the sedentary people at all times. Already at this time, houses and fields disturbed the self-regulation of the soil, so that former forests became steppe (cf. Metzner & Reichelt, 1997, p. 117).

(b) The invention of the product. In order to produce objects, the sedentary people dug open the soil and extracted material from it. By leading (ducere) it out of the ground (pro) and shaping it, products – things – were created. From house and soil – the secured possession – the sedentary began to expand their sphere of influence. With their surpluses they traded, developed crafts, created trade routes and markets and gradually inhabited the earth. The first great empires arose. The largest network of routes was the Roman Empire, which accelerated the transportation of weapons, goods and people, and continued to open up and seal the ground and create things.

The Technosphere – Destruction and Conversion of Resources

Humans are constantly changing the earth. The totality of all human interventions in the earth – the technosphere – consists of soil, water and air. This sphere grew enormously in thirteen century Europe due to the rise of crafts and trade. When mediaeval crafts were dissolved in the Renaissance, industry developed, the production form of capitalism, which optimized production. Since capital needs growth, productivity and mass production increased, resulting in a systematic exploitation of resources, as well as an increase in world population and world energy consumption.

Resources in the Anthropocene

The earth is being permanently transformed. By volcanic eruptions, meteorite impacts, earthquakes, or by animals such as beavers that greatly change their environment by feeding and dwelling. Every activity changes the Earth system.

Humans have the greatest potential to change resources. Initial effects produce fire, simple tools, agriculture, animal husbandry, and crafts. The Anthropocene as a new geological age starts at the latest with industry, probably not 70,000 years ago, as Yuval Noah Harari writes (2017, p. 117.), but rather at the time of sedentarization. But industrialization appears most significant, because the basic, bio and cultural resources were changed immensely by industrial nations and exploited mercilessly, leading to their destruction. The destructive action is fed by a mental attitude of perfectionism and hubris.

In the Anthropocene, soil is disrupted. Biotopes are closely interconnected communities of animals and plants, including climate, light, and geology. In dynamic ecosystems, small interventions can radically change everything. For monocultures and factory farming, wildlife is displaced and must seek new habitats. Then, as soil-dwelling species disappear, so do insects, birds, and mammals, and biodiversity declines, and as insects die, so does a vital resource: pollination. In "Japan, Argentina, Chile, New Zealand, and Italy," there are partly no bees anymore and hand pollination becomes necessary (Settele, 2020, p. 138f).

In the Anthropocene, the atmosphere is doubly stressed. It is warming up due to greenhouse gases such as water vapour, methane and carbon dioxide, and it is polluted by fine dust such as car tyre abrasion, by soot, fumes and smoke. This affects the lungs, heart and skin and can lead to disease, disability and death.

In the Anthropocene, waters are polluted by toxins from industry, artificial fertilizers from agriculture, air-polluted rain, toxins from garbage dumps that seep into groundwater, vast amounts of plastic in the world's oceans, and ocean acidification that dissolves outer layers of calcium from marine animals.

In the Anthropocene, food loses its quality through genetic manipulation and constant breeding. They may also contain microplastics, radioactive substances and drug residues. Such foods make people sick or deprive them of the pleasure of eating and thus of some of the joy of life.

In the Anthropocene, people are exposed to a wide variety of pollutants. Right down to their cells, they are connected to the world through their senses. Cells and senses need stimuli, just as muscles need movement, and people need communication and closeness. This requires trust – especially physical contact, which reduces stress and positively stimulates many areas of the brain (cf. Grunwald, 2017, p. 64). The Anthropocene is a busy, agitated time: permanent news in real time, cell phone ringing, constant readiness, no rest, and an incessant striving for fitness, beauty, and perfection. A constant self-optimization that causes dissatisfaction and mood disorders.

In the Anthropocene, people recognize themselves as exploiters and destroyers of their environment. In this self-knowledge and the simultaneous perception of their indifference, they can become strangers to themselves. The indifference to the destruction of their own livelihood lies in the lack of empathy and responsibility. Therefore, people can destroy soils and pollute water and air, handle resources irresponsibly and eat animals tortured in industrial mass production without any remorse. In these self-damages, the Anthropocene leaves people perplexed. For Bruno Latour, however, living in the Anthropocene means making an effort and redefining the political task (cf. Latour, 2017, p. 247).

Conservation and reconstruction of resources

(a) World domestic policy and world community. Since today everyone can recognize that they are contributing to the destruction of the world, the question of a global responsibility forces itself, which offers the possibility to raise the awareness of the global of present crises and to start a changed thinking and acting. The resource community has become fragile. A reflection of the ecological situation. Communities are losing what holds them together - common ground of trust and responsiblity. Creating that is a key resource project. Since resource depletion, climate change and species extinction stem from the mental attitude of representatives of Western cultures, a paradigm shift is required: politics, civil societies and each individual have a responsibility and must therefore cooperate in the conservation of resources - there is already acute resource scarcity. This can only be done globally and as a world domestic policy - a world domestic policy that develops concepts against the dangers of anthropogenic interventions and acts accordingly.

(b) World education and resource protection. The World Resources Council, an international scientific body of the UN watches over the fact the raw materials are treated sparingly, materials are recycled and things are repaired, and the G-20 states are working to transform the world economy into a resource-efficient economy.

Such ideals are already being practiced on a small scale, in private and in civil society: Promoting handycrafts and growing regional products, reducing plastic, moderating consumption, eating healthy, driving less, and getting away from the throw-away mentality. These ideals are also practiced on a large scale. Paradigm shifts are evident among business leaders and politicians. Wasting resources and climate change are recognized as a threat to one´s own existence and to the survival of humanity. Likewise, that the world situation springs from one´s own actions: cities like Paris are becoming greener, road traffic is being limited and electrified, bike lanes are being expanded, and urban gardening and underground growing like in London are being practices and supply chains controlled. Meanwhile, even financial markets consider circular economy and climate protection to be decisive.

What is happening on a large and small scale needs to be translated to the very big, the earth and the global community. This requires cooperation beween science, companies, politics and private individuals: What is needed are clever theories, responsible production, future-oriented policies, considered consumption and the use of renewable resources. This requires a new global value system that overcomes the sedulousness of industry. The logic of capital, market and growth still prevails and has penetrated consciousness, and every European carries capitalism within him or her. The practice of this logic leads to alienation, to the rationalization of ways of life as well as to the distance between people, who now have to learn the letting go. The non-doing. That does not mean renunciation, but a wanting that is not naive, but reflected and sustainable.

Education is one of the factors that determine whether it will be possible to use resource sensibly. This will only be possible in the context of a global political and cultural order. Therefore, for Christoph Wulf, "the development of an awareness of the need for a planetary world community" (2021, p. 465) is necessary, and a sense of belonging to this community must arise in as many people as possible (2021, p. 476).

If the planet is to remain habitable, non-renewable resources are to be left alone and steady economic growth is to be abandoned. The clearing and poisoning of the soil as well as the cutting down and burning of forests mus be prohibited, as well as the pollution of the atmosphere with greenhouse gases and fine dust, and excessive consumption. The hatred in digital media, the violence in the streets and the mere meaning must be discussed in order to be able to let go of them.

What is produced must be recyclable. From the outset, products are to be aligned with this. In order for everyone to still have the chance to "inscribe a better future in the enduring evidence of the earth's history in the rock" (Ellis, 2020, p. 225), international political cooperation is needed under a fair distribution of burdens within the framework of a political economy and a circular world economy.

References

Ellis, E. C. (2020). *Anthropocene. The age of man – An introduction.*
Grunwald, A. (2017). *Homo hapticus. Why we cannot live without the sense of touch.*
Harari, Y. N. (2017). *Homo deus. A history of tomorrow.*
Latour, B. (2017). *Battle for Gaia. Eight lectures on the new climate regime.*
Metzner, H., & Reichelt, G. (1997). *Habitat Earth. The special position of our inhabited planet.* European Academy for Environmental Studies (ed.).
Settele, J. (2020). *The triple crisis. Species extinction, climate change, pandemics.*
Wulf, C. (2021). Global citizenship education. Education towards a planetary world community in the Anthropocene. *Vierteljahresschrift für wissenschaftliche Pädagogik, 97.*

Hajo Eickhoff, Dr. Phil, works as a cultural historian, exhibition organizer and cultural and business consultant in Berlin. Studied philosophy, history and art history in Freiburg, Aachen and Berlin. His research focuses on postures, the inner design of the human being, the intelligence of cells, the visual arts and the structure of occidental cultural development.

Terraformation

Yves Citton

Abstract One usually speaks of 'terraforming' to discuss the possibility of altering the environment of another planet in order to make it sufficiently similar to the conditions that have allowed human beings to dwell on Earth. But the notion of 'terraformation' can also be used to reflect upon the various ways in which humans (and other animals) have actively transformed planet Earth in order to make it more suitable to their needs. The Anthropocene can thus be seen as a terraforming adventure that is about dramatically to fail: humankind was endowed with a marvellously liveable planet, and it is managing to mess it up almost beyond repair. This article discusses what can be gained by planning for an alternative terraformation of planet Earth—a move that calls for a further elaboration of the notion of 'planetarity'.

Science fiction writers have devoted volumes to imagining how, in the future, mankind could send pioneers to other planets or constellations that could be used as extraterrestrial colonies on which to settle human forms of life (see for instance Robinson, 1992–1996). The main challenge consists in devising and implementing planetary-scale transformations that can allow humans to get out of their spaceship or capsulary life, so that they can walk freely on the surface of their new dwelling, finding enough air, water and food, at proper temperature, to allow for migration in large numbers. This terraforming process usually involves operations that need to be planned and operated at a tremendous scale, with irreversible consequences for the guest environment.

Reading the daily news in the age of the Anthropocene—with ever more precise and worrying predictions about global warming, rising sea levels, carbon sequestration and solar geoengineering—triggers an uncanny resemblance with such fictional narratives. In 2019, media and design theorist Benjamin H. Bratton (2019) has launched a three-year program of investigation at the Strelka Institute, a think-tank located in Moscow, where two dozens of young researchers, coming from a wide array of disciplines and renewed each year, collaborate on projects intended to

Y. Citton (✉)
Université Paris 8, Paris, France

N. Wallenhorst, C. Wulf (eds.), *Handbook of the Anthropocene*,
https://doi.org/10.1007/978-3-031-25910-4_46

imagine what types of transformations will have to be performed on planet Earth to ensure that humans can continue to find it habitable.

A provocative and polemical manifesto entitled *The Terraforming 2019* stated the general principles and goals of the program. They can be summarized along the following lines. 1° Humankind has just a few more years (until 2030) to drastically and globally redesign its dwelling on Earth before (more) irreparable damage is done to our habitat. 2° Humankind has always artificially altered—i.e. 'terraformed'—its environment. 3° In order to remedy the self-destructive terraformation enacted over the last 150 years, a return to 'nature', as advocated by many thinkers of ecology, would be impossible, pointless and misled: what we need is more artificiality (technology, automation, computation, prostheticisation, surveillance) to counteract the damaging side-effects of the current artificialization. 4° We can no longer afford the ineffective demagogical processes of political representation which are currently failing our imperative need drastically to transform our productive infrastructure. 5° We need purposefully to design and develop an ambitious form of planetary-scale computation taking advantage of our ubiquitous sensing devices to subordinate our individualist desires to collective forms of planning capable of maintaining the habitability of our terrestrial environments. 6° Humanist culture, with its romantic cult of nature and its illusions of sovereign subjectivity, stands in the way of the type of collectively-designed terraforming necessary to prevent the ecocidal consequences of the current unplanned alteration of our earthly habitat.

With such a program, Bratton invites us to understand the name 'Anthropocene' on at least two levels. On a first level, it simply means that mankind, through its technological agency, has become a geological force capable of messing up its terrestrial dwelling. But it also means, on a second level, that a certain fetishist conception of the human—'humanism' as an obsolete ideology, dangerously fond of self-possessive individualist sovereignty—is the anthropocentrist cause for the self-destructive misuse of this technologically acquired geological force.

While deliberately provocative, Bratton's intervention in the Anthropocene debates resonates with a number of other (more conciliatory) philosophical voices that have recently intersected on the notion of 'planetarity'. Coming from the historical tradition of the Subaltern Studies that flourished in the Indian subcontinent over the last decades, Gayatri Spivak (1999) and Dipesh Chakrabarty (2019) have advocated for a distinction to be made between the 'global' and the 'planetary' (see also Gabrys, 2018). Simultaneously Isabelle Stengers (2009) and Bruno Latour (2017) reinvigorated the figure of Gaia, originally sketched by Lovelock and Margulis (1974). The resulting terminological landscape could be synthetically and somewhat brutally summarized as follows:

1. All human beings live in a 'WORLD' which makes sense for them insofar as it is substantiated by their specific cultural values, and there are as many worlds as there are human cultures, sometimes overlapping (with potentially conflictual frictions) in our multicultural urban environments.

2. Over the last five hundred years, these worlds have been progressively subsumed under a process of 'globalization' enforced by a conquering system of economic and financial valorization (capitalism): our current 'GLOBE' is organized along the lines of logistic containerization and along the flows of financial (high-speed) trading.

3. Economic value ultimately results from scarce resources and human labour which are always grounded on 'EARTH', understood as the juxtaposition and overlap of the countless territories and heterogeneous 'patches' (Tsing, 2015) which sustain our material, social and cultural life.

4. Along the expansion of European colonialism, the earth has been mapped, appropriated, contracted, traded, negotiated, governed by various types and levels of political agencies, from the commune to the nation-State, and all the way to our international organizations (EU, UN, IPCC, COP, etc.): the resulting 'COSMOS' is the proper stratum of geopolitics, understood simultaneously as a self-professed exercise in cosmopolitanism and as a cynical continuation of the colonial order.

5. The increasingly precise awareness of the geophysical and biological interdependencies that entangle together the various entities taken into account by the Earth System Science has been figured under the name of the Greek goddess 'GAIA', with the double and somehow contradictory result of stressing, on the one hand, the indifference of these entangled forces towards the narrow and fragile requirements of human welfare and, on the other hand, a spontaneous tendency of the system to stabilize around points of homeostasis and equilibria that almost make it look like an (anthropo-bio-morphic) organism.

6. Planetarity as such kicks in last, hammered in by the tough lessons of ozone depletion and climate change: we are now compelled to reconsider our terrestrial milieu with the same attitude we had toward other 'PLANETS': in terms of habitability, i.e., as environments potentially hostile to human dwelling—a premise that it would have previously seemed absurd to apply to the Earth, since it was a given that we could thrive on it.

By writing an *Introduction to Comparative Planetology*, Lukáš Likavčan (2019) adopts the emblematic point of view of a planetarity conceived as an intellectual exercise in terraformation. Because of the degradations it is subjected to by our thoughtless and unplanned climatic, chemical and biological alterations, the Earth needs to be reconsidered, not as 'our' home, but like any other planet: we are *de facto* doing comparative cost analyses of the possibilities, relative advantages and drawbacks of fixing it (through geoengineering, clean-ups, decontamination) or leaving it (in the fantasies of transhumanist dreamers and billionaire tourists).

In Likavčan's analysis, in order to steer our path away from the current ecocidal trajectory, an alternative terraformation would call for 1° an overcoming of the nation-state, 2° a re-dimensioning of our most relevant political institutions at the scale of the bioregions, 3° the institutionalization of a transnational legal accountability for the 'slow violence' unleashed by (fossil fuel) corporations, and 4° a reconsideration of our collective agency at the level of the species. Most importantly however, 5° the visions of a planetary-scale politics sketched by Bratton and Likavčan invite us to place centre-stage of any terraforming plans the crucial (but also deeply disturbing) political role played in them by our technological infrastructures: "Recognition of non-human agency in planetary infrastructures brings a situation in which we cannot be sure anymore whether we are the operators of technologies, or whether the technologies operate us. These reverse prostheticisations bring ontological redistributions leading to a new cosmogram of infrastructural space" (Likavčan, 2019, 106).

Even if it stresses the role of design and planning, even if it is in part elaborated in Moscow—not too far (spatially and intellectually) from where the Soviet Union attempted to program its economy through five-year plans—this conception of terraformation for the Anthropocene is anything but a return to the modern yet romantic fantasy of human mastery through control technologies. Artificiality is welcome, prostheticisation is ubiquitous, plans are desperately needed, geoengineering may be unavoidable—but our human pride and delusion about our effective level of agency must sober up: at the planetary scale and at the high level of complexity inherent to our terraforming infrastructures, we are bound to be simultaneously operating the technologies and operated by them. Such is our destiny as consciously terraforming entities in the Anthropocene: we must learn to cohabit on a planet and with technical infrastructures that are equally 'alien' to us—unmastered and extra-terrestrial.

References

Bratton, B. H. (2019). *The terraforming 2019*. Strelka Institute.
Chakrabarty, D. (2019). The planet: An emergent humanist category. *Critical Inquiry, 46-1*, 2–31.
Gabrys, J. (2018). "Becoming planetary", *e-flux architecture*. https://www.e-flux.com/architecture/accumulation/217051/becoming-planetary/
Latour, B. (2017). *Facing Gaia. Eight lectures on the new climatic regime*. Polity.
Likavčan, L. (2019). *Introduction to comparative planetology*. Strelka Press.
Lovelock, J. E., & Margulis, L. (1974). Atmospheric homeostasis by and for the biosphere: The Gaia hypothesis. *Tellus, 26*, 1–10.
Robinson, K. S. (1992–96). The Mars trilogy. : Spectra/Random House.
Spivak, G. C. (1999). *Imperatives to re-imagine the planet*. Passagen Verlag.
Stengers, I. (2009). *In catastrophic times: Resisting the coming barbarism* (p. 2015). Open Humanities Press.
Tsing, A. L. (2015). *The mushroom at the end of the world. On the possibility of life in the capitalist ruins*. Princeton University Press.

Yves Citton is fortunate enough to be paid to study and teach media and literature at the University of Paris 8 Vincennes-Saint Denis and to be co-director of the journal *Multitudes*. He is the author of a dozen book, including *Altermodernités des Lumières* (Seuil, 2022), *Faire avec. Conflits, coalitions, contagions* (Les Liens qui Libèrent, 2021), *Générations collapsonautes* (Seuil, 2020, in collaboration with Jacopo Rasmi), *Mediarchy* (Polity Press, 2019), *Contre-courants politiques* (Fayard, 2018), *The Ecology of Attention* (Polity Press, 2016), *Renverser l'insoutenable* (Seuil, 2012).

Waste

Isabelle Hajek

Abstract This entry addresses the recent flurry of citizens' and activists' initiatives to fight waste. It shows that they challenge the prevailing economic and technocratic approach to the problem by putting the question of waste and wastage in the context of social relationships, thereby reconnecting with the founding project of political ecology.

A surge of interest in fighting waste has been observed in many industrialized countries. Against the backdrop of an economic crisis and of rising environmental awareness, calls to eliminate waste are more and more frequently voiced, for instance in documentaries depicting the devastating impact of waste on natural and human environments. Waste prevention and recycling policies have themselves been strengthened. Food waste, in particular, has become a public policy concern, gradually broadened to include all objects. Yet, the prevailing interpretation of the impact of waste dates back to pre-Anthropocene times: its main terms emerged at turn of the 1960s, as European and American social critique focused on denouncing consumer society and raised environmental concerns. Emphasizing the waste problem was then far from a unanimously agreed choice in ecologist and environmental movements. In France, it was considered as a management problem, part of a political approach to ecology or "ecologism" (Gorz, 1977, 1992) which sought to deviate from the reformist, scientific vision of sociology, faulted for seeking to correct the workings of industrialized, urban societies without challenging their underlying principles. A foundational conflict of ecological critique – "reform or revolution?" (Gorz & Bosquet, 1978, p. 9) – played out around the issue of waste. Should the waste-based capitalist social model be monitored or combatted? Informed by a variety of strains of thought, from 1930s non-conformists to post-war existentialist Marxism, which updated the concept of alienation by emphasizing its quotidian

I. Hajek (✉)
University of Strasbourg, Strasbourg, France
e-mail: hajek@unistra.fr

dimension, and the situationist critique of consumer society disseminated in May 68, "ecologism" proposed a comprehensive political project. It intended to give individuals spaces for experiencing autonomy and sociality and to promote self-limitation at a time when the extension of economic rationality to all areas of social life steered them in the opposite direction, making them dependent on a "megamachine" by pushing an "industrial definition of existence" (Illich, 2004) and a "heteronomy" of needs and desires destined to yield what Jean Baudrillard (1970) called "productive" waste or "consummation". Under that approach, focusing on waste meant refusing to turn to the general causes enabling to consider their role as a nuisance within the system.

In the turbulent context of the 1970s, as developed countries pursued productivist goals and sought to anticipate urban pollution, the formalization of national waste policies was primarily backed by a technocratic elite of modernizers. The transnational rise of the reformist ecology defended by some critics of industrial society and by proponents of a scientific approach to urban ecology was used by this elite as an ideological framework to (re)consider the waste issue in terms that were no longer solely hygienic. Indeed, by the late 1950s, critiques had emerged in the United States to single out the excesses of a "throwaway society". In 1958, John Kenneth Galbraith, in *The Affluent Society* (1958), developed the concept of "revised sequence" and the idea that businesses force products on consumers. Vance Packard published *The Status Seekers* (1959) and *The Waste Makers* (1960), which denounced incentives to consume in the face of the impending exhaustion of natural resources; then, Ralph Nader, in *Unsafe at any Speed* (1965), took the automotive industry to task over the impact of its quest to maximize sales. A similar approach was also taken by Bertrand de Jouvenel, whose output challenged the messianic consumerism of Jean Fourastié (1949, 1951) and the then fashionable theses on the so-called classless civilization of abundance and leisure. Jouvenel criticized the conflation between the improvement of living standards and the increase of goods and services per individual, denounced the impact of urban sprawl and unchecked car use, and examined waste within the framework of a reflection on the economy's "negative flows", leading him to be one of the earliest advocates of recycling as a solution to the waste of raw materials of the "social metabolism" of industrial societies (de Jouvenel, 1976). Also, the development of a scientific urban ecology gave credence to the idea of "urban metabolic cycles", in an approach that increasingly took economic issues in stride – some ecologists now drew reverse parallels between energy or matter flows and monetary flows (Larrère, 1991). These analyses addressed the question of waste as part of a new – qualitative – phase of industrial and urban development, and inscribed it in a rationality that was compatible with the idea of a permanent "cycle" of fertility and nature. In the wake of the 1973 oil crisis, this interpretive framework took on a normative dimension with the formalization of waste policies in many countries, emphasizing the connections between, respectively, waste and pollution, the recovery/recycling/sorting of waste and savings on raw materials and energy, and the fight against waste and productivity. The ecological idea of cyclicity was used to legitimize an economics-based way of dealing with waste, resulting in an industrial, managerial model of waste processing that stresses their "reclamation" – a model that still largely prevails today, including in circular economy approaches.

While for some time, the failings of this model, which generates toxic waste itself, as evidenced by the crises pertaining to incineration, refocused criticism on the impact and management of that waste, the idea of waste has been experiencing a second life. Citizens' and activist initiatives have been flourishing (Anstett & Ortar, 2015; Goyon, 2016); from "dumpster diving" (salvaging from the waste containers of shops or restaurants) to "freeganism" (a lifestyle based on limited consumption focused on free and vegan products), "disco soups" (parties where soups made from discarded vegetables are cooked and eaten), "repair cafes" (where broken household devices are repaired for free) and the "zero waste" movement. These initiatives circulate between national spaces, in a very decentralized manner, draw inspiration from one another, and propose a variety of approaches to fighting waste: extending the working lives of devices, giving back power to social actors, passing on know-how, creating social links, fighting poverty, preserving resources. The methods they put forward to combat waste are at odds with the end-of-pipe technical and managerial situations that prevailed during the rise of consumer society. They combine an ecology of personal solutions, which emphasizes each individual's ability to alter their everyday conduct to reduce waste, if not deconsume – thus the popularity of "average" people such as Bea Johnson, the face of the Zero Waste movement, and the proliferation of blogs where individuals describe changes in their everyday lives – and actions, often by individual entrepreneurs, to create a waste-free economic supply (Hajek & Diestchy, 2018). Ultimately, this is not so much about developing a policy to promote appropriate behaviours under a productivist economy as a way of life invented by social actors who seek to enable it by themselves producing an economy that reflects their values, or to make the existing economy conform to their values in the name of a quest for self-fulfilment and for the alignment of individual values and practices in all areas of life. Far from pursuing private hedonism or the depoliticization of the waste problem, this activism is characterized by a tireless search for continuities between personal, private, domestic forms of investment and collective – political, economic, environmental – investments. They do so for instance by turning domestic efforts to reduce waste into activist work, by introducing activities previously conceived as self-production in the fight against waste, by repurposing discarded objects in various sectors and inventing a new semantics in connection to the process, or by creating venues for exchange and socialization, where knowledge and waste reduction and recovery practices are developed and shared. These activists operate as brokers of new anti-waste standards and values in various spheres of social life. They are also brokers of objects, who challenge the existing delegation to the technological and economic sector and reintroduce the question of what kinds of waste we produce, what we can do with it, and how we can prevent it. They develop the circulation of "leftovers" (Joulian et al., 2016; Benelli et al., 2017), painstakingly devising new uses for them, in a way that cannot be reduced to the ecological and economic dimensions, conceived as situated, often DIY action in targeted domains, rooted in plural social ties and finalities and in "instituted significations" (Castoriadis, 1975). In doing so, they are reconnecting with the opposition to a strict ecological determinism that characterized the founding project of political ecology, reminding us that waste, in a way that echoes Illich's (2004) demonstration on the concept of scarcity, is first and foremost the result of a social organization in need of a comprehensive anthropological and psychological reform.

References

Anstett, É., & Ortar, N. (2015). *La deuxième vie des objets. Recyclage et récupération dans les sociétés contemporaines*. Éditions PETRA.

Baudrillard, J. (1970). *La société de consommation*. Gallimard.

Benelli, N., Corteel, D., Debary, O., Florin, B., Le Lay, S., & Rétif, S. (2017). *Que faire des restes? Le réemploi dans les sociétés d'accumulation*. Presses de Sciences Po.

Castoriadis, C. (1975). *L'institution imaginaire de la société*. Points-Essais.

De Jouvenel, B. (1976). *La civilisation de puissance*. Fayard.

Fourastié, J. (1949). *Le grand espoir du 20ème siècle. Progrès technique, progrès économique, progrès social*. PUF.

Fourastié, J. (1951). *Machinisme et bien-être*. Minuit.

Galbraith, J. K. (1958). *The affluent society*. Houghton Mifflin.

Gorz, A. (1977). *Écologie et liberté*. Galilée.

Gorz, A. (1992). L'écologie politique, entre expertise et autolimitation. *Actuel Marx, 12*, 15–29.

Gorz, A., & Bosquet, M. (1978). *Écologie et politique*. Le Seuil.

Goyon, M. (2016). L'obsolescence déprogrammée: fablabs, makers et repair cafés. Prendre le parti des choses pour prendre le parti des hommes. *Techniques et Culture, 65–66*, 236–239.

Hajek, I., & Diestchy, M. (2018, décembre). RELGA. Étude sur les réseaux émergents de lutte contre le gaspillage. Les "passeurs" de la lutte contre le gaspillage. *ADEME: Rapport final*. https://librairie.ademe.fr/dechets-economie-circulaire/238-etude-sur-les-reseaux-emergents-de-lutte-contre-le-gaspillage-les-passeurs-de-la-lutte-contre-le-gaspillage.html. Accessed 22 June 2021.

Illich, I. (2004). L'histoire des besoins. In I. Illich (Ed.), *La perte des sens* (pp. 71–105). Fayard.

Joulian, F., Tastevin, Y.-P., & Furniss, J. (2016). Réparer le monde: excès, reste et innovation. *Techniques & Culture, 65–66*, 15–27.

Larrère, R. (1991). L'écologie, ou le geste d'exclusion de l'homme. In A. Roger & F. Guéry (Eds.), *Maîtres et protecteurs de la nature* (pp. 173–196). Champ Vallon.

Nader, R. (1965). *Unsafe at any speed*. Richard Grossman.

Packard, V. (1959). *The status seekers*. David MacKay.

Packard, V. (1960). *The waste makers*. David MacKay.

Isabelle Hajek is associate professor at the University of Strasbourg and affiliated with the laboratory Societies, Actors and Government in Europe (SAGE, UMR CNRS 7363). She is a specialist in urban ecology and environmental sociology. She notably published: Lutter contre le gaspillage. Réforme ou révolution. *Écologie & politique*, Ed. Le Bord de l'eau, 60, 2020; *Rethinking Nature. Challenging Disciplinary Boundaries* (eds., with Aurélie Choné and Philippe Hamman), Routledge, 2017; *Guide des Humanités environnementales* (with Aurélie Choné and Philippe Hamman), Ed. Septentrion, 2016; *De la ville durable à la nature en ville* (with Philippe Hamman and Jean-Pierre Lévy), Ed. Septentrion, 2015.

Section II
Epistemology, Sciences and Humanities

Nathanaël Wallenhorst and Christoph Wulf

The concept 'Anthropocene' marks the emergence of a new epoch, with new conceptions of nature, the world and humanity. In this process an important role is played by an expanding spectrum of different scientific disciplines. What are the basic suppositions of these sciences and how do they acquire their knowledge? What is it that creates certainties and convictions? How do existing insights gain new connections and new interpretations? How does existing knowledge lead to new kinds of knowledge and how that knowledge is used? Radical changes are on the horizon. In this process, sciences and philosophy play a role in the creation of new self-conceptions of humanity.

In the 18th and 19th centuries, newly emerging sciences began to advance the development of society. This process accelerated in the 20th and 21st centuries and contributed to the continued evolution of the Anthropocene, where humans have a considerable influence on the fate of the planet. Now it is the natural sciences and technology that have taken the lead. The sciences have become central forces that determine the relationship of humans to the world. In this process, new forms of knowledge have arisen, leading to the new term 'knowledge society'. As important as the sciences have become for our conceptions of the world and ourselves, their insights must not be allowed to obscure other characteristics of the globalized world, as created, for example, by neo-liberal capitalism.

The expansion of the sciences plays a central role in the Anthropocene. It has radically changed the living conditions of many people. At the same time these changes affect the type and quality of our knowledge. In all regions of the world,

N. Wallenhorst (✉)
Catholic University of the West, Angers, France
e-mail: nathanael.wallenhorst@uco.fr

C. Wulf
Freie Universität Berlin, Berlin, Germany
e-mail: christoph.wulf@fu-berlin.de

sciences play a role in both positive and negative developments in society
(Wallenhorst, 2021; Wulf, 2022; Wallenhorst & Wulf, 2022). In research into these,
the various paradigms, questions, methods, and techniques are of considerable
importance. This is particularly true of the sciences that have led to the emergence
of global technologies, industries and companies.

Can binding criteria be specified that can be used to determine what is considered scientific? Or does the historical and cultural nature of scientific research make
us question the validity of scientific criteria? In view of the two tendencies, on the
one hand the dynamics of globalization which tend to homogenize the world in the
Anthropocene and on the other, the tendency to emphasize its diversity, the question
of what can be considered scientific is of central importance, even if it cannot always
be answered clearly. Even where the sciences produce coherent, justifiable knowledge that is open to critical argument, the general criteria for scientific knowledge
are often culturally different (Michaels & Wulf, 2020).

Natural Sciences – Social Sciences – Humanities

What do we understand by 'sciences' in view of this situation? There is no simple
answer. In the English language, within scientific research a distinction is made
between Natural Sciences and Humanities (cf. Snow, 1963). This distinction also
plays an important role in epistemological discussion in Germany. There is however
a major difference, in that in Germany both the humanities and the natural sciences
are traditionally regarded as sciences. Although they have different roles and differ
in terms of content and method, both are recognized in society as sciences.

The social sciences have a place in between the natural sciences and the humanities. According to a common differentiation, the natural sciences have the task of
describing and explaining the world. It is demanded that science should be 'naturalistic', that is, that it should only deal with what is evident and measurable. That this
understanding of science is too narrow and therefore often harmful is clear. This has
led to the numerous unwanted negative effects of positivist knowledge. In the natural and technical sciences that predominate today there is a distinction between
Dilthey's 'spiritual sciences' (*Geisteswissenschaften*) and humanities in that they
attempt to understand the meaning of life. The focus is on hermeneutics, which
dates back to Friedrich Schleiermacher and his efforts to understand and interpret
meaning (Schleiermacher, 1977). Dilthey sees hermeneutics as an "empirical science of mental phenomena" or as a "science of the spiritual world" (Dilthey, 1988,
2008: Gadamer, 2002; Ricoeur, 1973).

Ensuing from extensive epistemological debates there is agreement that it is not
possible to uphold the idea of there being a clear distinction between natural sciences, social sciences and spiritual sciences or humanities. Instead of emphasising
the difference between natural and spiritual sciences it is more fruitful to attempt to
relate the different paradigms to each other so that they complement each other.
There is a consensus that both forms of research are important. But sometimes the

representatives of the quantitative paradigm insist that only they provide "true scientific" results. This is certainly true, although recently a comprehensive study on the reproducibility of 200 research findings in psychology published in *Science* demonstrated that the reproducibility of the findings could only be proved in one third of the studies (Ioannidis, 2005; Staddon, 2017). Since the representatives of the quantitative research paradigm insist that reproducibility guarantees the scientific character of the research findings, they now have a problem. According to the study, the reproducibility of the research results is limited. There must at least be an acknowledgement of historical and cultural differences. The rigid representatives of the quantitative paradigm in the social sciences only accept that qualitative research contributes to the development of hypotheses, which can then only be verified or shown to be scientifically false by using quantitative methods. According to Popper's Critical Rationalism, research produces scientific results if it uses one specific and normative research method (Popper, 2002). The application of this research method to different subjects assures the scientific character of the research findings.

Science and Methodology

If we understand methods as being paths to knowledge (from the Greek *methodos)*, then scientific knowledge in the Anthropocene is acquired via many different paths. Windelband's differentiation between 'nomothetic' natural sciences and 'ideographic' humanities is still relevant today (Windelband, 1924). In contrast to this, in the English-speaking world the Vienna Circle's idea of 'Unified Science' dominates. According to this, scientific knowledge should comply with certain features that concern the intersubjective validity of scientific knowledge and that are binding for all scientific methods. In his "Logic of Scientific Discovery", Karl Popper disputes such a dogmatic view, using his Falsification Principle to demonstrate that there is no scientific method that has succeeded in achieving knowledge that is valid once and for all (Popper, 2002). As a rule, methods and the knowledge acquired by means of them are only valid as long as they cannot be proved false. Therefore, in Popper's Critical Rationalism, 'medium-scope theories' are valid only if they cannot be falsified. Paul Feyerabend was critical of this, seeing methodological pluralism as "anything goes" (Lakatos, 1977).

The positivism dispute (Adorno et al., 1976) highlighted the complexity of methodological questions and the contradictory nature of the different positions in scientific theory. The Frankfurt School emphasized that it is not enough to limit scientific methods to 'medium-scope methods. The claim that following a universally accepted method ensures that the research and its results will be 'scientific' is not acceptable. Science also involves far-reaching theories which must be examined based on whether they are plausible and appropriate. Because of the complexity of the conditions, empirical falsification is often not possible. Wittgenstein too, in his later work, distances himself from the monism of Critical Rationalism with its demand for falsifiability. He demands diversity of methods and points out the limitations of scientific insights that are expressed in words. In view of this situation, there is still a certain tension in the humanities and social sciences today between methods that are becoming increasingly formalized and methods that are more flexible or open. It is therefore useful to see the natural sciences, social sciences, and the humanities as a *unitas multiplex* within which there are differences and similarities.

Paradigms of Research

Research in the individual scientific disciplines continues to play a central role in research on the Anthropocene. In addition, the scientific landscape has been enriched by the emergence of new developments. These involve new questions and goals as well as new methods and structures. The great challenges facing science today include questions of inter-, multi-, and trans-disciplinarity and issues of multiculturality connected with the developments on the planet. These issues are of particular importance in the context of network research (Latour, 1999, 2018), in which there is collaboration between numerous projects and different scientific paradigms (Kuhn, 1972; Wulf, 2003). Here, an attempt is made to overcome the limitations of research in a single discipline through interdisciplinary and international collaboration, thereby creating new forms of complex knowledge.

The humanities make a major contribution to integrating scientific knowledge into people's life contexts. In addition, we see how important the knowledge generated by the humanities is, so that people can live together despite their differences and the difficulties resulting from these. Knowledge of the humanities is required, for example, to deal constructively with violence, alterity, and sustainability. To meet the demands of the Anthropocene, there is a need for both extensive knowledge in the natural sciences and social sciences as well as the development of appropriate action and behaviour. Knowledge from the fields of the humanities and social sciences contributes significantly to the acquisition of the necessary knowledge and skills. What is important is their contribution to the transmission and evolution of culture and cultural identity and to coping with alterity.

The growth of scientific knowledge has required it to be structured in clear, transparent forms. This increases the importance of the individual scientific disciplines. Working methods are being developed which will cause lasting changes in individual disciplines. Interdisciplinary and transdisciplinary research transcends boundaries. The consideration of the "difference" of the other disciplines is a constitutive element of transdisciplinary research. This enables new themes, concepts, and methods to be developed that change and call into question established knowledge and research in the individual scientific fields. This results in increased diversity and greater complexity: the more radical the selection of subjects and methodological processes, the less predictable the results of the research.

Networking with researchers from other countries and cultures can create new forms of sustainable knowledge in the natural and social sciences and in the humanities. By focusing on scientific knowledge and relating this knowledge to existing historical, ethnographic, and aesthetic knowledge, the humanities, for example, contribute to a better understanding of the meaning of the results of scientific research. The high regard for the sciences and the interpretation of their meaning through the humanities often leads to the devaluation of other forms of cultural and social knowledge, which are of central importance for living together in society. These forms of knowledge include, for example, practical knowledge, body knowledge and aesthetic knowledge, knowledge of wisdom and other forms of tacit knowledge (Wulf & Baitello, 2018; Kraus et al., 2021; Kraus & Wulf, 2022).

Many of the issues are so complex that they cannot be explored in sufficient depth using the approaches of one single discipline. Inter- and transdisciplinary research helps to enhance the multi-dimensionality of the subjects of study, the methodological approach, and the investigation itself. In many cases, the different skills of various scientific fields can be combined to achieve the desired increase in complexity. The plurality of the scientific paradigms under consideration results in complex research that extends beyond the scope of single disciplines. Democratic societies need the humanities and their forms of knowledge to sustain people, to help them orient themselves in an increasingly complex world.

Diversity: Inter- and Transculturality

If the specialist disciplines and scientific paradigms are supplemented by viewpoints from other cultures, and if the research thereby becomes inter-, or transcultural in nature, everything becomes increasingly complex (Gil & Wulf, 2015). In view of humanity's shared responsibility for the future of the planet there is a need for research that develops diverging cultural perspectives on global processes. In this research, an important role is played by differences and how they are dealt with. It is not possible to have intercultural and transcultural research projects if cultural differences are not processed. 'Difference' acquires particular importance in the UNESCO convention for the protection of cultural diversity, which envisages cultural difference as a universal human right (UNESCO, 2005).

The global change we are experiencing today is a multidimensional process with economic, political, social, and cultural consequences. Something that is very important is our encounter with what is foreign to us and our recognition of how experiences of what is foreign are determined by our own cultural situation. Today there is a danger that people increasingly only encounter themselves on the planet.

In this process, it is above all these aspects of western culture that adversely affect our experience of what is foreign and of alterity: (1) a pronounced subjectivism, and (2) extreme rationalism, and (3) an excessive ethnocentrism (Wulf, 2016).

In our encounter with nature, with other people and their cultures and in much anthropological research, foreignness and alterity have played an important role. Mimetic and hermeneutic approaches help us to understand these concepts. They are particularly difficult in cases of hybrid cultural phenomena. The concept of hybridity is used to determine cultural contacts, not only in a dualistic and essentialist way but also to show that through these contacts, new transcultural identities can emerge with the help of a "*third space*" (Bhabha, 2004). This third space is liminal; it is a space of in-between-ness. In this liminal space, spaces are bypassed and restructured, and hierarchies and power relations are changed. It is crucial to know to what extent these processes and their results are determined by performative practices and in what way new transcultural hybrid forms are born in them. These forms are mixed forms in which elements from different systems and contexts change their character in a mimetic process and give rise to new, partly transcultural identities.

Amazement, radical questioning, philosophical criticism, and self-criticism play a key role when researching these increasingly important processes. These forms of philosophical reasoning cannot be adequately described as methods. They cannot be formalized, and their significance can only be gauged when they are used to examine phenomena, events, actions, and problems. Depending on their context, these types of philosophical reflection result in differing views and insights as well as an increase in the complexity of research. They help to ensure that we continue to regard the fundamental questions concerning the human condition as unanswered and to realize that there can be no definitive concept of human beings and the Anthropocene (Kamper & Wulf, 1989, 1994).

Sustainability as a Challenge

In research on the Anthropocene, sustainability, and the Sustainable Development Goals in particular play an important role. The issue of sustainability leads to the introduction of new subjects for research, new research areas and methods. In addition, there are also new ways of structuring transdisciplinary and transcultural research, and also there is a re-evaluation of the outcomes of existing research. For this to happen, there is a need for an investment of financial and personal resources. The focus on sustainability, that influences many areas of human knowledge and that is closely linked with peace and the reduction of violence, must not be allowed to place restrictions on scientific freedom. This is ensured by the fact that the initiation of large parts of this research came from science itself, as did its normative focus and subject matter.

Although these research projects and institutions can be inspired and promoted form outside the sciences, they must be independent of these outside interests and must follow scientific criteria and standards. It is only through this that the creativity and productivity of scientific research can be guaranteed, although the insights that may emerge and the impact of the work cannot always be envisaged beforehand.

It is also important to examine what is meant by sustainability. This is particularly important in those contexts where the goal is a reduction in violence towards nature and other people. What does the term sustainability encompass? Currently there is much to be said for a broad use of the term. Then the aim could be to re-interpret familiar issues in light of the contribution they make to furthering sustainability.

Regardless of their foundations, intentions, form or content, research projects contain a comparative element. Implicit or explicit comparison plays a particularly important role in interdisciplinary or intercultural studies (Dreher & Stegmaier, 2007). If you want to find out something about a subject you contact someone else and establish what is identical, different or similar and where there are deviations. To compare something, you have to see the similarities. In science, this process must undergo methodological control, to ensure that criteria of objectivity, reliability and validity are met. The key objective in the comparison is to find a *tertium quid* that is equidistant from the two positions compared and permits a 'fusion of horizons' (Gadamer). This can only be achieved by mimetic assimilation (Detienne, 2009). Here, historical, and comparative cultural aspects are important (Wulf, 2013; Wulf et al., 2011). These studies throw up many new epistemological questions to be addressed in the Anthropocene.

References

Adorno, T. W., Dahrendorf, R., Pilot, H., Albert, H., Habermas, J., & Popper, K. R. (1976). *The positivist dispute in German sociology*. Heinemann.

Bhabha, H. K. (2004). *The location of culture*. Routledge.

Detienne, M. (2009). *Comparer l'incomparable*. Seuil.

Dilthey, W. (1988). *Introduction to the human sciences. An attempt to lay a Foundation for the Study of society and history* (Ramon J. Betanzos, Trans.). Wayne State University Press.

Dilthey, W. (2008). *The construction of the historical world in the humanities*. Vandenhoeck & Ruprecht.

Dreher, J., & Stegmaier, P. (Eds.). (2007). *Zur Unüberwindbarkeit kultureller Differenz*. transcript.

Gadamer, H.-G. (2002). *Truth and method*. Continuum.

Gil, I. C., & Wulf, C. (Eds.). (2015). *Hazardous future: Disaster, representation and the assessment of risk*. De Gruyter.

Ioannidis, J. P. A. (2005). Why Most published research findings are false. *PLOS Medicine, 2*(8), e124.

Kamper, D., & Wulf, C. (1989). *Looking back at the end of the world*. Massachusetts Institute of Technology.

Kamper, D., & Wulf, C. (Eds.). (1994). *Anthropologie nach dem Tode des Menschen. Vervollkommnung und Unverbesserlichkeit*. Suhrkamp.

Kraus, A., & Wulf, C. (Eds.). (2022). *The palgrave handbook of embodiment and learning*. Palgrave Macmillan. (in preparation).

Kraus, A., Budde, J., Hietzge, M., & Wulf, C. (Eds.). (2021). *Handbuch Schweigendes Wissen*. [Handbook of silent knowledge]. Juventa Beltz.

Kuhn, T. S. (1972). *The structure of scientific revolution*. The University of Chicago Press.

Lakatos, I. (1977). *The methodology of scientific research Programmes: Philosophical papers* (Vol. 1). Cambridge University Press.

Latour, B. (1999). *Pandora's hope: Essays on the reality of science studies*. Harvard University Press.

Latour, B. (2018). *Down to earth: Politics in the new climatic regime*. Polity Press.

Michaels, A., & Wulf, C. (Eds.). (2020). *Science and Scientification in South Asia and Europe*. Routledge.

Popper, K. (2002). *The logic of scientific discovery* (2nd English ed.). Routledge Classics.

Ricoeur, P. (1973). Task of hermeneutics. *Philosophy Today, 17*(2), 112–128.

Schleiermacher, F. (1977). *Hermeneutics. The handwritten manuscripts* (James Duke and Jack Forstman, Trans.). Ed. Heinz Kimmerle. Scholars Press.

Snow, C. P. (1963). *The two cultures: And a second look: An expanded version of the two cultures and the scientific revolution*. Cambridge University Press.

Staddon, J. (2017). *Scientific method*. Routledge.

UNESCO. (2005). *The convention on the protection and promotion of the diversity of cultural expressions*. UNESCO.

Wallenhorst, N. (2021). *Mutation. L'aventure humaine ne fait que commencer*. Le Pommier.

Wallenhorst, N., & Wulf, C. (2022). *Humains – Dictionnaire d'anthropologie prospective*. Vrin.

Windelband, W. (1924). *Geschichte und Naturwissenschaft. Windelband, Wilhelm. Präludien. Aufsätze und Reden zur Philosophie und ihrer Geschichte* (Vol. 2, 9th ed.). Mohr.

Wulf, C. (2003). *Educational science. Hermeneutics, empirical research, critical theory*. Waxmann.

Wulf, C. (2013). *Anthropology. A continental perspective*. University of Chicago Press.

Wulf, C. (Ed.). (2016). *Exploring alterity in a globalized world*. Routledge.

Wulf, C. (2022). *Education as human knowledge in the Anthropocene. An anthropological perspective*. Routledge. Weinheim and Basel.

Wulf, C., & Baitello, N. (Eds.). (2018). *Sapientia Uma arqueologia de saberes esquecidos*. SESC.

Wulf, C., Suzuki, S., Zirfas, J., Kellermann, I., Inoue, Y., Ono, F., & Takenaka, N. (2011). *Das Glück der Familie: Ethnografische Studien in Deutschland und Japan* (p. 2013). Verlag für Sozialwissenschaften (Japanese edition).

Nathanaël Wallenhorst is Professor at the Catholic University of the West (UCO). He is Doctor of Educational Sciences and Doktor der Philosophie (first international co-supervision PhD), and Doctor of Environmental Sciences and Doctor in Political Science (second international co-supervision PhD). He is the author of twenty books on politics, education, and anthropology in the Anthropocene. Books (selection): *The Anthropocene decoded for humans* (Le Pommier, 2019, *in French*). *Education in the Anthropocene* (ed. with Pierron, Le Bord de l'eau 2019, *in French*). *The Truth about the Anthropocene* (Le Pommier, 2020, *in French*). *Mutation. The human adventure is just beginning* (Le Pommier, 2021, *in French*). *Who will save the planet?* (Actes Sud, 2022, *in French*). *Vortex. Facing the Anthropocene* (with Testot, Payot, 2023, *in French*). *Political education in the Anthropocene* (ed. with Hétier, Pierron and Wulf, Springer, 2023, *in English*). *A critical theory for the Anthropocene* (Springer, 2023, *in English*).

Christoph Wulf is Professor of Anthropology and Education and a member of the Interdisciplinary Centre for Historical Anthropology, the Collaborative Research Centre (SFB, 1999–2012) "Cultures of Performance," the Cluster of Excellence (2007–2012) "Languages of Emotion," and the Graduate School "InterArts" (2006–2015) at the Freie Universität Berlin. His books have been translated into 20 languages. For his research in anthropology and anthropology of education, he received the title *"professor honoris causa"* from the University of Bucharest. He is Vice-President of the German Commission for UNESCO. *Major research areas:* historical and cultural anthropology, educational anthropology, imagination, intercultural communication, mimesis, aesthetics, epistemology, Anthropocene. Research stays and invited professorships have included the following locations, among others: Stanford, Tokyo, Kyoto, Beijing, Shanghai, Mysore, Delhi, Paris, Lille, Modena, Amsterdam, Stockholm, Copenhagen, London, Vienna, Rome, Lisabon, Basel, Saint Petersburg, Moscow, Kazan, Sao Paulo.

Part V
The Anthropocene as an Interpretative Framework

Anthropocene Working Group

Jan Zalasiewicz, Colin Waters, Simon Turner, Mark Williams, and Martin J. Head

Abstract The Anthropocene Working Group of the Subcommission on Quaternary Stratigraphy, of the International Commission on Stratigraphy, has been active since 2009. Its primary role is to consider the Anthropocene as a potential formal addition to the Geological Time Scale. Unusual in composition because many members work in disciplines other than stratigraphic geology —the Anthropocene incorporates geological, historical, and instrumental records— it initially needed to establish whether the Anthropocene could be the basis of a valid chronostratigraphic unit. That task achieved, work then focused on gathering evidence from 12 sections in different geological settings around the world, to establish which of these might be proposed as a Global boundary Stratotype Section and Point (GSSP, or 'golden spike') to define the Anthropocene formally. The Group aims to submit a formal proposal on this in 2023.

The Anthropocene Working Group (AWG, sometimes known as the Working Group on the 'Anthropocene') is one of many working/task groups attached to the various subcommissions of the International Commission on Stratigraphy (ICS). The ICS maintains and develops the International Chronostratigraphic Chart, that serves as the basis for the Geological Time Scale (GTS). Each boundary working group focuses on one boundary, and therefore comprises experts in that part of the stratal record. Its initial task is to initiate, where necessary, and support separate groups researching individual sections that have outstanding potential to define the

J. Zalasiewicz (✉) · C. Waters · M. Williams
Geography, Geology and the Environment, University of Leicester, Leicester, UK
e-mail: jaz1@leicester.ac.uk; cw398@leicester.ac.uk; mri@leicester.ac.uk

S. Turner
Geography, University College London, London, UK
e-mail: simon.turner@ucl.ac.uk

M. J. Head
Earth Sciences, Brock University, St. Catharines, ON, Canada
e-mail: mjhead@brocku.ca

315

boundary in question. Formal units of the GTS are defined at their bases, each top demarcated by the base of the superjacent unit. Each unit is defined by a Global boundary Stratotype Section and Point (GSSP), or more colloquially a 'golden spike'. One or more research groups will submit a GSSP proposal to the boundary working group, which then weighs the evidence during a period of open discussion, and votes to decide upon a single candidate GSSP. The selected proposal is then advanced to the relevant subcommission, which in the case of the Anthropocene Working Group is the Subcommission on Quaternary Stratigraphy (SQS). After a period of further discussion and scrutiny, the subcommission itself votes on the proposal, and if successful the proposal is submitted to the full voting membership of the ICS for additional examination and voting. If the proposal is approved by the ICS, it must be ratified by the Executive Committee of the International Union of Geological Sciences to come into force. All committees in this process are international in composition and voting at every level requires a 60% supermajority: the GTS is hence conservative by design to maintain stability, and boundaries reflect international agreement that reinforces their legitimacy and acceptance (Head, 2019).

The AWG operates within this formal framework which until now has worked almost exclusively within the domain of geological stratigraphy. The units of the GTS have been of greatest relevance to the geological community and form a temporal backbone. However, in some respects this framework has had to develop new research approaches to suit the particular circumstances of the Anthropocene.

Firstly, whereas all other subcommission working groups consist exclusively of expert stratigraphers (although with different kinds of expertise—in paleontology, isotope geochemistry, radiometric dating, etc.), the AWG has widened its membership to reflect the overlap of geological time (classically represented by strata) with the various human timescales studied by archaeologists and historians, and also with recently observed and recorded aspects of Earth System change studied by biologists, oceanographers, geographers and so on, while an expert in international law reflects the Anthropocene's wider utility (Vidas, 2015). Therefore, the AWG was perforce originally set up, and continued as, the first truly multidisciplinary group within the ICS, with representatives of all these communities, not least the highly integrative Earth System science (ESS) community (Steffen et al., 2020) where the Anthropocene concept arose (Crutzen & Stoermer, 2000; Crutzen, 2002) and soon became widely used as a *de facto* framing concept (Zalasiewicz et al., 2021). More than half of the current 37 AWG membership are geological stratigraphers, the others representing additional disciplines.

A second major difference is that other ICS working groups deal with geological time units which, although not precisely defined and formally established, nevertheless usually have a long history of practical use. Geological communities are familiar with these units and with the strata that act as their basis. Deep time boundaries are generally supported by a detailed, mature framework of data and interpretation, and widely shared underlying concepts on which the tightly focused work of selecting, refining—or even radically changing the position of—a boundary can be built. With the Anthropocene, by contrast, there was no tradition among geologists of recognizing and distinguishing a coherent unit of time and strata in what was simply regarded as the latest part of the Holocene Epoch of the Quaternary Period—let alone one characterised by an array of overwhelming human impacts, and recognized using evidence such as concrete and plastics. While such ideas of a 'human epoch' (or era) had occasionally been mooted over the past few centuries, the reaction of the mainstream geological community was almost universally disparaging, contrasting the brief time and 'superficial' nature of human impact with the much longer time scales and larger-scale changes (such as the splitting of continents and the building of mountains) of Earth history.

Hence, the first task of the AWG was to build a geological description of the Anthropocene, if there was indeed one to build, because the initial study (Zalasiewicz et al., 2008), that led to the formation of the AWG (in 2009) by invitation from the SQS, was couched in very general terms. It was not then clear at all whether on closer examination the Anthropocene might dissolve, geologically at least, into an indefinable gradation from the Holocene. A particular problem was the location of the boundary; in Crutzen's launching of the term (*op. cit.*), the beginning of the Industrial Revolution had been mooted, but the resulting geological effects were gradual, and took place at different scales and times around the world, leaving no single clear isochronous signal recognizable in the stratal record.

The AWG initiated a series of wide-ranging surveys of the physical (lithostratigraphic), biological (biostratigraphic) and chemical (chemostratigraphic) signals within strata that might be ascribed to the Anthropocene, and whether these formed coherent patterns consistent with a potential chronostratigraphic unit. These were partly conducted by various combinations of AWG members, and partly by other researchers who were beginning to become interested in this question. The AWG compiled several thematic volumes (Williams et al., 2011; Waters et al., 2014) and individual papers (e.g. Waters et al., 2015). Work on these was conducted largely by email as the AWG then had no dedicated funding to support its work. AWG meetings in person were therefore rare, but nevertheless occasionally and fruitfully took place, notably in Berlin (2014) and Oslo (2016), supported by the Haus der Kulturen der Welt and Fridtjof Nansen Institute respectively, both being humanities-based institutions). Through all this, a body of work emerged that began to show clear patterns.

It became clear that the Anthropocene was significant at a geological scale, not just because of the magnitude of its impacts, but because of the irrevocable nature of at least some of them—effectively changing the course of Earth history for many thousands, indeed millions, of years into the future. Moreover, the Anthropocene took on considerably greater geological coherence when its beginning was placed in the mid-twentieth century (Zalasiewicz et al., 2015). This coincides with the beginning of the 'Great Acceleration' of population, industrialization and globalization initially recognized within the Earth System science community (Steffen et al., 2007), when an array of near-synchronous and more or less global signals, including such newly recognized ones as fly ash (Rose, 2015), nitrogen isotopes and diatoms (Wolfe et al., 2013) and microplastics (Zalasiewicz et al., 2016), were imprinted into strata. This led to a summary of the patterns identified to date (Waters et al., 2016) and a non-binding AWG vote carried out for the 2016 International Geological

Congress (Zalasiewicz et al., 2017a). The votes revealed majority AWG opinion affirming the geological reality of Anthropocene, and recommending its formalization as a new epoch/series beginning in the mid-twentieth century, and defined by a GSSP, with artificial radionuclides of the beginning of the 'bomb spike' the most indicated option among potential primary markers.

Work then began to shift emphasis from overall characterization of the Anthropocene, once its geological identity had been satisfactorily crystallized, to focusing on the search for potential GSSPs, within various geological environments with continuous, high-resolution stratigraphic archives, including lake, estuarine and anoxic marine sediments, peat, polar ice, speleothems and annually banded coral skeletons (Waters et al., 2018). Wider discussions have nevertheless continued, including analyses of the Anthropocene biosphere (Williams et al., 2016), responses to critiques of the Anthropocene (Zalasiewicz et al., 2017b), chronostratigraphy/Earth System science relations (Steffen et al., 2016), quantitative analyses of the energy outputs that drive Anthropocene impacts (Syvitski et al., 2020) and how the Anthropocene is viewed through different disciplinary lenses (Zalasiewicz et al., 2021). A detailed summation of the main research findings of the AWG in this first phase of work was published in a third edited volume (Zalasiewicz et al., 2019).

In 2020, the AWG entered a new formal cycle to reflect these changes, with Chair's position moving from JZ to CW, and the Secretary's role from CW (formerly MW) to ST (while JZ became Chair of the parent organization, the SQS). In addition, the AWG's structure changed to encompass voting members (with expertise to assess detailed stratigraphic evidence of GSSP candidate sites, commensurate with the overall SQS mandate) and non-voting members (to continue the wider exploration of the Anthropocene, and particularly how its geological application interfaces with use in other disciplines—an issue that is not present, or is generally negligible, for older geological time units). GSSP investigations received the generous and transformative sum of 800,000 euros, the initiative of the Haus der Kulturen der Welt in Berlin, as a collaborative endeavour with the AWG to explore the scientific, historical, and cultural construction of the Anthropocene.

Twelve candidate GSSP sites and other reference sections have been analyzed by a variety of research teams, in a range of environments worldwide. Work on these has been affected by the Covid-19 pandemic, but has now been completed, allowing publication of the results and detailed comparison of, and voting on, the sites to select a candidate Anthropocene stratotype. This then is aimed to be proposed, in 2023, to the SQS, ICS and IUGS executive to deliberate and vote on in turn. Normally, ICS working groups are disbanded after such a task is completed. In this case, given the amount of basic research still to do on the Anthropocene, the central role of the AWG within this study, and the strong working relationships developed within it, some components of the AWG might well continue in the future, even if informally, to help pursue this necessary work.

References

Crutzen, P. J. (2002). Geology of mankind. *Nature, 415*, 23. https://doi.org/10.1038/415023a

Crutzen, P. J., & Stoermer, E. F. (2000). The "Anthropocene.". *IGBP Global Change Newsletter, 41*, 17–18.

Head, M. J. (2019). Formal subdivision of the Quaternary System/Period: Present status and future directions. *Quaternary International, 500*, 32–51.

Rose, N. L. (2015). Spheroidal carbonaceous fly-ash particles provide a globally synchronous stratigraphic marker for the Anthropocene. *Environmental Science and Technology, 49*, 4155–4162. https://doi.org/10.1021/acs.est.5b00543

Steffen, W., Crutzen, P. J., & McNeill, J. R. (2007). The Anthropocene: Are humans now overwhelming the great forces of nature? *Ambio, 36*, 614–621. https://doi.org/10.1579/0044-7447(2007)36[614:TAAHNO]2.0.CO;2

Steffen, W., Leinfelder, R., Zalasiewicz, J., et al. (2016). Stratigraphic and earth system approaches in defining the Anthropocene. *Earth's Future, 8*, 324–345. https://doi.org/10.1002/2016EF000379

Steffen, W., Richardson, K., Rockström, J., et al. (2020). The emergence and evolution of earth system science. *Nature Reviews Earth & Environment 1*, 54–63. https://doi.org/10.1038/s43017-019-0005-6

Syvitski, J., Waters, C.N., Day, J., et al. (2020). Extraordinary human energy consumption and resultant geological impacts beginning around 1950 CE initiated the proposed Anthropocene Epoch. *Communications Earth & Environment 1*, 32. https://doi.org/10.1038/s43247-020-00029-y

Vidas, D. (2015). The Earth in the Anthropocene – And the World in the Holocene? *European Society of International law (ESIL) Reflections, 4*(6), 1–7.

Waters, C. N., Zalasiewicz, J. A., Williams, M., Ellis, M. A., & Snelling, A. M. (Eds.). (2014). *A Stratigraphical Basis for the Anthropocene*. Geological Society, Special publication 395.

Waters, C. N., Syvitski, J. P. M., Gałuszka, A., et al. (2015). Can nuclear weapons fallout mark the beginning of the Anthropocene epoch? *Bulletin of the Atomic Scientists, 71*(3), 46–57. https://doi.org/10.1177/0096340215581357

Waters, C. N., Zalasiewicz, J., Summerhayes, C., et al. (2016). The Anthropocene is functionally and stratigraphically distinct from the Holocene. *Science, 351*(6269), 137. https://doi.org/10.1126/science.aad2622

Waters, C. N., Zalasiewicz, J., Summerhayes, C., et al. (2018). A Global boundary Stratotype Section and Point (GSSP) for the Anthropocene Series: Where and how to look for a potential candidate. *Earth-Science Reviews, 178*, 379–429. https://doi.org/10.1016/j.earscirev.2017.12.016

Williams, M., Zalasiewicz, J., Haywood, A., & Ellis, M. (Eds.). (2011). The Anthropocene: A new epoch of geological time? *Philosophical Transactions of the Royal Society, 369A*, 833–1112.

Williams, M., Zalasiewicz, J., Waters, C. N., et al. (2016). The Anthropocene: A conspicuous stratigraphic signal of anthropogenic changes in production and consumption across the biosphere. *Earth's Future, 4*(3), 34–53. https://doi.org/10.1002/2015EF000339

Wolfe, A. P., Hobbs, W. O., Birks, H. H., et al. (2013). Stratigraphic expressions of the Holocene–Anthropocene transition revealed in sediments from remote lakes. *Earth-Science Reviews, 116*, 17–34. https://doi.org/10.1016/j.earscirev.2012.11.001

Zalasiewicz, J., Williams, M., Smith, A., et al. (2008). Are we now living in the Anthropocene? *GSA Today, 18*(2), 4–8. https://doi.org/10.1130/GSAT01802A.1

Zalasiewicz, J., Waters, C. N., Williams, M., et al. (2015). When did the Anthropocene begin? A mid-twentieth century boundary level is stratigraphically optimal. *Quaternary International, 383*, 196–203. https://doi.org/10.1016/j.quaint.2014.11.045

Zalasiewicz, J., Waters, C. N., Ivar Do Sul, J. A., et al. (2016). The geological cycle of plastics and their use as a stratigraphic indicator of the Anthropocene. *Anthropocene, 13*, 4–17. https://doi.org/10.1016/j.ancene.2016.01.002

Zalasiewicz, J., Waters, C. N., Summerhayes, C. P., et al. (2017a). The Working Group on the Anthropocene: Summary of evidence and interim recommendations. *Anthropocene, 19*, 55–60. https://doi.org/10.1016/j.ancene.2017.09.001

Zalasiewicz, J., Waters, C. N., Wolfe, A. P., et al. (2017b). Making the case for a formal Anthropocene: An analysis of ongoing critiques. *Newsletters on Stratigraphy, 50*, 205–226. https://doi.org/10.1127/nos/2017/0385

Zalasiewicz, J., Waters, C. N., Williams, M., & Summerhayes, C. (Eds.). (2019). *The Anthropocene as a Geological Time Unit: A Guide to the Scientific Evidence and Current Debate* (p. 361). Cambridge University Press.

Zalasiewicz, J., Waters, C. N., Ellis, E. C., et al. (2021). The Anthropocene: Comparing its meaning in geology (chronostratigraphy) with conceptual approaches arising in other disciplines. *Earth's Futures, 9*(3), e2020EF001782. https://doi.org/10.1029/2020EF001896

Jan Zalasiewicz is a geologist (now retired) at the University of Leicester, and a member (formerly Chair) of the Anthropocene Working Group. He is co-editor of *The Anthropocene as a geological time unit: a guide to the scientific evidence and current debate*, has been involved in other AWG publications, and with Julia Adeney Thomas and Mark Williams has co-written *The Anthropocene: A Multidisciplinary Approach*. His books for a general audience that *inter alia* explore Anthropocene concepts include *The Earth after Us* and (with Mark Williams) *The Goldilocks Planet* and *Ocean Worlds*.

Colin Waters is an honorary professor in the Department of Geography, Geology and the Environment at the University of Leicester, UK. He has been the chair of the Anthropocene Working Group since 2020, following 9 years as its secretary. He has a central role in coordinating activities of the Working Group members and searching for the "golden-spike" section as part of the formalisation process for recognising the Anthropocene as a new geological time unit. He also has specific interests in characterising the nature and scale of human modification of the landscape, particularly through the accumulation of novel materials (e.g. plastic and concrete) and artificial deposits. He retired in 2017 as a Principal Mapping Geologist at the British Geological Survey, where over nearly 30 years' service he specialised in geological mapping of the UK, Morocco and Mauritania and is also an expert on Carboniferous stratigraphy.

Simon Turner is a Senior Research Fellow in Geography at University College London and Secretary of the Anthropocene Working Group (AWG). His research background is in paleoenvironmental analysis, geochemistry and aquatic monitoring that record environmental change in lakes and wetlands. He has worked in multiple locations around the world in aquatic and coastal systems extracting sediments from the Pleistocene to the present that contain records of past disturbance, contamination and pollution. His most recent work has been serving as the scientific coordinator for the AWG and Haus der Kulturen der Welt collaborative project to define a Global boundary Stratotype Section and Point (GSSP) for the Anthropocene and explore its cultural and historical implications.

Mark Williams is a palaeobiologist at Leicester University. He studies the evolution of life on Earth over hundreds of millions of years and has been a long-time member (and former Secretary) of the Anthropocene Working Group. He has co-authored popular science books with Jan Zalasiewicz on climate change, ocean evolution, and the story of life on Earth, and with Jan and Julia Adeney Thomas he co-authored *The Anthropocene: A Multidisciplinary Approach*. His new book with Jan is called *The Cosmic Oasis: the remarkable story of Earth's biosphere* (for Oxford University Press).

Martin J. Head is a stratigrapher at Brock University. He is currently Vice-Chair of the International Subcommission on Quaternary Stratigraphy, having served as its Chair (2012–2020), and is concurrently Co-Convener of its Working Group on the Middle–Upper Pleistocene Subseries Boundary. He has been involved in the formalization of the Quaternary System, its Calabrian, Chibanian, Greenlandian, Northgrippian, and Meghalayan stages, and with the introduction of formal subseries for both the Pleistocene and Holocene series.

Anthropocene

Alexander Federau

Abstract The article presents the concept of the Anthropocene, a new geological epoch that follows the Holocene, characterized by human impact. While this scientific concept is rather new and still debated among geologists, the idea that humans have a significant and lasting impact on the planet is almost as old the discipline of geology. Various interpretations of the Anthropocene are then presented, and some philosophical consequences of the idea explored.

The Anthropocene is the proposal to define a new geological epoch that ends the Holocene, the interglacial episode that has lasted just over 10,000 years and whose primary characteristic was a high degree of climatic stability. Theorized in the early 2000s by climate chemist Paul Crutzen and biologist Eugene Stoermer, and then further developed with other global environmental change specialists and historians (Crutzen & Stoermer, 2000; Steffen et al., 2007, 2011), the Anthropocene is based on the recognition that current environmental conditions differ significantly from those of the Holocene. The reason for this is the emergence of a new geological force as a consequence of human activities and comparable to other geological forces such as volcanism or plate tectonics. The environmental impact of this force is established and lasting, and it is already stratigraphically detectable, which is of primary importance for a geological definition. The formal introduction of a new epoch in the geological time scale is a subject of discussion among geologists, and follows a formal process that takes several years. One point of controversy is the choice of stratigraphic criterion. When geologists dig, should they look for plastic, radioactive elements, the absence of remains of certain extinct species? A second debate concerns the date of entry into the Anthropocene (see Capitalocene): 1950, 1850, the seventeenth century or even prehistory are discussed. If the Anthropocene is a recent concept, coined to describe a contemporary situation, the idea of specifically characterizing an 'age of man' has been discussed in geology since the dawn of the discipline in the first half of the nineteenth century. Geologists then proposed

A. Federau (✉)
Mobilidée, Geneva, Switzerland

© The Author(s), under exclusive license to Springer Nature
Switzerland AG 2023
N. Wallenhorst, C. Wulf (eds.), *Handbook of the Anthropocene*,
https://doi.org/10.1007/978-3-031-25910-4_49

neologisms such as the Anthropic Period, the Psychozoic or the Anthropozoic Period to describe the contemporary geological epoch (Federau, 2017, 149–95). Used by several authors, the Anthropozoic enjoyed a certain popularity for a time thanks to the work of the Italian geologist and priest Antonio Stoppani. However, it was not until the beginning of the twentieth century that the most successful precursor to the Anthropocene idea emerged. Conceptualized by the Russian geochemist Vladimir Vernadsky, the French paleontologist and Jesuit priest Pierre Teilhard de Chardin and the French philosopher Édouard le Roy (Samson & Pitt, 1999), the noosphere describes a sphere of the mind that is added to the biosphere to transform it. These various precursor concepts have in common the progressive interpretation of the Earth's history. It presents itself as a growing sophistication, with several qualitative leaps: for example, the appearance of life came to modify a previously inert Earth, and recently the spirit came in turn to transform life. In these interpretations, humanity is in a way the culmination of evolution, its most important and qualitatively most complex element. The original creators of the Anthropocene took up this progressive narrative, even though it is tinted with a fundamental anthropological pessimism in Crutzen's work. They assert that the Anthropocene provokes a reflexive moment for humanity, which must become aware of its own power in order to tame it and adequately regulate the great biogeochemical cycles. If followed to the end, this logic culminates in the "good Anthropocene" hypothesis, the advent of humanity's golden age as the world's dominant species (Lynas, 2011; Asafu-Adjaye et al., 2015; Hamilton, 2015). However, apart from the authors mentioned above, no one is thinking of equating the Anthropocene with any form of utopia. Rather, most interpretations are the exact opposite of this optimism. Indeed, the overwhelming majority of scientific data shows that the entry into the Anthropocene represents a series of existential threats to humanity (Lenton & Schellnhuber, 2007; Rockström et al., 2009; Barnosky et al., 2012). By irreversibly and sustainably transforming the global environmental conditions that allowed agriculture and civilizations to flourish, geological force promises evils, not wonders: climate models predict increasingly extreme weather events, as well as ocean acidification that is deadly for many species. The simultaneous disappearance of nature, through the rapid loss of biodiversity and natural habitats, leads to even more severe problems, since if nature lives well without humanity, the reverse is not true. Another feature of the Anthropocene, the destruction of the ozone layer, also has no known benefit. The catalogue of destruction grouped under the concept of the Anthropocene, however, produces very diverse reactions. The first type, surprisingly common in the political world, is minimization, or outright denial. The latter is often defended by putting scientific facts on the same level as personal opinions, implying that everyone is free to believe it or not. For instance, the claim that coal is a "clean" source of energy is enough to make it true. A second type of reaction, fatalism, concludes from the teachings of the Anthropocene that the imminent collapse of societies is inevitable (Servigne & Stevens, 2015). This posture leads either to pessimistic quietism, survivalist activism or to ecological transition initiatives (Hopkins, 2008). Finally, a third type of reaction considers that, for humanity, progress is no longer about producing more and better, but about achieving societies and lifestyles that are sustainable for the environment and human societies. This calls for a fundamental reinterpretation of the place of human beings on Earth (Latour, 2015; Bourg, 2018), who should less seek to dominate nature than the consequences of their own actions. The aim is not to return to Holocene conditions, an unattainable goal, but to avoid and limit as far as possible the catastrophic effects of this undesirable "geological force". This anthropological reinterpretation, which has been worked on and thought about for at least half a century with the emergence of environmental ethics and ecological thinking, took on a new impetus with the Anthropocene. The reality it presents fundamentally contradicts two presuppositions of modern Western thought. First, that the separation of nature and society makes sense, in theory and in practice. Second, that infinite growth on a finite planet is possible. The first presupposition, the need to separate human beings from nature, runs throughout the history of Western thought. It is already present in Plato's thought, and it is an important element of Christianity (White Jr., 1967). It is reinforced with Descartes and modern thought. There are now countless criteria that distinguish us from other animals: language, tools, mastery of fire, consciousness of death, etc. Epistemologically, the separation is even institutionalized in the separation between the human and social sciences and the natural sciences. The Anthropocene, by showing the intimate entanglement between human actions and nature, puts this conception to rest. It intertwines natural and social destinies, and forces us to a new understanding of history (Chakrabarty, 2009). Our historical identity is once again linked to the history of Earth, as it is to our genealogy and social context. By making explicit the fragility of biogeochemical processes, the Anthropocene brings to the forefront that the Earth is our home, the only hospitable place in the known universe. This need to conserve our common home has direct implications for the idea of progress. In our liberal societies, this is nowadays often reduced to an imperative of economic growth. Yet the central feature of the Anthropocene is not the demonstration of human power, but the awareness of the finite nature of our planet. The Earth is not only limited geographically, a fact that has been known for centuries, but also in its ability to be resilient to the disruptions it is undergoing today. There is only one atmosphere, only one ozone layer. When a living species disappears, it is forever. The spread of non-biodegradable compounds such as plastics or toxics has consequences over thousands of years.

The economic and anthropological challenge is now to go beyond this imperative of growth (Raworth, 2017), even "green" growth, to reinvent an economy and a society that operates within planetary limits.

References

Asafu-Adjaye, J., Blomqvist, L., Brand, S., Brook, B., Defries, R., Ellis, E., Foreman, C., et al. (2015). An Ecomodernist Manifesto. http://www.ecomodernism.org. Accessed Apr 2015.

Barnosky, A. D., Hadly, E. A., Bascompte, J., Berlow, E. L., Brown, J. H., Fortelius, M., Getz, W. M., et al. (2012). Approaching a state shift in Earth's biosphere. *Nature, 486*(7401), 52–58.

Bourg, D. (2018). *Une nouvelle Terre*. Desclée De Brouwer.

Chakrabarty, D. (2009). The climate of history: Four theses. *Critical Inquiry, 35*(2), 197–222. https://doi.org/10.1086/596640

Crutzen, P. J., & Stoermer, E. F. (2000). The Anthropocene. *Global Change Newsletter, 41*(May), 17–18.

Federau, A. (2017). *Pour une philosophie de l'Anthropocène. L'écologie en questions*. Presses Universitaires de France.

Hamilton, C. (2015). The Theodicy of the 'Good Anthropocene'. *Environmental Humanities, 7*, 233–238.

Hopkins, R. (2008). *The transition handbook: From oil dependency to local resilience*. Green.

Latour, B. (2015). *Face à Gaïa: huit conférences sur le nouveau régime climatique. Les empêcheurs de penser en rond*. La Découverte.

Lenton, T. M., & Schellnhuber, H. J. (2007). Tipping the scales. *Nature Reports Climate Change, 1*, 97–98.

Lynas, M. (2011). *The god species: Saving the planet in the age of humans*. National Geographic.

Raworth, K. (2017). *Doughnut economics: Seven ways to think like a 21st-century economist*. Random House Business.

Rockström, J., Steffen, W., Noone, K., Åsa, P., Stuart, F., Chapin, I. I. I., Lambin, E., Lenton, T. M., et al. (2009). Planetary boundaries: Exploring the safe operating space for humanity. *Ecology and Society, 14*(2), 1–33.

Samson, P. R., & Pitt, D. (Eds.). (1999). *The biosphere and Noosphere reader: Global environment, society, and change*. Routledge.

Servigne, P., & Stevens, R. (2015). *Comment tout peut s'effondrer : petit manuel de collapsologie à l'usage des générations présentes*. Seuil.

Steffen, W., Crutzen, P. J., & McNeill, J. R. (2007). The Anthropocene: Are humans now overwhelming the great forces of nature? *Ambio: A Journal of the Human Environment, 36*(8), 614–621. https://doi.org/10.1579/0044-7447(2007)36[614:TAAHNO]2.0.CO;2

Steffen, W., Grinevald, J., Crutzen, P. J., & McNeill, J. R. (2011). The Anthropocene: Conceptual and historical perspectives. *Philosophical Transactions of the Royal Society A: Mathematical, Physical and Engineering Sciences, 369*(1938), 842–867.

White, L., Jr. (1967). The historical roots of our ecologic crisis. *Science, 155*(3767), 1203–1207.

Alexander Federau is Mobility Consultant at Mobilidée. PhD in Philosophy and Environmental Sciences. Author of an Anthropocene philosophy. Alexander is a specialist in environmental ethics and has worked on the concept of nature. His interest focuses on behavioural changes related to ecological transition.

Anthropocentrism

Bryan L. Moore

Abstract A key part of the warrant for naming our era the Anthropocene lies in the stratigraphic record, though the term has become important in many fields outside geology, including the social sciences and humanities as well as hard sciences. Human traces appear in the rocks in the late Pleistocene period and show acceleration relatively very recently, during the Industrial Revolution--the era of William Blake's "dark Satanic mills." Humans ramped up their alteration of the environment more radically in the middle of the twentieth century with the marked increase in the burning of fossil fuels (see Steffen, 2005). The term shares a prefix, *anthropos* ("human"), with anthropocentrism, which is more subjective but helpful in tracing the causes of the Anthropocene. This causality is particularly apparent for us living in the era of modern science, which has both enabled humans to transform ecosystems and to better understand our species' position locally and universally. Anthropocentrism undergirds the basis for trust in human technologies of convenience and profit, some of which operate counter to sound ecology.

Ecologists and environmental philosophers have identified variants of anthropocentrism and positions counter to it following the work of Aldo Leopold, who in his 1948 book *A Sand County Almanac* proposes a land ethic which "changes the role of *Homo sapiens* from conqueror of the land-community to plain member and citizen of it" (1966, p. 240). "Hubris" is, in some cases, a near equivalent to anthropocentrism, but the latter term carries a wider significance. Though it is a contentious idea among environmental philosophers, no alternative term has stuck (Curry 54). An ecocentric view, like the closely related biocentric one, holds that all members of land communities possess intrinsic value outside of their possible human value and are implicit in Leopold and sources before and after him. Of course, the large and fine details of ecocentrism/biocentrism call for scrutiny amidst contexts. Few, for example, would claim that a microbe holds the same value as an elephant. Yet to

B. L. Moore (✉)
Arkansas State University, Jonesboro, AR, USA
e-mail: bmoore@astate.edu

N. Wallenhorst, C. Wulf (eds.), *Handbook of the Anthropocene*,
https://doi.org/10.1007/978-3-031-25910-4_50

dismiss anthropocentrism and its related terms as too vague or unphilosophical a subject is problematic, dangerous for the planet, and common among proponents of unregulated industry.

Ecocentrism/biocentrism stand in opposition to anthropocentrism, a term that is not monolithic. One may hold a "strong" or "hard" anthropocentrism positing that humans have utter rights to the earth and its resources without constraints, ecological considerations being at best secondary. Perhaps few industrialists or elected officials with a good understanding of public relations would profess such a strong anthropocentrism outright, but numerous actions, including the eco-disaster of Exxon Valdez and the sheer bravado of mountaintop removal mining, show that such a view is not rare and is apparent over time across the globe, under governments of almost all types across centuries. A "soft" or "weak" anthropocentrism grants that although humans are more important than nonhuman nature, we should take large steps, where needed, to ensure the viability of our fellow species and ecosystems (see Oelschlaeger, 1991, Chap. 9).

Counter to statements by some (mostly strongly anthropocentric) promoters and agents of industry, to reject a strong anthropocentrism is not equivalent to stripping the rights of people to take care of themselves and their families or even to pursue (most?) pecuniary interests. Far from antihumanist or misanthropic, to hold a weak anthropocentric or anti-anthropocentric view is, on the contrary, a most humane and socially responsible position; this is especially true given the serious challenges presented by our current environmental crises. Most holding such views understand that these crises will continue to affect the poor and vulnerable more than the well-off and secure.

Practically speaking, perhaps a modest drawing down of our imprint on the planet (e.g., a carbon tax) would suffice to address many of our environmental emergencies. How such a drawing down might occur is a matter of contention (e.g., Sutoris, 2021), though, doubtless, such change will be the result of ecological-ethical education and political action. Yet a species that truly loves itself and acts as though it (we) were truly the most valuable event in the universe would, it seems, be more successful in acting in the interest of its own wellbeing. Instead, humans have not responded sufficiently to climate change, the destruction of habitats, the extinction of species, and other serious, interrelated environmental challenges. If humanity does not possess a death wish, through repeatedly injuring our fellow species, flora, bodies of water and land, we have acted as if it did, as if our new species were invulnerable.

There is, of course, no worldwide polling on anthropocentrism, but data shows that most people in the United States "personally worry about the quality of the environment" either a "great deal" or a "fair amount," while 24% do so "only a little" or "not at all." Almost all Gallup polls over more than 30 years indicate that most Americans favour environmental protection over economic growth (Gallup). It is likely that the apparent concern for the environment is centred in its relation to human conditions, not as an inherent good, but it may be, again, that a softer anthropocentrism is sufficient for positive change. Yet Americans, like those in most democracies around the world, often elect officials who support actions and legislation that do demonstrable harm to natural environments.

The roots of anthropocentrism extend to prehistory and are discernible in some of the foundational Western texts. The Bible is undoubtedly a source and justification for the human-centred alteration of the planet. In Genesis 1:28, for example, God tells Adam and Eve to "replenish the earth, and subdue it; and have dominion over the fish of the sea and over the fowl of the air, and over every living things that moveth upon the earth." Though many people of faith understand the natural world as the work and embodiment of God, some read the Bible as a divinely sanctioned license or even command for earthly plunder. The greatest Roman Catholic theologian, Thomas Aquinas, wrote that nonhuman animals are "ordered to man's use," yet he has been defended by some for his proto-ecological ideas (see Glacken, 230). Many among today's U.S. religious right ascribe to the view that God (not humans) oversees the planet's ecosystems, which "are robust, resilient, self-regulating, and self-correcting, admirably suited for human flourishing, and displaying His glory" (Cornwall). A subset of evangelical Christians, the Dominionists, see the destruction of "the old earth" as a precondition for the second coming of Christ and the future abode of Christians in the new heaven and earth as outlined in Revelations 21 (see Hendricks).

Anthropocentrism is not, however, based solely in religious ideas, and some explicitly religious figures central to environmentalism (e.g., John Muir) reject anthropocentrism. Many humanist writers, including Shakespeare and Marx, extoll humans and suggest that the universe was formed with humanity as its centre. The writings of Plato and Aristotle, though differing substantially, state teleological views, and many Stoics, including Posidonius, reinforce the notion that the universe exists for the sake of the gods and humans (Moore, 2017, pp. 50–51). In the early modern era, Francis Bacon established an ontological divide between humans and the rest of nature, and his "Great Instauration" set the world on the path toward a graded use of natural resources, including the burning of fossil fuels (see Merchant, 1980, Chap. 6; Oreskes, 36). Descartes also separated humans from the rest of nature, nonhuman animals being "automata" (Moore, 2008, p. 67), though he also helped overturn the Aristotelian physics dominant for two millennia. Kant writes that "man" is, practically speaking, the "end of nature" and "the titular lord of nature." Such designations are, he writes, analogical and means "reflective judgement"; humans should proceed *as if* we were the ultimate end of nature (Moore, 2017, p. 12).

In the East, Hinduism and Taoism contain much older nonanthropocentric traditions. The Hindu deity Shiva (Nataraja), for example, represents a complex, conflicted, constantly changing universe not out of step with modern science. The god dances on one foot before the cosmos--life and death, creation and destruction, and energy, mass, and space--while the demon of ignorance, Apasmara, lies subdued underfoot (A. Huxley, 2012). Yet much of the Eastern hemisphere faces environmental challenges as serious as those in the West.

Renaissance and Enlightenment writers, by definition, put humans in the centre of creation, but they also initiated thought that pushed back against the notion of a human telos. The philosophical roots extend to ancient writers who inspired these movements. Some of the pre-Socratics, including Empedocles (fifth century BCE), rejected anthropocentrism. Lucretius, the chief ancient Epicurean, wrote in his poem *De rerum natura*, "not for us and not by gods/Was this world made. There's too much wrong with it!" Anticipating James Lovelock's 1972 Gaia hypothesis, the first-century Stoic Seneca writes of the earth as a living creature in his *Natural Questions*, which presents a case against a human-centred universe (Moore, 2017, pp. 55–57). In the Middle Ages, church fathers evoke *contemptus mundi* (spiritual contempt for earthly things), though St. Francis veers closer to rejecting anthropocentrism. Sufism complicates what is in Islam a largely anthropocentric view. And, for many centuries, American Indians have viewed the natural environment much less anthropocentrically than the late-arriving Europeans.

Although few major philosophers critique anthropocentrism directly, many of them do so indirectly, anti-teleologically. An important source is Spinoza, who suggests that God and Nature are synonymous, an idea explored by Goethe, Coleridge, Wordsworth, Emerson, Thoreau, and others. For many neoclassical writers, the world is a machine, and humans are removed from and above nature. For romantics such as Wordsworth, the world is an organism of which humans are a part, and much romantic poetry responds negatively to an increasingly industrialized world.

Even as the turn from a Ptolemaic to a Copernican view of the cosmos took hold quickly among scientists after Galileo, this was not accompanied by a practical shift in how humans saw themselves on a large scale. Nor did a wide shift occur after Darwin's *On the Origin of Species* (1859), which demonstrates the close kinship of all species and further problematizes a human telos. As early twentieth-century poet T.E. Hulme wrote, "The change which Copernicus is supposed to have brought about is the exact contrary of the fact. Before Copernicus, man was not the centre of the world; after Copernicus he was." Similarly, American anthropologist Loren Eiseley writes that "our world view … is still Ptolemaic, though the sun is no longer believed to revolve around the earth" (Moore, 2017, p. 23, 239).

Anthropocentrism, ecocentrism, and their variants are important but underdiscussed in the history of ideas, from antiquity to the near present, though some journals and conferences are helping to trace these complicated roots. The field of posthumanist studies, inspired by Nietzsche and Heidegger, has grown dramatically in the early twenty-first century, and it often links with ecological themes. Many science fiction works dramatize worlds in which humans have little or no importance. This tradition began, arguably, in the dialogues of Lucian of Samasota (second century CE). The idea of a universe not centred around humans developed in the wake of Newton, including in the works of the chief French *philosophes* Voltaire ("Micromegas"—an early science fiction story), Rousseau, and Diderot. H.G. Wells's *The Time Machine* and George R. Stewart's *Earth Abides* explore worlds without humans, and this tradition continued into the twenty-first century in the speculative fiction of Margaret Atwood, Kim Stanley Robinson, and Paolo Bacigalupi and in the more realistic fiction of Cormac McCarthy, Leslie Marmon

Silko, and Richard Powers. Some writers, including the poet Robinson Jeffers, philosopher Peter Wessell Zapffe, and horror fiction writer H.P. Lovecraft (whose fiction is not centred on natural environments), place humans at the bottom of the chain of natural selection and judge our species as insignificant cosmically (see Moore, 2017).

A human-centred view should not be confused with a "love for humanity." An ingenious if very recent species, humans have transformed the planet in a short time span. While some see technology as central in drawing down our impact on natural systems, others are skeptical about what Kenneth Burke called the human technological entelechy to cure ills largely *created by* misapplied or unwise technologies (Rueckert, 1994, p. 199). We live in a highly technological age, but perhaps ecological balance would be better achieved by, for starters, reassessing our place on our home planet, being more conscientious stewards of our home, and drawing down our impact on it. Whatever the case, inaction is not an option. As Kim Stanley Robinson writes, "There is no alternative way; there is no planet B."

References

Blake, W. (2008). And did those feet in ancient time. In Norton (Ed.), *Blake's poetry and designs* (2nd ed., pp. 147–148).

Cornwall Alliance. (2000). https://cornwallalliance.org/landmark-documents/the-cornwall-declaration-on-environmental-stewardship/

Curry, P. (2011). *Ecological ethics: An introduction* (2nd ed.). Polity.

Gallup. (2021). https://news.gallup.com/poll/1615/environment.aspx

Glacken, C. J. (1967). *Traces on the Rhodian shore: Nature and culture in Western thought from ancient times to the end of the eighteenth century*. University of California Press.

Hendricks, S. (2005). *Divine destruction: Dominion theology and American environmental policy*.

Huxley, A. (2012). Aldous Huxley describes the dancing shiva image. YouTube. Accessed 16 Sept 2021.

Leopold, A. (1966). *A Sand County almanac with essays on conservation from Round River*. Ballantine.

Merchant, C. (1980). *The death of nature: Women, ecology, and the scientific revolution*. Harper.

Moore, B. L. (2008). *Ecology and literature: Ecocentric personification from antiquity to the twenty-first century*. Palgrave.

Moore, B. L. (2017). *Ecological literature and the critique of anthropocentrism*. Palgrave.

Oelschlaeger, M. (1991). *The idea of wilderness: From prehistory to the age of ecology*. Yale UP.

Oreskes, N., & Conway, E. M. (2014). *The collapse of Western civilization: A view from the future*.

Robinson, K. S. (2018, March 20). Empty half the earth of its humans. It's the only way to save the planet. The Guardian.com.

Rueckert, W. H. (1994). *Encounters with Kenneth burke*. University of Illinois Press.

Steffen, W., et al. (2005). *Global change and the earth system: A planet under pressure*. Springer.

Sutoris, P. (2021, August 15). Behavioral science won't fix the climate crisis. Salon.com.

Bryan L. Moore is a Professor of English at Arkansas State University in Jonesboro, Arkansas. He is the author of, among other works, *Ecology and Literature: Ecocentric Personification from Antiquity to the Twenty-first Century* (Palgrave, 2008) and *Ecological Literature and the Critique of Anthropocentrism* (Palgrave, 2017).

Anthropology

Christoph Wulf

Abstract In the Anthropocene, humans have become increasingly uncertain about who they are and how they should view themselves. The future of the planet, people and animals depends to a great extent on our conception of ourselves and our actions. Amid all the criticism of inadequate conceptions of humanity, what is certain is that we have nothing beyond ourselves. That is why extensive anthropological research is needed, to determine our limitations and capabilities. In the face of globalization, it is especially important to combine diachronic historical perspectives with synchronic cultural anthropological and philosophical perspectives and reflections. The dual historicity and culturality of the objects of research and of the researchers themselves is important, as is the development of transdisciplinary and transcultural perspectives and insights into the limits of human knowledge and action.

As the age of human impact on earth, the Anthropocene raises the urgent question of who human beings are and what conception we have of ourselves. Many traditional certainties have lost their validity. Uncertainty and doubt accompany the question of who we humans are and how we are to understand ourselves. In the Anthropocene, humans largely influence the fate of the planet, including its future and its survival. The question who humans are is of irrefutable importance. This is the case even though we know less and less whether and how we can answer this question. The fact that it can only be partially answered is a characteristic of human beings. We are a riddle to ourselves, which we cannot solve completely. This is undeniable. Nevertheless, we must not forget this question of who human beings are. Again and again, it leads to important answers. Anthropology, as the science of human beings, is a fundamental science. What does anthropology mean today and what does it embrace? There can be no unambiguous answer to this question either. Many answers are possible and necessary. In principle, all disciplines of the natural

C. Wulf (✉)
Freie Universität Berlin, Berlin, Germany
e-mail: christoph.wulf@fu-berlin.de

sciences, humanities and social sciences can contribute to anthropology. Their contribution becomes clearer the more they relate their research to questions about humans in the Anthropocene.

If we wish to put the epistemology of anthropology on a more serious footing, then we must examine, in a critical and constructive way, the anthropological paradigms that are internationally significant is indispensable. These are:

- evolution and hominization,
- philosophical anthropology, developed in Germany,
- historical anthropology and the history of mentalities, initiated by historians in France and taking its cue from the *Annales-School,*
- the North American tradition of cultural anthropology, and
- historical cultural anthropology.

To provide a framework for anthropology I suggest that we use the paradigm of historical cultural anthropology as a basis for further research (Wulf, 2013, 2016, 2022b). This paradigm integrates perspectives from the other four major anthropological paradigms and provides a basis for an adequate understanding of phenomena, processes, and institutions in a globalized world. Together with historical and philosophical methods, the ethnographic approach is one of the main methods of anthropology, conceived as historical and cultural anthropology.

Evolution and Hominization

The branch of anthropology that studies hominization stems from an attempt to fit the natural history of human beings into the framework of anthropology which aims to understand the human being. Yet the natural history of human evolution can be understood only when considered as part of social and cultural history. Its irreversibility, as well as that of the history of life itself, is grasped today because of material self-organization.

Hominization, the long process of evolution from Australopithecus to primitive human beings, can be understood as a multidimensional morphogenesis arising from the interplay between ecological, genetic, cerebral, social, and cultural factors. This process necessitated three types of change. The first were ecological changes which led to the expansion of the savannah and thus to an "open" biotope. Second, a genetic change took place in the highly developed primates, which were already walking upright. Third, there was a change in social self-reproduction due to the splitting off of young groups and the use of new territories. It was the new ecosystem—the savannah—which triggered the dialectic between the feet, hands, and brain and which became the source of technology and all other human developments. The process of hominization was intensified by a prolonged infancy or neoteny, incomplete development of the brain at birth, and prolonged childhood with longer affective ties between the generations, with the associated potentials for comprehensive cultural learning. The cerebralization, prolonged youth, and increased social and cultural complexity were mutually dependent. The complexity of the brain requires a corresponding sociocultural complexity. The creative potential of the brain can be expressed and develop only in a sociocultural environment that grows in parallel. This dialectic relationship means that humans have been cultural beings from the very beginning, i.e., their "natural" development is cultural.

The final stage of this process of hominization is, in fact, also a beginning. The human species, which has reached its completion in *Homo sapiens*, is a youthful and childlike species: our brilliant brains would be feeble organs without the apparatus of culture; all our capabilities need to be bottle-fed. Hominization was completed with the irreversible and fundamental creative incompleteness of human beings. The course of hominization also clearly illustrates that *Homo sapiens* and *Homo demens* are inseparably linked and the great achievements of humanity have their downside: the horrors and atrocities perpetrated by humans.

Philosophical Anthropology

While taking evolution into account in anthropology serves to highlight the shared lineage and mutual parentage of all forms of life and the long time-span of hominization as well as the general laws of evolution, philosophical anthropology turns its attention to the particular character of the human being.

The centrepieces of philosophical anthropology are the anthropological works of Max Scheler, Helmuth Plessner and Arnold Gehlen. In 1927, Max Scheler gave a lecture in Darmstadt entitled "Die Sonderstellung des Menschen," which was published in 1928 under the title *Die Stellung des Menschen im Kosmos* (*The Human Place in the Cosmos*) and is regarded as the beginning of philosophical anthropology. When Scheler died in the same year, he left no concrete preparatory material for the anthropological work he had intended to publish in 1929. The philosopher and biologist Helmuth Plessner, however, published his main anthropological work *Die Stufen des Organischen und der Mensch* (*Levels of Organic Being and Man*) in 1928. Despite their differences, Scheler's seminal article (2009) and Plessner's book (1928/2003) share the assumption that organic life is structured in levels. Arnold Gehlen's work (1988) took a different approach and focuses on humans as acting beings.

The preoccupation of this strand of anthropological thought was to understand the essence, the nature, of human beings in general. Within this framework, anthropology concentrated upon a comparison between "human being" and animal (Gehlen, 1988; Plessner, 1970, 1980–1985, 2003), with a view to distinguishing shared features and differences. To grasp the *conditio humana* philosophical reflections were brought to bear upon biological insights. It was thought that the conditions for the formation of the human species could be glimpsed in biological and above all morphological characteristics. This perspective has had two consequences:

a focus of anthropological reflection and research on the human body, and a generalizing discourse relating to *one* unique and unitary model of man.

Due to its focus on *the* human being as such, Philosophical anthropology fails to address the historical and cultural diversity of human beings in the plural. To investigate the diversity of human life is the aim of a branch of historical science that is oriented towards anthropological issues.

The Annales School

Anthropology underwent an additional development and refinement in a historical turn, which can be discerned in the historical treatments of anthropological topics of the *Annales School* and the history of mentalities that flowed from it (Burke, 1991; Ariès & Duby, 1985). Quite opposed to those who insist that social structures as well as the social actor's subjective experiences be rooted in a character common to all human beings, the practitioners of historical studies with an anthropological orientation inquire into the specifically historical and cultural character of each of these phenomena. This complements the new issues and the new methodological procedures used in the depiction and analysis of the history of events and the examination of structural and social history. Concentrating on anthropological issues brings into focus both historical structures of social reality and subjective moments of agency in social subjects; this focus is used for research on the basic conditions of human behavior. The studies carried out by Lucien Febvre and Marc Bloch in France are examples of the successful examination of anthropological issues in the field of history, in which historical knowledge arises from the disputed borders between events and narrative, reality and fiction, structural history and narrative historical writings. These works, which have since become classics of their genre, appeared at the same time as the works on philosophical anthropology in Germany. Fernand Braudel's study of the Mediterranean (Braudel, 1949), Emmanuel Leroy Ladurie's on the village of Montaillou (Leroy Ladurie, 1978), and Carlo Ginzburg's on the world of millers around 1600 (Ginzburg, 1980) may be cited as successful examples for this endeavor.

Historical anthropology investigates elementary situations and basic experiences of being human. It studies a basic stock of patterns of thought, feeling, and behavior that is anthropologically coherent, basic human phenomena and elementary human behavior, experiences, and basic situations, not to make statements about humans in general but to gain an understanding of the multidimensional conditions of life and experiences of real people in their respective historical contexts. This diversity of phenomena is paralleled by the multidimensionality and open-endedness of anthropological definitions and research paradigms. From the point of view of the historical sciences, the feelings, actions, and events under investigation can be understood only in terms of their historic uniqueness. It is this that lends them their dynamic nature and makes them subject to historical change.

Cultural Anthropology

Cultural anthropology does not view human beings as being "behind" (i.e., responsible for) the diversity of their historical and cultural characteristics but studies them within the context of these characteristics. Therefore, it is not sufficient to identify body, language, or imagination as universal cultural entities; they must be examined in the context of different cultures. It is this diversity of culture that enables us to draw conclusions about humans. Comparing culturally different forms of expression results in new ideas and calls some areas of accepted thinking into question. Ethnological research into the heterogeneity of cultures yields important results for cultural anthropology. These findings have had a lasting effect on the understanding of what is different in our own cultures. New developments have resulted in an expanded concept of culture in which both the disparities and the shared characteristics of different cultures play an important role. The globalization of politics, economics, and culture is resulting in the overlapping, blending, and assimilation of features that are global, national, regional, and local. This creates a need for new ways of examining different cultures.

Even though anthropology is the result of a process of philosophical and scientific evolution, it can no longer pretend, these days, that at the end of the day Europeans are the only human beings and act as though European human beings are the only possible yardstick. It is obvious, even in an era of globalization deeply marked in its content and form by Western culture, that different forms of human life exist today, influenced by various local, regional, and national cultures. The Anglo-Saxon tradition of cultural and social anthropology has turned its attention to this situation. Its accent lies on the social and cultural diversity of human life. Its research explains both to what extent cultural evolutions are heterogeneous and to what extent the profound diversity of human life remains unnoticed. It is precisely the analysis of foreign cultures that makes it plain to us how limited and problematic our understanding is. Thanks to the analysis of cultural manifestations drawn from heterogeneous cultures, anthropological inquiries make an important contribution to the elaboration and development of anthropology while its ethnographical methods oblige practitioners to draw upon historical sources. Besides creating a sensitivity for the strange and foreign character of other cultures, it creates a sensitivity for the strange and foreign in its own culture. The (self-)reflexive point of view adopted by cultural anthropology towards European cultures has contributed to a considerable evolution and advance of anthropological knowledge (see Geertz, 1993; Sahlins, 1976; Harris, 2001; Evans-Pritchard, 1981; Malinowski, 1922; Lévi-Strauss, 1992).

The issue of understanding the limits of our comprehension of different cultures becomes central. The ethnographic methods developed in social and cultural anthropology based on fieldwork and participant observation lead to forms of knowledge other than those gleaned from historical source interpretation and philosophical reasoning. They not only make us aware of what is different in other cultures but also what is different in our own culture. Therefore, the application of the anthropological perspective to the cultures of the world broadens and deepens the scope of anthropological research.

Historical Cultural Anthropology

The specific situation of human beings in the world, the comparison with animals or machines, is no longer anthropology's centre of interest. If the relationship between humans and animals is the issue today, then discussion centres less on the special status of human beings and more on how much we have in common with other living creatures and with primates in particular. The focus is also more on historical and cultural inquiries and on understanding the cultural diversity of social life. A particular and very pronounced interest in the study of current phenomena is noticeable. Expanding on historical anthropology, historical cultural anthropology touches upon the historical and cultural determination of culture and its manifestations and demands that their study and reflection consider ethnological and philosophical perspectives and questions. Committed to this task, historical cultural anthropology makes an important contribution to the self-comprehension and self-interpretation of cultures and societies in the Anthropocene. In this process of cultural understanding, research efforts rapidly run the risk of being unable to move beyond the level of their own initial insights. To safeguard against this risk, historical cultural anthropology needs to reflect upon its relation to power and knowledge, as well as to make efforts specifically aimed at bringing to light the involuntary and often unacknowledged normative implications of its own research.

Within this frame of reference, "historical cultural anthropology" also designates multiform transdisciplinary and transnational efforts to pursue the universal idea of an abstract anthropological norm and to continue analyzing diverse human phenomena. Historical cultural anthropology is the common denominator of history and the humanities. Nevertheless, it is not limited to a history of anthropology as a discipline nor to contributing to history from the perspective of an anthropological subdiscipline. It attempts, rather, to bring into an accord the historical and cultural determination of its perspectives and methods with the historical and cultural determination of its object of study. It can harness insights gleaned in the humanities to those yielded by a critique of anthropology based on the history of philosophy and bring both to fruition to create new perspectives and lines of inquiry out of a new consciousness of methodological problems. Historical anthropology is limited neither to certain spatial frames nor to particular epochs. Reflecting on its own historicity and its own cultural condition, it succeeds both in leaving behind the Eurocentrism of the humanities and moves from the interest in history to concern with current and future problems (Wulf, 2010, 2013, 2016; Wulf & Kamper, 2002; Wulf et al., 2010, 2011).

In view of this situation in anthropology, I suggest that we try to connect lines of thought from these different mainstreams and, where possible, to develop them into an anthropology that adequately accounts for the historicity and culturality of the researchers and their objects of study. Philosophical reflection can then help to render the results of this research fruitful for our understanding and definition of human beings. I have attempted to achieve this in comprehensive research on mimetic processes, which lasted more than twenty years (Gebauer & Wulf, 1995; Wulf, 2013, 183–198).

Epistemological Perspectives

As the influence of normative anthropologies has waned, anthropological research has begun to pay more attention to the body with its natality and mortality, which is both the product and the agent of its own socialization and enculturation (Wulf, 2013, 275–290; Kraus & Wulf, 2022). The lived human body is the result of multifarious mimetic processes that include not merely imitation, but an active acquisition of cultural knowledge. In these mimetic processes culture is produced, handed down, and transformed. Here the performativity of the body, how it is staged and enacted, plays an important role. Performativity is of particular significance for language processes, cultural performances, and aesthetics (Wulf, 2013, 199–214). If rituals, for example, are perceived and interpreted only as "texts," they lack a dimension that is associated with their material and bodily aspects. This led me to focus on the performative side of social practices and ritual acts in a large research project entitled "Berlin Study on Rituals and Gestures," in which I examined these aspects in detail, taking a close look at how social practices, rituals, and ritualization contribute to the performative formation of communities and how they shape educational processes and promote learning (Wulf et al., 2010, 2011; Wulf, 2013, 215–235). In these investigations, the material nature and sensory capacities of the body played a major role. The results revealed that human corporeality is shaped by language and imagination (Wulf, 2022a).

The fragmentation and "deterritorialization" of contemporary anthropology hold considerable potential for the development of new modes of anthropological reflection and research. They also provide an opportunity to free ourselves from outdated traditions of our discipline and to redefine the horizons of anthropology. In this redefinition, the perspectives that are emerging from globalization are becoming increasingly important. Among other things, they have engendered criticism of the neoliberal economic trends that are marginalizing the social market economy and of the associated tendency for many societies to become more and more alike. Today, giving anthropology a global orientation means to open it up for research in all societies and cultures of the world and to address the issue of what will be the most important conditions of human life in the future. The aim of anthropological research is to contribute to a better understanding and better explanations of human phenomena and problems in our globalized world and thus also to a better understanding between people. The lively debates on the historical involvement of sections of anthropology in colonialism and racism, the problem of representation, and the extent to which the other can "speak back" are evidence of efforts to broaden the horizons of anthropology and open it up for new tasks.

There is a *dual historicity and culturality* in anthropology that arises from the historicity and culturality of the different perspectives of anthropological researchers and from the historical and cultural character of the contents and subjects of the research. The historicity and culturality of the anthropologists themselves form the background against which the phenomena and structures that came into being in a different time or culture are perceived and investigated. New research questions and methodologies develop in a reciprocal relationship as researchers reflect upon this dual historicity and culturality. In anthropological research it is important to think of historicity and culturality as belonging together and not, for instance, to play culturality off against historicity.

The historical and cultural approach to anthropology that I present here employs both *diachronic* and *synchronic* methods to investigate human societies and cultures. In addition to anthropological issues and the hermeneutic and text-critical methods from the historical sciences that are applied diachronically, field research with its numerous qualitative and quantitative methods still plays an important role as a method of synchronous anthropological research. The interpretative and reflexive methods offer the possibility of lending expression to the individual and subjective perspectives.

One new challenge that anthropologists have long failed to address is how to define the *relationship between general insights and specific insights* relating to human beings as individuals and human beings in general. While in archaeology, biological anthropology, and linguistic anthropology it is permissible to make universal statements about human beings, in historical and cultural anthropological approaches the emphasis is more on being able to use hermeneutic methods to make complex statements on historic-cultural phenomena. These approaches are oriented towards the investigation and assurance of cultural diversity. However, even when we are concerned with cultural diversity, the question still arises as to what is common to all human beings.

In my view, the aim of anthropological research is not to reduce but to increase the complexity of our knowledge about human beings. This requires interpretation, reflection, and self-criticism, and an ongoing, philosophically inspired critique of anthropology that must include an examination of the fundamental limits of human self-interpretation. In analogy to a definition of God in theology, there is thus talk of the *Homo absconditus*. This term expresses the notion that anthropological insights and findings can only grasp the human condition in part, that is, from various perspectives and thus incompletely. Anthropological research and discovery is location-related and subject to historical and cultural change. Its starting point is a willingness to wonder or marvel that the world is as it is and not otherwise. Marveling (*thaumazein*) is the beginning of fascination with the mystery of the world and curiosity about the possibilities of anthropological knowledge. This aim implies scepticism towards all-encompassing and universal anthropological interpretations, such as those occasionally found in biological science, for example. Anthropology is not a single discipline. It touches on many different sciences and disciplines, including philosophy.

Anthropology cannot be regarded as a closed field of research. It is the result of the interplay between different sciences. Depending on the issue to be examined, the range of disciplines involved can be very different. The object and subject of anthropology can encompass the entire field of human culture in different historical areas and cultures. Anthropology presupposes a plurality of cultures and assumes that cultures are not closed systems; rather, they are dynamic, able to permeate each other, and they have an indeterminate future.

Anthropology can be understood as an academic attitude towards examining issues relating to different times and cultures. Anthropological research therefore can be found in many different disciplines, such as history, literature, linguistics, sociology, psychology, and the theory of education. However, the research frequently tends to transcend the boundaries of individual disciplines, thereby becoming transdisciplinary. This results in completely new scientific disciplines and issues that require new forms of scientific interaction and cooperation. Many different research methods are used in these processes. Historical-hermeneutical processes of text, image and music interpretation, qualitative social research methodologies, and philosophical reasoning are widely used, the latter being an approach that is difficult to categorize in terms of specific methodology. Some research makes use of artistic and literary materials, thereby transcending the traditional boundaries between science, literature, and art. A growing consciousness of the role of cultural traditions in the development of different research areas, subjects, and viewpoints has made the increasing trend towards crossing international cultural boundaries a central issue of anthropological research. In the light of globalization, this transnational approach to anthropology is becoming increasingly important. It provides the framework that nurtures a spirit of inquiry and a commitment to expanding our knowledge, which in turn lead to the development and testing of new research paradigms.

Since the Anthropocene is determined by the effects of human activity across our planet, the question as to how humans view and evaluate themselves and their actions, and how they project this view into the future is crucial for the fate of the planet and the future of humanity. However, openness, flexibility and imagination offer us the chance to create new relationships with nature and the world. Yet these will still be anthropogenic, even if people strive for new forms of action that are less violating to overcome the destructive anthropocentrism that is now widespread. International and transdisciplinary network thinking, humility, and a carefully considered ethical approach demonstrating the willingness to show solidarity are important prerequisites of anthropological research, which must place more importance on bringing together the views of the science, technology, and the humanities.

The conditions of the Anthropocene today require, for example, an opening of anthropology to the questions and perspectives of the earth sciences and the search for a new understanding of nature, the relationship between nature and culture, and the entanglement of nature and culture in humans. In the Anthropocene, there is hardly any area of nature that is not influenced by humans. At the same time, humans are shaped by a nature that they shape culturally. Nature and culture form an indissoluble connection in human beings, which leads to new questions and research tasks. The effects of human action can no longer be understood only as consequences of purposeful action by conscious actors. Rather, many effects of human action are unintentional and surprising for the actors. This situation results from the fact that humans are natural beings and cultural beings at the same time. They are part of nature, but do not merge into it and have the possibility of self-determined action. The consideration of this problem constellation is of special interest in anthropological research in the Anthropocene.

References

Ariès, P., & Duby, G. (1985). *Histoire de la vie privée* (Vol. 1–5). Seuil.

Braudel, F. (1949). *La Méditerranée et le monde méditerranéen à l'époque de Philippe II.* A. Colin.

Burke, P. (1991). *The French historical revolution: The Annales School, 1929–89.* Stanford University Press.

Evans-Pritchard, E. E. (1981). *A history of anthropological thought.* Faber and Faber.

Gebauer, G., & Wulf, C. (1995). *Mimesis. Culture, art, society.* California University Press.

Geertz, C. (1993). *The interpretation of cultures: Selected essays.* Basic Books.

Gehlen, A. (1988). *Man, his nature and place in the world.* Columbia University Press.

Ginzburg, C. (1980). *The cheese and the worms: The Cosmos of a sixteenth century Miller.* Johns Hopkins University Press.

Harris, M. (2001). *The rise of anthropological theory: A history of theories of cultures* (Rev. ed.). AltaMira Press.

Kraus, A., & Wulf, C. (Eds.). (2022). *Palgrave handbook of embodiment and learning.* Palgrave Macmillan.

LeRoy Ladurie, E. (1978). *Montaillou. Cathars and Catholics in a French Village, 1294–1324.* Scolar Press.

Lévi-Strauss, C. (1992). *Tristes tropiques.* Atheneum.

Malinowski, B. K. (1922). *Argonauts of the Western Pacific.* Routledge.

Plessner, H. (1970). *Laughing and crying: A study of the limits of human behavior.* Northwestern University Press.

Plessner, H. (1980–1985). *Gesammelte Schriften.* Ed. G. Dux, O. Marquard, and E. Ströker. Suhrkamp, 10 vols.

Plessner, H. (2003). *Die Stufen des Organischen und der Mensch. Einleitung in die philosophische Anthropologie.* Suhrkamp.

Sahlins, M. (1976). *Culture and practical reason.* University of Chicago Press.

Scheler, M. (2009). *The human place in the Cosmos.* Northwestern University Press.

Wulf, C. (Ed.). (2010). *Der Mensch und seine Kultur. Hundert Beiträge zu Problemen des menschlichen Lebens in Vergangenheit, Gegenwart und Zukunft.* Anaconda.

Wulf, C. (2013). *Anthropology. A continental perspective.* University of Chicago Press.

Wulf, C. (Ed.). (2016). *Exploring alterity in a globalized world.* Routledge.

Wulf, C. (2022a). *Human beings and their images. Imagination, mimesis, performativity.* Bloomsbury.

Wulf, C. (2022b). *Education as sustainable development in the anthropocene, an anthropological perspective.* Routledge.

Wulf, C., & Kamper, D. (Eds.). (2002). *Logik und Leidenschaft. Erträge Historischer Anthropologie.* Reimer.

Wulf, C., et al. (2010). *Ritual and identity. The staging and performing of rituals in the lives of young people.* Tufnell. (French edition 2004, German edition 2001).

Wulf, C., Suzuki, S., et al. (2011). *Das Glück der Familie. Ethnographische Studien in Deutschland und Japan.* Springer. (Japanese translation 2013).

Christoph Wulf is Professor of Anthropology and Education and a member of the Interdisciplinary Centre for Historical Anthropology, the Collaborative Research Centre (SFB, 1999–2012) "Cultures of Performance," the Cluster of Excellence (2007–2012) "Languages of Emotion," and the Graduate School "InterArts" (2006–2015) at the Freie Universität Berlin. His books have been translated into 20 languages. For his research in anthropology and anthropology of education, he received the title "*professor honoris causa*" from the University of Bucharest and the honorary membership of the German Society of Educational Research. He is Vice-President of the German Commission for UNESCO. *Major research areas:* historical and cultural anthropology, educational anthropology, imagination, intercultural communication, mimesis, aesthetics, epistemology, Anthropocene. Research stays and invited professorships have included the following locations, among others: Stanford, Tokyo, Kyoto, Beijing, Shanghai, Mysore, Delhi, Paris, Lille, Strasbourg, Modena, Amsterdam, Stockholm, Copenhagen, London, Vienna, Rome, Lisbon, Basel, Saint Petersburg, Moscow, Kazan, Sao Paulo.

Asian Anthropocene

Paul Jobin

Abstract Postcolonial perspectives on the Anthropocene aim at disentangle it from its current Western-centrism. Another reason to bring the focus away from Western approaches is that Asia—from South to East Asia—is now the world's biggest carbon-emitter, and a major habitat of biodiversity.

An overwhelming proportion of authors who discuss the Anthropocene, both advocates and opponents of the notion, are from Western Europe, North America, and Australia. Moreover, scholars from other regions have so far shown little enthusiasm for this debate. This discrepancy engenders empirical shortages and theoretical flaws, sometimes openly assumed (e.g., Corlett, 2013), but the domination of Western paradigms in the Anthropocene literature is more and more openly challenged (Malm & Hornborg, 2014; Davis & Todd, 2017; Marquardt, 2019; Simpson, 2020). A few scholars have therefore deemed it necessary to redefine the concept from the perspective of Africa (Hecht 2018), and Asia (Hudson, 2014; Baldwin & Erickson, 2020; Chatterjee, 2020; Simangan, 2019, 2020; Jobin et al., 2021).

As a concept aimed to define a very long period of time on a universal scale, the Anthropocene is not supposed to be confined within a particular place. From that angle, there should be nothing such as an African or an Asian Anthropocene. Yet these formulas have been used to reverse perspectives on the Anthropocene, and disentangle it from its current Western-centrism. This article focuses on the case of Asia, largely defined (South Asia, Southeast Asia to East Asia).

Hudson (2014) paved the way by identifying three research axes: the role of Asia in Anthropocene histories, the social and ecological vulnerabilities this epoch poses for Asia today, and how Asia addresses these global challenges. But as Simangan (2019, p. 565) aptly notes: "In a discourse saturated by universalising agenda, a regional level of analysis is an attempt to bridge global action and local capacity." Moreover, in an echo of *Provincializing Europe*—Chakrabarty's seminal book for

P. Jobin (✉)
Academia Sinica Taiwan, Taipei, France
e-mail: jobin@sinica.edu.tw

© The Author(s), under exclusive license to Springer Nature Switzerland AG 2023
N. Wallenhorst, C. Wulf (eds.), *Handbook of the Anthropocene*,
https://doi.org/10.1007/978-3-031-25910-4_52

subaltern studies—the historian of India Elisabeth Chatterjee (2020) invites researchers to "provincialize" the notion, and depart from the Western focus on the history of coal and oil to study other drivers of the Anthropocene, such as hydro-electricity, which has been instrumental in the modernization of Asian countries.

However, the geography of contemporary Asia is intertwined with the logic of asymmetrical world exchanges. The Organization for Economic Co-operation and Development (OECD), which includes Japan and South Korea, thus accounts for more than two-thirds of the world's gross domestic product, but less than 20 per cent of its population. What Alf Hornborg terms "time-space appropriation" and an "unequal ecological exchange" have meant a huge transfer of wealth and resources from the "rest of the world" to Europe and North America. As he further observes (2018), this transfer still operates to the advantage of economic alliances, such as the OECD, because while these countries import and consume merchandise from China, carbon emissions resulting from the production of these imports are attributed to China (see also Zhang et al., 2017).

Hamilton (2017) tackles the argument of unequal relations: not only China is now the world's biggest carbon-emitter, but as China reorients its economy toward domestic consumption, its share of emissions arising from export manufacture is declining; it is thus becoming harder to place all of the responsibility on its exports. At the 2015 climate conference in Paris, Chinese diplomats were compelled to give up this line of argument, which had sabotaged the negotiations at the 2009 Copenhagen conference. Following on Chakrabarty (2009), Hamilton (2017, p. 31) thus argues: "If the 'Anthropocene' was a Eurocentric idea when it was coined, it is now Sino-Americo-Eurocentric, and in a decade or two it will be Indo-Sino-Americo-Eurocentric". Or as Horn and Bergthaller (2020, p. 173) put it: "The old industrial nations of Europe and North America may have started the recent transformation of the Earth system, but they are no longer in the driver's seat. Today the Asian nations are as much a part of the problem—and they must be a part of the solution, if there is to be one."

Indeed, from 1998 to 2017, five of the top ten countries most affected by climate change were in Asia: Myanmar, the Philippines, Bangladesh, Pakistan, and Vietnam. Inversely, Taiwan, Japan, and South Korea rank in the bottom of those most on duty to reduce their carbon emissions. The Asia Pacific region as a whole is the highest contributor of greenhouse gas emissions to the atmosphere, with 40% of global emissions in 2015; this percentage is projected to increase until 2030, with 89% of Asia-Pacific's contribution coming from China, India, and Indonesia (Simangan, 2020).

Regarding the concrete consequences of what the Anthropocene means for Asia, a great deal of discussion has so far focused on climate change and its most immediate consequences, like rising sea levels or stronger typhoons. For instance, Jakarta has been proclaimed "the city of the Anthropocene" for its vulnerability to rising sea levels and the resilience of the *kampong*—its floating slums (Chandler, 2017). The Indonesian government is thus planning to transfer the capital to East Kalimantan (on the island of Borneo), with possibly detrimental effects for local indigenous populations and lush rainforests that are home to orang-utans and countless other animal species. Singapore is another city threatened by rising sea levels, with 35 per cent of its territory lying less than five meters above sea level; but from Lee Kuan Yew's vision of a "Garden City" to its iconic Supertrees, Singapore's techno-nature and green-washing policy reflect the firm intention of the city-state to become a champion of resilience in the Anthropocene (Schneider-Mayerson, 2017).

Along with rising seas, biodiversity loss in Southeast Asia is a major issue—if not the main issue—of the Asian Anthropocene. After the Amazon and the Congo, Southeast Asia is the world's third-largest zone of tropical forests and a concomitant repository of terrestrial biodiversity, now at the forefront of "mass destruction" and at so rapid a rate that current data quickly becomes obsolete (Zeng et al., 2018). Since the 1999s, mobilizations of local people and transnational networks against deforestation by agribusiness (such as oil palm) have continued unabated. Furthermore, the construction of large hydroelectric dams, and the expansion of mining and monocultures are almost inevitably accompanied by the displacement of entire communities, massive pollution of land and rivers, and a homogenization (and oversimplification) of human and natural ecology.

This violence is nothing new; it started during colonial times, was further aggravated by post-colonial regimes of European to Japanese domination, and has been described by an abundant literature. What is more specific to the Anthropocene paradigm is that departing from the naïve belief that brutal infrastructure projects and the expansion of agribusiness are "sustainable development", there is now a large consensus among international organizations that further destruction of the "cultural and natural heritage" must be avoided.

These considerations on the Anthropocene of Asia intertwine also with geopolitical concerns. As a mark of its ascendancy, China has undertaken a comprehensive global investment program, the Belt and Road Initiative (BRI), with the aim of fashioning a China-centred economic sphere, by developing and economically integrating the countries along the historic Silk Road, which could be the largest infrastructure project in human history, with possibly devastating consequences for biodiversity and a further increase in global warming emissions due to the export of coal-burning power plants.

Thus far publications on the Anthropocene in Asian vernacular languages have been very limited. The "indigenization" of the notion has started only recently. For instance, in Taiwan, the notion of the Anthropocene has prompted some academic discussions, though limited to a small circle of social science scholars and artists engaged in ecological issues (e.g., Chuang & Gong, 2020). The perspectives include various angles such as eco-feminism or land and custom rights of indigenous peoples. The theoretical frameworks include local literature as much as a long list of Western scholars—with some recurrent figures such as Bruno Latour, Dona Haraway and Anna Tsing.

In spite of the limited indigenization of the Anthropocene in Asian languages thus far, we must keep in mind that all over Asia environmental movements have played a significant role to curb the massive forces of destruction presented above (Jobin et al., 2021), with emblematic examples such as the litigations for climate justice initiated from the Philippines and Indonesia. It remains to be seen however to what extent can these social movements cope with the pace and scope of the ongoing destruction.

References

Baldwin, A., & Erickson, B. (2020). Introduction: Whiteness, coloniality, and the Anthropocene. *Environment and Planning D: Society and Space, 38*(1), 3–11.

Chakrabarty, D. (2009). The climate of history: Four theses. *Critical Enquiry, 35*, 197–222.

Chandler, D. (2017). Securing the Anthropocene? International policy experiments in digital hacktivism: A case study of Jakarta. *Security Dialogue, 48*(2), 113–130.

Chatterjee, E. (2020). The Asian Anthropocene: Electricity and fossil Developmentalism. *The Journal of Asian Studies, 79*(1), 3–24.

Corlett, R. (2013). Becoming Europe: Southeast Asia in the Anthropocene. *Elementa: Science of the Anthropocene, 1*, 000016.

Chuang, C.-M., & Gong, J.-J. (Eds.). (2020). "Yishu celiang, renleishi meixue pipan" [Artistic measuring, aesthetic critique in Anthropocene epoch]. *ACT: Art Critique of Taiwan, 80*, 4–128.

Davis, H., & Todd, Z. (2017). On the importance of a data, or, decolonizing the Anthropocene. *ACME: An International Journal for Critical Geographies, 16*(4), 761–780.

Hamilton, C. (2017). *Defiant earth: The fate of humans in the Anthropocene*. Allen & Unwin.

Hecht, G. (2018, February 6). The African Anthropocene. *Aeon* (online).

Horn, E., & Bergthaller, H. (2020). *The Anthropocene: Key issues for the humanities*. Routledge.

Hudson, M. (2014). Placing Asia in the Anthropocene: Histories, vulnerabilities, responses. *The Journal of Asian Studies, 73*(4), 941–962.

Jobin, P., Ho, M.-s., & Hsiao, H.-H. M. (2021). *Environmental movements and politics of the Asian Anthropocene*. ISEAS.

Malm, A., & Hornborg, A. (2014). The geology of mankind? A critique of the Anthropocene narrative. *The Anthropocene Review, 1*(1), 62–69.

Marquardt, J. (2019). Worlds apart? The Global South and the Anthropocene. In T. Hickmann, L. Partzsch, P. Pattberg, & S. Weiland (Eds.), *The Anthropocene debate and political science* (pp. 200–218). Routledge.

Schneider-Mayerson, M. (2017). Some islands will rise: Singapore in the Anthropocene. *Resilience, 4*(2–3), 166–184.

Simangan, D. (2019). Situating the Asia Pacific in the age of the Anthropocene. *Australian Journal of International Affairs, 73*(6), 564–584.

Simangan, D. (2020). Where is the Asia Pacific in mainstream international relations scholarship on the Anthropocene? *The Pacific Review, 34*, 724–746. https://doi.org/10.1080/0951274 8.2020.1732452

Simpson, M. (2020). The Anthropocene as colonial discourse. *Environment and Planning D: Society and Space, 38*(1), 53–71.

Zhang, Q., Jiang, X., Tong, D., et al. (2017). Transboundary health impacts of transported global air pollution and international trade. *Nature, 453*, 705–717.

Zeng, Z., Estes, L., Ziegler, A. D., Chen, A., et al. (2018). Highland cropland expansion and forest loss in Southeast Asia in the twenty-first century. *Nature Geoscience, 11*, 556–562.

Paul Jobin is Associate Research Fellow at the Institute of Sociology, Academia Sinica Taiwan. Previously, he was Associate Professor in the Department of East Asian Studies at the University of Paris. His research focuses on the socio-politics and geopolitics of environmental issues in Taiwan, Japan and East Asia. His recent publications include journal articles in *Environmental Sociology, EASTS, The Asia-Pacific Journal* and *Politique internationale,* book chapters in *Critical Zones* (MIT Press, coedited by Bruno Latour), *Repairing Environments* (Routledge, coedited by Laura Centemeri) and *Legacies of Fukushima* (Penn Press), and a coedited volume *Environmental Movements and Politics of the Asian Anthropocene* (ISEAS).

Dating Debate

Nathanaël Wallenhorst

Abstract This notice provides in-depth insight into the debate (based on stratigraphic findings) as to the exact date that Earth entered the Anthropocene Epoch. Generally, in Human and Social Sciences, the date of the boundary is paid less attention than the systemic shifts resulting from it. (There have been a number of publications on our surpassing of the planet's limits, and passing of the point of no return). This article explores each of the possible dates when stratigraphic data suggest the Anthropocene may begun. We propose a synthesis of the seven main possible dates for the entry into the Anthropocene.

Hypothesis 1: 400,000 Years Ago, with the Mastery of Fire

The earliest proposed date for the dawn of the Anthropocene is in the Stone Age (Doughty, 2013). The frequent use of fire in the late Pleistocene, around 400,000 years in Africa, is certainly the first way in which humans made a mark on their environment (Roebroeks & Villa, 2011). However, as the use of fire is always local, it does not give rise to a GSSP (global), and therefore cannot be used to mark the boundary (Lewis & Maslin, 2006, p. 173). Another event which indelibly marked the biosphere was the megafauna extinction, with half of large mammal species dying out. However, this event (between 50,000 and 12,000 years ago) cannot be used as a GSSP either, because the extinction was very uneven from one continent to another, and took place at different times (Lewis & Maslin, 2006, p. 174).

N. Wallenhorst (✉)
Catholic University of the West, Angers, France
e-mail: nathanael.wallenhorst@uco.fr

© The Author(s), under exclusive license to Springer Nature
Switzerland AG 2023
N. Wallenhorst, C. Wulf (eds.), *Handbook of the Anthropocene*,
https://doi.org/10.1007/978-3-031-25910-4_53

Hypothesis 2: Between 8000 and 5000 Years Ago, with the Development of Agriculture

A number of authors, such as American science journalist Michael Balter (2013), American-Belgian ecologist Jed O. Kaplan et al. (2011) and American palaeoclimatologist William F. Ruddiman (2003, 2013) and Ruddiman et al. (2014), believe the Anthropocene began with the development of agriculture and the resulting alteration of the atmosphere's chemical composition. As the climate stabilised at the dawn of the Holocene, agriculture was able to develop in various locations across the globe. It began continents some 10,000 years ago, in southeast Asia, northern China and South America (Lewis & Maslin, 2006, p. 174). Over time, agriculture led to natural vegetation being replaced by crop species. It also drove certain animal species to extinction, in favour of others which could be domesticated. Agriculture also altered the flow of biogeochemicals. Indeed, drawn from Lewis and Maslin (2006), p. 174), the atmospheric peak of methane (CH_4) in 5020 BP ('before present', with 'present' taken to be 1950) recorded in ice cores could represent a GSSP, because it may be due to rice agriculture in Asia and growing numbers of ruminants.

The hypothesis of an 'early Anthropocene' is supported by a number of findings: the alteration of the terrestrial habitat and terrestrial biotic shifts, microbiotic changes in the ocean and shifts in the makeup of the atmosphere, caused by agriculture (Zalasiewicz et al., 2014). Conceivably, the development of agriculture may have helped maintain the warmer temperatures of the Holocene, preventing a new Ice Age. For example, we see a spike in CO_2 levels 8000 years ago, and a similar one for methane 5000 years ago (MacFarling Meure et al., 2006). However, this hypothesis has yet to be fully confirmed. The unusually long interglacial period we are currently experiencing could also be due to non-anthropogenic changes in solar radiation and shifts in Earth's orbit, altering methane emissions in humid tropical zones (Steffen et al., 2011, p. 847; Lewis & Maslin, 2006, p. 174).

Two authors (Certini & Scalenghe, 2011) propose to date the start of the Anthropocene at 2000 years ago. This does not correspond exactly to the development of agriculture, but to the development of large civilisations which have left traces in soil cores. They consider the paedosphere (soil free from ice) to be 'the best indicator of the increasing human impact on the whole of the environment come up because it strongly reflects the increasing impact of the earliest civilisation on a large portion of the Earth's surface' (2011, p. 1269). Finally, they believe soil is a better stratigraphic indicator than atmospheric chemical alterations, in determining where to place a golden spike. Soils have undergone numerous modifications as the result of human activity, to fertilise crops. However, their proposal falls down, as they view the Anthropocene as the anthropisation of the planet, whereas in fact, the term refers to the wholesale alteration of Earth's system.

Hypothesis 3: 1610, when the Old World Met the New

Another event could come into play in defining the start of the Anthropocene – what Lewis and Maslin (2006) term the collision between the Old and the New World. To date, they are the only researchers to have suggested this as a GSSP. This event marks the start of organised human society all over the world, with shared food products, reshaping both plant and animal life. Above all, Europeans' arrival in America in 1492 heralded a sharp decline in world population. In 1492, America' population was estimated at between 54 and 61 million. Only 158 years later, in 1650, it had fallen to 6 million. This population slump in America was due to war, slavery, disease brought by Europeans, and famine. This marked decrease in human population led to a reduction in farmland, and woodland began to recover, reaching an estimated 50 million hectares of forest. This decreased atmospheric CO_2 by 7–10 ppm, which can be seen in Antarctic sediment cores from between 1570 and 1620. Lewis and Maslin estimate 1610 as an appropriate GSSP marker. Secondary stratotypes are associated with this approximate date, such as contact between biotopes from the Old and the New World (the fossil record shows that maize was introduced to Europe around this time), or the eruption of Huaynaputina, which can be seen in sediments from both poles and from the tropics. We see a drop in atmospheric methane, and the spread of Arctic ice. Two different ice cores, taken from the Antarctic, show this 7–10 ppm reduction in atmospheric CO_2, which exceeds the possible margin of error (between 1 and 2 ppm). In addition to being visible in the sediments and representing a promising potential GSSP, the meeting of the Old and New Worlds was a major event in human history, representing the start of globalisation – the true characteristic of the contemporary period.

One of the strengths of Lewis and Maslin's (2006) approach is that they analyse geological data, rather than human historical data. This is why the proposed GSSP is 1610, corresponding with the dip in atmospheric CO_2, as opposed to 1492. They reject the other possible dates for the start of the Anthropocene, as they are not based on a global marker. In their view, this date is the most promising candidate for a GSSP, in view of the radical and global shift it marks. As the meeting of Old and New Worlds also opened the door to the Industrial Revolution, the date is compatible with Crutzen and Stroemer's (2000) initial hypothesis.

Hypothesis 4: 1769 or 1800, with the Start
of the Industrial Revolution

In his earliest articles, Crutzen (Crutzen & Stroemer, 2000; Crutzen, 2002) aligns the onset of the Anthropocene with the Industrial Revolution. If asked for a specific date, he would choose 1769, which is when the steam engine was invented. In the eyes of a group of researchers including Crutzen, industrial technologies are to blame for our entry into the Anthropocene (Steffen et al., 2011; Zalasiewicz et al.,

2011). They also point out that the global effects of the Industrial Revolution on the environment have surpassed those of agriculture. In the view of Zalasiewicz et al. (2008), the April 1815 eruption of Mount Tambora, which caused a volcanic winter lasting a year in the northern hemisphere, could be a 'natural' stratigraphic marker for the onset of the Anthropocene, which corresponds to the dawn of the Industrial Revolution.

In recent years, there has been less focus on the Industrial Revolution as the leading candidate for the start of the Anthropocene, but when the concept was first proposed, it was the prime suspect. In 2007, Steffen, Crutzen and McNeil described the industrial era as the first stage of the Anthropocene. The Industrial Revolution began in Britain, with the invention of the steam engine (which emits CO_2 into the atmosphere). It truly transformed Britain in around 1850, and eventually spread to the rest of the world. The tricky aspect of using the Industrial Revolution as the proposed date of the onset of the Anthropocene is that it occurred at the beginning of the eighteenth Century in the United Kingdom; then, between 1820 and 1880, it took over the east coast of the United States and western Europe; and at the very end of the eighteenth Century, it conquered the rest of Europe and the USA. Other countries were not touched by the Industrial Revolution until the twentieth Century. Steffen, Grinevald, Crutzen and McNeil, in a 2011 article, proposed that the Anthropocene should be treated as a new epoch, dating from around 1800, with the beginning of the Industrial Revolution. Their position on the matter shifted, and in 2014, all four signed an article proposing 1945 as the start date of the Anthropocene from a stratigraphic viewpoint, with the placement of a golden spike (Zalasiewicz et al., 2014). Steffen et al. (2011) view atmospheric CO_2 concentration as a good indicator of the onset of the Anthropocene. However, this argument may be contested by certain stratigraphers paying close attention to the sediment record. In 1750, atmospheric CO_2 was 277 ppm; it rose to 279 ppm in 1775; then to 283 ppm in 1800; and to 284 ppm in 1825. These figures are within the range of variability determined for the Holocene – between 260 and 285 ppm. The maximum natural variation for the Holocene was reached in 1850, at 285 ppm; then by 1900, it had been substantially surpassed, at 296 ppm (Steffen et al., 2011, p. 848; Etheridge et al., 1998; Kleinen et al., 2010). In Lewis and Maslin's view, the trace metal pollution associated with the Industrial Revolution cannot be used as a GSSP, because it is local and diachronic (traces of the use of such metals can be seen between 8000 BP and the Industrial Revolution) (Lewis & Maslin, 2006, p. 175). Whilst a collection of markers can be found from 1800 in North America and northern Europe (including the rise in atmospheric CO_2 from the nineteenth Century onwards), none of them are clear indicators of a GSSP, in Lewis and Maslin's opinion.

Hypothesis 5: 1950, with the Start of the Great Acceleration

Since the mid-twentieth Century, there has been a huge rise in world population. This is particularly true of urban population, which soared from 730 million in 1945 to 3.7 billion city dwellers in 2014 (Zalasiewicz et al., 2014, p. 3). Though the term

'Great Acceleration' had not yet been coined, in 2004, Steffen et al. presented the curves showing the phenomenon. Then, in 2007, applying the term, they presented further plots illustrating evolving human activity on the planet. The Great Acceleration refers to an upswing in the production of manmade sediments. A number of authors also believe it can serve as a relevant stratigraphic marker (Waters et al., 2014; Holtgrieve et al., 2011; Wolfe et al., 2013; Zalasiewicz et al., 2014). Zalasiewicz et al. (2014, p. 4) identify a set of possible stratigraphic markers pertaining to the Great Acceleration: the disruption of the nitrogen cycle with the development of the Haber–Bosch process; technofossils, plastic waste and aluminium spread across the globe; the explosion of human deposits in the ground; the enormous increase in concentration of pollutants linked to the development of industrial activity; the passing of a significant milestone in anthropogenic changes to the biosphere, with numerous species dying out; the acceleration of hydrocarbon consumption, leading to an increase in atmospheric CO_2; fossil traces connected to oil drilling; oil spills in coastal areas; and the construction of massive dams, reclaiming land from river deltas.

When updating the graphs of the Great Acceleration, extending the plots to 2010, Steffen et al. (2016) took up the debate as to the date of onset of the Anthropocene. They state that the start of the Great Acceleration, in the 1950s, is the most promising candidate from the standpoint of Earth system science, as that is when the natural variability margins of the Holocene were surpassed, and evidence indicates that these disruptions of Earth's system were driven by human activity. From a strictly stratigraphic perspective, Zalasiewicz and Williams (2013) also view the Great Acceleration, of the 1950s, as the start of the Anthropocene.

Hypothesis 6: 1945 or 1964, with the Denotation of Nuclear Bombs

Another hypothesis that has been investigated as the possible start date of the Anthropocene is 16 July 1945 – the date of the world's first nuclear test (codenamed Trinity) at Alamogordo in New Mexico – or the radioactivity peak seen some 20 years later. After this first explosion, other nuclear detonations followed at an average of one per 9.6 days until 1988, altering the chemical makeup of the atmosphere (Zalasiewicz et al., 2014, p. 1). The radioactive peak can be seen in 1964, just after the implementation of the Partial Nuclear-Test-Ban Treaty, signed in Moscow on 5 August 1963. Lewis and Maslin (2006) and Masco (2010) regard the atomic explosions as a sensible point at which to situate the onset of the Anthropocene, and the atmospheric radiocarbon ($\Delta^{14}C$) peak provides an unambiguous GSSP. For this reason, they propose the 1964 peak as a possible GSSP to date the planet's entry into the Anthropocene. Lewis and Maslin (2006) propose to use a pine tree, 25 km east of Kraków in Poland, whose growth rings show a radiocarbon spike, as the GSSP. The advantage of Lewis and Maslin's proposed GSSP in 1964 is the same as for the proposal of 1610. Their findings are based on stratigraphic data, rather than on human history (though the stratigraphic data directly reflect human history, with a lag of a few years or decades). The benefit of choosing 1964 as the start date for the Anthropocene is that it connects to a set of markers showing human impact on Earth's system, which can be seen in the course of the Great Acceleration. On the other hand, whilst the nuclear explosions did leave a mark in the stratigraphic record, they did not radically alter the Earth system (though the power they unleashed would have had the potential to do so). The proposed date is approximately the same as the start of the Great Acceleration, but we wish to draw a distinction between the two events (which not all authors do). The atomic bomb detonations do not correspond to the acceleration of consumption and industrial production, but to the crossing of a techno-scientific threshold directly linked to humanity's capacity for self-destruction.

Crutzen gradually revised his proposal on the date of the onset of the Anthropocene. Originally, he situated it alongside the invention of the steam engine, giving the approximate date of 1800, with the start of the Industrial Revolution in Britain. He later joined other researchers in proposing the first nuclear bomb detonation as a GSSP, on the basis of more solid stratigraphic evidence.

Zalasiewicz et al. (2014, p. 6), in an article entitled 'When did the Anthropocene begin? A mid-twentieth century boundary level is stratigraphically optimal', specify that the significance of the Anthropocene lies in the 'finding of the earliest traces of human impact on the Earth system, and the extent and longevity of the anthropic changes made to the Earth system'. For this reason, they propose a GSSA (Global Standard Stratigraphic Age) rather than a GSSP, situating the dawn of the Anthropocene in 1945, with the Alamogordo nuclear test. The strength of this approach is that it opens the door to multidisciplinary research on the Anthropocene, rather than confining it to geology. Their argument seems meritorious, and unusual for geologists and Earth system scientists, who make up the majority of the article's 26 signatories.

The main advantage of using the fallout from nuclear bombs, from a stratigraphic point of view, is that it offers a clear, precipitous and widespread signature – and stratigraphers value signals such as this very highly (Waters et al., 2015).

Hypothesis 7: Somewhere in the Future

For Wolfe et al. (2013), there can be no doubt that human activity has altered elements of the Earth system; the traces can clearly be seen in ice cores from the ice caps – the glaciers covering the poles which are particularly important for climate stability. However, he feels the human impact currently visible on the ice caps is not sufficient to mark a new geological boundary between the Holocene and the Anthropocene, because the change so far could, potentially, be reversible within the next 10,000–100,000 years. He does believe, however, that human activity is likely to delay the next glacial formation by at least 100,000 years. Wolfe et al. (2013)

recognises the significant – and global – anthropogenic traces left by nuclear explosions. Whilst Wolfe is very reserved and conservative in his reasoning, he believes that a GSSP could be placed in the ice in Greenland or the Antarctic, but is aware that it is not ideal from a stratigraphic standpoint, given the impermanence of the ice caps. For Wolfe et al. (2013), it is evident that human beings have altered large parts of the Earth system. However, as it is not yet clear exactly what the Anthropocene will bring in the next few hundred or few thousand years, he feels it more sensible to allow future generations to decide on the date of the transition, when they are able to look at it with the benefit of hindsight.

Conclusion: Some Pointers on Using These Data in Teaching

It is important to be aware of the debate over stratigraphic data, but such knowledge ought not to be a goal *in itself*. It opens the door to reflection, and can be worked on in a variety of ways with students and pupils. This debate appears to be a helpful lens through which to analyse the different facets of the polemic concept of the Anthropocene. Though indubitably scientific, it is also a political concept (which will determine the future direction of the human adventure). It is also one that arouses powerful feelings, as can be seen from the academic articles on the subject – clearly, researchers cannot remain neutral on topics so disquieting as those stirred up by the Anthropocene (Wallenhorst, 2019).

This debate also provides a starting point for a critical examination of school curricula. In view of that examination, students (perhaps older students more than young pupils) could be encouraged to reflect on the socialisation they receive at school. What fundamental truths has school taught us which now appear in a new light, as we venture into the uncharted waters of the Anthropocene?

It is also sensible to examine the political agendas underlying each of the hypotheses for the dawn of the Anthropocene (pointing the finger at certain culprits, serving a particular moral outlook, opening or closing the door to the future).

Whilst there is vigorous debate among experts, we can easily examine the purely anthropological point underlying the various hypotheses. Ultimately, what is at stake in the way in which humans exist on Earth? Looking at the bigger picture, the figure of *homo oeconomicus*, seeking selfishly to maximise their own individual interests, is very much in the foreground.

The Anthropocene also provides the impetus to re-examine modern life, and the prominent dualism between nature and culture. The Anthropocene blurs the boundaries between the two. What is truly 'natural'? Can there be anything genuinely 'cultural' and 'social' that is not anchored in 'nature'? What can be said of our connections with the Earth itself, and with the rest of the living world?

Thus, this scientific debate can also help awaken rational thinking. The Earth keeps a permanent record of our presence and our activities. This much is obvious: it speaks *about* us – but does it speak *to* us… and were we to listen, what would it say?

References

Balter, M. (2013). Archaeologists say the "Anthropocene" is here – But it began long ago. *Science, 340*, 261–262.

Certini, G., & Scalenghe, R. (2011). Anthropogenic soils are the golden spikes for the Anthropocene. *The Holocene, (21)*8, 1269–1274.

Crutzen, P. J., & Stoermer, E. F. (2000). The Anthropocene. *Global Change, Newsletter, 41*, 17–18.

Crutzen, P. J. (2002). Geology of mankind: "The Anthropocene". *Nature, 415*, 23.

Doughty, C. E. (2013). Preindustrial human impacts on global and regional environment. *Annual Review of Environment and Resources, 38*, 503–527.

Etheridge, D. M. et al. (1998). Historical CO_2 records from the Law Dome DE08, DE08-2, and DSS ice cores. In Trends: a Compendium of Data on Global Change, Oak Ridge National Laboratory, U.S. Department of Energy, Oak Ridge.

Holtgrieve, G. W., et al. (2011). A coherent signature of anthropogenic nitrogen deposition to remote watersheds of the Northern Hemisphere. *Science, 334*, 1545–1548.

Kaplan, J. O., et al. (2011). Holocene carbon emissions as a result of anthropogenic land-cover change. *The Holocene, 21*(5), 775–791.

Kleinen, T. et al. (2010) Holocene carbon cycle dynamics. *Geophysical Research Letters, 37*, 1–5.

Lewis, S. L., & Maslin, M. A. (2006). Defining the Anthropocene. *Nature, 519*, 171–180.

MacFarling Meure, C., et al. (2006). Law Dome CO_2, CH_4 and N_2O ice core records extended to 2000 years BP. *Geophysical Research Letters, 33*, 1–4.

Masco, J. (2010). Bad weather: On planetary crisis. *Social Studies of Science, 40*(1), 7–40.

Roebroeks, W., & Villa, P. (2011). On the earliest evidence for habitual use of fire in Europe. *Proceedings of the National Academy of Science, USA, 108*(13), 5209–5214.

Ruddiman, W. F. (2003). The anthropogenic greenhouse era began thousands of years ago. *Climatic Change, 61*, 261–293.

Ruddiman, W. F. (2013). The Anthropocene. *The Annual Review of Earth and Planetary Sciences, 41*, 45–68.

Ruddiman, W. F., et al. (2014). Does pre-industrial warming double the anthropogenic total? *The Anthropocene Review, 1*, 1–7.

Steffen, W., et al. (2011). The Anthropocene: Conceptual and historical perspectives. *Philosophical Transactions of the Royal Society, 369*, 842–867.

Steffen, W., et al. (2016). Stratigraphic and Earth system approaches to defining the Anthropocene. *Earth's Future, 4*, 1–22.

Wallenhorst, N. (2019). *L'Anthropocène décodé pour les humains*. Le Pommier.

Waters, C. N., et al. (2014). Evidence for a stratigraphic basis for the anthropocene. In R. Rocha, J. Pais, J. Kullberg, & S. Finney (Eds.), *STRATI 2013, Springer Geology* (pp. 989–993). Springer.

Waters, C. N., et al. (2015). Can nuclear weapons fallout mark the beginning of the Anthropocene Epoch? *Bulletin of the Atomic Scientists, 71*(3), 46–57.

Wolfe, A. P., et al. (2013). Stratigraphic expressions of the Holocene–Anthropocene transition revealed in sediments from remote lakes. *Earth-Science Reviews, 116*, 17–34.

Zalasiewicz, J., et al. (2008). Are we now living in the Anthropocene?. *GSA Today, 18*(2), 4–8.

Zalasiewicz, J., et al. (2011). The Anthropocene: A new epoch of geological time? *Philosophical Transactions of the Royal Society, 369*, 835–841.

Zalasiewicz, J., et al. (2014). When did the Anthropocene begin? A mid-twentieth century boundary level is stratigraphically optimal. *Quaternary International, 30*, 1–8.

Zalasiewicz, J., & Williams, M. (2013). The anthropocene: A comparison with the Ordovician-Silurian boundary. In R. Lincei (Ed.), *Scienze Fisiche e Naturali* (pp. 3–11). Springer.

Nathanaël Wallenhorst is Professor at the Catholic University of the West (UCO). He is Doctor of Educational Sciences and Doktor der Philosophie (first international co-supervision PhD), and Doctor of Environmental Sciences and Doctor in Political Science (second international co-supervision PhD). He is the author of twenty books on politics, education, and anthropology in the Anthropocene. Books (selection):*The Anthropocene decoded for humans* (Le Pommier, 2019, *in French*). *Education in the Anthropocene (*ed. with Pierron, Le Bord de l'eau 2019, *in French*). *The Truth about the Anthropocene* (Le Pommier, 2020, *in French*). *Mutation. The human adventure is just beginning* (Le Pommier, 2021, *in French*). *Who will save the planet?* (Actes Sud, 2022, *in French*). *Vortex. Facing the Anthropocene* (with Testot, Payot, 2023, *in French*). *Political education in the Anthropocene* (ed. with Hétier, Pierron and Wulf, Springer, 2023, *in English*). *A critical theory for the Anthropocene* (Springer, 2023, *in English*).

Good Anthropocene

François Prouteau

Abstract Based on the scientific evidence of the alarming context of changing living conditions on Earth, this article analyses how the Anthropocene can be claimed to be good or bad. We explore the extent to which the use of the oxymoron 'good Anthropocene' is relevant or not for moving towards a positive future.

The triple fold concept of the Anthropocene is simultaneously scientific, political and ideological. This has an impact on whether we see it as good or not. These different points of view challenge us to reflect on our current situations, our different heritages and the future that is being written. Since the mid-twentieth century, societies have benefited from the global spread of technological knowledge and have become more economically interdependent (the Great Acceleration). The human population has grown rapidly, and societal and medical advances have extended life spans. At the same time, the risks and threats of disasters are increasing. Despite all of this, we hope to find the right path and the necessary tools for the human adventure to continue in the Anthropocene. Could the Anthropocene be good, "given that biodiversity and healthy ecosystems are requisite for sustained human life" (Jeanson et al., 2020) but their survival is more deeply threatened than it has been for the last 2.7 million years?

Before asking whether, from a moral or a political point of view, the Anthropocene is good or bad, we need to consider this new geological and biological epoch as a scientifically observable entity. Paul Crutzen (Nobel Prize in Chemistry in 1995) and biologist Eugene Stoermer, both affiliated with the International Geosphere-Biosphere, are the first scientists to refer to the Anthropocene in an official paper published in 2000. They used this neologism to define the new geological epoch in which anthropological impact has been central to geology and ecology since the second half of the eighteenth century. Such a starting date accords with James Watt´s invention of the steam engine but "alternative proposals can be made"

F. Prouteau (✉)
LIRFE – Catholic University of the West, Angers, France
e-mail: f.prouteau@iffeurope.org

(Crutzen & Stoermer, 2000). The new geological epoch called Anthropocene could be officially adopted by the International Commission on Stratigraphy (IGS) before the end of 2021. The IGS, who is the guardian of Earth's timeline formed the advisory Anthropocene Working Group (AWG) in 2009. The AWG is comprised of the paleobiogist Jan Zalasiewicz as chair and a team of sixteen members, about half being stratigraphers plus a mix of environmental scientists including Paul Crutzen, Will Steffen, Erle C. Ellis, an archaeologist, an historian of sciences, a writer and a law professor. The AWG endorsed the Great Acceleration (Steffen et al., 2015; Waters et al., 2016), which occurred in the mid-twentieth century, "as the main narrative explaining Earth's transition to the Anthropocene" (Ellis, 2018). There have been debates among scientists about designating a clear 'golden peak' or boundary marker for the beginning of the Anthropocene epoch (Zalasiewicz et al., 2019). Among the array of human signals that might be used as markers, anthropogenic radionuclides, for example ^{239}Pu, associated with thermonuclear explosion tests are the most promising (Syvitski et al., 2020), but potential secondary markers including plastic or carbon in a wide variety of sedimentary bodies, could be also retained.

Good Anthropocene? The answer is no if we examine a range of data that strongly underpins the gap in the trajectory of the Earth systems away from what they were during the Holocene, substantially and globally, and, potentially, irreversibly (Steffen et al., 2018). Since 2015, the concentration of CO_2 in the atmosphere has exceeded 400 ppm (parts per million, equivalent to 0.04% of the number of CO_2 molecules in the atmosphere) and is still rising (on average above 410 ppm during 2020); it has almost doubled in the last 30 years. According to studies of ice and marine sediment cores taken in 2019 by the Potsdam Institute for Climate Impact Research (PIK), a level above 400 ppm has never been reached. And we know that CO_2 is the most significant greenhouse gas in terms of emissions and lifetime (at least more than a century).

The Earth's average temperature has never exceeded the pre-industrial value by more than 2 °C since the beginning of the Quaternary period, just over 2.7 million years ago (Willeit et al., 2019). The current increase in the concentration of CO_2 is taking us in the wrong direction away from global warming of well below +2 °C compared to the so-called "reference period" 1850–1900. +2 °C was the target of the Paris Agreement on climate change which entered into force in November 2016, but since then, very few measures have been put in place. By 2100, global temperatures will rise by at least +3.5 °C if "Business as usual" policies remain. Temperatures will then be close to those of the Pliocene, one of the last periods of the Ternary Period, when trees grew in Antarctica and sea levels were 15 metres higher. Not to mention the acidification of these same oceans which, at +1.5 °C, would significantly affect marine biodiversity, fisheries and marine ecosystems, and the risks associated with the melting of 70% of the permafrost, to a depth of 3 or 4 metres, resulting in the release of hundreds of billions of tonnes of greenhouse gases and millions of viruses and bacteria that have been dormant until now.

It seems impossible to keep talking about a good Anthropocene when the Planet is becoming more and more uninhabitable for humanity. Already, with a temperature rise of 2 °C above pre-industrial levels, there are major risks of activating a domino effect: even higher temperatures and the overstepping of other planetary limits that will further destabilize the Earth system through the loss of biodiversity, chemical pollution, soil use, freshwater use. Very fast or very large climate changes can cause extensive disturbances (e.g., wildfires, insect attacks, droughts). Existing biomes (macro ecosystems with similar ecologic conditions) could be removed with risks to the positive feedback back into the climate system. Such factors irrevocably take the Earth System on a trajectory to a hotter state and in undesired, non-linear change at regional to global scales. For example, deforestation in the Amazon could not only irreversibly transform this region into a semi-arid savannah but could also reduce the availability of water resources in Asia. For The Intergovernmental Panel on Climate Change (IPCC) that was created to provide policymakers with regular scientific assessments, on the melting of the Greenland and West Antarctic ice sheets, warming above 2 °C, could be a tipping point, too. Exceeding +1.5 °C, which is increasingly likely by the end of the current decade, could lead to irreversible consequences for centuries. Scientists underline that while most life on Earth can recover from major climate change by evolving into new species and creating new ecosystems, Humanity cannot.

With this data in hand, how can we support the idea of a "good Anthropocene", especially when it comes from scientists who are aware of the degradation and threats caused by anthropogenic impact on the Earth system? One of these scientists is Erle C. Ellis, famous not only for his work that has enabled new ecosystem maps to be drawn up with the human factor at the centre (anthropic biomes or anthromes) but also for having used for the first time the expression of "good Anthropocene" (Hamilton, 2015). It is true that the Anthropocene has so far been a period of great acceleration in population growth, economic development and health care. In Ellis' very optimistic and technophile vision, the Anthropocene is not seen as a 'crisis'. This strange diagnosis or blind spot impedes learning from the past and analysing the reasons for the extremely critical and perilous situation for the future of humanity on Earth, as if nothing had happened.

We know that humanity is definitely a force for transforming the planet (Ellis et al., 2018). In the origin stories of Anthropocene inspired by the mythology of ancient Greeks, the god *Prometheus* plays a key role: he creates humans and "enables them to thrive by giving them fire stolen from the gods" (Ellis, 2018). A technocratic Promethean is using his powers in an unbridled way, with overbearing pride or presumption (*hubris*). How can we affirm that seeing the better future we have already created is enough "to imagine a good anthropocene (ibid.)?" How dare we? As we know, unlimited technological advances are linked to destructive *hubris*, and if business as usual continues, there will be no stopping the Earth's evolution towards a hotter, more polluted, less biodiverse and less humanly habitable state for millennia. *Homo Prometheus* plays with fire: from his point of view, it is by trusting in technology that we will succeed in solving the problems of the Earth system; in order to build a good Anthropocene, "creating that future will mean going beyond fears of transgressing natural limits" (Ellis, 2012).

More broadly, the good Anthropocene is animated by a techno utopia associated with The Breakthrough Institute. This Californian think tank published in 2015 *An*

Ecomodernist Manifesto written by "ecomodernists". Some of them are professors of economics, sustainable development, environment, applied physics, history or philosophy; others seem to be impressed by the comparison that the Anthropocene allows between the human species and God (Federau, 2017). They are all convinced "that knowledge and technology, applied with wisdom, might allow for a good, or even great, Anthropocene" (Ecomodernism, 2015) because humans have unbounded growing social, economic and technological powers. The reasons why the environment has been destroyed are never discussed, nor is there any question of changing the way of life. On the contrary, the authors embrace an optimistic view toward human capacities and the future, ultimately championing an "acceleration of technological progress" and "the active, assertive and aggressive participation of private sector entrepreneurs, markets, civil society and the state".

Many manifestos published since 2010 seem to tell us that a better Anthropocene is possible, that another world is desirable, but all of them support different theses in a war of ideas around the Anthropocene (Wallenhorst, 2021). The *Ecomodernist Manifesto* is the only one which encourages blind trust in the human genius inventing more powerful technologies without changing our relationship to the world. Contrary to the ecomodernists, the convivialists from the First Convivialist Manifesto (Internationale convivialiste, 2013) to the Second Convivialist Manifesto (Convivialist International, 2020) want to contribute to a post-neoliberal world that would put an end to the reign of individuals limited to their social bubble driven by the omnipotence of technology. What convivialists call "alternative humanity »" (AH) is subordinated to the absolute imperative of *hubris* control. Thus, the interjection "AH!" can also mean "*anti-hubris*" or "advance in humanity" (Prouteau, 2021) on the basis of five interrelated principles (Convivialist International, 2020): the interdependence of all living beings including humans with nature (common naturality); respect for humanity in the diversity of each of its members (common humanity); the greatest wealth is that which humans maintain in their relations with each other (common sociality); a legitimate policy and an equal freedom that allow each individual to develop his or her capacities, power to be and to act without harming that of others (legitimate individuation); commitment to the common good in a peaceful and deliberative rivalry that allows human beings to differentiate themselves, both ethically and politically (creative opposition). Convivialism calls for a genuine anthropological transformation without disregarding educational thinking; in this, it outlines, without saying so, the foundations of a better Anthropocene that is not only desirable but also concretely possible.

References

Jeanson, A. L., Soroye, P., et al. (2020). Twenty actions for a "good Anthropocene"—Perspectives from early-career conservation professionals. *Environmental Reviews, 28*(1), 99–108. https://doi.org/10.1139/er-2019-0021

Convivialist International. (2020). *The second convivialist manifesto: Towards a Post-Neoliberal World.* https://doi.org/10.1525/001c.12721

Crutzen, P. J. & Stoermer, E. F. (2000). The Anthropocene. *Global Change Newsletter.* n°41 [archive] (pp. 17–18). IGBP.

Ecomodernism. (2015). *Ecomodernist Manifesto.* www.ecomodernism.org

Ellis, E. (2012). *The Planet of no return. Human resilience on an Artificial Earth.* The Breakthrough Institute. https://thebreakthrough.org/journal/issue-2/the-planet-of-no-return

Ellis, E. C., Magliocca, N., Stevens, C. J., et al. (2018). Evolving the Anthropocene: Linking multi-level selection with long-term social–ecological change. *Sustainability Science, 13,* 119–128. https://doi.org/10.1007/s11625-017-0513-6

Ellis, E. C. (2018). *Anthropocene. A very short introduction.* Oxford University Press.

Federau, A. (2017). *Pour une philosophie de l'Anthropocène.* PUF.

Hamilton, C. (2015, April 21). The Technofix Is In: A critique of "An Ecomodernist Manifesto". *Earth Island Journal.* https://clivehamilton.com/the-technofix-is-in-a-critique-of-an-ecomodernist-manifesto

International Convivialiste. (2013). *Manifeste convivialiste. Déclaration d'interdépendance.* Le Bord de l'eau.

Prouteau, F. (2021). Un alterhumanisme pour l'Anthropocène. In *La Revue du Mauss* (n°57) (pp. 78–82). Le Bord de l'Eau.

Steffen, W., Rockrstrom, J., et al. (2018). Trajectories of the earth system in the Anthropocene. *Proceedings of the National Academy of Sciences of the United States of America, 115,* 8252–8259.

Steffen, W., Broadgate, W., Deutsch, L., Gaffney, O., & Ludwig, C. (2015). The trajectory of the Anthropocene: The Great Acceleration. *The Anthropocene Review.* https://doi.org/10.1177/2053019614564785

Syvitski, J., Waters, C. N., Day, J., et al. (2020). Extraordinary human energy consumption and resultant geological impacts beginning around 1950 CE initiated the proposed Anthropocene Epoch. *Communications Earth & Environment, 1,* 32. https://doi.org/10.1038/s43247-020-00029-y

Wallenhorst, N. (2021). *Mutation. L'aventure humaine ne fait que commencer.* Le Pommier.

Waters, C., et al. (2016). The Anthropocene is functionally and stratigraphically distinct from the Holocene. *Science, 351*(6269), aad2622. https://doi.org/10.1126/science.aad2622

Willeit, M., Ganopolski, A., et al. (2019). Mid-Pleistocene transition in glacial cycles explained by declining CO_2 and regolith removal. *Science Advances, 5*(4), eaav7337. https://doi.org/10.1126/sciadv.aav7337

Zalasiewicz, J., Waters, C., Williams, M., & Summerhayes, C. (2019). *The Anthropocene as a Geological Time Unit.* Cambridge University Press.

François Prouteau is a graduate engineer from the French Grande école Institute Mines-Telecom Atlantique with a PhD in Educational Sciences from the Lumière University Lyon 2. He is teacher and associated researcher at The Catholic University of the West (LIRFE Laboratory). His current research focuses on education facing contemporary anthropological and cultural challenges. He is cofounder of the Institutes of Formation Fondacio, of which he was the director for Europe from 1989 to 2009, and advisor for Africa, Asia and South America. He is currently the President of Fondacio.

Holocene

Michel Magny

Abstract The Holocene is an epoch defined by geologists as the last interglacial phase in the succession of glacial/interglacial phases which characterizes the Quaternary period. Its originality can be put in the so-called "Neolithic revolution" which corresponds (i) to the development of a new mode of subsistence based on agriculture and breeding, (ii) to an increasing human impact on environment, and (iii) to deep transformations of human societies. The Holocene epoch shows how the trajectories of the nature and human societies are tightly linked.

Investigations and observations by naturalists and geologists during the XIXth and the XXth centuries have given evidence of important climatic oscillations marked by successive advances and retreats of glacier tongues in Alpine valleys. Since 1970s, a systematic development of analysis of sediments accumulated in oceans, lacustrine basins and peat bogs as well as analysis of ice cores from Greenland and Antarctic ice sheets, allowed the astronomic theory of climate changes. Proposed for the first time in 1941 by the Serb astrophysicist Milutin Milanković, this theory stipulates that the alternance of cold glacial periods and warmer interglacial periods results from cyclic variations in the geometry of the Earth orbit and rotation under the influence of other planets in the solar system ("orbital forcing"). This influence results in a combination of 23,000, 41,000 and 100,000 yr cycles which modulate solar insolation on Earth surface and provoke cyclic climate changes. From −2.8 to circa − 1 million of years, the 41,000 yr cycle dominates with relatively limited magnitude of changes. Since − 1 million of years, the magnitude increases with a domination of the 100,000 yr cycle. Thus, the orbital forcing induces a marked glaciation period every 100,000 yr, separated from the following by a ca 10,000 to 20,000 yr-long interglacial phase characterized by a climatic optimum (insolation maximum). The Earth orbital conditions have produced the Last Glacial Maximum

M. Magny (✉)
CNRS-UMR 6249, Laboratoire Chrono-environnement, UFR Sciences et Techniques, Université de Franche-Comté, Besançon, France
e-mail: michel.magny@univ-fcomte.fr

© The Author(s), under exclusive license to Springer Nature Switzerland AG 2023
N. Wallenhorst, C. Wulf (eds.), *Handbook of the Anthropocene*,
https://doi.org/10.1007/978-3-031-25910-4_55

around −24,000 yr, while we benefit today from an interglacial climate phase that geologists have called *Holocene*. Moreover, the Milanković theory predicts that, *without other perturbation*, a new cold glaciation period should culminate in circa 100,000 years.

The count of annual layers observed in Greenland ice cores allows to date to circa −9700 yr changes in the isotopic composition of ice that mark the beginning of the Holocene interglacial (Mackay et al., 2003; Oldfield, 2005). The Greenland isotopic records show two successive trends in the climate history of Holocene in response to changes in insolation determined by orbital forcing. A first trend displays a temperature increase from −9700 to −7000 yr until a climate optimum; a second trend ("Neoglacial") shows a decrease in temperature which marks a progressive evolution towards colder climatic conditions, announcing a next future glaciation according to the Milanković theory. Superimposed on these general trends, the Greenland palaeoclimatic records as well as multiple other continental and oceanic records show the existence of secondary climate oscillations which punctuated the whole Holocene in response to different climate forcing factors: changes in the thermohaline oceanic circulation associated to sudden outbursts of large proglacial lakes during the deglaciation phase, cyclic changes in the solar irradiance, and volcanic eruptions (Mayewski et al., 2004; Magny, 2013). The effects of these secondary climatic oscillations vary depending on latitudes as shown by studies at European or global scales. Thus, a climate cooling over polar regions may correspond to a reinforcement of westerlies and associated lows over the mid-latitudes resulting in increasing precipitation, while over tropical latitudes the global decrease in temperature results in decreasing evaporation over oceans and more continental aridity. In 2018, referring to two major climatic events at ca −6200 and −2200 yr, and in agreement with the International Union of Geological Sciences, the International Sub-commission of the Quaternary Stratigraphy subdivided the Holocene in three phases: the Greenlandian Stage (from −9700 to −6200 yr), the Northgrippian Stage (from −6200 to −2200 yr) and the Meghalayan Stage (since −2200 yr).

However, the originality of the Holocene interglacial can be put in the irruption of a new factor now able to influence the trajectory of the Earth climate and ecosystems before determined only by natural factors, i.e. human activities. Formerly linked to hunting and gathering (Palaeolithic), the subsistence of human groups underwent important transformations during the Holocene with the emergence of a productive economy now based on agriculture and breeding. This so-called "Neolithic revolution" happened independently in different regions and to different dates; it firstly developed in the Middle East to ca −9000/−8000 yr, in China to ca −8000/−7000 yr, as well as to ca −8000 yr. in the Central America, to ca −5000 yr in the South America et to ca −3000 yr in Africa, before progressively extending to large parts of societies and terrestrial spaces (Guilaine, 2011). The Neolithic revolution rapidly became associated with an increasing impact of human societies on the environment (clearing and opening of forests, retreat or extinction of wild species, increasing number of domestic animals) resulting from an unprecedented demographic expansion of human populations: probably below 6 millions of individuals until the end of the Palaeolithic around −12,000 yr, the human populations reached 100 millions to ca −2000 yr and 250 millions to the beginning of the Christian era. The Neolithic revolution is also rapidly associated with the emergence of social inequalities and first forms of State with correlative apparition of first cities and the urban phenomenon, of slavery, of war, and of first "world-economies". This revolution also marks a new speeding of technological progresses particularly with the development of metallurgy and extractive activities which resulted in atmospheric pollution at a hemispheric scale as soon as the Antiquity. It also coincided with the development of a new imaginary; the former horizontal relationship between humans and other living organisms suggested by parietal pictures of the final Paleolithic (for instance Chauvet, Lascaux, Altamira) is now replaced by a new verticality incarnated by anthropomorphic divinities: the domination relations imposed by humans to nature echo the social stratifications via economic and political inequalities (Testart, 2012; Demoule, 2017; Scott, 2019).

These new types of relationship between humans and environment or between humans became still exacerbated by the development of the capitalist economy in Europe at the end of the Middle Ages and by the intellectual revolution associated with the Modernity. Referring to the Descartes methodology, the science and the technics establish a drastic separation between the man considered as unique subject and the nature reduced to simple object. At the end of the XVIIIth century, the industrial revolution that develops in Europe marks a new step in the allegiance of terrestrial ecosystems to the man becoming the undisputed master of their trajectory. To such an extent that the Dutch chemist P. Crutzen and the American biologist E. Stoermer consider in 2000 that the development of the industrial civilization now signs the end of the Holocene and that the Earth enters into a new epoch: the *Anthropocene* (Crutzen & Stoermer, 2000; Bonneuil & Fressoz, 2013; Wallenhorst, 2019). Today the human population overpasses 7.5 billions of individuals and attains a technological power which appears unlimited. An only species, *Homo sapiens*, is on the point of appropriating all terrestrial spaces and it threatens to ruin large parts of terrestrial ecosystems and to compromise its own future existence. Considered as a first absolute principle, the economy now overpasses the orbital forcing which regulated the Earth climate since 3 millions of years; it dictates its own capitalist frenetic tempo: in a contrary direction to that predicted by the Milanković theory, the climate warming induced by human activities has already put the global temperature to a level close to that of the Holocene optimum (Marcott et al., 2013).

Finally, Holocene shows that, even before in the Anthropocene and through a long *Palaeoanthropocene* (Foley et al., 2013), the trajectory of Earth ecosystems and that of societies appear to be tightly linked: the transformations underwent by the nature mirror those which shape societies since the beginning of the Neolithic revolution. In this sense, restoring the future of Earth ecosystems today threatened by collapse directly questions the predation imaginary which through the neoliberal ideology has imposed its yoke not only to the nature but also to societies (Magny, 2019).

References

Bonneuil, C., & Fressoz, J.-B. (2013). *L'événement anthropocène. La Terre, l'histoire et nous.* Le Seuil.

Crutzen, P., & Stoermer, E. (2000). The « Anthropocene ». *IGBP Newsletter, 41*, 17–18.

Demoule, J.-P. (2017). *Les dix millénaires oubliés qui ont fait l'histoire. Quand on inventa l'agriculture, la guerre et les chefs.* Fayard.

Foley, S., et al. (2013). The Palaeoanthropocene. The beginnings of anthropogenic environmental change. *Anthropocene, 3*, 83–88.

Guilaine, J. (2011). *Caïn, Abel, Ötzi. L'héritage néolithique.* Gallimard.

Mackay, A., Battarbee, R., Birks, J., & Oldfield, F. (2003). *Global change in the Holocene.* Arnold.

Magny, M. (2013). Orbital, ice-sheet, and possible solar forcing of Holocene lake-level fluctuations in West-Central Europe: A comment on Bleicher. *The Holocene, 23*, 1202–1212.

Magny, M. (2019). *Aux racines de l'Anthropocène. Une crise écologique reflet d'une crise de l'homme.* Le Bord de l'Eau.

Marcott, S., et al. (2013). A reconstruction of regional and global temperature for the past 11 300 years. *Science, 339*, 1198–1201.

Mayewski, P., et al. (2004). Holocene climate variability. *Quaternary Research, 62*, 243–255.

Oldfield, F. (2005). *Environmental change. Key issues and alternative approaches.* University Press.

Scott, J. (2019). *Homo domesticus. Une histoire profonde des premiers États.* La Découverte.

Testart, A. (2012). *Avant l'histoire. L'évolution des sociétés, de Lascaux à Carnac.* Gallimard.

Wallenhorst, N. (2019). *L'Anthropocène décodé pour les humains.* Le Pommier.

Michel Magny is emeritus Director of Research at CNRS (CNRS-UMR 6249, Laboratoire Chrono-environnement, University of Franche-Comté, Besançon, France). His main focuses are (i) on reconstruction of Lateglacial and Holocene climate changes and (ii) interactions between environment and on societies in west-central Europe.

Planet

J. Baird Callicott

Abstract Geocentric Greek astronomers called seven heavenly bodies visible to the naked eye (sun, moon, Mercury, Venus, Mars, Jupiter, and Saturn) πλανητες—wanderers—because unlike all the other αστρα, their relative positions were constantly changing. While the Romans changed the Greek names of the individual planets to those still in use, they retained the generic name, which now occurs in many European languages. In acentric modern astronomy a planet is a satellite of a star. In addition to the five actual planets known to the Greeks, Earth, Uranus, and Neptune also orbit the sun, a smallish star about half the distance from the centre of the Milky Way to its outer reaches. Earth is the only planet on which life is known for certain to exist. The search is on for concrete evidence of existing or formerly existing life on Mars (Dunbar B. et al. Perseverance Mars rover. NASA. https://www.nasa.gov/perseverance. Retrieved 06/16/2021, 2021). Technology for detecting the chemical signature of life in the atmospheres of some planets of other stars may become available in the near future (Fujii Y. et al. Astrobiology 18 (6): 739–778, 2018)—proving a point made by (Lovelock J. Gaia: A new look at life on earth. Oxford University Press, New York, 1979) that life is a whole-planet phenomenon. That is, the Earth does not just harbour or support life; it has a life of its own. The Earth is a living planet in relationship to which individual organisms are as ephemeral cells (Vernadsky VI. Am Sci 33:1–12, 1945).

In our galaxy alone, there may be as many as six billion Earth-like planets (Kunimoto & Matthews, 2020). That makes the odds pretty good that some of them are home to some forms of life (Kipping, 2020). Voyages of discovery or colonization to any such are, however, ruled out by the laws of physics. The nearest star to Earth, Proxima Centauri, is ~4.25 light years distant. At a maximum obtainable velocity of 10% of the speed of light, with time allowed for acceleration and deceleration, the trip to it would require ~100 years, several human generations, living in a huge

J. B. Callicott (✉)
University of North Texas, Denton, TX, USA
e-mail: JohnBaird.Callicott@unt.edu

© The Author(s), under exclusive license to Springer Nature
Switzerland AG 2023
N. Wallenhorst, C. Wulf (eds.), *Handbook of the Anthropocene*,
https://doi.org/10.1007/978-3-031-25910-4_56

self-sufficient vehicle, powered by an enormous amount of energy (Malik, 2002). And what if no habitable satellite orbited Proxima Centauri? Slightly more plausible fantasies of "terraforming" and colonizing Mars abound. But all such fantasies seem to bespeak a profound lack of species self-awareness, an ignorance of just what αυθρωπος is—an evolved species, and like all other extant species, exquisitely adapted to the precise environmental conditions of this planet. Adapted indeed to the precise conditions found on this planet during only the past 300,000 years—the late Pleistocene and Holocene conditions.

Which brings us to the Anthropocene, a controversial new "epoch" marked by the emergence of Homo sapiens as a force of change sufficiently powerful to alter the ecological and evolutionary trajectories of the planet. The Anthropocene Working Group of the Subcommision on Quaternary Stratigraphy recommended recognition of the Anthropocene epoch to the International Union of Geological Sciences, which so far has not done so. The Working Group also recommends a mid-twentieth-century Holocene-Anthropocene boundary, marked by globally distributed anthropogenic radionuclides, produced from 1945 to 1988 in atmospheric atomic-bomb detonations (Zalaziewicz et al., 2014). Crutzen and Stoermer (2000), who coined the name, argued for a late eighteenth-century boundary marked by increased atmospheric CO_2, caused by burning coal during the Industrial Revolution. Ruddiman (2003) argued that the Anthropocene began much earlier, some 8000 ybp, which is when anthropogenic carbonation of the atmosphere began, caused by clearing forests for settled agriculture during the Neolithic. I argue that the boundary should be older still (Callicott, 2018), marked by the wave of extinctions following the diaspora of *Homo sapiens* out of Africa, culminating with the anthropogenic extinction of at least154 species of large (> 44 kg) mammals most of which occurred 11,500 to 10,000 ybp in North and South America (Sandom et al., 2014). The temporal resolution of past chronostratigraphic boundaries is never as sharp as several decades or even several centuries. And the ossific evidence of anthropogenic extinctions became global (a requirement for a chromostratigraphical boundary) with the distribution of αυθρωπος throughout the Americas. And on the near side of the boundary, we find the relatively sudden appearance of bones of artificially selected animals. Moreover, many past chronostratigraphic boundaries are marked by mass extinction events; and that of the Holocene is still underway, spreading from large mammals to other vertebrate and invertebrate fauna and to plants (Ceballos et al., 2015).

Consider the near coincidence of the following phenomena: (i) the global spread of *Homo sapiens*; (ii) the extinction of the Pleistocene megafauna in the Western Hemisphere; (iii) the advent of the Holocene; (iv/a) the shift from foraging to settled agriculture and (iv/b) plant and animal domestication; and (v) the rise of civilizations. This makes the Anthropocene coeval with the Holocene; that is, the Anthropocene (human-recent) is a more fitting new name for the Holocene (wholly-recent).

Determining the Holocene-Anthropocene boundary is more than a chronostratigraphic issue; it is also a moral issue. If the Anthropocene is a recent successor of the Holocene, it represents a new normal, which global society must accept and to which it must adjust. "The sixth mass extinction, climate change … dude, get used to it … like we're in the Anthropocene now, man." The climate during the Holocene/Anthropocene was remarkably favourable to the human species. It enabled humans to abandon foraging for a living and develop settled agriculture, which started up almost as soon as the Holocene/Anthropocene kicked in full tilt. Settled agriculture enabled humans to live in cities. Living in cities fostered a division of labour. Accordingly, there emerged various specialized artisans, artists, commercial classes, priesthoods, and politicians. In short, we owe the existence of human civilization—with its graphic, poetic, and musical arts; its philosophies and sciences; its technologies; its polities; and its economies—to the Holocene/Anthropocene climate. This halcyon interval in the human tenure on the planet is indeed in great danger of being succeeded by another. The Tartarucene would be a good name for it, a hellish place of heat, drought, distorted polar vortices, drowning island nations and coastal cities, violent storms, floods, radical ecological dislocation, pandemics, famine, failed states, mass migration, the collapse of global civilization, population crash followed by remnant bands mercilessly ruled by warlords, a new and irreversible Dark Age. If the Holocene is (or was) the Anthropocene and the new age we are on the cusp of entering is named as it should be—the Tartarucene or the Erebucene—we might be less complacent about the need to preserve the Holocene/Anthropocene climate as a moral and existential imperative.

And as to the living planet: it's just fine. It has endured catastrophic change far greater than any $\alpha\nu\theta\rho\omega\pi o\varsigma$ might effect over the course of its 3.5 billion year lifespan, only to recover more vital and vigorous than before. And doubtless it will again if we cannot mitigate this little disruption—but no $\alpha\nu\theta\rho\omega\pi o\iota$ will be around to see that happen.

References

Callicott, J. B. (2018). Environmental ethics. In D. A. Della Sala & M. Goldstein (Eds.), *The encyclopedia of the Anthropocene* (Vol. 4, pp. 1–10). Elsevier.

Ceballos, G., Ehrlich, P. R., Barnosky, A. D., et al. (2015). Accelerated modern human-induced species losses: Entering the sixth mass extinction. *Science Advances, 1*(5), e1400253. https://doi.org/10.1126/sciadv.1400253

Crutzen, P., & Stoermer, E. (2000). The "Anthropocene". *Global Change Newsletter, 41*, 17–18.

Dunbar, B., & Greicius, T. (2021). *Perseverance Mars rover*. NASA. https://www.nasa.gov/perseverance. Retrieved 06/16/2021.

Fujii, Y., Angerhausen, D., Dietrick, R., et al. (2018). Exoplanet biosignatures: Observational prospects. *Astrobiology, 18*(6), 739–778.

Kipping, D. (2020). An objective Bayesian analysis of life's early start and our late arrival. *PANAS, 117*(22), 11995–12003.

Kunimoto, M., & Matthews, J. M. (2020). Searching the entirety of Kepler data. II. Occurrence rate estimates for FGK stars. *The Astronomical Journal, 159*(6), 248. https://doi.org/10.3847/1538-3881/ab88b0

Lovelock, J. (1979). *Gaia: A new look at life on earth*. Oxford University Press.

Malik, T. (2002). Sex and society abord the first starships. *Space*: https://web.archive.org/web/20090327233721/http://www.space.com/scienceastronomy/generalscience/star_voyage_020319-1.html. Retrieved 06/16/2021.

Ruddiman, W. F. (2003). The anthropogenic greenhouse began thousands of years ago. *Climate Change, 61*, 261–293.

Sandom, C., Faurby, S., Sandel, B., & Svenning, J.-C. (2014). Global late Quaternary extinctions linked to humans not climate change. *Proceedings of the Royal Society B, 281*(1787). https://doi.org/10.1098/rspb.2013.3254

Vernadsky, V. I. (1945). The biosphere and the noosphere. *American Scientist, 33*, 1–12.

Zalaziewicz, J., Waters, C. N., Williams, M., et al. (2014). When did the Anthropocene begin? A mid-twentieth century boundary is stratigraphically optimal. *Quaternary International, 30*, 1–8.

J. Baird Callicott is Distinguished Research Professor Emeritus and Regents Professor of Philosophy, ret., University of North Texas. Callicott is co-Editor-in-Chief of the *Encyclopedia of Environmental Ethics and Philosophy* and author or editor of many books and author of dozens of journal articles and book chapters in environmental philosophy. His writings have been translated into French, Spanish, German, Russian, Japanese, and Chinese. He is a leading contemporary exponent of Aldo Leopold's land ethic and has elaborated an Earth ethic, *Thinking Like a Planet*, in response to climate change. Most recently, he returned to his roots in classical scholarship with *Greek Natural Philosophy*.

Quaternary

Martin J. Head

Abstract The Quaternary System/Period represents the past 2.58 Ma on the Geological Time Scale, during which time distinctive biotic associations arose, the genus *Homo* diversified, and climates were strongly influenced by Northern Hemisphere glaciation. Although derived from a seventeenth century geological subdivision of northern Italy and having a long history of use, the term Quaternary was formalized only in 2009. It is defined by a Global boundary Stratotype Section and Point (GSSP) at Monte San Nicola in Sicily, Italy and is subdivided into the Pleistocene and Holocene series/epochs. All subdivisions of the Quaternary are defined using paleomagnetic reversals and/or distinctive climatic changes. The Holocene Series/Epoch began 11,700 years b2k (before 2000 CE) and extends to the present day. If the Anthropocene is officially recognised as a series/epoch, then it will terminate the Holocene in the mid-twentieth century thereby adding a third series/epoch to the Quaternary.

The Quaternary System/Period represents the past 2.58 Ma and is the final major unit of the Cenozoic Erathem/Era, superceding the Neogene System/Period. The Quaternary is currently subdivided into the Pleistocene and Holocene series/epochs. If the Anthropocene is officially recognised as a series/epoch, it will terminate the Holocene. The lower boundary of the Quaternary and those of its official subdivisions are each defined by a Global boundary Stratotype Section and Point (GSSP), approved by the International Commission on Stratigraphy and ratified by the Executive Committee of the International Union of Geological Sciences (IUGS EC), in conformity with Geological Time Scale requirements. The Anthropocene must therefore be defined by a GSSP at its base to be recognised as a formal subdivision of the Quaternary.

M. J. Head (✉)

Earth Sciences, Brock University, St. Catharines, ON, Canada

e-mail: mjhead@brocku.ca

© The Author(s), under exclusive license to Springer Nature Switzerland AG 2023

N. Wallenhorst, C. Wulf (eds.), *Handbook of the Anthropocene*, https://doi.org/10.1007/978-3-031-25910-4_57

373

Each unit on the Geological Time Scale represents time (geochronology) and the rock sediment, or other geological material formed during that time (chronostratigraphy), and is defined by the same GSSP. The geochronological rank-terms age, epoch, period, and era and their chronostratigraphic equivalents stage, series, system, and erathem reflect this duality. In simplifying the following account, which is largely summarized from Head and Gibbard (2015a) and Head (2019), only chronostratigraphic terms are cited.

Historically, the term Quaternary refers to the fourth (*Quarto*) and most recent "order" in an early (chrono-)lithostratigraphic classification of the rocks in northern Italy proposed by Veronese mining specialist Giovani Arduino in 1670 (Vaccari, 2006). However, it was Desnoyers in 1829 who coined the term *Quaternaire* or *tertiaires récens* for strata in France and in Italy that were younger than the "older Tertiary" of the Seine (Paris) Basin, although ironically he dismissed the *Quaternaire* in the same publication because of its transitional nature. de Serres in 1830 considered the *Quaternaire* synonymous with the term Diluvium, noting that humans were contemporaneous with these sediments. The Quaternary was finally established unequivocally by Henri Reboul in 1833 on the basis of its distinctive paleontological characteristics.

The Pleistocene was introduced by Charles Lyell in 1839 (Greek, *pleïstos*, most; and *kainos*, recent) as a substitute for his Newer Pliocene, but in 1863 he proposed abandoning the term because Forbes (1846) had popularized it not as originally intended but in place of Lyell's "Post-pliocene". By the time Lyell had finally conceded this new definition, in 1865, the Pleistocene was already being used widely in this way (Head, 2021). The Pleistocene meanwhile had been linked with Desnoyers' Quaternary by Lyell himself in 1839, and both terms had been associated with glaciation. The base of the Quaternary and that of the Pleistocene therefore became equated. The Quaternary is now characterized by the strong influence of Northern Hemisphere glaciation, and the genus *Homo* diversified and possibly first appeared during this time.

An increasing need for precision within the Geological Time Scale led to the realization that defining a unit by its lower boundary alone, with its upper boundary set by the base of the superjacent unit, was more effective than the traditional body stratotype concept that inevitably created gaps and overlaps at boundaries. During the 18th International Geological Congress in London in 1948, it was agreed to place the "Pliocene–Pleistocene (Tertiary–Quaternary) boundary … at the horizon of the first indication of climatic deterioration in the Italian Neogene succession." (King & Oakley, 1949). Accordingly, in 1985, the Pliocene–Pleistocene boundary was defined by a GSSP at Vrica in Italy, placed near a cooling event now dated to 1.80 Ma. However, more severe cooling lower in the record was disregarded and the issue of the Tertiary–Quaternary boundary was not addressed. Earlier cooling was subsequently acknowledged by the establishment of the Gelasian Stage, with a GSSP at Monte San Nicola, Sicily, Italy (Rio et al., 1998) and dated to 2.58 Ma. The Gelasian Stage was initially the highest stage of the Pliocene Series. When the Quaternary eventually came to be defined officially in 2009 (Gibbard & Head, 2010), it utilized the GSSP at Monte San Nicola. This required lowering the

Pliocene–Pleistocene boundary to the same level in compliance with the timescale hierarchy. The Vrica GSSP now defines only the Calabrian Stage which was ratified in 2011 (Cita et al., 2012) and forms the second stage of the Lower Pleistocene Subseries (Head et al., 2021).

The deposits at Monte San Nicola represent a slope-basin setting at a water depth of 500–1000 m. The GSSP defines the lower boundaries of the Quaternary System, Pleistocene Series, Lower Pleistocene Subseries, and Gelasian Stage (Head et al., 2021). It is placed at the base of a marly layer conformably overlying the sapropelic Nicola bed and has an astronomically calibrated age of 2.58 Ma. The Nicola bed corresponds to Mediterranean Precession-Related Cycle (MPRC) 250 and to Marine Isotope Stage (MIS) 103 based on correlation with the well-studied Singa section in Calabria, Italy. The Gauss–Matuyama paleomagnetic reversal is accepted as occurring in MIS 103 (Head, 2019). It is not precisely located at Monte San Nicola but a sedimentary succession at the Gardar Drift, North Atlantic provides an astronomically tuned directional midpoint age of 2.587 ± ≥5 Ma with a duration of 5 kyr. Although significant intensification of Northern Hemisphere glaciation occurred at 2.7 Ma, MIS 104 at ~2.6 Ma is a pronounced glacial cycle representing a climatic reorganization across the middle and higher latitudes of the Northern Hemisphere (Hennissen et al., 2014). Important climatic signals at ~2.6 Ma and immediately after, along with the Gauss–Matuyama paleomagnetic reversal, therefore aid global recognition of the base of the Quaternary.

Silty marls at Vrica, Calabria were deposited at a water depth exceeding 500 m. The GSSP defining the Calabrian Stage is placed at the base of the marl bed conformably overlying sapropelic bed 'e' which is assigned to MPRS 176 (Lourens et al., 1996). The GSSP has an astronomically calibrated age of 1.80 Ma (Cita et al., 2012) and coincides with the transition from MIS 65 to 64. It occurs ~8m below the observed top of the Olduvai Subchron, with diagenetic overprinting preventing a more precise placement of this polarity reversal (Roberts et al., 2010). An important climatic transition occurs near the base of the Calabrian, including intense cooling at ~2.06 Ma (MIS 78) in the Mediterranean followed by sustained lowering of sea-surface temperatures during both warm and cool climate cycles. Cooling at ~1.8 Ma in Siberia and elsewhere suggests an expansion of Eurasian glaciation from this time (Herbert et al., 2015) and low winter temperatures may account for the immigration of cold-adapted species into the Mediterranean Sea (Crippa et al., 2016).

The GSSP defining the Middle Pleistocene Subseries and Chibanian Stage, ratified in 2020, is placed within the Chiba section, Japan (Suganuma et al., 2021). The Chiba section represents muddy deposition along an open-ocean continental slope setting on the Pacific margin of Japan at a water depth in excess of 800–100 m. The GSSP loosely aproximates the midpoint of the Early–Middle Pleistocene Transition, a fundamental shift of the Earth's climate state between ~1.4 and 0.4 ka characterized by a progressive increase in the amplitude of climate oscillations, an evolving waveform, and a shift towards a quasi-100 ky frequency (Head & Gibbard, 2015b). The GSSP has an astronomically tuned age of age of 774.1 ka and is placed at the base of a regional lithostratigraphic marker, the Ontake-Byakubi-E (Byk-E) tephra bed. The primary guide to this boundary is the Matuyama–Brunhes paleomagnetic reversal which in the Chiba section has a directional midpoint 1.1 m above the GSSP. The Chiba composite section gives an astronomically dated directional midpoint of 772.0 ± 5.4 ka with a duration of up to 2 kyr for this reversal. The GSSP occurs immediately below the MIS 19c/b boundary which represents the inception of glaciation within this interglacial cycle. The base of the Chibanian Stage is therefore recognised globally by a paleomagnetic and numerous climatostratigraphic signals (Suganuma et al., 2021; Head, 2021).

The Chibanian is currently the only stage of the Middle Pleistocene Subseries although a second stage has been suggested with a lower boundary at ~430 ka coinciding with the onset of the "Mid-Brunhes Event". This event marks an increase in the amplitude of quasi-100 kyr glacial–interglacial cycles and represents a step change in Quaternary climate (Head, 2021).

The Upper Pleistocene Subseries, although ratified as a term with a provisional base at ~129 ka, has yet to be officially defined. Its lower boundary has long been equated with that of the Last Interglacial which approximates MIS 5e (Head, 2019; Head et al., 2021).

The term *holocène* (Greek *holos* and *kainos* meaning "entirely recent") was introduced by Paul Gervais in 1867–69 for the warm interval that follows the last glacial episode, effectively replacing Lyell's (1839) "Recent" and Forbes' (1846) "Postglacial". The Holocene Series was ratified by the IUGS EC in 2008 with a GSSP at a depth of 1492.45 m in the NGRIP2 Greenland ice core, and dated at 11,700 yr b2k (before 2000 CE) using annual layer counting with a maximum counting error of 99 yr (Walker et al., 2009). The GSSP represents rapid climatic warming at the end of the Younger Dryas/Greenland Stadial 1 cold phase, but is marked by an abrupt decline in deuterium excess values as the primary guide and which corresponds to an ocean surface temperature drop of 2–4 °C. This cooling seems to reflect a rapid (1–3 year) shift in the source of precipitation, as ice-free waters of the mid-Atlantic Ocean moved northwards in response to warming. Additional signals at the base of the Holocene include a short-term shift to higher $\delta^{18}O$ values.

The Holocene is subdivided into the Greenlandian, Northgrippian and Meghalayan stages and their corresponding subseries (Walker et al., 2018), with ratification by IUGS EC effected in 2018. The Greenlandian Stage and Lower Holocene Subseries are defined at their base by the GSSP in the NGRIP2 which also defines that of the Holocene Series (see above).

The Northgrippian Stage and Middle Holocene Subseries are defined by a GSSP in the NGRIP1 Greenland ice core at a depth of 1228.67 m, and dated at 8236 ± 47 yr b2k using annual ice-layer counting from the DYE-3 ice core. The GSSP marks a brief cooling episode known as the "8.2 ka climatic event" widely considered to reflect a disruption of the thermohaline circulation triggered by catastrophic release of meltwater from the Laurentide Ice Sheet into the North Atlantic. Abrupt cooling is registered in the NGRIP1 core by a conspicuous shift to more negative $\delta^{18}O$ and δD values and a decline in ice-core annual layer thickness. A double acidity peak at the GSSP allows it to be recognized precisely within the Greenland ice core record. The 8.2 ka event is near-global in extent and expressed in an array of climate proxies.

The Meghalayan Stage and Upper Holocene Subseries are defined by a GSSP located at a depth of 7.45 mm in speleothem KM-A from the Mawmluh Cave, State of Meghalaya, India (Walker et al., 2018; Head, 2019). The GSSP has a modelled age of 4250 ± 30 yr b2k, based on U/Th dating, and occurs at the approximate mid-point between two progressive shifts towards less negative $\delta^{18}O$ values. These shifts represent reductions in precipitation owing to a weakening of the monsoon across India and southeast Asia. The primary guide to this boundary is the "4.2 ka event" which lasted for two or three centuries and is expressed in many low- and mid-latitude regions as an aridification episode.

If the Anthropocene is ratified at the rank of series/epoch, with its base and that of its corresponding stage defined by a GSSP dated to the mid-twentieth century, as is currently preferred by the Anthropocene Working Group (Zalasiewicz et al., 2017; Head, 2019), the upper boundaries of the Holocene Series and Meghalayan Stage will be set by this GSSP.

References

Cita, M. B., Gibbard, P. L., Head, M. J., & The Subcommission on Quaternary Stratigraphy. (2012). Formal ratification of the base Calabrian Stage GSSP (Pleistocene Series, Quaternary System). *Episodes, 35*(3), 388–397.

Crippa, G., Angiolini, L., Bottini, C., Erba, E., Felletti, F., Frigerio, C., Hennissen, J. A. I., Leng, M. J., Petrizzo, M. R., Raffi, I., Raineri, G., & Stephenson, M. H. (2016). Seasonality fluctuations recorded in fossil bivalves during the early Pleistocene: Implications for climate change. *Palaeogeography, Palaeoclimatology, Palaeoecology, 446*, 234–251.

Forbes, E. (1846). On the connexion between the distribution of the existing fauna and flora of the British Isles, and the geological changes which have affected their area, especially during the epoch of the Northern Drift. *Memoirs of the Geological Survey of Great Britain, 1*, 336–432.

Gibbard, P. L., & Head, M. J. (2010). The newly-ratified definition of the Quaternary System/Period and redefinition of the Pleistocene Series/Epoch, and comparison of proposals advanced prior to formal ratification. *Episodes, 33*, 152–158.

Head, M. J. (2019). Formal subdivision of the Quaternary System/Period: Present status and future directions. *Quaternary International, 500*, 32–51.

Head, M. J. (2021). Review of the Early–Middle Pleistocene boundary and Marine Isotope Stage 19. *Progress in Earth and Planetary Science, 8*, 50.

Head, M. J., & Gibbard, P. L. (2015a). Formal subdivision of the Quaternary System/Period: Past, present, and future. *Quaternary International, 383*, 4–35.

Head, M. J., & Gibbard, P. L. (2015b). Early–Middle Pleistocene transitions: Linking terrestrial and marine realms. *Quaternary International, 389*, 7–46.

Head, M. J., Pillans, B., Zalasiewicz, J. A., and the ICS Subcommission on Quaternary Stratigraphy (2021). Formal ratification of subseries/subepochs for the Pleistocene Series/Epoch of the Quaternary System/Period. *Episodes 44*(3), 241–247. https://doi.org/10.18814/epiiugs/2020/020066

Hennissen, J. A. I., Head, M. J., De Schepper, S., & Groeneveld, J. (2014). Palynological evidence for a southward shift of the North Atlantic Current at ~2.6 Ma during the intensification of late Cenozoic Northern Hemisphere glaciation. *Paleoceanography, 28*. https://doi.org/10.1002/2013PA002543

Herbert, T. D., Ng, G., & Peterson, L. C. (2015). Evolution of Mediterranean sea surface temperatures 3.5–1.5 Ma: Regional and hemispheric influences. *Earth and Planetary Science Letters, 409*, 307–318.

King, W. B. R., & Oakley, K. P. (1949). Report of the temporary commission on the Plio-Pleistocene boundary, appointed 16th August 1948. In A. J. Butler (Ed.), *IGC, 18th session, Great Britain, 1948. Part I. General proceedings* (pp. 213–228). The Geological Society.

Lourens, L. J., Hilgen, F. J., Raffi, I., & Vergnaud-Grazzini, C. (1996). Early Pleistocene chronology of the Vrica section (Calabria, Italy). *Paleoceanography, 11*(6), 797–812.

Lyell, C. (1839). *Nouveaux éléments de géologie*. Pitois-Levrault et Cie, 648 pp.

Rio, D., Sprovieri, R., Castradori, D., & Di Stefano, E. (1998). The Gelasian Stage (Upper Pliocene): A new unit of the global standard chronostratigraphic scale. *Episodes, 21*, 82–87.

Roberts, A. P., Florindo, F., Larrasoaña, J. C., O'Regan, M. A., & Zhao, X. (2010). Complex polarity pattern at the former Plio–Pleistocene global stratotype section at Vrica (Italy): Remagnetization by magnetic iron sulphides. *Earth and Planetary Science Letters, 292*, 98–111.

Suganuma, Y., Okada, M., Head, M. J., Kameo, K., Haneda, Y., Hayashi, H., Irizuki, T., Itaki, T., Izumi, K., Kubota, Y., Nakazato, H., Nishida, N., Okuda, M., Satoguchi, Y., Simon, Q., & Takeshita, Y. (2021). Formal ratification of the Global Boundary Stratotype Section and Point (GSSP) for the Chibanian Stage and Middle Pleistocene Subseries of the Quaternary System: The Chiba Section, Japan. *Episodes 44*(3), 317–347. https://doi.org/10.18814/epiiugs/2020/020080

Vaccari, E. (2006). The "classification" of mountains in eighteenth century Italy and the lithostratigraphic theory of Giovanni Arduino (1714–1795). *Geological Society of America Special Paper, 411*, 157–177.

Walker, M., Johnsen, S., Rasmussen, S. O., Steffensen, J. P., Popp, T., Gibbard, P., Hoek, W., Lowe, J., Bjorck, S., Cwynar, L., Hughen, K., Kershaw, P., Kromer, B., Litt, T., Lowe, D. J., Nakagawa, T., Newnham, R., & Schwander, J. (2009). Formal definition and dating of the GSSP (Global Stratotype Section and Point) for the base of the Holocene using the Greenland NGRIP ice core and selected auxiliary records. *Journal of Quaternary Science, 24*, 3–17.

Walker, M., Head, M. J., Berkelhammer, M., Björck, S., Cheng, H., Cwynar, L., Fisher, D., Gkinis, V., Long, A., Lowe, J., Newnham, R., Rasmussen, S. O., & Weiss, H. (2018). Formal ratification of the subdivision of the Holocene Series/Epoch (Quaternary System/Period): Two new Global Boundary Stratotype Sections and Points (GSSPs) and three new stages/subseries. *Episodes, 41*(4), 213–223.

Zalasiewicz, J., Waters, C. N., Summerhayes, C. P., Wolfe, A. P., Barnosky, A. D., Cearreta, A., Ellis, E., Fairchild, I., Gałuszka, A., Haff, P., Hajdas, I., Head, M. J., Jeandel, C., Leinfelder, R., McNeill, J., Neal, C., Steffen, W., Syvitski, J., Wagreich, M., & Williams, M. (2017). The working group on the Anthropocene: Summary of evidence and interim recommendations. *The Anthropocene, 19*, 55–60.

Martin J. Head is a Professor of Earth Sciences at Brock University and currently vice-chair of the International Subcommission on Quaternary Stratigraphy (SQS), having served as its chair from 2012 to 2020. He was co-convener of the SQS Working Group on the Lower–Middle Pleistocene Subseries Boundary (2010–2020) and is co-convener of its Working Group on the Middle–Upper Pleistocene Subseries Boundary. He is a voting member of the SQS Anthropocene Working Group and of the International Subcommission on Stratigraphic Classification, and is a member of the INQUA Commission on Stratigraphy and Chronology.

Stratigraphy

Colin N. Waters, Jan Zalasiewicz, and Simon Turner

Abstract The Anthropocene is proposed as a stratigraphic unit of geological time commencing in the mid-twentieth century. Stratigraphy, though rooted in the analysis of ancient rocks, also provides diverse methodologies to investigate geological and archaeological archives in order to determine the relative timing, scale and diversity of human impact on the planet. Our footprint as regards biodiversity, climate change, sea-level rise, pollution, and the anthropogenic deposits that are the residues of our existence on the planet, are encompassed by stratigraphy. This provides a means of establishing how parameters, such as changing temperatures or atmospheric chemical composition, or documented historical events such as nuclear bomb-testing, are preserved immediately in geological archives, or show delay and dissipation before their ultimate registration in strata.

Stratigraphy is the approach by which geologists comprehend the relative ages of rocks, fossils and the processes that formed them over 4.6 billion years of Earth history. The science developed over the last three centuries, constructing a planetary history initially blind to the true age of the rocks being studied. Through detailed regional and global correlations it became possible to establish the relative order in which strata accumulated, magma formed igneous rocks and mountains grew by deformation of the Earth's crust. The wonder of stratigraphy is that it has enabled profound understanding of how Earth has evolved in times long before humans first walked its surface. Though the methodologies involved in stratigraphy were developed on deep-time rock records, they can be applied to the analysis of recent Earth history, spanning the time of human evolution. Thus stratigraphy provides an

C. N. Waters (✉) · J. Zalasiewicz
Geography, Geology and the Environment, University of Leicester, Leicester, UK
e-mail: cw398@leicester.ac.uk; jaz1@leicester.ac.uk

S. Turner
Geography, University College London, London, UK
e-mail: simon.turner@ucl.ac.uk

© The Author(s), under exclusive license to Springer Nature
Switzerland AG 2023
N. Wallenhorst, C. Wulf (eds.), *Handbook of the Anthropocene*,
https://doi.org/10.1007/978-3-031-25910-4_58

invaluable record that, together with archaeology and historical archives, gives a means of charting the scale of human impact over time.

To aid the process of correlation—matching up events in time across space—in geology, numerous stratigraphical approaches have developed (Rawson et al., 2002). These tools, including geochronology, chrono-, bio-, climato-, sequence-, chemo- and litho-stratigraphy help illuminate our profound influence on the workings of this planet as an integrated Earth System, and provide the evidence base for definition of the Anthropocene as a geological time unit.

The concept of a stratigraphic Anthropocene requires establishment of a formal time framework via geochronology (the subdivision of geological time and process) and complementary chronostratigraphy, which relates time to the rock record. Geological time units have specified, globally synchronous starts and ends (the end of one unit being defined by the start of the next), and are characterized by imprints of physical, chemical and biological properties during that interval of planetary evolution. It is via this process that the Anthropocene is being investigated currently as a potential epoch (of geochronology) and series (of chronostratigraphy) (Waters et al., 2016; Zalasiewicz et al., 2020).

Direct measurement of geological time relies heavily on radiometric dating methods based on known decay rates of unstable isotopes present in minerals or organic materials. For example, radiocarbon (^{14}C) dating, widely used in archaeology to date human occupation sites, has a useable range from 60,000–200 years. The stratigraphic Anthropocene is proposed to have commenced in the mid-twentieth century in response to a sudden 'Great Acceleration' in human population, energy consumption and economic development (Zalasiewicz et al., 2015; Syvitski et al., 2020). The Anthropocene is therefore too young for standard radiocarbon dating to be effective, but global perturbations of ^{14}C by atmospheric nuclear weapons testing is a key stratigraphic marker. With a useful dating range of 150 years, the most suitable radiometric technique for the Anthropocene is ^{210}Pb dating. When used in association with synthetic radionuclides and counting of seasonal layers formed in some environments, chronologies are produced of unprecedented precision in the geological record (Waters et al., 2018a). This permits opportunities to test the extent to which documented anthropogenic impacts, such as pollution events, are faithfully and permanently preserved in geological archives.

Formalization of the Anthropocene as a chronostratigraphic series requires recognition of a single reference horizon within a physical section, known as a Global boundary Stratotype Section and Point or 'golden spike' (Waters et al., 2018b). This will define the base of the Anthropocene by reference to a primary marker, in practice supported by a range of other signals. These markers will then be correlated globally to allow worldwide recognition of Anthropocene strata in diverse environments. Analysis of potential GSSPs by the Anthropocene Working Group of the Subcommission on Quaternary Stratigraphy is ongoing, with sections in marine, estuarine and lake sediments, peat, corals, glacial ice and speleothems being considered (Waters et al., 2023).

Biostratigraphy, the use of fossils as a means of correlation, provides primary markers for most chronostratigraphic units. The Anthropocene is a time of increased species extinction rates and the planet is now considered to be on a trajectory towards the sixth mass extinction event (when >75% of species are lost) of the last half billion years (Barnosky et al., 2011). The current early phase is associated with dramatically reduced wildlife populations due to human predation, habitat destruction and climate change; a counterpart is increased populations of domesticated species (Williams et al., 2018). Most dramatic are the translocations of non-native species and resulting unprecedented homogenization of biota across the planet through human intervention, both accidental and intentional. The rapid appearance/disappearance of species provides a tool for correlating sediments regionally and a means of checking the observations from the biological sciences on the timing of these ecological transformations.

Human production of novel artefacts, increasingly diverse in form and composition, can be treated in terms of technofossils, permitting even greater scope as a tool for precise correlation of strata (Zalasiewicz et al., 2014). Humans have created more than 200,000 new mineral-like compounds, many since 1950, concrete has become the most abundant artificial rock and since the mid-twentieth century fly ash particles and novel plastic polymers have become ubiquitous in sediments worldwide (Zalasiewicz et al., 2019a; Zalasiewicz et al., 2019b).

For geological time units over the last 2.6 million years (the Quaternary Period), correlations have largely relied upon climatostratigraphy. Proxy signals of repeated, often abrupt climate change in Quaternary strata reflect more than a hundred switches between cold glacial and warmer interglacial intervals; the latest of these, starting 11,700 years ago, is defined as the Holocene Epoch. The Anthropocene represents a time when a sudden trend of rapid global warming (1 °C since 1975), mainly caused by anthropogenic emissions of greenhouse gases through burning fossil fuels, contrasts with the preceding 7000 years of broadly stable climate (Summerhayes & Zalasiewicz, 2018). The resulting rise in sea level, through melting ice sheets and ocean thermal expansion, is still incipient (0.3 m over 160 years) but is accelerating (now ~4 mm/year). There is an inherent lag in sea level rise relative to elevated temperatures, and sea levels may rise for thousands of years after peak temperatures; the analogue of a thermal event 3 million years ago suggests Anthropocene sea levels may be on a trend towards a 10–20 m rise. Sea level rise may be analyzed in terms of sequence stratigraphy, which deals with the large-scale, three-dimensional arrangement of sedimentary strata, and will be most apparent on almost flat coastal areas. For major coastal deltas, sea level rise compounds the effect of subsidence caused by sediment starvation through upstream damming of major rivers, extraction of water and hydrocarbon resources from deposits, and simply by the weight of large urban centres. Combined, these factors will result in inundation, likely this century, and the stratigraphy of many coastal megacities (e.g. Dhaka, Jakarta and Shanghai) being incorporated into shallow marine sedimentary successions.

A growing trend in stratigraphy is the use of chemical patterns, especially isotopic ratios of particular elements, to identify stratal units. Such chemostratigraphy provides robust means of correlating 'hidden' relationships associated with large-scale changes, e.g. in climate, rates of weathering and global biomass. This is true too for the Anthropocene, where novel chemostratigraphic signals reflecting human pollution of the environment are both diverse and extensive (Zalasiewicz et al., 2019a). Stable carbon isotope ratios across the Earth's surface are being radically transformed by burning fossil fuels that are enriched in the lighter ^{12}C isotope. Similarly, stable nitrogen isotope ratios have been markedly perturbed, especially from increased use of artificial nitrogen-based agricultural fertilizers since the 1950s. Industrial emissions of trace metals and novel persistent organic pollutants provide further means of regional correlation. However, it is the radiogenic isotopes that form a 'bomb-spike' produced by the fallout from atmospheric detonations of thermonuclear devices in the 1950s–1960s, that are the most distinctive and globally synchronous signal in modern strata (Waters et al., 2015), marking the transition to this new epoch.

Lithostratigraphy deals with the description, definition and naming of bodies of stratified deposits in which the principle of superposition applies (i.e. that younger strata overlie older ones). The basic unit, the formation, represents strata of consistent lithological characteristics and stratigraphic position. It is such units that appear on most geological maps, providing information relevant to mineral resource extraction or land use planning. In the applied geosciences, knowing the composition of strata is more important than knowing their age. Lithostratigraphic units extend over local to regional scales and their boundaries are placed at lithological changes which may be of different ages in different places, reflecting environments that change through time and space. Modern 'natural' successions such as alluvium and peat are included in lithostratigraphic schemes, but it is only in recent times that geologists have started to consider anthropogenic (artificial) deposits in the same way. Prior to this, such deposits along with cultivated soils were purely of interest to archaeologists, who used their own stratigraphic techniques to analyze such human-generated accumulations. Such deposits may contain multiple internal surfaces representing complex cuts and reworking of the deposits, or prolonged gaps in accumulation evident as buried soil horizons. Successive appearances of distinctive artefacts and changes in types of plant and animal remains, and physico-chemical signals, also help archaeologists to understand how humans interacted with a site over time (Edgeworth et al., 2015). The basal boundary of such anthropogenic deposits can be markedly diachronous (time-transgressive) even within small areas. The composition of these deposits can be complex and heterogeneous on a local scale (typically more so than the variability in natural strata). Archaeologists cope with the complexity by detailed analysis of sequential excavations over small sites. Globally, artificial deposits of similar origins, e.g. landfill sites, industrial complexes, may differ widely in composition and age, reflecting social trends that can be diachronous around the world.

Geologists involved with mapping such inherently heterogeneous anthropogenic deposits have developed a morphogenetic classification approach; this permits large areas of artificial deposits and excavations to be mapped through recognition and interpretation of landscape features, often combined with searches of historical maps (Ford et al., 2014; Waters, 2018). This permits city-wide quantification of the mass of artificial deposits and allows separation of Anthropocene from Holocene deposits based on the kind of debris they contain (Terrington et al., 2018). The Anthropocene has also witnessed increased human penetration into the Earth's sub-surface through extraction of mineral resources or using the rock mass itself as a resource to live in, or for transport or waste storage (Waters et al., 2018c). Such deep structures may leave the longest record of our occupation of this planet, being pro-tected from the erosion that will recycle and redeposit many of the surface manifes-tations of the physically, chemically and biologically unique strata that we have made.

References

Barnosky, A. D., Matzke, N., Tomiya, S., et al. (2011). Has the Earth's sixth mass extinction already arrived? *Nature, 471*, 51–57. https://doi.org/10.1038/nature09678

Edgeworth, M., Richter, D. deB., Waters, C. N., et al. (2015). Diachronous beginnings of the Anthropocene: The lower bounding surface of anthropogenic deposits. *Anthropocene Review, 2*(1), 33–58. https://doi.org/10.1177/2053019614565394

Ford, J. R., Price, S. J., Cooper, A. H., & Waters, C. N. (2014). An assessment of lithostratigraphy for anthropogenic deposits. In C. N. Waters, J. A. Zalasiewicz, M. Williams, et al. (Eds.), *A Stratigraphical basis for the Anthropocene* (Vol. 395, pp. 55–89). Geological Society, London, Special Publication. https://doi.org/10.1144/SP395.12

Rawson, P. F., Allen, P. M., Brenchley, P. J., et al. (2002). *Stratigraphical Procedure. Geological Society Professional Handbook* (57 pp). The Geological Society, London.

Summerhayes, C. P., & Zalasiewicz, J. (2018). Global warming and the Anthropocene. *Geology Today, 34*(5), 194–200. https://doi.org/10.1111/gto.12247

Syvitski, J., Waters, C. N., Day, J., et al. (2020). Extraordinary human energy consumption and resul-tant geological impacts beginning around 1950 CE initiated the proposed Anthropocene Epoch. *Communications Earth & Environment, 1*, 32. https://doi.org/10.1038/s43247-020-00029-y

Terrington, R. L., Silva, É. C. N., Waters, C. N., et al. (2018). Quantifying anthropogenic modifica-tion of the shallow geosphere in central London, UK. *Geomorphology, 319*, 15–34. https://doi.org/10.1016/j.geomorph.2018.07.005

Waters, C. N. (2018). Artificial ground. In P. T. Bobrowsky & B. Marker (Eds.), *Encyclopedia of Engineering Geology* (pp. 30–44). Springer International Publishing. https://doi.org/10.1007/978-3-319-12127-7_21-1

Waters, C. N., Syvitski, J. P. M., Gałuszka, A., et al. (2015). Can nuclear weapons fallout mark the beginning of the Anthropocene Epoch? *Bulletin of the Atomic Scientists, 71*(3), 46–57. https://doi.org/10.1177/0096340215581357

Waters, C. N., Zalasiewicz, J., Summerhayes, C., et al. (2016). The Anthropocene is function-ally and stratigraphically distinct from the Holocene. *Science, 351*(6269), 137. https://doi.org/10.1126/science.aad2622

Waters, C. N., Fairchild, I. J., McCarthy, F. M. G., et al. (2018a). How to date natural archives of the Anthropocene. *Geology Today, 34*(5), 182–187. https://doi.org/10.1111/gto.12245

Waters, C. N., Zalasiewicz, J., Summerhayes, C., et al. (2018b). Global Boundary Stratotype Section and Point (GSSP) for the Anthropocene Series: Where and how to look for a potential candi-date. *Earth-Science Reviews, 178*, 379–429. https://doi.org/10.1016/j.earscirev.2017.12.016

Waters, C. N., Graham, C., Tapete, D., et al. (2018c). Recognising anthropogenic modification of the subsurface in the geological record. *Quarterly Journal of Engineering Geology and Hydrogeology, 52*(1), 83–98. https://doi.org/10.1144/qjegh2017-007

Waters, C. N., Turner, S. D., Zalasiewicz, J. & Head, M. J. (2023). Candidate sites and other reference sections for the Global boundary Stratotype Section and Point of the Anthropocene series. *Anthropocene Review*. Epub ahead of print 7 February 2023. https://doi.org/10.1177/20530196221136422

Williams, M., Zalasiewicz, J., Waters, C., et al. (2018). The palaeontological record of the Anthropocene. *Geology Today, 34*(5), 188–193. https://doi.org/10.1111/gto.12246

Zalasiewicz, J., Williams, M., Waters, C. N., et al. (2014). The technofossil record of humans. *Anthropocene Review, 1*(1), 34–43. https://doi.org/10.1177/2053019613514953

Zalasiewicz, J., Waters, C. N., Williams, M., et al. (2015). When did the Anthropocene begin? A mid-twentieth century boundary level is stratigraphically optimal. *Quaternary International, 383*, 196–203. https://doi.org/10.1016/j.quaint.2014.11.045

Zalasiewicz, J., Waters, C. N., Williams, M., & Summerhayes, C. (Eds.). (2019a). *The Anthropocene as a Geological Time Unit: A Guide to the Scientific Evidence and Current Debate* (361 pp). Cambridge University Press.

Zalasiewicz, J., Gabbott, S. E., & Waters, C. N. (2019b). Chapter 23: Plastic waste: How plastic has become part of the Earth's geological cycle. In T. M. Letcher & D. A. Vallero (Eds.), *Waste: A handbook for management* (2nd ed., pp. 443–452). Elsevier. ISBN: 9780128150603.

Zalasiewicz, J., Waters, C., & Williams, M. (2020). Chapter 31: The Anthropocene. In F. Gradstein, J. Ogg, M. Schmitz, & G. Ogg (Eds.), *A Geologic Time Scale 2020* (Vol. 2, pp. 1257–1280). Elsevier BV. ISBN: 978-0-12-824363-3.

Colin Waters is an honorary professor in the Department of Geography, Geology and the Environment at the University of Leicester, UK. He has been the chair of the Anthropocene Working Group since 2020, following 9 years as its secretary. He has a central role in coordinating activities of the Working Group members and searching for the "golden-spike" section as part of the formalisation process for recognising the Anthropocene as a new geological time unit. He also has specific interests in characterising the nature and scale of human modification of the landscape, particularly through the accumulation of novel materials (e.g. plastic and concrete) and artificial deposits. He retired in 2017 as a Principal Mapping Geologist at the British Geological Survey, where over nearly 30 years' service he specialised in geological mapping of the UK, Morocco and Mauritania and is also an expert on Carboniferous stratigraphy.

Jan Zalasiewicz is a geologist (now retired) at the University of Leicester, and a member (formerly Chair) of the Anthropocene Working Group. He is co-editor of *The Anthropocene as a Geological Time Unit: a Guide to the Scientific Evidence and Current Debate*, has been involved in other AWG publications, and with Julia Adeney Thomas and Mark Williams has co-written *The Anthropocene: A Multidisciplinary Approach*. His books for a general audience that *inter alia* explore Anthropocene concepts include *The Earth after Us* and (with Mark Williams) *The Goldilocks Planet* and *Ocean Worlds*.

Simon Turner is a Senior Research Fellow in Geography at University College London and Secretary of the Anthropocene Working Group (AWG). His research background is in paleoenvironmental analysis, geochemistry and aquatic monitoring that record environmental change in lakes and wetlands. He has worked in multiple locations around the world in aquatic and coastal systems extracting sediments from the Pleistocene to the Present that contain records of past disturbance, contamination and pollution. His most recent work has been serving as the scientific coordinator for the AWG and Haus der Kulturen der Welt collaborative project to define a global stratotype section and point (GSSP) for the Anthropocene and explore its cultural and historical implications.

Part VI
Complexity as a Paradigm for Understanding Reality

Co-existence

Jozef Keulartz

Abstract Co-existence is a major theme in the philosophical discourse on the Anthropocene. The best known movements within Anthropocene thinking are ecomodernism and posthumanism. Both movements distance themselves from the idea that nature and culture are strictly separate domains. In the Anthropocene, this separation no longer appears tenable; nature and culture have become inextricably entangled. The world consists exclusively of hybrids, compositions of both human and non-human entities. While the ecomodernists seem to aim for total dominance of human over non-human actors, posthumanists also want to grant non-human actors the right to representation in the public arena, in what Bruno Latour has called 'the Parliament of Things'. Ecomodernists and posthumanists are, however, of one mind in their criticism of the traditional nature movement, which believes it can return to a past when nature and culture were still separated.

Ecomodernism

The concept of the Anthropocene has sparked a fierce debate about the relevance of traditional nature policy and nature management. This debate erupted in 2004 in response to the essay *The Death of Environmentalism* by Ted Nordhaus and Michael Shellenberger, founders of the Breakthrough Institute. In this essay they express their conviction "that modern environmentalism, with all of its unexamined assumptions, outdated concepts and exhausted strategies, must die so that something new can live" (p. 10). The movement they initiated is known as 'ecomodernism'.

Ecomodernists distance themselves from the doom and gloom of the old nature movement, which sees the Anthropocene exclusively as a potential ecological disaster. Instead, they welcome this 'age of man' as a new step in the progress of humanity. In their view, humans are not a pest species, as traditional conservationists seem

J. Keulartz (✉)
Radboud University Nijmegen, Nijmegen, The Netherlands
e-mail: jozef.keulartz@wur.nl

N. Wallenhorst, C. Wulf (eds.), *Handbook of the Anthropocene*,
https://doi.org/10.1007/978-3-031-25910-4_59

to believe, but a 'God species' that, thanks to its infinite creativity and ingenuity, will have a magnificent green future. Ecomodernists have replaced the techno-phobia of the traditional nature movement by an outspoken techno-triumphalism.

Ecomodernists are quite optimistic; they believe that humanity can achieve a 'good Anthropocene' if only it puts its scientific-technological skills at the service of responsible planetary stewardship. With the 'inevitable domestication' of the earth, ecomodernists believe we must shift our focus from protecting biodiversity to promoting ecosystem services that are vital to the health and well-being of the population.

It is clear that ecomodernists, with regard to the co-existence of human and non-human actors, are firmly relying on the power of the Anthropos: there is not a single piece of nature left that does not show the finger and footprint of man, and if it were up to the ecomodernists, that imprint would only become more striking in the future. This humanization of nature will increasingly erase the individuality and otherness of all non-human life on earth.

Posthumanism

Posthumanism appears to counter ecomodernist progress optimism. It aims to make a radical break with the anthropocentric worldview, in which humans take centre stage; instead, it views humans as part of a complex and all-encompassing whole made up of both human and non-human entities.

Posthumanists agree with ecomodernists that nature – as a completely independent and autonomous domain – is 'dead': nature and culture are inseparably entangled. The world consists solely of hybrids, of what Latour, who was a senior fellow at the Breakthrough Institute in 2010, also calls 'naturecultures', assemblages of both human and non-human entities: the hole in the ozone layer, genetically modified crops, irrigation systems, the coronavirus, etc. Posthumanists therefore share the ecomodernists' criticism of the traditional nature movement that wants to return to a past when there was such a thing as 'pristine nature' – an outdated idea in their view.

But when it comes to dealing with hybrids, posthumanists hold a very different view from ecomodernists. While ecomodernists seem to aim for total dominance of human over non-human actors, posthumanists also want to grant non-human actors the right to representation in the public arena; they deserve a seat in what Latour has called 'the Parliament of Things'. This influential concept was introduced in *We have never been Modern* (1991), and further elaborated in the *Politics of Nature* (1999).

The first major pilot project experimenting with Latour's concept was a simulation of the twenty-first climate conference (COP21) that took place in Paris in December 2015, where the famous climate agreement was adopted. Over 200 students from universities around the world formed 41 delegations, representing not only human entities, such as national governments, NGOs and indigenous peoples, but also non-human entities, such as the ocean, the atmosphere, the Sahara, the Amazon and endangered species. Each delegation consisted of five people: an administrator, a scientist, an economist, a politician, and finally someone free to choose.

Latour initially placed a strong emphasis on the role of scientific expertise in interpreting the interests of non-human entities; this is reflected in the subtitle of his *Politics of Nature*: 'How to bring the sciences into democracy'. But in *Facing Gaia* (2017), he acknowledges that in exploring 'fictions' such as the Parliament of Things, there is also a role for "plays, exhibitions, art forms, poetry, and perhaps also rituals" (p. 283).

Co-existence or Separation

As we have previously seen, ecomodernists and posthumanists agree on two points. They make short work of the traditional dualism of culture and nature and only recognize the existence of hybrids. And they deny that it would be possible to return to a past in which human and non-human actors would exist separately. Nature conservation and nature restoration are therefore considered by both movements as obsolete activities. Nature conservation is impossible without human intervention; after all, it asks, according to Latour (2008), for "our constant care, our undivided attention, our costly instruments, our hundreds of thousands of scientists, our huge institutions, our careful funding."

Here we come up against what is known as the 'nature conservation paradox': if nature is seen as being completely independent of human intervention and control, then nature conservation and nature restoration are by definition impossible. Nature managers are often faced with a dilemma: they can never do it right – they are condemned if they do not intervene (because this would cause the loss of important natural values) but also if they do intervene (because this would diminish naturalness).

Not surprisingly, there has been considerable criticism of this post-environmental vision, in which self-appointed representatives of non-human entities collectively determine the fate of nature. According to some critics, denying the possibility of any disentanglement between humans and non-humans will result in the non-humans beings severely curtailed in their freedom to shape their own lives autonomously. As an alternative to what might be called an 'ecology of co-existence', these critics have argued for an 'ecology of separation', as the subtitle of a 2019 book by French philosopher Frédéric Neyrat reads. Another French philosopher, Virginie Maris, has pointed out that such an ecology is badly needed to protect wildlife from ongoing exploitation and destruction. In her 2018 book *La part Sauvage du Monde – Penser la Nature dans l'Anthropocène*, Maris states that the distinction between the natural and the artificial should not be understood as a dichotomy, but as a broad continuum of hybrid intermediate forms. Thinking in terms of boundaries thus gives way to thinking in terms of degrees: It is no longer a matter of 'one or the other' but of 'more or less'. This way, one can escape the dualism denounced by Anthropocene thinkers without falling into a counterproductive monism that merges the two domains of culture and nature into one.

On account of such a gradualist concept of nature, human intervention does not always and automatically lead to decreasing naturalness, but can also lead to increasing naturalness. This can be illuminated using an analogy drawn from the civilization theory of the German-British sociologist Norbert Elias (1897–1990). Elias differentiates between roughly two phases in the Western civilizing process. In the period from 1200 until 1750, people learned to repress spontaneous reactions. In the following period they were expected no longer to simply repress their feelings and emotions but to express them in a 'controlled' way. Elias called this form of self-restraint 'controlled decontrolling of emotional controls'. Likewise, one could speak of a 'controlled decontrolling of ecological controls' with respect to nature restoration (Keulartz, 2012). This formula makes it clear that given the dynamics of natural processes (such as grazing and predation) more room again is not an uncontrolled process, but requires new, less rigid and more flexible forms of nature management. If we want to loosen the reins and promote the ability of non-human actors to actively shape their own way of life and habitat, it will require more rather than less horsemanship.

Rewilding

A form of nature management that puts into practice the idea of a controlled disentanglement of nature and culture is rewilding. Whereas traditional restoration ecology aims to return an ecosystem back to some historical condition, rewilding is forward-looking rather than backward-looking: it examines the past not so much to recreate it but to learn from the past how to activate and maintain the ecological and evolutionary processes that are crucial for biodiversity conservation. Rewilders want to restore these natural processes through the creation of large protected core areas and the corridors between them, creating networks that allow for the spread and exchange of species.

This is urgently needed, as currently only 13% of the land surface and 5% of the oceans are protected. A draft text for the fifteenth biodiversity conference, to be held in the Chinese city of Kunming in the spring of 2022, calls for ensuring that 30% of the earth is protected by 2030. This plan, also known as '30 × 30', is seen as a good stepping stone to an even more ambitious goal: protecting half the planet by 2050. Calls to protect half the earth date back as far as the 1970s, but the concept has gained momentum in recent years thanks to the founding in 2009 of the *Nature Needs Half*-movement and the publication in 2016 of the book *Half Earth* by renowned ecologist Edward Wilson.

To initiate and stimulate natural processes, it is also necessary to make use of keystone species. These are usually large carnivores and herbivores that exert a strong influence on the structure of the landscape and on the composition of plant and animal species, but that have often become locally or globally extinct. Their return is crucial for restoring the evolutionary and ecological potential lost due to their disappearance. To realize this comeback, rewilders make use of a variety of techniques. Besides the familiar method of reintroducing animals in areas where populations have decreased dramatically or even gone extinct, rewilders also employ some more controversial methods including 'back breeding' to restore wild traits in domesticated species, 'taxon substitution' to replace extinct species by closely related species with similar roles within an ecosystem, and 'de-extinction' to bring extinct species back to life again using advanced biotechnological technologies such as cloning and gene editing.

Living Apart Together

The gradual disentanglement of nature and culture is meant to give wild animals back their freedom to shape their own lives autonomously. Of course, this will never lead to the complete separation of wildlife from our human culture; there will always remain a form of co-existence, one that can best be described as 'living apart together'.

References

Keulartz, J. (2012). The emergence of enlightened anthropocentrism in ecological restoration. *Nature and Culture, 7*(1), 48–71.

Latour, B. (1991). *We have never been modern.* Harvard University Press.

Latour, B. (2004). *Politics of nature. How to bring the sciences into democracy.* Harvard University Press.

Latour, B. (2008). 'It's development, stupid!' Or: How to modernize modernization. In J. Proctor (Ed.), *Postenvironmentalism* (pp. 1–13). MIT Press.

Latour, B. (2017). *Facing Gaia – Eight lectures on the new climatic regime.* Polity Press.

Maris, V. (2018). *La Part Sauvage du Monde – Penser la Nature dans l'Anthropocène.* Seuil.

Neyrat, F. (2019). *The unconstructable earth – An ecology of separation.* Fordham University Press.

Shellenberger, M., & Nordhaus, T. (2004). *The death of environmentalism: Global warming politics in a post-environmental world.* Breakthrough Institute.

Wilson, E. (2016). *Half-earth. Our planets fight for life.* Liveright Publishing Corporation.

Jozef Keulartz is emeritus Professor Environmental Philosophy at Radboud University Nijmegen, the Netherlands. He is also senior researcher at Wageningen University & Research Center, the Netherlands. His books include Die verkehrte Welt des Jürgen Habermas (Junius, 1995), and Struggle for Nature (Routledge, 1998). He is co-editor of Pragmatist Ethics for a Technological Culture (Kluwer, 2002), Legitimacy in European Nature Conservation Policy (Springer, 2008), New Visions of Nature (Springer, 2009), Environmental Aesthetics (Fordham, 2014), Old World and New World Perspectives in Environmental Philosophy (Springer, 2014), Animal Ethics in the Age of Humans (Springer, 2016), and Animals in our Midst (Springer, 2021).

Complexity

Jürgen Scheffran

Abstract Complexity emerged as a paradigm for a world that is increasingly difficult to understand. Concepts from natural sciences, notably chaos theory, self-organization and complex adaptive systems, diffused into the social sciences. The complexity-stability nexus is associated with crisis, robustness, resilience, sustainability and viability. Complexity science offers an integrative framework for the interaction of natural and social systems in the Anthropocene, including growth, climate change, compound risk, tipping points and cascades.

While complexity corresponds to the difficulty to describe, understand or explain something, there is no common definition but numerous meanings (Abraham et al., 1990). The complexity of a real-world object or system can be characterized by its "essential attributes" used for a description or model in an appropriate language which can be words, computer code, graphical or mathematical symbols. Besides the number of units needed to construct or decompose a system, complexity is indicated by the connectivity between them. In a network the nodes and links can be ordered in different ways, from hierarchical to star-like structures. Whether a model description is a good representation of reality depends on the quality of approximation and its errors. The challenge is to find a balance between a description that is too simple and misses the essence of reality or too complex and cannot be handled. According to Albert Einstein, everything should be made as simple as possible, but not simpler. This leads to the question when a representation is sufficiently complex to meet the needs, as expressed in adaptive complexity.

Complex adaptive system use the self-organization of their parts to form emerging patterns which together are more than the sum of the parts (Haken, 1977). Other features are the tendency to be unpredictable, generate surprises and react sensitively near bifurcation points. Uncertainties open possible futures, with risks and

J. Scheffran (✉)
Institute of Geography, University of Hamburg, Hamburg, Germany
e-mail: juergen.scheffran@uni-hamburg.de

393

opportunities, and the freedom of choosing alternative pathways. A famous image from chaos theory is the butterfly effect, supposing that small things (a butterfly) can have big effects (a storm). After a tiny virus initiated a pandemic through global chains, the sensitivity of our interconnected world became apparent.

This points to the relationship between the complexity of a system and its stability which means that a disturbance will not be magnified but dampened to have a small and disappearing impact. A stable system is able to preserve the essential attributes that determine its identity and complexity. While every system has limits of stability, on the edge to instability small deviations can lead to qualitative changes through death, break-up or transformation. The interplay between the complexity and stability of dynamic systems has been studied in ecosystem research since the 1970s and extended to social systems (May, 1972; Landi et al., 2018). While some claim that growing complexity of systems drives them to instability, others claim the opposite. In the real world naturally evolving systems such as biological organisms, ecosystems, societies and networks often tend to be robust, adaptive and stable to environmental conditions for a certain "lifetime" from birth to death. In contrast, random or inadequately constructed complex systems can be dysfunctional if their parts are not mutually fitting to or in conflict with each other, getting "out of control" to collapse or lose attributes essential for their identity and existence. Mutations harmful to organisms are identified as destabilizing in the evolutionary process while innovative mutations improve the organism's fitness and chance of survival. Specialized systems efficient for a particular purpose may be less flexible, adaptive and error-friendly and thus more vulnerable to changing conditions (Scheffran, 2015). Thus, over time unstable systems disappear and stable ones remain, at the cost of higher complexity to keep the system under control against natural variation.

A major driver towards instability is exponential population growth which continues until natural resources are depleted or coexistence among competing species is achieved. Evolutionary learning processes support a stable balance of mutually regulating mechanisms of magnification and dampening, to keep growth under control. Examples are rainforests and coral reefs, which due to their diversity are more stable to natural variations than specialized monocultures. When climate change and other human interventions push them out of their stability range, they are endangered. Similar problems also arise in organizations and technical systems when increasing efficiency and reduced redundancy reach a critical density of couplings which increases the likelihood of "normal accidents", as experienced in nuclear reactor accidents (Perrow, 1984).

A related concept is resilience which is the ability of a system to withstand, cope with or compensate for external shocks and surprises. A system is robust if it performs reasonably well across a wide range of plausible futures (Lempert et al., 2009). Broader is the concept of viability, seeking to stay within necessary boundaries to ensure livability with mechanisms to modify, regulate and control systemic change (Aubin & Saint-Pierre, 2007). In the guardrail concept (tolerable-windows approach), actions are to be avoided that threaten the existence of a system. Since control mechanisms add to complexity and may fail, such as control rods in nuclear reactors, keeping them as simple and as effective as possible contributes to stability. If limits are reached and resources scarce, social learning can help to find new strategies and innovations to ensure survival despite changing conditions.

Scientific disciplines investigate transitions from the simple to the complex (and vice versa): from nuclear particles to atoms (physics), from atoms to molecules (chemistry), from molecules to cells and ecosystems (biology), from consumers to firms (economy), from citizens to states (political science), from single agents to social networks (social science), from conflict to cooperation (peace and conflict studies), and from environmental destruction to sustainability (environmental sciences) (Mainzer, 2007). While much of the world is essentially stable and predictable most of the time, occasionally destabilizing processes accumulate and end the world as we know it through crisis events, followed by qualitatively new periods of history, as was experienced in major revolutions such as 1789 in France or the two world wars.

Complexity theory can contribute to understanding the wickedness of such transitions (Harrison, 2006). The 1980s that established complexity science ended the Cold War with a chaotic domino effect, following the fall of the Berlin Wall, German unification and the collapse of the socialist world order, an event of global and historic proportions (Scheffran, 2008). The structurally simple East-West conflict, with its modelling paradigms of linear dynamic systems, rational choice optimization and two-player games, was followed by a period of uncertainty, disorder and instability, represented by non-linear adaptive agent models, multi-scale decision-making and bounded rationality (Axelrod, 1997; Bendor & Scheffran, 2019). Complexity thinking is slowly making its way into international relations and policy-making (Bousquet & Curtis, 2011; Geyer & Pickering, 2011; Geller, 2011). Many actors and factors dynamically interact in today's fractal crisis and conflict landscape where events could have immediate impacts elsewhere, coupling local and global shifts to "normal emergencies".

Complexity science offers an integrative framework and opportunity for transdisciplinary and inclusive science in the Anthropocene, beyond a limited understanding of the interconnectedness emerging from the interaction of environmental and social phenomena (Balbo et al., 2020: 292). One of the most complex global challenges is anthropogenic climate change which threatens the rather stable climate during the Holocene since the last ice age. Multiple pathways connect climate change, natural resources, human security and societal stability (Scheffran, 2016). Changes in the climate system (such as variations in temperature and precipitation) together with other risks to planetary boundaries, such as biodiversity loss, land use or nitrogen pollution, affect the functioning of ecological systems and natural resources (e.g. soil, water, forests, biodiversity) which stress human needs and security (e.g. availability of water, food, energy, health and wealth). Depending on vulnerability human responses can affect societal stability, provoke tensions and social disruptions in regional hot spots, such as forced migration or violent conflict. The challenge is to develop adaptive strategies to avoid disaster risk and maintain stability despite unprecedented systemic changes. Multi-level crises are linked through long-distance teleconnections such as globalized trade and financial markets;

infrastructure and supply chains; media and social networks; communication and transportation systems; as well as resource flows and climate change (Scheffran, 2016).

Climate change is a global connector and risk multiplier, exceeding adaptive capacity and resilience of natural and social systems. Compound effects combine multiple stressors (Zscheischler et al., 2018), in particular weather extremes such as heavy rainfall, floods and storms damaging infrastructures, or droughts, heat waves and wildfires, associated with crop and vegetation losses, air pollution and human health problems. Complex interactions are represented in the water-food-energy nexus or the climate-conflict-migration nexus.

A tipping point is defined as a "threshold at which small quantitative changes in the system trigger a non-linear change process that is driven by system-internal feedback mechanisms and inevitably leads to a qualitatively different state of the system", characterized by multiple stable states, abruptness, feedbacks and irreversibility (Milkoreit et al., 2018: 9). Tipping elements of the climate system include melting of Greenland and West Antarctic ice sheets, release of methane from frozen soils, rain forest losses, weakening of the North Atlantic Current, or changes in the Asian monsoon. Their amplification in "Hothouse Earth" (Steffen et al., 2018) would have extreme and long-term consequences. Gradual warming can connect ecological and social tipping points, with negative or positive effects from local to global scales (Rodriguez-Lopez et al., 2019; Otto et al., 2020).

Beyond tipping points complex risk cascades can spread through networks (AghaKouchak et al., 2018), similar to a nuclear chain reaction beyond critical mass. Reactor accidents like in Chernobyl 1986 and Fukushima 2011 were followed by global cascades (Scheffran, 2016). In the COVID pandemic humanity became part of a global chain reaction. Climate-related cascading events include disasters and weather extremes, famines and epidemics, poverty and refugee movements, violent conflict and terrorist acts, which may trigger economic and political crises across regions.

Coping with complex crises requires humanitarian assistance and disaster response, regulation of markets and prices, resilience and sustainable peacebuilding in the context of a socio-ecological transformation that leverages positive tipping points (Otto et al., 2020; Thonicke et al., 2020). Protest movements such as Fridays for Future and the spread of innovations could build critical mass and momentum for system change based on self-organization in viable adaptive networks (Scheffran, 2015). The challenge is to move away from a negative nexus of problems to a nexus of solutions and synergies, using complexity as an opportunity.

References

Abraham, N. B., Albano, A. M., et al. (Eds.). (1990). *Measures of complexity and chaos*. Springer.
AghaKouchak, A., Huning, L. S., et al. (2018). How do natural hazards cascade to cause disasters? *Nature, 561*(7724), 458–460.

Aubin, J.-P., & Saint-Pierre, P. (2007). An introduction to viability theory and management of renewable resources. In J. Kropp & J. Scheffran (Eds.), *Decision making and Risk Management in Sustainability Science* (pp. 43–80). Nova Science.

Axelrod, R. M. (1997). *The complexity of cooperation: Agent-based models of competition and collaboration*. Princeton University Press.

Balbo, A., Rothe, D., & Scheffran, J. (2020). The Anthropocene: An opportunity for transdisciplinary and inclusive science? In M. Brzoska & J. Scheffran (Eds.), *Climate change, security risks, and violent conflicts* (pp. 287–205). Hamburg University Press.

BenDor, T. K., & Scheffran, J. (2019). *Agent-based modeling of environmental conflict and cooperation*. Taylor & Francis/CRC Press.

Bousquet, A., & Curtis, S. (2011). Beyond models and metaphors: Complexity theory, systems thinking and international relations. *Cambridge Review of International Affairs, 24*(1), 43–62.

Geller, A. (2011). The use of complexity-based models in international relations. *Cambridge Review of International Affairs, 24*(1), 63–80.

Geyer, R., & Pickering, S. (2011). Applying the tools of complexity to the international realm: From fitness landscapes to complexity cascades. *Cambridge Review of International Affairs, 24*(1), 5–26.

Haken, H. (1977). *Synergetics*. Springer.

Harrison, N. E. (Ed.). (2006). *Complexity in world politics: Concepts and methods of a new paradigm*. State University of New York Press.

Landi, P., Minoarivelo, H. O., et al. (2018). Complexity and stability of adaptive ecological networks: A survey of the theory in community ecology. In *Systems analysis approach for complex global challenges* (pp. 209–248). Springer.

Lempert, R., Scheffran, J., & Sprinz, D. F. (2009). Methods for long-term environmental policy challenges. *Global Environmental Politics, 9*(3), 106–133.

Mainzer, K. (2007). *Thinking in complexity: The computational dynamics of matter, mind, and mankind*. Springer.

May, R. M. (1972). Will a large complex system be stable? *Nature, 238*, 413–414.

Milkoreit, M., Hodbod, J., et al. (2018). Defining tipping points for social-ecological systems scholarship – An interdisciplinary literature review. *Environmental Research Letters, 13*, 1–12.

Otto, I. M., Donges, J. F., et al. (2020). Social tipping dynamics for stabilizing Earth's climate by 2050. *PNAS, 117*, 201900577.

Perrow, C. (1984). *Normal accidents: Living with high risk technologies*. Princeton University Press.

Rodriguez Lopez, M., Tielbörger, K., et al. (2019). A transdisciplinary approach to identifying transboundary tipping points in a contentious area: Experiences from across the Jordan River region. *Sustainability, 11*, 1184.

Scheffran, J. (2008). The complexity of security. *Complexity, 14*(1), 13–21.

Scheffran, J. (2015). Complexity and stability in human–environment interaction: The transformation from climate risk cascades to viable adaptive networks. In E. Kavalski (Ed.), *World politics at the edge of chaos* (pp. 229–252). University of New York Press.

Scheffran, J. (2016). From a climate of complexity to sustainable peace: Viability transformations and adaptive governance in the Anthropocene. In H. G. Brauch, U. Oswald-Spring, et al. (Eds.), *Handbook on sustainability transition and sustainable peace* (pp. 305–347). Springer.

Steffen, W., Rockström, J., et al. (2018). Trajectories of the earth system in the anthropocene. *PNAS, 115*(33), 8252–8259.

Thonicke, K., Bahn, M., et al. (2020). Advancing the understanding of adaptive capacity of social-ecological systems to absorb climate extremes. *Earth's Future, 8*, e2019EF001221.

Zscheischler, J., Westra, S., et al. (2018). Future climate risk from compound events. *Nature Climate Change, 8*, 469–477.

Jürgen Scheffran is Professor of Geography at Universität Hamburg and chair of the Research Group Climate Change and Security in the Center for Earth System Research and Sustainability and the Climate Excellence Cluster CLICCS. After his physics PhD he worked in research groups in environmental science, peace and conflict research at the universities of Marburg, Darmstadt, Paris and Illinois, as well as the Potsdam Institute for Climate Impact Research. Besides peer-reviewed journals, he authored or edited more than 30 books, including "*Handbook on Sustainability Transition and Sustainable Peace*" and "*Climate Change, Human Security and Violent Conflict*".

Dualism

Rosine Kelz

Abstract In the wide-ranging and diverse debate about the Anthropocene thesis, the status of the Human/Nature dualism is a divisive issue. After defining the term dualism, this chapter will outline diverging perspectives on the role it plays for understanding the Anthropocene. Dualistic thinking, some commentators argue, has been integral to bringing about and maintaining an unjust socio-economic global system which is closely intertwined with today's planetary environmental crises. Others highlight, however, that earth system science, which provides the aggregated data and models on which the Anthropocene thesis is founded, is part of a tradition of non-dualistic theories in the natural sciences, which, in turn, are connected to a broader non-dualist modern western culture. Nevertheless, popular narratives and tropes of the Anthropocene discourse often re-articulate dualistic narratives of human history and of Earth's future.

A dualism, Val Plumwood writes, is more than just a binary distinction. While the latter just highlights the differences between two groups or concepts, the former expresses a hierarchical relationship between two terms. The differences between them are exaggerated, and the dependency of the dominant term on the subordinated one is denied (Plumwood, 1993, p. 41). Moreover, for its stability, a dualism – like Human/Nature – depends on a system of interconnections with other dualisms – like Mind/Body, Human/Animal, Man/Woman. These sets of related pairs order western thought – they form a 'fault-line which runs through its entire conceptual system' (Ibid, p. 42). Highlighting the interconnections between philosophy and socio-political practices, Plumwood argues that dualisms are also 'major cultural expressions and justifications' of power relations in western societies. Which pairs take a central position within the network of dualisms therefore changes in correspondence with transformations in the socio-political order in the

R. Kelz (✉)
Institute of Intercultural and International Studies, University of Bremen, Bremen, Deutschland
e-mail: kelz1@uni-bremen.de

© The Author(s), under exclusive license to Springer Nature Switzerland AG 2023
N. Wallenhorst, C. Wulf (eds.), *Handbook of the Anthropocene*, https://doi.org/10.1007/978-3-031-25910-4_61

399

history of western societies. 'Nature', however, always had a central role to play as subordinated term. In Plumwood's analysis, Plato already established the groundwork for the modern Human/Nature dualism by constructing an ideal of the human as equivalent with mind or reason. This excludes 'the whole rich range of other human and non-human characteristics', designating them to the realm of nature. Mindless nature is then defined as 'ineluctably alien', which separates the human from animals and non-human nature (Ibid, p. 107). These ideas are made more explicit, however, in Cartesian thought which marks the Enlightenment (Ibid, p. 107). Descartes, Plumwood agrees with many commentators, views nature as an external realm to be controlled and exploited, where it is the role of humans 'to make ourselves … the masters and possessors of nature' (Descartes, 2006, p. 51). This separation of 'man' from nature, Plumwood writes, also provided the ideological background for European colonial expansion which 'from the fourteenth century onwards brings to the fore civilized/primitive as a variant of reason/nature' (Plumwood, 1993, p. 44).

Jason Moore (2017) draws on Plumwood when he identifies the early modern version of the human/nature dualism as closely interwoven with the rise of capitalism. Capitalism, for him, is the true cause of the Anthropocene – which Moore therefore argues would more aptly be called the 'Capitalocene'. For Moore, the exaggerated separation of 'man' from nature, pivots 'man' against a nature external to him. This makes it possible to treat nature as a cheap resource to be exploited. With the development of capitalism, he argues, the distinction between 'man' and nature became a 'real' abstraction: a thought construct which guides concerted action and therefore allowed for the 'great remaking of planetary life': 'The capitalist revolution, far from a narrowly economic process, was an epochal shift in the ways of earth-moving (mining, farming), state-making, mechanization and symbolic praxis. Not for nothing, the first thing every great European empire set about doing was not merely 'exploring', but mapping and cataloguing the globe as a potential storehouse of wealth' (Moore, 2017, p. 605). At the same time, Moore stresses, many human beings were denigrated into the realm of nature (Ibid, p. 606). Early capitalist accumulation depended on the devaluated or unpaid work and energy of 'women, nature, and colonies' (see also Mies, 1986, p. 77). To summarize, Moore stresses how dualistic thinking has enabled and justified an exploitative economic system, which has brought about the catastrophic planetary conditions which are now referred to as the Anthropocene.

To argue that all western thought and societal practices are dualistic, however, would itself be a problematic reduction of complexity. Contemporary earth system science, where the Anthropocene term first gained traction, highlights the interconnections between human systems and natural systems (see e.g. Steffen et al., 2007). When Clive Hamilton describes earth system science as an 'integrative meta-science of the whole planet as a unified, complex, evolving system beyond the sum of its parts' (Hamilton, 2016, p. 94), this formulation is reminiscent of mid-twentieth century systems theory. It builds on an understanding of Earth as a complex system producing and maintaining the conditions for life, a notion which was popularized in the 1970s by the Gaia concept proposed by James Lovelock and Lynn Margulis.

Including human systems into their frameworks, planetary systems approaches are themselves embedded within a longer tradition of monist, holistic and relational notions in modern western culture. Monism, for example, is a well-established tradition in modern European philosophy (see e.g. Weir, 2012). At times, European monists have drawn on Asian non-dualist traditions of thought – however, usually without seriously challenging the Eurocentric bias of modern western philosophy. Monism played a key role in the development of biology as a modern science during the nineteenth century. German biologist Ernst Haeckel, for instance, argued that the natural sciences had provided empirical verification of Baruch Spinoza's philosophical monism – the notion that mind and matter were two modes of a single substance. For Haeckel, who first coined the term 'ecology' for the study of the relationships between organisms, and between organisms and their environments (Cooper, 2003, p. 4), Darwinian evolution provided 'a master theory linking the multiplicity of biological life to the development of human consciousness and civilization as a single meaningful totality' (Weir, 2012, p. 1).

Philosophical and cultural critiques of modern western societies' exploitative and estranged relationship to nature already accompanied early European industrialization. Eighteenth century romantics inspired early nature conservation movements in the late 19th and early twentieth century. While in many of these critiques the binary distinction between nature and human culture was retained in some form, the hierarchical relationship between human and nature was called into question – the goal was a communion with awe-inspiring nature or the return to a more natural form of living (see Toepfer, 2020, p. 221). In the early twentieth century, critiques of human/nature dualism found manifold philosophical expressions. Theodor Adorno, for example, argues in a 1932 lecture against the idea of nature as 'the invariable backdrop against that history unfolds' (Weißpflug, 2019, p. 24), and stresses the transience of nature. Prefiguring later environmental thinkers like Plumwood, Adorno writes that the idea of nature/culture dualism is based on a fantasy of dominating nature, which denies humanity's interwovenness with the natural realm.

Despite its non-dualistic scientific aspirations, however, mainstream Anthropocene discourse is often criticized for being itself dualistic. This charge is levelled against two related targets. The first is directed against common narratives offered by earth system scientists, who aim to popularize and explain the Anthropocene thesis. While scientists highlight that humans are part of the greater planetary system, the dominant narrative of the Anthropocene, Jeremy Baskin (2015, p. 11) argues for instance, reinserts '"man" into nature only to re-elevate "him" above it'. The human species, Steffen et al. (2007) suggest, was able to emancipate itself from nature through the use of tools and technologies. Today, humanity has finally reached a stage where it will be able to 'dominate' the great forces of nature. The promise of technological control over nature is prominently expanded upon by ecomodernist and 'good' Anthropocene texts promoting the development of geo-engineering and other invasive technological solutions to the side-effects of humanity's 'immature' technologically enabled growth-spurt (for a critique see e.g. Neyrat, 2019). Such storylines not only re-establish the classical dualism's hierarchical relationship, they also homogenize and simplify both terms. As Baskin writes, many Anthropocene narratives normalize 'a certain portion of humanity as the "human" of the Anthropocene' (Baskin, 2015, p. 11). Instead of highlighting differences between countries, social groups and individuals, where only some have caused the majority of the environmental destructions of the Anthropocene, and have benefited from the extraction and exploitation of resources, dominant Anthropocene narratives reiterate a narrative of human progress modeled on the more affluent parts of western societies.

The second charge of dualism is directed against the popular imagery of the Anthropocene. In iconic media and artistic depictions of the Anthropocene, an anonymous humanity, visible only through their artefacts, overwhelms nature to the point of denying the existence of a non-human world. Stacy Alaimo argues that in the visual narratives of the Anthropocene humanity acts on an 'externalized' planet. (Alaimo, 2017, p. 90) The forgetting of how 'human agencies are entangled with those of nonhuman creatures and inhuman substances and systems', she writes, is reinforced in visual depictions of the Anthropocene. Aerial pictures of vast cityscapes, of Earth by night where only electric illumination remains visible, or maps of human transport and communication channels, all used to illustrate popular writings about the Anthropocene, make Earth into 'an eerily lifeless entity, devoid of other species, as if the sixth great extinction had already concluded' (Alaimo, 2017, p. 91). For Alaimo, 'in the dominant visual apparatus of the Anthropocene, the viewer enjoys a comfortable position outside the systems depicted'. These representations of the Anthropocene 'ask nothing from the human spectator ... they neither involve nor implore. The images make risk, harm, and suffering undetectable' (Alaimo, 2017, p. 92). Similarly, Baskin (2019, p. 157-8) comments on the 'Welcome to the Anthropocene' video, which opened the 2012 United Nations Conference on Sustainable Development, that it reinforces the prevalent 'space' view from nowhere, depicting an Anthropocene where humanity as a species is central, but humans and human societies, in all their complexity and multiplicity are absent. The popular imagery of the Anthropocene can thus be interpreted as a recent expression of a Cartesian rationalism where 'vision and thought' are 'funneled into a spectator's view of the world', rendering 'the viewer bodyless and placeless'. Earth, firmly in humanity's hands, is here seen from the universal perspective of 'a single disembodied, omniscient, and panopticonic eye.' (Warf, 2008, p. 53).

There are many prominent non-dualist approaches in the natural and social sciences and the humanities which aim to better grasp the interrelations between humans, other living beings, and their environments. However, as Plumwood's definition should remind us, to criticise dualism one does not have to abandon drawing careful distinctions. Instead, non-dualist thought in the Anthropocene needs to highlight the complex multiplicities of histories and forms of engagements with their environments which differentiate human societies and individuals across the planet. Acknowledging differences in socio-economic position and the differential situatedness of people in relations of power and political representation are key for a critical engagement with the global politics of the Anthropocene. Similarly, climate change mitigation and environmental protection policies need to take the complexities of various forms of interrelations between human and non-human processes into account, and explore these relationships from perspectives critical of the 'default' anthropocentrism Plumwood's criticism of the modern human/nature dualism highlights.

References

Alaimo, S. (2017). Your Shell on acid: Material immersion, anthropocene dissolves. In R. Grusin (Ed.), *Anthropocene feminism* (pp. 99–119). University of Minnesota Press.
Baskin, J. (2015). Paradigm dressed as epoch: The ideology of the anthropocene. *Environmental Values, 24*, 9–29. https://doi.org/10.3197/096327115X14183182353746
Baskin, J. (2019). Global justice and the anthropocene: Reproducing a development story. In F. Biermann & E. Lövbrand (Eds.), *Anthropocene encounters: New directions in green political thinking* (pp. 150–168). Cambridge University Press.
Cooper, G. (2003). *The science of the struggle for existence. On the foundations of ecology.* Cambridge University Press.
Descartes, R. (2006). *A discourse on the method of correctly conducting one's reason and seeking truth in the sciences.* Oxford University Press.
Hamilton, C. (2016). The Anthropocene as rupture. *The Anthropocene Review, 3*(2), 93–106. https://doi.org/10.1177/2053019616634741
Mies, M. (1986). *Patriarchy and accumulation on a world scale.* Zed.
Moore, J. (2017). The capitalocene, Part I: On the nature and origins of our ecological crisis. *The Journal of Peasant Studies, 44*, 594–630. https://doi.org/10.1080/03066150.2016.1235036
Neyrat, F. (2019). *The unconstructable earth. An ecology of separation.* Fordham University Press.
Plumwood, V. (1993). *Feminism and the mastery of nature.* Routledge.
Steffen, W., Crutzen, P., & McNeill, J. (2007). The anthropocene: Are humans now overwhelming the great forces of nature. *AMBIO: A Journal of the Human Environment, 36*(8), 614–621. https://doi.org/10.1579/0044-7447(2007)36[614:TAAHNO]2.0.CO;2
Toepfer, G. (2020). Artenschutz durch Gentechnik? Vom Dilemma zur Tragik des Naturschutzes im Anthropozän. *Natur und Landschaft, 95*(5), 220–225.
Warf, B. (2008). *Time-space compression.* Routledge.
Weir, T. (2012). The riddles of monism: An introductory essay. In T. Weir (Ed.), *Monism: Science, philosophy, religion, and the history of a worldview* (pp. 1–44). Palgrave.
Weißpflug, M. (2019). A natural history for the 21st century: Rethinking the anthropocene narrative with Arendt and Adorno. In T. Hickmann et al. (Eds.), *The anthropocene debate and political science* (pp. 15–30). Routledge.

Rosine Kelz teaches Political Theory at the University of Bremen. Prior to her position in Bremen Rosine was a Research Associate at the Institute for Advanced Sustainability Studies (IASS) and an Andrew W. Mellon Fellow in the Bio-Humanities at the University of Illinois. She holds a D.Phil in Politics (Theory) from the University of Oxford. Her research explores the role of temporality for notions of justice, democracy and sustainability, as well as philosophical and political questions related to new biotechnologies and nature conservation. Her publications include the article *"Politics of Time and Mourning in the Anthropocene"* co-authored with Henrike Knappe and published in the journal *Social Sciences*, and the book *"The non-sovereign self, responsibility and otherness"* (2016 Palgrave Macmillan).

Entropy

Anne Alombert

Abstract This article explains the meaning of the notion of entropy and its implications for the understanding of the Anthropocene epoch. Although the notion of
entropy belongs to the thermodynamic field, it also has some consequences in the
field of biology and human sciences: living organisms can be defined by their anti-
entropic tendencies towards organization and evolution, whereas human beings displace the play between entropy and anti-entropy through the production of artificial
organs, which can become very anthropic if they are not adopted through the practice of collective knowledge. According to Bernard Stiegler, in the current epoch,
the "disajustment" between technical evolution and social evolution provokes an
increase of entropy and anthropy, which has to be countered thanks to the valorization of anti-anthropic practices of different knowledge, at the source of cultural and
social evolution and diversity.

The concept of entropy appeared in the nineteenth century in the field of thermodynamic physics and was coined by Rudolf Clausius. The "entropy law", also known
as the "second law of thermodynamics" describes the irreversible dissipation of
energy in a closed system, through its degradation into heat. The transformation of
energy into heat means that energy changes from a usable state (where it can be
used by man) to an unusable state (when it can no longer be used). The transformation of the system does not correspond to an energy consumption strictly speaking,
but to a change of state of the energy: for example, "when a piece of coal is burned,
its chemical energy neither decreases nor increases", "but its initial free energy has
dissipated so much in the form of heat, smoke and ashes, that man can no longer use
it" (Georgescu-Roegen, 2006, p. 68). This degradation of energy corresponds to the
passage from a certain ordered structure (an improbable configuration) to a state of
dispersion and disorder (a more probable configuration): "free [usable] energy
implies a certain ordered structure comparable to that of a store where all the meats

A. Alombert (✉)
Université Catholique de Lille, Lille, France
e-mail: anne.alombert@univ-paris8.fr

© The Author(s), under exclusive license to Springer Nature
Switzerland AG 2023
N. Wallenhorst, C. Wulf (eds.), *Handbook of the Anthropocene*,
https://doi.org/10.1007/978-3-031-25910-4_62

are on one counter, the vegetables on another, etc. While bound energy [not usable] is energy scattered in disorder, like the same store after being hit by a tornado" (Georgescu-Roegen, 2006, p. 68). This is the reason why "entropy is also defined as a measure of disorder" (Georgescu-Roegen, 2006, p. 68).

In the 1950s, the theory of entropy was taken up in the field of information theory (in particular through the works of Claude Shannon (Shannon, 1948) and in cybernetics, in particular through the work of Norbert Wiener (Wiener, 1961). Despite referring to a completely different process than that which is described by the second law of thermodynamics, the name "entropy" was then used to describe the degree of uncertainty contained in a message, introducing a confusion into the signification of the concept.

Of greater importance, however, was the physicist Erwin Schrödinger's effort to mobilize the notion of entropy in biology, in order to show that life entails a kind of "negative entropy" or "negentropy", that is, a tendency that runs counter to the overall entropic process (Schrödinger, 1992). According to Schrödinger, organisms are endowed with a metabolic capacity to reduce the increase of entropy by exchanging matter and energy with the environment and thus maintaining their life. No living thing escapes the global entropic tendency, but through their organization and their evolution, organisms are engaged in a local and temporary struggle against the dissipation of energy. Hence Norbert Wiener refers to living organisms as "local and temporary islands of decreasing entropy in a world in which the entropy as a whole tends to increase" (Wiener, 1998, p. 36).

More recently, the philosophers and mathematicians Giuseppe Longo and Francis Bailly have developed a notion of "anti-entropy" to describe the specific characteristics of the living state of matter, that is, of biologically organized matter (Bailly & Longo, 2009): they argue that life, unlike the machine, can arise only as a dynamic and historical process capable of maintaining and differentiating the "multilevel entangled structure" of living organisms. The high level of correlation of parts and whole involved in biological organization amounts to an "extended critical situation" where the integration and regulation of the various levels of this organization is the crucial anti-entropic process of life.

To sum up, the production of entropy can be defined as a tendency towards disorganization, destructuring and disorder, which corresponds to a dissipation and degradation of energy: an entropic process is a process in which a system tends to exhaust its dynamic potentials, as well as its capacity for conservation or renewal, by dissipating its energy and gradually rejoining a state of inertia. On the contrary, anti-entropy refers to a tendency towards organization, structuration, diversification and the production of novelty or improbability – the development of a system that tends to self-conservation, renewal or transformation towards improvement. Although anti-entropy can never eliminate the inevitable increase of entropy, it can locally delay or defer this increase.

In his recent work based on his reading of Alfred Lotka (Lotka, 1945), Bernard Stiegler created the concepts of anthropy and anti-anthropy in order to designate the production of entropy and anti-entropy at the level of exosomatic life, that is to say of the technical, psychological and social life that is commonly known as human life (Stiegler, 2016; Stiegler et al., 2020). Indeed, according to Stiegler, the exosomatization process (the process of technical externalization) "displaces the play of entropy and negentropy" (Stiegler, 2016, p. 14): unlike the endosomatic or biological organs of living beings, which are always local and temporary producers of negentropy (organization and diversification), exosomatic or technical organs are ambivalent. On the one hand, they can accelerate the production of entropy (through the process of combustion and energy dissipation that technological production involves, and through industrial standardization that homogenizes and standardizes behaviours). On the other hand, exosomatic organs can produce new, improbable and singular (social, artistic, cultural and technical) forms of organization and diversification, provided that these technical organs are successfully adopted by humans, through the practise of collective knowledge through social organizations. The production of knowledge thus corresponds to an anti-anthropic process (organization and diversification on the psychic, technical and social levels).

In other words, if a living organism is able to organize itself to produce antientropy through its biological organization, by temporarily and locally delaying entropy, human beings can and must organize themselves on the anti-anthropic level, by practising knowledge and constituting social organizations, in order to postpone the anthropogenic effects inevitably entailed by the production of exosomatic or technical organs, which have now become industrial and digital. Social organizations, however, tend to themselves become anthropic: knowledge tends to rigidify (in the form of dogmas) and social institutions tend to close. Thus, anti-anthropy refers to the ability to refresh knowledge and institutions by transforming them diachronically, that is, by causing them to evolve or bifurcate towards new horizons.

Stiegler argues that the Anthropocene era results from an immeasurable acceleration in the evolution of exosomatic organs: what is truly new in the Anthropocene era stems from the production of technical organs that did not exist during the Holocene, and from the explosion of technical evolution that occurred with the industrial revolution and intensified once again in the second half of the twentieth century. Becoming more and more complex, these technical organs are transformed so quickly that the knowledge (know-hows, know how to live, theoretical knowledge) and social organizations required for their fruitful adoption does not have time to develop: a disadjustment occurs between the technical system and social systems.

Therefore, according to Stiegler, the current epoch of the Anthropocene can be characterized as an "Entropocene" (Stiegler et al., 2020, p. 12), insofar as this period corresponds to a massive increase in the rates of entropy at all possible levels: physical (dissipation of energy), biological (destruction of biodiversity) and psychosocial (destruction of cultural and social diversity). According to Stiegler, "the various disturbances afflicting the current stage of the Anthropocene *all* consist in an increase of thermodynamic entropy, as the dissipation of energy, biological entropy, as the reduction of biodiversity, and informational entropy, as the reduction of knowledge to data and computation – and, correspondingly, as loss of credit, as mistrust, as generalized mimetism and as the domination of what has been called the 'post-truth era'" (Stiegler et al., 2020, p. 37).

This is the reason why Stiegler argues that "in a context where the Anthropocene is reaching its limits, the economy must be redefined above all as *collective action in the struggle against entropy and against anthropy*" (Stiegler et al., 2020, p. 37). This implies to conceive and implement one or more economic models, based on the systemic valorization of anti-entropic production, that is, on the systemic valorization of the practices of various knowledge (technical knowledge, practical knowledge, existential knowledge, as well as the knowledge of how to do, how to live and how to conceive), serving to readjust the disadjustment between the digital technical system and the social systems characteristic of the Anthropocene.

References

Bailly, F., & Longo, G. (2009). Biological organization and anti-entropy. *Journal of Biological Systems, 17*, 63–96.

Georgescu-Roegen, N. (2006). La décroissance: entropie, écologie, économie. Ellébore-Sang de la terre. [Original edition 1979].

Lotka, A. (1945). The law of evolution as a maximal principle. *Human Biology, 17*, 167–194.

Schrödinger, E. (1992). *What is life?*, in *what is life?, with mind and matter and autobiographical sketches*. Cambridge University Press. [Original edition 1944].

Shannon, C. (1948). A mathematical theory of communication. *The Bell System Technical Journal, 27*, 379–423. 623–56.

Stiegler, B. (2016). *Automatic society, volume 1: The future of work* (Daniel Ross, Trans.). Polity Press. [Original edition 2015].

Stiegler, B., et al. (2020). *Bifurquer. Il n'y a pas d'alternative*. Les liens qui libèrent.

Wiener, N. (1961). Cybernetics, or control and communication in the animal and the machine, 2. MIT Press. [Original edition 1944].

Wiener, N. (1998). The human use of human beings. *Cybernetics and Society*. Free Association of Books. [Original edition 1950].

Anne Alombert is a teacher and researcher in philosophy at University Paris 8. Her research focuses on the relationships between knowledge and techniques, as well as on the anthropological issues of contemporary technological transformations. She is the author of a philosophy thesis carried out at Paris Nanterre University, which examines the question of the relationships between life, techniques and minds in the work of Gilbert Simondon and Jacques Derrida. She is co-author of the book *Bifurquer*, co-written with the philosopher Bernard Stiegler and the collective Internation.

Existence

Christian Arnsperger

Abstract This article argues that a main hidden driver of the Anthropocene is existential—namely the wholesale denial, in capitalist civilization, of human fragility and mortality. Mainstream economics, which unthinkingly validates the unboundedness of human wants and the necessity for open-ended growth, must give way to existential ecological economics—an approach that recognizes that capitalism, which clearly propels the overshoot of material flows, is itself a device for denying and repressing deep human fears about death.

In the nineteenth century, economics used to be coined the "sinister science." In the wake of Thomas Malthus and the classical economists, it was all about the unchecked growth of human population and human wants coming up against the limits of the Earth's capacity to provide—and losing (Kallis, 2019). The sinister character of economics came from its being, basically, an existentialist discipline: it portrayed the absurdity of the human condition, understood as an irrepressible drive for growth (fuelled, in Malthus's austere religious worldview, by original sin, unchecked lust leading to too many births, and general gluttony making human wants unlimited) frustrated by nature's biophysical limits, to the point of rendering humanity's presence on Earth an optional situation at the mercy of the biosphere's whims.

This centrality of limits mutated radically when economics became, during the first half of the twentieth century, predominantly neoclassical. Growth theory took centre stage as Cold War rivalries made national economies focus on output expansion and GDP-per-capita size (Arnsperger this volume). Individual wants were still considered unlimited: agents' preferences were modelled mathematically using properties such as "local non-satiation" and "monotonicity"—fancy technical words for the idea that more always feels better than less. What bounded people's consumption in this approach was the so-called price mechanism, which basically meant that markets had to function "perfectly" so that (*i*) each individual's budget

C. Arnsperger (✉)
University of Lausanne, Lausanne, Switzerland
e-mail: christian.arnsperger@unil.ch

constraint imposed a limit on that individual's consumption bundle and (*ii*) all individuals' budget constraints taken together led to a situation where on each separate market, the total amount all consumers want to purchase equals the total amount all producers want to sell.

The capital stock was often part of the model—especially in dynamic growth models where the economy was assumed to gradually converge on a steady state with constant population, aggregate consumption, technical progress and aggregate capital stock. Natural resources, however, were not. For all intents and purposes, neoclassical growth theory claimed—at least implicitly—to have superseded the "sinister science" by replacing ecology with technology, renaming nature's limited biophysical metabolisms "natural capital." Research and development as well as perfect market competition, both being eminently human institutions, replaced the often brutal regulation of the human population by nature. Techno-optimism, free-market fundamentalism and an imaginary of limitless possibility combined to promote "weak sustainability": while human wants are indeed unbounded, this is not in itself a bad thing; any shortfall in natural capital could be compensated for through with the appropriate combination of innovations in technological and human capital, thus making it possible to imagine an "unbounded" satisfaction of wants (Pettit, 2010) through "green" growth. Productivity enhancements and rebound effects could generate an open-ended path along which, *as a matter of principle*, the unlimited character of human wants no longer had to be seen as a problem.

Recently, degrowth thinkers have sought to invert the direction of the discussion by insisting on the fact that human wants are *not* unlimited and that this is a good thing because consciously limiting our wants is the gateway to new horizons of enjoyment and pleasure (Kallis, 2019; Soper, 2020). This is indeed the direction in which the conversation urgently needs to move. Human self-limitation has long been a centerpiece of political ecology, figuring prominently in the work of major thinkers such as André Gorz, Ivan Illich or Ernst Schumacher. "Basic needs" approaches (Max-Neef et al., 1991; Gough, 2017) have garnered renewed interest as applied researchers seek to recast the material foundations for post-growth cultures in the Anthropocene (Millward-Hopkins et al., 2020).

How can a sufficient number of us—especially in the wealthy regions of the world—transition psychologically and philosophically from a view of human wants as legitimately unlimited to one of positive self-limitation? Fostering self-awareness as well as empathy will be necessary. The rest of this article suggests that one important component of deeper self-awareness is *existential lucidity*: as a civilization, we need to re-actualize the "sinister science" by reuniting economics with ecology and by insisting on the necessity of *accepting existential finitude*. Both the biosphere and each of our biological organisms are finite, and this is part and parcel of existing as an earthling. Helping one another compassionately to accept this existential fact will be key if we are to avoid making the Anthropocene into a theatre of ecological and social decay and destruction.

The film *Planet of the Humans* (Gibbs, 2019) may be flawed when it comes to assessing renewable technologies, but it contains an existential gem lodged in a short interview with existential psychologist Sheldon Solomon. One of the main invisible drivers of capitalism and its relentless pressure towards destructive economic growth may have long been overlooked: Since the inception of the industrial revolution, Western as well as Westernized peoples appear to have collectively bought into the unconscious notion that capitalism's productivism and consumerism offer solace and meaning in the face of human frailness and mortality (Arnsperger, 2005). In other words, the Anthropocene can be seen as a vast attempt at denying finitude and death: Through cultural messaging and through socialization, we routinely behave in ways that overshoot the biosphere's finitude in order to be able to deny our awareness of our own finitude.

Solomon is a contemporary representative of a current in social psychology called "Terror Management Theory" (TMT). This approach is rooted in the pioneering work of anthropologist and psychologist Ernest Becker (1973, 1975). According to Becker, all cultures have at their inception our eternal grappling with the fundamental terror of mortality: Humans have always created "systems of existential heroism" whereby certain roles, functions or behaviours are seen as giving life meaning and as bestowing upon the individual some "cosmic significance" in the form of a transcendent legacy, be it religious or secular. In other words, at the heart of every culture is an individual quest for immortality. This project is doomed to failure, of course, at the biophysical level since any organism is subject to entropy, but it gets sublimated into the culture, where it can feed people's attitudes and aspirations with the unconscious notion that, symbolically speaking, negentropy could defeat decay and death.

TMT has grown into a significant body of theoretical and empirical research (Solomon et al., 2015). It has advanced into environmental psychology, investigating how death-denying tendencies lead a culture to value behaviours that deny biophysical limits and harm nonhuman life (Koole & Van den Berg, 2004; Vess & Arndt, 2008). Viewed from the angle of TMT, mainstream economics is both (*i*) decidedly part of the ideological infrastructure of the Anthropocene because it existentially justifies its denial of biophysical limits and (*ii*) based on theoretical assumptions (*homo economicus*, monotonic individual preferences, absence of resource constraints, etc.) which cohere with this denial (Arnsperger, 2023).

Neither a blanket denunciation of greed or a well-meaning call for "alternative hedonism" (in favour of self-limitation) will, *in and of themselves*, get to the root that first needs to be uncovered—namely, the invisible collusion of capitalism (as well as its old nemesis, productivist socialism) with our terrified denial of finitude and mortality. In their majority, degrowth activists who practice radically alternative ways of living have—whether consciously or unconsciously—taken on this terror and somehow made their peace with it. Pre-colonial, Indigenous cultures contain both practical and spiritual "resources" that can help us bear the burden of finitude in a non-destructive manner. By and large, however, Western and Westernized populations still live in the unconscious grip of capitalism as a device to deny human finitude and, therefore, the Earth's finitude as well.

To tackle the challenge of making this twofold denial conscious, we need more than an ecological economics that centres around the critique of capitalism

(Pirgmaier & Steinberger, 2019), as crucial as that is. We need to develop an *existential ecological economics* (Koller, 2021; Arnsperger, forthcoming) which has two crucial tasks: to show the way in which capitalism has become a mechanism for the denial of existential finitude, and to offers new, more ecologically and humanly congenial, ways to make peace with frailty and mortality.

References

Arnsperger, C. (2005). *Critique de l'existence capitaliste: Pour une éthique existentielle de l'économie*. Éditions du Cerf.

Arnsperger, C. (2023). *L'existence écologique: Critique existentielle de la croissance et anthropologie de la post-croissance*. Éditions du Seuil.

Arnsperger, C. (Forthcoming). *Existential ecological economics*. Routledge.

Arnsperger, C. (this volume). Growth. In N. Wallenhorst & C. Wulf (Eds.), *Handbook of the Anthropocene*. Springer.

Becker, E. (1973). *The denial of death*. Free Press.

Becker, E. (1975). *Escape from evil*. Free Press.

Gibbs, J. director. (2019). *Planet of the humans* (Documentary film). Rumble Media.

Gough, I. (2017). *Heat, greed and human need: Climate change, capitalism and sustainable well-being*. Edward Elgar.

Kallis, G. (2019). *Limits: Why Malthus was wrong and why environmentalists should care*. Princeton University Press.

Koller, S. (2021). Towards degrowth? Making peace with mortality to reconnect with (one's) nature: An ecopsychological proposal for a paradigm shift. *Environmental Values, 30*(3), 345–366. https://doi.org/10.3197/096327120X15916910310590

Koole, S., & Van den Berg, A. (2004). Paradise lost and reclaimed: A motivational analysis of human-nature relations. In J. Greenberg, S. Koole, & Pyszczynski (Eds.), *Handbook of exprimental existential psychology* (pp. 86–103). Guilford Press.

Max-Neef, M., Elizalde, A., & Hopenhayn, M. (1991). *Human-scale development*. Apex Press.

Millward-Hopkins, J., Steinberger, J., Rao, N., & Oswald, Y. (2020). Providing decent living with minimum energy: A global scenario. *Global Environmental Change, 65*, 102168.

Pettit, J. (2010). A defense of unbounded (but not unlimited) economic growth: The ethics of creating wealth and reducing poverty. *Journal of the Society of Christian Ethics, 30*(1), 183–204.

Pirgmaier, E., & Steinberger, J.K. (2019). Roots, riots and radical change—A road less travelled for ecological economics. *Sustainability, 11*. https://doi.org/10.3390/su11072001

Solomon, S., Greenberg, J., & Pyszczynski, T. (2015). *The worm at the core: On the role of death in life*. Random House.

Soper, K. (2020). *Post-growth living: For an alternative hedonism*. Verso.

Vess, M., & Arndt, J. (2008). The nature of death and the death of nature: The impact of mortality salience on environmental concern. *Journal of Research in Personality, 42*, 1376–1380.

Christian Arnsperger is professor of sustainability and economic anthropology at the University of Lausanne. He holds a PhD in economics from the University of Louvain (Belgium) and has also been active as a scientific advisor to the Alternative Bank Switzerland. He publishes widely on the critique of economic growth, the existential underpinnings of capitalism, and the links between money and sustainability. His latest books are *Écologie intégrale: Pour une société permacirculaire* (with Dominique Bourg, PUF, 2017) and *L'existence écologique: Critique existentielle de la croissance et anthropologie de l'après-croissance* (Seuil, 2023).

Gaia

Birgit Althans

Abstract This article aims to reconstruct James Lovelock's development of the Gaia Hypothesis, taking into account the narrative of Hesiod, found in adaptions of Lovelock & Margulis, Latour, Haraway and Stengers.

Gaia means 'the earth' in ancient Greek, and in Greek mythology, she is also regarded as the 'birth-giver' and embodiment of the earth. Hesiod's Theogony describes her as one of the first deities to emerge from the 'gaping abyss' of Chaos; she would also go on to found Olympus. Gaia, according to Hesiod, the 'broad-bosomed goddess', first gave birth to Uranos, the 'starry sky, that it might encompass her everywhere' and with him, she, as the 'solid ground', produced mountains, valleys, the sea, the Titans, the one-eyed Cyclopes and the hundred-armed Hecatoncheires. Gaia's children were hated by their father Uranos from the start and were pushed back by him into the womb, the bodily cavity of the earth (Gaia). Gaia thereupon created, 'because it became oppressively narrow to her', the 'grey ore of meteoric iron', forged from it a large sickle and tried to persuade her children to avenge the 'detestable outrages' of their father and to kill him. Out of fear, the children shied away from this, but her son, the 'great but devious' Kronos, uses the 'meteoric sickle' to cut off his father's genitals while pressing down his mother Gaia, 'full of lust [...] everywhere down.' Meanwhile, the blood spurting out of Uranos fertilises the 'broad earth', Gaia, again, who brings forth the giants, the Erinys and Nymphs. From the genitals, which had been thrown into the sea, came Aphrodite, purportedly born of sea foam (Hesiod, 2014: 14–15). Such is the mythological figure of Gaia, a narrative brimming with (self-)procreation and fertility, but also rape and revenge, whose genealogy extends into the god-figures of Olympus.

B. Althans (✉)
Kunstakademie Düsseldorf, Düsseldorf, Germany
e-mail: birgit.althans@kunstakademie-duesseldorf.de

© The Author(s), under exclusive license to Springer Nature
Switzerland AG 2023
N. Wallenhorst, C. Wulf (eds.), *Handbook of the Anthropocene*,
https://doi.org/10.1007/978-3-031-25910-4_64

In the 1970s, James Lovelock, founder of the Gaia hypothesis, resorted – incidentally on the advice of his village neighbour William Golding, the author of 'Lord of the Flies' – to the so monstrously described mythological figure of Gaia as a both life-giving and destructive 'Mother Earth' to illustrate his new perspective of the Earth as an active, living system that enables life in the Earth's biosphere. Lovelock had begun his career as a British scientist with academic degrees in medicine, chemistry and biophysics, conducting research at Harvard, Yale and NASA. For Lovelock, the extra-terrestrial view of Earth made possible by NASA in the 1960s was the spark. He turned the question 'Why is there no life on Mars?' into 'Why is there life on Earth?' (Bardi, 2021). From my point of view, Lovelock, in his adaptation of the mythological Gaia, needed a fictional character, similar to the writings of businessman and financial journalist Daniel Defoe at the beginning of the eighteenth century, to illustrate to his contemporaries the workings of a hitherto unknown system in which, through their behaviour, they all participated. If Defoe's Lady Credit, with her whims and sensitivities, embodied the unpredictability of the eighteenth-century financial market, Lovelock's adaptation of the Gaia figure illustrated the complex and highly sensitive interactions of microorganisms in the atmosphere and biosphere. For Lovelock, it is precisely this 'imbalance' of the earth as a 'living system', enabling life, that constitutes the vitality of the earth: 'Gaia is the planetary life system that includes everything influenced by and influencing the biota. The Gaia system shares with all living organisms the capacity for homeostasis – the regulation of the physical and chemical environment at a level that is favourable for life.' (Lovelock, 2000: 56) Ugo Bardi, professor of physical chemistry and member of the Club of Rome, emphasises in his introduction to the Gaia hypothesis that Lovelock's specific view of Earth was made possible because, as did Charles Darwin in the nineteenth century in his views on 'The Origin of Species', he observed an already-known system with different eyes. Unlike biologists, who are mostly concerned with theories of atmospheric composition and climate research, Lovelock, a chemist, recognised that the earth contained reactive gases that were not in chemical equilibrium, especially oxygen. Since nothing can remain out of equilibrium for long – 'No biosphere, no oxygen. And as a consequence, no oxygen, no biosphere.' (Bardi, 2021: 11) – and oxygen was present in the atmosphere, it had to be produced, according to the thesis, by an active process, by the photosynthetic activity of living beings. Lovelock developed from this the theory of the biosphere, in which oxygen is not only produced as a waste product in photosynthesis but, in the contingent course of the earth's history, developed through the encounter with other substances into an elementary prerequisite for life. Lovelock reconstructed this concept in evolutionary theory, especially in his collaboration with biologist Lynn Margulis and her work on microorganisms (Margulis & Sagan, 1997) as the interaction of highly divergent organisms that suddenly became atypically diverse.

'Oxygen is poisonous, it is mutagenic and probably carcinogenic, and it thus sets a limit to life spans. But its presence also opens abundant new opportunities for organisms. At the end of the Archean, the appearance of a little free oxygen would have worked wonders for those early ecosystems. Oxygen would have changed the environmental chemistry. The oxidation of atmospheric nitrogen to nitrates would have increased, as would the weathering of many rocks, particularly on the land surfaces. This would have made available nutrients that were previously scarce, and so allowed an increase in the abundance of life.' (Lovelock, 2000: 114)

Lovelock used his concept of homeostasis to describe how the biosphere and atmosphere interact to maintain a nearly constant concentration of oxygen. From these considerations arose Lovelock's and Margulis's view of the earth's biosphere as a single, interrelated system in which all components interact. Lovelock named this system, in reference to the Greek name for the earth, 'Gaia'. The problem? 'There is only one Gaia but Gaia is not One.' (Conway, 2016). So how can we think of Gaia?

Philosopher and sociologist Bruno Latour took up Lovelock's Gaia hypothesis and expanded it in six Clifford Lectures held in Edinburgh in 2013, republished in 2017 in the form of eight heavily revised chapters. Here, Latour examines Lovelock's descriptions of microorganisms interacting in the biosphere and atmosphere with reference to his own work on the 'agency' of microbes as related to the work of Louis Pasteur. He also reflects, including exchanges with Donna Haraway and Isabelle Stengers, on the best narratives for illustrating the problem of the Anthropocene. Latour expresses a fascination with Lovelock's writing, which (in a way, *avant la lettre*) fulfils both Felix Guattari's call in the 'Three Ecologies' (2014, org. 1989) for analyses of the 'hard sciences' to become more narrative, as well as Donna Haraway's view that scientific texts, especially those about the Anthropocene, should preferably be written as 'science fiction, speculative feminism, science fantasy, speculative fabulation, science fact' (Haraway, 2016: 10). The aim is to better relate differing perspectives to one another, but also, due to the urgency of the Anthropocene, to find the broadest possible reception. 'Lovelock's prose', as Latour finds, always reads a little like a detective story, except that the mystery to be solved is not triggered by the discovery of a dead body; on the contrary, it begins with the mystery of why a character has not been assassinated – at least not yet!" (Latour, 2017: 92) He applies this narrative pattern from an evolutionary theoretical perspective to a geochemical situation: 'Then we shall have to name the invisible protector that ensures the continuity of what ought to have disappeared billions of years ago, as it did on Mars and Venus. […] Carbon dioxide ought to be present in much larger quantities in the air? Where does it fall? Into the soil. By the intermediary of what agent? By the action of micro-organisms and vegetation. Now let us look to see whether these micro-organisms are up to the new role assigned to them.' (Latour, 2017: 92). Latour elucidates Lovelock's descriptions of the unexpected agency of hitherto little-noticed chemical actors using an illustration from Walt Disney: 'This staging has a cartoonish aspect, as if every time Lovelock touched some part of the décor with his magic wand, suddenly, as in a Disney version of "Sleeping Beauty", all the servants in the palace, until then passive and inert, awoke from their sleep, yawning, and began to move frenetically about – the dwarves and also the clock, the trees in the garden and also the knobs on the doors. The humblest accessories henceforth play a role, as if there were no more distinctions between the main characters and the extras. Everything that was a simple intermediary serving to transport a slim concatenation of causes and consequences becomes a mediator adding its own grain of salt to the story.' (Latour, 2017: ibid.). However, it always remains a natural science narrative. For Lovelock, everything located between the top of the upper atmosphere and the bottom of the sedimentary rock formations – what biochemists aptly call the 'critical zone' – turns out to be caught up in the same seething broth. 'The Earth's behavior is inexplicable without the addition of the work accomplished by living organisms, just as fermentation, for Pasteur, cannot be started without yeast.' (Latour, 2017: 93) Lovelock's equation of planet Earth with the Gaia figure attempts to make what Earth does no longer purely scientific, as movement in the universe, but rather 'to translate into a more or less comprehensible language the agency responsible for the fact that the Earth has a behavior' (Latour, 2017: 85), to increase our attention to the distribution of agency. Latour's adaptation of Gaia, inspired by Detienne's reading of Hesiod, describes her as a 'chthonic power, dark-skinned, dark-haired and somber' (Latour, 2017: 83), transforming her into a figure of the embodiment of a power 'before' the gods, an often terrible, sometimes benevolent power, read by Latour from a 'performative perspective', 'to replace what gods, concepts, objects, and things *are* by what they *do*.' (Latour, 2017: 86). In this, Latour supports Lovelock's position of not conceiving of Gaia as a 'superorganism' but rather as a 'structure' or 'assemblage', in which each part of the whole influences all others: 'If A modifies B, C, D, and X to benefit A's own survival, it is just as true that B, C, D, and X modify A in return. Animation is immediately propagated at all points.' (Latour, 2017: 99). And this presents the greatest complexity for scientific thinking, in that these non-intentional processes are not, as is usual, observable from the outside. The Gaia hypothesis teaches us that 'This time, we humans are not shocked to learn that the Earth no longer occupies the center and that it spins aimlessly around the Sun; no, if we are so profoundly shocked, it is on the contrary because we find ourselves at the center of its little universe, and because we are imprisoned in its minuscule local atmosphere.' (Latour, 2017: 80). Referring to the Anthropocene, Latour concludes that this situation needs to be politicized. Referencing the conservative philosopher Carl Schmitt, Latour sees the 'earthbound of the Anthropocene', who tell the Gaia story, 'at war' with the people who still deny the Anthropocene: 'Whereas Humans are defined as those who take the Earth, the Earthbound are taken by it [...] If the Humans and the Earthbound are at war, this could also happen to "their" scientists in conflict. Naturalist scientists – those who proudly assert that they are "of Nature" – are unfortunate figures, bound to disappear, disembodied, behind their knowledge, or to have souls, voices, and places, but at the risk of losing their authority.' (Latour, 2017: 251, 252). This position of Latour's has been criticized by the feminist theorists of the Anthropocene, Donna Haraway and Isabelle Stengers: 'In much of his writing, Latour develops the language and imagery of trials of strength; and in thinking about the Anthropocene and the Earthbound, he extends that metaphor to develop the difference between a police action, where peace is restored by an already existing order, and war or politics, where real enemies must be overcome to establish what will be.' (Haraway, 2016: 42) This is why, Haraway argues, Latour ultimately tells the Gaia hypothesis as a war narrative: 'Those trials—the war of the Earthbound with the Humans— would not be conducted with rockets and bombs; they would be conducted with every other imaginable resource and with no god trick from above to decide life and death, truth and error. But still, we are in the story of the hero and the first beautiful words and weapons, not in the story of the carrier bag. Anything not decided in the presence of the Authority is war; Science (singular and capitalized) is the Authority; the Authority conducts police actions. In contrast, sciences (always rooted in practices) are war. Therefore, in Latour's passionate speculative fabulation, such war is our only hope for real politics.' (Haraway, 2016: 42). Haraway does not consider this narrative to be helpful, and Isabelle Stengers holds a similar view. Haraway sees the confrontations of different groups with Gaia in the Anthropocene from the perspective of a cosmopolitically conceived 'ecology of practices', as a continuous interlocking – or entangling (Barad, 2007) and 'mutual seizure'. 'Cosmopolitics' in Stenger's sense does not aim at systematizing and unifying relations into an overarching whole but captures the concrete practices of 'mutual apprehension' through which entities situated at particular times and symbiotic milieus can form a common stock of practices: 'Ecology is, then, the science of multiplicities, disparate causalities, and unintentional creations of meaning.' (Stengers, 2010: 34).

Haraway holds Stengers' description of Gaia to be more complex than Latour's: 'Shaping her thinking about the times called Anthropocene and 'multi-faced Gaia' (Stengers' term) in companionable friction with Latour, Isabelle Stengers (2013) does not ask that we recompose ourselves to become able, perhaps, to "face Gaia".' (Haraway, 2016: 43). According to Haraway, Stengers, like Latour, evokes the name of Gaia in the same manner as did James Lovelock and Lynn Margulis, 'to name complex nonlinear couplings between processes that compose and sustain entwined but nonadditive subsystems as a partially cohering systemic whole. In this hypothesis, Gaia is autopoietic—self-forming, boundary maintaining, contingent, chapter two dynamic, and stable under some conditions but not others. Gaia is not reducible to the sum of its parts, but achieves finite systemic coherence in the face of perturbations within parameters that are themselves responsive to dynamic systemic processes.' (Haraway, 2016: 43f). According to Stengers' and Haraway's material feminist perspective: 'Earth/Gaia is maker and destroyer, not a resource to be exploited, or ward to be protected or nursing mother promising nourishment. Gaia is not a person but complex systemic phenomena that compose a living planet. Gaia's intrusion into our affairs is a radically materialist event that collects up multitudes. This intrusion threatens not life on earth itself—microbes will adapt, to put it mildly—but threatens the livability of earth for vast kinds, species, assemblages, and individuals in an "event" already under way called the Sixth Great Extinction.' (Haraway, 2016: 43). Haraway mentions that Gaia literally demands a fundamentally new way of thinking about our (damaged) planet and humankind: 'Gaia does not and could not care about human or other biological beings' intentions or desires or needs, but Gaia puts into question our very existence, we who have provoked its brutal mutation that threatens both human and nonhuman livable presents and futures. Gaia is not about a list of questions waiting for rational policies. Gaia is an intrusive event that undoes thinking as usual.' (Haraway, 2016: 44)

References

Barad, K. (2007). *Meeting the Universe Halfway*. Duke University Press.
Bardi, U. (2021). Gaia, eine Idee, die voranschreitet. In J. Lovelock (Ed.), *Das Gaia-Prinzip. Die Biographie unseres Planeten* (pp. 9–19). Oekom-Verlag.
Conway, P. (2016). Back down to earth: Reassembling Latour's Anthropocenic Geopolitics. *Global Discourse, 6*(1–2), 43–71. https://doi.org/10.1080/23269995.2015.1004247
Guattari, F. (2014). *Three ecologies*. Bloomsbury.
Haraway, D. (2016). *Staying with the Trouble. Making Kin in the Chthulucene*. Duke University Press.
Hesiod. (2014). *Theogonie* (R. Schrott, Trans.). Hanser.
Latour, B. (2017). *Facing Gaia. Eight lectures in the New Climatic Regime*. Polity Press.
Lovelock, J. (1991/2000). *Gaia: The practical science of planetary medicine*. Oxford University Press.
Margulis, L., & Sagan, D. (1986/1997). *Microcosmos: Four billion years of evolution from our microbial ancestors*. University of California Press.
Stengers, I. (2010). *Cosmopolitics I: The science wars*. University of Minnesota Press.
Stengers, I. (2013). In conversation with Heather Davis and Etienne Turpin. Matters of Cosmopolitics: On the provocations of Gaia. In E. Turpin (Ed.), *Architecture in the Anthropocene: Encounters among design, deep time, science and philosophy* (pp. 171–182). Open Humanities.

Birgit Althans is Professor for Pedagogy in the department of art-related sciences at The Art Academy Duesseldorf. Main Research in Historical and Pedagogical Anthropology; Gender and Cultural Studies; qualitative, cultural science oriented research in pedagogical and artistic fields of work (especially theater), cultural education in anthropocene.

History

Anna Echterhölter

Abstract For the first time in 4.54 billion years, Earth, flora, fauna, and climate are unrectifiably affected by the sum of human actions. As a result, historians are witnessing their own domain of expertise, the development of human society over time, transgress into the realm of nature. Clearly, the humanities are challenged by the epoch and state of the Anthropocene, which is more than just a matter of geology. In response, historians have taken decisive steps to develop more viable versions of recording history. These include: (a) reconsidering frameworks of historiography to accommodate nature amidst the history of human interactions, (b) paying attention to multiple dynamic scales, (c) minding and multiplying the categories of historical actor, and (d) viewing key topics such as environmental economics, energy and resource extraction through a critical, epistemological lens. It is time to adjust the patterns of justification and responsibility which lie in historical narratives.

It was no small provocation of historiography, when the French Annales historian Emmanuel Le Roy Ladurie put climate into the centre of his archival research and suggested to write "history without men" (Le Roy Ladurie, 1967). To investigate the history of natural environments (and not human society) in an isolated manner demanded an unexpected asceticism, resulting in an emphasis on nature changing perspectives of history. However, this is the opposite of what Anthropocene history claims to pursue. It is not primarily interested in the misogyny unleashed by a small Ice-Age. Anthropocene history is interested in how humans affect the Earth. A prominent example is the "orbis spike", the measurable decline in atmospheric CO_2 in the early seventeenth century. It is explained through the depopulation of the Americas though extractive capitalism and colonialism (Davis et al., 2019: 3).

A. Echterhölter (✉)
University of Vienna, Vienna, Austria
e-mail: anna.echterhoelter@univie.ac.at

© The Author(s), under exclusive license to Springer Nature Switzerland AG 2023
N. Wallenhorst, C. Wulf (eds.), *Handbook of the Anthropocene*,
https://doi.org/10.1007/978-3-031-25910-4_65

Historiography under the auspices of Anthropocene cannot be regarded without the role of humans. It has to recognize humans as a natural force (and important contributing factor). Additionally, it must encompass local and planetary scales in the face of extinction. Just as global historians, historians of Anthropocene grapple with an epistemic object of awkward planetary proportions. Thus, there are clear characteristics which constitute one special form of environmental history, alongside big history, social ecology, history of meteorology, approaches towards multispecies, energy humanities and others. All in all, Anthropocene history represents a form of unsettled historiography and remains doubtful of its framework. The new facts encourage to rethink conceptual matters and to break with certain notions, which were regarded highly for a long time (Robin & Steffen, 2007; Chakrabarty, 2009; Thomas, 2014; Otter et al., 2018; Höhler & Westermann, 2020). It makes the attempt to innovate and includes several methodological strategies, such as the following:

Anthropocene History as Unsettled Historiography

The challenges in the field of Anthropocene history begin with the unclear positioning of science and humanities. Anthropocene as a geological discourse was coined by a group of geologists stemming from earth systems science and chronostratigraphy. Anthropocene as a public and transdisciplinary research field was sustained by a world-wide "Knowledge-Action-Network" and a cooperation of scholars, artists and activists that even developed new textbooks (Horn & Bergthaller, 2020) and interdisciplinary research journals (*Anthropocene*, or *The Anthropocene Review*). Albeit that historians of science and technology have been accompanying the Anthropocene Working Group from its very origin in 2009, there was widespread doubt whether the humanities in general and history in particular could even contribute to this "geological" field. International policy discourse assigned climate change to data driven science and singular indicators, excluding social sciences and humanities as a whole (Robin & Steffen, 2007: 1698). However, the subject in question was soon revisited. Instead of succumbing to the technocratic narrations of the Anthropocene, the very constellation of human nature calls for the implication of historians, as they are closely involved with the cultural reality and socio-economic dynamics of the affected societies. As a result, they might be better equipped to reflect on the complex dynamics of nature and humans as well as natural politics (Bonneuil & Fressoz, 2016: Chap. 9).

What characterizes history under the auspices of Anthropocene is that conceptual distinctions which separate history and the sciences have to be reconsidered, as for example the difference between cultural and natural forces (Chakrabarty, 2009: 201; Arni, 2018). Nevertheless, the question of how history can accommodate both, human and non-human agency, society and nature, remains. New ground was broken in the 1920s, when the concept of the *biosphere* was developed. It added life to the geological strata of lithosphere, hydrosphere, and atmosphere. The concept devised by the Soviet geochemist Wladimir Iwanowitsch Vernadskij made all the difference, because it highlighted the unique ability of living organisms to interact with matter and to shape environments by their constant establishment of order (Rispoli & Grinevald, 2018). Gradually, new concepts began to accommodate the far-reaching concerns of human nature. To delineate human-Earth interactions further, Vernadskij introduced the *noosphere*. Borrowed from theology, this methodological tool captures the changes in the environment, which are effectuated by inventions of the human mind, like for instance chemical knowledge. With the introduction of the *technosphere* (Nelson et al., 2017), the planetary impact of human thought has today become a more comprehensive notion, telling the history of tools, machines and infrastructures, which form their own strata around the globe.

Scales and the Politics of Geological Time

The Anthropocene imposes a temporal order of geology onto the historian's craft. Admittedly, epoch-making, calendars, and chronology were never solely at the disposal of historians. At all times they had to share the craft of creating epochs, centuries and time reckoning systems with religions, rulers, astronomers and geologists. The discovery of the deep time of geology in the course of the eighteenth century is a case in point. Scientific knowledge helped phase out the use of the Biblical chronology of *anno mundi* in historiographical writing. Now another geological fact is making demands on historiography and its chronological order. Contrary to earlier scientific conceptualizations of time, the Anthropocene carries political connotations. It was consciously designed to hold the future over the present and to recognise the rights of future generations over those ruling today. Admittedly, these social and political implications disqualify the Anthropocene as geological era for some scientists, who believe that geology does deal with measurable geological facts, not with futurology. However, in view of geology's own past, one might contest these claims, since geology did include religious perspectives, for instance, not so long ago (Warde et al., 2017).

The intermingling of scientific and historical scales, of abstract ones and those burdened with politics and responsibilities, is even more common if we look beyond Europe. Better grounds for combining geological and historical, scientific and anthropological scales of time have been reported from Australia, where strong traditions of millennial and geological time reckoning have evolved. Thus, the intellectual climate was favourable "to combine human and planetary time in a single, multi-scalar framework" (Bashford, 2013; Otter et al., 2018: 571). Furthermore, abstract scientific time exported by colonialism, seizes to remain politically neutral. The introduction of the colonialist's geological time keeping in India in the nineteenth century is a well investigated case (Chakrabarti, 2020). A neutral, science-based chronology superseded Indian views of Antiquity, previous canons, cosmologies and historical accounts.

Very often the subject of scales goes beyond mere temporal issues. It attests to efforts of redistributing responsibilities within historical narratives (Höhler & Westermann, 2020), economic calculation (Warde et al., 2017) or commodity chains. Deborah Coen has shown that it can be very fruitful to consider dynamic scales. She defined the versatile scaling of local and global perspectives as the secret to the success of Alpine meteorology. What is more, she drew parallels between historical and natural scales, or better: the multiscale views of the liberal Habsburg empire and the first successful dynamic modelling of climate (Coen, 2018).

Actor Categories – Subaltern Animals, Black and White Social Histories

The place of humans in the Anthropocene is a controversial topic. Andreas Malm has ridiculed the idea that the Anthropocene is caused by the human species poignantly (Malm & Hornborg, 2014). While he and Alf Hornberg concede that no other species seems to throw the planet off balance, they refuse to see the whole species at fault. This diminishes the role played by capitalism while concealing the fact that the life and production styles of very few white males, who began to wield steam engines against colonized people and exploited nature without measure, are to blame. Moreover, while the risks are accumulating on a planetary scale, the species perspective blurs all access to precise understanding of the dynamics. Historical analysis may thus find more to analyse on the meso-scale, e.g. communal infrastructure planning, the space in which the majority of environmental protests and their specific reframing of scientific knowledge operate (Güttler, 2019). Gabrielle Hecht has pointed out that for a thorough understanding of the Anthropocene as a global phenomenon, the divergent actors' perspectives and economic dynamics from Africa are a necessity (Hecht, 2018). Black feminist theory has further decentred and tentatively provincialized the Anthropocene as a history of dispossession in which "white geology" is identified as a key driver (Yussof, 2019; Davis et al., 2019).

While motives of subaltern studies may have already been visible in this shift of perspective towards the Global South, they were further explored beyond species boundaries. There is a rising interest in animals, both wild and domestic, as disadvantaged actors, and even their non-human empires have been taken into consideration (Deb Roy & Sivasundaram, 2015; Bird Rose et al., 2017). While it is easy to agree that the tides must turn and the behaviour of the species ought to be broken down along the colour lines of race, class, and gender, the most appropriate point of view for larger developments remains elusive. Yet, it would be indispensable for historians to agree on actor's categories that reflect multispecies concerns and environmental justice in the long term. Too often historical narratives hinge on the interests of singular actors or institutions, rarely eclipsing the timespan of a generation.

Toward an Epistemology of Resources

While most issues surrounding the Anthropocene touch upon method, some key topics are also of renewed interest. With regards to the history of technology, it has been common to phrase historical epochs according to types of energy consumption, or to trace the impact of large technological systems based on their regulatory frameworks and "momentum"—such as roads, energy pipelines, or data infrastructures (cf. the research network "Tensions of Europe"). The world history of crucial materials is on the rise, as well as the history of resource thinking and phantasies of "cornucopianism", or perceived overabundance of energy (Albritton Jonsson, 2014). In many European languages, the term "resources" appears comparatively late in the nineteenth century, as a highly specific type of matter, firmly cornered by economic, geologic, and geopolitical interests. Under the auspices of modern economic theory, nature has become an exchangeable, yet finite, form of capital, increasingly accounted for in global resource surveys (Westermann, 2017). Yet, as far as classical economic theory is concerned, nature was never perceived as a factor to be reckoned with, despite the remarkable efforts of hetero-dox traditions like ecological economics (Martinez-Alier & Muradian, 2015). The history of resource economics, as another pertinent subfield of economic theory, is criticized of having too narrow of a scope, and for "the mismatch between the scales at which human and natural systems organize. These lead to failures in feedback, when, for instance, benefits accrue at one scale, but costs are carried by another" (Robin & Steffen, 2007: 1698).

While resources and energy are key topics and draw due attention, they are sometimes reduced to technical issues, such as the reduction of the future to questions of climate, the reduction of crisis responses to technological fixes and the translation of the social imaginary into the singular goal of survival according to path dependencies churned out by earth system modelling (Selcer, 2021). Anthropocene history, just as environmental or energy humanities, should not amount to an immanent history of economics or resources. It should strive to maintain the much larger societal outlook acculturated by history, while abandoning national and increasingly nationalist histories.

One way to avoid these technocratic pitfalls has been to look south for decolonized methods (Smith, 1999). Elisabeth Povinelli has taken counter knowledge and "radical alterity" around the globe as a starting point for the investigation of new epistemologies (Povinelli, 2001). As far as the current crisis emerged from global extraction regimes (also addressed to as "plantationocene"), modern Northern epistemologies may be too fatally implicated to serve as sole response. In analogy to the current extinction of plants and animals the extinction of thought systems is recorded in an increasing pace. The damage done by such epistemicides is summarised as a global impoverishment of divergent ontologies and outlooks on life (de Sousa Santos, 2014). The loss of diversity in ways to organise the relation of humans to nature is not just another case of salvage anthropology. It has a practical side. With these ways of perceiving of nature, rare patterns of justification and ways of sharing responsibilities in subsistence economies are lost, which may have proven fruitful for arriving at an outlook for multiple species living on a finite Earth. The comparative research program New Earth Histories in particular promotes a mundane and comparative outlook and teaches all historically occurring relations towards Earth on equal par (Bobbette et al., 2023). As the example of personal rights for rivers shows, ecological knowledge and legal tropes of non-industrialized societies are increasingly transformed into practical policies (Ludwig & Macnaghten, 2020).

References

Albritton Jonsson, F. (2014). The origins of Cornucopianism: A preliminary genealogy. *Critical Historical Studies, 1*(1), 151–168.

Arni, C. (2018). Nach der Kultur. Anthropologische Potentiale für eine rekursive Geschichtsschreibung. *Historische Anthropologie, 26*(2), 200–223. https://doi.org/10.7788/ha-2018-260206

Bashford, A. (2013). The Anthropocene is modern history: Reflections on climate and Australian deep time. *Australian Historical Studies, 44*(3), 341–349. https://doi.org/10.108 0/1031461X.2013.817454

Bird Rose, D., et al. (Eds.). (2017). *Extinction studies. Stories of time, death, and generations.* Columbia University Press.

Bobbette, A., Bashford, A., & Kern, E. (Eds.). (2023). *New earth histories.* University of Chicago Press. (Accepted for publication).

Bonneuil, C., & Fressoz, J.-B. (2016). *The shock of the Anthropocene. The earth, history and us.* Verso.

Chakrabarti, P. (2020). *Inscriptions of nature: Geology and the naturalization of antiquity.* Johns Hopkins University Press.

Chakrabarty, D. (2009). The climate of history: Four theses. *Critical Inquiry, 35*, 197–222.

Coen, D. (2018). *Climate in motion: Science, empire, and the problem of scale.* University of Chicago Press.

Davis, J., Moulton, A. A., Van Sant, L., & Williams, B. (2019). Anthropocene, Capitalocene, ... Plantationocene? A manifesto for ecological justice in an age of global crises. *Geography Compass, 13*, e12438. https://doi.org/10.1111/gec3.12438

de Sousa Santos, B. (2014). *Epistemologies of the south: Justice against Epistemicide.* Routledge.

Deb Roy, R., & Sivasundaram, S. (Eds.). (2015). Nonhuman empires. *Comparative Studies of South Asia, Africa and the Middle East, 35*(1), 66–75. https://doi.org/10.1215/1089201X-2876104

Güttler, N. (2019). Hungry for knowledge. Towards a meso-history of the environmental sciences. *Berichte zur Wissenschaftsgeschichte, 42*, 235–258.

Hecht, G. (2018). Interscalar vehicles for an African Anthropocene. On waste, temporality, and violence. *Cultural Anthropology, 33*(1), 109–141.

Höhler, S., & Westermann, A. (2020). Writing history in the Anthropocene. Scaling, accountability, and accumulation. *Geschichte und Gesellschaft, 46*, 579–605.

Horn, E., & Bergthaller, H. (2020). *The Anthropocene: Key issues for the humanities.* Routledge.

Le Roy Ladurie, E. (1967). *Histoire du climat depuis l'an mil.* Flammarion.

Ludwig, D., & Macnaghten, P. (2020). Traditional ecological knowledge in innovation governance: A framework for responsible and just innovation. *Journal of Responsible Innovation, 7*(1), 26–44.

Malm, A., & Hornborg, A. (2014). The geology of mankind? A critique of the Anthropocene narrative. *The Anthropocene Review, 1*(1), 62–69. https://doi.org/10.1177/2053019613516291

Martinez Alier, J., & Muradian, R. (Eds.). (2015). *Handbook of ecological economics*. Edward Elgar Publishing.

Nelson, S., Rosol, C., & Renn, J. (2017). Perspectives on the technosphere. Special issue of *The Anthropocene Review, 4*, 1.

Otter, C., Bashford, A., Brooke, J. L., Jonsson, F. A., & Kelly, J. M. (2018). Roundtable: The Anthropocene in British history. *Journal of British Studies, 57*, 568–596. https://doi.org/10.1017/jbr.2018.79

Povinelli, E. (2001). Radical worlds: The anthropology of incommensurability and inconceivability. *Annual Review of Anthropology, 30*, 319–334.

Rispoli, G., & Grinevald, J. (2018). Vladimir Vernadsky and the co-evoloution of the biosphere, the Noosphere, and the Technosphere. *Technosphere Magazine*.

Robin, L., & Steffen, W. (2007). History for the Anthropocene. *History Compass, 5*(5), 1694–1719.

Selcer, P. (2021). Anthropocene. *Encyclopedia of the History of Science*. https://doi.org/10.34758/zr3n-jj68.

Smith, L. T. (1999). *Decolonizing methodologies: Research and indigenous peoples*. Zed Books.

Thomas, J. A. (2014). History and biology in the Anthropocene. Problems of scale, problems of value. AHR roundtable. *American Historical Review, 119*, 1587–1607.

Warde, P., et al. (2017). Stratigraphy for the renaissance. Questions of expertise for "the environment" and "the Anthropocene". *The Anthropocene Review, 4*, 246–258.

Westermann, A. (2017). The end of gold? Monetary metals studied at the planetary and human scale during the classical gold standard era. In I. Borowy & M. Schmelzer (Eds.), *History of the future of economic growth. Historical roots of current debates on sustainable degrowth* (pp. 69–90). Routledge.

Yussof, K. (2019). *A billion black Anthropocenes or none*. University of Minnesota Press.

Anna Echterhölter is a Professor of History of Science at the University of Vienna. She has been a member of the Vienna Anthropocene Network and a co-convener of the Anthropocene Histories partnership seminar at the IHS in London. She participated in the Anthropocene Markers Group at the MPIWG and with Alexa Färber she pre-enacted an Austrian Climate Audit Court. Publications: 2022. Human-Mineral Classification: Taxonomy, Totemism, and the Technofossils of the Anthropocene https://www.anthropocene-curriculum.org; 2020. Quantifzierungsdenken im Westpazifik. Für eine Epistemologie der Ressourcen. *Nach Feierabend* 15: 125–144; 2020. Shells and Order. Questionnaires on Indigenous Law in German New Guinea. *JHoK* 17: 1–19.

I thank Sophie Page, Amanda Power, Sujit Sivasundaram for ongoing discussions in the "Anthropocene Histories" seminar, and John Sabapathy for bringing us all together.

Humanities

Rosi Braidotti and Hiltraud Casper-Hehne

Abstract The term 'Anthropocene' was proposed by the geological and natural sciences community to describe the current geological epoch and show the influence of human activity on the planetary ecosystem and its dynamics. This idea was taken up by Humanities scholars from a wide range of disciplines. It functions within the Humanities as a complex and multi-facetted notion that refers to the simultaneous occurrence of different environmental, technological and social transformations. This focus is particularly marked in the New Humanities, that call for renewed attention for the role of cultural, narrative and social issues in shaping collective responsibility for the future of the Earth. The exact meaning and empirical evidence supporting the Anthropocene however, are also met with criticism. We argue that the concept needs to be supplemented by more specific notions and practices, in order to avoid a growing sense of disciplinary segregation in the emergent areas of Humanities scholarship.

The term 'Anthropocene' has been proposed in 2002 by the scientific community as a way to describe the current geological epoch. It aims to show the measurable impact of human activity – notable technological developments and unchecked consumerism – in relation to the 'glocal' ecosystem and its dynamics. This has given rise to a rather polemical debate that continues to occupy the academic world and which has been exacerbated by the Covid-19 pandemic.

We think it is best therefore to approach the notion of the Anthropocene within the Humanities as a complex and multi-facetted idea – both a description of our historical condition and a methodological tool to navigate some of its contradictions. It refers to a complex phenomenon that points to the simultaneous occurrence

R. Braidotti (✉)
Distinguished University Professor, University of Utrecht, Utrecht, Netherlands
e-mail: r.braidotti@uu.nl

H. Casper-Hehne
Intercultural German Studies, University of Göttingen, Göttingen, Germany
e-mail: h.casper-hehne@phil.uni.goettingen.de

© The Author(s), under exclusive license to Springer Nature Switzerland AG 2023
N. Wallenhorst, C. Wulf (eds.), *Handbook of the Anthropocene*,
https://doi.org/10.1007/978-3-031-25910-4_66

of different – and internally contradictory – kinds of environmental, technological and social transformations. At the environmental level, we are witnessing the climate change crisis, the extinction of many species on a depleted planet struck by extreme weather conditions and new epidemics. At the technological level, the traditional understandings of the human have been redefined by the expansion of the life sciences and genomics, neural sciences and robotics, nanotechnologies, the new information technologies and the digital interconnections they construct. At the social level, the joint impact of those two phenomena is causing increasing polarizations and social injustices through the unequal distribution of wealth, prosperity and access to technology. According to Oxfam in 2020, the world's 22 wealthiest people owned more Wealth than the 4.6 billion poorest people.

This acute situation has also been described in terms of the posthuman convergence (Braidotti, 2013), with the Fourth Industrial Revolution meeting the Sixth Extinction. The Fourth Industrial Revolution (Schwab, 2015) involves the expansion of advanced technologies, but also their intrusion into the very fabric of living matter – both in humans and non-humans – through unprecedented advances in genomics and the life sciences. This means that the categorical divide between biology and technology, or nature and culture, has shifted significantly. The Sixth Extinction on the other hand refers to the endangered status of many species during the current geological era, as the result of human activity, frantic consumerism and technological intrusion (Kolbert, 2014). These are world-wide or planetary phenomena, which however acquire specific features in different contexts and thus call for multiple perspectives, rather than a mono-paradigmatic approach. They need to be addressed as intersecting phenomena, happening concurrently.

The Humanities in general and especially cultural studies argue that humans' relationship to nature needs some drastic revisions. "Our concept of nature is outdated. Nature is neither an obstacle nor a harmonious other, no longer a power that can be separated from or ambivalent towards human action. Man shapes nature. Humanity finds its expression in the history of the earth" (Scherer & Klingan, 2013: 2).

The COVID-19 pandemic is emblematic of the sharp contradictions of the era of the Anthropocene, in that it combines all the three aspects mentioned above. It highlights the negative effects of undue human interference in the lives of multiple species, and shows how less privileged social classes, marginalized genders and ethnic groups are disproportionately more exposed to the risks and dangers of this condition. The pandemic, however, has also shown and to a large extent increased our collective dependence on the very technologies that lie at the core of our consumeristic and energy-wasting culture. Both information, communication and bio-medical technologies have become all the more important as a result of the Covid-19 contagion.

Thus, the Anthropocene as a marker of this particular moment in history affects social and environmental ecologies, but also the social imaginaries, as well as individual psychological states and emotions. The emotional or affective dimension is of great importance, as it sets a social mood of pain and anxiety, uncertainty about the present and the future. It also imposes on us all the imperative to review established opinions and to question received notions and understandings of what it means to be human. These affective, ethical and even pastoral care aspects are highly relevant for the pedagogical practice of the Humanities.

The concept of the Anthropocene, which originated in the geological and natural sciences, has been taken up by Humanities scholars in a multiplicity of approaches, across a wide range of disciplines and academic practices. It provides a useful but not linear framework for research practices and broader intellectual debates about the current environmental, technological and social changes. Further, it helps us assess collectively how these material conditions affect our shared sense of humanity, the representations and values we can uphold today. This concept functions therefore like a theoretical navigational tool that assists Humanities scholars in the task of critical reflection on contemporary cultural and socio-economic formations. In this respect, the Anthropocene as a horizon of thought not only entails scientific, technical, social, economic and even cultural aspects, but also raises issues of representation, ethical values and participatory citizenship (Möllers, 2015: 122). The New Humanities in particular call to strengthen the role of the cultural narrative and social issues in shaping collective responsibility for the future of the Earth (Mauch & Trischler, 2013: 9).

Our argument is that we need to learn to address these complexities in a parallel and not compartmentalized fashion. We need to confront the contradictions not only intellectually, but also affectively and to do so in an affirmative, ethical manner. This conviction rests on the firm belief that the Anthropocene is not only a crisis for the Humanities, but also a great opportunity for the field to renew and update itself. We equally believe, however, that such a change requires the analysis and revision of set and established ideas, that is to say, a bit more conceptual creativity and methodological innovation.

However, there has also been much criticism of the concept, its exact meaning and range of applications. Paradoxically, the notion of the Anthropocene ends up actually highlighting the anthropocentric dimension, as it "evokes human-centredness" (Crist, 2013: 129).

Donna Haraway is even more explicit: "Please tell me that you share my anger, that in this moment of transdisciplinarity and multispecies everything, in this moment of beginning to get a glimmer of how truly richly complex the world is and always has been, someone has the unmitigated arrogance to name it the Anthropocene" (Haraway, 2016: 545). This designation also hides the relationship of humans to animals and plants and other cycles of the earth, that is to say, the importance of non-human factors and entities.

In other words, exclusive focus on the Anthropocene can result in too partial a picture. Braidotti has argued for instance (Braidotti, 2019) that because of its over-generic nature, the Anthropocene could not stand the pressure of the multiple approaches it generated and has thus become an "anthropomeme" (Macfarlane, 2016). That is to say it has generated a plethora of alternative but aligned notions, such as: 'Chthulucene' (Haraway, 2016), 'Capitalocene' (Moore, 2013), and Anthropobscene' (Parikka, 2015). And there are yet others: Plasticene, 'Plantationcene' (Tsing, 2015) and 'Misanthropocene' (Clover & Spahr, 2014).

Thus, just referring to the Anthropocene is not enough. Rather the concept needs to be supplemented by other notions and practices, in order to avoid a growing sense of disciplinary segregation in the emergent areas of Humanities scholarship. We need to keep complexity and systemic inter- and trans-disciplinarity in mind. The Anthropocene is not complete without an analysis of socio-political conditions and the economic disparities that it entails. Nor can it avoid a serious confrontation with its effects on identities and cultural belonging, and on the formation of subjectivity: what does it mean to do academic research as the planet all around us is dying?

By addressing social justice, ethical and political concerns at the core of the geo-centered discussions, the Humanities also raise questions of self-representation, that is to say the formation of social imaginaries about the current predicament and the societal challenges it throws our way. Most people's perception and understanding of the climate change crisis, for instance, but also of the threats and opportunities of the new technologies, is mediated by cultural, visual, literary and media representations, which constitute the core fields of enquiry of the Humanities. We would even dare to suggest that the Humanities contribute to bring the allegedly 'unrepresent-able' dimensions of the Anthropocene into public representation. We think that the construction of the social imaginary and the analysis of these forms of representation of our current predicament are the prerogative of the Humanities as a teaching and research field.

This specific function of the Humanities is especially important as the Anthropocene as a whole is a rather gloomy concept, that often causes a morose mood of pending disaster and inevitable apocalypse. The entertainment industry has been quick in commodifying this mood, turning the catastrophe into a highly profit-able genre: disaster movies, extinction series, 'morning after' visions. They are variations on what is becoming known as "Collapsogy" (Servigne & Stevens, 2020), that is to say, a terminal form of cultural pessimism. The future of the human – in this culturally specific inception – is now pre-occupying vast numbers of social commentators and philosophers (Fukuyama, 2002; Habermas, 2003; Sloterdijk, 2009; Pope, 2015). The mood is generally sombre, but the Humanities bring also more affirmative and generative scenarios. They can offer innovative and daring visions for the future, through the study of history, culture, the arts and the literary sources of speculative science fiction and other genres that investigate and design possible futures for our and other species. Comparative intercultural perspectives are then crucial in lifting the West out of its current gloom.

References

Braidotti, R. (2013). *The posthuman*. Polity Press.
Braidotti, R. (2019). *Posthuman knowledge*. Polity Press.
Clover, J., & Spahr, J. (2014). *Misanthropocene: 24 theses*. Commune Editions.
Crist, E. (2013). On the poverty of nomenclature. *Environmental Humanities, 3*, 129–147.
Fukuyama, F. (2002). *Our Posthuman future: Consequences of the biotechnological revolution*. Profile Books.

Habermas, J. (2003). *The future of human nature*. Polity Press.

Haraway, D. (2016). *Staying with the trouble: Making kin in the Chthulucene*. Duke University Press.

Kolbert, E. (2014). *The sixth extinction. The unnatural history*. Henry Holt and Company.

Macfarlane, R. (2016). Generation Anthropocene: How humans have altered the planet forever. *The Guardian*, 1 April. Available at: https://www.theguardian.com/books/2016/apr/01/generation-anthropocene-altered-planet-for-ever

Mauch, C., & Trischler, H. (2013). *Making tracks. Human and environmental histories*. Rachel Carson Center.

Möllers, N. (Hrsg.) (2015). *Willkommen im Anthropozän. Unsere Verantwortung für die Zukunft der Erde* (pp. 122–125). Deutsches Museum.

Moore, J. (2013). Anthropocene, Capitalocene, and the myth of industrialization. *World-ecological imaginations: Power and production in the web of life*, 16 June. Available at: https://jasonwmoore.wordpress.com/2013/06/16/anthropocene-capitalocene-the-myth-of-industrialization/

Oxfam. (2020). https://www.oxfam.org/en/press-releases/worlds-billionaires-have-more-wealth-46-billion-people

Parikka, J. (2015). *The Anthropocene*. University of Minnesota Press.

Pope, F. (2015). *Encyclical letter Laudato si': On care for our common home*. The Vatican Press.

Scherer, B., & Klingan, K. (2013). Das Anthropozän-Projekt. *Eine Eröffnung*. 10.-13. Januar 2012.

Schwab, K. (2015). The fourth industrial revolution. *Foreign Affairs*, 12 December.

Servigne, P., & Stevens, R. (2020). *Collapsology*. Polity Press.

Sloterdijk, P. (2009). Rules for the human zoo: A response to the letter on humanism. *Environment and Planning. Society and Space, 27*(1), 12–28.

Tsing, A. L. (2015). *The mushroom at the end of the world: On the possibility of life in capitalist ruins*. Princeton University Press.

Rosi Braidotti is distinguished University Professor at Utrecht; Honorary Degrees Helsinki, 2007, Linkoping, 2013; Fellow of the Australian Academy of the Humanities (*FAHA), 2009*; Member of the Academia Europaea *(MAE)*, 2014.; recipient Humboldt Research Award, 2021. Main publications: More Posthuman Glossary (Bloomsbury Academic) with Rosi Braidotti, Emily Jones, Goda Klumbyte (ed). *Nomadic Subjects* (2011a), and *Nomadic Theory*. (2011b), Columbia University Press. *The Posthuman*, 2013, *Posthuman Knowledge*, 2019 and *Posthuman Feminism*, 2021, Polity Press.

Hiltraud Casper-Hehne is distinguished University Professor at Göttingen; 2009–2021 Vice President for International Affairs at Göttingen University; since 2021 Director of the network NEH21: New European Humanities in the twenty-first Century; since 2020 member of the Executive Committee of the *European University* ENLIGHT. Main publications: Braidotti, Rosi/Casper-Hehne, Hiltraud/Ivkovic, Marjan/Oostveen, Daan F. (2021) (eds.): *Trajectories for the Humanities in the twenty-first Century*. Edinburgh Publishing House (forthcoming).

Hydro-Social Cycle

Marie-Eve Perrin

Abstract The concept of the hydro-social cycle developed by Linton and Budds allows us to theorise and analyse water-society relations on the one hand, but also to understand the social nature of water itself. Water, in fact, shapes and is shaped by social relations, structures and subjectivities. It is not simply the object of politics, but also internalises these political and social relations (Linton and Budds. Geoforum 57:170–80, 2014). In this respect, the case study on the subject of water in a mining landscape in northern New Caledonia shows the intertwining of the social, the hydric and the political in the extractivist daily life of miners in the Anthropocene era.

Evaporation, condensation, transpiration and precipitation: this is how the water cycle is commonly represented and taught. For Linton and Budds, however, this axiom is based on a certain vision of 'nature'. Nature is said to be detached from humans and their practices, external to culture. A nature/culture dichotomy that characterises the naturalistic way of being in the world on which modern Western society is based (Latour, 1991; Viveiros de Castro 2014; Descola, 2005). However, Bruno Latour, in *Nous n'avons jamais été modernes* (1991), argues that despite this 'Great Divide', in practice, the 'moderns' never cease to produce 'hybrid' objects, which belong to both, and which they refuse to think about (Demeulenaere, 2017).

It is at the heart of this critique that the concept of the hydro-social cycle is inscribed: an intrinsically hybrid cycle producing – and produced by – hybrid objects/subjects. Indeed, "the need to control water affects the organisation of society, which in turn affects the disposition of water, which in turn causes new forms of social organisation to emerge, and so on, in a cyclical process" (Linton & Budds, 2014). In this sense, water and society shape and reshape each other materially and discursively, across time and space. Moreover, to the extent that water embodies social relations, particular types of relations produce particular types of water. The

M.-E. Perrin (✉)
EHESS, Paris, France
e-mail: M.budna@iffeurope.org

433

following case study highlights several types of water and the interweaving of the subject of water in mining practices, imaginaries and representations.

When rainwater falls on the highest altitudes of the commune of Poum (New Caledonia), it encounters a mountain that has been exploited since the end of the nineteenth century for its ore – cobalt, chromium and then nickel – and flows into the Titch and Ponvio tribes, before flowing into the sea. The water then appears uneventful, its course seems peaceful and inexorably unchanged for ever. However, if we focus on the mine, we can see material and immaterial paths, both hydric and social: stretches of water make up the mining landscape, "watering machines" drive alongside large 100-tonne trucks, employees are dedicated exclusively to "water management", and water is thought of right from the design of the mining project.

The water reservoirs and sedimentation basins are dug by the miner to ensure social, technical and economic functions. The former is intended to water the roadways in order to prevent dust blowing and potential respiratory problems for drivers. If the drivers feel that the track is not watered enough, they are entitled to stop the vehicle despite the consequent decrease in mining productivity. In this sense, putting water to work conditions the putting of people to work and ultimately the achievement of the extractivist project. The interdependence of their activities in the extractivist system shows a first facet of the hybridity of water. As for the sedimentation basins, or decanters, they are designed to slow down the surface water. The process is simple: the height of the water and the surface of the structure are sufficiently large to slow down the flow and allow the coarse particles to fall by gravity to the bottom of the basin, and then the fines to settle. Thus, the sedimentation basin plays an essential pivotal role: while it slows the flow and effectively reduces the risk of erosion downstream, it also removes from the water the traces of erosion – natural and man-made – upstream of the houses. The decanter thus acts as a socio-hydric separator between the mine and the inhabitants at the foot of the massif. This is a crucial issue in a New Caledonian context where water made cloudy by mining activity has often been the subject of conflict between the inhabitants and the industrialist, both locally and in other communes in the territory.

For this reason, three employees of the mining company – named SLN- work full time on the so-called environmental work: a technician, an excavator driver and a truck driver. A brief description of a scene that could be described as anthropoceni-cal shows the immaterial presence of water in the daily life of these workers: the backhoe extends its bucket one metre above the ground, the rear hydraulic arm descends slightly while the bucket and the front arm close synchronously in the ground. The large arm rises, the excavator rotates on itself, and the bucket releases the mound of earth six metres away. A few minutes later, the technician is talking to the shovel operator on the site, right next to the rumbling machine. He points to positions in the landscape, gestures to the imaginary lines that need to be dug further. In fact, both interlocutors are talking about the path on which they want to invite the water to pass. The nature/culture dichotomy no longer makes sense. These humans commit their bodies, their time, their increased knowledge of the landscape and their imaginations, to build structures to invite water. In a sense, water, absent when they work, continues to flow in their representations. And when it is materially present, the paths it takes are neither natural nor man-made, they are hybrid. In contrast to the useful water for watering the slopes, the three agents represent water as a constraint, which has the power to sabotage mining productivity and mine/resident relations. It is an important player in mining decisions, and one in which the industrialist invests time, money, and wear and tear on his machinery.

Water is also an issue in pre-project meetings between the company, the town hall, the Northern Province and the local people. For example, during a communal mining commission (CMC) concerning a project to explore the plateau by drilling boreholes, a discussion about water began. When the representative of the Province, a Kanak and pro-independence executive, took the floor, he anchored his speech in the local landscape by referring nine times to the proper names of rivers, a cove and a forest: 'Titch', 'Ponvio', 'Pwané', 'Paevala'. Then he insists on one point: the Titch River must be preserved from the mining operation that is being extended. The Titch River rises to the north-east of the summit plateau being explored and joins the tribe of the same name below. It has an essential place in one of the tribe's founding myths, as it is said to have been the result of a request for water addressed to the mountain goblins by three tribesmen. Today, it is their main source of drinking water.

However, in response to the concern raised, the industrialist invites us to consider the "water resource" contained in the alluvial plain, located at the foot of the mining massif to the south, which could meet the tribe's drinking water needs instead of the Titch. In the space of a short dialogue, he managed to translate a socio-ecological entity with a proper name into H_2O water. A Latourian purification acted out live (Latour, 1991). In this sense, rivers, waterholes, creeks and groundwater are all spatial discretizations of the all-encompassing H_2O. What Jamie Linton calls a "modern abstraction"(Linton, 2010).

This readjustment of focus in the miner's discourse is not insignificant. In 2019, the establishment of a water catchment upstream of the Titch has led to the creation of three nested protection areas – immediate, close (PPR) and remote – within which no activity is authorised. Thus, in order to repeal the Titch catchment and its PPR, SLN donated two water boreholes -named C1 and C9 -, located in the plain, to the commune. The company and its subcontractor are also financing 75% of the project to connect the water supply network. In the long term, these works will cover 80% of the commune's drinking water needs – including the inhabited blocks – in a context where inequalities in access to drinking water have been the commune's major concern since its creation in 1977. Indeed, having become a public problem in the 1950s in a post-Brazzaville context of development of the French colonies, drinking water is still scarce in Poum. Through the retrocession of these boreholes by the private sector, H_2O water is moving from being a public problem to an opportunity for social acceptance of the mine. A kind of exchange between infrastructure and perimeters: in exchange for C1 and C9, the Titch and Ponvio catchments would be abrogated, the polygons representing the PEPs would be removed from the maps and would no longer constrain the limits of mining.

The choice of following water in practices and discourses, and in the infrastructures that constrain it, reveals a water cycle that is both hydrological and social, as well as different types of water that reconfigure the organisation of society on several scales: that of the mining activity and that of the commune. Firstly, water put to work allowed the improvement of the miners' working conditions and the maintenance of a certain rate of mining productivity. Then, over the course of the decades, the use of water under pressure modified SLN's way of working. The company learned to deal with environmental and social sensitivities by converting into internalities what was previously external to its activity. Thus, valleys, watercourses and the landscape are now part of the mining mindset from the design of the mining project to the shipping of the ore. Finally, the representation of water-H_2O, homogeneous and universal, has met that of the water of the Titch, locally anchored. A conflict of representations that is both scalar and ontological. This led to the retrocession of two water boreholes and a large financial contribution from the miner to the commune. In exchange, on the one hand, the mining activity is better accepted by the population, and on the other hand, in the long run, the mining project can extend upstream of the Titch. Here, it is the representation of water as a scarce element and a public problem that gives rise to new forms of economic, political, social and water organisations that combine the public and the private in the management of this common good.

For Leslie Canoye, the recycling of a public problem into an issue of social acceptance of the mine is a characteristic feature of the capitalist system, which recuperates criticism: the latter constrains it while at the same time fuelling it (Carnoye, 2017). This raises the question of the "compossibilities" of worlds in the Anthropocene era (Balaud & Chopot, 2021). Is negotiation with a system as resilient as the capitalist system not at the genesis of this era also called the Capitalocene (Bonneuil, 2017)? Balaud and Chopot invite us to stop the dialogue and to fight with and for the collective rather than the human. In New Caledonia, however, the issue is more complex. While in South America the struggle for indigenous rights is often synonymous with the struggle against extractivist projects, in Kanak country, Kanak pro-independence ideology converges on the idea of Kanak entrepreneurial mining development (Demmer, 2018). How can we understand the compromise that exists between the feeling of belonging to the land (Le Meur, 2010), a pillar of Kanak culture, and the exploitation of the land in the name of economic emancipation in a post-colonial context?

References

Balaud, L., & Chopot, A. (2021). *Nous Ne Sommes Pas Seuls Politique Des Soulèvements Terrestres*. Seuil.
Bonneuil, C. (2017). Capitalocène Réflexions Sur l'échange Écologique Inégal et Le Crime Climatique à l'âge de l'Anthropocène. *Association EcoRev, 44*(11), 52–60.

Carnoye, L. (2017). L'écologisme, Une Critique Du Capitalisme? L'école Française Des Conventions Au Risque de La Question Environnementale. *Revue de Philosophie Économique, 18*, 29–58. https://doi.org/10.3917/rpec.182.0029

Demeulenaere, E. (2017). L'Anthropologie Au-Delà de l'anthropos. Un Récit Par Les Marges de La Discipline. In *Humanités Environnementales. Enquêtes et Contre-Enquêtes*, Editions de la Sorbone (pp. 43–73). Hal archives-ouvertes.

Demmer, C. (2018). Nationalisme Minier, Secteur Nickel et Décolonisation En Nouvelle-Calédonie. *Cahiers Jaurès, 230*(4), 35–52. https://doi.org/10.3917/cj.230.0035

Descola, P. (2005). *Par-Delà La Nature et La Culture*. Gallimard.

Latour, B. (1991). *Nous n'avons Jamais Été Modernes Essai d'anthropologie Symétrique*. La Découverte.

Linton, J. (2010). *What is water? The history of a modern abstraction*. UBC Press.

Linton, J., & Budds, J. (2014). The Hydrosocial cycle: Defining and mobilizing a relational-dialectical approach to water. *Geoforum, 57*, 170–180. https://doi.org/10.1016/j.geoforum.2013.10.008

Le Meur, P. Y. (2010). La Terre En Nouvelle-Calédonie: Pollution, Appartenance et Propriété Intellectuelle. *Multitudes, 41*(2), 91–98. https://doi.org/10.3917/mult.041.0091

Viveiros de Castro, E. (2014). Perspectivisme et multinaturalisme en Amérique Indigène (Perspectivism and multinaturalism in indigenous America). *Journal Des Anthropologues, 138–139*, 161–181. https://doi.org/10.4000/jda.4512

Marie-Eve Perrin has worked on the links between mining, water and society in New Caledonia. A geological technician for a mining company in the past, she now holds a master's degree in Knowledge in Society, with a specialization in environmental studies, from EHESS Paris.

Interculturality

Johann Chalmel

Abstract The challenges of the Anthropocene are many and diverse. One of the areas that need to be worked on to meet them is the transformation of education. In a complex society, each citizen must develop the power to act while respecting ecosystems and planetary boundaries. In this respect, the "liquid" approach to interculturality, which takes into account the cultural identity of each individual in interaction, offers us an interesting framework for (re)thinking pedagogical methods that embrace complexity. Pedagogical methods developed on this basis lead learners to reflect on their place in society and the opportunities they have, while making the most of the complexity of the context in which they live.

The term interculturality has multiple meanings. It is difficult, if not impossible, to elaborate a universally accepted definition. The way J. Demorgon (2003, p. 59–60) suggests to address it, is to reflect on it through the prism of related notions: the transcultural, which characterizes the integration by a given culture of an element of another culture, and the multicultural, which presents cultures as distinct sets that do not evolve, even when they interact with each other. This conceptual framework allows us to rethink the intercultural as a dynamic focused on the moment of exchange between two cultures with the resulting co-construction. An intercultural approach makes it possible to relax the "rigidity" of multiculturalism by analysing the real impact of encounters (however small) on the cultures involved and to evaluate the transcultural elements with the mutations they undergo during the integration process. Interculturality is therefore concerned with moments of interaction in which one or more cultures evolve in ways that are often imperceptible due to the brevity, multiplicity, and complexity of exchanges. Moreover, the long process of integration and adaptation of an element by one or more cultures rarely allows it to be observed.

J. Chalmel (✉)
Haute-Alsace University, Mulhouse, France
e-mail: johann.chalmel@uha.fr

N. Wallenhorst, C. Wulf (eds.), *Handbook of the Anthropocene*,
https://doi.org/10.1007/978-3-031-25910-4_68

Interculturality is all the more difficult to define as it includes the notion of culture, which also has many meanings. When talking about the encounter or exchange between cultures, one must first of all agree on what level this concept should be placed. For Edgard Morin (2015, p. 4) the human condition is ternary: individual, society, species. Culture can therefore be perceived as individual, namely the identity that the individual has develops through his or her interaction experiences, collective when it refers to the practices and knowledge of a given community, or universal when it refers to the human species as a whole with its own characteristics.

Interculturality, with its focus on the dynamics of exchange, cannot be directly concerned with the universal culture that brings humans together as a whole (the interaction of the species with its environment, the earth system, cannot be considered as "intercultural"; culture is after all inherently specific to humanity) meaning that interculturality can be observed only on the first two levels of individual and society.

Historically, the notion of interculturality was born at the end of the 1950s, when the American occupiers wanted to develop communication with the Japanese occupied (Demorgon, 2003, p. 44). This would mark the genesis of intercultural communication, which would develop in the business world to facilitate international relations between large companies and specialists who would concentrate then mostly on the collective aspect of culture, often reducing it to the nationality of the individual who holds it. This approach left its mark on international management training with, for example, G. Hofstede, who conducted a famous survey involving 10,000 employees of the multinational IBM in the 1970s. He defined six cultural dimensions to characterize individuals, mainly according to their nationality and presented culture as "the collective programming of the mind that one acquires while growing up in a given country" (Hofstede et al., 2010, p. 600). The training courses created on this basis are based on one principle: by knowing the country of origin of my interlocutor I can adapt my behaviour according to the score of each dimension (https://www.hofstede-insights.com/product/compare-countries/).

In Europe, the notion of interculturality was approached differently starting in the late 1970s. As a response to the reception of immigrants, particularly in the classrooms, intercultural training focused first on the issue of language and then on extending the subject to the culture of the home country. That's how institutes such as the Franco-German Youth Office promote interculturality to enable cooperation (Demorgon, 2003, p. 47). Interculturality is mainly used as an educational tool, as evidenced by the first appearance of the term in 1998 in the Encyclopaedic Dictionary of Education and Training (Champy et al., 2005) in an article entitled "Intercultural Education" by M. Abdallah-Pretceille. She would change the understanding of interculturality by approaching it through the prism of cultural identity: "The intercultural approach (…) emphasises the processes and interactions that unite and define individuals and groups in relation to each other. (…) More than an additional variable, culture gives or restores a place to the subject, to the interactions, to the context, to the processes of analysis using the other as a mirror." (Pretceille, 2015).

Inspired by the work of Z. Bauman, F. Dervin (2012) follows this path by opposing solid approaches to interculturality "which ignore the context of interaction and the complexity of the individuals brought into contact, individuals who are reduced to cultural facts", with a "liquid" approach "which refuses the quasi-systematic equation between discourse and acts, therefore between "internal" or "external" descriptions of cultures or their members as "evidence" or veri-conditional arguments". This approach embraces the complexity of situations and individuals, leading the individuals themselves to continually question their own perception of Others, of contexts, but also of their way of acting, interacting and thinking.

In the context of crossing of planetary boundaries (Rockström et al., 2009), this capacity of individuals and groups to properly deal with complexity and uncertainty becomes crucial. According to N. Wallenhorst (2020) an educational paradigm shift is needed in order to achieve this: from "educating about" to "educating within" the Anthropocene era. Through this idea, the author pursues a twofold objective: to break out of the habitus of rationalist and segmented reasoning, with a strong emphasis on disciplinary knowledge (neglecting the interactions between them), and to move from an educational approach that is mainly focused on the personal development of the individual to the training of citizens capable of envisaging their development in relation to contextual imperatives (Wallenhorst, 2020). This paradigm shift requires the teaching of new methods to "(…) capture the mutual relationships and reciprocal influences between parts and whole in a complex world." (Morin, 2015).

For this reason, the individual cannot focus solely on his economic prosperity and the increase of his own comfort but must consider his or her context differently, not as the person responsible for the whole but as part of a whole, of an ecosystem. On the other hand, the rational thinking on which our modern societies are based opposes nature and culture in the way it functions and does not allow for this change of cosmology (Curnier, 2017). To achieve such a transformation, K. Wilber (2000) proposes that our actions, whatever the need to which they respond, should systematically pass through the filter of ethical questioning to ensure that the complexity of each situation is taken into account. His scheme leads the individual seeking to meet his or her economic needs to ask himself or herself whether he or she respects the others, the living, the ecosystems and the self. In this way, the collective and the individual, the subjective and the objective are being questioned in order to develop tailor-made solutions to an individual's problems while respecting the political, ecological and systemic challenges of his/her context. Based on this scheme, Curnier (2017) proposes to move from the rationalist way of thinking to an ecological humanism based on strong sustainability and distanced from hypermodernity, offering a social organization that respects planetary boundaries and simultaneously takes into account ecological, political and social temporalities.

The social transformation proposed by ecological humanism requires a profound change in our current cosmology and educational systems. Moving away from a rationalist education framework requires moving towards "a model that fosters the multiplicity of intelligence forms according to a global vision of thought that would make it possible, among other things, to overcome the oppositions of reflection-action and creativity-activity. The questioning of the relationship between man and nature could be based on an epistemology that strives for dialogue between natural sciences and human/social sciences, as well as making use of practical and traditional knowledge from different cultural contexts" (Curnier, 2017). The potential of such a design of education is vast and the pedagogical work associated with it is considerable.

This state-of-the-art in interculturality and education within Anthropocene leads us to the following hypothesis: pedagogies developed on the basis of the "liquid" approach to interculturality offer interesting avenues for reflection, with a vision of transforming education.

Chalmel et al. (2019) propose a framework adapted to intercultural accompaniment methods (NovaTris, Center for cross-border skills, ANR-IDFI-0005): each accompanier can develop his or her own method using theoretical invariants which ensure that the complexity of the other and its context are taken into account in the process of developing intercultural competence. We believe that it is possible to make use of and adapt these methods in the context of the educational design in favour of ecological humanism and, at the same time, to take advantage of intercultural dynamics in order to bring out innovations that take into account the diversity of learners.

First of all, intercultural training as mentioned above is part of the "educating within" proposed by Wallenhorst. The shift of the trainer from being in a position of authority to a position of accompaniment facilitates the consideration of the contexts of each student, who then is enabled to build individualised competences in the group, adapted to his/her values, needs and context.

To do this, the suggested activities initially guide the learner to understand who he or she is and what his or her context is (the famous "educating about"), in order to deconstruct it. The interaction with the "diverse diversities" (Dervin, 2012) of the group leads them, in a mirror game, to consider the potentials thus detected from new perspectives, better adapted to their context, their possibilities and their limits. This didactic process of conscientisation (Freire et al., 2001) seems to be an interesting path for development in order to give the learner a glimpse of the possibilities and hence escape from the immobility often observed when confronted with the Anthropocene's immensity of challenges.

Secondly, intercultural accompaniment requires the setting up of enabling environments (Oudet, 2012) for learners in intercultural training, whose "conversion factors" enable each individual to test his or her potential, to discover his or her power to act and get prepared to act in society with a sharpened knowledge of his or her capability and context. Experiential learning (Kolb, 1984) is in this case a key framework that trainers implement through active pedagogy activities that are collectively debriefed and individually formalised so that each person can discover his or her potential (internal and external resources) and concretely consider their use according to his or her real context with its possibilities and limits. This approach could, among others, integrate Wilber's (2000) scheme, to lead everyone to reflect on their capability (Sen, 2009) while respecting ecosystems and planetary limits.

The learner could then have the freedom to condition the choice whether or not to engage his or her capability in his or her real context to an eco-citizen ethic: "If we keep the idea of a major contribution of education to a goal of societal transformation and not only adaptation or improvement due to the urgency and importance to meet up challenges of the Anthropocene, then we must choose an emancipatory project of an autonomous, proactive, creative and committed subject: the author-subject"(Lange et al., 2019).

Finally, the "liquid" approach of the provided intercultural accompaniments makes use of the various fields of expertise and points of view of each person. The projects on which learners work together are thereby approached through a variety of disciplinary lenses and divergent opinions. The process of negotiation, joint reflection and co-construction leads everyone to consider the complexities as a requirement for the coherent resolution of the problem they work on. Such an approach could be applied in a training course "within Anthropocene", taking care to involve a heterogeneous audience from a variety of disciplines.

References

Chalmel, J., et al. (2019). Oser une nouvelle approche de l'interculturel—Exemple de la « pédagogie NovaTris ». In F. Dervin & N. Auger (Éds.), *Les nouvelles voix/voies de l'interculturel* (Le langage et l'homme), pp. (179–174). EME éditions.

Champy, P., et al. (2005). *Dictionnaire encyclopédique de l'éducation et de la formation* (3e éd. Entièrement revue et mise à jour). Retz.

Curnier, D. (2017). Éducation et durabilité forte : Considérations sur les fondements et les finalités de l'institution. *La Pensée écologique, N°, 1*(1), 252–271.

Demorgon, J. (2003). L'interculturel entre réception et invention. Contextes, médias, concepts. *Questions de communication, 4*, 43–70.

Dervin, F. (2012). *Impostures interculturelles*. L'Harmattan.

Freire, P., et al. (2001). *Pédagogie des opprimés : Suivi de Conscientisation et révolution*. La Découverte.

Hofstede, et al. (2010). *Cultures et organisations : Nos programmations mentales*. Pearson Education France.

Kolb, D. A. (1984). *Experiential learning: Experience as the source of learning and development*. Prentice-Hall.

Lange, J.-M., et al. (2019). Penser l'éducation au temps de l'anthropocène : Conditions de possibilités d'une culture de l'engagement. *Éducation et socialisation, 51*.

Morin, E. (2015). Les sept savoirs nécessaires à l'éducation du futur.

Oudet, S. F. (2012). Concevoir des environnements de travail capacitants : L'exemple d'un réseau réciproque d'échanges des savoirs. *Formation emploi. Revue française de sciences sociales, 119*, 7–27.

Pretceille, M. (2015). L'interculturel comme paradigme de transgression par rapport au culturalisme. *Voix Plurielles, 12*(2), 251–263.

Rockström, J., et al. (2009). A safe operating space for humanity. *Nature, 461*(7263), 472–475.

Sen, A. (2009). Un nouveau modèle économique : Développement, justice, liberté.

Wallenhorst, N. (2020). III. Quel paradigme pour éduquer en Anthropocène? In *Humains, animaux, nature* (p. 253–265).

Wilber, K. (2000). *A theory of everything: An integral vision for business, politics, science, and spirituality*. Shambhala.

Johann Chalmel is a junior professor at the University of Haute-Alsace (UHA) in the Chair in Interculturalities. He is a doctor in educational science, specialising in the themes of pedagogy, intangible cultural heritage and interculturality. He is the author of the book: Pedagogy, an intangible cultural heritage? (l'Harmattan, 2020, in French) and has contributed to the development of pedagogical methods of intercultural accompaniment at NovaTris, the centre for cross-border skills (ANR-IDFI-0005).

Narratives

Martin Bohle

Abstract Earth System Literacy offers a rich matrix of narratives about (planet) Earth, World and Humans when interwoven with cultural artefacts (e.g. arts, history). Anthropogenic global change locks them together in a unifying framework. By making a distinction between reproduction, work (*ergon*, Greek) and civicness, there are narratives that convey what characterises the human condition today, including the material, social and historic settings of the Anthropocene.

Humans shape their environment (niche), responding to the needs and insights that they filter through their worldviews and experiences. As part of an 'economy of knowledge' (the *"social processes pertaining to the production, preservation, accumulation, circulation, and appropriation of knowledge"* (Renn, 2020; p.429, p.151ff)), story-telling helps to analyse worldviews and experiences, including being a tool for making value statements, shaping individual and societal perspectives or handling abstract concepts, comparing observations, constructing thoughts, or assessing cultural and social contexts. Furthermore, sharing narratives is a social practice that, for example, consists of showing affective relations, describing different contexts, and spreading or challenging practices.

Traditional narratives about supernatural agents ruling people, the World and planet Earth have been replaced in contemporaneous Worlds by narratives casting traditions and conventions or scientifically validated knowledge. While historical and modern narratives differ on how humans intersect with planet Earth, both describe the human condition and contextualise facts within value-laden frames of reference (Salvatore et al., 2019).

Civic awareness of how contemporaneous people intersect with planet Earth requires a double bearing of Earth and society. Therefore, narratives must be about the planet and people. Also, they must be woven to tell about people-people

M. Bohle (✉)
International Association for Promoting Geoethics (IAPG), Rome, Italy

Ronin Institute, Montclair, NJ, USA
e-mail: martin.bohle@ronininstitute.org

interactions, including the specific cognitive and affective relations that dominate the urbanites' world (Bohle et al., 2017).

Contemporary societies bind the entire globe into a single social-ecological system, the Earth System. The instruments are global supply chains, an all-embracing division of labour, a planetary technosphere, and a shared knowledge system (Otto et al., 2020; Rosol et al., 2017). Cycles of matter, energy, and information link the socio-economic (sub)systems of the Earth System to the planet's physical and biological (sub)systems. People purposefully design (engineer) the economic intersections with planet Earth to meet their needs (e.g. food, shelter, health) and preferences (e.g. lifestyle). Many artisans, technicians, architects, engineers, scientists and artists use Earth Science knowledge to alter (natural) environments, enhance the human niche, or create artefacts. Various philosophical and cultural frameworks describe how planet Earth and World intersect, combining scientific understanding, economic-technological know-how, and socio-cultural endorsement or criticism (Fressoz, 2012; Purdy, 2015; Meiske, 2021).

Narratives in the incipient Anthropocene, the good, the bad and the ugly (Dalby, 2015), weave a dense fabric, thread by thread. They tell about the functioning of the Earth System, social relations, including relations of power, modes of production and consumption (Rosol et al., 2017; Dryzek & Pickering, 2019; Otto et al., 2020). They also describe how affluent groups of individuals (people, classes, nations) emerged as a *telluric* force (Mokyr, 2016; Reinhard, 2016).

The contemporary hegemonic culture of anthropogenic global change that tackles Nature as a cheap resource, and its marginalised alternatives, offer multiple storylines, whether told by scholars, citizens or folks. Although the stories differ, the shared subject is the same, how to relate planet Earth, the World and Humans. A prominent witness is the Anthropocene Working Group (Will, 2021), established by the international geological community to debate the geologist's professional story about (to be or not to be) the Anthropocene. The scholarly storylines of anthropogenic global change can be structured by epistemic communities, such as Earth sciences, engineering, sociology, economy, history, arts or philosophy. A broader set of storylines would resist such classification, and the scholarly discourses spill over in the public sphere (Sklair, 2020). Over the past two or three decades, a fabric of contemporary narratives of the Anthropocene emerged describing Earth as one entangled system, as different as the various yarns may be.

Perceiving Earth, World and Humans as a single system has been the cultural standard, for aeons, among indigenous people, non-Western cultures, or Europe's feudal cultures. For example, referring to the deep historical past, the Gilgamesh epic (George, 2000) tells the story of Uruk, one of the first cities built 5000 years ago in Mesopotamia. It describes how these city-dwellers perceive the experiential connections of (their) World and planet Earth. Faith-based Earth-centric perceptions were interwoven with society-centric views about people-people interactions, earthly gods and godly earthlings.

Contemporaneous people describe World and Earth differently from the inhabitants of Uruk, be it only because of the respective physical and mental artefacts at their disposal. Nevertheless, commonalities exist. Applying Hannah Arendt's works

(Arendt, 1958), the concepts of labour (*laborans*' struggle for biological and social reproduction at levels of subsistence or affluence), work (*homo-faber* building and operating the ergosphere (*ergon*, Greek, meaning "work"; see (Renn, 2020; [p. 382ff]) for the concept of a physical technosphere being enlarged by the related knowledge system to the ergosphere)), and civic activities (*zoo politikon's* political and cultural activity) apply to historical and contemporaneous settings. Subsequently, these concepts classify narratives into three broad categories, narratives of reproduction, work and civicness (governance), whichever role an individual, collective, or institutional human agent may assume.

Although *laborans'* impact on Earth, struggling for reproduction, has been noticeable for several thousand years (Lewis & Maslin, 2018), the contemporary patterns of the use of renewable and non-renewable resources have a new quality (Rosol et al., 2017). Quite recently, *homo-faber* has built a planetary technosphere conceived by *zoon-politikons* of European origin (Mokyr, 2016; Reinhard, 2016; Dryzek & Pickering, 2019). Contemporary humans labour to reproduce biologically and socially, design and operate the technosphere, and act as citizens. The notion 'Anthropocene' (i.e. the physical anthropogenic global changes, including the societal processes that led to them) describes the appropriation of biotic and abiotic resources for reproduction, enabled through an ergosphere, the planetary technosphere coupled with a hegemonic culture of norms and lifestyles. That is the essence of the human condition in the Anthropocene. The onset of the Anthropocene marks the end of the *laborans'* unintentional impact on Earth. Instead (Hamilton, 2017), it challenges the *zoon-politikons* regarding the endeavours of the *homo-fabers* (engineering the ergosphere) to enable lasting reproduction in the Anthropocene (for *laborans* and others).

To handle the intersections of Earth, World and Humans in the complex-adaptive dynamics of Anthropocene, citizens should benefit from Earth Systems Literacy (Wysession et al., 2012; Bohle, 2015). The citizens' insights should also dwell on the awareness of philosophical systems' appropriateness and sustainability of daily practices (which may have grown over centuries). Consequently, narratives in the Anthropocene should inform the governance of anthropogenic global change from the perspectives of *laborans*, *homo-faber*, and *zoon-politikon*.

As most humans live in cities, specific narratives that address urbanites are particularly needed. Densely packed, urban people live on a relatively small percentage of the Earth's habitable surface. *Homo-faber* is engineering urban environments to shelter people from hazards; to frame *laborans'* dependence on the natural pace of Earth System dynamics; to favour *zoon-politikon's* political and cultural experiences. Living a modern urban lifestyle, experiences of planet Earth are biased towards events that disrupt the functioning of the engineered structures. Hence, experiencing planet Earth is the exception, and experiencing the World is the rule. In the Anthropocene, the rule will likely be broken.

To summarise, Earth sciences have many societal bearings because this knowledge domain is relevant to economic development (e.g. production of goods, living conditions, individual well-being) and cultural value-setting (the functioning of the Earth System, including the evolution of life-bearing planets). Narratives about the human condition are empty without Earth System Literacy. When interwoven, for example, with arts and cultural history, the bearing of Earth System Literacy offers a matrix of Earth, World and Humans narratives in the Holocene and the Anthropocene(s).

References

Arendt, H. (1958). *The human condition*. The University of Chicago Press.

Bohle, M. (2015). Simple geoethics: An essay on daily earth science. *Geological Society, London, Special Publications, 419*(1), 5–12. https://doi.org/10.1144/SP419.3

Bohle, M., Sibilla, A., & Graells, R. C. I. (2017). A concept of society-earth-centric narratives. *Annals of Geophysics, 60*(7). https://doi.org/10.4401/ag-7358

Dalby, S. (2015). Framing the Anthropocene: The good, the bad and the ugly. *The Anthropocene Review, 3*(1), 1–19. https://doi.org/10.1177/2053019615618681

Dryzek, J. S., & Pickering, J. (2019). *The politics of the Anthropocene*. Oxford University Press.

George, A. (2000). *The Epic of Gilgamesh*. Penguin Classics, London, 228p.

Fressoz, J.-B. (2012). *L'Apocalypse Joyeuse – Une Histoire Du Risque Technologique*. Le Seuil.

Hamilton, C. (2017). *Defiant earth – The fate of humans in the Anthropocene*. Wiley, Polity Press.

Lewis, S. L., & Maslin, M. A. (2018). *The human planet – How we created the Anthropocene*. Penguin Random House.

Meiske, M. (2021). *Die Geburt Des Geoengineerings: Großbauprojekte in Der Frühphase Des Anthropozäns*. Wallstein Verlag.

Mokyr, J. (2016). *A culture of growth*. Princeton University Press. https://doi.org/10.1515/9781400882915

Otto, I. M., Wiedermann, M., Cremades, R., Donges, J. F., Auer, C., & Lucht, W. (2020). Human agency in the Anthropocene. *Ecological Economics, 167*(September 2018), 106463. https://doi.org/10.1016/j.ecolecon.2019.106463

Purdy, J. (2015). *After nature: A politics for the Anthropocene*. Havard University Press.

Reinhard, W. (2016). *Die Unterwerfung Der Welt – Globalgeschichte Der Europäischen Expansion 1415–2015*. Verlag C.H. Beck oHG.

Renn, J. (2020). *The evolution of knowledge – Rethinking science for the Anthropocene*. Princeton University Press.

Rosol, C., Nelson, S., & Renn, J. (2017). Introduction: In the machine room of the Anthropocene. *The Anthropocene Review, 4*(1), 2–8. https://doi.org/10.1177/2053019617701165

Salvatore, S., Rochira, A., & Kharlamov, N. (2019). The embodiment of cultural meanings. Symbolic universes as forms of life. In S. Salvatore, V. Fini, T. Mannarini, J. Valsiner, & G. A. Veltri (Eds.), *Symbolic universes in time of (post)crisis* (pp. 235–253). Springer. https://doi.org/10.1007/978-3-030-19497-0_8

Sklair, L. (2020). *The Anthropocene in global media: Neutralising the risk*. Routledge. https://doi.org/10.4324/9780429355202

Will, F. (2021). Evidenz Für Das Anthropozän -Wissensbildung Und Aushandlungsprozesse an Der Schnittstelle von Natur-, Geistes- Und Sozialwissenschaften. In C. Mausch & H. Trischler (Eds.), *Umwelt und*. Vandenhoeck & Ruprecht.

Wysession, M. E., LaDue, N., Budd, D. A., Campbell, K., Conklin, M., Kappel, E., Lewis, G., et al. (2012). Developing and applying a set of earth science literacy principles. *Journal of Geoscience Education, 60*(2), 95–99. https://doi.org/10.5408/11-248.1

Martin Bohle obtained a doctoral degree from the Ecole Polytechnique Fédéral de Lausanne in 1986. He published on aquatic systems and was an official of the European Commission (DG for Research and Innovation). Retired (2019), he is affiliated with the Ronin Institute for Independent Scholarship (Montclair, NJ, USA) and cooperates with the International Association for Promoting Geosethics (Rome, Italy). He publishes on the relationships of Earth sciences and society; for example, Bohle, M. & Marone, E. (eds.) (2021). Geo-societal Narratives – Contextualising Geosciences. https://doi.org/10.1007/978-3-030-79028-8.

NBIC

Edouard Kleinpeter

Abstract NBIC is an acronym that stands for Nanotechnology, Biotechnology, Information Sciences and applied Cognitive Sciences. It is usually associated with the idea of a putative convergence between these four fields of human technoscience. From the contemporary perspective, it encompasses two major meanings: as an epistemological view on scientific activity and as an agenda of how scientific research is to be pursued. In this article, we describe how this term emerged from a report published in 2003, before examining its philosophical foundations and underlying concepts, namely materialism, the notion of emergence borrowed from complex systems theory, and a certain analysis of scientific work. We then discuss how it relates to the modern practice of interdisciplinarity in science.

NBIC is an acronym that designates four fields of human technoscience: Nanotechnology, Biotechnology, Information Sciences and applied Cognitive Sciences. It is generally used in the expression "*NBIC Convergence*" which would thus designate the putative convergence between these aforementioned four fields. It originates from a report commissioned jointly by the US National Science Foundation and Department of Commerce and co-directed by the sociologist William Bainbridge and Mihail Roco, an engineer specialising in nanotechnology (Roco & Bainbridge, 2003). They expose a prospective vision of this convergence's opportunities which they identify in various domains of human enhancement: intelligence, physical capacities, health and longevity, communication, social life, education, etc. This report soon provided a scientific grounding for many transhumanist schools of thought and proved to be a science policy programme that is still influencing research funding nowadays in the US. As questionable as this report may be for its programmatic dimension and its (very) laudatory tone, it is nonetheless based on the observation of a real convergence, already elaborated at the end of the 1930's. At this time, the idea of a possible generalization of the informational machine,

E. Kleinpeter (✉)
Bordeaux University, Bordeaux, France
e-mail: edouard.kleinpeter@u-bordeaux.fr

from artificial control systems to biological and sociological systems began to take shape, prefiguring the emergence of the cybernetics paradigm in the 50's (Claverie, 2014). The centrality of concepts such as "entropy" or "code", as described in Erwin Schrödinger famous text *"What is Life?"* (Schrödinger, 1993) to the initiation of a convergence between biology and physics, has laid the foundations of a community of principles between knowledge fields previously distinct.

In a modern perspective, the development of interdisciplinarity on the one hand and of technologies (information processing, manipulation of objects on a very small scale, modelling and simulation of cognitive processes, functional imagery, Artificial Intelligence, etc.) on the other have led to the discovery, among the four aforementioned fields, of the efficiency of the method used by one to solve the problem of another, to the possibility of translating the questions of one into the language of another, etc. A cross-fertilization of knowledge and a mutual enrichment of paradigms then ensued. It led to the formulation of the possibility of a common ontology, thus causing the emergence of epistemologies getting rid of disciplinary boundaries and based firstly on the hypothesis of the transversal nature of the notion of information, most entities of the world being considered as systems that process, exchange and store information. Hence, for example, the "genetic code" metaphor entails a conception of living beings as transfer vectors of an information that flows and modifies itself independently of all eternity (Dawkins, 1999). Besides, the NBIC convergence heavily relies on the hypothesis of a strong materialism, every phenomenon of the world being reducible to interactions between material entities. It entails a gradation of the heuristic scale modelled on the physical scale: from the most explanatory (the atomic scale) to the most global (the social scale). One consequence is the centrality of the concept of emergence, which is the distinctive mark of the science of complex systems since the 80's. The general idea is that the material and informational interaction of a vast number of low-level entities will cause their auto-organization and lead to the emergence of higher-level phenomena that cannot be adequately explained with low-level heuristics. The canonical example is the emergence of high-level psychological phenomena such as, say, reflexive consciousness, morality or empathy, which would eventually emerge from electrical interaction between neurons in the brain. Modern technologies such as deep learning through convolutional neural network or spiking neural networks, based on the interdependence between artificial entities layers (programmes and/or circuits) which mimic the functioning of biological neurons, have led to the conception of machines that exhibit behaviours usually associated with human intelligence: visual recognition, data classification, playing (and winning) games based on reasoning such as Go and chess, etc. Their undeniable successes in fields that still remain very specific (general intelligence now seems to be unattainable), would tend to give credit to the idea of the possibility of the emergence of a form of specialized intelligence based on the interaction between low-level auto-organized entities. The third pillar of the NBIC convergence, alongside strong materialism and emergence, is based on a pragmatic analysis of technological progress. The four disciplinary fields subsumed under the acronym are indeed applied sciences with a strong engineering component. Nanotechnologies enable the manipulation of matter at the atom scale, paving the way to the conception of nano-robots which could, for example, act on living organisms from the inside. Biotechnologies and genetic sequencing technologies, such as the spectacular "molecular scissors" CRISPR-Cas9 (Hsu et al., 2014; Charpentier & Kaldy, 2015 for a popularized introduction), rendered accessible the deciphering and edition of DNA sequences. Progress in information sciences facilitate storage, manipulation and high-performance processing of information once it has been numerically encoded. Lastly, applied cognitive sciences, or "*cognitics*" to put the emphasis on their technological vocation (Claverie, 2005), build a unified approach to cognition, relying heavily on AI, experimental psychology and quantitative sociology.

The NBIC convergence, as a research agenda, aspires to produce a holistic understanding of the human being based on unifying concepts and a progression through the explanatory scale by a succession of emergences, all of that being backed up by an increasingly efficient technology. Cross-fertilization can be fruitful when it is performed with discernment. However, some technologists, Google's director of engineering Ray Kurzweil in the vanguard (Kurzweil, 2006), predict the advent of the "*Technological Singularity*" around 2050, that is, the emergence of an artificial intelligence superior to humans and to which we would be wise to delegate our individual and collective decisions. Among other hypotheses, this discourse relies on the reduction of intelligence to the ability of processing information on the one hand and, on the other, on the emergence of cognition from the informational interaction between a large number of small entities, disregarding their nature, indifferently biological or artificial. Moreover, the nesting of explanatory scales upon which the NBIC convergence project relies, collides with criticism of complex systems sciences, already formulated by Edgar Morin in 1982 (Morin, 1990). Hence Morin demonstrates that it does not suffice to think of the modalities of emergence of the complex from the simple, but also to understand how the complex retro-acts on the simple, thus enforcing the circularity on which Morin's "complex thought" is based. For example, if a high-level concept such as democracy undeniably emerges from a form of organization between individual entities (here, the citizens), its internalization by each of these entities will, in return, modify their behaviours and world views, which will then entail a redefinition of the concept, etc. Yet, this kind of retroaction constitutes a blind spot in the discourses, which is usually confined to a simplistic reductionism. Furthermore, the primacy given to the notion of information and the strong bias towards technology of the NBIC convergence tend to induce research practices relying heavily on the processing of "Big Data", exposing itself to recent epistemological critics. Some researchers indeed speak of a "science without theory" or "agnostic science" (Napoletani et al., 2011, 2014; Kitchin, 2014). For example, since the 1990s, microarray DNA chips enable researchers to measure the level of expression of certain genes in a cell. They are made of a matrix of DNA fragments (probes) with which complementary DNA strands extracted from the cell will hybridize. It then suffices to observe which strand has been hybridized to deduce which genes are present in the cell. Typically, by differentiating between a healthy and a diseased cell, it becomes possible to determine which genes are involved in the disease without (almost) any prior hypothesis. Moreover, thanks to the progress of machine learning, a new kind of scientific practice is then emerging, relying on the machines' ability to detect structures in huge and disparate data sets, or even to establish proto-laws based on measures, without any theoretical *a priori*. These authors stress the fact that it constitutes a paradigm shift, moving human intelligence from the observation of nature to the conception of tools and algorithms to process data. Generally speaking, the perspective of the NBIC convergence in the US takes place in a framework of human enhancement at the individual level. It clashes with a more sociological vision held, notably, by the European Union (Nordmann, 2004) that aims at putting convergent technologies at the service of an enhancement *for* (and not *of*) the human being.

References

Charpentier, E., & Kaldy, P. (2015). CRISPR-Cas9, l'outil qui révolutionne la génétique. *Pour la Science, 456*, 24–32.

Claverie, B. (2005). *Cognitique: Science et pratique des relations à la machine à penser.* L'Harmattan.

Claverie, B. (2014). *De la cybernétique aux NBIC: l'information et les machines vers le dépassement humain* (Vol. 68, pp. 95–101). Hermès, La Revue. CNRS Editions.

Dawkins, R. (1999). *Qu'est-ce que l'évolution ? Le Fleuve de la Vie.* Hachette Littérature.

Hsu, P. D., Lander, E. S., & Zhang, F. (2014). Development and applications of CRISPR-Cas9 for genome engineering. *Cell, 157*(6), 1262–1278.

Kitchin, R. (2014). Big data, new epistemologies and paradigm shifts. *Big Data and Society, 1*, 1–12.

Kurzweil, R. (2006). *The singularity is near: When humans transcend biology.* Penguin Books.

Morin, E. (1990). *Science avec conscience.* Seuil. (Original edition 1982).

Napoletani, D., Panza, M., & Struppa, D. (2011). Agnostic science. Towards a philosophy of data analysis. *Foundations of Science, 16*(1), 1–20.

Napoletani, D., Panza, M., & Struppa, D. (2014). Is big data enough? A reflection on the changing role of mathematics in applications. *Notices of the American Mathematical Society, 61*(5), 485–490.

Nordmann, A. (2004). *Converging technologies – Shaping the future of European societies.* Rapport de la Direction Générale de la Recherche et de l'Innovation de la Commission Européenne.

Roco, M. C., & Bainbridge, W. S. (2003). *Converging technologies for improving human performance. Nanotechnology, biotechnology, information technology and cognitive science. NSF-DOC sponsored report.* Kluwer.

Schrödinger, E. (1993). *Qu'est-ce que la vie?* Seuil. (Original edition 1944).

Edouard Kleinpeter is a research engineer for the French National Centre for Scientific Research (CNRS). After an engineering degree in Physics, he shifted towards scientific journalism before turning to academic research with a Ph D. thesis on science popularization. He then integrated an interdiciplinary laboratory and worked on philosophical foundations of human-machine interactions, namely of Human Enhancement and its affiliated ideologies (Transhumanism,

Posthumanism, …) under the direction of a senior philosopher, Jean-Michel Besnier, between 2012 and 2016. He then visited a second engineering school in Computer Science in order to acquire a better understanding of the inner structure of the machines and the technical details of the way we interact with them. In 2019, he joined the Research Group in Theoretical and Applied Economics (GREThA, UMR 5113, CNRS-University of Bordeaux), where he pursues his research on Human Enhancement with a focus on its economic aspects.

Networks

Iris Clemens

Abstract The Anthropocene as an intellectual concept and the Relational Approach, or thinking in networks, are tightly linked. One can say that the Anthropocene is *the* age of thinking in and about relations. Humans have perhaps never before thought so extensively about their connections to all their surroundings, be they other humans or animals, plants, substances such as CO2 or water, microbes, technology, fabrics, forests, temperatures, etc. For the European tradition, this thinking in relations is not self-evident.

In general, humankind is a species that thinks continuously about itself and nature, and invents theories about its species, their different forms of living, their bodies or their relations to other beings, gods, ancestors, things, etc. These emerging and changing theories and concepts are always rooted in the specific historical and cultural context, influenced extensively by the recent dominant technologies of the epoch, too. For the European (or Western), so-called modern image of the human being, the enlightenment idea of the intelligent, rational *individuum* is central. This idea of the individuum is based on a concept of a singular monad, an isolated entity. Ideally speaking, this individuum should be autonomous and equipped with the capability for self-determination. Many very diverse scientific disciplines such as educational sciences or economics still work with parts of this concept. In this idea of the monadic individuum with its singular, independent and unique I, we still find plenty of references to the Christian concept of human beings with the one soul somewhere *inside* as gift from a god, a core that cannot be transformed without destruction. However, we cannot find this concept of the human being in other socio-cultural contexts. It is not universal, but contingent on history and culture. As with any ideas and concepts, it is a product of a specific social evolution, in this case rooted in the European-North American context. Such cultural formations also

I. Clemens (✉)
University of Bayreuth, Bayreuth, Germany
e-mail: Iris.Clemens@uni-bayreuth.de

© The Author(s), under exclusive license to Springer Nature
Switzerland AG 2023
N. Wallenhorst, C. Wulf (eds.), *Handbook of the Anthropocene*,
https://doi.org/10.1007/978-3-031-25910-4_71

457

influence extensively the *perception* and feeling of human beings. We learn in our socio-cultural socialization to perceive ourselves in a specific way, e.g., as independent, isolated individuals, *or not*. However, the concepts of the human being as well as our perceptions of ourselves are changing considerably in the Anthropocene. The many trends in perspectives, summarized as Anthropocene, force the species, but especially humans in the *minority world* (Akkari & Dasen, 2008; formerly called 'the West'), to rethink their very condition and nature, as much as their position in the world because their behaviour leads to disasters which are so prominent for the description of the Anthropocene.

In the Anthropocene, the human being expands its boundaries in many regards, and due to these developments becomes used to thinking of itself and its surroundings as connected or *relational*. Concepts, explanations and perspectives in the minority world move away from the isolated, closed entity, e.g., the core monad of the individual or the national market or context, towards globally interwoven dynamics, in which humans are always and inescapably embedded, be it in a global perspective as a species or on an individual level as actors. This has consequences for the self-image of the species and the (self-) perception of humans. Due to the climate crisis, globalization, digitalization or the COVID-19 pandemic, to name only these, humankind has started changing its view of itself. It is no coincidence that the concept and idea of the network grew in importance at the same time. The human realizes the extent to which he/she is embedded in a web of essential relationships and therefore loses a centred position as the pride of creation or as a demiurge decoupled from nature. The term Anthropocene itself can be read as an expression of human chauvinism (Coccia, 2018), and humans become the focus as the enemy of their own species and of all others. Products, markets, behaviour, everything is connected globally to an endless variety of other places, actors etc., and this is presented in our daily lives. In the internet – this interconnection of digital networks – algorithms produce a new, previously unknown type of data (big data) and knowledge through data mining. Even machines can perform *deep learning* through these digital networks. But the global, interwoven connections can be even more dramatic, when the power of human species' behaviour stops the Gulf Stream or changes the climate in different ways. As consumers, the inhabitants of the minority world realize, due to global economic connections, that, with or without their consent, they are supporting child labour in other parts of the world or causing other species to become extinct through their hunger for meat or soya. Their way of life is even connected to the Arctic and its wild life, as microplastic is found there. Human beings learn that literally every move they make is connected globally and can have dramatic consequences. Humans can no longer ignore their constant interconnectedness. They are the last knot in an endlessly connected global net that is impossible to overlook or ignore any longer.

Of course, these links are not new, but the attention in the minority world to this embeddedness is a rather recent phenomenon. Especially the subsuming of human beings as components, among others, in an overall interrelation with no specific position (beside maybe the one of being the main problem) is new, e.g., in comparison to a Buddhistic perspective in which humans have never been excluded from other beings or nature in such an extraordinary way. Also in the social sciences and the humanities, humans are more and more conceptualized as being constituted through their relations and not as pre-given, stable entities (Emirbayer, 1997; White, 2008). The human being becomes the homo connectus (Clemens, 2021a, b). We can now think of intelligence not only as an individual attribute, but of swarm intelligence (Christakis & Fowler, 2009), and society is described as a network society (Castells, 2000). Importantly, in the relational approach, humans lose their supreme position in scientific theories to explain social phenomena (Barad, 2007). Humans have to cope with the assumption, that agency is no longer their privilege, and that in the parliament of things (Latour, 2004) there are many voices beside theirs (or at least there should be). The importance and explanatory potential of such dichotomies as subject-object and nature-culture are diminishing, and phenomena are more often conceptualized and explained across these lines in a relational way that connects rather than separates persons or actors. In New Materialism, matter is conceptualized as dynamic (Barad, 2007) and agentive. The relational approach, or network perspective, is geared towards patterns of relational structures and is based on the assumption that we can understand the world and all its phenomena only if we start from these relation structures, not from a priori artificially separated, independent entities.

In social sciences and the humanities, network perspectives or relational approaches have been increasing for some time. For network theory according, for example, to White (2008), the social itself persists out of nothing but relations webbed together to form social networks. The social *is* a network. Actors, identities, everything an action can be attributed to can only be identified and described in relation to other actors in a particular environment, and the position of the actor in this specific setting. Therefore, any description and analysis of social phenomena must start with relations, the in-between spaces, the social embeddedness of what is observed, and not with constant entities with ostensibly stable characteristics that act independently or autonomously similar to the concept of the individual. There are neither independent entities, nor actors, nor actions. We see here parallels to other traditions of thinking and philosophies such as Buddhism, where it is said: There is action, but no agent (Hiriyanna, 2005). Excluding a single actor and ascribing action exclusively to them is always an act of ascription done by the observer and says therefore more about the observer then about the observed (White, 2008). In everyday life, actors (or identities or knots/nodes) are inextricably related and are interdependently interwoven with other actors. Without being aware of it, we are continuously embedded in multiple networks with our diverse identities that appear and disappear relational to these networks and their specific formations. As a consequence, actors have many different identities, depending on the variety of networks to which they have access (ibid.). And as relations constitute us, Powell (2013) points out that our knowledge changes, when our relations change. Therefore, the emergence of knowledge has to be seen as relational, too (Clemens, 2021a). This is particularly interesting when we think of our relations to new technologies, as discussed above. The New Materialism following Barad (2007) and others, includes non-human actors in its consideration of relations. Therefore, thinking in networks challenges one of the greatest and most consequential hierarchies in the history of thinking in the minority world: that humans are somehow outside and/or above all other forms of existence. Consequently, within the Anthropocene biophilia may emerge, and the possibility arises that a biocentric concept (Crawford, 1982) known for a long time in parts of the majority world, will displace the anthropocentric perspective of the minority world.

This has consequences for the way we know, perceive, and conceptualize ourselves. As Breidbach (1997) has shown, our conceptualization of ourselves always correlates with the contemporary technical developments and dominating techniques. If humans think about themselves, they focus on the surrounding techniques (e.g., in the age of the steam engine, Sigmund Freud worked with the concept of 'pressure' to explain human behaviour and perception). According to Breidbach, the evolution of computer architectures, from hierarchical towards parallel processing machines, in the 1970s is crucial for our understanding of humans today. Meanwhile, we are surrounded by parallel processing computers at any moment of our lives and perceive this as normal. Through these media, we are relational in a way never seen before. Even the boundaries of our bodies are shifting fluidly, including technical devices of all kinds: smartphones, −glasses, patterns or clothing, etc. Conversely, these techniques can penetrate our bodies. Through the new media, parts of our selves are processed in parallel as a matter of course, when our mobile phone knows if we have a date tonight, while an investment app invests our money or orders what we will eat at the weekend, and another app knows whether we will take the train or a shared car, and yet another whether we have to do workouts when we arrive home because we haven't achieved our workload for today. All this is very common already: why should we look up a route before leaving the house, if media can tell us the way on the move? This development has only just started. Our interaction *with* and *through* the techniques massively influences our daily lives, our perception, self-perception and our knowledge. Parallel processing has become a daily phenomenon that includes us. Different aspects of ourselves were processed in parallel in many different ways, de-located and de-centralized due to recent digital techniques. And these technologies refer very naturally to emerging processes beyond us. We are a part of these processes, interwoven in this network of millions of global elements that interact in parallel and endlessly. These techniques, as well as the contemporary models and conceptualizations of the human being that are closely linked to them, influence our perceptions and feelings. The technology evolution and the ecological crises both point towards the relationality of our existence. Therefore, it is no coincidence that ideas such as the monadic self are becoming weaker, and thinking in relations and networks is now becoming more widespread. Network thinking is emerging recursively to these developments. We perceive our constant embeddedness in networks for better or worse.

References

Akkari, A., & Dasen, P. R. (2008). *Educational theories and practices from the majority world*. SAGE.

Barad, K. (2007). Agential realism: How material-discursive practices matter. In K. Barad (Ed.), *Meeting the universe halfway: Quantum physics and the entanglement of matter and meaning* (pp. 132–186). Duke University Press.

Breidbach, O. (1997). *Die Materialisierung des Ichs. Zur Geschichte der Hirnforschung im 19. und 20. Jahrhundert*. Suhrkamp.

Castells, M. (2000). Toward a sociology of the network society. *Contemporary Sociology, 29*(5), 693–699.

Christakis, N. A., & Fowler, J. H. (2009). *Connected. The surprising power of our social networks and how they shape our lives*. Little Brown.

Clemens, I. (2021a). The relationality of knowledge and postcolonial endeavours: Analysing the definition, emergence, and trading of knowledge(s) from a network theory perspective. In E. T. Woldegiorgis, I. Turner, & A. Brahima (Eds.), *Decolonisation of higher education in Africa: Perspectives from hybrid knowledge production* (pp. 100–117). Routledge.

Clemens, I. (2021b). Wissen und der homo connectus: Überlegungen zu einem Grundbegriff der Erziehungswis-senschaft aus einer relationalen Perspektive. In M. E. von Eschenbach & O. Schäffter (Eds.), *Denken in wechselseitiger Beziehung: Das Spectaculum relationaler Ansätze in der Erziehungswissenschaft* (pp. 50–70). Velbrück Wissenschaft.

Coccia, E. (2018). *The life of plants: A metaphysics of mixture*. John Wiley & Sons.

Crawford, S. C. (1982). *The evolution of Hindu ethical ideals* (Asian studies at Hawaii) (Vol. 28). University Press of Hawaii.

Emirbayer, M. (1997). Manifesto for a relational sociology. *American Journal of Sociology, 103*(2), 281–317.

Hiriyanna, M. (2005). *Outlines of Indian philosophy* (4th ed.). M/S Kavyalaya Publishers.

Latour, B. (2004). *Politics of nature. How to bring the sciences into democracy*. Harvard University Press.

Powell, C. (2013). Radical Relationism: A proposal. In C. Powell & F. Dépelteau (Eds.), *Conceptualizing relational sociology: Ontological and theoretical issues* (pp. 187–207). Palgrave Macmillan.

White, H. C. (2008). *Identity and control. How social formations emerge* (2nd ed.). Princeton University Press.

Iris Clemens holds the chair for Educational Science at the University of Bayreuth, Germany. She is one of the founder of the German Society for Network Research and a PI of the Cluster of Excellence Africa Multiple: Reconfiguring African Studies. Her main research interests are network theory, relational approaches and research, global relations of the emergence of knowledge and the global circulation of knowledge, decolonial perspectives on knowledge and education, culture sensitive views on epistemologies and indigenous knowledge, India.

New Humanities

Rosi Braidotti and Hiltraud Casper-Hehne

Abstract The New Humanities are a significant critical and creative response to the Anthropocene, especially the Environmental, Medical, Public and Digital Humanities. This chapter examines the rise of these new fields of research and teaching and assesses their salient features, theoretical assumptions and methodological innovations, as well as their institutional applications. The authors argue that far from being the symptom of crisis, the New Humanities are a sign of vitality and innovation and that their capital of knowledge and representation needs to be enlisted to public debates about the Anthropocene. Equally necessary is the inspiration provided by many generations of critical interdisciplinary fields, notably feminist, gender, decolonial and race studies. The article argues forcefully for a culture of mutual respect between the Humanities, Social sciences and Life sciences to support researchers and citizens through the ongoing transformations.

The Anthropocene addresses the humans' role as a geological force in shaping the Earth's current status and possible future. Because advanced economies bear a larger responsibility in polluting the environment, while poorer populations carry a disproportionate amount of negative climate change consequences, the conditions described as Anthropocene are neither politically nor ethically neutral. This multiplies the challenges in both assessing the present and facing the future, but the urgency of devising new models for actions is checked by the scale and complexity of the issues involved. This difficulty spells out a crucial social and cultural role for the New Humanities.

R. Braidotti (✉)
University of Utrecht, Utrecht, Netherlands
e-mail: r.braidotti@uu.nl

H. Casper-Hehne
Intercultural German Studies, University of Göttingen, Göttingen, Germany
e-mail: h.casper-hehne@phil.uni.goettingen.de

© The Author(s), under exclusive license to Springer Nature Switzerland AG 2023
N. Wallenhorst, C. Wulf (eds.), *Handbook of the Anthropocene*,
https://doi.org/10.1007/978-3-031-25910-4_72

Although the Anthropocene is still contentious in the scientific community, there is a consensus among Humanities scholars that this concept is too important to be left to natural scientists alone. What is urgently needed is a redefinition of the relationship between nature and culture and the complex interrelations between humans and the natural world. To paraphrase Haraway, to properly understand 'nature', we need the social and human sciences (Haraway et al., 2016: 548). The Humanities must therefore contribute their specific voices to this discussion (Braidotti et al., 2022a, b).

The Anthropocene challenges also the traditional divides between disciplines, moving beyond C.P Snow's (1998 [1959]) binary distinction between the Humanities and the Natural sciences, to encompass the "three cultures" of Humanities, Social sciences and Life Sciences (Kagan, 2009). This reshuffling has important implications for the periodisations that organize knowledge in the Humanities. Their history and historiography are usually narrated through the eyes and interests of humans and their knowledge about the natural world. An anthropocentric bias is constitutive of these disciplines.

In order to cope, the discipline of history in particular needs to develop an interdisciplinary combination of geological and socio-economic histories, which integrates planetary factors into the analysis of human cultural dynamics, shedding anthropocentric assumptions (Chakrabarty, 2009). This approach entails a change of both spatial and temporal scales. The scale of Anthropocene scholarship stretches over hundreds of thousands of years, defeating the traditional historical timeframes that organize academic knowledge, thereby collapsing the divide between human chronology and geological time.

Institutionally, this pushes together disciplinary categories which were until now kept apart – for instance the earth sciences, literature, philosophy and history, producing a mix of human-centred and post-anthropocentric knowledge practices, which grants the same cognitive attention to non-human as to human entities and actors (Braidotti, 2019). By extension, such an Anthropogenic turn cannot fail to question the assumptions of linear progress and evolution that sustain Humanism.

The New Humanities are a significant response to the Anthropocene, both at the level of research and publications as in teaching new curricula. They are, for instance, the Ecological or Environmental Humanities, sub-divided into Blue Humanities, which study seas and oceans, and the Green Humanities, which study the Earth. They are also known as the Sustainable, Resilient and Energy Humanities. Other significant instances are: the Medical or Health Humanities, also known as the Bio, Neural and Evolutionary Humanities. The rise of the Public Humanities has triggered the Civic or Community Humanities; the Translational Humanities; the Global or Greater Humanities. More neo-liberal variations are the Interactive Humanities and the Entrepreneurial Humanities. The Digital Humanities, which are also called the Computational, Informational and Data Humanities, are possibly the most powerful institutional developments of the last decades. And the list is open and continues to grow.

The energy, critical acumen and creative self-reinvention of the New Humanities developing around – and often against - Anthropocene discourses are a sign of great vitality and courage. Their creativity is so intense and diverse, that it requires specific tools to document and study it, such as specialized glossaries (Braidotti & Hlavajova, 2018; Braidotti, Jones, & Klumbyte, 2022b) and companions (Åsberg & Braidotti, 2018). They have also prompted meta-discursive analyses of the New Humanities, for instance: the Post Humanities (Wolfe, 2010); Inhuman Humanities (Grosz, 2011); Transformative Humanities (Epstein, 2012); the Nomadic Humanities (Stimpson, 2016) and the Critical Posthumanities (Braidotti, 2013, 2019).

What are their defining features? The first is an uncommon degree of social relevance, as they are attuned to the posthuman times and the Anthropocene sociopolitical concerns; they raise issues of community, control and security. They tend to be concrete and pragmatic discourses, with strong support from the corporate and financial sectors.

The New Humanities share a number of assumptions. Firstly, that the subject of knowledge is neither Man – the homo universalis of humanism – nor Anthropos – the dominant species. The knowing subject is no longer the liberal individual, but a more complex transversal ensemble which includes awareness of the environmental roots of all living beings and the importance of technological factors.

Thematically, the New Humanities deal with non-human objects/subjects of study and approach non-human agents as collaboratively linked to knowledge production and research practices. Today, the objects of research and enquiry of the Humanities include, on top of human diversity, animal studies, eco- and geo-criticism.

The strength of the New Humanities therefore is that they blur the distinction between bios – human life and zoe – non-human-life, asserting this diversity in a non-hierarchical manner. They acknowledge the differential intelligence of matter and the respective degrees of ability and creativity of all living entities, thereby reducing anthropocentric arrogance. (Braidotti, 2006, 2013). This approach positions terrestrial, planetary and cosmic concerns, as well as the conventional naturalized others, animals, plants and the technological apparatus, as serious agents and co-constructors of collective thinking and knowing.

The reference to neo-materialist philosophies, such as contemporary studies of Spinoza (Lloyd, 1994, 1996; Deleuze, 1988, 1990), supports the redefinition of the relationship between humans and non-humans, by stressing that we are all part of a common matter. By extension, there are no dualistic oppositions between minds-bodies, natures-cultures, *bios-zoe* but rather a continuum between entities which connect through the shared capacity to affect and be affected by each other.

This approach raises serious methodological issues as well: firstly, about defamiliarization as a research method. And secondly about the limitations of a social constructivist approach, which is based on dualistic oppositions that do not help to deal with the nature-culture continuum (Haraway, 1985) and the zoe/geo/techno-mediated milieus (Braidotti, 2021). This gives them a 'supra-disciplinary' character (Lykke, 2018) of systemic cross-hybridization.

The New Humanities are contiguous with, but not identical to 'cognitive capitalism' (Moulier-Boutang, 2012), in that they pursue a non-profit orientation, whereas most scientific research today is monetarized and integrated into financial capital.

They also engage with social and cultural movements, new kinds of economically productive practices and multiple curiosity-driven knowledge formations, that do not always coincide with the surplus-value profit motive.

We would want the Anthropocene Humanities to address all the three pillars of our present condition: environmental, technological and social, in a more inclusive manner than they have so far. They have to combine 'species thinking' and 'network thinking', within post-anthropocentric and decolonial re-configurations of what it means to be a subject of knowledge (Braidotti & Gilroy, 2016).

The Anthropocene as a horizon of critical thinking is a planetary phenomenon constructed by unequal post/de/neo-colonial power relations (Banerji & Paranjape, 2016). We think it urgent therefore to integrate in the New Humanities the critical discourses of feminism, environmentalism, posthumanism (Åsberg & Braidotti, 2018); de-colonial, inter-cultural, anti-racist, and Indigenous theories (Bleck et al., 2013; Danowski et al., 2017; Braidotti & Bignall, 2019). This is a way of counteracting the perpetuation of discrimination against the feminist, queer, migrant, poor, decolonial, diasporic, Black and otherwise-abled humans and their knowledge practices.

The New Humanities have risen to the challenge. Since Rob Nixon's seminal work on slow violence (2011), the missing links between postcolonial theories, Environmental Humanities and Indigenous epistemologies have been filled by De-colonial Transnational Environmental Justice Humanities. Similar changes have occurred in the Digital Humanities (Nakamura, 2002; Ponzanesi & Leurs, 2014), providing new platforms to re-think transnational spaces, such as the Postcolonial Digital Humanities project (http://dhpoco.org/) and the Decolonial Digital Humanities (Cardenas et al., 2021). The critique of Western imperialism and racism provides an added critical distance, an extra layer of dis-identification, which also expresses enduring care for Indigenous and First Nation people. The New Humanities show us the way to combine sophisticated theories with attention to the excluded and dehumanized humans, the earth and organic non-human others, as well as the technological apparatus. This is a crucial message.

We feel very strongly that we need to work towards a culture of mutual respect between the Life sciences – also known as 'hard' sciences – and the Humanities – which we see as 'subtle', rather than 'soft' sciences. We would like to push their respective complexity to explore multiple potential visions, while resisting the pitfalls of relativism. We can work together towards the actualization of shared, commonly held, community-oriented revisions of knowledge practices in the Anthropocene. The capital of knowledge of the academic Humanities, notably literature, music, poetry, science-fiction, cinema and media is vital, as are the original sources provided by many generations of critical "studies"; notably feminist, gender, decolonial and race studies.

The New Humanities are uniquely placed to encourage people to reflect upon and possibly change their living habits in order to confront the challenges of the Anthropocene. They can steer the process in an affirmative direction away from fear, conformism and flat consumerism, to broaden people's critical horizons. The Humanities today focus our attention on emergent areas of knowledge production, welcoming experimental modes of thinking, the quest for alternative values, and some perplexity before the great challenges of the day. These critical orientations are taken as indicators not only of sharp minds and well-trained intellects, but also of human ethical compassion and decency, as well as of active and discerning citizenship. In the broader sense, they express an unshakable commitment to the democratic exercise of social and cultural criticism. Such an approach requires both an on-going understanding of meanings, of the past, and of value as open processes that need to be constantly re-examined and contested. However, it also demands respect for the imagination, to see artistic and literary practices as major components of the capital of knowledge which the Humanities have built over centuries of subtle, complex and transdisciplinary study and transdisciplinary dialogues.

References

Åsberg, C., & Braidotti, R. (Eds.). (2018). *A feminist companion to the posthumanities*. Springer.

Banerji, D., & Paranjape, M. R. (Eds.). (2016). *Critical posthumanism and planetary futures*. Springer International Publishing.

Bleck, N., Dodds, K., & Williams, C. B. (2013). *Picturing transformations*. Figure 1 Publishing.

Braidotti, R. (2006). *Transpositions*. Polity Press.

Braidotti, R. (2013). *The Posthuman*. Polity Press.

Braidotti, R. (2019). *Posthuman knowledge*. Polity Press.

Braidotti, R. (2021). *Posthuman feminism*. Polity Press.

Braidotti, R., & Bignall, S. (Eds.). (2019). *Posthuman ecologies*. Rowman and Littlefields.

Braidotti, R., & Gilroy, P. (Eds.). (2016). *Contesting humanities*. Bloomsbury Academic.

Braidotti, R., & Hlavajova, M. (Eds.). (2018). *Posthuman Glossary*. Bloomsbury Academic.

Braidotti, R., Casper-Hehne, H., Ivkovic, M., & Oostveen, D. (Eds.). (2022a). *Roads ahead: Trajectories for the humanities in the 21st century*. Edinburgh University Press.

Braidotti, R., Jones, E., & Klumbyte, G. (Eds.). (2022b). *More posthuman glossary*. Bloomsbury Academic.

Cardenas, M., Beydon, N. F., & Kavaloski, A. (Eds.). (2021). www.hastac.org/forums/colonial-legacies-postcolonial-realities-and-decolonial-futures-digital-media

Chakrabarty, D. (2009). The climate of history: Four theses. *Critical Enquiry, 35*, 197–222.

Danowski, D., de Castro, V., & Eduardo. (2017). *The ends of the world*. Polity Press.

Deleuze, G. ([1970] 1988). *Spinoza: Practical philosophy*. City Lights Books.

Deleuze, G. ([1968] 1990). Expressionism in philosophy: Spinoza. Zone Books.

Epstein, M. (2012). *The transformative humanities: A manifesto*. Bloomsbury Academic.

Grosz, E. (2011). *Becoming undone: Darwinian reflections on life, politics, and art*. Duke University Press.

Haraway, D. (1985). A manifesto for cyborgs: Science, technology, and socialist feminism in the 1980s. *Socialist Review, 5*, 2.

Haraway, D., Ishikawa, N., Gilbert, S. F., Olwig, K., Tsing, A. L., & Bubandt, N. (2016). Anthropologists are talking – About the Anthropocene. *Ethnos, 81*(3), 535–564.

Kagan, J. (2009). *The three cultures: The natural sciences, social sciences and the humanities in the twenty-first century*. Cambridge University Press.

Lloyd, G. (1994). *Part of nature*. Cornell University Press.

Lloyd, G. (1996). *Spinoza and the ethics*. Routledge.

Lykke, N. (2018). Passionately posthuman: From feminist disidentifications to postdisciplinary posthumanities. In C. Asberg & R. Braidotti (Eds.), *A feminist companion to the Posthumanities*. Springer International Publishing.

Moulier-Boutang, Y. (2012). *Cognitive Capitalism*. Polity.

Nakamura, L. (2002). *Cybertypes*. Routledge.

Nixon, R. (2011). *Slow violence and the environmentalism of the poor*. Harvard University Press.

Ponzanesi, S., & Leurs, K. (2014). Introduction to the special issue: On digital crossings in Europe. *Crossings, Journal of Migration and Culture, 4*(1), 3–22.

Snow, C. P. (1998 [1959]). *The Two Cultures*. Cambridge: Cambridge University Press.

Stimpson, C. R. (2016). The nomadic humanities. *Los Angeles Review of Books*, 12 July.

Wolfe, C. (2010). *What is posthumanism?* University of Minnesota Press.

Rosi Braidotti is a distinguished University Professor at Utrecht; Honorary Degrees Helsinki, 2007, Linkoping, 2013; Fellow of the Australian Academy of the Humanities *(FAHA)*, 2009; Member of the Academia Europaea *(MAE)*, 2014; recipient Humboldt Research Award, 2021. Main publications: *Nomadic Subjects* (2011a), and *Nomadic Theory*. (2011b), Columbia University Press. *The Posthuman*, 2013, *Posthuman Knowledge*, 2019 and *Posthuman Feminism*, 2021, Polity Press. *More Posthuman Glossary*, 2022, Bloomsbury Academic, with Emily Jones and Goda Klumbyte (eds).

Hiltraud Casper-Hehne is distinguished University Professor at Göttingen; 2009–2021 Vice President for International Affairs at Göttingen University; since 2021 Director of the network NEH21: New European Humanities in the twenty-first Century; since 2020 member of the Executive Committee of the *European University* ENLIGHT. Main publications: Braidotti, Rosi/ Casper-Hehne, Hiltraud/Ivkovic, Marjan/Oostveen, Daan F. (2021) (eds.): *Trajectories for the Humanities in the twenty-first Century*. Edinburgh Publishing House (forthcoming).

Philosophy

J. Baird Callicott

Abstract Diogenes Laertius begins *Lives of the Eminent Philosophers* thus: "There are some who say that the study of philosophy had its beginning among the barbarians." He goes on to review possible claims on behalf of the Persians, Babylonians, Indians, "Druids," and Egyptians granting that each such peoples have wisdom traditions, but no true philosophy. Think what you may of Diogenes' blunt Greek chauvinism, there is, indeed, something peculiar and unique about Greek philosophy. It begins in the early sixth century BCE with speculation about the stuff of the world (water, air, fire); the forces moving that stuff around (Love and Strife, Mind); its beautiful-order ($\kappa o \sigma \mu o \varsigma$) and the law(s) governing the moving cosmos (Justice, Logos). It rapidly develops by way of a diachronic dialectical process. That is, succeeding philosophers respond to the speculations of their predecessors with, often implicit, but always incisive criticism followed by ever subtler speculation.

By the mid-fifth century BCE, Greek natural philosophy had achieved the atomic theory of matter and a very primitive theory of evolution by chance combination and natural selection. By the early fourth century BCE, Greek natural philosophers had discovered that the Earth is a sphere; and by the end of the third century BCE, Eratosthenes had calculated the actual angle (23.5°) of the spherical Earth's axial tilt; and his calculation of its circumference missed the actual figure by only −1.4 percent. Also, in the third century BCE, Aristarchus of Samos came up with a heliocentric cosmology and a correct ordering of the orbits of the planets (including the Earth) out from the sun. Gassendi and Newton adopted Greek atomism wholesale; Copernicus acknowledged the achievement of Aristarchus; and Darwin acknowledged Aristotle's report of Empedocles's theory of evolution (Callicott et al., 2018).

Greek natural philosophy dealt a mortal blow to Greek religion, as documented by Aristophanes's satire of philosophy in the *Clouds*. At least among the intelligentsia, the death of Zeus was implicitly acknowledged. (Remembering, however, the

J. B. Callicott (✉)
University of North Texas, Denton, TX, USA
e-mail: JohnBaird.Callicott@unt.edu

N. Wallenhorst, C. Wulf (eds.), *Handbook of the Anthropocene*,
https://doi.org/10.1007/978-3-031-25910-4_73

fate of Socrates, one must conclude that the gods were still very much alive in the imaginary of the rabble, whom Socrates's accusers, Meletus and Anytus, managed to rouse against him.) Because Zeus had established a moral order, sanctioned by his terrible thunderbolt, disbelief in the gods created a need for some other explanation of and inducement to adhere to ethics. To fill the breach moral philosophy got underway, initiated by Socrates, according to Diogenes. Socrates was only one among quite a few moral philosophers of the fifth century BCE. Most of the others, however, have been demeaned and dismissed as "sophists," thanks largely to Plato's characterization of them. Prominent among them were Protagoras, dean of the lot, Antiphon, and Thrasymachus.

The sophists all professed one or another variety of the social contract theory of ethics, crisply summarized by Plato toward the beginning of *Republic* II (358e-359b): To commit injustice is good and to suffer it bad, but the cost in suffering injustice is greater than the benefit of committing it. So, when tasting of both, those lacking the power to seize the one and escape the other contracted with one another neither to commit nor to suffer injustice. Plato sees three shortcomings in this theory: It pits artificial νομος (convention or law) against φυσις (nature, understood as unbridled self-interest); thus, to commit injustice in secret and evade retribution is tempting to everyone; and it does not appeal to those strong few (tyrants) who are able openly to commit injustice with impunity. Socrates (if we can trust Plato's portrait of him) argued that ανθρωπος is quintessentially a social species and that justice (and virtue more generally) is the healthy condition of the soul (and thus its natural state); and if a healthy body is a necessary condition of happiness, so much more so is a healthy soul. Aristotle elaborated on this theme of natural virtue, regarding happiness as realizing the full potential of human nature. Ανθρωπος is the rational animal; thus to govern one's animal life of desires and their satisfaction by reason is to achieve human excellence or virtue and, *eo ipso*, happiness. A third type of moral philosophy, hedonism, was introduced by Democritus, who is most celebrated for the atomic theory of matter, but hedonism is more usually associated with Epicurus. Pleasure is good and pain is bad; and the goal is to maximize the one and minimize the other. This moral philosophy, however, requires a careful analysis and assessment of pleasure and pain. Many pleasures—such as those of rich foods, strong drink, and promiscuous sex—are followed soon after or eventually by associated pains. Thus, the less costly pleasures of friendship, intellect, and spirit should instead be pursued. In addition to ethics, moral philosophy includes psychology, political theory, and epistemology (Bobonich, 2017).

After Rome came to dominate the Mediterranean basin, the tradition of philosophy begun by the Greeks lost its forward momentum. The individual brilliance of the natural and moral philosophers in classical antiquity gave way to more or less stagnant schools—Stoicism, Epicureanism, Skepticism, Neoplatonism—around the turn of the first millennium. It remained in that condition until Muslim philosophers gave it new life and momentum beginning in the ninth century. The creative philosophers of the "Islamic Golden Age" built on the sound foundations of Plato and

Aristotle, the former stimulating interest in mathematics and cosmology, the latter in logic and systematic philosophy (Leaman, 2002). After the twelfth century, the Aristotelianism preserved and cultivated by Ibn Suna (Avicenna) and Ibn Rushd (Averroes) was further developed by Christian philosophers, most notably by Thomas Acquinas. Much of the philosophical activity from then through the fourteenth century was, however, more a matter of recovering the extant works of the ancients, especially those of Aristotle and Plato, and translating them into Latin. Rapid forward momentum began to occur in the sixteenth century with the heliocentric astronomy of Copernicus and gained speed with that of Kepler and Galileo in the seventeenth along with the empirical epistemology of Bacon, the rational epistemology and cosmology of Descartes, and the ethical and political philosophy of Hobbes.

Natural philosophy began to be transformed into science with the introduction of the experimental method by Newton. The fragmentation of the natural sciences into "disciplines" happened gradually, but by the turn of the twentieth century, it had proliferated and become intrenched. And by then the social sciences—anthropology, psychology, sociology, economics—had begun to bud off from moral philosophy. In response, philosophy took a revolutionary turn in the early twentieth century. It transformed itself into a discipline in its own right and severely narrowed its scope to logic and the analysis of language in the Anglophere and to the description of immediate experience on the European continent. During the twentieth century, disciplinized philosophy became ever more isolated and inbred, preoccupied by arcane puzzles and obscure and often incomprehensible phenomenology. And the natural and social sciences became ever more specialized and fragmented into micro-disciplines.

If we are not now living in the Anthropocene, we are certainly living in the first century of the third millennium. Twentieth-century philosophy abandoned its long historical legacy and abdicated its traditional role of generalizing and synthesizing the many branches of knowledge into a coherent metaphysics and worldview. The current state of the sciences, especially quantum mechanics, relativity, cosmology, evolutionary biology, evolutionary psychology, and ecology provide the ingredients with which philosophers might rebuild a grand, all-inclusive narrative in the venerable manner of Plato, Aristotle, Descartes, Spinoza, and Kant. Furthermore, a return to its roots holds the only hope for philosophy to survive in an academic environment that demands accountability and return on investment. Philosophers can make important contributions in an academic environment that is also becoming ever more interdisciplinary and ever more fraught with ethical issues regarding representation, inclusion, social justice, the treatment of animals in research settings and conservation, and climate change. Nor are the sciences "value-free" and objective as scientists once pretended; and values are not the same as subjective preferences. Exposing, clarifying, and critically assessing the cryptic values lurking in the sciences is also an important contribution that philosophers can make to interdisciplinary research.

References

Bobonich, C. (Ed.). (2017). *The Cambridge companion to ancient ethics*. Cambridge University Press.

Callicott, J. B., van Buren, J., & Brown, K. W. (2018). *Greek natural philosophy: The Presocratics and their importance for environmental philosophy*. Cognella Academic Publishing.

Leaman, O. (2002). *An introduction to classical Islamic philosophy* (2nd ed.). Cambridge University Press.

J. Baird Callicott is Distinguished Research Professor Emeritus and Regents Professor of Philosophy, ret., University of North Texas. Callicott is co-Editor-in-Chief of the *Encyclopedia of Environmental Ethics and Philosophy* and author or editor of many books and author of dozens of journal articles and book chapters in environmental philosophy. His writings have been translated into French, Spanish, German, Russian, Japanese, and Chinese. He is a leading contemporary exponent of Aldo Leopold's land ethic and has elaborated an Earth ethic, *Thinking Like a Planet*, in response to climate change. Most recently, he returned to his roots in classical scholarship with *Greek Natural Philosophy*.

Science

Alice B. M. Vadrot

Abstract Science is both an indispensable source of evidence concerning global environmental change and a precondition for an accelerated destruction of nature. With the aggravation of climate change and the emergence of new environmental threats the role of science in supporting policy responses will increase in importance but will also continue to be contested. This chapter explains the Janus-faced nature of science in the Anthropocene and argues that future research is needed to implement a pluralism of knowledge and address global inequalities in science-making.

If we wish to understand how contemporary societies are addressing global environmental change, we must turn our attention to science, for science has significantly shaped the ways in which modern societies know and govern the natural world. We have developed scientific and technological means to extract, process, and use natural resources. Yet we also use science to map, monitor, measure, assess, model, and predict human impacts on terrestrial and marine ecosystems. The ability to systematically study the environment is a precondition for responding to climate change, biodiversity loss, chemical pollution, the depletion of the ozone layer, and other forms of environmental degradation. Indeed, it has contributed to the institutionalization of environmental politics on a global scale.

However, the increased political and public delegitimization of scientists and of expert bodies—as exemplified by climate denial—shows that environmental knowledge is a fought-over and contested object. Appeals to recognize non-scientific knowledge have anticipated reflexive, inclusive, and participatory approaches to the production of global environmental expertise; these could balance the many needs and interests associated with different knowledge forms. It is this ambiguity in the role of science in the Anthropocene that has led to increased scholarly interest in how science has been shaping both the relationship between society and nature and current attempts to tackle global environmental crises.

A. B. M. Vadrot (✉)
University of Vienna, Vienna, Austria
e-mail: alice.vadrot@univie.ac.at

© The Author(s), under exclusive license to Springer Nature
Switzerland AG 2023
N. Wallenhorst, C. Wulf (eds.), *Handbook of the Anthropocene*,
https://doi.org/10.1007/978-3-031-25910-4_74

Science as Empowerment

Much of the development and wealth of the industrial world resulted from, and continue to be based on the destructive, large-scale extraction of natural resources (e.g. industrial agriculture or deep-seabed mining), made possible by scientific technology, especially in the poorest parts of the world. Indeed, the application of scientific knowledge has often been viewed as a precondition for the ascendency of some states over others.

Francis Bacon, for instance, thought that scientific knowledge provided the opportunity to dominate nature (Bacon, N.O.I. Aph.3.). Similarly, Michel Foucault described scientific knowledge as a tool for hegemonically taking hold of the world. Thus, he argued that colonialism, international relations, and agriculture were at the origin of biology, since the development of scientific knowledge is closely tied to power and domination (Foucault & Colin, 1980). The principles constituting power and those constituting knowledge are similar insofar as they both designate and establish notions of control and regulation. Science equips us with capabilities, which is why, in Karl Mannheim's terms (1929), (scientific) knowledge is always a condition and a result of empowerment (*Befähigung*).

Speaking Truth to Power

Science can produce evidence on the damages inflicted by human activities on our planet, inform policymaking, and raise public awareness on the need for rapid action. The Fridays for Future movement, for instance, has heavily relied on scientific insights and regular reports by the Intergovernmental Panel for Climate Change (IPCC), which has significantly contributed to putting climate change high on the political agenda. In a similar vein, public and political attention to the impacts of biodiversity loss increased with the publication of the first global assessment report by the Intergovernmental Science-Policy Platform on Biodiversity and Ecosystem Services (IPBES), which stated that around one million plant and animal species were threatened with extinction—more than ever before in human history. As global expert bodies, the IPCC and IPBES both rely on experts producing regular assessments on the state of scientific knowledge regarding climate change and biodiversity loss, respectively.

Without the existence of scientific communities advocating for political action to combat these problems, several important intergovernmental agreements, such as the Convention on Biological Diversity (CBD) or the United Nations Framework Convention on Climate Change (UNFCCC) would not have been adopted. In order to explain the role played by science in the emergence of international environmental regimes, Peter Haas introduced the concept of "epistemic communities", arguing that a group of experts sharing the same values and problem understanding might succeed in transforming knowledge claims into a political agenda (Haas, 1992). Ideally, science for policy should be credible, objective, and value-free, which reso-nates with the idea of science "speaking truth to power" (Wildavsky, 1979).

This view implies that science is separate from power, and once taken up by poli-cymakers is subordinated to the dynamics and inherent logic of the political system: "Politicians do not want science, they want a justification for pre-existing political programs which are driven principally by political anticipations of gain" (Haas, 2005, 384–385). Thus, science can be instrumentalized to legitimize either political decisions or non-decisions. Non-decisions may be justified by referring to scientific uncertainty or a lack of evidence on the appropriateness of specific policy measures. However, emerging environmental issues, such as microplastic pollution—or, in the past, depletion of the ozone layer—often suffer from a lack of scientific data or knowledge that could underpin policymaking. Against this background, the "pre-cautionary principle" and the notion of "the best available science" have gained popularity to prevent any delay in political action.

Science Is Political

Although science may be instrumentalized by policymakers serving particular needs and interests, it is gradually being acknowledged that science itself embodies implicit values; these may shape the political responses and instruments thought to be appropriate to tackle a specific policy issue. This implies that science is political in its effects on decision-making. It may contribute to the delimitation of the discur-sive space within which certain environmental debates take place.

Increasingly, the social dimension of scientific knowledge production has been emphasized, including processes of co-production and the relationship between knowledge and power (Jasanoff & Martello, 2004). The institutions we have set up to respond to global environmental change are inherently connected to the ways in which we understand environmental problems (Jasanoff & Martello, 2004). From this perspective, scientific knowledge is embedded within a specific institutional, historical, social, and political context. As a product of social work, science depends on mediated rules, norms, and paradigms, and on its specific modes of organization. STS scholars, for instance, mention dynamics within the production of scientific knowledge at different levels and sites. This includes the study of scientific investi-gation in terms of practical work, for example in the laboratory (Latour, 1987), or the many ways in which science and society are co-produced and enable each other (Jasanoff, 1990). As a social product, science intentionally and/or unintentionally contributes to the construction and endorsement of ideas, values, paradigms, and beliefs. Thus, science is not truth or wisdom, but a social practice that requires ana-lytical attention.

For example, knowledge-oriented approaches in Political Ecology start from the assumption that the ways in which we know are inseparable from "[…] the forms of access and control over resources and their implications for environmental health and sustainable livelihoods" (Watts, 2000, 257; Peet et al., 2011, Escobar, 1998). Here, the relationship between science and policy is viewed as an epistemological struggle where particular representations of environmental problems compete for leadership in framing the issue, which means that certain knowledge-holders are favoured over others; some forms of knowledge are supported whereas others are marginalized. What a "meaningful" and "valid" understanding of the environment is, is always, including its scientific representation, contested and influences the way in which land or nature-society relations are organized.

The emergence of the "ecosystem services" approach provides a good example. Since 2005, this new notion has shaped many political discussions on how to protect biodiversity; it has contributed to a narrow, anthropogenic understanding of nature and the popularization of marked-based environmental instruments which, under specific circumstances, may even harm both the natural environment and its inhabitants. However, its resonance with political and economic thinking, along with its potential to solve trade-offs between different uses of land and resources by providing a tool that estimates the economic costs of environmental degradation, suppressed many critical voices. This has led to "epistemic selectivities" delimiting the forms of knowledge and understandings of nature that specific actors select to underpin their strategies in a political setting (Vadrot, 2014).

Epistemic selectivity implies that some knowledge forms, scientific concepts, and terminologies are favored over others; this, in turn, will shape how individual actors, groups, and institutions formulate their responses to global environmental change. This can have damaging consequences, such as an unequal distribution of the costs and benefits flowing from changes in our natural environment. It may also entail a democratic deficit whereby those suffering from the application of scientific knowledge do not take part in its development; hence they tend to view scientific processes as an "illegitimate and exploitative set of discursive practices" (Haas, 2005, 384). This is how science (analytically) becomes both an object and subject of political regulation and contestation.

Future Research Directions

The Janus-faced nature of science as both an indispensable source of evidence concerning global environmental change and a precondition for an accelerated destruction of nature is creating tensions; it implies that the kind of knowledge that should guide environmental policy needs to be reflected on and negotiated. Research is thus needed, firstly to understand the tensions related to the production and use of scientific knowledge in reaction to environmental damage and, secondly to carefully untangle values from facts. With the aggravation of climate change and the emergence of new environmental threats (e.g., microplastic pollution and ocean acidification), the role of science in supporting policy responses will increase in importance but will also continue to be contested. In this regard, practices of (de)legitimizing scientific knowledge in public discourse and policymaking should be examined, together with the rules determining the forms of knowledge used to justify political (non-)decisions.

While the pursuit of pluralism in the field of knowledge is nothing new—it was already postulated by philosophers of science in the 1950s—this abstract claim needs to be translated into practice. Several international expert bodies, such as the IPCC and IPBES, have already turned their attention to non-scientific knowledge, experimenting with participatory processes to ensure the inclusion of all knowledge forms relevant to global environmental change. Research analyzing and supporting such processes is needed; but it should also critically assess whether the knowledge base used for decision-making fulfils the criteria of credibility, saliency, and legitimacy without marginalizing non-scientific knowledge. While deeply rooted inequalities in science-making between the Global North and Global South have been demonstrated, including colonial patterns that reinforce the dominance of northern academia and epistemologies in the Global South, there are few indications on how to close persistent gaps and diversify the science base informing policy. This will require a fine-grained analysis of scientific fields, from the micro-level of knowledge products to their use by policymakers in the negotiation room (Vadrot, 2020).

References

Escobar, A. (1998). Whose knowledge, whose nature? Biodiversity, conservation, and the political ecology of social movements. *Journal of Political Ecology, 5*(1), 53–82.

Foucault, M., & Colin, G. (1980). *Power/knowledge: Selected interviews and other writings, 1972–1977*. Pantheo Books.

Haas, P. M. (1992). Introduction. Epistemic communities and international policy coordination. *International Organization, 46*(1), 1–35.

Haas, P. M. (2005). Science and international environmental governance. In P. Dauvergne (Ed.), *Handbook of global environmental politics*. Edward Elgar.

Jasanoff, S. (1990). *The fifth branch – Science advisers as policymakers*. Harvard University Press.

Jasanoff, S., & Martello, M. L. (Eds.). (2004). *Earthly politics. Local and global in environmental governance*. MIT Press.

Latour, B. (1987). *Science in action: How to follow scientists and engineers through society*. Harvard University Press.

Mannheim, K. (1929/1955). *Ideology and utopia: An introduction to the sociology of knowledge*. Harvest Books.

Peet, R., Robbins, P., & Watts, M. (Eds.). (2011). *Global political ecology*. Routledge.

Vadrot, A. B. M. (2014). *The politics of knowledge and global biodiversity*. Routledge.

Vadrot, A. B. M. (2020). Multilateralism as a "site" of struggle over environmental knowledge: The north-south divide. *Critical Policy Studies, 14*(2), 233–245.

Watts, M. (2000). Political ecology. In E. Sheppard & T. Barnes (Eds.), *A companion to economic geography*. Blackwell.

Wildavsky, A. (1979). *Speaking truth to power: The art and craft of policy analysis*. Little, Brown & Company.

Alice B. M. Vadrot is Associate Professor for International Relations and the Environmental at the Department of Political Science of the University of Vienna, Visiting Research Fellow at the Centre for Science and Policy (CSaP) of the University of Cambridge, member of the Young Academy of the Austrian Academy of Sciences, and Senior Fellow of the Earth System Governance Platform. Her work addresses the role of knowledge and science in global environmental politics. She has conducted extensive research on the Intergovernmental Platform on Biodiversity and Ecosystem Services (IPBES) and has developed the concept of "epistemic selectivities". Since 2018, she leads the ERC Starting grant project MARIPOLDATA, which develops and applies a new methodological approach for grounding the analysis of science-policy interrelations in ocean politics in empirical research.

System

Philippe Hertig

Abstract This short contribution aims to summarize the relevance and necessity of using the concept of system and the systemic approach to understand the major environmental and societal issues facing human societies in the context of the Anthropocene. After a brief historical review, the text proposes a definition of the concept of system and presents its general characteristics, refers to certain fields of application, and concludes with some reflections on the teaching and learning of systems thinking in schools.

The major challenges facing human societies are and will be characterized by complex interactions between multiple factors, some of which fall under the purview of the natural sciences, while others fall under the purview of the social sciences and refer to political and ethical issues, explicitly or implicitly related to value systems (Audigier et al., 2011; Audigier et al., 2015; Hertig, 2017a, b). In the context of the Anthropocene, considered as a global change reflecting a set of biophysical forcings, on a planetary scale, such as global warming and its systemic effects, the decline of biodiversity, the depletion of non-renewable resources (and of some renewable but overexploited resources), or the transformation of large biotic and abiotic systems (e.g., soils or oceans) and the modification of large biogeochemical cycles such as those of nitrogen or phosphorus, the impacts of these changes are occurring on a planetary scale, but also synchronously on all spatial scales (Lussault, 2022), while affecting human societies. Understanding these major societal and environmental issues requires the ability to decipher and understand multiscalar phenomena and processes, involving a multitude of interacting actors and factors, with feedback loops and recursivity. The classical analytical approach is powerless to shed light on such phenomena or processes, which can only be understood by mobilizing the systemic approach and the complexity paradigm as defined by Edgar Morin (2005, 2007).

P. Hertig (✉)
University of Teacher Education, State of Vaud, Lausanne, Switzerland
e-mail: philippe.hertig@hepl.ch

© The Author(s), under exclusive license to Springer Nature
Switzerland AG 2023
N. Wallenhorst, C. Wulf (eds.), *Handbook of the Anthropocene*,
https://doi.org/10.1007/978-3-031-25910-4_75

Influenced by cybernetics, communication theory and structuralism (Durand, 2010), systems theory emerged in the 1950s, thanks in particular to the work carried out under the aegis of the Society for the Advancement of General Systems Theory, founded in 1954 by the biologist Ludwig von Bertalanffy, the economist Kenneth Boulding, the mathematician and psychologist Anatol Rapoport, and the neurophysiologist and behavioral scientist Ralph W. Gerard. In particular, von Bertalanffy sought to define the general properties common to all systems, and he brought together the essence of his work and the reflections arising in the fertile crucible of the Society in a major work published in 1968 entitled *General System Theory: Foundations, Development, Applications*. Inspired by and based on the contributions of scientists from several disciplinary backgrounds, systemics today constitutes "a legacy of approaches and concepts, applicable or transposable to approaches that reject the reduction of the whole to the sum of its parts and the reduction of knowledge of the whole to the simple sum of knowledge of its parts" (Thibault, 2013a: 974) (original quotation in French, adapted by the author of this article). As a result, since the second half of the twentieth century, many areas of science and technology have benefited from the application of the principles of systems thinking.

Before discussing the general characteristics of a system, it is useful to note that while the rationalist and reductionist tradition of thought based on the principle of direct causality has held a central position for centuries in the West, this is not the case in other cultures which were and are still based on a holistic conception of the world. It should also be noted that in the West, well before the twentieth century, thinkers had already expressed reflections that questioned reductionism. For example, we can quote a famous fragment of Pascal's *Pensées*, published posthumously in 1669:

> *Mais les parties du monde ont toutes un tel rapport et un tel enchaînement l'une avec l'autre que je crois impossible de connaître l'une sans l'autre et sans le tout. (...) Toutes choses étant causées et causantes, aidées et aidantes, médiates et immédiates, et toutes s'entretenant par un lien naturel et insensible qui lie les plus éloignées et les plus différentes, je tiens impossible de connaître les parties sans connaître le tout non plus que de connaître le tout sans connaître particulièrement les parties.*[1]

There are countless definitions of the concept of system, which vary according to the scientific field in which they are rooted. However, in the context that interests us here, the concept is defined on the basis of the idea of an open system formalized in 1937 by von Bertalanffy, and all the definitions that have been proposed have points in common that can be synthesized as follows: a system is an organized set of elements and processes, linked together by interactions of a diverse nature that confer a certain autonomy on it, and that constitute a global unit. The essential property of

---

[1] "But the parts of the world are all so related and linked to one another that I believe it impossible to know one without the other and without the whole. (...) Since everything, then, is cause and effect, dependent and supporting, mediate and immediate, and all is held together by a natural though imperceptible chain which binds together things most distant and most different, I hold it equally impossible to know the parts without knowing the whole and to know the whole without knowing the parts in detail" (Pascal, 1958: 20).

any system is *emergence*, which can be illustrated by a very simple example: taken individually, the constituent parts of a bicycle are of little use in themselves; it is only by the adequate assembling of these parts and by the play of their interactions that the "bicycle" system becomes operational. The principle of emergence is reflected in the well-known statement that "the whole is greater than the sum of its parts".

Mobilizing a systemic approach to understand a social situation or a natural phenomenon implies being able to identify what "makes a system" in the case in question. To this end, an analytical approach is generally necessary at first. Such an approach is based on the characteristics and properties of the system, which can be translated by means of a few key concepts. The *elements* or *components* of the system are in relation with one another. These relationships are generally not limited to the simple causal action of one element on another (A –> B), but very often take the form of a double action (A <=> B) that can be qualified by means of the fundamental concept of *interaction* (Durand, 2010; Thibault, 2013a). Some interactions take the form of *feedback*, with *positive feedback* or *negative feedback loops* (concepts inherited from cybernetics) being at the heart of a system's *dynamics*. A system is a whole that cannot be reduced to the sum of its parts, and is therefore defined in its *globality*, which reveals emergent qualities that the components of the system do not possess when considered separately – here we find the essential property of *emergence* mentioned above. *Organization* is another central concept of the systemic approach: it accounts for both the arrangement of relations between the components of the system, which thus produces a new unit whose qualities are different from those of the elements that constitute it, and the processes by which matter, energy and information are assembled and put into action and form (Durand, 2010: 10). The *limit* or *boundary* of a system may be more or less clearly identifiable, and more or less permeable to the system's interactions with its *environment*. The systems approach also relies on the concept of *complexity* (which must be clearly distinguished from the notion of complication). A system is said to be complex when it is impossible to predict its behavior, even if it remains intelligible: random or chaotic processes, or new arrangements between elements can cause it to evolve in new ways (Thibault, 2013b: 214).

The physicist and philosopher of science Mario Bunge uses a basic definition of the concept of a system – "a complex object in which each part or component is related to at least one other component" (Bunge, 2020: 450)[2] – to propose a model for analyzing a system based on four concepts: composition, environment, structure, and mechanism (CESM analysis). The set of parts of the system constitutes its *composition*. The *environment* of a system – beyond the boundary of the system – defines all that acts on the system as a whole or on its components, and on which the system and its components also act. The *structure* of the system is characterized by the set of relationships that exist among the components of the system and between its components and certain elements of its environment. The *mechanism* of a system

---

[2] Translated from the French.

designates the processes that govern its functioning and that can vary according to the conditions of its interactions with its environment, or under the effect of the feedbacks that affect the system internally. Strictly speaking, a mechanism refers to a material system, and we use the more general notion of process to understand the dynamics of natural or social systems.

Among the many typologies of systems which exist in the literature, one of the most interesting is the one devised by Mario Bunge. It is based on the principle of emergence, each level emerging from the lower level, and thus accounts for the increasing complexity of the world: the first level is that of physical systems, followed by physical-chemical systems, which have their own activity; then come living systems, then social systems, human social systems, and finally systems produced by human activities, with or without artifacts, namely technical systems, semiotic systems, and conceptual systems (Bunge, 2020; Durand, 2010).

The systemic approach uses effective tools, including analogy and especially graphic representation and modeling. The power of graphic representation lies in its capacity to allow a global reading, but also the identification of a component or a group of elements, and the reading of the relations among the components of the system. It is also particularly suitable for reporting on feedback loops and for developing models, which can be cognitive, decision-making, normative or predictive (Durand, 2010: 60–61).

Articulated around the concept of system, the systemic approach constitutes a theoretical, conceptual and methodological corpus that is essential for understanding and reflecting on the world in the Anthropocene. The increase in interdependencies and interactions, and also the hazards and uncertainties that characterize this world, result in an increase in complexity that calls more than ever for the use of systems-thinking tools to ensure that our thought patterns are able to meet the ever more pressing challenges facing human societies (Durand, 2010: 118–119). The systems approach enhances the conceptual networks and methods that allow us to understand social, economic, geopolitical or environmental phenomena or problems as diverse as the family, the relations of a population with its built and natural environment, migratory movements, access to and management of resources, or pandemics. In geography, for example, it is common to consider a city as an open system whose relationships with its environment are studied, identifying the inputs and outputs of the system (among which are relationships with other cities or with the surrounding countryside), as well as feedback loops (Serre, 2015) – urban heat domes being one manifestation among others. Within the natural sciences, biology and ecology were among the first disciplines to make extensive use of systems thinking. Considering the Earth as a system is a key to understanding the changes taking place on a global scale (see the article *Earth Systems*). One cannot understand current global warming without referring to the climate system, a highly complex system characterized by interactions among several components or subsystems. The atmosphere, the hydrosphere, the cryosphere, the lithosphere, the pedosphere and the biosphere are the six subsystems that, together with the contribution of solar radiation, constitute the natural climate system – subsystems and a global system disturbed and unbalanced by interactions with the anthroposphere. The complex feedbacks generated by these imbalances are likely to result in "tipping cascades" of some elements once warming thresholds are reached and exceeded (Steffen et al., 2018).

"In a perspective of citizenship training, it is the responsibility of the school to give students the intellectual tools that will enable them to analyze complex problems and issues, to make choices, to take decisions and to act in a rational way" (Hertig, 2015: 126).[3] The first didactic challenge a teacher faces when he or she aspires for his or her students to gradually appropriate the tools of complex thinking is to make them understand that they must overcome the spontaneous mode of thinking that reduces any explanation to monocausality or to a chain of linear causalities (Hertig, 2018). Linear causal reasoning and binary reasoning are prevalent in common sense discourse and too often in the media as well. It is therefore essential that schools encourage students as early as possible to think in terms of systems without being satisfied with the "one cause -> one effect" scheme. This work must be initiated from the earliest stages of schooling by means of appropriate didactic approaches (Assaraf & Orion, 2005; Bollmann-Zuberbühler & Kunz, 2008) relating to objects of knowledge from the student's daily environment that are understood in terms of systems, while working on the basics of graphic representation (Hertig, 2017b). In the context of the Anthropocene, the concept of system, the systemic approach and complex thinking are essential tools that all students should appropriate, as they are for them both keys to the intelligibility of today's world and resources for thinking about the future world as responsible citizens (Hertig, 2017a, b, 2018). These are also crucial issues for initial and in-service teacher training.

References

Assaraf, O. B.-Z., & Orion, N. (2005). Development of system thinking skills in the context of earth system education. *Journal of Research in Science Teaching, 42*(5), 518–560.

Audigier, F., Fink, N., Freudiger, N., & Haeberli, P. (Eds.). (2011). *L'éducation en vue du développement durable: sciences sociales et élèves en débats.* (Cahiers de la Section des sciences de l'éducation n° 130). Université de Genève.

Audigier, F., Sgard, A., & Tutiaux-Guillon, N. (Eds.). (2015). *Sciences de la nature et sciences de la société dans une école en mutation. Fragmentations, recompositions, nouvelles alliances ?* De Boeck.

Bollmann-Zuberbühler, B., & Kunz, P. (2008). Ist systemisches Denken lehr- und lernbar? In U. Frischknecht-Tobler, U. Nagel, & H. Seybold (Eds.), *Systemdenken. Wie Kinder und Jugendliche komplexe Systeme verstehen lernen* (pp. 33–52). Verlag Pestalozzianum/PHZH.

Bunge, M. (2020). *Dictionnaire philosophique. Perspective humaniste et scientifique.* Editions Matériologiques. (Original edition 2003).

Durand, D. (2010). *La systémique.* PUF.

---

[3] Translated from the French.

Hertig, P. (2015). Approcher la complexité à l'Ecole: enjeux d'enseignements et d'apprentissages disciplinaires et interdisciplinaires. In F. Audigier, A. Sgard, & N. Tutiaux-Guillon (Eds.), *Sciences de la nature et sciences de la société dans une école en mutation. Fragmentations, recompositions, nouvelles alliances?* (pp. 125–137). De Boeck.

Hertig, P. (2017a). Des outils de pensée pour approcher la complexité. In J. Didier, Y.-C. Lequin, & D. Leuba (Eds.), *Devenir acteur dans une démocratie technique. Pour une didactique de la technologie* (pp. 155–170). Université technique de Belfort-Montbéliard.

Hertig, P. (2017b). Education à la complexité. In A. Barthes, J.-M. Lange, & N. Tutiaux-Guillon (Eds.), *Dictionnaire critique des enjeux et concepts des "Educations à"* (pp. 74–81). L'Harmattan.

Hertig, P. (2018). Géographie scolaire et pensée de la complexité. *L'Information géographique, 82*(3), 99–114.

Lussault, M. (2022). *L'Anthropocène: Un champ d'expérimentation pour la formation post-disciplinaire*. Public conference held at the University of Teacher Education, State of Vaud (HEP Vaud), March 24, 2022.

Morin, E. (2005). *Introduction à la pensée complexe*. Seuil.

Morin, E. (2007). Complexité restreinte, complexité générale. In J.-L. Le Moigne & E. Morin (Eds.), *Intelligence de la complexité. Epistémologie et pragmatique* (pp. 28–64). L'Aube.

Pascal, B. (1669 [posth.]). *Pensées [Fragments – 199-72, "Disproportion de l'homme"]*. Retrieved from https://www.ub.uni-freiburg.de/fileadmin/ub/referate/04/pascal/pensees.pdf (p. 39). Accessed 14 April 2022. Translation: *Pascal's Pensées* (W. F. Trotter, E. P. Dutton, Trans), 1958.

Serre, D. (2015). Concevoir la résilience urbaine: un défi face à des complexités. In J.-C. S. Lévy (Ed.), *Complexité et désordre. Eléments de réflexion* (pp. 113–133). Grenoble Science.

Steffen, W., Johan, R., Richardson, K., Lenton, T. M., et al. (2018). Trajectories of the earth system in the Anthropocene. *Proceedings of the National Academy of Sciences (PNAS), 115*(33), 8252–8259.

Thibault, S. (2013a). Système. In J. Lévy & M. Lussault (Eds.), *Dictionnaire de la géographie et de l'espace des sociétés* (pp. 973–975). Belin.

Thibault, S. (2013b). Complexité. In J. Lévy & M. Lussault (Eds.), *Dictionnaire de la géographie et de l'espace des sociétés* (pp. 214–216). Belin.

von Bertalanffy, L. (1968). *General system theory: Foundations, development, applications*. George Braziller.

Philippe Hertig, PhD, is a geographer and was full professor for geography didactics at the University of Teacher Education, State of Vaud, Lausanne, Switzerland, where he was head of the Department of Didactics of Human and Social Sciences (UER SHS) and of the International Research Lab on Education for Sustainable Development (LirEDD) until his retirement in December 2022. The main focus of his research concerns the learning processes of complex subject matters, the contributions of the social sciences to education for sustainable development, especially the links between geography and ESD, and educational and methodological approaches regarding the use of pictures and images in geography. In Switzerland he has also contributed to a large research and development project on climate change education (Climate Change Education and Science Outreach, CCESO).

Thought

Claus Leggewie and Frederic Hanusch ⓘ

Abstract This article examines thought in the Anthropocene. While thought has been attributed to the human species in particular, in the Anthropocene thought opens up as a category to the non-human. We define thought as an inter- and trans-disciplinary practice of thought collectives which aim to "think like a planet" that results in planetary knowledge ecologies.

Thought in the Anthropocene as a 'Denkstil'

The *Copernican Revolution* is seen as a profound replacement of the predominant (Ptolemaic) model of the Heavens – with the planet Earth as the centre of the universe – by a heliocentric model – with the Sun at the centre of a fragment of the universe, the solar system. This "revolution" of our world view was started in the sixteenth century by the astronomer Copernicus' in his treatise "De revolutionibus orbium coelestium" (1543). In "The Copernican Revolution" (1957) the philosopher Thomas Kuhn argued that Copernicus' discovery of the nature of the solar system not just provided better astronomical evidence but was also in tune with a disposition to change prevailing worldviews with modern ones. Three years later Thomas Kuhn published his famous study on "The Structure of Scientific Revolutions", which generally challenged the prevailing view of progress in science in which scientific progress was viewed as "development-by-accumulation" of accepted facts and theories. He hold that this was valid in periods of "normal science" but episodically interrupted by periods of revolutionary science leading to new paradigms. Science moves beyond the mere "puzzle-solving" of the previous paradigm, new paradigms change the rules of the game and direct research into new directions. The example at hand was the Copernican Revolution, and for Kuhn such

C. Leggewie (✉) · F. Hanusch
Justus-Liebig-University Giessen, Giessen, Germany
e-mail: claus.leggewie@zmi.uni-giessen.de; frederic.hanusch@zmi.uni-giessen.de

© The Author(s), under exclusive license to Springer Nature Switzerland AG 2023
N. Wallenhorst, C. Wulf (eds.), *Handbook of the Anthropocene*,
https://doi.org/10.1007/978-3-031-25910-4_76

a paradigm shift was a result of sociological factors, not a logically determinate procedure.

Even more appropriate for understanding thought in the Anthropocene seems to be a recourse to a concept brought up by the less known Polish physician and historian of science Ludwik Fleck. His monograph "Genesis and Development of a Scientific Fact" written in 1935 had virtually no impact until the English and German translations of this study and other works. Fleck wrote them under almost impossible conditions in the Lemberg ghetto and finished them as a Holocaust survivor in Lublin and Warsaw. More attention was granted in the shadow of Kuhn's template while Fleck is more radical in his sociological input. We refer to his main concepts of "thought-collective" and "thought-style" to describe what we call "planetary thinking" that comes about in the Anthropocene (Hanusch et al., 2021). A *Denkkollektiv* is a community of researchers who interact collectively towards the production or elaboration of knowledge using a shared framework of cultural customs and knowledge acquisition. Any scientific discovery is an interaction between the discovered phenomenon, the discoverer, and the existing pool of knowledge from which they draw, a prevalent *Denkstil*. By interacting socially in their production of knowledge, researchers produce common concepts and practices that they use to discuss and debate one another's ideas and discoveries. These common concepts and practices both provide a shared language of science within which they can communicate, and consequently limit the possibilities of thought that the researchers can have. Under the influence of a particular "thought style," the common concepts and practices previously mentioned, a community of researchers "shares joint attributes of problems and judgments considered as evident, joint methods to acquire knowledge, and agrees in the determination of meaningless questions." Within this conception of the production of scientific knowledge, the ideas, concepts, and theories of these researchers are permanently conditioned by their present "thought style" and cannot be considered independently of them. It has to be pointed out that those thought-collectives aren't sects and thought-styles not dogmas. Within a given thought-style there coexist many particular concepts, theories and models which are developed by esoteric and exoteric circles within the collective. That is why Fleck emphasized by the way that the scientific community is fundamentally democratic – in contrast to religious or metaphysical thought-collectives. The central thesis of Fleck and the larger Lemberg School in the 1930's was that discovery is a collective process. The inter- and transdisciplinary research community or in Fleck's words "thought collective" is a vivid expression of how various disciplinary accounts contribute to one "thought style".

Thinking Like a Planet

The range of those involved in the thought collective of planetary thinking in the Anthropocene is broad and ranges from sensual nature writing to formalized earth system sciences.

The genre of nature writing, whose beginnings can be traced back to the late eighteenth century, has been flourishing again of late. In addition to thick descriptions of landscapes, some authors now try to reconnect humanity to non-human nature by empathically writing from the point of view of animals, trees, fungi, rocks, etc. In extreme cases, this tentative science of nature goes so far as to try to emulate the behaviour of its objects, e.g., to really "be a beast" (Foster, 2016). To be sure, to adopt the cognitive and physiological processes of other lifeforms is an absurd endeavour—there are clear neuronal limits to such attempts. But since it reveals the historical change in human perceptions of nature, such writing is of value nonetheless. Forests, for instance, were interpreted over the centuries as kingdoms, cathedrals, or factories, whereas now they are seen as information networks, and people try to decipher the 'language' of animals (Meijer, 2019). In his now classic *Sand County Almanac*, the American environmentalist Aldo Leopold famously suggested that to be better able to protect ecosystems we should learn to "think like a mountain": "I know suspect that just as a deer herd lives in mortal fear of its wolves, so does a mountain live in mortal fear of its deer. And perhaps with better cause, for while a buck pulled down by wolves can be replaced in two or three years, a range pulled down by too many deer may fail of replacement in as many decades. So also now with cows. The cowman who cleans his range of wolves does not realize that he is taking over the wolf's job of trimming the herd to fit the range. He has not learned to think like a mountain. Hence, we have dustbowls, and rivers washing the future into the sea." (Leopold ([1949] 1987, 132; see also Callicott, 2013) Leopold's vivid call for a new representation of the world in our thought has found many followers, and his adage many analogues: from "thinking like a river" (Worster, 1994) to "thinking like a climate" (Knox, 2020) to "thinking like a mall." "[E]nvironmental thinking," the philosopher Steven Vogel (2015) argues, "would be better off if it dropped the concept of 'nature' altogether and spoke instead of the 'environment'—that is, the world that actually surrounds us." Instead of discounting such approaches as irrational and speculative, then, we ought to perhaps take more seriously what they point to: the various kinships, overlaps, and contradictions that characterize our existence on this planet.

On the opposite side to planetary thinking yet perfectly grounding its symbolic empathy is the attempt to make the planet's thought accessible to humans through sensors. These can be elicited, for instance, on the level of individual "earthlings"— such as trees that 'tweet' data about their growth and hydraulic functioning. Just like humans use pulse watches to learn more about their health condition, a tree can be equipped with a sensor and thus enabled to 'report' on its 'impressions'. The Canadian forest ecologist Suzanne Simard, for example, examined tree diseases occurring particularly often in newly reforested areas. Through DNA analyses of fungi she ascertained how trees communicate with each other and with their environment, namely, by means of so-called mycorrhizal networks underground (Simard et al., 2012; see also Simard, 2021). In addition, seismographic measurements, from the Earth's interior to interplanetary space via satellite, now form an entire information ecological infrastructure, allowing for a planetary monitoring that to some observers amounts to another Copernican Revolution (Schellnhuber, 1999). The data collected provide the basis for models of the Earth system which show how processes proceeding on different levels and entailing substance flows caused by animate as well as inanimate matter are intertwined in complex and precarious ways. The digital pervasion of the planet raises the question of its programmability. In her book *Program Earth: Environmental Sensing Technology and the Making of a Computational Planet* (2016), Jennifer Gabrys demonstrates that the cybernetic perspective of Earth system science—which can be translated into formats of "environmental computing" (see the Treewatch project above) — represents just one variety of planetary processes and relations and thus of ways to "think like a planet".

Planetary Knowledge Ecologies

With vastly diverse perspectives of those involved in the thought collective, the challenge lies in accounting for the multitude of views of the Anthropocene, which do not, however, lead to a multitude of worlds but to one world, a world that can be described very differently and in which very different behaviors are possible. Planetary knowledge neither relativistically aims at letting everyone live in their own world, nor monistically at creating a unified science or a dualism of scientific and non-scientific approaches to the world. It is pluralistic, or rather pluriversalist (Mitchell, 2003; Reiter, 2018). Planetary thinking conceives of the world as a variously entangled structure that corresponds to a non-linear idea of the causality of complex interdependencies and a multi-level system of inter- and transdisciplinary methods. The design of planetary ecologies of knowledge can be regarded as a kind of evidence-based activity in search of variability and as permanently composing shapes. This activity of designing embodies the genuinely human aspect in planetary knowledge. The shaping of knowledge is an act carried out not merely intellectually but as part of a comprehensive conduct of life. Planetary knowledge has a reflective or "diffractive" quality (Barad, 2007). We can thus better surmise what authors mean when they talk about ecologies of knowledge (Rahder, 2020; Akera, 2007), namely, an "ethico-onto-epistemic" frame (Barad, 2007) that is capable of integrating a great number of assumptions and explanations about the Earth as a planet—which change the more we know. The thinking and non-thinking parts of the planet can thus converge.

In their seminal book on Planetary Social Thought (2021), Nigel Clark and Bronislaw Szerszynski show what such an ecology of knowledge may look like. They use the phrase "planetary multiplicity" to explicate that planet Earth, because it has changed or was forced to do so, has taken various shapes—in fact, there are different Earths: The connections between humans and that permanently changing Earth, and the challenges arising from its evolutionary self-organization and disruptive variation, Clark and Szerszynski call "earthly multitudes." "[H]uman difference and social 'otherness,'" they hold, "are bound up with the capacity of the Earth to self-differentiate or become other to itself" (9). Which means that biological and cultural diversity, though not identical, are related.

For now, planetary thinking raises more issues than it solves. To engage in this mode of thought means to recognize that 'our' Earth really belongs to all, epistemologically, ontologically, and ethically. It means to acknowledge that human societies are deeply entangled with the life of an ever-changing planet—a planet that expands spatially from the Earth's core to interplanetary space, temporally from nanoseconds to geologic time scales, and materially from elementary particles to the dark matter of the universe. The "planet" is currently joining the "earth," the "globe," and the "world" as a fourth dimension, and in the planetary vocabulary, mytho-poetic, analytical-scientific, and ethical-normative semantics mix. Very different orders of knowledge thus contribute to planetary thinking, but they have one thing in common: the insight that everything is interdependent. Inspired by an "overview" (of Earth from outer space) as much as by an "ultraview" (of the universe from our solar system), planetary thinking combines anthropocentric with planetocentric perspectives. We would misconstrue the Anthropocene if we reduced it to a realization of the effects of human interference with the planet and to the development of yet another anthropocentric strategy to solve this problem. This would not do justice to our insight that planetary constellations are always bidirectional: planetary forces affect human societies and vice versa. Rather, planetary thinking requires us to expand our normative cosmopolitanism by way of a new global constitution and transform it into a comprehensive cosmopolitics. The partial emergency state that accompanied the Covid-19 pandemic has made it crystal clear that we need to adjust our priorities. The planet will survive climate change, species extinction, and other tipping points of the Earth system—the question is if, and how, we will.

References

Akera, A. (2007). Constructing a representation for an ecology of knowledge: Methodological ad-vances in the integration of knowledge and its various contexts. *Social Studies of Science, 37*(3), 413–441. https://doi.org/10.1177/0306312706070742

Barad, K. (2007). *Meeting the universe Halfway: Quantum physics and the Entanglement of matter and meaning*. Duke University Press.

Callicott, J. B. (2013). *Thinking like a planet: The land ethic and the earth ethic*. Oxford University Press.

Clark, N., & Szerszynski, B. (2020). *Planetary social thought: The Anthropocene challenge to the social sciences*. Polity.

Donges, J. F., Winkelmann, R., Lucht, W., Cornell, S. E., Dyke, J. G., Rockström, J., Heitzig, J., & Schellnhuber, H. J. (2017). Closing the loop: Reconnecting Hu-man dynamics to earth system science. *The Anthropocene Review, 4*(2), 151–157. https://doi.org/10.1177/2053019617725537

Fleck, L. (1935). Genesis and development of a scientific fact.

Foster, C. (2016). *Being a beast*. Profile Books.

Gabrys, J. (2016). *Program earth: Environmental sensing technology and the Making of a Computa-tional planet, electronic mediations*. University of Minnesota Press.

Hanusch, F., Leggewie, C., & Meyer, E. (2021). *Planetar denken – Ein Einstieg*. Transcript.

Knox, H. (2020). *Thinking like a climate: Governing a City in times of environmental change*. Duke University Press.

Leopold, A. [1949] (1987). A sand county almanac, and sketches here and there. Oxford University Press.

Meijer, E. (2019). *When animals speak: Toward an interspecies democracy*. New York Uni-versity Press.

Mitchell, S. (2003). *Biological complexity and integrative pluralism*. Cambridge University Press.

Purdy, J. (2017). Thinking like a mountain: On nature writing. n+1 29. https://nplusonemag.com/issue-29/reviews/thinking-like-a-mountain/

Rahder, M. (2020). *An ecology of knowledges: Fear, love, and Technoscience in Guatemalan Forest conservation*. Duke University Press.

Reiter, B. (Ed.). (2018). *Constructing the Pluriverse: The geopolitics of knowledge*. Duke Uni-versity Press.

Schellnhuber, H. J. (1999). 'Earth system' analysis and the second Copernican revolution. *Nature, 402*, C19–C23. https://doi.org/10.1038/35011515

Simard, S. W. (2021). *Finding the mother tree: Discovering the wisdom of the Forest*. Knopf.

Simard, S. W., Beiler, K. J., Bingham, M. A., Deslippe, J. L., Philip, L. J., & Teste, F. P. (2012). Mycorrhizal networks: Mechanisms, ecology and modelling. *Fungal Biology Reviews, 26*(1), 39–60. https://doi.org/10.1016/j.fbr.2012.01.001

Thomas Kuhn The structure of scientific revolutions.

Vogel, S. (2015). *Thinking like a mall: Environmental philosophy after the end of nature*. The MIT Press.

Worster, D. (1994). Thinking like a river. In *Wealth of nature: Environmental history and the eco-logical imagination by worster* (pp. 123–134). Oxford University Press.

Claus Leggewie is holder of the Ludwig Börne professorship and director of the "Panel on Planetary Thinking" at Giessen University, Germany. He currently is co-editor of the book series "Climate & Cultures" (Brill) and the "Routledge Global Cooperation Series" (Routledge). Earlier affiliations include visiting professorships at the University of Paris-Nanterre and New York University (Max Weber Chair). From 2008 to 2016, he was a member of the German Advisory Council on Global Change (WBGU).

Frederic Hanusch is scientific coordinator of the "Panel on Planetary Thinking" at Giessen University, Germany. His research is focused on the intersections of democracy and planetary change. He is currently working on a book entitled "Deep-Time Governance" (Cambridge University Press) to explore how planetary temporalities can be institutionalized in governance systems. In his previous book "Democracy and Climate Change" (Routledge), he analyzed why more democracy leads to better climate performance.

Transdisciplinary

Javier Collado Ruano ⓘD and Florent Pasquier

Abstract This essay focuses on regenerative perspectives that can transform how the Anthropocene is viewed globally. Because research on the Anthropocene is increasingly complex, we argue that a transdisciplinary approach will shift contemporary debate and provide a fuller array of effective long-term solutions. We argue that scientific knowledge must be integrated with factors such as ancestral wisdom, artistic expression, emotions, and spiritual worldviews. The following article highlights the destructive consequences of global change based on scientific data, it questions conventional economic vision, and examines how regenerative cultures promote human well-being, ethical awareness, and environmental justice.

Anthropocene and Global Change Within the Earth System

The term 'Anthropocene' informs current scientific, philosophical, and academic debate. According to Steffen et al. (2007), the Anthropocene began around 1800 with the onset of industrialization, the central feature of which was the enormously expanded use of fossil fuels. In recent times, concerns about the Anthropocene emphasize the drastic effects of human actions on the Earth System (Waters et al., 2016). The term 'global change' refers to multidimensional changes on a planetary scale – culturally, ecologically, geologically – that occur in the Earth System, encompassing problems such as biodiversity loss, climate change, land use change, shifts in the nitrogen and phosphorus cycles, global water use, ocean acidification,

J. C. Ruano (✉)
National University of Education in Ecuador (UNAE), Azogues, Ecuador
e-mail: javier.collado@unae.edu.ec

F. Pasquier
Sorbonne University, Paris, France

ozone depletion, chemical pollution, and atmospheric aerosols, among other inter-systemic problems (Bowman et al., 2009). The Earth as a meta-system, constituted by biophysical systems that interact with each other, functions adequately or inadequately based on the prevailing global environmental conditions. As a result, solutions cannot be found when examining countries independently from each other. Indeed, today's socioecological problems are wholly interdependent (Collado, 2018).

In the latest report of the Global Atmosphere Watch (GAW), a program established by the World Meteorological Organization (WMO) of the United Nations, the CO_2, CH_4, and N_2O fractions of the year 2019 in the atmosphere reached historical figures, the highest of the last 800,000 years. The values of more than 410 parts per million (ppm) of Carbon Dioxide (CO_2), 1877 parts per billion (ppb) of Methane (CH_4), and 332 ppb of Nitrous Oxide (N_2O) indicate that immediate change is needed to help our planet (WMO, 2020). As indicated by the Intergovernmental Panel on Climate Change (IPCC), created by the WMO and the UNEP in 1988, these levels of gases increased the average global temperature by 0.85 °C during the period from 1880 to 2012 (IPCC, 2014, p. 2). Through mathematical models based on the recorded data of the atmosphere, the biosphere, and the oceans, climatologists predict that the global average of the air temperature on the surface will increase by 1.4 °C to 5.8 °C for 2100 compared to 1990, and that the world average sea level will see "an increase of 0.09 to 0.88 meters by 2100" (IPCC, 2001, p. 3). While these forecasts are uncertain, the current records of global change have caused a great alert for politicians and environmentalists, highlighting the urgent need to create an Earth Democracy (Shiva, 2005).

The Mass Extinction of Species and the Loss of Biodiversity

For Leakey and Lewin (1996), human actions have caused the "sixth mass extinction". Scientists say that in 1989 one species disappeared per day, and that in 2000 this loss occurred every hour. Species may be headed toward extinction even more rapidly, at a rate of about 17,500 per year. Oberhuber (2004) estimated that between 10% and 38% of the existing species will have disappeared between 1990 and 2020. Decreased biodiversity, climate change, global warming, droughts, the acidification of the oceans, and the rise in sea level are just some examples of global change phenomena that show the acceleration caused by our anthropogenic actions. "The overconsumption of limited resources and the production of unlimited pollution are inconsistent with the continued functioning of the Earth's ecological system in a way that is compatible with the survival of human civilization" (Gore, 2013, p. 332). There is an urgent need to develop a complex, intercultural, transdisciplinary way of thinking (Nicolescu & Pasquier, 2019), in order to more fully account for transnational damages of an intersystemic nature.

What Social Consequences Will the Global Change of the Anthropocene Have?

The alteration of the Earth system by human actions has led scientists to project various consequences. For example, the global rise in sea levels could drastically affect large coastal metropolises, such as Tokyo, New York, Rio de Janeiro, Cape Town, Sydney, Tijuana or Barcelona (Kopp et al., 2016). Ecosystems will continue to suffer irreversible negative impacts and points of no return. Experts estimate that, in the next twenty years, some 1.8 billion people will suffer from chronic shortages of natural resources and will not be able to thrive. Contemporary agricultural activities exacerbate such problems by using excessive water. It is estimated that to produce a ton of cereals, about a thousand tons of water are required. Worldwide, agriculture uses about 70% of the fresh water available. In addition, the considerable geopolitical influence of China and India risks exponentially multiplying humankind's overuse of natural resources.

The Importance of Designing Regenerative Cultures

For Stiglitz et al. (2010), GDP as an economic indicator only takes into account the value of the goods and services produced, leaving aside fundamental questions such as how proliferation affects biodiversity and how ecosystems have inherent limits. The first step needed to address problematic aspects of the Anthropocene is to recognize the economy as a subsystem of the Earth System. Introducing the physical principles of thermodynamics into the economy allows fuller recognition of the capitalist economic system's irrationality (Martinez-Alier, 2003). Economic growth is based on the exploitation of natural resources. Since the Industrial Revolution, money has colonized life. We must therefore replace GDP as a measure of economic and social progress, as it destroys the planet's real capital, namely the capacity to support life. We must not accumulate an artificial, abstract, dead capital in the form of money, a measure of growth that lacks intrinsic value.

Fullerton (2015) questions the long-term viability of the dominant capitalist system, and proposes a regenerative economic system that is self-organizing and self-sustaining. He encourages an economic transition towards regenerative capitalism, which offers a new paradigm for the modern economy, one that leaves behind centuries of efforts to dominate and control nature, and particularly focuses on generating and distributing abundance (Pasquier, 2016).

Conventional economies have been shown to bring social inequality and ecological devastation, while green economic proposals are unfeasible because they increase the costs of production and, therefore, of retail sales. Sustainable economies and so-called responsible consumption that allows biophysical restoration have historically failed, because Western countries have globalized their imperial lifestyle.

According to environmental leader Müller (2020), it is time to rethink the concept of sustainable development, because it does not call into question economic growth. The concept of regenerative cultures can bridge gaps in terms of how we perceive, shape, and respond to economic growth. Collado et al. (2019) define regenerative cultures as the transdisciplinary dialogue of scientific knowledge and indigenous communities' ancestral practices. According to Wahl (2016, p. 264): "The creation of regenerative cultures is also rooted in a shift from seeing ourselves only as separate individuals, communities, nations, and species to understanding our deep interbeing as fundamentally interconnected expressions of life itself."

Humankind needs regenerative cultures as a new way of looking forward, beyond sustainable cultures. While sustainable development emphasizes minimizing the negative human impact on the planet, regenerative cultures focus on maximizing human beings' positive impact and fully respecting the Earth. Regenerative cultures can establish a new collective imaginary, based on what we continually learn from nature – tangibly as well as holistically – rather than on raw materials physically extracted from it (Benyus, 2009). They can change the ecocidal economic patterns that have destroyed nature for centuries. They establish instead eco-efficient economic patterns inspired by those of nature, whose beings and systems intrinsically follow instructive paths.

References

Benyus, J. (2009). *Biomimicry: Innovation inspired by nature*. Harper Perennial.
Bowman, D. M. J. S., Balch, J. K., Artaxo, P., Bond, W. J., Carlson, J. M., Cochrane, M. A., D'Antonio, C. M., Defries, R. S., Doyle, J. C., Harrison, S. P., Johnston, F. H., Keeley, J. E., Krawchuk, M. A., Kull, C. A., Marston, J. B., Moritz, M. A., Prentice, I. C., Roos, C. I., Scott, A. C., … & Pyne, S. J. (2009). Fire in the earth system. *Science, 324*(5926), 481–484. https://doi.org/10.1126/science.1163886.
Collado, J. (2018). Co-evolution in big history: A transdisciplinary and biomimetic approach to the sustainable development goals. *Social Evolution & History, 17*(2), 27–41.
Collado, J., Madroñero, M., & Álvarez, F. (2019). Training transdisciplinary educators: Intercultural learning and regenerative practices in Ecuador. *Studies in Philosophy and Education, 38*(2), 177–194.
Fullerton, J. (2015). *Regenerative capitalism. How universal principles and patterns will shape our new economy*. Capital Institute: The future of finance.
Gore, A. (2013). *The future. Six drivers of global change*. Random House.
IPCC. (2014). *Climate change 2014: Impacts, adaptation, and vulnerability*. Cambridge University Press.
IPCC (Intergovernmental Panel on Climate Change). (2001). *Climate change 2001: Impacts, adaptation, and vulnerability*. Cambridge University Press.
Kopp, R., et al. (2016). Temperature-driven global sea-level variability in the Common Era. En *Proceedings of the national academy of sciences of the United States of America* (pp. 1–8).
Leakey, R., & Lewin, R. (1996). *The sixth extinction: Biodiversity and its survival*. Phoenix Books.
Martinez-Alier, J. (2003). *The environmentalism of the poor. A study of ecological conflicts and valuation*. EE.

Müller, E. (2020). Regenerative development as natural solution for sustainability. In: Sarmiento, F., & Frolich, L. (coord.) *The elgar companion to geography, transdisciplinarity and sustainability*, (pp. 201–218).

Nicolescu, B., & Pasquier, F. (2019). To be or not to be transdisciplinary, that is the new question. So, how to be transdisciplinary? *Transdisciplinary Journal of Engineering & Science, 10.*

Oberhuber, T. (2004). *Camino de la sexta gran extinción*. En *Ecologista*, n° 41. Ecologistas en acción.

Pasquier, F. (2016). Dans nos sociétés de rentabilité, quelle place pour l'amour et l'amitié transpersonnels ? *Bulletin de l'Association Française du Transpersonnel*, 20–25.

Shiva, V. (2005). *Earth democracy. Justice, sustainability and peace.* South End Press.

Steffen, W., Crutzen, P., & McNeill, J. (2007). The anthropocene: Are humans now overwhelming the great forces of nature? *AMBIO, 36*(8), 614–621.

Stiglitz, J., Sen, A., & Fitoussi, J. (2010). *Mis-measuring our lives. Why GDP Doesn't add up. The report by the commission on the measurement of economic performance and social Progress.* The New Press.

Wahl, D. (2016). *Designing regenerative cultures.* Triarchy Press.

Waters, C., et al. (2016). The Anthropocene is functionally and stratigraphically distinct from the Holocene. *Science, 351*(6269).

World Meteorological Organization (WMO). (2020). *WMO greenhouse gas bulletin. The state of greenhouse gases in the atmosphere based on global observations through* 2019. N° 16.

Javier Collado Ruano is titular Professor in the National University of Education (UNAE) in Ecuador. PhD in Dissemination of Knowledge by the Federal University of Bahia (Brazil), and PhD in Philosophy by the University of Salamanca (Spain).

Florent Pasquier is Associate professor HDR at Sorbonne University, Costech Lab (Connaissance Organisation et Systèmes Techniques / Knowledge, Organization and Technical Systems, EA 2223), UTC. President of Ciret: International Center For Transdisciplinary Research, Paris. https://pasquierflorent.wixsite.com/pasquier

Part VII
The Paradigmatic Impasses of Technology?

Dystopia

Ariel Kyrou

Abstract From Mary Shelley's *Frankenstein* in 1818 to the current novels of so-called 'climate fiction' such as Paolo Bacigalupi's *The Water Knife*, via Aldous Huxley's and George Orwell's classics *Brave New World* and *1984*, science fiction dystopias are in essence critical works of the Anthropocene. This early criticism can be seen as an expression of our collective unconscious. But since the 1960s many such works have been consciously critical, functioning as indispensable environmental alerts.

In the opening pages of his novel *The Water Knife*, science fiction writer Paolo Bacigalupi describes "a family home of three generations made valueless because the suburb's water had gone dry and Phoenix (the capital of Arizona) wouldn't allow a hookup" (Bacigalupi, 2018, p. 42). A little further on, he introduces us to the flats and over-protected enclave of a privileged few, paying a fortune not to have to buy their water by the centilitre, at the fluctuating market price, from a pump under a sandstorm. The dystopia becomes clearer, with the lack of water causing a feeling like a violent and suffocating collapse of our so-called civilised world. The worst is reached with a scene in a hospital, after the discovery of a mass grave of climate refugees who wanted to flee from Texas to the north of the United States, killed by supposed smugglers from some mafia, who "took the money from their victims and just buried them in the desert". The young journalist at the heart of the story, though used to "statistics of people displaced by tornadoes, hurricanes or rising seas", is overwhelmed by "the bodies piled up trying to buy their way north to areas where water, work and hope were plentiful". The apocalypse deeply affected her, in painful empathy with the murdered refugees: "With all the dead bodies and the overwhelmed emergency staff, she felt as if the whole world was falling apart". (Bacigalupi, 2018, p. 158–161).

A. Kyrou (✉)
Moderne Multimédias, Paris, France
e-mail: ariel.kyrou@gmail.fr

Paolo Bacigalupi observes the collapse from the sewers of humanity. His pen goes into the rubble of the Anthropocene, or rather of its twin, the Capitalocene – for it is indeed the most mortifying capitalism that is transforming the Earth into an immense dry ruin, a physical and metaphysical graveyard of geology as well as of the flora, fauna, bodies and souls of humans. The author describes what would remain of his world, ours, alongside those left behind by our civilisation, close to the migrants who are drowning in the Mediterranean or trying to escape from camps, for example on the Greek island of Lesbos. Published in 2015, *The Water Knife* is a major work of a recent sub-genre of science fiction, of which the American writer Kim Stanley Robinson could be the godfather: "climate fiction" or "cli-fi", which is supposed to anticipate more directly than before the short-term consequences of future climate upheavals, via recognised scientific work and plots that are intended to be credible. In his works, Bacigalupi weaves together three of the four main threats to human existence, according to the American biologist and geographer Jared Diamond: global warming, the depletion of natural resources, social inequality and the resulting clashes – the only thing missing is the nuclear threat (Diamond, 2009). Paolo Bacigalupi paints, in his own words, "a visceral picture of what is likely to happen" on the basis of models, scientific data and an extrapolation of the social and political phenomena he observes (Kyrou, 2020). *The Water Knife* is a twenty-first century dystopia, but the novel's radical critique of the Anthropocene comes from afar.

Until Mary Shelley's *Frankenstein* in 1818 – to give a symbolic, if simplistic, date – the notions of the creation of life as well as the destruction of humanity were almost entirely a matter of divine grace or punishment. The great myths of our beliefs and religions of all times. But for decades they have vibrated at the heart of science fiction, a disparate genre that is only the free extrapolation of our dreams and nightmares. As Hiroshima and Chernobyl showed before we realised our multiple potential for self-destruction in the form of collapse, from global warming to the supposed domination of machines, man is turning into a sorcerer's apprentice. Armed with his new technological power, he tries his hand at making prostheses, simulating life, designing 'artificial intelligences' and creating clones or genetic mutants of plants or, why not, living beings. Science fiction, under this double prism of destruction and creation, is in essence the imagination of the Anthropocene. It tells the multiple, sometimes violent and often contradictory stories of a new geological and vital era in which human beings are the origin and the telluric force.

What then of that particular form of science fiction that we call 'dystopias', that is, negative utopias of our modern age civilisations, counter-utopias of our industrial and technological ages? From Mary Shelley's *Frankenstein*, which is perhaps the founding text, to Paolo Bacigalupi's *The Water Knife* and other works of our contemporary disasters, haven't the great literary or cinematographic dystopias been and still are for us today among the first major warnings of the suicidal drifts of the Anthropocene?

In the literal sense of the term, Aldous Huxley's *Brave New World* (1932) and George Orwell's *1984* (1949), the two most famous novels in this dystopian territory, do not have the Anthropocene as their theme. The word did not exist when they were written. They do not even have ecology as an issue. On the other hand, for today's reader, the Anthropocene seems to be their major collective unconscious. Their ghost, which was then unnamed.

The human beings of *Brave New World* are manufactured in laboratories, within 'incubation and conditioning centres'. Sexual reproduction is an object of disgust, from the low castes of the Deltas and Epsilons multitude, literally designed on an assembly line like industrial parts, to the Alphas 'programmed' to be the ruling class of the world state. Books and flowers, in particular, are to arouse an 'instinctive hatred' in these artificially shaped populations for their happiness and the smooth running of the economy that society is all about. For as London's DIC (Director of Incubation and Conditioning) says, primroses and landscapes "have one serious flaw: they are free. The love of nature provides no work for any factory. It was decided to abolish the love of nature, at least among the lower classes", but not the tendency to consume transport. As the Director concluded, "we condition the masses to hate the countryside, but at the same time we condition them to love all outdoor sports. At the same time, we make sure that all outdoor sports require the use of complicated equipment. So that we consume manufactured goods as well as transport." (Huxley, 1977, p. 41) The God of the Christians is gone. In his divine place sits the figure of Henry Ford. So the cross has been flattened to become a 'T', in the glory of the Model T Ford series.

The factory and the plantation are the models for the utopias that are the source of *1984* and *Brave New World* – even if they have become dystopias in the eyes of the reader. As in a caricature of the Anthropocene, the human or especially the non-human environment is merely a resource to be shamelessly exploited. Whether nature is damaged by industry or by war – which also contributes 'positively' to the running of the capitalist machine – is of no importance. At best, for the Ministry of Truth official Winston Smith, nature provides the nostalgic, Shakespearean romantic backdrop to his dreams of freedom and love without control or thought police, with a "clear, slow-moving stream forming under the willow trees, ponds in which golden fish swam" (Orwell, 2004, p. 38). In classical dystopias, fauna and flora, the agents of Gaia, are only loved for themselves and grasped in interdependence with humans by weak beings, savages or heretics to society.

What about apocalyptic or post-apocalyptic dystopias from before the concept of the Anthropocene appeared in this perspective? In René Barjavel's *Ravage*, published as a serial in 1943, the sudden disappearance of all forms of electricity on a June evening in 2052 sounds like a religious punishment: "Men released the terrible forces of nature that stood locked carefully. They thought they were in control. They named it Progress. This is an accelerated progress towards death" (Barjavel, 1943, p. 85). The 'great catastrophe' here is not a symbol of any ecological awareness, it is just "a flurry of fire and merciless evil, a manifestation of divine wrath against the pride of men" (Barjavel, 1943, p. 299).

Earth Abides, a novel by George R. Stewart published in 1949, the same year as *1984*, is less reactionary and more interesting in the perspective of the Anthropocene as we see it today. For the pandemic of unknown origin that has decimated the planet's population, resulting from some Covid-19 with much more deadly effects, is not a divine fatality. Better still, at the very beginning of the plot, the reader learns that the main character is preparing a thesis, *The Ecology of the Black Creek Region*, for which he is studying "the past and present relations between the men, plants and animals of this region" (Stewart, 2018, p. 23). The fate of the young white man, when he recreates a 'tribe' with an African-American woman, has a hint of the Adam and Eve myth, but the book is peppered with reflections on evolution, the biology of living species, and how the planet would react to a sudden and drastic collapse of the human population. The key, which would bring us closer to a critique of the Anthropocene rather than a religiously inspired call for a return to the cosmos and pre-industrial ways of life, lies in this ecological dimension, which is more prominent in *Le Monde enfin*, a 2006 dystopia by Jean-Pierre Andrevon very close du *Earth Abides*, but openly misanthropic, marking the revenge of the animal and plant world on the human race ravaged by a prion or a virus (Andrevon, 2010).

On the oh-so-dystopian theme of the end of the world, fiction resists the wear and tear of the years all the better because it remains open to interpretations, in particular those of a suicidal responsibility of all or part of humanity in its fatal fate. This is one of the explanations for the timelessness of J. G. Ballard's 1960s tetralogy: *The Wind from Nowhere* (1962), T*he Drowned World* (1962), *The Drought* (1965) and The *Crystal World* (1966). Without an immediately understandable cause, rather like Cormac McCarthy's *The Road* (2006), these novels foreshadow the environmental register of later fictions, especially that anticipating climate change. In the same way, in most of John Brunner's or Philip K. Dick's texts, the collapses – social and environmental in the case of the former, nuclear in the case of the latter – are less like rationally apprehensible, even avoidable, events than like permanent settings that are impossible to deny (Kyrou, 2021). They form the context, necessarily chaotic, with which beings have no other solution than to deal. The causes of the catastrophe are neutralised, forgotten, because they are a thousand times less crucial than the consequences, experienced daily by their characters, as we are today, living with the effects of global warming or the Covid-19 pandemic. The novel or the short story that stages the disaster then becomes a fable of all our apocalypses, past, present or future. The radical critique of the Anthropocene therefore sounds like an obvious science fiction feature. Thus, among many others, these two novels published in 1994: Robert Silverberg's *Hot Sky at Midnight*, which anticipates the consequences of the greenhouse effect and the disappearance of the ozone layer in the twenty-fourth century (Silverberg, 1999); and Bruce Sterling's *Heavy Weather*, whose furious tornadoes imagined for 2031 (Sterling, 2004) seem premonitory of those that devastated cities in Kentucky and Tennessee in the western United States in December 2021. It is in this spirit that we should see and revisit the movies to our environmental warnings, from Richard Fleisher's *Soylent Green* (1973) to Adam McKay's *Don't Look Up* on Netflix in November 2021.

But can the major works of this age, which is aware of our future in the Anthropocene era, from John Brunner to Ursula K. Le Guin, from Kim Stanley Robinson to a whole new generation of European authors, including Li-Cam and Catherine Dufour, be described as 'dystopias'? For them, dystopia is no longer something to be invented: it is there at the heart of our real world. And that is why, like the town of Precipice, a city of mutual aid and the common good at the heart of the hell of generalized control in John Brunner's *The Shockwave Rider* (1975), they willingly slide down the tracks of lucid utopias into the heart of their tragic dystopias (Brunner, 1977).

References

Andrevon, J.-P. (2010) [original edition 2006]. *Le Monde enfin*. Pocket.

Bacigalupi, P. (2018) [original edition *The Water Knive* 2015, French translation 2016]. *Water knife*. J'ai lu.

Barjavel, R. (1972) [original edition 1943], *Ravage*, Folio / Denoël.

Brunner, J. (1977) [original edition *The Shockwave Rider* 1975, French translation 1977]. *Sur l'onde de choc*. Robert Laffont.

Diamond, J. (2009) [original edition *Collapse: How Societies Choose to Fail or Succeed* 2005, French translation 2006]. *Effondrement: Comment les sociétés décident de leur disparition ou de leur survie*. Folio essais.

Huxley, A. (1977) [original edition *Brave New World* 1932, French translation 1932]. *Le Meilleur des mondes*. Presses Pocket.

Kyrou, A. (2020). *Dans les imaginaires du futur*. ActuSF.

Kyrou, A. (2021). *ABC dick*. ActuSF.

Orwell, G. (2004) [original edition 1949, French translation 1950]. *1984*. Ebooks libres et gratuits.

Silverberg, R. (1999) [original edition *Hot Sky at Midnight* 1994, French translation 1995]. *Ciel brûlant de minuit*. J'ai Lu.

Sterling, B. (2004) [original edition *Heavy Weather* 1994, French translation 1997]. *Gros temps*. Denoël.

Stewart, G. R. (2018) [original edition *Earth Abides* 1949, French translation 1951]. *La Terre demeure*. Fage éditions.

Ariel Kyrou a member of the editorial collective of the revue Multitudes, he recently published *Dans les imaginaires du futur* (ActuSF, 2020, in French), a book in which he questions our fictions and our "terrestrial and anarchist" utopias to better extricate us from the Anthropocene, as well as *ABC Dick, We live in the words of a science-fiction writer* (ActuSF, 2021, in French). He is the co-writer of the documentary film *Les mondes de Philip K. Dick* (2016, Nova Prod/Arte). Associate director of the company Moderne Multimédias, he is the editorial director of the Laboratoire des solidarités of Fondation Cognacq-Jay, and as such, editor-in-chief of the solidarum.org site. He also gives courses on "the imaginaries of today and tomorrow" and on the "sociology of current cultural practices" in Master 1 at the University of Versailles Saint-Quentin en Yvelines.

Genetic Engineering Revolution

Benjamin Gregg

Abstract Genetic engineering in general, and human genetic editing in particular, is revolutionizing humankind's self-understanding: an evolved organism taking ever greater control of its own evolution. This Anthropocenic phenomenon is deeply equivocal (Gregg B. Human genetic engineering: biotic justice in the anthropocene? In: DellaSala D, Goldstein M (eds) Encyclopedia of the Anthropocene, vol 4. Elsevier, Oxford, pp 351–359, 2018). While delivering humans from some risks, it renders them vulnerable to unintended consequences as well. Even in the face of seemingly intractable differences of opinion about how best to understand genetic manipulation culturally, and how to evaluate it ethically, political communities and international organizations alike must address its possible future legal regulation. Governments and other elite institutions would do well to include the general public, to the extent possible, in deliberations about regulation, even as genetic engineering is based on highly specialized knowledge.

The revolution of genetic engineering reaches back 10,000 years to when humans between the Tigris and Euphrates genetically modified plants and animals through artificial selection. Almost all the foods humans currently consume have genes produced through selective breeding, rendering the human species itself a genetically modified organism (for example, in adults' lactose tolerance). Human genetic engineering at the molecular level, whether as changes to the species' germline or to an individual person's genetic material, vastly increases the speed, power, and precision of the manipulation. Somatic editing modifies DNA in non-reproductive cells; germline editing modifies the nuclear DNA of human eggs, sperm, or embryos, thus affecting every cell of an unborn life and thereby the genes of all of its descendants.

B. Gregg (✉)
University of Texas at Austin, Austin, TX, USA
e-mail: bgregg@austin.utexas.edu

N. Wallenhorst, C. Wulf (eds.), *Handbook of the Anthropocene*,
https://doi.org/10.1007/978-3-031-25910-4_79

This revolution is normatively ambivalent along several dimensions: its distinct advantages for human life cannot be decoupled from its distinct risks to human life. Ambivalence marks (1) its complexities in application, (2) its intrinsic risks, (3) local and global disagreement on how best to evaluate genetic engineering, (4) unequal access to its therapeutic promises, and (5) its problematic economic dimensions. Thus ambivalence also marks the issues of (6) necessary legal regulation and (7) future public engagement.

1. Human genetic engineering is revolutionary as therapy that eliminates genetic defects that generate disease or disease resistance. It is also revolutionary as a means to enhancing an already healthy body. Yet this distinction between therapy or disease prevention, on the one hand, and enhancements beyond health, on the other, is contextual and easily collapses: therapy can sometimes also be enhancement. For example, somatic cell engineering to treat, say, cognitive decline during Alzheimer's, or cognitive disability in newborns with autism spectrum disorder, may also be considered enhancements delivered through somatic gene therapies. And biotechnology aimed at eliminating genetic disease may be deployed in an attempt to render persons, say, more intelligent, athletic, or longer-lived. The notion of health is indeterminate in this sense but in other ways as well. Thus a gene that contributes to one health problem may also provide benefits: a "gene that lowers an individual's risk of getting HIV increases the risk of succumbing to the West Nile virus," and a "gene that lowers the risk of type-1 diabetes brings a higher risk of Crohn's disease" (Baker, 2016: 270). Note that analysts disagree as to whether some conditions (such as autism or aging) are indeed health problems.

2. Human genetic engineering is inherently risky. Evaluation of those risks leads to ambivalent results. Engineering may recommend itself where it is shown to be safe (for example in treating such scourges as sickle cell anemia, cystic fibrosis, or Huntington's) and where alternatives are absent. Yet unknown possible adverse effects cannot be measured; risk-freedom may be impossible; and "we cannot confidently predict all the consequences, whether of introducing deleterious traits or by losing unanticipated benefits to retaining particular alleles," and downstream effects are intrinsically uncertain (Carroll & Charo, 2015: 5). The natural world is deeply unpredictable; evolution, for example, operates by means of genetic mutations, some of which confer survival advantages in some environments. Correspondingly, there are always uncertainties as to how particular gene edits may impact various components of a cell. Diseases caused by a single mutation affect fewer people and are easier to prevent, but many diseases are polygenic, including such prevalent conditions as schizophrenia and heart disease. Such complexity only increases the possibility of unintended consequences of gene editing, such as off-target effects, that is, gene deletion or mutation at sites other than the intended one. Editing at the one-cell stage but not over multiple embryonic cleavages can lead to cellular-disease inducing mosaicism, where the individual's body has two or more genetically different sets of cells.

Finally, unlike traditional medicines, gene edits cannot be withdrawn in the face of eventual adverse effects. Recruiting healthy persons for clinical trials may be unethical. Monitoring the safety of the engineering over the course of the patient's life may entail life-long surveillance, which may be unsustainable for ethical, legal, and economic reasons.

3. Human genetic engineering is a form of eugenics (defined as efforts to affect genetic heritage in health-affirming ways): liberal not coercive, individualistic not collectivistic, aimed at both increasing desirable traits as well as decreasing undesirable ones (Veit et al., 2021). Members within any political community, as well as across different communities, are unlikely ever to agree consensually on how best to define "desirable" and "undesirable"; what forms of genetic engineering to allow and which to prohibit; and for what reasons. Any given answer to any of these questions likely presupposes one or the other notion of human identity if not of human nature (Gregg, 2022). Should the transfer to humans of genes of artificial, laboratory-created genes, or genes from other species, be permitted? Does editing out deleterious traits from a family's genetic line imply anything about the genetic superiority or inferiority of that family or of other persons and groups? Such questions go to the social construct of human identity (Gregg, 2012a) which, at different stages of human life, is implicated in different ways. Consider several. First, in human genetic engineering, what moral status, and what legal status, should a political community accord to the human embryo? Second, germline engineering affects future persons who, at the point of engineering, cannot have had an opportunity to make a voluntary, informed decision on whether to take the risk of engineering the embryo or fetus from which they later developed. By what standards might a political community weigh genetic risks to the future adult against cultural risks of relevant parental child-rearing choices (including conventional parental decisions regarding education, religion, diet or lifestyle)? Third, what obligations does the generation performing the engineering have toward future persons (Gregg, 2021a)? Fourth, by what standard could a political community balance risk assessment with benefit assessment, for example when engineering may be beneficial but not yet adequately safe? Fifth, does genetic engineering toward preventing diseases and disabilities necessarily devalue the lives of persons currently living with those diseases and disabilities?

4. Unequal access to human genetic engineering is expectable along several dimensions (Gregg, 2012b). First, if somatic and germline editing become a regular part of healthcare, they may be unaffordable excect for the wealthy, thus exacerbating extant economic inequalities and possibly reinforcing the exclusion of many stakeholders from decision-making about the technology's development, deployment, and regulation. Second, access to genetic engineering also varies markedly among different countries and world regions. Third, medical systems in the West tend to regard the white middle class as the default population in ways that may sometimes miss many other groups, including persons with disabilities of all ethnicities, whose presence is wider than socio-economic classes.

From yet other perspectives, one might think that human genetic engineering is too accessible. Consider CRISPR technology (clustered regularly interspaced short palindromic repeats): low in cost and highly efficient in obtaining consistent results across different cell types and species, it can be mastered and deployed more easily and quickly than older gene editing tools. Widely available to persons committed to purchasing it (online, even), it constitutes a dangerous tool for rogue biohackers and others operating outside accredited laboratories, and seeking to exploit the vulnerabilities of various groups and individuals and who, through technical incompetence or otherwise, may do serious harm to those groups and individuals.

5. Evaluation of the various economic dimensions of genetic engineering may be ambivalent as well. Consider agriculture: because crops are the greatest source of nutrition in the world, genetic enhancement of nutritional value and resilience are desirable—yet allowing private corporate property rights for engineered seeds should give one pause (Moyn, 2018). Consider consumer freedom: parents purchasing the genetic engineering necessary to deliver their future children from genetic diseases may lead to a kind of commodification of the child, perhaps with embryos regarded as economic goods and some future persons reduced to designed-products. Consider healthcare systems: a one-time genetic therapy would be less expensive than the accumulated costs of conventional treatments over a lifetime. But might increased costs due to exorbitant prices charged by private companies raise healthcare costs overall? Is cost analysis even the right metric for evaluating genetic engineering, for example, by comparing the lower cost of preventing the birth of a child with cystic fibrosis relative to the cost of caring for person with cystic fibrosis over a lifetime?

6. Regulation confronts these myriad moral and practical ambiguities. CRISPR in particular will be more challenging than other forms of genetic engineering due to its low cost, efficiency, precision, and global availability, making moratoria or outright bans ineffective, all the more so in a world of medical tourism, rogue agents, and massive economic interests eager to profit from the bioengineering of humans. But all forms of human genetic engineering require regulation and accountability, from licensing to surveilling to taxing authorized users. The question is: Who should decide issues involving the legal regulation of genetic engineering, and how? Whereas eugenic movements of the nineteenth and twentieth centuries were conducted by the state, today individuals (parents above all) are more likely to drive demand. Possible regulatory entities include state administrative and advisory agencies, recommending legislation, ensuring legal compliance, licensing practitioners and institutions and research projects, and protecting the fundamental rights of patients; oversight agencies, such as the Federal Drug Administration or the European Medicines Agency, to evaluate research proposals, monitor treatments and conduct surveillance of treatment results; scientists, speaking as experts but also as citizens, interested in conveying their research to the general public, mindful of public concerns; professional bodies that can credential practitioners and educate publics, such as the American

Society for Transplantation and Cellular Therapy; advisory bodies that issue expert statements, guidelines, recommendations, committee reports and statements that propose parameters for human genetic engineering, such as the Nuffield Council on Bioethics, the National Institutes of Health, and the National Academies of Science; patients and patient advocacy groups, providing perspectives for other patients and their families in preparation for clinical trials; and international organizations, such as the World Health Organization, operating across legal systems and regulating cross-border activity involving genetic engineering, but also issuing guidelines and standards of conduct. To be sure: mostly certain Western countries with regulatory infrastructures have the capability to enforce such standards. Still, countries with no laws regulating human genetic engineering might rely on such elite international organizations.

7. Human genetic engineering also calls for informal regulation, especially by means of broad public engagement involving a variety of communities and stakeholders; plausible public education in the relevant sciences and technologies as well as in multiple ethical approaches, toward solicitinag informed input from the political community in discussing and devising policies for regulation, recommending legislation, developing licensing standards for qualified individuals and institutions in relevant research and deployment, and supporting compliance with existing laws and possible future legislation (Gregg, 2021b). Such efforts would diffuse accountability among multiple stakeholders at different levels of government and administration as well as in public spheres domestic and, perhaps one day, international.

References

Baker, B. (2016). The ethics of changing the human genome. *Bioscience, 66*, 267–273.

Carroll, D., & Charo, R. A. (2015). The societal opportunities and challenges of genome editing. *Genome Biology, 16*, 1–9.

Gregg, B. (2012a). *Human rights as social construction*. Cambridge University Press.

Gregg, B. (2012b). Genetic enhancement: A new dialectic of enlightenment? In D. Wetzel (Ed.), *Perspektiven der Aufklärung: Zwischen Mythos und Realität* (pp. 133–146). Verlag Wilhelm Fink.

Gregg, B. (2018). Human genetic engineering: Biotic justice in the anthropocene? In D. DellaSala & M. Goldstein (Eds.), *Encyclopedia of the Anthropocene* (Vol. 4, pp. 351–359). Elsevier.

Gregg, B. (2021a). Regulating genetic engineering guided by human dignity, not genetic essentialism. *Politics and the Life Sciences, 41*(1), 60–75.

Gregg, B. (2021b). Political bioethics. *Journal of Medicine and Philosophy: A Forum for Bioethics and Philosophy of Medicine, 47*(4), 516–529.

Gregg, B. (2022). *Creating human nature: the political challenges of genetic engineering*. Cambridge University Press.

Moyn, S. (2018). *Not enough: human rights in an unequal world*. Harvard University Press.

Veit, W., Anomaly, J., Agar, N., Singer, P., Fleischman, D., & Minerva, F. (2021). Can 'eugenics' be defended? *Monash Bioethics Review, 39*, 60–67.

Benjamin Gregg is Professor of Social and Political Theory at the University of Texas at Austin. Author of over 90 articles, reviews, and translations as well as five books: *Thick Moralities, Thin Politics*; *Coping in Politics with Indeterminate Norms*; *Human Rights as Social Construction*; *The Human Rights State*; and *Creating Human Nature: The Political Challenges of Genetic Engineering*. Fulbright Distinguished Chair in Public International Law at Lund University, Sweden, 2021–2022; Visiting Researcher, Centre for Biomedical Ethics, National University of Singapore, 2022. Visiting Researcher, Centre for Bioethics and Medical Humanities, Universitas Gadjah Mada, Yogyakarta, Indonesia, 2023.

Human Genome

Jérémy Choin and Lluis Quintana-Murci

Abstract Archaeology, palaeoanthropology and linguistics have provided valuable insights into human history. However, over the last two decades, fuelled by the advent of improved sequencing technologies, the study of the diversity of the human genome has greatly increased our understanding of the peopling history of the world. The information provided by these genomic data is highly complementary to these other disciplines as it offers another dimensionality for the study of population history. Here, we review how evolutionary and population genetic approaches have allowed the detailed reconstruction of the migratory history of our species, admixture episodes between modern human populations or between humans and now-extinct hominins, and events of human adaptation to new environments.

The release of the first draft of the human genome in 2001 marked the beginning of the genomic era and dramatically impacted the field of anthropological and population genetics (Lander et al., 2001). Shortly after, several consortia were built with the aim of characterizing the levels of human genetic diversity at the worldwide scale (Auton et al., 2015; Mallick et al., 2016). Today, we know that the human genome is composed of 3.2 billion nucleotides and hosts around 20,000 protein-coding genes. We also know that, among its more than 3 billion nucleotides, around four million are different between two randomly-chosen individuals. If we compare the genome of a random individual to the genome reference sequence, it differs at around 10,000 nucleotides that change the amino-acid sequence of the

J. Choin
Institut Pasteur, Université Paris Cité, CNRS UMR2000, Human Evolutionary Genetics Unit, Paris, France

L. Quintana-Murci (✉)
Institut Pasteur, Université Paris Cité, CNRS UMR2000, Human Evolutionary Genetics Unit, Paris, France

Chair of Human Genomics and Evolution, Collège de France, Paris, France
e-mail: quintana@pasteur.fr

© The Author(s), under exclusive license to Springer Nature Switzerland AG 2023
N. Wallenhorst, C. Wulf (eds.), *Handbook of the Anthropocene*,
https://doi.org/10.1007/978-3-031-25910-4_80

corresponding protein, it presents up to 300 genes with mutations that disrupt their function, and it harbours up to 100 mutations, at the heterozygote state, that have already been associated with a genetic disorder.

Mutations can be divided into three classes based on the number or the size of the modification: substitutions (point mutations), insertions and deletions (indels), and chromosomal rearrangements. We will focus here on point mutations, also known as single nucleotide polymorphisms (SNPs), because they are the most frequent in the human genome. Point mutations that change the final gene product, the protein amino-acid sequence, are known as 'non-synonymous mutations', while variants that does not change the protein sequence are known as 'synonymous mutations'. The rate of substitutions per site and per generation is expected to be 10^{-8}, but the mutation rate can go up to 10^{-5} substitutions per site and per generation depending on the genomic region (e.g. CpG sites) (Lipson et al., 2015).

Most mutations are neither beneficial nor deleterious; thus, they do not have any phenotypic consequence that alter the fitness of the carrier. The fate of these mutations, which are known as neutral mutations, in the population is driven by a stochastic process known as 'genetic drift'. Under genetic drift, the probability of a neutral mutation to be fixed in the population (or to be eliminated) depends on its initial frequency. In a population with a large effective population size, known as Ne (i.e., the number of individuals who contribute genetically to the next generation), the strength of the genetic drift is weak, leading to a stability of allelic frequencies during a long period of time. Conversely, in small populations (small Ne), for example those that have experienced reductions in population size following a bottleneck or a founder event, genetic drift is stronger and can cause sharp allele frequency variations from one generation to another.

Contrary to genetic drift, which can exacerbate the levels of population genetic differentiation, gene flow (i.e., the exchange of migrants between populations) reduces population differentiation. Thus, populations that are geographically close tend to be less genetically differentiated, because of their more recent divergence and/or the occurrence of gene flow (or admixture) between them; a process known as the 'isolation-by-distance model'. Hence, the levels of genetic differentiation between human groups depend on their demographic history, which affects the strength of genetic drift, as well as their levels of genetic isolation.

Over the past two decades, the study of the patterns of population genetic diversity, first using a handful of neutral genetic markers, such as the maternally-inherited mtDNA or the paternally-inherited Y-chromosome, and more recently using the information contained in the whole DNA sequence, has allowed the detailed reconstruction of the migratory history of human populations (Nielsen et al., 2017) as well as the different events of admixture between modern groups and between humans and archaic human forms, such as Neandertals and Denisovans (Choin et al., 2021; Vernot et al., 2016). Population genetic studies have traced back the origin of *Homo sapiens* to Africa around 200,000 years ago, and revealed that the peopling of the world started by a single, successful exit route from East Africa around 60,000 years ago. The out-of-Africa exodus was followed by a series of founder events associated with the peopling of Eurasia, Australia and Papua

New-Guinea around 60,000–40,000 years ago, the Americas around 35,000–15,000 years ago, and, more recently, the remote islands of Oceania around 3500–700 years ago.

The improvement of sequencing technologies has also allowed to access the genomes of fossil remains (i.e., ancient DNA), including those of extinct hominins (Liu et al., 2021). Ancient DNA studies indicate that the ancestors of present-day non-Africans interbred with Neanderthals, shortly after the Out-of-Africa exit around 50,000 years ago, with the genomes of all present-day non-Africans harbouring around 2% of Neanderthal genetic material. Likewise, there is increasing evidence to suggest that modern humans admixed with Denisovans, probably in Southeast Asia and at multiple, different times, with Denisovan DNA segments being today observed in present-day Australo-Papuans, constituting around 3.5% of their genomes, and to a lesser extent (< 1%), in East and Southeast Asian groups.

Despite most mutations of the human genome evolve neutrally, a fraction of them can evolve under the action of natural selection because they alter the fitness of the carriers by affecting phenotype variation. If the new phenotype confers a selective advantage in a given environment, the carriers of the advantageous mutation involved in the phenotype would have a higher survival probability and reproductive success (i.e., higher fitness). Consequently, the mutation would increase in the population by 'positive selection'. Conversely, deleterious mutations that decrease the fitness of the carriers, such as mutations in genes involved in basic developmental processes or functions related to innate immunity, will be purged from the population through the action of 'negative selection' (also known as 'purifying selection').

In this context, population genetics studies have increased our understanding of the genes and phenotypes that have participated in the adaptive history of humans (Fan et al., 2016). Probably the most iconic example of local adaptation is the lactase persistence trait in adulthood, which present strong signals of positive selection in the genomes of Europeans and East Africans (Tishkoff et al., 2007). There is also increasing evidence supporting that pathogen exposure has played a key role in the genetic adaptation of modern humans, including at the *DARC* gene, which confers resistance to *vivax* malaria in Africa, or at genes involved in the NF-κB signalling pathway, which confer resistance to cholera in the Bangladeshi population (Quintana-Murci, 2019). Furthermore, there are numerous examples of local adaptation to extreme environments (Ilardo & Nielsen, 2018). For example, signatures of positive selection have been identified at genes associated with the thyroid hormone signaling pathway in Aboriginal Australians, which could have contributed to their adaptation to the desert climate (Malaspinas et al., 2016), or in genes involved in the fatty acid metabolism in Greenlandic Inuits, which likely allowed them to adapt to the cold climate of the artic environment (Fumagalli et al., 2015). Other studies have shed light on how humans have adapted to the extreme conditions of the tropical rainforest (genes associated to stature), high altitude (genes involved in hypoxia-inducible factor pathway), or UV exposure (genes involved in skin pigmentation) (Fan et al., 2016).

These examples of positive selection follow the 'classic sweep' model, a process in which a new, strongly beneficial mutation appears and rapidly increases in frequency to ultimately reach fixation. However, there is compelling evidence to suggest that human genetic adaptation over the last 250,000 years involved a moderate number of classic sweeps, suggesting that other modes of natural selection occurred (Pritchard & Di Rienzo, 2010). For example, following a change in environmental pressures (e.g. the settlement of a new geographic region), a fraction of previously-neutral standing variation can indeed become advantageous, because they confer a specific advantage in the new environment (e.g., different climate, new pathogens, varying nutritional resources). Thus, the frequency of these mutations would no longer be driven by genetic drift alone but also by the action of natural selection. This mode of selection is known as 'selection on standing variation'. As these mutations already exist in the population, the adaptive process will be much faster than under the classic sweep model. Another selection regime that has gained considerable interest over the past few years is that of 'polygenic adaptation'. Contrarily to the classic sweep model, polygenic adaptation involves the action of weak positive selection that targets multiple genomic regions simultaneously, all of them associated with the same adaptive phenotype, for example, human height.

Each mode of natural selection leaves specific molecular signatures in the genome, which can be detected by an always increasing number of statistical tests (Vitti et al., 2013). For example, positive selection can be detected with several metrics that are based on different aspects of the data; the site frequency spectrum, the degree of genetic differentiation between populations, and the levels of extended haplotype homozygosity. Under the classic sweep model, we expect a selected allele to be highly frequent in a specific population and be carried by haplotypes conserved over long genomic distances, due to hitchhiking effects. However, for complex traits or diseases, many genetic variants can be involved in the same phenotype owing to the polygenic basis of such phenotypes, with each of these variants having a small contribution to the variance of the trait. Consequently, under a model of polygenic adaptation, we expect only a subtle shift in the allelic frequencies of the adaptive variants in a specific population.

More recently, gene flow (or admixture) between human populations, a phenomenon that has been pervasive during human history, has been increasingly documented as having played a key role in the rapid genetic adaptation of humans. Beneficial mutations can be indeed transmitted from one 'donor' population to a 'receiver' population via gene flow, a process known as 'adaptive admixture'. For example, it has been shown that the lactase persistence trait in eastern Bantu-speaking groups was acquired through gene flow from east Africans pastoralists (Patin et al., 2017). Similarly, the malaria-protective *DARC* variant has been acquired by Pakistani admixed individuals via gene flow with sub-Saharan Africans (Laso-Jadart et al., 2017). Interestingly, advantageous mutations can also be acquired through admixture with archaic hominins, a process known as 'adaptive introgression' (Dannemann & Racimo, 2018). While most mutations transmitted by Neandertals and Denisovans were deleterious to modern humans, and were thus purged from their genomes, a fraction of them were beneficial, increased in frequency and contributed to modern human adaptation. For example, signatures of adaptive introgression have been observed at genes related to metabolism, skin pigmentation, and immunity to infection, in particular in genes encoding for human proteins that interact with viruses.

In sum, more than two decades after the release of the first draft of the human genome, population genetic studies have provided insights into the degree of human genome diversity at an unprecedented level of resolution. In doing so, they have greatly increased our knowledge on the peopling history of the world, the levels of admixture between modern human populations and between humans and extinct human forms, and the different ways in which populations have adapted to the different environments they encountered during their dispersals around the world.

References

Auton, A., Brooks, L. D., Durbin, R. M., Garrison, E. P., Kang, H. M., Korbel, J. O., Marchini, J. L., McCarthy, S., McVean, G. A., & Abecasis, G. R. (2015). A global reference for human genetic variation. *Nature, 526*, 68–74.

Choin, J., Mendoza-Revilla, J., Arauna, L. R., Cuadros-Espinoza, S., Cassar, O., Larena, M., Ko, A. M., Harmant, C., Laurent, R., Verdu, P., Laval, G., Boland, A., Olaso, R., Deleuze, J. F., Valentin, F., Ko, Y. C., Jakobsson, M., Gessain, A., Excoffier, L., Stoneking, M., Patin, E., & Quintana-Murci, L. (2021). Genomic insights into population history and biological adaptation in Oceania. *Nature, 592*, 583–589.

Dannemann, M., & Racimo, F. (2018). Something old, something borrowed: Admixture and adaptation in human evolution. *Current Opinion in Genetics & Development, 53*, 1–8.

Fan, S., Hansen, M. E., Lo, Y., & Tishkoff, S. A. (2016). Going global by adapting local: A review of recent human adaptation. *Science, 354*, 54–59.

Fumagalli, M., Moltke, I., Grarup, N., Racimo, F., Bjerregaard, P., Jorgensen, M. E., Korneliussen, T. S., Gerbault, P., Skotte, L., Linneberg, A., Christensen, C., Brandslund, I., Jorgensen, T., Huerta-Sanchez, E., Schmidt, E. B., Pedersen, O., Hansen, T., Albrechtsen, A., & Nielsen, R. (2015). Greenlandic Inuit show genetic signatures of diet and climate adaptation. *Science, 349*, 1343–1347.

Ilardo, M., & Nielsen, R. (2018). Human adaptation to extreme environmental conditions. *Current Opinion in Genetics & Development, 53*, 77–82.

Lander, E. S., et al. (2001). Initial sequencing and analysis of the human genome. *Nature, 409*, 860–921.

Laso-Jadart, R., Harmant, C., Quach, H., Zidane, N., Tyler-Smith, C., Mehdi, Q., Ayub, Q., Quintana-Murci, L., & Patin, E. (2017). The genetic legacy of the Indian Ocean slave trade: Recent admixture and post-admixture selection in the Makranis of Pakistan. *American Journal of Human Genetics, 101*, 977–984.

Lipson, M., Loh, P. R., Sankararaman, S., Patterson, N., Berger, B., & Reich, D. (2015). Calibrating the human mutation rate via ancestral recombination density in diploid genomes. *PLoS Genetics, 11*, e1005550.

Liu, Y., Mao, X., Krause, J., & Fu, Q. (2021). Insights into human history from the first decade of ancient human genomics. *Science, 373*, 1479–1484.

Malaspinas, A. S., et al. (2016). A genomic history of aboriginal Australia. *Nature, 538*, 207–214.

Mallick, S., et al. (2016). The Simons genome diversity project: 300 genomes from 142 diverse populations. *Nature, 538*, 201–206.

Nielsen, R., Akey, J. M., Jakobsson, M., Pritchard, J. K., Tishkoff, S., & Willerslev, E. (2017). Tracing the peopling of the world through genomics. *Nature, 541*, 302–310.

Patin, E., Lopez, M., Grollemund, R., Verdu, P., Harmant, C., Quach, H., Laval, G., Perry, G. H., Barreiro, L. B., Froment, A., Heyer, E., Massougbodji, A., Fortes-Lima, C., Migot-Nabias, F., Bellis, G., Dugoujon, J. M., Pereira, J. B., Fernandes, V., Pereira, L., Van der Veen, L., Mouguiama-Daouda, P., Bustamante, C. D., Hombert, J. M., & Quintana-Murci, L. (2017). Dispersals and genetic adaptation of Bantu-speaking populations in Africa and North America. *Science, 356*, 543–546.

Pritchard, J. K., & Di Rienzo, A. (2010). Adaptation – not by sweeps alone. *Nature Reviews Genetics, 11*, 665–667.

Quintana-Murci, L. (2019). Human immunology through the lens of evolutionary genetics. *Cell, 177*, 184–199.

Tishkoff, S. A., Reed, F. A., Ranciaro, A., Voight, B. F., Babbitt, C. C., Silverman, J. S., Powell, K., Mortensen, H. M., Hirbo, J. B., Osman, M., Ibrahim, M., Omar, S. A., Lema, G., Nyambo, T. B., Ghori, J., Bumpstead, S., Pritchard, J. K., Wray, G. A., & Deloukas, P. (2007). Convergent adaptation of human lactase persistence in Africa and Europe. *Nature Genetics, 39*, 31–40.

Vernot, B., Tucci, S., Kelso, J., Schraiber, J. G., Wolf, A. B., Gittelman, R. M., Dannemann, M., Grote, S., McCoy, R. C., Norton, H., Scheinfeldt, L. B., Merriwether, D. A., Koki, G., Friedlaender, J. S., Wakefield, J., Paabo, S., & Akey, J. M. (2016). Excavating Neandertal and Denisovan DNA from the genomes of Melanesian individuals. *Science, 352*, 235–239.

Vitti, J. J., Grossman, S. R., & Sabeti, P. C. (2013). Detecting natural selection in genomic data. *Annual Review of Genetics, 47*, 97–120.

Jérémy Choin is Doctor in Population Genetics by the University of Paris. He obtained his Undergraduate degree in Ecology and Evolution from the University of Toulouse and his Master degree in Genomics from the University of Paris. His research focuses on the peopling history of the South Pacific using a combination of theoretical population genetics and empirical genomic approaches. His main expertise is to reconstruct the demographic past of human populations using massive genomic datasets and state-of-the-art computational methods.

Lluis Quintana-Murci is Professor at the Collège de France and Institut Pasteur, and is Doctor in Population Genetics by the University of Pavia (Italy). He heads the Unit of Human Evolutionary Genetics in the Institut Pasteur since 2007. His research focuses on demographic and adaptive inference in humans, using genomic approaches. In particular, he is interested in how pathogens have exerted selective pressures on the human genome. He has co-authored over 200 publications and published 12 book chapters. He is the author of the book *Le peuple des humains* (Odile Jacob, 2021, *in French*).

Industrial Revolution

François Jarrige and Thomas Le Roux

Abstract This article examines the close links between notions of the Anthropocene and the Industrial Revolution with a view to interpreting historical times which marked the shifting of the world into a new economic regime shaped by the increasing extraction of resources and the significant increase in pollutant discharges into the environment. The age of the Anthropocene forces us to rethink how the industrialization of the world happened, without concealing its negative aspects. This enables us to reconsider our view of progress and reconnect with what is really happening.

The expression "Industrial Revolution", which first appeared among French merchants in 1797, was coined by nineteenth-century economists. It was popularized in France by Adolphe Blanqui in 1837, in his *Histoire de l'économie française*, and taken up by Friedrich Engels in the 1840s. It was to be disseminated in the community of historians following the publication of *Lectures on The Industrial Revolution in England* by the British Arnold Toynbee (1884). By analogy with political revolutions, the expression aimed to mark the break experienced by contemporaries in the socio-economic order; gradually, it conveyed the idea of economic growth, ushering in an era of progress, enrichment and emancipation from natural constraints. Among twentieth-century historians, it is a convenient label to describe the entry of Europe into a new economic and energy regime between the end of the eighteenth century and the middle of the nineteenth century, with the ruptures produced by technical change, and in particular the steam engine, textile mechanisation and the use of coal (Landes, 1969). The concept is also used as a

F. Jarrige (✉)
Burgundy University, Dijon, France
e-mail: francois.jarrige@u-bourgogne.fr

T. Le Roux
CNRS-EHESS-CRH, Paris, France

© The Author(s), under exclusive license to Springer Nature Switzerland AG 2023
N. Wallenhorst, C. Wulf (eds.), *Handbook of the Anthropocene*,
https://doi.org/10.1007/978-3-031-25910-4_81

reading grid to characterise subsequent developments. We thus speak of the second industrial revolution with the advent of electricity, then of the third industrial revolution with the digital advent.

It is thus as much a pedagogical figure as a discourse. Nevertheless, the immense historiography and the malleability of this concept have made it possible to qualify the principle. The linear narrative centred on the study of growth and the primacy of technology has given way to greater attention to the plurality of paths and forms of work organisation inscribed in dense and ancient social ties. The model of the "Industrial Revolution" has thus been succeeded by that of "proto-industrialization" or "soft industrialization" which characterised certain regions, far from the British model (Woronoff, 1994; Horn, 2006). In addition, some historians have put forward the idea of an "industrious revolution", by pointing out the role of demand, popular aspiration for consumption and the intensification of work, without giving technical systems a preponderant role (de Vries, 2008; Verley, 1997), while others have insisted more broadly on the cultural roots of the British Enlightenment (Mokyr, 2010) or the high level of wages (Allen, 2009). This reflection on the processes has also shed light on the uneven development of industrialization, both from a geographical and chronological point of view, leading, in a transnational and Marxist approach, to examine the relations of power and domination or unequal exchange (Malm, 2016; Hornborg, 2013).

Though marginalised for some time in historical accounts, the concept of Industrial Revolution regained its centrality with the beginning of the twenty-first century and the worsening of the climate crisis, as it was propelled as the starting point of the Anthropocene by Paul Joseph Crutzen, Nobel Prize in chemistry (Bonneuil & Fressoz, 2016). According to him, it was indeed the development of James Watt's steam engine at the end of the eighteenth century that inaugurated this new era in which human societies became a geological force capable of modifying the major balances of the biosphere. Fuelled by their contemporary concerns, many historians are currently reinterpreting the notion of Industrial Revolution as a fundamental rupture: they no longer promote an imaginary sense of progress and emancipation, but rather pay greater attention to the destruction and unprecedented transformations of the world's physical environments.

The notions of Anthropocene and Industrial Revolution therefore maintain close links to interpret historical times, and to represent the shifting of the world into a new economic regime shaped by the increasing extraction of resources and the significant increase in pollutant discharges into the environment. However, these links are hotly debated and discussed. For example, some Marxist-inspired scholars are wary of the notion of Anthropocene, which they consider too encompassing, and prefer to speak of the Capitalocene to better establish the link with class relations and the rise of industrial capitalism in the eighteenth century (Moore, 2015). Furthermore, the historical questions on the impacts of industrialization on the environment draw their sources from the questioning of the productivist system, of which there are regular manifestations during the twentieth century, themselves largely consecutive to various forms of social dislocations (Polanyi, 1944). The approach began to flourish in the 1970s with the birth of political ecology, then environmental history (Clapp, 1994). At the turn of the 2000s, even before the Anthropocene became a framework for reading, ecological factors and environmental externalities became decisive elements in explaining the unequal development of societies in several major narratives (Davis, 2001). While regaining its centrality, the concept of Industrial Revolution was deprived of its evolutionist and determinist vision with the marginalisation of classic explanatory factors such as technical progress or factory concentration.

This focus on environmental factors illustrates the growing concerns about the effects of the ecological crisis of the Anthropocene. Historians such as Rudolf Sieferle (2001), Anthony Wrigley (2010) and Kander et al. (2013) propose to re-read the industrialization of Europe through the prism of energy metric. The Industrial Revolution is therefore interpreted as the transition from an ancient organic economy based on wood, muscular power and hydraulic energy to a mineral economy based on coal, this "underground forest" which allowed Western Europe to escape the constraint of land availability for plants production, making long-term economic growth possible. In his great fresco of global history on the reasons for the great divergence between Europe and Asia, Kenneth Pomeranz (2000) also places environmental issues and coal at the heart of the story. He shows that in the eighteenth century the Yangtze River Delta region in China and England had similar economic characteristics and faced the same Malthusian ceiling, due to lack of space and new raw materials. But unlike the Yangtze, England managed to overcome its ecological limitations and constraints by taking advantage of cheap, abundant and accessible coal, coupled with the massive agricultural resources of the American and West Indian colonies of its empire. Coal, combined with the strength of the Atlantic maritime empire that England had built, allowed it to make up for land shortages and to industrialize with an exogenous source of energy. Coal energy and associated technologies are therefore resurfacing to explain industrialization.

Interest in the environmental consequences of the Industrial Revolution has long been limited among historians. For example, many studies have been devoted to the chemical industry without a single line about the gigantic pollution it caused. But historiography has been paying increasing attention to the issue of pollution and risks, for several decades in the United States and Great Britain, and more recently on the European continent (Melosi, 1980; Luckin, 1986; Mosley, 2001). While it is true that smog was invented in England at the beginning of the twentieth century, problems related to water or air pollution have not ceased to generate debates and controversies on the continent since the beginning of industrialization. At the beginning of the nineteenth century, for example, following the first factory settlements, Paris was sick of its pollution, which prompted the police prefect to create a Health Council, made up of experts and responsible for visiting the workshops. In 1810, a law on the pollution of harmful establishments, the first one in the world, was passed: nestled in the decades of the first industrialization, this law heralded a paradigm shift towards the authorities' support for industrial progress, the latter seeking to legitimise the presence of industry in the city by asserting its harmlessness and proposing to remedy the nuisances by resorting to technical improvements, despite the harmful consequences on the environment, which were not unknown (Le Roux, 2011). There is no doubt that the irruption of the Anthropocene in the social sciences helps to propose new interpretations of the Industrial Revolution, without hiding its setbacks (Jarrige & Le Roux, 2020), in order to get rid of the imagination of progress and reconnect with our biological materiality.

References

Allen, R. C. (2009). *The British industrial revolution in global perspective*. Cambridge University Press.
Bonneuil, C., & Fressoz, J.-B. (2016). *The shock of the Anthropocene. The earth, history and us*. Verso.
Clapp, B. W. (1994). *An environmental history of Britain since the industrial revolution*. Longman.
Davis, M. (2001). *Late Victorian holocausts: El Niño famines and the making of the third world*. Verso.
de Vries, J. (2008). *The industrious revolution. Consumer behavior and the household economy, 1650 to the present*. Cambridge University Press.
Horn, J. (2006). *The path not taken. French industrialization in the age of revolution, 1750–1830*. The MIT Press.
Hornborg, A. (2013). *Global ecology and unequal exchange*. Routledge.
Jarrige, F., & Le Roux, T. (2020). *The contamination of the earth: A history of pollutions in the industrial age*. MIT Press.
Kander, A., Malanima, P., & Warde, P. (2013). *Power to the people. Energy in Europe over the last five centuries*. Princeton University Press.
Landes, D. (1969). *The unbound Prometheus. Technological change and industrial development in Western Europe from 1750 to the present*. Cambridge, Cambridge University Press.
Luckin, B. (1986). *Pollution and control: A social history of the Thames in the nineteenth century*. Hilger.
Malm, A. (2016). *Fossil capital. The rise of steam power and the roots of global warming*. Verso.
Melosi, M. (Ed.). (1980). *Pollution and reform in American cities, 1870–1930*. University of Texas Press.
Mokyr, J. (2010). *The enlightened economy: An economic history of Britain (1700–1850)*. Yale University Press.
Moore, J. (2015). *Capitalism in the web of life: Ecology and the accumulation of capital*. Verso.
Mosley, S. (2001). *The chimney of the world: A history of smoke pollution in Victorian and Edwardian Manchester*. White Horse Press.
Polanyi, K. (1944). *The great transformation: The political and economic origins of our time*. Beacon Press.
Pomeranz, K. (2000). *The great divergence: Europe, China, and the making of the modern world economy*. Princenton University Press.
Le Roux, T. (2011). *Le laboratoire des pollutions industrielles à Paris, 1770–1830*. Albin Michel.
Sieferle, R. P. (2001). *The subterranean Forest. Energy systems and the industrial revolution*. The White Horse Press.
Toynbee, A. (1884). *Lectures on the industrial revolution in England*. Rivingtons.
Verley, P. (1997). *L'Échelle du monde. Essai sur l'industrialisation de l'Occident*. Gallimard.
Woronoff, D. (1994). *Histoire de l'industrie en France du XVIᵉ siècle à nos jours*. Le Seuil.
Wrigley, E. A. (2010). *Energy and the English industrial revolution*. Cambridge University Press.

François Jarrige is Doctor in History and Associate Professor at the Burgundy University (France), in the laboratory LIR3S (UMR CNRS-UB). He studies the social and ecological aspects of industrialization, he is the author of *Au temps des tueuses de bras* (PUR, 2009, in French); *Technocritiques. Du refus des machines à la contestations des technosciences* (La découverte, 2014, in French); and (with T. Le Roux) *The Contamination of the Earth. A History of Pollutions in the Industrial Age* (MIT Press, 2020).

Thomas Le Roux is Associate Professor of Research in History at the French National Center for Scientific Research (CNRS), doctor in History and director of the Center for Historical Research (CRH, Centre de recherches historiques) of the EHESS. Specialist of the history of industry and pollution, he is the author of *Le laboratoire des pollutions industrielles. Paris, 1770-1830* (Paris, 2011, in French); (with F. Jarrige) *The Contamination of the Earth. A History of Pollutions in the Industrial Age* (MIT Press, 2020) and (dir. with Béatrice Delaurenti) *Cultures of Contagion* (MIT Press, 2021).

Synthetic Biology

Joshua Wodak

Abstract This article examines the practice of synthetic biology, and its relevance in terms of technoscientific responses to the Anthropocene. The article offers a brief history of the practice, and contextualises synthetic biology within the wider field of controversial 'technofixes' that are proposed for ameliorating the existential challenge that is the Anthropocene. In particular, it explores the emerging relationship between conservation biology and synthetic biology, whereby the latter is put forward as an integral means for intervening in ecosystems and evolution itself.

The Anthropocene marks a key catalyst for the unfolding Mass Extinction Event, being the sixth such event since the evolution of multicellular life 570 million years ago. While the recognition of human-induced destruction of the more-than-human living world is commonly limited to larger flora and fauna, the destruction is also registered at the microbial scale. Writing in the journal *Anthropocene*, microbiologists Michael Gillings and Ian Paulsen make such a case for why some of the most telling impacts are at the microbial, rather than human, scale. Having documented how human actions have influenced microbial evolution from the origins of the Holocene through to the advent of the Anthropocene, they illustrate why this largely imperceptible realm is so relevant to current global biophysical change. And further, why this realm is so relevant to the tenuous prospects for the future of humans: "Microbial evolution is currently keeping pace with the environmental changes wrought by humanity. It remains to be seen whether organisms with longer generation times, smaller populations and larger sizes can do the same" (Gillings and Paulsen, 2014, p. 1).

Reading between the lines of the hedged language that is so typical of scientific synopses of the Anthropocene: microbes were the sole lifeforms for the first three billion years of life on this planet. At a macro-evolutionary scale, microbial life will weather this 'storm' just as they have all those Mass Extinction Events that came

J. Wodak (✉)
Western Sydney University, Sydney, NSW, Australia
e-mail: j.wodak@westernsydney.edu.au

© The Author(s), under exclusive license to Springer Nature
Switzerland AG 2023
N. Wallenhorst, C. Wulf (eds.), *Handbook of the Anthropocene*,
https://doi.org/10.1007/978-3-031-25910-4_82

before. The same cannot be said for larger organisms: say bipedal apes who propagate one-at-a-time with generational intervals measured in decades. In the same article, *Microbiology of the Anthropocene*, Gillings and Paulsen put forward a measured case for why they have both moved from being microbiologists, who work via observation-and-analysis, to synthetic biologists, who no longer seek to just observe and analyse, but now to engineer and design microbes (2014, pp. 3–8).

Synthetic Biology is a diverse field of practices that have rapidly risen in prominence, since first emerging in the early 2000's. Defined by the Royal Society as "the design and construction of novel artificial biological pathways, organisms or devices, or the redesign of existing natural biological systems" (2010) the short history of synthetic biology has run in parallel with the contemporary coinage of the Anthropocene, by Paul Crutzen in 2000. Indeed, in a 2010 blog post co-authored with Christian Schwägerl on *Living in the Anthropocene: Toward a New Global Ethos*, Crutzen made an explicit connection between synthetic biology and the Anthropocene, remarking that Albeit clumsily, we are taking control of Nature's realm, from climate to DNA. We humans are becoming the dominant force for change on Earth. A long-held religious and philosophical idea — humans as the masters of planet Earth — has turned into a stark reality… While driving uncountable numbers of species to extinction, we create new life forms through gene technology, and, soon, through synthetic biology (Crutzen & Schwägerl, 2011).

So, while the Anthropocene is signified by human-induced extinction on a planetary-wide and geological scale, synthetic biology appears to signify human-induced evolution, albeit on the microbial scale, and predominantly used for perpetuating so-called human civilisation rather than benefiting the more-than-human living world (Wodak, 2020).

The field aims to synthesise microbial organisms into 'biofactories', whereby their metabolism is directed towards making medicines, bioplastics, or biofuels. This applies to microbes that have already been co-opted for human benefit since the proverbial dawn of civilisation, such as yeast for beer and bread, through to currently non-existent but seriously proposed new-to-nature microbes notionally designed to fulfil specific desired outcomes. While the field is highly diverse, it is unified by a promissory zeal, which maintains that successful synthesis will allow for the prodigious productivity of the microbial world to be harnessed into biomanufacturing (such as swapping biofuels for fossil fuels, or bioplastic for petroleum-derived plastic), producing lower biophysical impacts.

This promissory zeal goes back a lot further than the decade since Crutzen and Schwägerl speculated that "we [will] create new life forms … soon, through synthetic biology." The "soon" has yet to appear in any domain outside of the laboratory, where so-called synthetic lifeforms are ephemeral and deliberately 'designed' to not propagate outside of laboratory-controlled conditions. Setting aside the dearth of real-world biomanufacturing outcomes to date, synthetic biology has more recently gained growing support for use in conservation. Such proposed conservation applications create a vexing connection between synthetic biology and the Anthropocene: whereby the former is posited as being key to responding to the existential challenges of the latter.

The potential relationship between these strange-bedfellows is unintentionally expressed by evolutionary biologist E. O. Wilson in the opening gambit to his book on *The Future of Life*. Here Wilson argues that "the race is now on between the techno-scientific forces that are destroying the living environment and those that can be harnessed to save it" (2003, xii). Contra Wilson, what if harnessing techno-science – exemplified by synthetic biology – may become part of the "race" between "destroying the living environment" and "forces … to save it"? Given that biophysical environments are changing orders of magnitude faster than multitudes of species can adapt to, synthetic biology has recently been put forth as offering 'last-ditch' conservation in this "race" against time, by intervening at the genomic level.

Conservationist synthetic biology only first appeared in 2013, in a framing paper by the Wildlife Conservation Society, titled 'How Will Synthetic Biology and Conservation Shape the Future of Nature?'. In it the authors echo Crutzen and Schwägerl's promissory rhetoric that synthetic biology will "soon" allow us to "create new life forms", as they argue that synthetic biology could provide conservationists with more effective methods of conservation, including the creation of new tools that can help to gather and process field samples affordably or to monitor for the presence of particular threats – be they pathogens or chemicals. Likewise, synthetic biology could be used to reintroduce lost genetic variation into extant, but diminished and threatened populations (Redford, 2013, p. 14).

Such visions have subsequently been amplified in a series of papers, reports and conferences which seek to develop conservationist synthetic biology (Coleman & Goold, 2019; Piaggio et al., 2017; Redford et al., 2013, 2019).

What these promissory visions share is the idea that synthetic biology will "soon" be able to help conserve imperilled species *in situ* rather than the present-day confines of the *ex situ* laboratory. Proposals range from enhancing the resistance of the American chestnut tree to a fungal blight brought over a century ago in transplanted trees from Japan; enhancing the thermal tolerance of coral; synthesising the blood of the Horseshoe crab so that it is no longer harvested for its blood to be used for human vaccination manufacture; making malarial mosquitoes all male so that they cannot breed, to reduce malaria, and likewise making invasive rodents on secluded islands all male so that they cannot breed, to reduce island extinctions caused by the rodents. How such promise may translate into practice is an area of rapidly growing research, given that the science and technology does not exist for most of the proposals to be able to be realised, yet the timeframes available to do so are rapidly diminishing, as per Wilson's "race."

Research into this arena extends far beyond technoscientific mindsets, to incorporate critical perspectives from the humanities and social sciences (Preston, 2018; Wodak, 2021). These perspectives seek to shed light on yet another lurking danger that stems from positing synthetic biology as an existential response to the Anthropocene. Namely, if the former was to prove efficacious in relation to that latter, it could lend ill-conceived support to the so-called 'Good Anthropocene' hypothesis (Hamilton, 2016), or other related proposals to use technofixes to ameliorate the fuller consequences of the Sixth Extinction Event.

In this vein, synthetic biology belongs to what Jamie Lorimer terms "the 'dream of mastery' in his book *Wildlife in the Anthropocene: Conservation after Nature*. He defines this as presenting "the Anthropocene as an economic and scientific opportunity necessitating more modernization", and contrasts it to "the 'dream of naturalism'" which is "based on a return to some premodern or even prehistorical state revealed through a valorisation of traditional/indigenous knowledge" (Lorimer, 2015, pp. 2–3). Further, Lorimer is critical of both 'dreams', as he argues that "they share a totalizing and anthropocentric belief in the power of science and technology to either destroy or manage the earth" (2015, p. 3). Both dreams, he writes, neglect our persistent vulnerabilities to the earth's unruly geopower manifest in earthquakes, tsunamis, and other geological hazards. Second, [they] downplays the biopower and resilience of life itself, which continues to elude Promethean aspirations for planetary management and will no doubt survive even the most extreme scenario for a warming world" (2015, p. 3).

If the Anthropocene has cleaved environmentalism into these two opposing "dreams", then there is a great deal at stake for returning to a sobering reality. This is the reality of Wilson's "race", during which proposals are increasingly in momentum and hubris for moving from "albeit clumsily … taking control of Nature's realm" to intentionally taking 'control' in the not-too-distant future. At stake here is an ontological condition that undermines any simplistic grounds for supporting or objecting to such proposed synthetic biology usage. Because, if it is actually time to use synthetic biology for conservation, then Anthropocene ecology and evolution will bear the mark of human intention, as a counterpoint to the current unintentional desecration caused by those "techno-scientific forces that are destroying the living environment." Therein, the Anthropocene will not merely be registered in the strata as a boundary layer of mass extinction, which will persist through the earth's strata over geological time scales. For it will also be registered in the human-induced evolution of microbes, and, thus, human-inflected evolution of whatever larger lifeforms that become successful beneficiaries of bioengineered genomes.

This is simply because species subject to any human intervention that successfully confers evolutionary benefits will carry discrete genotypic marks in their human-inflected phylogenesis. Not in the domain of synthetic yeast co-opted into doing the bidding of so-called civilisation, but in the flora and fauna bearing synthesised genes that have increased their capacity to adapt to prevailing biophysical conditions. In the instance of species' directly subjected to synthetic biology, their evolution would be ontologically unique. It is high time, it would seem, that engagement with the Anthropocene turn toward the microbial realm, where our marks of intention and intervention will play out with planetary-wide repercussions for that which is susceptible to the unfolding mass extinction event.

References

Coleman, M., & Goold, H. (2019). Harnessing synthetic biology for kelp Forest conservation. *Journal of Phycology, 55*, 745–751.

Crutzen, P., & Schwägerl, C. (2011). Living in the Anthropocene: Toward a new global ethos. *Yale E360*. Available online: https://e360.yale.edu/features/living_in_the_anthropocene_toward_a_new_global_ethos. Accessed 9 July 2021.

Gillings, M., & Paulsen, I. (2014). Microbiology of the Anthropocene. *Anthropocene, 5*, 1–8.

Hamilton, C. (2016). The theodicy of the 'Good Anthropocene. *Environmental Humanities, 7*(1), 233–238.

Lorimer, J. (2015). *Wildlife in the Anthropocene: Conservation after nature*. University of Minnesota Press.

Piaggio, A., Segelbacher, G., Seddon, P., Alphey, L., Bennett, E., Carlson, R., Friedman, R. M., Kanavy, D., Phelan, R., Redford, K., Rosales, M., Slobodian, L., & Wheeler, K. (2017). Is it time for synthetic biodiversity conservation? *Trends in Ecology & Evolution, 32*, 97–107.

Preston, C. (2018). *The synthetic age: Outdesigning evolution, resurrecting species, and reengineering our world*. MIT Press.

Redford, K. (Ed.). (2013). *How will synthetic biology and conservation shape the future of nature? A framing paper*. Wildlife Conservation Society.

Redford, K., Adams, W., & Mace, G. (2013). Synthetic biology and conservation of nature: Wicked problems and wicked solutions. *PLoS Biology, 11*, e1001530.

Redford, K., Brooks, T., Macfarlane, N., & Adams, J. (Eds.). (2019). *Genetic Frontiers for conservation: An assessment of synthetic biology and biodiversity conservation: Technical assessment*. IUCN, International Union for Conservation of Nature.

The Royal Society. (2010). *Synthetic Biology*. https://royalsociety.org/topics-policy/projects/synthetic-biology/. Accessed 9 July 2021.

Wilson, E. (2003). *The future of life*. Vintage.

Wodak, J. (2020). (Human-inflected) Evolution in an age of (human-induced) extinction: Synthetic biology meets the Anthropocene. *Humanities, 9*(4), 1–17.

Wodak, J. (2021). Drawing a line in the sand: Bioengineering as conservation in the face of extinction debt. *Queensland Review, 28*(2), 24–36.

Dr. Joshua Wodak works at the intersection of the Environmental Humanities and Science & Technology Studies. His research addresses the socio-cultural dimensions of the climate crisis and the Anthropocene, with a focus on the ethics and efficacy of conservation through technoscience, including Synthetic Biology, Assisted Evolution, and Climate Engineering. He is currently a Senior Research Fellow at the Institute for Culture and Society, Western Sydney University; a Chief Investigator at the Australian Research Council Centre for Excellence in Synthetic Biology; and an Adjunct Senior Lecturer, School of Biological, Earth, and Environmental Sciences, UNSW. His recently completed book, Petrified: Living During a Rupture of Life on Earth, about what kinds of conservation experimentation we should be considering in response to the unfolding Sixth Extinction Event, will be published in 2023.

Technology

Pascal Marin

Abstract The harmful consequences of the headlong technological rush for humankind and their natural environment are becoming visible in the context of the Anthropocene. They also reveal how technology is much more than a matter of machines and technical means, as Martin Heidegger had well established in 1953 in *The Question Concerning Technology*. Will humanity be able to free itself from its quest for omnipotence over nature, in order to rediscover the meaning of the technical know-how that we have received since the paleoanthropological beginnings of our strange creative and thinking nature?

Any and all human actions go hand in hand with representations, and are imbued with meaning. All the mental activity behind an action (which the world never sees, and of which the person themselves is often barely aware or completely unaware) takes the act 'by the reins' and determines the action. Thus, technology – which, in order to accomplish various tasks in our personal and social lives, transforms the natural environment and populates it with artefacts – is more complex than the technological resources that must be built and deployed. Technological activity develops and implements its mechanical reality on the basis of a mental conception of the universe, which pre-dates any action and guides those actions. The distinction between the reality of technological artefacts and the 'imagination, the realm of forms' (Malraux, 1977, p. 179), which organises and drives technological action, is central to Martin Heidegger's thinking in *The Question Concerning Technology*. There is one central theme running through the philosopher's argument: the distinction, discussed in detail in Heidegger's text, between technology in the form of machines, and what Heidegger defines as the 'essence of technology' – i.e. the way in which it is present in our thoughts; that mental representation which, here, we speak of in terms of conception and imagination. 'Technology is not the same thing as the essence of technology', declares Heidegger; 'the essence of technology is not

P. Marin (✉)
Catholic University of Lyon, Lyon, France
e-mail: pmarin@univ-catholyon.fr

© The Author(s), under exclusive license to Springer Nature
Switzerland AG 2023
N. Wallenhorst, C. Wulf (eds.), *Handbook of the Anthropocene*,
https://doi.org/10.1007/978-3-031-25910-4_83

529

anything technological' (Heidegger, 1953, p. 10). In other words, the issues raised by technology go further than just technological objects that must be understood, exploited and mastered. This goes against technocratic thinking, whereby the issue is viewed solely as a problem that can be solved by a technological solution. On the other hand, in essence, technology in its modern guise aims to take full possession of nature. The archetypal maxim of this goal, which is inherent in today's science and technology, is that put forward by Descartes in *Discourse on the Method* whereby, by 'knowing the power and actions of fire, water, air, the stars, the heavens and all the other bodies which surround us, as distinctly as we know the various crafts of our artisans, we might also apply them in the same way to all the uses to which they are adapted, and thus render ourselves the lords and possessors of nature' (Descartes, 1953, p. 168). Beyond a doubt, these words embody the very soul of techno-science. As Descartes describes, modern technology continuously parallels the creative arts of the past, artisans and their crafts. However, the power which modern technology would gain from mastery of the natural elements, and the power those elements hold, set it clearly apart from our ancestors' *technè*, which did not have the same aim of domination. Yet while our contemporaries may seek to render themselves 'the lords and possessors of nature', in spite of the formidable achievements of scientific discovery and its technological applications, it must be recognised that the quest has failed. The looming climate emergency demonstrates that nature (of which our knowledge was supposed to make us masters) is beyond our control. More disquieting still, we can now quantify the power that this conception of technology has over our minds, provided we can recognise the degree to which technology affects our own sense of self. Without even appealing to our thinking and spirituality, the conception of technology first corrupts the sense of natural existence, of corporeal life. The most manifest symptom of this loss of awareness is the rise in cultural power of the ideas of Man-machine hybrids, between transhumanism and humanoid robots. These hybrids may be fictional, as yet, but they have sprung from real human minds. They are currently breaking down our sense of responsibility, awareness of others, and respect for life. Indeed, in these concepts of humans denatured by technology and humanised machines, life has disappeared, and the vessel of life, the body, is simply disposed of. Yet of course, it must be noted that this concept has no grounding in reality. It is, we believe, a sort of optical illusion, born of the fact that modern science does not perceive life, as such, because it studies reality through the lens of matter. Yet for both biologists and physicists, 'there is no "living matter". There is matter which makes up living beings, and that matter has no particular property that is not possessed by matter which makes up inert substances' (F. Jacob, 2000, p. 24). This materialistic view of human beings, which foreshadows their machinisation through technology and scientific knowledge, has deep roots. It is grounded in anthropology which has, from the very outset, mirrored the rising importance, in western society, of science and technological advancement: dualism. Yet this is an odd sort of anthropology: its purpose is not to reflect on Man's nature, but rather to justify our ability to know about the world. This ability (known as the soul or spirit) is viewed as being separate from the body, the latter being grounded in the pluralistic physical reality, utterly unaware of the universal.

The very word 'dualism', therefore, is misleading. It incorporates a monism of the spirit or soul, which condemns the body, the very vessel of life, to obscurity. These ideas of Man-machine hybrids, which are currently infiltrating our mindscapes under the influence of the media, radicalise and reveal the meaning of our history: 'In philosophy, dualism indicates disregard for the body. With cognitive techniques, it constitutes efforts to rid ourselves of the body entirely' (Besnier, 2009, p. 71). The concept of the Man-machine, which lies behind such efforts, is thus absurd. The fantastical nature of a non-corporeal existence is proof enough of the absurdity of dualist concepts. Who, though, subscribes to this conclusion in today's world? Of course, any conscious person has the ability to understand that their life cannot be reduced to a machine. To awaken this knowledge, however, they need to draw on resources other than those provided by their dominant culture. Indeed, that culture is involved in the increasing use of even more technology to keep control of the technology which is already beyond our grasp. How can we dispel the vain and dangerous idea of technological dominion over nature?

In *Une éthique pour la nature* (Ethics for Nature), German philosopher Hans Jonas sagely remarks that it is not possible to halt the 'technological quest'. To do so would no longer be desirable, because it is only through technological advancements that some of the harmful effects of technological development can be rectified. He concludes that only fear is cause for hope – a variation on the theme of 'But where danger is, grows the saving power also', advanced by Friedrich Hölderlin, whom Heidegger cites in the closing pages of *The Question Concerning Technology*: 'indeed, we are forced to live in the shadow of impending calamity', declares Jonas. 'Yet in awareness of that threat, paradoxically, lies the only glimmer of hope: that very awareness will keep the voice of responsibility alive' (Jonas, 2000, p. 154). However, what the philosopher does not explain is how we can achieve awareness of the threat. It has now been irrefutably shown that the idea of progress is no longer widely popular. Whilst, until recently, it was omnipresent in politicians' discourse, today it is conspicuous by its absence (Klein, 2017). The mood is rather pessimistic in western societies. However, in our view, in this context of disenchantment, the idea of all-powerful technology may not actually have lost any ground. Indications that it remains popular are to be found in the fascination with ideas of human–machine conversion and, in particular, all things relating to the world of robots and artificial intelligence. Think, for example, of the hugely successful publications of futurist Yuval Noah Harari, a thinker who leaves no room for doubt as to the scope of his predictions, if we give credit to the ending of his *Homo Deus. A Brief History of Tomorrow*. When, as he says, the 'all-knowing supercomputer [has] managed to conquer the entire galaxy', then '*homo sapiens* will vanish' (Harari, 2017, p. 409). In a way, while consideration of the climate crisis, along with the difficult economic and geopolitical context, hamstrings the idea of progress, on the other hand, the idea of 'techno-prophets', to use Dominique Lecourt's word (2003, p. 12) – of salvation through technology – is gaining traction. In this context of collective quasi-forclusion of human beings' sense, we believe Martin Heidegger is correct when he says that peril is not yet to come – it is here, it is now. 'What is dangerous is not technology', he says. 'The threat to man does not come, in the first instance, from potentially lethal machines and the apparatus of technology. The actual threat has already affected man in his essence [… It] threatens man with the possibility that it could be denied to him […] to experience the call of a more primal truth' (Heidegger, 1953, p. 45). However, in order to hear that call, we at least need our ears: we need to listen to discourse which can break through the dumb conceptions of seductive power. What opens us up to listening, in particular, is the experience of a body that has been left behind by the accelerated motion of technology: a body experiencing *burn-out*, which does not cope well with acceleration. The body, in order to live once more and catch its breath, fall back into its own rhythm, engages in the practice of 'deliberate deceleration', according to the analyses of Hartmut Rosa, a philosopher critical of acceleration (Rosa, 2014). Humans, having reclaimed their bodies and their ears, can then begin listening to all (philosophers and scientists) who have understood the vacuity of the technolatrous ideals of the Man-machine. It is worth heeding the words of Gilbert Simondon, saying that 'there is no such thing as a robot; that a robot is no more a machine than a statue is a living being' (Simondon, 1958, p. 10), and Steven Pinker, saying that in his view, Man 'will always be the same primate' (Pinker, 2000) – a primate enamoured of his own power.

References

Besnier, J.-M. (2009). *Demain les posthumains. Le futur a-t-il encore besoin de nous ?* Hachette littératures.

Descartes, R. (1953). *Œuvres et lettres*. Éditions Gallimard/Pléiade.

Harari, Y.-N. (2017). *Homo deus. Une brève histoire de l'avenir*. Albin Michel.

Heidegger, M. (1953). La question de la technique. In: *Essais et conférences* (pp. 9–48). Éditions Gallimard.

Jacob, F. (2000). Qu'est-ce que la vie ? In: *Université de tous les savoirs*. Éditions Odile Jacob.

Jonas, H. (2000). *Une éthique pour la nature*. Éditions Desclée de Brouwer.

Klein, E. (2017). *Sauvons le progrès*. La Tour d'Aigues.

Lecourt, D. (2003). *Humain post humain*. Puf.

Malraux, A. (1977). *L'homme précaire et la littérature*. Éditions Gallimard.

Pinker, S. (2000). Dans mille ans, l'homme sera toujours le même primate. *Courrier international*, 2000/09/07.

Rosa, H. (2014). *Aliénation et accélération. Vers une théorie critique de la modernité tardive*. Éditions La Découverte.

Simondon, G. (1958). *Du mode d'existence des objets techniques*. Éditions Aubier.

Pascal Marin is Professor at the Faculty of Philosophy of the Catholic University of Lyon. He is a member of the editorial boards of *La Revue des Sciences philosophiques et théologiques*, of the review *Sémiotique et Bible* and of the review *Théophilyon*. Specialized in the fields of hermeneutics and language, his research focuses at this time on the conditions of possibility of a philosophical anthropology in the context of artificial intelligence and transhumanism. He recently coordinated the dossier of the Revue *Théophilyon, Le naturalisme et ses critiques*, Tome XXVI, 2021–1 and published *Le robot et la pensée. Contre-philosophie de l'homme-machine*, Cerf, 2019.

Technoscientific Materialism

Edouard Kleinpeter

Abstract In this article, we analyse a concept forged by the encounter between an eighteenth century philosophical monism, Materialism, and the modern intertwining of science and technology, subsumed under the notion of "technoscience". In particular, we show how Materialism has been enrolled as the main philosophical foundation of technoscience, since this found in the materialist monad a "natural" substantiation of a certain conception of science and innovation. We first briefly introduce both terms of the dyad separately before discussing how they relate to one another in the contemporary discourse about scientific progress, while taking a small detour via the writings of the science fiction author Iain M. Banks.

Materialism is an eighteenth century philosophical school of thought which advocated metaphysics based on the idea that matter was the sole universal component of reality. It established a simple ontology resting on the materialistic monad claim: everything is matter or emerges from the interaction between materialistic entities. Thus, it opposed Cartesian dualism which postulated the existence of two distinct substances, mind and matter, and Spiritualism which stated the primacy of the mind. From its encounter with nineteenth century's naturalistic sciences a form of scientific materialism emerged, discarded as "vulgar" by Engels since it did not take into account historical and dialectic dimensions. Again, it was based on the materialistic monad and the affirmation of the superiority of the natural sciences when it comes to accessing the (material) truth about the world. During the 70's and 80's, materialism was incorporated into the field of the Theories of the Mind. In life sciences first, where reductionist theories were developed, consisting in explaining high-level cognitive phenomena, such as consciousness, by the physiological interactions and regulation mechanisms occurring in the brain (Changeux, 1983). Philosophers notably reacted to these issues by pinpointing the blind spots of a purely physical acceptance of cognition which is unable to explain, among other things, the fact that

E. Kleinpeter (✉)
Bordeaux University, Bordeaux, France
e-mail: edouard.kleinpeter@u-bordeaux.fr

intentional properties can be realized by different physical properties while maintaining their nomological characteristics (Jacob, 1993). During the last decade, scientific materialism met with renewed interest, in particular, to stress what Marc Silberstein calls its "pluralistic unity": as a regulatory principle to develop a scientific understanding of the world, as an unification principle of the sciences (from biology to society), as a heuristic principle to assess the assertability of scientific statements and, lastly, as the best possible ontology for natural sciences, indeed endorsed by the vast majority of scientists (Silberstein, 2013). In parallel, a certain interest in the materiality of cognitive processes was developed in psychoanalytical literature to oppose the purely informational theories based on the brain-computer metaphor, even if the neurosciences discarded it long ago (Dehaene, 2014). Here, the stress is put on the embodiment of cognitive mechanisms: thought is then understood as the product of the interaction between the brain, the body, the environment and the history of the individual (Benasayag, 2016), the latter two continuously retroacting to modify the states of the former two. Henceforth, machines (computers, namely), deprived from these interactions since they are unable to evolve in their materiality and are stuck in the rigidity of their code, could not *de facto* be conscious, nor have feelings in the same fashion as biological beings do (Chapouthier & Kaplan, 2011).

The Belgian philosopher Gilbert Hottois undertook to build the genealogy of the term "technoscience", while developing his own conception in his thesis published in 1979 and continued in his later works (Hottois, 2013; Hottois, 2015). He distinguishes three meanings of the term: a vision of scientifico-industrial politics, a societal approach, and a definition based on its transgressive nature related to the technophilic/technophobic duality of a distant future prospective, mostly through science fiction writings. The first one refers to the "*technoscience agencies*" which were created by the US government just after World War 2 (NASA, NIH, NSF, etc.). It reflects the fact that most American bureaucracies did not (and still do not) make any distinction between science and technology, both being at the service of progress and the leadership of the nation at the global scale. This conception of a science driven by the techno-industrial progress is still dominant in current discourses about science politics in most countries, through related notions such as "technological transfer" or "innovation". It stirs much criticism of "Big Science" as an avatar of techno-capitalism, which accuses it of forsaking the objectivity and rationality of the scientific approach in the name of the pursuit of public or private financing, thus giving up on the theoretical and grounding aspects of research as a quest for knowledge. The second acceptance of the term refers to the reality of technoscientific activity on a daily basis (Hottois, 2015, *op. Cit.*), as it is practised in the laboratories. It considers science as an activity that is essentially social on the one hand, marked by conflicting interests, norms and values and, on the other hand, as semiotic, symbolic and textual (Latour, 1989; Latour, 1988). Nonetheless, contrary to the previous case, it does not imply that scientific productions could be reduced to this construction process, their objectivity being guaranteed by social control (Bourdieu, 2001). The third meaning, developed by Hottois himself, aims at integrating both this social and economic immanence of technoscience and its transgressive and operative nature in the perspective of surpassing human boundaries (Hottois, 2013, *op. cit.*). He bases his reasoning on the speculative technoscientific imaginary at work in science fiction literature, on the exploration of philosophical questions it triggers, which he claims have been scarcely addressed by the philosophy of the first third quarters of the twentieth century: the extinction or mutation of mankind (thus prefiguring the contemporary debates about trans/post-humanism), the radical strangeness of the distant future, the introduction of morality and ethics into science, etc.

In this context, how can one define technoscientific materialism? The expression can be found, once again, in the writings of Gilbert Hottois (Hottois, 2013, *op. cit.*) when he analyses the works of the science fiction writer Iain M. Banks in his *The Culture* series. Banks describes an eponymous utopian civilization constituted of transhumans, aliens and artificial intelligences who have colonized the Milky Way. The technocosmos in which they live grants them an unlimited abundance of goods and services, to the point that they have eradicated illness and death. Hottois qualifies this civilization as post-modern and post-metaphysical. Hence, the technoscientific materialism on which it is founded is pragmatic above all: it is a 'working hypothesis and way of life that no reality has invalidated and which has extraordinarily flourished' (*ibid.*, p. 205, my translation). In the modern context, this opportunistic dimension of a non-metaphysical materialism resonates with the abolition of the great divides of modernity which has been endorsed by the technoscientific project since its origins. The distinctions between living and inert, thought and matter, natural and artificial, lose their conceptual and heuristic validity. Hence, if the technoscientific utopia aspires to this flat, world which offers no relief, then "common" materialism seems to be the most appropriate philosophy. Furthermore, if nature was considered as a limiting factor for deploying human technologies (their objective being, precisely, to extract us from our natural state, hence in opposition towards nature), it is now reduced by technoscience to an assembly of basic blocks actuated by principles allegedly simple enough to be manipulated, replicated and modified *ad libitum*. As Bernadette Bensaude-Vincent states: '*The technological undertaking [has transformed] the negotiation with our limits into a limitless exploration of possibilities*' (Bensaude-Vincent, 2009, p. 146, my translation). The proposed convergence project between the four fields of nanotechnology, biotechnology, information sciences and cognitive sciences (NBIC) harbours the seed of the advent of a new, better, enhanced human being, which echoes with the transgressive potential found in Hottois's definition of technoscience. Mainly, it is based on the hypothesis of the existence of a unifying principle making it possible to consider such a convergence. Yet, materialism provides this *a priori* simple, monolithic and universal principle. But it remains that there is no foundational or metaphysical purposes here: the aim is not to apply a discourse onto the world by enforcing the materialist monad, but rather, on the one hand, to conveniently subsume the whole technoscientific project under a sane rationality while evacuating all essentialist considerations regarding the status of the human being, its relationship with technology and nature and, on the other hand, to assert the potentiality of an hybridization between biology and technology since they both share the same materiality. Paradoxically, it is also worth mentioning that technoscience fosters a certain disdain towards the technical objects themselves and the modalities of their access to existence, an individuation process accurately described by Gilbert Simondon (Simondon, 1958). Indeed, according to the technoscientific conception, a good technical object is the one that integrates into our daily lives so well that it makes itself invisible. The whole process involved in its conception (realization of the ingenuity of its inventors, integration of constraints from the environment, etc.) is whisked off and replaced by an incremental process of micro-innovation, usually driven by socio-economical motives, which will soon replace it by a better "2.0 version". The same goes for the human being who awaits the advent of the transhuman…

References

Benasayag, M. (2016). *Cerveau augmenté, homme diminué*. La Découverte.
Bensaude-Vincent, B. (2009). *Les Vertiges de la technoscience. Façonner le monde atome par atome*. La Découverte.
Bourdieu, P. (2001). *Science de la science et réflexivité*. Paris Raisons d'agir.
Changeux, J.-P. (1983). *L'homme neuronal*. Arthème Fayard.
Chapouthier, G., & Kaplan, F. (2011). *L'homme, l'animal et la machine*. CNRS Editions.
Dehaene, S. (2014). *Le Code de la conscience*. Odile Jacob.
Hottois, G. (2013). *Généalogies philosophique, politique et imaginaire de la technoscience*. Librairie Philosophique J. Vrin.
Hottois, G. (2015). *Technoscience*. In: Hottois, G., Missa, J.-N., & Perbal, L (dir.), *Encyclopédie du trans/posthumanisme. L'humain et ses préfixes* (pp. 455-465). Librairie Philosophique J. Vrin.
Jacob, P. (1993). *Sens commun, psychologie cognitive et philosophie de la psychologie: Croyances, matérialisme et externalisme*. *L'Année psychologique, 93*, 59–83.
Latour, B. (1988). *La Vie de laboratoire*. La Découverte.
Latour, B. (1989). *La science en action*. La Découverte.
Silberstein, M. (2013). *L'unité plurielle du matérialisme*. In: Silberstein, Marc (dir.), *Matériaux philosophiques et scientifiques pour un matérialisme contemporain, vol. 1* (pp. 7-27). Editions Matériologiques.
Simondon, G. (2012). Du mode d'existence des objets techniques. Aubier. (Original edition 1958).

Edouard Kleinpeter is a research engineer for the French National Centre for Scientific Research (CNRS). After an engineering degree in Physics, he shifted towards scientific journalism before turning to academic research with a Ph D. thesis on science popularization. He then integrated an interdiciplinary laboratory and worked on philosophical groundings of human-machine interactions and, namely, of Human Enhancement and its affiliated ideologies (Transhumanism, Posthumanism,…) under the direction of a senior philosopher, Jean-Michel Besnier, between 2012 and 2016. He then passed through a second engineering school in Computer Science in order to acquire a better understanding of the inner structure of the machines and the technical details of the way we interact with them. In 2019, he joined the Research Group in Theoretical and Applied Economics (GREThA, UMR 5113, CNRS-University of Bordeaux), where he pursues his research on Human Enhancement with a focus on its economic aspects.

Technosphere

Peter K. Haff

Abstract The technosphere, the interlinked network of the world's humans and technological artefacts, is the defining structure of the Anthropocene. The technosphere is undesigned, autonomous and possesses agency. Where influenced by human knowledge, its future behaviour is unpredictable, although constrained by generic principles of organization. Humans face a fundamental dilemma in the conflict between (i) the technosphere's increasing rate of energy consumption, required to support discovery and application of the new knowledge essential for improvement and maintenance of human well-being, and (ii) the fact that no system can sustain an accelerating regime of energy use indefinitely.

The technosphere concept arose from an attempt to describe in physical terms the nature of a new geological Epoch, the Anthropocene (Haff, 2014a). Massive, energetic, and globally ubiquitous, the technosphere is a product of the continuing process of planetary evolution. It is an emerging Earth sphere, taking its place alongside the classical spheres of air (atmosphere), water (hydrosphere), rock (lithosphere) and living matter (biosphere).

Like the other spheres, the technosphere is a physical system. Here 'system' means 'dynamical system', an energy-consuming collection of entities (parts) identified and recognized on the basis of their mutually organized motion (behaviour).

Examples of the contents of the technosphere include most of the world's humans, domestic animals and plants, agricultural soils, transistors, cell phones, computers, computer networks, legal treatises, artwork, medical offices, buildings, schools, corporations, political parties, government bureaucracies, armies, and infrastructure, as well as the world's communication, transportation, educational, health and financial systems, and its nation states.

The emergence of the technosphere represents a major transition in the development of Earth through time. Its imprints and waste products mark the beginning of

P. K. Haff (✉)
Duke University, Durham, NC, USA
e-mail: pkhaff@gmail.com

the Anthropocene Epoch and constitute the Anthropocene's most distinctive stratigraphy (Zalasiewicz et al., 2014).

Among the most important questions for humans regarding the technosphere is not simply "what is it?" but rather "what does it mean for humans that they themselves are among its components?" The answer is counterintuitive, calling into question the view that civilization and technology are human constructs.

This physical view of the technosphere has been criticized as incompatible with sustainable development discourse founded on humanitarian principles (e.g., Donges et al., 2017). However, it is the detached perspective afforded by paying attention to physical necessity that defines the possible scope of those principles and identifies limits on achievability of desired goals.

Because patterns of civilized life across the globe are enabled by the actions of the technosphere, the future of the new sphere will determine the prospects for human well-being in the Anthropocene.

The path to understanding this novel phenomenon is obscured by the fact that humans do not stand outside the system but are integral parts of it. The individual human's perspective on the technosphere tends to be restricted in scope, magnifying the importance of that which is "up close" and affects the individual directly. One consequence of this condition is a failure to credit the power and reach of larger-scale forces that are out of sight but mould human experience and behaviour.

Because the technosphere has grown up around them, and because their ideas, skills and intentions were clearly essential to its emergence, humans tend to view the technosphere as a human construct.

However, the technosphere was created not by human actions alone, but by positive feedback processes whose scale, complexity and even existence, like Adam Smith's Invisible Hand, lay beyond the knowledge or comprehension of most humans.

The technosphere grew organically as opportunity permitted, with only piecemeal planning, into a globe-spanning, interlinked composite of billions of humans and trillions of technological artefacts.

Given its complexity, emergent (undesigned) origin and its ad hoc growth pattern, the technosphere's ability to maintain metabolic activity over years and at global scale in the absence of human understanding of the underlying dynamics, suggests that the system has at its command more modes of behaviour than putative human controllers are aware of or have the means to regulate.

In this picture (Ashby, 1957), the technosphere, like the four classical Earth spheres, is effectively autonomous, i.e., largely beyond human control (Haff, 2014b). The emergence of technospheric autonomy marked a technological tipping point in Earth history, the effects of which cannot now be undone.

Although control is lacking, the technosphere is responsive to and in turn affects the actions of its human components. These action-reaction loops are the sinews of technospheric dynamics (Haff, 2014b). The participation of knowledge-producing humans in positive feedback processes is a principal source of innovation in the Anthropocene and so of improvements in human well-being.

Equation-of-motion approaches like those used to predict some aspects of climate change are not sufficient to anticipate the evolution of the technosphere, a system which lacks such governing equations. A new framework is necessary to help prepare humans for dealing with the future of this recently emerged sphere. A similar situation prevailed in the nineteenth century where in order for humans to understand the behaviour of the biosphere (which is also devoid of an equation of motion) a novel reconception of the nature of change through time was required—the principles of Darwinian evolution.

In the twenty-first century, progress has been made in understanding technospheric dynamics thermodynamically (e.g., Garrett, 2011, 2014), and, alternatively, through elucidation of generic, or regulative, rules of behaviour that apply to all dynamical systems (Haff, 2014b). These rules include:

The Rule of Agency

That every system, including the technosphere, has an intrinsic purpose, namely, to ensure its own survival (Haff, 2016). The presence of intrinsic purpose, an emergent physical property, does not imply consciousness or intentionality.

System survival requires energy consumption, a task implemented according to two additional regulative rules:

The Rule of Performance

That the parts of a system act to support the intrinsic purpose of that system. As parts of the technosphere, humans and other components are obligate participants in the process of securing and using the energy and material resources that underpin survival of the system.

The Rule of Provision

That a system acts to ensure that its parts follow the Rule of Performance. As the host system for humans and technological artefacts, the technosphere enables, and also compels, constrains, or incentivizes humans and other components to help secure and process the resources essential to its survival.

Compulsion and incentivization are specific mechanisms supporting a positive feedback process that acts to increase technospheric energy consumption. However, metabolic speedup in emergent systems like the technosphere is more an expectation than an exception. In the presence of abundant energy reserves, there are more ways for the system to increase energy use than there are to maintain it at a fixed level.

The state of chronically increasing energy consumption by the technosphere presents humans with an ironic and difficult-to-resolve conflict, one which might be called the Fundamental Dilemma of the Anthropocene, namely:

(i) An increasing rate of global energy consumption is essential to human well-being: Improvement in the human condition (e.g., creation of a global health network) and preparing for and responding to potential and recent disasters that can threaten civilizational stability (e.g., effects of climate change, pandemics, Carrington-type Event, asteroid impact) require continuing increases in energy use beyond the base metabolic rate of the technosphere.

(ii) However, a state of chronically increasing global energy consumption is unsustainable: Increasing energy consumption, even when used to improve human well-being, opens the door to increasing frequency of unwanted side-effects. These can spread and intensify at rates that threaten to outpace the speed of effective human response. Recent examples include spread of fake news and techniques of mass surveillance, dispersal of novel pathogens and cyber weapons, and the destructive effects of climate change. If the rate of occurrence and intensity of such phenomena continue to increase, eventually biologically-limited humans will be unable to adjust to changing conditions fast enough to follow the Rule of Performance. At this point the technosphere would lose its internal coherence and, in consequence, the ability to improve and sustain human well-being.

This dilemma might be addressed (Haff, 2019) in the short run by making use of the fact that both humans and the technosphere possess the same intrinsic purpose common to all physical systems—to survive. For example, human support for increases in overall energy consumption might be the anthropic compromise offered in return for technosphere acquiescence in focusing energy flows into sectors such as medicine or atmospheric carbon capture where growth seems compatible with increasing human well-being.

In the long run, avoiding the negative consequences of growth in technospheric energy consumption while retaining acceleration-enabled benefits of resilience and innovation will require a radical reconceptualization of possible technospheric futures. The apparent haven of a steady-state technosphere is not a solution, not because of lack of human commitment, but because the human-technological artifice is a precariously-balanced structure that will require energy-demanding innovation if it is to maintain and increase human well-being in a turbulent, unpredictable world.

Summarizing: The technosphere is the emergent stage on which human desires and actions will play out, and human fate be realized, during the opening years of the Anthropocene. The future of *Homo sapiens* depends on the human ability (i) to discover and use new tools for framing the dynamics of the technosphere, (ii) to acknowledge that the technosphere has agency and is unpredictable where its behaviour depends on future knowledge, (iii) to relinquish the view that humans are the sole architects of the technosphere, (iv) to abandoned the idea of control over an innovating technosphere, (v) to accept the need for increasing technospheric energy consumption, (vi) to compromise with this new world system in a way that both humanity and the technosphere can, at least in the short run, achieve their most fundamental goal, survival, and finally (vii) to strategize novel reconfigurations of the human-technological enterprise that might be compatible with human prosperity over the long run.

References

Ashby, W. R. (1957). *An introduction to cybernetics*. Chapman and Hall.

Donges, J. F., Lucht, W., et al. (2017). The technosphere in earth system analysis: A coevolutionary perspective. *The Anthropocene Review, 4*, 23–33.

Garrett, T. J. (2011). Are there basic physical constraints on future anthropogenic emissions of carbon dioxide? *Climatic Change, 104*, 437–455.

Garrett, T. J. (2014). Long-run evolution of the global economy: 1. Physical basis. *Earth's Future, 2*, 127–151.

Haff, P. K. (2014a). Technology as a geological phenomenon: Implications for human well-being. In C. N. Waters, et al (Eds.), *A Stratigraphical Basis for the Anthropocene. Geological Society London Special Publications 395*, 301–309.

Haff, P. (2014b). Humans and technology in the Anthropocene: Six rules. *The Anthropocene Review, 1*, 126–130.

Haff, P. K. (2016). Purpose in the Anthropocene: Dynamical role and physical basis. *Anthropocene, 16*, 54–60.

Haff, P. K. (2019). The technosphere and its relation to the Anthropocene. Chap. 4 In J. Zalasiewicz et al. (Eds.), *The Anthropocene as a Geological Time Unit*. Cambridge University Press.

Zalasiewicz, J., Williams, M., Waters, C. N., Barnosky, A. D., & Haff, P. (2014). The technofossil record of humans. *The Anthropocene Review, 1*, 34–43.

Peter K. Haff, PhD in Physics, is Professor Emeritus of Geology and of Civil and Environmental Engineering at Duke University. He is currently working on the emergence and behavior of Earth's newest sphere and its implications for future human well-being.

Utopia

Bertrand Bergier

Abstract Utopia evades any definition or attempt at general categorization (Riot-Sarcey M. L'utopie en questions. Presses Universitaires de Vincennes, Saint Denis, 2001). In light of the multiplicity of texts produced by utopians, a "bibliography (…) would occupy a whole volume on its own," notes the sociologist Desroche. Faced with this proliferation, he proposes systems of classification that take into account, in a transformative perspective, several dimensions: familial, educational, economic, ergonomic, ecological, andragogical, political-scientific or even theological-political (Desroche H. Histoire des Mœurs. Tome III. Thèmes et systèmes culturels. Gallimard, Paris, 1993, p.85).

Among these innumerable accounts, some, boasting the best form of government, have legitimized domination and, in doing so, possible repression. This legitimization has been at work from the original utopias.

Thomas More's utopia (1516), denouncing the societal and political shortcomings of sixteenth-century English society, describes a substitute society, ideal, egalitarian, perfectly regulated, claiming to adhere to nature and not having recourse to any transcendence. Unproductiveness becomes reprehensible, which "grants Utopians the natural right to annex any land they need that is mismanaged by the natives" (Morgan, 1996).

Bacon's utopia (1627), New Atlantis, combines social progress with the unlimited progress of science. The intellectual elite derives its laws from the observation of nature. It eludes all democratic control, emancipating itself from political and religious power alike, and reigns sovereign over society. Like More, the philosopher-lawyer turns utopia into a historical thrust permitting one to think a better world, if not the best of worlds, is possible.

Translation: Jessee Kilgore

B. Bergier (✉)
Catholic University of the West, Angers, France
e-mail: bertrand.bergier@uco.fr

N. Wallenhorst, C. Wulf (eds.), *Handbook of the Anthropocene*,
https://doi.org/10.1007/978-3-031-25910-4_86

Condorcet's utopia (1793), the *Sketch for a Historical Picture of the Progress of the Human Mind*, is a remathematized extension of the *New Atlantis* (Desroche, 1993, P. 114). The perfection of human faculties is unlimited. The future looks bright. It is that of a scientific program to master both nature and societal mechanisms. Reason must triumph, even if in the face of irrational reactions, it must permit—by rationally justifying it—dictatorship.

Saint-Simon's utopia, like Condorcet's, does not oppose reality but, unlike the latter, intends to take into consideration power struggles within history and the functioning of the social body. To draw up the table of the laws of social change, he combines the human and physical sciences, and emphasizes a "science of production, that is, the science which has for its object the order of things most favourable to all kinds of production" (Gurvitch, 1965, p.60). In fact, he adds the myth of work to the myth of progress. The figure of the entrepreneur, rationally organizing work, becomes central to reforming the social body. He and his peers are the new masters of society: "The scholars show what the laws of social hygiene are; then, from among the measures they propose as a result of these laws, the industrials choose" (Durkheim, 2010, p. 150).

The constitutive and interdependent dimensions of these "rational utopias" (Metzger, 2001) are the imagination and the contestation. They oppose the conservative opinion which speaks in favour of the established order (Mannheim, 2015; Kupiec, 2006). The critique of existing society is based on the description of an ideal, progressive society drawing from science (Ruyer, 1988/1950), or more broadly, reason (Talmon, 1957), including market-minded reason. Thus, the techno-financial utopia (Mattelart, 1999) boasts an all-powerful, limitless technique (Latour, 2017), and a market, touted as a providential, self-regulating space, bringing about a global community and leading humanity towards a brighter tomorrow. Business and free trade are glorified, as they pacify the world and fulfill the quest for a space without borders. The local, the national, and the international are no longer compartmentalized. Commercial deterritorialization prevails. The community is a community of consumption, a community of consumers. The language is that of the "universals of communication": "(…) Computer science, marketing, design, and advertising, all the disciplines of communication [seize hold] of the word concept itself" (Deleuze & Guattari, 1994, p. 10). And so the professionals who work in these disciplines become designers. The enterprise vaunts its cosmic vocation and takes over for politics in the construction of social bonds. Management becomes the technical version of politics.

Not only do "utopias want to install reason in the realm of the imagination," (Baczko, 1989, p. 16) but they want to see this coherent imaginary embodied. Utopia becomes a method of action (Pisani, 2005) to deal with the problems confronting the "making of society together" in the here and now. In his work on "historical utopias," Riot-Sarcey (1998) invites us to conceive the reality of utopia, to grasp it as "a concrete act of imagining (…) that enables us to change the world, to change social relations" (Mathieu, 2008, p. 96). Social hope is meant to be active. The prospect of excellence is not an illusion. "We will not say: that is impossible because it is too beautiful. On the contrary, we will say: that is too beautiful not to be possible" (Breton cited by Mathieu, 2008, p. 95). Both to be realized and being realized elsewhere, utopia is not only the dismissal of a trial (the mistrial of a trial), it is at the same time the dismissal of a mistrial (the mistrial of a mistrial) in the sense that it refuses to take refuge in the realm of ideas and appeals to acts, to practice (Rancière, 2001, p. 66).

However, this detailed plan of action targeting the progress of humanity, hardly tolerating contestation, is even obsessed with "non-contradiction" (Ricoeur, 1997/1986, P. 231). Any criticism of a rational utopia is, by definition, perceived as irrational. It calls for the reeducation of dissidents, if not their elimination. Ultimately, it is only when they cling onto the implementation of the model, manifesting a frantic concern over the smallest details of its organization (Deléage, 2008), that utopias can become "instruments of repression" (Sargent, 2006, p. 14) and turn into dystopias.

Inverted twin of utopia, dystopia delivers a gruesome message. "Night has fallen on the dystopian cities and seems to last indefinitely" (Godin, 2010, p. 61). Nature is irreversibly polluted by individuals tearing each other apart. There where utopia manifested humanism, dystopia expresses the inhumane, the barbarity of man become wolf to man. It brings about the reign of inequalities and insecurities.

However, dystopia is not reducible to a nightmarish future, to a negative utopia. Put another way, utopia and dystopia have some features in common.

Their anticipatory temporalities act on the future. They are opposed to the fascination with presentism and the pervasive nostalgia of repeated commemorations. Both refuse the "hereafter" in the construction of their future; they define themselves as the product of human will (Mannheim, 2015; Kupiec, 2006). They both are thought systems in advance of history which differentiate themselves from fiction in that they claim an imaginary, that is not only likely, but already at work, a near future, a time that is already ours. Utopia tends towards that which is insufficiently present, dystopia towards that which is already excessively present. They are not opposed to the real, but bitterly question it and ask us: "How do we define ourselves? What do we want to do with ourselves?" They are matters for deliberation for each of us, both singularly and collectively. While the utopian summons a hope in action, the dystopian is not devoid of it: while warning us of the dire consequences of our actions, they maintain the hope that we can change (Sargent, 2006).

One and the other manage the enclosure of a global society. Monolithic, they produce the indisputable. While dystopia proceeds from a manifest political will of omnipotence and terror likely to strike anyone at any moment, it exacerbates the implications of utopia to the point of absurdity, by revealing its interventionist, if not totalitarian dimension: in other words, a utopia that ignores what is foreign and that does not permit opposition within itself, a utopia that sacrifices individual liberty on the altars of equality and security, dissolving the individual in an undifferentiated community. In More's utopia (1516), the Utopian who oversteps the bounds of his province is treated as a criminal. In the case of recidivism, he is reduced to the condition of a slave. In Nicholas' utopia (1579), a bloodthirsty repression is supposed to impose honesty: a corrupt judge exposes himself to having his leg amputated with a saw. Utopian perfection proves to be deadly. Dystopia reveals its hidden defects (Godin, 2010).

Utopia and dystopia can succeed one another. The twentieth century saw the utopia of communism transform into its own dystopia: with the fall of the Wall came "the utopia of free trade, which for many has already become its own dystopia" (Sargent, 2006, p.11). The somber criticism of a technological onslaught associated with market reasoning threatening not only the social order but, more broadly, the natural order, appears in numerous ecological dystopias "while later inspiring new forms of social project," of new ecological utopias that can be termed ecotopic (Deléage, 2008, p.33). Ecotopia manifests an ecological hope, designing a future for humanity that escapes the mortal dangers of resource depletion and pollution. But, as with positivist utopias, one cannot exclude "[the threat of] an authoritarian and conformist order, perhaps even some kind of "eco-fascist" regime" (Kumar, 2000, p. 262).

Whether they promote downsizing, praise conviviality, or even the exit from a society of consumption, ecological utopias – those of Kropotkine, Bargavel, Callenbach (Paquot, 2018, p. 35) – like all utopias, are to be read in context, in this instance that of ecosystem imbalances and irreversible damages. Although utopias are potentially dangerous, we must nevertheless choose them (Sargent, 2005) because they oppose the tyranny of fatalism. Thus, ecological utopias – and the myriads of microprojects that implement them– aim to build a future free from "the powerful forces of the globalization of productivism bolstered by those of big data and the digital age" (Paquot, 2018, p. 111). They revalue sobriety, the encounter, a "relationship with the living world concerned with *best* at the expense of *most*" (ibid).

References

Baczko, B. (1989). *Utopian Lights*. Paragon House.
Deléage, J.-P. (2008). Utopies et dystopies écologiques. *Écologie et politique, 37*, 33–43.
Deleuze, G., & Guattari, F. (1994). *What is Philosophy?* Verso.
Desroche, H. (1993). Humanismes et utopies. In J. Poirier (Ed.), *Histoire des Mœurs. Tome III. Thèmes et systèmes culturels* (pp. 78–134). Gallimard.
Durkheim, E. (2010). Socialism and Saint-Simon. Routledge. (Original edition 1959).
Godin, C. (2010). Sens de la contre-utopie. *Cités, 42*, 61–68.
Gurvitch, G. (1965). *Claude-Henri de Saint-Simon, La physiologie sociale. Œuvres choisies par Georges Gurvitch*. PUF.
Kumar, K. (2000). Utopia and Anti-Utopia in the Twentieth Century. In R. Schaer, G. Claeys, & L. T. Sargent (Eds.), *Utopia: The Search for the Ideal Society in the Western World* (pp. 251–267). The New York Public Library.
Kupiec, A. (2006). Karl Mannheim, l'utopie et le temps. Brève anthologie. *Mouvements, 45–46*, 87–97.
Latour, B. (2017). *Facing Gaia: Eight Lectures on the New Climatic Regime*. Polity Press.
Mannheim, K. (2015). *Ideology and Utopia: An Introduction to the Sociology of Knowledge*. Martino Publishing.
Mathieu, N. (2008). L'utopie féminine: faire de tous les lieux une maison. *Ecologie & politique, 37*, 93–101.

Mattelart, A. (1999). *Histoire de l'utopie planétaire. De la cité prophétique à la société globale.* La Découverte.

Metzger, J.-L. (2001). Management réformateur et utopie rationnelle. *Cahiers Internationaux de Sociologie, 111,* 233–259.

Morgan, N. (1996). L'utopie de Thomas More revisitée. *Futuribles, 205,* 5–19.

Paquot, T. (2018). *Utopies et utopistes.* La Découverte.

Pisani, E. (2005). *A Personal View of the World: Utopia as Method.* Legas.

Rancière, J. (2001). Sens et usages de l'utopie. In M. Riot-Sarcey (Ed.), *L'utopie en questions* (pp. 65–80). Presses Universitaires de Vincennes.

Ricoeur, P. (1997). *L'idéologie et l'utopie.* Seuil. (Original edition 1986).

Riot-Sarcey, M. (1998). *Le réel de l'utopie. Essai sur le politique au XIX siècle.* Albin Michel.

Ruyer, R. (1988). *L'utopie et les utopies.* Gérard Monfort. (Original edition 1950).

Sargent, L. T. (2005). The Necessity of Utopian Thinking: A Cross-National Perspective. In R. Jörn, M. Fehr, & T. Rieger (Eds.), *Thinking Utopia: Steps Into Other Worlds* (pp. 1–14). essay, Berghahn Books.

Sargent, L. T. (2006). In Defense of Utopia. *Diogenes, 53,* 11–17.

Talmon, J. L. (1957). *Utopianism and Politics.* Conservative Political Centre.

Bertrand Bergier Professor at the Catholic University of the West (Angers), associate professor at the Université of Sherbrooke (CERTA), and scientific collaborator at CREN at the University of Nantes.

Part VIII
The Acceptance of Limits, Containing and Salutary

Climate Ethics

Michel Bourban

Abstract This article outlines key developments in the philosophical literature on climate change. The allocation of the costs and benefits of greenhouse gas emissions between states and individual duties of climate justice are two major topics that climate ethics scholars have discussed by drawing on deontological theory, consequentialism, and virtue ethics. This article explores the connections between ethics and climate justice to present these two topics. In addition, it introduces three emerging sub-fields in climate ethics: the ethics of climate engineering, non-anthropocentric climate ethics, and the ethics of procreation. It concludes that as the remaining global carbon budget dwindles, radical lifestyle changes become more and more pressing and should move to the forefront of the debate.

Climate ethics is a field of research in applied ethics and political theory. It started to develop in the early 1990s with Dale Jamieson's reflection on the inadequacy of our conventional value system to deal with global environmental problems (Jamieson, 1992) and Henry Shue's distinction between subsistence emissions and luxury emissions (Shue, 1993). Since then, climate ethics has followed two main objectives: justifying collective agents' duties of climate justice (burden-sharing justice) and clarifying individual agents' responsibilities in the fight against climate change (individual climate ethics).

A first major branch of climate ethics is burden-sharing justice. It arises out of analytical political philosophy and focuses on principles of distributive justice, more specifically on the fair allocation of the costs and benefits of greenhouse gas (GHG) emissions between states. The three main normative ethics (deontological theory, consequentialism, and virtue ethics) have been used to justify the three major duties of climate justice held by states: reduce their GHG emissions, help vulnerable populations to adapt to adverse climate impacts, and compensate – as far as possible – for losses and damages (Bourban, 2018).

The original version of the chapter has been revised. A correction to this chapter can be found at https://doi.org/10.1007/978-3-031-25910-4_279

M. Bourban (✉)
University of Twente, Enschede, Netherlands
e-mail: m.bourban@utwente.nl

Since principles of climate justice generally come from Kantian-Rawlsian-inspired theories, deontological ethics is very influential in climate ethics. In this perspective, Simon Caney (2009, 2010) has developed a human-rights approach complemented by a "hybrid view" that reconciles the polluter pays principle (major historical and current GHG emitters ought to pay) with the ability to pay principle (the wealthiest ought to pay) in order to justify the obligation for developed countries to take the lead in the fight against climate change.

Consequentialism has also greatly influenced climate justice scholars. Since the disruption of the climate system has adverse effects on vulnerable people and populations, such as sea-level rise, more frequent and intense storms and hurricanes, and severe droughts and forest fires, it is possible to morally assess the consequences of state actions (and inaction). In relation to this, Peter Singer (2016) explains that the polluter pays principle and the principle of equal distribution of emission rights can be based on a calculation of utility that aims to ensure the greatest net happiness for the greatest number.

Turning now to virtue ethics, Stephen Gardiner (2011) explains that political inertia in the face of climate change is an ethical failure resulting, in particular, from the problem of "moral corruption". Moral and political agents are subverting moral language and arguments to serve their own interests, leading to attitudes of complacency and procrastination. Climate change also challenges our traditional normative categories, such as harm and responsibility. Dale Jamieson (2007) therefore calls for a shift in the value system of Western societies through an educational program that promotes "green virtues", such as humility, temperance, and respect for nature.

Considerations on attitudes, values, and character traits lead us to individual climate ethics, the second major branch of climate ethics, which arises out of applied ethics and environmental ethics. Here, the focus is more on individual behaviour and values and what individuals can do in the absence of ambitious and effective climate policies. Faced with the cumulative delay of decades of political inertia, individual climate ethics has taken off since 2010. In particular, it seeks to instil a sense of duty to reduce one's individual carbon footprint and to promote and support collective action against climate change (Fragnière, 2016).

Climate change is the result of the actions of a very large number of agents who are dispersed in both space and time. Although individual emissions are virtually harmless taken separately, once they are added to past and present anthropogenic emissions into the atmosphere, they can contribute to causing harm to people living far away, and to creating harmful circumstances for people who will live in the future. No individual action is the sole cause of harm; however, each high-emitting activity potentially contributes to the emergence of a harmful effect. This is the reason why Dale Jamieson (1992) observes that "Today we face the possibility that the global environment may be destroyed, yet no one will be responsible". In such circumstances, how can we assign responsibility for fighting climate change to individuals?

A first line of argument avoids relying on consequences by focusing on intentions. It rests on the integrity of moral agents (Hourdequin, 2010), which is morally relevant both for deontological theorists and virtue ethicists. Even if the effects of a reduction in individual GHG emissions are imperceptible, acting virtuously still matters. An action with a low or imperceptible impact but arising from virtuous motives, such as temperance or benevolence, should be pursued, because it reveals something about our integrity as moral agents, about the coherence between our ethical commitments and our behaviour. Integrity contributes to reducing the moral dissociation between what we should do and our everyday actions (Bourban & Broussois, 2020a).

Following a consequentialist logic based on the harm principle, other climate ethics theorists have tried to calculate the concrete impact of individual emissions. While John Nolt (2011) estimates that the lifetime emissions of an average American would result in the serious suffering and/or death of two people over the next millennium, John Broome (2012) calculates that the monetary value of the harm caused by the lifetime emissions of a westerner ranges between $19,000 and $65,000. While these figures are still debated, they show that one possible way to justify the duty to reduce one's individual carbon footprint is that it prevents harm to others.

Even if individual emissions have harmful impacts, the victims of these impacts remain so distant in space and/or time that it is difficult for the person contributing to the harm to realize it. It is indeed quite counter-intuitive to feel responsible for a state of the world to which one contributes in an infinitesimal way, or to which one contributes without knowing it. Anthropology challenges ethics on this point. We tend to be concerned only with the effects of our actions that are close and visible, and to ignore those that are dispersed in space and time and beyond our perception. What kind of measure would be suited to a situation where everyone contributes to disrupting the climate system, without anyone feeling responsible for it?

In such a situation, it may be tempting to bypass individual behavioural changes and focus instead on technological innovation. Indeed, climate engineering is an increasingly favoured measure in scientific, economic, and political circles. It represents an intentional and large-scale manipulation of the climate system designed to counter global warming or offset some of its effects. There are two main climate engineering methods: carbon dioxide removal techniques, the purpose of which is to reduce the levels of already emitted carbon dioxide in the atmosphere, and solar radiation management techniques, which endeavour to reduce the solar radiation reaching the Earth's surface.

The emerging sub-field of the ethics of climate engineering is rapidly expanding in critical response to the popularity of climate engineering projects (Gardiner, 2010; Preston, 2013). Its main purpose is to highlight the issues of justice, ethics, and governance raised by climate engineering projects. Because of their potential side effects on poor and marginalized populations, these projects could exacerbate the climate injustices that the global poor are already suffering. Another challenge is to find institutions adapted to the complex task of governing climate engineering. Can a governance system ensure that once in place, a climate engineering program is not suddenly stopped due to political unrest or technological malfunction, leading to a quick increase in global temperatures (the 'termination problem')? Or can it ensure that once in place, a climate engineering program would stop once it has achieved its end, despite the risks of sociotechnical lock-in (the 'phasedown problem')? Another and perhaps even more important problem is ethical. Climate engineering is based on a Promethean logic of controlling and dominating nature through technical means, a logic which is itself at the root of the climate problem. It would be more virtuous, if not wiser, to change our relationship to others and to the natural world, rather than trying to adapt planetary boundaries to our high-emitting lifestyles (Bourban & Rochel, 2021).

Changing our relationship to the world would mean changing our habits in terms of mobility, diet, and family size. Since its emergence in the early 1990s, climate ethics has focused on transforming our lifestyles through rehabilitated old values, such as temperance, and through finding new values, such as mindfulness (Jamieson, 1992, 2007). In addition to technological innovations and institutional reforms, adopting more sustainable individual behaviours that respect our natural environment, other human beings, and other living beings is crucial to contributing to the fight against climate change.

How can we change our ways of thinking, acting, and consuming? Two other emerging sub-fields of climate ethics are particularly relevant here. The first is non-anthropocentric climate ethics. Although most climate ethicists adopt an anthropocentric approach, Clare Palmer (2011), Elizabeth Cripps (2013) and Katie McShane (2016) have recently highlighted our responsibility for climate impacts on ecosystems, other species, and non-human animals. This approach could be further extended by moving from the consequences of climate change to its causes, especially industrial livestock farming. Such an approach to climate ethics would converge with animal ethics and environmental ethics to encourage individuals to reduce or simply stop their consumption of animal-based products (Bourban & Broussois, 2020b).

Another new sub-field of climate ethics deals with individuals' procreative choices, which also have a significant carbon footprint (or a 'carbon legacy'). Philosophers have explored three types of population growth reduction measures to help combat climate change: education and empowerment, which lead to voluntary measures to reduce family size (Cafaro, 2012); negative and positive incentives, which encourage individuals to have small families (Hickey et al., 2016); and coercive measures, which legally sanction parents who have too many children (Conly, 2016). These political measures have varying degrees of effectiveness, feasibility, and ethical acceptability, but in our context of scarce global carbon budget and overpopulation, a small family ethics becomes more and more compelling (Rieder, 2016; Bourban, 2019).

Today, we have to philosophize in a context of climate emergency. The carbon budget left to humanity before triggering tipping points and crossing critical thresholds in the climate system is scarce and rapidly shrinking. Since betting on climate engineering projects in the future may turn out to lead to a dangerous anthropogenic interference with the climate system, the very situation we should try to avoid, a crucial task is to focus on education, communication, and incentivization to convince or at least 'nudge' individuals to adopt low-emitting lifestyles. Ethical and democratic debates on controversial solutions, such as changing our diets, banning ridiculously large houses, yachts, and cars, and having fewer children should therefore gain momentum in the coming years in order to help avoid the worst-case scenarios.

References

Bourban, M. (2018). *Penser la justice climatique. Devoirs et politiques.* PUF.

Bourban, M. (2019). Croissance démographique et changement climatique: repenser nos politiques dans le cadre des limites planétaires. *La Pensée écologique, 3*(1), 19–37. https://doi.org/10.3917/lpe.003.0019

Bourban, M., & Broussois, L. (2020a). The Most good we can do or the best person we can be? *Ethics, Policy & Environment, 23*(2), 159–179. https://doi.org/10.1080/21550085.2020.1848175

Bourban, M., & Broussois, L. (2020b). Nouvelles convergences entre éthique environnementale et éthique animale: Vers une éthique climatique non anthropocentriste. *VertigO – la revue électronique en sciences de l'environnement, 32*, 1–29. https://doi.org/10.4000/vertigo.26893

Bourban, M., & Rochel, J. (2021). Synergies in innovation: Lessons learnt from innovation ethics for responsible innovation. *Philosophy & Technology, 34*(2), 373–394. https://doi.org/10.1007/s13347-020-00392-w

Broome, J. (2012). *Climate matters: Ethics in a warming world.* W. W. Norton.

Cafaro, P. (2012). Climate ethics and population policy. *WIREs Climate Change, 3*(1), 45–61. https://doi.org/10.1002/wcc.153

Caney, S. (2009). Climate change, human rights and moral thresholds. In S. Humphreys (Ed.), *Human Rights and Climate Change* (pp. 69–90). Cambridge University Press.

Caney, S. (2010). Climate change and the duties of the advantaged. *Critical Review of International Social and Political Philosophy, 13*(1), 203–228. https://doi.org/10.1080/13698230903326331

Conly, S. (2016). *One child: Do we have a right to more?* Oxford University Press.

Cripps, E. (2013). *Climate change and the moral agent: Individual duties in an interdependent world.* Oxford University Press.

Fragnière, A. (2016). Climate change and individual duties. *WIREs Climate Change, 7*(6), 798–814. https://doi.org/10.1002/wcc.422

Gardiner, S. M. (2010). Is "arming the future" with geoengineering really the lesser evil?: Some doubts about the ethics of intentionally manipulating the climate system. In S. M. Gardiner, S. Caney, D. Jamieson, & H. Shue (Eds.), *Climate ethics: Essential readings.* Oxford University Press.

Gardiner, S. M. (2011). *A perfect moral storm: The ethical tragedy of climate change.* Oxford University Press.

Hickey, C., Rieder, T. N., & Earl, J. (2016). Population engineering and the fight against climate change. *Social Theory and Practice, 42*(4), 845–870. https://doi.org/10.5840/soctheorpract201642430

Hourdequin, M. (2010). Climate, collective action and individual ethical obligations. *Environmental Values, 19*(4), 443–464. https://doi.org/10.3197/096327110X531552

Jamieson, D. (1992). Ethics, public policy, and global warming. *Science, Technology, & Human Values, 17*(2), 139–153. https://doi.org/10.1177/016224399201700201

Jamieson, D. (2007). When Utilitarians should be virtue theorists. *Utilitas, 19*(2), 160–183. https://doi.org/10.1017/S0953820807002452

McShane, K. (2016). Anthropocentrism in climate ethics and policy. *Midwest Studies In Philosophy, 40*(1), 189–204. https://doi.org/10.1111/misp.12055

Nolt, J. (2011). How harmful are the average American's greenhouse gas emissions? *Ethics, Policy & Environment, 14*(1), 3–10. https://doi.org/10.1080/21550085.2011.561584

Palmer, C. (2011). Does nature matter? The place of the nonhuman in the ethics of climate change. In D. G. Arnold (Ed.), *The ethics of global climate Change* (pp. 272–291). Cambridge University Press.

Preston, C. J. (2013). Ethics and geoengineering: reviewing the moral issues raised by solar radiation management and carbon dioxide removal. *WIREs Climate Change, 4*(1), 23–37. https://doi.org/10.1002/wcc.198

Rieder, T. N. (2016). *Toward a small family ethic: How overpopulation and climate Change are affecting the morality of procreation*. Springer.

Shue, H. (1993). Subsistence emissions and luxury emissions. *Law & Policy, 15*(1), 39–60. https://doi.org/10.1111/j.1467-9930.1993.tb00093.x

Singer, P. (2016). *One world now: The ethics of globalization*. Yale University Press.

Michel Bourban is a Postdoctoral Researcher at the University of Twente. He works on the project "Cosmopolitan Citizenship and Ecological Citizenship", funded by a SNSF research grant. He holds a PhD in Philosophy from the University of Lausanne and Paris-Sorbonne University (Paris IV). His work on climate justice, climate ethics, ecological citizenship, effective altruism, and innovation ethics has been published in journals such as *Ethics, Policy & Environment*, *Philosophy & Technology*, and *De Ethica*, and in edited volumes published by Springer, Routledge, Elsevier, and Göttingen University Press. He has also published two books on climate justice and climate ethics with the French publishers PUF and Vrin.

Climate Justice

Marie Toussaint

Abstract The climate crisis, the collapse of biodiversity and, more generally, the overstepping of planetary boundaries, call for profound civilizational change. This change cannot take place without the transformation of our law, which is an essential lever for being in harmony with the living. If climate justice uses existing standards, it is esentially to change the law through jurisprudence. By seeking the courage of judges through ambitious rulings, this mobilization aims to reform current environmental legislation through jurisprudence, to rethink our way of creating society and to legislate our interactions and relationships with the living. *In fine*, the challenge is to transform our social contract, which, for too long, has been based on an anthropocentric vision of law, into a "natural contract" that would integrate fundamental human rights and the rights of living organisms.

The Climate Justice Movement, Mobilizing the Law to Hold Polluters Accountable and Protect Human Rights

From the inhabitants of the Maldives threatened by rising sea levels, to the indigenous populations driven away from their territory by deforestation and forest fires, to the inhabitants of working-class neighbourhoods whose health is directly affected by atmospheric pollution and heat domes: human beings are not equal in the face of climate disruption, and some are already violently suffering from its consequences. These inequalities are a reflection of economic, social and cultural injustices, which primarily affect minorities and the most vulnerable. The concept of climate justice seeks above all to shed light on, analyse and denounce these environmental inequalities by using the law. This is why climate justice litigation aims not only to hold States and companies accountable by forcing them to take ambitious actions to fight and adapt our territories to the climate crisis, but also to ensure the effectiveness and universality of our fundamental right to a healthy environment, to clean air, to

M. Toussaint (✉)
European Parliament, Strasbourg, France

N. Wallenhorst, C. Wulf (eds.), *Handbook of the Anthropocene*,
https://doi.org/10.1007/978-3-031-25910-4_88

unpolluted soils, to living ecosystems. Climate litigation therefore strives to ensure compliance with international and national texts that set targets for the reduction of greenhouse gas emissions, but also to protect the fundamental rights intrinsically linked to the protection of life and the climate. Although there are many constitutions and texts setting environmental objectives, these standards are very poorly applied. This ineffectiveness of environmental law is today counterbalanced by the emergence of jurisprudence that can give the texts their full scope and contribute to the reclaiming of democratic sovereignty in the face of the global law emerging from arbitration tribunals and international agreements and contracts.

These climate litigation cases have become an essential means of action used by civil society, brought by citizens and supported by environmental organizations. Their reasoning is simple: if governments do not take the current climate emergency seriously, civil society and citizens should find the courage of judges who, by relying on an ambitious interpretation of national and international law in force, confront decision-makers with their responsibility to take drastic measures to reduce greenhouse gas emissions. While the objective is the same, climate justice has many facets and continues to diversify: directed against States or companies, brought before national or regional jurisdictions, and based on international texts, national implementation laws or a set of texts whose binding value is often varies. This global movement, which began in the United States, has seen a sharp increase in climate cases since the beginning of the 2010s, from just a few cases per year to nearly 144 disputes in 2020. The United States still has the most climate actions today with some emblematic cases, such as the Juliana case initiated in 2015 by the NGO Our Children's Trust. In Europe, cases have also flourished in recent years (in France, Italy, Belgium, Great Britain, the Netherlands, etc.), and some have already been successful, such as in the Netherlands with the recent success of the Urgenda association.

In addition to States, corporations are also targeted by these climate actions. Citizens are calling on them to apply and respect their due diligence regarding the impact of their activities on human rights and the environment. In France, the law on due diligence of multinational companies was the basis for two recent litigation cases against the multinational Total and then the Casino group. In the Netherlands, Shell, the world's ninth largest CO_2 emitting company, was ordered to reduce its emissions by 45% by 2030 compared to 2019 levels, following an action led by MilieuDefensie and over 17,000 co-plaintiffs. Companies are also trying to defend themselves through private justice and arbitration, which is becoming increasingly important in environmental matters. Examples are unfortunately numerous: the German multinationals RWE and Uniper claiming billions of euros in compensation following the Netherlands' decision to phase out coal by 2030, or the British company Rockopper asking for 350 million euros to Italy following its refusal to grant it a drilling concession in the Adriatic Sea. Recent rulings, such as the decision recognizing the illegality of the Energy Charter Treaty in disputes between polluting industries and States, are proof of a growing awareness of the need for State sovereignty in the preservation of the world's natural commons. Through such decisions, judges are making it possible to move from a solitary national sovereignty to a sovereignty of solidarity, as advocated by Mireille Delmas Marty.

However, this approach to climate justice does not allow for a break with the anthropocentric vision of the law because it is based on interdependence between environmental and human rights violations. Consequently, compensation for environmental degradation is conditional on the demonstration of a violation of human rights and a prejudice suffered by the claimant. Christopher Stone's famous 1972 article, "*Should Trees Be Allowed to Plead*", still makes sense today. In parallel with the climate justice movement, another wave of litigation is arriving straight from Latin America, calling for compensation for the damage suffered directly by ecosystems. If climate change litigation is considerably changing the situation with regard to the obligations of States and the rights of citizens, these lawsuits initiated to protect the living could radically change our relations with ecosystems by making them real subjects of rights. Wetlands in Florida, the Atrato River in the Amazon, or the Whanganui River in New Zealand are examples among others to whom the judge has granted a legal status and with it the right to exist, to live, to regenerate at a natural pace, but also to take legal action to defend their rights and the designation of legal guardians to ensure their effectiveness.

Acting as a real counter-power to an executive disconnected from reality and the climate emergency, this movement asking for justice for the climate and for the living does not stop in the courts. Beyond their legal success, these lawsuits constitute a real lever for climate and environmental policies that are equal to the ecological crisis, but also a change of legal paradigm that is indispensable in the Anthropocene era.

The Legal Revolution Needed to Protect the Living

Behind their technical nature, climate disputes also have a political and democratic role. The media coverage of recent cases, such as the "Affaire du siècle" in France, which was supported by more than 2.3 million people, has made it possible to highlight the political risks of climate inaction, at the same time as the legal risks are becoming clearer. In the Netherlands, after being ordered by the Supreme Court in The Hague to reduce its emissions by 25% by the end of 2020 compared to their level in 1990, the government was forced to revise its climate policy drastically. The legislators have even taken up some of the proposals made by the association behind the appeal, particularly on carbon-free mobility. Faced with the legal risk, some companies are also experiencing significant internal mobilization, including from certain shareholders of the Carbon majors who are asking for more action in favour of the climate. Last May, an ExxonMobil activist fund, Engine No. 1, succeeded in electing two of its representatives to the Supervisory Board, calling on the multinational to focus on accelerating rather than delaying the transition. In the same week, a Chevron shareholder's proposal to reduce the company's scope 3 emissions, which include energy products, was approved by a majority of investors.

Nevertheless, despite this constant legal pressure, environmental and climate legislation adopted at national and European level continues to fall far short of the
objectives set out in the Paris Agreement. The Climate Law in France, the European
Green New Deal and many other first-generation climate laws are sorely lacking in
ambition and are decried by scientists who have been sounding the alarm for several
decades already. Pressure from lobbies and the short timeframe favoured by governments guided by the sole indicator of economic growth are all reasons that drain
these laws of their substance and take us further away from the climate objectives.
It is therefore essential to change the laws and this transition cannot be achieved
without changing our founding texts. In order to place the issue of life at the heart
of our democratic system, constitutions at the national level and binding treaties at
the regional or international level must embody this paradigm shift. The adoption of
an ecological Constitution would make it possible to set up the preservation of the
environment, biodiversity and climate, among the major stakes of our century and
to call into question bills that destroy the climate and living organisms.

The climate cannot survive without the fundamental transformation of rules in
the European system. A European Environmental Treaty is also essential to ensure
that all policies conducted at Union level respond to the ecological emergency and
are conditional on compliance with environmental rules. We must put the general
interest, that of the living, present and future generations, and human rights, back at
the top of the priorities, and therefore of the hierarchy of European standards. We
should make the climate, biodiversity and planetary boundaries the compass of a
European Union that has rediscovered its values, its rationality and its power to act.
At international level, the idea of a binding Global Environment Pact recognising
the universal right to a healthy environment has come up against the inherent limits
of global governance but will certainly return to the forefront of the international
stage during the next major international events.

Because of its scope and speed, the climate and living justice movement has
already played, and will have to play in the coming years, a considerable role in
influencing public and private policies. It is also bringing a new discourse into the
public debate: an ecocentric approach to law based on new legal concepts and a
profound civilizational transformation.

Conclusion

We have perhaps never had so much need for the imaginative forces of law so dear
to Mireille Delmas-Marty. What is at stake in the twenty-first century is the survival
of humanity on a planet, which has become hostile. The law, our social pact, must
be one of the tools mobilized to save the planet. The protection of fundamental
rights, the repair of damaged or threatened ecosystems, the reduction of environmental and climate inequalities are all objectives supported by the climate justice
movement, which must be carried out. The challenge is to create innovative legal

tools to protect natural commons, whether they are global or territorial, in order to re-establish the ecological solidarity that is essential to confront the twenty-first century's ills.

References

CJEU. (2021, September 2). Case C741/19 Republic of Moldova v Komstroy. ECLI:EU:C:2021:655.

Climate Change Litigation Databases. http://climatecasechart.com/climate-change-litigation/

Delmas-Marty, M. (2020). Let's take advantage of the pandemic to make peace with the earth. *Le Monde Tribune*. https://www.lemonde.fr/idees/article/2020/03/17/mireille-delmas-marty-profitons-de-la-pandemie-pour-faire-la-paix-avec-la-terre_6033344_3232.html

Hague District Court. (2021, May 26). *Milieudefensie et al. v. Royal Dutch Shell plc.* ECLI:NL:RBDHA:2021:5339.

High Council for the Climate. (2021). *Annual report – Strengthening mitigation, engaging in adaptation*. https://www.hautconseilclimat.fr/publications/rapport-annuel-2021-renforcer-lattenuation-engager-ladaptation/

Intergovernmental Panel on Climate Change. (2021). *AR6, climate change 2021: The physical science basis*. https://www.ipcc.ch/report/sixth-assessment-report-working-group-i/

Our business. (2020). *First climate litigation against a multinational oil company in France: 14 local authorities and 5 associations sue Total for breach of duty of care*. https://notreaffaire-atous.org/actions/polleurs-payeurs/

Our business. (2021). *Deforestation and human rights abuses in the Amazon: Indigenous peoples' representatives and international NGOs take casino to court*. https://notreaffaireatous.org/actions/polleurs-payeurs/

Serres, M. (1990). *The natural contract*. François Bourin, in-8 paperback, 191 pp.

Marie Toussaint is a French environmental activist, a jurist in international environmental law, co-founder of the association *Notre affaire à tous* and at the origin of the *l'Affaire du siècle* campaign, the climate action supported by more than 2.2 million people in France. She has been elected MEP on the Greens/EFA group in May 2019 and sits on the ITRE (energy, industry), ENVI (environment) and JURI (legal affairs) committees where she fights, among other things, for divestment from fossil fuels, environmental justice, recognition of the rights of nature and the crime of ecocide.

Earth Jurisprudence

Cormac Cullinan

Abstract The ecological destruction that has characterised the Anthropocene to date, has been facilitated and legitimised by legal and governance systems that entrench a "colonial" relationship between humans and Earth. These systems are based on anthropocentric beliefs that humans are separate from, and superior to, other aspects of Nature and have the right to manage and exploit Nature. Earth jurisprudence on the other hand, is a philosophy of law and governance rooted in the recognition that humans are an integral part of a complex living community ("the Earth Community"). Since human wellbeing is derived from the Earth Community, it cannot be sustained at the expense of that community. Consequently, in order for contemporary societies to become ecologically viable, humans must rediscover their roles within the Earth Community, and reorient their legal and governance system to focus on the primary purpose of guiding people to co-exist harmoniously within the Earth Community.

The Anthropocene has been characterised by massive ecological degradation and destruction. Human activities have caused the emission of vast quantities of greenhouse gasses into the atmosphere causing global warming, climate change and ocean acidification, and initiating the 6th period of mass extinction. Extinction rate of species have increased more than 100-fold, and between 1970 and 2016 the sizes of monitored populations of mammals, birds, amphibians, reptiles, and fish declined by an average of 68% (IPBES, 2019). Dramatic declines in insect populations are also being reported from around the world.

The Intergovernmental Science-Policy Platform on Biodiversity and Ecosystem Services (IPBES) has advised that arresting this decline and beginning the process of restoring biological diversity can now only be achieved by means of "transformative change" which it defines "as a fundamental, system-wide reorganization across

C. Cullinan (✉)
Wild Law Institute, Cape Town, South Africa
e-mail: Cormac@greencounsel.co.za

technological, economic and social factors, including paradigms, goals and values."
(IPBES, 2019).

The human cultures and civilisations primarily responsible for this destruction are characterised by deeply anthropocentric worldviews. From this perspective, human beings are separate from Nature and can transcend its laws (human exceptionalism), and the planet is a collection of "natural resources" which (certain) humans are entitled to deal with as they will. As Berry has pointed out, these cultures "think of the universe as a collection of objects rather than as a communion of subjects" (Berry, 1999, p.16). The shift from seeing the world as a machine to understanding it as a network or system that occurred in science, has not yet occurred in law (Capra & Matthei, 2015, p.4)

In order to emphasise that Earth jurisprudence is concerned primarily with maintaining healthy interrelationships (or "communion" in Berry's terms) between the myriad of diverse beings (including human beings) that constitutes Earth, this chapter uses the term "Earth Community" to refer to Earth, and "being" or "ecological being" to refer to any entity which has come into being as a consequence of the creation and evolution of this planet, and are consequently members of the Earth Community.

Despite significant differences between the human cultures that dominate the world, they share a conception of humans as rulers or managers of other-than-human beings, and tend to relate to them in a manner that is analogous to how colonizers treated the people and lands which they colonized. This "eco-colonial" attitude is reflected in the legal systems of these cultures, which define all non-human aspects of Nature as property or "natural resources" and legitimise ecological degradation by recognising the rights of individual humans and juristic persons (e.g. corporations and the state) to determine what happens to that property. This provides the foundation for economic systems in which the state promotes economic growth and international trade by authorising ecologically damaging activities instead of encouraging humans to prioritise the best interests of the Earth Community (which would also benefit humans).

There is no scientific evidence for an evolutionary discontinuity that would justify the belief that humans are separate from the Earth Community that brought us into being and continues to sustain us. It is also clear that humanity does not have the wisdom and powers that would be necessary to manage the complex living community within which we are embedded - if indeed that were even possible. On the contrary, it is now clear that the ecological destruction caused by eco-colonial cultures means that they cannot persist (i.e. are unsustainable) and must either collapse or transform. This raises the fundamental question of how societal governance systems would have to be transformed in order to enable a society to abandon an ecologically unviable trajectory and transform itself into a society which has an enduring place within the Earth Community by virtue of the fact that its presence within that community is mutually beneficial.

Earth Jurisprudences

An Earth jurisprudence is a philosophy of law and governance devised by a particular human culture or community to provide a philosophical basis for the development and implementation of a system of human governance that seeks to guide humans to contribute to the integrity, healthy functioning, beauty and on-going evolution of the Earth Community (Cullinan, 2002, 2011).

The diversity of human cultures and ecosystems must inevitably produce a range of Earth jurisprudences, each adapted to the specific conditions within which it evolved and reflecting a particular human community's understanding of how to regulate itself as part of the Earth Community. This is evident in the great variety of governance systems developed by Indigenous Peoples, which nevertheless exhibit common features. These typically include the understanding that humans are required to live in accordance with a pre-existing system of order, reverence for "Mother Earth" (or other expressions for the whole community), and an emphasis on: respect for other beings, personal and collective responsibility, and reciprocity.

An Earth jurisprudence would typically:

recognize that by virtue of their existence, all members of the Earth Community (i.e. all beings) have the fundamental right to exist and to be free to play their unique role within that community;

regard these inherent, fundamental rights as inalienable and recognise that human have a duty to respect the rights of other-than-human beings, and to seek to co-exist harmoniously with them;

regard as illegitimate and "unlawful" any human acts or laws that infringe upon these rights because they violate the fundamental relationships and principles that constitute the Earth Community;

provide effective legal remedies to protect these fundamental rights if they are violated by human acts;

advocate restorative justice (which focuses on restoring damaged relationships and ecological health) rather than punishment (retribution); and

seek to resolve competing rights on the basis of what is best for the Earth Community as a whole and thereby contribute to maintaining a dynamic balance between the rights of humans and those of other members of that community.

The Great Jurisprudence

The diverse Earth jurisprudences are united by their common commitment to understanding the "Great Jurisprudence" of the Universe and developing laws and cultural practices that guide people to act in accordance with it. The Great Jurisprudence refers to the inherent characteristics of the Universe which determine how it is structured and ordered, and how different aspects relate to one another and behave. In other words, it comprises the fundamental relationships and principles that constitute the Earth Community, and in that sense is analogous to the constitution of that community. As Thomas Berry put it: "the Universe is the primary law- giver" (Berry, 1999, p. 64). Consequently the Great Jurisprudence is the primary source of law and provides a logical premise for the development of law by diverse human societies (Cullinan, 2002; Burden, 2011). A society's Earth jurisprudence and the laws that give effect to it, may be understood as a bio-culturally specific application of the Great Jurisprudence.

The Great Jurisprudence includes *a priori* characteristics, principles or "laws of Nature" that influence the functioning of the Earth system regardless of whether not we know of their existence. For example, the molecular structure of greenhouse gasses enables them to absorb infrared radiation which causes them to vibrate and warm the atmosphere around them. The structure of these molecules and how they interact with infrared radiation can be characterised as inherent and predictable characteristics of the Earth system, analogous to laws.

The Great Jurisprudence also encompasses the tendency of a system to behave in a particular way under certain conditions. These characteristics are insufficiently deterministic to be characterised as a "law" of Nature, but are nevertheless enable us to make reasonably accurate predictions about the system. For example many organisms seek to maintain a dynamic internal equilibrium ("homeostasis") and will attempt to counteract perturbations in order to restore equilibrium until it is no longer possible to do so and the organism dies.

Many ever-changing natural systems use negative feedback mechanisms to maintain a degree of stability and prevent accelerating change. On the contrary, most legal systems do not prevent runaway climate change because they do not incorporate sufficiently powerful negative feedback mechanisms to regulate greenhouse gas emitting activities more stringently as atmospheric concentrations of greenhouse gasses increase. This is an example of how the failure of human societies to align their legal systems with the Great Jurisprudence has harmful consequences for the Earth Community.

Rights of Nature

Recognition of the rights of Nature or Mother Earth is a technique used to give effect to Earth jurisprudence. Advocates of rights of Nature argue that if legal systems recognise that human beings have inherent, inalienable human rights simply by virtue of their existence, it is necessary to recognise analogous rights for all beings that constitute the Earth Community. If this is done, legal systems can be used to promote ecological health by resolving conflicts between humans and other aspects of Nature according to what is in the best interests of the Earth Community as a whole.

Applying the rights of Nature approach involves: (a) expanding the class of legal/juridical persons to include other ecological beings (i.e. members of the Earth Community); (b) recognizing that those ecological beings have inherent and inalienable fundamental rights like human rights; (c) imposing legal duties on human beings and human institutions (including artificial juristic persons such as corporations and governments) to refrain from infringing upon the rights of those ecological beings without adequate justification; and (d) establishing legal mechanisms for enforcing compliance with those duties.

These rights are conceived of as being inherent, inalienable, limited by the rights of other beings (at least to the extent necessary to maintain the integrity, balance and health of the communities within which that being exists), and differentiated in the sense that some are common to all beings while other are specific to a particular species or kind. The universal rights common to all beings are conceived of as including the rights: to exist, to habitat or a place to be, and to participate in the evolution of the Earth Community.

Application and Future Prospects

The application of the Earth jurisprudence approach is steadily gaining momentum as observed by Boyd (2017) and as evidenced by the annual reports of the United Nations Harmony with Nature Programme (http://www.harmonywithnatureun.org) which tracks the development and implementation of measures to promote living in harmony with Nature throughout the world (Harmony with Nature, United Nations (n.d.)). Key milestones include: the adoption in Ecuador in September 2008 of a new constitution that expressly recognised the rights of Nature; and the adoption in Bolivia April 2010 by a World People's Conference of approximately 35,000 people of a Universal Declaration of the Rights of Mother Earth (UDRME). Since then legislators in several countries have enacted laws that recognised rivers, forest and mountains (among other ecological beings) as legal subjects, and courts have done the same in many countries such as Bangladesh, Colombia, Ecuador, and India.

The UDRME articulates not only the fundamental rights of all ecological beings, but also specific human duties to respect, protect, conserve and restore the integrity of vital ecological cycles and processes, and to take proactive measures to reform legal and economic systems that promote harmony with Nature.

The emergence of new Earth jurisprudences is an aspect of a wider global cultural shift in our understanding of the Universe and the place and role of humanity within the ecological systems of Earth. It is reflects changing cultural norms in the face of the realities of the Anthropocene, and is likely to gather strength as the ecological crises intensify and reveal that most legal systems are not only ineffective in preventing the harm, they are complicit in driving and legitimising it.

References

Berry, T. (1999). *The great work: Our way into the future*. Harmony/Bell Tower.

Boyd, D. R. (2017). *The rights of nature. A legal revolution that could save the world*. ECW Press.

Burden, P. (Ed.). (2011). *Exploring wild law: The philosophy of earth jurisprudence*. Wakefield Press.

Capra, F., & Mattei, U. (2015). *The ecology of law: Towards a legal system in tune with nature and community*. Berrett-Koehler Publishers.

Cullinan, C. (2002). *Wild law: Governing people for Earth*. SiberInk.

Cullinan, C. (2011). *Wild law: A manifesto for earth justice* (2nd ed.). Green Book.

Harmony with Nature, United Nations (n.d.). See http://www.harmonywithnatureun.org Accessed 7 Nov 2021.

IPBES. (2019). *Summary for policymakers of the global assessment report on biodiversity and ecosystem services of the Intergovernmental Science-Policy Platform on Biodiversity and Ecosystem Services*. IPBES secretariat.

Cormac Cullinan is a practising environmental lawyer, author, speaker and advocate for the rights of Nature and ecological sustainability. He is a director of the Wild Law Institute and of the specialist environmental law firm Cullinan & Associates Inc. His ground-breaking book Wild Law (2002) pioneered Earth jurisprudence and has played a significant role in informing and inspiring a growing international movement. Cormac led the drafting of the Universal Declaration of the Rights of Mother Earth (proclaimed on 22 April 2010 in Bolivia), is a founder and Executive Committee member of the Global Alliance for Rights of Nature, and is a judge of the International Rights of Nature Tribunal.

Earth Overshoot Day

Mathis Wackernagel and David Lin

Abstract Earth Overshoot Day marks the day of a year when humanity's demand on ecosystems' productivity exceeds what Earth can regenerate in that year.

From January 1st to Earth Overshoot Day, humanity has demanded as much as the biosphere can regenerate in the entire year. In 2021, Earth Overshoot Day 2022 fell on July 28, based on estimates extending on the National Footprint and Biocapacity Accounts (called "nowcasting") (Lin et al., 2018, 2021; York University et al., 2022). These accounts estimate the →*ecological footprint* and →*biocapacity* of every country and the world.

Global →*overshoot* is the difference between humanity's ecological footprint and the planet's biocapacity. Therefore, Earth Overshoot Day is computed by dividing the planet's biocapacity (the amount of ecological resources Earth is able to generate that year), by humanity's Ecological Footprint (humanity's demand for that year), and multiplying by the number of days in that year:

$$\left(Earth's \, Biocapacity \, / \, Humanity's \, Ecological \, Footprint \right) \times 365 = Earth \, Overshoot \, Day$$

The number of days from January 1 to Earth Overshoot Day corresponds to the portion of the demand that Earth is able to regenerate, while the remaining days, from Earth Overshoot Day to December 31, correspond to the portion that humanity lives off the depletion of our planet's biological assets. This portion reflects global overshoot.

Earth Overshoot Day is hosted and calculated by Global Footprint Network, an international research organization that provides decision-makers with a menu of

M. Wackernagel (✉) · D. Lin
Global Footprint Network, Geneva, Switzerland
e-mail: mathis@footprintnetwork.org; david.lin@footprintnetwork.org;
https://www.footprintnetwork.org

N. Wallenhorst, C. Wulf (eds.), *Handbook of the Anthropocene*,
https://doi.org/10.1007/978-3-031-25910-4_90

tools to help the human economy operate within Earth's ecological limits. Its dedicated website (www.overshootday.org) features media responses from around the world. The strongest media resonance over the past 10 years has been in France and Germany, with major news outlets dedicating substantive reports to the day. For instance, on 2017's Earth Overshoot Day, "La Libération" dedicated its initial five pages to that day. In 2021, over 6000 internet articles referred to Earth Overshoot Day (or its equivalent in other languages), which is estimated to have generated over 7 billion media impressions (Meltwater.com search October 2021). Earth Overshoot Day 2022 had a similar reach.

Overshoot Days can also be calculated for each country – they represent the day Earth Overshoot Day would be if all the world consumed at the rate of people in that country. The dates for each country are also published on the Earth Overshoot Day website atovershootday.org/newsroom/country-overshoot-days/. Note that countries with a per person rate of consumption that is less than global biocapacity per person do not have a country overshoot day.

Countries' demand can also be compared to their own territorial biocapacity. This reveals for which portion of the year a country's own biocapacity would suffice to provide for that country's demand. Since biocapacity is unevenly distributed among countries, this ratio allows us to determine the country's ecological deficit day, which is different from the country's overshoot day. For instance, some city states like Qatar or Singapore have very little biocapacity available domestically and would not even have enough biocapacity for a month per year if they had to rely solely on their own ecological resources. However, one country's overshoot day and deficit day are similar: Germany has about the same biocapacity per person as world average. This may come as a surprise because the population density in Germany is much higher than world average. However, its productivity is also higher than world average, and as a result, the number of global hectares (i.e., average productive hectares) per person in Germany and the world are currently within just a few percentage points of each other (Global Footprint Network, 2021).

To make Earth Overshoot Day actionable, Global Footprint Network introduced the challenge to "Move the Date." This proposal is based on the observation that by delaying the date of Earth Overshoot Day by 6 days each year starting in 2022, humanity's Footprint would be smaller than the planet's biocapacity before 2050. Given the accumulated ecological debt (i.e., the sum of the annual ecological deficits), with a big portion of this debt being the excessive greenhouse gases in the atmosphere, a faster reduction may be necessary. The "1.5° trajectories" as outlined by IPCC would correspond to delaying Earth Overshoot Day by 10 days each year for the next 10 years. Also note: Given that humanity is in competition for biocapacity with wild animals, allotting the entire biocapacity to human use is not compatible with biodiversity preservation. E.O. Wilson suggested that humanity use ½ Earth, which, if well managed, enables 85% of the world's biodiversity to be maintained (Wilson, 2016; Half Earth Project, 2021).

The proposal to "Move the Date" of Earth Overshoot Day emphasizes the benefits of action. Moving the date makes the world more resource secure and therefore more robust rather than conveying a sense of diminishing people's lives.

Overemphasizing the need to reduce resource use may amplify people's sense of "suffering and sacrifice", easily shifting the emotional focus towards fear of scarcity. This deep-rooted fear becomes a barrier to embracing the need to address overshoot. Similarly, when working on personal weight loss, health coaches do not emphasize the feeling of hunger and the pain of exercises, but the yearning for health and vitality which can be regained by embracing a healthy diet and exercise regime. To highlight the opportunities to gain resource security (and therefore eliminate overshoot by design, rather than disaster), Global Footprint Network introduced the platform "Power of Possibility" (www.overshootday.org/pop), which presents a multitude of complementary, economically viable ways to increase human resource security. It demonstrates the abundance of possibilities that are available, with the intent to create a sense of agency in and desire for the sustainability transformation.

Some have criticized Earth Overshoot Day, including Richardson (2019), typically questioning the relevance of the underlying National Footprint and Biocapacity Accounts (York University et al., 2022). Global Footprint Network has responded to such criticisms (Global Footprint Network, 2021), including that of Richardson (Wackernagel & Lin, 2019). In those responses, Global Footprint Network researchers point out that many criticisms are misrepresentations of what the Ecological Footprint accounts are set up to measure (Wackernagel & Lin, 2019). While there clearly is room for methodological and data improvement in Ecological Footprint accounting, as acknowledged by Global Footprint Network and the York University Footprint Initiative, Wackernagel and Lin also emphasize that, if anything, the accounts provide an overly conservative estimate of actual global overshoot (Wackernagel et al., 2019, 2021). Global Footprint Network also provides a detailed discussion of common criticisms on its website, along with a detailed handbook discussing past criticism (Global Footprint Network, 2021).

References

Global Footprint Network research team. (2021). *Ecological footprint accounting: Limitations and criticisms, handbook, Global Footprint Network, Oakland*, available from Global Footprint Network website at https://www.footprintnetwork.org/our-work/ecological-footprint/limitations-and-criticisms/

Global Footprint Network, York University, FoDaFo. (2021). *Data platform*. https://data.footprint-network.org

Half Earth Project. (2021). *Why Half?* https://www.half-earthproject.org/discover-half-earth/#why-half. Accessed Nov 2021

Lin, D., Hanscom, L., Murthy, A., Galli, A., Evans, M., Neill, E., Mancini, M. S., Martindill, J., Medouar, F.-Z., Huang, S., & Wackernagel, M. (2018). Ecological footprint accounting for countries: Updates and results of the National Footprint Accounts, 2012–2018. *Resources, 7*(3), 58. https://www.mdpi.com/2079-9276/7/3/58

Lin, D., Wambersie, L., & Wackernagel, M. (2021). *Estimating the date of earth overshoot day 2021*. Global Footprint Network. https://www.overshootday.org/content/uploads/2021/07/Earth-Overshoot-Day-2021-Nowcast-Report.pdf

Meltwater. (2021). *Media search engine*. www.meltwater.com

Richardson, R. C. (2019). *Resource depletion is a serious problem, but 'footprint' estimates don't tell us much about it*. GreenBiz. https://www.greenbiz.com/article/resource-depletion-serious-problem-footprint-estimates-dont-tell-us-much-about-it

Wackernagel, M., & Lin, D. (2019). *Ecological footprint accounting and its critics*. GreenBiz. https://www.greenbiz.com/article/ecological-footprint-accounting-and-its-critics. Accessed Nov 2021.

Wackernagel, M., Lin, D., Evans, M., Hanscom, L., & Raven, P. (2019). Defying the footprint Oracle: Implications of country resource trends. *Sustainability, 11*(7), 2164. https://www.mdpi.com/2071-1050/11/7/2164

Wackernagel, M., Hanscom, L., Jayasinghe, P., Lin, D., Murthy, A., Neill, E., & Raven, P. (2021). The importance of resource security for poverty eradication. *Nature Sustainability, 4*, 731–738. https://doi.org/10.1038/s41893-021-00708-4

Wilson, E. O. (2016). *Half-earth: Our Planet's fight for life*. Liveright.

York University Ecological Footprint Initiative & Global Footprint Network. National Footprint and Biocapacity Accounts. (2022 edition). *Produced for the Footprint Data Foundation and distributed by Global Footprint Network*. Available online at: https://data.footprintnetwork.org

Dr. Mathis Wackernagel created the *footprint* concept in the early 1990s with Prof. William E. Rees. The carbon footprint has become the most popular variant. In 2003, he founded Global Footprint Network, a sustainability think-tank, making planetary constraints relevant to decision-making. Its largest engagement campaign is its annual Earth Overshoot Day. Mathis's honors include the 2018 World Sustainability Award, the 2015 IAIA Global Environment Award, and the 2012 Blue Planet Prize.

Dr. David Lin leads Global Footprint Network's research team, and contributes to the production, development, and improvement of the National Footprint and Biocapacity Accounts. Prior to joining Global Footprint Network, David earned his Ph.D. and worked as a post-doctoral researcher in the Systems Ecology Laboratory at the University of Texas at El Paso. His research focused on integrating models of ecosystem function with land cover change analysis in Arctic ecosystems. David is a native of California, and holds a BS in ecology, behavior, and evolution from the University of California, Los Angeles.

Earth System Law

Louis J. Kotzé and Rakhyun E. Kim

Abstract The existing body of international environmental law has been created in the context of a relatively stable and harmonious Holocene epoch. This assumed regulatory premise of Holocene stability and harmony has resulted in a collection of international environmental law norms that are unable to sufficiently address the governance challenges emanating from within the context of the Anthropocene's complex, unstable, unpredictable, and intertwined earth system. Earth system law has recently been proposed as an alternative vision for international environmental law in the Anthropocene. Earth system law is intended to serve as an imaginative framework that can guide innovative questions regarding the difficulties posed to international environmental law in responding to the complex challenges of earth system governance, and as a roadmap for international environmental law to better address these challenges on an appropriate planetary level in the Anthropocene.

The rapidly deteriorating state of the earth system is now clearly evident through the lens of the Anthropocene (Steffen et al., 2016). The Anthropocene trope evidences progressively invasive and destructive patterns of human mastery on planet Earth that lie buried beneath and within centuries-old layers of corporate greed, political opportunism, appropriation, slavery, colonialism, imperialism, patriarchy, ecocide, and systemically entrenched entitlement that feeds off an increasingly vulnerable and uneven living order (Fineman & Grear, 2013). Law generally, and international environmental law in particular, have played a central role in facilitating these and other drivers of the Anthropocene; and have been unable to effectively prevent, halt and/or minimize continuing earth system decay and to offer meaningful solutions to

L. J. Kotzé (✉)
Faculty of Law, North-West University, Potchefstroom, South Africa
e-mail: louis.kotze@nwu.ac.za

R. E. Kim
Copernicus Institute of Sustainable Development, Utrecht University,
Utrecht, The Netherlands
e-mail: r.kim@uu.nl

573

the Anthropocene's deepening socio-ecological crisis in a way that can also safe-guard planetary integrity (Kotzé, 2019a).

Several studies reveal international environmental law's perceived regulatory gaps (e.g., the fact that it does not yet regulate potentially harmful new climate technologies such as solar geoengineering (Biermann et al., 2022); its structural complicity in creating the conditions that drive the Anthropocene (Kim & Mackey, 2014; Kotzé, 2019b); its inability to address socio-ecological damage and resultant climate and other inter- and intra-species injustices in an inter and intra-generational way (Grear, 2014); and its lack of regulatory ambition, evidenced in particular by its lack of binding ecological norms (French & Kotzé, 2019; Kim & Bosselmann, 2013; Bridgewater et al., 2014). Moreover, the architecture of international environmental law, including its core assumptions, orientation, operation, and objectives, are not commensurate with the most recent understanding of the governance challenges that arise from within the context of a complex and interlinked earth system (Cardesa-Salzmann & Cocciolo, 2019; Viñuales, 2018). A recent study, for example, highlights how international environmental law struggles to grapple with the coordination of planetary boundaries and the many complex planetary scale governance challenges emanating from interacting planetary boundaries (French & Kotzé, 2021). Clearly, international environmental law must change if it is to remain relevant, useful and effective in pursuit of global sustainability in the Anthropocene (Biber, 2017).

Scholars have been exploring alternative visions for international environmental law in the Anthropocene, which they hope will be more fully able to legally respond to the prevailing Anthropocene reality and complex earth system governance challenges (e.g., Kotzé, 2017; Webster & Mai, 2021). While terms such as 'Earth-centred law' (Bosselmann, 2016) and 'planetary boundaries law' (Chapron et al., 2017) have been suggested, the growing epistemic project of earth system law holds out considerable potential to re-imagine and craft "next-generation international environmental law" for the Anthropocene (Kim, 2021), or *Lex Anthropocenae* (Kotzé & French, 2018). Earth system law also has the potential to redirect much needed and renewed attention to the critically important role of law "as a technology of social organisation [that] has been neglected in the otherwise highly technology-focused accounts by natural and social scientists of the drivers of the Anthropocene" (Viñuales, 2018: 2). In other words, earth system law reminds us that the Anthropocene predicament is not only interesting for and a concern of earth system science. This predicament, and ways to maintain planetary integrity in the face of unprecedented earth system decay, are also very relevant for the social sciences where law is a "purposeful vehicle for shaping [human] behaviour to achieve desired ends" (Hadfield & Weingast, 2012: 473).

The concept of earth system law was first proposed in a 2019 publication (Kotzé & Kim, 2019); an endeavour that Biermann (2021: 12) describes as an "attempt to chart a new legal field". Although the discourse on earth system law is all but mature, interest in this proposal is growing, as the emerging scholarship shows (e.g., du Toit et al., 2021; Gellers, 2021; Kotzé et al., 2022; Kim & Kotzé, 2021; Kotzé & Kim, 2020; Mai & Boulot, 2021; Kim, 2021; Leach, 2023; Petersmann, 2021;

van Asselt, 2021). In essence, the project of earth system law offers an alternative framing for international environmental law to facilitate the type of transformations and governance interventions that are in step with a continuously transforming earth system, and that are required to address the socio-ecological crisis of the Anthropocene and the multiple governance challenges arising from this crisis (Kim et al., 2022). Because of the intimate disciplinary, conceptual and practical links between law and governance, earth system law situates such an alternative framing within the context of "earth system governance", which is defined as "organised human responses to earth system transformation, in particular the institutions and agents that cause global environmental change and the institutions, at all levels, that are created to steer human development in a way that secures a 'safe' co-evolution with natural processes" (Biermann, 2007: 328).

Earth system law's principal objective is to align international environmental law with an earth system perspective. It does so by prompting lawyers and policymakers to discard assumptions of one-dimensional Holocene-nested linearity, predictability, simplicity and harmony on which international environmental law rests; and instead to embrace an alternative understanding of the role and contribution of international environmental law in governing complex, non-linear, interconnected, multi-scalar and unpredictable earth system governance challenges that arise in the Anthropocene. Earth system law therefore encourages law, lawyers and other social actors to grapple more deliberately with the natural science aspects of the earth system and to translate these into the social science domain in a way that also meaningfully embraces "earth system governmentality" (Lövbrand et al., 2009). To this end, earth system law seeks to facilitate a deeper understanding of international environmental law's (in)ability and potential contribution to respond to the multiple governance challenges and implications flowing from earth system thinking. In terms of such a description, earth system law is not so much a new body of law (such as human rights law or trade law that focuses on a specific issue) as it is a vision or imaginary of what international environmental law could become for the purpose of facilitating the legal aspects of earth system governance in the Anthropocene.

Earth system law is defined as an innovative legal imaginary that is rooted in the Anthropocene's planetary context and its perceived socio-ecological crisis. Earth system law is aligned with, and responsive to, the earth system's functional, spatial and temporal complexities; and the multiple earth system science and social science-based governance challenges arising from a no-analogue state in which the earth system operates. Earth system law seeks to respond to the earth system's instability and unpredictability and its governance challenges through a continuous norm development process that drives meaningful transformations as well as intra, inter- and trans-disciplinary learning and deliberation (Kotzé et al., 2022). To this end, and in pursuit of desirable planetary futures, earth system law potentially offers: an intra-, inter- and trans-disciplinary analytical framework to better understand and respond to the legal dimensions of earth system governance; the normative foundations to govern the full spectrum of earth system relationships in a way that promotes planetary integrity and justice in their fullest sense; and the legal means to facilitate transformative earth system governance for long-term sustainability (Kotzé & Kim, 2020).

In an *analytical* sense, earth system law offers a framework to critique the current deficiencies of international environmental law, and to reimagine international environmental law; to open up the closures of earth system science for lawyers, while illuminating the juridical aspects of earth system governance for earth system scientists; to reveal the regulatory implications of earth system thinking for law; and to serve as a new crosscutting theme of scientific enquiry for scholars working in the area of global sustainability. In a *normative* sense, earth system law offers a framework to design better and more ambitious legal rules to respond more effectively to the type of planetary governance challenges that the dynamic and complex earth system presents. The transformative dimension of earth system law involves both reforming existing international environmental law alongside the governance demands of a complex earth system (internal transformations), as well as pursuing initiatives that are fully embedded in an earth system law paradigm that can trigger and steer societal transformation towards planetary integrity and justice (external transformations).

References

Biber, E. (2017). Law in the Anthropocene epoch. *The Georgetown Law Journal, 106*, 1–68.

Biermann, F. (2007). 'Earth system governance' as a crosscutting theme of global change research. *Global Environmental Change, 17*, 326–337.

Biermann, F. (2021). The future of 'environmental' policy in the Anthropocene: Time for a paradigm shift. *Environmental Politics, 30*, 61–80.

Biermann, F., Oomen, J., Gupta, A., Ali, S. H., Conca, K., Hajer, M. A., Kashwan, P., Kotzé, L. J., Leach, M., Messner, D., Okereke, C., Persson, Å., Potočnik, J., Schlosberg, D., Scobie, M., & VandeVeer, S. D. (2022). Solar geoengineering: The case for an international non-use agreement. *WIREs Climate Change, 13*(3), e754. https://doi.org/10.1002/wcc.754

Bosselmann, K. (2016). Shifting the legal paradigm: Earth-centered law and governance. In P. Magalhães, W. Steffen, K. Bosselmann, A. Aragaõ, & V. Soromenho-Marques (Eds.), *The safe operating space treaty: A new approach to managing our use of the earth system.* Cambridge Scholars Publishing.

Bridgewater, P., Rakhyun E., Klaus, K., Bosselmann. (2014/2016). Ecological Integrity: A Relevant Concept for International Environmental Law in the Anthropocene? *Yearbook of International Environmental Law 25*(1), 61–78. https://doi.org/10.1093/yiel/yvv059

Cardesa-Salzmann, A., & Cocciolo, E. (2019). Global governance, sustainability and the earth system: Critical reflections on the role of global law. *Transnational Environmental Law, 8*, 437–461.

Chapron, G., Epstein, Y., Trouwborst, A., & López-Bao, J. (2017). Bolster legal boundaries to stay within planetary boundaries. *Nature Ecology and Evolution, 1*, 1–5.

du Toit, L., Lopez Porras, G., & Kotzé, L. J. (2021). Guiding environmental law's transformation into earth system law through the telecoupling framework. *European Energy and Environmental Law Review, 30*, 104–113.

Fineman, M. A., & Grear, A. (Eds.). (2013). *Vulnerability: Reflections on a new ethical foundation for law and politics.* Routledge.

French, D., & Kotzé, L. J. (2019). 'Towards a global pact for the environment': International environmental law's factual, technical and (unmentionable) normative gaps. *Review of European, Comparative and International Environmental Law, 28*, 25–32.

French, D., & Kotzé, L. J. (Eds.). (2021). *Research handbook on law, governance and planetary boundaries*. Edward Elgar.

Gellers, J. C. (2021). Earth system law and the legal status of nonhumans in the Anthropocene. *Earth System Governance, 7*, 100083.

Grear, A. (2014). Towards 'climate justice'? A critical reflection on legal subjectivity and climate *in*justice: Warning signals, patterned hierarchies, directions for future law and policy. *Journal of Human Rights and the Environment, 5*, 103–133.

Hadfield, G., & Weingast, B. (2012). What is law? A coordination model of the characteristics of a legal order. *Journal of Legal Analysis, 4*, 471–514.

Kim, R. E. (2021). Taming Gaia 2.0: Earth system law in the ruptured Anthropocene. *The Anthropocene Review, 9*, 1–14.

Kim, R. E., & Kotzé, L. J. (2021). Planetary boundaries at the intersection of earth system law, science and governance: A state of the art review. *Review of European, Comparative and International Environmental Law, 30*, 3–15.

Kim, R. E., et al. (2022). Law systems and planet earth: Editorial. *Earth System Governance, 11*, 100127. https://doi.org/10.1016/j.esg.2021.100127

Kim, & Bosselmann, (2013). International Environmental Law in the Anthropocene: Towards a Purposive System of Multilateral Environmental Agreements *Abstract*. *Transnational Environmental Law 2*(2), 285–309. https://doi.org/10.1017/S2047102513000149

Kim, R. E., & Mackey, B. (2014). International environmental law as a complex adaptive system. *International Environmental Agreements: Politics Law and Economics 14*(1), 5–24. https://doi.org/10.1007/s10784-013-9225-2.

Kotzé, L. J. (Ed.). (2017). *Enviornmental law and governance for the Anthropocene*. Hart Publishing.

Kotzé, L. J. (2019a). International environmental law and the Anthropocene's energy dilemma. *Environmental and Planning Law Journal, 36*, 437–458.

Kotzé, L. J. (2019b). International environmental law's lack of normative ambition: An opportunity for the global pact for the environment? *Journal for European Environmental and Planning Law, 16*, 213–238.

Kotzé, L. J., & French, D. (2018). A critique of the global pact for the environment: A stillborn initiative or the foundation for *Lex Anthropocenae*? *International Environmental Agreements: Politics, Law and Economics, 18*, 811–838.

Kotzé, L. J., & Kim, R. E. (2019). Earth system law: The juridical dimensions of earth system governance. *Earth System Governance, 1*, 100003.

Kotzé, L. J., & Kim, R. E. (2020). Exploring the analytical, normative and transformative dimensions of earth system law. *Environmental Policy and Law, 50*, 457–470.

Kotzé, L. J., Kim, R. E., Blanchard, C., Gellers, J. C., Holley, C., Petersmann, M., van Asselt, H., Biermann, F., & Hurlbert, M. (2022). Earth system law: Exploring new frontiers in legal science. *Earth System Governance, 11*, 100126.

Leach M. C. (2023). Seeking 'Systems' in Earth System Law: Boundaries identity and purpose in an emergent field. *Earth System Governance, 15*, 100162. https://doi.org/10.1016/j.esg.2022.100162

Lövbrand, E., Stripple, J., & Wiman, B. (2009). Earth system governmentality: Reflections on science in the Anthropocene. *Global Environmental Change, 19*, 7–13.

Mai, L., & Boulot, E. (2021). Harnessing the transformative potential of earth system law: From theory to practice. *Earth System Governance, 7*, 100103.

Petersmann, M.-C. (2021). Sympoietic thinking and earth system law: The earth, its subjects and the law. *Earth System Governance, 9*, 100114.

Steffen, W., et al. (2016). Stratigraphic and earth system approaches in defining the Anthropocene. *Earth's Future, 8*, 324–345.

van Asselt, H. (2021). Governing fossil fuel production in the age of climate disruption: Towards an international law of 'leaving it in the ground'. *Earth System Governance, 9*, 100118.

Viñuales, J. (2018). The organisation of the Anthropocene in our hands? *International Legal Theory and Practice, 1*, 1–81.

Webster, E., & Mai, L. (Eds.). (2021). *Transnational environmental law in the Anthropocene: Reflections on the role of law in times of planetary change.* Routledge.

Louis J. Kotzé is Research Professor at the Faculty of Law, North-West University, South Africa; and Senior Professorial Fellow in Earth System Law, University of Lincoln, UK. He is a Senior Fellow of the Earth System Governance Network; Member of the Scientific Steering Committee of the Network; and co-convenor of the Network's Taskforce on Earth System Law. He is associate editor of the journal *Earth System Governance* (Elsevier) and has published extensively in the areas of global environmental constitutionalism, law and the Anthropocene, and earth system law. He currently serves as Klaus Töpfer Sustainability Fellow at the Potsdam Institute for Advanced Sustainability Studies.

Rakhyun E. Kim is Assistant Professor of Global Environmental Governance at the Copernicus Institute of Sustainable Development, Utrecht University in the Netherlands. He directs a five-year research programme on the complex dynamics of 'problem shifting' between international environmental treaty regimes, supported by a 1.5 million euro 'Starting Grant' from the European Research Council. Kim is a Senior Research Fellow of the Earth System Governance Project, where he co-convenes the Taskforce on Earth System Law.

Ecocide

Valérie Cabanes

Abstract The destruction of the Earth system by industrial technologies is jeopardizing the living conditions of present and future generations, humans and non-humans. Along with political, diplomatic and economic initiatives, international law has a role to play in transforming our relationship with the natural world, shifting that relationship from one of harm to one of harmony. Individuals, but also legal entities that are actively responsible for this destruction, must be prosecuted when their decisions undermine the integrity of life on Earth and the safety of the planet. To meet that challenge, law has to shift from anthropocentrism to ecocentrism and recognize an international crime: Ecocide.

The deterioration of living conditions on Earth and the accelerating destruction of the planet's ecosystems make it urgent to adopt innovative and binding measures to control human activity, particularly industrial activity. The current economic system, with its unsustainable patterns of consumption and production, has continuously altered the dynamics and functioning of the entire Earth system to an extent unprecedented in human history. Humanity is now standing at a crossroads. The scientific evidence points to the conclusion that the emission of greenhouse gases and the destruction of ecosystems and their biodiversity at current rates, as well as the crossing of each of the planetary boundaries will have catastrophic consequences for the Earth's system including human species. It is becoming necessary to reaffirm the supremacy of human rights over trade law on the one hand, but also to recognize that our fundamental rights are conditioned by respect for higher standards defined by biological laws. If the conditions of life itself are threatened, how can we hope to guarantee humanity's right to water, food, health and even shelter? For several decades now, a number of actors have been trying to

V. Cabanes (✉)
Notre affaire à tous, Paris, France
e-mail: valerie@endecocide.eu

© The Author(s), under exclusive license to Springer Nature
Switzerland AG 2023
N. Wallenhorst, C. Wulf (eds.), *Handbook of the Anthropocene*,
https://doi.org/10.1007/978-3-031-25910-4_92

579

have the intrinsic value of nature and the rights of ecosystems to exist, thrive and regenerate recognized and defended in court. But when damages done to the Earth system are severe, it seems necessary to broaden the range of the most serious international crimes by recognizing a fifth crime against peace and security of mankind: the crime of ecocide. By destroying the ecosystems on which we depend, we are destroying the foundations of our civilization and mortgaging the living conditions of all future generations. This is no less serious than war crimes, crimes against humanity, or the crimes of genocide or aggression, already recognized in international criminal law.

The intent to make ecocide an international crime isn't new. The work has been inspired by earlier efforts, in 1945, to forge definitions of new international crimes, including 'genocide' and 'crimes against humanity'. Ecocide draws from both terms, in form and substance. The word ecocide combines the Greek '*oïkos*', meaning house/home (and later understood to mean habitat/environment), with the Latin '*occidere*', meaning to kill. This draws on the approach taken by the Polish jurist Rafael Lemkin, who invented the word 'genocide' in November 1944. In 1972, the Swedish Prime Minister, Olof Palme, gave a speech at the UN Conference on the Human Environment, in which he described destruction brought about by indiscriminate bombing, by large-scale use of bulldozers and herbicides as an outrage sometimes described as 'ecocide', which requires urgent international attention. But even before that biologist Arthur W. Galston used the word 'ecocide' at the 1970 Conference on War and National Responsibility in Washington, D.C. to characterize the use of Agent Orange by the US Army during the Vietnam War. In Vietnam, voices were raised as early as 1968 to describe the Vietnamese ecocide as a war against a land and the unborn, in order to recall that the acts of war committed by the Americans went beyond the definition of crimes established at the Nuremberg trials. Its consequences affected not only civilians but also future generations.

Several attempts at formalizing ecocide as an international crime have emerged since then. Richard Falk, a professor of international law at Princeton, proposed in 1973 to raise ecocide to the same level as genocide by drafting an international convention to be submitted to the UN member states. Then in 1985, the "Whitaker Report" presented to the UN Sub-Commission on the Prevention and Punishment of the Crime of Genocide, clearly recommended the inclusion of ecocide as an autonomous crime alongside genocide and ethnocide or cultural genocide. But the final text of the Rome Statute founding the International Criminal Court (ICC) in 1998 finally retained deliberate and serious damage to the environment (Article 8.2.b.iv) as a war crime but did not mention it in peacetime.

Following this decision, which was too limited in scope, lawyers as Lynn Berat, Polly Higgins, Laurent Neyret and I began to advocate for the recognition of ecocide in peacetime. The Stop Ecocide Foundation took up the challenge in 2017 and started diplomatic lobbying. Vanuatu and the Maldives called for consideration of adding the crime of 'ecocide' to the Rome Statute at the 18th Meeting of the ICC

Assembly of States Parties in 2019, followed by Belgium a year later. In late 2020, the Stop Ecocide Foundation convened an Independent Expert Panel for the Legal Definition of Ecocide, which I joined. It comprises twelve lawyers from around the world, with a balance of backgrounds, and expertise in international, criminal, environmental and climate law. We have worked together for 6 months, charged with preparing a practical and effective definition of the crime of 'ecocide'. The Panel was assisted by outside experts and a public consultation that brought together hundreds of ideas from legal, economic, political, youth, faith and indigenous perspectives from around the globe. In developing its work, the Panel has sought to draw, as far as possible, upon existing precedents and authorities in international treaty and customary law, as well as the practice of international courts and tribunals. Particular regard has been placed on practice in international criminal law and the approaches reflected in the Rome Statute.

The Ecocide definition unveiled on June 22, 2021 marks a big first step in the global campaign's efforts to prevent future environmental disasters. The panel defines ecocide as "unlawful or wanton acts committed with knowledge that there is a substantial likelihood of severe and either widespread or long-term damage to the environment being caused by those acts." This definition of ecocide offers the States Parties to the Rome Statute the opportunity to meet current challenges. It offers a new and practical legal tool. Each term chosen has been clearly defined to understand how the crime would be qualified:

> "Wanton" means with reckless disregard for damage which would be clearly excessive in relation to the social and economic benefits anticipated;
>
> "Severe" means damage which involves very serious adverse changes, disruption or harm to any element of the environment, including grave impacts on human life or natural, cultural or economic resources;
>
> "Widespread" means damage which extends beyond a limited geographic area, crosses state boundaries, or is suffered by an entire ecosystem or species or a large number of human beings;
>
> "Long-term" means damage which is irreversible or which cannot be redressed through natural recovery within a reasonable period of time;
>
> "Environment" means the Earth, its biosphere, cryosphere, lithosphere, hydrosphere and atmosphere, as well as outer space.

The introduction of the qualifier 'unlawful' captures environmentally harmful acts that are already prohibited in law. The Panel considered narrowing this to unlawful under international law. However, this was felt to be too narrow. International environmental law contains obligations for States in treaties and customary law but relatively few absolute prohibitions, and it leaves the bulk of the protection to be formulated at the national level, through national laws. While the lawfulness of an act under the relevant national law cannot be relied on to permit acts that are unlawful under international law, there is no reason that national illegality – in particular as a matter of domestic criminal law - should not be part of an international law definition.

The Panel recognizes that defining the 'environment' (or 'natural environment'), has proved to be challenging for international law. To date, there is no single agreed definition of these terms. Available definitions vary with regard to their scope, content, and approach. One possible approach was to leave the term 'environment' undefined, as the International Law Commission has done in the topic 'Protection of the environment in relation to Armed Conflicts.' One benefit of this approach is that, as human understanding of the environment changes, the evolution in knowledge could be taken into account for the purposes of this crime. The Panel decided to adopt a different approach, taking into account that the criminal law may require greater clarity and specificity than might be the case in general environmental law context. For our limited purposes, therefore, the term has been defined as Earth System Science does, encompassing the Earth and the interactions between its five main spheres which interact in complex ways, but also the outer space. The five Earth spheres are the lithosphere (the brittle upper portion of the mantle as well as the crust), the biosphere (all ecosystems and living beings); the hydrosphere (the combined mass of water found on, under, and above the surface); the atmosphere (the layer of gases retained by Earth's gravity that surrounds the planet); the cryosphere (Earth's surface where water is in solid form, including sea ice, lake ice, river ice, snow cover, glaciers, ice caps, ice sheets, and frozen ground (which includes permafrost). Unlike the existing four international crimes, ecocide would be the only crime in which human harm is not a prerequisite for prosecution, moreover the Earth system becomes a subject of law.

The International Criminal Court is asked to rule independently by firmly applying the principle of universal jurisdiction, according to a common higher interest placed above States, with jurisdiction possible on any national territory when vital ecosystems are threatened. While it will become a crime in the countries where it is ratified, ratifying nations may arrest non-nationals on their own soil for ecocide crimes committed elsewhere. This means citizens of countries that are not members of the ICC could still be prosecuted. But for the ICC to adopt the ecocide definition and amend the Rome Statute it would take a few steps. One of the International Criminal Court's 123-member countries (which do not include the U.S., China or India) would have to submit a definition to the United Nations secretary-general. The proposal must then be voted on by a majority of members of the ICC at the annual assembly in December in order to be considered. Between the formal proposal of an ecocide crime by a member country and ratification, however, an amendment process could take years. Once the final text for an amendment is discussed and agreed upon, two-thirds of member countries must vote in favour. When the vote is ratified in General Assembly of the ICC states parties, it must be enforced in countries a year later.

Valérie Cabanes is a legal expert in international law, specializing in human rights. She is a member of the Executive Committee of the Global Alliance for the Rights of Nature since 2015 and one of the experts of the UN Harmony with Nature initiative network since 2016. She launched in 2013 a European citizen initiative on the crime of ecocide, co-founded in 2015 Notre Affaire à Tous at the initiative of a climate case against the French government and in 2019, Wild Legal, a school training in the doctrine of the Rights of Nature. She contributed to the drafting of a Universal Declaration of Humankind Rights commissioned by the French President, François Hollande in 2015 and joined in 2020 the expert drafting panel on the legal definition of "ecocide" convened by the Stop Ecocide Foundation. She is the author of *Homo Natura, en harmonie avec le vivant* (Paris: Buchet/Chastel, 2017) and *Un nouveau droit pour la Terre. Pour en finir avec l'écocide*, (Paris: Seuil, 2016) translated in English under the title *Rights for Planet Earth, End to crimes against nature* (Delhi: Natraj Publishers, 2018).

Ecological Footprint

Mathis Wackernagel and David Lin

Abstract The ecological footprint is a resource accounting approach that tracks human demand on ecosystems, or more precisely on the planet's *biocapacity*. It allows us to determine the size of economies, from households to cities, countries, or humanity as a whole, compared to the size of ecosystems, a region or the entire planet.

Ecological footprint accounting allows researchers to compare how much nature there is to how much nature is needed to support human activities. In other words, such accounting tracks people's or activities' competition for what ecosystems can regenerate. Since regeneration occurs on the biologically productive surface areas of our planet, this metric tracks how much mutually exclusive, biologically-productive area is necessary to regenerate people's demand for nature's products and services (Borucke et al., 2013; Lin et al., 2018; Wackernagel & Beyers, 2019; Wackernagel et al., 2019, 2021).

To map human dependence on biocapacity, ecological footprint accounting is based on two principles (Wackernagel et al., 2019).

Additivity: Given that human life competes for biologically productive surfaces, these surface areas can be added up. The ecological footprint (or footprint), therefore, adds up all human demands on nature that compete for biologically productive space. These spaces are added up to the extent they mutually exclude each other, for instance if they provide for biological resources at the expense of accommodating urban infrastructure or absorbing excess carbon from fossil fuel burning. In other words, surfaces that serve multiple human demands or provide multiple crops are counted only once.

Equivalence: Since not every biologically productive surface area is of equal biological productivity, areas are scaled proportionally to their biological

M. Wackernagel (✉) · D. Lin
Global Footprint Network, Geneva, Switzerland
e-mail: mathis@footprintnetwork.org; david.lin@footprintnetwork.org;
https://www.footprintnetwork.org

© The Author(s), under exclusive license to Springer Nature
Switzerland AG 2023
N. Wallenhorst, C. Wulf (eds.), *Handbook of the Anthropocene*,
https://doi.org/10.1007/978-3-031-25910-4_93

productivity (Wackernagel et al., 2018). Global hectares, which are the measurement unit for ecological footprint accounting, are biologically productive hectares with world average productivity. This is necessary since the planet's surface areas vary considerably in productivity given latitude, rain, soil conditions, solar orientation, slope, and other parameters. The global hectare serves as an "ecological currency" that enables analysts to compare bioproductivity over time and geographies.

These principles make the footprint comparable to the available biologically productive space (biocapacity).

Human demands on nature that compete for biocapacity include:

food, fibre, timber,
space for roads and structures,
energy production (from hydropower to biomass), and
waste absorption, incl. CO_2 from fossil fuel or cement production.

Since wild animal species also compete for biocapacity, some biocapacity would need to be left for non-human uses. E.O. Wilson postulates to use just half of Earth to maintain about 85% of the planet's pre-industrialization biological diversity (Wilson, 2016; Half Earth Project, 2021). The question is which half: Global Footprint Network suggests that it makes practical sense to translate using half Earth into using half the biocapacity of this planet. Applying the biocapacity lens takes into account the fact that wild species require productive areas to thrive, not just icefields or deserts (just protecting half of the planet's lowest productivity areas would do little for biodiversity preservation). By postulating as the goal to set aside half the biocapacity for wild species would also allow for some flexibility. There would be choice whether productive areas are entirely left for other species, or whether more area is used by people, but at a lower level of intensity.

The →carbon footprint is an integral part of the ecological footprint. In ecological footprint accounts, carbon emissions are expressed in global hectares, but often, carbon footprints are also measured in tonnes per year. There is a direct translation from tonnes into global hectares in any given year, since the conversion describes how many biologically hectares with average productivity it takes to sequester excess emissions (Mancini et al., 2016).

In contrast, the →water footprint, inspired by the ecological footprint, tracks embodied (or 'virtual') water. Virtual water in a product means accounting for the water that was needed to produce product. Water footprint (as defined by the Water Footprint Network) is not a direct part of the Ecological Footprint because those water footprints answer a different research question, one of absolute water use. Water use in Ecological Footprint accounting is measured in terms of the biocapacity used or compromised by that water use. In other words, it takes the ecological context into account. For example, if the diversion of water reduces irrigation potential elsewhere, this loss in productivity would be counted as the ecological footprint of that water diversion. Also, the energy needed to build, maintain, and operate the water distribution system and to then treat the wastewater is included in the ecological footprint of water.

Ecological Footprint accounts are applicable at any scale. It started with country and city analyses developed by Mathis Wackernagel and William E. Rees in the early 1990s at the University of British Columbia, Vancouver, but was soon applied to other scales (Wackernagel & Rees, 1995). The most developed accounts are the National Footprint and Biocapacity Accounts, based on UN statistics. They use UN statistics; using those statistics was chosen to not pick data arbitrarily, even though the data set may be limited, and as a result lead to underestimates of ecological footprints. The first National Footprint and Biocapacity Accounts started in 1997 at the then Center for Sustainability Studies at the Universidad Anáhuac de Xalapa in Veracruz, Mexico. From 2003 onwards, they were produced and further developed by Global Footprint Network. Since 2019, they are produced annually by an independent entity, the Footprint Data Foundation, in collaboration with York University, Toronto. Latest results show that human demand in 2018 exceeded the planet's regeneration by at least 75% (York University et al., 2022).

For sub-global ecological footprint calculations, one needs to distinguish between the consumption footprint (how much is embedded in final consumption) versus the production footprint (how much is embedded in producing everything in the region, independent of where it is finally consumed). The difference between the two is the footprint of net-imports (production + net-imports = consumption). More details on sub-national assessments are available in the Ecological Footprint Standards, established in 2009 by Global Footprint Network in collaboration with the primary footprint practitioners at that time (footprintstandards.org).

The benefits of a biological perspective, as inherent in →planetary boundaries or ecological footprint accounting, build on the insight that the biosphere's power to regenerate has become too small compared to human demand, leading to all major environmental challenges including climate change, biodiversity loss, water scarcity, deforestation, desertification, or ocean acidification. This perspective carries three main advantages:

1. A biological approach is comprehensive: From the perspective of biological regeneration, all human pressures on water, climate, biodiversity, food, energy, etc. are considered as parts of the same overall dynamic, rather than as aggregated as one rather than counted as various separate, competing issues. This one dynamic is simply that we are caught in a human competition for regeneration. This perspective shows that, given global competition for renewable, biological resources, these resources have become more fragile and limiting than "non-renewable (or non-biological)" resources. Rather than addressing environmental issues as separate from each other, and thereby pegging one against another one or shifting one pressure from one to another domain. By presenting all these pressures as one overarching phenomenon, (competition for regeneration), rather than as one at the cost of another, the challenge becomes more solvable. Therefore, this biological perspective also bridges conservation and climate change.

2. Biological metrics are intuitive and understandable. Very few people relate to 2 °C, ppm, or tons of carbon. But even primary school kids understand biological balances ("how much do we take compared to how much ecosystems regenerate?").

Results of ecological footprint accounts use easy-to-understand units of measurement: number of Earths, number of Germanies Germany uses, the date of Earth Overshoot Days, or global hectares.

3. A biological approach emphasizes resource security and economic benefits ("skin in the game"). Perhaps most importantly: In contrast to the 'carbon-only' view, a biological approach makes obvious that the company, city, or country is a subsystem of the biosphere. Therefore, any human entity is part of, or embedded in, its biological context. It is inextricably dependent on the biological resources. Such a view helps emphasize resource security as the central concern; it makes obvious the risk each country faces when not prepared for the predictable future of climate change and biological resource constraints. Resource security thinking builds the bridge to recognizing economic self-interest. A comprehensive biological approach therefore paints climate action as necessary rather than noble. In contrast, the current more electro-mechanically informed climate debate is mostly based on a "noble argument" ("your action is your responsibility to humanity and the future"), leading to timid action. Given the slow rate at which infrastructure can be retrofitted or upgraded, a country or city's economic robustness depend on aggressive climate action, preparing itself for an inevitable carbon-free future. Boosting one's resource security becomes economically critical (while also benefiting humanity's sustainability).

Ecological footprint accounting has not been without its critics. Some researchers have misinterpreted ecological footprint accounting as a 'social theory' or a policy guideline, and therefore reject its results. In reality, footprint accounts are merely a metric documenting competing uses of biocapacity (Grazi et al., 2007). Similarly, Newman (2006) has argued that the ecological footprint concept carries an anti-urban bias, as it does not consider the opportunities created by urban growth. He argues that the ecological footprint of densely populated areas, such as a cities or small countries with a comparatively large population (e.g., Paris and Qatar), may lead to the perception that these populations are "parasitic". In contrast, ecological footprint accounts don't judge or interpret, but rather document: they track the resource dependence of cities—like a fuel gauge documents a car's fuel availability, or a camera captures an image of the object in front of its lens. Newman questions the metric because these communities have little intrinsic biocapacity, and instead must rely upon large hinterlands. Ecological Footprint researchers would state that footprints of cities only reveal how much biocapacity is needed to maintain that city. This information empowers city managers to make good decisions and keep cities resilient and operational in a world of resource constraints. Others claim that ecological footprint accounting denies the benefits of trade (Stiglitz et al., 2009). In reality, the accounts merely provide descriptive results that others then can interpret to what extent the described trade flows are beneficial or not. Therefore, ecological footprint accounts neither prove nor deny the benefits of trade – they just document the biocapacity flows associated with trade.

Global Footprint Network has collected the criticisms and discusses them in detail on their website (Global Footprint Network, 2021).

References

Borucke, M., Moore, D., Cranston, G., Gracey, K., Iha, K., Larson, J., Lazarus, E., Morales, J., Wackernagel, M., & Gall, A. (2013). Accounting for demand and supply of the Biosphere's regenerative capacity: The National Footprint Accounts' underlying methodology and framework. *Ecological Indicators, 24*, 518–533. https://doi.org/10.1016/j.ecolind.2012.08.005

Global Footprint Network. (2021). *Limitations and Criticism, Global Footprint Network website.* https://www.footprintnetwork.org/our-work/ecological-footprint/limitations-and-criticisms/

Grazi, F., van den Bergh, J. C. J. M., & Rietveld, P. (2007). Spatial welfare economics versus ecological footprint: Modeling agglomeration, externalities and trade. *Environmental and Resource Economics, 38*, 135–153.

Half Earth Project. (2021). *Why half?* https://www.half-earthproject.org/discover-half-earth/#why-half. Accessed Nov 2021

Lin, D., Hanscom, L., Murthy, A., Galli, A., Evans, M., Neill, E., Mancini, M. S., Martindill, J., Medouar, F.-Z., Huang, S., & Wackernagel, M. (2018). Ecological footprint accounting for countries: Updates and results of the National Footprint Accounts, 2012–2018. *Resources, 7*(3), 58. https://www.mdpi.com/2079-9276/7/3/58.

Mancini, M. S., Galli, A., Niccolucci, V., Lin, D., Bastianoni, S., Wackernagel, M., & Marchettini, N. (2016). Ecological footprint: Refining the carbon footprint calculation. *Ecological Indicators, 61*, 390–403.

Newman, P. (2006). The environmental impact of cities. *Environment and Urbanization., 18*(2), 275–295. https://doi.org/10.1177/0956247806069599. ISSN 0956-2478.

Stiglitz, J., Sen, A., & Fitoussi, J.-P. (2009). *The Measurement of Economic Performance and social progress revisited*, OFCE, N° 2009–33, december 2009, Centre de recherche en économie de Sciences Po. https://www.researchgate.net/publication/239807212.

Wackernagel, M., & Ree, W. E. (1995). *Our ecological footprint: Reducing human impact on the Earth.* New Society Publishers. Gabriola Island.

Wackernagel, M., & Beyers, D. (2019). *Ecological footprint: Managing the biocapacity budget*, New Society Publishers. Gabriola Island. https://www.footprintnetwork.org/2019/09/04/18187/

Wackernagel, M., Galli, A., Hanscom, L., Lin, D., Mailhes, L., & Drummond, T. (2018). Chapter 16: Ecological footprint accounts: Principles. In S. Bell & S. Morse (Eds.), *Routledge handbook of sustainability indictors* (pp. 244–264). Routledge International Handbooks; Routledge.

Wackernagel, M., Lin, D., Evans, M., Hanscom, L., & Raven, P. (2019). Defying the footprint Oracle: Implications of country resource trends. *Sustainability, 11*(7), 2164. https://www.mdpi.com/2071-1050/11/7/2164

Wackernagel, M., Hanscom, L., Jayasinghe, P., Lin, D., Murthy, A., Neill, E., & Raven, P. (2021). The importance of resource security for poverty eradication. *Nature Sustainability, 4*, 731–738. https://doi.org/10.1038/s41893-021-00708-4

Wilson, E. O. (2016). *Half-earth: Our Planet's fight for life.* Liveright.

York University Ecological Footprint Initiative & Global Footprint Network. National Footprint and Biocapacity Accounts. (2022 edition). *Produced for the Footprint Data Foundation and distributed by Global Footprint Network.* Key results available online at: https://data.footprintnetwork.org. More information at www.fodafo.org and https://footprint.info.york.ca

Dr. Mathis Wackernagel created the *footprint* concept in the early 1990s with Prof. William E. Rees. The carbon footprint has become the most popular variant. In 2003, he founded Global Footprint Network, a sustainability think-tank, making planetary constraints relevant to decision-making. Its largest engagement campaign is its annual Earth Overshoot Day. Mathis's honors include the 2018 World Sustainability Award, the 2015 IAIA Global Environment Award, and the 2012 Blue Planet Prize.

Dr. David Lin leads Global Footprint Network's research team, and contributes to the production, development, and improvement of the National Footprint and Biocapacity Accounts. Prior to joining Global Footprint Network, David earned his Ph.D. and worked as a post-doctoral researcher in the Systems Ecology Laboratory at the University of Texas at El Paso. His research focused on integrating models of ecosystem function with land cover change analysis in Arctic ecosystems. David is a native of California, and holds a BS in ecology, behaviour, and evolution from the University of California, Los Angeles.

Environmental Justice

Wolfgang Sachs

Abstract This article addresses some of the power asymmetries which divide mankind, the so-called "we" in the discourse about the Anthropocene. Humanity is beset with long-running cleavages between Northern and Southern countries, just as between the wealthy and the poor across countries. Power is wielded environmentally, too. First, responsibility for greenhouse gas emissions around the world is highly skewed towards the Global North, while impacts of climate change on society and biosphere fall disproportionately onto the Global South. Second, human rights of indigenous, rural and fishing communities are in danger when mining, terrestrial and marine resources are exploited. Third, unequal trade relations in terms of raw materials, land, fisheries and gene patents continue to exist from colonial times until today. And fourth, restoring justice for possible victims is an arduous process without a world government and judiciary. Structural power is largely unaccountable.

In the historical accounts of the Anthropocene, the industrial revolution is regarded as a landmark, launching the fossil age. But the very same event was also something of a big bang that catapulted parts of humanity into a world of rising incomes and economic growth, while the majority of world population stood still (Milanovic, 2016, 2). Since then Western Europe and North America have sprinted forward, only to be caught up by East Asia in the past 30 years. As a result, a huge income inequality arose between countries: just by being born in the United States rather than in the Congo, a person enjoys an income 93 times higher (ibid, 133). However, most theories of justice have a handicap: they are devised for territorially bounded societies. In a globalized world, however, the class differences between the haves and have-nots across countries become ever more evident. Still, the dimensions of resource justice are valid in a transnational world: justice as redistribution, recognition, reciprocity, and as redress (Sachs & Santarius, 2007, pp.119).

W. Sachs (✉)
Wuppertal Institute of Climate, Environment, Energy, Berlin, Germany
e-mail: wolfgang.sachs@wupperinst.org

N. Wallenhorst, C. Wulf (eds.), *Handbook of the Anthropocene*,
https://doi.org/10.1007/978-3-031-25910-4_94

Justice as redistribution aims at overcoming inequality. Inequality of what? Income inequality is largely reflected in carbon inequality. In this perspective, inequality runs the whole gamut of climate change, from mining over emissions to impacts. Looking at the world population by income class and their share of CO_2 emissions, a huge disparity emerges: In 2015, half of the world population, those who earned income, caused a staggering 93% of CO_2 emissions, while the other half of the poor accounted for only 7% (Kartha et al., 2020, 6). (Other researchers arrive at similar but slightly different figures: Hubarek et al., 2017: well-off 85%, the poor half 15%; Chancel & Piketty, 2015: well-off 87%, the poor half 13%). What a striking gap! Of the global emissions of the middle/high-income earners, 35.9% come from North America and Europe, 24.8% from China, 13.6% from the rest of Asia including India, 13.3% from the Middle East and Russia/Central Asia, 3.5% from Latin America and 1.7% from Africa (ibid. 2020, 11). In contrast, the other half of the world population, the one with minor emissions, is mostly found in India, China, Africa and Latin America. Equally, in 2015 the top 10% of the income pyramid emitted 45% of global emissions, while the other 55% of emissions were distributed among the remaining 90% (Chancel & Piketty, 2015, 10). The distribution of carbon emissions follows the economic division of the world. Air travel, real estate, and steaks set the tone in the global upper class (Ivanova & Wood, 2020), while used cars, washing machines, and air-conditioning are common in the middle class. And then there is the class of have-nots, who have to be content with standing in packed buses, suffering malnutrition, and sharing outhouses. Yet, equality is a treacherous concept in times of limits. Both securing human dignity and maintaining the planetary boundaries claim priority over equality.

Furthermore, impacts of global warming, such as drought, storms, rising sea level, melting of glaciers, are not distributed equally. To be sure, no country, no classes of people will be able to shield from the impacts. But they will not affect everyone equally — not in the same way, not at the same time, not at the same magnitude (UNDP, 2019, 175). People of all over the world will face the consequences in terms of human health, agricultural yield, changes in flora and fauna, but in spite of this, the poorer countries and poorer people will be hit earliest and hardest. Some countries could quite literally disappear. Children, women, and elderly people are the most vulnerable, in particular in economically weak regions. Inequality breeds injustice, when the situation of the victim is permanent, and injustices are cumulative. Besides, global warming has already led to more inequality of nations, somehow wiping out the progress made in human development (ibid. 183; Diffenbaugh & Burke, 2019). In a nutshell, climate change is a classical showpiece for shifting environmental loads to other countries and to other people. Those who are the least responsible, carry the heaviest burden.

Justice as recognition aims at equal dignity, not equal distribution (Fraser & Honneth, 2003). People desire to be treated with decency and respect, they strive to have full rights in society. On the contrary, people hate to be discriminated and disenfranchised. Due recognition is not just a courtesy, but a vital human need. Accordingly, it is also a collective need for all those who are treated condescendingly in the prevailing hierarchy – for women and persons of color, for Muslims as well as Mayas. The human rights charter builds on that, identifying people as a moral community, whose members hold equal and inalienable rights preceding the jurisdiction of a nation-state. People are citizens in a transnational legal space, this has been the revolution of human rights.

For herders in the savannas of Tanzania, for fisherfolk along the coast of Ghana, for peasants in the Andes, for rice farmers in Vietnam, for indigenous communities all over the globe, to have free access to land, rivers and oceans is a matter of survival and self-esteem. This regards, roughly speaking, a quarter of the world population. But the territories of life (ICCA Consortium, 2021) are under threat. At least since the time of colonialism fertile land and mineral deposits have always attracted the "omnivores" (Gadgil & Guha, 1995) domestically and globally, undermining livelihoods of the "ecosystem people". Resource extraction continues, enforcing quite often the displacement and dispossession of inhabitants. For instance, there are approximately 17,000 large-scale mining sites in 171 countries, mostly managed by international corporations (IPBES, 2019, 28). And the number of dams has increased rapidly in the past 50 years, there are now about 50,000 large dams (higher than 15 m) worldwide (ibid, 30). Moreover, the expansion of agriculture in the tropics has smashed precious ecosystems and removed forest-dwellers for cattle-ranching in Latin America and for plantations in South-Asia. Soy, palm oil and beef are the commodities with the largest embedded tropical deforestation imported into the EU, followed by wood products, cocoa and coffee (WWF, 2021, 5). The Environmental Justice Atlas (ejatlas.org) tells the story of about 3500 cases, albeit not all of them on human rights violations (Temper et al., 2015). Even in nature conversation areas, the original inhabitants were driven away, despite the fact that they are mostly caretakers of the natural world (Boyd & Keene, 2021).

It is therefore not surprising, that a further dimension of justice is neglected: the reciprocity of ecological trade relations between nations. With globalization, consumers are largely separated from producers through global commodity chains. For example, lithium is mined in Bolivia and shipped into South Korea to make batteries which are then exported to German automakers as equipment for electric cars destined for the US market. As the links in the commodity chains are widespread over the globe, so are the ecological impacts of resource depletion, manufacturing and consumption. Trade relations are unequal in ecological terms, if nations are asymmetrically affected by the consequences of resource exploitation and waste sinks. This is largely the case between the Global North and the Global South, the power of the former imposes an inferior position on the latter.

However, this is crucial when the finitude of nature is becoming more and more apparent. An undeclared war is on: Who owns the remaining biosphere? After Europe and North America have removed in the past centuries an immense amount of raw materials together with flora and fauna from theirs territories, the resource hunger of the global upper-middle class is now largely directed toward the Global South. As a global input-output study shows (Dorninger et al., 2021), high-income countries net appropriate a disproportionately large share of materials, energy, land, and labor through international trade. This unequal distribution grew from 1990 until the 2008 global financial crisis, stabilizing thereafter (ibid. 5). With regard to biodiversity, rich economies have more than 50% of their biodiversity footprint outside of their territorial boundaries, mostly in middle-income and poor countries. Those, in turn, are in danger to lose a large part (35–60%) of domestic species because of the production and export of coffee, tea, sugar, fish, textiles or manufactured items in international commodity chains (Wiedmann & Lenzen, 2018, 317).

Finally, justice needs to account for the dimension of redress. High-income countries (and classes) are seldom held accountable for violating global injustice, be it historically or in the present. The deeper reason for this failure that international law is based, except for human rights law, on the sovereignty of states. However, issues of climate, biodiversity or ocean governance go beyond state sovereignty, they involve the Earth as global commons. Up to now, the global environment is still largely regarded as *res nullius,* waiting to be exploited by private or state interests. In contrast, states should also act as trustees for the common natural assets and for the future generations (Bosselmann, 2017). Multilateral policymaking is the arena of the never-ending tug-of-war between trusteeship and sovereignty.

The Global South, however, suffers a double injustice: to be humiliated and to be without legal remedies. More precisely, there are tiny attempts to achieve redress through international and human rights law, but they are both rudimentary and fragmentary. For instance, in the UN Climate Convention the principle of "common but differentiated responsibilities" is laid down, as are the compensation payments for historical emissions as well as damages and losses. Furthermore, under the UN Biodiversity Convention fair and equitable sharing of genetic resources is proclaimed, as well as the security of indigenous territories in the UN Declaration on the Rights of Indigenous Peoples. But these agreements are full of holes, even mostly lacking in implementation. In sum, they do not live at all up to the injustice done. No wonder that the cry for justice is so popular with the actors of the civil society. They attempt to hold the governments accountable, for instance, increasingly bringing litigation before the courts. They appear to be inspired by the British philosopher C.S.Lewis, who once said about technology, "What we call Man's power over Nature turns out to be a power exercised by some men over other men with Nature as its instrument" (2001, 55).

References

Bosselmann, K. (2017). Governing the global commons: A 'planetary boundaries' approach. *The Political Quarterly, 13*(1), 37–42.

Boyd, D. S., & Keene, S. (2021). *Human rights-based approaches to conserving Biodiversity.* A Policy Brief from the UN Special Rapporteur on Human Rights and the Environment.

Chancel, L., & Piketty, T. (2015). *Carbon and inequality: From Kyoto to Paris.* School of Economics.

Diffenbaugh, N. S., & Burke, M. (2019). Global warming has increased economic inequality. *PNAS, 116*(20), 9808–9813.

Dorninger, C., et al. (2021). Global patterns of ecologically unequal exchange: Implications for sustainability in the 21st century. *Ecological Economics, 179*, 106824.

Gadgil, M., & Guha, R. (1995). *Ecology and equity. The use and abuse of nature in contemporary India*. Routledge.

Fraser, N., & Honneth, A. (2003). *Recognition or redistribution? A political-philosophical exchange*. Verso.

Hubarek, K., et al. (2017). Poverty eradication in carbon-constrained world. *Nature Communications, 8*, 912.

ICCA Consortium. (2021). *Territories of life*. Report www.iccaconsortium.org

Intergovernmental Science-Policy Platform on Biodiversity and Ecosystem Services (IPBES). (2019). *Summary for policymakers of global assessment report*. IPBES.

Ivanova, D., & Wood, R. (2020). The unequal distribution of household carbon footprints in Europe and its link to sustainability. *Global Sustainability, 3*, 1–12.

Kartha, S., et al. (2020). *The carbon inequality era*. Report of the Stockholm Environment Institute/Oxfam.

Lewis, C. S. (2001). *The abolition of man*. Harper One. [Original edition 1943].

Milanovic, B. (2016). *Global inequality: A new approach for the age of globalization*. Harvard University Press.

Sachs, W., & Santarius, T. (Eds.). (2007). *Fair future. Resource conflicts, security and global justice*. Zed Books.

Temper, L., del Bene, D., & Martinez-Alier, J. (2015). Mapping the frontiers and front lines of global environmental justice: The EJAtlas. *Journal of Political Ecology, 22*, 255–278.

United Nations Development Programme (UNDP). (2019). *Human Development Report 2019*. UNDP.

Wiedmann, T., & Lenzen, M. (2018). Environmental and social footprints of international trade. *Nature Geoscience, 11*, 314–321.

World Wide Fund for Nature (WWF). (2021). *Stepping up? The continuing impact of EU consumption on nature worldwide*. WWF.

Wolfgang Sachs is a former research director of Wuppertal Institute of Climate, Environment, Energy in Germany. He was a university lecturer in the United States, Italy, England, in addition to Germany. He published, among other books, *For Love of the Automobile. Looking Back into the History of Our Desires*. Berkeley, CA: University of California Press, 1992; (Ed.) *Development Dictionary. A Guide to Knowledge as Power*. London: Zed, 2019 (orig. 1992); *Planet Dialectics: Explorations in Environment and Development*. London: Zed Books, 2015 (orig. 1999)

Ethics of Justice

Daphnée Valbrun

Abstract Since antiquity, ethics and justice have been two essential themes in philosophy. From Socratic ethics (470–399 BC), whose main foundation is the objective good and virtue, to Aristotelian morality (384–322 BC), which advocates happiness as the supreme goal of human life, to Foucauldian ethics, which is oriented towards concern for the self, we see that ethics have always been at the heart of the great philosophical debates. Through their reflections, the philosophers always tried to articulate in the same movement of thought ethics and justice, because for them, ethical reflection needs to be supported by justice without which it risks to be denatured. However, the plurality of meanings given to ethics makes it difficult to attempt to define the concept and to apply the concept of ethical justice. Taking into account this ambiguity and in view of the importance of ethics in the practice of law, one wonders how to reconcile ethics and justice? This is the problem identified through this article.

Justice and ethics are two concepts that seem to differ in many ways (such as living together, respect for oneself and for others) but offer a unique perspective in our society because both orient their relationship towards humans. One regulates the society and the other postulates the knowledge of how to live in society. These two concepts seem to contain real points of similarity through their implementation. Does this complementarity exclude a certain ambiguity in their conciliation?

First of all, ethics refer to a plurality of meanings depending on the disciplinary field in which the word is used. For Gagnon and Lewis, it 'concerns the will to live together as well as the conceptions of good and evil, of the just and the unjust, which must guide the relationships between humans' (Gagnon et al., 2008, p. 81). This living together is based on individual and collective values. Thus, ethics appears as a penetrating view that makes it possible to become aware of human responsibility in a given situation without necessarily imposing a constraint. This is what makes it possible to say that ethics has an aim that is of the order of the

D. Valbrun (✉)
Catholic University of the West, Angers, France

relationship between oneself and the other, which prioritizes personal well-being and that of the other. To bring out this relational dimension, Moreau emphasizes that ethics 'is rooted in praxis, in the action and decisions that each person makes to guide his or her life' (Moreau, 2012, p. 34). For him, ethics is the set of values that guide the life and social action of each citizen. This is why ethics is very important because it consolidates the civicism advocated in European societies. Moreover, it characterizes not only a principle of self-respect and respect for others, but also a principle of autonomy.

Unlike ethics, justice has a different scope. It is concerned more specifically with the relationship of order that governs relations between humans in a social and political organization. According to Gagnon, it consists of assigning 'what is due to each person in the political and social hierarchy' (Gagnon et al., 2008, p. 81). Justice is then associated with everything that conforms to the law, everything that is just or legal. It refers to 'positive laws established by the authority of the city, so that it excludes from the term all implicit social norms (Morrisson, 1995, p. 331). Thus, only laws established by legitimate authorities are considered legal. He excludes values and social norms altogether, because it appears that our values are personal and may not be beneficial to all. Given the Socratic point of view, we can deduce that the idea of justice refers to the logic that to be just is not only to conform to laws but also to conform to the spirit of justice. Moreover, justice is the goal of all politics, insofar as it aims at establishing a true and anonymous equality, which does not take into account either the social situation or the personality of individuals. This is also the position of Ricoeur, who affirms that justice concerns the relationship to the other as a stranger, as a distant other, as each person. Its purpose is to establish the 'right distance between all humans, a middle ground between the too little distance that is characteristic of many dreams of emotional fusion and the excess of distance that is maintained by arrogance, contempt, and hatred of the stranger, the unknown' (Ricœur, 1995, p. 72). According to Ricoeur, it is important to have the capacity to recognize oneself through the other in order to treat him or her at his or her true value.

Today, in our society, justice is considered everything that is legally written and that does not affect the moral value of individuals. Everything that defends the moral integrity and the right of individuals participates in justice. Considering, this fundamental principle of Socratic ethics, namely: 'Virtues are good and that the good is beneficial' (Morrisson, 1995, p. 331), we can deduce that justice is a virtue and all virtues should be good. The same is true for social values, if they are beneficial to society and do not violate the law. They can be considered legal or just. This leads us to say that 'ethics and justice are complementary and are united under a common thought' (Gagnon et al., 2008, p. 81). In view of the relationship between ethics and justice, justice is said to be ethical when it is thought of primarily as a relationship with oneself. This means that justice in the ethical sense must allow us to keep our integrity and our human dignity above all.

From the point of view of legal practice, ethical justice aims to open up avenues of reflection that allow for a broader look at 'the act of judging in a perspective where a positive approach to judicial judgment that focuses primarily on the project of 'saying the law' can find the way to a possible articulation with reflections that are interested in the project of doing justice' (Landheer-Cieslak, 2012, p. 43). Thus, justice has a close relationship with ethics insofar as it guides social actions so that all citizens are treated fairly. Is it not this indestructible link between justice and ethics that Merlier attempts to show when he argues: 'Justice is the imperative that governs the three ethical principles of social action which are autonomy, benevolence and equity' (Merlier, 2013, p. 21). He clarifies that 'the justice imperative is situated above the ethical principles, which are subordinate and relative to it' (Ibid p. 21). Showing the complementarity of the two concepts, he advocates that ethics is below justice in order to be able to establish principles of legitimate justice. Is it because individuals are selfish by nature as Rawls (1971) postulates in his theory of justice? Or, can we say that our selfishness drives us to such an extent that it prevents us from being impartial and acting ethically?

It is important that all citizens are invited, in the manner of Rawls, to have 'the veil of ignorance' (Rawls, 1971) in order to put themselves in the position of choosing the principles of justice without knowing their future position in society. All citizens of a social and political organization must be sensitized through citizenship education to adopt ethics as a reflexive act that allows them to become aware of and act justly. Also, through the judicial institutions, it is especially necessary to question the competences and the professional ethics of the men of law because justice must be an ethical virtue. As a virtue, it must be complemented by ethics in order to have a society based on equality and social justice. 'Judicial institutions, as well as any other institution, should participate in the project of doing justice, treating all citizens fairly, and finding the right distance between all human beings while knotting the *nexus* between the self, the neighbor, and the distant through a just sharing' (Landheer-Cieslak, 2012, p. 8). This way of acting contributes to the living together and the well-being of each citizen of a social sphere. In the environmental sphere, for example, ethical justice is of paramount importance because it offers us the possibility of adapting our actions for better environmental protection. This environmental ethical justice can be distributive in the sense that it allows us to put our selfishness aside and act ethically to protect future generations. It can also be corrective in the sense that we must take symbolic actions (plant more trees, avoid all that causes pollution) or material actions (pay the damages caused by the exploitation of the wealth of underdeveloped countries) in view of a pacified planet, based on equal rights and protection for all.

In sum, it is important to think about ethical justice in order to shed light on the way our institutions administer responsibilities and promote the benefits of collective effort, especially in the environmental field. Ethical justice is useful for studying the principles that guide our major social, economic, and especially environmental tradeoffs. Thus, thinking about ethical justice is ideal for building a society with citizen's equal in rights. Thus, ethics and justice are complementary, on the one hand, in defining the relations between citizens of the same social organization and, on the other hand, in the implementation of laws protecting these citizens against the abuses of authorities. Moreover, the problem that arises today through the reconciliation of these two concepts is to be able to separate legal law and personal ethics, because this can in many cases risk encroaching on our own values in the reading of legal texts.

References

Gagnon, B., Lewis, N., & Ferrari, S. (2008). Environnement et pauvreté: regards croisés entre l'éthique et la justice environnementales. *Écologie & politique, 35*(1), 79–90.

Landheer-Cieslak, C. (2012). Paul Ricœur et l'éthique du jugement judiciaire: Quelles relations entre justice et sollicitude ? *Revue interdisciplinaire d'études juridiques, 68*(1), 1–47.

Merlier, P. (2013). *La justice et les trois principes éthique: Philosophie et éthique en travail social, sous la direction de Merlier Philippe.* Presses de l'EHESP.

Moreau, D. (2012). *L'éthique professionnelle des enseignants: Enjeux, structures et problèmes.* L'Harmattan.

Morrisson, D. (1995). Xenophon's Socrates on the just and the lawful. *Ancient Philosophy, 15,* 329–347.

Rawls, J. (1971). *Théorie de la justice.* Seuil.

Ricœur, P. (1995). *Le Juste 2, supra,* note 2.

Daphnée Valbrun is in her first-year of PhD in Educational Sciences at the Catholic University of the West and Strasbourg University. She is a lecturer in the department of Education Sciences at Catholic University of West. Her main research topics are the anthropocene, citizenship, educational pedagogies, autonomy and experience in education.

Finitude

Pascal Marin

Abstract After a millennial history of converting the energy of finitude into technocratic power over nature, revelations of the Anthropocene in today's world invite us to rediscover the meaning and the ethical fruitfulness lying within that which limits human power: our vulnerability, our mortality. But to recultivate this in a poetics of finitude requires the criticism and the deconstruction of the representations of the man-machine. In a senseless headlong rush they claim to liberate human beings definitively from their finite existence.

Finitude – humans' *Sein-Zum-Tode* (Heidegger, 1993/1927, 'Being-toward-death'), awareness of their own mortality – is no longer widely reflected in our society, in the ways in which we live our lives and plan for the future. 'In industrial and post-industrial societies, […] death has been forgotten' (Dastur, 2007); further than this, though, futurists from the transhumanist movement are now gaining prominence. They hope to use technology to free humans from the finitude of our mortal body, which they blame for all ills – disease, aging, and ultimately death. In their view, the living mind and the marvel of technology mean that nothing at all is owed to the mortal flesh. However, as much today as it ever has been, death is 'an immense mystery – a great question-mark which we carry with us in our very souls' (de Hennezel, 1995, p. 13). Some evidence of this can be found in the importance of care of the dead in the social organisation that is at the root of culture itself. Paleoanthropologists interpret the vestiges of funeral rites, burials and cremations as undeniable indicators which set human beings apart within the primate family (Maureille, 2004, p. 9). However, for early Man, born into a reality in which everything appears animated and eternally alive, the death of his contemporaries belies the omnipresence of life, darkening his world. The vigilant consciousness, driven by the awareness of death, thus becomes the perfect medium for the 'awakening of the questioning mind' (Jonas, 2001, p. 20). Hegel paints Man's constant journey toward

P. Marin (✉)
Catholic University of Lyon, Lyon, France
e-mail: pmarin@univ-catholyon.fr

death as the cradle of negativity (Kojève, 1947): the magic cauldron in which continuously simmer ideals, essences, images, scientific explanations and inventions of technical engineering. It is the constant awareness of 'the nothing' (what Heidegger terms *Das Nichts* – oblivion) which gives humans their ability to think about the realities of the world in virtual terms. At the very dawn of western civilisation, though, long before Hegel, the Greeks – labelling themselves *thnètoï* (mortals) – explicitly formed the link between our 'mortality' and the formidable achievements of human industry (Arendt, 1972). However, in addition to fear of death and awe of mortals' creative arts, these thinkers identified another human emotion: the guilty conscience of a progeny who feels unworthy of the Mother Nature which produced him. Man's own mortality is a contradiction of natural liveness. His plans for the future fly in the face of simple existence without specific goals or targets, forever new and forever the same, in perpetual cycles of regeneration: 'When Sophocles (in the celebrated chorus of Antigone) says that there is nothing more wonderful and frightening than Man, to illustrate the point he cites Man's deeds, which are a violation against nature because they disrupt what would, in the absence of mortals, be the eternal equilibrium of the self-contained eternity' (Arendt, 1972, p. 60). Here, Sophocles expresses a wisdom born of finitude, fully cognisant of the 'otherness' of the world. The thinker and poet witnesses a primitive form of humans, still aware of their debt to nature. Whilst Man may, to some extent, go against nature through his acts, he still reveres and respects it, knowing that his power over nature can only grow in proportion to his respect for it. However, today, such wisdom is dying out. Over the centuries since Sophocles' time, the original meaning of finitude has gradually been obscured, forgotten and allowed to decay. This loss of awareness is, seemingly, the price of power – the power of science, and the power of technology to transform nature. Anaximander, Heraclitus, Aristotle and other philosophers and physicists marked the beginning of an effort to study the world in its entirety and the laws which hold it together. However, in so doing, they embarked on a path which has led the human race to lose sight of themselves and their own finitude. This change of direction is evident in fragment 50 of Heraclitus, who exclaims '*ouk emou*' – 'Not me, my words, my language, my body, but the world!' We see the other face of this movement today, having come to the other end of the centuries-long history of converting the drive born of finitude into phenomenal technocratic power over nature. The process is not only ceaseless progress of science and technology. More than that, it has another side: the quelling of the idea of finitude, and therefore loss of sense of nature and the balances within it (the only environment in which we can, ourselves, survive). We must recognise the profound link between these two aspects of our existence. The sense of mortality, awareness of the fragility of life and the vast natural living environment which harbours all life on Earth, is grounded in respect: the sentiment behind all ethical discussions. Witnessing the glowing ember of an ancient wisdom that combines a sense of finitude and respect for nature, Montaigne (1962, p. 546) coined the maxim that in all human affairs, "tis a mortal Hand that presents it to us, 'tis a mortal Hand that accepts it'. Soon after Montaigne, though, came Descartes, with the idea that new science could make his contemporaries 'masters and possessors of nature' (Descartes, 1953,

p. 168) – the same Descartes who viewed death as a breakdown of the machine body (Chazal, 2013, p. 28) and believed that the human race would soon be free of it. Thus, the notice of his own death in the Antwerp Gazette was immensely ironic: 'he died in Sweden, a madman insisting that he could live however long he chose to' (Wolff, 2010, p. 264, note 3). Today, as we face the undeniable symptoms of a major ecological emergency, we are reminded of the wisdom of Sophocles and Montaigne, keeping a true sense of humanity at the very heart of our society. How, though, is it possible to encourage people to embrace their own finitude in today's highly technological culture, which is becoming ever further removed from precisely the thing which makes us human?

If we agree with Heidegger (1953, p. 285) that 'More primordial than man is the finitude of the Dasein in him', then not only is any effort to overcome that finitude vain and doomed to failure; it is self-contradicting: in precisely the goal it aims to achieve, it draws inspiration from the very finitude which it rejects. Such an attempt will end, ultimately, only in chaos and destruction. Rather, it is essential to encourage people to act in concert with their own finitude, though never ceasing the quest for scientific solutions to make that finitude easier to bear. Indeed, we believe it is important to resist both the utopia of transhumanism and the counter-utopia of a return to a world before science/technology-driven development. The crux, here, is moderation; the goal is to restore a balance between nature and culture. A critical factor in striking this balance is the adoption of a true sense of finitude within our culture. In order to achieve such a balance, with the sense of Man and the sense of nature growing in harmony, we must critically engage in deconstructing the ideals of identification of Man and machine, and more broadly, the reduction of nature to something material. Indeed, in today's world, as reality is examined through a materialistic lens, the differences between Man, animal and machine are disappearing from view. In particular, life is no longer a part of the knowledge base, as the criterion of materiality draws no distinction between the animate and the inanimate. Thus, while the differences between the animate and the inanimate, between living beings and machines, between humans and other living things of course subsist, we are losing sight of them, both epistemologically and culturally, in a culture dominated by objective material sciences. This critical view is taken by both academics and philosophers, decrying any attempt to reduce reality to only what we know of it. The academics include, for example, Albert Einstein, who believed that the *hic et nunc* of present reality cannot be included in the equations of physics and that 'there is something essentially of the Now which is beyond the grasp of science' (Klein citing Einstein, 2007, p. 92). The philosophers notably include Edmund Husserl, who uses phenomenology as a way of combating such 'naturalism', under which a being is merely an object of study for natural sciences (Husserl, 1955, p. 57). These battles over the meaning of reality continue to rage today, in numerous arenas. On one side are those who only wish to know what exact, experimental science can reveal about Man and nature. On the other are those who cultivate 'negative certainties' (Marion, 2010), holding that Man and nature cannot be boiled down to scientific knowledge and technological power. However, this struggle would be moot and powerless were it not for the fact that it is waged through culture and given widespread media coverage, which has power over people's minds and can guide communal actions. Thus, a sense of finitude must be accompanied by a poetic conception of it, in the same way as, in ontological reductionism, the ideals of the Man-machine in a machinised nature are supported by poetic, romanticised ideas of salvation through technology. Thus, Raphaël Liogier (2010, p. 26) said of the transhumanists' utopian idea: 'If we do not want this religion, this eschatology, this end, then we must construct another, […] similarly poetic, […], desirable, and ultimately more likely to encourage people to dream than to encourage them to think'. Here, we see two poetic approaches set against one another, but the two are not equal. As a transhumanist approach effectively reduces life to a machine, it holds nothing but death for human beings. On the other hand, the poetic view of finitude, by cultivating the sense of mortality, safeguards living conditions for the future.

References

Arendt, H. (1972). *La crise de la culture*. Éditions Gallimard.
Chazal, G. (2013). *Philosophie de la machine. Néo-mécanisme et post-humanisme*. Editions Universitaires.
Dastur, F. (2007). *La mort. Essai sur la finitude* (2e éd. augmentée). PUF.
de Hennezel, M. (1995). *La mort intime*. Robert Laffont.
Descartes, R. (1953). *Œuvres et lettres*. Éditions Gallimard/Pléiade.
Heidegger, M. (1953). *Kant et le problème de la métaphysique*. Éditions Gallimard.
Heidegger, M. (1993/1927). *Sein und Zeit* (17e Auflage ed.). Niemeyer Verlag.
Husserl, E. (1955). *La philosophie comme science rigoureuse*. PUF.
Jonas, H. (2001). La vie, la mort et le corps dans la théorie de l'être. In H. Jonas (Ed.), *Le phénomène de la vie. Vers une biologie philosophique* (pp. 19–44). De Bœck Université.
Klein, E. (2007). *Le facteur temps ne sonne jamais deux fois*. Éditions Flammarion.
Kojève, A. (1947). L'idée de la mort dans la philosophie de Hegel. In A. Kojève (Ed.), *Introduction à la lecture de Hegel*. Éditions Gallimard.
Liogier, R. (2010). *Introduction & La vie rêvée de l'homme. De l'humain, nature et artifices. La pensée de midi, 2010/1*, N° 30, 9–27.
Marion, J.-L. (2010). *Certitudes négatives*. Bernard Grasset.
Maureille, B. (2004). *Les origines de la culture. Les premières sépultures*. Éditions Le Pommier.
Montaigne. (1962). *Les Essais. Œuvres complètes*. Éditions Gallimard.
Wolff, F. (2010). *Notre humanité. D'Aristote aux neurosciences*. Fayard.

Pascal Marin is Professor at the Faculty of Philosophy of the Catholic University of Lyon. He is a member of the editorial boards of *La Revue des Sciences philosophiques et théologiques*, of the review *Sémiotique et Bible* and of the review *Théophilyon*. Specialized in the fields of hermeneutics and language, his research focuses at this time on the conditions of possibility of a philosophical anthropology in the context of artificial intelligence and transhumanism. He recently coordinated the dossier of the Revue *Théophilyon, Le naturalisme et ses critiques*, Tome XXVI, 2021-1 and published *Le robot et la pensée. Contre-philosophie de l'homme-machine*, Cerf, 2019.

Intergenerational Justice

Tracey Skillington

Abstract The impact of climate change is felt in every corner of the world today. Yet, rather than take responsibility for accelerating rates of natural resource destruction and subject them to more stringent forms of regulatory control, the tendency to date has been towards continuing with practices that knowingly jeopardize future life and development. Obligations to 'bequeath to future generations an Earth which will not one day be irreversibly damaged by human activity' are openly compromised (UNESCO Declaration on the responsibilities of the present generations towards future generations, Article 4, (1997). http://portal.unesco.org/en/ev.php-URL_ID=13178 & URL_DO=DO_TOPIC & URL_SECTION=201.html#:~:text=Article%204%20%2D%20Preservation%20of%20life%20on%20Earth & text=Each%20generation%20inheriting%20the%20Earth,not%20harm%20life%20on%20Earth. Accessed 29 June 2021). How are we to make sense of the 'chronological injustice' of this arrangement? Can 'interaction problems' between distant generations continue to be hailed as sufficient justification for not honouring constitutionally grounded commitments to equal protection? The discussion below considers some of the main arguments raised in relation to these concerns.

Today, the life support systems of this planet are in serious decline. Carbon, nitrogen and phosphorus cycles have been so badly affected by human activity that the Earth can no longer be considered the product of natural geological forces. Instead, Homo sapiens or, at least, a peculiar breed of humans (i.e., 'industrialised humans') have become the principal agents of planetary change. In an influential piece published in 2002, scientist Paul J. Crutzen describes how the cumulative destructive tendencies of these change agents accelerate the advancement the Anthropocene, a geological age characterized by multiple global crises, including the depredation of natural biogeochemical cycles, rapid loss of biodiversity, rising temperatures and ecosystem collapse. Inherently complex, these processes of change are also quite

T. Skillington (✉)
Department of Sociology & Criminology, University College Cork, Cork, Ireland
e-mail: t.skillington@ucc.ie

© The Author(s), under exclusive license to Springer Nature
Switzerland AG 2023
N. Wallenhorst, C. Wulf (eds.), *Handbook of the Anthropocene*,
https://doi.org/10.1007/978-3-031-25910-4_97

605

simply the product of human interference. Yet rather than take responsibility for this destruction as 'a common concern' (UNFCCC, 1992) and honour commitments to a partnership model of climate justice, to date, the tendency has been to continue with practices of environmental 'dis-saving' (Rawls, 2001: 118). That is, ongoing depletions of aggregate resource reserves at rates insufficient to stabilize climate conditions and preserve a safe living environment for the future. How are we to explain the rational basis of these actions?

In spite of unprecedented knowledge of geological crisis, the tendency still is to prioritize the immediate 'advantages' of largescale resource depletions over their more remote impacts (i.e., a discounting of future environmental conditions). When applied consistently, 'intertemporal discounting' results in a bias being displayed across multiple policy sectors towards the interests of the present in justice reasoning. Carbon intensive pathways thus continue largely unchallenged while the risks they pose to the future are deemed distant concerns (Weisbach & Sunstein, 2009) affecting subjects who, in not yet being born, lack an identity and, for many, a legitimate claim to justice (e.g., expectations that demand for gas and oil will continue to rise over the next 15 to 20 years, especially amongst emerging economies, see, BP Energy Outlook, 2020). There are many unresolved issues attached to a discounting approach, not least the way it imposes cumulative harms on future generations. As critics point out, current policy-determining generations do not leave the environmental and political stage all at once. Rather, they continue to share connections with those that follow, as much as those who came before. For instance, in terms of their contribution to GHG pollution, the effects of which reach far beyond the here and now and impose a negative forcing effect on future climate conditions (NASA, 2017). The destruction created by cumulative emissions into the Earth's atmosphere almost certainly cannot be reversed, leading to an inevitable overlap in the ecological fate of past, present, and future generations. Whilst we may assume technology will offer future generations better ways of coping with the dire effects of this cumulative destruction, we cannot assume that the knowledge they accrue will be reasonable compensation for what were once avoidable harms inflicted upon them (e.g., the decision of the European Investment Bank to continue to invest in new gas projects in spite of commitments to phase out the financial backing of fossil fuels by 2022. See Taylor, 2021).

Unlike the type of debates that confront each generation on inequalities prevailing between world regions (horizontal inequalities), the intergenerational context constitutes a different kind of collective action problem (i.e. one reflecting both horizontal and vertical inequalities). Climate justice campaigners call for a justice contract that more consciously embraces multiple generations. Moves have already been made in several states to incorporate the interests and, occasionally, the rights of present and future generations to a safe and sustainable environment into their constitutions as a legally binding commitment. For instance, Article 7 of the constitution of the Plurinational State of Bolivia refers to all citizens' fundamental right 'to enjoy a healthy environment, ecologically well balanced and appropriate to her wellbeing, while keeping in mind the rights of future generations, while Article 24 of the constitution of South Africa affirms everyone's 'right…to have the environment protected for the benefit of present and future generations'. Sceptics, however, point to the impracticalities of actualising such commitments, highlighting, in particular, how a lack of 'interaction' across generations, or truly reciprocal relations of exchange between equals (Darwall, 2002, p. 1) prevents the type of cooperation needed to make such a contract feasible in real terms. Typically, cooperation encompasses an agreed 'give and take' between co-existing parties with one giving to the other something and receiving something in return. On the surface, this type of exchange model would appear to create difficulties for a truly intergenerational model of climate justice, especially when further future generations are taken into consideration (i.e., those not yet born). Traditionally, contract theories of justice take as their primary focus the issue of distribution – how to distribute goods produced by shared productive endeavours – but how are we to understand 'shared productive endeavours' across generations who do not occupy the same time frame?

According to Rawls, it would be unfair to expect present generations to make sacrifices in terms of living standards (e.g., commit to zero carbon living) for the benefit of future peoples given that the latter cannot reciprocate to the advantage of those presently living. For Rawls (1999), such a scenario would give rise to a type of 'chronological unfairness' that is unjustified. Due to the limitations imposed by 'interaction problems' between distant generations, a contract model of intergenerational justice is said not practically feasible chiefly because questions of distributive justice are not applicable in the intergenerational context. But can we assume this continues to be the case? Equally, a reverse argument could be made in relation to the issue of chronological unfairness – those that come later will suffer disproportionately from the ecological losses incurred by the actions and decisions of their predecessors. Only when justice is determined from a future, and not just a purely present vantage point, can a contract model of intergenerational justice be considered procedurally, as well as morally justified, even if practically difficult to implement. However, if we continue on the current global emissions trajectory and further exhaust the capacities of the atmosphere to absorb excessive GHG without detrimental consequences, it will no longer be possible to claim that the foundational principles upon which an international order of liberal democratic states is built remain true or justified (including a Lockean principle advocating that 'enough and as good' is left for those that follow). It will also no longer be possible to defend our appropriation of what remains of the global carbon sink on the grounds that we the peoples act in ways that are fair, democratic and reasonable. A new model of justice, therefore, is required, one capable of expanding the range of relevant subjects of allocated justice. Amongst those most articulate on these and related concerns are global youth (i.e., members of the world's majority) whose strike for climate rallies around the world in 2019 and early 2020 captured widespread public attention and prompted UN Secretary-General, Antonio Guterres, speaking at the Climate Action Summit in September 2019, to encourage world leaders to consider more carefully the urgency of the issues youth raise, including emerging evidence proving that no world region is presently on track to meet the terms of the Paris Agreement (e.g., see Climate Action Tracker, 2021). Current rates of depletion of fossil fuels, biodiversity, arable lands and more, in gravely affecting the welfare of future generations, affect not only their interests but also their capacity to exercise constitutionally grounded rights to health, development, and equal protection (Skillington, 2019a, 2019b). As the science of climate change grows more precise, so too does knowledge of its impact on the ecological circumstances of the future. For those in favour of extending rights to future generations of the Anthropocene (even a category of 'group rights' relevant to generations not yet born), the capacity to predict more accurately future scenarios makes a long-term rights approach to justice both ethically reasonable and practically necessary.

References

BP Energy Outlook. (2020). https://www.bp.com/content/dam/bp/business-sites/en/global/corporate/pdfs/energy-economics/energy-outlook/bp-energy-outlook-2020.pdf. Accessed 22 June 2021.

Climate Action Tracker. (2021). *Climate summit momentum: Paris commitments improved warming estimate to 2.4°C*. https://climateactiontracker.org/documents/853/CAT_2021-05-04_Briefing_Global-Update_Climate-Summit-Momentum.pdf. Accessed 22 June 2021.

Crutzen, P. (2002). Geology of mankind. *Nature, 415*, 23.

Darwall, S. (2002). *Contractarianism/Contractualism*. Wiley Blackwell.

NASA. (2017). *Short-lived greenhouse gases cause centuries of sea level rise*. https://www.climatechange.ie/short-lived-greenhouse-gases-cause-centuries-of-sea-level-rise/. Accessed 21 June 2021.

Rawls, J. (1999). *A theory of justice*. Harvard University Press.

Rawls, J. (2001). *Law of peoples*. Harvard University Press.

Skillington, T. (2019a). Changing perspectives on natural resource justice, human rights and intergenerational justice. *The International Journal of Human Rights, 23*(4), 615–637.

Skillington, T. (2019b). *Climate change and intergenerational justice*. Routledge.

Taylor, K. (2021, March 4). Not quite over yet: EIB spent €890 million on fossil gas since phase out, activists say. *Euractiv*. https://www.euractiv.com/section/energy-environment/news/not-quite-over-yet-eib-spent-e890-million-on-fossil-gas-during-phase-out-activists-say/. Accessed 21 June 2021.

United Nations Framework Convention on Climate Change. (1992). https://unfccc.int/resource/docs/convkp/conveng.pdf. Accessed 23 June 2021.

Weisbach, D., & Sunstein, C. R. (2009). Climate change and discounting the future: A guide for the perplexed. *Yale Law and Policy Review, 27*, 433–457.

Tracey Skillington is Director of the BA (Sociology), Department of Sociology & Criminology, University College Cork, Ireland. Recent monographs include *Climate Justice & Human Rights* (Palgrave), *Climate Change & Intergenerational Justice* (Routledge) and forthcoming, *A Critical Theory of Climate Trauma* (Routledge). Her publications have appeared in many journals over the years, including the *European Journal of Social Theory*, the *British Journal of Sociology*, the *International Journal of Human Rights, Distinktion: Journal of Social Theory*, S*ociology*, the *Irish Journal of Sociology* and *Sustainable Development*.

International Law

Davor Vidas

Abstract International law has developed as a system of legal rules and principles regulating relations between States. At its core is the principle of the sovereign equality of States and their sovereignty over territory within their own boundaries. Whereas international relations are continually exposed to political change, international law aims to work towards facilitating legal stability in order to prevent or reduce conflicts and offer peaceful settlement of international disputes. The Anthropocene introduces unprecedented new challenges for international law: changes at the Earth System level, not only at the political level. Today's international law is a system of legal rules resting on foundations that evolved under the circumstances of the late Holocene, assumed to be ever-lasting. These changes will soon create a serious gap between international law as it is currently understood, and Earth System conditions that have already moved beyond the Holocene.

International law is a system of legal rules and principles applicable to a wide range of relations between States and other subjects recognized in the international community (Andrassy, 1976; Crawford, 2019; Jennings & Watts, 2012). Whereas various national laws regulate relations *within* any of the current 200 or so States, international law is a body of rules that applies to relations *between* those States. Also other types of entities may possess international rights and duties (such as international organizations, and sometimes also individuals), yet all those subjects are related, indirectly or directly, to States. Although States are not the sole subjects of international law, they are still its principal subjects.

International law provides the legal criteria for the creation of States and the recognition of their statehood (Montevideo Convention, 1933; Crawford, 2006). International law regulates how boundaries are delimited between States, and how maritime areas (like the territorial seas, exclusive economic zones, and the continental shelves) are delineated, so to determine to where State sovereignty or certain

D. Vidas (✉)
Fridtjof Nansen Institute, Lysaker, Norway
e-mail: dvidas@fni.no

© The Author(s), under exclusive license to Springer Nature Switzerland AG 2023
N. Wallenhorst, C. Wulf (eds.), *Handbook of the Anthropocene*,
https://doi.org/10.1007/978-3-031-25910-4_98

sovereign rights extend—and where these stop, and the freedoms of the high seas apply; and the content of rights and duties within the various spaces (UN Convention on the Law of the Sea, 1982). Rules of international law concerning the acquisition of territory and the delimitation of boundaries between States apply equally to *all* States (Jennings, 2017). Moreover, diplomatic and consular relations between them are subject to the rules of international law (Vienna Convention on Diplomatic Relations, 1961; Vienna Convention on Consular Relations, 1963), as are the dispute settlement mechanisms and procedures.

One element remains a central feature: *political change.* International relations are continually exposed to political change—an inherent feature of the (world) politics itself—whereas the objective of international law as a legal system is to work towards facilitating stability in those relations. The aim is to prevent or reduce conflicts, by offering options for peaceful settlement of international disputes. Achieving this objective has proven difficult ever since the emergence of international law—for obvious reasons.

A key feature of international law, and one which distinguishes it from other legal systems, is the sovereignty of its principal subjects: the States. Each of them possesses supreme authority within its own jurisdiction; there is no legal authority or power hierarchically above a State. Moreover, each State has its own territory over which it exercises sovereignty. Indeed, "the function of international law, at the most basic level, is to secure the coexistence of sovereign States" (Lowe, 2015, p. 2). This is reflected in the basic principles of international law.

First, international law is based on the principle of sovereign equality of States—a principle fundamental to the United Nations and reflected in Article 2(1) of the UN Charter (1945). In fact, however, States differ profoundly: their territories can be as massive as the Russian Federation's 17 million km^2 or as small as Monaco's less than 2 km^2; and their populations may exceed a billion, as is the case with China and India, or barely number ten thousand, as with Nauru and Tuvalu. In their political influence, economic development, military power, and many other aspects, States constitute a highly heterogeneous group in the sphere of international relations—but, under international law, each State is a sovereign entity, and all sovereign entities are legally equal. Under international law, any given State counts as only one party to an international treaty, and only one member of an international organization.

Second, consent by States is among core ingredients of international law. Due to the lack of a legislative process as in the national legal systems of individual States, international treaties are negotiated by the States themselves—and States become bound by any treaty only upon their consent (Vienna Convention on the Law of Treaties, 1969). Moreover, a State must agree to the jurisdiction of an international court or arbitral tribunal to be involved in the process of dispute settlement. Indeed, a State also freely decides whether to become a member of an international organization such as the UN.

International law promotes fundamental principles upon which legal rules in relations between States are based. However, this is not an abstract system detached from political, economic, or other relevant contexts. A clear example is the UN

Charter, which, while guaranteeing one seat as well as an equal vote to all its (currently 193) member States in the General Assembly, grants five powers—China, France, Russia, the UK, and the USA—permanent membership on the Security Council and the effective right of veto over Council decisions on matters of substance.

Beyond the changing political, economic, and social context, another type of change not inherent in the system of international law has emerged recently: the increasingly rapid change of Earth System conditions. Those have long been perceived as basically stable, generally unchanging from generation to generation, as witnessed for millennia. Indeed, changes have occurred occasionally, like shifting river courses (and thus international law rules on border determination in cases where inter-state boundaries follow rivers), or the sudden creation of volcanic islands (and thus the rules on rights acquired over these)—but all such natural changes lie well within the overall envelope of general stability that has characterized much of the Holocene Epoch for the past 11,700 years or so.

Of course, there were no States 11,700 years ago, no international borders between them, no exclusive jurisdictions or sovereignty—these abstract products of our imagination have found expression through legal principles and rules only in recent centuries. What we today take for granted is, in fact, a product of a relatively recent past, a past safely nested within the envelope of environmental stability of the Late Holocene. It is on those foundations that many social constructs—including international law—have developed.

To date, international law has focused on the interplay unfolding between the overall objective of legal stability, to which international law aims to contribute, and change as a political fact which it must constantly face. The sphere of that interplay is the World in its political composition, and the related legal principles and rules. The geological context of the Earth, which concerns the interplay between naturally dominant forces that determine the functioning of the Earth System, has remained beyond the focus of change as understood in international law.

This is where the question of the implications of the Anthropocene for international law arises. Resting on concepts of statehood and sovereignty, international law is to a large extent territorially based—and that very basis will soon be undergoing change, due not solely to political reasons as such (where, for instance, one sovereign may sometimes replace the previous one, on the same territory, or a part of it) but owing to the profoundly changing natural conditions of our planet, whereby a sovereign State may find itself permanently left without any territory—as in the case of several low-lying island States in consequence of sea-level rise. The types of changes characteristic of the Anthropocene may become particularly acute for core elements of international law that regulate the territorial divisions of sovereignty and jurisdiction, and related rights to the maritime areas. These reflect the overall stability of environmental conditions in the late Holocene, such as the generally stable sea levels of the past six to seven thousand years. In other words, political implications will be triggered by what have been seen as "non-political" reasons deemed irrelevant in the sphere of international law.

International law thus faces a horizon of change fundamentally different from any since its inception—and this change is underway at two different levels, not one. Changes at the level of the politically conceived World will increasingly interact with changes at the level of the geologically understood Earth, instead of relying on stability there, as in recent centuries. This will make achieving the overarching objectives of international law—facilitating stability in relations between States, avoidance of conflicts between them, and maintaining the peace—an even more daunting task.

A systemic challenge for international law is set to emerge when the changes that are already affecting the Earth System (albeit not yet manifest to the extent of impacting our political systems) will call into question the factual basis of the territorial divisions currently in force, impacting on cross-boundary movements of populations—and ultimately challenging the criteria for statehood as set by international law. Sea-level rise will be the first case to involve all these prospects – and, given recent scientific findings, this may occur within the coming decades.

The scope of the Anthropocene, of course, is far broader than climate change and sea-level rise. However, this issue-area is fundamental to the Earth System, affecting many of its other processes. Further, its effects will increasingly concern such core concepts of international law as statehood and are directly related to the rights of the affected populations.

Achieving a forward-oriented international law of the Anthropocene faces a further problem: our understanding of what international law is does not—and in itself cannot—incorporate the new context of the Anthropocene. Today's international law is a "system of legal rules resting on foundations that evolved under the circumstances of the late Holocene, assumed to be ever-lasting" (Vidas et al., 2015). This will soon create a serious gap between international law as currently understood, and Earth System conditions that have already moved beyond the Holocene. To remain relevant and suited to its functions, international law will have to bridge that gap.

This gap concerns the history of international law as much as its future. At the most general level, the linkages between the development of international law and the emergence of the Anthropocene are twofold. First, there is a linkage of origin: the ideological foundations expressed through the enabling principles and rules contained in core parts of international law—such as the early seventeenth century introduction of the "freedom of the seas" ideology (Grotius, 1609) in the interest of unhindered international trade (and the resultant profits), or the rules regarding acquisition of overseas territories during several centuries following their "discovery" (Anghie, 2005)—facilitated the emergence of forces that led to levels and patterns of development exerting ever-greater human impacts on the Earth System (Vidas, 2011). Second, there are linkages as to the consequences of changes in the Earth System, such as climate change and sea-level rise (Vidas, 2014). How can international law evolve to be able to embrace these consequences, while remaining relevant for the regulation of relations between States and other (perhaps also new) subjects of international law? – That is a key question to which all those who make international law and shape its content will need to keep responding, so that international law may retain its relevance and function also in the Anthropocene.

References

Andrassy, J. (1976). *Međunarodno pravo (International Law)* (6th ed.). Školska knjiga.
Anghie, A. (2005). *Imperialism, sovereignty and the making of international law*. Cambridge University Press.
Crawford, J. (2006). *The creation of states in international law* (2nd ed.). Clarendon Press.
Crawford, J. (2019). *Brownlie's principles of public international law* (9th ed.). Oxford University Press.
Grotius, H. (1609). *Mare Liberum, sive de jure, quod Batavis competit ad Indicana commercial, dissertation*. Elzevir.
Jennings, S. R. (2017). *The Acquisition of Territory in international law* (2nd ed.). Manchester University Press.
Jennings, S. R., & Watts, S. A. (2012). *Oppenheim's international law, Vol. I: Peace* (9th ed.). Longman.
Lowe, V. (2015). *International law: A very short introduction*. Oxford University Press.
Montevideo Convention on Rights and Duties of States. (1933). *League of Nations Treaty Series, 165*, 19.
United Nations Charter. (1945). *United Nations Treaty Series, 1*, 16.
United Nations Convention on the Law of the Sea. (1982). *United Nations Treaty Series, 1833*, 3.
Vidas, D. (2011). The Anthropocene and the international law of the sea. *Philosophical Transactions of the Royal Society – A, 369*(1938), 909–925.
Vidas, D. (2014). Sea-level rise and international law: At the convergence of two epochs. *Climate Law, 4*, 70–84.
Vidas, D., Zalasiewicz, J., & Williams, M. (2015). What is the Anthropocene – And why is it relevant for international law? *Yearbook of International Environmental Law, 25*, 3–23.
Vienna Convention on Consular Relations. (1963). *United Nations Treaty Series, 596*, 261.
Vienna Convention on Diplomatic Relations. (1961). *United Nations Treaty Series, 500*, 95.
Vienna Convention on the Law of Treaties. (1969). *United Nations Treaty Series, 1155*, 331.

Davor Vidas is Research Professor in International Law at the Fridtjof Nansen Institute, Norway, and Honorary Visiting Professor in the School of Geography, Geology and the Environment at the University of Leicester, UK. He is the Chair of the Committee on International Law and Sea Level Rise (International Law Association, 2012–2024), and is an advisory member of the Anthropocene Working Group. Professor Vidas is the founding co-Editor-in-Chief of the short monograph series *Brill Research Perspectives in the Law of the Sea*. Among his recent publications is *International Law and Sea Level Rise* (Brill: 2019).

Just Transition

Alexandre Berthe and Pascale Turquet

Abstract This article analyzes the emergence of the concept of just transition (JT) in the academic and institutional literature in recent years. The concept, put forward at COP 26, can cover different meanings that we analyze in this text. From a general point of view, JT means ensuring that ecological or energy transitions take place in a just way, i.e., by giving an adequate place to each actor in these transitions. The article proposes an overview of the visions of JT in order to understand how actors, industrial firms, unions and politicians position themselves around this new issue.

JT is now at the heart of the international dialogue on climate change. At COP 26, 17 parties, including governments of developed countries, such as France and the United States, signed an international declaration on JT. This declaration is based on the will to encourage low- and middle-income countries to move towards a world without net carbon emissions, by means of a new program based on six axes: support for workers in the transition to new jobs, support for social dialogue, the development of long-term economic strategies, the promotion of local, decent and inclusive jobs and finally, the inclusion of these JT efforts in national reports relating to the subject (COP26, 2021). This declaration is in line with the work previously carried out within the framework of the International Labour Organization (ILO), namely, the report on JT published in 2015 (ILO, 2015). This was based, in particular, on the need to support qualitative green jobs in the future, using the slogan, "no jobs on a dead planet".

A. Berthe (✉)
LiRIS, University Rennes, Rennes, France

LIED, Université de Paris, Paris, France
e-mail: alexandre.berthe@univ-rennes2.fr

P. Turquet
LiRIS, University Rennes, Rennes, France
e-mail: pascale.turquet@univ-rennes2.fr

Today, the country cited most often as having embarked on a JT, in partnership with international organizations and developed countries, is South Africa. This partnership was renewed in November 2021 with new commitments around a just energy transition in the country. The partnership focused, in particular, on the reorientation of the country, so as to exploit new energy resources and limit the dependence of the South African economy on the coal sector. The question now arises of extending this perspective to other developed and developing regions, such as Southeast Asia (Berthe et al., 2022).

To start from a definition that is as comprehensive as possible, the JT can translate into a process of shifting socioeconomic systems, in particular the energy sector, towards production and consumption methods that are more virtuous from an environmental perspective in a fair way. This last point means to ensure that the various populations, within their work activity or not, would not be negatively affected. It should be noted that the concept of JT immediately offers a vision for the future, since the term chosen is that of "transition", thus assimilating to a desired and desirable sidestep, unlike terms such as collapse, decline or progress. Furthermore, the term "just" also conveys the idea that the proposed transition is able to meet the criteria of justice - distributional and procedural justice (involving the participation of citizens and stakeholders) - which are still to be defined. The term chosen is, therefore, quite broad since it does not orientate on the proposed vision of justice, unlike other concepts that could have been used, such as those of equality or inclusiveness. Likewise, when it is used to describe an industrial evolution, for example, the shift from a coal-fired power station to the use of green energies, this term makes invisible those in positions of authority being called upon to define the contours of the transformation, the existence of winners and losers in this transformation and any potential debates on the real environmental benefits of the proposed transition.

From this initial definition, the literature will, therefore, grasp this concept in various ways and is divided, in particular, in relation to three major subjects: the actors at the heart of the implementation of the transition, the more or less significant extension to consider a variety of transition issues and the more or less transformative degree of the visions of justice conveyed.

Regarding the first point, the concept of JT originally emerged from the American and international labour trade union movement and was based on the desire for workers to take their place in the dialogue regarding the closure of coal- and oil-dependent activities to ensure that they would not be liable for the cost of this transition to a new economy (Rosemberg, 2010). As described by Newell and Mulvaney (2013), this concept was developed in response to new regulations to prevent pollution and the resulting plant closures. In such a context, JT aimed to take appropriate measures to protect jobs. This concept was taken up by the UK's Trade Union Congress, as well as the South African trade unions. In 2016, the International Trade Union Confederation and partners established the Just Transition Centre (JTC), which seeks to ensure the participation of labour in the social dialogue, focusing on a JT to a low carbon world (see JTC website). According to the JTC, such a plan must provide better and decent jobs, social protection, more training opportunities and greater job security for workers affected by global warming and climate change policies, extending the concept to social rights and professional careers. JTC follows the ILO Guiding Principles for a Just Transition (2015). Key policy areas are very wide-ranging, including macroeconomic, labour, industrial and sectoral, social protection policies, etc.

Emphasis is placed on social dialogue and tripartism policies. However, social dialogue and tripartism are certainly not the norm everywhere, in a context where the rate of unionization tends to decline. Furthermore, in many developed countries, social bargaining focuses on wages, work conditions, *etc.* and the choice of the productive combination is left to the employers. JT requires that employees are involved in this choice and consequently, in the in-depth transformation of the wage relationship. In the Global South, where JT is seen as a means of fostering social justice, tackling poverty eradication and safeguarding the environment (Just transition Centre and Union to Union, 2020), ambitions go far beyond the capital-labour relationship and aim to tackle development challenges, resting on civil society, including NGOs and foundations. However, recent research on the link between social and environmental policies in Southeast Asia (Berthe et al., 2022) shows that the question rarely arises and that no analysis regarding welfare systems in general, including social and environmental protection within the same approach, has been developed in the region to date.

The question at the heart of JT is whether the modes of transition rely on a bottom-up approach when defining transition policies and not only on a top-down approach. This origin of JT in the trade union movement also directly inscribed the concept from the perspective of the capital/labour relationship in the new modes of decarbonized production. A broader vision of the actors involved is developing today, considering that everyone, regardless of their position in production, must be involved in the JT process. This raises the question of how this concept can be linked to issues of climate, environmental and energy justice (Heffron & McCauley, 2018). The vision allows for more debate to be included, particularly in the context of climate change, which concerns everyone, but presupposes a vision in which we all face the same challenges which can lead to the removal of the political force behind the concept, by distancing us from the question of production, the power relations associated with it and the possible divergence of interests and values.

Regarding the second point, the question of JT originally concerned the potential transition from carbon-based energy to green energy. In this context, JT involves production issues and, therefore, the question of defining new industrial policies in developing countries that are compatible with the social and ecological objectives that have been set. The transition considered can also correspond to any transition, ecological or energy ones, the objective of which is to allow humanity to live within planetary boundaries. This definition is, therefore, close to the perspective of sustainable development, as defined in the Brundtland report: "Sustainable development is development that meets the needs of the present without compromising the ability of future generations to meet their own needs" (WCED, 1987). This vision of sustainable development has often been represented by three spheres, environmental, social and économic, which have had to converge to allow for sustainable development. JT is, therefore, linked to the question of 'liveability' in sustainable development, i.e., the fact of taking into account both the social and environmental dimensions. The question that subsequently arises is the following: how does JT deal with this third economic variable which is abandoned in this concept? Does this mean that growth is assumed as a prerequisite or that the growth paradigm is abandoned? For example, in the case of the ILO (2015), the paradigm is kept, since in this text it is said that governments, in consultation with social partners, should "align economic growth with social and environmental objectives". It is ultimately the question of the radicality of the change associated with JT that is raised.

In relation to the third point, the more or less transformative character of the JT joins the debates around the eco-social state in the thinking constructed by Gough (2013). His three-stage classification of the social-ecological state allows us to understand the extent to which the social issue can modify the prospects for transition. Firstly, social policy as compensation means that social policy could take into account the fact that environmental policies have differentiated social consequences. Secondly, social policy as a co-benefit is to focus on "win-win" strategies, i.e., policies that have positive impacts on both social and environmental dimensions. Finally, eco-social policies take into account context where "win-win" strategies are not enough because of the need to reduce pressure on the environment. Eco-social policies then look at the identification of scenarios that can achieve ecologically beneficial and socially just impacts (Gough, 2013). This leads to the necessity to rethink our vision of consumption and to envision the possibility of social policies without growth. Concerning JT, the question arises as to whether the JT leads to social compensation, to the search for win-win strategies or to a complete redefinition of our development models.

In view of this discussion on the concept of JT, it is easy to see the difficulty facing the academic literature on this theme, as it attempts to position itself between the need to have a precise and unique interdisciplinary concept, and the need to produce a concept that can be appropriated by actors to construct a common narrative in the dialogues between international actors or between populations affected by these transitions. From this perspective of science for action, the question that arises is to what extent can the concept of JT accompany the transitions that are necessary in a context of extreme tensions on many planetary boundaries? This raises the question of the performativity of the concept of JT, which, when attached to a political measure, can immediately conduct to legitimize this policy. In this context, a very diluted concept could enable the promotion of transition policies but would not necessarily enable the selection of the most relevant or desirable policies.

References

Berthe, A., Turquet, P., & Huynh, T. P. L. (2022). Just transition in Southeast Asia: Exploring the links between social protection and environmental policies. *AFD research papers.* Agence Française de Développement.

COP26. (2021). Supporting the Conditions for a Just Transition Internationally. https://ukcop26.org/supporting-the-conditions-for-a-just-transition-internationally/

Gough, I. (2013). Climate change, social policy, and global governance. *Journal of International and Comparative Social Policy, 29*(3), 185–203.

Heffron, R. J., & McCauley, D. (2018). What is the 'just transition'? *Geoforum, 88*, 74–77.

ILO. (2015). *Guidelines for a just transition towards environmentally sustainable economies and societies for all*. International Labour Organization.

Just transition Centre and Union to Union. (2020). *Just Transition in the international development cooperation context report*.

Newell, P., & Mulvaney, D. (2013). The political economy of the just transition. *The Geographical Journal, 179*, 132–140.

Rosemberg, A. (2010). Building a just transition: The linkages between climate change and employment. *International Journal of Labour Research, 2*, 125.

WCED (World Commission on Environment and Development). (1987). *Our common future*. Oxford University Press.

Alexandre Berthe is associate professor at the University of Rennes 2 (LiRIS). He is a specialist in ecological and agricultural economics. His work focuses on territorial dynamics in the context of ecological and energy transitions, especially focusing on biogas production in rural territories and on the inequality-environment nexus.

Pascale Turquet is professor at the University of Rennes 2 (LiRIS). She is a specialist in social protection policies. Her work focuses on health insurance, private insurance and social protection funding, both in France and Europe. More recently, she has taken an interest in studying the extension of social security in Southeast Asia.

Limitations

Jean-Yves Robin and Catherine Nafti

Abstract Any entry into the human condition is not without conditions, it is one of the postulates of psychoanalytic anthropology. The first of them refers to learning about limits. However, this education in otherness requires that authority figures, weakened for several decades, may be able to invest the regulatory function devolved to them. This is an essential issue because it is the only way to stem the hubris that has tipped humanity into a new era, the Anthropocene.

If there is one notion borrowed from the common language and particularly popular in the field of social sciences, it is that of limits. Its Latin etymology (limitem) refers to the idea of boundaries, borders, demarcation lines or thresholds that need or need not being crossed. Here are gathered many metaphors making us think of human condition. For example, in certain circumstances it is urgent to wait, not to go through different stages too quickly, to be patient, as it is the case in scuba diving when it comes to gradually surfacing. It is even vital to respect these decompression thresholds to avoid a major incident. Similarly, any meeting, either diplomatic or romantic, requires that we do not rush negotiations, games of seduction or exchanges; in other words, it is essential to respect certain silences, certain "proximal development zones", to take into account certain preliminaries, certain rituals because both sides consider them essential. These formulas undoubtedly deserve some clarification, they are the basis of our argumentation. The entry into human condition cannot be made without the internalization of a certain number of prohibitions which are used like dykes endorsed by the subject; they allow us to live together, they contain the devastating impulse of a deceiving imaginary which remains inexorably under the influence of the omnipotence (Lebrun, 1997, 2010). At least, it is one of the postulates of psychoanalytical anthropology.

However, if we look at the history of science, it must be said that since the triumph of modernity and hypermodernity, man has never ceased to inexorably push

J.-Y. Robin (✉) · C. Nafti
Catholic University of the West, Angers, France
e-mail: jean-yves.robin@uco.fr; catherine.nafti@uco.fr

N. Wallenhorst, C. Wulf (eds.), *Handbook of the Anthropocene*,
https://doi.org/10.1007/978-3-031-25910-4_100

621

back the limits. He set foot on the moon in 1969, which is an undeniable scientific achievement. Over the past 15 years, researchers have succeeded in decoding many of the characteristics of the human genome. In other words, science has made huge strides forward. So why refrain from leading researches that a priori pursue the project of improving human condition? Such laudable intentions are quite respectable, but are they not also the promise of some disappointments and disenchantments? Still willing to cross demarcation lines, Homo Deus no longer manages to contain himself (Harari, 2017). He is in a way overwhelmed by his unconscious which keeps inviting him to take the path of hubris or excess. By doing so, a precious balance could be upset, simply think of global warming, species extinction, biological invasions and the threats to biodiversity today.

The purpose of this notice is not to list all ecological threats. It is only a question of mentioning certain facts that suggest man has entered a new era, characterized by the disintegration of borders. Thus, the distinction between public and private spaces has become increasingly blurred over time, particularly as a result of digital technology. From now on, it is possible, via social networks, to show parts of our intimate life without any restraint. The introduction of this plot of our "extimity" is in many ways a danger (Tisseron, 2002). Films, photos, comments, selfies can then circulate on the net; they are not systematically compromising but so they can become. Here is the experience of a student in information and communication. At the end of his Master's degree, his application was accepted by the HR manager of a multinational company. There was only a test left: the interview, the great oral in a way, which would decide between the three remaining candidates who would successfully pass the first stage of the selection. How bewildered this Master's student was when he discovered that the company's communications director had been able to consult his Facebook page without any difficulty, and had found a photo of this young man on which he was drunk and naked! From the very start of the interview, he would face a polite but firm refusal which put a definitive end to his project of professional integration within the headquarters of this company: "Sir, when you know how to control your corporate image, come back and see us!" A formula that was similar to a flat no. In this kind of situation, this young candidate could only hope for one thing: benefit from the right to be forgotten.

This disintegration of boundaries can also be seen on the management side. Many managers spend a third of their time every day dealing with their emails. This "infobesity" (Sauvajol-Rialland, 2013) is growing to such an extent that it contributes to the despecialization of time and space. In these conditions, the conciliation between private and professional life is becoming increasingly problematic, precisely because the boundaries between these two fields have broken down over time, because of all the digital means that make it possible to be connected 24 h a day, 7 days a week. This situation can induce a certain number of effects. The executive becomes a stranger to himself and is unable to resonate with his environment, here and now. This "bowing-down" syndrome does not encourage dialogue and exchanges when attention, concentration and sight remain essentially captured by a screen.

In addition, studies show that a confrontation with screens disturbs sleep, internal clocks and circadian rhythms. Why should we be surprised then, as Hartmut Rosa (2010) shows it, that adults have lost an average of 2 h of sleep over the past 20 years or so? Finally, in France, a study conducted in 2015 among 776 young people from the 6th to the 3rd grade in the Paris region shows that 23% of middle school students say they sleep or fall asleep in class, 58% admit that it's difficult for them to get up in the morning, 15% send SMS messages at night, 11% connect with social networks during the same period and 74% immediately check and use their digital tablet or their smartphone if they wake up in the middle of the night. Finally, several studies reveal that the exposure time to screens in France has been increasing steadily over the past 5 years. For example, in 2010, it was 3 h 20 min per day for 15–19-year-olds, in 2015, it was 7 h 54 min (all the data in this paragraph are borrowed from Bihouix & Mauvilly, 2016).

And what about those most extreme cases that no longer distinguish day from night? There are one million of them in Japan, they are called hikikomori (Sauvajol-Rialland, 2013, p. 116). Most of them are young, they spend most of their time in their rooms. They stop all social activities and only consult their computers. Their sleep is reduced to an average of 5 h per night and only urgent needs keep them temporarily away from their screens. From the subject's side, this form of addiction reveals a very problematic relationship with the object. As the Lacanian psychoanalytical anthropology argues, the object that creates desire is not the one that satisfies it. This is why these young people in a situation of digital dependence are similar to these workers compelled to repetition in pursuit of a satisfaction that will never happen.

This disintegration of the boundaries between staff and professionals, private and public, day and night, has become a major political issue in some societies. Indeed, in certain circumstances, digital technology serves security ideals, as it is the case in the People's Republic of China. Within the last 10 years, a growing number of cctv cameras have been developing in urban areas of the Middle Kingdom, directly connected to computers. All the movements of any citizen can then be recorded; if necessary, a facial recognition process makes it possible to identify, almost immediately, a particular individual. For example, when walking across a street without using the zebra crossing, anyone can therefore, if necessary, not only be identified but also sanctioned. This will result in the loss of a few points. In fact, each citizen is awarded between 350 and 950 points. This is what is commonly referred to, under a relatively "sanitized" terminology, as the social credit system. This figure varies for each individual according to criteria related to whether or not they adopt behaviours qualified as "citizens". The fact of benefiting from an excellent score allows you to obtain many advantages: authorization to travel abroad, prime rate bank loans, etc....

Constantly pushing the boundaries, this project, marked by the seal of hypermodernity, establishes a process of deregulation. It is not only democracy that can be threatened, as the previous example suggests, but it is also the characteristics of the human condition that are themselves being challenged. Indeed, any entry into human condition is not unconditional (Lebrun, 2010). The most essential among them goes through learning the "symbolic castration". Yet everything contributes, particularly in the West, to do without this educational modality. It is about listening, hearing, being empathetic, accompanying without giving the impression of leading. The emergence of the "structural mother" is conducive to the development of "limitless states" that are only regulated by prevention measures, coming from the outside, as if the subject no longer had the possibility to internalize a certain number of prohibitions. Thus, as soon as a person invests "a so-called exceptional status" and tries to remind someone of the rules, he or she is more or less threatened or disqualified. Doctors, social workers, constables, police officers, teachers, educators, school heads – and this list is not exhaustive – are regularly questioned, arrested and sometimes even brutalized. It is therefore the question of authority and what makes a third party in the relationship that is being asked. Indeed, which authority can we rely on today when any speech, any offer of service becomes less and less audible and therefore admissible? And yet, the regulation of our production modes and growth marked by the seal of possession remains an imperative necessity. This will inevitably lead to painful decisions at local, regional and global levels. Let's hope that they are not taken by a minority that could lead us into an "Iron Age".

References

Bihouix, P., & Mauvilly, K. (2016). *Le désastre de l'école numérique – plaidoyer pour une école sans écrans*. Seuil.
Harari, Y.-N. (2017). *Homo deus – Une brève histoire de l'avenir*. Albin-Michel.
Lebrun, J.-P. (1997). *Un monde sans limite – Essai pour une clinique psychanalytique du social*. Erès.
Lebrun, J.-P. (2010). *La condition humaine n'est pas sans conditions*. Denoël.
Rosa, H. (2010). Accélération. Une critique sociale du temps. La Découverte. french translation.
Sauvajol-Rialland, C. (2013). *Infobésité – Comprendre et maîtriser la déferlante d'informations*. Vuibert.
Tisseron, S. (2002). *L'intimité surexposée*. Hachette.

Jean-Yves Robin is a psychosociologist. He is a Professor at the Catholic University of West. He is the author of several books on managers and executives. His latest book: Chefs d'établissement – Le burn-out n'est pas une fatalité ! (Le Bord de l'eau, 2022).

Catherine Nafti is a sociologist. She is a lecturer at the Catholic University of West. She is the author of several books on the relationship to knowledge. She is the editor of Actor and Knowledge.

Overshoot

Mathis Wackernagel and David Lin

Abstract Overshoot is the state in which human demand, within a given time period, exceeds the amount ecosystems regenerate in the same time period. Overshoot is possible because the accumulated stock of ecosystems can be depleted. The accumulation of greenhouse gases in the atmosphere is one of the most prominent symptoms of overshoot. But global depletion is not a long-term option. Therefore, global overshoot will inevitably end.

Overshoot is the state in which human demand, within a given time period, exceeds the amount ecosystems regenerate in the same time period (Meadows et al., 1972a; Catton, 1980). It means that demand is larger than what ecosystems regenerate. Humanity has become a depleting force of the biosphere due to global overshoot, and is now a major force shaping the planet's condition. Hence the size of global overshoot is a core parameter of the →*Anthropocene.*

Overshoot is possible because ecosystems' accumulated stocks can be depleted. Examples of overshoot include cutting trees faster than the forest can renew them, or overgrazing pastures. Symptoms of current global overshoot include deforestation, overfishing, groundwater depletion, soil loss, and accumulation of greenhouse gases in the atmosphere. As global depletion is not a long-term option given the limited size of accumulated stocks that can be depleted and limited capacities of waste sinks that can be filled, global overshoot will inevitably end. The question is only whether it ends by design or disaster.

Given that most environmental ills, such as climate change, biodiversity loss, deforestation, freshwater scarcity, and desertification, are ultimately driven by humanity's excess use of the biosphere, it is surprising that overshoot is barely recognized in the public discussion as the key dynamic driving ecological depletion. For instance, the media service Meltwater (2023) finds that throughout all of 2022,

M. Wackernagel (✉) · D. Lin
Global Footprint Network, Geneva, Switzerland
e-mail: mathis@footprintnetwork.org; david.lin@footprintnetwork.org
https://www.footprintnetwork.org

N. Wallenhorst, C. Wulf (eds.), *Handbook of the Anthropocene,*
https://doi.org/10.1007/978-3-031-25910-4_101

625

only 476 news articles on the web mentioned "ecological overshoot". One application of the concept, →Earth Overshoot Day, was referenced in 10 times as many articles: 4770 during the same period. "SDGs" or "sustainable development goals" which are just one, potential insufficient (Wackernagel et al. 2017), strategy to counteract overshoot, was found in 1.46 million articles in that year. "Climate change", which is merely one symptom of overshoot, appeared in 6.43 million articles. This demonstrates that the phenomenon of ecological overshoot as a key driver of environmental trends is vastly underestimated and underappreciated in academia, policy circles, and among the public.

Given its implications for resource security, conflict, and economic stability, ecological overshoot may be one of the most dominant risks to human wellbeing in the twenty-first century. This risk is only surpassed by the self-imposed threat of ignoring it, thereby staying underprepared.

Overshoot, or the difference between what ecosystems can provide and the amount of resources being used, can be measured in different ways: →*Net Primary Productivity* (biomass accumulation over a year) is compared to HANPP or human appropriation of net primary productivity (Vitousek et al., 1986; Haberl et al., 2007), but it is challenging to define a sustainable level of appropriation; →*carrying capacity*, or the number of animal units that can be sustained by a habitat, is compared to actual populations; or →*biocapacity* (regeneration measured in area which is scaled proportionally to its potential net primary productivity) compared to →*ecological footprints*. →*Planetary boundaries* are closely related to overshoot, but measure components of it, not the whole (Steffen et al., 2015). Planetary boundaries indicate conditions that, if breached, would lead to global overshoot. For each of the nine identified dimensions, thresholds are identified which, if exceeded, can lead to the triggering of irreversible shifts in ecosystems.

The dynamics of overshoot have been modelled by various researchers, including in early "predator and prey models" by Alfred Lotka and Vito Volterra in the 1920s (Brauer & Castillo-Chavez, 2000). A prominent computer-based study of global overshoot and its long-term implications was "Limits to Growth" published in 1972 (Meadows et al., 1972b). Others have modelled carrying capacity as a logistical function that asymptotically reaches carrying capacity. However, in the real world, demand can exceed regeneration if feedbacks are weak or delayed, leading to large overshoot. For instance, the feedback of climate change on emissions comes with large time delays, thereby not benefiting from self-corrective forces. In contrast, if feedbacks are strong and relatively fast, even light overshoot might cause some oscillations in the amounts ecosystems can provide to its hosts. If ecological feedback is weak or time delayed, the most likely scenario of overshoot is a gradual (as with forests) or rapid (as with fisheries) reduction of the amount that hosts can harvest. The extent of such a crash following unmanaged overshoot depends on how depleted stocks already are and how much such depletion reduces regeneration rates.

Global overshoot can be observed for carbon emissions alone. For instance, the Intergovernmental Panel on Climate Change (IPCC)'s 2014 report states that a concentration of greenhouse gases in the atmosphere of 450 ppm CO_2 equivalent gives humanity a 66% chance to comply with the Paris Agreement's 2-degree Celsius (2 °C) goal (IPCC, 2014). 450 ppm of a gas means that there is an occurrence of 450 molecules of this gas per million molecules of air (ppm stands for parts per million). CO_2 equivalent is a measure of all relevant greenhouse gases (minus water vapour) expressed in the amount of CO2 that would produce the equivalent amount of warming over a 100-year period. In contrast to those 450 ppm $CO_{2equivalent}$, the National Oceanic and Atmospheric Administration of the United States Department of Commerce (or NOAA) reports that in 2020, our planet's atmosphere already reached a greenhouse gas concentration of 504 ppm CO_2 equivalent (Butler & Montzka, 2021). This indicates that the greenhouse gas dimension alone is already in severe overshoot. This also illustrates the relationship between deficit spending (annual emissions) and the accumulated debt (greenhouse gas concentration in the atmosphere), run somewhat parallel to financial deficit and debt. As with finances, low levels of ecological debt are tolerable, but as ecological deficit spending continues, the debt becomes so large that its impact becomes difficult to manage. For instance, increasing the carbon debt by augmenting greenhouse gas concentrations from 300 to 305 ppm may not have significant consequences, but adding 5 ppm at the current level of greenhouse gas concentrations radically increases the likelihood of runaway climate change (Steffen et al., 2015; Randers & Goluke, 2020).

Taking all competing demands on regeneration into account makes the picture even clearer. Based on UN statistics, adding up those competing demands for regeneration and comparing them to what the planet's ecosystems can regenerate, the National Footprint and Biocapacity Accounts conclude that by 2018, human demand exceeded planetary regeneration by 75% (York University et al., 2022). These →*ecological footprint*-based accounts also estimate the ratio for each country. Also, for countries, it is possible to run a Footprint that is larger than the country's biocapacity. Three mechanisms allow for such a country deficit:

1. Net-importing biocapacity from elsewhere (this means that the ecological footprint embodied in imports exceeds the ecological footprint embodied in exports);
2. Using the global commons (such as in the case of fishing international waters or emitting greenhouse into the global atmosphere); and
3. Overusing or overshooting one's own territorial biocapacity (overharvesting forests or fish stocks, etc.). This would be local overshoot: use of the local ecosystems beyond their ability to regenerate.

At the global level, the ecological deficit is identical to overshoot because the globe as a whole does not trade, and cannot use an extraterritorial "global commons". Therefore, the difference between humanity's ecological footprint and the planet's biocapacity is both the global deficit as well as global overshoot.

Overshoot is maintained by liquidating stocks of ecological resources and accumulating waste, such as greenhouse gases in the atmosphere. Continuous deficit spending accumulates in an ever-larger debt, and one that may not be easily reversible.

To increase the prominence and recognition of overshoot risks, Global Footprint Network runs an annual campaign called *"→Earth Overshoot Day"*. From January 1 to that day, humanity has on aggregate used as much from the biosphere as the biosphere can regenerate in the entire year. According to the media search facility Meltwater.com, Earth Overshoot Day generated over 7 billion media impressions in 2022.

References

Brauer, F., & Castillo-Chavez, C. (2000). *Mathematical models in population biology and epidemiology.* Springer.

Butler, J. H., & Montzka, S. A. (2021). *The NOAA annual greenhouse gas index (AGGI).* NOAA Earth System Research Laboratory. https://gml.noaa.gov/aggi/aggi.html

Catton, W. J. (1980). *Overshoot: The ecological basis of revolutionary change.* University of Illinois Press.

Haberl, H., Heinz Erb, K., Krausmann, F., Gaube, V., Bondeau, A., Plutzar, C., Gingrich, S., Lucht, W., & Fischer-Kowalski, M. (2007). Quantifying and mapping the human appropriation of net primary production in earth's terrestrial ecosystems. *Proceedings of the National Academy of Sciences, 104*(31), 12942–12947. https://doi.org/10.1073/pnas.0704243104

Intergovernmental Panel on Climate Change. (2014). *Climate change 2014: Synthesis report. Contribution of working groups I, II and III to the fifth assessment report of the Intergovernmental Panel on Climate Change* (p. 151). Intergovernmental Panel on Climate Change. https://www.ipcc.ch/site/assets/uploads/2018/02/AR5_SYR_FINAL_SPM.pdf

Meadows, D. H., Meadows, D. L., Randers, J., & Behrens III, W. W. (1972a). *The limits to growth.* Universe Books.

Meadows, D. H., Meadows, D. L., Randers, J., & Behrens III, W. W. (1972b). *The limits to growth: A report for the club of rome's project on the predicament of mankind.* Potomac Associates book. https://www.library.dartmouth.edu/digital/digital-collections/limits-growth

Meltwater. (2023). *Media search engine.* meltwater.com. Accessed January 2023.

Randers, J., & Goluke, U. (2020). An earth system model shows self-sustained thawing of permafrost even if all man-made GHG emissions stop in 2020. *Scientific Reports, 10*, 18456. (2020). https://doi.org/10.1038/s41598-020-75481-z

Steffen, W., Richardson, K., Rockström, J., Cornell, S. E., Fetzer, I., Bennett, E. M., Biggs, R., Carpenter, S. R., de Vries, W., de Wit, C. A., et al. (2015). Planetary boundaries: Guiding human development on a changing planet. *Science, 347*, 1259855.

Vitousek, P. M., Ehrlich, P. R., Ehrlich, A. H., & Matson, P. A. (1986). Human appropriation of the products of photosynthesis. *Bioscience, 36*, 368–373. http://www.jstor.org/stable/1310258

Wackernagel, M., Laurel H., & David, L. (2017). Making the sustainable development goals consistent with sustainability. *Frontiers in Energy Research, 5*, 18. https://doi.org/10.3389/fenrg.2017.00018. http://journal.frontiersin.org/article/10.3389/fenrg.2017.00018/full

York University, Footprint Data Foundation, Global Footprint Network. (2022). *National footprint and biocapacity accounts 2022 edition.* https://data.footprintnetwork.org

Dr. Mathis Wackernagel created the *footprint* concept in the early 1990s with Prof. William E. Rees. The carbon footprint has become the most popular variant. In 2003, he founded Global Footprint Network, a sustainability think-tank, making planetary constraints relevant to decision-making. Its largest engagement campaign is its annual Earth Overshoot Day. Mathis's honors include the 2018 World Sustainability Award, the 2015 IAIA Global Environment Award, and the 2012 Blue Planet Prize.

Dr. David Lin leads Global Footprint Network's research team, and contributes to the production, development, and improvement of the National Footprint and Biocapacity Accounts. Prior to joining Global Footprint Network, David earned his Ph.D. and worked as a post-doctoral researcher in the Systems Ecology Laboratory at the University of Texas at El Paso. His research focused on integrating models of ecosystem function with land cover change analysis in Arctic ecosystems. David is a native of California, and holds a BS in ecology, behavior, and evolution from the University of California, Los Angeles.

Sobriety

Bruno Villalba

Abstract Sobriety has long been a personal ethical choice, as well as a theological imperative. The idea of sobriety hinges on individuals' decisions to renounce the superficial (material goods and ostentation), to focus instead on the essential: developing a mind which is composed and at peace. There have also been political drives for sobriety, largely because of the interweaving of religion and political power. The dawn of the Anthropocene brings sobriety, and its political role, into a completely new light. A sober mindset tells us that we must keep our needs, desires and behaviours proportional to what the planet is able to cope with; in that sense, sobriety becomes a policy of turning one's back on the ceaseless desire to accumulate possessions. This relates to all aspects of social activity (work, energy, leisure, transport, food and much more), globally. The anthropological fundaments of modern existence (individual freedom through material wellbeing) are called into question, with a view to establishing a proportionate relationship with the Earth system.

In Greek, *sophrosunè*, and in Latin, *sobrietas*, the etymology of sobriety refers to *moderation*, as a sensible measure. It is a form of frugality, but one which does not entail a complete and utter renouncement of the material mindset which shapes the world today. Above all, the aim of sobriety is to prevent *hubris* – disproportionate behaviour, intemperance and excess which, according to Aristotle (1999), leads to 'deregulation', the creation of an imbalance in the human body. It is primarily a question of temperance, which is a sign of personal 'moral virtue': by avoiding excess, the individual maintains self-control. By extension, this perception can also apply to the body politic of the people, as is the case in the democratic model. Political regimes should not quest for material wealth at all costs – to do so would be hubris. Instead, they should aim to regulate such wealth sensibly (Aristotle uses the term 'liberality' to denote this idea), which entails moderation in production and balance in the distribution of wealth.

B. Villalba (✉)
AgroParisTech, Printemps, Paris-Saclay, France
e-mail: bruno.villalba@agroparistech.fr

Constructing a Moral Sense of Moderation

The stoicists (Zeno of Citium, Seneca, Epictetus and other leading figures) champion the ideal of a form of frugality, indicative of the wise man's ability to reach a moderate relationship with the world and its excessive wealth (luxury, etc.). The wise man, they teach, focuses on developing internal moral resources – in particular, those which will allow him to come to terms with his own mortality. Stoicism aims to achieve ataraxia – tranquillity of the soul; inner peace. Epicureans (Epicurus, Lucretius, etc.) favour the principle of balance, in opposition to all forms of excess, including excess in the pursuit of moderation. What is important, in their view, is to build self-sufficiency. Indeed, self-sufficiency is the principle pursued by both these schools of philosophical thought, with individuals being endowed with a personal ethical sense of moderation, based on self-knowledge and bringing inner peace.

Echoes of this quest for internal moderation can be heard in the words of certain modern philosophers, such as Jean-Jacques Rousseau (1992). Rousseau rails against what is useless, ostentatious luxury and the harmful imitation thereof which leads society down a rabbit hole, on a never-ending and ludicrous quest, reducing the interest which is paid to public assets and civic virtue.

Monotheistic religions (such as Judaism, Islam and Christianity), along with Buddhism, Taoism and Hinduism frequently call for sobriety. From this point of view, however, sobriety is defined by recognising the omnipotence of the supreme being or entity and, through humility, subjugating oneself to the deity's authority. For example, in Christianity, sobriety represents the desire to focus on the essential: prayer, in the quest to save one's immortal soul. 'The end of all things is near. Therefore be alert and of sober mind so that you may pray' (*First Epistle of Peter*, 4:7) Evangelical poverty is portrayed as a virtue, because it marks compliance with the message of God: the faithful adjust their practices to the divine precepts. Self-denial proves the casting aside of worldly goods and leads Christians to form a closer relationship with other human beings, and with all God's creatures (including animals, according to St Francis of Assisi). Other thinkers, too, embrace the spiritual dimension, questioning the relationship we have with materiality in the modern world. For example, Ralph Waldo Emerson and David H. Thoreau advocate a frugal existence, to rid oneself of the superficiality characteristic of the western world. Recently, Pope Francis, in his encyclical *Laudato si'. On care for our common home* (2015), reaffirms the need to develop 'Christian spirituality [which] proposes growth through sobriety, and the capacity to be happy with little' (The English version uses the term *moderation*). The Christian faith includes 'our sister, Mother Earth', in the name of the shared suffering of humans and the planet. Thus, the encyclical underlines the need to associate the principle of sufficiency with ecological matters, in order to comply with God's wishes.

Finally, as Marshall Sahlins (1972) points out, sobriety is not a purely western moral stance, because the earliest peoples developed the principles of collective self-limitation, aimed at controlling humans' impact on the world, to ensure their own survival.

These approaches (both philosophical and religious) were developed in the context of a *certain* world – a world designed without limits, subject in its entirety to the infinite will of God or of reason. That is to say, the idea of moderation did not arise in the context of limited material resources, marked by the prospect of irreversible change, such as we face today.

Abundance and Freedom

In the modern era, the morality of self-restraint tends to be downplayed. Thanks to the explosion of scientific knowledge and the development of technological might, in modern society, it is possible to produce material goods on an ever-growing scale. The democratic system champions individual freedom. This is not just a simple ethical stance: the concept is constructed on the basis of the accumulation of rights for one and all (political rights, social rights, the right to recogntion, and so forth). These rights ought to lead to constant improvement in material living conditions. The growth of wellbeing for all individuals is a tangible marker of freedom. In implementing this creed, however, liberal democracy allows the development of a right to have one's needs met, and then a right to have one's desires met. That objective can be achieved through industrial productivism. Liberal democratic regimes, just as Marxist regimes once did, continue to negotiate the terms of a social contract based on the constant accumulation of material possessions. The cornucopia is now a reality (Cotgrove, 1982).

This leads to a 'Great Acceleration' (Steffen et al., 2011, 842–867) of the anthropic pressures on the Earth system. World population is constantly growing, and with it, so too are the needs of that population (in terms of energy, food, etc.). What does it matter if the satisfaction of desires is very unequally distributed – if not everyone *yet* has access to optimum comfort (Shove, 2003, 395–418)? The promise of plenty is still felt, and thanks to technological innovation, markets and social policies of redistribution, 1 day, everyone should be able to access that level of living comfort.

Sobriety and the Limits of the Planet

In vain, theorists have tried, and continue to try, to demonstrate the impossibility of reconciling this idea of limitless abundance with the very real limits of the material world. Some advocate sufficiency as the means of a more balanced existence (Erich Fromm, Paul R. Ehrlich, Anne H. Erlich and Duane Elgin, among others). Others seek to rein in our current course of action, in the interests of social justice (including Murray Bookchin and André Gorz). Yet more thinkers seek the conditions to bring about a shared ethos of convivial frugality (Jacques Ellul, Ivan Illich, etc.), and happy coexistence (Pierre Rabhi), which allow humanity and nature to exist harmoniously (Dominique Bourg).

Today, though, in the newly dawned Anthropocene, the nature of sobriety has changed. It is no longer simply a question of catering for an internal need; rather, it is a matter of adapting the way we live in order to cope with irreversible ecological imbalances. According to the 2015 Paris Climate Agreements, we need to decarbonise our economies. In order to do so, we must reduce humanity's carbon footprint to 2 tonnes CO_2 equivalent per capita per year, by 2050. To succeed, we have to reduce the footprint by 7–10% each year for 30 years, and then become totally carbon neutral between 2050 and 2100. Up until now, we have mainly looked at 'green growth' policies, technological innovations (such as nuclear power), transitioning to renewable energies, more efficient means of resource production, etc. Overall, though, these decoupling solutions have not achieved their aims, and merely shift the problem (causing harmful knock-on effects) or delay the implementation of the necessary adaptations. Whilst they do indeed help to reduce carbon emissions, they do not solve the anthropic causes of the constant pressure on the Earth system. The simple fact is that technological efficiency does nothing to rectify the burgeoning needs of the human population, and alternative forms of energy merely shift the balance of our dependency on non-renewable resources.

IPCC's report published on 4 April 2022, for the first time, has an entire chapter given over to how to change individuals' behaviour, calling for demand-management policies ('deep demand reduction, low demand scenarios, reduced demand, demand-side options, and demand-side measures', Sect. 5.3.3). The report focuses on lifestyle choices, institutions and cultural norms, and essentially emphasises the tryptic *Avoid* (consume less of something), *Shift* (substitute one type of consumption for another) and *Improve* (make an existing mode of consumption greener). However, the terms *sobriety*, *soberness*, *restraint* and *sufficiency* do not appear in the Summary for Policymakers, in Chap. 5 of the report entitled 'Demand, services and social aspects of mitigation', or in the *Glossary* (*Annex* VII). More than anything else, the report advocates constructing policies of 'deep demand-side reductions incorporating socio-cultural change and the cascade effects' (Sect. 5.3.3). The objective remains to bring about a 'transition toward high well-being and low-carbon demand societies' (Sect. 5.4), with emphasis being placed on the services provided rather than on the quantity of primary energy available. Although we can plainly see the urgent need to implement measures, and quickly, the report stops short of calling for a sober attitude among humans as a means of achieving that goal. Sobriety has yet to become political capital.

Politicising Sobriety

We find ourselves facing a profound contradiction: ought we to continue, or even ramp up, development-driven policies in the name of social justice and the free market … or ought we, on the other hand, to attempt to control – or, even better, reduce – the negative impact that we are having on the Earth system? In the quest for intragenerational balance, we run the risk of perpetuating certain economic policies, and a certain way of living, which are continually chipping away at the planet's finite, non-renewable resources. We also need to pay attention to intergenerational balance, so that future generations may enjoy ranges of choices equivalent to those which we, ourselves, enjoy. However, our ability to act is significantly constrained by the limits of the planet: no longer is it a matter of answering the call for *the right to have rights* (the political foundation of liberal democracy), but instead, of measuring *what we can still access* (Heinberg, 2007), fairly and sustainably. Indeed, we must begin to spotlight a policy of sobriety which will allow cohesion among human societies, living ecosystems and the Earth system to endure in the long term.

In order to achieve this, it is necessary to enter into a debate as to the terms, objectives and ends of the policies developed on the basis of the principle of sobriety (Princen, 2005). What are the fundamental needs and standards of wellbeing common to everyone, which are compatible with what the planet can sustain (O'Neill et al., 2018, 88–95)? What must we give up in order to keep what is necessary – what is essential? How can we fairly distribute this renouncement of the superfluous, to enable as many people as possible to attain a certain level of comfort? How much is enough (Spengler, 2016, 921–940; Villalba et al. 2018)? Such a debate could help to overcome the distaste for sobriety: it would no longer be seen as reactionary (a step backwards towards more 'traditional' morals), technophobic or inequitable. Whilst the idea of having less may seem counterintuitive and thus attract some resistance, sobriety is actually an inoffensive doctrine, for a variety of reasons. Firstly, it offers quick solutions by which we can begin to address the ecological disasters we face (everyone can easily reduce their ecological impact by assessing the consequences of their behaviour as consumers and professionals). Such an approach would, above all, highlight the extent of individuals' freedom to make decisions. Every individual can determine their own ability to break free of the consumerist trends which aggravate social frustration (Günther Anders, Ivan Illich). Thus, we can examine our own relationship with productivity and the social function of work (André Gorz). Social trends may finally break free from the continuous acceleration (Hartmut Rosa) of our way of life. Sobriety-driven policies would open up vast ranges of potential strategies to adapt to the complex situations we face. Thus, they could potentially free us up to take action in a way unparalleled by the technology-based options (which actually reduce our civic abilities to take action). Sobriety policies would have a different impact on matters of social inequality, because they go beyond the framework of technological redistribution of fairness (*more for more people*), reframing the conditions for a fair situation that takes account of the ecological limits (Alcott, 2008, 770–786). Finally (and this is the most important aspect, all too often overlooked), sobriety-based policies would help shift the focus of political debate, integrating the perspective of the living world, soil, and biodiversity. Sobriety forces us to re-examine our anthropocentric perspective, which currently dominates all over the world, and ensures that non-humans are afforded a duly significant place in the discussions. This eco-centric stance is a logical consequence of a worldview no longer based solely on the wants and needs of humans, but on the complex web of interactions between humans and other terrestrial beings.

Of course, sobriety-based policies can only be developed through democratic debate. The task will involve a great deal of political juggling between irreversible ecological impacts and social inequality. As part of the debate, we can examine and discuss complex potential means of action (such as quotas, rationing, environmental taxes, and so forth). In addition, the negotiations will need to take place at different levels of the decision-making hierarchy, from local to international.

Sobriety is a transition which appears to offer the least drastic way possible of moving from one type of world (one of abundance without consequence) to another, and adapting our societies to exist in a more constrained world. It is both a method and a narrative. As a method, it calls into question the viability of the transitional solutions that are deployed, taking account of the consequences of our decisions in terms of fairness. Thus, sobriety is a social and political process of coordination, of negotiation, aimed at establishing equitable division of the effort required to reduce the consequences of our social practices on ecological balances. As a narrative, it calls into question the very foundations of our political apparatus (institutional dimensions, means of production, individual identity, scientific perspectives, etc.) by looking at them stakly in relation to the horrific reality of ecological meltdown, with no form of sugar-coating. However, it opens up new prospects for individual emancipation and a more balanced, sustainable relationship with the Earth system.

References

Alcott, B. (2008). The sufficiency strategy: Would rich-world frugality lower environmental impact? *Ecological Economics, 64*, 770–786.

Aristotle. (1999). Moderation. In *Nicomachean ethics* (Vol. 13, pp. 1117b–1118b). Batoche Books Kitchener. https://biblehub.com/niv/1_peter/4.htm

Cotgrove, S. (1982). *Catastrophe or cornucopia: The environment, politics and the future*. John Wiley and Sons Publishing.

Francis, P. (2015). *Laudato Si': On care for our common home*, Ed. Our Sunday Visitor.

Heinberg, R. (2007). *Peak everything: Waking up to the century of decline in Earth's resources*. New Society Publishers.

Marshall, S. (1972). *Stone Age Economics*. Tavistock.

O'Neill, D. W., Fanning, A. L., Lamb, W. F., et al. (2018). A good life for all within planetary boundaries. *Nature Sustainability, 1*, 88–95.

Princen, T. (2005). *The logic of sufficiency*. MIT Press.

Rousseau, J.-J. (1992). *Discourse on the origin of inequality*. Hackett Publishing Company.

Shove, E. (2003). Converging conventions of comfort, cleanliness and convenience. *Journal of Consumer Policy, 26*(4), 395–418.

Spengler, L. (2016). Two types of 'enough': sufficiency as minimum and maximum. *Environmental Politics, 25*(5), 921–940.

Steffen, W., Grinevald, J., Crutzen, P., & McNeill, J. (2011). The Anthropocene: Conceptual and historical perspectives. *Philosophical Transactions of the Royal Society, 369*(1938), 842–867.

Villalba B., & Semal, L. (Eds.). (2018). *Energy sobriety. Material constraint, social equity and institutional perspectives*, Ed. Quæ (in French).

Bruno Villalba is Professor of Political Science, at AgroParisTech (Paris) and Member of Printemps UVSQ (CNRS UMR 8085). His areas of research are Political Ecology, Sustainable Development and collapse studies. His research focuses on environmental political theory, notably through analysis of the capacity of the democratic system to reformulate its goals based on environmental constraints. He is the author or co-author of over ten books on politics and ecology. Books (selection): Collapsologists and their enemies (ed. Le Pommier, 2021, in French); Energy Sobriety. Material constraint, social equity and institutional perspectives (with L. Semal, ed. Quæ, 2018, in French) and Political ecology in France (Paris, La Découverte, 2022, in French).

Part IX
The Refusal of Limits, Illusory and Destructive

Capitalocene

Alexander Federau

Abstract The Capitalocene is a critical alternative to the concept of the Anthropocene, conceived to replace it. The article discusses the reasons why scholars from the social sciences and the humanities defend this alternative version of the "human epoch", and dismiss the narrative that comes with the Anthropocene. This terminological quarrel expresses a tension between two postures that are frequently expressed in political and academic debates on environmental issues.

The Capitalocene is a critical alternative to the concept of the Anthropocene (see Anthropocene), which is designed to replace it. It describes the same scientific reality as the Anthropocene, but differs from it by blaming capitalism for the current ecological situation. Whereas the Anthropocene presents damage to the biosphere as a consequence of "human activities", the Capitalocene describes an "age of capital", and argues about the responsibility of the "capitalist system of production". The opposition between these two terms is not simply a terminological quarrel, but expresses a tension between two postures that are frequently expressed in political and academic debates on environmental issues. The first posture thinks of environmental damage in general terms based on the opposition between human beings and nature. For our own good and our own expansion, human beings must "unfortunately" and involuntarily destroy part of nature. This posture can be tinted with optimism - we dominate - or pessimism - human beings cannot help but "stain their nest". The second posture does not give credit to such a general presentation and focuses on concrete cases. For example, it is not humanity that is responsible for the sinking of the oil tanker Erika in 1999 in Brittany and the pollution that followed, but the companies Total and RINA. If humanity as a whole cannot be blamed for maritime pollution, can it be blamed when the processes extend over hundreds or thousands of years, as in the case of the Anthropocene? Yes, according to its main promoters. Its main designer, climate chemist Paul Crutzen, supported by former director of the International Geosphere-Biosphere Programme (IGBP) Will Steffen and historians Jacques Grinevald and

A. Federau (✉)
Mobilidée, Geneva, Switzerland

John McNeill, have repeatedly described the entry into the Anthropocene as a consequence of humanity's growing energy needs. (Steffen et al., 2007, 2011). Their energy story is a series of stages, each with an ever-increasing environmental impact. It begins with the mastery of fire, the first technique to have had a significant impact on the environment. Then they move directly to the steam engine, then to the atom, and end up speculating on a total control of large biogeochemical cycles using geoengineering techniques. This story proposes the steam engine as the "official" beginning of the Anthropocene, not because of its direct link to the birth of capitalism, but because it is a pivotal moment in the history of energy. This synthesis of human history has been strongly criticized by historians (Bonneuil & Fressoz, 2013; Malm & Hornborg, 2014; Campagne 2017), for its historical weakness and the misleading interpretation that emerges from it. According to these scholars, the Anthropocene narrative is historically weak because, by presenting humanity as a homogenous block, it ignores the socio-historical context that led to some key decisions. It is fallacious because it amalgamates historical, economic, sociological and political processes under the heading of "human activities". From there, responsibility is diluted collectively and can only be attributed to human nature. What these historians show is that this simplification is historically untenable and politically dangerous.

It is indeed questionable to attribute responsibility to all mankind, since historical and current responsibilities are very unequal. An inhabitant of Qatar emits a thousand times as much greenhouse gas (GHG) as a Malian, Nepalese or Tanzanian. Similarly, the majority of historical fossil emissions have been emitted by developed countries over the last two centuries. The United States alone is responsible for 27% of historical fossil emissions. At the same time, Great Britain emitted more than 5% of GHGs while India is responsible for only 3% of the total (Boden et al., 2010). In the interpretation of historian Andreas Malm (Malm, 2015, 2016, 2017), capitalist elites used fossil fuels to enslave the working class and colonize part of the world. Far from being a subject of consensus, these economic decisions were the source of social conflict and were made against the population. For example, although coal was known in India for centuries and was used on a small scale, it was not until the arrival of the British that it was systematically and extensively exploited. According to journalist and essayist Naomi Klein, the majority of humanity does not want to resort to hydrocarbons, as the establishment of large oil and gas groups is met with universal hostility (Klein, 2015). Moreover, the unlimited nature of capital accumulation explains the progressive intensification of environmental destruction and its current severity (Fischbach, 2009; Patel et al., 2018). While this argument convinces many social scientists, the narrative of the Capitalocene in turn raises some difficulties. For example, how valid is this analysis in the case of China? Now the world's largest emitter of GHGs, soon to be the largest historical emitter, can this country be assimilated to the capitalist system, given its current role in the global market system and the influx of foreign capital? And what about other communist countries? It is well known that the ecological balance sheet of the planned economies of the USSR and the Eastern countries of the twentieth century is not more glorious than that of the capitalist economies. These socialist economies were as much based on fossil fuels as their capitalist counterparts. Their environmental record was just as bad as that of Western countries.

Thus, if the analysis offered by the Capitalocene is convincing, it is also distinctive. The Capitalocene is a synecdoche in the sense that, while capitalism does indeed bear responsibility for the entry into the Anthropocene, it shares it with Marxist communism. What they have in common is that both are systems which make the growth of production their primary objective. However, all productivist thinking, whether capitalist or socialist, begins by postulating the existence of a planet and nature made up of infinite resources to be exploited. By definition, its goal is infinite growth, through the never-ending quest to maximize the productivity of its economy. The logic of productivism can only be overcome by negating its assumptions. This is the innovative contribution of ecological thinking, which reminds us *a contrario* that planet Earth is a fragile and finite planet, and that infinite growth on a finite planet is impossible. In this interpretation, the major teaching of the Capitalocene is not only an indictment of a mode of production, but above all a warning of the vital importance of conserving a hospitable planet for the flourishing of humanity. By defusing the deceptive and paralyzing discourse on the responsibility of the human species, the Capitalocene requires and promotes political action. Yet if a sustainable society can only exist if it is equitable, it will not be achieved by starting from the socialist recipes of the twentieth century, but by innovative structural changes that preserve planetary limits.

References

Boden, T. A., Marland, G., & Andres, R. J. (2010). *Global, regional, and national fossil-fuel CO_2 emissions*. Oak ridge national laboratory. http://cdiac.ornl.gov/trends/emis/overview_2007.html

Bonneuil, C., & Fressoz, J.-B. (2013). *L'événement Anthropocène: La Terre, l'histoire et Nous*. Seuil.

Campagne, A. (2017). *Le capitalocène: aux racines historiques du dérèglement climatique*. Éditions Divergences.

Fischbach, F. (2009). *Sans Objet: Capitalisme, Subjectivité, Aliénation. Problèmes et Controverses*. Vrin.

Klein, N. (2015). *This changes everything: Capitalism vs. the climate*. Penguin Books.

Malm, A. (2015). The Anthropocene myth. *Jacobin*. March 30, 2015. http://jacobinmag.com/2015/03/anthropocene-capitalism-climate-change/

Malm, A. (2016). *Fossil capital: The rise of steam power and the roots of global warming*. Verso.

Malm, A. (2017). *L'anthropocène contre l'histoire: Le réchauffement climatique à l'ère du capital* (Etienne Dobenesque, Trans.). La Fabrique.

Malm, A., & Hornborg, A. (2014). The geology of mankind? A critique of the Anthropocene narrative. *The Anthropocene Review, 1*, 62–69.

Patel, R., Moore, J. W., & Vesperini, P. (2018). *Comment notre monde est devenu cheap: une histoire inquiète de l'humanité*. Flammarion.

Steffen, W., Crutzen, P. J., & McNeill, J. R. (2007). The Anthropocene: Are humans now overwhelming the great forces of nature? *Ambio: A Journal of the Human Environment, 36*(8), 614–621. https://doi.org/10.1579/0044-7447(2007)36[614:TAAHNO]2.0.CO;2

Steffen, W., Grinevald, J., Crutzen, P. J., & McNeill, J. R. (2011). The Anthropocene: Conceptual and historical perspectives. *Philosophical Transactions of the Royal Society A: Mathematical, Physical and Engineering Sciences, 369*(1938), 842–867.

Alexander Federau is Mobility Consultant at Mobilidée. PhD in Philosophy and Environmental Sciences. Author of an Anthropocene philosophy. Alexander is a specialist in environmental ethics and has worked on the concept of nature. His interest focuses on behavioural changes related to ecological transition.

Denialism

Mikael Karlsson

Abstract This article describes the phenomenon of science denialism with a focus on the denial of climate science. It is evident from much research that science denial causes serious problems in coping with environmental change in the Anthropocene. However, while the reasons and characteristics of denialism have been detailed, straightforward answers on how to counteract science denial are missing. The article still offers some ideas in this respect.

When Swedish Television reviewed the well-received film *Don't look up* by interviewing an astronomer about the probability of a giant comet colliding with Earth, completely missing the film's unsubtle allegory of climate science denial in Trump's America, the unconscious denial of denialism reached a new level. This misunderstanding eloquently illustrates a tragic dilemma of the Anthropocene: while science and its technological extensions have enabled humanity to be a telluric force, widespread denial of science seriously impedes transformations that would allow societies to embark on a sustainable route.

This article describes the phenomenon of denialism and in particular science denial related to some of the most pressing planetary challenges in the Anthropocene. It moreover highlights some of the ideas put forward by researchers on how to cope with denialism, particularly when aiming for environmental societal transformation.

A number of terms and concepts are used in the research on various forms of misrepresentations of science. "Scepticism" is one such term, yet with a problematic connotation since it signifies a genuine scientific attitude and tradition. Science denial, by contrast, is the dogmatic opposition to the self-critical and reflective mindset that researchers – and preferably also citizens – should have. In many ways, denialism is a type of pseudoscience, concerned with issues in the scientific domain,

M. Karlsson (✉)
Climate Change Leadership, Uppsala University, Uppsala, Sweden
e-mail: mikael.karlsson@geo.uu.se

© The Author(s), under exclusive license to Springer Nature
Switzerland AG 2023
N. Wallenhorst, C. Wulf (eds.), *Handbook of the Anthropocene*,
https://doi.org/10.1007/978-3-031-25910-4_104

but suffering from lack of evidence and reliability, but denialism implies confronting science in contrast with pseudotheoretical promotion, such as astrology (Hansson, 2017). At the tip of this iceberg of problems is the deliberate, organised disinformation campaigns by denialists, a far more insidious phenomenon than whether laymen are unsure of whom to trust on complex matters. In the context of the Anthropocene, organised forms of denial delays the achievement of various sustainable development goals (Karlsson & Gilek, 2020). The "merchants of doubt" have unfortunately been quite effective (Oreskes & Conway, 2010). Today, denialism is perhaps most conspicuous in relation to science on climate change and climate governance (Edvardsson Björnberg et al., 2017).

Climate science denial takes different forms; *literally* by rejecting the evidence (IPCC, 2021) that global warming takes place, mainly due to human activities; *interpretative*, by accepting that such changes take place but refuting the evidence (IPCC, 2022a) on their negative consequences; and *implicatory*, by opposing the implications that reasonably follow the evidence in relation to adopted policy objectives (Cohen, 2001; Rahmstorf, 2004), for example by telling fables on solutions with little or no support in the science on climate mitigation (IPCC, 2022b). Science denialism reaches further than the climate crisis though, and targets other environmental risks such as emissions of hazardous chemicals (Karlsson, 2019), as well as other scientific issues, from the harm on health of tobacco (Oreskes & Conway, 2010) to the wonders of vaccines (Pierri et al., 2022), not to mention the horrific denial of the Holocaust (UN, 2022).

The strategies used by deniers fill a large palette. Common methods include communication of conspiracy theories, reliance on fake and often self-appointed experts, presentation of selectively picked or isolated data masquerading as comprehensive evidence, placing impossible expectations of what research and researchers can deliver, and defence of outright logical fallacies (Diethelm & McKee, 2009). It is not uncommon that both research and researchers are harshly criticised on nonsubstantial ground, sometimes even being threatened.

The foundations of science denial in the environmental field are well studied (Edvardsson Björnberg et al., 2017). The explanations presented include various psychological (Stoknes, 2014) and social factors, among the latter the elite cues hypothesis (McCright & Dunlap, 2011; Lewandowsky, 2021). At the bottom though, lies a combination of individualistic, conservative and often evangelical values and worldviews (Gauchat, 2015; Lewandowsky, 2021). Denialism is moreover pumped up by a well-oiled "denial machine" in some countries, most notably the U.S. (Dunlap, 2013), and amplified by the so-called media balancing principle, according to which an opponent to a statement ideally should always be given airtime, even when the messenger presents a case that is completely incompatible with the scientific consensus (Boykoff, 2013).

What can be done to counteract denialism, then? On that question, science is less conclusive (Edvardsson Björnberg et al., 2017). In the climate case, besides a common call for more education and improved communication – from prebunking to debunking misinformation and disinformation – studies point in different directions and the intentions behind counteracting measures vary widely, from science advocacy and enhancing climate literacy to supporting mitigation and opposing vested interests (Mendy & Karlsson, 2022). Among the strategies proposed for countering science denial, some researchers consider presenting emotional narratives (Rode et al., 2021) and using deliberative approaches and processes (Dryzek & Lo, 2015) as promising, whereas others instead recommend to frame messages in terms of security and economy (O'Sullivan & Emmelhainz, 2014), or to focus on describing co-benefits when advocating climate mitigation strategies (Bain et al., 2016; Karlsson et al., 2020). Part of the answer is thus that the context in question matters, even though the jury is out on when a specific measure might work and how.

At the same time, there are positive experiences of how to deal with other forms of denialism, and from confronting pseudoscientific theory promotion (Hansson, 2017). A general recommendation is to refuse to accept the agenda that is pushed forward by denialists, and instead seek to reveal deniers' motives, funding and activities – and to contrast that with a clear description of how research is conducted and what signifies scientific knowledge. Here though, care must be taken to avoid blaming the victims of misinformation and disinformation, and to never picture science as flawless and free from values (Hansson, 2020), since that is simply not the case. In parallel, it is of key importance to continue developing, identifying and communicating scientific consensus. Considering these insights and recommendations, denialism may after all be counteracted.

Zooming out, however, the question remains unanswered about how much humanity as a collective has really understood. Perhaps the largest denial of all is that of the Anthropocene itself: that while there is general worldwide acknowledgement of the precarious situation in which humanity has placed itself, considering the planetary crisis, with unprecedented rates of species extinction and global temperature rise, a metaimplicatory denial of these realities seems obvious and persistent. From this perspective humanity does not seem to "look up" at all, making the Swedish Television misinterpretation of the referred film quite telling for present times.

Nevertheless, although science denial endures and causes critical delay in environmental goal achievement, and perhaps even obscures the planetary dilemma humanity is confronted with at large, it is clear that the solid scientific picture of the climate crisis – and of environmental problems in general – has gained more and more recognition around the world in recent decades. With this recognition, new policies are developed and nothing actually speaks against the idea that humanity eventually will realise and cope with the challenges of Anthropocene.

References

Bain, P. G., Milfont, T. L., Kashima, Y., Bilewicz, M., Doron, G., Garðarsdóttir, R. B., et al. (2016). Co-benefits of addressing climate change can motivate action around the world. *Nature Climate Change, 6*, 154–157.

Boykoff, M. T. (2013). Public enemy no. 1? Understanding media representations of outlier views on climate change. *The American Behavioral Scientist, 57*, 796–817.

Cohen, S. (2001). *States of Denial: Knowing about atrocities and suffering*. Polity Press.

Diethelm, P., & McKee, M. (2009). Denialism: What is it and how should scientists respond. *European Journal of Public Health, 19*, 2–4.

Dunlap, R. E. (2013). Climate change skepticism and denial: An introduction. *The American Behavioral Scientist, 57*, 691–698.

Dryzek, J. S., & Lo, A. Y. (2015). Reason and rhetoric in climate communication. *Environmental Politics, 24*, 1–16.

Edvardsson Björnberg, K., Karlsson, M., Hansson, S. O., & Gilek, M. (2017). Climate and environmental science denial. A review of the scientific literature published in 1990–2015. *Journal of Cleaner Production, 167*, 229–241.

Gauchat, G. (2015). The political context of science in the United States: Public acceptance of evidence-based policy and science funding. *Social Forces, 94*, 723–746.

Hansson, S. O. (2017). Science denial as a form of pseudoscience. *Studies in History and Philosophy of Science, 63*, 39–47.

Hansson, S. O. (2020). How not to defend science. A Decalogue for science defenders. *Disputatio. Philosophical Research Bulletin, 9*, 13.

IPCC. (2021). *The pyhisical science basis. AR6 WGI*. IPCC.

IPCC. (2022a). *Impacts, adaptation and vulnerability. AR6 WGII*. IPCC.

IPCC. (2022b). *Mitigation of climate change. AR6 WGIII*. IPCC.

Karlsson, M. (2019). Chemicals Denial—A challenge to science and policy. *Sustainability, 11*, 4785.

Karlsson, M., & Gilek, M. (2020). Mind the gap: Coping with delay in environmental governance. *Ambio, 49*(5), 1067–1075.

Karlsson, M., Alfredsson, E., & Westling, N. (2020). Climate policy co-benefits: A review. *Climate Policy, 20*, 292–316.

Lewandowsky, S. (2021). Climate change disinformation and how to combat it. *Annual Review of Public Health, 42*, 1–21.

McCright, A. M., & Dunlap, R. E. (2011). The politicization of climate change and polarization in the American Public's views of global warming, 2001–2010. *The Sociological Quarterly, 52*, 155–194.

Mendy, L., & Karlsson, M. (2022). Towards a schematic of responses to climate science denial – A review. Paper to the Nordic Environmental Social Science Research Conference (NESS), Göteborg 7–9 June 2022.

Oreskes, N., & Conway, E. M. (2010). *Merchants of doubt: How a handful of scientists obscured the truth on issues from tobacco smoke to global warming*. Bloomsbury Press.

O'Sullivan, T. M., & Emmelhainz, R. (2014). Reframing the climate change debate to better leverage policy change: An analysis of public opinion and political psychology. *Journal of Homeland Security and Emergency Management, 11*, 317–336.

Pierri, F., et al. (2022). Online misinformation is linked to early COVID-19 vaccination hesitancy and refusal. *Scientific Reports*. https://doi.org/10.1038/s41598-022-10070-w

Rahmstorf, S. (2004). *The climate sceptics*. Potsdam Institute for Climate Impact Research. Available from: http://www.pik-potsdam.de/~stefan/Publications/Other/rahmstorf_climate_sceptics_2004.pdf

Rode, J. B., Dent, A. L., Benedict, C. N., Brosnahan, D. B., Martinez R. L., & Ditto P. H. (2021). Influencing climate change attitudes in the United States: A systematic review and meta-analysis. *Journal of Environmental Psychology, 76*, 101623.

Stoknes, P. E. (2014). Rethinking climate communications and the "psychological climate paradox". *Energy Research and Social Science, 1*, 161–170.

UN. (2022). *Holocaust denial*. UN General Assembly A/76/L.30.

Mikael Karlsson is Associate Professor in Environmental Science and research leader in Climate Change Leadership at Uppsala University, Sweden. Dr. Karlsson has worked with environmental issues for 30 years. His present research concerns climate, energy and biodiversity governance, with a focus on science-policy interactions, including science denial and decision-making. Dr. Karlsson is frequently consulted as expert by the Swedish government and the European Commission. He was previously President of Sweden's and Europe's largest environmental NGOs for over a decade.

Doughnut

Christian Arnsperger and Julia K. Steinberger

Abstract This article presents the basic building blocks, as well as the main implications, of Kate Raworth's "Doughnut economics," arguing that it is an essential tool for navigating the Anthropocene and for understanding what variables, both ecological and social, need to be adjusted and by how much. We show that Raworth's "growth agnosticism" is not as problematic as it might appear, and we offer some elements of reflection on whether "the Doughnut" leads to post-capitalism.

One of the hallmarks of the Anthropocene is the persistent breaching of vital thresholds of the Earth system by human activity. This has been recognized for quite some time (see e.g. Meadows et al., 1972), but at the start of the twenty-first century the most widespread way, by far, of representing this breach of thresholds is the so-called "planetary boundaries" model. Devised by scientific ecologist Johan Rockström and his colleagues (Rockström et al., 2009; Steffen et al., 2015), it allows to visualize in one disk-shaped graph the extent to which certain variables, measuring crucial Earth-system processes that affect the habitability of the Earth system—climate change, change in biosphere integrity, stratospheric ozone depletion, ocean acidification, biochemical flows, land-system change, freshwater use, atmospheric aerosol loading and chemical pollution—are being, or have already been, pushed structurally beyond viable threshold values. Since each process is connected to and affected by the other processes, these nine variables are a model for the *system* formed by the biosphere as it functions to support life on planet Earth. The central area of the disk, where all variables remain below their respective critical thresholds, is called by Rockström and his colleagues "the safe operating space for humanity," i.e., the space in which human activities would be safe for both non-human and human life, from the ecological and material-flows perspective.

C. Arnsperger (✉) · J. K. Steinberger
University of Lausanne, Lausanne, Switzerland
e-mail: christian.arnsperger@unil.ch; julia.steinberger@unil.ch

N. Wallenhorst, C. Wulf (eds.), *Handbook of the Anthropocene*,
https://doi.org/10.1007/978-3-031-25910-4_105

651

This focus on the biogeochemical aspects of our planet's habitability is, of course, extremely important. It provides and makes graphically visible what Eduard Suess and Vladimir Vernadsky called the biosphere: there is an *ecological ceiling* "beyond" which life on Earth becomes impossible. It has led Earth scientists such as Rockström to squarely acknowledge that human activities, which in the Anthropocene have been driven by a relentless quest for economic growth and the planetary expansion of resource extraction, are literally "bankrupting Nature" (Wijkman & Rockström, 2012). Such a graphic formulation is strongly congruent with the long-standing claims of ecological economics and of "degrowth" thinkers: As long as both economists and decisionmakers neglect the crucial planetary-scale information about physical flows coming from Earth science, they will grossly underestimate the massive damages and social costs caused by growth-oriented human activities (Arnsperger, 2023).

In direct response to the biogeochemical knowledge about planetary boundaries, Wijkman and Rockström (2012) call for a circular economy and for limiting economic growth. However, their outlook on systemic economic change remains rather simplistic, and they end up advocating for little more than green growth and absolute decoupling. In other words, they remain rather squarely within the mainstream faith that the economy can be made innocuous for the biosphere through appropriate technological means.

As emphasized by Pirgmaier and Steinberger (2019), the fact that many geoscientists as well as ecological economists position themselves with respect to economic growth as the core problem of the sphere of human activity, and therefore look mainly at geophysical flows and their necessary limitation, leads them to view "the economy" as an abstract, disembodied entity. In actual fact, the Anthropocene's growth dynamics has been driven mainly by one very specific system: capitalism. It is capitalism, with its globalizing force, its colonialist drive, its unabated extractivity and its entrenched tendency towards ever-widening inequality (Bellamy Foster & Burkett, 2016; Moore, 2016; Hickel, 2018), that has generated the growth that those focusing on geophysical planetary boundaries rightly find so problematic. In other words, truly caring about planetary boundaries and how to remain within a "safe operating space for humanity" requires critiquing capitalism and looking for post-capitalist alternatives.

This realization brings with it a crucial implication: not only has capitalism driven humanity as a whole, but mainly the wealthy, Western and Westernized portion of humanity, above the *ecological ceiling*—it has also pushed humanity as a whole, but mainly the poorer or impoverished, colonized and exploited portion of humanity, below a *social foundation*. In other words, in the Anthropocene "the economy" has failed humanity on both counts: we face ecological overshoot together with social shortfall.

This fact has long been a staple of ecological economics and its emphasis on ecological scale and fair distribution, but (again) at the start of the twenty-first century the most widespread way, by far, of representing this twofold failure is the so-called "Doughnut" model. It has enjoyed stunning success and has become one of the most popular graphics in the contemporary discussion on how to create an economy that is both environmentally sound and socially equitable. Created by former Oxfam economist Kate Raworth (2012, 2017), it builds critically on Rockström's planetary boundaries and adds a number of social, economic, political and cultural variables: access to sufficient and good-quality energy, water and food; access to basic health care and education; income and gender equality as well as the availability of fulfilling work; the ability to live in a peaceful, equitable and just society as well as to exercise one's political voice; and access to decent housing as well as sufficient social networks.

For each of these variables (a number of which coincide with time-honored United Nations criteria, as set out e.g. in the Human Development Index or the Sustainable Development Goals), indicators are constructed and critical thresholds are put forward. Raworth does not claim that these thresholds are indisputable; they are subject to debate in their own specific way, just like the geophysical thresholds computed by Rockström and his colleagues. Once the thresholds are established, they define the social foundation below which a human existence cannot be considered decent and minimally fulfilling. Between the ecological ceiling and the social foundation, there is what Raworth calls the "safe and just space for humanity," which because of its specific shape has become widely known as "the Doughnut."

The task of any economy is to function in such a way that it allows the human community in question—whether humanity as a whole (Rockström & Gaffney, 2021) or, for instance, a single country (Stratford & O'Neill, 2020) or city (Doughnut Economics Action Lab, 2020)—to live permanently within the Doughnut. In order to do this, both the ecological overshoot and the social shortfall must be remedied at the same time, which is a particularly apt way of fundamentally characterizing the notion of degrowth. As Raworth has emphasized, what is needed is not merely absolute decoupling—which capitalist "green growth" is in any case incapable of attaining—but "sufficiently absolute" decoupling that makes us go back inside the Doughnut and remain there forever.

The Doughnut model can be seen as a diagnostic tool as well as a therapeutic device: it allows to visualize what needs to be done and by how much different variables need to be adjusted downward (for the ecological thresholds) or upward (for the social ones). It does not *per se* prescribe a specific alternative economic model, and Raworth herself has argued that Doughnut economics is not anti-growth, but merely "growth agnostic": it rejects the centrality of economic growth as a measure of a society's performance, but does not reject the occasional need for growth in certain sectors or over certain periods, if this actually helps to stay within the Doughnut.

Is the Doughnut model post-capitalist? This is difficult to say on the sole basis of Raworth's own statements. Some have criticized her for it (Teicher, 2021), to no avail in our opinion: The alternative vision of economics and of human agency that Doughnut economics advances (for instance, a radical change in how humans interact socially and how they seek and find existential meaning) do, indeed, point in the direction of post-capitalism. A number of recent research papers have investigated the potential for the Doughnut model to offer insights and guidelines as to how reorganize our societies and economies. Using indicators designed to measure a "safe and just" development space, O'Neill et al. (2018) have shown that no country meets basic needs for its citizens at a globally sustainable level of resource use. They argue that strategies to improve physical and social provisioning systems, with a focus on sufficiency and equity, have the potential to move nations towards sustainability. With the help of another variant of Raworth's approach, Millward-Hopkins et al. (2020) have demonstrated that global final energy consumption in 2050 could be reduced to the levels of the 1960s despite a tripling of population, but that this would require a massive rollout of advanced technologies across all sectors, as well as radical demand-side changes in order to reduce consumption to levels of sufficiency. It is highly doubtful that these changes would be, in any shape or form, compatible with capitalism.

References

Arnsperger, C. (2023). Growth. In N. Wallenhorst and Ch. Wulf (Eds.), *Handbook of the anthropocene*. Springer.

Doughnut Economics Action Lab. (2020). *The Amsterdam City doughnut: A tool for transformative action*. Available at https://www.kateraworth.com/wp/wp-content/uploads/2020/04/20200406-AMS-portrait-EN-Single-page-web-420x210mm.pdf

Bellamy Foster, J., & Burkett, P. (2016). *Marx and the earth: An anti-critique*. Haymarket.

Hickel, J. (2018). *The divide: A brief guide to global inequality and its solutions*. Penguin.

Meadows, D., Meadows, D., Randers, J., & Behrens, W. (1972). *The limits to growth*. Potomac Associates.

Millward-Hopkins, J., Steinberger, J.K., Rao, N., & Oswald, Y. (2020). Providing decent living with minimum energy: A global scenario. *Global Environmental Change, 65*, 102168.

Moore, J. (Ed.). (2016). *Anthropocene or Capitalocene? Nature, history and the crisis of capitalism*. PM Press.

O'Neill, D., Fanning, A., Lamb, W., & Steinberger, J.K. (2018). A good life for all within planetary boundaries. *Nature Sustainability, 1*(2), 88–95.

Pirgmaier, E., & Steinberger, J.K. (2019). Roots, riots and radical change—A road less travelled for ecological economics. *Sustainability, 11*. https://doi.org/10.3390/su11072001

Raworth, K. (2012). *A safe and just space for humanity: Can we live within the doughnut?* Oxfam Discussion Paper. Available at https://www-cdn.oxfam.org/s3fs-public/file_attachments/dp-a-safe-and-just-space-for-humanity-130212-en_5.pdf

Raworth, K. (2017). *Doughnut economics: Seven ways to think like a 21st-century economist*. Penguin.

Rockström, J., & Gaffney, O. (2021). *Breaking boundaries: The science of our planet*. Dorling Kindersley.

Rockström, J., Steffen, W., Noone, K., Persson, Å., Chapin, S., et al. (2009). Planetary boundaries: Exploring the safe operating space for humanity. *Ecology and Society, 14*(2), 32.

Steffen, W., Richardson, K., Rockström, J., et al. (2015). Planetary boundaries: Guiding human development on a changing planet. *Science, 347*, 1259855.

Stratford, B., & O'Neill, D. (2020). *The UK's path to a doughnut-shaped recovery*. University of Leeds. Available at https://goodlife.leeds.ac.uk/doughnut-shaped-recovery

Teicher, J. (2021, April 24). Doughnut economics has a hole at its core. Jacobin. Available at https://www.jacobinmag.com/2021/09/doughnut-economics-raworth-amsterdam-capitalism-socialism

Wijkman, A., & Rockström, J. (2012). *Bankrupting nature: Denying our planetary boundaries*. Earthscan.

Christian Arnsperger is professor of sustainability and economic anthropology at the University of Lausanne. He holds a PhD in economics from the University of Louvain (Belgium) and has also been active as a scientific advisor to the Alternative Bank Switzerland. He publishes widely on the critique of economic growth, the existential underpinnings of capitalism, and the links between money and sustainability. His latest books are *Écologie intégrale: Pour une société permacirculaire* (with Dominique Bourg, PUF, 2017) and *L'existence écologique: Critique existentielle de la croissance et anthropologie de l'après-croissance* (Seuil, 2023).

Julia K. Steinberger is professor of ecological economics at the University of Lausanne and is an author of the sixth Assessment Report of the Intergovernmental Panel on Climate Change (IPCC). She holds a PhD in physics from the Massachusetts Institute of Technology (USA). She publishes widely on the complex systemic relationships between the use of resources and the performance of societies. Her latest publications as co-author include "Scientists' warning on affluence" (*Nature Communications*, 2020) and "Socio-economic conditions for satisfying human needs at low energy use: an international analysis of social provisioning" (*Global Environmental Change*, 2021).

Earth Stewardship

Agnès Sinaï

Abstract The notion of earth stewardship implies the idea of holistic management of ecosystems. This approach ranges from technologies such as geo-engineering to local and bioregional management of soil and resources. The idea of earth stewardship implies a reflection on the interdependencies between humans and all living things. How can we best conserve the earth?

In 2015, Will Steffen et al. published in the journal *Ambio* an article on the ways to ensure planetary stewardship. According to them, the advent of the Anthropocene suggests that we need to fundamentally change our relationship with the planet we inhabit. Many approaches could be taken, ranging from geoengineered solutions that deliberately manipulate parts of the Earth system, to becoming active stewards of our own survival system. The Anthropocene is a reminder that the Holocene, during which complex human societies developed, was a stable and accommodating environment and is the only state of the earth system that we know for sure can support contemporary society. There is an urgent need to achieve effective planetary stewardship. As we move further into the Anthropocene, we risk leading the Earth system on a course to more hostile states from which we cannot easily return. The article calls for the mobilization of the concept of socio-ecological systems as the human enterprise is now a fully coupled and interacting component of the earth system itself.

For the co-authors of this article, all specialists in the issues of the Anthropocene, the magnitude, speed and complexity of the challenges of the twenty-first century suggest that responses based on marginal changes in the current trajectory of human enterprise risk leading to the collapse of large segments of the human population or of contemporary globalized society as a whole. More transformational approaches may be needed. Geoengineering and reducing human pressure on the Earth system at its source represent the end points of the spectrum in terms of philosophies, ethics and strategies.

A. Sinaï (✉)
Sciences Po, Paris, France
e-mail: Agnes.sinai@sciencespo.fr

657

At the other end of the spectrum, the planetary boundaries approach proposed by Johan Rockström et al. (2009) and the Stockholm Resilience Center in 2009 attempts to define a "safe operational space" for humanity by analyzing the intrinsic dynamics of the Earth system and identifying the tipping points and critical processes on a global scale beyond which humanity must not go. The fundamental principle behind the Planetary Boundaries approach is that a Holocene-like state of the earth system is the only one that provides an environment conducive to human development.

Is it already too late to return to a Holocene world that may already be lost? Is the Anthropocene a one-way journey for humanity towards an uncertain future, a new state of the earth system, more threatening and very different from the present one? The earth stewardship approach is about reconnecting with the feeling of interdependence that binds us to the rest of life. Well-being and plenitude can only be achieved in a community that is extended to include all living beings in an amplified sensitivity to what surrounds us. Refusing the idea of linear progress in favour of the diversity of knowledge and intrinsic values recognized in nature, practising a form of self-government, achieving one's freedom through extending one's self to all entities, would also lead us to the "good" Anthropocene.

According to the International Geosphere Biosphere Program (IGBP), humanity is already managing the planet, but in an unconnected and haphazard way driven ultimately by individual and group needs and desires. As a result of the innumerable human activities that perturb and transform the global environment, the Earth System is being pushed beyond its natural operating domain. Many of these global changes are accelerating as the consumption-based Western way of life becomes more widely adopted by a rapidly growing world population. The management challenges to achieve a sustainable future are unprecedented. The word management or the term planetary management often raise concern. It is important to differentiate between attempts to manage the functioning of the Earth System (e.g., geo-engineering proposals) and attempts to manage human activities at the global scale (e.g., the Kyoto Protocol) so as to lessen their impact on the Earth System and thus to allow it to function in a more natural mode.

Stewardship entails a more rational long-term view by people of their place in nature and therefore the ways in which they manage and otherwise interact with ecosystems to watch and understand the land and use it respectfully forever, in the words of Canadian First Nations people. Stewardship requires effective governance at scales ranging from individuals to the globe. People have historically interacted with nature most directly at local scales, providing opportunities to learn from the consequences of their actions. However, the largescale environmental changes resulting from widespread unsustainable behaviour indicate a need for a human-Earth-system perspective that informs governance at local-to-global scales to foster learning and sustainable behaviour (Ostrom et al., 1999).

Earth stewardship therefore raises the question of scale (Paquot, 2020). Permaculture has taken an interest in this by proposing forms of creative energy descent at the level of communities and local communities. Care of earth is the fundamental motive for permaculture and there are two ways of looking at it. One could be called environmentalism. The other, ecology. Environmentalism sees humans as essentially separate from the rest of nature. Ecology sees humans as part of nature. It holds that all beings have rights, not just humans. Hence we need to leave as much of the Earth's surface as possible undisturbed by human activity. Since permaculture can enable us to meet our needs on a small area of land, it can help us towards this aim (Whitefield, 2004).

On the territorial level, earth stewardship can take the form of post-urban bioregions (Magnaghi, Sinai, 2017) capable of proposing new morphologies of human settlements placed under the sign of frugality and shared responsibility for the use of vital commons like water, farmland and energy. Bioregionalism is defined as first knowing the land on which we live and its resources. And its limits: limits of resources, place, carrying capacity of its lands and waters. Rather than making the organization of human regions dependent on extrinsic material resources, rather than seeing the regions exporting their wealth to distant shores and coming under distant head offices, the challenge is to imagine what could be done if each region used its funds, its reserves and its talents by limiting itself to the capacity of the territory and its own ecological constraints.

Attempts to manage the functioning of the Earth System itself are fraught with difficulties, perhaps insurmountable. Systems theory suggests that complex systems can never be managed; they can only be perturbed and the outcomes observed. Furthermore, many of these outcomes will likely be unpredictable, even with a vast amount of information on the system, leading to unintended and potentially severe consequences. This property of complex systems is manifest in the Earth System, for instance, as the instrinsic level of unpredictability in climate.

Humans and their societies and institutions are embedded in and are an integral part of the Earth System. Humans cannot fully stand outside the human-environment system as they attempt to analyse it, and thus cannot be in position to manage the Earth System in any objective fashion. Humans cannot be dispassionate observers and objective managers. Locally-controlled conservation produces better outcomes because it fosters active and collective stewardship of the environment. Such approaches can establish a shared vision for the landscape communities inhabit and mobilise people to preserve, restore and defend it while adapting to any threats or changes.

References

Dawson, N. M., Coolsaet, B., Sterling, E. J., Loveridge, R., Gross-Camp, N. D., Wongbusarakum, S., Sangha, K. K., Scherl, L. M., Phuong Phan, H., Zafra-Calvo, N., Lavey, W. G., Byakagaba, P., Idrobo, C. J., Chenet, A., Bennett, N. J., Mansourian, S., & Rosado, F. J. (2021). The role of indigenous peoples and local communities in effective and equitable conservation. *Ecology and Society, 26*(3), 19. https://doi.org/10.5751/ES-12625-260319

International Geosphere-Biosphere Programme (IGBP). (2004). *Global change and the earth system. A planet under pressure* (p. 285).

Ostrom, E., Burger, J., Field, C. B., Norgaard, R. B., & Policansky, D. (1999). Revisiting the commons: Local lessons, global challenges. *Science, 284*(5412), 278–282. https://doi.org/10.1126/science.284.5412.278

Paquot, T. (2020). *Mesure et démesure des villes*. CNRS éditions.

Rockström, J., Steffen, W., Noone, K., Persson, Å., Chapin, F. S., III, Lambin, E. F., Lenton, T. M., Scheffer, M., et al. (2009). A safe operating space for humanity. *Nature, 461*, 472–475.

Sinaï, A. (2017). Pour un aménagement permaculturel des territoires. In A. Sinaï & M. Szuba (Eds.), *Gouverner la décroissance*. Presses de Sciences Po.

Whitefield, P. (2004). *The Earth care manual*. Permanent publications.

Young, O., & Steffen, W. (2009). The earth system: Sustaining planetary life support systems. In F. S. Chapin III, G. P. Kofinas, & C. Folke (Eds.), *Principles of ecosystem stewardship: Resilience-based natural resource management in a changing world* (pp. 295–315). Springer.

Agnès Sinaï is a professor at Sciences Po, environmental journalist and author, founder of the Momentum Institute, a network for reflection on Anthropocene policies. Doctor in spatial planning and town planning (PhD, Université de Paris Est). Co-author of *Le Grand Paris after the collapse* (Wildproject, 2020) and of various books, including *Small treatise on local resilience* (with Pablo Servigne, Raphaël Stevens, Hugo Carton, ECLM éditions, 2015), *Walter Benjamin facing the storm of progress* (Le Passager clandestin, 2016) (all these publications are in French).

Emotional Commodisation

Simon Mallard

Abstract When we define 'emotion', we first define it as a moral issue. A further defining characteristic is the spurring of an individual to action, having taken stock of a given situation. Emotion stems from discord between perceived reality and the social norms governing the situation, its subjects or objects. One possibility is to view emotions as a body's actions towards others and itself (Dumouchel, Paul, Émotions. Essai sur le corps et le social. Le Plessis Robinson, Institut Synthelabo, 1999).

Goffman (1973), and later Hochschild (1983), described the emotional work inherent to social relations – especially to professions involving social interaction. Actors play a role, modulating their verbal and bodily expressions to serve the expectations of the interlocutor (member of the public), and thereby sculpt what is socially accepted behaviour. Here, the term 'work' refers to the measured, controlled, conscious decisions involved in displaying one's emotions and using them as tools – in other words, the strategic facet of emotions. Armed with this information, we can examine the social norms, professional rules, values, etc., governing the 'production' of emotions.

The display of emotions is governed by implicit *'feeling rules'* and *'display rules'*, which control the precursors to action. Feeling rules are the guiding principles and socially shared values in a given context – governing which emotions can, or should, be expressed to others. Under the influence of feeling rules, humans learn to associate a feeling they experience with a situation. Display rules define the meaning that the actor attaches to the sentiments and the emotions displayed. These rules refer to social codes, and 'govern' the commoditisation of emotion and the socio-affective.

Here, emotion is associated with a commodity – i.e. with goods and trade. In order to shift from the object (emotion) to the commodity (emotional commodity),

S. Mallard (✉)
Université Catholique de l'Ouest, Angers, France
e-mail: simon.mallard@uco.fr

a certain process must take place; this process is known as emotional commoditisation (i.e. commoditisation of feelings). In addition to the emotional work discussed above, a broader, denser economic importance is attached to social interactions and purchases. This also indicates that – at least to a degree, and on all levels –emotions are coming to occupy an important place within our capitalist society. Illouz (2019) describes emotional commodities as 'nodes'. These phenomena link consumerism with emotions – that is, they link the physical and the intangible aspects of products.

It is widely known that there is an emotional part inherent in consumer commodities, which create an environment imbued with emotion. Goods evoke emotions, and those emotional experiences, in turn, become *bona fide* commodities in the capitalist structure. Commodities are consumed because they cater for people's desires, and fit into their life plans; emotions help penetrate the shell of an individual's subjective judgement. Consumerist one-upmanship is dictated by novelty as a condition for enjoyment (Freud, 1920/2010).

In the twenty-first century, we have witnessed the intensification of the commercialisation that comes hand in hand with normative and instrumental approaches. This trend heralds new social relationships in a changing world, with significant emotional work involved, and rampant infiltration by emotional commodities. Consumption is now guided by emotional behaviour; our consumerism is steered by the emotions which the consumer items evoke. People use emotion as a tool to help sell goods, but the reverse is also true – emotion itself becomes a saleable good. In this new market, a huge amount of work is invested in shaping emotions in order to profit from so-called *soft skills* (emotional intelligence). From this standpoint, every one of us could be said to be managers of our own little 'business', operating in a highly competitive market.

To explain the rules in force and the production, sculpting and delivery of emotions, Illouz (2019) points to three ideals that are key in shaping individual identity: authenticity, intimacy and self-control. An individual must (1) appear genuine, (2) endeavour to be kind and fair in their relations with others, and (3) exercise self-control. In this context, emotion is used as a tool in the service of capitalism.

There is an ever-growing body of emotional guidance (such as top-10 lists), relaying sensational and emotional experiences, telling individuals which experiences to consume and which to avoid. Thus, our behaviour, and the ways in which we think, act, feel and, indeed, live, are increasingly being shaped by external forces. The subtle domination of emotional commodities can be seen in our consumer practices. Examples are numerous (Illouz, 2019).

The increasingly frenetic pace of life (in the political, economic, social and professional spheres, and so on) means people have to make a life choice: they must choose between accomplishment and losing themselves. Humans in today's 'hypermodern' (Aubert, 2018) and 'fluid' (Bauman, 2006) society are described as 'frivolous' (Melman, 2009), 'perverse' (Enriquez, 2006), 'bipolar' and 'tired' (Ehrenberg, 2000). This condition is attributed to information overload and lack of markers. The boundaries between external and internal, between intimate and social, between private and work life, between oneself and others are blurred by the complex, global, interconnected world in which we live. In this changing reality, emotions are markers or tools that can be used to sell something, be it to oneself or to others. Thus, each individual becomes a consumer/producer in a hyper-capitalist world. Inevitably, then, one must think about consumption habits, civilian life and the social and economic models which govern this new world.

'Where there's a will, there's a way!', 'Be yourself!' 'Become who you are!' These are just a few of the slogans upon which this new world is founded. People are indoctrinated into thinking they are in full control of their work, personal life, emotions, etc. Yet this outlook fails to take account of organisations, institutions and groups which exert an influence. It also assumes that there is no individual or collective subconscious at work.

If we are guided solely by instrumental rationality, we may define this tendency toward the emotional as a case of objectification (Nussbaum, 1995): humans are treated like objects; it should also be noted that, from certain perspectives, people may even view themselves as objects. However, if we view emotion as being *apparatus* – that is, 'an arrangement of [ABC] in support of [XY]' (Caron, 2010) – then the apparatus can be reconfigured (see Albero, 2010) by using different rationalities: existential, axiological or epistemic. As a set of social practices, the apparatus is driven by social systems and actors; within it, individuals tread the line between submission and resistance.

Illouz's idea is bold and provocative, but appears to indicate a powerful trend toward more psycho-analytical approaches (such as that of Melman, 2009) or sociological ones (such as that of Aubert, 2018). 'Emotions become the repository of truth and existential experience. They define the subject's own [version of] truth' (Illouz, 2019, p. 335). We may, of course, passively accept that our lives are governed entirely by emotions, but passive acceptance is not the only course of action.

Emotions tell us nothing about what *is* – about truth, or about the world – but instead only about what we deem relevant. They help shape our perception, weaving together the past, present and future. They teach us about our own learning styles and the ways in which we can act both *in* and *on* the world, because they change the way in which we interact with objects. This more philosophical approach examines the intimate link between values and emotions (Tappolet, 2000; Livet, 2002).

As we saw earlier, emotions can be viewed as a body's actions towards others and itself (Dumouchel, 1999). This definition emphasises both the innate and the acquired aspects – i.e. the biological, psychological and social aspects – of emotion. As mentioned previously, more weight tends to be attached to the economic dimension than to the others.

Notwithstanding the general trend in our hypermodern society, we ought not to overlook the various realities of more marginal, vulnerable populations, or those who choose an alternative way of life. Furthermore, it is important to include certain living environments – rural settings, suburbia, so-called third places, etc. (which are sometimes ignored by researchers) – in our analysis.

In terms of methodology, to complement Illouz's sociological approach, microsociological case studies could be conducted, to gain as detailed a view as possible of the subjects. Besides being a movement linked to emotion, a gesture holds meaning both for the person making it and for those perceiving it; there is a degree of interpretation required on the part of both, and this interpretative filter must be taken into account. Such a view should illuminate the various aspects of this phenomenon through a grounded, comprehensive or inductive approach. Ethnographic approaches (Kaplan, 2019), which are increasingly popular today, seem particularly well suited to analyse and grasp the full complexity of an emotional gesture – its physical, cognitive, psychological, social, symbolic and operative scope, as well as its intangible aspects.

Finally, for social and professional practices in particular, by studying emotions, we should be able to examine issues relating to human ontology and citizenship. Such examination should come, hand in hand, with critical reflection on the notion of individual autonomy, to help shape a genuine drive for emancipation (Eneau, 2016).

References

Albero, B. (2010). La formation en tant que dispositif: du terme au concept. In B. Charlier & F. Henri (Eds.), *La technologie de l'éducation: recherches, pratiques et perspectives* (pp. 47–59). Presses Universitaires de France.

Aubert, N. (2018). @ *la recherche du temps: Individus hyperconnectés, société accélérée: tensions et transformations*. Toulouse.

Bauman, Z. (2006). *La vie liquide* (C. Rosson, Trans.). Le Rouergue – Chambon.

Caron, P.-A. (2010). Du dispositif pensé par les enseignants au dispositif construit par l'informaticien: la place de la modélisation. In Gerard Leclercq et Renata Varga (Ed.), *Dispositifs de formation, quand le numérique s'en mêle. Une approche pluridisciplinaire*. Hermès/Lavoisier.

Dumouchel, P. (1999). *Émotions. Essai sur le corps et le social*. Institut Synthelabo.

Ehrenberg, A. (2000). *La Fatigue d'être soi. Dépression et société*. Odile Jacob.

Eneau, J. (2016). Autoformation, autonomisation et émancipation: de quelques problématiques de recherche en formation d'adultes. *Recherches & éducations, 16*, 21–38.

Enriquez, E. (2006). L'idéal type de l'individu hypermoderne: l'individu pervers. In N. Aubert (Ed.), *L'individu hypermoderne*. Erès.

Freud, S. (1920/2010). *Au-delà du principe de plaisir*. Payot.

Goffman, E. (1973). *La mise en scène de la vie quotidienne. 1. La présentation de soi* (A. Kihm Trans.). (vol. 1). Les Éditions de Minuit.

Hochschild, A. (1983). *The managed heart: Commercialization of human feeling*. University of California Press.

Illouz, E. (Ed.). (2019). *Les marchandises émotionnelles*. Premier parallèle.

Kaplan, D. (2019). Les cartes sexuelles de Tel-Aviv. Création d'ambiance, sexualité récréative et atmosphère urbaine. In E. Illouz (Ed.), *Les marchandises émotionnelles* (pp. 177–204). Premier Parallèle.

Livet, P. (2002). *Émotions et rationalité morale*. Presses Universitaires de France.

Melman, C. (2009). *La nouvelle économie psychique: la façon de penser et de jouir aujourd'hui*. Érès.

Nussbaum, M. (1995). Objectification. *Philosophy & Public Affairs, 24*(4), 249–291.

Tappolet, C. (2000). *Émotions et valeurs*. Presses Universitaires de France.

Simon Mallard is Associate Professor at the Université Catholique de l'Ouest. He is also research associate at Centre de Recherche sur l'Éducation, les Apprentissages et la Didactique (EA, 3875, Univ Rennes 2).

Eurocentrism

Rosi Braidotti and Hiltraud Casper-Hehne

Abstract The term Anthropocene was criticized for its tendency to simplify space and time, through over-universalizing language. Climate change narratives are overwhelmingly white, Eurocentric and masculine. The wealthy minority of the world's population is responsible for the effects of the Anthropocene, while the consequences fall disproportionately on the poorer sections of humanity. The Anthropocene glosses over these differences, ascribing the Anthropocene to "The Human" as a universal category. This is presented in a classically Enlightenment, humanist fashion, as a single anthropocentric entity. A critique of the excluding aspects of humanism is consequently crucial to Anthropocene research and to critical posthuman thought. Intercultural, post-colonial and de-colonial scholarship from the humanities, feminist and gender theories, and ecological thought are necessary to assess the discriminatory aspects of the humanist legacy and make sure they are not continued in the Anthropocene.

The term Anthropocene has been criticized for its tendency to simplify space and time, through over-universalizing language. Climate change narratives and scenarios, theories of the Anthropocene and posthuman scholarship are overwhelmingly white, Eurocentric and masculine (Mertens & Ulstein, 2020). This is ironic to say the least, given that the wealthy minority of the world's population is largely responsible for the effects of the Anthropocene, while the disastrous consequences fall disproportionately on the poorer sections of humanity.

The Anthropocene is a term that glosses over these differences, rather than deepening our understanding of them. It relies on assumptions, terminology and a value system based on scientific rationality as defined in the age of Enlightenment by a

R. Braidotti (✉)
University of Utrecht, Utrecht, The Netherlands
e-mail: r.braidotti@uu.nl

H. Casper-Hehne
Intercultural German Studies, University of Göttingen, Göttingen, Germany
e-mail: h.casper-hehne@phil.uni.goettingen.de

© The Author(s), under exclusive license to Springer Nature Switzerland AG 2023
N. Wallenhorst, C. Wulf (eds.), *Handbook of the Anthropocene*,
https://doi.org/10.1007/978-3-031-25910-4_108

Western worldview. The term Anthropocene is ascribed a uniform meaning and presented as a universal category, in keeping with a classical Enlightenment, humanist approach. This approach combines the humanist values of continuity of space and time – hence a unitary vision of the human subject – with an implicit assumption of anthropocentrism, that makes humans the dominant and most evolved species. Both as 'the Man of Reason" (Lloyd, 1996) and as *homo sapiens*, the human is expected to rule as the supreme entity.

Braidotti (2013, 2019) has argued consequently that a critique of the exclusionary aspects of Humanism is crucial to research on the Anthropocene and critical posthuman thought. The version of humanism at play in the posthuman convergence is drawn from the European Renaissance ideal of the human as "the measure of all things". This idea makes 'Man' coincide with universal reason, self-regulating rational judgment, moral self-improvement and enlightened governance. This image is a marker of European culture and society, and for the scientific and technological activities it privileges and it thus applies to groups and nations as well as individuals. It results in presenting Europeans as the motor of human evolution and the progress of human civilization through science and technology. All other, non-European, cultures and all non-human species are organized in a descending hierarchical scale, whereby being 'other than' or 'different from' 'Man', is negatively coded as being 'worth less than' 'Man'. This is not just a symbolic exclusion, but also a deeply and violently material one, which affects the women and LBGTQ+ people (the sexualized others), Black and Indigenous people (the racialized others) and the animals, plants and earth-entities (naturalized others). Their social and symbolic value is denied and their material survival compromised. This civilizational model was further instrumental to the project of Western modernity and the colonial ideology of European expansion (Braidotti, 2006; Weheliye, 2014; Wynter & McKittrick, 2015). It shows that the human is not a neutral category, but one that is marked by powerful differences within. It also indicates that Europe is not just a geo-political location, but rather a universal attribute of human consciousness that can transfer its quality to any suitable subjects, provided they comply with the required norms.

It is not surprising therefore that the sense of panic triggered by mainstream Western discussions of climate change, the Anthropocene and species extinction, has been criticized by Black, decolonial and especially Indigenous critical thinkers, LBGTQ+ people and feminists. They have stressed the extent to which the First Nation people have already experienced the devastation of their lands and cultures through the violence of colonization (Clarke, 2008; Whyte, 2017; Todd, 2015). As Danowski and Viveiros de Castro poignantly put it: "for the native people of the Americas, the end of the world already happened – five centuries ago. To be exact, it began on October 12, 1492" (2017: 105). Black and indigenous thinkers, like many non-Western feminists and philosophers, consequently call for a more critical account of the climate change emergency, one that allows for decolonial perspectives and anti-racist approaches.

The wealth and rigour of inter−/de−/post-colonial theories, race and Indigenous philosophies, feminist theories of transnational justice and radical ecology, are central to the critique of Eurocentrism. The scholarship is far too vast for us to do justice to it in this essay. What we want to argue here is that the concept of the Anthropocene still very much reflects a Western scientific cosmology, and that this bias needs to be examined critically. The critique of Eurocentrism within the Anthropocene debate combines two lines of enquiry, that target humanism and anthropocentrism respectively and draw attention to their multiple intersections (Braidotti, 2013, 2019, 2021). The double charge is that the residual humanism of the Anthropocene flattens out the staggering differences in power and access that mark our current era of simultaneous advances in technological development and even more advanced environmental degradation. Furthermore, its unexamined anthropocentrism fails to recognize the existence, sentient nature and know-how of multiple other species. These blind spots structure the Anthropocene and limit its scientific and cultural impact. We want to argue instead that the earth is not a uniform, closed space with the same homogeneous relationships between nature and culture, or between humans and their environment. An approach like that is biased, in that it describes only one kind of humanity, as well as one kind of nature, which create their opposites through their distinction (Baskin, 2015: 155). Eurocentrism on this scale is not just a cultural bias, but rather a structural error of judgment, that needs to be corrected by introducing heterogeneity, hybridity and complexity.

Within the debates aimed at assessing the human responsibility for the Anthropocene, this critical line unfolds in more polemical directions. The all-inclusive and personified idea of 'humanity', for instance, is taken as a problematic construct that conveniently conceals political questions about specific responsibility for the environmental disaster (Oppermann & Iovino, 2017). Climate change narratives and scenarios circulate largely within the global North, as a way of processing of the bad conscience of the West, which is primarily responsible for destroying the planet (Manemann, 2014: 44).

Given that the understanding and logic of the Anthropocene idea are based on a Western worldview (Danowski et al., 2014: 3) that separated nature from culture, bodies from minds, humans form non-humans and organized all differences hierarchically (Descola, 2013), it then follows that "the Anthropocene is a product of Western humans" (Morton, 2014: 261). The discourse refuses to question the power relations that organize human domination – and dehumanize many humans in the process – and instead tend to look for solutions by falling back on strategies based on technologies and management (Crist, 2013). This strategic approach reiterates the faith in progress, growth and development, conveying an ideal of humanist perfectibility and linear historical evolution, which are central to the Enlightenment and the Western project of modernity. Grown historically from this specific Western nucleus of values, it was applied globally, also through the enslavement of vast sections of the human population and the systematic exploitation of natural resources. This combination of genocide and ecocide has been criticized by ecofeminists and Indigenous thinkers for decades (Mies & Shiva, 1993; Plumwood, 1993; Rose, 2004; Shiva, 1997).

Intercultural post-colonial and de-colonial scholarship from the humanities, feminist and gender theories, and ecological thinking are consequently necessary to analyse these problematic and unsettling aspects of the Anthropocene debate. A decolonial approach exposes the narrative of Western modernity as assuming the positive impact of technologically driven industrial and extraction economies. That also implies that nature is the provider of endless raw material and resources for humans to use at their will. Anthropocene-oriented scholars question this presumptuous attitude, joining forces with eco-feminist and environmentalist critiques of human exceptionalism (Plumwood, 2002). Theriault for instance asks whether humanity is seen as the planetary engineer with inexhaustible resourcefulness, or as a starved global parasite destroying its own host (Theriault, 2014: 2). These scholars encourage us to see humans as knowledgeable guardians who protect the planet, and not as the sovereign rulers of creation. The idea of humans ruling the rest of the world on the Christian model seems to be at play here (Rickards, 2015: 285). As Mannemann explains, "the Anthropocene vision of the world gardener threatens to tip over into that of the superman" (2014: 89). The Anthropocene in this respect is the return of nature and its intrusion into the dominant, Western, Christian, capitalist civilizational matrix (Danowski & Viveiros de Castro, 2017).

These lines of critical enquiry converge on the crucial question of what we mean when we say, for instance: "we humans live in the Anthropocene". Who is the "we" that composes this humanity? Whose pain and anxiety is being expressed here? This generic language obscures the fact that the effects of the Anthropocene are caused by only a minority of the world's population, but are uniformly and globally ascribed to an entity called "The Human" (Baskin, 2015: 16). Universalizing statements are all the more objectionable considering that almost 50% of global emissions are produced by the wealthiest 10% of the world's population, yet the impact of these emissions is harming the poorest first and worst, especially in the global South (Nixon, 2011).

Critical thinkers, especially within posthumanism, have made one thing clear: the project of Western industrial modernity, which deployed a fossil fuel-led model of extractive economies and colonial economic growth and assumed the unlimited exploitation of 'natural' resources, has shown its limitations (Braidotti, 2019). This capitalist, colonial and patriarchal model, which is at the root of ecological depletion, is inextricably linked to the oppression of women and LBGTQ+ people; the colonial appropriation of land, culture and resources from Indigenous people, their enslavement and their dehumanization. These are not accidental problems, but structural components of a global design of domination that got us to where we are today. To become aware of such historical responsibilities – at long last – should be greeted as an opportunity to develop adequate knowledge of the abuses of the Western world and to atone ethically and politically for them. This is a task for the Humanities in the twenty-first century.

In summary, the term Anthropocene has so far not been clearly defined, and is subject to several kinds of criticisms, but it nevertheless has played a crucial role in shaping contemporary debates about environmental responsibility, sustainability, and the transformations in the status of the human. Relatedly, it further sheds light on the distribution of wealth and power. It has generated new perspectives that have already been applied across the human and social sciences and stimulated lively discussion across disciplines. These problematic aspects are part of a Humanist and anthropocentric legacy that was used in the past to justify – even through scientific language – discriminations, hierarchies, and exclusions. Anthropocene and posthuman scholarship must therefore make sure that these patterns of discrimination are not continued, but are instead exposed, examined and applied to further research in the New Humanities.

References

Baskin, J. (2015). Paradigm dressed as epoch. The ideology of the Anthropocene. *Environmental Values, 24*(1), 9–29.

Braidotti, R. (2006). *Transpositions: On nomadic ethics*. Polity Press.

Braidotti, R. (2013). *The Posthuman*. Polity Press.

Braidotti, R. (2019). *Posthuman knowledge*. Polity Press.

Braidotti, R. (2021). *Posthuman feminism*. Polity Press.

Clarke, A. (2008). *Supersizing the mind. Embodiment, action and cognitive extension*. Oxford University Press.

Crist, E. (2013). On the poverty of nomenclature. *Environmental Humanities, 3*, 129–147.

Danowski, D., & Viveiros de Castro, E. (2017). *The ends of the world*. Polity Press.

Danowski, D., Viveiros de Castro, E., & Latour, B. (2014). Position paper. *The Thousands Names of Gaia. From the Anthropocene to the Age of the Earth*. https://thethousandnamesofgaia.files.wordpress.com/2014/07/position-paper-ingl-para-site.pdf

Descola, P. (2013). *Beyond nature and culture*. University of Chicago Press.

Lloyd, G. (1996). *Spinoza and the ethics*. Routledge.

Manemann, J. (2014). *Kritik des Anthropozäns*. Transkript.

Mertens, M., & Ulstein, G. (2020, November 26). Decolonizing the Cli-Fi corpus. *Collateral. Online Journal for Cross-Cultural*. Close Reading. http://www.collateral-journal.com/

Mies, M., & Shiva, V. (1993). *Ecofeminism*. Zed Books.

Morton, T. (2014). How I Learned to Stop Worrying and Love the Term Anthropocene. In *Cambridge Journal of Postcolonial Literary Inquiry, 1*(2), 257–264.

Nixon, R. (2011). *Slow violence: Environmentalism of the poor*. Harvard University Press.

Oppermann, S., & Iovino, S. (2017). *Environmental humanities. Voices from the Anthropocene*. Rowman and Littlefields.

Plumwood, V. (1993). *Feminism and the mastery of nature*. Routledge.

Plumwood, V. (2002). *Environmental Culture*. Routledge.

Rickards, L. A. (2015). Metaphor and the Anthropocene. Presenting humans as a geological force. *Geographical Research, 53*(3), 280–287. https://www.hkw.de/media/de/texte/pdf/2013_2/programm_6/anthropozaen/booklet_anthropozaen_eine_eroeffnung.pdf

Rose, D. B. (2004). *Reports from a wild country*. University of New South Wales Press.

Shiva, V. (1997). *Biopiracy. The plunder of nature and knowledge*. South End Press.

Theriault, N. (2014). Rezension zu Crist, Eilen. 2013: On the poverty of our nomenclature. *Environmental Humanities, 3*, 129–147. https://inhabitingtheanthropocene.com/2014/11/24/on-the-pverty-of-our-nomenclature/

Todd, Z. (2015). Indigenizing the Anthropocene. In H. Davis & E. Turpin (Eds.), *Art in the Anthropocene: Encounters among aesthetics, politics, environments and epistemologies* (pp. 241–254). Open Humanities Press.

Weheliye, A. (2014). *Habeas viscus*. Duke University Press.

Whyte, K. P. (2017). Indigenous climate change studies: Indigenizing futures, decolonizing the Anthropocene. *English Language Notes, 55*(1–2), 153–162.

Wynter, S., & McKittrick, K. (2015). *Sylvia Wynter. On being human as Praxis*. Duke University Press.

Rosi Braidotti is distinguished University Professor at Utrecht; Honorary Degrees Helsinki, 2007, Linkoping, 2013; Fellow of the Australian Academy of the Humanities *(FAHA),* 2009; Member of the Academia Europaea *(MAE),* 2014.; recipient Humboldt Research Award, 2021. Main publications: *Nomadic Subjects* (2011a), and *Nomadic Theory* (2011b), Columbia University Press. *The Posthuman,* 2013, *Posthuman Knowledge,* 2019 and *Posthuman Feminism,* 2021, Polity Press.

Hiltraud Casper-Hehne is distinguished University Professor at Göttingen; 2009–2021 Vice President for International Affairs at Göttingen University; since 2021 Director of the network NEH21: New European Humanities in the Twenty-First Century; since 2020 member of the Executive Committee of the *European University* ENLIGHT. Main publications: Braidotti, Rosi/ Casper-Hehne, Hiltraud/Ivkovic, Marjan/Oostveen, Daan F. (2021) (eds.): *Trajectories for the Humanities in the Twenty-First Century.* Edinburgh Publishing House (forthcoming).

Neoliberalism

Renaud Hétier

Abstract The neoliberal policies exercised in large parts of the world pose a certain problem. It is interesting to think about the place of neoliberalism in the history of capitalism, the driving factors behind it, and obviously, its effects. A critical analysis is required in order to understand the paradox whereby the system's effects are enormously damaging, but no viable alternatives emerge to replace that system, whilst ordinary citizens, largely dispossessed of their power to act, do not rise up in revolt. When we reflect on this state of affairs, we cannot help but think of the question of evil, and our tendency to voluntarily ignore it.

The approach to the concept of neoliberalism in this article is not strictly a technical one, for two reasons. Firstly, the concept does not refer to a strictly 'objective' phenomenon, but a nebulous one, combining aspects of the ideological, the economical and the political. Secondly, and importantly, to take a 'neutral' approach would be a dereliction of the duty to engage in critical reflection. Readers most likely already have an idea of what the term 'neoliberalism' covers. In addition, the doctrine cannot be 'objectively', given that above all, it is a partly dark ideology, in the sense that its true intentions are often concealed, and its effects are enormous. It is impossible to say anything meaningful about neoliberalism without dissecting it and examining its mechanical workings, which is the only way to truly understand it. As a term, 'neoliberalism' does not convey the same concrete concept as does 'capitalism', as a global, historical system (Bihr, 2018). Neoliberal leanings appear to be the next chapter in a story: the intensification and acceleration of the potential effects of capitalism. However, neoliberalism is a breakaway movement whose speed and violence are unprecedented. One of the benefits of the concept is that it includes the idea of 'liberalism', and therefore, the idea of liberty somewhere in the background. The term can be linked to a doctrine, but also to political policy, or to a particular

R. Hétier (✉)
UCO, Angers, France
e-mail: Renaud.hetier@uco.fr

© The Author(s), under exclusive license to Springer Nature
Switzerland AG 2023
N. Wallenhorst, C. Wulf (eds.), *Handbook of the Anthropocene*,
https://doi.org/10.1007/978-3-031-25910-4_109

period in the history of capitalism (Harvey, 2005). These three possibilities are interlinked:

the *doctrine* that is publicly displayed (greater economic freedom and greater individual freedom) and the clandestine doctrine that underlies it (the quest to destroy all forms of freedom that do not fit in with the logic of profiteering);

the *policy* which must also be unfolded and examined in full (making sacrifices in the name of lower cost and less 'control' (notably control over movement of capital), and at the same time, increasing the number of ways in which one can interfere, control, and monitor);

the *historical* moment, paradoxical in itself, whereby neoliberalism is apparently the very embodiment of capitalism in its purest form, and a true departure from (the polar opposite, indeed, of) liberalism, in that an ever-smaller number of individuals enjoy a sort of unrestricted freedom, while the masses lose more and more and more.

More specifically, on a financial level primarily, neoliberalism aims to take away a number of constraints (relating to State control, worker protection and workers' rights, market regulation, etc.), with the aim of allowing capital to circulate more freely, supply and demand for labour to increase, and growth to take off. Secondarily, this liberation of economic power, which is supposed to liberate businesses, could produce more wealth, more jobs, and thus benefit everybody. However, it is difficult to know whether this secondary benefit (which we find, for example, in the idea of 'trickle down') is the genuine goal for neoliberal thinkers, or whether it is merely a means of justifying a 'policy' which, in fact, helps maximise its proponents' profits (and disguises the fact, which actually results in worsening inequality). What is undeniable is that, be it true or false, the promise of neoliberalism has 'worked' – it has largely won the battle for hearts and minds, and garnered majority support. This invites questions about how this doctrine serves a twofold expectation: the expectation of material pleasure, although it is only virtual; and the expectation of 'individual freedom' – freedom for individual enterprise (Foucault); human liberty that is measured by economic success. We shall revisit this idea later on. One might legitimately wonder whether we have reached the (historic) point at which the ideal of freedom can be superposed upon that of material pleasure. Of course, the spread of this ideology may also awaken and sharpen the instincts that were already present, and dormant to a greater or lesser degree, and thus, reinforce itself (in much the same way as a drug dealer makes a new client dependent on the product, until that dependency ultimately becomes the client's problem (Rahnema, 2003)). Certainly, any ideology does something similar, but neoliberalism appears to have particularly effective tools with which to do it: rendering its promises extremely *material* and tangible (rather than dangling the carrots of justice, equality, etc.), contributing to a formidable 'construction of consent' (Harvey, 2005) in individuals who have long been 'prepared' (Hétier, 2021), and who have lost all else (genuine freedom, hope, and dignity).

At a theoretical level, neoliberalism, as it arose in the twentieth Century, was based upon Keynesian liberalism. For Keynes, who was writing in the interim period between the two World Wars, the limitations of conventional liberalism, based on the principle of '*laisser faire*', had become apparent (notably in the 1929 stock market crash). The State needed to act in order to support and shore up the economy – an idea which is apparent in that of the 'Welfare State'. The Second World War proved the relevance of Keynes' theory afresh: with the ability of the great democracies (the USA and Great Britain) to firstly mobilise their economies to support the war effort on a massive scale (Jorion & Burnand-Galpin, 2020), and then after the war, to restart the ordinary economy through mass public investment (notably with the Marshall Plan). Neoliberalism is not a homogeneous and coherent school of thought. Instead of being a positive 'doctrine', it is primarily a rejection of conventional liberalism, and actually seems to be a *reaction* to Keynesianism. Within neoliberalism, there are four discernible main streams (Hoang, 2021): supply-side economics, monetarism, neo-Walrasianism, and neo-institutionalism. These various sectors are interoperable, to varying degrees. In this article, we shall focus on the idea which has spread to almost all corners of the globe – that of the supply economy: the market's selectivity among competing businesses, which is supposed to stimulate better performance, rather than State funding and the heavy taxation which comes with it.

Overall, the desire to 'liberate' the economy could be understood from two different standpoints. Firstly, by advocating that the State refrain from regulating the flow of money (monetarism), and supporting supply-side economics (i.e. production by companies) rather than demand-side economics (which would help buoy up wages), the aim is to counteract inflation and joblessness (Hayek and Mont Pelerin Society), increase growth and sustain freedom through a market-based economy (Friedman). To a degree, from the 1980s on, this strategy has worked (if we ignore the social impacts – worsening inequality – and environmental ones – the precipitation of the planet into the Anthropocene). Then, neoliberalism (apparently) supports a model of political freedom (Friedman, 1962), raising the spectre of masterminding (and dictatorship) by the Soviets. Today, it is clear that neoliberalism has not helped make democracy stronger. It undermines the political and economic alternatives, firstly, and considerably weakens all opponents and hurdles to the logic of unbridled profit, utterly blind to any and all consequences, secondly.

Historically, Thatcher's Britain and Reagan's USA, in the 1980s, are the flagship models of the triumph of this doctrine, with the violent social breakdown and the attendant deterioration in democracy, along with an unprecedented rise in inequality (today, a true explosion). However, it is important not to underestimate the events in Chile, because firstly, they pre-date those in Britain and the US, and secondly, they reveal how far neoliberalism can go if no credible resistance remains. Thus, Chile served as the testing ground for the USA in this area, in pared down circumstances, as the experiment was unrestrained by democracy, and as the country was run by a violent dictatorship, it was completely 'free' of any 'resistance'. One of the greatest achievements – if not indeed the single greatest achievement – of the doctrine is that it has become associated not just with freedom, but with movement, progress, with the very nature of human civilisation. Therefore, any resistance to it can be presented as a simple hurdle – archaic and regressive – which robs any alternative of the vestige of legitimacy as soon as it begins to emerge. Britain's swerve toward neoliberalism in the 1980s was not exactly bloodsoaked. Nor has the world's subsequent slide into neoliberalism been. (It must be noted, though, that this 'coup' has not been entirely bloodless: protests such as industrial action have, on occasion, been repressed with horrific violence). However, these events do confirm what those in Chile indicated. In neoliberalism, politics and economics merge into one. That is, politics is exploited to serve economic goals (only those political actions which contribute to the economy are implemented), and ultimately, politics (at least, in the democratic sense of the word) is entirely subsumed. This may indicate why the greatest success of neoliberalism, in China, has not emerged from a democratic society. (China's leaders, united on this point, view democracy as a mere obstacle to efficiency, and also, more cynically, as a luxury that is secondary to the needs of the people, reduced to access to material pleasure through the consumer society). The 'liberty' that the proponents of neoliberalism so passionately vaunt is decidedly not an authentic form of freedom: not democratic, nor moral, nor spiritual.

In 1979, Michel Foucault was a visionary, developing his theory of what he calls 'biopower'. Even at this stage, the philosopher pointed to the 'naturalness' of phenomena, saying that in these contexts, the State could justifiably take a back seat (paraphrasing Rousseau, Foucault holds that we must 'allow nature to run its course'). The aim of liberal governmentality – which does not apply here – is to seek maximum rationality (the greatest possible efficiency for the least possible cost). In this context, it is important to reiterate that neoliberalism is not an evolved or more advanced form of liberalism. It is a matter of economic *'laisser faire'*, ignoring all risk of consequences, surrendering to the 'blind hand of the market', in the hope that it will 'naturally' regulate itself. It publicly vaunts the principle of total freedom whilst, with one hand, dismantling all forms of State regulation, and with the other, interfering to serve the interests of elites with public money. The final piece of the puzzle, though, according to Foucault, is that neoliberalism ultimately spreads its requirements into domains other than just the economy, extending into all areas of society. (One obvious example, in today's world, is that of a hospital, which is supposed to be 'profitable' in the view of those in charge, although it is a caregiving, rather than a producing, institution). Foucault quite rightly points to a 'political economy' (rather than economic politics), confirming that the two spheres have merged – or rather, that politics has been subsumed by neoliberal economics. It is a particularly warped act to pretend to champion 'liberty' (i.e. economic liberty, which is supposed to bring all other forms of freedom with it) in order to *destroy* liberty: to discredit and uproot any resistance (which is dismissed as 'backward-looking'), interpret protest as civil disturbance, and thus justify the obsession with security (for which there is never any shortage of funding), with the aim of 'monitoring and punishing'.

Pursuing the analysis still further, and examine the triumph of neoliberalism – which has torn down all opposition, destroyed all political alternatives, trampled over the rights won in the social struggles of the twentieth Century, undermined all democratic institutions and, of course, steered the State away from its role as defender of the people – from a historical and anthropological perspective, we

cannot fail to mention the unprecedented affront to liberty and solidarity that individualism represents. The irrepressible surge in individualism takes place in a context of enormous wealth, which provides fertile ground for it. Also, though, it occurs
in societies that are vast and complex, in which it is not easy to feel the spirit of
'community', in a culture which prioritises individual rights (and rejects all authority, including the authority of knowledge); in an industrialised, globalised economy
where all individuals – even from childhood – feel the need to have everything for
themselves and lack nothing, rather than sharing, and prioritising sharing over sati-
ety. Ultimately, notwithstanding its harmful consequences, neoliberalism offers a
particularly violent form of hope. Most incredibly, its extreme rationality (in which
every action can be reduced to a calculation) is actually based on fundamental irrationality. Like in a lottery, almost everybody loses, but almost everybody continues
to play the game, in the fantastic hope of a staggering gain. What gain might that be?
Certainly, they hope to gain happiness, but a form of happiness that is confused with
material comfort, access to unlimited consumption fed by limitless supply and, if
possible, luxury (inspiring figures are, less and less nowadays, moral heroes, and
increasingly, are 'successful' individuals). Undeniably, there are many poor people.
After centuries of capitalism and decades of neoliberalism, though, who has the
mental and spiritual strength to eschew wealth when it 'falls from the sky'? Who
still has the strength, not through 'failure' but through choice, to prefer a certain
poverty, associated with a form of dignity (not desiring a bigger house, not wishing
to build a private swimming pool, not seeking a 4 × 4 or an SUV, not jetting off
around the world, and ultimately, not wanting *even more*)?

All sorts of factors contribute to this dire situation. The traditions and values of
simplicity, sobriety and poverty (which must not be confused with penury (Rahnema,
2003)) have been brushed off, and utterly crushed in some instances. All alternatives
to materialism (spirituality, intellectualism, art for non-commercial purposes, ennui,
etc.) have been devalued, bringing a host of mental, social and environmental consequences. Non-monetary solidarity has become a thing of the past. Finally, the idea
of citizenship has faded into the background, and a State that is certainly *neo* but by
no means liberal (China) has triumphed economically. Neoliberalism has injected a
profound evil into the world, and it has spread like a virulent disease. Here, then, is
the point: what neoliberalism has liberated is evil. The whole history of civilisation
is a struggle to contain the antisocial tendencies that can arise in individuals. Today,
though, those tendencies are allowed to spread unchecked, all in the name of growth
and profit (Dufour, 2019). The impressive conjurer's trick is to condemn us all to
continue to wage war on one another – in the spirit of 'every man for himself' –
through financial competition. The further we advance, the more we seem to move
away from the barbarism of the past, where bloodshed was commonplace, and the
suffering of other people a spectacle. We appear to have moved past the archaic
violence that inhabits us all. In the western world, we grant extended rights to an
ever-growing number of people, and are increasingly recognising the right to be
different. On the surface, the economic war appears perfectly clean: the major enterprise which suddenly delocalises its operations, leaving its workers in the dirt; the
State withdrawing financial support from associations; the consumer who wilfully

closes their eyes to what their goods cost to workers on the other side of the world ... can all blindly ignore the harm that they are doing, because it is happening elsewhere, far away, and they are not *physically* getting their hands dirty. A prime example of this can be seen in a smartphone: in our society, made up of individuals who are highly trained and very highly informed, who has the *moral fortitude* to resist a 'better-performing' smartphone, which endangers the lives of children in the mines of the Congo (Lebrun, 2020)? In our view, childhood is sacred, and anyone who harms a child is now seen as a monster. However, that conviction wavers, provided the harm is not happening under our noses, and when our own material enjoyment is at stake. Unlike what Kant suggests in *What is Enlightenment?*, to know is not enough. Knowing means nothing, if we *do not feel* the evil we are doing. This is neoliberalism in a nutshell: the promise of pleasure, for which the trade-off – the evil, the suffering of others – becomes sufficiently indirect, sufficiently invisible, sufficiently imperceptible, that we no longer lose sleep over it.

References

Bihr, A. (2018). *Le premier âge du capitalisme. Tome 1. L'expansion européenne*. Page 2/Syllepse.
Dufour, D.-R. (2019). *Baise ton prochain. Une histoire souterraine du capitalisme*. Actes sud.
Friedman, M. (1962). *Capitalism and freedom*. University of Chicago Press.
Harvey, D. (2005). *A brief history of neoliberalism*. OUP Oxford.
Hétier, R. (2021). *L'humanité contre l'Anthropocène. Résister aux effondrements*. PUF. Foreword by Dominique Bourg.
Hoang, N. (2021). *Néolibéralisme*. Encyclopedia universalis.
Jorion, P., & Burnand-Galpin, V. (2020). *Comment sauver le genre humain*. Fayard.
Lebrun, F. (2020). *On achève bien les enfants. Écrans et barbarie numérique*. Le Bord de l'eau.
Rahnema, M. (2003). *Quand la misère chasse la pauvreté*. Acte Sud.

Renaud Hétier is a Professor at the Catholic University of the West (UCO). He holds a Doctorate in Educational Sciences. He has authored books on education, and anthropology in the Anthropocene. Books (selection): *Humanity versus the Anthropocene* (PUF, 2021, *in French*); *Cultivating attention and care in education* (PURH, 2020, in French); *Creating an educational space with fairy tales* (Chronique sociale, 2017, *in French*); Education between presence and mediation (L'Harmattan, 2017, *in French*).

Plantationocene

Yves Citton

Abstract This article discusses the debates that pushed several scholars and activists to suggest alternatives to the name 'Anthropocene'. It first states some of the reasons that have attracted criticism of this choice of word. It then surveys some of the alternatives that have been proposed, before focusing on one of them, 'Plantationocene', which is presented at greater length as the least unsatisfactory label for our age.

As this *Handbook* demonstrates, the emergence of the term 'Anthropocene' marks a major turn in the way English and Globish speakers have considered human history. One can read this turn as an assignation of responsibility—a gesture which, during the second half of the twentieth century, was anticipated by thinkers like Gunther Anders (1956), Ivan Illich (1973) or Hans Jonas (1979), who differently stressed the large scale unintended and worrying consequences of technical developments occurring at the time. Even if the word emerged from debates among geologists, the more popular framing of the issue it addresses tends to depict Mankind (*anthropos*, *Mensch*) as having unleashed the uncontrolled Power of a Technicity (*technè*, *Gestell*) that is now coming back to haunt us, in the most spectacular form of environmental destructions and climate change. The main *dramatis personae* on this AnthropoScene are therefore the Earth (the victim, wounded and scared by thoughtless human actions), Technology (the runaway adjuvant, turned unmanageable crook), Mankind (the innocent protagonist, surprised to find itself in the position of the culprit), and Science (the tentative saviour, in need of funding to better understand what had gone wrong and to develop innovative green tech in charge of cleaning up the crime scene).

Quite early on, this general plot came under attack. Rather than attempting to retrace the editorial history of the ambiguities, criticisms and defences raised around the 'Anthropocene' (see Wallenhorst, 2019, 2020), it may suffice, in order to explain the proposition of the alternative term 'Plantationocene', to understand what types

Y. Citton (✉)
Université Paris 8, Saint-Denis, France

N. Wallenhorst, C. Wulf (eds.), *Handbook of the Anthropocene*,
https://doi.org/10.1007/978-3-031-25910-4_110

of reproaches have emerged against this scenario, and what alternative scripts have been advanced as more realistic, more fair, and/or more empowering.

A first and most common set of attacks have come from voices located in, or sympathetic with, the global South. The character of Mankind (*anthropos*) seems much too inclusive: it tends to dilute the specific responsibility of Western colonizing nations and people in the wounds suffered by the Earth. Describing our age under the label 'Anthropocene' fits suspiciously well with the attempt made by rich Western governments to ask "all the nations of the world" and "all the people on Earth" to make shared sacrifices in order to prevent a catastrophic worsening of the damage done by greenhouse gas emissions. In consideration of the wrecking of our common (and only liveable) planet, however, any *anthropos* was certainly not born equal to any other. Because of their differently affluent and wasteful lifestyles, an average US resident emits eight times more CO_2 than an average resident of India, and one hundred times more than a resident of Madagascar or Niger (World Bank, 2020). In light of the historical record, as well as in light of current ecocidal impacts, authors like media archaeologist Jussi Parikka (2015) proposed to rename our age the 'Anthrobscene' to account for the North/South injustice of this dominant framing.

From then on, the story morphs into a whodunit. Who is to blame for the ecocide? Who bears the responsibility—and who should be made to shoulder the cost—of the damage done and of the repair to come? Clearly not the people of Niger or Madagascar, but who else? A likely suspect is denounced by Jairus Victor Grove (2019) in proposing the word 'Eurocene'. Seen from a decolonial perspective, the crime scene leaves little doubt: populations and cultures originating in Western Europe have progressively extended their wars and military conquests over the whole planet. They have developed various forms of technologies—material (combustion engine) and managerial (finance)—which have spread worldwide, following their military expansion. One very peculiar 'praxis of being human' (Wynter, 2015), which can be identified with *homo oeconomicus*, has become a hegemonic anthropologic model, and its extractivist relations to its surround are directly responsible for the damage. This story deserves to be written in black and white—and the devastations of the Anthropocene are a crime of whiteness.

Here again, though, the label used to denounce the culprit may be too inclusive. Can equal blame be put on a small family farmer in Southern Italy and on a London financial trader? What the European colonial armies have spread is not so much a genome or a culture, but rather an economic and financial model: capitalism. To say we live in the 'Capitalocene' puts frontstage, not mankind as a whole, not a subsection of it, but a socio-economic dynamic geared towards and driven by the limitless accumulation of what Karl Marx conceptualized as 'the capital'. More precisely, the plot now revolves around a certain "way of organizing nature—as a multispecies, situated, capitalist world-ecology" (Moore, 2016, 6). Even if this way of organizing nature has originated in Europe, it has a logics and a drive of its own, and it is this very specific narrative logic which answers the whodunit question.

The crime at hand may not be that easy to solve, however. Critics point towards the rather dismal ecological record of the Soviet Union—hardly a champion of capitalism—to suggest that the detectives should pursue their investigation. And the new proposals have flourished over the past years—like 'Manthropocene': look at the gender inequalities in decision-making instances, and you will probably find a good correlation between the uses of technology that have damaged our environment and the proportion of tie-wearing bipeds.

The advantage of 'Plantationocene' (Tsing, 2015; Haraway, 2016; Perry & Hopes, 2022) may be that it manages to superpose, compound and articulate most of the partial alternative stories we briefly surveyed. It goes back in time, not to an original sin common to most human population (like the Neolithic Revolution), but to a relatively recent episode (directly aligned with the Eurocene story): the establishment of colonial plantations (sugar, coffee, cotton) in the Atlantic triangle from the seventeenth century on. In Marxist and decolonial terms, the slave plantation was a horrendous way to accumulate capital by brutalizing, displacing and exploiting (to death) populations extracted from Africa and brought to the Caribbean, as well as Brazil and North America. But the plantation was also a place of experimentation for industrial modes of production, where Technology (another major character in our plot) was to play a major role. Finally, nothing could describe more accurately a plantation than "a way of organizing nature": its standardized mode of settlement consisted in destroying whatever forms of biodiversity could inhabit the place, in order to establish a clean slate (*tabula rasa*) upon which to rationalize strictly scalable operations under a regime of monoculture.

From a decolonial point of view (La Paperson, 2017, 6–12), the slave plantation, in its historical invention, was the epitome of the 'settlement', in its purest and most abstract form: 'Settlers' conceived production through an appropriation that would not only instrumentalize, brutalize and dehumanize (black) 'Slave' populations, but also burn, destroy and negate the preexisting ecosystem and 'Native' people, by reducing a living milieu to a set of potential 'resources' which could be artificially recombined in order to be 'exploited' in strict alignment with the Settlers' interests. Opting to label our age 'Plantationocene' invites us to read the past four centuries—and the present—through this decolonial lens, as a struggle between these same three racialized classes of *anthropoi*, who are indeed in very different positions within the misnamed 'Anthropocene': COP negotiations as well as forest exploitation in the Amazon reenact the same conflicts between the calculations of Settlers' interests, the displacement (forced migrations, incarceration, exploitation) of black/Slave people, and the negation-extermination of indigenous/Native people. We tell quite a different story about our past, our present and our future, whether we choose to cast Settlers, Slaves and Natives, or whether we stick with the Earth, Mankind, Technology and Science plot.

Today's agro-industrial plantations have not made slave labour obsolete: working conditions in the Southern regions of Italy, Spain or California (relying on illegal migrant workers) are barely more respectful of human dignity than the historical plantation. But even in the cases when farm employees are decently paid and treated, another crucial feature of the plantation model continues to define them: monoculture. Anna Tsing (2017) has brilliantly analyzed the ecological dementia of this 'way of organizing nature': by negating the biodiversity that ensured a certain resilience to the previous ecosystem, monoculture exposes our living milieus to the onslaught of uncontrollable contagions. Monoculture still persists in countless projects of tree plantations proudly displayed worldwide as 'green' and 'eco-friendly'. Insisting on calling our age 'Plantationocene' should remind us that some cures maybe just as poisonous as the disease, because they carry the disease in their very monocultural formula.

According to philosophers-poets Stefano Harney and Fred Moten (2021), we have been and continue to be living in a regime of 'Plantocracy', which is the main culprit for the 'ge(n)ocide' that has been going on for more than three centuries (on Black and Indigenous people), with the new perspective soon to engulf wide sections of the Settlers' class itself in its horrors. If this recasting of the AnthropoScene in the *dramatis personae* of the Plantationocene keeps the Earth as the crime scene, it does not only split Mankind in three antagonist classes. It also urges us to reconsider the many ways in which Technology and Science have made themselves the accomplices of the ge(n)ocide.

The ecological dementia of the monocultural plantation is a terribly rational form of productive economy, supported by loads of awfully scientific studies. While the Anthropocene depicts Mankind and Technology as the guilty parties in the wreckage of the Earth—and Science as the latter-day savior—acknowledging the Plantationocene induces the uncomfortable splitting our identities as scholars. The history of male domination in Science (aligning this part of the story with the Manthropocene plot), as well as the firm anchorage of scientific activities and institutions in the Settlers' agenda, certainly invite us to ask what can be our future roles as students and academics in our collective emancipation from the ecocidal regimes of power that doom the future of the Plantationocene. Rewriting the plot is a necessary first step to reconsidering and altering its (un)happy ending. May our anxieties about 'the end of the world' lead to putting an end to Plantocracy.

References

Anders, G. (1956). *Die Antiquiertheit des Menschen*. Beck.

Grove, J. V. (2019). *Savage ecology. War and geopolitics at the end of the world*. Duke University Press.

Haraway, D. (2016). *Staying with the trouble. Making kin in the Chthulucene*. Durham.

Harney, S., & Moten, F. (2021). *All incomplete*. Minor Composition.

Illich, I. (1973). *Tools for conviviality*. Harper & Row.

Jonas, H. (1979). *The imperative of responsibility: In search of ethics for the technological age*. University of Chicago Press, 1984.

Moore, J. W. (2016). *Anthropocene or Capitalocene? Nature, history, and the crisis of capitalism*. PM Books.

Paperson, L. (2017). *A third university is possible*. University of Mnnesota Press.

Parikka, J. (2015). *The Anthrobscene*. University of Minnesota Press.

Perry, L., & Hopes, A. (Eds.). (2022). The Plantationocene series. Plantation worlds, past and present. *EdgeEffect*. https://edgeeffects.net/plantationocene-series-plantation-worlds/

Tsing, A. L. (2015). *The mushroom at the end of the world. On the possibility of life in the capitalist ruins*. Princeton University Press.

Tsing, A. (2017). A threat to Holocene resurgence is a threat to livability. In M. Brightman & J. Lewis (Eds.), *The anthropology of sustainability* (pp. 51–65). Palgrave Macmillan.

Wallenhorst, N. (2019). *L'Anthropocène décodé pour les humains*. Le Pommier.

Wallenhorst, N. (2020). *La vérité sur l'Anthropocène*. Le Pommier.

World Bank. (2020). *CO_2 emissions (metric tons per capita)*. https://data.worldbank.org/indicator/EN.ATM.CO2E.PC (consulted 12/31/2021).

Wynter, S. (2015). In K. McKittrick (Ed.), *Sylvia Wynter: On being human as praxis*. Duke university Press.

Yves Citton is fortunate enough to be paid to study and teach media and literature at the University of Paris 8 Vincennes-Saint Denis and to be co-director of the journal *Multitudes*. He is the author of a dozen book, including *Altermodernités des Lumières* (Seuil, 2022), *Faire avec. Conflits, coalitions, contagions* (Les Liens qui Libèrent, 2021), *Générations collapsonautes* (Seuil, 2020, in collaboration with Jacopo Rasmi), *Mediarchy* (Polity Press, 2019), *Contre-courants politiques* (Fayard, 2018), *The Ecology of Attention* (Polity Press, 2016), *Renverser l'insoutenable* (Seuil, 2012).

Postcolonialism

Fred Poché

Abstract This article presents the philosophical influences of postcolonial studies and their object through its protagonists, namely E.W. Said, A. Nandy, G.C. Spivak and H. K. Bhabha. The author also explains the importance of this research for coexistence or "living together" in our multicultural societies.

Postcolonial criticism deconstructs the representations and symbolic forms that served as the basis for the colonial project. In *Orientalism* (1978), the founding father of postcolonial studies, E. Said (1935–2003), describes the way in which Western scholars and poets have, since the eighteenth century, constructed the image of a mythical and obscure Orient, an antithesis of the Enlightenment, suitable for justifying its colonization. For the American-Palestinian intellectual, far from being a universal science, *Orientalism* has more value as a sign of European and Atlantic power over the East than as a truthful discourse on it. This deconstruction unmasks, then, the power of falsification of the colonial language. It allows us to measure that what passed for European humanism appeared each time, in the colonies, in the form of duplicity, double language and distortion of reality (Mbembe, 2006). Said questions the academic knowledge constructed by the West about the rest of the world. He insists on the ideological dimension of this view and sees Orientalism as the source of American anti-Arab prejudice.

Said relies on the French philosopher Michel Foucault to work on the notions of "representations" and "discourses", rather than on the objective conditions of colonial reality. His work thus opened up a field of study attentive to the relationship between knowledge and power: colonial discourse. Said's various works constitute an essential point of support for postcolonial studies. *Orientalism* (Said, 1978) functions as a method for understanding the colonial period, but also the normative orientations and representations that continue to structure the relationships of Western countries with other societies and cultures. Said's work allows us to give full scope

F. Poché (✉)
Catholic University of the West, Angers, France

© The Author(s), under exclusive license to Springer Nature Switzerland AG 2023
N. Wallenhorst, C. Wulf (eds.), *Handbook of the Anthropocene*,
https://doi.org/10.1007/978-3-031-25910-4_111

to the analyses of discourse found in particular in the structuralist or post-structuralist current (Foucault, Lacan, Derrida, Deleuze). In this sense, post-colonial research is linked to the notion of representation. Indeed, in the discourse of colonialism, the representation of the Other is based on the immutable fixity of its characteristics. Now, this fixity implies that otherness is thought of as atemporal, "outside of history and progress, enclosed in an original and authentic dimension that precedes civilization, by definition Western" (Matheron et al., 2007).

In the Indian context, A. Nandy (1937 –) widens the breach opened by Said. In a book entitled The Intimate Enemy, Ashis Nandy emphasizes that colonialism was, above all, a psychic affair. As such, the anti-colonialist struggle constituted, according to him, both a material and a mental war. For the Indian psychologist, the resistance to colonialism and its corollary, nationalism, were forced to operate in the terms previously defined by the West. Long before the others, it was Nandy who facilitated Frantz Fanon's passage to India. At the same time, he introduced psychoanalysis into the postcolonial discourse while opening a dialogue with the philosophers of the Frankfurt School: Adorno and Horkheimer.

It is common to say that two researchers in particular, along with E.W. Said, form the "Holy Trinity" of postcolonialism: Gayatri Chakravorty Spivak (1942–) and Homi Bhabha (1949 –). Spivak's work traces the entire experience of post-colonial migration from India to the United States. This important author presents herself as a post-colonial intellectual caught between the socialist ideals of the national independence movement and the legacy of the colonial education system. By virtue of her training, the literary theorist can be seen as the heir to a colonial education policy implemented in India by the British Empire since the nineteenth century. It should be noted that the education policy encouraged, at the time, the training of the Indian middle classes so that they would integrate the values of British culture.

According to Spivak, the teaching of English literature in colonial India provided an insidious but effective pathway for the "civilizing mission" of imperialism; hence her concern to develop research showing this ideological function of English literature in the colonial context. In "Three Women's Texts and the Critique of Imperialism" (1985), the philosopher emphasizes the impossibility of reading nineteenth century British literature without remembering that imperialism, understood as the "social mission", was part of the cultural representation of England for the English. Finally, G. C. Spivak helped bring Derrida to the attention of the United States by translating and prefacing: On Grammatology (Spivak, 1974). She is, notably, the author of a text commented on around the world: "Can the subaltern Speak?" (Spivak, 1988).

Similarly, Homi Bhabha is one of the most important and influential thinkers in the postcolonial critical movement. The author of a widely commented work around the world, he comes from a small community, the Parsis (from Parashika: people of Perce), a faith derived from Zoroastrianism. Bhabha works with open borders on the colonial question and its extensions. He is interested in the history of art, languages and civilizations of South Asia, as well as human rights. Particularly marked by the writings of Edward Said, our author is in critical dialogue with feminist theorists (Kristeva, Cixous…) as well as post-structuralist philosophers such as Foucault, Deleuze and Derrida.

Beyond a certain heterogeneity, postcolonialism can be summarized in four main axes: a reflection from the crisis of representation, both aesthetic (avant-garde art) and political (the crisis of representative democracy); a critique of essentialist and totalizing thought, in this case, of the notion of identity understood as a fixed reality; a concern to decentralize the subject as it appears in the Cartesian tradition; and a questioning of any form of binarism. Bhabha, precisely, attacks Western production and the realization of certain binary oppositions (centre/margin, civilized/savage, enlightened/ignorant). His work reveals a fragmented terrain where the objects, like the sources of study, are particularly heterogeneous. Thus, we come across, in his writings, texts of English missionaries of the nineteenth century, novels of Toni Morrison, or of Nadine Gordimer, the reflections of Heidegger, Lacan or Derrida; or still that of Edward Said, of poets of the Caribbean, of Frantz Fanon and of Walter Benjamin. This division of the textual corpus naturally follows from conscious choices. Indeed, for Bhabha, the analysis of cultural difference requires the shaping of a transformed interdisciplinary discourse. We could speak in this regard of an undisciplined epistemology. Moreover, the author of *The location of culture* (1994) invents new terms to define the changing relationships between the various poles of the postcolonial world: ambivalence, hybridity, sly civility, mimicry, third space. These concepts describe the ways in which colonized peoples have resisted the power of the colonizer.

A Challenge for Coexistence

Relatively recently in France, contrary to the Anglo-Saxon world, research concerning the situation of the suburbs deserves to be developed. Indeed, the problems of discrimination, the quest for recognition, and the question of "wounded memories" of colonization all call into question the republican model of integration. Basically, what the young Sartre reproached liberal or bourgeois anthropology for, namely the fact of defining human being once and for all and of removing the perception of collective realities, remains relevant and promising. In fact, the contemporary tendency towards tension, or even the denial of the traumas of colonization and its present effects, requires a new impulse to think of diversity as richness. In this sense, creatingwhat is common to us, at a time of generalized fragmentation, should not be centred on a quest for what is similar, but the sharing of diversity.

From Postcolonialto "Decolonial". Postcolonial Theories Developed During the 1980s and 1990s in Anglo-Saxon Thought

Centred on the colonial legacies of the nineteenth and twentieth centuries in India, Australia, Africa and the Middle East, these completely ignored Latin America. Yet these countries have been emancipated since the dawn of the nineteenth century.

Some so-called "decolonial" currents emphasizethat the postcolonial critique of Eurocentrism remains Eurocentric because it is limited "to the legacies of the North European empires of the 19th century" (Boidin, 2009). More precisely, then, it is proposed that we not only question slavery and colonization, but that we go further back in history and question capitalism, in this case up until 1492. The emergence of this economic system "is thus not perceived by decolonials as being related to the "industrial revolution" of the eighteenth century, but as arising from the invasion of America, the driving of Muslims and Jews out of Spain and the ensuing process of ethnic cleansing (limpieza de sangre)" (Amselle, 2018).

Culture, Class, Race, Gender

The future of postcolonial, or "decolonial", issues is not just a matter for the academic field. Indeed, research conducted in various parts of the world should be able to further inspire social and political practices. Moreover, it seems necessary to cross the cultural question with the wounded memories and the questions relating to the problems of class, race or gender. In an era marked by identity-based tensions, such attention should allow for the implementation of a real politics of difference (C. West, 1993). Indeed, as the African-American philosopher Cornel West points out, racial and gender identities are unintelligible without reference to class. The racial explanation, alone, cannot therefore account for the positionof people of immigrant background who are sometimes confronted with double, or even triple, oppression.

References

Amselle, J. L. (2018). Au-delà et en deçà du postcolonialisme. Hommage à Yambo Ouologuem (1940–2017). *Les temps modernes, 3*(699), 76–84.
Boidin, C. (2009). Études décoloniales et postcoloniales dans les débats français. *Philosophie de la libération et tournant décolonial, Cahiers des Amériques latines*, n°62: 129–140.
Matheron, F., Corsani, A., Degoutin, C., & Zapperi, G. (2007). Narrations postcoloniales. *Association Multitudes, 2*(29), 15–22.
Mbembe, A. (2006 décembre). Qu'est-ce que la pensée postcoloniale? (Propos recueillis par O. Mongin, N. Lempereur et J. L Schegel), *Esprit*: 117–133.
Said, E. W. L'Orientalisme. *L'Orient créé par l'Occident*. Cerf. (Original edition 1978, French translation 2004)
Spivak, G. C. (1974). Translator's preface. In J. Derrida. *Of grammatology* (G. C. Spivak, Trans.). (pp. IX–LXXXVII). The Johns Hopkins University Press.

Spivak, G. C. (1988). «Can subaltern speak ?», dans Cary Nelson et Lawrence Grossberg (dir.), *Marxism and the interprétation of culture*, pp. 271–313. Champaign, University of Illinois Press.

West, C. (1993). *Beyond eurocentrism and multiculturalism volume two. Prophetic reflexions. Notes on races and power in America*. Common courage press.

Fred Poché is a professor of contemporary philosophy at the Faculty of Human and Social Sciences of the Catholic University of the West Angers, France, as well as a member of the RPPsy "Research in Psychopathology and Psychoanalysis", component of Angers. Author of about twenty books, he is the winner of the Jean Finot prize of Political and *Moral* Sciences *Acasdemy* for his book: *Blessures Intimes, blessures sociales* (2008).

Posthumanism

Mads Rosendahl Thomsen and Jacob Wamberg

Abstract This article provides an overview of various positions in the heterogenous theoretical field termed "posthumanism." Posthumanism may be understood
as both an umbrella category for post-anthropocentric movements in general, from
New Materialism to Anthropocene theories, and as a more specialized field that
focuses on the prospects of deliberate technological modifications of the human
body, often termed "the posthuman." In the latter, body-centered and cyborgian
approach, posthumanism supplements the typical Anthropocene focus on technology's involuntary destructive effects on the planet's environments.

As a commonly-used term, posthumanism is only a few years older than the
Anthropocene—there are few mentions of either prior to 1990—and like theories on
the Anthropocene, it refers to several positions that often conflict with each other.
As an umbrella term, posthumanism may be understood as both an overarching and
a supplementary category, when speaking of Anthropocene theories. In its most
comprehensive sense, "posthumanism" is simply a synonym for "post-
anthropocentrism," and encompasses theories related to the Anthropocene, together
with domains such as New Materialism and theories of technological modification
of the human body, often called "the posthuman" (Thomsen & Wamberg, 2020).
Here, posthumanism would typically cover all sorts of deconstructions of human
exceptionalism, in an attempt to confer agency, rights, and dignity on other entities,
from other large animals, to all biological life, to cyborgs, to machines, to inorganic
matter at large. Indeed, when linked to earlier genealogies, posthumanism has the
potential to become a hyper-intersectionalist node for all those hermeneutics of suspicion that, in an internal critique of the humanities, expose the series of Others that
the humanist subject has had to suppress in order to manifest its alleged autonomy,
including Marxism (the proletarian Other suppressed by the middle and upper
classes), psychoanalysis (the subconscious Other suppressed by consciousness),

M. R. Thomsen (✉) · J. Wamberg
Aarhus University, Aarhus, Denmark
e-mail: madsrt@cc.au.dk; kunjw@cc.au.dk

689

N. Wallenhorst, C. Wulf (eds.), *Handbook of the Anthropocene*,
https://doi.org/10.1007/978-3-031-25910-4_112

feminism (the female Other suppressed by the male subject), deconstruction (the Others of intertextuality and fluctuating signification suppressed by the transcendental signified), postcolonialism (the non-Western Other suppressed by the West), and disability studies (the corporeal Other suppressed by the normative body).

Perhaps more usually, posthumanism also specifically addresses theories of the posthuman, that is, the technologically-modified human body. Here, posthumanism becomes a supplementary category with respect to Anthropocene theories, since both domains concern nature, modified by technology. However, whereas Anthropocene theories are typically oriented towards technology's unintended and destructive effects on the environment—global warming, species extinction, exhaustion of natural resources, pollution—posthumanism's focus is the human body intentionally modified by technology. Thus, the Anthropocene's conflicted narrative of humanity being at its most powerful while dooming itself by its own actions has parallels in two of the more widespread uses of "posthumanism" in this simultaneously technology- and body-oriented sense.

First, we encounter the overly optimistic theory of transhumanism that focuses on the technological potential of modifying the human species to a point where it becomes a new species (More & Vita-More, 2013). In a radicalization of certain strands of Enlightenment thinking (Thomsen and Wamberg, pp. 25–46), the human being is considered as being in such technological control of its own destiny that it can indefinitely upgrade aspects such as cognition, emotion, physical abilities, and lifespan, perhaps to the point of being able to upload its mind to more durable media, and thereby become immortal. Proponents of this substrate-independent progressionist thinking, such as Ray Kurzweil (2005), assume that technological acceleration will pursue an exponential curve to a certain point, the so-called Singularity, where change will be so pervasive and rapid that it can no longer be perceived or modelled. This transhumanist, utopian narrative of accelerating progress due to the intended effects of technology is obviously in an utterly ironic contrast to the Anthropocene dystopian narrative of accelerating planetary disaster caused by the unintended effects of technology, and the simultaneous existence of these two narratives that allegedly concern the same planetary future is essentially paradoxical. Nevertheless, they have a strangely intertwined relationship, the Anthropocene disaster emerging directly as the suppressed, dark shadow of the transhumanist desire, if not greed, for progress.

This entanglement becomes especially obvious if stated in terms of complexity theory. As emphasized by transhumanist philosopher Max More, civilization's transhumanist progress may be seen as technology's take-over of biological evolution's main drive, the tendency to counter the otherwise universal trend of increasing entropy and disorganization, and instead produce increasing complexity, what often is called negative entropy, or negentropy, but what More terms "extropy" (More, 2003). This is the same ultimate grand narrative of technology's role in evolution that is expressed by the otherwise arch-sceptic of grand narratives, postmodern philosopher Jean-François Lyotard (2001 [1993]). In a so-called postmodern fable, Lyotard speculates about what will happen to the evolutionary drive towards rising complexity, when, in the distant future, the sun will expand into a nova (actually a red giant) and swallow the whole solar system (actually, only its inner parts). His answer is that humans are only a temporary shell for evolution's rising complexity, which will therefore seek a more durable, cyborg medium, a better performing system, in which to escape the sun's feverish death.

What is notable in the Anthropocene update on this postmodern transhumanist fable is that the earth's destruction by heat is not only imagined as a remote phenomenon in time, caused by the distant sun. It is happening right now, right here, due to the suppressed shadow of capitalism's greed for accelerating complexity, the excessive amounts of rising negentropy's own exhaled entropy, which, in the form of excess carbon dioxide and other gases, causes global warming. Therefore, Anthropocene present and transhumanist future manifest two sides of the same coin. This is indeed the central point of the late work of philosopher of technology, Bernard Stiegler (2018). Here, planetary entropy emerges as simply the repressed remnants of so-called anthropological culture, capitalist computerized industry, culminating in transhumanism, so the Anthropocene simply amounts to what Stiegler terms the "Entropocene." Therefore, entering Stiegler's hoped-for Neganthropocene, the epoch following the Anthropocene bifurcation that would otherwise lead to an environmental apocalypse, cannot be qualified other than by caring more about negentropy, and reducing entropy (Stiegler, 2018, pp. 45 and 51).

However, transhumanist progressionist desires and their entropic discontents comprise only an extremist position in the posthumanist landscape. Most obviously, this is countered by what is often termed Critical Posthumanism (Hayles, 1999; Herbrechter, 2013 [2009]; Braidotti, 2013; Wolfe, 2010; Haraway, 1991 [1985]). Although rarely involved with truly ecological forms of complexity theory, including concerns about entropy, Critical Posthumanism is in tune with Anthropocene theories in its decentering of the human mind and body, now typically informed by various forms of deconstruction and systems theory, from cybernetics to Niklas Luhmann's sociology. Facing the convergence of computer and biotechnologies that have the cyborg, the amalgamation of machine and biological organism, as its most condensed hybrid (Haraway, 1991 [1985]), these positions are governed by a more material perspective on both mind and information. The systems thinking that typically informs Critical Posthumanism assumes some sort of self-organization that runs through all levels of matter—from inorganic entities, to organisms, to machinic information, to human semiotic meaning—and Critical Posthumanism is thus disconnected from the control of the individual Cartesian mind that still lingers as the ideal of transhumanism. Nevertheless, it often supposes a complete cyborgian integration of machines and organisms, which still seems to conflict with biological life. Life still refuses to be seamlessly absorbable by general theories of self-organized systems, just as the mind does not seem to be reducible to machine-based information. Perhaps Critical Posthumanism still suffers from a methodological gap between social constructivism and materialism, and, on a more ontological level, between the concepts of mind and matter. These gaps seem to only become deeper when we move into those more broadly posthumanist theories that wholly foreground material entities at the expense of semiotic ones, such as Bruno Latour's

Actor-Network-Theory (ANT), ironically derived from semiotic structuralism (Latour, 2007), or New Materialisms, such as Speculative Realism (Harman, 2018).

Although one may think that the posthumanist dissolution of human exceptionalism, and by extension, the individuum/surroundings dichotomy, would prevent it, for historical reasons there still seems to exist a considerable gap in theoretical focus between environmental destruction (Anthropocene thinking about surroundings) and bodily construction (posthumanist thinking about the individual body, even when hybridized). This is probably due to the circumstance that theoreticians concerned with the Anthropocene, appalled by the environmental disaster caused by the capitalist misuse of technology, are distinguished by a more or less Luddite disillusionment with the positive potential of technology. Therefore, the environmental parallel to the cyborg, the technological modification of the environment known as geoengineering, is mostly denigrated by Anthropocene theoreticians, not only in its techno-optimistic ("trans-Anthropocene") versions, but also in its pragmatic role as mere damage control, ("Critically Anthropocene") techno-fatalism (Kolbert, 2021, p. 200). Here, the path taken by Donna Haraway, pioneer of posthumanist cyborg theory, is illuminating, since, although she once suggested an inevitable techno-infiltration, as far as human and other individual species are concerned (Haraway, 1991 [1985]), she now, focusing on the environment, minimizes technology's constructive capacity, or just constructive givenness, and considers its potential intervention in the environment as "a comic faith in," indeed a "touching silliness about technofixes" (Haraway, 2016, p. 3). Even when James Lovelock, the theoretician of planet Earth as a comprehensive cybernetic system, Gaia seeks to integrate Gaia with technology-oriented posthumanism in its narrower sense, his envisaged cyborgian sphere of a thoroughly artificial intelligence seems somewhat disconnected from the now indeterminately-healed planetary environment in which it is embedded.

The narrower focus on the body in technology-oriented posthumanism also seems to comprise a blind spot in theories on the Anthropocene that ought to be more thoroughly addressed. If the implied grand narrative of the Anthropocene is that humanity changes everything on the planet, it seems odd how little attention has been paid to how humanity might alter itself, even on a pragmatic level. One does not have to buy into transhumanist fantasies to recognize a number of developments that change the human condition more directly than changes in the environment do. Human–machine interaction is transforming the way humans perceive the world and handle information. Cyborgification, even when this term is not used, is becoming more advanced and normalized, particularly for therapeutic purposes. Gene modification, particularly through CRISPR-Cas9 techniques, is being developed at a speed such that moral and legal restraints, rather than technical capacity, determine its development. Nonetheless, the degree to which it is possible to prevent experiments that seek to alter and improve humans is open to question, even if "improvement" remains a contested term connected to progressionism. Therefore, at the very least, to paint a realistic picture of what the (post)human condition presently involves, the analytical lens of the Anthropocene should be combined with one that gauges the effects on humanity directly caused by deliberate human acts. Ignoring the potential developments humans may undergo, as the spectrum of technologies keeps expanding, blocks the paths to a new, balanced view of a planetary future that is not centered on the human, but which may hardly disregard the deliberate effects of technology on the human body.

Apart from decentering cognition, both Anthropocene and Critically Posthumanist thinking encompass an ethical decentering that criticizes the greater value assigned to human life. However, although all posthumanist observers would principally agree on the necessity of expanding rights to non-human entities, there is no consensus of how far this should be taken in practice. Although some animal rights are more or less universally agreed upon, conferring rights on more diffuse entities, such as ecosystems, or the planet as a whole, quickly becomes difficult to define. Politically and juridically, we still seem surprisingly bound to the asymmetry of rights that are based in human-centered systems.

In all variants of posthumanism and Anthropocene theories there lurks, almost per definition, a certain strand of anti-humanist thinking. Posthumanism, even transhumanism, implies that the time of the human being as we know it, is past, and the very idea of naming a geological period after the human species also suggests that there will be a time after humans, perhaps after their suicide through planetary devastation. Nevertheless, how exactly this anti-humanism manifests varies quite a lot. Does antihumanism mean that the human species is over and done with, or that a certain humanist way of thinking and behaving is becoming obsolete, so the human species remains, only in new discursive and artifactual contexts—as has been the case in earlier periods, perhaps throughout most or all of human history? In his deservedly influential work, *What is Posthumanism?*, Cary Wolfe stresses that he is neither antihumanist nor transhumanist, but that the "post" should be understood in the same sense as in "the postmodern," that is, as indicating something that both precedes and follows humanism (p. xv). If even so moderate a posthumanism may help to steer the human species in a direction that will also benefit its own condition by leading it to better care for the environment and change its behavior in ways that respect other living beings, it need not to be so shameful to remain human, after all.

References

Braidotti, R. (2013). *The Posthuman*. Polity.

Haraway, D. J. (1991 [1985]). A cyborg manifesto: Science, technology, and socialist feminism in the late twentieth century. In Haraway, *Simians, cyborgs and women: The reinvention of nature* (pp. 149–181). Free Association Books.

Haraway, D. J. (2016). *Staying with the trouble: Making Kin in the Chthulucene*. Duke University Press.

Harman, G. (2018). *Speculative realism: An introduction*. Polity.

Hayles, N. K. (1999). *How we became Posthuman: Virtual bodies in cybernetics, literature and informatics*. University of Chicago Press.

Herbrechter, S. (2013 [2009]). *Posthumanism: A critical analysis*. Bloomsbury.

Kolbert, E. (2021). *Under a white sky: The nature of the future*. The Bodley Head.

Kurzweil, R. (2005). *The singularity is near: When humans transcend biology*. Penguin Books.

Latour, B. (2007). *Reassembling the social: An introduction to actor-network theory*. Oxford University Press.

Lovelock, J. (1979). *Gaia: A new look at life on earth*. Oxford University Press.

Lovelock, J., with Appleyard, B. (2019). *Novacene: The coming age of hyperintellegence*. Allen Lane, Penguin Books.

Lyotard, J.-F. (2001 [1993]). A Postmodern Fable. (G. Van Den Abbeele, Trans.). In S. Malpas (Ed.), *Postmodern debates* (pp. 12–21). Palgrave.

More, M. (2003). Principles of extropy (Version 3.11): An evolving framework of values and standards for continuously improving the human condition. Extropy Institute, on https://web.archive.org/web/20131015142449/http://extropy.org/principles.htm

More, M., & Vita-More, N. (2013). *The transhumanist reader*. John Wiley.

Stiegler, B. (2018). *The negenthropocene* (Ed. & Trans.), and introduction D. Ross. Open Humanities Press.

Thomsen, M. R., & Wamberg, J. (2020). *The Bloomsbury Handbook of Posthumanism*. Bloomsbury Academic.

Wolfe, C. (2010). *What is Posthumanism?* University of Minnesota Press.

Mads Rosendahl Thomsen is Professor of Comparative Literature at Aarhus University, Denmark. His research interests are centred on world literature, posthumanism, and digital humanities. He is the author of *Mapping World Literature: International Canonization and Transnational Literatures* (2008), *The New Human in Literature: Posthuman Visions of Changes in Body, Mind and Society after 1900* (2013), a co-author with Stefan Helgesson of *Literature and the World* (2019), and the editor of 14 books, including *World Literature: A Reader* (2012) and *The Bloomsbury Handbook of Posthumanism* (2020 with Jacob Wamberg).

Jacob Wamberg is Professor of Art History at Aarhus University, Denmark. He works on a paleo-futurist mapping of the visual arts, syncretically fusing areas such as complexity theory, philosophy of technology and posthumanism. His publications include *Landscape as World Picture: Tracing Cultural Evolution in Images* (2009 [2005]), *Totalitarian Art and Modernity* (2010, ed. with Mikkel Bolt Rasmussen), *Art, Technology and Nature: Renaissance to Postmodernity* (2015, ed. with Camilla Skovbjerg Paldam), and *The Bloomsbury Handbook of Posthumanism* (2020, ed. with Mads Rosendahl Thomsen).

Transhumanism

David Doat and Gabriel Dorthe

Abstract This chapter provides an overview of the transhumanist movement, its origins, its main figures and its main positions. It then highlights the fact that transhumanism shows little concern for environmental issues, as it is mostly focused on individual bodies, health and longevity. Finally, this chapter examines how transhumanists activists or related academics address contemporary ecological disasters, focusing on the human engineering hypothesis first, and then the "good Anthropocene" and its connections with some aspects of the debate on solar geoengineering.

The British biologist and first director of UNESCO Julian Huxley first coined the term "transhumanism" at his William Alanson White Memorial lecture in Washington, DC, in April 1951, to support the idea that "the human species can, if it chooses, transcend itself…. We need a name for this new belief. […] I believe in transhumanism: once there are enough people who can really say that, the human species will be on the threshold of a new kind of existence" (Huxley, 1951). This belief resonates with an American public imbued with technological utopianism, which has religious and philosophical roots since the nineteenth century (Noble, 1999). The transhumanist movement in the United States derives from this widely distributed idea that technology could and should drive humanity's fulfilment.

One of the first authors to take up the word "transhumanism" after Huxley – who might have taken it up himself from the French engineer Jean Coutrot (Dard & Moatti, 2016) – was the futurist Fereidoun M. Esfandiary (better known as FM-2030). In his essays *UpWingers: A Futurist Manifesto* (1973) and *Are You a Transhuman? Monitoring and Stimulating Your Personal Rate of Growth in a Rapidly Changing World* (1989), he advocates for the advent of a "transitional human", taking its

D. Doat (✉)
Catholic University of Lille, ETHICS EA 7446, Lille, France
e-mail: David.Doat@univ-catholille.fr

G. Dorthe
Catholic University of Lille, ETHICS EA 7446, Lille, France

Harvard STS & RIFS, Potsdam, Germany

© The Author(s), under exclusive license to Springer Nature
Switzerland AG 2023
N. Wallenhorst, C. Wulf (eds.), *Handbook of the Anthropocene*,
https://doi.org/10.1007/978-3-031-25910-4_113

evolution in charge by means of science and technology, exploring the path to a radically new post-human condition and post-humanist culture. Along with FM-2030, a small group of authors and activists shape transhumanists ideas in the countercultural context of the 1970s California, such as Robert Ettinger for cryonics, Timothy Leary with psychology and experiments with psychedelics, Eric Drexler on nanotechnology, or the designer Natasha-Vita More (born Nancie Clarke).

In 1988, Max More (born O'Connor), a young British philosopher, moved to the Californian technophile and futurist milieu in the hope of realizing his libertarian and immortalist aspirations. He later founded the Extropy Institute, a think tank that laid the foundations of the first philosophy of transhumanism, named extropianism. Investing the new means of communication provided by the emerging Internet, many entrepreneurs and future figures of the Silicon Valley, academics such as the philosophers David Pearce, Anders Sandberg and Nick Bostrom, as well as various supporters, enthusiastically joined this nascent movement.

The sociology of the transhumanist movement was already structured into three groups of actors, that would help developing it up to today: (1) industrialists and entrepreneurs of neoliberal or even libertarian obedience; (2) activists who gather in associations in a dozen countries, more and more leaning towards the technoprogressist branch of the movement (especially in Europe) that advocates for an equal access to technology and public funding for research on healthy longevity; and (3) academics working in bioethics or sociology at Oxford University (UK) and in some US universities who tend to distance themselves from More's extropianism, root transhumanism into the utilitarian tradition, and moderate its technosolutionist optimism by an exploration of existential risks.

In 1998, Nick Bostrom and David Pearce founded the World Transhumanist Association (known today as Humanity+), and spearheaded the writing of the first Transhumanist Declaration, which was slightly updated over the years, in order to provide the movement with a structured core of values. In an extended explanation of these general principles, transhumanism is defined as "(1) the intellectual and cultural movement that affirms the possibility and desirability of fundamentally improving the human condition through applied reason, especially by developing and making widely available technologies to eliminate aging and to greatly enhance human intellectual, physical, and psychological capacities; (2) the study of the ramifications, promises, and potential dangers of technologies that will enable the overcoming of fundamental human limitations, and the related study of the ethical matters involved in developing and using such technologies" (Transhumanist FAQ, 2003). This definition of transhumanism allows taking into consideration the issues of accessibility of technologies and the involved risks for humans, society and nature. But above all, it sanctions the convergent ideological commitment of all transhumanists (extropianists, utilitarians, technoprogressists) to the modification and enhancement of the human body, and the elimination of its deficiencies and natural shortcomings.

At its core, transhumanism is not so much concerned with the ecological conditions of enhancement nor with sustainable relations between humans and non-humans or the preservation of natural equilibriums. It focuses on the means for enhancing or transcending the human body, and the instrumentalization of natural resources, deemed as unlimited, for the purposes of humanity and its trans- or post-human destiny. Yet, even if transhumanism is not historically nor doctrinally interested in ecological issues, transhumanist scholars and actors of the movement at large happen to address some of them.

The responsibility of human beings in the degradation of their natural environment due to the imperfection of their nature, the risks of collapse of the possibility of existence of the human species on earth, or in the long term of the cosmos itself (Big Crunch), are seriously taken into consideration in the transhumanist literature (Bostrom, 2013; Vidal, 2014; Cotton-Barratt et al., 2020). Facing these issues, transhumanists traditionally argue for a techno-scientific enhancement of humans; some of them claim for a wiser use of technology for the accomplishment of transhumanists ideals. So doing, they stick with a highly technosolutionist and anthropocentric perspective. Nevertheless, in relation to the concept of the Anthropocene, some recent transhumanist statements are worth mentioning, as potential signs of an emerging ecological awareness in the movement.

In 2012, three philosophers from Oxford and New York University, S. Matthew Liao, Rebecca Roache and Anders Sandberg came up with an idea that has given rise to much criticism. Sandberg is an early transhumanist who, among other things, had been taking an active role in the elaboration of the first versions of the Transhumanist Declaration in the late 1990s. Deeming mitigation of greenhouse gases and adaptation to climate change as being too little too late, Liao, Roache and Sandberg imagine human engineering measures, such as making humans smaller, or pharmacologically induced meat intolerance (Liao et al., 2012). In an interview with one of the authors of this chapter in 2019, Sandberg cheerfully admitted that this article was a piece of academic provocation that was not meant to be taken at face value. This paper nevertheless reflects a typical feature of the transhumanist thought: when it imagines solutions to counteract global warming, it tends to focus on technological interventions on individual bodies.

In 2020, the technoprogressist French Transhumanist Association released the Viridian Manifesto, which assumes that "A technoprogressive future will only be possible in the decades ahead given an indefinitely sustainable relationship with the environment." To that end, the text envisions radical transitions induced by renewable energy, publicly funded research on sustainable development, recycling and decrease of pollution through regulation of over-consumption and advertisement, 3D printing, effective use of AI, or neurosciences (in line with the above-mentioned arguments). Healthy life extension is also justified in this perspective, assuming that living longer would make humans more respectful of their environment (Technoprog 2020). These claims seem to pave the way for a more foundational and structured ecological sensibility in the transhumanist thought than what had featured the utilitarian and libertarian strands. Since the early days of the transhumanist thought, for example in Esfandiary's work, environmental issues have barely been mentioned, except in calls for more precise and powerful technologies that might be able to clean up the pollution already generated. More concerned about social justice and equality, the technoprogressist branch of transhumanism now seeks to address these issues in more depth.

The ecomodernist movement is a source of inspiration and influence for transhumanist thinkers. It stands for a "good Anthropocene", taking the term at face value and promoting an extended stewardship of humans on the planet. *An Ecomodernist Manifesto*, published in 2015 by a group of eighteen scientists and intellectuals led by the founders of The Breakthrough Institute advocates for a wise use of knowledge and technology in order to protect nature, help striving human well-being and stabilize the climate (Shellenberger et al., 2015).

While it is not explicitly mentioned in the manifesto, this latter perspective roughly calls for geoengineering, the "deliberate large-scale manipulation of the planetary environment to counteract anthropogenic climate change" (Shepherd, 2009). One of the signatories is none other than one of the most prominent advocates for carbon capture and storage and solar geoengineering: David Keith, professor of applied physics and public policy at Harvard University. Across the Atlantic, the Oxford principles for the governance of geoengineering have been published in 2009 by an interdisciplinary team of scholars that include one of the early figures of transhumanist bioethics, Julian Savulescu (Rayner et al., 2013). The *Viridian Manifesto* also mentions this controversial technological scheme, accounting that it should be used "only when we know that the means envisaged are reversible without new technology." Back in 2014, Kris Notaro, who was then the Institute for Ethics & Emerging Technologies' Managing Director (an online think tank co-founded by major transhumanist thinkers), was framing geoengineering as a human right that should be guaranteed under the auspices of the UN's Universal Declaration of Human Rights (Notaro, 2014).

These ecomodernist ideas are progressively taken up by major figures of the transhumanist movement. In the wake of the French Viridian Manifesto, James Hughes, the main inspirational figure of technoprogressism, has published a plea for strong connections between socialism and ecology. Calling for reinvigorating the ideal of progress for all, he promotes, among other things, the wise use of genetically modified crops, geoengineering and nuclear power, all but highly contested issues for environmental thinkers and activists (Hughes, 2021). Adopting a broader perspective, David Wood, chair of the London Futurists and prominent figure of the transhumanist movement, calls his fellow transhumanists "to challenge and reorient the public narrative" about the future, in order to "take wise advantage of the remarkable capabilities of twenty-first century science and technology." The expected acceleration of technological progress is meant to bring abundance and freedom to all humans (Wood, 2019).

This brief overview of the transhumanist movement's positions towards the Anthropocene and its challenges, and its attempts to accommodate transhumanism and ecology, shows a steady optimistic vision of the future despite ecological threats. The movement remains constrained by its classical technosolutionism and anthropocentrism. Technology is seen as morally neutral, and ultimately called to solve human problems and transcend short-term perspectives as well as political divides. In this respect, transhumanism fits in many of the ambiguities of the concept of the Anthropocene.

References

Bostrom, N. (2013). Existential risk prevention as global priority. *Global Policy, 4*, 15–31. https://doi.org/10.1111/1758-5899.12002

Cotton-Barratt, O., Daniel, M., & Sandberg, A. (2020). Defence in depth against human extinction: Prevention, response, resilience, and why they all matter. *Global Policy, 11*, 271–282. https://doi.org/10.1111/1758-5899.12786

Dard, O., & Moatti, A. (2016). Aux origines du mot "transhumanisme". *Futuribles, 413*.

French Transhumanist Association Technoprog. (2020). *Viridian manifesto: Technoprogressive and ecological proposals*, https://transhumanistes.com/viridian-manifesto-technoprogressive-and-ecological-proposals. Accessed 19 Oct 2021.

Hughes, J. J. (2021, March 16). EcoSocialism and the Technoprogressive perspective. *Medium* (blog). https://medium.com/institute-for-ethics-and-emerging-technologies/ecosocialism-and-the-technoprogressive-perspective-671098733fef. Accessed 19 Oct 2021.

Huxley, J. (1951). Knowledge, morality, and Destiny: I. *Psychiatry, 14*, 129–140. https://doi.org/10.1080/00332747.1951.11022818

Liao, S. M., Sandberg, A., & Roache, R. (2012). Human engineering and climate change. *Ethics, Policy & Environment, 15*(2), 206–221. https://doi.org/10.1080/21550085.2012.685574

More, M. (2013). A letter to mother nature. In M. More & N. Vita-More (Eds.), *The transhumanist reader* (pp. 449–450). John Wiley & Sons, Ltd.. https://doi.org/10.1002/9781118555927.ch41

Noble, D. W. (1999). *The religion of technology: The divinity of man and the Spirit of invention*. Penguin Books.

Notaro, K. (2014, July 10). Geoengineering as a human right. *IEET – Institute for Ethics and Emerging Technologies* (blog). www.homepages.ed.ac.uk/shs/Climatechange/Geo-politics/Geoengineering%20as%20a%20Human%20Right.htm. Accessed 19 Oct 2021.

Rayner, S., Heyward, C., Kruger, T., Pidgeon, N., Redgwell, C., & Savulescu, J. (2013). The Oxford principles. *Climatic Change, 121*(3), 499–512. https://doi.org/10.1007/s10584-012-0675-2

Shellenberger, M., & Nordhaus, T. et al., (2015). An Ecomodernist Manifesto, *The Breakthrough Institute*, www.ecomodernism.org/manifesto-english. Accessed 19 Oct 2021.

Shepherd, J. (2009). *Geoengineering the Climate: Science, Governance and Uncertainty*. The Royal Society. https://royalsociety.org/topics-policy/publications/2009/geoengineering-climate. Accessed 23 Feb 2023.

Transhumanist FAQ. (2003). www.nickbostrom.com/views/transhumanist.pdf. Accessed 15 Oct 2021.

Vidal, C. (2014). Cosmological immortality: How to eliminate aging on a universal scale. *Current Aging Science, 7*, 3–8.

Wood, D. (2019). *Sustainable superabundance: A universal transhumanist invitation*. Delta.

David Doat is an associate professor in philosophy at the ETHICS Laboratory of the Catholic University of Lille. His work focuses on human enhancement and vulnerability in contemporary philosophical, scientific, trans- and posthumanist literature. He has published and edited several books, such as "Transhumanisme: quel avenir pour l'humanité?" (2021, Paris, Cavalier Bleu), and "L'homme augmenté en Europe: rêve et cauchemar de l'entre-deux-guerres" (2021, Paris, Hermann).

Gabriel Dorthe is a postdoctoral researcher co-affiliated with the Program on Science, Technology and Society at Harvard and the Research Institute for Sustainability Potsdam. He is an associate researcher at the ETHICS Laboratory of the Catholic University of Lille. He holds a PhD in philosophy and environmental humanities (University of Paris I and University of Lausanne). His dissertation provides an ethnographic account of the transhumanist movement and argues that it is foremost a form of public engagement towards technoscientific promises.

Section III
Human Beings: Bridging Nature and Culture

Nathanaël Wallenhorst and Christoph Wulf

The scientific data emerging in recent decades are evidence of how human activity has disrupted the climate and the balance of life. Thus, the Anthropocene is marked, among other phenomena, by collapses in ecosystems. The current geological problem is that we are launching a trajectory which is very rapidly taking us away from the glacial –interglacial cycle which controlled the functioning of the Earth system for the last million years. If human activity has become the prime geological force governing the very functioning of the Earth system, the main consequence of our actions is to have "forced" (to use the sacred geological expression) numerous subsystems of the Earth system which is at the edge of a new threshold where humans will no longer be the prime geological force. The time is coming when the Earth system will run the risk of entering a transition period that is independent of human actions. In other words, in the Anthropocene, contrary even to the etymology of the term, human actions are not the prime geological force, apart from during a short period (not even a century if see the Anthropocene starting in 1950).

We should also link up our biological knowledge of living things and the biogeochemical knowledge of the Anthropocene with our philosophical tradition. The articles in this section re-examine certain basic elements of the human adventure (culture, religion, art, etc.) and shed new light on what the Anthropocene reveals about our humanity – we understand the extent to which our humanity is dependent on our relationship with the non-human (living as well as cosmic). In fact, the encounter between the long history of the Earth and the short history of the human adventure makes us question what we are. "Something" deep is happening in the relationship between what is cultural and what is natural, impacting directly on the

N. Wallenhorst (✉)
Catholic University of the West, Angers, France
e-mail: nathanael.wallenhorst@uco.fr

C. Wulf
Freie Universität Berlin, Berlin, Germany
e-mail: christoph.wulf@fu-berlin.de

way in which we, as humans, are going to continue to live sharing a terrestrial space with "other inhabitants of the Earth", i.e. non-humans.

Thus, the importance of the concept of the Anthropocene lies fundamentally in the way in which it comes to reconfigure the nature/culture or nature/society dualism that is central in the development of western thinking. Now we can no longer think within the confines of this dualism, since it is clear that a fundamental idea of humanity, of humanity as a geological force (the idea at the heart of the Anthropocene) has now been slowly establishing itself for almost two centuries. In fact, it is important here to note that the question of the impact of human activity on the biosphere as a whole largely precedes the coining of the concept of the Anthropocene. Awareness of the impact of human activity on the climate or the Earth as a whole has a long history, and this idea has been developed by several scientists over two centuries. In the concept of the Holocene itself (the previous geological epoch) we find suggestions of the idea of the global impact of humanity. Initially, this concept involves the idea of the presence of humans on the Earth. In fact the term Holocene, the etymological meaning of which is "entirely recent" was popularised by the French geologist Paul Gervais, who picked up and changed slightly the term "recent epoch", proposed by the Scottish geologist Charles Lyell (1833), who established that the end of the glacial period coincided with the development of human civilisations (Lewis & Maslin, 2015, 172). After dating the start of the geological Quaternary Period as the time when *Homo habilis* appeared, the concept of the Holocene marks the second time that the human race is involved in geological dating. The Anthropocene will be the third occurrence of humanity playing a role in geological dating.

In 1778 the French naturalist, Count Georges-Louis Leclerc de Buffon published an article in *Des époques de la nature*, saying that the whole face of the Earth today "bears the imprint of the power of man". In particular, this makes humanity capable of changing the climate by impacting their environment – "Man can change the influences of the climate in which he lives, and finally set the temperature at the point that suits him" (Buffon, 1778, 237). A few centuries earlier, in July 1494, Christopher Columbus already sensed the capacity of humanity to affect the climate based on the management of deforestation - he is pleased that humans are able to control the climate. After that, between 1830 and 1833, the British geologist Charles Lyell defined the contemporary epoch in *Principles of Geology* as the "human epoch"; 20 years later the English geologist and philosopher William Whewell wrote that "the Human Epoch of the earth's history is different from all the preceding Epochs" (1853, 88). In 1854, the Welsh geologist and theologian Thomas Jenkyn was the first person to identify a geological epoch marked by humanity (1854) which he defines as "the human epoch". Two years later the English intellectual William Adams described his geological epoch based on his intuition of there being traces of human activity in sediments: "The Modern or Human epoch is illustrated by alluvial deposits, which are the effects of atmospheric and other more powerful causes: these still continue in operation, imbedding remains of man and inorganic matter." (1856, 247) About a dozen years later, in 1865, the Irish Reverend Samuel Haughton published his *Manual of Geology,* in which he defined the

Anthropozoic epoch as the "epoch in which we live" (138). From the 1880 edition of his manual, Federau (2016, 64) picks up on one of the consequences that Haughton draws from this geological power: "humans should be at the 'head of the system of life', because of their spiritual nature and their power to progress indefinitely [578–579]". In 1863 the American geologist James Dwight Dana also published a *Manual of Geology* in which he referred to "the Age of Mind and the Era of Man" (130). Then the Italian geologist and priest Antonio Stoppani also described his contemporary period in the same way (1873), using the expression "Anthropozoic Era" (1873, vol. 2, 732). Next, the Russian geologist Aleksei Pavlov (1854–1929) towards the end of his life (the late 1920s), used the expression "Anthropocene" or "Anthropozoic Era" when speaking of his epoch, as mentioned by Vernadsky or Shantser (1973, 140).

In the second edition of *Man and Nature* in 1874 (10 years after the first edition) which the American diplomat and ecologist George Perkins Marsh entitles *The earth as modified by human action*, Marsh refers to the works of Stoppani and his idea of the Anthropozoic Era and concludes that there is an intrinsic incompatibility between industrial societies and the balance of nature (Federau, 2016, 61). In fact, for Marsh "The Earth is fast becoming an unfit home for its noblest inhabitants" (Marsh, 1874, 44). The works of Marsh had quite an impact – although in 1833 Lyell considered that human beings were capable of transforming geography but only played a minor role, after reading Marsh's book 30 years later he changed his mind and modified his assertions (Federau, 2016, 60–62). In *La pensée écologique – Une anthologie*, we also learn that in 1915 the German geologist Ernst Fischer considered "man to be a geological factor" (Bourg et Fragnière, 2014, 137), with the capacity to change the climate. In 1922 in his work *Man as a geological agent*, the British geologist Robert Lionel Sherlock developed this same idea of humanity as a geological force. All these precursors analysed the impact of humanity on the Earth system one or two centuries before it became a theory in the years just after 2000.

Next, at the beginning of the twentieth century, the Ukrainian geochemist and naturalist Vladimir Vernadsky developed the concept of the biosphere, taking up the ideas of the Austrian geologist Eduard Suess who invented the word and for whom it simply meant the space which shelters life in Earth. Vernadsky went on to make this concept more complex. He put forward the idea of there being a relationship between the biosphere and human intelligence by way of the concept of the noosphere (from the Greek for spirit), a concept which has at its base the power of the human spirit. For Vernadsky (1945), the spirit is able to affect its future by altering its environment. Vernadsky does not interpret the geological power of humanity in a religious or spiritual way, but he has great confidence in science being able to good care of the planet (Federau, 2016, p. 71). He has a profound belief in progress – and yet is aware at the same time of humanity's capacity to destroy itself. We can identify this idea of the noosphere as being the predecessor of the concept of the Anthropocene (Steffen et al., 2011). In developing the idea of the biosphere (and then that of the noosphere) which he studies from a biogeochemical perspective also introducing a holistic element, Vernadsky made a seminal contribution to the development of Earth system science, within which the concept of the Anthropocene emerged in 2000.

After that, the evolution of the idea of humanity being a geological force was interrupted by two world wars, only to reappear in 1955 at a conference at the University of Princeton. This was then published by the American geographer William M. Thomas (1956) as *Man's role in changing the face of the earth* (Steffen et al., 2011, p. 844; Robin et al., 2014). Finally, more recently, as spotted by Steffen et al. (2011), a similar term, *"Anthrocene",* was used in 1992 by the American science journalist Andrew C. Revkin, "Perhaps earth scientists of the future will name this new post-Holocene era for its causative element—for us. We are entering an age that might someday be referred to as, say, the Anthrocene. After all, it is a geological age of our own making." (1992, p. 55). Other terms have been used to refer to the global importance of human activity (without referring to the geological power of humanity). For example, Michael Samways, a South African entomologist, invented the term "Homogenocene" to denote the world becoming more uniform, and the American biologist Michael Soulé referred to the cotemporary period as the "Catastrophozoic" era (Kolbert, 2015).

As the history of the idea of humanity as a geological force shows, i.e. this encounter of nature and culture, it is now important to be aware of the need to go beyond this dualism (Wallenhorst, 2019, 2020, 2021, 2022; Wallenhorst & Wulf, 2022; Liebau et al., 2003; Gil & Wulf, 2015; Wulf, 2022, 2013). The work of the contemporary German biologist and philosopher Andreas Weber illustrates this particularly well. His thinking fits into the framework of an anthropology of immersion in nature, through his biological interpretation of life and his reading of the geoscientific works on the Anthropocene. He develops a particularly radical way of thinking which breaks with one of the paradigms of Promethean modernity: it is marked by the distinction between a silent, non-human world and a human world which alone is capable of speech. For Weber it is essential to discover, or rediscover, our relationship with the non-human. "We should conserve nature because it is all that we are not." (Weber, 2007, 294) Andreas Weber, by maintaining the singularity and uniqueness of the subject, shifts the boundary between subject and world by showing that the subject is also the world. He bases his theoretical framework on the sharing relationship between world and subject, which is necessary and vital, from breathing to food. Breathing, for example, is regarded as sharing with the biosphere – ecosystems being dependant purely on exchange. It is not only that we incorporate the elements that surround us, but following this same very simple principle of biophysical exchanges, Andreas Weber goes further: we become transformed as we return to what surrounds us. Indeed, we are the Earth.

References

Adams, W. H. D. (1856). *The history, topography, and antiquities of the Isle of Wight*. Smith Elder.
Bourg, D., & Fragnière, A. (2014). *La pensée écologique – Une anthologie*. PUF.
Buffon, G. L. L. (199, (ed. or. 1778)). *Les époques de la nature*. Diderot éditions.

Federau, A. (2016). *Philosophie de l'Anthropocène – Interprétation et épistémologie*. Thèse de doctorat en cotutelle de l'Université de Lausanne et de l'Université de Bourgogne.

Gil, I. C., & Wulf, C. (Eds.). (2015). *Hazardous future: Disaster, representation and the assessment of risk*. De Gruyter.

Lewis, S. L., & Maslin, M. A. (2015). Defining the Anthropocene. *Nature, 519*, 171–180.

Liebau, E., Peskoller, H., & Wulf, C. (Eds.). (2003). *Natur. Pädagogisch-anthropologische Perspektiven*. Beltz.

Lyell, C. (1990, ed. or. 1830–1833). *Principles of geology* (Vol. I, II, III). University of Chicago Press.

Marsh, G. P. (1970, ed. or. 1874). *The earth as modified by human action: A new edition of "man and nature"*. Arno Press.

Revkin, A. C. (1992). *Global warming: Understanding the forecast*. American Museum of Natural History, Environmental Defense Fund, Abbeville Press.

Robin, L., et al. (2014). Three galleries of Anthropocene. *The Anthropocene Review, 1*(3), 207–224.

Shantser, E. V. (1973). The anthropogenic system (period). In *The great soviet encyclopedia* (Vol. 2, pp. 139–144). Macmillan.

Steffen, W., et al. (2011). The Anthropocene: Conceptual and historical perspectives. *Philosophical Transactions of the Royal Society, 369*, 842–867.

Stoppani, A. (1873). *Corso di geologia, Geologia Stratigrafica* (Vol. 2). Bernardoni Brigola Editori.

Wallenhorst, N. (2019). *L'Anthropocène décodé pour les humains*. Le Pommier.

Wallenhorst, N. (2020). *La vérité sur l'anthropocène*. Le Pommier.

Wallenhorst, N. (2021). *Mutation. L'aventure humaine ne fait que commencer*. Le Pommier.

Wallenhorst, N. (2022). *Qui sauvera la planète?* Arles.

Wallenhorst, N., & Wulf, C. (Eds.). (2022). *Humains. Un dictionnaire d'anthropologie prospective*. Vrin.

Weber, A. (2007). *Alles fühlt: Mensch, Natur und die Revolution der Lebenswissenschaften* (p. 294). Berlin Verlag.

Whewell, W. (1853). *Of the plurality of worlds*. Library of Alexandria.

Wulf, C. (2013). *Anthropology. A continental perspective*. University of Chicago Press.

Wulf, C. (2022). *Education as human knowledge in the Anthropocene. An anthropological perspective*. Routledge.

Nathanaël Wallenhorst is Professor at the Catholic University of the West (UCO). He is Doctor of Educational Sciences and Doktor der Philosophie (first international co-supervision PhD), and Doctor of Environmental Sciences and Doctor in Political Science (second international co-supervision PhD). He is the author of twenty books on politics, education, and anthropology in the Anthropocene. Books (selection): *The Anthropocene decoded for humans* (Le Pommier, 2019, *in French*). *Education in the Anthropocene* (ed. with Pierron, Le Bord de l'eau 2019, *in French*). *The Truth about the Anthropocene* (Le Pommier, 2020, *in French*). *Mutation. The human adventure is just beginning* (Le Pommier, 2021, *in French*). *Who will save the planet?* (Actes Sud, 2022, *in French*). *Vortex. Facing the Anthropocene* (with Testot, Payot, 2023, *in French*). *Political education in the Anthropocene* (ed. with Hétier, Pierron and Wulf, Springer, 2023, *in English*). *A critical theory for the Anthropocene* (Springer, 2023, *in English*).

Christoph Wulf is Professor of Anthropology and Education and a member of the Interdisciplinary Centre for Historical Anthropology, the Collaborative Research Centre (SFB, 1999–2012) "Cultures of Performance," the Cluster of Excellence (2007–2012) "Languages of Emotion," and the Graduate School "InterArts" (2006–2015) at the Freie Universität Berlin. His books have been translated into 20 languages. For his research in anthropology and anthropology of education, he received the title *"professor honoris causa"* from the University of Bucharest. He is Vice-President of the German Commission for UNESCO. *Major research areas:* historical and cultural anthropology, educational anthropology, imagination, intercultural communication, mimesis, aesthetics, epistemology, Anthropocene. Research stays and invited professorships have included the following locations, among others: Stanford, Tokyo, Kyoto, Beijing, Shanghai, Mysore, Delhi, Paris, Lille, Modena, Amsterdam, Stockholm, Copenhagen, London, Vienna, Rome, Lisabon, Basel, Saint Petersburg, Moscow, Kazan, Sao Paulo.

Part X
Humanity as Birth

Anthropological Mutations

Jean-Louis Genard

Abstract This article takes up and develops Michel Foucault's intuition (The order of things: an archaeology of the human sciences. Pantheon Books, New York (original French edition 1966), 1970) that modern anthropology, born in the seventeen and eighteenth centuries, would have envisaged the human being as an "empirical-transcendental subject", thinking of humans as being torn between responsibility and determinism, between capacity and incapacity, between autonomy and heteronomy… The 'disjunctive' conception of these anthropological coordinates, which was dominant in the nineteenth century, where there was a harsh opposition between those who were capable and those who were not, depriving the latter (the poor, women, the insane, servants, workers…) of rights granted to the former, has gradually been replaced by a 'conjunctive' conception, as a response to numerous social struggles, seeing humans as being situated on an 'anthropological continuum', considering each person as being both capable and incapable, certainly fragile and vulnerable, but always resilient as well. This is an anthropological continuum that now extends beyond humans alone, to animals and anti-speciesism on the one hand, and to robots and transhumanism on the other.

One specificity of Western modernity is that it has attributed an autonomous will to humankind, conceived of and established as a responsible subject (see the entry on 'Responsibility'). Another specificity of this same modernity is undoubtedly its grounding in the development of science and the fact that not only has it made the human a responsible 'subject', but also an 'object' of knowledge. Foucault, drawing inspiration from Kantian dualism, understood this perfectly and theorised that man was 'born' with Western modernity, as an 'empirical-transcendental double' (1966),

---

Translation by Gail Ann Fagen.

J.-L. Genard (✉)
Université libre de Bruxelles, Brussels, Belgium
e-mail: jean.louis.genard@ulb.be

as constituted at the heart of this antinomy between freedom and determinism that Kant had formulated (1999).

Primo, 'Disjunctive' Anthropology, 'Conjunctive' Anthropology. Nevertheless, what Foucault did not see was that two interpretations, two accentuations can be drawn from this antinomy, this 'empirical-transcendental double'. They can be termed 'disjunctive' (humans are one OR the other, free OR pre-determined, capable OR incapable, autonomous OR heteronomous, active OR passive, etc.) and 'conjunctive' (humans are always free AND pre-determined, capable AND incapable, and so on). Without a doubt, these two interpretations have always co-existed. Yet, historically, when we analyse the forms of their co-existence since the nineteenth century, the progressive shift from a disjunctive predominance towards a conjunctive predominance becomes evident.

In the nineteenth century, disjunction prevailed. This led to a separation among humans: those who were capable on one side and, on the other, those who were not: the insane, women, all who were not autonomous economically, servants, etc. Capacity and incapacity were largely considered as statuses. This status was expressed, among others, in significant differences under the law: those who were 'incapable' were deprived of fundamental rights, especially the right to vote, relegating them to second-rank citizenship.

With time, an effect notably of so many social struggles – by workers, women, etc., but also evolutions in the way people were represented, in particular knowledge gained in the area of mental health, and so on, the anthropology moved from a disjunctive to a conjunctive predominance. Henceforth, even if disjunction persists for some (young children, for example), we can say that overall, rather than being one or the other, once and for all, each human is situated along what could be called an *anthropological continuum.*

Secondo, from Disjunction to Conjunction. From the 1960s, Robert Castel evoked the rise in the USA of 'therapy for the normal' (1981). Anti-psychiatry triumphed, strongly calling into question the institutions hitherto responsible for confining the insane. There are other multiple examples. Humans, every one of us, are always in-between. We are always confronted with the risk of falling into heteronomy, but each still has the resources, potentials to regain control over the situation. The words for this have changed: 'fragile' and 'vulnerable' on one side, but, on the other, always endowed with resources, 'potentials' and 'resilience' (Genard, 2009). Since the 1980s, these terms have steadily increased in importance and have gradually become ubiquitous both in the specialist vocabulary and in search engines. Anthropological and psychological knowledge also confirms this: humans actually are endowed with largely unexploited potentials. The old terminology adapts to new semantics. For example, people who used to be called 'gifted' are now described as having 'high potential'. The oppositions between capacity and incapacity, between active and passive have become blurred. They are no longer statuses that seal one's fate once and for all. The question is now one of 'empowering', 'activating', 'rendering responsible'. The question is also one of transforming vulnerability, trauma or what was seen a handicap, and turning it into a resource, an opportunity (Genard, 2015). The book by Boris Cyrulnick, *Un merveilleux malheur* ("A marvellous misfortune") (1999) made the author famous. The sale of pharmaceuticals to cure hyper- and hypo-activity is soaring (Ehrenberg, 2013). Post-traumatic stress, at first limited to the military, has become commonplace. Books on management are riding the wave and recommend using our vulnerabilities as a strength. Health conditions or social propensities, once interpreted solely as disabilities are now seen as different lifestyles, rich in potential. Highlighting Asperger's sheds a positive light on autism. Far from reduced to a to a mental health disorder, schizophrenia reveals unexpected potentialities that just need to be encouraged, especially in capacity for dreaming (Ehrenberg, 2017).

Tertio, the Policy of Rendering Responsible. The welfare state has undergone considerable transformations. In relation to the services it provides, citizens can no longer claim unconditional entitlement. This unconditionality, following the growth of neo-liberalism, is considered an infringement of the requirement for responsibility. The aim is now to 'render citizens responsible' in their relationship with these benefits underwritten by social welfare. The strategies include making people participate financially in services that used to be free of charge, impressing on them that social rights must be 'deserved', counselling people, empowering them, coaching them, and so on. And then if they do not take their lives in hand, deprive them of access to social welfare. Exclusions multiply, poverty grows...

Nevertheless, among us are those who are 'truly' vulnerable, 'truly' fragile, for whom these strategies to assume responsibility and take one's life in hand would be indecent. In relation to these subjects, there is the obligation for care which, in this context, appears as a counterpart, in a way compensating for responsibility as a policy. Alongside the ethics of care (Tronto, 1993), towards those who cannot, who no longer can, parallel instruments come into place, in continuity with traditional welfare, which can be termed care 'policies', the landscape of this new assistance henceforth called humanitarian. And here is where we see the other facets of responsibility. If pressure to assume responsibility is conjugated in the first person (I) as the ability to start off, as an initiative (Arendt, 1958), care nevertheless sees responsibility in the second person, as an obligation, not to answer *to* the other, but *for* the Other (Thou) and in plural (We) as a collective responsibility (Genard, 2017).

Quarto, an Anthropological Continuum Beyond the Frontiers of Humanity: Transhumanism and Anti-Speciesism. Moving forward, positions along the anthropological axis ranging from heteronomy to autonomy, from incapacity to capacity, become mobile and fluctuating. What takes shape, nevertheless, beyond this fluctuation that until now has remained largely within human bounds, is an extension of the continuum beyond these bounds.

On the one hand, there is transhumanism, an over-empowerment where the question is no longer merely to mobilise, enrich and reveal the potential inside us, but also to 'enhance' them through technical means. The question is no longer that of 'high potential', the new name for 'gifted' but, in a way, one of seeing humans through the lenses of deficiency, lack... To make disability, and its overcoming, the horizon of an approach to humanity.

And then there is anti-speciesism. To begin with, from the human perspective, it is a matter of refreshing the view of hitherto unimagined potentialities that were dismissed or minimised. Potentialities that no longer constitute a human exception, but which humans share with all living beings, with varying degrees of proximity, of course, but at the very least without any discontinuity in this continuum. Genetic heritages differ less than we imagined, showing that humans are close not only to the great apes, but far beyond. These proximities also hold promise for undreamed of medical exploits such as organ grafts. They also show the way to forms of intelligence, especially infra-propositional, that we share with animals (Ferry, 2004). But this perspective mainly opens us to a common sensitivity, the capacity for suffering, thus a vulnerability, that must be respected.

These new anthropological coordinates thus hold the extraordinary particularity of breaking through the human boundaries that modernity set for us (Genard, 2020). On the one hand we are pondering humankind via the dual horizon of our technological 'augmentation', the cyborg, the prosthetic man. On the other side, we are exchanging the, as yet highly fragile, horizon of a 'common humanity' for that of a 'community of beings' where the millennia-old human exception may well appear itself as an avatar, the illusion of a being uprooted from the living world.

References

Arendt, H. (1958). *The human condition*. University of Chicago Press.
Castel, R. (1981). *Les Gestion des risques, de l'anti-psychiatrie à l'après-psychanalyse*. Minuit.
Cyrulnick, B. (1999). *Un merveilleux malheur*. Odile Jacob.
Ehrenberg, A. (2013). La santé mentale ou l'union du mal individuel et du mal commun. *SociologieS*, Grands résumés, L'Ombre portée: l'individualité à l'épreuve de la dépression. http://journals.openedition.org/sociologies/4505. Accessed 20 June 2020.
Ehrenberg, A. (2017). *La mécanique des passions. Cerveau, comportement, société*. Odile Jacob.
Ferry, J.-M. (2004). *Les grammaires de l'intelligence*. Cerf, Passages.
Foucault, M. (1970). *The order of things: An archaeology of the human sciences*. Pantheon Books. [Original French edition 1966].
Genard, J.-L. (2009). Une réflexion sur l'anthropologie de la fragilité, de la vulnérabilité et de la souffrance. In T. Périlleux & J. Cultiaux (Eds.), *Destins politiques de la souffrance, Intervention sociale, justice, travail* (pp. 27–46). Eres.
Genard, J.-L. (2015). L'humain sous l'horizon de l'incapacité. *Recherches Sociologiques et Anthropologiques, 46-1*, 129–146. https://doi.org/10.4000/rsa.1424
Genard, J.-L. (2017). La place du care dans la réorientation des théories critiques. In M. de Nanteuil & L. Merla (Eds.), *Travail et care comme expériences politiques* (pp. 21–35). Presses universitaires de Louvain. Sciences politiques et sociales.
Genard, J.-L. (2020). Un bouleversement radical de nos repères anthropologiques et des conditions de la moralité: le déclin ou la fin de l'exception humaine ? *SociologieS*, Dossiers, La société morale. http://journals.openedition.org/sociologies/13202. Accessed 17 Jan 2020.
Kant, E. (1999). *Critique of Pure Reason (The Cambridge Edition of the Works of Immanuel Kant)* (P. Guyer & A. W. Wood, Trans. and Ed.). Cambridge University Press. [Original German edition 1781].
Tronto, J. (1993). *Moral boundaries: A political argument for an ethic of care*. Routledge.

Jean-Louis Genard is a philosopher, doctor in sociology, professor at the Université libre de Bruxelles, co-editor-in-chief of the journal *SociologieS*. He has published numerous books, including *La Grammaire de la responsabilité* (Cerf, 2000), *Action publique et subjectivité* (with F. Cantelli, LGDJ, 2007), *L'éthique de la recherche en sociologie* (with M. Roca i Escoda, 2019), as well as numerous articles on anthropological transformations and developments, responsibility, capacitation and empowerment, ethics, public policies and epistemological and ethical issues specific to the human sciences.

Birth

Frédéric Spinhirny

Abstract At a time when the birth rate is becoming a political issue again, in Europe, and in France particularly, my contribution aims to go beyond acknowledging the simple fact that the demographic crisis is linked to the pandemic. Indeed, the problem is certainly not a recent one, so that giving birth to a new relevance though reconsidering our lifestyles has become an urgent need. For if some people fear the Great Replacement or the ecological disaster, is it not even more terrifying to envisage that human beings may be replaced with…nothing?

At a pandemic-stricken moment when people lament the excess death rate resulting from the COVID-19 crisis more than anything else, pondering about ways to have people give life is crucial. In fact, there is also an unexpected consequence of the pandemic on the birth rate, when pertinent observations or sociological analyses of crisis phenomena predicted an increase in births after periods of lockdown. Without prejudging the future, the early months of 2021 delivered a stinging denial: our hopes have been buried as a « baby crash » is looming. In Europe, figures speak volumes: the fertility rate reached 1.84 births per woman in France in 2020, compared with 1.86 births per woman in 2019. The number of babies born during the year, that is roughly 740,000 (down 13,000 on 2019) is the lowest recorded since the end of WW2. In France, a mere 53,900 babies were brought into the world during the first month of 2021 which represents an unprecedented decline since 1975 and the post-baby boom period. However, France remains first among European countries. Spain's and Germany's rates, and especially Italy's are generating concerns and rekindling reflections on how to increase the birth rate. Throughout the world most developed countries are concerned, with Japan and Russia in the lead. But the collapse in the number of births started long ago and it questions our relationship to the world.

Nowadays, it is commonplace to hear that our world may collapse. So why have children when the great catastrophe is about to take place? Indeed, many an author

F. Spinhirny (✉)
Hospital Manager and Philosopher, Paris, France

has enabled the general public to become aware of the reasons for the disintegration or for the disappearance of a civilization: *Collapse,* by Jared Diamond; *The Collapse of Complex Societies,* by Joseph A. Tainter; and more recently, *Another End of the World is Possible,* by Pablo Servigne, Raphaël Stevens and Gauthier Chapelle. If ecology plays a major role in these collapsology works, it is established that both human action and political decisions do lead to the end of the world as we know it today. Therefore the aforementioned works are permeated with the idea that making responsible choices is called for above all. Few are those who seriously bring up the question of demography and specifically of birth policy. We sometimes hear that bringing a child into the world in a toxic environment – an environement condemned to destruction – or simply in the context of a political regime that is harmful to human rights would be irresponsible. Of course, an existential contradiction hangs over those individuals who proclaim they do not want to have children, as they argue such a personal choice can be interpreted as an ethic orientation or action for the public good.

This line of reflection already guided Hans Jonas after the WW2 disaster in *The Imperative of Responsibility: In Search of an Ethics for the Technicological Age.* If the philosopher questions the introduction of a right to be born, he immediately notes that evidence of the asymmetrical nature of the relationship to the child can already be found in terms of the very obligation of care, that is the parents' responsibility to their new-born, who, as we know both by instinct and thanks to science, wouldn't survive without protection and food. Thus it is incumbent upon us to base a responsibility on this original fragility, this precariousness, this incompleteness, this unfinishedness. Jonas describes in general « an obligation to the future » which morally establishes our duty to have children, i.e. paradoxically an obligation to and respect for what does not (yet) exist and so does not have rights (yet). Therefore contemporary ethics must, for moral reasons, be based on the fact that posterity is the fundamental issue of the twentieth century. Therefore men are needed on earth in order to perpetuate the very possibility of Man. And from this principle another obligation follows: that of having new individuals be born into a properly human world that can accommodate them, making sure they are not in a position to turn against their creators and hold them accountable for the state the world is in. But, if everything is falling apart, what is a liveable world today? In the context of his conversation with Hannah Arendt, Jonas asks the following topical question: « If politics ought to transform the world into a decent abode for man, then what would a decent abode for Man be? ». In theory, we perfectly understand that our duty consists in having children. But given that ecological resources are likely to become exhausted, isn't it contradictory, since there can be no birth if living on Earth is impossible? Is multitude compatible with the finiteness of our unique planet?

Today, there is undoubtedly a close link between demography and living conditions on Earth. Eventhough it does not necessarily depend on the number of inhabitants on Earth, we are aware that lifestyles do have the most significant impact on the resources of the planet. A debate on the need to reduce the world population through a policy aiming at having the birth rate decrease, inherited from Malthus, is still topical. However, such a policy would contradict the essence of birth as the very possibility to give life to individuals who are in a position to change the course of the world. Seriously considering the option of radical behavioural change, under pain of preventing the world from preserving life, is preferable. Such issues as food waste, that could feed the large malnutrition-stricken part of the population, the limitation of polluting collective means of transport, and the education of women in developed countries may be more suitable priorities.

If it appears that our housing environment on Earth conditions the emergence of human life, we must also ponder the consequences of the evolution of society itself on having children. In other words, we could well imagine that once the climate crisis has passed, once ecological practices have become sufficiently rooted in our culture to ward off the spectre of desolation, there will still be ways of living or living conditions that are unsuitable for the increase in births. Axel Honneth and Byung Chul Han have shown the entanglement between intimacy and performance, and between emotion and work processes. Moreover, our behaviour is gradually shaped around the success of our individual actions, leaving little room for the other, who is seen as a constraint. Today, it is complicated to find one's place in an environment where techniques and increasingly formalized lifestyles shape one's behaviour. Evoking well-being, friendliness, innovation, originality or recognition in a world that concretely limits behavioural scope and inevitably brings about defections and an alienation from the world. Unlike the human being born to other human beings and shaped by them, the ideal-type individual becomes the self-made entrepreneur whose life reference is work and personal development the driving force. This is the self-made man/woman strictly speaking, the one who has shaped himself/herself, the one who was born to no one and needs no one to be born. It is a powerful modern mythological representation where the other does not exist. It's an illusion of the modern Man who wishes to deny his birth, to forget the openness of the origin in order to become « master and possessor of nature » and to instrumentalize the world in his own way and for his very own success. Harmut Rosa's conclusion is that these new desires which characterize our late modernity inevitably lead to alienation from the self and from the others, to an erosion of the ability to become attached, and to a difficulty in establishing a sincere relationship with others that is not predicated on efficiency.

So what can birth philosophy do in the face of this situation? What future is a child born into a world in ruins likely to have? In our world, the diversification of life choices is not, after all, synonymous with happiness but rather with an existential difficulty: that of fulfiling oneself as a unique human being who is tuned to their environment. This is Harmut Rosa's source of reflection on resonance as an authentic relationship with oneself and the world. For the German philosopher a different kind of being-in-the-world can exist, but only through a genuine individual, cultural and political transformation. If the soul doesn't go back to the Anthropocene Epoch, the world cannot be transformed so that it can welcome our children. Encouraging people to have children for the latter to roam the world quite on their own, through fake mediation, screens for instance, or to instil practices which don't connect them to others and instead push them to predation or to an extreme use of the world's available resources into them, does not constitute a pertinent approach.

However, it must be asserted that birth is the revelation of a mutual relationship, which is miles away from the feeling of self-sufficiency or autarky. Its symbol links individual upheaval and the construction of a liveable world for Man. The birth rate raises profound questions we need to be able to answer in order to give children every opportunity to take responsibility for the world we pass on to them: will I give them the means to preserve the world? Will they have the ability to transform it? Finally, what have I done with the promise revealed through birth, the promise to regenerate the world around me indefinitely, to change it, to act and invent something radically new? For Corine Pelluchon, following Arendt's reflection on birth rate, the fertility of which she acknowledges, birth undeniably refers to what we do with our freedom in the world. The philosopher transforms birth rate into one of the bases of her Ethics of Consideration and specifies that we must choose the political organization which is bound to guarantee that each birth is both continuity and novelty, the transmission of reference points and an increase in the capacity to change the world. Birth contains all the dimensions of environmental metamorphosis, which can change any perspective, within itself. In *Repairing the World,* Corine Pelluchon emphasizes that being born alive, naked, fragile and dependent is a humility lesson we must remember in order to come into communion with the living, nature and animals included.

It is therefore a question of taking the birth issue seriously, as if we had to prepare for it, just as death is widely studied by the philosophical tradition. Now I put forward the hypothesis that birth continues to shape the way we live and that we are beings-for-birth. Today, it is urgent to hear its call, which comes from the origins, whispering that we remain free to choose what we really want to do and responsible for what we really want to be. Birth constitutes a total commitment, which is first intimate, then political. At the end of the day, nothing is more topical than birth at a time when every country is counting its dead, these victims of the pandemic or of its consequences. Reclaiming birth is an urgent need if we want to take care if the world in its social, ecological and political dimensions. Above all, it appears as generosity, as a gift: the mother gives a child to the world, transmits the ability to engender again and thereby triggers a virtuous cycle of accountability the Convivialists hold dear: giving, receiving and giving back. In the midst of catastrophism, the radical nature of birth remains: giving birth partakes of the construction of a world open to plurality where other solutions to life exist; and the evocative power of birth is not confined to childbirth itself, for it is a reclaiming of oneself that can take place at any moment. Finally, let us keep in mind the idea that until something brand new has come into the world, we are not aware of its possibility.

Frédéric Spinhirny is hospital manager and Gestions Hospitalières review chief editor. He is also a philosopher and the author of some 50 articles on management. In addition he has published 5 philosophy essays with Editions Sens & Tonka: *Eloge de la dépense*, 2015; *L'homme sans politique*, 2017; *Hôpital et modernité*, 2018 and, with Editions Payot: *Naître et s'engager au monde. Pour une philosophie de la naissance*, May 2020; Les caractères aujourd'hui. Ce qui résiste et ce qui cède, November 2022.

Body

Christoph Wulf

Abstract Since nature and culture are inseparably fused in the human body, the central, dynamic force of the Anthropocene that shapes the structures of our planet also finds concrete form in the human body. Thus, the exploration of the body makes an important contribution to the exploration of the Anthropocene. It is where micro, mezzo and macro perspectives unite, the individual and the universal overlap and a new understanding of the *unitas multiplex* of the planet becomes possible.

The human body comes into existence and is formed in a reciprocal interchange between nature and culture. It is an example of many processes that are characteristic of the Anthropocene (Wulf, 2022b). Just as culture influences the development of our bodies even before birth, in the Anthropocene it is nature and life on our planet that is shaped and altered by humans and their cultures. The traditional distinction between nature and culture as two quite distinct realms is no longer valid. Instead, to gain a new understanding of what it is to be human, it is a question of recognizing that nature is being changed by the effects of culture. Instead of assuming the existence of a bipolar relationship between nature and culture we must assume a multipolar systemic relationship in which all the elements relate to each other in several different ways. This results in a complex understanding of the body, nature and culture that requires more international, transdisciplinary research. Following the decline of normative anthropology, anthropological research paid more attention to the body. There is a problem here, however, insofar as the human body is not accessible as a whole but only partially. This gives rise to different forms of expression, description, and representation. A look back at hominization, history and at other cultures shows that the human body is perceived, experienced, and interpreted in very different ways. The wide range of different viewpoints that have developed illustrates the degree to which in the Anthropocene the social and cultural dissemination of body images is linked to power, the economy, and biopolitics.

C. Wulf (✉)
Freie Universität Berlin, Berlin, Germany
e-mail: christoph.wulf@fu-berlin.de

N. Wallenhorst, C. Wulf (eds.), *Handbook of the Anthropocene*,
https://doi.org/10.1007/978-3-031-25910-4_116

Research on *evolution* and *hominization* clearly illustrates the temporality and genesis of the human body, as well as its relationship to the history of life on Earth (Wuketits, 2005; Schrenk & Mueller, 2009; Leroi-Gourhan, 1993). The human body is the result of an irreversible evolutionary process dating back to the beginnings of life on Earth, which is understood to have developed because of material self-organization. The human body shares its origins with all known species and genera and is related to these in varying degrees; its development is a consequence both of this relationship and of the diverging paths of development over the course of evolution. Its form and development were heavily influenced by the urge for self-preservation and the human powers of innovation, adaptation, and specialization. The evolution of the body was largely driven by genetic recombination and natural selection, as well as the subtle interdependencies between internal and external selection.

Whereas research into evolution, hominization (Morin, 1973), genetics (Doudna & Sternberg, 2018), and the brain strives to obtain general insights into the human body, *philosophical anthropology* seeks to understand the particular characteristics of the human body that distinguish it from those of other animals, that is, from the bodies of the other primates, and to study the significance of these differences for further human self-understanding. Plessner (1981) and (Scheler, 2009) both put forward a structure of life divided into different stages. Unlike plants, both animals and humans have a centre that enables them to move in space. The centric position of animals and humans allows them to face an objectively structured environment and to act spontaneously. In contrast to the bodies of animals, the human body allows humans to distance themselves from the environment and to adopt an *ex-centric* position. This gives rise to three states of being: the human body is characterized by the mode of *having a body* in which it experiences an external world; consequently, it is also characterized by the mode of *being a body*, under which it experiences its soul and inner life. Ultimately this enables humans to take up an imaginary point of view outside of themselves, allowing them to perceive these two other modes and their interrelationship. Corresponding to this structure of the human body, the world is experienced as an outside world, an inside world, and a coexistent world [*Mitwelt*].

In contrast to this theory, Gehlen (1988) assumes that the human body is by nature deficient, and this forces human beings (as invalids of their higher powers) to act to overcome their deficiencies. In Gehlen's view it is largely neoteny (or the extrauterine year), reduced instinct, excess of drives, and relief [*Entlastung*] from and openness to the world that characterize human corporeality. According to the thesis of neoteny, humans retain the fetal stage, early birth, slow bodily development, and a long childhood and adolescence. Residual instinct and excess of drives are equally important corporeal characteristics. The hiatus between stimulus and response makes learning possible and enables humans to adapt to heterogeneous biotopes. Excess of drives, arising from neoteny and instinct reduction, enables the broad spectrum of human behavioural patterns to develop. Because of excess of drives, discipline and domestication are required, processes in which rituals and institutions play a central role. Relief helps to coordinate perception and motion. Behavioural patterns are practiced, become automatic, and can be performed without reflection. Habits are formed, creating continuity, and liberating energy for new activities. Whereas the bodies of animals are attuned to a specific environment and have specialized organs for dealing with these habitats, the human body "has world"; that is, it has unspecialized organs that can adapt to very different conditions. In Gehlen's view, the unspecialized nature of the human body enables humans to be open to the world.

Where this view of the body concentrates on the general conditions of the human body and its development, other anthropological concepts of the body concentrate on the historical and cultural nature of the human body. These concepts assume that the body changes as part of a historical and cultural process (Kamper & Wulf, 1982; Feher, 1989; Benthien & Wulf, 2001; Wulf & Kamper, 2002). The methods of research employed and the results produced differ accordingly. *Historical anthropology* examines the human body over the course of time (Burke, 1991; Braudel, 1949; Ginzburg, 1980; Ariès & Duby, 1985). Researchers in this field investigate how collectively held attitudes prevalent during a specific historical era can give rise to specific bodily feelings and sensations. The historical character of feeling and thinking, as well as of collective memory, becomes apparent (Corbin et al., 2005). The conception of time in the present era differs from that of the Middle Ages, for example, in that nowadays the discrepancy between world time and individual time and the acceleration of time play a significant role. As the concept of time alters, the experience of space also changes. Different and partially contradictory conceptions of time and space simultaneously shape life in the modern era. The historical nature of elementary situations and basic experiences and the historical character of attitudes to death, love, and work become apparent (Scarry, 1985; Butler, 1993; Wulf, 2016). Subjectivity appears as the result of historic and cultural processes such as adjusting, distancing, and disciplining. Even sexuality and birth, childhood, youth, old age, and clothing reveal their historical nature, clearly illustrating that the human body appears only in concrete historical forms and that its historical nature must therefore be researched if we are to understand its specific character at any historical juncture.

Historical anthropological research has the aim of identifying historical changes in humans' relationship to their bodies and of examining and evaluating these changes. It focuses on images and concepts of the body as well as bodily practices and the use of the senses. Many of the methods and procedures used in this field are based on research performed for example by Norbert Elias (1978), and Michel Foucault (1977). These works form the basis for our attempt to analyze how the body and its senses are changing today. The aim here is to estimate what influence the acceleration of time, the ubiquity of the new media, and the greater use of computers have on human perception and on the way we relate to our bodies, and to determine what part increasing abstraction and imagination within society play in this.

In his reconstruction of the process of civilization in Europe, Norbert Elias (1978) showed that the human body is becoming increasingly disciplined. The controls exercised affect eating habits, social behaviour, and emotional life. Greater shame and embarrassment thresholds play a central role here. Humans are establishing a greater distance from their bodies. Modern humans are becoming increasingly distanced from each other, the world, and themselves, resulting in self-control and self-discipline, which can be employed as strategies to attain self-perfection. A more extensive inner world is being developed, consisting of susceptibilities, feelings, moral principles, and sets of values. This development occurs through practical acts and social activities, which create examples and models. The transformation to a body exercising a greater degree of control and under the dominance of rational thinking emerges during mimetic acquisition in the form of regularly repeated exercises and imitation, following guidelines and instructions, and control and correction. Gradually the closed and hermetic bodies (*corpus clausum*) of modern humans emerged.

Foucault's (1977) analysis points in the same direction. In contrast to Elias however, he places greater emphasis on the controlling and disciplining power of institutions in whose spaces the roles of society and individuals interlink, a process in which the activity of the body and the constructive side of the subjects play an important part. Foucault describes power as a subcutaneous body politic, which he does not see as only suppressing human beings, but also as advancing them, and "producing" them as individuals. What looks like humanization—more lenient punishments, the introduction of psychology into the penal system, understanding the culprit and the crime—is, in fact, a more subtle means of control, heralding a new era of civilization. The aim of this is the *control*, *discipline*, and *standardization* of the body, its gestures, and its behaviour. The exemplary development of such processes can be seen in the temporal and spatial structures of prisons, schools, and the military, where the body's ability to learn enables it to be subjected to the "microphysics of power" and to be colonized politically.

Whereas historical research into the body takes a *diachronous* view of the human body, focusing on its historical particularity, cultural anthropology has developed a *synchronous* perspective. What applies to the historical nature of the human body applies analogously to its culturality. Its accent lies on the social and cultural diversity of human life. Its research explains both to what extent cultural evolutions are heterogeneous and to what extent the profound diversity of human life remains unnoticed. It is precisely the analysis of the body in foreign cultures that makes it plain to us how limited and problematic our understanding is (Michaels & Wulf, 2013, 2014; Wulf, 2022a). Thanks to the analysis of cultural manifestations drawn from heterogeneous cultures, anthropological inquiries make an important contribution to the elaboration and development of anthropology. Besides creating a sensitivity for the strange and foreign character of other cultures, it creates a sensitivity for the strange and foreign in its own culture. The (self-)reflexive point of view adopted by cultural anthropology towards European cultures has contributed to a considerable evolution and advance of anthropological knowledge (Geertz, 1993; Sahlins, 1976; Harris, 2001; Evans-Pritchard, 1981; Malinowski, 1922; Mead, 1980; Lévi-Strauss, 1999). The issue of understanding the limits of our comprehension of different cultures becomes central. The ethnographic methods developed in social and cultural anthropology based on fieldwork and participant observation lead to forms of knowledge other than those gleaned from historical source interpretation and philosophical reasoning. They not only make us aware of what is different in other cultures but also what is different in our own culture. Therefore, the application of the anthropological perspective to the cultures of the world broadens and deepens the scope of anthropological research. A clear example of this is to be found in the Berlin Study of Rituals and Gestures, which researched the importance of rituals in the education of adolescents and children over a period of 12 years. This took place in the four most important socialization fields: family, school, peer group and media. The research into the ritual practices in the four socialization fields showed the importance of mimetic and ritual processes for the development of children's social competence and the importance of silent, body-based knowledge for the socialization, development and education of children and adolescents (Wulf et al., 2010; Wulf, 2022a, b, c; Kraus et al., 2021; Kraus & Wulf, 2022). Despite collective historical and cultural experiences, every human body is unique. Even in cases where the collective character of attitudes and opinions is emphasized in studies of mentalities, the processes of socialization and enculturation are different for every person.

The heterogeneous nature of cultural influences results in the development of bodies with very different cultural characteristics; their variety and scope are irrevocable. Only by reducing alterity to what we own and know from experience can we create the impression that it is possible to transcend cultural differences without any loss of complexity. On closer examination, however, it becomes apparent that such a reduction ignores the specific features of a cultural anthropological view of the body, which can only be found by studying and illustrating alterity and corporeal difference (Lakoff & Johnson, 1999). What is to be regarded as a cultural difference in terms of the body depends on the prerequisites of the research task and is constituted in a reciprocal relationship between the cultural frame of reference applied and the related perceptions of alterity. The task is to increase the complexity of our knowledge of the human body by identifying cultural diversity.

Since nature and culture are inseparably fused in the human body, the central, dynamic force of the Anthropocene that shapes the structures of our planet also finds concrete form in the human body. Thus, the exploration of the body makes an important contribution to the exploration of the Anthropocene. It is where micro, mezzo and macro perspectives unite, the individual and the universal overlap and a new understanding of the *unitas multiplex* of the planet becomes possible.

References

Ariès, P., & Duby, G. (1985). *Histoire de la vie privée* (Vols. 1–5). Seuil.
Benthien, C., & Wulf, C. (Eds.). (2001). *Körperteile: Eine kulturelle Anatomie*. Rowohlt.
Braudel, F. (1949). *La Méditerranée et le monde méditerranéen à l'époque de Philippe II*. A. Colin.

Burke, P. (1991). *The French historical revolution: The Annales School, 1929–89*. Stanford University Press.

Butler, J. (1993). *Bodies that matter: On the discursive limits of "sex"*. Routledge.

Corbin, A., Courtine, J.-J., & Vigarello, G. (Eds.). (2005). *Histoire du corps* (Vols. 1–3). Éditions du Seuil.

Doudna, J. A., & Sternberg, S. H. (2018). *A crack in creation*. Random House Children's Books.

Elias, N. (1978). *The civilizing process*. Urizen Books.

Evans-Pritchard, E. (1981). *A history of anthropological thought*. Faber and Faber.

Feher, M. (Ed.) (1989). *Fragments for a history of the human body* (3 Vols). Zone.

Foucault, M. (1977). *Discipline and punish. The birth of the prison*. Pantheon Books.

Geertz, C. (1993). *The interpretation of cultures: Selected essays*. Basic Books.

Gehlen, A. (1988). *Man. His nature and place in the world*. Columbia University Press.

Ginzburg, C. (1980). *The cheese and the Worms: The cosmos of a sixteenth century miller*. Johns Hopkins University Press.

Harris, M. (2001). *The rise of anthropological theory: A history of theories of cultures*. (Rev. ed). Alta Mira Press.

Kamper, D., & Wulf, C. (Eds.). (1982) *Die Wiederkehr des Körpers*. (4th ed., 1992). Suhrkamp.

Kraus, A., & Wulf, C. (Eds.). (2022). *Palgrave handbook of embodiment and learning*. Palgrave Macmillan.

Kraus, A., Budde, J., Hietzge, M., & Wulf, C. (Eds.). (2021). *Handbuch Schweigendes Wissen* (2th ed.). Beltz Junventa.

Lakoff, G., & Johnson, M. (1999). *Philosophy in the flesh: The embodied mind and its challenge to Western thought*. Basic Books.

Leroi-Gourhan, A. (1993). *Gesture and speech*. MIT Press.

Lévi-Strauss, C. (1999). *Tristes tropiques*. Atheneum.

Malinowski, B. (1922). *Argonauts of the Western Pacific*. Routledge.

Mead, M. (1980). *Sex and temperament in three primitive societies*. Morrow Quill Paperbacks.

Michaels, A., & Wulf, C. (Eds.). (2013). *Emotions in rituals and performances*. Routledge.

Michaels, A., & Wulf, C. (Eds.). (2014). *Exploring the senses*. Routledge.

Morin, E. (1973). *Le paradigme perdu: La nature humaine*. Édition Seuil.

Plessner, H. (1981). Die Stufen des Organischen und der Mensch. In G. Dux, O. Marquard, & E. Ströker (Eds.), *Gesammelte Schriften* (Vol. 4). Suhrkamp.

Sahlins, M. (1976). *Culture and practical reason*. University of Chicago Press.

Scarry, E. (1985). *The body in pain: The making and unmaking of the world*. Oxford University Press.

Scheler, M. (2009). *The human place in the cosmos*. Northwestern University Press.

Schrenk, F., & Mueller, S. (2009). *The Neanderthals*. Routledge.

Wuketits, F. M. (2005). *The evolution of living systems*. Wiley-VCH.

Wulf , C. et al. 2010. *Ritual and Identity: The Staging and Performing of Rituals in the Lives of Young People*. Tufnell Press.

Wulf, C. (Ed.). (2016). *Exploring Alterity in the Globalized World*. Routledge.

Wulf, C. (2022a). *Human beings and their images. Imagination, mimesis, performativity*. Bloomsbury.

Wulf, C. (2022b). *Education as sustainable development in the Anthropocene, an anthropological perspective*. Routledge.

Wulf, C. (2022c). Embodiment through mimetic learning. In A. Kraus & C. Wulf (Eds.). *Palgrave handbook of embodiment and learning*, Palgrave Macmillan.

Wulf, C., & Kamper, D. (Eds.). (2002). *Logik und Leidenschaft. Erträge Historischer Anthropologie*. Reimer.

Christoph Wulf is Professor of Anthropology and Education and a member of the Interdisciplinary Centre for Historical Anthropology, the Collaborative Research Centre (SFB, 1999–2012) "Cultures of Performance," the Cluster of Excellence (2007–2012) "Languages of Emotion," and the Graduate School "InterArts" (2006–2015) at the Freie Universität Berlin. His books have been translated into 20 languages. For his research in anthropology and anthropology of education, he received the title *"professor honoris causa"* from the University of Bucharest and the honorary membership of the German Society of Educational Research. He is Vice-President of the German Commission for UNESCO. *Major research areas:* historical and cultural anthropology, educational anthropology, imagination, intercultural communication, mimesis, aesthetics, epistemology, Anthropocene. Research stays and invited professorships have included the following locations, among others: Stanford, Tokyo, Kyoto, Beijing, Shanghai, Mysore, Delhi, Paris, Lille, Strasbourg, Modena, Amsterdam, Stockholm, Copenhagen, London, Vienna, Rome, Basel, Saint Petersburg, Moscow, Kazan, Sao Paulo.

Community (A Case Study)

Claire Meunier Kjetland and Frédérique Brossard Børhaug

Abstract We explore the notion of community by discussing a concrete experience of creating community as we argue that this abstract notion cannot be applicable without being experienced in practice. More specifically, we explore in this contribution the concept of community through the concrete work of planting trees. Based on the pedagogical innovation from the French NGO LIKEN, we discuss experiences of resonance with trees and convivialist community building in higher education.

Why explore the notion of community through a concrete example of pedagogical innovation? We believe that theoretical notions must be confronted in practice in one or another way to see if they prove to be relevant. Furthermore, in academic texts, readers are used to cognitively visualising concepts but less is done on visualising in other ways, for instance by fostering feelings (surprise, empathy, dismissal etc.). This contribution seeks to expand the more traditional visualisation process by including a broader understanding of community and what creating community means. As such, community is not only a theoretical notion, but an anthropological question of being in the world. This contribution is one example dealing with this anthropological question, how can we be in the world and how can we educate community building in the Anthropocene. We are aware that this contribution will surprise some of the readers, it is part of our educational endeavour to question our humanity in dialog with nature and promote new ways of combining the abstract and the concrete also within research discourse.

Looking at the etymology of the term 'community', it first relates to a number of people who inhabit the same area, thus physically living together and secondly, a number of people who aim to form a reciprocal community so they can act together

C. M. Kjetland (✉)
Charity for Environmental Preservation, LIKEN, Pau, France

F. B. Børhaug
VID Specialized University, Stavanger, Norway
e-mail: frederique.brossardborhaug@vid.no

© The Author(s), under exclusive license to Springer Nature
Switzerland AG 2023
N. Wallenhorst, C. Wulf (eds.), *Handbook of the Anthropocene*,
https://doi.org/10.1007/978-3-031-25910-4_117

and constitute a fellowship (community | Search Online Etymology Dictionary (ety-monline.com)). In the second denotation of the term, the bounds to everyone is thus based on a common will and a moral responsibility to create community. However, how to create community with non-human inhabitants sharing the same physical space of residence? In this section, we share concrete experiences about creating community with trees through a French NGO's work of tree planting (LIKEN) at the university of Pau in France. More specifically through LIKEN's pedagogical innovation, we reflect on how encounters with trees in higher educational settings can foster resonance (Wallenhorst, 2021b) and convivialist community experiences (Convivialist International, 2020a, b).

LIKEN is a charity for environmental preservation, founded in 2015 and located in the southwest of France. Their associative projects link together art, nature and the active participation of all kinds of public of different ages and social backgrounds. One key mission is to restore over-urbanised areas with trees.

A concrete example of LIKEN's work is situated at the university campus of Pau in south-west France where an ongoing experimental project initiated in 2018, promotes student participation for revitalizing the campus with trees. As such, students participate in creating and implementing solutions for the planting area and student community garden. The pedagogical innovation includes the use of diverse materials provided by nature close by and from the compost site for leaves, cut grass and crushed branches, it develops an autonomous rainwater harvesting system, plants fruit trees and bushes aiming to provide food for students, creates flower meadows for preserving insects and helping pollination, and it organizes dead hedges constructed from cut branches for providing shelter for insects and small animals. The pedagogical work helps students to learn about co-creating an island of freshness and better food resilience on the campus; it also builds a closer relationship with nature in academic studies when these pedagogical innovations are included in teaching and fosters a sense of ownership of the campus' area.

So, to what extent does this example of LIKEN activities participate in creating community? We believe that cultivating encounters with trees in higher educational settings may enlarge the anthropocentric instrumental view of nature and promote experiences of resonance with trees going beyond the sole welfare of human beings and foster convivialist community experiences.

Reflecting on a renewed way of becoming human, the Convivialist manifesto proposes five key principles for creating community together (Convivialist International, 2020a, b). Firstly, it stipulates the principle of common naturality that is that humans are interdependent with nature, not standing above nature. Secondly, the principle of common humanity expresses that there is only one humanity that must be respected by all beyond any concrete differences of gender, culture etc. Thirdly, the principle of common sociality appreciates that the greatest wealth is building relationships with the human and non-human other. Fourthly, the principle of legitimate individuation argues that everyone contributes to community building in deep accordance with her or his singular individuality. Finally, the principle of creative opposition implies that creating community, is learning to encounter opposition in a non-destructive way in accordance with the above-mentioned principles.

Furthermore, these five principles are subordinate to a metaprinciple, the absolute imperative of hubris control, that is a desire for omnipotence, excess, and never-ending possession, a desire that is not only present within any community, but also lays deep within everyone (2020a, p. 42–45). The five principles and the metaprinciple represent an existential educational question and LIKEN's work might help us to reflect on pedagogical innovation in higher educational settings for building convivialist communities creating bonds with trees, other non-humans and with humans.

For instance, in the light of the principle of common sociality, it is valuable to study how trees are living interdependently as an ecosystem. Through LIKEN, students learn in concrete actions about the trees' life cycle and represent more theoretically how trees are unifying and creating entities with other entities (mushrooms, birds, etc.). Students also understand better their common naturality with nature; through deeper awareness of the specific local surroundings and closer cooperation with nature, they learn the capacity for enhancing the art of living together, thus counteracting the human posture of mastery. Furthermore, as the student groups and the members of LIKEN also come from various backgrounds, age and life experiences, it thus offers a valuable opportunity for sharing personal and context-based life experiences with trees (emotions, cultural points of view, roles, future visions etc.) as they are highly welcomed in all pedagogical activities.

Furthermore, LIKEN's pedagogical activities represent an opportunity to reflect on the metaprinciple of 'no hubris', as trees do not exceed their capacity of living; they extract only from nature what is needed for their condition and give back during their whole life cycle and even after, from the very seed to lying dead with roots and trunks providing further habitats for other entities (mushrooms, animals etc.). The gift of death thus opens for new life cycles and enriches human metaphysical considerations about giving and finiteness in life. In addition, planting trees and taking care of the biodiversity help students re-learn the cycle of nature, adapting and evolving with its continuous changes. For instance, students wait for the right time for trees to be planted and adjust their calendar accordingly to nature, not human needs. Counteracting global warming locally, they reflect on how to personally contribute, sustaining the principal of singularity legitimation– each individual in their own way has an important role to play for the common good. In other words, it creates better opportunities for convivialist community building and it opens for new kinds of experiences beyond academic rational knowledge that implies experiences of resonance.

How to become human in a new way? Wallenhorst (2021a) suggests that inhabiting earth must be profoundly changed; he welcomes an existential mutation which deepens further Arendt's quest of human condition (p. 159–176) and opens for experiences of resonance (p. 177). Thus, the human condition in the Anthropocene questions not only the democratic organization of human societies but interrogates fundamentally its continuity (p. 161). In the global neoliberal society leading to ever-growing acceleration, accumulation, and alienation, relearning the capacity of listening to the world that small children have from birth but lose it in their education (Hétier, 2021), is crucial. As the world speaks to humans but humans do not listen having lost their capacity of being together with the world, the world becomes muted. Many humans then learn mostly to take the world and use it instrumentally for their sole interests, they do not care for it (p. 177).

Therefore, an important educational question is how to relearn the capacity of resonance with the world? One possible approach is promoting convivialist experiences as other ways of valuing our existence and acting in the world. Many French citizens aspire to other ways of living with the world; these aspirations must be strengthened by educative actions deviating the desire of hubris to nature caring needs (Wallenhorst, 2021b, p. 65–67). In the work of LIKEN, many students express their desire to be part of changing the world, to belong to a more convivialist place, awakened by the beauty of the world when belonging to the community of trees. This in-between arousing from the encounter of humans with trees makes the students more human, they become one with the trees.

However, experiences of resonance must also lead to experiences of resistance as the socio-political order homogenizes and controls the world's plurality and its polyphonic experiences of living (p. 73–4). It supposes resistance to leaders and socio-political regimes who do not want to resonate with the world. Thus, higher education does not only imply pedagogical actions of caring for nature's needs; it also implies political actions of resistance in education. In Norwegian higher education for instance, academics not only request new types of teaching and research practices but also a more fundamental refoundation of modes of governance and systems of evaluation, rewards and regulations hindering the necessary flexibility and creativity (UiB, 2020, p. 19). These systemic practices must complete the pedagogical innovation as LIKEN does on its level in its concrete actions.

Grasping the existential finitude and the beauty of inhabiting the same space by creating community with trees is one valuable opportunity among others to experience conviviality and resonance, that is experiencing oneself in deep solidarity with one's inner self, other humans and non-human entities. Because our existence is experienced first through senses, sensitivity and openness, the relation with the world goes beyond hard sciences, or profit and consumption. Arts and poetry open another and necessary way to relate to nature, a landscape, a forest, the sky, a tree (Lamarre, 2021). We thus wish to close this contribution by a short poem attempting to better learn to think, fight and dream (Wallenhorst, 2021a).

References

Convivialist International. (2020a). *Second manifeste convivialiste: Pour un monde post-libéral.* Actes Sud.

Convivialist International. (2020b). The second Convivialist Manifesto: Towards a post-neoliberal world. *Civic Sociology, 1*(1), 12721. Retrieved from: THE SECOND CONVIVIALIST MANIFESTO: Towards a Post-Neoliberal World | Civic Sociology | University of California Press (ucpress.edu)

Hétier, R. (2021). Le temps de la résonance. In N. Wallenhorst (Ed.), *Résistance, résonance: Apprendre à changer le monde avec Harmunt Rosa* (pp. 17–40). Le Pommier.

Lamarre, J. M. (2021). Habiter poétiquement le monde, habiter politiquement le monde? In N. Wallenhorst (Ed.), *Résistance, résonance: Apprendre à changer le monde avec Harmunt Rosa* (pp. 105–123). Le Pommier.

UiB. (2020). *SDG – Quality in higher education: Developing a platform for sharing of ideas and practices within the universities.* https://www.uib.no/sites/w3.uib.no/files/attachments/sdg_-_quality_in_higher_education_-_report_feb_2020.pdf

Wallenhorst, N. (2021a). *Mutation: L'aventure humaine ne fait que commencer.* Le Pommier.

Wallenhorst, N. (2021b). Apprendre la résonance. In N. Wallenhorst (Ed.), *Résistance, résonance: Apprendre à changer le monde avec Harmunt Rosa* (pp. 63–84). Le Pommier.

Claire Meunier Kjetland holds a MA in French at the University of Bergen, Norway, and a MA in Town and Country Planning at Pau University, France. She is a former teacher at upper secondary school, she taught 20 years in Norway French studies, Classic Geography, Social Geography, Physical Geography (Oceanography, Climate Change and Natural Resources etc.). She is the president and co-leader of LIKEN, a charity for environmental preservation since 2020. LIKEN promotes art, nature and civic participation through numerous pedagogical activities with all kinds of public in the Southern part of France. She also writes poetry. Selected publications: *Enchanté* (2006/2012, Coed. with Liautaud and Hønsi) (in French), *Rendez-vous* (2006/2012), school manuals in French for upper secondary schooling.

Frédérique Brossard Børhaug is Professor of Education at VID Specialized University, Stavanger, Norway. She holds a Ph.D. in education from the University of Oslo, Norway. She conducts research on intercultural ethics based on the philosophy of Emmanuel Levinas, anti-racist Education in French and Norwegian multicultural school contexts, on the *Human Development and Capability Approach* (*HDCA*), on inclusion of minority youth, on V*a*KE-didactics (*Values and Knowledge Education*) in intercultural teaching, and on education in the Anthropocene. Publications (selection): (2020), *Educate in the Anthropocene: A Norwegian perspective. Annales des Mines, Revue Responsabilité et Environnement, 101* (in French); (2021) Missing links between intercultural education and anthropogenic climate change? *Intercultural Education, 32*(4); (2022); *Climate change and ecological relation building at school* (Fagforlaget with S. M. Gamlem, in Norwegian).

Emotions

Sabine Seichter

Abstract For a long time the importance of emotions was disregarded. The current "emotional turn" demonstrates the function of emotions, both for the development of individual identity and for social coexistence. The notion of a humane global society demands, in addition to the securing of material, technical and financial resources, the capacity for empathy and compassion. Furthermore, it calls for a holistic view of the individual to be acknowledged with dignity. Emotional education, which should aim to empower individuals to act ethically, is a profoundly moral and ethical challenge which can only succeed through the interaction of thinking and feeling.

Human beings are designed to relate to each other. From birth to death they are involved in relationships. In other words, humans cannot live outside of social and community relationships. The existential connection between "I" and "you" is, in anthropological terms, part of the fundamental definition of what it is to be human. This was expressed somewhat formulaically by Martin Buber in his philosophy of dialogue ("Through the Thou a person becomes I" 1958, p. 39). A conclusion made by all human sciences, from neuroscience to psychology, sociology to philosophy, is that the development of individual identities takes place primarily through face-to-face relationships. This scientific conviction is based on the anthropological fact that the individual is not a causally induced or mechanically functioning, passive object, but must instead be understood as an interactive and responsive actor who creates and develops his personality in relationship with others. The development of personal identity, according to this anthropologically significant observation, thus takes place within the tensions of a complex web of sociality, cognition and emotion (see Seichter, 2015).

S. Seichter (✉)
Fachbereich Erziehungswissenschaft, Universität Salzburg, Salzburg, Austria
e-mail: sabine.seichter@plus.ac.at

© The Author(s), under exclusive license to Springer Nature
Switzerland AG 2023
N. Wallenhorst, C. Wulf (eds.), *Handbook of the Anthropocene*,
https://doi.org/10.1007/978-3-031-25910-4_118

While there is, and has long been, consensus over the social and cognitive dimensions of individual personal development, in the context of anthropological analysis the sphere of the emotions has for a long time been neglected, devalued and even regarded as taboo. In educational practice it was not unusual (and this remains the case) for feelings to be strictly controlled and even suppressed. Education tended and still tends to be characterized by "emotional coldness". In the scientific sphere, the one-sided, empirical-rationalist or empirical-scientific focus in education, which was dominant throughout the twentieth century, led to the separation of reason and emotion. The consequence of this separation was that emotions were devalued in academic anthropological discourse (Seichter, 2017a, b).

The fact that the importance of emotions is currently (once again) the subject of vigorous debate internationally is primarily due to the globally respected study, *Upheavals of Thought. The Intelligence of Emotions*, by North American philosopher Martha C. Nussbaum (2001). Through this work Nussbaum overcomes the dualistic opposition of cognition and emotion or thinking and feeling and shows that, just as applies to cognitions, emotions also always include reason-based judgment and rational evaluation. Emotions represent a different form of cognitive mental state, they have an "evaluative, representational" element (Döring, 2009, p. 16), they appear intentional, convey knowledge about the world and express this, for instance, through gestures and facial expressions (see Wulf, 2021). Emotions move individuals to act, lead to evaluations and judgments about oneself and others and are thus of existential importance.

Particularly in the light of a holistic view of the individual, which is gaining prominence with the contemporary tendency towards the reification of the individual as an "alternative concept" in anthropology, emotions also play a central role that cannot be overestimated. This thus results in anthropological concepts which consider the individual as a whole "person" rather than as an instrumentally rational "object". They recognize and understand the individual in interaction with other people and attach great importance to the individual's relational interconnectedness.

Personalistic ways of thinking have been significantly shaped by Western philosophy and Western European educational thinking and their various approaches continue to be similarly shaped today. What these personalistic approaches have in common is their recourse to human dignity, recognition of personal alterity and heterogeneity and their immutable interdependence in the social sphere (see Böhm, 1994). In terms of discourse history, personalistic thinking is articulated in opposition to "perceived depersonalizing elements in Enlightenment rationalism, pantheism, Hegelian absolute idealism, individualism as well as collectivism in politics, and materialist, psychological, and evolutionary determinism. In its various strains, personalism always underscores the centrality of the person as the primary locus of investigation [...]." (Williams & Bengtsson, 2014).

Looking at the anthropological transformations of the twenty-first century, the focus is increasingly on the reformulations of the human being in terms of how persons relate to themselves and to the world around them, on a general level, and there is thus greater consideration of the emotional at the specific level.

Human emotions are undergoing a historical and cultural transformation. They can be understood as products of religious, moral, socio-cultural and also economic changes, and they are generated and constantly modified through the practices of social coexistence.

In her sociological studies the Israeli sociologist, Eva Illouz, is able to demonstrate clearly how practices around emotions have undergone a lasting transformation in the context of modern digital technologies. The essence of her thesis is that, in the close and mutually dependent connection between capitalism and consumption, it is not only the concept of the emotional that is changing but that of the individual themselves, together with their emotions. Following the analyses Illouz undertook (1997) on the historically and culturally determined transformation of emotions, it becomes very clear that emotions have increasingly become economic products or, more precisely, goods. Based on a postmodern critique of capitalism following Karl Marx, and from the modern perspective of the consumption of goods, Illouz sees emotions (and especially sexuality) as products of modern consumption and this is crucially taking place within an increasingly digitalized world. According to Illouz (2019), emotional relationships are being reshaped more and more in economic terms. This analysis is illustrated particularly vividly by the examples of social media platforms and numerous online dating apps (see Seichter, 2020).

The primary change in an increasingly virtual world is the dwindling significance of "real" face-to-face relationships. This transformation is also influencing approaches to feelings and emotional practices. For example, IT developers see "emotional artificial intelligence" as a very important component of AI systems (see Misselhorn, 2021). So-called "social robots" as used in areas such as care-giving, treatment and education, have special cameras, microphones and sensors which are designed to recognize and evaluate the emotions of the individuals with whom they interact. The belief in this rapidly growing field of IT is that emotion recognition is a pivotal factor in the "intelligent" functioning of AI. Algorithmic emotion recognition by AI systems (as already incorporated into smart watches and other wearables), is one of the fastest growing areas within AI development. Currently, this artificial empathy in AI systems (still) functions on the basis of causal stimulus-response relationships and conforms to a highly reductionist understanding of behavioral theories. What social robots are not (as yet) capable of is their "own" expression of emotions based on the (human) interplay of language, gestures, facial expressions and demeanor. It is also questionable whether social robots will in future be able to perform such actions as expressing empathy or making moral judgments. The gulf between technical capabilities and ethical desirability currently still appears very wide indeed.

In the early twenty-first century, these technical transformations – and this above all concerns anthropological, ethical and pedagogical considerations – must be regarded as a challenge to reflect anew on the existential meaning of emotionality. Against the background of a personalistic conception of the individual, as briefly outlined above, interaction with other people (not just with machines!) in the social (and not just the virtual!) sphere is not simply an added extra, rather it is the existential basis for the development of personal identity. In this context emotions assume – in the wider sense – a political dimension (see Nussbaum, 2013). A liberal, postmodern society which is based on the dignity of every individual, on inclusion and social justice, and which thereby seeks to alleviate inequalities, poverty and need, depends on solidarity. The notion of a humane global society demands, in addition to the securing of material, technical and financial resources, the capacity for empathy and compassion. Furthermore, it calls for a holistic view of the individual to be acknowledged with dignity. Emotional education, which should aim to empower individuals to act ethically, is a profoundly moral and ethical challenge which can only succeed through the interaction of thinking and feeling.

If we again look to Martha Nussbaum here, emotions are central to establishing a humane social order. "Emotions: Being able to have attachments to things and people outside ourselves; to love those who love and care for us, to grieve at their absence; in general, to love, to grieve, to experience longing, gratitude, and justified anger. Not having one's emotional development blighted by fear and anxiety. (Supporting this capability means supporting forms of human association that can be shown to be crucial in their development.)." (Nussbaum, 2011, pp. 33–34).

Emotions such as guilt, embarrassment, shame, pride, envy, love, affection, empathy and sympathy are not simply elements of social relationships at a general level, they are the very basis of social action at a specific level. Social coexistence is thereby always based on and led by emotions. Emotions enable us to change our social perspectives, to appreciate other realities and to put ourselves in other situations and in other people's positions. Emotions can provide the moral compass for a democratic, relational and simultaneously "successful" coexistence.

As early as the 1920s Martin Buber coined the term "inclusion" in his philosophy of dialogue. He used it to express the importance, when in dialogue and interaction with others, of not only taking into account oneself and one's own emotions, but of always including those of the other individual or individuals. This means cultivating one's practical powers of discernment which should never consist purely of rational judgment but should always include an emotional evaluation.

References

Böhm, W. (1994). *Theory and practice and the education of the person*. Organization of American States.

Buber, M. (1958). *I and thou* (R. G. Smith Trans.). Scribner Classics.

Döring, S. A. (2009). *Philosophie der Gefühle*. Suhrkamp.

Illouz, E. (1997). *Consuming the romantic utopia: Love and the cultural contradictions of capitalism*. University of California Press.

Illouz, E. (2019). *The end of love: A sociology of negative relations*. Oxford University Press.

Misselhorn, C. (2021). *Künstliche Intelligenz und Empathie. Vom Leben mit Emotionserkennung, Sexrobotern & Co*. Reclam.

Nussbaum, M. C. (2001). *Upheavals of thought. The intelligence of emotions*. Cambridge University Press.

Nussbaum, M. C. (2011). *Creating capabilities. The human development approach*. The Belknap Press of Harvard University Press.

Nussbaum, M. C. (2013). *Political emotions. Why love matters for justice*. Harvard University Press.

Seichter, S. (2015). Science de l'éducation entre unicité et multiplicité. *Rassegna di Pedagogia, 1-2*, 27–39.

Seichter, S. (2017a). Educación y relación interpersonal. *Rassegna di Pedagogia, 3-4*, 13–21.

Seichter, S. (2017b). *Pädagogische Liebe. Erfindung, Blütezeit, Verschwinden eines pädagogischen Deutungsmusters*. Schöningh.

Seichter, S. (2020). Emotionen: Zwischen Romantisierung und Technologisierung. *Paragrana, 1*, 157–165.

Williams, T. D., & Bengtsson, J. O. (2014). *Personalism. The Standford encyclopedia of philosophy*. https://plato.stanford.edu/entries/personalism. Accessed 1 July 2016.

Wulf, C. (2021). Emotion and imagination: Perspectives in educational anthropology. *International Journal of African Studies, 1*, 45–53.

Sabine Seichter Dr. phil. habil. has been a professor for general education at the University of Salzburg since 2014. Her main areas of work are history and theory of education and ›Bildung‹, historical-cultural and personalistic conceptions of pedagogical anthropology, history and anthropology of childhood. Her publications appeared in several editions, have been widely received and are already considered as standard works in educational science: "Erziehung und Ernährung" ("Upbringing and Nutrition") 2020, 3rd ed.; "Erziehung an der Mutterbrust "("Upbringing at the Mother's Breast") 2020, 2nd ed.; "Projekt Erziehung" ("Project Education") 2019, 6th ed.; "Wörterbuch der Pädagogik" ("Dictionary of Pedagogy") 2018, 17th ed.; "Das "normale" Kind. Einblicke in die Geschichte der schwarzen Pädagogik" ("The 'Normal' Child. Insights into the History of Dark Pedagogy") 2020.

Enlivenment

Nathanaël Wallenhorst

Abstract *Enlivenment* is a term proposed by the contemporary German biologist and philosopher Andreas Weber, whose work follows on from the Frankfurt School's Critical Theory. The term notably features in two of Weber's works: the first, co-authored with fellow German philosopher Hildegard Kurt, *Lebendigkeit sei! Für eine Politik des Lebens. Ein Manifest für das Anthropozän* (2015) (*Towards Cultures of Aliveness: Politics and Poetics in a Postdualistic Age – an Anthropocene Manifesto*), and the second, *Enlivenment. Eine Kultur des Lebens. Versuch einer Poetik für das Anthropozän* (2016) (*Enlivenment. Towards a Poetics for the Anthropocene*).

The term *Enlivenment* contains a tacit reference to the Age of Enlightenment. Its proponents believe that humanity needs to advance from the *Enlightenment* that brought us the Industrial Revolution, to a state of *Enlivenment*, which is needed in the Anthropocene. We need to build upon the Enlightenment as a society, and rediscover an appreciation of the living world. The proposed intellectual shift is from enlightenment to vitalisation. In each of Weber's works in German, the term *Enlivenment* is left in English – undoubtedly to highlight the parallel with *Enlightenment* and the importance of the paradigmatic shift – both in how we think and how our societies are organised. The term refers to a form of dynamics that is intrinsic to living things, which must be allowed expression, and allowed to spread, in order for humanity to embark the 'adventure' our uncertain future holds. Weber's German rendering of *Enlivenment* is *Verlebendigung*, which carries notes of encouragement, cheer, enchantment or birth, and the giving of life.

Enlivenment is based on what we share with other beings: the very fact of living and feeling. The notion also marks the end of the idea that humanity and nature are separate. The argument is that humans have long falsely believed they hold dominion over the living matter of which all life is composed. Today, everything has

N. Wallenhorst (✉)
Catholic University of the West, Angers, France
e-mail: nathanael.wallenhorst@uco.fr

become an element of culture, and we have developed the mentality that we are superior to nature. Weber decries the idea of a world based on mechanisms that are designed for efficiency. Rather, his concept of the world, echoed by Kurt in their joint *manifesto*, is of 'a meshwork of mutually transformative and meaningful relationships, which are experienced by subjects' (Weber & Kurt, 2015, p. 11). This approach 'is starting to find root in the current revolution of biological thinking, similar to the revolutions in physics roughly 100 years ago through relativity theory and quantum physics' (p. 11). In fact: 'Humans and nature are one, because creative imagination and feeling expression are natural forces' (p. 11).

Above all, Enlivenment celebrates aliveness. Life carries uncontrollable, subversive power and creativity, which cannot be contained or manipulated. In addition, the aliveness of others (be they human or non-human) and of our environment renders our own existence possible. In this notion, the emphasis is not on the exceptionality of humanity in relation to the rest of the living world, but our solidarity with the whole fabric of living things. Essentially, life is beyond our control, as it should be, according to Weber's philosophy of Enlivenment. It is by fully embracing our place within the living world that we will be able to survive the Anthropocene. This power of Enlivenment and solidarity among living things can offer hope for humanity. As such, it is essential that we do not view ourselves as being removed from that 'fabric' of all living things. Enlivenment refers to a concept of life as a creative practice. This contrasts with the ideas of technicists, who view life as a tool – a technique which must be understood in order to be exploited. The idea of Enlivenment, notably expounded in the *Manifesto*, focuses on the need to re-examine how we think about humanity: we should think of it in terms of its biological connection with the tapestry of life.

Weber refers to Gary Snyder, the American poet and ecophilosopher, and his view of the wild as a process beyond human control. We cannot make Earth better by controlling it, but by contributing and playing our part, we can make a positive difference. This is an important paradigmatic shift: the human adventure has a role to play, and can choose to participate more fully in the fabric of solidarity that is the living world. Participation is possible, but is poetic, and is the antithesis of control.

Enlivenment is a type of second Enlightenment ("*Aufklärung 2.0*") (Weber, 2016, p. 25). Andreas Weber critiques the Age of Enlightenment and the ideology of death which pervades it, whereby everything around us is considered inert matter. It should, however, be remembered that the Age of Enlightenment encompasses multiple discoveries – a fact which Weber appears to overlook). Weber follows in the footsteps of Max Horkheimer and Theodor W. Adorno in their critique, *Dialectic of Enlightenment (Dialektik der Aufklärung)*. Horkheimer and Adorno point out that the ideology of Enlightenment, while it has liberated many, has also resulted in disastrous totalitarian systems. However, the authors propose no alternative concept. This is what Weber aims to provide, in the concept of Enlivenment. In that effort, Weber is directly engaged with Hartmut Rosa's approach, which is also an extension of the Critical Theory. Rosa's concept of resonance is one alternative – a type of positive political proposition. Enlivenment is a type of "corrective" concept. Enlivenment and resonance are particularly closely related. For example, in his original German text, Weber frequently uses the term '*Verbindung*', which could be translated as relationship, link or connection. Enlivenment includes a type of biological foundation that the concept of resonance lacks (in addition, for example, to the mirror neurons that Rosa cites). Furthermore, Enlivenment is equally as radical as convivialism. The aim behind both concepts is to break down *hubris* (or at least to contain it, under the 'lid' of conviviality).

The concept of Enlivenment differs from that of sustainable development ('*Nachhaltigkeit*') in its radical nature. Seeing how the Anthropocene has become synonymous with ecomodernism or transhumanism, Andreas Weber proposes an alternative: the poetics of vitality. Enlivenment is a critique of the neo-Darwinian idea of biological optimisation, and of the neoliberal quest for economic efficiency (Weber, 2016, p. 45). Indeed, these two paradigms may sometimes seem sufficient to identify the knowledge we now have of how the world works. However, they entirely overlook the commonality and sharing, a) amongst humans, and b) between humans and the rest of the living world. They also omit the idea of cooperative dynamics. Weber demonstrates that the anthropological concepts based on competitiveness and individual competition, which form the basis for our biological functions (Darwinism) and economic activities (liberalism), are connected to a particular way of viewing reality. He puts forward a different frame of reference. The world is not a free-for-all or an out-and-out war, with all creatures looking out for their own interests alone. Weber aims to supplant the bioliberal principles underlying our scientific, political and educational decisions, replacing them with the dynamics and principles of Enlivenment. By becoming aware of our own aliveness, we should begin to be able to forge a connection with the natural world, and with other living organisms. Education in the Anthropocene needs to guide people towards resonance. Also, we need to steer learners towards Enlivenment (experiential enlivenment, so they are aware of their own place as part of the living world, and cognitive enlivenment, to help them break free of the dominant Cartesian rationalism). The term *Enlivenment* is quite deliberately linked to *Enlightenment*. The aim is not to replace rational thought and empirical observation with poetic fancy, but rather, to weave different rationales together. The idea is for scientists, politicians and society as a whole to again begin taking an interest in the sensitive aspect of both human and nonhuman existences.

Enlivenment is the cornerstone of a policy of civilisation that can come about through establishing a culture of vitality, to help us survive the Anthropocene. Thus, Enlivenment counters the prevailing ideas about nature which tend to shape political positions. Firstly, nature is not efficient. Quite the contrary – it is continually wasteful: fish, amphibians and insects have to lay millions of eggs in order for just a few to reach maturity. Another example of this inefficiency is the fact that warm-blooded animals expend 90% of their energy simply to keep their metabolism running. Secondly, the biosphere is not growing. The biomass of the biosphere is in balance (with only very slight variations). Thus, nature is characterised by a steady state. Thirdly, no new species has ever appeared as a result of competition for resources. Rather, it is new cooperation and symbioses (or simply chance) that allow new lifeforms to emerge. Fourthly, nature provides resources enough for everyone; the prime example being solar energy, which is abundant enough for all living things. Thus, we can conclude that, through symbioses and cooperation, all species should be able to coexist. Fifthly, the concept of ownership has no place in the biosphere. The body itself is not the property of the creature which inhabits it, as it must interact with its environment, and is characterised by exchanges of matter (Weber, 2016, pp. 55–57). In light of the above observations, Andreas Weber developed biopoetics as a model of living relationships.

What is particularly interesting in Weber's philosophical and political thinking is that the dawn of the Anthropocene does not represent the death of hope. On the contrary, the vitality of living things means we can embrace revitalisation and renewed solidarity, and our existences will become political capital. The *Ecomodernist Manifesto* (2013) celebrates the power of technology, and human beings' incredible ability to master their environment. However, in Weber's view, the opposite is true: Earth, and the living macro-ecosystem it hosts, show us the way to survive the Anthropocene. We are not masters of the earth; rather, the earth – and the web of life which inhabits it – is master of us. From this perspective, Enlivenment breaks with all forms of anthropocentrism.

References

Weber, A. (2016). *Enlivenment. Eine Kultur des Lebens. Versuch einer Poetik für das Anthropozän.* Matthes und Seitz.
Weber, A., & Kurt, H. (2015). *Lebendigkeit sei ! Für eine Politik des Lebens. Ein Manifest für das Anthropozän.* Think Oya.

Nathanaël Wallenhorst is Professor at the Catholic University of the West (UCO). He is Doctor of Educational Sciences and Doktor der Philosophie (first international co-supervision PhD), and Doctor of Environmental Sciences and Doctor in Political Science (second international co-supervision PhD). He is the author of twenty books on politics, education, and anthropology in the Anthropocene. Books (selection): *The Anthropocene decoded for humans* (Le Pommier, 2019, *in French*). *Education in the Anthropocene* (ed. with Pierron, Le Bord de l'eau 2019, *in French*). *The Truth about the Anthropocene* (Le Pommier, 2020, *in French*). *Mutation. The human adventure is just beginning* (Le Pommier, 2021, *in French*). *Who will save the planet?* (Actes Sud, 2022, *in French*). *Vortex. Facing the Anthropocene* (with Testot, Payot, 2023, *in French*). *Political education in the Anthropocene* (ed. with Hétier, Pierron and Wulf, Springer, 2023, *in English*). *A critical theory for the Anthropocene* (Springer, 2023, *in English*).

Evolution and Communication

Sebastián Agudelo

Abstract We propose an interpretation of processes of evolution that share an analogical structure with processes of communication in human beings regarding language development.

Biological evolution is also the evolution of communication, if we consider that all living beings require the means to deal with, pass on and incorporate information on different levels: between DNA and RNA, among cells and organs, between members of the same species and with members of other species. These means of communication need to be efficient and allow, follow or facilitate their speciation process. If we take as starting point a simple definition of evolution as the process of biological transformation of how organisms cope with genetic and environmental information, in order to favour their survival, the relationship between evolution and communication appears clearer. However, we need to differentiate between information and communication, as the language use attributes to the latter an intentional and active component in which sender and recipient interact through a channel and with the former is understood that the sender does not necessarily have an intention and the recipient has no influence on the content of the message. The notion of information was clarified by Shannon (1948) as a set of discrete symbols and functions (code) given in a temporal unit, where transmission results in the decrease of uncertainty for a recipient. He related it to the second law of thermodynamics by opposing it to entropy. If the tendency of particles of matter is to cool down and lose their order, that of living beings is to process information in order to achieve a greater organisation. The mathematical conception of information had the advantage of dispensing with the entities that produce the message and its content, so that it increases both the comprehension of the communication systems of living organisms and the development of communication technologies according to physico-chemical principles. But, it also disregarded the problems of asymmetry of power

S. Agudelo (✉)
Independent Philosopher, Berlin, Germany

N. Wallenhorst, C. Wulf (eds.), *Handbook of the Anthropocene*,
https://doi.org/10.1007/978-3-031-25910-4_120

and control that it poses for societies through reification of human beings (Fuchs, 2009). The introduction of the notion of feedback in the field of what will henceforth be known as cybernetics by Wiener (2019) or the idea of structural closed cycles in the experimental medicine of V. von Weizsäcker (1950), is to prove decisive in the stimulation of a dynamic approach to communication and action control schemes in machines and animals. There was a new conception of interaction without necessarily resorting to intentionality, preferring instead to draw on the irritability of supports, i.e. their ability to react and embody new data. The long-standing debate between idealists and materialists, vitalists and mechanists, will be resolved, at least partially, with the aid of technological models that distance themselves from all prior ontological commitment. However, despite the advantage of these models in making available new analogies and empirical methods of addressing newly biological questions, the technological mentality driven by its relative success deeply permeated the field of natural sciences to the point of claiming a certain homology of dynamics and temporally discarding the self-organization capacity of life. This will later be rescued by the postulates of autopoiesis only to be challenged again by AI scientists. Up to this point, the problems of cognition, perception, and structure of nature were considered from a perspective that was still influenced by the technomorphist conception through which nature has been understood since the beginnings of modernity and which Kant, in a visual paradigm of knowledge, had already pointed out as the basic form of our understanding (Kant, 1986). Consider, for example, the apparent structure of DNA. It looks like a double helix of identical information that is recorded on the basis of four different components, adenine (A), cytosine (C), guanine (G) and thymine (T), which are translated into protein structures according to their combination. DNA is self-replicating, and it is on the basis of this natural possibility that new engineering technologies, such as CRISPR, have been developed allowing scientists to cut and paste already existing parts of DNA at will. DNA seems to function like a simplified alphabet, whose visible and more or less intelligible sentences make up the realm of the living, whose biological language can be interpreted under our arbitrary language.

The metaphoric structure of human knowledge is perhaps one of the most underestimated problems of science (Lakoff & Johnson, 2008), whose ideal of unveiling the ultimate structure of reality is constantly carried over into new ways of communicating with and about it. In this sense, communication is a system of the metaphors that best suit the general understanding that we have of the reality that living organisms endure and create, including that of biological evolution and the evolution of communicating systems in the continuation and overcoming of biological processes.

Evidence of communication processes can be found across all the different living kingdoms (animal, plant, fungi, protist and monera), indicating that they form analogical and homological structures necessary for the processes of in-formation of biological systems. Here we limit ourselves to an interpretation of their evolution in human beings in order to support a conception of communication as a process of expression and interaction that can shed some light on our understanding of the configuration of reality encompassing both the biological and the technological evolution.

The evolution of communication is correlative to the evolution of the supports and the means of communication. In biological beings, these correspond mainly to living bodies, but are not limited to them, since communication implies an externalization and therefore a relation to inorganic bodies. Paleoanthropological research has shown, in close collaboration with neurology, archeology and anatomy, that the evolution of humans is closely linked to the organization of activities conducive to survival that depend on mobility, perception, motivation and action. According to A. Leroi-Gourhan (1993), since the origin of vertebrates, a "living mechanics" has pushed them to find anatomical solutions between the dichotomy of physical and anatomical pressures and a tendency towards freedom of movement and interaction. Bipedalism allowed the transformation of the two poles of the anterior field, the hand and the mouth. The former was freed from locomotion and took over the grasping and preparation role that belonged to the latter, which in turn became specialised for speech. Technical skills are so deeply related to communication skills that, at the neurological level, the movement of the lips and tongue is controlled by the same pyramidal areas that control the movement of the fingers. This correlation is palpable in the ordinary behavior of accompanying speech with hand movements or manipulating objects while thinking aloud. Just as humans can use their hands to speak, they can also use their mouths as a grasping instrument. Moreover, the invention of writing is, from the point of view of neuroanatomical evolution, a logical step in the development of the hand and face pole by rearranging the means of expression. But writing is only a later visible and more memorable achievement that is situated in an external support emancipated from the brain and constitutes a technique of communication. The role of the hands in the early evolution of human communication has been reconsidered and given greater importance by the studies of sign languages and their recognition as fully functioning natural languages (Stokoe et al., 1995).

A main axis of this discussion corresponds to the comparison of animal and human communication systems, where studies of chimpanzees have received special attention. Since Darwin, linguists have tried to find out if anything can be added to the chimpanzee cognition that controls calls, gestures, parades and technical skills to make them communicate thoroughly with humans. According to this perspective, the difference of expression between high-order primates and human beings is only a matter of degrees of intelligence and not an entirely new configuration of it. During the 1960s and 1970s, linguists and trainers tried to teach chimpanzees and gorillas first to speak, then, when it was considered that their limited ability to express themselves verbally derived primarily from their vocal anatomy, to sign. Primates have a very good visual memory, on which they rely in their natural habitats to find food, detect dangers, and interpret acts and gestures of their conspecifics. But this has shown to be no better guarantee for signing than their ability to produce a certain range of meaningful sounds was for speaking. Besides the difficulties that they have to articulate words finely, either with their hands or mouth, and the fact that their motivation to interact with humans relies largely on food stimuli, their semantic achievements fall short of comprehending abstractions, and the lack of use of any type of syntax arises as a definite ontological barrier to the production of human-like language. The sought-after continuity between primate communication systems and human language seems to be one chapter more in the discussion on the continuity of the evolutionary lineage between primitive primates and early humans, which kept nineteenth century naturalists on tenterhooks and gave free rein to a fanciful cultural iconography of the ape-man. Although this image proved false, it returned under the guise of evolutionary linguistics as the effort to demonstrate the transition from a primate form of communication to language. This enterprise rested on a correct premise that mere upright anatomy does not suffice to establish the originality of human beings, but it must be related to brain and language development. The question of whether humans or language came first had been dismissed in the eighteenth century by W. von Humboldt as metaphysical and unsolvable (von Humboldt, 2020). But, if the phylogenetical thesis alone led to a dead end, the ontogenetic research still had a say. A breakthrough came with S.J. Gould in the 1970s, when he provided a first synthesis of two seemingly irreconcilable conceptions by linking Haeckle's recapitulatory theory with Bolk's theory of retardation (Gould, 1977). He then built with Eldredge a case against Darwinian gradualism by proposing the theory of punctuated equilibrium. This theory accounts for the paleontological record, according to which large periods of phylogenetical stability are sporadically interrupted by short times of intense species variability and emergence (Gould & Eldredge, 1977). These insights have been considered to support Chomsky's hypothesis of the emergence of an ad hoc language acquisition device (Chomsky, 2014), which is still held by a large group of enthusiasts, even though it has not yet been attributed to an organ. The problem lies probably in the view that language is the expression of a single disposition rather than a whole system made up of many parts, some of which existed before humans and were reconfigured in them. This was the position embraced earlier by Herder, who saw that the sphere of perception and action of humans is not determined by already given objects, but must be constructed in interaction with them. The activity of the senses is mediated by the production of sounds to discriminate, qualify, fix, and render available what the eye sees and the hand grasps even in the absence of external stimuli. He anticipated cybernetics when he stated that words are a self-sufficient reality that closes a cognitive cycle where they are simultaneously emitted and received by their producers. He did not introduce new elements into the cognitive equation of language production, but proposed a reorganization of what already existed in the zoological realm. The outcome was not a difference of degrees between human language and animal communication systems, but one of essence. Such a conception lies in the vicinity of another important contribution of Gould, together with Lewontin, concerning the restructuration of organs and the new functions that anatomical and physiological by-products of evolution can eventually assume with success (Gould & Lewontin, 2020). In Herder, the senses of sight and touch transfer information to each other by translating stimuli into an acoustic channel. Objects appear to the eye differently from the way they are felt by the hand, but although these perceptions are not the same, they relate to each other through a third element alien to both, which is sound (Herder, 2012). Sound transposes these perceptions in words that say what is seen and what is touched, and which in a predication enclose what they present and stand for. Words represent what the senses present. Language is thus metaphorical and analogical, and not only in its rhetorical use. It is already so in its structure, where the senses constantly transfer information among them to make a common sense even if they belong to completely different supports and channels.

References

Chomsky, N. (2014). *Aspects of the theory of syntax*. The MIT Press.
Fuchs, C. (2009). Towards a critical theory of information. *TripleC: Communication, Capitalism & Critique. Open Access Journal for a Global Sustainable Information Society, 7*(2), 243–292.
Gould, S. J. (1977). *Ontogeny and phylogeny*. Belknap.
Gould, S. J., & Eldredge, N. (1977). Punctuated equilibria: The tempo and mode of evolution reconsidered. *Paleobiology, 3*(2), 115–151.
Gould, S. J., & Lewontin, R. C. (2020). The spandrels of San Marco and the Panglossian paradigm: A critique of the Adaptationist program. In *Shaping Entrepreneurship Research* (pp. 204–221). Routledge.
Herder, J. G. (1772/2012). *Treatise on the origin of language*. Cambridge University Press.
Kant. (1986). *Critique of judgement*. Hackett Publishing.
Lakoff, G., & Johnson, M. (2008). *Metaphors we live by*. University of Chicago Press.
Leroi-Gourhan, A. (1993). *Gesture and speech*. The MIT Press.
Shannon, C. E. (1948). A mathematical theory of communication. *The Bell System Technical Journal, 27*(3), 379–423.
Stokoe, W. C., Armstrong, D. F., & Wilcox, S. E. (1995). *Gesture and the nature of language*. Cambridge University Press.
von Humboldt, W. (2020). *Ueber das vergleichende Sprachstudium in Beziehung auf die verschiedenen Epochen der Sprachentwicklung* (Vol. Band 3, pp. 241–268). De Gruyter.
Von Weizsäcker, V. (1950). *Der Gelstaltkreis. Theorie der Einheit von Wahrnehmen und Bewegen*. Kohlhammer.
Wiener, N. (2019). *Cybernetics or control and communication in the animal and the machine*. The MIT Press.

Sebastián Agudelo is an independent researcher and translator based in Berlin. He holds a PhD in Philosophy from the University of Paris VIII and his work is mainly concerned with problems of language and technological development from an evolutionary perspective. He uses and reflects on approaches from Historical and Philosophical Anthropology, Pragmatism, Critical Theory and Transcultural Aesthetics.

Feminism and Gender

Birgit Althans and Gabrielle Ivinson

Abstract The term Anthropocene refers to the impact of human activities on the earth that interfere with climate, land and biosphere that sustain plant, animal and human life at unprecedented scales of intensity that set a course towards species extinction. The Industrial Revolution and the aftermath of this over two centuries is cited as a turning point in levels of pollution. The term Capitolocene refers to the interconnection between the ecological state and the capitalist condition, and alludes to a time in which economic productivity is reliant on the extraction of coal, oil and minerals and the ensuing environmental degradation of terrain, rivers and atmospheres. Across time human, nature, territory, labour and politics inter-connect yet in different ways. In this section we are interested in how feminists have taken up and developed the concept Anthropocene.

First there has been a shift to understand world systems beyond the terms of positive scientific methods towards integration across humanities, arts and scientific thinking. Feminist scholars have long questioned the human as homogeneous and have been interested in, for example, how women and girls have historically not enjoyed the same rights to citizenship, possession and access to land as men; disparities in who benefits from capitalism; the deep north-south geopolitical divide related to unequal profits (Chakrabarty, 2009); a more geocentric understanding of planetary life (Braidotti, 2012), and a recognition that humans are not the only important actors on earth. They place an emphasis on 'ongoing multispecies stories and practices of becoming-with other beings' which Haraway refers to as the Chthulucene (Haraway, 2016, 55). Feminists have also been concerned with the agency of matter and rejecting Cartesian binaries such as alive-dead, and specifically the notion that matter as inert and 'HuMAN' minds are agentive/relational (Bennett, 2010). They reject linear cause-effect rationality by 'queering concepts of time and space' (Barad, 2007) to emphasise that what happens in one space impacts another place

B. Althans (✉) · G. Ivinson
Kunstakademie Düsseldorf, Düsseldorf, Germany
e-mail: birgit.althans@kunstakademie-duesseldorf.de

© The Author(s), under exclusive license to Springer Nature
Switzerland AG 2023
N. Wallenhorst, C. Wulf (eds.), *Handbook of the Anthropocene*,
https://doi.org/10.1007/978-3-031-25910-4_121

and that pasts, presents and futures are entwined. Heather Davis argues that the term Anthropocene has the capacity to remind us of our limits as human beings, always indebted to others and 'that our being is tied to the rocks and other-than-humans that compose us' (cited in Braidotti & Hlavajova, 2014, 63).

Donna Haraway is a feminist biologist who takes a critical stance towards the term 'Anthropocene' to describe the state of the planet. Her preferred term is 'Chthulucene' from the Greek *khthon* meaning heritance or remembering and *kainos* meaning time of being (Haraway, 2016, 2). While she recognizes we live in times of urgency on the verge of mass death, extinction and the onrush of disaster, she urges us think of humans interconnectedly living with all inhabitants of the earth such as 'cnidarians, spiders, raccoons, squid, and jellyfish' in networks, webs and lines of existence rather than thinking of separate spheres of inhabitation (ibid. 32). She rejects the autopoietic notion of feedback loops in systems theory and instead uses a tentacular metaphor to stress a sense of porous tissues, open edges and contact zones. She warns that in the onrushing catastrophe there is an unprecedented 'looking away' (Haraway, 2016, 52). She compares this collective attitude of looking away to Hannah Arendt's analysis of the Nazi war criminal Adolf Eichmann who displayed an "inability to think" (Haraway, 2016, 52). In that surrender of thinking lay the "banality of evil that could make the disaster of the Anthropocene, with its ramped-up genocides and speciescides, come true." (Haraway, 2016, 36). What Eichmann was not able to do, according to Arendt, was to live with others in a world he was interested in. She states, he "could not entangle, could not track the lines of living and dying, could not cultivate response-ability" (ibid.). Instead, he undertook rational-bureaucratic thinking. "Function mattered, duty mattered, but the world did not matter for Eichmann" (Haraway, 2016, 36). She argues that this inability to perceive a living world, "…..was not an emotional lack, a lack of compassion, although surely that was true of Eichmann, but a deeper surrender to what I would call immateriality, inconsequentiality, or, in Arendt's and also my idiom, thoughtlessness. … There was no way the world could become for Eichmann and his heirs—us?—a "matter of care". (Haraway, 2016, 36).

The Eichmann example reinforces the need for a shift away from thinking that places Western, male humans at the centre of things. Haraway refers here to the Greek etymology of *anthrōpos* "man, human", literally "he who has the face of a man." (Online Etymology Dictionary). Haraway also cites Bruno Latour who wrote, "Some are readying themselves to live as Earthbound in the Anthropocene; others decided to remain as Humans in the Holocene" (Latour, 2013 cited in Haraway, 2016, 58). Both Haraway and Isabelle Stengers (2014) refer to Bruno Latour's strategic cosmopolitican perspective of the Anthropocene. For Latour, the historical earth is a biotope of past, present as well as future entities, "an entity that is composed of multiple, reciprocally linked, but ungoverned self-advancing processes." (Latour, 2013, 8.) Insofar as ecological conflicts are always geopolitically motivated, any cosmopolitics or political ecology for him requires a willingness to negotiate and the use of diplomacy. Latour draws in part on Carl Schmitt's "political theology," when he sees humans in history and the earthbound in the Anthropocene engaged in trials of strength where there is no referee who/which can establish what is/was/will be. For Haraway "The question of whom to think-with is immensely material." (2016, 43.) Her strategic perspective comes from her emphasis on compositional bonds between human and non-human 'mortal earthlings in thick copresence' (Haraway, 2016, 4) which is always situated, entangled and worldly.

Haraway refers here to Anna Tsing's book *The mushroom at the end of the world*, which provides a rich empirical example of interconnectedness with other species. Tsing describes the way fungal bodies extend themselves below the forest floor in 'nets and skeins, binding mineral soils long before producing mushrooms' (Tsing, 2015, viii). She portrays the 'arts of living on a damaged planet' and the possibility of life in Capitalist ruins (Tsing, 2015; Tsing et al., 2017).

> If a rush of troubled stories is the best way to tell contaminated diversity, then it's time to make that rush part of our knowledge practices … Matsutake's willingness to emerge in blasted landscapes allows us to explore the ruins that have become our collective home. To follow Matsutake guides us to possibilities of coexistence within environmental disturbance. This is not an excuse for further human damage. Still, Matsutake shows one kind of collaborative survival. (Tsing, 2015, 34)

Haraway makes links as well with the concept of care – for all 'entanglements' in the Anthroprocene – in Maria Puig de la Bellacasa's work on "Matters of Care" (2011, 2017). The entanglements concern 'human-soil-affections'. While her dystopian vision of anthropocenic soils invokes yet another objectification of a natural resource brought to exhaustion by deadly human-centred capitalist production she also points to more relational futures. She refers to research on current practices, material involvements and stories emerging from scientific accounts, community involvement and artistic manifestations. Emerging motifs are of renewed imaginaries of entanglements of human-soil matter, which feed into each other. She draws attention to soil's aliveness: "In a sense we are unique moist packages of animated soil". These are the alluring words of Francis D. Hole, a professor of soil science renowned for encouraging love for the soil and understanding of its vital importance. Affirming humans as being soil entangles them in substantial commonness." (Puig de la Bellacasa, 2019, 391).

Her sense of human-soil entanglements places the focus on intensifying intimate interdependency. "These new involvements with soil's aliveness open up a sense of earthy connectedness that animates and re-affects material worlds and a sense of more than human community in those involved. "(Puig de la Bellacasa, 2019, 391).

Kathryn Yusoff (2018) in her book *A Billion Black Anthropocenes or None* draws attention to the notion of the Anthropocene as a European, liberal concern based on a White Geology which fails to recognise race. She reminds us that, 'Imperialism and ongoing (settler) colonialisms have been final words for as long as they have been in existence (2018, xiii). When the 'Anthropocene' is seen as the threat to the planet, she argues, it is actually seen as a threat to liberal white communities. The current concern named 'Anthropocene' can only be recognised as 'the' stage of planetary emergency' if the billions of extinctions of people, cultures and ways of living and being that supported and continue to support capitalism are forgotten. She argues that imperial capitalism worked by removing black, brown and indigenous peoples from their lands, taking possession and remorselessly extracting minerals, while eradicating ecosystems and ways of life. There have been a billion Black extinctions. The world has already come to an end for so many communities, black, brown or white.

Yusoff questions what she calls the origin stories that underlie the term Anthropocene. Three spikes are often specified; 1610 colonial expansion, 1800s Victorian Industrial Revolution and 1950s the Great Acceleration. Along with other Black scholars she argues that the enforced enslavement of people from Africa to sugar cane plantations and mines in the New World (the Americas) from the fifteenth century onwards founded capitalism. When slavery was abolished in Britain in 1833, large sums of money equivalent to £20 million were given to powerful White Men in 'compensation' for losses relating to their involvement in the slave trade. This excessive wealth enabled them to invest in railways, canals, mines, cotton factories and warehouses, creating the industrial revolution. Accordingly, slavery is not in the prehistory of capital, it is foundational to it and its afterlives reverberate in current capitalist times.

For Yusoff race is central to the historical geographies of extraction. Her trenchant critique is that imperial Capitalism was predicated on categorising Black persons, not only as Other, but as inhuman. Human and its others were bifurcated to create two different species (Fanon, 1963). Binaries such as dead/alive; inert/agentive; human/inhuman are sedimented in scientific disciplines such as geology. Extracted from land, culture and spiritual life, diasporic slaves were dispossessed of their land and humanity, beaten, tortured and rendered inhuman. Both Black persons and inhuman earth were coded in the 'indifferent register of matter' (Yusoff, 2018, 5) and subjugated to extraction; Black bodies for their labour and earth for minerals. By classifying Black as inhuman, flesh became a commodity that could be weighed and counted, bought and sold. In plantations Black flesh became the energy that powered capitalist productivity. 'Coloniality cuts across earth and flesh, (ibid, 12). The 'slave-sugar-coal nexus' enriched Britain (ibid. 43).

In Yussof's argumentation racial environmentalism persists. Those who are first in line to suffer from climate change and pollution, whose ecosystems have been destroyed by mono crops and mono industries and who have been subjected to Western scientific experimentation with nuclear devices are excluded from major 'scientific' debates. Rosi Braidotti agrees with this view and states that there is an urgent need to de-segregate the different and highly specialized spheres of knowledge production, especially the case for perspectives developed from decolonial, black and race studies that are often marginalized. (Braidotti & Hlavajova, 2014, 5). Yusoff suggests we need a more speculative geology, a *geology-in-the-making*: a future oriented social geology (*ibid.* 60) that recognises the everyday enactments of racial living and a far more revolutionary shift in the geographies of the Anthropocene where co-operative human efforts can have a place and can take place (McKittrick, 2011, 960).

The end of the world has already happened for some subjects (Césaire, 1972/2000). Yusoff, (2018, 63) thinking with Saidiya Hartman, (2003) states, 'The world must end for another relation to the earth to begin.' In the ruins of these worlds new imaginaries are growing. In the afterlives of slavery peoples 'embodied experience of power located in the earth [is] the basis of knowledge and the affirmation of a more exorbitant world or planitarity' (Yusoff, 2018, 39). Accordingly, Blackness becomes a prerequisite for the possibility of imagining 'living and breathing again' for others (ibid. 12) paving the way for a poetic undercommons (Moten, 2016) and more relational inhuman subjectivities. We end with a new beginning as described by Rosi Braidotti, "We define our era as the Anthropocene, by which we understand the geological time when humans are having a lasting and negative effect upon the planet's systems. ... We believe that new notions and terms are needed to address the constituencies and configurations of the present and to map future directions'. This moment is marking the 'end of the self referential arrogance of a dominant Eurocentric notion of the human and the opening up of new perspectives" (Braidotti & Hlavajova, 2014, 1, 3).

References

Barad, K. (2007). *Meeting the universe halfway: Quantum physics and the entanglement of matter and meaning*. Duke University Press.

Bennett, J. (2010). *Vibrant matter. A political ecology of things*. Duke University Press.

Braidotti, R. (2012). *Nomadic theory. The portable Rosi Braidotti*. Columbia University Press.

Braidotti, & Hlavajova (Eds.). (2014). *Posthuman glossary*. Bloomsbury Publishing.

Césaire, A. (1972/2000). *Discourse on colonialism*. Monthly Review Press.

Chakrabarty, D. (2009). The climate of history: Four theses. *Critical Inquiry, 35*, 197–222.

Fanon, F. (1963). *The wretched of the earth*. Groove Press.

Haraway, D. (2016). *Staying with the trouble. Making Kin in the Chtulucene*. Duke University Press.

Hartman, S., & Wilderson, F. (2003). The position of the Unthought. *Qui Parle, 13*(2), 183–201.

Latour, B. (2013, September 23). *"War and Peace in an Age of Ecological Conflicts."* Lecture for the Peter Wall Institute, Vancouver, BC, Canada. Video and abstract at http://www.bruno-latour.fr/node/527. Accessed 7 Aug 2015.

McKittrick, K. (2011). On plantations, prisons, and a black sense of place. *Social and Cultural Geography, 12*(8), 947–963.

Moten, F. (2016). *A poetics of the Undercommmons*. Sputnik and Fizzle.

Puig de la Bellacasa, M. (2011). Matters of Care in Technoscience: Assembling neglected things. *Social Studies of Science, 41*(1), 85–106.

Puig de la Bellacasa, M. (2017). *Matters of care: Speculative ethics in more than human worlds*. University of Minnesota Press.

Puig de la Bellacasa, M. (2019). Re-animating soils: Transforming human–soil affections through science, culture and community. *The Sociological Review Monographs, 67*(2), 391–407.

Stengers, I., & in conversation with Heather Davis and Etienne Turpin. (2014). Matters of cosmopolitics: On the provocations of Gaia. In E. Turpin (Ed.), *Architecture in the Anthropocene: Encounters among design, deep time, science and philosophy* (pp. 171–182). Open Humanities.

Tsing, A. (2015). *The mushroom at the end of the world: On the possibility of life in capitalist ruins*. Princeton University Press.

Tsing, A., Swanson, H., Gan, E., & Bubandt, N. (Eds.). (2017). *Arts of living on a damaged planet: Ghosts and monsters of the Anthropocene*. University of Minnesota Press.

Yusoff, K. (2018). *A billion black Anthropocenes or none*. University of Minnesota.

Birgit Althans is Professor for Pedagogy in the department of art-related sciences at Art Academy Duesseldorf. Main Research in Historical and Pedagogical Anthropology; Gender and Cultural Studies; qualitative, cultural science oriented research in pedagogical and artistic fields of work (especially theatre), cultural education in anthroprocene.

Gabrielle Ivinson is Professor of Education and Community at Manchester Metropolitan University, UK. She develops arts-based methods to attune to what lies beyond the spoken word, enabling difficult issues to be surfaced by drawing on post personal and new materialist concepts. She is a co-editor of Material Feminisms and Education (with Carol Taylor, 2015).

Homo Oeconomicus, Homo Collectivus and Homo Religatus

Nathanaël Wallenhorst

Abstract This article is an extension to 'The Human Condition in the Anthropocene' and 'The Human Adventure' (cross-reference). The former suggests that the dawn of the Anthropocene should bring about a shift in our understanding of humanity. The latter explores the potential relevance of the notion of the 'human adventure', as opposed to the human race. This notion emphasises the uncertainty of the future, the possibility of altering the characteristics on which we are contingent, and the need for political anthropology that acknowledges the 'adventure' of living. This article expands on the conceptualisation of the human adventure, dividing it into *Homo oeconomicus*, *Homo collectivus* and *Homo religatus*. Each of these concepts refers to one of the human adventure's three dimensions: *hubris*, the world and coexistence – hence, this article should be read directly after "The Human Adventure".

To begin with, let us look at *hubris* and the drive for profit behind *Homo oeconomicus*.

The rampant capitalism and consumerism born of the Industrial Revolution go hand in hand with burgeoning individualisation. Today's individuals are consumers, seeking to maximise their own interests. The capitalist economy has become all encompassing, and hegemonic (Peyrelevade, 2005). Consequently, individuals today are indistinguishable from *Homo oeconomicus*, characterised by an omnipresent drive for profit and consumption. *Homo oeconomicus* is a creation of the Industrial Revolution, but also the manifestation of the utilitarian paradigm. The limitations of utilitarianism have been demonstrated by a raft of anti-utilitarian writers (Caillé et al., 2018). *Homo oeconomicus* is the self-interested individual so plainly described by eighteenth-Century Scottish economist Adam Smith in *The Wealth of Nations*, with his well-known 'invisible hand' metaphor. Smith contends that the pursuit of individual interests benefits the collective. This utilitarian logic

N. Wallenhorst (✉)
Catholic University of the West, Angers, France
e-mail: nathanael.wallenhorst@uco.fr

was built upon and further explored by British writers Jeremy Bentham (1748–1832) and John Stuart Mill (1806–1873), and has since been seized upon by a host of so-called 'classic' liberal economists.

The limitations of *homo oeconomicus* are considerable, and are regularly pointed out in the literature: *Homo oeconomicus* 'has neither "altruism" nor "responsibility" for others and for future generations' (Faber et al., 2002, p. 324). In a way, *Homo oeconomicus* is an extension of Descartes' 'subject', one step removed from his environment, which he attempts to understand and to exploit (Flahault, 2005, p. 378). However, the rationality of *Homo oeconomicus* is relative. In the view of economist Christian Arnsperger, he is incapable of reflecting on what is behind his rationality, and is 'deaf and blind to his own worries and existential anguish' (Arnsperger, 2011). Thus, the dominant anthropological model today, based on *Homo oeconomicus*, is that of 'economic growth rooted in the psychology of lack' (Arnsperger, 2010, p. 25). The logic of accumulation at the heart of capitalism causes those who have too much to develop an illusory sense of immortality. Individuals fear the future, which will inevitably bring suffering, frailty and, ultimately, death; and they fight tooth and nail against that 'horizon of existential finitude' (Arnsperger, 2010, p. 24) through capitalism. We could even go so far as to say, as Arnsperger does, that the terror of frailty and death, which *Homo oeconomicus* cannot comprehend, is at the root of today's economic crisis.

Hannah Arendt shows that, in contemporary society, the role of political action is assumed by the acts of manufacture and consumption. Thus, economics has triumphed over politics: labour has been emancipated, with the *animal laborans* occupying public space – though in reality, this is a private space that is exposed to public view (Arendt, 1958/1983, p. 150). Useful articles created to serve, and artisanal products, have become consumer goods. An item such as a dress or a chair may be consumed and disposed of almost as quickly as would food (Arendt, 1958/1983, p. 140). *Homo oeconomicus* occupies the space between humans.

Secondly, we shall look at the awareness of the world and the logic of responsibility driving *Homo collectivus*.

While the rationale behind *Homo oeconomicus* has become practically hegemonic in today's society, it is at odds with another anthropological component, marked by collective rather than individual interest. In the academic literature, the anthropological model underlying the maximisation of collective interests is defined as *Homo politicus* (the term *politicus* was included in the sixteenth-Century Latin *Dictionarium* by Robert Estienne, referring to 'civility' or '*civilis*' – Demonet, 2005). *Homo politicus* is an alternative concept for human behaviour, differing markedly from *Homo oeconomicus*. The concept of *Homo politicus* 'is founded on political philosophy and hinges on human interest for justice for the good of the community' (Faber et al., 2002, p. 324). *Homo politicus* refers to responsibility, which is the polar opposite of the individualism that is so rife today. *Homo politicus* champions responsibility and the civic desire to maximise 'social wellbeing' (Nyborg, 2000, p. 306). Faber, Petersen & Schiller define *Homo politicus* as 'human behaviour which tries to consider what is best for society' (Faber et al., 2002, p. 328). In the eyes of British historian John Greville Agard Pocock, *Homo politicus* is linked to Aristotle's *zôon politikon*, affirming its existence and virtue, meaning that man is in the [Aristotelian] City and that is his rightful place. *Homo politicus* is the *Homo* of the world and concern for others besides oneself – it is in stark contrast to *hubris*.

Homo politicus is important in this understanding of the human adventure, which it helps define, without essentialisation. Human beings are constituted and defined by a range of external factors, rather than by ones inherent to their person. However, it may be difficult to conceive of *Homo politicus*, in that it is partly linked to the *Homo religatus* – the third anthropological component that we intend to study. *Homo politicus* does not exist except in connection with other humans. Indeed, 'Politics is based on the fact of human plurality. (…). Because philosophy and theology are always concerned with *man*, because all their pronouncements would be correct if there were only one or two men or only identical men, they have found no valid philosophical answer to the question: What is politics? (…) [there] is the assumption that there is something political in *man* that belongs to his essence. This simply is not so; man is apolitical. Politics arises *between men*, and so quite *outside of man*' (Arendt, 1993/1995, pp. 31–32).

There appear to be two preconditions for politics to arise 'between men': the presence of a pluralist collective and the sharing of human existence. The term *politicus*, which refers primarily to the first condition, is ambiguous in relation to the politics we are attempting to define here. The main component of *Homo politicus*, as defined above, is the capacity to focus on the interests of the collective in a very broad sense – that broad collective cannot simply be reduced to the specific group to which the individual belongs. This anthropological subset – the individual that can focus on the interests of a collective bigger than his own group – can be called *Homo collectivus*. The term *collectivus* has nothing whatsoever to do with Marxist collectivism or communist totalitarianism, in which the pluralistic nature of the collective is suppressed. Rather, *Homo collectivus* indicates the capacity for individuals to understand the interests of a vast, pluralist group, based on the logic of responsibility.

Homo collectivus offers a form of resistance to the hegemonic *Homo oeconomicus*, which is manifested, for example, in the thinking of Jonas and Arendt. *Homo collectivus*'s relationship with the environment is one of sustainability and durability, but founded on solidarity with the rest of the biosphere, which represents a space for political coexistence. In part, *Homo collectivus* is linked with *Homo sustinens* (Siebenhüner, 2000), and the anthropological characteristics of *Homo reciprocus* are also fundamental for sustainability (Faber et al., 2002, p. 332). *Homo collectivus* is responsible not only for other humans, but for everything in the biosphere, which becomes an *agora* (Cabanes, 2017; Weber & Kurt, 2015; Pelluchon, 2017; Servigne & Chapelle, 2017).

Finally, let us look at coexistence and the sense of hospitality driving *Homo religatus*.

Homo religatus refers to a human who, rather than existing in essence, in his own right, exists specifically in relation to others. The concept includes the same relationship to the universe outside of oneself as does *Homo collectivus*: that world is viewed from the perspective of human plurality for *Homo collectivus*, while *Homo religatus* relates to a singular 'other'. *Homo religatus* is the human of relationships, interconnected with others, whose coexistence (i.e. shared existence) precedes existence. The word *existence* comes from the Latin *ex stance*, meaning standing outside. It refers to the idea that we can only live by being in the world (Boutinet, 1990). Existence is the common condition which all members of the human adventure share.

From the time of Aristotle, it has been understood that *man* is made human by having a place in society. One of the fundamental paradigms in our understanding of human beings is that they are relational: our existence is preceded by that of others. Hence, existing equates to existing alongside others, and 'self-interest is inseparable from the (positive or negative) interest paid to others' (Flahault, 2008, p. 315). This logic of coexistence requires us to fit within a certain clearly bounded framework, in which each individual has their place. For coexistence to take precedence over existence, certain limitations must be imposed, which runs counter to the limitless cycle of desire and response that is the cornerstone of economics. Indeed, one of the features of the modern era is denial about the finitude of our environment, and consequently, denial about our own human finitude.

In his latest book, French anthropologist François Flahault continues to demonstrate that the individual does not precede society (2018). This is a crucial point, because as a corollary thereof, economics is the cornerstone of societies. Drawing on a series of contemporary scientific works – particularly ones in the field of biology, focusing on the principle of 'coevolution' – Flahault constructs a post-Promethean anthropology that flies in the face of work such as Israeli historian Juval Noah Harari's *Homo Deus* (2015/2017), which has enjoyed widespread success. Humanity did not shape itself; and no more does each individual shape themselves without outside intervention. Against the modern, individual-centred utilitarian ideology, Flahault demonstrates the importance of relationships, and the need for joint reflection about the living world, the environment, the organism and its biotope (also see Delannoy, 2017). This is a drastic paradigmatic shift away from the modern concepts of individuals and subjects.

References

Arendt, H. (1983) [original edition 1958, French translation 1961]. *Condition de l'homme moderne*. Calmann-Lévy.

Arendt, H. (1995) [original edition 1993]. *Qu'est-ce que la politique ?* Seuil.

Arnsperger, C. (2010). Changer d'existence économique : enjeux anthropologiques de la transition du capitalisme au post-capitalisme. *Revue d'éthique et de théologie morale, 258*, 23–50.

Arnsperger, C. (2011). Dépasser le capitalisme, mais par étapes. *Revue Projet, 324–325*, 73–81.

Boutinet, J.-P. (1990). *Anthropologie du projet*. PUF.

Cabanes, V. (2017). *Homo natura – En harmonie avec le vivant*. Buchet Chastel.

Caillé, A., Chanial, P., Dufoix, S., & Vandenberghe, F. (Eds.). (2018). *Des Sciences sociales à la Science sociale – Fondements antiutilitaristes*. Le Bord de l'Eau.

Delannoy, I. (2017). *L'économie symbiotique – Régénérer la planète, l'économie et la société.* Actes sud.

Demonet M.-L. (2005). Quelques avatars du mot politique (XIVème-XVIIème siècles). *Langage et société, 113*, 33–61.

Faber, M., Petersen, T., & Schiller, J. (2002). Homo oeconomicus and homo politicus in ecological economics. *Ecological Economics, 40*, 323–333.

Flahault, F. (2005). Vers une nouvelle pensée sociale. *Revue du MAUSS, 26*, 377–382.

Flahault, F. (2008). Comment l'homme peut-il être à la fois égoïste, bon et méchant ? *Revue du MAUSS, 31*, 307–317.

Flahault, F. (2018). *L'homme, une espèce déboussolée – Anthropologie générale à l'âge de l'écologie.* Fayard.

Harari, Y. N. (2017) [original edition 2015]. *Homo deus, une brève histoire de l'avenir.* Albin Michel.

Nyborg, K. (2000). Homo economicus and homo politicus: Interpretation and aggregation of environmental values. *Journal of Economic Behavior & Organization, 42*, 305–322.

Pelluchon, C. (2017). *Manifeste animaliste – Politiser la cause animale.* Alma.

Peyrelevade, J. (2005). *Le capitalisme total.* Seuil.

Servigne, P., & Chapelle, G. (2017). *L'entraide – L'autre loi de la jungle.* Les Liens qui Libèrent.

Siebenhüner, B. (2000). Homo sustinens als Menschenbild für eine nachhaltige Ökonomie. *sowi-online, 1*, 1–13.

Weber, A., & Kurt, H. (2015). *Lebendigkeit sei! Für eine Politik des Lebens. Ein Manifest für das Anthropozän.* Think Oya.

Nathanaël Wallenhorst is Professor at the Catholic University of the West (UCO). He is Doctor of Educational Sciences and Doktor der Philosophie (first international co-supervision PhD), and Doctor of Environmental Sciences and Doctor in Political Science (second international co-supervision PhD). He is the author of twenty books on politics, education, and anthropology in the Anthropocene. Books (selection): *The Anthropocene decoded for humans* (Le Pommier, 2019, *in French*). *Educate in Anthropocene* (ed. with Pierron, Le Bord de l'eau 2019, *in French*). *The Truth about the Anthropocene* (Le Pommier, 2020, *in French*). *Mutation. The human adventure is just beginning* (Le Pommier, 2021, *in French*). *Who will save the planet?* (Actes Sud, 2022, *in French*). *Vortex. Facing the Anthropocene* (with Testot, Payot, 2023, *in French*). *Political education in the Anthropocene* (ed. with Hétier, Pierron and Wulf, Springer, 2023, *in English*). *A critical theory for the Anthropocene* (Springer, 2023, *in English*).

Hospitality

Benjamin Boudou

Abstract This article reconstructs the different meanings of hospitality and articulates a politics of hospitality that would enable care and home-making practices for displaced persons.

Hospitality generally refers to the generous reception of strangers and an ethical principle of openness and benevolence. It is a practice (the action of welcoming, hosting and caring for strangers at "home") and a reason (strangers should be welcomed in the name of hospitality). The concept of hospitality thus has a descriptive and a normative side, along with individual-private connotations (welcoming home) as well as political-public ones (a virtue of institutions). At the state level, hospitality may be simply understood as a duty or principle of mutual aid, i.e., a natural duty of states to help, under certain conditions, anyone in need, regardless of political status (Walzer, 1983).

To better determine the political content of hospitality, we must first distinguish six major types of questions that have problematized hospitality. The first type is historical. Studies addressing this type of question describe precise determinations of hospitality practices, generally based on theological recommendations in tension with contemporary secular norms and institutions (Heal, 1990; Cavallar, 2002). The second type of question relates to phenomenology. Hospitality names a problem in phenomenology, namely the distinction between the familiar and the foreign. It is a question of understanding how the familiar is constituted and what the modes of access to the alien are. Hospitality generally takes the form of a metaphor to describe the process of familiarization, that is, the construction of a scale of familiarity with the world around us (Waldenfels, 2011). The third type of question is ethical and is mainly represented by the scholarship following the work of Jacques Derrida (Derrida & Dufourmontelle, 2000). Hospitality is another name for ethics as responsibility towards the Other. But hospitality can only be experienced through an action

B. Boudou (✉)
University of Rennes, Rennes, France
e-mail: benjamin.boudou@univ-rennes.fr

761

that necessarily falls short of its absolute standard. Hospitality can only be achieved by betraying its ideal of unconditional openness to the Other. The fourth type of question relates to political philosophy. By situating hospitality within the more general issue of the ethics of migration, this literature attempts to make hospitality a relevant concept for contemporary political thought (Boudou, 2021; Bulley, 2017). Paul Ricœur (2010) articulates the political stakes of hospitality (particularly the distribution of membership) with its ethical dimension (respect due to foreigners) and Seyla Benhabib (2004) borrows from Kant and Arendt to conceptualize a political hospitality as the basis of a cosmopolitan right to justification. The fifth line of questioning is sociological. It addresses the tensions between membership and hospitality, i.e., the mutual conflicts of integration between community and newcomers (Stavo-Debauge, 2017). The final type of question is anthropological. It focuses on gestures of hospitality from rituals in traditional societies (Pitt-Rivers, 1977) to the material and spatial modalities of reception and daily interactions between guests and hosts (Candea & Da Col, 2012; Vandevoordt, 2017).

These different approaches draw a composite image of hospitality. This image lies between its local, domestic practices and its metaphorical use (where it refers to ethics itself or cosmopolitanism); between a descriptive use and an abstract conceptualization; between a common-sense welcome, a political virtue and a legal norm. To define what a politics of hospitality means, we should consider this plurality of points of view, establish a common context and some normative coherence. The goal is to avoid confusing hospitality with other related concepts such as altruism, solidarity or fraternity.

It is first necessary to recall the context in which hospitality has known a strong revival: although it is not a new phenomenon, migration has recently become an unprecedented political issue. While the number of migrants around the world is stable (relative to the world population), the diversification of mobility, its various motivations and the policies that organize it, have transformed this universal anthropological fact into an issue (and for some a threat) that needs to be addressed. The politicization of migration and its problematization in public discourse has associated the figure of the migrant and welcoming practices as a triple threat to the identity and sustainability of Western societies: (i) an identity threat against the maintenance of a national character, customs and traditions, (ii) an economic threat from free riders, i.e., migrants who supposedly enjoy the benefits of the welfare state without having contributed to it, (iii) a political threat, i.e.,the loss of a nation-state's ability to decide about its migration policies because of the pressure from human rights standards and international organizations managing migration.

These framings are factually problematic and normatively questionable. Social scientists, political theorists, lawyers and activists have responded with counter-discourses. On the one hand, they aimed to bring into the political debate the results of the social sciences that rationalize and relativize intellectual biases and feelings of urgency, invasion, or loss of status associated with migration. On the other hand, the normative literature has questioned the value of border control by reassessing the weight of individual freedom of movement, the coherence of the idea of sovereignty as a unilateral right to enclose a territory, the scope of the principles of justice and the nature of the responsibility of states towards foreigners.

These approaches have found, in the concept of hospitality, an opportunity to synthesize a set of norms, values and practices (individual and collective). Hospitality would signal the urgency of thinking differently about the porosity of borders, the experience of exile and migration, the virtue and duty of welcoming the stranger.

The idea of a politics of hospitality thus refers to a particular way of governing foreigners that is associated with the deconstruction of negative prejudices about migrants and migration, the perception of what a political community is and of its duties foreigners. Historically, a politics of hospitality meant regulating contact with foreigners, without guaranteeing their integration. It was based on a strict distinction between the inside and the outside of the community, the quasi-sacred duty to welcome newcomers, but also to control them (Boudou, 2020).

Contemporary uses of hospitality in the political sphere can be conceptualized in two ways. Firstly, a politics of hospitality refers to a systematic questioning of a society's political, legal and administrative shortcomings in terms of refuge and immigration. A politics of hospitality is a critical discourse against inhospitality, that is, the inhumane, unjust and dangerous treatment of foreigners. Hospitality involves a strict respect for human dignity, a guarantee of the protection and fulfilment of people's rights, a sense of responsibility towards vulnerability partly caused by the conditions of hospitality itself, and an ambition to facilitate mobility, protection and integration.

From a more conceptual point of view, a politics of hospitality is based on a particular obligation which is not entirely limited to the legal and administrative framework of asylum and migrants' human rights. Hospitality today describes domestic reception practices (Gerbier-Aublanc, 2018), the policies of sanctuary cities (Boudou, 2019), the duty of states towards refugees and emergency humanitarian reception (DeBono, 2019; Rozakou, 2012).

I suggest defining hospitality as a principle of care for the displaced that relieves the distress caused by border crossings (Boudou, 2020, 2021). Hospitality alleviates ordinary obstacles that prevent a functional life in a new environment and allows for home-making practices. It is triggered by a sense of vulnerability in a context of displacement. Displacement involves specific harms, related to the loss of home. These harms are material (the destruction of, or eviction from, one's home); emotional (the deterioration of the sense of ease, familiarity, and attachment to one's place); and political (the impediment to safety, privacy, and autonomy). A politics of hospitality prevents and controls these harms that produce private and public forms of domination. It is not, as such, a legal duty. The obligation is imperfect because there is not a general right to hospitality, but various rights depending on whether it is a question of access to a territory, to protection or to social or political rights. It is collective because it is not the responsibility of anyone in particular, but if no one practices it, injustice persists.

This makes any form of hospitality necessarily controversial. It can be mobilized to denounce and repair the injustices committed at the borders, but also to reinforce a homogenizing and paternalistic vision of the national community, which is thus described as a household generously offering hospitality. It thus always presupposes precise justifications of the actions to be carried out, of the meaning of hospitality that is mobilized, of its limits regarding individual rights or, more radically, of its transformative potential of law and politics.

References

Benhabib, S. (2004). *The rights of others: Aliens, residents, and citizens*. Cambridge University Press.

Boudou, B. (2019). Hospitality in sanctuary cities. In S. Meagher, S. Noll, & J. Biehl (Eds.), *The Routledge handbook of philosophy of the City* (pp. 279–290). Routledge.

Boudou, B. (2020). Migration and the duty of hospitality. *Transitions: Journal of Transient Migration, 4*(2), 257–274.

Boudou, B. (2021). Beyond the welcoming rhetoric: Hospitality as a principle of care for the displaced. *Essays in Philosophy, 22*(1/2), 85–101.

Bulley, D. (2017). *Migration, ethics and power: Spaces of hospitality in international politics*. Sage.

Candea, M., & Da Col, G. (2012). The return to hospitality. *Journal of the Royal Anthropological Institute, 18*(s1), 1–19.

Cavallar, G. (2002). *The rights of strangers: Theories of international hospitality, the global community and political justice since Vitoria*. Ashgate.

DeBono, D. (2019). Plastic hospitality: The empty signifier at the EU's Mediterranean border. *Migration Studies, 7*(3), 340–361.

Derrida, J., & Dufourmontelle, A. (2000). *Of hospitality*. Trans. R. Bowlby. Stanford University Press.

Gerbier-Aublanc, M. (2018). Un migrant chez soi. *Esprit, 446*, 122–129.

Heal, F. (1990). *Hospitality in early modern England*. Clarendon Press.

Pitt-Rivers, J. (1977). *The fate of Shechem or the politics of sex: Essays in the anthropology of the Mediterranean*. Cambridge University Press.

Ricœur, P. (2010). Being a Stranger. *Theory, Culture & Society, 27*(5), 37–78.

Rozakou, K. (2012). The biopolitics of hospitality in Greece: Humanitarianism and the management of refugees. *American Ethnologist, 39*(3), 563–577.

Stavo-Debauge, J. (2017). *Qu'est-ce que l'hospitalité? Recevoir l'étranger à la communauté*. Liber.

Vandevoordt, R. (2017). The politics of food and hospitality: How Syrian refugees in Belgium create a home in hostile environments. *Journal of Refugee Studies, 30*(4), 605–621.

Waldenfels, B. (2011). *Phenomenology of the alien: Basic concepts*. Northwestern University Press.

Walzer, M. (1983). *Spheres of justice: A defense of pluralism and equality*. Basic Books.

Benjamin Boudou is professor of political science at the University of Rennes. He is the editor of the French journal of political theory *Raisons Politiques* and the author of *Politics of hospitality: A conceptual genealogy* (CNRS Éditions, 2017, in French) and *The dilemma of borders: Ethics and politics of immigration* (Éditions de l'EHESS, 2018, in French).

Human Adventure

Nathanaël Wallenhorst

Abstract The concept of human nature refers to the essence of our species, and to contingency on the present. Meanwhile, the emerging idea of the 'human adventure' emphasises the uncertainty of the future, and the possibility of altering the characteristics on which we are contingent. (Notably, we could modify the earth's system, or even alter humans' own biological make-up). This article is an extension to 'The Human Condition in the Anthropocene', based on Hannah Arendt's anthropological model developed in *The human condition* (1958). That model defines *animal laborans* as driven by work, *Homo faber* as driven by creativity, and *zôon politikon* as driven by action. Here, from the standpoint of humans in the Anthropocene, we explore these three categories that make up Arendt's idea of the human condition. We propose to build upon the model by speaking of the *human adventure*.

Firstly, over the past few decades, there have been a wealth of publications on the dominance of *Homo oeconomicus*. The category has become widely accepted, and built on by multiple authors. *Homo oeconomicus* is an extension to Arendt's *animal laborans*, except more emphasis is placed on consumption than on work. Arendt perspicaciously noted that any work produced is automatically destined for consumption, and therefore, such product had no longer-term existence. In the 60 years since the publication of *The human condition*, consumerism has gone from strength to strength, and today, *Homo oeconomicus* (the consumer) is certainly dominant over *animal laborans*.

Secondly, Arendt's *Homo faber* is also ripe for re-examination. One of the strengths of Arendt's model is that it acknowledges the overlap between the private and public spheres, in *Homo faber*'s creations. *Homo faber*, whose products are not destined for consumption, helps build a world shared by all humanity. Today's world is marked by considerable socioeconomic acceleration since the 1950s, as

N. Wallenhorst (✉)
Catholic University of the West, Angers, France
e-mail: nathanael.wallenhorst@uco.fr

© The Author(s), under exclusive license to Springer Nature Switzerland AG 2023
N. Wallenhorst, C. Wulf (eds.), *Handbook of the Anthropocene*,
https://doi.org/10.1007/978-3-031-25910-4_124

attested by Steffen et al.'s 'Great Acceleration' curves (2004, 2015) and Hartmut Rosa's analyses (2010, 2012). In that context, the work that humans produce largely contributes to the global spread of consumerism. The same is true of nanotechnology, biotechnology, information technology and cognitive science (NBIC), exploited by web-based businesses; progress in aerospatial engineering; humankind's desire to colonise Mars – these efforts are driven not by an association of nations, but by relatively young multi-billion-dollar industries; or even art designed for consumption (Menger, 2003). Against this backdrop, can it really still be claimed that *Homo faber* helps build a shared world through creations not directly designed for immediate consumption?

Thirdly, Arendt's action-based *zôon politikon* is worth refining. From Chap. 5 of *The Human Condition*, it becomes clear that we must distinguish between two components that facilitate action: the primacy of collective interests over individual ones, and the shared existence of humankind. In Arendt's view, the world and life refer to two separate fundamental components of the human condition. Life is the intervening time between birth and death – between entering the world and leaving it. Life refers to the economic sphere, and the necessity of consumption-driven perpetual motion. The world, for its part, is political; it exists before and persists after life. The concept refers to a type of permanence, and always to extremely long periods of time.

Fourthly, the opposing categories of 'life' and 'world' ought to be re-examined, in the context of the Anthropocene. The revised anthropological approach employed throughout this compendium consolidates the idea of 'the world' (that political space in which cooperation emerges). However, the 'world' is not in opposition to 'life', because the shared fabric of the living world makes up the body politic (and the world). The political world exists only because of interactions (both organic and chemical) between living beings. The political world is merely the consequence of exchanges between living beings – and also interactions with non-living, non-organic things. The work of contemporary German philosopher and biologist, Andreas Weber, is of particular interest when considering the biological and living materiality of the (political) world. Weber continually demonstrates the profound link between the paradigms of the world and of life. The world is characterised by its aliveness (2016, 2017; Weber & Kurt, 2015) (this idea is reminiscent of Corine Pelluchon's approach in *Les nourritures*, 2015). Thus, reality is not merely objective, but creative and expressive above all (and therefore unpredictable and uncontrollable), because it is living. Even above all in the Anthropocene, movement is possible, and so, therefore, is hope, because the primary characteristic of our reality is its aliveness. Aliveness, in the Weberian sense, includes Arendt's political component, based on the relationships within plurality, and the novelty allowed by natality. Such aliveness is central element in Weber's thinking and his way of looking at the world, and is characterised by its expansiveness. However, this expansiveness is not that of insatiable capitalism, characterised by the constant need for 'more'. For example, the paradigm of life is not viewed in reference to consumption (unlike in Arendt's model). The expansiveness is that of life, rather than that of an individual. Thus, rather than individuals having the opportunity to develop and expand, *life* can develop and expand by being shared. Here, coexistence precedes existence: 'I am because you are' (Weber, 2017, p. 10). In Weber's view, when we share, rather than impoverishing us or taking something away from us, it actually enriches us.

Arendt's distinction between life and world is no longer of interest in our time, marked by the Anthropocene. Indeed, in today's world, we need to strengthen our roots in the living universe we all share, and break away from the destructive hegemony of capitalism. We need to think of politics in contrast to economic hegemony, which is rooted in the aliveness of the living universe. Thus, the category is limited to express that economic hegemony. We need to come up with another category, indicating that destructive domination of economic-based logics – destructive because it is unlimited. Here, we propose to identify it as part of *hubris*. The choice of this term to represent this component, that stands in opposition to the political, is quite deliberate: destruction is a fundamental anthropological component which must be reined in, otherwise the very survival of the human adventure is threatened. Also, *hubris*, which is a fundamental part of our nature, cannot be easily overcome.

The table below offers an overview of the three dimensions of the human adventure. These categories are interlinked. Chemical interactions facilitate the political world; in turn, the political may encourage coexistence, amongst humans and between humans and non-humans. This helps 'living' relationships to operate more fluidly (allowing everyone to live without the threat of one species dominating others). These categories are to be considered in the context of peaceful coexistence.

Sphere	Private sphere Economical	Public sphere Political	
Aspects of the human adventure	*Hubris*	World	Coexistence
Driving logic	Profit	Responsibility	Hospitality
Central hub/pole	Individual	Collective	Others
Anthropological model	*Homo oeconomicus*	*Homo collectivus*	*Homo religatus*
Biosphere	Anthromes	Agora	Milieu

The three dimensions of the human adventure

Once *hubris* is accepted as one of the three components of the human adventure, the category of life again has its place in the biosphere, giving these three dimensions materiality and power. Life, as a biosphere, refers to the whole of the planet's surface, looking onto the cosmos and receiving sunlight. There are three aspects to the biosphere which imbues the human adventure with materiality.

Firstly, *anthromes* (anthropogenic biomes) are areas of the earth's surface that have been appropriated and profoundly altered by humans to suit their own purposes (leading to the destruction of animal habitats). Today, a huge proportion of the biosphere has been anthropised, which risks causing major systemic disruption.

Secondly, the *agora* is an arena for coming together and exchanging words, giving rise to subsequent action. The biosphere can be viewed as an agora, in that it is the place where all political subjects, or entities, that inhabit the earth (humans, animals, plants, minerals, etc.) encounter one another, and have various types of exchanges. If we speak of the biosphere as an agora, then we must also discuss the interactions of which life is made up. This is the responsibility of humans.

The third dimension of the biosphere is the so-called *milieu*. The emphasis, in the notion of milieu, is on how spaces bear the indelible hallmarks of the representations and relationships that take place there (Berque, 2016).

We hold that the human adventure (the etymological roots of which are outlined in the article 'The Human Condition in the Anthropocene') is the intersection between three contradictory axes of human behaviour. One such axis is individualism, characterised *Homo oeconomicus*'s neoliberal drive to maximise individual interests. However, there are two other axes which do not fit with individualism: firstly, collectivism, marked by *Homo collectivus*'s logic of responsibility for the group; and secondly, altruism, marked by *Homo religatus*'s hospitality and drive for coexistence with others.

Homo oeconomicus is an integral part of the human adventure. In today's world, the hegemony of *Homo oeconomicus* is a problem. Fundamentally, the model we propose here opposes *Homo oeconomicus* with the idea of *post-Promethean political togetherness*. That idea determines the structure of the three-dimensional discussion in this compendium. The counterweight to *Homo oeconomicus*' individualism, rather than an individual driven by other philosophies, is simply a void: a relational space with nobody at its centre. Two anthropological figures are needed in order for the aforementioned post-Promethean political togetherness to emerge: the group-oriented *Homo collectivus* and the *Homo religatus*, capable of hospitality towards others. The post-Promethean political togetherness is on a par with Arendt's concept of action: it can only come to pass through humans acting together. These contradictory logics driving *Homo oeconomicus*, *Homo collectivus* and *Homo religatus* represent conflict between mercantile goods (*Homo oeconomicus*), collective goods (*Homo collectivus*) and the common good (*Homo religatus*), as discussed by Flahault (2013). These three anthropological characteristics are discussed in depth in '*Homo oeconomicus, Homo collectivus, Homo religatus*'. One of the major issues of the twenty-first Century – which the dawn of the Anthropocene brings into sharp focus – is the urgent need to break away from capitalism and mercantile economics. In order to do so, we need a new concept of what human beings are.

References

Arendt, H. (1958). *Condition de l'homme moderne*. Calmann-Lévy. [original edition 1958, French translation 1961].

Berque, A. (2016). La relation perceptive en mésologie: du cercle fonctionnel d'Uexküll à la trajection paysagère. *Revue du MAUSS, 47*, 87–104.

Flahault, F. (2013). Pour une conception renouvelée du bien commun. *Études, 418*, 773–783.

Menger, P.-M. (2003). *Portrait de l'artiste en travailleur, Métamorphoses du capitalisme*. Seuil.

Pelluchon, C. (2015). *Les nourritures*. Seuil.

Rosa, H. (2013). *Accélération. Une critique sociale du temps*. La découverte, French translation. [original edition 2010].

Rosa, H. (2014). *Aliénation et accélération – Vers une théorie critique de la modernité tardive*. La Découverte, French translation. [original edition 2012].

Steffen, W., et al. (2004). Global change and the earth system. A planet under pressure. In *The IGBP book series*. Springer.

Steffen, W., et al. (2015). The trajectory of the Anthropocene: The great acceleration. *The Anthropocene Review, 2*(1), 81–98.

Weber, A. (2016). *Enlivenment. Eine Kultur des Lebens. Versuch einer Poetik für das Anthropozän*. Matthes und Seitz.

Weber, A. (2017). *Sein und Teilen – Eine Praxis schöpferischer Existenz*. Transcript Verlag.

Weber, A., & Kurt, H. (2015). *Lebendigkeit sei! Für eine Politik des Lebens. Ein Manifest für das Anthropozän*. Think Oya.

Nathanaël Wallenhorst is Professor at the Catholic University of the West (UCO). He is Doctor of Educational Sciences and Doktor der Philosophie (first international co-supervision PhD), and Doctor of Environmental Sciences and Doctor in Political Science (second international co-supervision PhD). He is the author of twenty books on politics, education, and anthropology in the Anthropocene. Books (selection): *The Anthropocene decoded for humans* (Le Pommier, 2019, *in French). Educate in Anthropocene (*ed. with Pierron, Le Bord de l'eau 2019, *in French). The Truth about the Anthropocene* (Le Pommier, 2020, *in French). Mutation. The human adventure is just beginning* (Le Pommier, 2021, *in French). Who will save the planet?* (Actes Sud, 2022, *in French). Vortex. Facing the Anthropocene* (with Testot, Payot, 2023, *in French). Political education in the Anthropocene* (ed. with Hétier, Pierron and Wulf, Springer, 2023, *in English). A critical theory for the Anthropocene* (Springer, 2023, *in English).

Human Condition

Nathanaël Wallenhorst

Abstract This article examines the human condition in the Anthropocene, based on the definition given by Hannah Arendt in *The Human Condition* (1958). That book was particularly fitting in the age of modernism. Today, however, the dawn of the Anthropocene appears to necessitate a shift in the way in which we view humanity. This article proposes the notion of the 'human adventure', rooted in the whole adventure of living beings on this planet, and highlighting the uncertain fate that humanity currently faces.

Arendt wrote *The Human Condition* in the wake of her three volumes on *The Origins of Totalitarianism*. After addressing the 'banality of evil', she wrote a book that is seminal in anthropology, attempting to define the absolute essence of the human condition. This work includes a number of opposing terms. The human condition is marked by *vita activa* (active life; the term serves as the title for the German translation of *The Human Condition*). *Vita activa* is differentiated from *vita contemplativa* (contemplative life), the latter being the subject of her final work, *The Life of the Mind*, which was unfinished at the time of her sudden death. Two further terms which are mutually opposing, but complementary, form the backbone of *The Human Condition*: the private sphere (economic) and the public sphere (political). These two concepts frame Arendt's critique of the modern age, in which the private sphere appears to dominate the public sphere. Among other causes, this reflects the dominance of interest over disinterested selflessness. Against this dichotomy between the private and public spheres, Arendt constructed a three-pronged anthropological model of humans' relationship to the world: the labour of the *animal laborans*, the creativity of the *Homo faber*, and the action of the *zôon politikon*.

N. Wallenhorst (✉)
Catholic University of the West, Angers, France
e-mail: nathanael.wallenhorst@uco.fr

N. Wallenhorst, C. Wulf (eds.), *Handbook of the Anthropocene*,
https://doi.org/10.1007/978-3-031-25910-4_125

One of Arendt's criticisms of the modern age is that '[creative] work has become [fruitless and unending] labour' (1958, p. 142). The difference between labour and work is the permanence of the product of work (an enduring creation). By lasting, work contributes to the creation of a world, in which all humanity shares. Arendt decries the fact that we have become a consumer society: instead of using the enduring products of work, we consume the fleeting and disposable products of labour. Thus, instead of building a world in which we can all live, we are depleting natural resources and destroying our world – the one and only world in which humanity can exist (Arendt, 1958, p. 31).

Arendt's critique of the modern age is particularly interesting for our purposes. Through her lens, we can see how the hegemony of economics, unrestrained by political regulation, has led us to the jaws of the Anthropocene. The third aspect of Arendt's model – action – is somewhat enigmatic. In Arendtian thinking, there is no greater human activity than cooperative action through dialogue. Action and words are the two facets of the supreme political activity. Arendt has a very high opinion of the actions of which humans are capable. She presents action as part of the solution to the dominance of the private sphere over the public. She seems to have drawn inspiration for her concept of 'action' from the roots of the American Revolution – the bloody conflict was preceded by a period of ideas stimulated by both French and British thinkers, so the seeds were sown in a democratic, non-violent fashion. Arendt's concept of action is a beacon of hope, the counterpart to the darkness of totalitarianism. In Arendt's own words, action is a miracle. Its power lies in its fleetingness. Action endures only when human beings work together. An action is an event in the public sphere, born of our shared existence (the awareness of which is at the heart of pluralism). Action has been lacking throughout the modern era, and that lack of action has led us, inescapably, into the Anthropocene. Today, then, is the notion of the human condition still apt and applicable, despite the fundamental importance of action, facilitating the emergence of the political (which has been in such short supply in the modern era)? Now that we are in the Anthropocene, ought we to re-examine how we think about humanity?

The geoscientific research on the beginning of the Anthropocene published over the past two decades reframes the notion of the human species in a particularly naturalist fashion. The defining feature of naturalism is that, instead of individuals with free will, it views us as a united species with a simple history. This simplistic approach appears always to underlie research in earth sciences on the Anthropocene. It is linked partly with the notion of human nature and the naturalisation of technology in the guise of progress (Bonneuil, 2014; Federau, 2017; Wallenhorst, 2019, 2020). The notion of humanity is contingent upon that of the human condition. Now more than ever, as we face the Anthropocene, we need a clear understanding of that contingency, to counter the anthropological oversimplification in many geoscientific works. For her part, Hannah Arendt, given her experience in detention camps and her first-hand knowledge of the Nazis' bid to radically transform humanity, was particularly circumspect in regard to human nature, preferring to focus on the dimension contingent on the human condition (Arendt, 1948, p. 277). However, thinking about humanity without reference to the idea of human nature is problematic – it does not acknowledge the influence of an external force (nature) on humanity. For our part, we believe that the living world is an essential element, which must be taken into consideration when thinking about humanity in the Anthropocene. Recognising our place in the living world could hold the very key to humanity's survival. Without a solid grounding in the unified fabric that is the living world, we cannot hope for long-term survival in the biosphere (Cabanes, 2017; Weber, 2016, 2017; Weber & Kurt, 2015; Flahault, 2018; Pelluchon, 2015; Servigne & Chapelle, 2017). The vast-scale destruction of biodiversity going on today, and the conception of humans as being in some way removed from the rest of the living world, are among the major problems on which the Anthropocene shines a light. It is important that our view of humanity be based on reference to that living world.

Another major feature of the Anthropocene will have a long-lasting impact on the human condition. We face a future marked by uncertainty (more so now than at any other time in human history): an 'adventure' in a very real sense. Here, the term 'human adventure' could refer simultaneously to such awareness of humanity's place in the living world, and to the degree of uncertainty as to humanity's future on the planet. The idea of adventure is representative of the threat we face, and emphasises that we share a common destiny, in relation to the uncertainty the future holds. The whole of humanity is united in facing our uncertain fate, though we are still conscious of the social, cultural and political differences between us. If we take this view, then the definition of humanity is no longer intrinsic and static; rather, we are defined by what we can become, and what we make of ourselves within our environment.

The term 'adventure' derives from the Latin *adventura*, meaning what is to come, and *advenire*, meaning to happen or occur. Other derivatives of *advenire* include *advena* (foreigner) and *adventicius* (what comes from outside, what occurs). The current use of the term 'adventure' includes the idea of uncertainty and danger, in connection with the environment in which the adventure is happening. Adventure contains intrigue: an event with an unknown outcome. Today, we face uncertainty as to what will become of humanity, given the point of no return marked by the Anthropocene. An adventure is an experiment (from the Greek *experior*, to put to the test). The word 'human' has the same etymological root as *humus*: the earth. *Humus* imbues us with *humanitas*, human nature. Human nature comes partly from the soil, as we are affected by the soil we tread. In biological terms, we come from the earth, and the matter that makes us up is part of a never-ending cycle that includes humanity and our planet. Humans are defined, in part, by our relationship with the earth – with *humus*. The word *humus* itself refers to the layer of natural compost on the earth's surface resulting from decomposition of organic matter. The product of death (decomposition) gives rise to life (germination). Human nature is also partly connected with the humility that comes of knowing our own existence is finite. Both *humus* and *humilitas* define *humanitas*. The terms 'human' and 'terrestrial' come from two different Latin roots, with the same meaning: Earth.

As humans, we need to learn to deploy the power we hold in a different way, and the first step is by acknowledging our own finitude, and that of our environment. The concept of the 'human adventure' includes what we are in the process of becoming. The anthropological bases of that notion share much with the work of the anthropologist François Flahault, who breaks with the modern essentialisation of the individual (Flahault, 2018, p. 38). We believe it would be interesting to think of humanity, in the Anthropocene, in relation to the idea of the human adventure rather than that of the human condition. 'Human nature' refers to an essence; the 'human condition' is contingent on present circumstances. The idea of the 'human adventure' includes the uncertainty of the future, and the possibility of altering the characteristics on which that future is contingent (notably modifying the earth's systems, or even, potentially, altering humanity's own biological make-up).

The Anthropocene represents a new beginning for humankind, because the world – so important to Arendt's description of the human condition – is under threat. Indeed, for Arendt, the world is a crucially important element of the human condition. Firstly, it accommodates human creations, and measured against the timescale of the world, humans are able to appreciate the fleeting nature of their own existence. Secondly, the world is the shared space in which action takes place. The world has existed for millions of years before we were born, and will survive long after we are gone. It is perennial. In speaking of the world, Arendt uses the metaphor of a tent pitched on the surface of the planet, accommodating all humanity. If we extend that metaphor, we can clearly see that, as the earth is constantly shifting, the tent will be shaken, and will require a degree of flexibility in order to adapt and remain standing. The Anthropocene alters the human condition. In this new geological period, the notion of the human adventure that we propose (defined in greater detail in the article so titled) becomes appropriate.

References

Arendt, H. (2002). *Les origines du totalitarisme. Trois tomes.* Seuil. (Original edition 1948).
Arendt, H. (1983). *Condition de l'homme moderne.* Calmann-Lévy. (Original edition 1958, French translation 1961).
Bonneuil, C. (2014). L'Anthropocène et ses lectures politiques. *Les Possibles, 3,* 1–7.
Cabanes, V. (2017). *Homo natura – En harmonie avec le vivant.* Buchet Chastel.
Federau, A. (2017). *Pour une philosophie de l'Anthropocène.* PUF.
Flahault, F. (2018). *L'homme, une espèce déboussolée – Anthropologie générale à l'âge de l'écologie.* Fayard.
Pelluchon, C. (2015). *Les nourritures.* Seuil.
Servigne, P., & Chapelle, G. (2017). *L'entraide – L'autre loi de la jungle.* Les Liens qui Libèrent.
Wallenhorst, N. (2019). *L'Anthropocène décodé pour les humains.* Le Pommier.
Wallenhorst, N. (2020). *La vérité sur l'Anthropocène.* Le Pommier.
Weber, A. (2016). *Enlivenment. Eine Kultur des Lebens. Versuch einer Poetik für das Anthropozän.* Matthes und Seitz.
Weber, A. (2017). *Sein und Teilen – Eine Praxis schöpferischer Existenz.* Transcript Verlag.
Weber, A., & Kurt, H. (2015). *Lebendigkeit sei! Für eine Politik des Lebens. Ein Manifest für das Anthropozän.* Think Oya.

Nathanaël Wallenhorst is Professor at the Catholic University of the West (UCO). He is Doctor of Educational Sciences and Doktor der Philosophie (first international co-supervision PhD), and Doctor of Environmental Sciences and Doctor in Political Science (second international co-supervision PhD). He is the author of twenty books on politics, education, and anthropology in the Anthropocene. Books (selection): *The Anthropocene decoded for humans* (Le Pommier, 2019, *in French*). *Educate in Anthropocene (*ed. with Pierron, Le Bord de l'eau 2019, *in French*). *The Truth about the Anthropocene* (Le Pommier, 2020, *in French*). *Mutation. The human adventure is just beginning* (Le Pommier, 2021, *in French*). *Who will save the planet?* (Actes Sud, 2022, *in French*). *Vortex. Facing the Anthropocene* (with Testot, Payot, 2023, *in French*). *Political education in the Anthropocene* (ed. with Hétier, Pierron and Wulf, Springer, 2023, *in English*). *A critical theory for the Anthropocene* (Springer, 2023, *in English*).

Human Existence

Daphnée Valbrun

Abstract This article is a philosophical reflection seeking to determine the importance of nature in the lives of human beings. Also, it questions certain prejudicial effects of human action on nature. Indeed, considering the close relations between the human being and nature, it is quite natural that we are called to take care of it, because, the future of humanity is dependent on the way we treat nature. On the other hand, the future of nature depends on how human beings use it. Hence the importance of managing our actions that impact the environment for a better cohabitation between human beings and nature.

The term of existence has multiple meanings. It comes from the Latin *ex-sistere* which means to come out of, to manifest. Some philosophers like Thomas Aquin (1271) allot to the existence a direction having report with the essence, that is, what makes a thing be. However, for other philosophers like Heidegger (1986), Levinas, the existence would be being out of being. This outside of being is the capacity to be able to think in the sense of René Descartes, *cogito ergo sum*. Can we say that thinking is the only thing that can determine an existence? For Heidegger, it is a capacity of each individual because the nature is simply there but does not exist according to him. This characteristic of the human, should it not mobilize a sensitivity as Rousseau advocates it towards all the other existing that is to say our fellow men and the there as for example the nature. What relation exists it between existing (the man) and the there (the nature)?

'To question man's relations with nature is to presuppose that he is capable of reflecting on these relations, of thinking them differently and of acting accordingly' (Flahault, 2013, p.125). Man as *homo faber*, has a very strong predominance over nature due to the fact that he is considered a being capable of thinking and making tools in order to meet these needs. As *homo rationae* and *homo faber*, he must reflect on the relationship he has with nature? For Flahault: "by using these two words, *human and nature*, we admit from the outset that nature is one thing and that

D. Valbrun (✉)
Catholic University of the West, Angers, France

man is another" (Ibid, p.125). The author shows through this quote that nature and human are two different things because of the very essence and definition of these two terms. Therefore, one wonders, does this difference, imply a total separation of the man in the nature? Taking into account the difference that Flahault makes between the man and the nature, it seems important to us to specify in what consists the environment and/or the nature, then to show the relation that exists between them.

The term *nature* is often misleading. In many cases, the meaning given to this concept refers much more to the notion of natural law. In antiquity (from Aristotle to the Stoics), this notion of nature played an extremely important role in the formation of the mind. For example, Stoic wisdom consisted in getting as close as possible to nature in order to acquire the knowledge that it contains. For the metaphysicians, this nature or the environment is made to be tamed by the man and would be comparable to 'the work conceived by a divine architect who assigned to him his form and his end' (Cf. Serres, 2004, p.69). What is this conceived end? Can we not consider this end as the condition of existence of the man? At least, could we say following René Descartes that the finality of the human is to master the nature?

For Serres, nature is 'acquired the status of an entity' (Serres, 2004, p.69). Also, he adds, 'the nature began to mean the whole of the essential properties of a being or a thing; to define thus their nature' (Ibid, p.69). This definition of the author is very enlightening and refers to a purely ontological dimension. Indeed, for Serres, the nature is what constitutes the essence of a thing or in clear term what makes that a thing is. Seeking to determine an exact definition of nature, Naim Gesbertb, raises a completely different dimension of it. For him, nature is 'the object of the system of legal relations of the surroundings at the heart of which man is situated and in which he lives'. The latter therefore has a relationship with our environment and is for the author, an open and interactive whole made up of everything in the earth's space such as air, water, etc. and which determines the conditions of existence and knowledge for man. This vision of the nature developed by Naim-Gesbertb joins the stoic perception of the nature because of the fact that it constitutes a source of knowledge. Well, more for him, the relation who exists between the man and the nature is inseparable. Therefore, whether it is philosophical or theological, nature is related to the social environment of the man or to everything that surrounds us. This means that we are led to collaborate with it throughout our existence.

If Naim-Gesbert brings out the side of collaboration between man and nature, for Descartes, Man has a relationship of power towards nature. Man would be a master, delegated by a divine power that grants him control over nature. On the other hand, Rousseau believes that this role of master makes the man lose all his essence, because it is the object of its denaturation. The Cartesian perception is completely different from that of Rousseau, because, the dimension of essence found in Rousseau is not present or at least, it is little in Descartes. Therefore, this idea of nature as Descartes conceives it is a work of God that must be controlled by man. Is this way of understanding nature without impact on it?

In response to this question, Serres gives the following answer: 'Since the first living beings began to colonize the face of the Earth, constantly evolving, leaving behind them more fossil species than we will ever know contemporary ones' (Serres,

2004, p.69). This quote attests to the seriousness of our actions on the planet and shows how the power we think we have over nature is harmful to humans as well as their contemporaries. This perception of dominated and dominant makes that today humans are at war with nature and consider it inferior. As a subaltern, man feels no pity and is only accountable to God. Consequently, by seeking to master nature, man tends more and more to enter into war against it in the idea of being victorious, but in reality he only destroys himself. For only the protection of nature will allow him to save himself and make a place for future generations.

Seeking to determine human place in nature, Flahault asks the question, and 'Does humanity believe it dominates the earth'? (Flahault, 2013, p.128). To this question, Pascal responds by making clear the relationship that exists between man and nature and locates humans limits vis-à-vis it. He affirms: 'In nature, man is nothingness with respect to the infinite, a whole with respect to nothingness, a medium between nothing and everything' (Blaise, 1925, p1106). Presumably, the author shows that nature is symbolized by the infinite and is represented by a force, from which man can institute a limited order that transcends, protects, nurtures, educates and sources pleasure etc. Taking into account these different important functions of nature in the human adventure, man must be its guardian because, thinking he has a certain power over nature, he misuses it and destroys himself. For it is impossible for man to live outside of nature. Therefore, if man does not exist outside of his natural environment, and if his cohabitation with nature is essential for his survival, it is obvious that an environmental ethic becomes quite indispensable to man.

From this fact, we understand that nature is part of the social environment of the man and the man undoubtedly cannot exist outside of this environment. In this logic, this one can be considered as a social and political space that favors the conditions of the life of the man in society. Thus, the power of the man on the nature must be controlled by a right and logical rationality. For nature is endowed with a purpose of its own. Probably, this finality is that of cohabitation between them. The man as a thinking being of the nature must put his egoism aside and has for duty to protect it not to destroy it if he wants to facilitate his adventure.

This can only be done through education and a surpassing of oneself because man's behavior is often driven by interests or money, and to meet his personal needs, makes him enter into a power relationship with nature. He confiscates it for sometimes useless needs, without taking into account the impact of his actions on himself and his fellow men. As a result, not only are our natural resources depleted and guarded by a small minority, but this also has serious consequences for the environment. Nature should be perceived by humans as the primary support for existence, and must be considered as a human social envelope that conditions, limits and protects human groups. Thus, it is important that humans take care of it if they want to lead a serene existence. This must begin with an awareness of his place as a cohabitant with nature and that it is important for man to adapt his actions in order to live in harmony with it.

From this reflection, we can deduce that the human adventure is complicated and that of the generations to come risks to be it even more by our selfishness and our greed. Wallenhorst believes that: 'the entry into the Anthropocene directly and precipitously compromises the sustainability of the human adventure' (Wallenhorst, 2020, p. 46). He invites us to readjust our actions towards nature in order to improve our living conditions and also to facilitate cohabitation because it is impossible for man to live outside of it. A protected nature is a condition for a good human adventure based on social, ecological and political well-being. It is a source of wealth, food, education and pleasure. The seriousness of the current situation demands that we act quickly and efficiently. A way is proposed to alleviate and allow us to live more serenely this human adventure of which Wallenhorst speaks, this solution is none other than an ethics of the environment through education.

References

Blaise, P. (1925). *Pensées*, Brunschwicg n° 72, Bibliothèque de la Pléiade, n° 84.
Flahault, F. (2013). L'homme fait-il partie de la nature ? *Revue du MAUSS, 42*(2), 125–128.
Heidegger, M. (1986). Etre et temps. Paris Gallimard.
Serres, M. (2004). Le concept de Nature. *Études, 400*(1), 67–73.
Wallenhorst, N. (2020). Quel type de citoyenneté en Anthropocène ? *Le Télémaque, 58*(2), 45–58.

Daphnée Valbrun is in her first-year of PhD in Educational Sciences at the Catholic University of the West and Strasbourg University. She is a lecturer in the department of Education Sciences at Catholic University of West. Her main research topics are the anthropocene, citizenship, educational pedagogies, autonomy and experience in education.

Natural and Cultural Heritage

Christoph Wulf

Abstract Culture is not the only thing that is passed on from one generation to the next and changed in the process. Nature is also a heritage that each generation can use and shape and that is passed on from generation to generation. In the Anthropocene, a clear distinction between nature and culture is hardly possible. At present, there is almost no area of nature that is not influenced by humans. Many of these influences are destructive. They have changed nature in such a way that life on the planet is endangered. In his own interest, therefore, the human being is striving to correct this situation. The common heritage of nature and culture has its origins in the past and is distinguished by its significance for the present and the future. It has a cross-generational and cross-cultural significance for individuals and communities. Humanity has become a telluric power. Therefore, it has a responsibility for its actions and the destiny of the planet. This responsibility raises complex ethical issues in which difficult trade-offs and clear choices must be made. There is rightly talk of global ethical responsibility. This includes dealing with values such as peace, non-violence, human dignity, freedom, justice, and sustainability. In the Anthropocene, the validity of these values must be extended so that it also determines the actions of humans towards other living beings. Climate and plant life also require difficult trade-offs and decisions. We are all dwarfs on the shoulders of giants, i.e. our possibilities for life and development are based on the actions of previous generations. Each generation has a responsibility to understand, shape, and make fruitful for future generations the natural and cultural heritage and their mutual entanglements (Resina, J. R., & Wulf, C. (Eds.). *Repetition, recurrence, returns: How cultural renewal works*. Lexington Books, Roman & Littlefield, 2019; Gebauer, G., & Wulf, C. (1995). *Mimesis. Culture, Art, Society*. University of California Press).

C. Wulf (✉)
Freie Universität Berlin, Berlin, Germany
e-mail: christoph.wulf@fu-berlin.de

N. Wallenhorst, C. Wulf (eds.), *Handbook of the Anthropocene*,
https://doi.org/10.1007/978-3-031-25910-4_127

When talking about a common natural and cultural heritage and its universal value, one must simultaneously consider its multiformity, diversity, and alterity. Both perspectives must be related to each other, intertwined and communicated to the next generations (Wulf, 2013). It is necessary to understand the importance of the commonality of heritage and responsibility for the planet. Climate protection and the sustainable use of resources make this clear. The concept of heritage implies careful handling, i.e. attention, protection, care and sustainable use. Heritage should also be preserved for future generations in order to ensure their life development. It contributes to a humane and sustainable development of life:

The world's heritage contains both tangible and intangible assets that are equally important to individuals, communities and humanity as a whole and have intergenerational significance. Heritage is an open generic term for all those tangible and intangible assets inherited from a past to which individuals, communities, or humanity as a whole attach salient importance. What is heritage for a community emerges as a consequence of a complex dynamic of determination and selection as well as of proving, updating, transformation and appropriation through practice and interpretative approaches.

Natural and cultural heritage is an anchor point for our rapidly changing society. Its interpretation makes it clear that the features and consequences of modern society are by no means without alternative. Studies show that in a changing world under pressure to homogenize, people soon perceive new conditions as "normal" and no longer question development processes. The encounter with heritage allows the creeping processes of "shifting baselines" to become visible and allows us to experience the present in an overarching historical context and to perceive the open formability (contingency) of our concrete life worlds (Wallenhorst & Wulf, 2022, 2023). This is a political empowerment and all the more relevant the more people are afraid of not being able to consciously shape their lifeworlds.

Unlike artifacts, localized monuments, and small ecosystems, landscapes face particular challenges as heritage. Around the world, the small remnants of nature not yet altered by humans are being destroyed on a dramatic scale. The same is true for the old and "real" cultural landscapes. They are being replaced by new (uniform, interchangeable) landscapes, not least because the concept of cultural landscapes as heritage has not yet been sufficiently developed conceptually and politically. Comprehensive global and local social transformation processes (e.g., infrastructure development, demographic change, migration, rural exodus, urbanization) are increasing pressure on natural areas, traditional cultural landscapes, and global and local human-generated transformation processes (e.g., loss of biodiversity, climate change).

When talking about common heritage, we need to understand how important natural and cultural heritage is, how much it is endangered and how easily it can be destroyed. The common heritage offers an opportunity to learn and to educate (Wallenhorst, 2021; Wulf, 2022b, c). The encounter and confrontation with the respective specific heritage and the heritage of humanity is the task of education (Resina & Wulf, 2019). In this context, UNESCO's value-based heritage concept focuses on communicating heritage in its significance for humanity, international understanding, sustainable development, openness to the world, global citizenship, and peace. Young people play an important role in defining, developing, interpreting, preserving, and communicating heritage. They develop the meaning of heritage in the future and therefore need to be involved in a participatory way as soon as possible. This allows them to develop their own perceptions and approaches to heritage, to relate to it freely, and to engage critically and productively with traditional and existing approaches.

Programmes for the Care of the World Heritage

UNESCO has developed several programmes for the conservation and promotion of natural and cultural heritage, with different emphases, to contribute to the preservation of heritage. These make up an important part of UNESCO's work to preserve the common heritage.

The best known of these is the *World Heritage Promotion Programme*, which began with a convention in 1972 and has since recognized 1154 World Heritage sites in 167 countries around the world. These World Heritage sites are outstanding testimonies to past cultures and unique natural landscapes. They are sensory testimonies to the diversity and dignity of cultures. What they have in common is their high universal value, their significance not only for national or local communities, but for humanity as a whole. The protection and sustainable preservation of these sites is therefore the responsibility of the entire international community. Examples of World Heritage Sites are the Great-Barrier Reef in Australia, the Serengeti National Park in Tanzania, Manchu Picchu in Peru, the Acropolis in Greece (UNESCO, 1972).

Whether dance, theatre, music, customs, festivals, or handicrafts - *Intangible Cultural Heritage* (ICH) is alive and is supported by human knowledge and skills. It is an expression of creativity, conveys continuity and identity, shapes social coexistence and contributes to sustainable development. More than 500 forms of intangible cultural heritage are listed on the international UNESCO lists, and more than 130 are on the German list of intangible cultural heritage. The 2003 Convention is the basis for both. Examples are organ building and organ music from Germany, yoga in India. Rumbas from Cuba (UNESCO, 2003, 2005).

Since 1992, the *World Documentary Heritage* has contained important testimonies of cultural turning points in history. There are more than 400 documents from 21 countries that are part of UNESCO's Memory of the World Programme. They are of exceptional value, raise awareness of the significance of historical events and developments, and serve as sources of knowledge for shaping present and future societies. They are safeguarded and made accessible in archives, libraries and museums. These testimonies include the Gutenberg Bible, Beethoven's Ninth Symphony, and the colonial archives of Benin, Senegal, and Tanzania (UNESCO, 1992).

UNESCO Global *Geoparks* (UGGp) are regions with important fossil sites, caves, mines, or rock formations. UNESCO's works with geoparks began in 2001. 2015 the 195 Member States of UNESCO ratified the Creation of a new label, the UNESCO Global geoparks. It offers the opportunity to better understand planet Earth and the conditions of life by following in the footsteps of the past. Currently, 169 geoparks in 44 countries have been designated worldwide. These geoparks are model regions for sustainable development. They work on viable future options for a region's landscape and address global societal challenges, such as the finite nature of natural (especially geological) resources and climate change. Examples include the Bergstrasse-Odenwald, the Swabian Alps, the German-Polish Muskauer Faltenbogen / Łuk Mużakowa (UNESCO, 2015a).

UNESCO identifies with its 727 *Biosphere Reserves* worldwide model regions and places of learning for sustainable development, in 131 countries and makes clear how in a concrete landscape sustainable development can succeed and nature conservation and economy can be brought together. More than 275 million people worldwide live in these biosphere reserves. In Argentina the delta of the Paraná, in Ethiopia Lake Fana, in Brazil Central Amazon belong to the Bioreserves (UNESCO today, 2007).

Education for the Sustainable Development of the Natural and Cultural Heritage

For UNESCO, the 2030 Agenda provides an excellent opportunity to work on a comprehensive values-based approach to heritage. The future of humanity depends significantly on success in these areas (Wallenhorst, 2021; Wulf, 2022a, b; Wallenhorst & Wulf, 2022, 2023). The goal of sustainable development is the realization of a continuous process of transformation throughout society that should lead to preserving the quality of life of the present generation while at the same time securing the choices of future generations to shape their lives. Sustainable development is now a recognized way to improve individual future opportunities, societal prosperity, economic growth, and environmental sustainability. Education for sustainability aims to empower people to creatively shape an ecologically compatible, economically efficient, and socially just environment, considering the international perspective (UNESCO, 2015b).

Sustainability is a regulative idea for dealing with the natural and cultural heritage, which can only be realized approximately. Education for sustainability is an important prerequisite for the gradual realization of sustainability. It addresses the individual, whose sensitivity and readiness for responsibility it wants to promote, and communities, without whose commitment its realization is not possible. Considering individual and social differences, it starts with existing structures and develops young people's creative competence in this area. The aim is to develop the ability to shape one's own life and living space in terms of sustainability. This requires learning in concrete problem constellations, working out their contexts and initiating participatory action. Education for sustainability also implies a reflexive critical understanding of education and a willingness to participate in the corresponding individual and social learning processes. To this end, it is necessary to develop standards for education for sustainable development that do justice to the multi-perspectivity of sustainability. Education for sustainable development should contribute to the establishment of social justice between nations, cultures, and world regions and between generations. Global responsibility and political participation are central principles of sustainability. Education for sustainability aims to develop a sustainability-oriented capacity for action that enables people to creatively shape the tasks required in this area.

To realize a sustainable approach to the natural and cultural heritage, a changed relationship to the external and internal nature of human beings is required (Kraus & Wulf, 2022; Wulf 2022c. It requires a transformation of capitalism and the development of new forms of economic production and cooperation, as well as a reduction of the colonization of nature in the interest of new forms of coexistence between humans and nature (Moore, 2016; Glikson, 2017; Haraway, 2016; Escobar, 2018). Sustainable management of natural and cultural heritage requires not only education for sustainability. It additionally requires education for global citizenship.

Global citizenship refers to belonging to a planetary community into which we are born and in which we have the task of helping to shape (Bernecker & Grätz, 2018; Wulf, 2021; Antweiler, 2011; Beck, 2004). This refers not only to our relationship with other people, but also to our relationship with animals and plants. Education for a sustainable use of natural and cultural heritage and education for a world community are at the heart of a new view of humanity and the world that embraces social participation and cultural sharing. It obligates the individual to the world community and the world community to the individual. On the one hand, global citizenship or world community implies far-reaching idealistic objectives; on the other hand, it is limited by the narrow boundaries of state interests and constraints. In the definition of the term, there is an explicit reference to sentiment. Thus, it says: global citizenship "is the feeling of belonging to a large human community. It particularly emphasizes interdependence politically, economically, socially, and culturally, and the interactions between the local, the national, and the global." Defining the term as a sense of belonging recognizes that the term contains both the same and different elements in different regions of the world. All people have feelings of belonging to communities. However, how these are articulated varies. In addition to biological and individual differences, historical and cultural differences play an important role. The simultaneity of the non-simultaneous between countries and regions, states and their political systems determines the quality and intensity of feelings, thus also the feeling of belonging to the world community. The formation to a world community connects desires and imaginations, rational cognitions and differing feelings and releases energies of action and behaviour, which are of central importance for dealing with the common natural and cultural heritage (Wulf, 2016, 2022a, b, c). The goal is, based on corresponding changes, to rediscover nature as a co-environment, to respect it and to preserve it (Meyer-Abich, 1990).

References

Antweiler, C. (2011). *Mensch und Weltkultur. Für einen realistischen Kosmopolitismus*. Transcript.

Beck, U. (2004). *Der kosmopolitische Blick*. Suhrkamp.

Bernecker, R., & Grätz, R. (Eds.). (2018). *Global citizenship education*. Steidl.

Escobar, A. (2018). *Designs for the pluriverse. Radical interdependence, autonomy and the making of worlds*. Duke University Press.

Glikson, A. Y. (2017). *The plutocene: Blueprints for a post-Anthropocene greenhouse earth* (Modern Approaches in Solid Earth Sciences, 13). Springer.

Gebauer, G., & Wulf, C. (1995). *Mimesis. Culture, Art, Society*. University of California Press.

Haraway, D. (2016). *Staying with the trouble. Making kin in the Chtulucne*. Duke University Press.

Kraus, A., & Wulf, C. (Eds.). (2022). *The Palgrave handbook of embodiment and learning*. Palgrave Macmillan.

Meyer-Abich, K. (1990). *Aufstand für die Natur. Von der Umwelt zur Mitwelt*. Hanser.

Moore, J. W. (Ed.). (2016). *Anthropocene or Capitalocene? Nature, history, and the crisis of capitalism*. PM Press.

Resina, J. R., & Wulf, C. (Eds.). (2019). *Repetition, recurrence, returns: How cultural renewal works*. Lexington Books, Roman & Littlefield.

UNESCO. (1972). *The world heritage convention*. UNESCO.

UNESCO. (1992). *Memory of the world*. UNESCO.

UNESCO. (2003). *Convention for the safeguarding of intangible cultural heritage*. UNESCO.

UNESCO. (2005). In Deutsche UNESCO Kommission (Ed.), *Übereinkunft über Schutz und Förderung der Vielfalt kultureller Ausdrucksformen*. DUK.

UNESCO. (2015a). *The UNESCO global parks*. UNESCO.

UNESCO. (2015b). Rethinking education. In *Towards a global common good?* UNESCO.

UNESCO today. *Journal of the German UNESCO Commission for UNESCO* (2007). German Commission for UNESCO, vol 2/2007.

Wallenhorst, N. (2021), Mutation. *L'aventure humaine ne fait que commencer.*

Wallenhorst, N., & Wulf, C. (Eds.). (2022). *Dictionnaire d'anthropologie prospective*. Vrin.

Wallenhorst, N., & Wulf, C. (Eds.). (2023). *Handbook of the Anthropocene*. Springer Nature.

Wulf, C. (2013). *Anthropology, a continental perspective*. The University of Chicago Press.

Wulf, C. (Ed.). (2016). *Exploring alterity in a globalized world*. Routledge.

Wulf, C. (2021). Global Citizenship Education. Bildung zu einer planetarischen Weltgemeinschaft im Anthropozän. *Vierteljahreszeitschrift für Wissenschaftliche Pädagogik, 97*, 463–479.

Wulf, C. (2022a). *Human beings and their images. Imagination, mimesis, performativity*. Bloomsbury.

Wulf, C. (2022b). *Education as human knowledge in the Anthropocene. An anthropological perspective*. Routledge.

Wulf, C. (2022c). Embodiment through mimetic Learning. In A. Kraus, & Wulf, C. (Eds.). *The Palgrave handbook of embodiment and learning* (pp. 39–59). Palgrave Macmillan.

Christoph Wulf is Professor of Anthropology and Education and a member of the Interdisciplinary Centre for Historical Anthropology, the Collaborative Research Centre (SFB, 1999–2012) "Cultures of Performance," the Cluster of Excellence (2007–2012) "Languages of Emotion," and the Graduate School "InterArts" (2006–2015) at the Freie Universität Berlin. His books have been translated into 20 languages. For his research in anthropology and anthropology of education, he received the title *"professor honoris causa"* from the University of Bucharest and the honorary membership of the German Society of Educational Research. He is Vice-President of the German Commission for UNESCO. *Major research areas:* historical and cultural anthropology, educational anthropology, imagination, intercultural communication, mimesis, aesthetics, epistemology, Anthropocene. Research stays and invited professorships have included the following locations, among others: Stanford, Tokyo, Kyoto, Beijing, Shanghai, Mysore, Delhi, Paris, Lille, Strasbourg, Modena, Amsterdam, Stockholm, Copenhagen, London, Vienna, Rome, Basel, Saint Petersburg, Moscow, Kazan, Sao Paulo.

Otherness

Muriel Briançon

Abstract This article examines the term of otherness, which is an essential root of Humanism. Indeed, humanity cannot be thought outside of relation to others. However, this many faceted term is still often used in a general sense, seeing otherness merely in terms of human differences and forgetting its other meanings. This article proposes three complementary dimensions of 'otherness', in order to better understand what it involves, namely how to live with others, with ourselves and with uncertainty.

The future of the human race will involve a new way of relating to others, a new dimension of self-awareness, and a new way of relating to knowledge. Underlying this threefold relation is the notion of otherness. Otherness is, at once, an "educational problem" and a "fundamental problem – perhaps the most fundamental problem of the 21st Century)" (Radford, 2009, p. 11). The notion has become unavoidable in the human and social sciences, and the term "otherness" is being used increasingly and in a burgeoning range of ways.

What sort of otherness are we talking about? The Quebec researchers Pierre Ouellet and Simon Harel, in their article from 2007, ask: *Quel Autre?* ("Which Other?"). Plainly, this is the essential question. Ouellet and Harel's patchwork wonderfully illustrates the disparate uses of this unclassifiable term. In France, over the past 10 years, the perception of otherness in the humanities has evolved from one anchored in the person of others (which, logically, leads to reflection on human rights, secularity, citizenship, disability, immigrant integration, tolerance and racism) to a many-faced, extremely fragmented form of otherness which many publications – such as *Figures de l'Autre* (Ardoino & Bertin, 2010) – have tried and failed to coalesce into a coherent concept. Anywhere there is difference, foreignness, barriers and boundaries, we now speak of otherness. In education, the notion is employed in widely varying ways: in *L'Altérité* (Groux & Porcher, 2003), no fewer

M. Briançon (✉)
Catholic University of the Westa, La Réunion, France
e-mail: mbrianco@uco.fr

© The Author(s), under exclusive license to Springer Nature Switzerland AG 2023
N. Wallenhorst, C. Wulf (eds.), *Handbook of the Anthropocene*,
https://doi.org/10.1007/978-3-031-25910-4_128

than a hundred different terms are used in the attempt to illustrate it. D. Jodelet (2005) observes that "the matter of otherness is debated in a hugely broad intellectual space, running the gamut from philosophy, morality and legality to the human and social sciences" (p. 24).

This inopportune and diverse use of the term "otherness" may be due to polysemy, to the varied etymology of the term, and its forgotten historical meanings. "Otherness" is an Anglo-Saxon word for the French alterité. This is a term borrowed by philosophers from the Low Latin alteritas (in the mid-fourth Century). It derives from the Latin alter, which gives us the French words "autre" (other), "autrui" (other people) and "altruisme" (altruism, care for other people). The word came to mean, at once, a "difference by way of change", "diversity" and "alteration". It appears to have fallen out of use, only to resurface in the seventeenth Century, with its more modern meaning: "the character or property of that which is other". It became widely used in philosophy in the early nineteenth Century, and relatively recently, has been adopted specifically in application to human relationships. Thus, it is unsurprising that the *Dictionnaire encyclopédique de l'éducation et de la formation* (Encyclopaedic Dictionary of Education and Training) offers no direct definition for "Altérité", instead referring the reader to the entry for "Intercultural (Pedagogy)" (Abdallah-Pretceille, 1994, p. 556): otherness is mentioned only twice in the volume – firstly in support of the definition of "the intercultural as a possible means of analysing and understanding issues connected to otherness and cultural diversity", and later as one of the tensions inviting the use of the intercultural approach (Groux, 2002), with the binomial "sameness/otherness". A decade later, otherness was finally included in the *Dictionnaire de l'altérité et des relations interculturelles* (Dictionary of otherness and intercultural relations), denoting "a quality or essence – the essence of being other", leading to "a dramaturgy of relations with others" (Rey, 2003, p. 4). Should otherness be reduced, though, to a set of differences between human beings?

François Jullien (2012), the first in France to hold a chair in Otherness at the Collège d'Études Mondiales, views otherness as the philosophical topic par excellence and indeed the most essential tool of philosophy. Its profound meaning, then, is to be found in the *Vocabulaire technique et critique de la philosophie* (Lalande, 1968, p. 39 – Technical and Critical Lexicon of Philosophy), which reveals a far more ancient Greek etymology: ἑτερότης (heterotes). Plato, in *The Sophist*, appears to have discovered this Form (idea) by accident (Cordero, 2005), indicating difference with respect to a given referent. Thus, from a logical standpoint, otherness is a symmetrical and intransitive relation: the negation of sameness. The Greek root hetero (as opposed to homo) is found in the composition of a plethora of scientific words containing the idea of difference in shape, nature, origin, between individuals, species or elements. However, the Greek dictionary defines the noun ἑτερότης: (1) difference of type or fundamental difference; (2) lack of unity, disunity. The adjective ἕτερος has six different uses: (1) one of the two, when speaking about dual organs; (2) the second, in a list; (3) other which is similar or analogous, other of the same type; (4) other in general; (5) different, contrary, opposite, other than, different by comparison; (6) other than the accepted norm – i.e. euphemistically bad (Bailly,

1901, p. 368). Aside from denoting a simple difference between human beings, otherness also refers to other and forgotten meanings.

Being a socially acute question (SAQ) in education at present, otherness is a social matter; a learning objective; a reason to re-examine teaching practices; a media object; and a source of controversy: is otherness a myth or a reality? Can it be boiled down to a difference from an external referent? Is it possible to define otherness? Can otherness be known and named? With respect to this latter, the first objection "relates to the very possibility of conceiving of the appearance of otherness as such", "the emergence of something which is 'other'" – "something which does not originate with us" in a knowledge system that is conditioned, conveyed through a medium, circular, enclosed and delimited (Bonoli, 2007, p. 3). The second objection "relates to the possibility of explicitly naming otherness when it does manifest itself" in a vernacular language which will, inevitably, oversimplify it (Ibid, p. 4). With this in mind, should otherness remain unknown and inexpressible?

In these circumstances, with a view to allowing humanity to move forward beyond the Anthropocene epoch, is it possible to teach, educate and train people in how to approach otherness (and if so, how)? This is a superhuman task, because to do so prerequires us to think of, and verbally express, something which cannot be thought of or verbally expressed. This seemingly vain quest to understand otherness gives rise to a three-dimensional, paradoxical concept of it.

Indeed, our proposed solution (Briançon, 2019) is to think paradoxically of otherness in terms of three mutually complementary, inextricably linked dimensions:

Firstly, external otherness relates primarily to the person in front of you. An "other" is not simply an alter ego: "he is what I, myself, am not" (Lévinas, 1946, p. 75). In philosophy, the relationship with other people, oscillating between intersubjective conflict (Sartre, 1943) and encounter in relation to the "I and Thou" pair (Buber, 1969), is wonderfully illuminated by Lévinas, in whose view, "the absolute Other is the other Person. [...] The absence of a shared territory makes the Other into a Foreigner: a Foreigner causing consternation in one's own territory" (Lévinas, 1961, p. 28). The issue of the presence/absence of an other person characterises absolute otherness, which comes into play in the ethical concept of the Face (face-to-face) and in the mysterious à-venir (yet to come) (Lévinas, 1946). External otherness teaches us the sense of ethics, responsibility and the overcoming of ego.

Secondly, should we not look for this "other" in ourselves rather than in other people? Internal otherness lies in the disquieting strangeness of the subject himself: "I is an other", said Rimbaud. The reflexive subject experiences concern and is troubled by contact with an other person (Ardoino), discovers his own uncontrollable emotions, the fragility of his identity over time, and the yawning gap between his idem identity and his ipse identity (Ricœur), the alienation of his conscious mind and the dialectic between his for-self and for-others (Hegel), not to mention the uncharted waters of his subconscious (Freud) and the object a (the first letter of the French word "autre", a point of resistance to theoretical thinking, a non-response, an absence, a consuming void (Lacan). Internal otherness teaches us reflection, self-acceptance and inner peace.

Thirdly, otherness can be seen as epistemological – it is an issue of knowledge – the Limit of human knowledge. It is the Other with a capital O: a conceptual and discursive aporia. This infigurable, inexpressible unknown is viewed as a tangible thing in western thinking, and as a void in eastern thinking. The term originates earlier than Plato's discovery of otherness as difference. Epistemological otherness is indeed the avatar of the Grecian concept of non-being (μή ἐόν), which the pre-Socratic philosopher Parmenides of Elea, the proponent of being (ἐόν), defines as non-existent, inconceivable and inexpressible, and condemns in his Poem, *On Nature* (Dumont, 1988, 28 B I DK). The conception of this Other, removed from being, thought and language which, for 2500 years, has been excluded by a long-forgotten philosophical choice conditioning the western way of thinking, leads to a positive reconsideration of our relationship with non-knowledge, transcendence and spirituality.

The three forms of otherness appear to differ widely. With this in mind, is it suitable to regroup them into a single conceptual meta-category? Ricœur concluded *Oneself as Another* (1990) with the statement: "only a discourse separate to oneself […] is suitable for the meta-category of otherness, with the danger that the otherness will cease to be by becoming the same as itself …" (p. 410). Whilst this is a legitimate stance for a philosopher, tirelessly wondering and debating, it is not for a teacher, whose goals are to transmit knowledge, inculcate and educate. Provided we do not lose sight of its heuristic and indefinite nature, the three-dimensional concept of Otherness becomes a new focus for research, a new tool to produce knowledge and a new topic to teach in order to renew Man's human beings' relationship with others, with himself and with the unknown, in the Anthropocene age.

It is a provocative, paradoxical and transgressive concept (because its epistemological form represents the absence of content). In spite of this – or, strictly speaking, because of it – it is teaching (enseignant) in the etymological sense (derived from insignire). Indeed, it highlights the otherness of a sign: it shows and signposts phenomena and a process which would, otherwise, remain unseen; in so doing, it transforms the way in which we exist in the world. It teaches us to live with others, to live with ourselves, and to live with what is by accepting uncertainty. It is precisely because Otherness is extremely instructive that it must be taught.

References

Abdallah-Pretceille, M. (1994). Interculturelle (Pédagogie). In P. Champy & C. Etévé (Eds.), *Dictionnaire encyclopédique de l'éducation et de la formation* (pp. 556–561). Nathan.
Ardoino, J., & Bertin, G. (2010). *Figures de l'Autre. Imaginaires de l'altérité et de l'altération.* Téraèdre.
Bailly, A. (1901). *Abrégé du dictionnaire Grec Français.* Hachette.
Bonoli, L. (2007). La connaissance de l'altérité culturelle. Expérience et réaction à l'inadéquation de nos attentes de sens. *Le Portique* [online]: 5.

Briançon, M. 2019. *Le sens de l'altérité en éducation. Enjeux, Formes, Processus, Pensées et Transferts*. Préface de Marie-Louise Martinez. Postface d'Eirick Prairat. Editions ISTE, Collection "Education", edited by Angéla Barthes and Gérard Boudesseul.

Buber, M. (1969). *Je et Tu*. Préface de G. Bachelard. Aubier. (Original edition 1923).

Cordero, N. L. (2005). Du non-être à l'autre. La découverte de l'altérité dans Le Sophiste de Platon. *Revue Philosophique de la France et de l'Etranger, 2*, 175–189.

Dumont, J.-P. (1988). *Les présocratiques* (Published by Jean-Paul Dumont in collaboration with Daniel Delattre and Jean-Louis Poirier. Translated texts collected by H. Diels in Fragmente der Vorsokratiker, Berlin, 1903). Gallimard, La Pléïade. (Original edition 1903).

Groux, D. (2002). *Pour une éducation à l'altérité. Actes de la Journée d'études sur l'éducation à l'altérité*. L'Harmattan.

Groux, D., & Porcher, L. (2003). *L'Altérité*. L'Harmattan, collection Cents mots pour.

Jodelet, D. (2005). Formes et figures de l'altérité. In M. Sanchez-Mazas & L. Licata (Eds.), *L'Autre: Regards psychosociaux* (pp. 23–47). Les Presses de l'Université de Grenoble.

Jullien, F. (2012). *Petite introduction: De l'altérité*, introduction à la Leçon inaugurale de la Chaire sur l'Altérité, Collège d'Etudes mondiales. http://www.college-etudesmondiales.org/fr/content/lecart-et-lentre.

Lalande, A. (1968). *Vocabulaire technique et critique de la philosophie*. PUF.

Lévinas, E. (1946/1947). *Le temps et l'autre*. PUF.

Lévinas, E. (1971). *Totalité et infini. Essai sur l'extériorité*. Martinus Nijhoff. (Original edition 1961).

Ouellet, P., & Harel, S. (2007). *Quel Autre? L'altérité en question*. VLB éditeur.

Radford, L. (2009). L'altérité comme problème éducatif. *Actes de la 15e Journée Sciences et Savoirs*, Université Laurentienne, 11–27.

Rey, J.-F. (2003). Altérité. In G. Ferréol & G. Jucquois (Eds.), *Dictionnaire de l'altérité et des relations interculturelles* (pp. 4–7). Armand Colin.

Ricœur, P. (1990). *Soi-même comme un autre*. Seuil.

Sartre, J.-P. (1943). *L'être et le Néant. Essai d'ontologie phénoménologique*. Gallimard.

Muriel Briançon is Associate Professor at the Catholic University of the West (UCO). She is Doctor of Educational Sciences and the author of five books on education and philosophy of education. Books (selection): *Those pupils in scholar difficulties which say they are first curious for the teacher* (L'Harmattan, 2011, in French). *The teaching otherness* (Publibook, 2012, in French). *The Meaning of Otherness in Education. Stakes, Forms, Process, Thoughts and Transfers* (ISTE & WILEY, 2019, in French).

Plasticity

François Prouteau

Abstract Sciences, through their current progress, show that the plasticity of life is so important in human beings that it can be considered as a constitutive characteristic of their identity. The robustness-vulnerability dynamic at the core of this plasticity, particularly developed in human beings and inherent to their complexity, calls for an ethical reflection on the use of techno-sciences.

The term plasticity is borrowed from the Greek word *plastikos* meaning malleable, in the feminine "plastikê, with the same meaning as *tekhnê*, the art of modelling figures and creating forms" (Demoustier, 1844). Its etymology is derived from ancient Greek *plassein*, "to shape" or "to model" a figure or a body, sometimes with art (aesthetic dimension). This notion of plasticity is present in philosophical anthropology, through the image of wax being modelled (Plato, *The Laws I. 5, 2017*) or being flexible (Montaigne, *Essays II. 12, 2018*).

Rousseau points out how, by nature, a child's brain has "this flexibility" "which makes it suitable for receiving all sorts of impressions" and "everything around it is the book in which, without thinking about it, it continually enriches its memory". The author of *Emile, or Education* (Rousseau, 1921) develops, in this sense, the notion of perfectibility, and Condorcet (1796) sees no end to the perfection of human abilities. A few years later, Hegel highlighted the concept of plasticity, from its historical beginning in art, to the development of subjectivity. He uses the term "plastic" to "characterise the fundamental mode of being human", according to Catherine Malabou (2018).

The Greek terms *plassein* and *anthropos* are side by side in *The Republic* (377c) where Plato (2000) urges women to shape the soul like a piece of wax, by words, rather than the body of their newborn child by massage. The proximity of the two Greek terms is also found in the biblical account of creation (*Genesis* 2:7) where *anthropos* receives from God a body (*plassein*) "shaped" from clay and a *psyche* is

F. Prouteau (✉)
LIRFE – Catholic University of the West, Angers, France
e-mail: f.prouteau@iffeurope.org

© The Author(s), under exclusive license to Springer Nature Switzerland AG 2023
N. Wallenhorst, C. Wulf (eds.), *Handbook of the Anthropocene*,
https://doi.org/10.1007/978-3-031-25910-4_129

breathed into him. The prologue of John in the Gospels shows the action of the divine *logos* which is like "the moving structure of the moving universe" (Grosjean, 1991): the plastic weave of the subjects and the world are woven in the same language and in continual transformation.

Biology shows that human beings manifest their capacities for evolution and interaction with environments through epigenetics. Epigenetics suggests how much the biological environment but also the psychological environment are essentials in gene expression, as observed scientifically, to varying degrees, throughout human development. The little human being is endowed with an enormous potential, for learning like nothing else, thanks to the extreme plasticity received before his birth, as a legacy of evolution. Brain imaging has shown us today that the brain's highways are formed during the 3rd month of pregnancy, due to certain genes that contribute to giving the brain a global architecture with specialized regions. Learning is linked to changes in the synaptic connections between neurons: these are activated and adjusted each time we learn, more so in children, then with diminished capacity at all stages of life. "More than *Homo sapiens*, we are *Homo docens* – because everything we know about the world, we have learned mostly from our environment or from our surroundings" (Dehaene, 2018). Learning to learn is certainly a key educational activity to harness anthropological plasticity for human growth.

Through the advances in nanobiotechnologies and NBICs (nanotechnologies, biotechnologies, information technology and cognitive sciences), scientists also highlight the ability of the nervous system to change its structure and function (neuronal plasticity) in reaction to the environment. Brain imaging has confirmed the main distribution of brain activities specific to each hemisphere, while highlighting a form of collaboration between hemispheres, or even migration from more specific areas from one hemisphere to another. For example, in children, several cases of almost total removal of the right hemisphere for therapeutic reasons show that the left hemisphere can develop, depending on environmental stimuli, activity zones that were thought to be reserved for the other hemisphere. In addition, various scientific studies led by teams of researchers show that regular meditation has a physiological effect on the brain (Lutz, 2017).

What is called neuroplasticity – in other words, the ability of structure of the brain to be modified through real experiences – can be observed in the cortex, as well in the cerebellum. Composed of more than half of the neurons in the brain, in humans, compared to other primates, the cerebellum is of greater significance than the cortex. It is known that learning and movement coordination is linked to the development of new networks in the cerebellum in dialogue with other structures in the brain, thus making it possible to link physical sensations to emotions and cognitive processes.

Scientists try to understand how to solve the "stability–plasticity dilemma" as a constraint for artificial and biological neural systems: learning "requires plasticity for the integration of new knowledge, but also stability in order to prevent the forgetting of previous knowledge" (Mermillod et al., 2013); they observe that "humans can rapidly learn enormous amounts of new information, on their own, throughout life, and can integrate all this information into unified conscious experiences that cohere into a sense of self" (Grossberg, 2019).

Other scientific knowledge in the animal world and more generally in the whole of nature, also demonstrates the plasticity of life, which allows us to see evolution and the laws of life differently. We discover that the dynamic balances between all the forces of selection at the different levels of life allow life to create ever more novelty, diversity and therefore adaptation to the environment: we are intertwined. In biology, different forms of life (plants, fungi, bacteria…) live in symbiosis with the human species. Mutual aid reaches exceptional levels in human beings, with a high sensitivity to external conditions thanks to epigenetics and culture. The immaturity and vulnerability of young humans at birth and in their early years forced our ancestors to develop collective strategies to provide them with a safe environment. The role of parents and family, not to mention solidarity within the tribe and cooperation between the males in the group, played a key role. This situation at birth would thus have led human beings to become better able to distinguish between individuals who could help them and those who could harm them, and also to develop the ability to put themselves in the other's shoes and pay attention to what they think. The immaturity and fragility of the baby cause the human being to become an expert in otherness in order to live. This makes it possible to better understand the paradox of human evolution: this extreme vulnerability of the human being at birth is in fact its strength. This strength is a bonding of its external plasticity in relation to the environment, and its internal plasticity. The human genome has benefited over millions of years from the knowledge of ancestral generations and has given a common structure to the human brain. The latter is able both to project abstract hypotheses onto the outside world and to be hypersensitive to signals from the environment, thus permanently producing an interaction between the integral subject – body, soul, spirit – and the world, and a joint transformation of both.

Human beings exist as they are today due to our plasticity, endowed with the plasticity of life that has developed due to dedicated thought and action on and in our world, and whilst not forgetting the contributions and interactions of the earth's eco-system, without which humans cannot exist. The plasticity of life can be characterised by the dynamism between robustness (invariance, rigidity or maintenance of coherence) and vulnerability (transformation or malleability) with a capacity to adapt to the environment.

Vulnerability and adaptation to the environment are two characteristics of living organisms that culminate in humans. The culmination of these is enhanced by their anthropological plasticity, providing capacities both prodigious and limited: of evolution and learning; of experiencing failures, maturing and resilience; of "empathy, harmony and solidarity between different humans" (Magnin). All the sciences – human and social sciences, life and earth sciences – are mobilized in the study of humans' plasticity. Thierry Magnin also showed how contemporary research and epistemology invite us to work with the "incompleteness of reason", "to learn interdisciplinary dialogue" and to enter "the mystery of knowledge".

Obviously, this position is opposed to a reductionism that would consider all life and especially the human being, as a game of mechanics where spare parts and connections can be repaired, changed or even increased at leisure, by increasingly efficient artefacts. With regard to humans, discoveries about brain function have fuelled impressive advances in computing, data analytics and machine learning, inspired by what we know about the brain, as little as that is today. Meanwhile, these algorithms are already challenging *Homo sapiens* because undeniably, such machines in many fields learn better than we do in areas where we thought we were, a short time ago, superior (translation of foreign languages for example) and invincible (game of Go). The next step from there is to replace human beings by smarter robots, that supporters of transhumanist current thought would use to take advantage for themselves. These same trends also have high hopes of the possibilities opened up by the technology commonly known as "DNA scissors" to qualify the CRISPR-Case 9 method (Clustered Regularly Interspaced Short Palindromic Repeats Associated Protein 9) for targeted genome modification. If this technique enables, in particular, replacement of a defective gene that can cause serious pathologies – highly commendable, per se provided precautionary principles are applied – some see it as a tool potentially capable of enabling a technological power to create a human being "liberated" from biological laws and its belonging to the living world. A simplistic and crude vision that fails to recognise the plasticity of the human being and his eco-systemic complexity. To suggest that we would benefit from the sustainable subjection of the brain to a computer is in total contradiction with brain plasticity.

Anthropological plasticity goes hand in hand with the complexity of the world, because the human subject and the world are closely interwoven, in fact intertwined (plexus). On the one hand, the world in its complexity modifies the human being by encouraging its adaptability and developing its learning capacities, and therefore its plasticity. Conversely, the plasticity of human beings through education allows them to acquire new talents that generate multiple improvements and complexity in the evolution of societies. The help of the senses and the plasticity of the body allow us to experience the world individually (Wulf, 2016), in the interactions and changes brought about by the impacts and responses of the world on our bodies, and vice versa (Polyani, 1969). Anthropological plasticity also manifests itself in the plasticity of language, whose practices (rhetoric, poetics, hermeneutics), with their totalizing, competing and complementary aims, constantly open up new identities and ways of being in the world, expanding the imaginary, discovering new dimensions of reality (Ricœur, 1999). Language has its own "plasticity" which uses a way of saying things, chooses a certain number of them rather than others, and regulates the elements. So, it is with the "philosophical *logos*" in search of speaking truthfully, not pure rhetoric but "that other thing which is both a technique and an ethic, which is both an art [*technē*] and a morality, and which we call *parrhēsia*" (Foucault, 2001).

Taking into account the human subject endowed with plasticity, both in his body and in his language in relation to the world and to others, human and non-human, requires systemic and complex thinking in an ethical context (Morin, 2013) in order to avoid any reductionist representation of the *anthropos* and to accompany the ongoing anthropological metamorphoses (Giorgini, 2020).

References

Christoph, W. (2016). *On historical anthropology: An introduction*. At SSRN: https://ssrn.com/abstract=3706830

Condorcet, M. de [1795] (1796). *Outlines of an historical view of the progress of the human mind*. M. Carey. https://oll.liberty.fund.org

Montaigne, M. de. [1580] (2018). *The complete essays*. Oxford World's Classics.

Dehaene, S. (2018). *Apprendre !* Odile Jacob.

Demoustier, A. C. (1844). *Manuel lexique philologique, didactique et polytechnique*. Ladrange.

Foucault, M. (2001). In A. Fontana & F. Gros (Eds.), *L'Herméneutique du sujet. Cours au Collège de France 1981–1982*. Seuil.

Giorgini, P. (2020). *La crise de la joie*. Bayard.

Grosjean, J. (1991). *L'ironie christique*. Gallimard.

Grossberg, S. (2019). The resonant brain: How attentive conscious seeing regulates action sequences that interact with attentive cognitive learning, recognition, and prediction. *Atten Percept Psychophys, 81*, 2237–2264. https://doi.org/10.3758/s13414-019-01789-2

Lutz, A. (2017). Postface. In: Zellner Keller Brigitte, Maskens Claude, Attala Jackie (dir), *Pleine conscience pour les seniors. Programme en 8 mois. Approche cognitive*, De Boeck Supérieur. https://www.cairn.info/pleine-conscience-pour-les-seniors%2D%2D9782807312128-page-317.htm

Malabou, C. (2018). Hegel nous invite à nous transformer en permanence. In. *Philosophie magazine*. N°123. Oct. 2018.

Martial, M., Aurélia, B., & Patrick, B. (2013). The stability-plasticity dilemma: Investigating the continuum from catastrophic forgetting to age-limited learning effects, 2013. *Frontiers in Psychology, 4*, 504. https://doi.org/10.3389/fpsyg.2013.00504

Morin, E. (2013). *Mes philosophes*. Fayard.

Nathanaël, W., François, P., Dominique, C. (dir.). (2018). *Éduquer l'homme augmenté*. Le Bord de l'eau.

Plato. [427 347 BC] (2000). *The republic*. CUP.

Plato [427 347 BC] (2017). *The laws*. CUP.

Polanyi, M. (1969). *Knowing and being*. Chicago University Press.

Ricœur, P. (1999). *Lectures 2, La contrée des philosophes*. Seuil.

Rousseau, J.-J. [1762] (1921). *Emile, or Education*. (Barbara Foxley, M.A. Trans.). J.M. Dent and Sons\E.P. Dutton. https://oll.liberty.fund.org

François Prouteau is a graduate engineer from the French Grande école Institute Mines-Telecom Atlantique with a PhD in Educational Sciences from the Lumière University Lyon 2. He is teacher and associated researcher at The Catholic University of the West (LIRFE Laboratory). His current research focuses on education facing contemporary anthropological and cultural challenges. He is cofounder of the Institutes of Formation Fondacio, of which he was the director for Europe from 1989 to 2009, and advisor for Africa, Asia and South America. He is currently the President of Fondacio.

Sufficiency

Wolfgang Sachs

Abstract The article reviews the concept of 'sufficiency' as an antidote to the expansive modernity. The notion, appreciated by the most time-honoured wisdom traditions of the world, appears to be of decisive significance for human action in confronting the Anthropocene. As a political response, sufficiency runs counter to the imperative of escalation in speed, distance, and volume in goods and services that rules (post)industrial societies. It promotes the over-all aim of living well within ecological and social limits, rejecting an unduly technical and economic optimism. The entry scans the efforts of sufficiency in energy, mobility, buildings, land, food, and data. Sufficiency is regarded as a strategy of resistance to the advent of the Anthropocene, long overdue and by now crucial for the flourishing of humans and other life forms.

In the present context, sufficiency (from Latin *sufficere*: 'to have enough strength or capacity, to be adequate') means to strive for a modest but adequate scale of living. It presents a continuous search for the right balance (Ingleby & Randalls, 2019) between the two extremes of deficiency and excess (Linz, 2004). According to the ethics of Aristotle as well as the teachings of Buddha and Confucius, the golden middle way leads to virtue in the personal and social life (Gottwald et al., 2016). Consequently, the pursuit of sufficiency recognizes two boundaries, a lower as well as an upper one, a floor as well as a ceiling (Spengler, 2016). To live by insufficient means may result in deficiency in health, food, housing, participation and a host of other human rights (cf. UNDP, 2007) - the destiny of roughly half of the world population. But to live by more means than enough may result in excess, for instance, in real estate, up-scale lifestyle, investments and a host of other items of affluence. However, the well-to-do, that half of the world population that leaves 4/5 of the human footprint on the earth, are largely responsible for the great acceleration of the emergence of the Anthropocene from 1950 onwards. Strategies of sufficiency,

W. Sachs (✉)
Wuppertal Institute of Climate, Environment, Energy, Berlin, Germany
e-mail: wolfgang.sachs@wupperinst.org

therefore, will involve the rich world rather than the poor. Wealth alleviation is more important than poverty alleviation. Remember the famous saying of Mohandas Gandhi, according to whom the world has enough for everyone's need, but not enough for everyone's greed. For that reason, this entry will focus on the second meaning of sufficiency, on reducing excess.

To live without apparent limits has been the hallmark of the industrial civilization, but to live well within limits is regarded to be the distinguishing trait of a resilient civilization. Pushing the frontiers of space and time has been the glory of science and technology, just as the accumulation of goods and services has been the triumph of economics. That type of modernity has failed, so the protagonists of sufficiency plead for a more modest society. They join the long line, reaching back to the eighteenth century, of social critics opposing the rise of the commercial society and the mechanical worldview, scholars and activists like Antonio Genovese (1712–1769), Alexander von Humboldt (1769–1859), Henry David Thoreau (1817–1862), William Morris (1834–1896) or Rabindranath Tagore (1861–1941), among many others.

However, it seems that today this counter-melody to uni-linear progress comes finally to the fore, confronting the predicament of the Anthropocene. Human agency in this epoch finds itself in a paradoxical situation (Horn & Bergthaller, 2019, 72): On the one hand, humanity gained an immense force through all kinds of technologies, strong economies, and rational mindsets that built up during centuries. On the other hand, these very same traits come along with all sorts of unintended consequences, social as well as environmental ones, leading to a profound loss of control over the techno-sphere with its infrastructures, energy regimes, and modes of production and consumption. Against this background, the protagonists of sufficiency intend to wrestle back control over aspects of the techno-sphere, just as the eco-modernists intent to do. However, eco-modernists (Asafu-Adjaye et al., 2015) put all their eggs into the technology basket, favouring artificial intelligence, geo-engineering and genome editing. They raise the flag of no limits for innovation, revolutionizing the means for achieving orthodox goals, namely domination and economic growth. In contrast, the proponents of sufficiency put forth different goals, supporting common good, justice and wellbeing in frugality. They want to subordinate and direct technical means towards these goals, welcoming limits in case a technology ruins the natural and social world. So while the proponents of eco-modernism pursue the ambition to systematically govern the Anthropocene, the proponents of sufficiency urge for an orderly retreat from this gloomy epoch.

Among eco-philosophers, there has been a significant tradition of speaking about sufficiency, in particular during the early decades of the environment movement. For instance, Ernst F. Schumacher suggested a Buddhist economy (1973), Ivan Illich acclaimed convivial tools (1973), Arne Næss coined the concept of deep ecology (1973) and Thomas Berry praised Earth-centred values (1999). With the turn of environmental thinking to more management and economics, the paradigm of resource efficiency took over, while the paradigm of resource sufficiency fell almost into oblivion, although the weakness of the former became quickly visible (Sachs, 1999; Princen, 2005). Even though the book 'Limits to Growth' became a standard reference, the talk about limits turned unfashionable during the period of neoliberal-ism. Yet, research and debate about sufficiency gained new momentum during the last decade, particularly in connection to discourses on post-growth and degrowth economics.

To carve out further characteristics of the concept, one could distinguish between efficiency, consistency, and sufficiency in resource use. Efficiency refers to a strat-egy of minimizing the resource use for a given goal, consistency refers to a change of the quality of the resource flow in order to render it compatible with natural flows, and sufficiency refers to a transformation of the goals of resource use, com-plementing the two other strategies. While efficiency and consistency are about doing things right, sufficiency is about doing the right things. Yet, unless there is no sufficiency, volume effects run danger to cancel out the desired reduction potential of efficiency and consistency. Likewise, nothing could be gained, if the sheer scale of nature-compatible resource flows would damage local and foreign ecosystems.

Sufficiency concerns the demand side and the supply side of the economy. On both sides, intelligent moderation is called for. Whereas modernity has been ruled by the imperative of escalation according to the motto 'bigger, faster, further and more', the politics of sufficiency (Paech, 2012; Schneidewind & Zahrnt, 2016) breaks with this logic, asserting that at times the opposite motto might be desirable and necessary: 'smaller, slower, shorter and less'. However, there is no royal road to sufficiency, unless welcoming collective caps on production and consumption in land, vehicles, airplanes, meat, and data. For instance, the sufficiency of buildings (Bierwirth & Thomas, 2019) offers the prospect of housing comfortably many peo-ple on limited land, combining smaller and flexible apartments with community spaces and public gardens. Ending soil sealing, incidentally, would contribute to the greatest achievement of sufficiency: leaving half of the earth protected for the diver-sity of plant and animals. Furthermore, transport sufficiency promises a life in live-able cities with less cars, giving more space to public transport as well as sharing of cars, bikes and scooters. Moreover, a truly resource-light mobility system will blun-der, unless all cars by design are eventually downsized in speed and weight, just as the jet-traffic might be largely replaced by comfortable airships and blimps. As to shorter distances, the move to economic decentralization will allow for the estab-lishment of regional value chains, supported by a distributed economy where local producers link electronically to suppliers and customers (Manzini, 2015). With regard to material sufficiency (Toulouse & Attali, 2018), whose appeal is to be inde-pendent of too many gadgets, people would use the appliances and products pru-dently, avoid over-size, be attentive to durability, and emphasize reuse and recycling. Moreover, downshifting the scale of living, a zero-option for buying products and services is always possible. Equally, for excessive food consumption. Caps for meat consumption would benefit agro-ecology, just as a ban of bottom trawling in the oceans would strengthen marine wild life. Finally, personal autonomy as well as democracy require data sufficiency (Lange & Santarius, 2020). Which limits for connectivity? Not only protection from fake-news or a preservation of privacy would be indispensable for the internet in a democracy, but also a suppression of monopolies and an exit from the marketing-driven internet, providing room for open-source applications and platform cooperatives. What is more, digital devices and services consume a huge amount of energy and raw materials. Data sobriety (Ferreboeuf et al., 2019) strive to install the least powerful devices possible, and changing them as seldom as possible, while refraining as much as possible from resource-consuming services, such as streaming films or blockchain technology. To put it into a nutshell, sufficiency aims at a double dividend: reducing resource consumption on the one hand and nurturing well-being, democracy and beauty on the other.

Expansive modernity, however, has engendered a particular understanding of freedom. Producers demand economic freedom and customers claim consumer sovereignty. Both refer to the world of goods and services, a subcategory of human pursuit of freedom. Since antiquity, freedom has meant freedom from oppression by other human beings. And according to Immanuel Kant, freedom finds its limit at another person's freedom. Yet with the advent of the Anthropocene, freedom has become threatened by a multitude of man-made natural disasters. The maximization of individual freedom, including the liberty of producers and consumers, is endangering the collective freedom of the present and future generations, not to mention the right of existence of a myriad of non-human life forms. Consequently, production and consumption in today's world must be infused by freedom and constraint, by rights and responsibilities (Fuchs et al., 2021, 32). It seems as if the narrative of liberal societies needs to be profoundly rewritten (Muller & Huppenbauer, 2016). This narrative had its origins in the four cardinal virtues, courage, prudence, justice and temperance. All the virtues had found a secular correlate, namely freedom, precautionary principle and social justice, except one: temperance. Sufficiency is the new meaning of temperance.

References

Asafu-Adjaye, J., et al. (2015). *An Ecomoderinst Manifesto*. The Breakthrough Institute.

Berry, T. (1999). *The great work: Our way into the future*. Bell Tower.

Bierwirth, A., & Thomas, S. (2019). *Energy sufficiency in buildings*. European Council of Energy Efficiency.

Ferreboeuf, H., et al. (2019). *Lean ICT: Towards digital sobriety*. The Shift Project.

Fuchs, D., Sahakian, M., Gumbert, T., Di Giulio, A., Maniates, M., Lorek, S., & Graf, A. (2021). *Consumption corridors living a good life within sustainable limits*. Routledge.

Gottwald, F.-T., Malunt, B. M., & Mayer-Tasch, P. C. (Eds.). (2016). *Die unerschöpfliche Kraft des Einfachen*. Springer.

Horn, E., & Bergthaller, H. (2019). *The Anthropocene. Key issues for the humanities*. Routledge.

Illich, I. (2001). *Tools for conviviality*. Marion Boyars. (Original edition 1973).

Ingleby, M., & Randalls, S. (Eds.). (2019). *Just enough. The history, culture and politics of sufficiency*. Palgrave Macmillan.

Lange, S., & Santarius, T. (2020). *Smart green world? Making digitalization work for sustainability*. Routledge.

Linz, M. (2012). *Weder Mangel noch Übermaß. Warum Suffizienz unentbehrlich ist*. oekom. (Original edition 2004).

Manzini, E. (2015). *Design, when everybody designs. An introduction to design for social innovation*. The MIT Press.

Muller, A., & Huppenbauer, M. (2016). Sufficiency, Liberal societies and environmental policy in the face of planetary boundaries. *Gaia, 25*, 105–109.

Næss, A. (1973). Shallow and the deep. Long-range ecology movements: A summary. *Inquiry, 16*, 95–100.

Paech, N. (2012). *Liberation from excess. The road to a post-growth economy*. oekom.

Princen, T. (2005). *The logic or sufficiency*. The MIT Press.

Sachs, W. (2015). *Planet dialectics: Explorations in environment and development*. Zed Books. (Original edition 1999).

Schneidewind, U., & Zahrnt, A. (2016). *The politics of sufficiency: Making it easier to live the good life*. oekom.

Schumacher, E. F. (2010). *Small is beautiful: Economics as if people mattered*. HarperCollins. (Original edition 1973).

Spengler, L. (2016). Two types of 'enough': Sufficiency as minimum and maximum. *Environmental Politics, 25*, 921–940.

Toulouse, E., & Attali, S. (2018). *Energy sufficiency in products*. European Council of Energy Efficiency.

United Nations Development Programme. (2007). *Thailand human development report 2007: Sufficiency economy and human development*. UNDP.

Wolfgang Sachs is a former research director of Wuppertal Institute of Climate, Environment, Energy in Germany. He was a university lecturer in the United States, Italy, England, in addition to Germany. He published, among other books, *For Love of the Automobile. Looking Back into the History of Our Desires*. Berkeley, CA: University of California Press, 1992; (Ed.) *Development Dictionary. A Guide to Knowledge as Power*. London: Zed, 1992; *Fair Future. Resource Conflicts, Security and Global Justice*. London: Zed, 2007 (with Tilman Santarius).

Part XI
Humanity as Temporality

Acceleration

Florian-Alexis Bongiraud

Abstract For most of us, in particular when we live in urban areas, each day seems to be a race against time. Acceleration is a normative frame that disturbs not only our daily life but also the ecosystem of planet earth. Hartmut Rosa's theoretical work enlightens the reflection regarding the link between acceleration as a social phenomenon and the era of the Anthropocene. Combined with other critical theories, Rosa's analyses offer an opportunity to define a post growth world that exceeds the negative impact of social acceleration.

In 2010, Hartmut Rosa, a German sociologist, inheritor of Frankfurt critical theory, wrote *Social Acceleration: A new theory of Modernity* in order to explain a significant experience of late modernity: the increase of speed in all the fields of our social life. "The pace, speed, duration and sequence of our activities and practices are almost never determined by us as individuals but rather almost always prescribed by the temporal patterns and synchronization requirements of society" (Rosa, 2011, p. 9). When we struggle with answering our emails, making sure we will spend a limited amount of time in a sports club to maintain our health or multiply the number of friends in the social networks without meeting them in a coffee place because we feel a lack of time, we experience social acceleration.

Why do we pursue the goal of increasing the amount of activities, friends and success until we reach the feeling of *burn out*? Social acceleration is intimately related to the age of Anthropocene because the modern societies impose the standards of growth, expansion and competition. In particular since the seventies, the capitalist dynamic "transforms the increase of production and productivity (…) into a systemic imperative (…)."(Rosa, 2011, p. 200). In the education, professional and personal fields, individuals aim at succeed challenges in a social structure of competition (Rosa, 2018, Chap. 4). As a result, we reinforce the volume of goals we would like to reach and compress the time available in our daily routine.

F.-A. Bongiraud (✉)
St Michel de Picpus High School, Paris, France
e-mail: fbongiraud@saintmicheldepicpus.fr

N. Wallenhorst, C. Wulf (eds.), *Handbook of the Anthropocene*,
https://doi.org/10.1007/978-3-031-25910-4_131

807

Based on a physical notion, Hartmut Rosa defines acceleration as a quantitative increase by unity of time (2018, Chap. 4). In particular since the industrial revolutions, modern societies are based on an energization logic which is stabilized to maintain the cultural and social structures of capitalism. The modern individuals must cope with three kind of social accelerations. We can consider each of them as a social standard incorporated and integrated through our lifespan in various fields such as school education, sport and leisure, work and family life.

First, technological – technical acceleration is an "intentional acceleration process driven by a goal." (Rosa, 2012) that interfere within the economy. Rosa holds that innovations increase productivity and the feeling that the present is more intense and reduced. As explained by Vostal (2018), there is no doubt that technical changes have impacted consumption and production patterns since the industrial revolution.

The innovative means of transports illustrate this change. If you want to reach Berlin from Paris, you will need 11 hours by car, around 9 hours by train and less than 2 hours by plane. If you would prefer using the plane rather than the train, you would reduce the time needed to travel but the footprint would be twenty one times higher. As an example, the modernization of transports networks in a city like Shanghai is so fast that Rosa considers "we cannot fix Shanghai in a picture, because the city never lets frozen nor immobilize." (Rosa, 2018b, p. 71). This permanent change in urban areas of China involves an irreversible destruction of natural resources and fields.

The second form of acceleration identified by Rosa (2012) is social change that relates with the early modern promise as defined by the philosophers of the age of Enlightenment during the eighteenth century. Social structures, knowledge, principles and values, internal organisation of social groups change permanently with an increasing speed. It is a disturbing experience for individuals that are forced to consider changes and reversibility as a social standard to pursue. Consumption patterns illustrate this form of acceleration. Although we can observe recently a change in ways of consuming, with organic foods or *buycott* actions for products with a low impact on the environment, the consumption is still central dynamic of the modern societies described by Rosa as "a behavior of maniac increase" (Rosa, 2018a, b, p. 40). It requires to maintain a dynamic stability of growth. For instance, the surface of the earth's expanse used by human activity intensively has grown from 10% to 25–30% (Wallenhorst, 2019).

Finally, Rosa associates the acceleration of lifestyle with a "temporal famine" Rosa (2012) that we experiment when at the end of the day we feel that we lacked of time. However, 1 day is still composed of 24 hours; but the density and layering of activities increase in the social fields. Rosa distinguishes a subjective dimension of this form of acceleration with the feeling of time pressure with an objective dimension that we can identify with the time reduction devoted to basic human activities such as fooding and sleeping. This way of life leads us to decrease our well-being and increase our consumption of goods and services that also contribute to increase the ecological footprint and disturb the ecosystem.

The combination of each form of acceleration, technological acceleration, acceleration of socialchange and acceleration of pace of life, feeds a vicious circle, named *acceleration spiral* (Rosa, 2011), that lead us to overexploit natural resources, influence deeply the climatic regime and block the earth regeneration.

The "great acceleration" became emblematic of this new geological era named Anthropocene (Wallenhorst, 2019), born in the mid 20th according to the scientist community, characterized by the impact of human activities on climate change and biosphere. In the book *The truth about the Anthropocene* (2020), Nathanaël Wallenhorst underlines the outcomes of a research realized by American scientists, including the geochemist Will Steffen. Looking at the evolution of several indicators, some related to consumptions (paper consumption, fertilizer consumption,…) and others related to global warming (carbon emissions, ocean acidification,…), they observe a significant correlation from 1950 till 2010. Named the Great Acceleration, this trend results of the overexploitation of natural resources, such as wood, water, that disturbs the entire ecosystem. Wallenhorst (2020) compares human economic activities impact on biosphere with a clock mechanism for which we would have removed a cogwheel.

When we look for solutions regarding social acceleration, we instinctively think about deceleration, which means slowing down our lifestyles, the economic growth and level of consumption. If most of individuals reduce the number of travels, the amount of meat they eat or the time spent to produce in their jobs, they may contribute to reduce growth, the foot carbon and the increasing risk of global warming. However, Rosa (2012) underlines that intentional deceleration is generally a way to reinforce acceleration in the future. For example, we take a gap year to reduce our stress level in order to be more efficient and resourceful when we will get back to work. That is the reason why, according to Rosa (Rosa, 2018a, b), resonance, the human ability to create a sensitive and mutual relation with others and the world, is a better answer to struggle with the alienation triggered by acceleration.

Resonance involves an ability to move (emotion) and be moved (affection). For example, I undertake to build a wood house (emotion) but I make sure that I don't waste primary resources because I am touched by the natural environnement (affection). However, the search for resonance in the modern world is generally misguided. In order to cope with structural competition related to social acceleration, human beings focus on experiencing resonance oasis, such as a pleasant atmosphere in our home which required to buy candles, pillows or perfumes, influenced by the internalized standards of self reification. Then, they move away from a genuine assimilation experience of the world. When Rosa deals with the environmental issue, he also shows that modern individuals trend to perceive nature as a resonance oasis by opposition with the tough reality of the professional world. In this perspective, the global warming threat can be interpreted as a collective fear to loose a resonance sphere.

At the end of his masterful reflection, Rosa reaches the conclusion that we need a political and cultural U turn to enter in a post growth society: "it is not the *access to things*, but the *quality* of the relationship to the world which must become the political and individual action standard" (Rosa, 2018a, p. 501). It requires a radical institutional change, for example by establishing a minimum universal income that would help to escape from social anxiety and allow a new relationship with the world.

According to the convivalists (2020), an active network of critical contemporary searchers, resonance is an answer to social acceleration in the era of the Anthropocene, but it is insufficient. There is a political challenge to deeply transform the social structures and institutions, it requires an ability to resist. According to the french philosopher Frederic Worms (2021), there is a genuine danger in our societies when emergency becomes a paradigm that crushes all the fields in social life. Indeed, social acceleration contributes to persuade the individuals and politicians that everything must be solved as soon as possible and it may become difficult to identify the priority emergencies. A positive manner to consider emergency is a vertical approach, an emergency with several layers, based on the political will to interfere with the main dangers for mankind and the earth planet, such as pandemics and climate change (Worms, 2021). And it involves conflicts and oppositions in a deliberative democracy based on rational discussions between equal citizens (Habermas, 1997) if we want "to give hope to move from the modern to the contemporary." (Latour, 2017).

References

Habermas, J. (1997). *Droit et démocratie*, trad. fr. C. Bouchindhomme et R. Rochlitz. Gallimard.
Latour, B. (2017). *Où atterrir ? Comment s'orienter en politique*. La Découverte.
Les convivialistes. (2020). *Résistance, Résonance. Apprendre à changer le monde avec Hartmut Rosa*. Le Pommier.
Rosa, H. (2011). *Accélération: une critique sociale du temps*. La Découverte.
Rosa, H. (2012). *Aliénation et accélération. Vers une théorie critique de la modernité tardive*. Editions La Découverte Poche.
Rosa, H. (2018a). *Résonance. Une sociologie de la relation au monde*. Editions La Découverte.
Rosa, H. (2018b). *Remède à l'accélération. Impressions d'un voyage en Chine et autres textes sur la Résonance*. Flammarion.
Vostal, F. (2018). Towards a social theory of acceleration: Time, modernity, critique. *Revue européenne des sciences sociales, 52*, 235–249. https://doi.org/10.4000/ress.2893
Wallenhorst, N. (2019). *L'Anthropocène décodé pour les humains*. Le Pommier.
Wallenhorst, N. (2020). *La vérité sur l'Anthropocène*. Le Pommier.
Worms, F. (2021). *Vivre en temps réel*. Bayard Editions.

Florian-Alexis Bongiraud is a French teacher in economics and social sciences (sociology and political sciences) at the high school Saint Michel de Picpus located in Paris. He also works as a tutor in Paris for trainee teachers. Supervised by Nathanael Wallenhorst, he will start in October 2021 a Phd in the educational sciences field in relation with the analytical framework of Hartmut Rosa to study it application of resonance experience for current and future high school students.

Catastrophe

Bruno Villalba

Abstract This article discusses how the notion of 'catastrophe' has evolved over time. Having examined the origins of the notion, it shows the transition from the desire to dominate nature in order to mitigate danger (the danger of a natural disaster), to enacting rational disaster-management strategies; then to implementing policies aimed at protection, prevention and risk management. However, in the Anthropocene, the nature of catastrophe has changed. Catastrophes are now characterised by threats on a planetary scale, systemic threats relating to social and eco-systems, and to long-term sustainability. The damage wrought is transcendental, affecting technology (chemical, nuclear, etc.), society (inequality) and the environment (climate, biodiversity, etc.). All these issues are converging, meaning we are now on the edge of a catastrophe such as the world has never seen.

A 'catastrophe' is a sudden and dramatic incident, which disrupts the normal course of events, and profoundly shifts the way in which society is organised. Such events are often the result of natural phenomena (e.g. volcanic eruptions, tsunamis, violent storms, meteor strikes, etc.). However, historically, these events have only been classified as catastrophes if humans have been affected. That is, events are labelled catastrophes, or disasters, when they have claimed lives and wreaked massive damage (to homes, infrastructure, etc.). Sometimes, the term may refer to the destruction caused by an enemy power. Now, though, in the Anthropocene, with burgeoning technological prowess (extractivism techniques, chemistry, nuclear technology, etc.), the causes of catastrophes are shifting (now, typically, they are caused by human activity). In addition, the timescales are different (a catastrophe today may well have a permanent, irreversible impact). The geographic area affected is broader than ever: potentially, they may be truly global. Finally, their consequences (for example, a nuclear holocaust) will impact both humans and non-humans (the collapse of biodiversity).

B. Villalba (✉)
AgroParisTech, Printemps, Paris-Saclay, France
e-mail: bruno.villalba@agroparistech.fr

The etymology of catastrophe is rooted in Greek (*katastrophê*, meaning a major upset) and Latin (*catastropha*). In Greek tragedy, the catastrophe is the decisive event which brings about the disastrous upheaval. It allows the audience to understand the deeper meaning behind the intrigue, and the reasons for the dire turn of events in the plot. Monotheistic religions attach very specific meaning to catastrophe. It is viewed as the expression of a deity's explicit will to teach humanity an unforgettable lesson (the Genesis Flood narrative, the destruction of Sodom and Gomorrah, etc.). The objective in these narratives is to change human behaviour in order to avoid future punishment. Thus, the catastrophe has meaning, as punishment for a transgression, and is intended to re-establish normal order in keeping with Providence. As rational knowledge (mathematics, physics, and Earth sciences) has developed, we have been able to move away from this fatalistic view of natural disasters, or the punitive perception of them, building up an objective understanding of the causes of catastrophes.

Particularly from the latter half of the nineteenth Century, political authorities have begun taking more direct action, to offer a political response to danger (notably, the danger of natural disasters), with policies aimed at *protection, prevention* and finally *risk management.*

Protection entails acceptance that a catastrophic event (a proven, known risk) is possible, though it may be unpredictable (due to lack of tools capable of gauging the situation and predicting outcomes). Protective measures aim to mitigate the effects of a catastrophe by using technology to reduce the extent of the damage (e.g. construction standards to combat fire, etc.).

Prevention comes into play when we are better able to quantify the risk of an event (i.e. its probability) and implement protocols to limit the likelihood of its occurrence (e.g. vaccination against epidemics, insurance policies, etc.). It is based on the development of experimental science (engineering, medicine, etc.), which produces new knowledge about a natural hazard, and then industrial science, which offers solutions to mitigate its effects.

Finally, *risk management* represents a rational, measured view of the existence of the danger, but also the advent of preventative measures (a speculative gain) so as not to hamper modern technical operations. Prevention and risk management encourage the development of bureaucratic administration, to oversee the management of the threat, but also liberal economic activity, to profit from the uncertainty (insurance dynamics, etc.).

This rationality offers the illusion that we can effectively control the risk of catastrophe, through technical power, rational territorial organisation, and collaboration between the State and the private sector. However, the rise of the thermo-industrial society, and its obsession with security (notably in military matters), has led to a drastic shift in the concept of catastrophe. The definition has been extended; the causes of catastrophes are now endogenous to human societies (demographic pressures, military and civilian nuclear technology, zoonotic diseases, etc.), and they now affect the entire planet (runaway climate change, the collapse of biodiversity, etc.). Worse still, the rational thinking upon which our preventative measures are founded may lead some to wreak a catastrophe of horrific proportions: the planned extermination of an entire sector of humanity. (The Hebrew term *Shoah*, שואה, means 'catastrophe', and is used specifically to refer to the Nazi Holocaust) (Arendt, 1963). Though our societies have devised strategies of resistance against such catastrophes, history shows that they have not been prevented from ever recurring. The deliberate elimination (total or partial) of a national, ethnic or religious group has indeed happened again multiple times during the twentieth Century (the same fate befalling the Armenians, the Khmers, the Tutsis and the Bosniak Muslims, for example). In the age when nuclear weapons are industrially produced and distributed, we must inevitably wonder about our ability to constantly monitor and control their use – particularly when we consider that they have indeed been deployed in war, not once but twice, regardless of the moral rectitude of that decision. We have suddenly been plunged into what is referred to as the 'end times' – the clock is ticking, counting down until nuclear weapons are deployed once again. 'In "the end times" means in that age when, every day, we could bring about the end of the world. – "Definitively" means that, for all the time that is left to us, it will always be "the end times". No longer can another time come about: the end times will end only with our end' (Anders, 2006: 116; see also Van Dijk, 2000).

In addition, by complexifying their organisation (industrial production, territorial administration, systematic globalisation of trade, etc.), our 'risk societies' are increasingly bringing about their own catastrophic situations. Catastrophes arise through the cumulative effect of all our social practices (consumption, transport, interconnection, acceleration, etc.), and affect all individuals (risk individualisation) (Beck, 1992).

Finally, when the ecological tipping points are passed (in terms of climate, biodiversity, disruption of the nitrogen and phosphorus cycles – Rockström et al., 2009), it causes irreversible imbalances, with multiple consequences for human society (climate migration, economic pressure, etc.), as well as for ecosystems and living species. Reports from international experts (IPCC, IPBES) regularly confirm the disastrous prospects we face. Nowadays, catastrophes arise from the cumulative effects of ecological, social (mainly rampant inequality) and technological threats.

Such 'transcendental harm' (Bourg, 2013) is testimony to the runaway consequences of today's catastrophic situations. No longer can we speak merely of *risks* (which are hypothetical, accidental, and refer to one-off events), but of *threats*. Such threats are characterised by their unpredictability, systemic nature, global extent, repeatability (for example, consider the Chernobyl and Fukushima disasters), but above all, their irreversibility. In addition, the problems we face require very long-term management. Thus, the dangers with which we are now confronted spark an unprecedented form of violence (where there is no discernible enemy, as is the case with runaway climate change), and a multifaceted threat (in the interconnections that form between the various types of threat). In addition, they are beyond our ability to comprehend: these dangers – be it the nuclear threat, the climate crisis or the devastating loss of biodiversity – are of 'unimaginable dimensions' (Welzer, 2017).

Various approaches have been put forward to address such imbalances. One is to improve technical means of risk management, in order to minimise the effects of any failing (broadening bureaucratic prerogative, particularly in terms of security measures; extending technical experts' freedom to take action; risk management; etc.). It is a matter of regulating uncertainty by reducing the scope for it. Another approach is to increase human interaction with the living world, to limit the risks (geo-engineering, transhumanism, etc.). Finally, there is a trend toward developing a strategy to adapt to these threats; such adaptation would require a shift in how we envisage the future of human society, and even our own place in the world. Hans Jonas (1985) proposes that we should develop an 'ethic of responsibility' that is compatible with our technophile civilisation. He sets out an ethical framework of responsibility, which would preserve an authentically human identity, and affirm the responsibility we bear for our shared future. This involves early management of the potential risk posed by technical innovations, on the basis of advances in scientific knowledge; we must examine the conditions in which new technologies should be used, and whether they can be useful from a social and ecological standpoint. Jean-Pierre Dupuy (2015) proposes to increase the ontological weight of catastrophes (what he terms 'enlightened catastrophism'). In so doing, we should be able to more fully understand the likelihood of their occurring, and thus make provisions for these eventualities as we move towards the future. Making people aware of an impending catastrophe ought to lead to a change in behaviour.

However, the implementation of these logical proposals is hampered by denial, minimisation, avoidance and inertia in decision-making, and in our perceptions of catastrophes. There is still too wide a chasm between the way in which we perceive catastrophe and our appreciation of how likely it is to occur.

References

Anders, G. ([1972] 2006). La menace nucléaire. *Considérations radicales sur l'âge atomique*. Le Serpent à Plumes.

Arendt, H. (1963). *Eichmann in Jerusalem: A report on the banality of evil*. The Viking Press.

Beck, U. (1992). *Risk society: Towards a new modernity*. SAGE Publications Ltd.

Bourg, D. (2013). Dommages transcendantaux. In D. Bourg, J. Pierre-Benoit, & K. Alain (Eds.), *Du risque à la menace. Penser la catastrophe*. PUF.

Dupuy, J.-P. (2015). *A short treatise on the metaphysics of tsunamis* (Studies in violence, mimesis & culture). Michigan State University Press.

Jonas, H. (1985). *The imperative of responsibility: In search of an ethics for the technological age*. University of Chicago Press.

Rockström, J., et al. (2009). Planetary Boundaries: Exploring the Safe Operating Space for Humanity. *Ecology and Society, 14*(2), 32.

Van Dijk, P. (2000). *Anthropology in the age of technology, the philosophical contribution of Günther Anders*. Brill.

Welzer, H. (2017). Climate wars: What people will be killed for in the 21st century. *Polity*.

Bruno Villalba is Professor of Political Science, at AgroParisTech (Paris) and Membre of Printemps UVSQ (CNRS UMR 8085). His areas of research are Political Ecology, Sustainable Development and collapse studies. His research focuses on environmental political theory, notably through analysis of the capacity of the democratic system to reformulate its goals based on environmental constraints. He is the author or co-author of over ten books on politics and ecology. Books (selection): Collapsologists and their enemies (ed. Le Pommier, 2021, in French); Energy Sobriety. Material constraint, social equity and institutional perspectives (with L. Semal, ed. Quæ, 2018, in French) and Political ecology in France (Paris, La Découverte, 2022, in French).

Future

Fabrice Flipo

Abstract There is a distinction in French between "*avenir*" (literally: "to-come" or "forthcoming") and "*futur*"; the second one differs from the first in that the latter is more distant and indeterminate, designating an era that the livings of another era will not experience. To confuse the two in an anglicism, for the French linguistic authorities. "*Avenir*" is therefore a particularly close form of future: it is an imaginable, familiar future, structured on roles and trajectories that are already known or not very different from those we can observe. Thus we can imagine a student or a child as a fireman, lawyer, mathematician or mechanic. "*Futur*" is more indeterminate, and therefore difficult to imagine. The relative monotony of science fiction narratives illustrate this fact: always robots, flying cars, spaceships and obstacles similar to fairy tales or great classical literary works – a hero, a quest, obstacles, Good and Evil. William Gibson's *Neuromancer*, for example, features a drug-addicted main character whose main quality is to be able to break through the fire-walls of the digital matrix, like a burglar. This future is remote because it is bristling with unknown objects, resulting in exotic names and surprising possibilities, such as the regeneration of organs or the ability to replace them with robotic prostheses. The "*futur*" is also distinguished by its more narrowly temporal dimension; it designates what has not yet factually occurred. It has a finalised dimension: it designates what must have a history, because it is desirable, as opposed to what will not. Thus "Un futur sans avenir" (a future without future), by the group Oblomoff, which intends to show why the future promised by scientists is neither desirable, nor even possible, and therefore will shortly disappear; or the punks' *No Future*. The nuance can also be heard in Bill Joy's famous text, Why the Future Doesn't Need Us (2000), in which he explains that "Our most powerful 21st-century technologies – robotics, genetic engineering, and nanotechnology – are threatening to make humans an endangered species" (subtitle), and therefore, in essence, why they should have no

F. Flipo (✉)
Institut des Mines de Paris, Paris, France
e-mail: fabrice.flipo@imt-bs.eu

817

N. Wallenhorst, C. Wulf (eds.), *Handbook of the Anthropocene*,
https://doi.org/10.1007/978-3-031-25910-4_133

future, which leads him to call for a moratorium. In contrast, the future is central to emancipatory thinking, insofar as, as thinkers as different as Bloch (1976) and Whitehead (1995) suggest, we aspire to harmony.

Becoming refers to time. Heidegger (1985) makes historicity one of the existentials of Dasein, of being-there; this means that written or told history presupposes an understanding of the passing of time, of the future and the past, as a succession, in the same way as being-with, anchored in culture. For Heidegger, in an analysis that may seem dated and "speciesist", the human being is the only one to exist, that is, to "ek-sist", to stand outside of itself, in temporal displacement with itself, inevitably perceiving what it has been, or what it could be, and is not yet, in displacement with what it is, or thinks it is. We discover ourselves primitively anchored in a world, moral and material, from which we come, and into which we are thrown (§58). To project, to aim, is to discover possibilities (§65), it is also to constitute ourselves. Heidegger evokes the delay of consciousness on what constitutes it; Derrida (1967) draws from this the notion of trace, a fragile sign indicating our own passage, in an uncertain way, explaining the indefinite character of any search of origin. Husserl distinguishes three types of "protentions" (anticipations of the future) and "retentions" (perceptions of the past): primary retention or immediate recollection (what I have just done, in the unity of action); secondary retention or distant, selective and constructed recollection (memories of youth or holidays); and tertiary retention, which is considered to be specific to Man because it depends on tools and more specifically on inscriptions such as texts. For Husserl, the human being constitutes time, in the sense that it is the retention by consciousness of the present that has become past (for example, a note in a melody) and the protention of the future that has not yet happened that makes time conscious. A temporality that we discover as the historical dimension of our being-there, Heidegger adds.

Hegel (1993) distinguishes three conceptions of history. The first is the simple movement of the world, an irrepressible constitutive fact of the human condition, and even of life, insofar as historical consciousness does not make history itself: it introduces a reflexivity or "specularity" on the pre-existing phenomenon. The second is history as progress in a given direction. Many authors, from Saint-Simon to Bergson (1907), Teilhard de Chardin (1955) and Stiegler (2018), via Leroi-Gourhan (1964), have characterised modern societies by their break with the cyclical conception (eternal return), in favour of a linear evolution whose direction would be that of a complexification, both in the order of life and in that of technology. This linear scale opens up a discourse of development, of 'advance' and 'delay', most often taking 'primitive societies' (Durkheim, 2008) as the starting point in the human order, and the most complex societies as the natural and (pre)destined 'omega point' (Teilhard de Chardin, 1955). The link with the economic order is obvious: what mainly progresses is trade and division of labour, the 'productive forces' (Marx, 1993) give rise to ever more specialists and specialities, including the sciences. Other authors, such as Joseph Tainter (2014) or the resilience theorists

(Gunderson and Holling, 2001), believe, on the contrary, that it is complexity itself that is cyclical, with periods of 'collapse' (simplification of systems) succeeding periods of 'capitalisation' (increase in the number of stabilised links). Hegel's third conception is history as the development of reason, the fulfilment of which he saw in the Europe of his time. And indeed some ways of writing or telling history go in this direction, such as the histories of techniques by Maurice Daumas (1962–1979) and Bertrand Gille (1978), for whom progress in the order of the increasing use of nature is accompanied by bulletins of victory, or the histories of the world focused on the West. Others, such as the Frankfurt School and later the postmodern (Lyotard, 1979), postcolonial (Saïd, 2005) and post-Marxist (Laclau and Mouffe, 2009) currents, will underline the ideological and illusory nature of this vision. Against an excessive and invasive universalism, eliminating democracy and minorities, these currents will also insist in a Nietzschean way on the link between past and future: become who you are. With the linguistic turn, the notion of narrative becomes central: who speaks, about what, in whose name, to whom, and to produce what effects?

Does the "*futur*" have an "*avenir*"? In the age of the Anthropocene, this question is being asked more than ever. The progress of foresight since the founding report of MIT to the Club of Rome (1972) may be the sign of a growing uncertainty about the future, which must be taken in hand and no longer expected as the natural result of the necessities put in place. Several narratives coexist. The dominant one continues to see the future in the form of a triumph of technology: it is Elon Musk, who has been explaining, in a very traditional way, that the next frontier is space. The second seeks to change this dynamic, which takes neither ecology nor social dimensions into account. The theories of complexity (Morin, 1977–2004) are also partly part of this attempt. Degrowth advocates seek to halt the growing complexity, to move towards low tech, flexible rather than rigid links, and adaptive technical systems rather than macro technical systems. The mobilising force of narratives and 'empty signifiers' (Laclau and Mouffe) is at the centre of their concerns, with the aim of getting individuals to project themselves outside inherited institutional destinies. Biology proposes visions close to the last two conceptions, by underlining the importance of epigenetics or through the concepts of 'punctuated equilibrium' (Gould) or 'panarchy'. Different political ideas have different relationships to historicity. Conservatism makes instrumental use of history as a factor of social order: the future is the past, and it is therefore in the past that we must seek our place in the future. Conservatism even goes so far as to take humanity out of history in favour of a supernatural (religious) order in order to achieve absolute submission. Liberalism allows for a greater degree of indeterminacy and debate about ends, which it admits are plural, and therefore, for conservatives, a factor of discord and chaos. Liberalism, however, is averse to prophecies and visions of the future that are too far removed from the established order (Popper). Emancipatory trends such as socialism or ecology, on the other hand, have a close relationship with the future, which they conceive in terms of a break with the present and the past. Duty-to-be largely prevails over being; the possibilities are open. There is a form of Pascalian wager here: there is no proof that the established order cannot be entirely changed, and there is no lack of historical examples. Today, there are three main visions of the future: salvation through technology, a change in lifestyles, and a chosen or imposed degrowth (Flipo, 2014).

References

Bergson, H. (1907). *L'évolution créatrice*. Alcan.

Bloch, E. (1976/1954). *Le principe espérance* (tome 1). Gallimard.

Daumas, M. (1962–1979). *Histoire générale des techniques. 5 tomes*. PUF.

Derrida, J. (1967). *La grammatologie*. Editions de Minuit.

Durkheim, E. (2008/1912). *Les formes élémentaires de la vie religieuse*. PUF.

Flipo, F. (2014). Les trois conceptions du développement durable, *Développement Durable et Territoire*, Vol 5, n°3, 5 décembre 2014. http://developpementdurable.revues.org/10493

Gille, B. (1978). *Histoire des techniques*. Pléiade.

Gunderson, L., & Holling, C. S. (2001). *Panarchy: Understanding transformations in systems of humans and nature*. Island Press.

Hegel, G. W. F. (1993). *La raison dans l'histoire (1822)*, UGE 10–18.

Heidegger, M. (1985). *Être et temps (1927)*. Gallimard.

Laclau, E., & Mouffe, C. (2009/2001). *Hégémonie et stratégie socialiste*. Les solitaires intempestifs.

Leroi-Gourhan, A. (1964). *Le geste et la parole*. Albin Michel.

Lyotard, J.-F. (1979). *La Condition postmoderne: rapport sur le savoir*. Minuit.

Marx, K. (1993). *Le capital (1867)*. PUF.

Morin, E. (1977–2004). *La méthode – 6 tomes*. Seuil.

Saïd, E. (2005/1978). *Orientalisme*. Seuil.

Stephen Jay Gould. (2006). *La Structure de la théorie de l'évolution*. Gallimard.

Stiegler, B. (2018). *La technique et le temps*. Fayard.

Tainter, J. (2014/1990). *L'effondrement des sociétés complexes*. Le retour aux sources.

Teilhard de Chardin, P. (1955). *Le phénomène humain*.

Whitehead, A. N. (1995). *Processus et réalité* (1929). Gallimard.

Fabrice Flipo works at Institut Mines-Télécom, Professor of political philosophy and philosophy of science and technology, researcher at the Laboratory of Social and Political Change (University of Paris). Author of, among others: Nature et politique (Amsterdam, 2014), Réenchanter le monde (Le Croquant, 2017), Écologie autoritaire (ISTE, 2018).

Great Acceleration

J. R. McNeill

Abstract This chapter reviews the origins and meaning of the concept The Great Acceleration. In the sense now current in global change science and Earth System science, and increasingly current in environmental history, the term was coined in 2005. It refers to the rapid expansion of various socio-economic driving forces, including energy use, population growth and economic growth, and the resulting rapid alteration of key indicators of the Earth System such as climate, atmospheric chemistry, ocean chemistry, and biodiversity.

'The Great Acceleration' is a term coined in 2005 in Dahlem, outside Berlin, at a workshop devoted to understanding the human-driven changes to the global environment on various time scales (Hibbard et al., 2007). It refers both to the enormous and unprecedented rate of growth in energy use, human population, the global economy since World War II, and other anthropogenic forces; and to the simultaneous enormous and unprecedented scale, scope, and pace of anthropogenic environmental change. That environmental change attained sufficient intensity to alter some of the Earth's governing biogeochemical systems such as the carbon cycle, the nitrogen cycle, the water cycle, and the climate, which together account for a large share of the Earth System. The term, by design, echoes the title of Karl Polanyi's 1944 book *The Great Transformation*. The parallel between, on the one hand, Polanyi's linked transitions in law, politics, social behavior, technology, and the organization of economic production (mainly the factory system) in Britain in the eighteenth and nineteenth centuries and, on the other hand, the linked transitions in the trajectory of the Earth System and socio-political-economic systems inspired the specific choice of words.

The earth science and global change science community came to adopt the term, thanks largely to the work of Will Steffen and colleagues in the International Geosphere-Biosphere Program (Steffen et al., 2007). To a lesser degree, the term

J. R. McNeill (✉)
Georgetown University, Washington, DC, USA
e-mail: mcneillj@georgetown.edu

N. Wallenhorst, C. Wulf (eds.), *Handbook of the Anthropocene*,
https://doi.org/10.1007/978-3-031-25910-4_134

found favor with social scientists and historians concerned with modern environmental change (e.g. McNeill & Engelke, 2014). Unlike the term and concept of the Anthropocene, the Great Acceleration has not attracted widespread criticism except from a mathematical perspective (Nielsen, 2021).

The argument for a great acceleration in global population, energy use, and economic production after 1945 (or 1950) is irrefutable. Human population reached a billion about 1820; then 2.5 billion by 1950, 6.1 billion by 2000 and will likely attain 8 billion by 2023. Human energy consumption, although known with less precision than population size, shows a similar ascent: 41 exajoules (EJ) in 1900, 100 EJ by 1950, then 377 EJ in 2000 and 525 EJ in 2020. In the years 1900–1950, the size of the world economy expanded about fourfold, but between 1950 and 2000 by another eightfold, and it has doubled once more since 2000. All these trajectories are unprecedented, indeed never remotely approached, in the prior demographic, energy, or economic histories of humankind. One can find broadly similar trajectories in global freshwater use and the production of cement, plastic, iron, salt and many other commodities (Syvitski et al., 2020).

As a result of this Great Acceleration in the socio-economic systems of humankind since 1950, the Earth System has undergone sharp changes that in many cases are unprecedented in the Holocene (the last 11,700 years), and in some cases in millions of years. Average global temperature has climbed, especially since 1980, at a rate not seen in at least 10,000 years and perhaps for much longer than that. Sea surface temperatures have risen as well, almost 1 degree C since 1950. Sea level is now almost 15 cm higher than in 1950. Atmospheric methane (a powerful greenhouse gas) concentrations have almost doubled since 1950, from 1162 parts per billion (ppb) to 1889 ppb. Carbon emissions have sextupled (1950–2020), producing carbon concentrations a quarter again as large as those of 70 years ago. The size of dam reservoirs has increased 20-fold in the same span. Ocean acidity has increased by about one quarter since 1950. Tropical forest area has declined, due to timber cutting and agricultural expansion, by at least half a billion hectares since 1950, equivalent to about half the surface area of China or the USA. By any measure, biodiversity is declining at rates that eclipse anything outside the five great extinction events on the history of life on Earth (Syvitski et al., 2020; IGBP, 2015; McNeill & Engelke, 2014).

The Great Acceleration therefore amounts to a great destabilization of the Earth System. The Holocene, after the exit from the last glaciation, offered remarkably stable conditions with respect to climate. That proved a favorable circumstance for the development of domestication, farming, complex society, cities, states and all the familiar master institutions of the human community. The significance of The Great Acceleration, therefore, includes the possibility that the Earth System will escape the envelope of (minor) variability that has characterized the Holocene. It might overtop certain unknown thresholds, create positive feedback loops that push it – perhaps irretrievably on any time scale relevant to the human career on Earth – into unfamiliar states. That, of course, remains to be seen and may well be avoidable if the human species can manage to stop The Great Acceleration (see below).

Should the Earth System escape the bounds of the Holocene, human communities will be hard-pressed to adapt and, of course, some much more so than others.

People inhabiting semi-arid lands (likely to become more arid) or those living in seacoast deltas (likely to be submerged by rising sea levels) will have few or no agreeable options. Migration will likely appear to them as the least bad possibility.

What is the relationship between The Great Acceleration and the Anthropocene? If one adopts the definition of the Anthropocene favored by the Anthropocene Working Group, the body charged with making a recommendation to the International Union of the Geological Sciences as to whether or not the geological time scale should be revised so as to include the Anthropocene, then the Great Acceleration and the Anthropocene began at the same time: the middle of the twentieth century. (Zalaciewicz et al., 2015). To date, they are contemporaneous. But not for long. The Great Acceleration will end, indeed is already ending, while the Anthropocene will last for millennia at least.

The strongest driving forces of the Great Acceleration, mounting fossil fuel use and global population growth, are unlikely to endure for another century. The rate of growth of both have declined since the 1960s. With respect to fossil fuel use, that decline has been choppy and brief returns to rapid growth, in 2002–2008 for example, have occurred. But the secular trend is one of slowly declining rate of growth, and now that solar and wind power are cheaper per kilowatt hour than electricity generated with oil, coal, or natural gas, and likely to become cheaper still, the odds are strong that fossil fuel use will grow more slowly with each passing year, and eventually cease. After that point – peak fossil fuel – growth will be negative and use will diminish, although how quickly is anyone's guess.

With respect to population, the growth rate today is about half what it was at its peak in 1968–72. Demographers expect it to decline further, and typically – but not always – predict that it will cease in another 50 or 70 years (Adam, 2021), after which growth will be negative and total population will decline. So the Great Acceleration has begun to slow, and will surely end. The Anthropocene will endure because the human impact on the Earth's governing systems will endure. Greenhouse gases already emitted will affect climate and ocean acidity for millennia into the future. Moreover, the human imprint on the biosphere, in the form of reduced biodiversity, will last far into the future, as long as the rate of the creation of new species lags far behind the rate of anthropogenic extinctions. And of course those species now extinct, and those soon to become so, will never return. The plastics, asphalt, and concrete now scattered around the planet will last for centuries or millennia. So the Anthropocene will last long after The Great Acceleration has been relegated to the dustbin of history.

References

Adam, D. (2021). How far will global population rise? *Nature* www.nature.com/articles/d41586-021-02522-6

Hibbard, K., et al. (2007). Decadal-scale interactions of humans and the environment. In R. Costanza, L. J. Graumlich, & W. Steffen (Eds.), *Sustainability or collapse? An integrated history and future o fpeople on earth* (pp. 341–375). MIT Press.

IGBP [International Geosphere-Biosphere Program]. (2015). igbp.net/globalchange/greataccelera
    tion.4.1b8ae20512db692f2a680001630.html
McNeill, J. R., & Engelke, P. (2014). *The great acceleration*. Harvard University Press.
Nielsen, R. W. (2021). Anthropogenic data question the concept of the Anthropocene as a new
    geological epoch. *Episodes: Journal of International Geoscience*. https://www.episodes.org/
    journal/view.html?doi=10.18814/epiiugs/2021/021020
Steffen, W., Crutzen, P. J., & McNeill, J. R. (2007). The Anthropocene: Are humans now over-
    whelming the great forces of nature? *Ambio, 36*, 614–621.
Syvitski, J., Waters, C. N., Day, J., et al. (2020). Extraordinary human energy consumption and
    resultant geological impacts beginning around 1950 CE initiated the proposed Anthropocene
    epoch. *Nature Communications: Earth & Environment, 1*. https://doi.org/10.1038/
    s43247-020-00029-y
Zalaciewicz, J., et al. (2015). When did the Anthropocene begin? A mid-twentieth century bound-
    ary level is stratigraphically optimal. *Quaternary International, 383*, 196–203.

J. R. McNeill, University Professor at Georgetown University, has authored or edited 23 books, including *Something New Under the Sun* (2000), listed by the London *Times* among the 10 best science books ever written (despite being a history book); and *Mosquito Empires* (2010), which won the Beveridge Prize from the American Historical Association; and most recently *The Webs of Humankind* (2020), 2 vols. In 2018, he received the Heineken Award for History from the Royal Netherlands Academy of Arts and Sciences. He is a former president of both the American Society for Environmental History and the American Historical Association; and an elected member of both the American Academy of Arts and Sciences and the Academia Europaea.

Kairos

Emmanuel Nal

Abstract This article examines the *kairos*, which is both the Greek notion and mythological figure of opportunity. Everyone is looking for different opportunities during the life, and this study is an occasion to think about timeliness and appropriateness, as paths to success. This article also links kairos with serendipity, and describe the components of a kairos intelligency, to enlight the modern stakes of this notion.

Acting at the right time, understanding what is best to do, to realize it in time: the difficulty of reconciling a measured action, adapted to a context, with the right moment can seem so wonderful that it could be considered as a happy coincidence. But yet, "The Greeks have a name for this coincidence of human action and time, which makes the time favourable and the action good: it is the καιρός, the favourable opportunity, the opportune time" (Aubenque, 1963, p. 96–97). However, the term suggests as many translations as it offers nuances throughout its history: it connotes at first, in Homer's works, "the decisive place", even the "flaw" and it is only later that it assumes a temporal value with the meaning of "the favourable moment", another qualitative nuance of time distinct from *chronos* and *aiôn*. By taking the meaning of "opportunity", it operates more than a lexical evolution since it integrates the idea of effective action (whether it is a technical, rhetorical, artistic, strategic or moral action), inscribing it in a space-time (in the right place, at the right time), according to a desired purpose (Balibar et al., 2013). The difficulty of limiting the kairos to a single definition is in itself eloquent; it must be conceived as a *notion* which manifests the complexity of its object and raises its epistemological – and major – problem: what can we know about this opportunity and more generally about its timeliness and appropriateness? The constitution of a science requires a constancy of the phenomenon studied, but the opportunity is unstable, it escapes from any model because of its relativity: first, the opportunity I perceive is not

E. Nal (✉)
University of Upper Alsace, Mulhouse, France
e-mail: emmanuel.nal@uha.fr

necessarily one for others; moreover, according to the circumstances, I will not be sensitive to the same forms of opportunities in my own life.

In the *Charmide*, Plato (2004) encounters a similar problem in trying to bring wisdom back to an essence – in other words, to a unique quality – and all attempts prove insufficient. Restraint and prudence may be relevant ways to act, but not in all cases, not in a universal way (if I refrain from acting when my friend needs to be saved, this would not be wise or appropriate). Ever-changing, the kairos asks us to stay attentive to beings, to temporalities, to the possibilities that open or obtain, according to the moments, each thing bearing its own measure (Pindare, 1970). This concern to stay attentive to occasion, called "καιροφυλακία" in ancient Greek, is illustrated, for example, by the character Cyrano de Bergerac (Act IV, scene 10 of the eponymous play): in the imminence of a battle, leaving him a faint hope of survival, he thinks this moment is the last possibility to confess his love to Roxane, but learning of Christian's death (Roxane's husband) he realizes that "It's over, I can never say it again!" out of respect for the memory of the deceased. In order to hope to grasp the kairos, it is necessary to be ready, by careful attention and by the ability to be "inconstant in a good way" knowing how to judiciously embody opposites. In the *Eudemian Ethics*, Aristotle (1984) denounces those who are inconsistent in a bad way "ὅλως ἀνώμαλοι κακῶς" (1234b): when the danger is not there, they boast, and when it happens, they flee. To be inconstant in a good way is therefore, according to the circumstances, to know how to be lively or being able to postpone the action. In Hellenic culture, the kairos is at the center of many recommendations; Hesiod, a Greek poet of the eighth century BC, does not hesitate to write, in *Works and Days*, that "In all things, kairos is what is the best" (2014, verse 694) and in *The Game of the Seven Sages*, composed around 390 AD, the Latin poet Ausone attributes to Pittacos of Lesbos, one of the mythical Sages of Greece in the sixth century BC, the maxim "Γίγνωσκε καιρόν" ("Get to know the kairos!"). However, an important question is how can we follow such prescriptions if we do not know what this kairos is?

The epistemological problem turns into a pedagogical problem, and this is perhaps what explains the recourse to allegory, with the appearance of the mythological figure of Kairos. He's the last son of Zeus, and he wanders around flying, the only lock of hair on his forehead offering the possibility, to the one who knows how to do it, to "seize the opportunity by the hair". In the most famous representation of the young god by the sculptor Lysippus (fourth century BC), his hand holds a razor on the thread of which rests the balance arm of a pair of scales, of which he tilts one of the pans by placing his finger upon it. The reference is already humanistic, since it shows that opportunity emancipates from any predestination, for two reasons. On the one hand, the race run by Kairos does not follow any given path, he does not seek to encounter anyone and does not let himself easily be captured; it is therefore a merit which belongs entirely to the man able to perceive his passage and to manage to seize it – in time. On the other hand, his influence on the pans of the scales (symbol of Destiny in Ancient Greece) reminds us that "opportunity is something we must learn to use" (Jankélévitch, 1980, p. 119) and that man is responsible for the "career of an occasion" (*Ibid.*). The last sentence of the epigram composed by Posidippe (third century BC), where a traveller has a dialogue with the Kairos statue by Lysippus, confirms the pedagogical nature of the allegory, when the statue declares that it is there "For you, a foreigner, [the sculptor] has placed me at the entrance to instruct you" ("*διδασκαλίην*") (Trédé, 1992, p. 78). In what Ricœur calls "the odyssey of freedom throughout the world of works", man must discover how to move from the virtuality of his projects, which presupposes a certain belief in his power of realization, to their inscription in "real history" (Ricœur, 1990). In this immemorial quest for meaning, there are sometimes *kairoi* in the event, but how can we recognize what we have never seen? How can we achieve in action that balance of right measure, space and time which seems to guarantee its effectiveness, and for which we are asked so often, without ever being taught how to achieve it?

Focused on various future events, the man of the twenty-first century lives in times that are still not his. In addition to the phenomenon of "acceleration" described by A. Rosa, the era requires him to be "the hero of his project" (Boutinet, 2015), or at least to be "open to new opportunities", and also to know how to manage periods of crisis (a particularly expected competence), to anticipate situations in order to make the most of them – and we are not far from being considered to be at fault if we have not mastered these qualities proper to "the Kairic man" (Moutsopoulos, 1991) – who is quite different from the opportunist. Whether it is the temporalities associated with individual life journeys, as well as those of the planet with its environmental issues, discourses exacerbate "the not-yet" and the "never-more", to the point that the hic et nunc action, characteristic of the kairos to be grasped, is increasingly promoted and even dramatized – "after, it will be too late". How, and in what forms can the kairos help to meet human needs in the light of contemporary demands? The areas likely to promote the development of a "kairos intelligence" can be divided into at least four realms (Mencacci, 2002; Nal, 2012, 2017, 2019), such as the ability to invest the possibilities and perceive the semiotics of meaning – signs that there is something here and now that is worth trying.

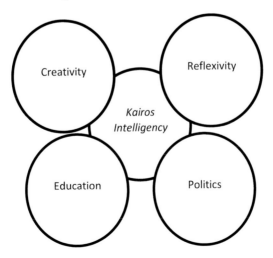

In addition to the mystery that surrounds inspiration and the appearance of the occasion, both require the subject's readiness to what comes: how can I perceive interest in encounters that I might make if I am not even open to the interest that they can present? The intelligence of the kairos encourages the attentive reception of the idea that comes and the possibility offered, opening up to a sensitivity of the potential of the situations (Jullien, 1997).

It is a condition for serendipity – this art of discovering what one does not necessarily seek, which is at the origin of many discoveries (Bourcier & Van Andel, 2010). The creative process also implies a "why not?" attitude, a "permission" that we must give to ourselves, through which one becomes an author and which is indispensable because one must be determined to seize the opportunity. What we call reflexivity is the ability to take an analytical look at our personal experience as well as our practices. The process of autobiographical narrative or life story, makes it possible to consider *a posteriori* the pivotal moments of existence and to become aware of what one has been able to achieve: it is an encouragement in transitional periods (occupational reorientation, diseases, etc.), a reminder of what our power to act can achieve and a supervision that helps to reconstitute a certain frame of existence – one of the meanings of the term derived "καῖρος" is precisely the "weft thread". From an educational perspective, it is a question of thinking of problem situations in order to make learners able to seize the opportunities of their autonomy (see in particular Mencacci, 2002); the educator is also the one who knows how to maliciously compel the kairos to work for him. "In fact, the opportunity is a grace that sometimes needs to be helped in a sneak way and therefore is not entirely gracious. (…) This is how everything can become an opportunity for a verve consciousness, capable of fertilizing chance and making it operative" writes Jankélévitch (1980, p. 128), and one might imagine that the training of educators could include an awareness-raising element/exercise– and maybe some form of training – in searching for these opportunities. From a political point of view, the kairos can emerge from a concerted effort and put into perspective the question of what a "collective opportunity" can be. Habermas has shown the interest but also the inherent difficulties of what can arise during an exercise of "ethical discussion"; but if the kairos is marked by the effective coincidence of a moment, a place and an action, it can also assert itself as a matrix of history when it marks the convergence of a collective capable of agreeing on a project of society (Ackerman, 1998).

References

Ackerman, B. (1998). *Au nom du peuple. Les fondements de la démocratie américaine.* Calmann-Lévy.

Aristote. (1984). *Ethique à Eudème.* Vrin.

Aubenque, P. (1963). *La prudence chez Aristote.* PUF.

Balibar, F., Büttgen, P., Cléro, J.-P., Collette, J., et Cassin, B. (2013). Moment, instant, occasion. *Trivium*, 15.

Bourcier, D., et Van Andel, P. (dir.). (2010). *La sérendipité. Le hasard heureux.* Hermann.

Boutinet, J.-P. (2015). *Grammaire des conduites à projet*. PUF.

Hésiode. (2014). *Théogonie – Les Travaux et les Jours – Le Bouclier*. Les Belles Lettres.

Jankélévitch, V. (1980). *Le Je-ne-sais-quoi et le Presque-rien. I. La manière et l'occasion*. Seuil.

Jullien, F. (1997). *Traité de l'efficacité*. Grasset.

Mencacci, N. (2002). Une approche de la pensée Mètis – la ruse –, et de l'intelligence du kaïros – l'occasion, l'à-propos –, chez le formateur qui agit en situation. Paris: 6ème Biennale Internationale des chercheurs et des praticiens de l'éducation et de la formation.

Moutsopoulos, E. (1991). *Kaïros, la mise et l'enjeu*. Vrin.

Nal, E. (2012). *Kaïros, l'irruption du sens au cœur de la complexité humaine: le joueur, le médiateur, le stratège*. Thèse de doctorat, Université de Paris-8, Saint-Denis.

Nal, E. (2015). Le kaïros dans les Histoires de vie. Chemins de formation au fil du temps, 19, 93–104.

Nal, E. (2017). Temps opportuns en éducation: perspectives sur le « bon » moment. Fribourg; Colloque international SSRE « Les temps de l'éducation et de la formation ».

Nal, E. (2019). Du kaïros à la médiation, petite réflexion sur l'occasion et le bon moment d'agir. *Tiers, 24*, 23–36.

Pindare. (1970). *Olympiques*. Les Belles Lettres.

Platon. (2004). *Charmide*. GF-Flammarion.

Ricœur, P. (1990). Avant la loi morale, l'éthique. *Encyclopaedia Universalis, Supplément I*, 62–66.

Trédé, M. (1992). *Kaïros, l'à-propos et l'occasion (le mot et la notion, d'Homère à la fin du IVe siècle avant J.-C.)*. Klincksieck.

Emmanuel Nal is a Philosopher, Senior Lecturer in Educational Sciences at the University of Upper Alsace. Researcher at LISEC (UR 2310), Associate researcher at MAPP (EA 2626). His research focuses on perception and ways of seizing opportunities, transformative time-spaces (heterotopias, thresholds, transitional spaces) and invention processes. These studies allow him to consider pedagogical practices and accompanying postures.

Mobility

Philippe Hamman

Abstract This entry addresses social and spatial mobility in contemporary societies, and in doing so examines a wide array of concrete practices (commuting, residential mobility, travel for leisure and also migrations) on an individual and collective level. Access to mobility is subject to social stratification, and spatial mobility is often disconnected from social mobility. For the future, in the face of intensifying globalization, four processes have special implications in terms of the societal stakes of mobility, namely, the relationship to living environments, space-times, sustainability and borders.

In contemporary Western societies, mobility is commonly associated with values of freedom and emancipation, reduced spacetime, and social fluidity (Kaufmann et al., 2004). For policy-makers, it is also an indicator of socio-economic dynamism, facilitating changes of spaces and statuses. This is mobility as social mobility, generally understood by sociology as a double movement of the individual in social space: from social and familial background to diploma, and then from diploma to occupation (Pfefferkorn, 2017), although we know that class membership and social recognition are not derived solely from formal qualifications and paid labour (Van den Berg, 2011). Hence the existence of a "right to mobility" linking mobility to the notion of citizenship, for accessing other rights – work, housing, education, health, etc. (Cresswell, 2006). This postulate is however based on a doubly reductive equation. First, individual behaviours are normatively attributed to a rational actor model wherein economic benefits and time spent on tasks are optimized, leading to the perception that spatial mobility is indispensable for self-fulfilment. Second, different forms of mobility are usually studied separately: commutes (journeys between home and work, etc.), residential mobility (changes in accommodation and living settings), travel for leisure (tourism) and migrations (on various scales and with various degrees of finality).

P. Hamman (✉)
University of Strasbourg, Strasbourg, France
e-mail: phamman@unistra.fr

Concurrently, there is the question of differences in access to these mobilities – in other words, socio-spatial inequalities, which we know to be cumulative, not compartmentalized (Pfefferkorn, 2017; Warwick-Booth, 2018). Members of privileged groups can select the location of their residence (as is illustrated by the tight-knit world of bourgeois neighbourhoods), of their workplace (see the example of long-distance high-speed train commutes: Öhman & Lindgren, 2003) and of their leisure activities (various destinations made "closer" by air travel). Conversely, mobility is often imposed on working-class individuals (asked to move to attend a training session, find a job, etc.), who regardless sometimes find themselves locked in suburban housing estates. According to Jacques Donzelot (2006), mobility is a structuring factor of the "three-speed city", which is fragmented by processes of marginalization (confinement in "sensitive" suburban neighbourhoods), periurbanization (the middle-class residents of suburban housing developments who commute by car) and gentrification (with the "Bobos" or "bourgeois-bohemians" moving into historic city centres and more generally the embourgeoisement of working-class neighbourhoods and even peri-urban or rural areas).

Mechanically associating mobility and the capital-endowed classes vs. sedentariness and the poor would be misguided. Indeed, wealthier groups have been known to root themselves in specific places, like the endogamous residents of US gated communities (Vesselinov et al., 2007), and the French working classes have been found to opt for "autochthonous social capital" (Retière, 1994) instead of paying the financial (car, gas), familial and social costs of mobility. The possession of such roots can also appear as a condition of mobility, between urban and rural spaces, metropolis and hinterland, between a residential space and a favoured "environment" that is difficult to leave, and amenities and work opportunities pursued elsewhere. One can move to take root: in such cases, spatial mobility is disconnected from social mobility. Inhabiting, as an anthropological fact, and being mobile are not mutually exclusive. For instance, residential multilocality (i.e., residing in several places) renews relationships to territoriality, in owners of secondary residences, commuters, students and children in joint custody (Weichhart, 2015). The increase in social, spatial and virtual forms of mobility (the latter facilitated by information technologies) does not result in an a-territorial "global citizenship". Alain Tarrius (2010) has shown this in his research on "circulatory territories", spaces of mobility where an informal economy develops that is both globalized (as in the trans-Mediterranean migration route) and localized (in some urban neighbourhoods).

By the 1920s, Chicago School sociologists had already made a distinction between mobility and displacements. Roderick D. McKenzie (1927) introduced a distinction between "mobility", which changes the individual's identity (change of residence, change of employment), and "fluidity", which actually relates to stability (going on an errand, commuting to and from work). Based on the observation that one can travel without being mobile, Vincent Kaufmann (2011: Chapter 2) identified three dimensions of mobility, which are not necessarily interlinked: the "field of possibilities", a societal context that can be more or less favourable for actors; the aptitude to move (in geographical, economic and social space) or "motility", at the crossroads between supply accessibility, skills and actors' appropriations; and movement or displacements in space. These give way to three concrete situations: "displacement + mobility", i.e. jointly spatial and social mobility; "frequent displacements + low mobility", like the corporate CEOs who circle the world but always remain in the same social universe (Sassen, 2001); "no displacements + mobility", especially thanks to widespread Internet use.

What does this entail for the future in the face of growing globalization? Four processes arguably have special implications in terms of the societal stakes of mobility, namely, the relationships to living environments, spacetimes, sustainability and borders.

Along with the development of a planetary urban system, the association between city and mobility has been strengthened, including by internationalized networks of "global cities" (Sassen, 2001). On another level, the role of cars in suburban sprawl has often been pointed out. Jacques Donzelot (2006) has noted a "forced hypermobility" of car-dependent residents to combine life in individual, low-rise suburban housing, a source of social distinction and proximity to "nature", with work, requiring increasingly long commutes, as well as the children's school and urban services.

This ties in with the problem of speed in a world characterized by receding physical distances thanks to the potentials of spacetimes – with cars, high-speed trains or planes – and the simultaneity of communication technologies. Combined with mobility, acceleration has become a prized social value (Rosa, 2010), leading to the formulation of a "right to speed" in a growth-oriented economy, which has in turn been challenged by the practices of the "slow" movement, promoting slower paces of life.

Indeed, mobility cannot escape calls for sustainability. This is true in periurban areas, which are a focus of recurring debates on the "dense city", although the carbon footprint of periurban residents is not necessarily higher than that of their urban counterparts if one considers not only commutes but also long-distance mobility and travel for leisure. This is also true for the so-called "sustainable", "soft" and "shared" mobility policies that encourage a shift from car use to public transit, biking and walking. The comeback of trams in major cities evidences the social implications of mobility, including accessibility for the elderly and disabled (low floors, etc.), for underprivileged categories (should social pricing be implemented on the basis of status – student or unemployed – or income?), impact on urbanities and social bonds, if not local development (connections between city centres and marginalized peripheral areas), and costs (serving less busy routes with dedicated bus lanes) (Hamman, 2015).

Lastly, does mobility mean the end of borders, especially national ones? Although John Urry (1999) claims we have entered a "post-societal era", organized around horizontal networks, mobilities and fluidities, dynamics of reconfiguration and hybridization are visible. This is exemplified by the 1.5 million cross-border workers in the EU-28 (2019 Annual Report on intra-EU Labor Mobility). They are defined by their residence and employment in two adjacent national spaces, between which they commute on a daily or weekly basis (for instance between Alsace and Bade-Württemberg or Lorraine and Luxembourg). Like the stranger in the work of

Georg Simmel (1950 [1908]), they have ties to a (local, professional) group and are not simply "passing through" like tourists, but they were not originally a part of that group (as they came from the neighbouring country for a job), nor are they fully a part of it (depending on the modalities of European coordination, they depend on one country or the other), or at all times (they go back and forth between home and their workplaces), meaning that they embody peer-to-peer relationships with the other, beyond the physical distances they travel. In the process, in-between territories materialize, structured both by the constraints (such as avoiding double taxation) and opportunities (such as securing a higher wage) inherent in these flows (Hamman, 2008, 2019).

This article received support from the Maison Interuniversitaire des Sciences de l'Homme – Alsace (MISHA) and the Excellence Initiative of the University of Strasbourg, France. The author wishes to thank Jean-Yves Bart for translating from the original French.

References

Cresswell, T. (2006). The right to mobility: The production of mobility in the courtroom. *Antipode, 38*(4), 735–754.

Donzelot, J. (2006). The three-speed City: Marginalisation, periurbanisation, gentrification. In B. Stiftel, V. Watson, & A. Henri (Eds.), *Dialogues in urban and regional planning* (Vol. 2, pp. 103–126). Routledge.

Hamman, P. (2008). Legal expertise and cross-border workers' rights: Action group skills facing European integration. *International Journal of Urban and Regional Research, 32*(4), 860–881.

Hamman, P. (2015). Negotiation and social transactions in urban policies: The case of the tramway projects in France. *Urban Research and Practice, 8*(2), 196–217.

Hamman, P. (2019). Borders and cross-border cooperation in Europe from a sociological perspective. In J. Beck (Ed.), *Transdisciplinary discourses on cross-border cooperation in Europe* (pp. 123–145). Peter Lang.

Kaufmann, V. (2011). *Rethinking the City: Urban dynamics and motility*. EPFL Press.

Kaufmann, V., Bergman, M. M., & Joye, D. (2004). Motility: Mobility as capital. *International Journal of Urban and Regional Research, 28*(4), 745–756.

McKenzie, R. D. (1927). Spatial distance and community organization pattern. *Social Forces, 5*(4), 623–627.

Öhman, M., & Lindgren, U. (2003). Who are the long-distance commuters? Patterns and driving forces in Sweden. *Cybergeo: European Journal of Geography*. https://doi.org/10.4000/cybergeo.4118

Pfefferkorn, R. (2017). The sociological discourse on inequality and social class in France. In M. C. Jones & D. Hornsby (Eds.), *Language and social structure in urban France* (pp. 35–45). Routledge.

Retière, J.-N. (1994). *Identités ouvrières. Histoire sociale d'un fief ouvrier en Bretagne. 1909–1990*. L'Harmattan.

Rosa, H. (2010). *Alienation and acceleration. Towards a critical theory of late-modern temporality*. NSU Press.

Sassen, S. (2001). *The Global City*. Princeton University Press. (Original edition 1991).

Simmel, G. (1950). translated and edited by Wolff, Kurt H. *The Sociology of Georg Simmel*. Glencoe, Ill.: The Free Press, Part 5: The Stranger (pp. 402–408). (Original edition 1908).

Tarrius, A. (2010). 'Circulatory territories' and the urban stages of Transmigrants. *Regards croisés sur l'économie, 8*, 63–70.

Urry, J. (1999). *Sociology beyond societies. Mobilities for the twenty-first century*. Routledge.

Van den Berg, M. (2011). Subjective social mobility: Definitions and expectations of 'moving up' of poor Moroccan women in the Netherlands. *International Sociology, 26*(4), 503–523.

Vesselinov, E., Cazessus, M., & Falk, W. (2007). Gated communities and spatial inequality. *Journal of Urban Affairs, 29*(2), 109–127.

Warwick-Booth, L. (2018). *Social inequality*. Sage. (Original edition 2013).

Weichhart, P. (2015). Residential multi-locality: In search of theoretical frameworks. *Tijdschrift voor Economische en Sociale Geografie, 106*(4), 378–391.

Philippe Hamman is a Professor of urban and environmental sociology, affiliated with the Institute for urbanism and regional planning (IUAR) and the laboratory Societies, Actors and Government in Europe (SAGE) at the University of Strasbourg. His recent publications include: *Cross-border Renewable Energy Transitions: Lessons from Europe's Upper Rhine Region* (ed., Routledge 2021). *Sustainability Governance and Hierarchy* (ed., Routledge 2019). *Sustainability Research in the Upper Rhine Region. Concepts and Case Studies* (ed. with S. Vuilleumier, Presses Universitaires de Strasbourg 2019). *Rethinking Nature. Challenging Disciplinary Boundaries* (ed. with A. Choné, I. Hajek, Routledge 2017).

Progress

Jörg Zirfas

Abstract The term "progress" is a new concept of the Enlightenment. It is an optimistic term that describes the development towards a better future, especially against the background of technical, industrial and scientific innovations. The history of Europe has been dominated by the "myth of progress" (Sting S (1991). Der Mythos des Fortschreitens. Zur Geschichte der Subjektbildung.Reimer, Berlin) since the Enlightenment at least. Progress means, first of all, in a formal sense, perceiving things in a specific way, i.e. interpreting change in a specific way, taking a specific direction in doing so, emphasizing a specific future, and ultimately seeking to realize a specific goal in the future, although progress is not identical with the achievement of the goal (Spaemann R (1983) Unter welchen Umständen kann man noch von Fortschritt sprechen? In Philosophische essays, 130–150. Reclam, Stuttgart).

Historically, the self-empowerment of the human being as *homo faber*, the consequent (capitalist) instrumentalization of nature, a history not oriented towards transcendence, and the depotentiation of God by deism make modern progress possible. "Progress" appears as a compensatory result of dwindling religious expectations of salvation, ontological assumptions of contingency, and related anthropological uncertainties (cf. Ritter, 1972). In this respect, the meaning of progress has, on the one hand, to do with the disappearance of the metaphysical systems of security and order of religion and, on the other, with an increased awareness of contingency of a new age whose self-understanding has to do with an open, accelerated future (cf. Burgen et al., 1997). Progress envisages undeniably a process into the future, which makes it almost impossible to differentiate between true or false, meaningful or senseless (cf. Spaemann, 1983). In this respect, it is an imaginary scheme of development (cf. Sting, 1991). However, where future projections approve of the deficits of experience, they become ideological weapons. Progress is one of the "grand

J. Zirfas (✉)
University of Cologne, Cologne, Germany
e-mail: joerg.zirfas@uni-koeln.de

© The Author(s), under exclusive license to Springer Nature
Switzerland AG 2023
N. Wallenhorst, C. Wulf (eds.), *Handbook of the Anthropocene*,
https://doi.org/10.1007/978-3-031-25910-4_137

narratives" (Lyotard) of modernity, which, because it has the function of orientating and legitimizing, it does not seem to be possible to abandon (cf. Zirfas, 2015).

Progress becomes a universal and legitimizing concept in the philosophy of history, a party and action concept, an ideological concept denoting movement towards, a concept of improvement, a linear, temporal and planning concept, and finally a moving concept of acceleration (cf. Kosselleck, 1975, p. 352 f.). Progressive thinking assumes an open future and thus switches from *profectus* as limited perfection to *progressus* as unlimited perfectibility. Progress only knows its own limits, or rather it no longer knows any limits, because it is the movement of shifting limits (Zirfas, 2021). The ubiquity of progress thinking knows no limits – historical, cultural, technical or economic etc. and no anthropological limits. Therefore, it never reaches its destination. It is possible to speak of a "dynamized utopia" (cf. Strasser, 2015).

The modern idea of progress is initially determined by the reduction of infant mortality, life extension, medical progress, literacy, but then also significantly by scientific and technical changes as well as by a capitalist logic of growth. These generate and accompany the specific restlessness of progress, which cannot let the things of this world stand on their own (Konersmann, 2015). With the rapid development of technology, it is less and less the human being or the human race and its body of knowledge, but technological process itself that becomes an important model of progress (cf. Ran, 2013a, 2013b). This is also obvious, because finite life is definitely too short for infinite progress, so one has to go beyond humans. This currently means machines, Artificial Intelligence, trans- and posthumanism (cf. Loh, 2020).

Progress has for a long time focused on human beings themselves as objects. The self-empowerment of humans does not stop at themselves. If only because humans are a factor of uncertainty for progress, it seems logical to develop them towards health, intelligence, freedom from error, predictability, effectiveness and efficiency, etc. Subordinating progress to the perspectives of domination, economics and technology – rather than ecology, social questions, peace or morality – degrades people to become means to progressive ends. And the question is whether progress destroys the foundations for the humane on the one hand, and the humane itself on the other. The desacralization of nature, which degraded humans to being the rulers of nature and thus in turn to the material of their self-empowerment (der Pot & Jacob, 1985), appears here as a central problem of progress.

At the latest at the beginning of the twentieth century, with World War 1, the model of progress becomes problematic despite the prominence of vulgar Darwinist, Marxist-historical-philosophical, nationalist, and colonialist discourses. Walter Benjamin, for example, points out that progress is a single catastrophe that "ceaselessly heaps rubble upon rubble" (Benjamin, 1984, p. 161), and Max Horkheimer notes that the "progress of technical means [is] accompanied by a process of dehumanization. Progress threatens to nullify the goal it is meant to realize – the idea of man" (Horkheimer, 1967, p. 13). Progress is the sign of a human being who has subjugated nature – with the effect that their own nature must now also progress. In the loss of nature as the Other of humanity, the latter seem to carry out their own destruction: a thesis that can probably be substantiated most sensuously with a view to the Anthropocene.

The English term "sustainability" (durability) expresses – perhaps even more strongly than the German term "Nachhaltigkeit" – that one's own lifestyle (sustainability-wise) should not be limited to present-related, short-term possibilities of freedom and satisfaction. In this context, there is no question of always considering every situation involving choices and decisions in the light of future possibilities, but rather of generally keeping a future perspective in mind when making important political, social and ecological decisions.

However, the term sustainability gets caught up in the time-theoretical problem of a multiplicity of futures. Occasionally, it refers to a *past* future, i.e. a future that is determined by the past, for example, in debates about nature conservation or species protection, in which the 18th and 19th centuries are often seen as the golden age of a "natural" balance between humans and nature, which in turn then has to function as a regulative model for the future. What is meant, however, can also be a *present* future, i.e. a future whose scope for action is essentially outlined by the present, as is expressed, for example, in the debates about nuclear power or even new debt, which will limit the ecological and economic scope of future generations considerably.

Occasionally, one also finds an indication that due to the negative self-defeating or self-defying prophecies of the future, other forms of the future, and thus self-refuting prophecies, occur, since people now behave other than prophesied. In this respect, a global negative assent can be assumed with regard to conceivable global catastrophes, such as those associated with the consequences of global warming, food and water shortages, nuclear catastrophes, structural poverty and terrorist dictatorships. These seem at least to open up a progress-oriented perspective, even if the direction of positive further development of cultures cannot be determined by them. Moreover, there is an implicit contradiction between development and sustainability: for those who promise development to present generations will often violate the imperative of sustainability, because in many cases (see mineral resources) it is not possible to operate sustainably.

Thus, finally, there is a future that is truly open to surprises and *newness*, which cannot begin as radically new in an emphatic sense if it is conceived merely as progress and therefore as a space to either perfect or eliminate what already exists. This perspective pleads for openness, connections, and possibilities of development other than simply a logic of progress.

The question of whether progress can still be a concept of the future, when the vast majority of humanity has still not come to enjoy it in the sense of satisfying their basic needs of food, housing, income, or, taking it a step further, has still to experience a substantial improvement in the quality of life, also arises with regard to its capitalist logic (Capitalocene). Progress follows a capitalistic logic when the development and dynamics of societies are measured in terms of increase of capital, which constantly restages and reorganizes the interrelationship of economic processes, social orders, and individual technologies of self. "Property has become a means of production; money, through investment in land, buildings, and machinery, has become capital, whose only goal is self-exploitation and self-aggrandizement, but whose condition is the exploitation of 'living labour,'which now itself becomes a tradable commodity" (Radtke, 2020, p. 418). The logic of this economy abstracts and objectifies goods and services, and thus abstracts them from their subjective contexts of origin: everything is transformed into seemingly inexhaustible sources of resources and commodities – people, societies, nature – that can be mastered, optimized, exploited, and marketed (cf. Patel & Moore, 2017).

A rejection of progress is a rejection of a capitalist and technological concept of constant accumulation and exponential growth, which is ultimately expressed one-dimensionally in numbers. Since the development of technologies is also often ultimately subjected to a capital valuation, problems of the Anthropocene can probably not be managed with a technological logic of increase and market-based instruments. Postcapitalism, postextractivism and postprogress appear as concepts that question technological and economic growth and want to overcome the exploitation and destruction logic of capitalism in order to ensure the development of people and their possibilities of action and life as well as the development of the natural resources of life (cf. Adorno, 2000).

What would be real, substantial progress? A real improvement in the quality of life, a good life for all? The following aspects will probably be decisive above all: an exit from the capitalist logic, an overcoming of the linear quantitative increase thinking, a different resource-oriented attitude to nature and a relational anthropology. Real progress consists in saving humanity from an anthropology of sovereignty.

References

Adorno, T. W. (2000). Fortschritt. In *Philosophie und Gesellschaft* (pp. 94–118). Reclam.
Benjamin, W. (1984). Über den Begriff der Geschichte. In *Allegorien kultureller Erfahrung* (pp. 156–169). Reclam.
Burgen, A., McLaughlin, P., & Mittelstraß, J. (Eds.). (1997). *The idea of Progress*. Walter de Gruyter.
der Pot, V., & Jacob, J. H. (1985). *Die Bewertung des technischen Fortschritts. Eine systematische Übersicht. 2. Bände*. Van Gorcum.
Horkheimer, M. (1967). *Zur Kritik der instrumentellen Vernunft*. Fischer.
Konersmann, R. (2015). *Die Unruhe der Welt*. Fischer.
Kosselleck, R. (1975). Fortschritt. In: *Geschichtliche Grundbegriffe*, hg. O. Brunner, W. Konze, und R. Koselleck, (pp. 351–423). Klett-Cotta.
Loh, J. (2020). *Trans- und Posthumanismus zur Einführung*. 3. Aufl. Junius.
Patel, R., & Moore, J. W. (2017). *A history of the world in seven cheap things*. University of California Press.
Radtke, F.-O.. (2020). Kapitalismus. In *Handbuch Erziehungs- und Bildungsphilosophie*, hg. von G. Weiß, und J. Zirfas (pp. 417–430). Springer VS.
Ran, B. (Ed.). (2013a). *Global perspectives on technological innovation*. Information Age Publishing.
Ran, B. (Ed.). (2013b). *The dark side of technological innovation*. Information Age Publishing.

Ritter, J. (1972). Fortschritt. In *Historisches Wörterbuch der Philosophie*, hg. J. Ritter, 1031–1059. Schwalbe.

Spaemann, R. (1983). Unter welchen Umständen kann man noch von Fortschritt sprechen? In *Philosophische essays* (pp. 130–150). Reclam.

Sting, S. (1991). *Der Mythos des Fortschreitens. Zur Geschichte der Subjektbildung.* Reimer.

Strasser, J. (2015). *Das Drama des Fortschritts.* Dietz.

Zirfas, J. (2015). Ohne Gewähr oder: Die unsichere Zukunft. *Paragrana.* Internationale Zeitschrift für Historische Anthropologie. Bd. 24. Heft 1: Unsicherheit: 26–38.

Zirfas, J. (2021). Die Grenzen des Fortschritts. *Paragrana. Internationale Zeitschrift für Historische Anthropologie.* Band 29. Heft 2: Religion als Perfektion.

Dr. Jörg Zirfas, Professor of Anthropology and Philosophy of Education at the University of Cologne. Head of the commission "Educational anthropology" (DGfE) and of the "Society of Historical Anthropology" at the Free University of Berlin. Historical and educational anthropology, philosophy and psychoanalysis of education, educational ethnography, aesthetics, culture and education.

Time

Renaud Hétier

Abstract Humans, inescapably aware of their own finitude, face a double-edged problem: not having enough time, yet also missing out on the time that is available. For them, both time and death are unthinkable. Thus, they project their experience of time into a figurative space: the imagination (Durand). Today, however, the imagination collides with realism and materialism: stripped of any illusions as to their own finitude, modern individuals seek to intensify their own lives – to accelerate them (Rosa); to live "more" within the limited time available. Some might do so by withdrawing into hedonism; some might throw themselves headlong into technology or even transhumanism. Somewhere between these two extremes, though, we all inevitably make our way towards the ever-advancing future: the future of the generations who will succeed us.

How is our relationship with time structured? The experience of any period of time is measured, apparently, in relation to our own lives. Human beings are unique among all animals in their awareness that they will ultimately die, and they experience life in relation to that finitude. Thus, the experience of time, as Heidegger (1985) argued, is an experience of angst. Death anxiety, when viewed clearly (rather than being obscured by such-and-such a fear), is also what gives meaning to life. Indeed, the risk is that, although human life is limited, we could forget to live before we die, and let our allotted time slip away. Rousseau gives a singularly heavy formulation of the problem, saying: 'it is cruel indeed to die before having begun to live' (1966, 96). Human beings are afflicted by such thoughts even from childhood (Yalom, 1980), and a long life by no means guarantees one the sense of being fully alive. It should also be noted that children themselves, when faced with the prospect of their own imminent death (in the case of incurable diseases), endeavour to live what life they have left to the full (Hétier, 2012).

R. Hétier (✉)
UCO, Angers, France
e-mail: Renaud.hetier@uco.fr

© The Author(s), under exclusive license to Springer Nature Switzerland AG 2023
N. Wallenhorst, C. Wulf (eds.), *Handbook of the Anthropocene*,
https://doi.org/10.1007/978-3-031-25910-4_138

How is our concept of time structured? In *The Anthropological Structures of the Imaginary*, Gilbert Durand also notes the decisive rejection of death: 'the whole representation is constructed in rebellion against lack of time – in particular, the representation in the purest form of anti-destiny: the fantastic function, including memory, is merely an incident. The task of the mind is insubordination against existence and death, and the fantastic function manifests as the master of that revolt' (Durand, 1984, 468). Durand distinguishes three 'dominant reflexes' (47), three main 'gestures' (55): one of 'position', hinging on verticality; one of nutrition/digestion; and one of 'sexual rhythm'. These three 'sensorimotor structures' (48) support imaginary structures: schizomorphic/heroic, synthetic/dramatic and mystical/antiphrasic structures. The first structure belongs to the 'diurnal regime', and the other two to the 'nocturnal regime'. In the diurnal regime, verticality determines the dichotomy between top and bottom; its visual aspect correlates light and darkness; and its moral facet divides good and evil. In the nocturnal regime, the oppositions become weaker: 'the antidote to time is no longer sought at the superhuman level of transcendence and purity of soul, but in the warm and comforting intimacy of substance or in the rhythmic constancy of phenomena and events. The heroic regime of antithesis will be supplanted by the plenary regime of euphemism' (220). Whereas, in the diurnal regime and the heroic structure, it was a question of 'distinguishing', in the nocturnal regime, for the synthetic structure, it is a question of 'linking' and, for the mystical structure, a question of 'merging'. Generally speaking, imagination renders time as a spatial object ('the imagination flies immediately in space' (462)), and facilitates the 'reversibility that representation allows' (465).

For the same, universal death anxiety, therefore, there are multiple potential solutions, which may coexist or intersect. What is the dominant position in our lives today? What is our experience of time? The sense of fragility and the fleeting nature of life is perhaps less acute, because our time investment is based on our individual experiences. On this subject, we can point to the lessening place of transcendent figures, such as that of religion (and with it, the prospect of eternal life) (Marcel, 1985), that of intergenerational transmission (and therewith, the possibility that our own existence will carry on in some way, by passing on certain knowledge, values, a trade or a skill set), or that of institutions and their representatives. Against this backdrop of loss of transcendence, there is still the possibility of projecting our essences in the form of creations and objects (again, individual), which can stand the test of time and possibly survive after we are gone. Undeniably, this would account for the huge investment in material goods in our consumer society. Digital technologies, notably, by means of self-presentation devices such as personal profiles, allow us to create and leave traces of ourselves.

We can interpret that investment in light of the dominant reflexes of 'posture' (with the importance attached to sight) and 'digestion' (with the importance of contact). A certain paradox (between diurnal and nocturnal) takes shape here, crossing distance and proximity. The importance attached to the image is connected to the spatialisation of the imaginary, which helps combat the sense of irreversibility. The explosion in the number of images not only consumed, but also produced and manipulated, reveals the fragility of this solution: an image, when it becomes a trace of the past, an archive, a memory, merely emphasises the passing of time which it was intended to negate – that is why new images are constantly having to be produced. Thus, our digital existence creates a peculiar relationship with time: while it does leave traces, it is continually overwriting those traces with new ones, similar to a palimpsest which is constantly being renewed and updated. Thus, *Homo numericus* is a being of (re)presentation and of the present.

If the past tends to be obscured in the eddy of this vast palimpsest, what can we say of the future? The concern over the future now goes far beyond individuals' deaths. In the Anthropocene, we are threatened by the disappearance of an entire world, of humanity, of life itself. The individual nature of the experience of time, mentioned above, is now giving way to awareness of our shared fate. At the very least, people are concerned for their children or their grandchildren: 'With every child that is born into this world, humanity begins anew in the face of mortality, and from this standpoint, it is the survival of humanity, as well, which is at issue here' (Jonas, 1990, 258). What is the solution? A technological leap forward on the Promethean and heroic mode is more than doubtful. The (mystical) retreat becomes uncertain in the time of global upheaval which has left no room for sanctuary (although certain people are dreaming of colonising Mars as a new home for humanity). Only the cyclic solution remains, specific to the synthetic imaginary. The advantage of the cyclic representation is that it brings back what has passed (think of the seasons, for example). From that point of view, the future becomes predictable, as a mere reiteration of a phenomenon that has already been experienced. However, it is easy to see the impasse here – it is precisely the natural cycles, which support a kind of stable representation of time, that are affected by the Anthropocene (Federeau, 2017, 46–104). Thus, 'the promise of modern technology has become a threat' (Jonas, 1990, 15).

What kind of representation of time do we need to get through the Anthropocene? This question can be split into two: firstly, what kind of concept of time is possible in the Anthropocene Epoch? Secondly, what kind of concept of time is necessary to weather the oncoming storm of the Anthropocene?

We have seen the different imaginaries relating to time tested anew. The only thing which is certain is that there is no unique, definitive solution. In fact, the history of life, and of the human adventure, appear to depend on particular conditions which have not always all been met and which, even regardless of the impact of human activities, will not always be met in future. Our relationship to time in physical and zoological terms, which is more or less informed, presents us with that limitation, which can give rise to a type of abandonment. Human responsibility seems to be overwhelmed by a fate which we cannot control. It is possible, therefore, that humans will suffer the effects of the Anthropocene with a degree of fatalism, viewing them simply as part of that inescapable destiny.

Against this backdrop, the main risk is of a certain retreat – what Durand calls 'mystical' conciliation. At the same time, such retreat may offer refuge (in overinvestment in private life and entertainment), and a halfway solution (in the demultiplication of initiatives at the micro scale – that of the individual, the family, the flat share, etc.). We know that to slow the climate emergency and the advance of extinction of species will also require large- and medium-scale initiatives, which brings the political (rather than simply domestic) into play. The heroic imaginary, which envisages bypassing the problems posed by the Anthropocene – notably through transhumanism – has less traction and, as previously indicated, is unlikely to succeed (Wallenhorst et al., 2018). Remember that the goal of transhumanism is nothing less than to conquer death. Here, it is worth adding to what we said about the synthetic structure: cyclical repetition, by producing (or, in this case, reproducing), works 'little by little' to bring about change and novelty. Thus, what is produced represents progress. On this subject, Durand points to the production of fire: 'time is no longer conquered by the simple assurance of return and repetition, but because a definitive "product" is created by the combination of opposites – because "progress" is made that justifies fate itself; because irreversibility itself is harnessed, and becomes promising' (Durand, 1984, 390).

In our view, the difficulty lies in identifying the type of progress that is desirable. The idea of humanity's emancipation through technological progress has seemingly misfired. The passing of time appears to be a vector of inevitable degradation of the world and, in a way, or regression – one which is irreversible (on a scale applying to all humanity). No longer is there global 'promise' – only specific technical and technological progress. The question then arises of whether we should join forces with Utopian ideals, striving to free ourselves from the certainty of annihilation. 'Without a Utopia, we are lost' (Bregman, 2017, 34). Indubitably, the future cannot be made attractive through a negative lens – such as the very necessary reduction in consumption. We must be able to continue to create, and in fact, to engage in new, sustainable cycles of activity. It has become untenable to reduce our thinking about time to the scale of the individual. In other words, we need to find new forms of transcendence, focused on the future. We need to think in terms of intergenerational solidarity – which is, in itself, cyclical (every new generation brings renewed life); we must attenuate the anthropic effects on natural cycles, and finally, collectively invest in recycling. As we can see, it is the cyclical imaginary that can be drawn upon, with its verbal theme of 'connecting', to open up a new perspective on time.

References

Bregman, R. (2017). *Utopies réalistes*. Seuil.
Durand, G. (1984 [1969]). *Les structures anthropologiques de l'imaginaire*. Dunod. Available in English, 1999, (M. Sankey & J. Hatten. Trans.). *The anthropological structures of the imaginary*. Boombana Publications.
Federeau, A. (2017). *Pour une philosophie de l'anthropocène*. PUF.
Heidegger, M. (1985 [1927]). Être et temps. Authentica.
Hétier, R. (2012). La mort à portée d'un enfant. In A. Kerlan & L. Loeffel (Eds.), *Repenser l'enfance* (pp. 315–329). Hermann.
Jonas, H. (1990 [1979]). Le principe responsabilité. Flammarion.
Marcel, G. (1985). *Le désenchantement du monde – Une histoire politique de la religion*. Gallimard.
Mustière, P., & Fabre, M. (Eds.). (2013). *Jules Verne. Science, crises et utopies*. Coiffard.

Rosa, H. (2014). *Aliénation et accélération – Vers une théorie critique de la modernité tardive*. La Découverte Poche.

Rousseau, J.-J. (1966 [1752]). Émile ou de l'Éducation. GF Flammarion.

Wallenhorst, N., Coatanéa, D., & Prouteau, F. (Eds.). (2018). *Éduquer l'homme augmenté – Vers un avenir postprométhéen*. Le Bord de l'eau, col. "Documents".

Yalom, I. (1980). *Existential psychotherapy*. Basic Books.

Renaud Hétier is Professor at the Catholic University of the West (UCO). He is Doctor of Educational Sciences. He is the author of books on education, and anthropology in the Anthropocene. Books (selection): *Humanity versus Anthropocene* (PUF, 2021, *in French*); *Cultivating attention and care in education* (PURH, 2020, in French); *Create an educational space with the fairy tales* (Chronique sociale, 2017, *in French*); Education between presence and mediation (L'Harmattan, 2017, *in French*).

Tipping Points

Timothy M. Lenton

Abstract This article introduces tipping points in the Anthropocene. It argues that to minimise the risk from negative climate tipping points, positive social tipping points need to be identified and triggered.

A tipping point is where a small change makes a big difference to a system. To take a simple example, when you lean back on a chair you get to a point where a tiny nudge one way or the other will determine whether you return back upright or end up sprawled on your back on the floor. This is a form of strong 'non-linearity' – output is *not* proportional to input. Tipping points can happen in all sorts of complex systems thanks to what scientists and engineers call positive feedback loops: A small initial change has consequences that feedback to amplify it. Then the now larger change goes around the loop and gets further amplified, and so on. For example, as the Arctic sea-ice melts it exposes a much darker ocean surface that absorbs more sunlight, warming things up and melting more sea-ice, which warms things further, and so on. Usually, such feedbacks are constrained: the initial change gets amplified but not enough for the whole thing to 'runaway' out of control. One unit of change goes around the loop and gives back an additional say 0.1 units of change, which goes around the loop again and gives back 0.01 units, and so on. However, sometimes the feedback gets strong enough to propel change from one state to another. In this case, one unit of change creates an additional unit (or more) of change, which becomes self-propelling. The point where that happens – if it happens – is the tipping point.

The corresponding mathematical concept of a 'bifurcation' was introduced by Henri Poincaré in 1885, and the language of 'tip[ping] point' first emerged in a social science context in the 1950s (Grodzins, 1957). Meteorologists started studying bifurcations in the climate in the 1960s, and by the late 1980s it was clear that the Earth had undergone repeated abrupt climate changes in the recent past, and that

T. M. Lenton (✉)
Global Systems Institute, University of Exeter, Exeter, UK
e-mail: t.m.lenton@exeter.ac.uk

849

humans were driving global warming in the present. As Wallace ('Wally') Broecker put it in 1998: "the climate system is an angry beast and we are poking it with sticks". But it was not until Malcolm Gladwell popularised the term 'tipping point' (Gladwell, 2000) that the language of 'tipping points' started to be used in the climate context. A key reason was that parts of the climate had started to change far more abruptly than climate modellers had forecast – notably the shrinking Arctic sea-ice (Stroeve et al., 2007). Added to that, abrupt and/or irreversible changes, such as the dieback of the Amazon rainforest, started to appear in some climate models' future projections (Cox et al., 2000). Putting these sources of evidence together with understanding of feedbacks, a shortlist of potential 'tipping elements' in the climate system was drawn up – large sub-systems that could be pushed past a tipping point by human activities, leading to major impacts (Lenton et al., 2008).

The last decade has seen increasing evidence of self-propelling change in several of these tipping elements (Lenton et al., 2019). The Greenland ice sheet is melting at an accelerating rate, partly because as the top of the ice sheet drops in altitude, this warms it up further, accelerating the melting. Observational evidence has revealed that part of the West Antarctic ice sheet may have started to irreversibly retreat. Part of the East Antarctic ice sheet is showing the same dynamics. Over centuries, the resulting melting could raise sea levels by over 10 m – noting that today one billion people live within 10 m of sea level. In the ocean, the Atlantic meridional overturning circulation has weakened 15% since the 1950s, risking plunging Europe into a climate like the 'Little Ice Age'. On land, the Amazon rainforest supports itself by recycling its own rainfall and suppressing fires – but increasing droughts and deforestation risk tipping it into a dry, fiery savannah state, releasing carbon to the atmosphere. The circumpolar-Arctic permafrost is thawing and collapsing abruptly releasing carbon into the atmosphere. The list goes on.

Tipping point changes are inherently harder to predict than linear changes, but if they occur, they cause large, often abrupt, and often irreversible damages. In any sane risk assessment, one needs to pay disproportionate attention to such high impact events, even if they are deeply uncertain. Climate change is the mother of all risk management problems – one where the risk could become existential – and our assessment of tipping point risk keeps escalating. In less than 20 years the Intergovernmental Panel on Climate Change's assessment of tipping points has shifted from 'unlikely until we reach 4°C global warming' to 'we are in the danger zone now at just over 1°C global warming' (Lenton et al., 2019). Worse still these tipping elements are coupled such that tipping one can make tipping the next more likely. For example, Arctic warming and meltwater pouring off Greenland risk stopping the sinking of surface water in the seas around Greenland that drives the Atlantic meridional overturning circulation. In the past, we know that weakening of the overturning circulation disrupted the monsoons in West Africa and India, and warmed the Southern Ocean, threatening the Antarctic ice sheets. In the worst case, this could lead to 'domino dynamics' – where tipping one thing tips the next – producing unstoppable climate change to a 'Hothouse Earth' state (Steffen et al., 2018). Thankfully, we are not at that point yet and we are not even sure if it is possible – but, crucially, we cannot rule it out.

This is a strange kind of progress. Modernism has brought us to the point where it falsifies its own assumption – that the world is a gigantic, predictable, clockwork machine that responds proportionately to our interventions. Instead, the more organic and volatile climate system – Broecker's angry beast being poked with sticks – appears to be rebelling (Latour, 2017). It may even be that some recent extreme weather events – notably the North American heatwave of late June-mid July 2021 that burned the Canadian town of Lytton to the ground – are an example of 'flickering' towards an alternative climate state that is destined to become the new normal.

Society is rightly demanding to know; how can we avoid (further) climate tipping points? The apparently simple answer is enshrined in the 2015 Paris Agreement; strive to limit global warming to 1.5 °C. The problem is we have left it very late to collectively realise that and still be able to do something about it. To have an even chance of limiting global warming to 1.5 °C, all human-caused greenhouse gas emissions must be halved by 2030 and stopped by mid-century. In 2020, those emissions dropped around 5% thanks to worldwide lockdowns in response to the COVID-19 pandemic. Alas, in 2021 they rebounded to where they were in 2019. Instead, they need to be dropping at least 7% *every* year from now on. Deforestation must stop this decade. Fossil fuel burning must be halved this decade, and stop, worldwide, by mid-century. Food systems must be transformed. Where greenhouse gas emissions cannot (or will not) be stopped, they must be counterbalanced by deliberate and permanent greenhouse gas removal from the atmosphere. This needs a global transformation.

From a modernist, linear-thinking perspective, such an abrupt social transformation looks impossible. Neoclassical economics, in particular, assumes there is just one stable state of the economy involving eternal economic growth, environmental damages are just a 'marginal' externality, and (crucially) mitigating greenhouse gas emissions will get more expensive the more of it we do (Nordhaus & Boyer, 2000). That would be a counsel of despair, if it were true – but as this is the ideology that got us into this mess, it is surely time to question it. Encouragingly, there have been abrupt transformations of society, technology and our relationship with ecology in the past – propelled by well-known positive feedback loops. Now, the only way we can hope to achieve the necessary scale and rate of transformation is to leverage those feedbacks to find and trigger some positive tipping points.

Greta Thunberg and her followers have provided one startling example. When Greta started her school strike in front of the Swedish parliament that made it easier for the next person to join her, who made it slightly easier for the next person, and so on (Kuran, 1989). Before we knew it, hundreds of thousands of us were marching on the streets demanding political action to tackle climate change. However, as Greta continually reminds us, we need tipping points of action not just political rhetoric. Happily, there are examples of positive tipping points in action starting to emerge. The rapid uptake of electric vehicles in Norway is one, and the shutting down of coal power generation in the UK since 2012 is another (Sharpe & Lenton, 2021). In each case, a mix of vision, happy accident and (limited) foresight has been at play.

Now we need leaders – from all over society – who can 'think in systems' (Meadows, 2008) and seize the opportunity that self-reinforcing dynamics provide. We need to learn by doing. To achieve economies of scale. To trigger social contagion. These are all well-understood reinforcing feedbacks, with the potential to propel 'upward-scaling tipping cascades' (Sharpe & Lenton, 2021). The only novelty is the deliberate intention of trying to trigger them towards a collective goal of transformation. This will inevitably be messy and somewhat unpredictable – because we are acting within a complex, evolving, networked reality that we cannot fully comprehend.

For many this may trigger a deep fear of setting in motion self-propelling processes that, by definition, are hard to control. Those who cling to modernism often seem to cling to the desire to control its consequences. However, to declare simultaneously that climate change was "unforeseen", but now that we have seen it, the same pattern of thinking and action that caused the problem is going to solve it, sounds delusional. Instead, we need to place some trust in feedback processes that we cannot readily control. As a fell runner, I relish descending scree slopes. Once you are hurtling downhill, you cannot control the flow of rocks moving underneath you – but you can develop the skill to ride the wave. There is some jeopardy – learning-by-doing can be painful – but it is also exhilarating. In the same spirit, if we commit to transformation, we must embrace the evolving complexity and learn to ride it.

Positive tipping points are already starting to interact in a good way (Sharpe & Lenton, 2021): The electric vehicle revolution is propelling ever-cheaper battery technology. That cheap energy storage capacity helps solve the critical challenge with renewable power – balancing supply and demand. Ever cheaper renewable power generation is in turn feeding back to make electric transportation even more desirable. Positive tipping points are also starting to extend beyond society and technology into the ecological realm. Witness the movement against single-use plastics inspired by concern for marine life.

Positive feedbacks can always operate in either direction and as a new way of living takes off an old way will suffer accelerating decline. That is causing many economists and the financial sector to fret about 'transition risk'. Fossil fuels that were being banked on as assets instead need to be stranded right where they are – in the ground. Happily, the new economy that replaces the old must grow and generate employment. Thus, the risks involved in transformation are transient, whereas the climate risk we need to avoid is irreversible.

Whilst climate tipping points are a key aspect of the climate crisis, the existence of positive social tipping points can give us back a sense of autonomy to tackle that crisis. All too easily we blame politicians or bankers or someone else with power for not solving the problem for us. Those same actors look at each other seemingly paralysed by the complexity of the situation – but any (and all) of us could be the trigger of tipping points that unlock positive global change.

References

Cox, P. M., Betts, R. A., Jones, C. D., Spall, S. A., & Totterdell, I. J. (2000). Acceleration of global warming due to carbon-cycle feedbacks in a coupled climate model. *Nature, 408*, 184–187.

Gladwell, M. (2000). *The tipping point: How little things can make a big difference*. Little Brown.

Grodzins, M. (1957). Metropolitan segregation. *Scientific American, 197*, 33–41.

Kuran, T. (1989). Sparks and prairie fires: A theory of unanticipated political revolution. *Public Choice, 61*, 41–74.

Latour, B. (2017). *Facing Gaia: Eight lectures on the new climatic regime*. Polity.

Lenton, T. M., Held, H., Kriegler, E., Hall, J., Lucht, W., Rahmstorf, S., & Schellnhuber, H. J. (2008). Tipping elements in the Earth's climate system. *Proceedings of the National Academy of Sciences, 105*, 1786–1793.

Lenton, T. M., Rockstrom, J., Gaffney, O., Rahmstorf, S., Richardson, K., Steffen, W., & Schellnhuber, H. J. (2019). Climate tipping points – Too risky to bet against. *Nature, 575*, 592–595.

Meadows, D. H. (2008). *Thinking in systems: A primer*. Chelsea Green Publishing.

Nordhaus, W. D., & Boyer, J. (2000). *Warming the world. Models of global warming*. MIT Press.

Sharpe, S., & Lenton, T. M. (2021). Upward-scaling tipping cascades to meet climate goals: Plausible grounds for hope. *Climate Policy, 21*, 421–433.

Steffen, W., Rockström, J., Richardson, K., Lenton, T. M., Folke, C., Liverman, D., Summerhayes, C. P., Barnosky, A. D., Cornell, S. E., Crucifix, M., Donges, J. F., Fetzer, I., Lade, S. J., Scheffer, M., Winkelmann, R., & Schellnhuber, H. J. (2018). Trajectories of the earth system in the Anthropocene. *Proceedings of the National Academy of Sciences, 115*, 8252–8259.

Stroeve, J., Holland, M. M., Meier, W., Scambos, T., & Serreze, M. (2007). Arctic Sea ice decline: Faster than forecast. *Geophysical Research Letters, 34*, L09501.

Tim Lenton is founding Director of the Global Systems Institute and Chair in Climate Change and Earth System Science at the University of Exeter. He specialises in modelling life's coupling to the Earth system, climate dynamics, and associated tipping points. His books 'Revolutions that made the Earth' (with Andrew Watson) and 'Earth System Science: A Very Short Introduction' have popularised a new scientific view of our planetary home. Tim is renowned for his work identifying climate tipping points and co-authored the 'Planetary Boundaries' framework. He is a member of the Earth Commission, an ISI Highly Cited Researcher, and in the Reuters 'Hot List' of the world's top climate scientists.

Part XII
Humanity as Symbolisation

Aesthetics

Mariagrazia Portera

Abstract This article examines the ideas of 'aesthetic pleasure' and 'aesthetic appreciation of nature' in the Anthropocene. In the framework of the current ecological crisis, the anthropogenic roots of which are today beyond dispute, are the aesthetic categories of 'beautiful', 'sublime', 'majestic' etc. still appropriate to describe our experience of nature? Can a landscape – or an animal or a plant – which have undergone changes and modifications due to climate change (a human-induced phenomenon) still be considered beautiful? More generally, is the idea of nature as an object of aesthetic contemplation still legitimate in the Anthropocene?

In the eighteenth century, particularly in England and Scotland, philosophers such as Edmund Burke, Joseph Addison, Francis Hutcheson provided relevant insights into the role and value of nature as an object of aesthetic contemplation, in the frame – Burke, in particular – of an in-depth discussion of the aesthetic category of the 'sublime'. The same did Kant in his *Critique of the Power of Judgement* (1790): the sight of a starry sky at night, the terrifying power of a volcano in eruption, the immensity of the ocean are, according to Kant, all experiences of natural phenomena of highest relevance to humans. Indeed, through them, we can get a grasp on our deepest moral force and intrinsic dignity as human beings. In opposition to Burke's empirical-physiological aesthetics, however, Kant's concept of the sublime is transcendentally orientated: it does not exist *in* natural phenomena, rather only in the subject; our aesthetic judgements about the sublimity of nature (and, at least to a certain extent, also about its beauty) reveal more of the status, of the moral values and cognitive functioning of the judging individual than about the object judged.

What happened after Kant – who may be considered the last 'philosopher of (the aesthetics of) nature' – is that post-Kantian aestheticians, paradigmatically those inspired by Hegel, took Kant's transcendental-subjective stance about nature, so to speak, to extremes. There is no point in aesthetically approaching natural

M. Portera (✉)
University of Florence, Florence, Italy
e-mail: mariagrazia.portera@unifi.it

© The Author(s), under exclusive license to Springer Nature
Switzerland AG 2023
N. Wallenhorst, C. Wulf (eds.), *Handbook of the Anthropocene*,
https://doi.org/10.1007/978-3-031-25910-4_140

phenomena, they argue, since what matters is what happens in the arts – in the 'spirit' and its transforming artistic-creative power, not in nature. As Hegel wrote in the *Introduction* to his *Aesthetics*, the beauty of art is higher than nature, since 'the beauty of art is beauty born of the spirit and born again'. Such an approach resulted in the neglect of natural beauty for almost 150 years after Kant and Hegel. The idea of natural phenomena as legitimate sources of aesthetic contemplation was rejected, as it appeared to be the vestige of an old-fashioned way of conceiving aesthetics. It was only in the mid-60 of the twentieth century that things started slowly to change, and nature and its aesthetic appreciation came back to the fore. Indeed, what we call today environmental aesthetics is a relatively recent subfield in aesthetics that formally emerged in the 70s and 80s of the twentieth century in the Anglo-American world as a reaction against the exclusive focus on the arts, as famously argued by Ronald Hepburn (1966) in his seminal essay *Contemporary Aesthetics and the Neglect of Natural Beauty*.

According to Hepburn, 'although some important features of art-experience are unattainable in nature, that by no means entitles the aesthetician to confine his studies to art': nature, understood as 'all objects that are not human artefacts' (Hepburn, 1966, p. 299), can be a source of aesthetic pleasure as much as art. Initiated by Hepburn, a wide movement of reassessment of the aesthetic value of landscapes, flowers, trees, nonhuman animals etc. flourished in the second half of the twentieth century, with an impressive increase of publications over the last few years. Scholars working in environmental aesthetics, however, do not seem to agree on how the aesthetic appreciation of nature should be understood.

For instance, since 1981 Allen Carlson has been elaborating what we may call a 'cognitivist aesthetics of nature': Carlson argues that the more we know scientifically about nature, the more we aesthetically appreciate it; aesthetics depends to such an extent on common sense and scientific knowledge that the only way for an appropriate aesthetic appreciation of nature is under the guidance of science (Carlson, 2000). Ecologist Gordon Orians and psychologist Stephen Kaplan put forward, between 1980 and 1990, an evolutionarily oriented environmental aesthetics, arguing that the real meaning of natural beauty consists in its evolutionary value. Throughout their evolutionary history, human beings have developed aesthetic preferences for those landscapes and natural settings that provided more adaptive advantages and chances of survival; 'beautiful', in this view, is another word for 'adaptive' (Orians & Heerwagen, 1986; Kaplan & Kaplan, 1989. Unlike Carlson, Arnold Berleant (1992), another leading figure in the development of environmental aesthetics, sees a clear difference between science and aesthetics in their approach to nature. In several essays and books, he sketches out a 'participatory model' of aesthetic appreciation, or 'aesthetics of engagement', focused on the notions of embodiment and sensory immersion and radically alternative to the scientific analysis of nature. Inspired by Husserl's and Merleau-Ponty's phenomenology, Berleant holds that the aesthetic perceiver of nature, far from being a passive recipient of environmental stimuli, actively and multisensorially interacts with the environment in which it is embedded.

Recently, philosophers Ted Toadvine (2010), Emily Brady (2017, see also Brady & Phemister, 2012) and Timothy Morton (2007) have discussed the idea of 'natural beauty' in the context of a broader reflection on today's ecological crisis. We live today in the Anthropocene, a new phase in the history of the Earth in which humans have become geological agents able to transform with their pervasive activities – and to harm – the structure of the planet itself. Today, there is virtually no landscape, animal, tree or flower, i.e., those natural things traditionally assumed as objects of aesthetic contemplation, which does not bear with itself the trace of some anthropogenic transformation (see Brook & Prior, 2018). In such a context, are the aesthetic categories of 'beautiful', 'sublime', 'majestic' etc. still appropriate to the Anthropocene nature? Can a landscape – or an animal or a plant – which have undergone massive changes and modifications due to the ecological crisis (undoubtedly, a human-induced phenomenon) still be considered beautiful? More in general, is the idea of nature as an object of aesthetic contemplation legitimate in the Anthropocene? Philosopher Timothy Morton has argued that, for truly effective environmentalism, we should get rid once and for all of the idea of 'natural beauty': 'Putting something called Nature on a pedestal and admiring it from afar does for the environment what patriarchy does for the figure of Woman. It is a paradoxical act of sadistic admiration' (Morton, 2007, p. 5).

All in all, the question of aesthetic appreciation in the Anthropocene appears to be deeply interwoven with moral issues. Let us consider the following examples: because of the rise in temperatures, many northern countries of Europe have in the last few years started to enjoy longer and warmer summers, with inevitable changes in the range of animal and vegetal species that populate the regions. But is, in this case, the use of the term 'enjoy' – even 'aesthetically enjoy' – still appropriate? Because of water acidification, due to anthropogenic modifications of the climate, the Great coral reef is notoriously bleaching: but if someone, diving among coral reefs for the first time, finds the contrast between the luminous white of the corals and the deep blue of the sea aesthetically appealing or 'beautiful', how should we consider this aesthetic judgement? Is it legitimate or not? Emily Brady, in a recent paper (Brady, 2017; see also Wilke, 2013), emphasizes the moral implications of valuing today's nature from an aesthetic standpoint, but she also contests that these moral pressures necessarily lead 'to finding less aesthetic value in the world'. According to the American environmental philosopher, 'the proper answer to that moral challenge will be to understand the ways that aesthetic experience can be educative, enabling us to grasp losses and gains in aesthetic value'. In other words, we should focus on the potential of the arts and aesthetic experience in terms of environmental education, while at the same time recognizing and respecting the relative autonomy of aesthetics from the domain of moral issues. The bleaching coral reefs can still appear beautiful, but since there is no 'pure' or 'mere' aesthetic experience, i.e. totally devoid of knowledge and cognitive information, to become aware that the corals' beauty is the result of a human-induced climate alteration may give the aesthetic experience a tragically connotated aesthetic touch that may encourage people to take action.

Greta Thunberg, the Swedish young climate activist internationally known for having initiated in 2018 the 'Friday school strikes for climate' movement, in January 2019 told the economists and politicians gathered at the Davos World Economic Forum that she did not want them to be hopeful about the chances to fix the current climate crisis, rather be in panic: in front of the current state of nature and, more in general, in front of the current state of the Earth ecosystem 'I want you to panic. I want you to feel the fear I feel every day. And then I want you to act'. Thunberg's words can be a powerful description of the specific character of the aesthetic appreciation of nature in the Anthropocene: on the one hand, as Brady stressed, nature, however humanly manipulated and altered, is still able to induce aesthetic experiences of enchantment, on the other hand, these experiences are imbued with a sense of fear, precarity, discomfort, which nevertheless is important and worthwhile because it can induce actions of contrast and opposition to the ecological crisis. The concepts of the 'weird' and the 'eerie', as recently described by Mark Fischer (2017), promise to be, in this sense, two of the most important aesthetic categories of the 'Anthropocene': they convey a sense of precarious enjoyment, of discomforting fascination, a sense of perturbation which is not completely unpleasant. The Anthropocene nature is beautiful but also endangered to such an extent that its beauty is somehow dreadful. In the specific case of the eerie, this notion conveys a sense of 'detachment from the urgencies of every day', which is similar to the disinterestedness that characterizes the contemplation of the beauty of nature but at the same time, it is slightly discomforting and precarious, urging, therefore, the aesthetic perceiver to move, take an interest in, take action, instead of being satisfied with pure contemplative appreciation.

References

Berleant, A. (1992). *The aesthetics of environment*. Temple University Press.
Brady, E. (2017). Climate change and future aesthetics. In A. Elliott, J. Cullis, & V. Damoran (Eds.), *Climate change and the humanities* (Historical, philosophical and interdisciplinary approaches to the contemporary environmental crisis) (pp. 201–220). Palgrave Macmillan.
Brady, E., & Phemister, P. (Eds.). (2012). *Human-environment relations: Transformative values in theory and practice*. Springer.
Brook, I., & Prior, J. (2018). *Between nature and culture: The aesthetics of modified environments*. Rowman and Littlefield.
Carlson, A. (2000). *Aesthetics and the environment: The appreciation of nature, art and architecture*. Routledge.
Fischer, M. (2017). *The weird and the eerie*. Repeater.
Hepburn, R. W. (1966). Contemporary aesthetics and the neglect of natural beauty. In B. Williams & A. Montefiore (Eds.), *British analytical philosophy*. Routledge and Kegan Paul.
Kaplan, R., & Kaplan, S. (1989). *The experience of nature: A psychological perspective*. Cambridge University Press.
Morton, T. (2007). *Ecology without nature. Rethinking environmental aesthetics*. Harvard University Press.

Orians, G. H., & Heerwagen, J. H. (1986). An ecological and evolutionary approach to landscape aesthetics. In E. C. Penning-Rowsell & D. Lowenthal (Eds.), *Landscape meanings and values*. Allen and Unwin.

Toadvine, T. (2010). Ecological aesthetics. In L. Embree & H. R. Sepp (Eds.), *Handbook of phenomenological aesthetics* (pp. 85–91). Springer.

Wilke, S. (2013). Anthropocenic poetics: Ethics and aesthetics in a new geological age. In H. Trischler (Ed.), *Anthropocene: Exploring the future of the age of humans* (Vol. 3, pp. 67–74). RCC Perspectives.

Mariagrazia Portera is currently a Research Fellow at the University of Florence, Department of Literature and Philosophy, after being a post-doctoral fellow at IASH, Institute for Advanced Studies in the Humanities, University of Edinburgh, at the Free University of Berlin, at the University of Zagreb and at the Centre for Advanced Studies South-East Europe, University of Rijeka. She holds a PhD in Philosophy from the University of Florence. Her areas of expertise are the history of Aesthetics between the 18-century and the 19-century in Germany, the history of Darwinism, experimental aesthetics, and environmental and evolutionary aesthetics. She has authored more than 60 publications (articles, book chapters, books) and co-authored book chapters and papers with biologists, geneticists, philosophers of biology.

Art

Bill Gilbert

Abstract This essay addresses the question of what role Art has in the Anthropocene. It begins with a discussion of the contemporary definitions of these two terms and then identifies several potential avenues for Art to pursue in the uncertainty of the unfolding geologic epoch.

The terms Art and Anthropocene occupy a broad and complex mix of territories in popular culture. The word Art comes from the Latin *ars* or *artis*, which is to say that it is a concept that comes out of western civilization. Other cultural traditions do not include the idea of Art as a distinct entity or practice. The term Art is then the product of a particular (western) cultural mindset driven by the need to categorize and differentiate distinct aspects of human experience.

If we turn to the Webster dictionary, the first definition of art is

1. *The disposition or modification of things by human skill, to answer the purpose intended. In this sense art stands opposed to nature.*

Art is defined here as belonging to culture as a human invention separate from and "opposed to nature". If this definition in fact still holds, it may prove problematic in finding a role for Art in the proposed new geologic epoch. T.J. Demos and others argue that helping to eliminate this false distinction between our species and the rest of nature is a primary task for artists in the Anthropocene. (Demos, 2009, p. 20)

2. *A system of rules, serving to facilitate the performance of certain actions; opposed to science, or to speculative principles; as the art of building or engraving. Arts are divided into useful or mechanic, and liberal or polite.*

Having set Art in opposition to nature it, Webster posits a further opposition to science and then divides the practice between craft and fine art. Within the

B. Gilbert (✉)
University of New Mexico, Albuquerque, NM, USA
e-mail: billgilbert@cybermesa.com

community of art practitioners in recent decades the rigid boundaries between art and science have dissolved with interdisciplinary collaborations such as Hans Haake *Rhinewater Purification Plant (Kastner, p 141)*, 1972 or Newton and Helen Mayer Harrison's *Sava River Project*, 1989 (Heartney, 2014, p. 79) and the lines between craft and fine art have been repeatedly challenged as in Judy Chicago's *The Dinner Party*, 1974–1979, (Gerhard, 2013, p. 4). These developments can be seen as a part of a larger trajectory in Art focused on infiltrating all aspects of human experience. In a move towards a conception typical in non-western cultures, Art is changing from a distinct practice separate from and, in many cases, in opposition to other recognized disciplines to a mycelial presence, a omnipresent connective tissue in daily life under rubrics such as Social Practice, Relational Aesthetics, and System Theories.

The fine art and craft definition offered by Wikipedia may reflect a more up to date understanding of the term Art in popular culture.

> *"Art is a diverse range of (products of) human activities involving creative imagination to express technical proficiency, beauty, emotional power, or conceptual ideas. There is no generally agreed definition of what constitutes art, and ideas have changed over time."*

This definition focuses on creativity as the essential quality whether it is expressed in technical, emotional, conceptual or aesthetic expressions. *"ideas have changed over time"* may well be the crux of the matter. In that sense, Art in the Anthropocene is a moving target whose role will be neither singular nor easily pigeonholed. It will necessarily evolve and adapt in response to the rapidly changing environmental and social reality of the Anthropocene.

The term Anthropocene is a combination of the Greek: *anthropo*, meaning 'human' and *cene* meaning 'new'. The term, first coined in science disciplines, asserts that humans have altered the ecology of the planet over a sufficiently extended period as to qualify as a new geological era. (Overy, 2016) There is an inherent species-centric element to this idea, an advanced level of human hubris. It can be seen as a continuation of the concept of Dominion (Genesis, 1:28) in Christian theology that assigns to humans control over and responsibility for all of God's creations. The term Anthropocene thus magnifies our sense of self-importance as master of all things while also offering the assertion that our runaway consumption of the earth's resources has left an indelible mark. For those burdened with a sense of ethics, this leads to a reckoning with the responsibility humans have to acknowledge the error in our behaviours, reverse the deleterious effects of our actions and take the necessary steps to bring our species back into a sustainable web of life on this planet.

It can be argued that this is quite simply a matter of survival. On one level, humans will have to make significant systemic changes in our production and consumption of carbon to avoid the worst effects of global warming. On a deeper level, the future of our species is dependent on our ability to make fundamental changes in our relationship to the ecological web of the planet. Can we as humans reimagine ourselves as one part of a complex web rather than as the unchallenged, apex predator species having 'Dominion' over all others?

While scientists will necessarily take the lead in addressing the technological issues, it falls to humanists to foment a change in human consciousness. Art has a role to play in this regard. Art has a long record of serving as a mirror presenting back to culture the assumptions that underscore our behaviours and Art, as a human practice, can provide examples of actionable pathways to reconnection. György Kepes in *Arts of the Environment* (Kepes, 1972) makes the case that creative imagination and artistic sensibility are themselves survival tools. The question becomes how to proceed. Several opportunities for Art present themselves and each follows from a certain perspective on the conception of the Anthropocene and how it might play out.

Sigue, No Mas

While many artists are responding to climate change and the implications for an Anthropocene epoch, we should not underestimate the inertia of the status quo. Our discussion of the possibilities for Art in the Anthropocene should perhaps begin with an acknowledgment the role art will continue to fill in contemporary western cultures as the provider of diversions, escapism and entertainment. Given the collective inability of the 'developed' world to acknowledge the end of the path we are currently following, let alone take the steps necessary to avert the unfolding ecological disaster, it seems fair to assume that art that dazzles and distracts, that embraces excess such as *Balloon Dog* 1994–2000, Jeff Koons, *For the Love of God,* 2007, Damian Hirst will continue to be in high demand as we ride global warming to oblivion.

It's Over

One of the key aspects of the Anthropocene to date is the advent of the sixth extinction (Kolbert, 2014). While not specifically part of the geologic record defining the epoch, it will certainly be more evident to humans than the thin layer of radioactive dust and carbon residue we have deposited. If in fact the Anthropocene marks the exit of the human species from the earth's ecology, one role for Art may be post extinction communication. Just as the Lascaux cave paintings serve as a voice from a past culture, it may now be time for artists to think about leaving a record of our species for our successors. This process has in fact already begun. The US Department of Energy commissioned the creation of a marker for our nuclear waste repository in Southern New Mexico, the Waste Isolation Pilot Project (Meier, 2016), that would theoretically communicate the danger of the site tens of thousands of years in the future. Fundamental to the endeavour is the understanding that written languages will not survive/be intelligible in that time frame. The search then turns to the visual language of signs, semiotics. The proposed *Landscape of Thorns* by

Michael Brill and Safdar Abidi fits easily within a traditional definition of Art as an aesthetic object, as a sculpture created to express threat and invoke fear.

A new nuclear repository is being constructed in France for the waste from fifty-eight reactors. Once the facility in Bure is full it will also need signage that can communicate the site's danger far into the future. The government sponsored a competition in which Alexis Pandellé's winning proposal, *Prométhéé oublié (Forgotten Prometheus)* evoking a scar on the ground, a suggestion of a wound that will never heal updates the early Land Art of Michael Heizer and Robert Smithson. Second place was awarded to Stéfane Perraud and Aram Kebabdjian for a forest of trees genetically modified to grow an alarming blue as a warning of the danger underground that clearly references Mel Chin's early Environmental Art work *Revival Field*.

Finland is currently building the Onkalo ("hiding place") spent nuclear repository, and the Äspö Hard Rock Laboratory is underway in Sweden. The other nuclear nation states will undoubtedly need to prepare their own repositories in the near future. The effort to sign these nuclear waste repositories provides the opportunity for Art to experiment with how and what to communicate in a post human epoch. Surely there is far more in the record of human existence that our nuclear heritage. The question is what should be addressed.

This Too Shall Pass

A more hopeful possibility is that Art can play a role in the process of adaptation required for our species to survive the Anthropocene. Any question about the advent of climate change is over. We are now faced with simultaneously limiting the further global warming and adapting to the new ecological reality. The change necessary in human mindset is fundamental. The amount of remediation required for past abuses is immense. Perhaps Art has a role in helping us move from the current obsession with creating ever greater wealth to allow ever greater consumption by extracting ever greater human and material resources from the less powerful regions of the planet towards a new value system that priorities beauty and balance. As Art in western cultures has evolved through postmodernism, the frame around art as a distinct practice has dissolved. It is now possible to see Art as being a component of all human activity, whether it is a question of material production or social interaction. There are precedents for this in non-western cultures. The Navaho concept of Hózhó, the Beauty Way, serves as one example in which aesthetics equates with living in a holistic environment of beauty, balance, and well-being in harmony with the rest of creation. There is a role for Art in the Anthropocene in nurturing this worldview in capitalist cultures.

In the Arts, the focus on engaging with environmental and social issues has found expression in many different avenues (Brown, 2014). While the forms of Art practice have proliferated exponentially, several shared aspects have emerged.

The dominant model of the solo artist is being balanced by increased focus on collaborations and collectives; a move from the I model to the We.

The context for art is being transferred from the white cube of the gallery to the world at large.

The line between artist and audience blurs as more and more projects are based in co-generation with community.

The focus in art works moves from semiotics and metaphor to actualization.

The evolution of Environmental Art into Eco-Art is marked by an understanding that the human species is a part of nature. Consideration of environmental issues, therefore, must now be matched with a social component. As T.J. Demos suggests, the focus on a sustainable ecology must include a commitment to social justice (Demos, 2009, p. 19). Artists have embraced this ecological perspective in projects that model adaptive behaviours by simultaneously building environmental and social resilience. Recent efforts by artists to work across disciplines with scientists, to merge left and right brain responses to environmental issues, NSF Antarctic Artists and Writers Program, Cape Farewell, can be traced back to early experiments in cross disciplinary collaboration such as Maurice Tuchman's *Art & Technology* project at LACMA. While some artists work to build community resilience based in agricultural focus (The Land/The Rice Field, 1998-), (la r.O.n.c.e, 2013-), (SeedBroadcast, 2011-) others choose an urban or suburban focus (Black Lunch Table, 2005), (Fritz Haeg, (*Edible Estates*, 2005-). Many are moving beyond the metaphoric or symbolic language of art to engage in post gallery projects based in remediation (Mary Mattingly (*Waterpod*, 2006-), Simparch (*Clean Livin'*, 2003-), Beata Tsosie-Peña & Kaitlin Bryson, (*Many Hands*, 2021), Buster Simpson (*Whole Flow*, 2009), Jackie Brookner (*The Fargo Project*, 2010-). These artists and many others are crafting a role for Art in the Anthropocene as an active agent in the public sphere charting a course for a correction in human behaviours while participating in the daily practice of building beauty and balance.

References

Brown, A. (2014). *Art and ecology now*. Thames and Hudson.

Demos, T. J. (2009). The politics of sustainability: Contemporary art and Ecology. In *Radical nature: Art and architecture for a changing planet 1969–2009* (pp. 17–29). Barbican Art Gallery.

Gerhard, J. (2013). *The dinner party; Judy Chicago and the power of popular feminism 1970–2007*. University of Georgia Press.

Heartney, E. (2014, February). Art for the Anthropocene. In *Art in America* (pp. 76–82).

Kepes, G. (1972). *Arts of the environment*. Braziller.

Kolbert, E. (2014). *The sixth extinction: An unnatural history*. Henry Holt & Company.

Meier, A. (2016, July 21). *A nuclear warning designed to last 10,000 Years*. https://hyperallergic.com/312318/a-nuclear-warning-designed-to-last-10000-years/

Multiple Authors. (1611). *The Holy Bible: Old and new testaments*. King James Version, Church of England.

Overy, N. (2016). *13 ways of thinking about the Anthropocene*. https://neilovery.com/2016/09/07/13-ways-of-thinking-about-the-anthropocene/?fbclid=IwAR2CINv5zT-h2CMKUVNEGaQkICJ0uniSW4kTaTdQHyEecj88XmlWrJBKRGw

Bill Gilbert is Emeritus Professor of Art and Ecology and Lannan Endowed Chair at the University of New Mexico. Gilbert is the founder the Land Arts of the American West program and the Art & Ecology Area in the Department of Art and Art History. He has authored two books; 2009, *Land Arts of the American West*, University of Texas Press and 2019, *Arts Programming for the Anthropocene; art in community and environment*, Routledge Press and numerous essays and articles on the Indigenous ceramics of the Americas. Gilbert's multimedia installations have been exhibited internationally since 1981.

Culture

Rolf Elberfeld

Abstract From the point of view of the concept of culture, the age of the Anthropocene means that the old distinction between nature and culture is collapsing. Today, there is less and less wild and "untouched" nature on earth that could be cultivated by humans, but nature itself has become a cultural product. "Natural nature" is so exposed to human cultivation that the distinction between nature and culture is disappearing – with effects that cannot be foreseen today. This article shows the emergence and shifts of the aforementioned linguistic distinction in European intellectual history.

The word *culture* is derived from Latin. Its basic verbal form *colere* in Latin means "to cultivate, to take care, to till the field". The verb *colere* expresses an actively formative relationship of human beings to organic nature, which unfolded above all through the emergence of agriculture. Even though humans used tools and domesticated animals before the age of arable farming, these practices, together with the invention of agriculture, formed a new way of life for humans that fundamentally changed their relationship with nature. With farming, which has emerged in various places around the world since the tenth millennium BCE, humans increasingly began to influence and "cultivate" the natural growth of plants (Zohary & Hopf, 2000). Within this new way of life, the distinction between a wild nature on one side, and the cultivating practices on the other, related to the fields, the plants and animals, emerged. Through this new practice, humans began to alter and subjugate "natural nature" in a more permanent and systematic way.

The history of the Latin term *colere* and *cultura* was preceded in Europe by another one, which is connected above all with the ancient Greek words *meletē* and *epimeleia*. Both words basically mean "care, attention" and thus correspond to the Latin *colere*. The word *meletē* occurs for the first time in European literature in

R. Elberfeld (✉)
University of Hildesheim, Hildesheim, Germany
e-mail: elberfeld@uni-hildesheim.de

Hesiod, who defines the nature of humans in his book *Works and Days* (Hesiod, 2018) through the caring work on a farm. According to Hesiod, the people who care for the fields, plants and animals in a good way are "just" because they follow the natural order. In ancient Greece, it takes until the fifth century BCE for the idea of care and cultivation to be applied to the "soul" of human beings. Socrates, Xenophon, Isocrates, Plato and others speak in their texts of "care/cultivation of the soul" (epimeleia tēs psychēs) as an essential purpose of philosophizing (Hadot, 2002). By relating the concept of care and nurture to the human soul, an influential and long-standing discourse on education and cultivation of the human being developed in Europe. What is to be cared for and cultivated in human being is the good natural dispositions in human soul. The innate nature of human being is to be developed through cultivation.

In Latin, the reflections on the cultivation of fields, plants and animals, as well as the cultivation of human souls are associated with the word *cultura* (cultivation). The oldest Latin evidence for the word *cultura* is found in Cato the Elder's (234–149) book *De agri cultura* (Cato, 1998). Cicero's central definition of philosophy, "*cultura animi autem philosophia est*" ("Philosophy is the care/cultivation of the soul," Cicero, 1989, Book II), demonstrates the application of *cultura* to the human soul adopted from Greek philosophy. His writings coined the distinction between *natura* and *cultura* in Latin, which is central to European intellectual history since the sixteenth century, while in the Roman and medieval period, this distinction was interpreted in various ways without becoming a central and guiding distinction.

The word *cultura*, which in its grammatical form is a *future participle active* of the verb *colere*, expresses the cultivation and processing of something that will lead to a *development* and *improvement* of the thing *in the future*. It is this positive expectation of progress for a future development that made the word *cultura* a leading term of intellectual development in modern Europe. With the book *Cultura ingeniorum* (1604) by Antonio Possevino (1533/4–1611), not only the word *cultura* became particularly prominent, but also the Latin word *ingenium*, which meant the natural endowments of humans. In connection with the European debates about the dignity of human beings since Petrarch, the developmental potential of humans through cultivation moved not only to the center of philosophy, but of the entire European culture, at the latest in the seventeenth century. Accordingly, Immanuel Kant, in the Age of Enlightenment, sees the cultivation and perfection of human being as the very purpose of human existence itself (Kant, 1996, Part II). Debates about the cultivation of human beings have been intensified in particular by European expansion since 1492, in the context of which European people have developed a global claim to dominance in political, cultural, and technical terms. The idea and practice of cultivating humans in mental as well as physical respects led to a general self-conception of its superiority in Europe over all other people in the world from the eighteenth century at latest.

In the eighteenth century in Europe, the focus was on the cultivation of the person in all spheres of life. In addition to the cultivation of the soul, there was also the cultivation of law, literature, history, the arts, the mind, etc. From the nineteenth century onwards, with the industrial revolution, the cultivation of material objects and various techniques increasingly came to the forefront of academic attention. In the middle of the nineteenth century, cultural science (Kulturwissenschaft) emerged in the German-speaking world. In this science, Gustav Klemm examined all material products produced by people worldwide (Klemm, 1854–1855). The word *culture* in this period is understood less and less as a process of cultivation, but rather as the totality of all products that have been created through human practice. In this context, not only material objects, plants, and animals are conceived as products of human cultivation, but also humans themselves, so that the narrative of cultivation and culture encompasses all areas of nature. Through human interventions and practices, *all of nature has become the object of human cultivation*, as evidenced by an infinite variety of products mediated by culture. This multiplicity of products is increasingly perceived in the late nineteenth century as a multiplicity of *cultures* – a plural that first appears in the mid19th century (Elberfeld, 2012, p. 280–302) – and develops its full semantic force in the twentieth century.

With the innovations in natural sciences, medicine, and engineering from the end of the nineteenth century onwards, the paradigm of cultivation, which started from the distinction between natural potentials and their cultivation by humans, was replaced by the paradigm of the producibility of nature–material, plants, animals, human beings. This paradigm tries to question the given nature of natural potentials, so that nature itself should become a cultural product altogether through technical interventions. This connection can be seen in relation to humans in the debates about eugenics, a word coined in England in 1883 (Galton, 1883). In the context of Galton's theory of the improvement of human nature through selective breeding and technical means, racial theories emerged in various sciences that contributed significantly to the preparation of the ideology of the Nazism, which sought to achieve an improvement of human nature through the "destruction of unworthy life" and the "improved breeding of special races." During this period, however, comprehensive visions emerged not only with regard to human nature – as, for example, in Nietzsche's vision of the *Übermensch* (Overman) – but also with regard to the producibility and influenceability of nature as a whole, which only gained new horizons of realizability at the end of the twentieth century through the digital revolution and developments in genetic medicine.

However, at the beginning of the twentieth century, Georg Simmel reflected on the "tragedy of culture" (Simmel, 1968), which, in his view, consists in the fact that people have created an objective culture for themselves throughout the centuries that is so complicated and comprehensive that no individual person can understand and see through it within the framework of his or her subjective cultivation. Simmel sees a central problem in the fact that the growing scope of objective culture increasingly overwhelms people, so that people become victims of the cultural complexity they have produced. Simmel's analysis clearly shows a shift in the distinction between nature and culture in favor of objective culture. The human being has transformed itself into a being in nature, that dominates nature itself through culture.

In this sense, humans have not only become the dominant being on earth because they influence and alter the climate in a previously unimaginable way. Rather, humans today are trying to decode all of nature – including themselves – in order to intervene in a cultivating and manufacturing way from within. This is testified not only by the debates on transhumanism (More & Vita-More, 2013), but also by all the technical possibilities, actual and unforeseen, that comprehensively change the nature of material substances, plants, animals and humans. From the perspective of the concept of culture, the age of the Anthropocene can thus mean that nature itself has increasingly become culture and that the distinction between nature and culture has dissolved in favor of culture.

References

Cato. (1998). *On farming* (A. Dalby, Trans.). Prospect Books.
Cicero. (1989). *Tusculan disputations* (J. E. King, Trans.). Harvard University Press.
Elberfeld, R. (2012). *Sprache und Sprachen. Eine philosophische Grundorientierung*. Alber.
Galton, F. (1883). *Inquiries into human faculty and its development*. Macmillan.
Hadot, P. (2002). *What is ancient philosophy? Translated by Michael Chase*. Harvard University Press.
Hesiod. (2018). *Works and days* (A. E. Stallings, Trans.). Penguin Classics.
Kant, I. (1996). *The metaphysics of morals* (M. J. Gregor, Trans.). Cambridge University Press.
Klemm, G. (1854–1855). *Allgemeine Culturwissenschaft*. Romberg.
More, M., & Vita-More, N. (Eds.). (2013). *The transhumanist reader: Classical and contemporary essays on the science, technology, and philosophy of the human future*. John Wiley & Sons.
Simmel, G. (1968). *The conflict in modern culture, and other essays* (Transl., with an introduction by K. Peter Etzkorn). Teachers College Press.
Zohary, D., & Hopf, M. (2000). *Domestication of plants in the Old World. The origin and spread of cultivated plants in West Asia, Europe and the Nile Valley* (3rd ed.). Oxford University Press.

Rolf Elberfeld is Full Professor of Philosophy at the University of Hildesheim/Germany. He studied Philosophy, Japanology, Sinology, History of Religion in Wuerzburg, Bonn and Kyoto/Japan. His fields of research are Intercultural Philosophy, Philosophy of Culture, Comparative Aethetics/Ethics, Phenomenology, Japanese Philosophy, Philosophy of Language and the body. He is the author of the books "Kitaro Nishida (1870–1945). Moderne japanische Philosophie und die Frage nach der Interkulturalität" (1999); "Phänomenologie der Zeit im Buddhismus. Methoden interkulturellen Philosophierens" (2004); "Sprache und Sprache. Eine philosophische Grundorientierung" (2012); "Philosophieren in einer globalisierten Welt. Auf dem Weg zu einer transformativen Phänomenologie" (2017); "Dekoloniales Philosophieren. Versuch über philosophische Verantwortung und Kritik im Horizont der europäischen Expansion" (2021).

Humour

Florent Trocquenet-Lopez

Abstract Let us dispel any terminological ambiguity from the outset, for it is an almost systematic tendency in all studies on *humorous intent* to deploy, for the sake of precision, a typology differentiating between "humour", "irony", "comedy", "laughter", "satire", "sarcasm", "witty", "burlesque", "farce", etc. In so doing, we run the risk of losing sight of the underlying problem with all of these phenomena, which is their *intentionality*. It was in the seventeenth century that the English borrowed from the French their word *humeur*, that the French had inherited from a physiological and psychological tradition dating back to Hippocratic-Galenician medicine, which largely shaped the medical theories of the classical era (Arikha N. Passions and tempers: a history of the humours. Ecco, New York, 2007). From the word *humeur*, first designating a bodily fluid whose "temperament" regulates the character of an individual, the English forged the word "*humour,*" which the French in turn reimported. *Humour* was eventually understood as the transmutation of a physiological character, attested by the very origin of the term, into a *style*, therefore a *strategy*, which Montesquieu, in a fragment of his *Pensées* written in 1743, defines as "the joke in the joke", or "the manner of comic force" (Montesquieu CL. Pensées. Robert Laffont, Paris, 1991). We will therefore speak here of humour rather than of *laughter*, in order to grasp the intention rather than its effect, since humour can play with laughter and even manifest it without necessarily provoking it, either because the humourist fails to do so or because he or she does not necessarily aim to achieve it: someone who laughs at their own jokes is not necessarily funny; or someone who laughs at someone may not want to make anyone else laugh but themselves.

In any case, humour as an intention poses an ethical problem: the question of whether or not it is humanly desirable, and under what conditions. Bergson strangely evades this question, stating only that "the function of laughter is to intimidate by humiliating" and that it "would not succeed" if man did not have "a small fund of

F. Trocquenet-Lopez (✉)
Ecole normale supérieure des lettres et sciences humaines, Lyon, France

wickedness or at least malice in him" (Bergson, 2001). The philosopher concludes: "Perhaps it would be better if we did not go too deeply into this point. We would find nothing very flattering for us" (*ibid.*). Since 1900, when Bergson's seminal work *Le Rire* was published, the innumerable studies on humour have only partly illuminated its darkness, essentially considering only the *extrinsic* dimension. Historical studies have been devoted to the manifestations, instrumentalization and repressions of humour in the age of totalitarianism, or to the possibility of practicing humour *a posteriori* when considering tragic events in history, especially the Shoah (one thinks here in particular of the work of Loeterwein & Strauss-Hiva, 2009), but also of much of the numerous analyses of Jewish humour, such as that of Judith Stora-Sandor (1984). A second, more anthropological field of research is divided between, on the one hand, work that extends Freud's psychoanalytical approach to humour as a "tendentious" practice (Freud, 2007) and, from a more contemporary perspective, considers the potential thaumaturgical effects of humour, as do, for example, Caroline Simonds and Bernie Warren (Simonds & Warren, 2001), and, on the other hand, ethnological or sociological studies that consider the different practices of "humour policing and policy", including the analysis of humour as a means of education (Calame-Griaule, 1965).

However, since the early 1980s, a new historical context has given rise to a critical discourse about humour that never became a systemic thinking, and which now needs to be deepened and updated. For today we are perhaps living in the twilight of what Gilles Lipovestky called "the humorous society", in which he saw the beginning of "the phase of the liquidation of laughter" (Lipovestky, 1983). This new "era" is marked by a domination of the humorous posture in the media field, and the emergence of humour as dogma, both in "one-man shows", which have multiplied considerately, and in talk shows, which have developed in parallel to the latter, and even perpendicularly, if we consider that many hosts or guests of these shows perform in one-man or one-woman shows (Leroux & Riutort, 2014). "When there is no longer any seriousness, can there still be laughter?" asked Georges Minois some 20 years later, on the threshold of the twenty-first century (Minois, 2000). David Le Breton proposes the term *humorism* to define this period when "taking oneself seriously is a tireless source of mockery and contempt, a sign of arrogance in our hedonistic societies", where "thinking without laughing is excluded because its seriousness breaks the soft consensus of the cool attitude" (Le Breton, 2018).

But this is neglecting the crisis that this humorism seems to be experiencing today, as was witnessed by recent polemics around comments made by humourists and on which they have had to account, and which are proof of a weakening of their power (for the year 2018 alone, we might think of Tex, who was expelled from France 2 for his comments on violence against women, or of Jean-Marie Bigard, who had to cancel a tour in the Var after his jokes on rape). Not least for this reason, Léa Lutaud titled a page in the "Enquête" section of the newspaper *Le Figaro* as follows: "Comedians don't make people laugh anymore" (Lutaud, 2019), and furthermore noted in particular the "distance" between three great figures of the French comedy scene: Dany Boon, Gad Elmaleh and Jamel Debbouze. All three experienced a sudden lack of success with the public.

What do humorism and its recent crisis reveal about the anthropological foundations of humour, and what ethical perspectives can such a crisis offer for human
progress? Far from seeking to determine a code of good conduct for humour, which
would be absurd, we shall focus here on understanding how humour, as practised in
the "humorous society", appears morbid for the social body and, ultimately,
for itself.

The conditions for the possibility of humour do not, of course, relate to content
that would be in keeping with its "essence". As David Le Breton reminds us: "There
is no such thing as the essence of laughter, let alone the essence of comedy" (Le
Breton, 2018). However, this does not mean that we should give up trying to identify major ethical principles, in the sense that Mark Hunyadi gives to the word in the
context of what he calls the "Great Ethics", i.e. principles that make it possible to
understand, and therefore judge, human "lifestyles" by inquiring about their "meaning" (Hunyadi, 2015). All the more that the humorous "way of life" is itself one of
the conditions of possibility of any society: no society can exist without humour.
Correlatively, any repression of humour in a society compromises the development
of the individuals who make it up. It is indeed a virtuous circle, which contemporary
humorism breaks. Finally, these ethical principles of humour can only be linked to
the situation of humorous enunciation, which, it should be remembered, always
involves three agents: the one who laughs, the one he or she makes laugh, and the
one who laughs at something or somebody (each of these agents can be embodied
by individuals or collectives).

It seems to us that three major ethical principles of humour can be formulated:
the *principle of ambiguity*, the *principle of risk*, and the *principle of inclusion*. Let
us take Figaro as our guide, the character of Beaumarchais, one of the most beautiful incarnations, in the French comic repertoire, of the ethical issues of humour.

"I hurry to laugh at everything, for fear of having to weep at it": this maxim from
Figaro in scene 2 of act I of *The Barber of Seville* (Beaumarchais de, 1980) could be
an excellent formulation of the *principle of ambiguity*. We have proposed
(Trocquenet-Lopez, 2018) to approach humour like a *pharmakon* in the sense that
Socrates understood it in a famous passage from the *Phaedra* devoted to the ambiguity of writing. The Greek word *pharmakon*, which designates both a "remedy"
and a "poison", attained the power of a concept in Plato's work. As an ambivalent
substance, humour oscillates between the poison of the often cruel reminder of a
social norm and the "good remedy, the Socratic irony [which] worries the intestinal
organization of self-indulgence," as Derrida notes in his commentary on the *Phaedra*
(Derrida, 1972). It is this substantially ambivalent character of humour that
Jankélévitch sums up in a fruitful oxymoron: "humorous seriousness" (Jankélévitch,
2017). However, two perils threaten, in humorism, the implementation of the *principle of ambiguity*. Firstly, as we have said, the *a priori* banishment of all forms of
seriousness, but also the conformism that animates "talk shows" where the plurality
of voices is only a decoy, concealing a monochrome of dominant opinions - "imper-
tinence" and "insolence" set as a norm as in the "spirit" of the TV channel "Canal+"
(Laborde, 2013), without any "counter-public" to qualify or contradict the jokes
expressed (La Sala Urbain, 2013). But to this lack of any antagonism, which

guarantees the *principle of ambiguity*, is added, more seriously, the practice of cynicism, which transforms the humorous *pharmakon* into a real poison: it denies the part of *lying true* that is at the root of any social bond. We have shown (Trocquenet-Lopez, 2018), based on the study of the *Tartuffe*, that hypocrisy is paradoxically, either to a certain extent, the guarantee of the cohesion of society, or the expression of our authentic desire to preserve it. Cynicism dissolves all forms of hypocrisy by making cruelty, on television or on social networks, a new (anti)social norm.

"Quivering, I try hard", bravely says Figaro, who calls himself "a speaker according to the danger" in his famous monologue in scene 3 of act V (Beaumarchais de, 1980). This states the second great ethical principle of healthy humour: the *principle of risk*. The humourist must expose himself, take responsibility and assume the uncertainty of his enterprise. Humour both distils and dispels anxiety, in a skilfully cathartic process. But this excludes any position of domination that corrupts humour into a pure principle of "distinction" (Bourdieu, 1979). However, humorism, especially as it appears on television, puts the laugher in a doubly dominant position: he or she is the enunciator of a discourse relayed by the television medium, which has the scientifically proven characteristic of placing the spectator in a situation of passivity (Postaire, 2013), and furthermore it is instituted as the undisputed arbiter of ridicule, by virtue of the system of talk shows centred around an undisputed and indisputable star presenter (La Sala Urbain, 2013; Trocquenet-Lopez, 2018).

This *principle of risk* leads to the third and last major ethical principle of healthy humour: the *principle of inclusion*. "Again I say my cheerfulness, without knowing if it is mine more than the rest, or even which self I am taking care of," Figaro confides to the audience in his monologue (Beaumarchais de, 1980): humour, for it is not the prerogative of anyone, or of any particular subject, it is everyone's business. The agents that have been identified (the one that laughs, the one he or she laughs at, and the one he or she makes laugh), must therefore enter into a perpetually dynamic relationship. The laugher must be ready to laugh at himself or herself, to incite the one he or she makes laugh to imitate him or her, and thus to enlighten the *facticity* (in the Sartrean sense) of the self, and of human existence: to dissipate any "spirit of seriousness", and to conjure up by humour the anxiety of our lack of being.

References

Arikha, N. (2007). *Passions and tempers: A history of the Humours*. Ecco.
Beaumarchais de, P. C. (1980 [1784]). The marriage of Figaro. In J.P. Beaumarchais (Ed.), *Théâtre. The Sacristan, The Barber of Seville, The marriage of Figaro*. Garnier.
Bergson, H. (2001 [1900]). *Le Rire*. PUF, "Quadrige".
Bourdieu, P. (1979). *La distinction*. Éditions de Minuit.
Calame-Griaule, G. (1965). *Ethnologie et langage. La parole chez les Dogons*. Gallimard.
Derrida, J. (1972 [1968]). La Pharmacie de Platon. In *La Dissémination*. Le Seuil.
Freud, S. (2007 [1905]). *Le Mot d'esprit et ses rapports avec l'inconscient*. M. Bonaparte & M. Nathan. Classiques des sciences sociales.

Hunyadi, M. (2015). La Tyrannie des modes de vie. In *Sur le paradoxe moral de notre temps*. Le Bord de l'eau.

Jankélévitch, V. (2017 [1963]). *L'Aventure, l'Ennui, le Sérieux*. Flammarion, GF.

La Sala Urbain, S. (2013). Le Petit Journal ou la séduction de l'infotainment. *Television*, 2013/1 n°4. CNRS Editions (pp. 105–123)

Laborde, B. (2013). Présidentielles 2012: quand la campagne s'invite au Grand Journal de Canal+. *Television*, 2013/1 No. 4. CNRS Editions (pp. 43–60).

Le Breton, D. (2018). *Rire. Une anthropologie du rieur*. Métaillé.

Leroux, P., & Riutort, P. (2014). Passer à la télé: Analyser la présence des professionnels de la politique au sein des émissions conversationnelles. *Réseaux, 187*(5), 51–77.

Lipovestky, G. (1983). *L'Ère du vide*. Gallimard.

Loeterwein, A., & Strauss-Hiva, C. (2009). *Rire, mémoire, Shoah*. Édition de l'Éclat.

Lutaud, L. (2019). *Les humoristes ne font plus rire*. Le Figaro, 13 July 2019, 26.

Minois, G. (2000). *Histoire du rire et de la dérision*. Fayard.

Montesquieu C. L. (1991 [1726–1755]). *Pensées*. Robert Laffont.

Postaire, E. (dir.). (2013). *L'Enfant et les écrans. Un avis de l'Académie des sciences*. Le Pommier, Institut de France.

Simonds, C., & Warren, B. (Eds.). (2001). *Le Rire médecin. Journal du docteur Girafe*. Albin Michel.

Stora-Sandor, J. (1984). *L'Humour juif dans la littérature. De job à Woody Allen*. PUF.

Trocquenet-Lopez, F. (2018). Prendre l'humour au sérieux. Éléments pour une éthique du rire. *Revue française d'éthique appliquée*, 2018-2/n°6 (pp. 92–106).

Florent Trocquenet-Lopez, a former student of the École normale supérieure des lettres et sciences humaines, a teacher of literature in preparatory classes at the Lycée Jean-Pierre Vernant in Sèvres, Florent Trocquenet-Lopez is also a writer (he published the novel Le Voisin in 2009 with L'Harmattan, and short stories in periodicals). He co-authored, with Véronique Anglard, La Nature, published in 2015 by Dunod. His current research focuses on the ethics of humour and the notion of "reality".

Imagination

Christoph Wulf

Abstract In the context of the Anthropocene the imagination has three main tasks. Firstly, it is the task of imagination to help to recall historical and cultural facts from history and other cultures. The imagination helps to transcend distance and to make things that are far away from us in time or space part of our personal experience that influence the way we act. To understand the situation of the Anthropocene we need to make things come alive through our imagination. Without the imagination it is not possible for us to understand historical developments, for our emotions to be stirred by them and for us to be challenged to act in new ways. Secondly, the imagination can qualify historical and cultural developments by making us aware that all developments could also have evolved differently. The imagination can combine diverse facts in a variety of ways; it makes new connections between individual factors and produces new ways of looking at things. As a result, we become aware of alternatives. Thirdly, the imagination helps to develop new scenarios which provide a framework for the destructive effects of the Anthropocene to be corrected. For this to happen new ways of living and acting are imagined that contain utopian elements geared towards sustainability. These serve as examples of good scenarios and practices.

The imagination can design alternatives to the present that have sustainability at their core. (Wulf, 2013, 251–274: Wulf, 2022a). Because of this, in the Anthropocene the imagination takes on great importance for our lives on an individual, social and global level (Wallenhorst, 2021; Wulf, 2022b). It assists in the development of possible ways of acting and living that will alleviate the destructive conditions of the modern world and the Anthropocene (Wallenhorst & Wulf, 2022, 2023). The imagination can design scenarios of new forms of life and create new vistas for a life that is less violent. From "The Limits to Growth", a 1972 publication of the Club of Rome (Meadows et al., 2004) and the Brundtland Report (1987), we know how

C. Wulf (✉)
Freie Universität Berlin, Berlin, Germany
e-mail: christoph.wulf@fu-berlin.de

threatening the situation created by humans is and what direction our thinking and actions must now take if we are to arrest the destruction of the planet. This was reconfirmed by the development and acceptance of the 17 Sustainable Development Goals (SDG) by the Community of States in 2015 and the global climate conferences that ensued. We need alternatives to the way we currently live and act that are based on sustainability. In order to achieve this, it is essential for the imagination to be more highly regarded and valued and for it to be promoted in society. A framework must be created which permits exploration and offers the space to imagine and think along different lines. Any curbs on imagination or thought are detrimental to this. Fundamental changes are needed in education, changes that aim to develop the imagination. It is a question less of regurgitating known facts and more of exploring unknown areas and going beyond what we already know as we confront what is not known.

In the context of the Anthropocene the imagination has three main tasks. Firstly, it is the task of imagination to help to recall historical and cultural facts from history and other cultures. The imagination helps to transcend distance and to make things that are far away from us in time or space part of our personal experience that influence the way we act. For us to understand the situation of the Anthropocene we need to make things come alive through our imagination. Without the imagination it is not possible for us to understand historical developments, for our emotions to be stirred by them and for us to be challenged to act in new ways. Secondly, the imagination can qualify historical and cultural developments by making us aware that all developments could also have evolved differently. The imagination can combine diverse facts in a variety of ways; it makes new connections between individual factors and produces new ways of looking at things. As a result, we become aware of alternatives. Thirdly, the imagination helps to develop new scenarios which provide a framework for the destructive effects of the Anthropocene to be corrected. For this to happen new ways of living and acting are imagined that contain utopian elements geared towards sustainability. These serve as examples of good scenarios and practices.

The imagination is not only important in the way we design scenarios of a sustainable life but also in how we can put them into practice. It is closely connected with language but also with that silent, practical knowledge that cannot be pinned down by language (Kraus, 2021). The imagination is an important condition for knowledge that is rooted in the body and movement (Kraus & Wulf, 2022; Wulf, 2022a, c). Many of our hopes for diminishing the anthropogenic, destructive conditions of the Anthropocene are placed in the creative power of the imagination, which we expect to be able to design and put into practice alternatives to what we have now. The extent to which this is possible, in view of the complexity of the way we live in the Anthropocene, is an open question, to which we must find the answer if we are to ensure the future of the planet (Wulf, 2022b).

Imagination, no less than language, is a *conditio humana*, part of our human condition, which is firmly rooted in the human body. Human action is not predetermined by instinct but shaped by the imagination. With the aid of imagination, past, present, and future are closely interwoven. The imagination creates the human world – social and cultural, symbolic, and imaginary. It makes history and culture possible and thus also historical and cultural diversity. It creates the world of images, and imaginaries, and is involved in the creation of practices of the body, such as dances, rituals, and gestures. Staging and performing these practices requires more than an awareness of them – they must be incorporated and become part of a practical, body-based, implicit knowledge, the dynamic nature of which makes social and cultural changes and shaping possible. Of central importance here are mimetic processes that are rooted in the imagination. It is here that the cultural learning takes place that generates social and cultural identity, which is an essential requirement for well-being and happiness (Gebauer & Wulf, 1995, 1998; Wulf, 2022a, c).

Individuals, communities, and cultures all generate imaginaries with the aid of the imagination. As a concept, the imaginary can be understood as an inner world of images that includes sounds, touch sensations, smells, and tastes. Human beings need the imaginary to be able to perceive the world in a way that is shaped by history and culture. The imagination remembers and produces, combines, and projects images. It creates reality. At the same time, it uses reality to generate images. The images of the imagination have a dynamic that structures perception, memory, and the future. Links are forged between the images that follow the dialectical and rhythmic movements of the imagination. Not only everyday life, but also literature, art and the performative arts have an inexhaustible reservoir of images. Some of them appear to be stable and not easily changed, while others are subject to historical and cultural changes. The imagination has a symbolizing dynamic which constantly generates new meanings, using images. People use these images that imagination has created to interpret the world (Blacking, 1977; Boehm, 1994; Huppauf & Wulf, 2009; Belting, 2011; Wulf, 2022a).

The imagination has a strong performative power which stages and enacts social and cultural actions. The imagination uses this power to create the imaginary, which contains remembered images from the past, images of the present and imagined images of the future. Mimetic movements can capture the character of the images. The images are absorbed into the imaginary. As part of our mental worlds, they represent the outside world. Which images, structures and models become part of the imaginary is determined by many different factors. The presence and absence of the outside world are inextricably interlaced in these images. Images that emerge from the imaginary are transferred to new contexts by the imagination. Networks of images are formed with which we capture the world, and which determine how we perceive it. The aim is to develop an individual and social imaginary that has as its focus the values and images of a sustainable world and a sustainable life, and which guides people's emotions and actions accordingly. Examples of images of the individual and collective imaginary that are relevant here are images of our planet in the vastness of the universe, images of destroyed landscapes, anthropogenic catastrophes. All of these images become part of the imaginary and urge us to act in a way that is based on the values of sustainability.

In addition to the images of nature a large part of the imaginary is made up of performative images from our social lives (Kress et al., 2021). The power structures of social conditions and societal structures are represented in these images. Many of these processes begin in childhood and take place unconsciously. We already learn to perceive social constellations and arrangements at this early age. These early experiences of perceiving the images that result from them play an important and indispensable role in how we grasp the world. All our perceptions are influenced in some way by historical and cultural schemata and mental images that we build up over the course of our lives, enabling us to grasp what we see. We see social actions and relate to them as we perceive them. They thus take on significance for us. When other people direct their actions towards us, the impulse to establish a relationship comes from them and they expect us to respond. Whatever happens, a relationship is formed, and the images of our imaginary are an important prerequisite of this relationship. We enter into an action game and act in relation to what is expected of us in this social arrangement, be it that we respond to it, modify it or oppose it. Our actions are mimetic, less in virtue of their similarity than in virtue of the corresponding reactions that they engender. Once we have engaged in an action game, we perceive the actions of the other participants, and our own actions relate to them mimetically.

Therefore, with the help of imagination we make our endangered world into part of our (mental) inner world. Here it is not only the iconic and acoustic images and the images of touch that become part of our imaginaries, but also the values, norms, and forms of action that they contain (Wulf, 2013). They influence our perception of the outside world. With the help of the imagination, situations that are distant both in time and space become part of our imaginaries and thus also of our bodies. The images, their values and norms, their emotional character, influence how we act. They can make us act in a way that promotes sustainability. It, therefore, is important to develop an imaginary that is based on the attainment of sustainability if we are to act in a sustainable way. In the context of Global Citizenship Education, it has an important role to play (Bernecker & Grätz, 2018; Wulf, 2021).

References

Belting, H. (2011). *An anthropology of images*. Princeton University Press.
Bernecker, R., & Grätz, R. (Eds.). (2018). *Global citizenship education*. Steidl.
Blacking, J. (Ed.). (1977). *The anthropology of the body*. Academic.
Boehm, G. (Ed.). (1994). *Was ist ein Bild?* Wilhelm Fink.
Brundtland Commission. (1987). *Our Common Future*. Oxford University Press.
Gebauer, G., & Wulf, C. (1995). *Mimesis. Culture, art, society*. University of California Press.
Gebauer, G., & Wulf, C. (1998). *Spiel, Ritual, Geste. Mimetisches Handeln in der sozialen Welt*. Rowohlt.
Huppauf, B., & Wulf, C. (Eds.). (2009). *Dynamics and performativity of imagination. The Image between the Visible and the Invisible*. Routledge.
Kraus, A., Budde, J., Hietzge, M., & Wulf, C. (Eds.) (2021). 2.edn. *Handbuch Schweigendes Wissen*. Beltz Juventa.
Kraus, A., & Wulf, C. (Eds.). (2022). *Handbook of embodiment and learning*. Palgrave Macmillan.
Kress, G., Selander, S., Säljö, R., & Wulf, C. (Eds.). (2021). *Learning as social practice. Beyond Education as an Individual Enterprise*. Routledge.

Meadows, H. D., Randers, J., & Meadows, D. (2004). *The limits of growth*. Taylor & Francis.

Wallenhorst, N. (2021). *Mutation. L'aventure humaine ne fait que commencer*. Le Pommier.

Wallenhorst, N., & Wulf, C. (Eds.). (2022). *Humains – un dictionnaire d'anthropologie prospective*. Vrin.

Wallenhorst, N., & Wulf, C. (Eds.). (2023). *Handbook of the Anthropocene*. Springer Nature.

Wulf, C. (2013). *Anthropology. A continental perspective*. The University of Chicago Press.

Wulf, C. (2021). Bildung zu einer planetarischen Weltgemeinschaft im Anthropozän. *Vierteljahreszeitschrift für Wissenschaftliche Pädagogik, 97*(2021), 263–479.

Wulf, C. (2022a). Humans and their images. In *Imagination, mimesis, performativity*. Bloomsbury.

Wulf, C. (2022b). *Education as human knowledge in the Anthropocene. An anthropological perspective*. Routledge.

Wulf, C. (2022c). Embodiment through mimetic learning. In A. Kraus & C. Wulf (Eds.), *The Palgrave handbook of embodiment and learning*. Palgrave macmillan.

Christoph Wulf is Professor of Anthropology and Education and a member of the Interdisciplinary Centre for Historical Anthropology, the Collaborative Research Centre (SFB, 1999–2012) "Cultures ofPerformance," the Cluster of Excellence (2007–2012) "Languages of Emotion," and the Graduate School "InterArts" (2006–2015) at the Freie Universität Berlin. His books have been translated into 20 languages. For his research in anthropology and anthropology of education, he received the title "professor honoris causa" from the University of Bucharest and the honorary membership of the German Society of Educational Science. He is Vice-President of the German Commission for UNESCO. Major research areas: historical and cultural anthropology, educational anthropology, imagination, intercultural communication, mimesis, aesthetics, epistemology, Anthropocene. Research stays and invited professorships have included the following locations: Stanford, Tokyo, Kyoto, Beijing, Shanghai, Mysore, Delhi, Paris, Lille, Strasbourg, Modena, Amsterdam, Stockholm, Copenhe, Basel, Saint Petersburg, Moscow, Kazan, Sao Paulo.

Museum

Hartwig Lüdtke

Abstract Museums are the institutions that acquire, preserve, research, and display the material evidence of human activities in the Anthropocene epoch. These include not only works of art, but also objects of daily use such as clothing, domestic furnishings, and tools and equipment used in production and manufacturing. Objects from a religious context also find their place in museum collections. As an educational institution, the museum is therefore precisely the right place to address the social debate about future developments and the various options available to human societies. It would of course be unreasonable to believe that museums could provide clear-cut answers to such complex questions as these, but they can certainly illuminate the many facets of highly topical issues, and offer points of reference enabling us to draw our own conclusions.

Over the course of the last 20 years, an increasingly intensive social debate has developed surrounding the term 'Anthropocene'. It would, however, be incorrect to assume that this issue has only recently been discovered. In fact, discussions of a geological epoch strongly influenced by human beings, called the Anthropozoic, have appeared for many decades in the relevant scientific fields, as many key reference works reflect (Filip, 1966, 37).

In many cases, however, the exact definition of the time period identified as the Anthropocene remains unclear to this day. Broadly speaking, three different points of view deserve consideration:

On the one hand, the term can be used to designate the period of geological history characterized by the presence of hominids. This would encompass approximately the last 3–4 million years. While humans already existed in the world during part of this phase, for much of it their way of life remained largely integrated into

H. Lüdtke (✉)
Mannheim's TECHNOSEUM – Landesmuseum für Technik und Arbeit,
TECHNOSEUM Stiftung, Mannheim, Germany
e-mail: hartwig.luedtke@technoseum.de

the natural conditions of their surroundings, and, as hunter-gatherers, they predominantly followed the opportunities offered by the natural environment, without significantly altering it.

On the other hand, one can make a case for beginning the Anthropocene at the start of the Neolithic Revolution, that is, with the development of agriculture and animal husbandry, which led humans to become sedentary. A number of other aspects characterize this period, which encompasses approximately the last 10,000 years. These include not only the emergence of larger population clusters in the form of early cities, but also the corresponding establishment of a consciousness of territorial ownership and territorial belonging. These in turn set the stage for armed conflicts and warfare, which have, over time, evolved to a scale that now allows for the possibility of nuclear confrontation and the extinction of life on this planet. More important for our inquiry, however, is the fact that humans began actively intervening in and changing the natural world during the emerging Neolithic period. These interventions included early irrigation systems as well as the clearing of forests in order to gain more arable land or to harvest appropriate building materials. In the course of the Iron and Bronze Ages the deforestation of large areas intensified, as this was the only means of obtaining the fuel required for smelting. This often large-scale deforestation set off processes of soil erosion, which in turn changed the environment and the landscape. From this perspective, much evidence speaks for assigning this entire time period to the Anthropocene.

Finally, however, the debates also take a closer look at the most recent phase of approximately the past 200 years, beginning with the Industrial Revolution and characterized by a dramatic increase in the consumption of natural resources and by the resulting changes in and impacts on the natural environment.

There are enough perfectly plausible arguments to support any one of these definitions of the term 'Anthropocene' as outlined above. It is not, however, our present concern to settle on one definition over another, as this bears little relevance to the keyword 'museum' in the context of the Anthropocene. As outlined at the beginning, museums collect and preserve the material evidence of human activity, regardless of whether we take a broader or narrower temporal view of the period.

> A museum is a non-profit-making permanent institution in the service of society and of its development, open to the public, which acquires, conserves, researches, communicates and exhibits, for purposes of study, education and enjoyment, the tangible and intangible evidence of people and their environment.

This wording represents the internationally accepted and currently valid definition of the museum as set out in the Code of Ethics of the International Council of Museums or ICOM (ICOM, 2006, 14). The ICOM, to which approximately 50,000 museums and individuals from around the world belong today, very intentionally agreed on this definition, which is universal in scope and applies to all museums, regardless of theme or geographic region. According to this definition, the core purpose of museums is therefore collecting items that are either objects of the natural world or human-made artefacts. Building on these collections, the goal is to research the objects themselves, and the structures and contexts they represent, and finally to present and convey both objects and findings to a broad public in the spirit of scholarly communication in the service of societal development.

The term 'museum' derives from the ancient Greek 'mouseion', which was used in the centuries before the Common Era to designate sites that on the one hand served cult purposes, but were, on the other, also places of research and learning. Of these, the *mouseion* in Alexandria, along with its extensive library, has enjoyed the most lasting fame. Today's museums carry on this fundamental mission of research and the transmission of knowledge, even though no direct line of descent links the ancient institutions to those of the modern era. The modern museum as we know it today developed in the late eighteenth and early nineteenth century, and ultimately stands as a product of the Enlightenment (Hochreiter, 1994, 181). The conversion, in the wake of the French Revolution, of the former Palais du Louvre into what is now a world-renowned museum forms a prominent milestone in the evolution of the museum and its increased accessibility to the general public. In many cases, museum collections are built on the art collections of the royal and princely courts of Europe, but at the same time they are also founded on collections compiled at universities for the purpose of scientific observation and study.

The first step in analysing the respective collections of objects is to carefully examine various details. Objects of one kind are grouped with others like them, creating a form of order. Following this organization of individual items, the next step entails an ordering of systems, culminating in a view of the multifaceted history of the world in general and of human beings in particular which emerges from the encyclopaedic structuring of these ordered and subordered systems. By extension, such structuring also charts our own cognizance of human activity and human intervention in the Anthropocene.

The nineteenth century saw a shift towards increasing specialization and a refinement of academic disciplines, leading to the rise of separate museums and museum types based on groups of materials and object contexts. As a consequence there emerged distinct museums or collections of art, archaeology, natural history, or technology. Interestingly, as early as the nineteenth century the idea also emerged that these museum collections should by no means only represent the past, but also examine the present day, and use the insights thus gained as a foundation for learning about and inquiring into the future. A good example of this in the context of Germany is the foundation of a 'Reichspostmuseum' in Berlin in 1871 (a museum of the history of communication), whose founding charter explicitly states that its collections were to be preserved for the instruction of *future* technicians and engineers (Katalog, 1897, VI).

Over the course of the twentieth century, the museum, like so many other institutions, was drawn into the maelstrom of ideological appropriation and instrumentalization. Museum artefacts were used to support notions of national identity while scholars traced supposed genealogies linking their epoch with past eras and alleged forebears. This phenomenon was particularly pronounced in many German museums in the period between 1933 and 1945 (Bouresh, 1996). This is not the place to address this aspect in greater detail, but it is worth noting that the fundamental danger persists to this day of museum collections being interpreted in a one-sided and limited manner, and being appropriated for political purposes. A territorial conflict between neighbouring countries, for example, can tempt the parties involved to use evidence of supposedly older artefacts in the ground to legitimize their current territorial claims.

In the modern world following the Second World War, museums of all disciplines have increasingly viewed themselves as institutions of research and public education. In many cases, this self-perception as a 'place of learning' has gone hand in hand with a considerable increase in the number of museums themselves and an increase in the year-on-year number of visitors to them. As in the past, collections remain central to the museum, but the importance of scholarly communication and educational programming is steadily increasing. Museums can play a particularly important role in lifelong learning, because they provide equal opportunities for people of all ages to engage with various subjects in history, art, natural history, or technology, and to gain new insights. In this sense, museums form a special kind of 'third place', the term used in current debates to identify precisely those educational opportunities offered outside the confines of formal education.

In recent years, 'entertainment value' has followed hot on the heels of 'educational value', giving rise to the new term 'edutainment' as a melding of the two. The idea behind this is that the specific experience of 'visiting a museum', which also includes a social component, allows people to gain new knowledge and new insights in a playful way. This development goes hand in hand with an expansion of participatory opportunities in many museums, which allow the visitors themselves to become actively involved in how and what they experience. Visiting a museum no longer means only looking at exhibited objects, but instead involves direct interaction with hands-on installations, for example, or participation in small workshops. The concept of participation and public engagement in museum work, however, goes a step further, and can even describe the process of actively involving the respective members of, for example, a local urban community in the curatorial planning of museum exhibitions. What subjects should be presented and debated in the museum? Who can and would like to contribute their experiences or their own objects to this debate and, if necessary, to the presentation? Museums are now posing these and similar questions to visitors, opening themselves up to a new kind of dialogue.

Interestingly, from one decade to the next we continue to observe an increase in the number of visitors to most museums, large and small. This only seemingly contradicts the fact that in other areas our world is becoming ever more digital and virtual. Our desire for authentic experiences is quite obviously growing in reciprocal response to the ever-increasing virtuality of the environments around us. During the television broadcast of a football match, for instance, viewers can follow and judge individual scenarios much better and more clearly than in a large stadium, where the seats are far from the playing field. Nevertheless, this hardly means that fans no longer flock to the stadium to experience the game live. The same applies to music: here again, in many cases, we can enjoy canned music quite comfortably via speakers in the home, while relaxing on our sofa; nevertheless, live concert performances in all genres, from classical to pop, remain hugely popular. We can say the same of a visit to a museum: here again we could acquire new knowledge conveniently and efficiently just by reading books or browsing online from the comfort of our own homes. The growing influx of visitors to museums, however, reflects precisely this interest in the authentic experience, in this case the first-hand encounter with the original object as a product of artistic creation or as direct material evidence of history. This 'aura' of the original has always determined what makes a museum special (Glaser, 1990).

In the debate surrounding the term 'Anthropocene', the question always arises of how humans will contend with the epoch, and how they will shape its further course (Möllers, 2015). In recent years, museums of natural history in particular have explored precisely this issue, presenting not only the history of the Earth, the history of life and biodiversity, but also the historical development of climate change, while also addressing future questions in order to draw attention to aspects of sustainable development (Koster, 2016). They probe subjects such as the relationship between humans and the environment, and to this extent these museums see themselves as educators for sustainability. Museums of science and technology also play a specific role with regard to the questions of how human activity has shaped the world during the Anthropocene, and which new, still undeveloped technologies humans can potentially use to influence and shape further development in a positive way (Gold & Lüdtke, 2012). The major issues of our age – such as climate change, global energy resources, and feeding a growing world population – can only be solved in the long term if we not only take into account the necessary social and behavioural changes, but also integrate new, resource-saving technologies into the package of solutions. This can involve the creation of resource-efficient infrastructure, but at the other end of the spectrum may even encompass geoengineering.

In conclusion, contemporary museums provide three different perspectives:

Firstly, they offer glimpses into other, earlier epochs for respective regions, illuminating the environmental situation as well as the living conditions and social circumstances of the people there during specific time periods. Museums allow us to address the question of what changes have occurred between a given point in time and our current era, and based on this, prompt us to explore what specific factors shape our situation today. In conjunction, they also highlight positive or negative developments and trends. After all, the central question is always how human activities and interventions have changed the world.

Secondly, museums provide a window onto the current conditions in other, often distant regions of the world. Here again museums illuminate the questions of what different social situations prevail, and what different approaches to managing natural resources and overcoming future challenges might exist.

Thirdly and finally, museums offer a glimpse into what can at least be imagined in the future. How will the future shape out? What are the different possible options and how much do they depend on active human intervention and decision-making? More than ever before, we must also address the question of how the relationship between humans and the environment will continue to evolve and present itself in the future.

Therefore, if we seek to discuss the definition of the Anthropocene and its history up to this point, and if we wish to explore the different options for human agency, we should turn to museums. They preserve not only the actual physical relics of human activities, but also the material evidence of immaterial cultural traditions, which are considered an essential facet of human life in the Anthropocene: musical instruments, theatre puppets, tools of various crafts, and liturgical objects. This evidence can also include the representation of abstract phenomena, as illustrated by the example of the term 'work' or 'labour' (Lüdtke, 2017). Building on this, museums offer individual opportunities for lifelong learning, while also providing a venue where society as a whole can engage in the public debates that will play a role in future developments. The exhibition format enables the museum to cut complex topics into bite-size portions, make phenomena more tangible, and in this way promote scholarly communication in the best sense of the word. Presenting scholarly findings to a broad public is, after all, an indispensable prerequisite for evidence-based decision-making in a democratic society. Only based on this foundation can we realistically create models for future developments and also define the individual steps necessary to bring these developments to fruition. We can ultimately only imagine and build the future structures of states, societies, and economic systems if we recognize and understand the evolutions that have taken place so far and that have been shaped by human beings. In this context, whether we define the Anthropocene as an epoch of 3 million years or of 200 proves irrelevant to the burning questions of our time.

References

Bouresh, B. (1996). Die Neuordnung des Rheinischen Landesmuseums Bonn 1930–1939. *Brauweiler*.

Filip, J. (1966). *Enzyklopädisches Handbuch zur Ur- und Frühgeschichte Europas* (Vol. 1). Berlin, Cologne, Mainz.

Glaser, H. (1990). Aura, Museen, Aufhebung. Kultur im Zeitalter der technischen Reproduzierbarkeit. In: *Preis, Achim, Stamm, Karl, Zehnder, Frank-Günter, Das Museum – Die Entwicklung in den 80er Jahren. Festschrift für Hugo Borger zum 65* (pp. 141–150). Geburtstag.

Gold, H., & Lüdtke, H. (2012). Die Technikmuseen. In: *Denkschrift zur Lage der Museen* (herausgegeben von B. Graf und V. Rodekamp) (pp. 367–380).

Hochreiter, W. (1994) *Vom Musentempel zum Lernort – zur Sozialgeschichte deutscher Museen 1800–1914*. Darmstadt.

ICOM (International Council of Museums). (2006). *ICOM Code of Ethics for Museums*, Paris.

Katalog. (1897). *Katalog des Reichspostmuseums*. Berlin.

Koster, E. (2016). From Apollo into the Anthropocene. The odyssey of nature and science museums in an external responsibility context. In B. L. Murphy (Ed.), *Museums, ethics and cultural heritage*. Routledge.

Lüdtke, Hartwig (2017). Die Darstellung von Arbeit im Museum – nicht einfach, aber spannend. In: Von der Weltausstellung zum Science Lab. *Handel – Industrie – Museum*. (ICOM Deutschland. Beiträge zur Museologie. Band 6). Berlin, pp. 111–119.

Möllers, N. (2015). Museums and the Anthropocene: Reconfiguring time, space and human experience. In N. Möllers, C. Schwägerl, & H. Trischler (Eds.), *Welcome to the Anthropocene: The earth in our hands*. Deutsches Museum Verlag. (pp. 108–112).

Prof. Dr. Hartwig Lüdtke (b. 1954), director (since 2006) of Mannheim's TECHNOSEUM – Landesmuseum für Technik und Arbeit, and chair of the TECHNOSEUM Stiftung. University study in archaeology, excavations and research in northern Europe. Visiting researcher in Norway. Director of Wikinger Museum Haithabu. Director of Rheinisches Landesmuseum Bonn. Curator of Museumsstiftung Post- und Telekommunikation. Publications in medieval archaeology and museology. Vice president (since 2014) of the German Commission for UNESCO.

Pantheism

Arianne Conty

Abstract Though reputed to be one of the most feared heresies of the Christian tradition, due to its deconstruction of the dichotomies of the sacred and the profane, pantheism is well-represented across world traditions, and has remained the religious position of some of the greatest minds of the Western tradition. From the Advanta Vedanta tradition in India to Taoism in China, and from the Stoics of ancient Greece to Spinoza, Hegel and Einstein, understanding the cosmos itself to be divine constitutes one of the most longstanding religious traditions of human history. Indeed, the reversal of values that pantheism enacts has led it to thrive in the Anthropocene Age, precisely because it re-enchants the material world and the mortal and fragile bodies that inhabit it. Instead of a transcendent deity interpreted as existing somehow outside the cosmos, Pantheism renders the immanent divine, thereby re-enchanting the immanent world of nature and of material kinship and inter-dependent unity. This article will differentiate pantheism from atheism and theism, clarify the meaning of monism that it pre-supposes, explain its different typologies (dual-aspect, ontological and teleological) and explain how the reversal of values intrinsic to pantheism can contribute toward the creation of a new paradigm that can respond to the Anthropocene Age.

From the Greek roots *pan* (all) and *theos* (God), the defining trait of pantheism is understanding the cosmos itself to be divine. The Stanford Encyclopedia of Philosophy defines pantheism, as the "view that God is identical with the cosmos, the view that there exists nothing which is outside of God, or else negatively as the rejection of any view that considers God as distinct from the universe."[1] The

---

[1] https://plato.stanford.edu/entries/pantheism/

A. Conty (✉)
American University of Sharjah, College of Arts and Science, Sharjah, UAE
e-mail: aconty@aus.edu

N. Wallenhorst, C. Wulf (eds.), *Handbook of the Anthropocene*,
https://doi.org/10.1007/978-3-031-25910-4_146

Routledge Encyclopedia of Philosophy similarly defines pantheism as "the view that Deity and Cosmos are identical" (Yandell, 1998: 202). Because pantheism understands the cosmos as constituting a monistic unity that is divine or sacred, individual entities are understood as modes or attributes of this totality, this unity. Pantheists thus deny that the divine can transcend or exist separately from the substance of the cosmos, and they also deny that the divine could be a transcendent person, all persons requiring bodies made of matter. The divine cannot transcend the cosmos, nor have an autonomous existence separate from the matter of the cosmos. The divine is immanent in the world.

Pantheism is present in many world traditions, such as the Advaita Vedanta school of Hinduism. Shankara, the most celebrated philosopher of the Advaita Vedanta school, held that the entire cosmos was one divine principle, called *Brahman*. Ignorance (*avidya*) keeps humans from understanding this divine unity, and it is thus the goal of philosophy to overcome the distorted perception of the world as constituted of individual intrinsic essences (*maya*), in order to reach the divine totality. The self (*purusha*) is capable of overcoming the ego (*ahamkara,* the principle of I, me, mine) in order to realize its true nature as *Atman,* the universal soul that is a reflection of and hence one with the divine totality that is *Brahman.*

The Taoist tradition in China can also be considered pantheistic, in that it understands the nature of reality in a monistic manner as the *Tao,* the one and all of which each individual being is a part. As a mysterious, numinous unity, the Tao sustains all things, and though inaccessible to rational thought, each person must seek to live in accord with its laws, which are the laws of nature, the way of all things. Thus the Tao master Lao Tzu wrote of the Tao in verse 25 of the Tao Te Ching (sixth century BCE):

> There is something formless and perfect.
> Before the universe was born.
> It is serene, empty.
> Solitary. Unchanging.
> Infinite. Eternally present.
> It is the mother of the universe.
> For lack of a better name,
> I call it the Tao. (Mitchell, 2006: 25)

Many polytheistic traditions, such as that of ancient Greece, can also be considered consonant with Pantheism, to the extent that they hold that the gods and goddesses that inhabit the cosmos are all manifestations of a single unity, or energy, or absolute totality. Pantheism is also present in many mystical traditions, including Jewish, Islamic and Christian mysticism. Though pantheism was considered a heresy by dogmatic Christianity, it has a long history in the West, and is the religious view of some of the greatest minds of the Western tradition, from the Presocratics, Giordano Bruno, Spinoza and Hegel, to Goethe, Ralph Waldo Emerson, Walt Whitman, Carl Sagan, and Einstein.

Pantheism represents a monistic position, over and against forms of dualism typical of monotheism in particular. In pantheism, the cosmos is a sacred totality or unity, manifest in myriad distinct forms. Such monism most often takes the form of Substance monism (Spinoza), but ontological monism (Ibn Arabi) and teleological monism (Hegel) are also well represented. Substance monism, most famously exposed in the philosophy of Spinoza, holds that the cosmos is absolute Substance, which can be apprehended in two distinct modes, thought or extension. While distinct, both thought and extension are modes of the same absolute Substance, which is infinite and all-encompassing. Spinoza's substance monism is called dual-aspect pantheism, because substance can be apprehended in two different modes, as thought and as extension.

From an ontological perspective, all that exists IS, and thus Being is understood as the absolute totality, the nature of the all. This position is most famously exposed in the philosophy of Heidegger, but it takes its most accomplished religious form in the mysticism of Ibn Arabi, for whom God is understood as absolute being, and absolute unity (*tawhid*), and thus as permeating all that exists as the unity of all being. Such an interpretation is called ontological pantheism.

Finally, teleological monism finds its best advocate in the philosophy of Friedrich Hegel, for whom the dialectical struggle of history reaches its teleological goal at the end of time, when self-consciousness finally recognizes its absolute identity with Geist, Spirit or God, and Geist is recognized as all and everything. Interpreting the entire universe as Geist, or Spirit, and understanding matter as a mere attribute of Spirit is called idealistic pantheism.

Today the most common form of religious monism is naturalistic pantheism, a position adopted in ancient Greece by the Stoics, and today by the Gaia movement as well as by many environmentalists and scientists. Unlike dual-aspect monism, such an interpretation understands the earth as a material body, and energy or mind as a quality or effect of organized matter. Treating the earth as a living self-regulating organism that seeks equilibrium, as formulated by James Lovelock and Lynn Margulis in the Gaia hypothesis, allows for a scientific form of monism that requires no belief in transcendent or personable deities.

If God is not distinct from the universe, since the divine and the universe are one, we will need to ask how such a view is different from atheism. Indeed, when Spinoza wrote *Deus sive natura,* "God or nature," his pantheistic philosophy was equated with atheism, since God seems to add nothing to the universe in itself. Yet as infinite substance, the universe as God is more than its many attributes, and it is this infinitude that best captures the religious dimension of Spinoza's thought and that inspired his religious awe. A naturalistic cosmos that is entirely immanent requires no reverence toward the cosmos, and tends to avoid thinking of the cosmos as a monistic unity, typical of pantheism. Yet for many Gaians, the earth itself, named after the Greek goddess of the earth, Gaia, warrants our veneration as a living and volitional organism seeking to maintain the best conditions for all of life. The difference between pantheism and atheism thus lies in the way one relates to the world, which for pantheists is a unity capable of inspiring wonder, thankfulness, veneration and grace.

Herself a pantheist, philosopher Mary-Jane Rubenstein expresses her beliefs as follows:

> Try and penetrate with our limited means the secrets of nature and you will find that, behind all the discernible concatenations, there remains something subtle, intangible and inexplicable. Veneration for this force beyond anything that we can comprehend is my religion. To that extent I am, in point of fact, religious. (Rubenstein, 2018: 162)

Scientists such as Einstein and Complexity theorist Stuart Kauffman also espouse this view, by honoring the "ceaseless creativity in the natural universe" as "so worthy of awe, gratitude, and respect, that it is God enough for many of us" (Kauffman, 2010: xi; 6). Einstein's words summarize quite well this naturalistic form of pantheism that is most common today:

> Try and penetrate with our limited means the secrets of nature and you will find that, behind all the discernible concatenations, there remains something subtle, intangible and inexplicable. Veneration for this force beyond anything that we can comprehend is my religion. (Kessler, 1971: 157)

The pantheistic position can most clearly be differentiated from theism in its refusal to attribute transcendence and personality to the divine. The closest theistic position to pantheism is that of panentheism, which agrees with pantheism that God is in the cosmos as the cosmos, while at the same time holding that God can somehow remain at the same time transcendent and thus separate from the cosmos. Such a separation implies dualism, which is anathema to most pantheists. It is this adamant refusal of dualism that earned pantheists the reputation as one of the most hated and feared heresies in the Christian tradition. Such a conflation between body and mind, world and god, put into question the entire structure of monotheistic dogma, inspiring what Mary-Jane Rubenstein has called the "monstrous." We will follow Rubenstein in understanding these dichotomies as structurally intrinsic to Modern culture since the distinction between mind and body, God and cosmos branches into the other dichotomies so typical of Modern ideology: transcendent/immanent, reason/emotion, eternal/mortal, human/animal, male/female ... By rejecting this first division, pantheism rejects all the others, making them redundant, hence its danger and the "monstrosity" of its risk.

As Rubenstein points out, the first terms of these dichotomies are those associated with the disembodied divine in Western metaphysics, understood as worthy of veneration, over and against the material earth, associated with the second terms of these dichotomies (Rubenstein, 2018: 3). By collapsing the dichotomy between God and world, the privilege of "spirit, masculinity, reason, light and humanity" over "matter, femininity, passion, darkness, and animal-vegetal-minerality" is also undone (Rubenstein, 2018: 11–12, citing Jantzen, *Becoming Divine,* 267).

Even more monstrous for the dogmatic monotheist, once the cosmos and all of its creatures are understood as part of God, humanism can no longer be understood in the exclusionary terms of anthropocentrism, in which other animals are denied a soul and treated exclusively as means to human ends. In pantheism, the nematode and the panther are parts of the divine to the same extent as the human animal, and are thus worthy of the same solicitude.

In her book *Environmental Culture: The Ecological Crisis of Reason,* Environmental philosopher Val Plumwood traces a continuous line between such remoteness from the natural world of mortal bodies, and a similar remoteness the male elite established between themselves and women, slaves, and colonized others, who were deemed lacking in human reason, and thus part of exploitable nature (Plumwood, 2001: 21). It thus becomes impossible to separate anthropocentrism from other problematic centrisms, such as androcentrism, ethnocentrism and eurocentrism. All of these centrisms use a binary structure to exclude human and nonhuman others by isolating values to the realm of the non-material mind or soul alone. Pantheism can thus be considered a valuable religious position consonant with contemporary post-colonial, feminist, animal-rights and environmental struggles to overcome such biases.

Just as tantrism is known to reverse the boundaries between the sacred and the profane in the Hindu and Buddhist traditions, so pantheism enacts the same reversal in the Western tradition, rehabilitating and valorizing precisely those traits deemed inferior in Western Christian culture. As earthly and immanent, the divine no longer inspires escape from the earth to a transcendent spiritual dimension, where death is overcome. Rather, the earth itself is heaven enough, and mortality is understood as essential to the regeneration of life, and the seasonal cycles of budding, flowering, decaying and dying are celebrated as the nature of the real.

Because pantheism is concordant with a scientific worldview while allowing for a spiritual attitude towards the cosmos, understood as a monistic unity that gives meaning to all of its constituent parts, pantheism's popularity today should not surprise us. Honoring the earth as Gaia and the cosmos as a mysterious force beyond our comprehension are attitudes that could inspire a paradigmatic shift in our understanding of matter and the life that it enables. Instead of understanding the earth as a source of exploitable resources waiting to be exploited, the norm in Western dualism that posits the divine as superior and opposed to the material world, pantheism could help us understand the material earth and all the material and mortal bodies that inhabit it as worthy of veneration and respect.

References

Kauffman, S. (2010). *Reinventing the sacred: A new view of science, reason and religion.* Basic Books.

Kessler, H. (1971). *The diaries of a cosmopolitan.* Weidenfeld & Nicholson.

Plumwood, V. (2001). *Environmental culture: The ecological crisis of reason.* Routledge Press.

Rubenstein, M.-J. (2018). *Pantheologies: Gods, worlds, monsters.* Colombia University Press.

Tzu, L. (1991/2006). *The Tao Te Ching* (Stephen Mitchell, Trans.). Harper Perennial.

Yandell, K. (1998). Pantheism. In E. Craig (Ed.), *Routledge encyclopedia of philosophy.* Routledge.

Arianne Conty is a continental philosopher working as associate professor of philosophy at the American University of Sharjah. Her research is in the fields of the philosophy of nature, the philosophy of technology, and the philosophy of religion. She has published over twenty articles in reputable peer-reviewed journals, and has just completed a book entitled *Grounding God: Religious Responses to the Anthropocene Age* with SUNY Press. Her recent articles on the Anthropocene are "Animism in the Anthropocene," 2021, *Theory, Culture & Society*, forthcoming; "Religion in the Anthropocene," 2021, *Environmental Values*; "Panpsychism: A Response to the Anthropocene Age," 2021, *Journal of Speculative Philosophy*; "The Politics of Nature: New Materialist Responses to the Anthropocene," 2018, *Theory, Culture & Society*; "How to Differentiate a Macintosh from a Mongoose: Technological and Political Agency in the Age of the Anthropocene," 2017, *Research in Philosophy and Technology*; "Who is to Interpret the Anthropocene? Nature and Culture in the Academy," 2016, *La Deleuziana*.

Relatedness

Sébastien Claeys

Abstract This article examines the concept of "relatedness", created by Roger Clausse and theorized in the 1970s by the sociologist Marcel Bolle de Bal. This concept is a way of looking at the crisis of connecting with others and of relating to the world in which we live. In this crisis, we are no longer able to relate to each other, nor to ourselves or to our environment. Furthermore we are not able to connect to a techno-scientific world that is changing at great speed, and which does not offer the possibility of synchronizing with a system of common social, political and legal norms. In this context of widespread "disrelatedness", the notion of "relatedness" emerges, not only as a tool for sociological analysis, but also as a tool for social transformation. It invites us to rethink the creation and perpetuation of links in a holistic manner, engaging at the same time with our environmental, social and mental ecology.

In order to speak of relatedness, we need to clarify the concept of "society of disrelatedness" (Bolle de Bal, 2003). According to this concept, modernity is characterized by a powerful movement of various forms of division. It brings the separation between man and nature, which institutes a predatory relationship with the environment (ecological disrelatedness). It also sets an extreme division of labour in the factory, separating men and women from their knowledge (psychological disrelatedness). Further, modernity fragments knowledge between distinct disciplines (epistemological disrelatedness), and also destructures traditional and intergenerational social ties (social disrelatedness). An additional division is given by the advent of methodological individualism and *homo œconomicus* in the field of social sciences (Généreux, 2006), taking away our capacity to think about collective dynamics (political disrelatedness). Analysing the centrifugal mechanics of these forms of 'disrelatedness' constitutes a central field of study in sociology. All combined, these "disrelatednesses" today take the form of a profound and protean crisis

S. Claeys (✉)
Sorbonne Université, Espace éthique Île-de-France, Paris, France
e-mail: sebastien.claeys@aphp.fr

N. Wallenhorst, C. Wulf (eds.), *Handbook of the Anthropocene*,
https://doi.org/10.1007/978-3-031-25910-4_147

899

of mediation with various manifestations. It manifests itself in citizens' mistrust against scientific discourse; as well as against media. These types of mistrust are demonstrated, in particular, by the rise of conspiracy theories. Major political crises can be another manifestation, which were analysed by Colin Crouch as a post-democratic turning point (Crouch, 2013). For Crouch, this turning point can be characterised by the questioning of institutions and elected representatives at all levels – whether within political parties, trade unions, or the National Assembly. A further manifestation takes the form of labour crises, accompanied by a loss of meaning linked to the excessive standardisation of tasks (Graeber, 2018).

This long inventory has been extensively detailed and studied in the field of social sciences. What the notion of relatedness now allows us to do, by following the path of "complex thinking" opened by the French philosopher Edgar Morin, is to think together, and in dynamic ways, these different phenomena. For we are not dealing with a succession of sectoral crises, but, more globally, with a crisis of connecting with, and relating to, the world in which we live (Rosa, 2018; Pelluchon, 2018). In other words, we are no longer able to relate to each other, nor ourselves, or our environment. We are not able to connect either to a techno-scientific world that is changing at great speed, and which does not allow us to synchronize with a system of common social, political and legal norms (Stiegler, 2016).

In this context of widespread "disrelatedness", the notion of relatedness ('reliance' in French, a neologism, not to be confused with the English term "reliance") appears not only as a tool for sociological analysis, but also as a tool for social transformation. It invites us to rethink the creation and perpetuation of links in a holistic manner, engaging at the same time with our environmental, social and mental ecology (Guattari, 1989). However, the concept of "relatedness" should not be seen as the ideal of an original "connectedness" ('liance' in French), nor as golden age to which we should return. Indeed, there is a difference between the concepts of "relatedness" and that of the idea of an "original connectedness" (between the 'reliance' and the 'liance' in French): "relatedness" does not presuppose an initial, instantaneous, nor consubstantial bond, born without any mediation, be it with a person's mother, their family or their community. By its very nature, any link can only exist through mediation, and even through re-mediation. It is important to note here, that the concept of "relatedness" emerges as the analysis of a remediation. In the work of Roger Clausse, this concept appears as tool to understand the psychosocial need to relate to others through newspapers, radio and television broadcasts (Clausse, 1963). The author then analyses these channels of information as means for readers and listeners to break their own isolation, to feel that they belong to a social group, and to enter into "communion" with their fellow citizens at moments of significant national or international events, broadcasted live. Maurice Lambilliotte completes this notion and broadens its meaning from a more ontological perspective by considering "relatedness" not only as a way of coming out of one's solitude, but also as "the state of feeling connected" to a higher entity that goes beyond us (Lambilliotte, 1968).

The notion did not flourish in the sociology of media, but was taken up and theorized in the 1970s by the sociologist Marcel Bolle de Bal. The researcher applied it in a study on "The aspirations of social relatedness" of the Belgian population, carried in 1975. He based his study on his experiences as a teacher, during which he noticed that students attached more value to the connections that had been established between them than to the educational content that had been transmitted to them (Bolle de Bal, 1996a). Bolle de Bal therefore focused, in his study, on "social relatedness", taken as the connections between a person and a system or subsystem to which they belong. The "social relatedness" was for him both "the act of connecting or relating" and "the result of this act" i.e. "the state of relatedness." The linkages between the two notions should be considered from a dynamic perspective. This led philosopher Edgar Morin later to say, when re-appropriating the notion, that "'being connected' is passive, 'connecting' is a way of participating, 'relatedness' is activating" (Morin, 2004).

The notions of 'linkages' and 'connections', stemming from the wider notion of 'relatedness' can serve as a valuable tool for understanding recent, fluid and non-fixed, phenomena of mediation which seem to have bypassed official institutions thus far. 'Connections' as processes can be understood as forms of institutions, better able to grasp the nature of movements such as "Nuit debout" in 2016 (the "Night, Standing Up" movement in Paris inspired by "Occupy Wall Street") or "Les gilets jaunes" in 2018 (the Yellow vests movement in France). These movements have often been subject to two contradictory political, philosophical and sociological analytical perspectives, seeking to describe them. One perspective would see in these movements, able to escape from any institutionalisation, the beating heart of politics. A second one would, instead, see in them movements with no link with existing institutions such as political parties and trade unions. In other words, they would only be than vain revolts with no future, unable to represent and to carry a distinct public voice. To escape this dichotomy, it might be relevant to use the concept of "relatedness", which allows us to consider these movements as an emerging institutional process. The matrix proposed by Marcel Bolle de Bal makes it possible to more accurately analyse these movements by distinguishing three main elements. It would be necessary to separate the "process of relatedness", by which mediations are instituted between actors; the "structure of relatedness", describing the institutional system linking actors together; and the "bond of relatedness", which analyses the result of these mediations (Bolle de Bal, 1996b). Under this framework, these fluid and non-fixed movements would be seen as examples of institutions. As a corollary, this change in perspective would as well allow seeing seemingly fixed, and traditional, institutions as also flowing and ever evolving entities. In this way, the concept of "relatedness" makes it possible to avoid two pitfalls: the one arguing for the beating and always moving heart, unable of entering the realm of the ideal-type of fixed and traditional institutions; and the other imagining sacrosanct traditional institutions, set in stone and deprived of any potential change (Claeys & Trocquenet-Lopez, 2021).

The notion of 'relatedness' is further used, also from a sociological perspective, by Michel Maffesoli to think about 'postmodern' social ties. For him, these ties would be based on imaginations; "tribes" of affinities; shared emotions and links with the environment (Maffesoli, 2017). According to the sociologist, those types of "postmodern" ties are more fragile than traditional social ties, because they are based, instead, on ephemeral collective trends or events (Maffesoli, 2007).

The notion of relatedness has gone beyond the field of sociology and into wider, plural and multidisciplinary fields of thoughts. Marcel Bolle de Bal had already proposed various types of links (Bolle de Bal, 2003): those "between a person and natural elements" (cosmic relatedness); "between a person and the human species" (ontological relatedness); "between a person and the various instances of his personality" (psychological relatedness); and, finally, "between a person and another social actor, whether individual [...] or collective" (social relatedness). Edgar Moring further adds to these categories that of the links existing between forms of knowledge (Morin, 1996), making it possible to tie together what seems scattered in order to acquire a more profound understanding of the world (intellectual relatedness). Linking these different types of relatedness, however heterogeneous they may be, seems today to be a key to enter into the "complex thinking" of contemporary philosophy (Claeys, 2018). Other thinkers have made allusions to the ideas conveyed by the notion of "relatedness" without using the term itself. Hartmut Rosa (2018), Corine Pelluchon (2018) and Félix Guattari (1989), to name but a few, each evoke, with their own perspectives, the need to think about the following relationships together, in one same dynamic and as if they were intimately linked with each other: that existing between a person and themselves (identity); their environment (corporeality); and others (solidarity). From this "generalized relatedness" between types of knowledge; between sciences and citizens; but also between citizens themselves, Edgar Morin derives an ethical imperative. As he indicates: "Every ethical act, let us repeat, is in fact an act of relatedness: relatedness with others, relatedness with those we hold dear, relatedness with the community, relatedness with humanity and, ultimately, it is an incorporation into the cosmic relatedness" (Morin, 2004). This ethical imperative is particularly helpful for reflecting on the new approaches to bring about empowerment through knowledge, such as those that have emerged from the notions of experiential knowledge and expert patients in the field of chronic diseases and disability. In these cases, asserting one's knowledge, sharing it and seeing it valued appears as an act of relatedness, of recognition and of integration into the social body.

This hybridization between different disciplinary fields turns the notion of "relatedness" into one, which might be more fluid and, sometimes even more contested, than other already established concepts such as those of "relationship", "belonging", or "community". Nonetheless, it also makes "relatedness" a notion allowing for rich multidisciplinary perspectives in various theoretical fields as a new way of understanding links in the broader sense. This richness is further enhanced in the field of practices, as "relatedness" further serves as a political horizon and as a method for understanding, but also renewing, broken social links. The diversity of approaches to this pivotal concept (from sociology to psychology, or from philosophy to cultural studies, etc.) makes it possible to re-appropriate it in the medico-social, educational and cultural fields as means of emancipation and social recognition.

References

Bolle de Bal, M. (1996a). Reliances autour de la reliance. In M. Bolle de Bal (Ed.), *Voyages au cœur des sciences humaines. De la reliance. Tome 1. Reliance et théories*. L'Harmattan.

Bolle de Bal, M. (1996b). La reliance ou la médiatisation du lien social: la dimension sociologique d'un concept charnière. In M. Bolle de Bal (Ed.), *Voyages au cœur des sciences humaines. De la reliance. Tome 1. Reliance et théories*. L'Harmattan.

Bolle de Bal, M. (2003). Reliance, déliance, liance: émergence de trois notions sociologiques, *Sociétés*, n°80, pp. 99–131.

Claeys, S. (2018). *De disruption à prosommateur: 40 mots-clés pour le monde de demain*. Le Pommier.

Claeys, S., & Trocquenet-Lopez, F. (2021). *Démocratiser nos institutions. Pour un réveil de la création politique* (Revue du MAUSS, n°57). Le Bord de l'eau.

Clausse, R. (1963). *Les Nouvelles*. Éditions de l'Institut de Sociologie de l'ULB.

Crouch, C. (2013). *Post-democracy*. Diaphanes éditions. [Original edition 2004].

Généreux, J. (2006). *La Dissociété*. Seuil.

Graeber, D. (2018). *Bullshit Jobs* (É. Roy, Trans.). Les Liens qui Libèrent.

Guattari, F. (1989). *Les Trois Écologies*. Galilée.

Lambilliotte, M. (1968). *L'homme relié. L'aventure de ma conscience*. Société Générale d'Édition.

Maffesoli, M. (2007). *Le réenchantement du monde. Une éthique pour notre temps*. La Table Ronde.

Maffesoli, M. (2017). *Ecosophie*. Les éditions du Cerf.

Morin, E. (1996). Vers une théorie de la relance généralisée? In M. Bolle de Bal (Ed.), *Voyages au cœur des sciences humaines. De la reliance. Tome 1. Reliance et théories*. L'Harmattan.

Morin, E. (2004). *La Méthode VI. Éthique*. Seuil.

Pelluchon, C. (2018). *Ethique de la considération*. Seuil.

Rosa, H. (2018). *Résonance. Une sociologie de la relation au monde* (S. Zilberfarb, Trans., with the collaboration of Sarah Raquillet). La Découverte.

Stiegler, B. (2016). *La disruption. Comment ne pas devenir fous?* Les Liens qui Libèrent.

Sébastien Claeys, Adjuncte professor at Sorbonne University (director of the Master "editorial consulting and content management"), Head of communication and public debate at the Espace de réflexion éthique de la région Île-de-France (Greater Paris center of ethics), columnist for the magazine *Socialter* and author of the book *From disruption to prosumer: 40 key words for tomorrow's world* (Le Pommier, 2018, *in French*).

Religion

Lisa H. Sideris

Abstract The Anthropocene concept, notwithstanding its secular-scientific trappings, exhibits religious and spiritual genealogies, aspirations, and implications. Discourse and debate surrounding the Anthropocene and its significance can often be read as competing normative claims about the status of humans vis-à-vis some cosmic, agential, or teleological force in the life beyond. These not-quite-secular elements of the Anthropocene make an especially interesting, and somewhat surprising, appearance in scholarly and popular discourse advocating a "good Anthropocene" future in which humans collectively attain wisdom and power to manage Earth and its future evolution.

The Anthropocene is widely regarded as a secular-scientific diagnosis of the present state of the Earth. The central claim that humans have become a geological force actively transforming the entire globe became legible in the wake of Earth System Science (ESS), a transdisciplinary field of research that emerged in the 1980s to establish a unified understanding of global change. This shift toward a holistic understanding of our planet as a system of interworking spheres, such as the lithosphere, biosphere, hydrosphere, atmosphere, and others, enabled researchers to understand and model climate systems and other dynamic feedbacks. An ongoing task of ESS is to study how human dynamics interact with, and potentially disrupt, these complex systems. In this sense, the Anthropocene concept begins with and belongs to Earth System Science (Hamilton, 2017, n.p.).

But whatever its debt to scientific advances in systems thinking, the Anthropocene concept is also beholden to older philosophical trends, some of which bear the imprint of theological commitments, or blur the lines between the religious and the secular. Anthropocene narratives, and in particular, normative visions of a so-called "good" Anthropocene that prophesy a dawning epoch of boundless human ingenuity, techno-prowess, and steady progress toward planetary management

L. H. Sideris (✉)
University of California Santa Barbara, Santa Barbara, CA, USA
e-mail: lsideris@ucsb.edu

905

(Asafu-Adjaye et al., 2015), overlap in important ways with quasi-religious conceptions of the Earth and humanity's role within it. The various religious antecedents, affinities, and implications of the Anthropocene concept are discussed below. The focus here is largely, though not exclusively, on religious dimensions of "good Anthropocene" advocacy. While apocalyptic predictions and end-times framings of the Anthropocene (what might be termed the "bad Anthropocene") undoubtedly spring from religious roots, the religiosity that infuses optimistic imaginaries of the Anthropocene may not be immediately apparent. Good Anthropocene narratives may be more compelling than narratives of environmental decline. Precisely for that reason, they warrant scrutiny of their religious entanglements.

Gaia

Many Earth System researchers acknowledge debts to the Gaia hypothesis, a holistic model of the Earth system first proposed in the early 1970s by chemist James Lovelock and co-developed with microbiologist Lynn Margulis. Interpretations of Gaia have proliferated and diverged over the years (Latour, 2017), but Lovelock's original proposal advanced the idea of Earth as a complex, self-regulating system akin to a super-organism (Lovelock, 1979). Lovelock's christening of this system "Gaia" recalls the ancient Greek goddess of Earth (Gaea), the mother of all life, suggesting an animistic or pagan perspective that personifies or possibly deifies the planet (Ruse, 2013). A provocateur among scientists, Lovelock sometimes distanced himself from spiritual interpretations of Gaia, while at other times seeming to welcome these nuances. With or without overt associations with a providential Earth agent, Gaia theory can be read as imputing teleological tendencies to the Earth system, suggesting that our planet's self-regulatory processes operate in a purposeful, goal-oriented manner. Earth does not merely have life; it *is* alive (Lovelock, 1979).

Perceptions of the Earth system as possessing will or agency align, in broad strokes, with interpretations of the Anthropocene as the arrival of a new and livelier—or deadlier—kind of planet. The very stability of the Earth and its component systems cannot be taken for granted in the Anthropocene. We now find ourselves threatened by an unpredictable and unruly planet. Earth is no longer an inert backdrop to the human drama, if ever it was. Nor is it a solicitous mother. Earth is an ornery beast, a sleeping giant, awakened by humanity's reckless provocation (Hamilton, 2017). Lovelock's own musings on Gaia, particularly in his later career, exude a gleeful misanthropy, tinged at times with fire-and-brimstone rhetoric: we are sinners in the hands of an angry Gaia (Lovelock, 2006). Most of us, he predicts, won't survive.

The Noosphere

But competing visions of the Anthropocene suggest the reverse arrangement: Earth is in our hands. These draw from familiar elements--intentionality, agency, and teleology--assumed to reside within natural and human systems, but their portrait of the human-Earth dynamic is rather different. If Earth-as-Gaia helped to reinstate nature as the central, and possibly erratic, character in our modern Anthropocene drama, other teleological interpretations of natural processes give humans pride of place in the Anthropocene. (In keeping with his reputation as an iconoclast, Lovelock is notoriously difficult to pin down. His recent work predicts that humans will be overtaken by a godlike machine intelligence that will subject us to its cyborg will. In a perverse way, this too is a "good Anthropocene" vision, as Lovelock awaits this "Novacene" epoch with great relish and optimism.) These accounts of humanity's role in relation to Earth are not new but they have undergone a revival in the Anthropocene. In the 1920s a group of European and Russian intellectuals began to speak of a phenomenon they dubbed the noosphere, a thinking layer and emerging sphere of the Earth (from the Greek, nous, meaning mind). The French Jesuit priest Teilhard de Chardin developed the noosphere concept in conjunction with French mathematician Édouard Le Roy and the Russian geochemist Vladimir Vernadsky (Samson & Pitt, 1999). Vernadsky is best remembered today for advancing the concept of the biosphere to describe Earth's living layer. The biosphere pointed to the capacity of life itself—above all, *homo sapiens*—to shape and transform the planet. Just as Earth's lithosphere, its solid outer crust, gave rise to the biosphere, the biosphere was now generating another vibrant layer—a planetary brain—through the collective thought and activity of the human species.

In the quasi-spiritual concept of the noosphere, we can detect some initial gestures toward the good Anthropocene (Sideris, 2016). For de Chardin, Vernadsky, and others, the noosphere marked the next critical stage in Earth's evolutionary unfolding, wherein humanity begins to remake the biosphere in its own image. This noospheric transition inspired optimism and excitement about the future, much as visions of the good Anthropocene do today. These ideas made it possible to conceive of the human species as evolving into a telluric yet complex and intelligent force that can radically, and beneficently (if self-interestedly), alter the planet. The noosphere, therefore, provides a framework for interpreting the Anthropocene, even if those who first deployed the concept could not have foreseen complex global changes, such as climate disruption, that we now associate with the Anthropocene.

Some scholars have argued forcefully for distinguishing the noosphere concept as it was articulated by Vernadsky and de Chardin, respectively (Hamilton & Grinevald, 2015). Vernadsky held to a materialist understanding of the noosphere in which humanity remains part of Earth processes, whereas de Chardin's fusion of theology with science led him to interpret the noosphere in metaphysical terms. On de Chardin's account, the noosphere—the culmination of complex human consciousness and accumulated information—advances toward a cosmic climax called the Omega Point, a kind of mystical union with God. De Chardin's belief that humans collectively achieve a state of self-awareness of our role in purposeful evolutionary processes shows affinities with calls for creating a good Anthropocene. These optimistic assessments are today issued by "ecomodernist" thinkers and certain science popularizers, who may or may not explicitly recognize de Chardin's influence (Asafu-Adjaye et al., 2015; Grinspoon, 2016). Like early noosphere enthusiasts, these thinkers propose the Anthropocene as a milestone of collective awakening to our outsize planetary powers. Our task is to embrace humanity's arrival at this pivotal moment in Earth's geological history, to seize the Anthropocene as an opportunity to remake the planet, in creative and propitious ways.

The claim that Vernadsky's noosphere concept was the materialist, secular counterpart to de Chardin's mystical-Christian version is open to debate. A handful of Anthropocene theorists explicitly invoke Vernadsky, and likeminded cosmic visionaries, as sources of their own philosophizing about humanity's role in the cosmos, and the future of Earth in the Anthropocene. For example, some astrobiologists interpret Vernadsky's noosphere concept to mean that humans evolve not only to transform the Earth but transcend it. This moment of transcendence is the Anthropocene by another name (Grinspoon, 2016). But what does it mean to transcend Earth? For astrobiologists, who study the origin and evolution of life in the universe, transcending Earth might entail physically leaving the planet, as humans evolve to explore and eventually colonize (or create) new worlds in space—otherworldly ambitions of a quite literal sort.

But transcendent aspirations with a decidedly numinous flavour swirl around Vernadsky's philosophy and that of his fellow "cosmists." These thinkers include Nikolai Fedorov (1829–1903), a Russian orthodox philosopher whose cosmic religion, like de Chardin's, inspired today's transhumanist movements, and Konstantin Tsiolkovsky (1857–1935), the father of Russian rocketry who, again, like de Chardin, believed in a universe suffused with consciousness at all levels (with humans as its highest expression). The cosmists' ambitions, collectively, included space flight and colonization, time travel and time reversal, human life extension and immortality, and the resurrection or reanimation (through cryonics or other technologies) of all living matter. They posited that humans were still in their early stages of evolution, and that maturity lay in our transformation into a collective force capable of redirecting cosmic evolution.

These (not quite secular) ideas have discernible influence on contemporary treatments of the Anthropocene. Astrobiologist David Grinspoon, for example, cites the cosmists as inspiration for his vision of "earth in human hands," and echoes them in prophesying the arrival of a "mature" Anthropocene, when humans attain the aggregate power and wisdom to oversee the planet and strike out for new worlds (Grinspoon, 2016).

Anthropocene Mythmaking

It should by now be apparent that these treatments of the Anthropocene reach toward theological pronouncements, not only in carrying forward religious or spiritual legacies but in their portrait of the human creature as engaged in an epic struggle with—or as the dramatic fulfilment of—some divine-like agency or force. The proposal, for example, that Anthropocene humans are evolving along a progressive continuum toward omniscience or omnicompetence is an exercise in theological anthropology. More specifically, we might read these narratives as theodicies—or better still, an anthropodies—in which humans providentially usher the planet into a state of flourishing, thus fulfilling an order that resides within all things, but awaits completion by one supremely complex species (Hamilton, 2016). Seen in this light, anthropocenic events like climate change are merely obstacles to be overcome, by-products of our ascent to a higher, and seemingly foreordained, state.

Contrary to their claims to novelty, Anthropocene narratives come to us clothed in the traditional, if tattered, raiment of myth and prophecy. They diagnose our current travails and foretell their resolution in a future world that humans will make. If humanity has fallen from a state of grace to which the good Anthropocene restores us, our fall is not the black mark of sin, on this account, but merely a function of our erstwhile ignorance and immaturity. This is a tale of blindness giving way to enlightenment, reminiscent of prophetic discourses that proclaim an "advent" (Bonneuill & Fressoz, 2017, 74). Having arrived at this self-reflexive moment in our species' evolution, we can chart a path forward with conscious intentionality. Failing that, and failing our planet, we can chart a course for new worlds beyond.

References

Asafu-Adjaye, J., et al. (2015, April). *An Ecomodernist Manifesto*. Breakthrough Institute.

Bonneuill, C., & Fressoz, J.-B. (2017). *The shock of the Anthropocene. The earth, history, and us*. Verso.

Grinspoon, D. (2016). *Earth in human hands: Shaping our planet's future*. Grand Central Publishing.

Hamilton, C. (2016, August 17). The Anthropocene belongs to earth system science. *The Conversation*.

Hamilton, C. (2017). *Defiant earth: The fate of humans in the Anthropocene*. Polity.

Hamilton, C., & Grinevald, J. (2015). Was the Anthropocene anticipated? *The Anthropocene Review, 2*(1), 59–72.

Latour, B. (2017). *Facing Gaia: Eight lectures on the new climatic regime*. Polity.

Lovelock, J. (1979). *Gaia: A new look at life on earth*. Oxford University Press.

Lovelock, J. (2006). *The revenge of Gaia: Earth's climate crisis and the fate of humanity*. Basic Books.

Ruse, M. (2013). *The Gaia hypothesis: Science on a Pagan Planet*. University of Chicago Press.

Samson, P. R., & Pitt, D. (Eds.). (1999). *The biosphere and noosphere reader: Global environment, society, and change*. Routledge.

Sideris, L. H. (2016). Anthropocene convergences: A report from the Field. "Whose Anthropocene? Revisiting Dipesh Chakrabarty's 'Four Theses.'" Edited by Robert Emmett and Thomas Lekan, *RCC Perspectives: Transformations in Environment and Society*. No. 2, 89–96.

Lisa H. Sideris is a professor in the Environmental Studies Program at University of California Santa Barbara, with research interests in environmental thought and ethics at the intersection of science and religion. She is author of *Environmental Ethics, Ecological Theology, and Natural Selection* (Columbia University Press, 2003), co-editor of a collection of interdisciplinary essays on Rachel Carson's life and work, *Rachel Carson: Legacy and Challenge* (SUNY Press, 2008), and author of *Consecrating Science: Wonder, Knowledge, and the Natural World* (University of California Press, 2017).

Sejahtera

Zainal Abidin Sanusi, Wan Zahidah Wan Zulkifle, and Dzulkifli Abdul Razak

Abstract The Anthropocene era has witnessed an acceleration of human activities that undeniably have a great impact on the ecological system. While these activities have contributed greatly to the betterment of society i.e. alleviating poverty, wider access to quality education and advancement of technology, these activities have a great impact in their destruction of the biosphere and Mother Nature. The development and advancement of technology comes with detrimental effects to the environment and the impacts are clearly observed now more than ever. The human capital-innovation-technology nexus remains very much the driver where ecological and human dimensions are not spared in the quest for material and mechanical prowess. In other words, it is economically biased which brought about an imbalance between the three aspects of Profit, Planet and People, otherwise dubbed as three Ps, collectively representing Sustainable Development which now seems to be distorted.

More recently, the context expanded by embracing the agenda of Sustainable Development Goals (SDGs) from 2016 until 2030, encompassing 17 goals with overarching targets of five Ps, where the Partnership and Peace make up the additional two Ps. The platform for SDGs was launched in New York in September 2015, slightly ahead of the 2016 World Economic Forum with the theme of Fourth Industrial Revolution (4IR) in Davos. Not only do SDGs act as a common global platform as endorsed by the United Nations General Assembly, but it is also a crucial bridge that connects to the unfinished Millennium Development Goals (MDGs) for the period 2000–2015. During the period, it encapsulates the United Nations Decade on Education for Sustainable Development (ESD) from 2005 to 2014. Yet

Z. A. Sanusi (✉) · W. Z. Wan Zulkifle
Sejahtera Centre for Sustainability and Humanity, IIUM, Kuala Lumpur, Malaysia
e-mail: zainalsanusi@iium.edu.my; zahidahzulkifle@iium.edu.my

D. Abdul Razak
International Islamic University Malaysia, Kuala Lumpur, Malaysia
e-mail: dzulrazak51@gmail.com

911

the world continues to be crises-susceptible which has been the source of concern over at least three decades; even more so of late given the frequency and severity of the crises, unlike during the traditional/tribal past experienced by most indigenous communities worldwide.

Arguably therefore, the threat of Anthropocene from the indigenous perspective calls for a more comprehensive and holistic framework through the Whole Person Transformation approach imbued with values of a 'complete person' as per Learning to Be to be fully integrated. In practically all cultures that we know of, there seems to be a 'specific term' that is dearly adhered to as its guiding light. Be it in the East or West, these words are encompassing with common themes and indeed values attached to them that resonate to one another. The importance of having a local indigenous concept to explain and integrate sustainable development agenda is very critical to ensure the sustainability of the agenda itself. « Unless we are able to translate our words into a language that can reach the minds and hearts of people young and old, we shall not be able to undertake the extensive social changes needed to correct the course of development. - Gro Harlem Brundtland" (Gadotti, 2008, p. 21).

Sejahtera as Local Indigenous Concept for Sustainable Development

Historical traces of the concept of development in the context of human-nature relations prove that sustainability is not a new concept that only emerged in the 1980s following the well-acknowledged Brundtland Report. Arguably, sustainability is an ancient concept in many indigenous traditions that has been overtaken and lost in the drive toward modern unsustainable development. The result is that development becomes purely a physical venture and no longer focuses on building 'collaborative relationships' between humans, the community, the environment, and the 'creator' as an enduring lifestyle. In doing so, the fine state of balance is severely offset by a hefty price tag for future generations. Here is where the Malay indigenous perspective on development through the concept of sejahtera becomes very relevant.

Sejahtera is one of such indigenous concepts that is rooted in the Malay Archipelago or Nusantara which is also intimately linked to Islamic worldview in a variety of ways. It is widely used culturally as a form of general greeting to convey the message of peace and harmony. Based on the standard Malay language dictionary, "sejahtera" is defined as safe and prosperous, happy and peaceful, safe and protected from disaster such as distress and disturbance (Dewan, 2005).

Sejahtera is not easily rendered into other languages because of its comprehensive and multi-layered meaning and nuances. It underscores that indigenous knowledge and wisdom have had their own uniqueness, strength, and relevance for the local community over the years (Razak et al., 2018). Within the context of anthropocene, Sejahtera can be described as a balanced lifestyle summarised by at least ten different elements neatly woven into the acronym SPICES, namely spiritual, physio-psychological, intellectual, cognitive, cultural, ethical, emotional, ecological, economic, and societal dimensions. All the components must co-exist in a balanced manner not only to achieve an overall state of well-being that is lasting (sustainable) over generations.

Central to the sejahtera concept is the 'heart' that represents each human being's inner self which is interconnected with the outer six elements that circling an individual. A person is considered being in sejahtera when he or she can keep the six elements balanced and in harmony between each other. A sejahtera person then will have the capability to keep the outermost four elements in balance and harmony guided by the self-sejahtera state particularly at decisions making and interactions process.

The uniqueness of the concept makes its translation to other languages not only difficult, if not impossible, but it may be misleading. Hence to understand it from a one-dimensional perspective, like the economy is not only to miss the whole point but also give a very distorted conception. What would be most sorely missed is the qualitative-cum-intangible aspects that are today's major concern. Health, for example, is not just about the absence of disease or illness that may be quantifiable in a reductionist way, but also universally recognised as the state of emotions, sans "physical" diseases, that could lead to a situation contrary to the concept sejahtera (such as depression, stress, violence) without any clear signs and symptoms until perhaps it is too late to deal with. This is well illustrated by the pandemic where wellbeing and wellness seem to override health per se.

Sejahtera may often be translated as 'well-being' or even 'prosperity,' but its inherent meaning is much more than that. In fact, it is 'beyond well-being.' It is human-centric in that it spans the macrocosmic-microcosmic nexus. It is macrocosmic because it relates humans to the external environment – nature and fellow beings, including other species. It is microcosmic because it embraces the 'self' and the inner (esoteric) dimensions, including spiritual consciousness. In short, the embodiment of sejahtera goes beyond the conventional three Ps of Planet, People, and Profit. Although each aspect can be individually targeted and developed, for example sejahtera ekonomi (economic well-being), it is only when expanded into the 'socio-ecological' dimension within SPICES that all elements are harmoniously blended and nurtured. That makes it a holistic endpoint for a sustainable future. In reality, however, it is a comprehensive, multilayered concept that embraces deeper and richer nuances. This rendering of sejahtera can easily link to the principle of 'thinking global and acting local' in finely balanced and harmonious ways. The biophysical, socio-emotional, and intellectual-spiritual needs are holistically interconnected to realise the holistic balance aligned with one's true nature.

Around the globe, similar concepts exist in many cultures. In Japanese culture, ikigai (pronounced Ick-ee-guy) brings the meaning "a reason for being" which is usually used to indicate the source of value in one's life or the things that make one's life worthwhile. Similarly, ubuntu (pronounced oo-BOON-too), a Nguni Bantu term meaning "humanity" is often rendered as "I am because we are," or "humanity towards others." In Xhosa (umntu ngumntu ngabantu), it is referred to as "the belief in a universal bond of sharing that connects all humanity". Lagom (pronounced LAH-gum) on the other hand, is a Swedish word that implies "not too little, not too much" or "just right." It represents the art of living a balanced, slower, fuss-free life "in moderation", "in balance", "perfect-simple", and "suitable". Whereas, hygge (pronounced hue-guh) is a Danish word used to acknowledge a feeling or moment. All these concepts convey similar concern on how the relationship should be established between man and nature. In all these concepts, the most important implication is how the relationship between man and nature should be governed. Thus, in the context of sejahtera this relationship takes an in-depth meaning, taking the cultural context and nuances into account. It implies a collaborative relationship in particular embraces compassion, empathy, and the uncompromising spirit of oneness transcending differences and bitterness, bringing about the much-needed close relationship, coexistence, and interdependency.

In a nutshell the element of sejahtera must be fully understood, internalised and practised because it is the fountainhead of good values/virtues that are innately human (and divine too) leading to a righteous and balanced way of life in nurturing the human person. This WPT approach, leading to Whole-Institution-Transformation, and eventually the Whole-Community-Transformation approaches.

Sejahtera, Sustainability and Anthropocentrism

According to Magni (2017), by exploring indigenous knowledge in greater depth, a society can better understand and benefit the sustainable development agenda especially in the current era of anthropocene. The components that make up sejahtera are the same traits needed to cater to the millions who are under urgent threat of global warming and climate change. The unprecedented occurrence of crisis after crisis cannot be handled effectively without nurturing the relationship that binds people via a set of common values and ethics. The world is getting highly complex, dynamic, and interdependent; therefore isolated, compartmentalised, independent, and conventional linear approaches are most likely to fail (because they are unsustainable). Instead, constructive relationships between the microcosm and the macrocosm as illustrated in the concept of sejahtera are very essential for sustainability.

According to Gadotti (2008, p. 29), sustainability is closely linked with well-being and a sustainable life is a lifestyle that promotes well-being for everyone in harmony (dynamic balance). Therefore, this Malaysian indigenous concept of sejahtera is well-fit with the sustainability notion, and Malaysian individuals who have achieved the sejahtera state would be in harmony and balanced to contribute for sustainable development. For example, the opening statement of Malaysian National Education Philosophy (NEP) is itself indicating a sustainable aspiration of the Malaysian education system to continuously develop the state of sejahtera at individual level (Zakaria, 2000). According to Abdul Razak (2021), the NEP implies a personal development process towards an achievement of individual 'quality of life' that is both sustainable and balanced that is embodied in sejahtera as a way of life. As such, it directly addresses the 'anthropocentric' factor that is the major cause of the global crises of today. In practical terms, it brought to life the practices of trusteeship, responsibility, harmony, and balance beyond that of ownership and growth in constructing 'better' Education for Sustainable Development (ESD) that is more naturally inclined without much imposition from the outside.

In conclusion, we are back full circle, by coming back to Learning to Be as a prerequisite to humanise education where the very essence of being human can be restored and experienced again based on values, involving ethics, authenticity and integrity, all subsumed by Learning to Be and Learning to Become. In the face of the Anthropocene, humanity must be better prepared with no one left behind, building more robust resiliency and the ability to cope by being holistic as subsumed by Learning to Be, including the domain of sejahtera as well as the wider traditional nexus of human heritage worldwide. And more recently Learning to Become, as the fifth pillar introduced by UNESCO, as the lead agency, in engaging "a global conversation as well as a report on the futures of (humanised) education, drawing on the diverse and fruitful ways of learning practised around the world, resolutely forward-looking, yet grounded in human rights at the service of the dignity of all." The latest (fifth) pillar can therefore be construed as a "disruption" to the conventional model which would otherwise remain unsustainable and irrelevant to the future, if not altogether obsolete to ESD. Devoid of indigenous counterpoints to the Anthropocene, the catastrophic material/mechanical/monetary-type of behaviours will be hard to constrain from the irrepressible rise of the anthropogenic tendency "towards unbounded freedom, insensitive to the needs of any time beyond our own" as "the sharpest manifestation of a sort of disregard for the natural world." (Hétier & Wallenhorst, 2021, p.51). In so doing, the forecasted degradation of the humans and its community as defined by the Anthropocene will be a reality.

References

Dewan, K. (2005). *Kuala Lumpur: Dewan Bahasa dan Pustaka* (4th ed.).

Gadotti, M. (2008). What we need to learn to save the planet. *Journal of Education for Sustainable Development, 2*(1), 21–30. https://doi.org/10.1177/097340820800200108

Hétier, R., & Wallenhorst, N. (2021). The COVID-19 pandemic: A reflection of the human adventure in the Anthropocene. *Paragrana, 30*(2), 41–52. https://doi.org/10.1515/para-2021-0023

Magni, G. (2017). Indigenous knowledge and implications for the sustainable development agenda. *European Journal of Education, 52*(4), 437–447. https://doi.org/10.1111/ejed.12238

Razak, D. A. (2021). *Essay on Sejahtera: Concept, principle and practice.* IIUM Press.

Razak, D. A., Khaw, N. R., Baharom, Z., Mutalib, M. A., & Salleh, H. M. (2018). Decolonising the paradigm of sustainable development through the traditional concept of sejahtera. Essay. In *Academia and communities – Engaging for change* (pp. 210–219). United Nations University, Institute for the Advanced Study of Sustainability (UNU-IAS).

Zakaria, H. A. (2000). Educational development and reformation in the Malaysian education system: Challenges in the new millennium. *Journal of Southeast Asian Education, 1*(1), 113–133.

Zainal Abidin Sanusi is currently Director, Sejahtera Centre for Sustainability and Humanity, International Islamic University Malaysia while serving as an Associate Professor at the Department of Political Science, Kulliyyah of Islamic Revealed Knowledge and Human Science. He has a Ph.D. in International Studies from Waseda University and spent a one-year postdoctoral study at UN University Institute of Advanced Studies doing research on multilateral environmental agreement. His current research interests are Sustainability Science, Education for Sustainable Development and Public Policy Analysis.

Wan Zahidah Wan Zulkifle is Assistant Director at Sejahtera Centre for Sustainability and Humanity, International Islamic University Malaysia. She received her Bachelor of Medicine and Surgery from USIM and Master of Human Sciences in Psychology (Industrial/Organisational) from IIUM. Her current work focuses on Sustainability-related Publication and Award, Education for Sustainable Development (ESD), and community empowerment with a special focus on health and wellbeing.

Dzulkifli Abdul Razak is currently the Rector of the International Islamic University Malaysia. He was the Vice Chancellor of USM from 2000–2011. He is the immediate past president of International Association of Universities, a Paris-based UNESCO-affiliated organisation. Dzul was the first Asian recipient of the prestigious 2017 Gilbert Medal. He is a Fellow of the Academy of Sciences Malaysia (FASc), World Academy of Art and Science (FWAAS) and the World Academy of Islamic Management (FWAIM). From February 2021, Dzul serves as an Expert for the Futures of Higher Education Project at UNESCO's Institute for Higher Education (IESALC) based in Caracas.

Sounds

Quentin Arnoux

Abstract This article proposes to set up an acoustic framework for the Anthropocene in order to develop an alternative and sensory narrative. Listening carefully to this new geological era can help us understand the issues and phenomena at work. Sounds can also be at the centre of a reflection on the ways to position the human being in the Anthropocene, as a part of an acoustic community.

In the early 1960s, the American biologist Rachel Carson published the book *Silent Spring* (1962) in which she criticizes the vices of intensive agriculture and the deleterious effects of pesticides on living organisms. This book is at the foundation of the environmental movement and proposes, by calling attention to sounds – the silence of a spring where no bird sings –, to open an alternative and sensory consideration of the Anthropocene in which sounds(−scapes) are interpreted as ecological indicators, enabling us to take stock of this new geological era, marked by the impact of the human activities on nature.

Claimed to be the source of objectivity and rationality, sight occupies a predominant place in Western knowledge (Attali, 2007). However, the world is not only to be seen, it is also to be heard. Within a thin atmospheric layer, the Earth's climate and ecology – 'a speck of sound within a silent universe' (Huxley & Koch, 1964, p. 15) – resonate and express themselves in an infinite number of acoustic phenomena and voices. It is only at the end of the 1970s, in connection with the technological advances in the field of sound recording, that sounds have had more importance within scientific research, notably under the impulse of the Canadian composer and environmentalist R. Murray Schafer. In *The Tuning of the World* (1977), the author lays the foundations of the concept of soundscape, later borrowed and shaped by numerous disciplines in the field of sound studies. Within this nebula of scientific disciplines, bioacoustics and ecoacoustics focus on the sounds of nature and share the premise that a sonic consideration of the environment can lead to a deeper understanding of life, interactions within ecosystems, global phenomena of climate

Q. Arnoux (✉)
Department of Environmental Humanities, University of Lausanne, Neuchâtel, Switzerland

© The Author(s), under exclusive license to Springer Nature Switzerland AG 2023
N. Wallenhorst, C. Wulf (eds.), *Handbook of the Anthropocene*,
https://doi.org/10.1007/978-3-031-25910-4_150

change and biodiversity loss, and can guide us towards an effective protection of nature (Krause, 2013; Sueur & Farina, 2015).

Climate change is accompanied by various sound transformations, particularly related to the disruption of seasons, the warming of temperatures, and the extreme weather phenomena. Melting glaciers, floods, hurricanes, monsoons, and droughts are all acoustic phenomena that characterize the current and future climatic upheavals of the Earth system. The loss of biodiversity, another planetary boundary, is manifested through the silencing of nature. The loss of acoustic diversity and density that affects many natural ecosystems is nothing other than the audible symptom of the damage suffered by living organisms – animals and plants. By disappearing or migrating to other habitats, animal species leave audible voids in the biophony, the acoustic layer within which all living organisms express themselves (Krause, 2016). Thus, a degraded or impacted habitat will have a less rich and organized soundscape than a habitat possessing an ecologically structured vitality. Plant biodiversity also participates in the acoustic architecture of ecosystems and accelerated deforestation in many tropical regions is transforming natural soundscapes (Deshays, 2017). Listening to soundscapes brings a deep understanding of the environmental upheavals at work and reveals their rapidity and globality. In this, they are 'acoustic compasses' capable of guiding us on a changing planet (Krause, 2013).

Sounds can also be placed at the centre of an alternative narrative of the Anthropocene and enable us to understand the relationship between the human being and nature through the lens of sound. If soundscapes reflect the conditions that produce them and provide information about the evolution and trends of a society (Schafer, 2010, p. 28), they can, to a large extent, be analysed as the acoustic outcome of modern Western ontology (Arnoux, 2021). In the nineteenth century, the Industrial Revolution concretized the progressive, growth-oriented ideal of modern Western civilization and led to the advent of the machine. A symbol of progress, this technical tool deeply transformed the soundscapes by projecting sonorities of new intensities. In industry, the increase of sound levels became synonymous with speed, efficiency, prosperity, and the use of new energy resources – coal, oil and then electricity – allowed modern Western civilization to go ever further, ever faster and in an ever greater cacophony. The silence of the forests and countryside were pierced by the steam engine (Thoreau, 2017) and cities became new kinds of sound environments. From the second part of the twentieth century onwards, the 'great acceleration' of human activities translated into a rapid and abrupt crescendo. The exponential curves of world demography, international tourism, and motor vehicles are all examples of a profound transformation of the world's acoustics. Below the surface of seas and oceans, on land, in the air, and in space, the increase of human activities translated into a growing tumult. The increase of human sounds contrasts with the silencing of natural ones and animal voices. The exponential curve of the urban population plunges human beings into their own acoustic reality, far from the sounds of nature, and the curve showing the increasing use of fertilizers in agriculture only perpetuates the fears of a mute world raised by Rachel Carson 60 years ago.

Sounds become noise when their power, density, or pervasiveness overwhelm the acoustic space. Today, human noise can be heard in most ecosystems, and natural orchestrations – which leave a large space for silence – are becoming rare (Krause, 1992; Hempton & Grossmann, 2009). As sentient beings, many animals have an acoustic perception of the world and adapt to the sounds of their environment. Recent studies in the field of bioacoustics have demonstrated the acoustic adaptation that some cetaceans or passerines must exhibit in order for their communication systems to outperform and not fall prey to human noise pollution: 'speaking' louder, in higher frequencies, and repeating sound signals so that they can be heard (Slabbekoorn et al., 2010). In a very different time frame compared to evolution, it is the adaptive capacities of living organisms that are being tested, and there is no indication that their elasticity is infinite. This vulnerability to noise pollution – which human beings share with the rest of living organisms – could be constitutive of a new morality and ethics towards nature (Pelluchon, 2018).

Sounds enable us to reflect on concrete ways on how to live and be a part of the Anthropocene. It is a listening capacity that we must recover, as sounds offer a differentiated 'reading' and a profound experience of the world. There should be a call for sounds to actively contribute to the debates concerning the state of the world in which we live and the one that we wish to leave to the future generations. The preservation of silent zones, silence as a principle of ecodesign, reflection on the acoustics of cities, and the sound implications of urban spread are all multidisciplinary examples of the consideration of sound during the development of human societies. The 'acoustic future' of the Anthropocene could be linked to the decrease of human activities, in other words, an inversion of the exponential curves that characterize the recent history of Western civilization. Such a civilizational turn would lead to notable acoustic changes: motors slowed down, sound volumes diminished. The sound metamorphoses would be more radical if the predictions of a collapse of industrial society were to come true in the coming decades. In this, collapsology proclaims, implicitly, a world of industrial silence.

The positioning of the human being in the Anthropocene cannot be done without the questioning and the deconstruction of the Western ontology, of its ideals and paradigms. The human being is not external and superior to nature. On the contrary, we draw our anthropological identity from those who express themselves and what resonates at our side. What would remain of our cultures, languages, arts, traditions, and beliefs if many non-human voices and presences continue to disappear? It is an 'anthropological void' that is to be feared if springs become silent and the voices of nature disappear (Arnoux, 2021). It is a relational ontology that we must adopt, a new representation of the world in which the human being is a participant of the biosphere's harmony – literally and figuratively. An ontology that could draw inspiration from indigenous oral cultures that listen, respond, and inscribe themselves in the expressions of nature (Abram, 2013). An ontology, finally, that would redefine myths, imaginaries and ideals, and that would see the return of the human being in a shared audible world, as the holder of a dialect among others, but constitutive of the same discourse, the one from the acoustic community, voice of the biosphere.

References

Abram, D. (2013). *Comment la terre s'est tue. Pour une écologie des sens* (D. Demorcy & I. Stengers, Trans.). La Découverte. The spell of the sensuous. *Perception and language in a more-than human world.* Vintage Books. (Original edition: Abram, D. (1997)).

Arnoux, Q. (2021). *Écouter l'Anthropocène. Pour une écologie et une éthique des paysages sonores.* Le Bord de L'eau.

Attali, J. (2007). *Bruits. Essai sur l'économie politique de la musique.* Le Livre de Poche. (Original edition 1977).

Carson, R. (1962). *Silent Spring.* Houghton Mifflin.

Deshays, D. (2017). Le paysage sonore: une congélation du vivant ? In J. Mottet (Ed.), *La forêt sonore. De l'esthétique à l'écologie* (pp. 7–22). Champ Vallon.

Hempton, G., & Grossmann, J. (2009). *One square inch of silence: One man's search for natural silence in a Noisy world.* Simon & Schuster.

Huxley, J., & Koch, L. (1964). *Animal language. How animals communicate.* Grosset & Dunlap. (Original edition 1938).

Krause, B. (1992). The habitat niche hypothesis: A hidden symphony of animal sounds. *Literary Review, 36*(1), 40–45.

Krause, B. (2013). *Le grand orchestre animal* (T. Piélat, Trans.). Flammarion. *The great animal orchestra. Finding the origins of music in the world's wild places.* Little, Brown and Company. (Original edition: Krause, B. (2012)).

Krause, B. (2016). *Chansons animales et cacophonie humaine. Manifeste pour la sauvegarde des paysages sonores naturels* (A. Prat-Giral, Trans.). Actes Sud. *Voices of the wild: Animal songs, human din, and the call to save natural soundscapes.* Yale University Press. (Original edition: Krause, B. (2015)).

Pelluchon, C. (2018). *Éthique de la considération.* Seuil.

Schafer, R. M. (2010). *Le Paysage Sonore. Le monde comme musique* (S. Gleize, Trans.). Wildproject. *The tuning of the world.* Random House. (Original edition: Schafer, R. M. (1977)).

Slabbekoorn, H., Bouton, N., van Opzeeland, I., Coers, A., ten Cate, C., & Popper, A. N. (2010). A noisy spring: The impact of globally rising underwater sound levels on fish. *Trends in Ecology & Evolution, 25*(7), 419–427. https://doi.org/10.1016/j.tree.2010.04.005

Sueur, J., & Farina, A. (2015). Ecoacoustics: The ecological investigation and interpretation of environmental sound. *Biosemiotics, 8,* 493–502. https://doi.org/10.1007/s12304-015-9248-x

Thoreau, H. D. (2017). *Walden ou la Vie dans les bois* (B. Matthieussent, Trans.). Le mot et le Reste. *Walden; or, life in the woods.* Ticknor and Fields. (Original edition: Thoreau, H. D. (1854)).

Quentin Arnoux holds a master's degree in Environmental Humanities from the University of Lausanne (Switzerland). He is the author of a sound-book on the sounds of the Anthropocene: *Listening to the Anthropocene. For an ecology and an ethic of soundscapes* (Le Bord de L'eau, 2021, *in French*).

Symbolic Boundaries

Claus Leggewie

Abstract This essay reflects upon the symbolic meanings and the polyinterpretability of real and imagined boundaries.

A Border of What?

When we talk about borders—or, more broadly, boundaries—in the planetary context, we should familiarize ourselves with their various, material as well as symbolical, meanings: the boundaries of the Earth system (Rockström et al., 2009; Steffen et al., 2015), the "limits of growth" as detected by the Club of Rome (Meadows et al. 1972), borders as basic elements in international politics, boundaries between scientific disciplines (Frodeman et al., 2010) or between age groups (Mannheim, 1952), and finally those lines we are accustomed to draw between nature and culture (Mortenson, 2011). But what are boundaries anyway, how are they drawn, where do they run? (https://plato.stanford.edu/entries/boundary) Wittgenstein (1922) once said that the limits of our language amount to the limits of our world. Accordingly, an etymological and semantic discussion of the concept may contribute to our understanding of *planetary boundaries* as well, a phrase that hails from Earth system science and is currently entering everyday language. Are they "real", are they "imagined"? The sociologist Georg Simmel once declared: "The boundary is not a spatial fact with sociological effects, but a sociological fact that is formed spatially." (https://journals.sagepub.com/doi/10.1177/0263276407084470) Is this also the case with physical boundaries and true for 400 parts per million in the atmosphere? Objectivist and constructivist views of the world mingle, clash or converge here.

C. Leggewie (✉)
Justus-Liebig-University Giessen, Giessen, Germany
e-mail: claus.leggewie@zmi.uni-giessen.de

921

Planetary boundaries demarcate areas where the resources and reproductive capacities of planet Earth are limited, i.e., not infinitely progressing. With all other pools of resources – water, energy, information – this is taken for granted; with respect to Earth: apparently not. The assumption of an infinite progression seems logical when we project the use of planetary resources into the Universe, which many consider to be itself infinite and thus elusive to intellectual inquiry, a sphere of spirituality or metaphysics. To engage in planetary thinking, however, means to acknowledge the infinity of the Universe and derive from that very infinity the finiteness of the planet. As much as the existence of *Homo sapiens* had a beginning, it will have an end, and according to Anthropocene theory, humanity is now massively accelerating this suicidal endeavour.

Manifold Meanings of Borders

To return to the question at hand: What are boundaries? there is an early definition in Euclid's *Elements*: "A boundary is that which is the extremity of something" (*Elements* Bk I, Df 13). Most borders, by their very definition, create binary distinctions between a here and a there, and we could draw a line from ancient Greek mathematics to the late twentieth-century systems theory of Niklas Luhmann (1995) who was intent on showing that each inclusion entails an exclusion. Functional differentiation means separation: within or without, North or South Korea. In the real world of course, things are not as black and white, not even in Korea. Boundaries are open, moveable, malleable, hazy, fuzzy. The boundary of Earth is not that of its surface or of the anthroposphere alone. Instead, the boundedness of our planet's resources is fundamentally dependent on the state of its atmosphere, of an 'out there' with certain temporal as well as spatial conditions. Not just things have an extremity; events, too, come with a duration, with an end. Temporal and spatial boundaries may coincide, differ, work against each other.

Most commonly, boundaries, a term that goes back to processes of colonization and war, are defined (and thus literally 'limited') as territorial or national borders. *Granica*, the Old Slavic word for border, which was adopted as a loanword in German (*Grenze*) and Dutch (*grens*), designated the terrain to be marked by a line, a fence, or a wall. Also consider the medieval marches, borderlands whose land-*marked* borders were easily crossed; though in Swiss German, *übermarchen* still means to exaggerate. *Border*—from Old French *bordeure*, the band along the edge of a shield—was used in its current sense from the late fourteenth century onwards to denote the same thing, namely zones separating the own territory from what was foreign (Abdulafia, 2002). This later developed into borders in a less physical sense: between civilization and wilderness (or civility and barbarism), between *Heimat* (home) and *Un-heim-lichkeit* (the unfamiliar and thus vaguely uncanny). Beyond the border, then, lies the unknown, the dangerous, the fascinating, the anarchic.

Borders may appear 'natural' but in general they are created (demarcated), subsequently institutionalized (managed), and eventually perpetuated, adjusted, or removed. They are social and political constructions, imposed by powerful elites. And there are typical borderlands: zones of transition and hybridity where the territorially excluded may well coexist with the inner life of a region. When it comes to the constructedness of borders, we should recall that passport inspections and checkpoints controlled by the police or the military were only introduced in Europe in the mid-eighteenth century and that the border wall—*the* symbol of the 'German twentieth century'—was a rather unusual phenomenon. In the Middle Ages, walls mainly surrounded cities, defining who and what belonged. Spheres of influence outside the city walls were loosely staked out by border stones or other boundary markers, but travellers were free to pass these. Only when feudal overlords began expressing their mutual power and dependency relationships in spatial terms, did borders really take on a legal meaning. The often life-threatening flipside of inclusion now dramatically came to light in the exclusion of minorities, particularly of Jews and nomadic communities, from the territories of nation states. Equally problematic is the revisionist appeal to people who live as ethnic minorities in other 'cultural spheres', with the goal of incorporating such enclaves into the own territory. This corresponds to the geopolitical tradition, popular in postimperial contexts, to obliterate border lines stipulated in treaties (for example, that of Versailles) by establishing buffer zones.

The other English word for border, *frontier*, comes from the military realm where it specifically referred to the front lines of armies and the outposts built in such areas. Political borders often concur with natural boundaries: hard-to-surmount mountain crests, wide rivers, deserts, woodlands, etc., which also tend to roughly coincide with linguistic and cultural borders. By contrast, language regions, cultural areas, and trade routes get torn in two, potentially causing long-lasting territorial conflicts, when political borders are determined on the drawing board, with purely artificial factors, such as geographic coordinates, as criteria. Unlike officially protected and fortified borders, so-called green borders are permeable, either by design or due to negligence. The drawing up or opening of a border is always a highly symbolic act that represents and at the same time enforces the sovereignty and monopoly on violence of nation states. Signs that read *Caution! No trespassing!* or *Do not proceed beyond this point!* underline this executive power. Throughout history, territorial disputes have given rise to armed interstate conflicts, so-called border wars. And since geodetic point positioning is all but impossible in such 'terrain,' the demarcation of water bodies and airspace constitutes a particular challenge. Currently, there are still border disputes, for example, in the Caucasus, the Persian Gulf region, and Northern Ireland.

This brings us to the problem of borders' porosity and penetrability. Clearly, while borders separate two entities, they also connect them (to a degree not the case with more remote other entities). Are boundaries, in Leonardo da Vinci's words, "neither air nor water"? Or a twilight as in the paintings of the Impressionists? Is a boundary two-, one-, or even zero-dimensional? Is it at all 'real' or just a linguistic convention represented by symbolic markers? Which lines separate the North Sea from the rest of the Atlantic Ocean? Where runs the border between Asia and Europe within Eurasia? Where does the Sahara begin, a cumulus cloud, the Alpine foothills? When did the American and French Revolutions start, or the Copernican and Industrial Revolutions? These are but a few possible mereological questions.

In 1893, the historian Frederick Jackson Turner first published his seminal essay "The Significance of the Frontier in American History" in which he claimed that the United States' moving, open frontier had had the effect of promoting a unique American brand of democracy: it did not evolve through European-style military or missionary movements, based on a feudal social structure, Turner argued, but through the settlers' gradual westward advance along a frontier between civilization and wilderness. This topos has been transferred to expansions of all sorts—not least to the progression of the "electronic frontier" since the 1990s—and today, *space* is regarded as the 'last frontier,' with other planets potentially being terraformed in the future.

The topos of the open frontier could also be applied to a narrative of economic, political, and cultural globalization. Despite all inequalities between territorial blind spots and regional clusters, and although it is still grounded in nation states, globalization can be read as a success story of boundary dissolution. Meanwhile, the proliferating field of "Border Studies" within International Relations and Cultural Studies focuses less on borders as such than on their permanent transgression, their constructedness, and their effects on migration and cultural hybridization.

The Climate Threshold as a Generational Boundary

In conclusion, I would like to address the 1.5 or 2 degree 'threshold,' another version of a boundary that was defined as a result of the recognition and consideration of planetary boundaries. First introduced by the American economist William D. Nordhaus as part of a cost-utility analysis, the 2-degree threshold was taken up and turned into a normative goal for the mitigation of climate change by the IPCC in its assessment reports of the 1990s. It was also adopted by the German Advisory Council on Global Change (WBGU) in its political counselling activities, and in 2010, the target was incorporated into the United Nations Framework Convention on Climate Change. Since then, it has been further underpinned by empirical studies and was lowered to 1.5 degrees in view of the critical situation of low-lying coastal and small island states, organized in the AOSIS. The new goal of "pursu[ing] efforts to limit the temperature increase to 1.5 °C above pre-industrial levels" became part of the Paris Agreement in 2015.

Mike Hulme doubted the universality; ambiguity; doubtful achievability; and questionable legitimacy of the benchmark. A threshold is, first and foremost, a low obstacle that is easily crossed and overlooked. But as the German word "Schwellenangst" illustrates—denoting the fear one sometimes feels when entering a new place or situation—thresholds have symbolic meaning as well; one also thinks of glass ceilings and other discouraging barriers. And 0.5 degrees difference makes a real difference, as is shown in a recent projection of the IPCC. Arctic summers without ice would not happen once within a decade, but ten times; not fourteen but 37 percent of the world's population would be exposed to extreme heat waves every 5 years; the annual fishing catch would not decrease by 1.5, but by 3 million tons; and the proportion of plants and insects that would be reduced by half within 5 years would double or triple. The 0.5 degree difference is as real as a two degree increase in body temperature.

For this reason, the 1.5-degree target was quickly institutionalized by being incorporated into all kinds of conventions and laws. The merely symbolic stop sign was thus converted into a concrete politics of resolute decarbonization, with climate protection goals being radicalized along the way. In both its genesis and impact, then, this boundary/threshold/target is a hybrid of a symbolic marking on the one hand, and empirical evidence, obtained from physical data, on the other. It is based on various boundary or threshold values, combining historical measurements of temperature developments from 1880 to 2017 (relative to the mean value of 1951 to 1980) with cumulated carbon dioxide emissions and concentrations, so as to arrive at a remaining CO_2 budget, measured in billion tonnes. This in turn leads to an emissions curve which drops the more steeply the longer it takes for it to reach the tipping point.

When you cross *this* boundary, you do not get captured or shot by border guards on the spot – and life goes on. However, seemingly isolated extreme events, taking place all over the Earth and each concerning the lives of a relatively small group of people, add up to form the total picture of a mushrooming catastrophe. This dynamic has not been curbed, and yet the hybrid, empirical-symbolic construct that is the 1.5-degree target has resulted in specific measures on the national level as well as in comprehensive transnational plans, which do make a limitation of global warming seem realistic. Not just within democracies, the debates on climate protection goals are of course characterized by all sorts of compromises. It is a constant negotiation, as Hans-Joachim Schellnhuber put it, between what is "scientifically" and what is "economically desirable"—to which we must add what is politically feasible: in a world community made up of almost 200 states, an ideal circular and sustainable economy will hardly ever be realized.

To conclude: *Limits to growth, planetary boundaries, climat sans frontières*: since we recognized that the notion of unlimited progress was but an illusion, such boundary concepts appear to most of us like metaphors of self-denial, of prohibition, of restricted options. In a different light yet, self-limitation may bring forth new freedoms from old constraints, constraints imposed on us by the logics of growth and discounting. Climate change and species extinction have **no** boundaries—if we do not raise barriers to ensure that the future of coming generations is no longer looking so gloomy. The most critical boundary of the planet thus appears to be a temporal one: the often paper-thin line between age cohorts or generations whose life-defining biographical experiences tend to diverge like mountain valleys separated by just a steep ridge.

References

Abdulafia, David u.a. (Hg.). (2002). *Medieval Frontiers: Concepts and practices*, Aldershot u.a.

Frodeman, R., Klein, J. T., & Mitcham, C. (2010). *Oxford handbook of interdisciplinarity*. Oxford University Press.

Luhmann, N. (1995). Inklusion und Exklusion. In: Luhmann, N. (Hrsg.) *Soziologische Aufklärung* Bd. 6 Opladen, S. 237–264.

Mannheim, K. (1952). The problem of generations. In P. Kecskemeti (Ed.), *Essays on the sociology of knowledge: Collected works* (Vol. 5, pp. 276–322). Routledge.

Meadows, D. et al. (1972). *The limits to growth*. Universe Books.

Mortenson, E. (2011). Bridging the nature/culture divide. *Topia, 33*, 254–257.

Rockström, J., Steffen, W., Noone, K., Persson, Å., et al. (2009). A safe operating space for humanity. *Nature, 461*, 472–475. https://doi.org/10.1038/461472a

Steffen, W., Richardson, K., Rockström, J., Cornell, S. E., et al. (2015). Planetary boundaries: Guiding human development on a changing planet. *Science, 347*(736), 1259855.

Wittgenstein, L. (1922). *Tractatus logico-philosophicus*.

Claus Leggewie is holder of the Ludwig Börne professorship and director of the "Panel on Planetary Thinking" at Giessen University, Germany. He currently is co-editor of the book series "Climate & Cultures" (Brill) and the "Routledge Global Cooperation Series" (Routledge). Earlier affiliations include visiting professorships at the University of Paris-Nanterre and New York University (Max Weber Chair). From 2008 to 2016, he was a member of the German Advisory Council on Global Change (WBGU).

Trust

Inka Bormann

Abstract Starting from the assumption of a fundamental vulnerability on the part of humankind, this article deals with trust as an anthropological universal under the conditions of the Anthropocene. Human beings depend on cooperation in order to be viable, and trust facilitates such cooperation. Trust is also needed for the purpose of meeting the challenges posed by the Anthropocene. Accepting our planetary boundaries and the massive societal transformation this necessitates, the article posits that trust in the capabilities of humankind to shape its future is both inevitable and a challenge when it comes to dealing responsibly with the uncertainties of the Anthropocene.

Human beings come into the world as entities needing protection and guidance because of their "weakness, defencelessness, helplessness, openness and exposure" (Misztal, 2011: 364), and as such they depend on the care and cooperation of others. This dependency on "another's goodwill" (Baier, 1986: 235) is interconnected with vulnerability. However, trust helps us to avoid reflecting on our own vulnerability and to remain capable of acting. It can be seen as an anthropological universal that facilitates living together.

Vulnerability is central in trust relationships at both the level of interpersonal interaction and interaction with institutions (Misztal, 2011). In trust relationships, vulnerability results from the fact that the recipient of the trust (trustee) is free as to whether or to what extent to consider the positive expectation of the party doing the trusting (trustor) that he or she will not be harmed (Rousseau et al., 1998). Understood as the willingness to become vulnerable, trust relies on interactions that lead, according to the quality of prior experience, to a propensity to trust on the part of the trusting party. Apart from such a propensity to trust, the perceived trustworthiness is crucial. It is rooted in the characteristics of the recipient of the trust, i.e. his/her competence, benevolence and integrity (Mayer et al., 1995) as

I. Bormann (✉)
Freie Universität Berlin, Berlin, Germany
e-mail: inka.bormann@fu-berlin.de

© The Author(s), under exclusive license to Springer Nature
Switzerland AG 2023
N. Wallenhorst, C. Wulf (eds.), *Handbook of the Anthropocene*,
https://doi.org/10.1007/978-3-031-25910-4_152

perceived by the trusting party. Whereas the perception of trustworthiness is in general pivotal for the development of trust and thus ultimately trusting behaviour, these characteristics may differ at the levels of interpersonal and institutional trust (Bachmann, 2018). In terms of interpersonal trust, for example, the level of perceived care, interest and honesty is essential. Institutional trust can be distinguished according to the target of the trust. It can on the one hand include trust in other individuals because of institutional frameworks in which interaction take place, based on routines, shared values and common norms; on the other hand, trust in institutions can rest upon their legitimacy and the guiding principles which they apply and operate efficiently and effectively (Lepsius, 2017).

Human vulnerability is rooted not only in being dependent on the cooperation and support of other people, but also in being exposed to the influences of nature. Both of these causes of vulnerability are subject to regulations that help to contain the basic human emotion of anxiety (Wulf, 2015). To curb the potential risk of being deceived and exploited by others because of one's faith in their trustworthiness, contracts are put into force to regulate (sometimes sanction) our behaviour. Technologies for mastering and cultivating nature have been created to contain the human's fear of nature and the power of its elements (Böhme, 2010: 38). However, from an institutionalist point of view, both forms of regulation are incomplete. As far as (symbolic, moral or legally secured) contracts are concerned, although they contribute to mutual predictability and therefore stimulate mutual trust, they always involve room for interpretation and fail to exclude the possibility that a partner in the interaction may violate the contract (Baurmann, 2002). And when it comes to technologies, it has been proven repeatedly that due to incomplete knowledge about their consequences, for example, technologies do not necessarily lead to unrestricted improvements, but can generate negative side effects which are temporally and spatially displaced (Beck, 1986). In this respect, in spite of the contracts and technologies which have been put in place to shape the world according to humankind's desires and needs, trust remains indispensable – it is a substantial ingredient of our humanity and a basic foundation of the social order which provides us with orientation. In other words, in terms of dealing with the inescapable vulnerability of the human race, trust is "a mode of intersubjective structuring of human relations with the world" (Endreß, 2001: 162).

Trust does not seem to be entirely rational because of this inescapable vulnerability – not even in the age of the knowledge society in which the Anthropocene was not only not prevented but, on the contrary, virtually unleashed. Even Simmel (1908/1992: 263) holds that trust is an intermediate state between knowledge and ignorance, with trust not being required when there is complete knowledge and unreasonable in the case of total ignorance. He identified three modes of trust, involving different foundations: faith, emotion and knowledge. As sociohistorical considerations of social change suggest, as social orders have changed, so have the foundations of trust. Whereas in traditional societies, social order and societal cohesion tended to be based on faith, personal relationships and familiarity, in complex modern societies the social order tends to rely increasingly on secularized, demoralized, and depersonalized structures as well as having a dependency on expert systems, all of which hamper familiarity and emotional ties but at the same time call for trust – the creation of which is at the same time challenged (Frevert, 2013; Misztal, 1996; Giddens, 1990).

This seems particularly true in the Anthropocene (Crutzen, 2002), a label for a human epoch which expresses the inseparable relationship of humankind with its environment, a relationship whose development has played a significant role in shaping this environment, which is increasingly suffering under the manifold distortions created by the globalized world. Based on an obviously low acceptance of the fact that the planetary boundaries (Rockström et al., 2009) will be exceeded in the foreseeable future (Lenton et al., 2019), the conditions of life on earth are rapidly changing. Habitat destruction and the loss of biodiversity, the consequences of climate change, hunger, droughts and floods, even in previously climatically temperate regions of the world, are only a few of the severe issues caused or accelerated by humankind, accompanied by the resulting social, economic and cultural upheavals. Neither hope nor blind trust therefore seems appropriate or sensible, given the overwhelming amount of knowledge about the earth's development, constantly approaching interconnected and mutually re-enforcing tipping points. In the Anthropocene, humankind is thus more than ever responsible for taking action.

Understood as an unprecedented era in which the developments on the earth are regarded as caused by humankind, hardly reversible, interconnected and threatening to life on earth, a great societal transformation seems to be indispensable when it comes to dealing with the consequences of the Anthropocene. But both the effects in the Anthropocene and the great transformation itself may cause anxiety, due to the complexity of the simultaneous and interconnected actions that will be required of all states and each individual citizen wishing to make a significant contribution – if not to fundamentally changing the situation, at least to slowing down the development. Even if it sounds contradictory: in terms of reducing complexity and maintaining a capability to act (Luhmann, 2014), trust in effective and efficient action to contain the consequences of the Anthropocene is required, even though there is an unmissable knowledge "gap" between knowing and acting (Kollmuss & Agyeman, 2002) as well as an apparent gap concerning what should be regulated and how. Because they have not only produced solutions but also contributed to the emergence of the Anthropocene, a reflected trust in technologies and their consequences seems reasonable. To this end, different modes of trust seem to be equally pivotal. Because social order is massively affected by the consequences of the Anthropocene, it would seem particularly important to have trust in the ability of legitimate institutions to constitute a sustainable societal order and to operate effectively and efficiently on the basis of shared democratic values for a sustainable future. At the same time, interpersonal trust seems more important than ever, because of the increasing heterogeneity resulting from global, climate change induced migration, for example, or because of the collaboration which is required across professions and regions to achieve the great transformation.

In other words, the cognitive and calculative capabilities of humankind – i.e. the ability to reflect on human behaviour and its consequences – do not appear to have contributed to shaping a sustainable future for all. Instead, humankind has increased its vulnerability. (Blind) trust in these particular human characteristics alone, therefore, does not seem advisable. Nevertheless, with regard to the consequences of the Anthropocene, it would seem to be reasonable – in combination with a portion of scepticism about technological solutions – to place a certain amount of trust in the interpersonal achievements of humankind as a species which has the capacity to both shape its future and to develop and maintain the social order in the great transformation. This will be necessary to ensure that humankind's anthropogenic state of nature is regulated in a reasonable and responsible way, for the good of all the citizens of this world.

References

Bachmann, R. (2018). Institutions and trust. In R. Searle, A.-M. Nienaber, & S. B. Sitkin (Eds.), *The Routledge companion to trust* (pp. 218–228). Routledge.

Baier, A. (1986). Trust and Antitrust. *Ethics and Education* (January), 231–260.

Baurmann, M. (2002). Vertrauen und Anerkennung: wie weiche Anreize ein Vertrauen in Institutionen fördern können [Trust and recognition: how soft incentives can foster a sense of trust in institutions]. In A. Maurer & M. Schmid (Eds.), *Neuer Institutionalismus: zur soziologischen Erklärung von Organisation, Moral und Vertrauen* (pp. 106–132). Campus.

Böhme, W. (2010). Elemente – Feuer Wasser Erde Luft [Elements – Fire water earth air]. In C. Wulf (Ed.), *Der Mensch und seine Kultur* (pp. 17–46). Anaconda.

Beck, U. (1986). *Risikogesellschaft. Auf dem Weg in eine andere Moderne* [Risk society - on the way to a different modernity]. Suhrkamp.

Crutzen, P. J. (2002). The "anthropocene". *Journal de Physique IV, 12*(10), 1–5.

Endreß, M. (2001). Vertrauen und Vertrautheit – Phänomenologisch-anthropologische Grundlegung [trust and familiarity – phenomenological-anthropological groundwork]. In M. Hartmann & C. Offe (Eds.), *Vertrauen. Grundlage sozialen Zusammenhalts* (pp. 161–204). Campus.

Frevert, U. (2013). *Vertrauensfragen. Eine Obsession der Moderne*. [Trust issues. An obsession of modernity]. Beck.

Giddens, A. (1990). *The consequences of modernity*. Polity Press.

Kollmuss, A., & Agyeman, J. (2002). Mind the gap: Why do people act environmentally and what are the barriers to pro-environmental behavior? *Environmental Education Research, 8*(3), 239–260.

Lenton, T. M., Rockström, J., Gaffney, O., Rahmstorf, S., Richardson, K., Steffen, W., & Schellnhuber, H. J. (2019). Climate tipping points — Too risky to bet against. *Nature, 575*(7784), 592–595.

Lepsius, M. R. (2017). Trust in institutions. In M. Rainer Lepsius & C. Wendt (Eds.), *Max weber and institutional theory* (Vol. 45, pp. 79–87). Springer.

Luhmann, N. (2014). *Vertrauen: Ein Mechanismus der Reduktion sozialer Komplexität*. [Trust: A mechanism of social complexity reduction] (5th ed.). UTB.

Mayer, R. C., Davis, J. H., & Schoorman, F. D. (1995). An integrative model of organizational trust. *The Academy of Management Review, 20*(3), 709–734.

Misztal, B. (1996). *Trust in modern societies. The search for the bases of social order: Significance, scope and limits of the drive towards global uniformity*. Polity Press.

Misztal, B. (2011). Trust: Acceptance of, precaution against and cause of vulnerability. *Comparative Sociology, 10*(3), 358–379.

Rockström, J., et al. (2009). Planetary boundaries: Exploring the safe operating space for human-ity. *Ecology and Society, 14*(2), 32.

Rousseau, D., Sitkin, S. B., Burt, R. S., & Camerer, C. (1998). Not so different after all: A cross-discipline view of trust. *Academy of Management Review, 23*(3), 393–404.

Simmel, G. (1908/1992). *Soziologie. Untersuchungen über die Formen der Vergesellschaftung.* [Sociology. Studies on the forms of socialization] (Vol. 11). Suhrkamp.

Wulf, C. (2015). Emotion. In C. Wulf & J. Zirfas (Eds.), *Handbuch Pädagogische Anthropologie.* [Handbook of pedagogical anthropology] (pp. 113–125). SpringerVS.

Inka Bormann is Professor of General Education at the Freie Universität Berlin. In her research she focuses on trust in educational settings, education for sustainable development and educational governance as well as the interplay of these phenomena. Her most recent articles on trust in edu-cational settings include, for example, *COVID-19 and its effects: On the risk of social inequality through digitalization and the loss of trust in three European education systems.* European Educational Research Journal (2021, with K. Brøgger, M. Pol and B. Lazarova); *A comprehensive view of trust in education: Conclusions from a systematic literature review.* Review of Education 9 (1), 124–158 (2021, with S. Niedlich, A. Kallfaß, S. Pohle); *Uncertainty and trust. Complementary elements of pedagogical interactions and their institutional over-formation.* Paragrana, 24(1), 151–163 (2015, in German); *Transformations of the thematization of trust in education.* In: S. Bartmann, S., M. Fabel-Lamla, M., N. Pfaff, & N. Welter (Eds.): Trust in educational research. Opladen: Budrich, 101–123 (2014, in German).

Part XIII
Humanity as Creation

Anxiety

Konrad Oexle and Thomas Reuster

Abstract Civilization has been viewed as a system to cope with anxiety. It is warranted, therefore, to look into the role of anxiety in the Anthropocene, as cause or effect and for better or worse.

On July 16, 1969, Neil Armstrong and his crew were launched into the sky on top of 3000 tons of explosive fuel. They knew the risks. Life insurances being unaffordable, they had signed space-themed envelopes as financial security for their wives. Four days later, approaching Moon's surface, the Lunar Module's computer kept issuing program alarms. Armstrong's heart rate rose to 156 bpm. The planned touchdown corridor was miles away. At Mission Control in Houston "a bunch of guys were about to turn blue". However, Armstrong did not panic. While he was not naturally free of fear but had needed "some years to sort of circumvent that concern", that is, "the uneasiness of facing the reality of death" (Armstrong, 2011), he had previously mastered nearly catastrophic situations as a pilot. Thus, although he had only 18 s of fuel left to land, he dodged boulders until he found a suited spot for a safe touchdown.

Man landing on the Moon is a paradigmatic event of the Anthropocene, highlighting the modern endeavour to conquer and control the world. Anxiety was not a leading emotion of those who realized President Kennedy's decision "to go to the Moon … and do the other things, not because they are easy, but because they are hard" (Kennedy, 1962). Instead, they followed a "risk-reward equation – the kind of balance that you always make" (Armstrong, 2011).

The original version of the chapter has been revised. A correction to this chapter can be found at https://doi.org/10.1007/978-3-031-25910-4_281

K. Oexle (✉)
Institute of Neurogenomics, Helmholtz Munich, Oberschleißheim, Germany
e-mail: konrad.oexle@helmholtz-muenchen.de

T. Reuster
Technical University Dresden, Dresden, Germany
e-mail: thomas.reuster@tu-dresden.de

However, that equation may not have accounted for all variables. At the time when man first walked on the Moon, the optimistic sixties were about to turn into the alarmed seventies. In 1972, the Club of Rome warned of the "limits to growth". Indeed, space travel contributed to this change of perspective. Earth's photograph as a tiny, vulnerable habitat surrounded by endless darkness strongly influenced the emerging environmental movement. On this "Spaceship Earth" the "cowboy economy", which mistakenly assumes infinite resources and space for waste disposal, needs to be replaced by a cyclical economy. This novel view on the limited global capacities matched Foucault's (1967) 'spatial turn' in the cultural sciences, including his assertion that "the anxiety of our era has to do fundamentally with space", that is, spatial relations (Höhler, 2015).

The space program thus related to both progress and its incipient crisis, with recognizing and seizing opportunities versus emphasizing self-inflicted threats and shying away from risks. Lunar voyages themselves were soon abandoned, actually, and space trafficking developed its own waste problem in the Earth orbit. Generally, the perceived ecological threats have increased dramatically, so that in Davos 2019 Greta Thunberg, the founder of the Fridays for Future youth movement, yelled at the world economy leaders, "I don't want you to be hopeful, I want you to panic!" (Thunberg, 2019).

In view of such stress, it is warranted to ask about the forms, functions, and consequences of anxiety in the Anthropocene. Various thinkers (Koch, 2013) regarded human civilization altogether as a system for coping with anxiety which, according to Hobbes, is fear of dying in an uncivilized war of all against all, while for Freud it is guilt due to instinctual drives resulting in their cultural sublimation. Anxiety may be deeply rooted in human nature as a universal feature of development and consciousness. However, anxiety certainly has cultural, historical and situation-dependent specifics, both in terms of quantity and quality. Europeans who crossed the ocean fleeing from poverty suffered from another anxiety than Africans who were sold into slavery. On the other hand, both may have experienced similar forms of anxiety, being removed from home and exposed to the open sea. We cannot be certain as we never can be certain about the feelings of other minds and other cultures. While paying attention to this limit, anxiety is discussed here in generalizing terms, presupposing that readers can understand the specific use of the notion even if it is not explicated in detail. Of note, the target to be understood is moving. For instance, anxiety often is practically synonym with fear but for Kierkegaard and Heidegger its existential bearing on emancipation and authenticity, respectively, referred to realizing the fundamental homelessness of being in the world rather than to the fear of a concrete circumstance.

As a plausible beginning of the Anthropocene, Lewis and Maslin (2015) proposed the daring but devastating conquest of the New World by the Old World in Early modernity. This encounter reduced the population of the Americas by about 90%, resulting in massive vegetation growth on abandoned land and causing a historic minimum of CO_2 at 1610 according to Antarctic ice core records. Early modernity also saw an important development in terms of anxiety: The monk Martin Luther recognized that he is not able to free himself from sin in order to overcome his fear of the Last Judgement and reasoned that the salvation from sin is only subject to the grace of God who cannot be manipulated. According to Max Weber, this shift in the theology of justification, most pronounced in Calvinism, was a major force in the upcoming capitalism, i.e., the economic mechanism that accelerated the Anthropocene: Labor force now could be redirected from providing for salvation after death to secular diligence whose success, incidentally, would be evidence of justifying endowment (Bongardt, 2013).

Luther gave his essay on justification by divine grace the title, "On the Freedom of a Christian". Individual freedom further gained importance in the eve of the Enlightenment while the relevance of the metaphysical diminished. Mankind was called upon to emerge "from its self-incurred tutelage". This appeared to be quite possible because reality was considered to be well ordered, culminating in Hegel's claim, "All what is real, is reasonable, and all what is reasonable, is real". However, at the same time doubts arose about the inherent rationality of the world. The terrible earthquake of Lisbon in 1755 challenged the optimistic stance, and Hegel's fellow Schelling in his later work emphasized the dark drives in human motivation and the anxiety that the destructive might overcome the ordered (Schulz, 1965).

While the fear of God had overwhelmed any other anxiety, that is, "ne timeas, modo timeas" ("fear not, provided you fear") in Pascal's words, the enlightened mankind lost that option of coping with anxiety. The technological application of reason certainly promised and achieved huge gains in wealth and live opportunities. However, in a "dizziness of freedom" the emancipated individual experienced an analogously increased anxiety of not grasping anything and finally losing everything. The existentialists thence responded by turning the perspective. For them, the anxiety of the mortal existence was the chance to gain authenticity and the experience of freedom inspired responsibility and courage to be (Bähr, 2013; Bongardt, 2013; Schulz, 1965).

But how should he act, the pragmatic existentialist, as embodied by the physician Rieux in Camus's Plague? Albeit in vain, Rieux applies the available scientific knowledge against the natural evil that threatens human lives. This allocation of evil, however, has become doubtful. Nature is now seen as the victim of technologically expanding humanity. Fear of natural dangers is being supplanted by concern for nature in its vulnerability. Or at least most of the time: recent viral epidemics have taught us once again, that nature is two-faced, being Medea no less than Gaia (Ward, 2009). Thus, the view on nature and the corresponding anxieties are incongruent and contested (Oexle, 2021a). Some doubt the scientific forecasts of catastrophes or their anthropogenic origin and fear the detrimental results of erroneous and authoritarian eco-politics, some fear the catastrophes and see hope only in radical (self-)restriction of man so that nature can return to its assumed equilibrium, and some think that the catastrophes can be averted only by scientific, i.e., geoengineering intervention. Anxieties about the interventions may soon outgrow the anxieties about the forecasts (Oexle, 2021a).

Anxieties concerning man-made catastrophes do not only refer to ecology, of course. Wars and especially the holocaust in the twentieth century undermined the trust in technocracy's ability to prevail irrational political movements. On the contrary, they emerged all the stronger as totalitarianisms in mass culture. Various authors including Adorno and Horkheimer and Arendt addressed this context. Hermann Broch most strongly emphasized the relief of individual anxiety provided by mass hysteria (Balke 2013; Koch, 2013).

Hence, for better or worse, anxiety is a driver as well as an effect in the Anthropocene. All the while, it is instable historically. Orwellian worries in West Germany about the 1980s census appear almost ridiculous in retrospect, while current fears of increasing commercial or governmental surveillance seem quite justified. This historical variability relates to anxiety's constitutive characteristic of being unsecure about its object. When we see a threat in concrete terms, the respective anxiety is correspondingly replaced by counteraction or relief. Anxiety therefore relates to action and politics. And the number of true or imagined threats has been inflated by the perpetually forecasting, self-reflexive modern "risk society". Evaluating threats and fears, allocating the appropriate means to counter them, and keeping those at bay who thrive on causing and communicating anxieties, threatens to become unmanageable (just to add one more self-reflexive anxiety to the list here). Beyond just preventing Hobbes' war of all against all, "biopolitics" now also has to manage the bodies whose individual and mutual (infectious) risks have become modifiable (Balke, 2013). These risks will further gain in weight the longer the bodies live. And if their management accesses novel molecular or virtual spaces such as genetics and data processing, new anxieties arise.

Even the scientific and intellectual discourse has been invaded by anxiety since an ever increasing number of diverging identities feel unsafe when their foundational assumptions are discussed. This is not a promising development. Rampant anxiety communication may indicate that society's functional subsystems such as science are about to fail. Then, as Luhmann emphasized, there is hardly a way back since "scientific attempts to explain the complicated structure of risk and safety issues just fuels anxiety with new alarms." (Oexle, 2021b).

"But where the danger is, also grows the saving power". Potentially, the fear about the breakdown of the "communication community" may be able to domesticate all other anxieties in the sense of Pascal's "ne timeas, modo timeas", because the requirement of successful communication must be primary since it is the prerequisite of all successful actions including its own criticism (Apel, 1973). However, as mentioned above, fear is not a constructive advisor. Brandom (2013) has reminded of Hegel's early suggestion that constructing conceptual norms of our communication requires "hermeneutics of magnanimity and trust".

References

Apel, K.-O. (1973). *Transformation der Philosophie*. Suhrkamp.
Armstrong, N. (2011). *An audience with Neil Armstrong*. YouTube. https://www.youtube.com/watch?v=KJzOIh2eHqQ
Balke, F. (2013). Politik der Angst. In L. Koch (Ed.), *Angst. Ein interdisziplinäres Handbuch* (pp. 80–93). Metzler.

Bähr, A. (2013). *Furcht und Furchtlosigkeit. Göttliche Gewalt und Selbstkonstitution im 17. Jahrhundert* (p. 55). V&R unipress.

Bongardt, M. (2013). Theologie der Angst. In L. Koch (Ed.), *Angst. Ein interdisziplinäres Handbuch* (pp. 20–30). Metzler.

Brandom, R. (2013). *Reason, genealogy, and the hermeneutics of magnanimity*. Howison lecture at the University of California/Berkeley. Video available at http://www.uctv.tv/shows/Reason-Genealogy-and-the-Hermeneutics-of-Magnanimity-with-Robert-Brandom-25074

Höhler, S. (2015). *Spaceship earth in the environmental age, 1960–1990*. Pickering & Chatto Publishers.

Kennedy, J. F. (1962, September 12). *Speech at Rice University*. https://er.jsc.nasa.gov/seh/ricetalk.htm

Koch, L. (2013). Einleitung: Angst als Gegenstand kulturwissenschaftlicher Forschung. In L. Koch (Ed.), *Angst. Ein interdisziplinäres Handbuch* (pp. 5–20). Metzler.

Lewis, S. L., & Maslin, M. A. (2015). Defining the Anthropocene. *Nature, 519*, 171–180.

Oexle, K. (2021a). Does nature think about the future? In M. Molls et al. (Eds.), *Science, reason, and responsibility* (pp. 42–45). TUM. University Press.

Oexle, K. (2021b). Vulnerability is a talent in the ecological crisis. In P. A. Wilderer, M. Grambow, M. Molls, & K. Oexle (Eds.), *Strategies for sustainability of the earth system* (pp. 99–102). Springer.

Schulz, W. (1965). Das Problem der Angst in der neueren Philosophie. In Hoimar v. Ditfurth (Ed.), *Aspekte der Angst* (pp. 1–13). Kindler.

Thunberg, G. (2019). *Our house is on fire!* YouTube. https://www.youtube.com/watch?v=M7dVF9xylaw

Ward, P. (2009). *The medea hypothesis: Is life on earth ultimately self-destructive?* University Press.

Konrad Oexle heads the Neurogenetic Systems Analysis Group of the Institute of Neurogenomics at the Helmholtz Center Munich. He is Associate Professor at the Technical University Munich and has published on applied and basic science.

Thomas Reuster studied philosophy and medicine. Until 2020 he headed a psychiatric clinic. He lectures at the Technical University Dresden teaching medical ethics and has published on psychosocial therapies and on anthropological psychiatry and phenomenology.

Artificial Intelligence

Shoko Suzuki

Abstract In various fields related to our daily lives, such as transportation, finance, logistics, corporate management, medical care, nursing care, and education, AI technology is built into the devices around us. For a new stage of technological civilization with Artificial Intelligence, it is necessary to reconsider what is human and what can we do as humans.

As the social implementation of Artificial Intelligence (AI) and the Internet of Things (IoT) has begun, the dissemination of intelligence-related technologies has progressed further, and society is starting to change its appearance remarkably. In various fields related to our daily lives, such as transportation, finance, logistics, corporate management, medical care, nursing care, and education, AI technology is built into the devices around us. As a system that functions via connection and cooperation rapidly forms, a period of change is coming that will affect the rules and values of society, as well as people's daily lives, including their working styles, lifestyles and communication.

When it occurs, data in physical space (i.e., the real world or real life) will be collected more and more quickly and in increasingly large quantities by information terminals embedded in all things, and will be accumulated in cyberspace. In addition, the analysed data will be fed back into the real world, leading to the creation of a "Cyber-Physical System" (CPS) at various levels in the community, formed through interaction between the two spaces (Floridi, 2014; Mayer-Schönberger & Cukier, 2013).

As large-scale projects involving industry, government, academia, and the private sector are developing and applying new technologies that bring about enormous economic effects and social reforms, concerns are spreading about the

S. Suzuki (✉)
Kyoto University, Kyoto, Japan

The Center for Advanced Integrated Intelligence, National Research Institute Riken, Saitama, Japan
e-mail: shoko.suzuki.ue@riken.jp

negative impact of new technologies, such as AI technology, on human society. At the same time, there is growing concern that the significance of human existence is diminished. The real question now is how to interact with technological civilizations that encompass all of human society from research and development of technology to its utilization (Dreyfus, 1978; Bostrom, 2014).

Of course, it is said that, as a result of starting to walk on two feet, human beings released their hands and brain to invent tools. In the early stages of human development, humankind, like other animals, was in the natural environment and through its interaction with this natural environment, humankind created a secondary environment, which can be referred to as a human and natural system, or ecosystem. It is an accumulation of technologies (technai) in a broad sense, including knowledge, skills and ways of using tools, and it can be regarded as a technological civilization in which a group of external mechanisms incorporating the accumulation of skill and experiences of technology and the mechanism of transmission and transmission of technology constitutes a system.

Humans build certain relationships with tools through the use of tools. The technique of using aids or instruments for supplementation and strengthening is cultivated through the relationship with the instruments, and humans use the instruments as if they were part of their own body. In order to utilize tools and machines, it is necessary to establish a collaborative system to understand the mechanisms necessary for them to function—in other words, the "language" of the tools and machines. In this process, humans reconfigure their own abilities by acquiring the knowledge, skills, and capacity to communicate with tools and machines (Stiegler, 1994; Hayles, 1999).

As technological civilizations continue to renew themselves, there will be a need for discussions that highlight the actual people experiencing constant change and transformation in the course of these developments. At the same time, human beings have acted on the natural environment while being influenced by it, and in the process of adaptation have created a technological civilization, also called the secondary environment, coming to be at the top of all other organisms. This also means that we should take responsibility, as we have the ability to determine the future of the global environment. The new challenge for humanity presented by the Anthropocene is nothing less than to find a way for human beings, through technological civilization, to live in well-being not only with living beings but also with all inanimate objects on the planet (Wulf, 2020).

It is important to build a human-centred world. However, questions about what it means to be human and how humans should live have always been asked repeatedly from the era of ancient Greece and ancient China to today. The question of how humankind should be viewed could be described as integral to the question of how humans should live in this world.

Humans have always maintained an attitude of questioning their own existence. That very attitude can be regarded as the essence of a human being. It is necessary for humans to control themselves so that certain fixed views and biased thoughts do not lead them to fall into the type of human-centred or human-oriented philosophy in which the superiority of humans alone is asserted, without regard for things that live all around in nature. It is necessary to constantly look back on the civilizations that have been built up by humans and the cultures that have been fostered in order to discourage ourselves from human arrogance. We need to reflect on our own deeds so that human behaviour does not become the sort of human-centred human orientation that benefits only humankind.

The idea of human centricity, which is common to all discussions addressing the vision of the future society to be brought about by AI technology, can be understood to incorporate the expectation that humans alone bear the right and responsibility to decide the future of the Earth as, in the past and in the future, unlike other living things, they have been responsible for technological civilization and have been in a position of superiority compared to other living things. If, however, the tendency to feel fear or panic at the advent of AI-dominated societies were to gradually become more and more intense in the future due to a desire to prevent a world in which humans are used by AI, an attempt based on human-centricity would concentrate on simply competitively comparing AI with humans in terms of their functions, or on strategies to avoid surrendering human leadership to AI.

What does it mean to be human? Where did humans come from and where will they go? Until now, humans have continued to question their own horizons by querying their own existence and significance. It goes without saying that questioning what makes us human is important not only for us to live as self-aware beings who consider their own origins and destination, but also to always continue asking whether we ought to be responsible for the future of the global environment.

In the same way that Darwin stated that it is the ability to deal with change that should be honed, it is necessary to reaffirm that humankind has survived to date because it changed itself flexibly according to the circumstances in response to changes in the environment. We also need to look back on the past from the viewpoint of what has been lost and what has been acquired during humankind's process of change, and to look ahead to humankind's future.

Now, humans again find themselves in a situation in which they are forced to question what it is that makes them human. This time, as the AI developed by human beings gains momentum toward surpassing human capabilities, we must look for what is uniquely significant about human existence that sets us apart from AI. In order to draw a boundary between humans and animals on the one hand and between humans and AI on the other hand, a redefinition of humanity is becoming necessary (Berberich et al., 2020; Suzuki, 2020).

Nevertheless, to the extent that we are human, continuing to question the very reason for the existence of humans, or in other words, the task of ascertaining our horizons, must not be forgotten, as it comprises the fundamental questioning attitude that underlies all academic disciplines. Especially in the academic system, which has been increasingly divided into specialties since the latter half of the nineteenth century and has now become fragmented, the approach of understanding humans as a whole has become extremely weak. As long as science, technology, civilization, and culture are all generated for humans by humans, the current situation in which questions about human existence are becoming rare may be evidence of the decline in not only the power to verify and reflect on what has been created by humans, but also in the very driving force toward the future that will be created.

Because, even if AI developed by people with a "low level of human qualities," so to speak, were capable of completing tasks that exceed those of humans in terms of each function itself, the ability to examine the future world with a view to pursuing human happiness or realizing a sustainable society, would be impossible to achieve with only a low level of human qualities. This applies not only to assessing the pros and cons of development itself but also to adjusting development methods and development speeds, as well as forecasting unexpected ripple effects and unexpected situations. It is the human qualities and virtue of the developer that determines the quality of AI. In that sense, it could be said that AI serves as a mirror for humans.

References

Berberich, N., Nishida, T., & Suzuki, S. (2020). Harmonizing artificial intelligence for social good. *Philosophy & Technology, 33*, 613–638. https://doi.org/10.1007/s13347-020-00421-8

Bostrom, N. (2014). *Superintelligence. Paths, dangers, strategies*. Oxford University Press.

Dreyfus, H. (1978). *What computers can't do: The limits of artificial intelligence*. Harper Collins.

Floridi, L. (2014). *The fourth revolution – How the infosphere is reshaping human reality*. Oxford University Press.

Hayles, K. N. (1999). *How we became Posthuman: Virtual bodies in cybernetics, literature, and informatics*. The University of Chicago Press.

Mayer-Schönberger, Viktor/Cukier, Kenneth. 2013. Big Data: A revolution that will transform how we live, work and think. Houghton Mifflin Harcourt Publishing

Stiegler, B. (1994). *La technique etle temps, Tome 1: La faute d'Épiméthée*. Èditions Galilée.

Suzuki, S. (2020). Redefining Humanity in the Era of AI- Technical Civilization. Wulf, Christoph and Jörg Zirfas (Hg.) Den Menschen neu denken. Paragrana: Internationale Zeitschrift für Historische Anthropologie. *Paragrana – Zeitschrift für Internationale Zeitschrift für Historische Anthropologie, 29/1*, 83–93.

Wulf, Chr. (2020). Den Menschen neu denken im Anthropozän. Wulf, Christoph/Jörg Zirfas (Hg.) Den Menschen neu denken. *Paragrana: Internationale Zeitschrift für Historische Anthropologie, 29/1*, 13–35.

Shoko Suzuki is Professor in Educational Philosophy at the Kyoto University and Principal Investigator of the Center for Advanced Integrated Intelligence, National Research Institute Riken in Japan. 2005–2019 she was a member of the academic council of Japan and 2009/2010 a guest Professor in Free University of Berlin in Germany. Books (selection): Pandemien im Anthropozän, Paragrana Internationale Zeitschrift für Historische Anthropologie 21/2 (ed. with Wulf, De Gruyter Verlag 2021). Auf dem Weg des Lebens –West- östliche Meditation (with Wulf, Logos Verlag 2013, in German). Takt in modern Education (Waxmann 2010).

Digitalization/Digital Transformation

Benjamin Jörissen

Abstract From the perspective of the Antropocene, digitalization appears as a complex phenomenon. It consumes material and energy and produces toxic waste, while at the same time providing spaces and networks for change, serving as an informational, communicational, social, and cultural infrastructure that drives globally networked activism and policy-oriented discourses. This article discusses the need to move away from 'solutionist' (Morozov) approaches to resilience-oriented strategies that promote responsive, responsible, and empowering approaches to digitalization.

In the "critical zone" that humans inhabit on this planet, silicon is the second most common chemical element after oxygen. From the most abundant down to the most "rare earths", the era of silicon-based computing literally turns earth(s) into cyberspaces and metaverses. The relationship between the materialities of digital calculating devices and the current state of the Anthropocene is complex and multifaceted, just like the stacked, interacting layers of technology that have finally led to today's global entanglement of human-human-machines-interactions, which drive not only communications but also the material logistics of life-sustaining systems, and thus directly or indirectly affect most of the current global state of being. "Digitalization" in this comprehensive sense contributes to the current reshaping of the material, energetic, and infrastructural, and also the sensorical, communicational, social and cultural face of the planet. It is a profoundly cultural process, especially if one acknowledges that the notion of "culture" cannot be separated from its material tiers. As the diagnosis of an "Anthropocene" calls for much more than "global consciousness", namely a planetary sense and sensitization (Spivak, 2002), the inherent moment of infrastructural globalization of a literally "worldwide" web is obviously entangled with the philosophical, aesthetical, medial and technological processes,

B. Jörissen (✉)
UNESCO Chair in Arts and Culture in Education, Friedrich-Alexander-Universität Erlangen-Nürnberg, Erlangen, Germany
e-mail: benjamin.joerissen@fau.de

moving towards an imagery and imagination of a "whole" planet earth (Nitzke & Pethes, 2017). The ways in which digital culture has contributed and will continue to contribute to notions of a "whole earth" in the past but also in the future are crucial in this regard: as a set of commodifiable resources (Zuboff, 2019), as an (allegedly) humanistic data-interspace (Schmidt & Cohen, 2013), or as a space signified by curatorial care (Turner, 2008) and critical data practices (Gabrys, 2018).

The Anthropocene perspective is bound to change classic ways of conceptualization, not only, but particularly with regard to digitality and digitalization. As it places an emphasis on the material, energetic, and environmental implications of human practice, it introduces "material subjects" that may appear as a contamination of theory and theorizing. For classical conceptualizations of digitality as a phenomenon of information and symbolicity, problems of physicality are either not at all theoretically defined (cf. Negroponte, 1996) or appear as a mere technical reverse side of an otherwise purely immaterial information sphere (cf. the issue of the "technological gambit" as discussed by Floridi, 2014, p. 213). In contrast to such positions, media philosophy has discussed at great length (e.g. Groys, 2012) the disappearance of materiality in processes of mediatization, critically pointing out that the invisibility implied in the phenomenal process tends to cause an "immaterial bias" of (media) theory itself.

Newer theoretical studies of digitality embrace its materiality (Dourish, 2017) and even incorporate findings from the field of information physics, particularly with regard to the significant energy costs caused by the strict requirement of digital information processing to control problems of noise and entropy (Cubitt, 2016). In this respect, digitalization as a substitution of materiality (be it of "analog" media or other physical materialities) and transmutation of the underlying informational-physical structure has to be closely scrutinized: not only with regard to environmental issues – from toxic waste to significant and increasing amounts of energy consumption in digitalized economies –, but also with regard to issues of inclusion versus exclusion of material properties in the process of digital "translation" to commodified and proprietary data structures, (e.g. socio-environmental) data goods, and compression algorithms.

The almost exclusively profit-oriented "cybernetic ecology of data" (Cubitt, 2020, p. 253) that is being built up today is thus not only "profiting from [ecological, BJ] debt by colonising the future" (ibid.), but also fails to pursue the "complete model of a world which it is constrained to exclude" (ibid.). Because "digitization" (as this process of translation of materialities into machine-readable digits) and "digitalization" (as the global, cultural-environmental process mentioned above) interact in such ways that only the digitized data connects to the digital infrastructures - and as such is part of a new, digitally driven "world making" -, the "uncountable" aspects of materiality or material life that either are not digitizable in themselves or that are assessed to be not worthy of having dedicated (meta-) data formats defined, fall outside the new, exclusively digitally-based technical representation systems, at least on the level of automated information processing and automated decision making (such as in Artificial Intelligence). The ways and means in which the digitization of the planet, as well as the digitalization of the cultural worlds, is taking place, are thus highly political issues, even if they are hardly ever discussed as such in public.

The ongoing global expansion and advancement of physical digital infrastructures should be considered a later aspect of the Anthropocene in its own right, as it literally adds a material layer of information to the surface and orbit of the planet (Blum, 2019). The consumption of digital devices, whose underlying physical-economical design is essentially characterized by planned obsolescence (Slade, 2007) and thus connected to increasing amounts of toxic digital waste (Gabrys, 2013), is tied to these global infrastructures and cannot be conceptualized separately from physical digital globalization. As the involved complexities are far too big for even a rough estimation of the environmental outcome of this "technological gambit", the dynamics of the expansion of digital technology and culture introduces a high level of contingency.

However, the ecological aspects of a world-wide digit(al)ization cannot be separated from the global-historical process that is "globalization". Digitality not only consumes material and energy, but also serves as an informational, communicational, social, and cultural infrastructure that also drives political and policy-oriented discourses and enables, for example, globally networked activism (della Porta, 2017; Merrill et al., 2020).

In regard to the kind of resilience that is required to cope with, adapt to and enable change in the face of the human impact on the planet, digitality thus remains a complex agent. Following Brown (2015), resilience can be understood as a "property of individuals, households, communities and social ecological systems" that helps with "maintaining identity or functioning - but also undergoing change, actively engaging in change or adapting to change" (ibid., p. 8). The fact that in times of the most intensive global interconnections of ecological, political and cultural challenges, digitalization can hardly be dispensed with – and cannot even be wanted to disappear – indicates the importance of digital means and infrastructures for new forms of cultural resilience. Two mostly, although not completely, opposite approaches are to be distinguished here.

1. The technological approach of so-called "smart" technologization follows a strategy of "saving nature with bits and bytes" by means of more digitalization (Lange & Santarius, 2020, p. 11), understanding sustainability as a management problem that operates alongside binary patterns such as opportunities and challenges (Lohrmann & Osburg, 2017, p. 37). Technological approaches such as these, however, are quite susceptible to "solutionist" logics – particularly with regard to industrial digital innovation (Morozov, 2013) – that tend to reduce complex social problems to defined problems with definite, computable solutions (ibid., p. 5), reducing the roots of that problem to a handful of easily identifiable and controllable factors (ibid., p. 322). The solutionist bet on improvement draws on and perpetuates classical western constructions of technological "progress" as societal progress. The fact that the solutionist chain of problem definition, feasibility analysis and implementation follows precisely the conception of progress that has led to the current state of global affairs in the first place - originating in the age of the European Enlightenment, tied to permanent technical innovation, and thus requir-

ing constant economic growth leads us to a closer examination of the historical and cultural (self-)reflexivity of the respective approaches.

2. Practices of (post-) digital cultural resilience differ vastly from solutionist approaches to digital technology, as "digitality" appears rather not as a collection of possible means to problems of an inherently technical nature, but as a structural agent of mostly unplanned and initially even involuntary cultural change. The cultural approach – although it may include advanced technical knowledge in the process –, begins with the question of how sustainability and resilience can be understood as inherently cultural concepts (Jörissen, 2022; Jörissen et al., 2023; Jörissen et al., in print), leading to corresponding understandings of the relationship between ontology, time, and change that may well differ from the rationalist standard model in the wake of European Enlightenment.

Again following Brown's (2015) analyses, three interrelated core elements of resilience-building processes can be identified (cf. Brown 2015, p. 185–198). The first is *rootedness*, a sense of belonging that is based on economic, environmental, and social participation. It is "more than place, partly because people are increasingly mobile, so it is about a more fluid set of attachments and multifaceted identities" (ibid, p. 197). Then, *resourcefulness* is the capability to connect inherited and local (practical and theoretical) knowledge to new elements and resources that become accessible through new and wider social networks (ibid). Finally, *resistance* is not to be understood as a simply oppositional practice, but as "a potential site for change and the means through which individuals change social processes and structures and build alternatives" (ibid, p. 194).

With regard to digitalization, these core elements may provide valuable guidance in order to identify (existing and possible) practices of (post-) digital cultural resilience-building. Such perspectives identify processes where digitalization and digitality a) support and enable (new) forms of rootedness, b) provide the means to build up new and extended networks, and c) help to enable sustainable resisting structures for change. Digitization and the networked distribution of material and immaterial cultural heritage, to focus on this important example, help to create autonomous and living archival practices that serve as "alliances against dissipation and loss, but also against enclosure, privatization and thematization of archives" (Anand, 2016, p. 78–79). The form of the archive as public and collective digital 'random access' databases releases archival knowledge from a hegemonial narrative order that "has been an internal privilege of governmental agencies"and its secrecies (Ernst, 2016, p. 14) and creates "a new kind of dynamic memory […] (co-)produced by online users for their own needs" (Ernst, 2013, p. 95). Digital archival practices and movements correspond to post-digital diasporas (Gopinath, 2018, Tan et al., 2016), to new collective endeavours of sustainable living (Beel et al., 2017), and also to the data-based new sensories of a digital planetarism (Gabrys, 2018).

References

Anand, S. (2016). 10 theses on the archive. In P. Tan, Ö. Çelikaslan, & A. Sen (Eds.), *Autonomous archiving* (pp. 79–94). dpr-barcelona.

Beel, D. E., Wallace, C. D., Webster, G., Nguyen, H., Tait, E., Macleod, M., & Mellish, C. (2017). Cultural resilience: The production of rural community heritage, digital archives and the role of volunteers. *Journal of Rural Studies, 54*, 459–468. https://doi.org/10.1016/j.jrurstud.2015.05.002

Blum, A. (2019). *Tubes: A journey to the Center of the Internet with a new introduction by the author*. HarperCollins.

Brown, K. (2015). Resilience, Development and Global Change. Routledge.

Cubitt, S. (2016). *Finite media: Environmental implications of digital technologies*. Duke University Press.

Cubitt, S. (2020). *Anecdotal evidence: Ecocritique from Hollywood to the mass image*. Oxford University Press.

della Porta, D. (2017). *Global diffusion of protest: Riding the protest wave in the neoliberal crisis*. Amsterdam University Press.

Dourish, P. (2017). *The stuff of bits: An essay on the materialities of information*. MIT Press.

Ernst, W. (2013). *Digital memory and the archive*. University of Minnesota Press.

Ernst, W. (2016). Radically de-historicising the archive. Decolonising archival memory from the supremacy of historical discourse. In L'Internationale Online (Ed.), *Decolonizing archives* (pp. 9–16). L'Internationale Online.

Floridi, L. (2014). *The fourth revolution: How the Infosphere is reshaping human reality*. OUP Oxford.

Gabrys, J. (2013). *Digital rubbish: A natural history of electronics*. University of Michigan Press.

Gabrys, J. (2018). *Becoming planetary*. e-flux.

Gopinath, G. (2018). *Unruly visions: The aesthetic practices of queer diaspora*. Duke University Press.

Groys, B. (2012). *Under suspicion: A phenomenology of media*. Columbia University Press.

Jörissen, B. (2022). Digitale sympoiesis und kulturelle resilienz. *Vierteljahrsschrift für wissenschaftliche Pädagogik, 98*(4), 474–489. https://doi.org/10.30965/25890581-09703066

Jörissen, B., Unterberg, L., & Klepacki, T. (Hrsg.). (2023). Cultural sustainability and arts education. *International Perspectives on the Aesthetics of Transformation*. Springer. https://doi.org/10.1007/978-981-19-3915-0

Jörissen, B., Bolden, B., Bresler, L., Engel, J., Jeanneret, N., Svendler Nielsen, C., Scheunpflug, A., & Wagner, E. (Eds.). (in print). Arts Education in Transformative Times: Changes, Challenges, and Chances, (IJRCAAE Vol. 0) (Vol. 0). Waxmann.

Lange, S., & Santarius, T. (2020). *Smart green world?: Making digitalization work for sustainability*. Routledge.

Lohrmann, C., & Osburg, T. (Eds.). (2017). Sustainability in a digital world: New opportunities through new technologies. In *CSR, sustainability, ethics & governance* (1st ed.). Springer International Publishing: Imprint: Springer. https://doi.org/10.1007/978-3-319-54603-2.

Merrill, S., Keightley, E., & Daphi, P. (2020). *Social movements, cultural memory and digital media: Mobilising mediated remembrance*. Springer Nature.

Morozov, E. (2013). *To save everything, click here: The folly of technological solutionism*. PublicAffairs.

Negroponte, N. (1996). *Being digital*. Vintage Books.

Nitzke, S., & Pethes, N. (Eds.). (2017). *Imagining earth: Concepts of wholeness in cultural constructions of our home planet, culture & theory*. Transcript.

Schmidt, E., & Cohen, J. (2013). *The new digital age: Reshaping the future of people, nations and business (kindle-edition)*. Knopf.

Slade, G. (2007). *Made to break: Technology and obsolescence in America*. Harvard University Press.

Spivak, G. C. (2002). Imperatives to re-imagine the planet. *Aut Aut, 312,* 72–87.

Tan, P., Çelikaslan, Ö., & Sen, A. (Eds.). (2016). *Autonomous archiving.* dpr-barcelona.

Turner, F. (2008). *From counterculture to cyberculture: Stewart brand, the whole earth network, and the rise of digital utopianism.* University of Chicago Press.

Zuboff, S. (2019). *The age of surveillance capitalism: The fight for the future at the new frontier of power.* Profile Books.

Benjamin Jörissen (Dr. phil.) is Full Professor of Education with a Special Focus on Culture and Aesthetics and UNESCO-Chair in Arts and Culture in Education at the Friedrich-Alexander Universität Erlangen-Nürnberg (FAU). His research focuses on educational theory and on empirical research in cultural, arts, and aesthetic education, especially with regard to UNESCO-related issues and the transformation of education in a post-digital culture. Selected publications: Jörissen, B. et al. (Eds.). (2023). Cultural Sustainability. Arts Education Research and the Aesthetics of Transformation. Springer (2023); Jörissen, B. et al. (Eds.). (2022). Ästhetik—Digitalität—Macht. MedienPädagogik. (2022); Jörissen et al. (Eds.). (2023). Digitalisierung in der Kulturellen Bildung: Erträge gegenwärtiger Forschung. kopaed. (in print); Jörissen et al. (Eds.). (2023). Arts Education in Transformative Times: Changes, Challenges, and Chances (= IJRCAAE Vol. 0). Waxmann (in print).

Inertia

Bruno Villalba

Abstract Inertia describes the ability of a movement to persist, after the application of the initial driving force. The earth's biophysical cycles have inertia of their own, while social systems have a particular dynamic, giving rise to other forms of inertia. Modern society has pushed back against natural inertial constraints, but our activities have also established other, disquieting inertia mechanisms (e.g. nuclear inertia). The Anthropocene raises questions about the compatibility of these inertial cycles; the timescales on which they operate are orders of magnitude apart, so it is no easy task to reconcile them. Today, we have begun to see clashes between the inertial mechanisms of the different systems, and urgently need to redefine the timescales that policymakers consider, so new policy can address and mitigate these clashes.

When considering ecological matters, one must, inevitably, also think about time (Adam, 2000). Barbara Adam continuously tries to reconcile the timescales on which ecological systems operate (the rhythms of the climate, the carbon cycle, the unpredictability of the living world, etc.) with those on which society runs (the pace of development, the conditions for wellbeing, and so on). Quite rightly, she focuses on the sustainability of the interplay between the two phenomena (for the safety of future generations, both human and non-human). Where her work differs from her predecessors' is that she takes account of temporal tipping points (crises, catastrophes) which could, potentially, threaten humanity's very survival. In this, Adam's view of time is in conflict with that of modern society. Society today tends to view itself as being able to wrest free from the influence of certain physical and social realities (the laws of physics and social traditions) which have, in the past, been seen as immutable inertial barriers.

B. Villalba (✉)
AgroParisTech, Printemps, Paris-Sacaly, France
e-mail: bruno.villalba@agroparistech.fr

951

Inertia describes a movement's ability to persist after the initial driving force has been applied. It may be viewed purely as the result of a physical principle, contingent upon the determinism of the laws of mechanics. In the absence of any outside interference, a moving body tends to remain in its original state of motion. However, inertia cannot be considered the absence of a reaction due to lack of activity. Quite the contrary, it describes the dynamic that is unique to each body.

Our planet is governed by the physical laws of the climate mechanism, thermodynamics, etc. The cycles on which these phenomena operate may be extremely long (tens of thousands of years, for geological phenomena) or shorter (only a few hundred years for oceanic circulatory systems). The differences in their duration aside, all such mechanisms are affected by inertia: a system is set into motion, it has an effect, it is prolonged, or else it ultimately comes to rest (it is finite). All these cycles interact with one another, which makes it much more difficult to assess the general inertia of the Earth system.

Human history also displays inertia. Here, we refer to the ability of an *idea* to persist, as the result of the notion which birthed it. It is certainly true that there is no universally shared anthropological concept of time. However, numerous theories draw upon the idea that we are shaped by specific factors, which have led us to form a certain collective imaginary concerning our role in the planet's evolution. Thus, we have developed models of relating to the world, which serve the particular goals of different groups of people. Viewed through this lens, history becomes logical and meaningful. The action of human groups results from an initial driving force, which imbues the movement with dynamism, with collective beliefs justifying the actions.

However, the western view of time has gradually gained wider acceptance. We have constructed the concept of 'clock time' founded on Newtonian principles, based on a linear view of the passage of time (Adam, 1998). According to this view, social time is an independent construct, entirely separate from the timescales of the natural world. There is an overwhelming focus on the present, with barely any consideration of nature's long inertial cycles. As a result, we have slowly developed what Hartog (2016) terms a 'regime of historicity', which determines the way in which time is recorded, imposing a certain order upon how it is perceived. This becomes a framework for long-term thinking which shapes our mentality – both individual and collective. The industrial and political revolution (which brought about individualistic liberal democracy) has led to a concept of time with two facets: *presentism* and *acceleration*. Presentism means that our experience of time is focused on here and now, with emphasis being placed on the *indefinite continuity* of the present. Acceleration involves *amplifying our immediate use of time*, increasing the possibilities for use of different slices of time (work time, leisure time and social interaction time) by increasing the capability and availability of technological tools (from the pocket watch to advanced digital technology). This focus on the endogenous conditions of maintaining this social motion (primarily through technological innovations) leads us to downplay the natural exogenous constraints which could threaten that evolution. Thus, the inertial mechanisms to which our ancestors have been subject are tamed, with individuals gaining greater autonomy. Individuals construct their relationships with the world on the basis of these two forms of emancipation: ecological timescales are subrogated to human developmental goals, while social timescales are transformed to allow for the perception of a constant opportunity of personal development. However, the imperative to make progress (as time passes) has caused us to lose sight of the inertial margin to which we, ourselves, are subject.

Nevertheless, the dawn of the Anthropocene calls into question whether it is appropriate to dissociate ourselves in this way from the timescales of the natural world (the biophysical cycles of the Earth system, the unpredictability of ecosystems and populations, etc.). The Anthropocene shines a light on the temporal disturbances resulting from human activities (notably, socio-technical development), which interfere with our presentist view of social time. We now face a situation never before seen in history, where the Earth system's inertia could critically impact the evolution of the inertia of social time in the modern era. Human action has given rise to its own types of socio-technical inertia, which interfere with the inertia of natural cycles. We are at a decisive moment in history, and need to come up with a policy to regulate time, which takes account of a particular conception of time (presentism, which tends to downplay the inertial aspects of natural cycles), and of major tipping points which threaten irreversible harm (the inability to adapt, leading to either the elimination, or a complete transformation – if indeed such a transformation is possible – of nature). This threat of irreversibility raises questions as to the very capacity of our social systems to ensure their own survival (Steffen, 2015). However, the inertia and feedback mechanisms of the Earth system are slow, which preserves the illusion that the changes currently being wrought are not as drastic as they actually are, and that we are still able to control their impact. The timescale over which these inertial phenomena manifest themselves makes it more difficult to address them here and now. Public policy is drawn up from a short-termist point of view (looking no further than 30 years, at most, into the future). How, then, are we to devise policy that takes account of inertial stresses which will increase the pressure on future generations?

In addition to natural inertia (and worsening impact of human activity), we must consider the inertia of technical threats (such as nuclear and chemical weaponry), and the inertia of political views (a *continuist* vision of time, owing to the ability to control risks through technological innovation, which is perceived as infinite). In the latter case, we need to fully take account of the inertia of institutional structures and of public action, which stem from the long historical practice of distancing ourselves from nature (McNeill, 2001). As time passes, it becomes increasingly difficult for policy to change direction, because previous policy choices shape current organisational rules, and determine the path of future solutions. This phenomenon, known as *path dependency*, leads only marginal changes to be made, as we seek to make use of tried and tested, past practices. Thus, the passage of time reinforces the existing concepts and institutions, making it difficult to bring about change. For example, it is enormously complicated to act against the role of oil in our society, because the use of such fossil fuels is woven into the very way in which our political and social systems function, the infrastructure upon which those systems are built, and our daily lives. We can see the 'lock-in through interdependency' to which

Pierson (2004) refers. The institutional inertia of the world of research also leads to veritable conservatism in academic circles. Self-referent mechanisms represent a barrier to inventive fundamental research, in the interests of academic objectivity. Such objectivity is founded on an existing knowledge base, which often takes little consideration of ecological issues. We also need to take account of behavioural inertia, because our way of life, by definition and by its structure, has an important inertial element to it. We have developed, and continue to perpetuate, a model of normalised comfort which should be of concern to everyone today, and which should certainly concern future generations. We envisage a continuous and extensive chain of improvements to individual comfort (Shove, 2003). However, this approach has ecological consequences which cannot be ignored (Hoekstra & Wiedmann, 2014).

Finally, socio-technical inertia also has a profound impact on social systems' ability to evolve, and may cause major shifts in natural inertial systems. The manufacture of certain tools (in particular, nuclear weapons, and nuclear waste from civilian activities) inevitably creates inertial mechanisms which operate over tens of thousands of years. In 2019, France's *Agence nationale pour la gestion des déchets radioactifs* (National Radioactive Waste Management Bureau) had a store of 1.67 million cubic metres of radioactive waste. 3% of that waste accounted for 99% of the radioactivity, but the materials would remain radioactive for hundreds of thousands of years. Even now, nowhere in the world are there facilities for the permanent storage of such highly radioactive waste. The nuclear threat (from both civilian and military quarters) leads to an inertial situation never before seen, because it brings with it the risk of an irreversible and radical shift (for this reason, some authors, including Waters et al., 2015, suggest that the 'militarocene' would be a more appropriate label for the new age in which we live). We have entered the so-called 'end times', characterised by the risk of the permanent erasure of all life on Earth: 'However far generations to come stretch into the future, and wherever they go to escape it, the bomb will always be one step behind them.' (Anders, 2002: 342). In addition, these threats create the need to maintain enormous long-term investment and bear huge risk-management costs. This will put pressure on States' financial resources – all the more so against the backdrop of tightening budgets, which will mean priorities need to be defined. The consequences of such human inventions over the very long term will go beyond what political and academic organisations, in vulnerable human societies, are capable of regulating.

Age-old forms of inertia are now resonating with more recent inertial mechanisms caused by humans. Certainly, these different types of inertia are characterised by different rhythms: our CO_2 emissions will continue to have a tangible impact on the climate for 5000 years to come, with average global temperatures continuing to climb past the end of this century. The inertia of the oceans and the atmospheric system lead to gradual transformations of the natural regulatory mechanisms; but the sudden rupture that a nuclear detonation represents could bring about a complete shift of social and ecological systems. Thus, we must deal with the cumulative effects of systemic inertia in multiple systems, creating ever more unpredictable situations, and further increasing the risk of irreversible damage. Now, within a limited window, political actors must attempt to address these conflicting times-cales, to preserve the possibility for human societies and the natural environment to co-evolve.

References

Adam, B. (1998). *Timescapes of modernity. The Environment and Invisible Hazards*. Routledge.
Adam, B. (2000). Time and the environment. In R. Michael & W. Graham (Eds.), *The international handbook of Environmental Sociology*. Edward Edgar Publishing Limited.
Anders, G. (2002). *L'Obsolescence de l'homme*, Paris, éd. Encyclopédie des Nuisances.
Hartog, F. (2016). *Regimes of historicity: Presentism and experiences of time*. Columbia University Press.
Hoekstra, A. Y., & Wiedmann, T. O. (2014). Humanity's unsustainable environmental footprint. *Science, 344*(6188), 1114–1117.
McNeill, J. (2001). *Something new under the sun: An environmental history of the twentieth-century world*. W. W. Norton & Company.
Pierson, P. (2004). *Politics in time*. Cambridge University Press.
Shove, E. (2003). *Comfort, cleanliness and convenience: The social organization of normality*. Berg.
Steffen, W. (Ed.). (2015). Planetary boundaries: Guiding human development on a changing planet. *Science, 347*(6223), 1259855.
Waters, C. N., James, P., Syvitski, M., Gałuszka, A., Hancock, G. J., Zalasiewicz, J., Cearreta, A., Grinevald, J., Jeandel, C., & McNeill, J. R. (2015). Colin Summerhayes & Anthony Barnosky, Can nuclear weapons fallout mark the beginning of the Anthropocene Epoch? *Bulletin of the Atomic Scientists, 71*(3), 46–57.

Bruno Villalba is Professor of Political Science, at AgroParisTech (Paris) and Member of Printemps UVSQ (CNRS UMR 8085). His areas of research are Political Ecology, Sustainable Development and collapse studies. His research focuses on environmental political theory, notably through analysis of the capacity of the democratic system to reformulate its goals based on environmental constraints. He is the author or co-author of over ten books on politics and ecology. Books (selection): Collapsologists and their enemies (ed. Le Pommier, 2021, in French); Energy Sobriety. Material constraint, social equity and institutional perspectives (with L. Semal, ed. Quæ, 2018, in French) and Political ecology in France (Paris, La Découverte, 2022, in French).

Machine

Martina Heßler

Abstract Since industrialization, machines have shaped and transformed human life. Currently, however, there is much talk of a new machine age. A new machine ethic is being discussed. But what characterizes these new machines? And what does it mean for the concept of the Anthropocene? The article differentiates between three categories of machines: the mechanical machine, the so-called trans-classical machine (computer, information-processing machine), and, finally, AI as a new category of machine. Each category of machines is related to historically different human-nature and human-machine relationships.

The philosopher Hans Blumenberg drew his readers' attention to "a colourful series of imaginations: apparatuses, vehicles, power units, and storage aggregates, instruments of manual and automatic function, lines, switches, signals, etc." He continued: It is a "universe of things (…), the classification of which has often been attempted with little satisfaction". (Blumenberg, 1981: 10, my own translation). Blumenberg observed that these things are usually classified as technology (cf. technology). However, he wrote, an attempt at defining the term technology will fill countless pages. The same applies to the concept of the machine. Above all, the relationship of the two terms is not clear-cut. When do we talk about machines, and when about technology? Would technology be the more comprehensive term under which to subsume the machine? Has there been a time in which one of the terms is more in vogue than the other?

Currently, there is much talk of machines and a new machine age (cf. Brynjolfsson & McAfee, 2016). Furthermore, a new machine ethic is being discussed (Misselhorn, 2018). However, most people would hardly classify many of the things surrounding us as machines: Is a cell phone, an automobile, or a laser printer a machine?

The reason for the current renaissance of the concept of machine are the successes of AI-research, in particular, of machine learning, which carries the word

M. Heßler (✉)
TU Darmstadt, Darmstadt, Germany
e-mail: hessler@pg.tu-darmstadt.de

© The Author(s), under exclusive license to Springer Nature   957
Switzerland AG 2023
N. Wallenhorst, C. Wulf (eds.), *Handbook of the Anthropocene*,
https://doi.org/10.1007/978-3-031-25910-4_157

machine in its term. However, these smart machines, now pervasive in everyday human life, are a novel category of machines. AI is fundamentally different from the classical machine and also from the so-called trans-classical machine, i.e., the computer or information-processing machine.

Machines, however, are not only physical objects like the apparatuses, devices or automata mentioned at the beginning. The concept of machine often serves as a metaphor: a metaphor for certain forms of government, for certain organizational characteristics of a society or of institutions, or for humans devoid of self-will, to give just a few examples. The concept of machine has also been discussed in relation to terms such as human or nature. Over the centuries and up to the present day, these terms have been repeatedly negotiated and reinterpreted in relation to each other. Over the past decades, they have been reconceptualized. Concepts running counter to dichotomous thinking, such as the hybridization of humans and nature, have been created.

Since the days of the industrialization, machines have permanently changed how humans deal with nature as well as how people live and produce things. Machines were used as a means of controlling and transforming nature. From the age of industrialization onwards until the second half of the twentieth century, mechanical machines performed work or generated energy, creating the category of the classical machine. This category dominated the concept of the machine until the second half of the twentieth century.

The steam engine in particular was seen as a symbol of the industrialization as well as its trigger or even as its cause. Research has corrected this deterministic view and placed the development and impact of the steam engine within the economic, political, and cultural contexts that made industrialization possible and pushed it forward. However, debates about the Anthropocene brought the steam engine into view again as a symbol of industrialization and lifestyle characterized by energy consumption. The steam engine and the railroad are indispensable "actors" in several lasting and massive transformation processes: energy production became independent of a fixed location, the production process itself underwent changes, and consequently, the way of life in industrialized societies started to look vastly different from the late 18th and early nineteenth century on. This enormous push of machine use in the Western world has been associated with the starting point of the Anthropocene (cf. Trischler & Will, 2019: 80–83). Even if historical research emphasizes that mechanization was neither a universal nor a linear process, the importance of machines for the transformation of man's environment as well as the human-nature relationship is beyond question.

In his book "The Machine in the Garden," Leo Marx described how American writers, e.g., Herman Melville, Henry David Thoreau, or Marc Twain, made the antagonism between rural ideals and the transformation of landscapes by machines a preeminent topic of theirs. Here, alienation by mechanization and seeing it as a blight on a pastoral landscape (Marx, 1964) was described.

Nature and machine were seen as opposites, with the overarching concept of the machine invading nature and transforming it. This way of thinking can still be found today, e.g., when wind turbines are constructed, which for some are a symbol of ecological technology but for others are a blight or contribute to the destruction of nature. However, in the American society of the nineteenth century, machines were not exclusively seen in damning contrast to nature. Machines also became an object of awe, exactly as nature had been before. They became a "technological sublime" (Nye, 1994).

Arguing against the strictly dichotomous concepts of nature and culture, or nature and machines, many scholars have emphasized the hybridization of nature and machines, describing landscapes as "organic machines" (White, 1995), or using the concept of "environ-tech" (Pritchard, 2014) to analyse the inescapable need for technological transformation of the environment.

However, looking with a critical or uneasy view at machines that cut through, conquered, and transformed the landscapes corresponded to the perspectives voiced by the critique of culture movement in the first half of the twentieth century. According to this perspective, it was necessary to defend the organic against the mechanization, as e.g., Siegfried Giedion or Lewis Mumford did. In his book "Mechanization Takes Command" (1948), Giedion examined what happens when mechanization meets the organic; and his book turned into a defence of the organic against the mechanical. With the concept of the "mega-machine", Lewis Mumford formulated the principles of a machine that, as he emphasized, was not bound to an actual physical machine, but rather was designed to exist within social structures. He used the term "mega-machine" to describe the integration of humans into a hierarchical organization in which humans become interchangeable while functioning as a part of a system the purposes of which are imposed by who or what is on its outside. According to Mumford, the "mega-machine" leads to the suppression of the living (Mumford, 1967/1970).

These critical voices took issue with the classical machine that had transformed society, the economy, the way of life, and nature in the 19th and early twentieth century: a rule-governed, efficient, standardized machine that always ran the same way, often interpreted in opposition to humans, in opposition to culture, and in opposition to nature.

In the middle of the twentieth century, a new concept of the machine emerged through cybernetics and computers. This did not only change the concept of the machine, but also the debates and perspectives on the machine. The debate about the transformation of nature took a back seat to the question of the transformation of humans.

In the 1960s, the philosopher Gotthard Günther spoke of the "trans-classical machine", which represented a new category of machines. Such a machine no longer performed work as the "classic Archimedean" machine did, but processed information (Günther, 1963: 183f.). Similarly, Max Bense emphasized that the "mathematical machines [...], occasionally also called thinking machines" (Bense, 1955: 7), represented a "new state of being of technology" (Bense, 1955: 8). Louis Couffignal stated with regard to the "thinking machine" that machines were "of the most diverse kind" which is why machines had to be classified using different categories (Couffignal, 1955: 13). The development of the computer, cybernetics, and artificial intelligence research served as the background to these new shifts regarding the concept of machine during the second half of the twentieth century. The computer evoked the image of a thinking machine, an information machine, or a symbolic machine (Krämer, 1988). Now the comparison with the human brain, which was to be imitated in a machine-like way, became central.

Within cybernetics, machines were designed and built that were among the first adaptive and learning machines, blurring and dissolving the clear distinction between the organic and the technical by putting the concept of information at the core (Cordeschi, 2002; Pickering, 2010; Müggenburg, 2018).

Whereas the concept of machine during the 19th and early twentieth century has born connotations of shaping and controlling the human environment, the cyborg (cf. Cyborg) as well as bio-, gene- and nanotechnology refer to changes in the human body. While genetics and biotechnologies are called *technology*, nanotechnology itself uses terms like nano*machines* or molecular *machines*. However, with these possibilities of interventions in "human nature", the human body became an object of design and machine transformation in a new way: either through gene modification technologies or through becoming a cyborg, i.e., a human who opts to get technological implants in order to modify their body – mostly with the goal to increase its capabilities.

Current AI, and machine learning in particular, changed the concept of machine again (Heßler, 2021). AI is designed to learn and then evolve through its learning processes. Unlike rule-governed machines, an example of which is the trans-classical machine, AI's behaviour is not predictable, and its results are not always reproducible. What distinguishes the machine as a machine is no longer the standardized, the fixed. Rather, AIs develop differently; they acquire an individual biography that depends on their human counterpart and the context of their usage. It is the everyday behaviour, the everyday use through individuals that permanently changes the development of AI. As a by-now integral part of everyday life, humans only notice these new digital and AI-machines if they do not work. Scholars therefore coined terms like informal or invisible technologies (Kaminski & Gelhard, 2014). These smart machines were regarded as part of the human ecology (Hörl, 2016; Haff, 2014). Finally, artificial neural networks represent a black box for AI developers: AI's behaviour is not always comprehensible. With these characteristics, AI seems to become more human-like, thus challenging human self-understanding and the human-machine relationship once again.

At the same time, digitization and AI are seen as opportunities for ensuring environmental sustainability, energy efficiency and conscientious use of resources. For example, the smart grid is supposed to efficiently manage energy consumption according to demand. Smart, flexible, individualizable machines and their technological logic are placed in the service of "nature". Nevertheless, the level of consumption of energy and resources used for digitization and AI goes against the idea of green technologies. It presents an unsolved problem, which is currently met with little public debate (Ensmenger, 2012).

In the context of the Anthropocene discussion, the new smart machines raise various questions that are highly relevant for the future. First, the question of the coexistence of people, things, technology, plants, substances, organisms, which is currently being discussed in theoretical concepts such as new materialism (cf. robotics). Second, the question of how much the human body could and should be transformed technically. Third, the question of how smart machines can be used to establish a relationship with nature that is counteracted neither by fantasies of domination and control nor by a level of consumption of energy and resources that is unsustainable. These questions make it clear how much not only machines transform the human and non-human environment but also how, conversely, the human way of life determines the development, function, and use of machines.

References

Bense, M. (1955). Vorwort. In L. Couffignal (Ed.), *Denkmaschinen*. Gustav Kilpper Verlag.

Blumenberg, H. (1981). *Wirklichkeiten in denen wir leben*. Reclam.

Brynjolfsson, E., & McAfee, A. (2016). *The second machine age: Work, Progress, and prosperity in a time of brilliant technologies*. Norton & Company.

Cordeschi, R. (2002). *The discovery of the artificial: Behavior, mind and machines before and beyond cybernetics*. Springer.

Couffignal, L. (1955). *Denkmaschinen*. Gustav Kilpper Verlag.

Ensmenger, N. (2012). The digital construction of technology: Rethinking the history of computers in society. *Technology and Culture, 53*(4), 753–776.

Giedion, S. (1948). *Mechanization takes command*. Oxford University Press.

Günther, G. (1963). *Das Bewusstsein der Maschinen*. Agis.

Haff, P. (2014). Humans and technology in the Anthropocene: Six rules. *The Anthropocene Review, 1*(2), 126–136.

Heßler, M. (2021). Artificial intelligence: A new category of machine or the humanisation of the machine. In Y. Keskintepe & A. Woschech (Eds.), *Artificial intelligence. Machine learning human dreams*. Wallstein.

Hörl, E. (2016). Die Ökologisierung des Denkens. *Zeitschrift für Medienwissenschaft, 8*(14), 33–45.

Kaminski, A., & Gelhard, A. (Eds.). (2014). *Zur Philosophie informeller Technisierung*. Wissenschaftliche Buchgesellschaft.

Krämer, S. (1988). *Symbolische Maschinen. Die Idee der Formalisierung in geschichtlichem Abriß*. Wissenschaftliche Buchgesellschaft.

Marx, L. (1964). *The machine in the garden: Technology and the pastoral ideal in America*. Oxford University Press.

Misselhorn, C. (2018). *Grundfragen der Maschinenethik*. Reclam.

Müggenburg, J. K. (2018). *Lebhafte Artefakte: Heinz von Foerster und die Maschinen des Biological Computer Laboratory*. Konstanz University Press.

Mumford, L. (1967/1970). *The myth of the machine*. Harcourt Brace.

Nye, D. (1994). *American technological sublime*. MIT-Press.

Pickering, A. (2010). *The cybernetic brain: Sketches of another future*. University of Chicago Press.

Pritchard, S. (2014). Toward an environmental history of technology. In A. C. Isenberg (Ed.), *The Oxford handbook of environmental history*. Oxford University Press.

Trischler, H., & Will, F. (2019). Die Provokationen des Anthropozäns. In M. Heßler & H. Weber (Eds.), *Provokationen der Technikgeschichte: Zum Reflexionszwang historischer Forschung* (pp. 69–105). Ferdinand Schöningh.

White, R. (1995). *The organic machine: The remaking of the Columbia River*. Macmillan.

Martina Heßler is Professor of the History of Technology at the Technical University of Darmstadt. Her research focuses on the history of technology from the Early Modern Period on. She is author of several books, e.g., *Kulturgeschichte der Technik,* Frankfurt 2012, Technikemotionen 2020, and *Handbuch Technikanthropologie* (together with Kevin Liggieri) 2020. Currently, she is working on the history of the man-machine-relationship.

Media

Leslie Sklair

Abstract Compared with media coverage of climate change there is relatively sparse coverage of the Anthropocene. At the time of writing, only one book has been published on how the print media all over the world report the Anthropocene. There has, however, been a modest number of published research articles on the topic, mostly accessing the online content of print media, and a growing interest in how the Anthropocene is dealt with in social media of various types.

There is an impressive amount of research on how climate change and global warming are reported in the media all over the world (see for example, Boykoff, 2011; McNatt et al., 2019). However, there is very little research on how the Anthropocene, is reported in the media. According to the University of Colorado research project MeCCO, led by Professor Max Boykoff, which monitors around 70 media sources in 38 countries, over 400,000 items on climate change and global warming were identified between 2004 and 2018 (http://sciencepolicy.colorado.edu/media_coverage). In 2016 the present author began to co-ordinate the Anthropocene Media Project (AMP). MeCCo contains valuable brief descriptions of global media sources, many of which also appear in AMP.

Assisted by a team of volunteer researchers from around the world AMP began to document and analyse how global media report the Anthropocene (see Sklair ed. 2021, hereafter referred to as AMP). The central role of the *New York Times* and, in particular, the 'dot.earth' column by Andy Revkin, in the coverage of the Anthropocene in the USA and abroad is difficult to exaggerate. Revkin began with a blog in 2007 which moved to the opinion pages of the *Times* in 2010, publishing many influential debates with Anthropocene notables until 2016 when Revkin moved on (see AMP, pp. 61–64). Around the same time Steven Corneliussen, a media analyst for the American Institute of Physics, was following Anthropocene coverage in three national newspapers in the USA and occasionally other sources,

L. Sklair (✉)
London School of Economics and Political Science, London, UK
e-mail: l.sklair@lse.ac.uk

writing up his findings in the popular science magazine *Physics Today* (see also the recent research of Hétier & Wallenhorst, 2021). Useful as these articles are, their coverage is limited.

AMP methodology was based on searches for mentions of the Anthropocene in all newspapers, magazines, and other news sites available online in almost all countries in the world (social media and media behind paywalls were excluded). The data base for the ensuing book was over 4000 items mentioning the term 'Anthropocene' (often in translation) from over 2000 publications in about 140 countries (AMP, Appendix 1). Three main research questions are highlighted. How likely is it that ordinary readers with no special interest in the topic would come across reference to the Anthropocene and what would it be telling them? How do different types of media (national/regional, big city papers, small local papers, general and special-interest magazines, news sites) report the Anthropocene to their readers? To what extent are media in different parts of the world 'provincializing' the Anthropocene by creating their own national, regional and/or ethnic narratives in contrast to uncritically replicating Western (Eurocentric and US-centric) narratives?

As is well-documented, the term 'Anthropocene' emerged publicly in 2000 in a relatively obscure geological newsletter and then in a short article in the prestigious journal *Nature* (Crutzen, 2002). Media coverage was sparse during the following decade although some media in Latin America, Western Europe, and Central/East Europe picked up the term early, and by 2006 all regions showed some Anthropocene items. Also worth noting is that total results for any 1 year between 2000 and the end of 2017 reached 100 items in only three regions – North America, Western Europe, and Central/Eastern Europe. The highest number of Anthropocene articles in all regions occurred in the year 2016, probably due to the outreach of the Anthropocene Working Group (AWG) which resulted in substantial global media reporting of the 35th International Geological Congress in Cape Town (see AMP, p.5 and *passim*).

The average numbers of articles mentioning the Anthropocene per source per year for each region are striking. Counting only sources with at least one 'Anthropocene' item (non-zero sources), the arithmetic average is about five items per source between 2000 and 2018. This means that any reader could be expected to encounter an article or item mentioning the Anthropocene once every three to 4 years. In Africa, Asia South, and the Middle East articles mentioning the Anthropocene occurred less than once per year. Even in the higher-scoring regions, North Asia, Latin America, North America, Western Europe, East/Central Europe, and Oceania (boosted by the large number of items from Australian media) Anthropocene articles appeared at a rate of about three per year.

Only 15 of the 2000+ newspapers and magazines searched mentioned the Anthropocene twice or more a year on average (that is at least 36 items between 2000 and the end of 2017). These publications are, in descending order of coverage, *New York Times*, *Süddeutsche Zeitung*, *Frankfurter Allgemeine Zeitung*, *Le Monde*, *Focus* (Germany), *Wired*, *Canberra Times*, *Morgenbladet* (Norway), *Slate* (USA), *Telegraph* (UK), *Discover* (USA), *Daily Mail* (UK), *Independent* (UK), *The Hindu*, and *The Guardian*.

Three foundational articles were selected to illustrate quality media coverage of the Anthropocene from *Frankfurter Allgemeine Zeitung* in 2000, *Le Monde* and *The Financial Times* both in 2004. These seemed to establish a sort of template for how quality media around the world approach the difficult idea of the Anthropocene (see AMP pp. 160–68). While these three foundational articles set the tone for most of the lengthy coverage in quality media, most references to the Anthropocene were quite short, some simply mentions of headlines like the 'Age of Man' or the 'Age of humans', frequently the shorthand for the Anthropocene in the mass media. There was also significant attention in media of all types to Anthropocene-related creative arts events (AMP, chapter "Living") and to a lesser degree reporting of the ways in which social science and environmental humanities scholars discussed the Anthropocene (ibid, chapter "Life"; for example Macfarlane (2016), and many articles in *The Conversation*, online). Also notable in publicising the concept was the headline "Welcome to the Anthropocene" in *The Economist* magazine (see Editors, 2011).

More than half of the 4000+ media items on the Anthropocene were randomly selected for classification. Three broad and sometimes overlapping narratives, extracted from the media coverage, point towards three different ways of framing the Anthropocene story. The first, the 'Neutral frame' was characterised by descriptive reporting, mostly passing mention of the Anthropocene, disagreements among scientists, especially over starting dates, and/or the presentation of the Anthropocene as a continuation of natural processes. The second, the optimistic 'good anthropocene frame' comprises a broad spectrum of mutually reinforcing narratives, ranging from the moderately to the definitely alarming. All recognise that the planet and humanity itself may be in some danger, and that we cannot ignore the warning signs. However, if we are clever enough, we can save ourselves and the planet with technological fixes (notably geoengineering) and other strategies that present opportunities for industry, science, and technology. The recognition of the need for change typically includes renewable energy, population control, conservation, sustainability, and resilience but stops short of any radical challenge to the economic and political status quo. This master narrative provides a more-or-less sophisticated optimistic 'good' Anthropocene frame. The third narrative (frequently hinted at but rarely elaborated) argues that human survival is at risk and that we cannot go on living and consuming as we do now, forcing the conclusion that we must strive to change our way of life, for example by bringing capitalism to an end and creating new types of global governance or state-less societies, religious or spiritual renewal, whether these are realistic or not. This is labelled the 'pessimistic radical change frame'. The fact that many articles combine two (occasionally all three) of these narratives is certainly a challenge for those tasked with categorizing media coverage of the Anthropocene (as well as climate change). For example, proposals to introduce electric cars might appear radical to some but, if they do not challenge all modes of transportation, they can be considered 'business as usual', which now has its own acronym in the literature (BAU). To a greater or lesser extent, combinations of the first two narratives become versions of what can be labelled 'reassurance narratives' and all these can be reconciled with what has been labelled the 'good'

Anthropocene (see Revkin, 2016, and AMP, pp. 27–28 and passim). Neutralizing and reassurance narratives together account for at least 99% of all items found in AMP. There are a few stories about the views of scientists, for example Barman (2010) reports the views of the late Frank Fenner, an eminent Australian biologist, that the human race itself is at risk of extinction. Some scholars have written specifically about the possibility of human extinction due to the Anthropocene, and their books have been reviewed in print media all over the world, for example Clive Hamilton (2017) and Roy Scranton (2015), reported in *The Australian*'s 'Books of the Year' section (19 December 2015).

In conclusion, it is important to remember that the Anthropocene is not a stable concept, Chwałczyk (2020) identifies 80–90 different versions of it. Most of those who write about it in the media are not Earth scientists. Journalists reporting the Anthropocene have a very difficult job (see Morton, 2016). There appears to be a growing interest in the Anthropocene in social media and other forms (for example, de Grosbois, 2013; Esposito, 2014; Gärdebo et al., 2017; Payne, 2017; Bergillos, 2021), probably the next big advance in this research field.

References

Barman, B. (2010). The extinction factor. *Daily Star* [BanglaDesh] 2 July.
Bergillos, I. (2021). Approaches to the Anthropocene from communication and media studies. *Social Science, 10*(10), 1–12. https://doi.org/10.3390/socsci10100365
Boykoff, M. (2011). *Who speaks for the climate? Making sense of media reporting on climate change*. Cambridge University Press.
Chwałczyk, F. (2020). Around the Anthropocene in eighty names — Considering the Urbanocene proposition. *Sustainability, 12*(11), 44–58. https://doi.org/10.3390/su12114458
Crutzen, P. (2002). Geology of mankind. *Nature, 415*(January), 23.
de Grosbois, A. M. (2013). Examining the Anthropocene with online media. *Environment and Earth Science, 68*, 607–608. https://doi.org/10.1007/s12665-012-2061-9
Editors. (2011). Welcome to the Anthropocene. *The economist* (28 May).
Esposito, P. (2014). *Mediated sustenance: The online news media discourse of sustainability, food, and agriculture in the Anthropocene* (State University of New York College of Environmental Science and Forestry, ProQuest Dissertations Publishing, 2014. 1568937).
Gärdebo, J., et al. (2017). Introduction to social Media in the Anthropocene. *Resilience: A journal of the environmental humanities, 5*(1), 1–17.
Hamilton, C. (2017). *Defiant earth: The fate of humans in the Anthropocene*. Polity Press.
Hétier, R., & Wallenhorst, N. (2021). From the idea of humanity as the power of nature to the representation of the Anthropocene in the media. *Revue française des sciences de l'Information et de la communication, 21*. https://doi.org/10.4000/rfsic.10116
Macfarlane, R. (2016). Generation Anthropocene: How humans have altered the planet for ever. *Guardian* (1 April).
McNatt, M. B., et al. (2019). Anthropocene communications: Cultural politics and media representations of climate change. In S. Davoudi et al. (Eds.), *The Routledge companion to environmental planning*. Routledge, chap 2.11.
Morton, T. (2016). The first draft of the future: Journalism in the 'age of the Anthropocene'. In J. P. Marshall & L. Connor (Eds.), *Environmental change and the World's futures: Ecologies, ontologies and mythologies*. Routledge, chap 3.

Payne, T. (2017). Dramatizing the Anthropocene through social media: The spatiotemporal coordinates of Hydrocitizens. *Resilience: A journal of the environmental humanities, 5*(1), 100–134.

Revkin, A. (2016). Building a "good" Anthropocene from the bottom up. *New York times* (6 October).

Scranton, R. (2015). *Learning to die in the Anthropocene: reflections on the end of a civilization.* City Lights.

Sklair, L. (Ed.). (2021). *The Anthropocene in global media: Neutralizing the risk.* Routledge.

Leslie Sklair is Emeritus Professor of Sociology at the London School of Economics. He has lectured all over the world and his publications have been translated into more than ten languages.

Resilience

Ortwin Renn

Abstract Policy advice for dealing with major crises has focussed on two concepts: resilience and sustainability. The article introduces the term resilience and explains its application in different disciplines. Furthermore, it explores the relationship between resilience and sustainability, illustrates the various concepts that are associated with each term and suggests an integrative approach that is based on the ideal of maintaining critical services for reaching humane living conditions for present and future generations based on fair distribution rules and inclusive governance processes.

The concept of resilience has been used in many disciplines for different notions of being able to respond adequately when the system is under stress. It has been widely applied in ecological research and denotes the resistance of natural ecosystems to cope with stressors (Holling, 1973). In the health sciences resilience is understood as the capacity of humans or animals to resist stressors that could jeopardize the physical or mental health of those exposed (Hanefeld et al., 2018). In the field of engineering and infrastructure planning, resilience is focused on the ability and capacity of systems to resist shocks and to have the capability to deal and recover from threatening events (Rose, 2007; Jackson & Ferris, 2017). This idea of resistance and recovery can also be applied to social systems (Review in Norris et al., 2008; Adger, 2000). The main emphasis here is on organizational learning and institutional preparedness to cope with stress and disaster. The US Department of Homeland Security (DHS) uses this definition: "Resilience is the ability of systems, infrastructures, government, business, and citizenry to resist, absorb, and recover from or adapt to an adverse occurrence that may cause harm, destruction, or loss [that is] of national significance (cited after Longstaff et al., 2010: 19). Hutter (2011) added to this analysis the ability of systems to respond flexibly and effectively when a system is under high stress from unexpected crisis. Pulling from an interdisciplinary body of theoretical and policy-oriented literature, Longstaff et al. (2010) regard

O. Renn (✉)
Research Institute for Sustainability - Helmholtz Center Potsdam (RIFS), Potsdam, Germany

resilience as a function of resource robustness and adaptive capacity. Finally, the International Risk Governance Council (IRGC, 2018) depicts resilience as a normative goal for risk management systems to deal with highly uncertain events or processes (surprises). It is seen as a property of risk-absorbing systems to withstand stress (objective resilience) but also the confidence of risk management actors to be able to master crisis situations (subjective resilience).

In the aftermath of the Corona Crisis and during the intense discussions on climate change and its implications, the relationship between resilience and sustainability has become a major topic of academic and political debate (Renn, 2020). For exploring this relationship, it is useful to distinguish three major concepts relating specifically to the role of resilience in the quest for sustainability (Reid & Botterill, 2013). The first school claims that resilience is independent of sustainability; resilience refers to a continuation of a desired service, such as electricity, fresh water or healthcare, in times of stress (Jackson & Ferris, 2017). Whether such a resilient strategy of designing more robust systems for vital services meets the goals of sustainability, such as environmental quality, circular economy or peaceful resolution of conflicts, is not included in this understanding.

The second concept basically claims that reliance is a part of sustainability. A major advocate of this view is the Stockholm Resilience Institute (Folke et al., 2016). The main claim here is that resilience is a way to sustain the functionality of services that are crucial to meeting human needs. So, both sustainability and resilience have "something to maintain" according to this understanding of the terms. The main goal is to sustain a development that evolves within the ecological and societal planetary boundaries.

The third concept takes a more radical view: resilience is seen as being in conflict with sustainability. Sustaining the status quo and making it more resilient defies the concept of sustainability (van der Leeuw, 2020, p. 81ff.). Resilience in this understanding is directed towards maintaining the given structures rather than engaging in a dynamic change system where sustainability is a dynamic goal to guide change not to preserve stability.

Given these three concepts of resilience with respect to sustainability, the answer to the question of commonality lies in the phrase "maintaining or sustaining something". This common feature may be key to a more coherent understanding of the relationship between the two terms (Faber et al., 2020). While resilience does not specify the type of services und functions that need to be maintained or quickly recovered (other than that they are critical or crucial for human societies), sustainability adds purpose to the services: they should be directed towards humane living conditions. These include respecting the boundaries of natural ecosystems and resources, meeting basic needs of all human beings, and ensuring peaceful means of conflict resolution (Robertson, 2017, p. 3–4).

Given this overlap, the combination of resilience and sustainability could be framed as a requirement that the *conditions for providing ecological services, for ensuring economic well-being and enabling quality of life with its essential social services are to be maintained over time in full awareness of multiple constraints such as present and future resource capacities/limitations (planetary boundaries) and of the occurrence of unexpected or unlikely stressors* (Renn, 2020). Meeting this goal does not imply sustaining current practices but rather fostering technical and social innovations that make it more likely that these humane conditions can be sustained and extended from the present to future generations.

Based on this reasoning, one can derive three major lessons from the resilience literature that could help to make societies more resilient once a crisis has occurred (such as Corona) or might occur (such as blackout of all electricity generating stations). First, investments in infrastructure and technical as well as social services can help to reduce vulnerabilities and make institutions better prepared for potential stressors. These investments should secure those goods and services that enable humane living conditions for the present and future generations, in particular healthcare, attractive working opportunities, clean environment and peaceful co-existence with fellow humans and other creatures. The UN-SDGs may act as a normative guide for selecting the targets for such investment programs.

Secondly, all measures to boost the economy and modernize the infrastructure need to consider the distribution of impacts among different target groups and populations (WEF, 2020). Who will gain from the measure, who is likely to lose? Policymakers are well advised to look for measures that are fair to all people affected. Alternatively, compensation measures are required if negative distributional effects cannot be avoided.

Thirdly, decisions about policy priorities, about distributing scarce resources and about prudent strategies between conflicting goals require the inclusion of major stakeholders, in particular those, who represent groups that suffer the most. Policy makers are well advised to provide a clear (ethical) justification for the necessary trade-offs that they assign between conflicting goals. Most prominent is the decision to design a middle ground between efficiency for reducing cost and resilience providing redundant, diversified and flexible means to sustain necessary services to society. It may be wise to determine this middle ground in view of the regional context conditions, but it is mandatory to come up with an ethically convincible rationale for making such trade-offs (Khoo & Lantos, 2020). It should be transparent of how much reduction in efficiency (increased costs) is worth how much gain in security of supply and services. Both, an inclusive approach to making such painful decisions and a convincing rationale for justifying the trade-offs is not only a prerequisite for a sustainable policymaking process, it is also a major condition for avoiding political polarization and public outrage.

The combination of resilience and sustainability provides effective guidance for managing global crises. It can and should be applied to design policies for many types of global threats such as climate and environmental changes, collapse of financial or economic systems, or growing inequalities. Investing in resilience for sustainable infrastructures, providing access to sustainable services to all people and implementing an inclusive, participatory governance approach are promising strategies to cope with sudden as well as slowly developing threats.

References

Adger, W. N. (2000). Social and ecological resilience: Are they related? *Progress in Human Geography, 24*(3), 347–364.

Faber, M. H., Miraglia, S., Qin, J., & Stewart; M.G. (2020). Bridging resilience and sustainability – Decision analysis for design and management of infrastructure systems. *Sustainable and Resilient Infrastructure, 5*(1–2), 102–124. https://doi.org/10.1080/23789689.2017.1417348

Folke, C., Biggs, R., Norström, A. V., Reyers, B., & Rockström, J. (2016). Social-ecological resilience and biosphere-based sustainability science. *Ecology and Society, 21*(3), https://www.jstor.org/stable/10.2307/26269981

Hanefeld, J., Mayhew, S., Legido-Quigley, H., Martineau, F., Karanikolos, M., Blanchet, K., Liverani, M., Mokuwa, E. Y., McKay, G., & Balabanova, D. (2018). Towards an understanding of resilience: Responding to health systems shocks. *Health Policy and Planning, 33*(3), 355–367. https://doi.org/10.1093/heapol/czx183

Holling, C. S. (1973). Resilience and stability of ecological systems. *Annual Review of Ecology and Systematics, 4*(1), 1–23.

Hutter, G. (2011). Planning for risk reduction and organizing for resilience in the context of natural hazards. In B. Müller (Ed.), *Urban regional resilience: How do cities and regions deal with change?* (pp. 101–111). Springer.

IRGC. (2018). *Resource guide on resilience* (Vol. 2). EPFL International Risk Governance Center. https://doi.org/10.5075/epfl-irgc-262527

Jackson, S., & Ferris, L. J. (2017). Designing resilient systems. In I. Linkow & J. M. Pala Oliviera (Eds.), *Resilience and risk* (pp. 121–144). Springer.

Khoo, E. J., & Lantos, J. D. (2020). Lessons learned from the COVID-19 pandemic. *Acta Paediatrica, 2*. https://doi.org/10.1111/apa.15307

Longstaff, P. H., Armstrong, N. J., Perrin, K., Parker, W. M., & Hidek, M. A. (2010). Building resilient communities: A preliminary framework for assessment. *Homeland Security Affairs, VI*(3), 1–23.

Norris, F. H., Stevens, S. P., Pfefferbaum, B., Wyche, K. E., & Pfefferbaum, R. L. (2008). Community resilience as a metaphor, theory, set of capabilities, and strategy for disaster readiness. *American Journal of Community Psychology, 41*, 127–150.

Reid, R., & Botterill, L.-C. (2013). The multiple meanings of 'Resilience': An overview of the literature. *Australian Journal of Public Administration, 72*(1), 31–40.

Renn, O. (2020). The call for sustainable and resilient policies in the COVID-19 crisis: How can they be interpreted and implemented? *Sustainability, 12*(6466). https://doi.org/10.3390/su12166466

Robertson, M. (2017). *Sustainability. Principles and practice* (2nd ed.). New York.

Rose, A. (2007). Economic resilience to natural and man-made disasters: Multidisciplinary origins and contextual dimensions. *Environmental Hazards, 7*, 383–398.

van der Leeuw, S. (2020). *Social sustainability, past and future. Undoing unintended consequences for the earth's survival.* Cambridge University Press.

WEF, World Economic Forum. (2020). *Coronavirus: A pandemic in the age of inequality.* www.weforum.org/agenda/2020/03/coronavirus-pandemic-inequality-among-workers. Accessed 4 Nov 2021.

Ortwin Renn is the forrmer scientific director at the Reserach Institute for Sustainability - Helmholtz Center Potsdam (RIFS) (Germany) and retired professor for environmental sociology and technology assessment at the University of Stuttgart. He also directs the non-*profit company DIALOGIK,* a research institute for the investigation of communication and participation processes. Renn is *Adjunct Professor for "Integrated Risk Analysis"* at Stavanger University (Norway), *Honorary Professor* at the Technical University Munich and Affiliate Professor for *"Risk Governance"* at Beijing Normal University. His research interests include risk governance (analysis perception, communication), stakeholder and public involvement in environmental decision making, transformation processes in economics, politics and society and sustainable development.

Risk

Philipp Seitzer and Jörg Zirfas

Abstract Risk is a modern mode of perceiving and reflecting uncertainty and insecurity and acting accordingly. In this sense, it is the constant companion of modern civilization and its achievements. The concept of risk in modernity emphasizes on the one hand the decisions – of individuals, of states or of the world society – but on the other hand, it also points to the fact that through these decisions the scope for decision-making and thus the possibilities for action become smaller. This concept seems adequate to describe the human-nature relationship in the Anthropocene insofar as it expresses a way of dealing with something that suggests control and predictability and that largely eludes control and calculation at the same time.

The concept of risk was first established in the mid-sixteen century in the field of commerce and insurance, to which its use remained limited until well into the modern era. In this context, it conveys a double connotation: "on the one hand – as hazard – the damage to be anticipated in the event of an unsatisfactory outcome of a trade, and on the other hand – as risk – the realization of the uncertainty of an expected outcome of the trade" (Rammstedt, 1992, p. 1046, own translation). This ambivalence between hazard and risk is one of the constants that still remains in its current usage. Today, the term risk is associated with many other terms such as uncertainty, insecurity, probability, contingency, uncontrollability, openness, fear, calculation, imputability, responsibility, reasonableness, etc. (cf. Wulf & Zirfas, 2015). Risk is thus a modern mode of perceiving and reflecting uncertainty and

---

The discourse on the Anthropocene implicitly refers to the assumption of a radical change in the relationship between humans and nature. The increasing and changing human influence on the global ecosystem is not only empirically evident, but also noticeable in theoretical terms. (Toepfer, 2013). The concept of risk plays a key role in describing the novelty of this relationship and in theoretically questioning its possibilities for interpretation and design.

P. Seitzer · J. Zirfas (✉)
University of Cologne, Cologne, Germany
e-mail: joerg.zirfas@uni-koeln.de

insecurity and acting accordingly. In this sense, it is the constant companion of modern civilization and its achievements (cf. Beck, 1986; Luhmann, 1991; Gross, 1994; Schmidt-Semisch, 2013; Beck, 2017). In conceptual terms, "risk" implies a considerable semantic diversity: Firstly, the term refers to a probable event that can be not only negative (damage) but also neutral or positive (opportunity); secondly, it refers to foreseeable future hazards and catastrophes that may threaten people (known risks); and thirdly, it refers to an incalculable future with its incalculable consequences (unknown or uncertain risks). Risks result from transforming uncertainties and hazards into decisions (Beck, 2017, p. 202).

If one follows a distinction made by Niklas Luhmann, hazards can be causally attributed to the possible damage caused by others or the environment, while risks refer to the possible damage caused by one's own decisions (Luhmann, 1990). "Risks are attributed to decisions, hazards are attributed externally" (ibid., p. 117, own translation). Unlike hazards, which exist independently of people, risks are dependent on individual intentions to act and decisions. In this sense, a lightning represents a hazard, not installing a lightning rod, a risk. It is important to note that risks do not increase with the effects that individual decisions are proven to bring about. They also do not exclusively correspond with the number of decisions that someone actually makes. However, they do increase with the imputability of effects to decisions (cf. ibid., p. 13). Accordingly, risks can never be eliminated, since every decision in the future can lead to a damage, which in turn can be traced back to decisions (or omitted decisions) of the past. Since modern times, decisions made by oneself are always decisions at one's own risk, while hazards refer to possible damage that may not have been caused by one's own decision – but can be caused by the decisions of others. In the discrepancy between the decision-makers and those affected, there is an indication of a discrepancy that ultimately becomes effective as a power imbalance. Risk polarizes, excludes and stigmatizes, it separates the "we of the decision-makers" from the "we of the living side-effects". Accordingly, "risk is another word for power and domination" (Beck, 2017, p. 255f., own translation), because riskiness is socially distributed quite differently between risk-causers and victims.

This already reveals another risk, which could be called a meta-risk: every decision of the decision-maker may not be rationally plausible for those affected in order to convince them of the harmlessness of the risk. Thus, residents of nuclear power plants may consider the risks unacceptable, despite scientific or political assurances. In this respect, the meta-risk arises from the range of affectedness: On the one hand, by the fact that, despite comprehensive protection, there are still more affected people than thought; and on the other hand, by the fact that even people who were thought not to be affected declare themselves to be so: „The spread of orientation to risks thus has serious consequences for the forms of solidarity still possible in modern society" (Esposito, 1997a, p. 162, own translation).

In addition, intentions and decisions have far-reaching consequences and can mean a dissolution of boundaries in at least three respects: Firstly, in temporal terms, since a decision can reach far into the future and thus also affect future generations; secondly, in issue-related terms, since taking risks in one area of life can have consequences in completely different areas; and thirdly, in social terms, when decisions can no longer be clearly attributed or all people are potentially affected. Wolfgang Bonß (1995, p. 80) calls these delimitations „second-order hazards". These are „transformed risks that become visible to the extent that the results of a risk system not only invalidate the boundaries of that system, but lead to a new initial situation that can no longer be described as a risk situation" (own translation).

The aforementioned dual aspect of hazard and risk is accompanied by a „tension between inevitable fate and personal responsibility. Only when the future is seen as at least partially under human control, it is possible to avoid hazards and mitigate their consequences" (Dreyer & Renn, 2010, p. 67, own translation). The concept of loss of control only makes sense when basic controllability of the uncontrolled is assumed. On the other hand, the risk would not be a risk if it were completely controllable. A risk differs from the completely incalculable nature of hazards in that it is at least calculable in its incalculability. A further field of tension opens up in the temporality of risk. A consciousness that becomes aware of a risk must be temporally structured in that it is directed towards the future or, better, from the future to the present and vice versa.

The concept of risk in modernity emphasizes on the one hand the decisions – of individuals, of states or of the world society – but on the other hand, it also points to the fact that through these decisions the scope for decision-making and thus the possibilities for action become smaller. The term describes not so much the possibility to act in the face of an anticipated future, but rather the necessity to act in the face of a future that has already occurred. On the one hand, the extent and simultaneous occurrence of many different risks has now taken on a magnitude that reaches far beyond individual and national spheres; the development of the atomic bomb can be regarded as the period in which this global awareness of risk took hold (cf. Anders, 1982). So-called global risks, such as those associated with the consequences of global warming, food and water shortages, but also nuclear catastrophes and terrorist dictatorships, are „in principle omnipresent", i.e., temporally, spatially, and socially unconstrainable, „incalculable", i.e., only limited and in parts not knowable (Beck, 2017, p. 103f.). On the other hand, it is becoming increasingly obvious that existing institutional compensation, protection, and prevention systems are not sufficient to counter the presumed risks. Furthermore, the (scientific) expert systems do not seem to be able to make exact calculations of the respective risks, which is why the question of the cultural perception of risks can be addressed and answered very differently. In this respect, risks are not only perspectives on the world, but also mirrors of the self. And fourth, risk analyses can contribute not only to the prevention of risks but also to their emergence, or prevention and control systems in turn unfold unexpected side effects.

In pre-Enlightenment times, the human relationship to nature was largely characterized by mythological and theological figures, so that natural phenomena such as earthquakes, floods, volcanic eruptions, droughts and drought periods still confronted people as more or less incalculable hazards and threats. It was only with the scientific revolution at the end of the eighteenth century that these contingencies increasingly took on the character of calculable risks. However, the transformation of natural events from a hazard to a risk progresses not primarily with the actual ability of humans to find plausible explanations for natural phenomena, to make accurate forecasts, and to effectively bring the course of events under their own control and domination. Rather, it is the degree to which humans attribute these abilities to themselves that paves the way for this transformation. At the same time, the concept of risk implies that the variables that can now be calculated continue to lead to certain unforeseeable effects and events.

The crucial point in the Anthropocene is not that natural disasters are anthropogenic in origin, but the knowledge about their impact: It is not that humans produce risks through their behavior, but the fact that they know about it, that leads to the need to rethink their behavior towards nature. The human-nature relationship in the Anthropocene is thus characterized by the fact that humans themselves become a risk factor for „nature, of which they are a part and without which they cannot survive" (Toepfer, 2013, p. 31, own translation). At the same time, the Anthropocene era is characterized by increasing complexity, which considerably complicates the assessment of potential or actual effects and correlations. Risks today are certainties of uncertainties (Bonß, 2010).

At present, risks increasingly escape the sphere of influence of those who are affected by them. They transgress in at least three interlocking ways: Firstly, the consequences of a centuries-old practice of shifting decisions and risks into the future become more and more apparent. They force people to assume responsibility for the consequences of decisions that are only to be expected in the near or distant future. The imperative of Hans Jonas' ethics for technological civilization proves particularly resonant here, urging us to always act in such a way „that the effects of your actions are compatible with the permanence of real human life on earth" (1997, p. 36, own translation). Secondly, this temporal extension of risk is joined by a spatial one: as knowledge of long-term consequences and global ecosystem interactions grows, so does awareness of the global consequences of local actions (cf. Toepfer, 2013).

In addition to the spatio-temporal transgression of risk and liability, we can thirdly speak of an expansion of the individual sphere of responsibility as a result of the collectivization of risks, which is expressed in the abstract talk of „mankind" as the decisive influencing factor. Through such ways of speaking, a highly questionable generic "we" is taken into responsibility, behind which the actual possibilities of influence of single individuals as well as the power imbalance between different decision-making instances and collectives remain veiled. It is true that each individual contributes to the causation of global risks through his or her consumption behavior. However, these behaviors only become risks when they are practiced collectively. In addition, the collectivization of risks in the talk of mankind as a factor of influence undercuts the systematic logic of the global "externalization society" (cf. Lessenich, 2016), which systematically exports all risks and side effects of affluent societies of the global North to the global South. The disparity between those who make decisions and those who are affected by them is thus geographically consolidated.

These temporal, spatial and generic transgressions challenge the principles of risk management by placing considerable strain on the organised practice of controlling nature in the Anthropocene. From a humanities perspective, however, an even more fundamental critique of the concept of the Anthropocene and the human-nature relationship it suggests was put forward from the perspective of the humanities. By talking about the Anthropocene – so the argument goes – the actual cause of man's overexploitation of nature would not be eliminated, but – on the contrary – even confirmed. The emphatic accentuation of human influence on the planet as an epochal caesura in the chronology of the earth turns out to be a further climax of the same reasoning that has led to this crisis in the first place. The earth thus continues to appear controllable, as "a planet managed for the production of resources and governed for the containment of risks" (Crist, 2013, p. 139). The concept of risk thus seems adequate to describe the human-nature relationship in the Anthropocene insofar as it expresses a way of dealing with something that suggests control and predictability and that largely eludes control and calculation at the same time (cf. Esposito, 1997b).

References

Anders, G. (1982). *Die Antiquiertheit des Menschen*. Beck.

Beck, U. (1986). *Risikogesellschaft. Auf dem Weg in eine andere Moderne*. Suhrkamp.

Beck, U. (2017). *Weltrisikogesellschaft. Auf der Suche nach der verlorenen Sicherheit* (5. Aufl.). Suhrkamp.

Bonß, W. (1995). *Vom Risiko. Unsicherheit und Ungewissheit in der Moderne*. Hamburger Edition.

Bonß, W. (2010). (Un-)Sicherheit als Problem der Moderne. In H. Münkler, et al. (hg.), *Handeln unter Risiko. Gestaltungsansätze zwischen Wagnis und Vorsorge* (pp. 33–63). Transcript.

Crist, E. (2013). On the poverty of our nomenclature. *Enviromental Humanities, 3*, 129–147.

Dreyer, M., und Renn, O. (2010). Vom Risikomanagement zu Risk Governance: Neue Steuerungsmodelle zur Handhabung komplexer Risiken. In H. Münkler, et al. (hg.), *Handeln unter Risiko. Gestaltungsansätze zwischen Wagnis und Vorsorge* (pp. 65–82). Transcript.

Esposito, E. (1997a). Risiko/Gefahr. In C. Baraldi, G. Corsi, und E. Esposito (hg.), *GLU. Glossar zu Niklas Luhmanns Theorie sozialer Systeme* (pp. 160–162). Suhrkamp.

Esposito, E. (1997b). Risiko und Computer. Das Problem der Kontrolle des Mangels der Kontrolle. In T. Hijikata und A. Nassehi (hg.), *Riskante Strategien. Beiträge zur Soziologie des Risikos* (pp. 93–108). VS Verlag für Sozialwissenschaften.

Gross, P. (1994). *Die Multioptionsgesellschaft*. Suhrkamp.

Jonas, H. (1997). *Das Prinzip Verantwortung. Versuch einer Ethik für die technologische Zivilisation*. Suhrkamp.

Lessenich, S. (2016). *Neben uns die Sintflut. Die Externalisierungsgesellschaft und ihr Preis,* hg. Bundeszentrale für politische Bildung. Carl Hanser.

Luhmann, N. (1990). Risiko und Gefahr. In Ders., *Soziologische Aufklärung 5. Konstruktivistische Perspektiven*, 131–169. Leske + Budrich.

Luhmann, N. (1991). *Soziologie des Risikos*. de Gruyter.

Rammstedt, O. (1992). Risiko. In J. Ritter und K. Gründer (hg.), *Historisches Wörterbuch der Philosophie* (Band 8, pp. 1045–1050). WBG.

Schmidt-Semisch, H. (2013). Risiko. In U. Bröckling, S. Krasmann, und T. Lemke (hg.), *Glossar der Gegenwart* (5. Aufl., pp. 222–227). Suhrkamp.

Toepfer, K. (2013). Nachhaltigkeit im Anthropozän. In *Nova Acta Leopoldina NF 117,* Nr. 398, 3–40.
Wulf, C., und Zirfas, J. Hg. (2015). *Paragrana. Internationale Zeitschrift für Historische Anthropologie. Band 24. Heft 1: Unsicherheit.* de Gruyter.

Philippp Seitzer works as a research associate in the field of special education at the University of Cologne and at the Ludwigsburg University of Education. He researches and publishes primarily on the topics of phenomenological perspectives for education, vulnerability as pedacogical reflection category, anthropological and ethical issues, especially in the context of mental disabilities, theoretical issues in the context of mental development and problems related to a science theory for special education.

Dr. Jörg Zirfas, Professor of Anthropology and Philosophy of Education at the University of Cologne. Head of the commission "Educational anthropology" (DGfE) and of the "Society of Historical Anthropology" at the Free University of Berlin. Historical and educational anthropology, philosophy and psychoanalysis of education, educational ethnography, aesthetics, culture and education.

Robot

Martina Heßler

Abstract Robots have become increasingly present in the human environment. As this happens, the question of the relationship of humans and robots is becoming more urgent. Also, much more broadly, the question of the relationship of robots to organic things, to the living, to nature itself, is coming to the fore. Even though the term robot only became common at the beginning of the twentieth century, the concept of an artificial being is a very old one, dating back to the ancient world. Over this long period of time the notion of what constituted a robot, including how it was technically constructed, underwent multiple changes. Five distinct narratives accompany the figure of the robot and its position in the human world. These narratives range from robots as servants to concepts of a post-biological world, and the so-called "Novocene" (Lovelock, Novacene: The Coming Age of Hyperintelligence. Penguin Books, Milton Keynes, 2019) in which humans and artificial intelligences live side-by-side. This article outlines the history of the robot and its accompanying narratives, and it raises the question of the position of robots in a web of diverse entities.

The play "Rossum's Universal Robots" ("R.U.R.") represents a kind of origin myth for the term robot. In this play, written by Karel Capek and performed in 1921, the term robot was used for the first time to describe artificial, human-like beings. After its introduction, the term spread rapidly. The term robot was derived from the Czech word for serfdom. R.U.R. frames one of the central narratives associated with robots to this day: Robots as servants to humans – albeit servants that could possibly incite a rebellion leading to the extinction of humankind.

While robots in the twentieth century were usually thought of as mechanical machines, the robots in Capek's play are artificial-organic beings. They undermine the dichotomy of nature and culture, of the artificial and the organic, and thus point to questions that are crucial for the Anthropocene: Which status should artificial beings, like robots, have? What should their relationship to the organic, to humans, plants, and animals look like? This also addresses the fuzziness of the term robot as

M. Heßler (✉)
TU Darmstadt, Darmstadt, Germany
e-mail: hessler@pg.tu-darmstadt.de

© The Author(s), under exclusive license to Springer Nature
Switzerland AG 2023
N. Wallenhorst, C. Wulf (eds.), *Handbook of the Anthropocene*,
https://doi.org/10.1007/978-3-031-25910-4_161

979

well as its kinship to concepts such as animate statues, artificial humans, to the golem, homunculi, androids, humanoids, and automata (Völker, 1994; Abnet, 2020). As John Jordan summarized, robots are "extremely difficult to define" (Jordan, 2016: 23 cf. various definitions pp. 23–27). Initial definitions in the context of early industrial robots emphasized that the robot was a "reprogrammable, multifunctional manipulator" (cited in Nocks, 2008: ix). Jordan refers to a broad definition from 2005 that highlights the historical transformation of robots: It now refers to a robot as "a machine that senses, thinks, and acts" (Jordan, 2016: 16). In terms of recent development, robots' ability to learn and to interpret and simulate human emotions should be added. At the same time, new questions of differentiation arise, e.g., are virtual AI applications such as robo-advisors, chatbots, avatars, and personal assistants also robots? In contrast to virtual applications, robots have a physical body and can perform actions.

The history of robots begins long before the term was coined in the early twentieth century. In the ancient world we find animate statues or early augmentative technologies (Mayor, 2018). In Egypt, the dead were buried in tombs accompanied by images of servants. During the Middle Kingdom (around 2100 B.C.E.), the dead were given so-called shabti, "a stylized, mummiform figurine" to accompany them in their graves as symbols of their servants (Nocks, 2008: 4–5). This is but to name only a few examples from ancient history.

Especially since the nineteenth century, science fiction, stories, toys and robots displayed at fairs made human imagination familiar with robots as mechanical beings. In the 1960s, workers in the industrial world made their first acquaintance with robots. Robots took over simple tasks previously performed by unskilled workers – feeding workpieces into mechanical machines and removing them again, moving parts from one conveyor belt to another, packaging and stacking finished products.

These first robots could only follow a fixed sequence programmed into them. This limited their field of action to repetitive, standardized tasks. Especially since the 1980s, robotics has been working extensively on making use of robots in a complex and dynamic environment. Robots now had to learn to perceive, see, and react appropriately to situations, to navigate autonomously, and – with their deployment in social fields – to interact with humans. This meant that they had to understand human communication and emotions (cf. the overview in: Nocks, 2008 and for detailed information Siciliano & Khatib, 2008).

For a long time, however, for most people robots remained rather a phenomenon from the world of science fiction. Hardly any other technology has been so strongly influenced by myths and science fiction. It was not until the 1990s, but especially since the turn of the millennium, that robots began to be utilized in many areas of human life, be it as service robots guiding customers in stores, as vacuuming or mowing robots performing household tasks, as surgical robots, or even as sex robots or deployed in war zones. Another field of application are robot pets such as the Tamagotchi or the robot dog Aibo, or educational robots.

As they became more and more deployed in complex human environments, expectations set for robots and their uses changed. However, five central narratives, which can only be separated analytically, have shaped the history of robots up to the present.

First, this is the notion of robots as servants to humans, a notion which, as shown, can be traced back to antiquity. In the late eighteenth century, white middle-class Americans amused themselves by giving orders to Native American-looking automata. In many cases, they were meant to serve male fantasies of control and taming, reinforcing racial and gender stereotypes. The term automaton also became a metaphor for persons without free will, intelligence, or self-control, a metaphor for people subordinate to the will of others. This typically meant women and slaves (Abnet, 2020).

The notion of the automaton servant, or, since the twentieth century, the robot obeying commands, has dominated narratives and science fiction. It defines their presentation at trade fairs but also their legitimations in the world of work or in the home: Robots were supposed to do the boring, exhausting or dangerous jobs. However, narratives about the robot used to continuously change between the inferior and obedient servant and the robot as a competitor that becomes superior to humans and might eventually replace them, thus reversing the hierarchy of master and servant. This narrative of the robot servant, and the fears associated with it, corresponds to the anthropocentric modernity and the idea of technology as a tool for humans to dominate and control the world.

Second, robots have also been envisaged as friend, partner or companion. It is often pointed out that the robot as friend or as friendly helper has a long tradition in Japan (Wagner, 2013; Jordan, 2016). In the Western world this conception can be found as early as the first half of the twentieth century. For the post-World War II period, Isaac Asimov was central in spreading the image of friendly and helping robots (Abnet, 2020). Starting in the 1990s, artificial companions and social robots have become physical objects in real life. Robots have also been established as partners in the world of work since the early years of the new millennium, as indicated by the term cobot (collaborative robot). Robots are now supposed to perform activities in close cooperation with humans. To mark a shift in the relationship between robots and humans, Sherry Turkle used the term "robotic moment" (Turkle, 2011). She noted that in the twenty-first century, humans are increasingly willing to accept robots as companions. Nevertheless, especially in the Western world, this is accompanied by an intense social and ethical debate about the relationship between humans and robots (Tzafestas, 2016; Loh, 2019).

Closely linked to this narrative, and not always distinguishable from it, is the third narrative that proposes that robots are better than humans and might even be something like an improved version of a human. This idea can be traced back to the Pygmalion myth, the idea of creating an artificial ideal woman. Short stories in the nineteenth century fantasized about the ideal mechanical husband or wife. Robots in the nursing field, it is argued today, do not get angry, are not impatient or, worse, malicious. Sex robots, it is argued, do not get bitchy, they are always available. This, too, provokes intense ethical debates about how to treat robots. Is violence against robots a crime? Or is sexist behaviour a problem for women and for society? (with regard to sex robots, see: Richardson, 2022). The roboticist Moravec is even convinced that a post-biological world would be a better world (Moravec, 1999).

Fourth, robots also fit into the narrative of technology as a means of enhancing human capabilities. Robots explore the planet Mars and Earth's oceans; they are sent into damaged nuclear power plants. By being able to be deployed into extreme environments, they have become part of a frontier idea: They are a means of conquering worlds inaccessible to humans, thus expanding the sphere of human life.

Fifth, robots give rise to the question of the distinction between the artificial and the living. Robots have been seen as the opposite of humans, especially in the world of work – mechanical and non-living. However, they have also served as attempts to artificially create the living. Currently, they simulate the living (movement, social interaction, emotions) in order to exist in a human-oriented society. Not only does the idea of life-like machines go back as far as antiquity. Then, the creation of machines imitating life indeed happened, see for example the dove of Archytas. Automata of the eighteenth century simulated humans and animals (Riskin, 2003). In the 1980s, an Artificial Life research project began to work on the emergence of "a new class of organisms" to create digital life (cited in Weber, 2003).

These five narratives are not to be understood as a historical sequence though historical shifts can be observed, e.g., since the 1990s in the direction of the robot as partner. All five narratives are anthropocentric narratives. They look at robots, their functions and tasks from the perspective of humans where robots are considered objects developed for use by humans. Particularly, the efficacy of the concept of the robot as servant is evident, because even if robots are constructed as companion or cobot, they are regarded as assistants and friendly helpers to humans. Isaac Asimov's "robot laws", which have become ubiquitous, are consistent with this viewpoint. However, recent fields of research, such as robot ethics, go beyond it. They also ask whether robots have rights, whether violence against robots is a crime, whether robots can be held criminally responsible. Thus, they point to a historic change in the position of robots and to new human-robot relations (Tzafestas, 2016; Loh, 2019).

These more recent considerations correspond to post-anthropocentric and anti-humanist theoretical concepts. Such concepts postulate a decentralization of humans and criticize the idea of their prominent position. In terms of a flat ontology, they see humans in a web of relations to other entities (Braidotti, 2017). These theoretical designs also criticize the notion of the Anthropocene and counter it, for example, with terms such as Chthulucene (Haraway, 2016) or Capitalocene (Moore, 2016). In this way of thinking, robots would be entities that stand in relation to other entities, including humans.

James Lovelock also shifted away the term Anthropocene and substituted it by "Novocene" (Lovelock, 2019). Using evolutionary logic and placing the concept of information at the centre, he predicts the development of artificially intelligent beings that are far superior to humans. However, he does not link this development to the classic narrative of human extinction. Thus, he is not focused, like Hans Moravec, Ray Kurzweil, and the Artificial Life researchers are, on digital rather than biological life. In Lovelock's vision, the alternative to the human-driven Anthropocene is not a society of artificially created life and artificial intelligences. Rather, he believes in the co-existence of these beings with humans, partly because they themselves are dependent on humans.

In the future, it is to be expected that robots will increasingly be present in many areas of human life. However, this does not only give rise to the necessity of shaping the human-robot relationship and of dealing with a robot ethics. It also raises questions about resources and energy consumption. Since robots already are and increasingly will become an even greater part of the mechanical-organic environment, they must do so in a resource-conserving and sustainable manner.

Fundamentally, robots do not only raise questions about the future of humans in the Anthropocene. They also raise questions about the relationship of humans to each other, and to other beings. Further, robots as energy- and resources-consuming beings will also be representative of the relationship of humans to the organic and to nature.

References

Abnet, D. A. (2020). *The American robot: A cultural history*. University of Chicago Press.

Braidotti, R. (2017). Becoming-world together: On the crisis of human. In M. Kries & A. Klein (Eds.), *Hello, robot: Design between humane and machine, catalogue* (pp. 238–253). Vitra Design Museum.

Haraway, D. (2016). *Staying with the trouble: Making kin in the Chthulucene*. Academic.

Jordan, J. (2016). *Robots*. MIT-Press.

Loh, J. (2019). *Roboterethik: Eine Einführung*. Suhrkamp.

Lovelock, J. (2019). *Novacene: The coming age of hyperintelligence*. Penguin Books.

Mayor, A. (2018). *Gods and robots: Myths, machines and ancient dreams of technology*. Princeton University Press.

Moore, J. W. (2016). *Anthropocene or Capitalocene? Nature, history and the crisis of capitalism*. PM Press.

Moravec, H. (1999). *Computer übernehmen die Macht*. Hoffmann und Campe.

Nocks, L. (2008). *The robot: The life story of a technology*. John Hopkins.

Richardson, K. (2022). *Sex robots: The end of love*. Wiley & Sons.

Riskin, J. (2003). Eighteenth-century wetware. *Representations, 83*(1), 97–125.

Siciliano, B., & Khatib, O. (2008). *Springer handbook of robotics*. Springer-Verlag Berlin Heidelberg. https://doi.org/10.1007/978-3-540-30301-5

Turkle, S. (2011). *Alone together: Why we expect more from technology and less from each other*. Basic Book.

Tzafestas, S. G. (2016). *Roboethics a navigating overview*. Springer.

Völker, K. (1994). *Künstliche Menschen: Über Golems, Homununculi, Androiden und lebende Statuen*. Frankfurt am Main.

Wagner, C. (2013). *Robotopia Nipponica: Recherchen zur Akzeptanz von Robotern in Japan*. Tectum Verlag.

Weber, J. (2003). Artificial Life Forschung und neuere Robotik. *FIFF-Kommunikation-Bioinformatik, 1*, 41–45.

Martina Heßler is Professor of the History of Technology at the Technical University of Darmstadt. Her research focuses on the history of technology from the Early Modern Period on. She is author of several books, e.g., *Kulturgeschichte der Technik*, Frankfurt 2012, Technikemotionen 2020, and *Handbuch Technikanthropologie* (together with Kevin Liggieri) 2020. Currently, she is working on the history of the man-machine-relationship.

Space

Merle Hummrich and Juliane Engel

Abstract Questions about the Anthropocene raise sensibility for critical approaches to theories of spaces in the context of globalization – and in so doing, overcoming the binary logic of culture, society, and nature. In this perspective, processes of space formation emerge through "(structured) configurations of social goods and people in places. Spaces come into being through action by synthesizing objects and people and arranging them relationally. In this process, actions are performed in pre-arranged spaces and the way they occur in everyday action refers back to institutionalized configurations and spatial structures." (Löw (2000) Raumsoziologie. Suhrkamp, Frankfurt a.M., p. 204) It is crucial to investigate the relationship between culture and nature and to ask questions that address how power and domination operate. In short, it is about discourses on how the planetary dimension – understood as a new critical relationality that also goes beyond Eurocentric epistemologies – of nature and culture can affect how research methods evolve. The following questions are central to this inquiry: How does this dimension transform issues that are discussed on a local or national level? What new aspects are revealed when the glocal is reconceptualized through the planetary dimension? Which role does the digital play in these processes? How does peoples' relationship to what they understand as "nature" (that which surrounds and permeates them) change, and how does this bring about changes in behavior? When addressing these questions, we will also consider their intersections with the modalities of the world of machines, robotics, artificial intelligence, and genetics. In terms of educational science, therefore, we can ask which processes of subject formation in spaces that are conceived of as planetary expand the current discourse. Presently, "Smart environments are expanding beyond smart cities to encompass many different milieus, from smart forests to smart oceans and smart agriculture. Tech companies, environmental researchers, and state actors are participating in developing and expanding these digital systems, often aiming to address urgent environmental issues. But these ambitions also become apparent as emerging planetary modes of governance, the

M. Hummrich (✉) · J. Engel
Goethe-University of Frankfurt, Frankfurt, Germany
e-mail: m.hummrich@em.uni-frankfurt.de; j.engel@em.uni-frankfurt.de

© The Author(s), under exclusive license to Springer Nature
Switzerland AG 2023
N. Wallenhorst, C. Wulf (eds.), *Handbook of the Anthropocene*,
https://doi.org/10.1007/978-3-031-25910-4_162

social–political effects of which have yet to be adequately assessed." (Gabrys, 2020, p. 7).

In this light, while spatial notions of order and the connections to territorial, bodily, physical, and corporeal conditions continue to be relevant, we have seen a rise in prominence of perspectives on spaces of power, space, and social structure, as well as space as a structure of relational ordering, which points to the interplay of inclusion and exclusion: "As a continuation of this dialogue, I provide the philosophical background behind the recent rise of posthumanist or 'new materialist' ecopolitics and argue why and how they can offer important insights for planning theory and practice. I lay out specifically how planners would execute this 'posthumanist normativity' in their everyday planning practices, focusing on three lessons that could be directly applicable." (Jon, 2020, p. 392).

In this context, two current research topics allow for scientific perspectives to unfold that systematically challenge universalist assumptions about the (symbolic) order of space: (1) postcolonial critique, understood as a possibility to analyze transnational entanglements and their powerful structuring of space; (2) postdigital approaches that refer to new structures of spatial arrangement in which virtual and real spaces interpenetrate each other.

Beginning with a short review on perspectives on space in the Anthropocene, this article presents these two current developments concerning the critique of Western modernity they offer through the example of education processes. The article concludes by reflecting on the ambivalence of spatial analysis as a whole: On the one hand, an analysis of space is indispensable for a critique of the power structures that inform spatial orders in processes of education, but on the other hand that analysis itself is entangled in the spatial arrangement determined by power relations within the Anthropocene.

Short Historical Review

Since the Enlightenment, there have been significant developments in the discourse on space. While in classical antiquity, space was principally the object of territorial disputes and therefore was understood as physical space, and metaphysical spatial orders were relegated to the realm of the religious (e.g. heaven and hell), secularization has significantly changed the way we think about space (Löw, 2000). eighteenth-century thinkers Kant, Newton, and Leibniz all considered space to be a condition of action (cf. Johansson, 2004) and hence the material basis of human action. Simmel (1908 [1999]) developed his perspective of space following this tradition, considering it a form that is 'in itself ineffective', "meaning that boundaries provide special configurations for experience and interaction." (Fearon, 2004). This assumption was the foundational thought of the Chicago School around Robert E. Park

(1925 [1967]) in the first quarter of the twentieth century. Park's concept of local, regional, and social spaces as boundaries of action can then be combined with the notion of environment (space) and conditions of social inequality. Park's concept has been criticized for its essentialism, such as its belief in the social heritage of racial temperaments, which he considered spatial boundaries.

The concept of space as a condition of action can hence be seen as an expression of social order. It builds on the notion of naturally occurring boundaries and therefore on the assumption of "natural" gender-related or ethnic differences. The criticism of this concept of space as essentialist forms the background for the development of the notion of space as relational order (Johansson, 2004), which considers space an expression of action, in the sense that architectures, boundaries, and things created by humans bear traces of action (Löw, 2000). Therefore, the social order of a space does not only precede action (in the sense of a condition of action), but is actively produced by action (Bourdieu, 1985). In this relational concept of space, the social spatiality of action is understood as embedded in the social space. Thus, e.g., school is thematized *as* a social space that is itself embedded *in* a social space (the neighborhood, the region, the state) (Hummrich, 2022). Spaces generate distributions that, in a society characterized by a strict social order, are generally unequal, or distributions in which some groups of people are favored over others. Therefore, spaces often form the object of social conflicts. Possibilities of how to use money, credentials, rank, or association are decisive in enforcing arrangements; and conversely, the possibility of utilizing spaces can become a resource." (Löw, 2016, pp. 232–233).

Now, while concepts of social order continue to highlight the limited nature of action through space and in space, discussing the Anthropocene brings forth another dimension of space: the dissolution of boundaries. The conquest of outer space and its tourist marketing (Spector & Higham, 2019) point to a special relationship between humans and space in the Anthropocene: On the one hand, the permanent relativization of boundaries of thought and action, on the other hand, an ever-widening gap in accessibility to social privileges. Meanwhile, the fact that humans are traveling to space illustrates how much the Earth's geological and ecological system is being affected by humans (Crutzen & Stoermer, 2000 [2013]) and the role of spatial concepts in the differentiation of nature and culture, which is part and parcel of the assumed universal right and entitlement to conquer other spaces. Ultimately, the meaning of space in the Anthropocene shows itself to be an interplay of limitations and dissolution that is related to ideas of participation and exclusion: The traces of humans (e.g. plastic in oceans, expanding landfills, rising carbon dioxide emissions) are found all over the planet (Waters et al., 2016; Horn & Bergthaller, 2019) and their consequences, therefore, affect humanity as a whole. However, only a few can access privileges such as luxury products, higher education, world and space travel, etc., while it is often the most marginalized people who are most affected by the consequences of environmental destruction and exploitation of resources.

Research Perspectives

Space and Postcolonial Order

The basic idea of postcolonialism is that imperialist structures did not simply vanish when formerly colonized areas became independent nations, but continue to be inscribed in thought and action. This is seen, for example, in the distinction between "the Orient" and "the West", an idea that still traces a spatial-geographical boundary and mobilizes certain assumptions about the respective inhabitants of these areas. (cf. Said, 1978 [1985]). This is also where the issue of education is particularly relevant: around the world, the history of the West and its idea of *reason* has become an integral part of the educational canon, which has been thoroughly shaped by hyper-focusing on the dominant culture. The differentiation between the West and the rest (Hall, 1992) has thus been solidified through this canon's distinction between what counts and knowledge and what doesn't, what is worth passing on via education and what isn't. The postcolonial diagnosis we reference here reflects the dialectic of postcolonial critique: it employs — indeed it must employ — "Western" colonial tools in order to be heard while critiquing the imperialism inherent in modern theorizing (cf. Spivak, 1999).

According to the Portuguese researcher Giuliani (2020), research on the Anthropocene is also bound up in this way of thinking. The logic of the Anthropocene and its epistemological foundation can be seen as ontologies and logic of othering of the Western "we" against the catastrophic tendencies coming from the South (such as mass migration and the threats it poses to Western education systems). Consequently, the spatial order of the postcolonial remains unaffected by theories of the Anthropocene, both in terms of epistemology and practical action. This concerns, among other things, discourses on education and its spatial implications and is reflected in the global distribution of educational capital.

The EFA Monitoring Report of 2015 discusses education as a human right – yet in over 30% of the countries included in the study, not all all children and young people have universal access to education: Children are not enrolled in school, teachers do not have adequate training, children and youth cannot pursue or achieve degrees (cf EFA Monitoring Report, 2015). UNESCO's 2020 Global Education Monitoring Report (2020) shows that school enrollment rates are much higher in rich Western countries than in the poorer countries of the Global South. This alone constitutes a global educational catastrophe in itself. In debates about the Anthropocene however, the catastrophe of education is mainly framed in the context of mass migration from the Global South to the Global North. Therefore, discourse about the Anthropocene reproduces the power structures of transnationalization that distinguish between internationalization on the one hand, which considers mobility (within the North) as an advantage for education, and interculturalization on the other hand, which considers mobility (from South to North) as a disadvantage. Yet another perspective considers the creative and commercial potential afforded by referencing the Anthropocene (Moore, 2015). The author tells of a case of sustainable tourism in the Bahamas, offering an example for a practical reflection on emerging practices of sustainability and postcolonial tourism. This kind of action can also be seen as a coping mechanism, as also highlighted by DeLoughrey (2019), who has been looking at how artists in the postcolonial South address the effects of colonialism and climate change in their work.

Reflecting upon space and the Anthropocene forces us to examine postcolonial power relations, revealing a critical perspective on the spatial structure in which the discourse on the Anthropocene is embedded, e.g. how spaces are differentiated as spaces of nature or culture. This differentiation also informs the distinction between primitive peoples belonging to nature and highly developed peoples, the former being constructed as located in the global South, the latter in the global North. On the other hand, the Anthropocene itself must be recognized as an expression of dominant discourses that construct spatial distinctions that are also governed by postcolonial structures of power, but within which more and more critiques of how inequality is spatialized are being raised (e.g., in the Fridays For Future, Extinction Rebellion, and Black Lives Matter movements).

Space and Postdigitality

The Anthropocene is interesting to postdigital thinkers because it discusses how human activity has been impacting the planet, addresses the precariousness of life, the boundaries of humanity, and emphasize the imagination needed to transcend these issues (cf. Hood & Tesar, 2019). In this context, the notion of postdigital space means that the boundaries between non-digital and digital space have become blurred. Theoretical perspectives on space and the Anthropocene open up new possibilities of knowledge production about dissolved spaces of action beyond the human visions of what is possible that have brought about the Anthropocene.It follows that a postdigital perspective would critique theories of the Anthropocene for the "human-centric positioning (…), where to be human remains a distinct, elevated entity." (Hood & Tesar, 2019, p. 307).

The postdigital perspective, however, allows not only for a critique of this perspective (i.e. human-centered thinking), but also the spatial structures associated with it. Moreover, the authors assume that the tension between human-centered anthropocentric perspectives and the fluid boundary between the human and non-human will become the fundamental experience for future educational processes and future experiences of growing up in general (Hood & Tesar, 2019, p. 308). "It is shown that this [Kantian] conception of emancipation sets the realm of autonomous beings humans over the realm of heteronomous beings. Accounts of the 'humanisation of nature' are analyzed as incomplete attempts to overcome this dualism. It is argued that the root of this incompleteness lies in the application of analytical rather than dialectical reason to the human–nature interaction. The Anthropocene is presented as the geological–historical moment when at the same time as nature is being humanised, humans are being made aware of themselves as animal." (Dobson, 2021).

In the post-digital age, nature, culture, and their interpenetration can be reassessed and discussed as a common legacy of humankind. The philosopher and biologist Andreas Weber suggests rethinking this complex as „Indigenialität "("indigeniality"), which, in a postcolonial or decolonial reading, can also be understood as "unlearning" (Spivak, 2012) of Eurocentric relationalities between the self and the world. In an interview, Weber explained that he considers human beings to be situated within the network of life; hence, for him, *indigenality* means recognizing oneself as an active part of a meaningful whole and acting in such a way that one's own quality of life enhances that of the whole. He cautions against drawing strict distinctions between the human, human culture, mind, and language on one side versus that which belongs to nature on the other. This separation, Weber says, can no longer be upheld. (Miller, 2019). The kind of current situations that Weber calls *western relationalities* can be understood as relations of domination.

Conclusion

The explanations above show that theories that emphasize the space of the Anthropocene allow for an analysis of how its parts are arranged with regard to the interrelatedness of geological, environmental, economic, and social aspects. The Anthropocene theory thus emphasizes how closely nature, culture, and society are interwoven. For educational science, this interweaving means that rather than focusing on the differentiation processes created by spatial orders, the focus lies on the resulting interfaces, i.e. on the spatial impact of humans not only on our planet but also in outer space. It seems appropriate, therefore, to challenge the theory of the Anthropocene by engaging with two concepts of space that expose the contradictions inherent in assuming the oneness of humanity (and how it harms the planet). This is not meant as a minimizing strategy, rather the intention is to highlight different positions of rationality and participation: South vs. North, West vs. rest – the boundaries are fluid.

> While for some scientists the term [Anthropocene] is a matter of the impact humans will leave on Earth in the deep geological future and is thus a matter of geology, for other Earth systems scientists such as Owen Gaffney and Will Steffen, the Anthropocene represents 'humanity's effect on the Earth cross[ing] a tipping point' and the Earth's shifting into a new domain of operation, out of the Holocene's safe operating space and into a question mark, with Earth heading into 'planetary terra incognita.' In this sense, the Anthropocene offers a name for a time of profound transformation. (Wakefield, 2020, p. 20)

In this light, we aim to analyze different dimensions of violence from a planetary perspective to avoid the binary opposition between nature and culture – for an analysis of conflicts between the human and the non-human must focus on how they are entangled. In that sense, the concept of the Anthropocene needs to be made operational beyond a mere critique of the destructive impact of industrialization on the environment (a critique that is also restricted too often). This critique should include all the spheres of society and culture that are driven by the notion of continuous "progress" and "evolution" of humankind, based on relations of exploitation. In her philosophical analysis of new forms of protest, Eva von Redecker, by advancing the notions of "nurturing" as opposed to "dominating", of "regeneration" instead of "exhaustion", and of "participation" rather than "exploitation", stresses the mediality of planetary thinking and action (von Redecker, 2020). What the Anthropocene calls for is not new ways of thinking, but new ways of practicing knowledge. In the Anthropocene, media play a crucial role in defining the epistemological framework that allows us to understand, simulate, and respond to phenomena like climate change (Schrickel & Stürmer, 2016). They might be an early phase in developing a new critical lens through which to question regimes of power and reflect on new planetary relations between the self and the world. Analyzing the relationship between nature and culture is crucial for developing new perspectives on theories of education.

In order to engage with these traces and to explore what they mean for educational science and the challenges they pose for education, we are involved in international and interdisciplinary collaborative efforts that allow for a broad range of expert scholars to collectively articulate these future-oriented questions.

References

Bourdieu, P. (1985). The social space and the genesis of groups. *Theory and Society, 14*, 723–744.

Crutzen, P.-J., & Stoermer, E. F. (2000). The anthropocene. How can we live in a world where there is no life without people. In L. Robbin, S. Sorlin, & P. Warde (Eds.), *The future of nature* (pp. 479–490). Yale University Press.

DeLoughrey, E. M. (2019). *Allegories of the Anthropocene*. Duke University Press.

Dobson, A. (2021). *Emancipation in the Anthropocene: Taking the dialectic seriously* (https://journals.sagepub.com/doi/10.1177/13684310211028148).

Fearon, D. (2004). *Georg Simmel. Sociology of space*. https://escholarship.org/content/qt7s73860q/qt7s73860q.pdf

Gabrys, J. (2020). Smart forests and data practices. From the internet of trees to planetary governance. *Big Data and Society*, 1–10. https://journals.sagepub.com/doi/pdf/10.1177/2053951720904871 [03/19/2023]

Giuliani, G. (2020). *Monsters, catastrophes and the Anthropocene. A postcolonial critique*. Routledge.

Global Education Monitoring Report Team. (2015). EFA Monitoring Report. https://en.unesco.org/gem-report/taxonomy/term/199 Zugegriffen am 03.01.2022

Global Education Monitoring Report Team. (2020). *Global education monitoring report. Inclusion and education*. All means all. http://hdl.voced.edu.au/10707/553248 Zugegriffen am 03.02.2022

Hall, S. (1992). The West and the Rest: Discourse and Power. In S. Hall (Ed.), *Essential essays, vol 2. Identity and diaspora* (pp. 141–184). Duke University Press.

Hood, N., & Tesar, M. (2019). Postdigital childhoods in the time of Anthropocene. *Postdigital Science and Education, 1*, 307–310.

Horn, E., & Bergthaller, H. (2019). *Anthropozän*. Junius.

Hummrich, M. (2022). Schule und Raum. Inklusion und Exklusion als Prozessdimensionen sozialer Differenzierung – In: Die deutsche Schule *114*(1), 22–33.

Johansson, I. (2004). *Ontological investigations: An inquiry into the categories of nature, man and society*. De Gruyter.

Jon, I. (2020). Deciphering posthumanism: Why and how it matters to urban planning in the Anthropocene. *Planning Theory, 19*, 392–420.

Löw, M. (2000). *Raumsoziologie*. Suhrkamp.

Löw, M. (2016). The sociology of space: Materiality, social structures, and action. In *Cultural Sociology*. Palgrave Macmillan.

Miller, S. (2019). *Philosoph Andreas Weber über „Indigenialität "– Hin zum Einklang mit der Natur* [Philosopher Andreas Weber on "indigeniality" – Towards harmony with nature]. Deutschlandfunk Kultur. https://www.deutschlandfunkkultur.de/philosoph-andreas-weber-ueber-indigenialitaet-hin-zum.2162.de.html?dram:article_id=459879 Zugegriffen am 03.02.2022

Moore, A. (2015). Tourism in the anthropocene park? New analytic possibilities. *International Journal of Tourism Anthropology, 4*, 186–200.

Park, R. E., Burgess, E. W., & McKenzie, R. D. (1925 [1967]). *The City*. The University of Chicago Press.

von Redecker, E. (2020). *Revolution für das Leben. Philosophie der neuen Protestformen* [Revolution for life. Philosophy of the new forms of protest]. Fischer Ver-lag.

Said, E. (1978 [1985]). *Orientalism*. Penguin Books.

Schrickel, I., & Stürmer, M. (2016). *Medienökologien fürs Anthropozän* [Media Ecologies for the Anthropocene]. *Zeitschrift für Medienwissenschaft, 14*, 180–185.

Simmel, G. (1908 [1999]). *Soziologie*. Suhrkamp Verlag.

Spector, S., & Higham, J. E. S. (2019). Space tourism, the anthropocene, and sustainability. *Tourism Social Science Series, 25*, 245–262.

Spivak, G. C. (1999). *A Critique of postcolonial reason. Toward a history of the vanishing present*. Harvard University Press.

Spivak, G. C. (2012). *An Aesthetic Education in the era of globalization*. Harvard University Press.

Wakefield, S. (2020). *Anthropocene back loop – Experimentation in unsafe operating space* (https://ubffm.hds.hebis.de/Record/HEB489914195)

Waters, C. N., Zalasiewicz, J., Summerhayes, C., Barnosky, A. D., Poirier, C., Gałuszka, A., Cearreta, A., Edgeworth, M., Ellis, E. C., Ellis, M., Jeandel, C., Leinfelder, R., McNeill, J. R., Richter, D. d., Steffen, W., Syvitski, J., Vidas, D., Wagreich, M., Williams, M., Zhisheng, A., & Wolfe, A. P. (2016). The Anthropocene is functionally and stratigraphically distinct from the Holocene. *Science, 351*. https://pubmed.ncbi.nlm.nih.gov/26744408/ [3/2/2022]

Dr. Merle Hummrich holds a Professorship für Educational Science with particular focus on Youth and School at Goethe-Universität of Frankfurt. Her work concentrates on: Qualitative Research Methods, Methodologies of educational research, Youth Research, School Cultures, Migration Societies, Transnationalization, Comparative Education.

Juliane Engel holds a Professorship for Educational Science with a particular focus on schools and cultural transformation. Her work concentrates on the systematic and empirical study of educational processes. These are contextualized against the background of social transformation dynamics, such as processes of cultural pluralization and (post-)digitality, and are questioned in terms of power theory. The focus of analysis is on the materiality and mediality of processes of subjectivation. She is particularly interested in discourses and practices of marginalization and minorization as well as the negotiation of vulnerability.

Virtuality

Jan Jagodzinski

Abstract What can the arts 'do' in the Anthropocene? The shift toward immanence, materiality, sensation and performance by what appears to be a new
Kunstwollen, directly addresses such a question. I develop the notion of the cosmic-
eco-artisan, influenced by the work of Deleuze and Guattari to further address such
a question.

The question of the place of art in the Anthropocene raises many questions as to
what its function ought to be. Heather Davis and Etienne Turpin have already
addressed a broad range of responses by artists in their *Art in the Anthropocene*
(2014), which presents an overview of the many concerns that surround the name
Anthropos. There is no recognition of indigeneity, for instance, which is said to
foster relationships of human and the more-than-human. The lack of theoretical
'ground' by which a clear direction is charted to curb greenhouse gas emissions,
e-waste and plastic pollution of the oceans is left to the swirl of network media
reports and capital interests that promote a 'good Anthropocene' for further growth
based on green industries and the convergence of nano-bio-info-cogno (NBIC)
technologies that promise the development of new materials based on biomimetic
design principles, channelling the now well-established hybridity of natureculture.

Artistic 'representations' of nature that continue to follow any residual modern
and postmodern principals such as 'form follows function,' now seem dated; that is,
structuralism and hylomorphism have been replaced by hylozoic accounts of matter
of one form or another, ranging from transcendent teleological positions, wherein
'life' or 'spirit' enter matter from 'without' and into matter by some exterior force
(God, Allah, the Great Creator, Intelligent Design) as is paradigmatic of organized
world-religions, to immanent positions, initially formulated by Baruch Spinoza in
his *Ethics* (1677), further 'scientifically' elaborated by Henri Bergson's *élan vital* as
argued in his *Creative Evolution* (1911) where matter is said to have degrees of

J. Jagodzinski (✉)
University of Alberta, Edmonton, AB, Canada
e-mail: jan.jagodzinski@ualberta.ca

N. Wallenhorst, C. Wulf (eds.), *Handbook of the Anthropocene*,
https://doi.org/10.1007/978-3-031-25910-4_163

self-organization. Various forms of panpsychism have been embraced by artists of every persuasion as 'materialism' has become prominent given that the Anthropocene was theorized initially as one immense open-system as Gaia by James Lovelock (1979); further elaborated on the global stage by prominent theoreticians such as Bruno Latour (2017) and Isabelle Stengers (2015) to the point where the anthropogenic productive labour of our species homo sapiens sapiens has been confirmed by a host of interdisciplinary scientists as being a significant force in the current phase change of the Earth. Artistically, we see adumbrations of these developments by ecofeminist artists charted by the writings of Lucy Lippard, for instance in *Overlay* (1983), while Nicolas Bourriaud's series of exhibitions and books, also attempted to chart artistic changes that began to take the Anthropocene into account through his concepts such as 'relational art,' 'postproduction,' 'the radicant,' 'exform' and especially his recent *Inclusions: Aesthetics of the Capitalocene* (2021).

Given this backdrop, the following entry develops what I take to be the 'cosmic artisan' as rudimentary introduced by Gilles Deleuze and Félix Guattari in their second volume, *Thousand Plateaus* (1987) in their two volume explorations of capitalism and schizoanalysis. Deleuze and Guattari's philosophy of immanence has been widely embraced by artists in their concerns with materialism, with digital design, with creativity in makerspaces, and generally within the creative performing arts. Félix Guattari's *Three Ecologies* (2000) went on to provide impetus for forwarding the 'posture' of the cosmic artisan by developing an ethico-aesthetic paradigm of ecosophy where three domains of ecology were theorized that would impact our species response to the current climate change, demanding a fundamental change in the economics of globalized capitalism. These three domains – environment ecology (nature), social ecology (relations with more-than-human) and mental ecology (subjectivity) would require a 'transversal' approach where all three domains are affected simultaneously to produce transformative change.

Affect theory, as seminally developed by Brain Massumi's *Parables of the Virtual* (2002), extrapolated from his translation of their *Thousand Plateaus*, has now become recognized as a fundamental understanding of the nonconscious neuronal embodiment that affects the psyche in terms of attitudes and beliefs, leading to possible changes in behaviour and action, fundamental for ecological social and environmental change. I develop the term 'cosmo–eco-artisan' to identify artists in their *singularity*, who attempt, through their artistic performances, to initiate a transversal affect/effect across these three ecologies to make some impact to at least stave off the possible extinction of our species. They are 'cosmic' in the sense that such artists concern themselves with the materiality that they use, including their quantum dimensions where BNIC technologies for design and change can be harnessed in ways that the most productive ecologically sensitive projects have embraced biomemisis to enable new visions of possibility. In this sense, Jennifer Gabry's *Program Earth: Environmental Sensing Technology and the Making of a Computational Planet* (2016) offers one direction for such a challenge provided, yet again, that the nonhuman organic world of plants and animals is not exploited for capitalist gains as Elizabeth Johnson (2017) has carefully shown. Cosmic refers to the invisible forces of the Earth that are in play through the electromagnetic spectrum of which only a limited part of its wave spectrum is registered 'naturally' by our senses. Cosmic artisans attempt to explore other 'invisible' wave functions that also affect/ effect our well-being and attunement with the Outside. Infrared cameras, X-ray and sonar apparati, and the new development of digitalized pixel colour play and filtration begin to open new dimensions of experiencing the Outside that help change reified visions. This applies especially to sound artists who play with Earth magmatic resonances that are changing – perhaps alarmingly so as the climate changes and begins to affect diurnal cycles of various species, much like electric-electronic technologies has changed our species cycles since the dawn of the twentieth century, enabling the productivity of a 24/7 workweek to our detriment.

I have, on other occasions, attempted to articulate the *performative* projects that 'cosmic-eco-artisans' have initiated (Jagodzinski, 2018); they are 'performative' in the sense that the question is raised as to what their affective apparatuses can 'do' to spectators and participants that are involved should engagement as an encounter take place where there is indeed and affective awakening, disturbance, and shock when it come to the Anthropocene problematic. 'Assemblages' (*agencements* in the lexicon of Deleuze and Guattari, 1987) are formed that are shaped by the forces of desire that circulate in the composition that becomes actualized by the cosmic artisan(s). Cosmic-eco-artisans are engaged especially with Earthly elements: air, water, land, minerals. They raise questions of ethics as profoundly explored by Amada Boetzkes, in her *The Ethics of Earth Art* (2010). I have maintained elsewhere (Jagodzinski, 2010) that cosmic-eco-artisans harness the technologies of *Lassen* rather than *Macht*. I borrow these terms from Krzysztof Ziarek's *The Force of Art* (2004) where his own appropriation of Heideggerian technological thought between *techne* and *poiesis* is developed in relation to an avant-garde, which, on another occasion I have articulated as being 'without authority' (Jagodzinski, 2019). There is no appeal to any transcendent telos or grand narrative. *Macht* and *Lassen* technologies are two ways of relating to an 'Outside,' by this I am referring to 'gaps,' 'spaces,' and times that are not captured and already categorized by common sense as being the way thing are: as 'reality.' The suggestion is made that *Lassen* technologies allow a gap to open (decoherence in the quantum sense) through an event or encounter with the art performance, whereas *Macht* technologies close this gap: no encounter takes place. Such technologies survey and reify 'reality.'

The enfoldment of an event which leads to transformative 'becoming' requires *poiesis*, a 'letting-be' that goes 'beyond measure' and de-mobilizes power. This becomes a way to de-anthropocentralize the *Macht* position of our species. It requires the cosmic artisan to become sensitive to the 'thinking' of material that is being exchanged. The plane of composition here involves the 'anorganic life of things' as forces that constitute their 'micro-brains' in the lexicon of Deleuze and Guattari. The *poiesis* of *Lassen* technologies enhances, enables, and grants the forces in play in ways that the relation formed is not calculable in terms of power increases, and cannot be measured or stated objectively. Cosmic-eco art attempts to perform such a gesture. *Lassen* technological forces are more like *mediators*. The intensification of the emergent 'phenomenon' is *singularly* other. It remains beyond normative forms of relation: it enhances the margin of alterity and exceeds available forms of relations. Appropriately, it occurs in the *middle voice*, between the active and the passive, performing an operation of mutation in the operation of making that desists or withdrawals from power. For Ziarek this is conceived as a certain 'force-work' that enables singularities beyond established categories, social depictions, cultural, aesthetic and so on. Vectors so created do not intensify, extend and amplify flows of power, but gather together a power-free (*machtlos*) momentum.

I end this entry with one such example that fits well with Guattari's transversal call on the three ecologies, an exemplary cosmic-eco-artistic performance. It is staged by Danish artist Tue Greenfort in 2007 called *Diffuse Einträge* [Diffuse Entries], a sculpture installation for Skulptur Projekte Münster 07 edition. The 'sculpture' consisted of a high-pressure liquid manure spreader spuing water taken from Lake Aa, Westphalia, Germany; an artificial recreational reservoir southwest of the center of Münster fed by the river Aa. The sculpture was, in effect, a mobile pressurized fountain. The lake is overgrown with blue-green algae making it a hazard to swim in. In short, it is contaminated, not only to humans but to birds and fish. The Cyanobacteria that proliferate there are toxic. Due to the process of eutrophication, the intensively farmed Münsterland region, where cows and pig farms are in abundance, phosphates enter the river from fertilizers and liquid manure that flow into the lake. The meat industry in the region is subsidized by EU subsidies and it has a powerful lobby and influence on the municipality. To 'protect' the meat industry and the many specialty products manufactured in the farmlands of Westphalia, a cosmetic solution was found to reduce the level of phosphates in the lake by adding Iron (III)-Chloride into the water. This was the first time this chemical (usually used in water system cleaning plants) was used in open waters – both in the river Aa and the lake to reduce the smell and the algae. The chemical itself is hazardous to health. To keep EU subsides, the source of the pollution was not mentioned but covered over, caused by "diffused entries" as Greenfort found out from the researcher who had developed the chemical solution.

Greenfort's 'sculpture' intensifies the smell of manure by adding Iron (III)-Chloride into the water as it forcefully pumps its water out, attempting to disrupt and bring to attention the irony of the cosmetic solution. Unlike the usual gesture to buy and install sculpture pieces from this seventh edition of the project, Greenfort's sculpture is an anti-form. Its affects are offensive to the city's decision. Greenfort intervenes in the romanticized landscape of peace and relaxation that this recreational lake promotes by causing an affront to the visitors to the lake, the affect created by the smell bringing together the ethico-political issues between state, municipality, and the meat industrial lobby. Greenfort exposes the invisibility of the 'causes' of the established aesthetic that was to preserve Lake Aa as recreational area by directly intervening in the *agencement* via his sculptural apparatus. The 'diffuse entries' are concretized and exposed. The more difficult question of such an ecosophical intervention into the political, economic and aesthetic dimensions is whether the actualized intervention would change the established state of affairs, or it becomes yet another interesting foray into nihilism? Yet, its affect has been felt even when such an answer is not immediately forthcoming.

References

Bergson, H. (1998). *Creative evolution*. Trans. Arthur Mitchell. Dover. (Original edition 1911).

Boetzkes, A. (2010). *The ethics of earth art*. Minnesota University Press.

Bourriaud, N. (2021). *Inclusions: Aesthetics of the Capitalocene.* Trans. Denyse Beaulieu. Sternberg Press. (Original edition 2021).

Davis, H., & Turpin, E. (2014). *Art in the Anthropocene: Encounters among aesthetics, politics, environments and epistemologies*. Open Humanities Press.

Deleuze, G., & Guattari, F. (1987). *A thousand plateaus. Vol. 1: Capitalism and schizophrenia*. Trans. Brian Massumi. University of Minnesota Press. (Original edition 1980).

Gabry, J. (2016). *Program earth: Environmental sensing technology and the making of a computational planet*. Minnesota University Press.

Guattari, F. (2000). *Three Ecologies*. Trans. Ian Pindar and Paul Sutton. Athlone. (Original edition 1989).

Jagodzinski, J. (2010). *Art and education in an era of designer capitalism: Deconstructing the oral eye*. Palgrave Macmillan.

Jagodzinski, J. (2018). From artist to the cosmic artisan. In C. Naughton, G. Biesta, & D. R. Cole (Eds.), *Art, artists and pedagogy: Philosophy and the arts in education* (pp. 84–97). Routledge.

Jagodzinski, J. (2019). An avant-garde "without authority": The Posthuman cosmic artisan in the Anthropocene. In P. de Assis & P. Giudici (Eds.), *Aberrant nuptials: Deleuze and artistic research 2* (pp. 233–254). Leuven University Press.

Johnson, R. E. (2017). At the limits of species being: Sensing the Anthropocene. *South Atlantic Quarterly, 16*(2), 275–292.

Latour, B. (2017). *Facing Gaia: Eight lectures on the new climate regime*. Polity.

Lovelock, J. (1979). *Gaia: A new look at life on earth*. Oxford University Press.

Massumi, B. (2002). *Parables of the virtual: Movement, affect, sensation*. Duke University Press.

Spinoza, B. (1996). *Ethics*. Intro. Stuart Hampshire, Trans. Edwin M. Curley. Penquin Classics. (Original Latin 1677).

Stengers, I. (2015). *In catastrophic times: Resisting the coming barbarism*. Trans. Andrew Goffey. Open Humanities Press (Meson Press). (Original Fr. edition 2009).

Jan Jagodzinski is Professor of Visual and Media Education at the University of Alberta, Canada. He is the author of 22 books and numerous chapters and articles. Latest non-edited book, *Pedagogical explorations in a posthuman age: Essays on designer capitalism, eco-aestheticism, visual and popular culture as West-East meet* (Palgrave-Springer, 2020). He is the editor of of the book series Palgrave Studies in Educational Futures.

Part XIV
Humanity as Justice

Animal Citizenship

Bruno Villalba

Abstract For as long as humans have been on the planet, their interactions with animals have been based on a continuous relationship – sometimes characterised by conflict, but sometimes by mutual benefit. Whilst the predominant relationship between humans and animals is one of exploitation, with humans using animals for their own benefit, animals have also been the subject of numerous symbolic relationships. For centuries, religions and scientific knowledge have minimised animals' identity. The advent of animal ethics, and the contributions of ethology, law and animal welfare activism, have profoundly altered the relationship between humans and animals. It is now recognised that humans and animals share the same fate, menaced in equal measure by the threat of ecological breakdown. In this light, it is important to examine the political status of animals.

Palaeolithic man (and woman) was responsible for the extinction of large numbers of megafauna species at that time. Palaeontological data show that these species died out around the same time that their continents were colonised by humans (Brook & Bowman David, 2004). Hunting drove animal populations into decline, caused degradation of animal habitats, and the permanent loss of food sources available to animals as a result of agricultural practices (for example, the use of fire).

In addition, the evolution of the living world (spatial distribution and number of species) is dependent on climate shifts (Barnosky et al., 2004). For this reason, palaeolithic humans constructed a complex symbolic system of ideas about (Marc & Marie-Christine, 2016). Thus, since the very dawn of our species, interactions between humans and animals have been shaped by a continuous relationship – sometimes one of conflict; at other times, one of mutual benefit. The coexistence of man and beast, in an extremely complex and fragile relationship, has also been shaped by ecological shifts, and alterations in the ways in which the two interacted.

B. Villalba (✉)
AgroParisTech, Printemps, Paris-Saclay, France
e-mail: bruno.villalba@agroparistech.fr

The dawn of the Anthropocene, bringing, as it does, an acceleration in the processes of social and ecological change, means that we urgently need to redefine the interactions and the aims of the constantly shifting web that links humans and other animals (Will et al., 2007).

Intensive Use of Animals As Tools

Thousands of years of Aristotelian, Christian and Cartesian thinking have led to the deep-rooted belief that there is a clear and insurmountable divide between humans and beasts. Animals are thought of as creatures defined by what they *lack*: the power of reasoning, altruism, intent, independence, and imagination – with no sense of time or concept of their place in the world. Ultimately, dualist western thinking, based on the idea of a separation between humans and the natural world, has taken root as the way in which we understand and regulate our relations with animals. However, in the eyes of the ancient Greeks, gods and humans alike are very close to animals – the boundaries between them are porous, and the qualities of the one may easily be passed to the other. Even today, for many peoples, their relationship with animals is based on proximity, without any fundamental separation. Thus, in the ontology of the animists, non-humans have the same internal essence as do humans, though they may be different in body (Philippe, 2014). Dualism also impacts a broad swathe of humanity itself. Animalisation (the idea that a person is close to animal) forces certain colonised peoples onto the fringes of human society (this is racialism). Thus, the idea of being animal becomes profoundly repellent! Everyone has their place in a hierarchy based on social status, colour and species (amongst other discriminating factors). In any case, the anthropomorphic culture which is so typical of our society makes it easy to distance ourselves from animals. Animals are seen as serving human goals, helping humanity to break free from the constraints imposed by the natural world. Animals greatly increase humans' working capabilities, keep humans safe, help them to conquer new lands and thus to plan for the future. Gods and monotheistic religions reinforce this instrumentalisation of animals, because in their thinking, humans enjoy a singular superior status (as the creation of the gods, the spark of God), which affords them a privileged position in the hierarchy of species – distinct from, and superior to, other species. Humans, too, forever reinforcing the idea of animals as being little more than tools – for example, through classification and selection/destruction of the living world, and through drawing a distinction between wild and domesticated animals (Charles, 1868). Nevertheless, humanity is constantly creating symbolic representations of animals – even those which we are destroying: images of animals which glorify their strength, independence, craftiness, and so forth.

A Step Toward Animal Emancipation

From the seventeenth Century onwards, and more so from the eighteenth Century onwards, political philosophy has been examining the boundaries of otherness: the otherness of citizens, children, the wild and the wilderness, and animals. Jean-Jacques Rousseau looks at animals' place on the political spectrum, and in so doing, steps away from the purely corporeal view of animals which Descartes popularised. Rousseau takes account of animals' sensitivity and feelings. However, he does not set out to define humans and animals as being equal. As animals are subject to the laws of nature, and in view of their limited capacity for reasoning, they are incapable of bettering themselves; humans, meanwhile, seen as having the ability to break free of the laws of nature. What is unique to humans is historicity – i.e. their capacity to improve themselves in the knowledge of what has gone before. Animals do not have this capacity. The above notwithstanding, Rousseau believes that animals are a model of balance – the balance that humans have lost by interfering with their own natural state. Thus, animals are celebrated as moral beings, Rousseau believes that they should be included in the community of rights. Unfortunately, this idea of his gained little traction.

In the British philosopher Jeremy Bentham's utilitarian approach, a balance must be struck between the interests of beings that are capable of having interests, regardless of whether they belong to one species or another. Utilitarianism holds that human actions should be judged on the basis of their consequences. The question which needs to be asked, in the utilitarian view, is not *are animals capable of rational thought?* or *can they speak?* but rather, *are they capable of suffering?* (*An Introduction to the Principles of Morals and Legislation*, 1789). As they clearly can suffer as the result of our actions, morality dictates that we should alter our relationship with animals.

Three centuries later, *animal ethics* emerged in the 1970s. Subscribers to this school of philosophical thought set out to examine humans' moral responsibility towards animals, on an individual basis. Animal ethics includes a number of subsidiary schools of thought, with two main approaches. The first is welfarism, represented by one of its first major contributors, Peter Singer (*Animal liberation*, 1975); the second is the theory of rights, formulated largely by Tom Regan (*The case for animal rights*, 1983). In Singer's view, our current moral attitudes toward animals are based on anthropocentric prejudices which are equally unjust as racism or sexism. Thus, equal consideration ought to be given to the interests of beings capable of suffering, be it physical or emotional. Singer advocates the principle of equal consideration of interests in favour of all sentient creatures, believing we should radically alter our attitude toward animals, aiming to ensure their *welfare*. Regan's theory of animal rights is based on the observation that, to all living beings, whether or not they are capable of rational thought, their own life matters (thus, all living beings are *equal from the perspective of livingness*). This shared biological existence overrides the differences between species on the basis of cognitive ability.

Thus, such differences can no longer serve as justification for different levels of entitlements to rights, or the belief that humans alone have moral capacity. Since then, philosophical thinking has delved into the complexity of the relations between animals and human worlds (Corine, 2019). Animal ethics has helped legitimise movements in favour of animals (anti-speciesist theories, the concept of animal welfare, veganism, etc.). These movements, in turn, have reshaped public perceptions of animals.

During the latter half of the twentieth Century, ethology – the study of animal behaviour – experienced a major shift in the way in which animals' social interactions are observed and explained. *What Would Animals Say If We Asked the Right Questions?*, asks the philosopher Vinciane Despret (2016). This change of direction was the result of a significant epistemological shift, driven notably by a number of trendsetters (such as Konrad Lorenz), but particularly by researchers Jane Goodall and Thelma Rowell, who changed the way in which observations are selected and interpreted. Paying closer attention to the power dynamics, these women refined the vague notions that scientists had, up until that point, projected onto primate social lives: the ideas of hierarchy, competition, the alpha male and his harem, an 'every animal for itself' war for existence, etc. For example, they demonstrated that primate hierarchy is not established purely coercively, and that, when a decision is to be made on a collective action, there is often a process of negotiation between group members. Ethologists, whilst they rarely seem to 'ask the right questions' of animals, and have stepped away from the reductionist presuppositions of behaviourists, can, even with their skewed human perspective, observe rich and complex realities in the animal kingdom (Despret, 2016). Ethologists have reached the conclusion that animals have mental awareness of life, and have intellectual capabilities that have been demonstrated by numerous observations. This branch of science has also proven that the use of language and moral awareness cannot be used as the basis for so clear-cut a distinction between humans and animals, whereby animals are perceived as being inferior. Intelligence, consciousness, language, culture, politics, affection, suffering, empathy, the capacity for socialisation and learning, etc., can no longer be viewed as the exclusive preserve of humans, because examples of these qualities can be found in many animals (de Waal, 2017). The *Cambridge Declaration on Consciousness* (2012) concludes that non-human animals have consciousness similar to that of humans. Thus stripped of what it was believed made us unique, humans were forced, once more, to accept our role as an integral part of the living world.

Animal rights reflect this shift. Little by little, animals' status within the world has changed. At European level, regulations have been developed which are more respectful of the wellbeing and humane treatment of animals raised for human consumption, and stricter controls on animal testing (both for scientific and for commercial purposes). The de-animalisation of animals in the food industry has been examined increasingly. For domestic animals as well, there has been an improvement in the status they are afforded. Since 2006, in Britain, they have had legal status in their own right. In France, animals have gone from being classed as 'movable property' (in the Civil Code of 1815); then to 'sentient beings' (in 1976) whose owner must house them in conditions compatible with the biological needs of their species; then, in 2015, to 'living beings with sensitivity' (this classification precludes animals from being treated as movable property). The same is not true, however, of wild animals, and the cruellest practices, such as the *corrida* of bulls, hunting for sport, cockfighting, ritual slaughter or certain forms of fishing or live-stock rearing. In 2021, a law was passed to combat mistreatment of animals, but in spite of that move, there seems to be no question of endowing animals with actual rights.

A Step Towards Animal Citizenship... Perhaps

The changing ways in which animals are perceived (in terms of their cognitive and relational capabilities), their legal status, and their role in society, have profoundly altered the relationship between humans and animals. However, worldwide, we kill somewhere between 60 and 140 billion animals for food each year – and meat-eating is still on the rise. The wild living world is collapsing at an ever-increasing rate (according to reports from the Intergovernmental Science-Policy Platform on Biodiversity and Ecosystem Services, IPBES; Elizabeth, 2015), and genetic diversity among domesticated animals is being lost. In addition, intensive livestock farming accounts for 15% of greenhouse gas emissions, consumes 45% of water reserved for the production of animal feed alone, and 60% of available land worldwide. However, in the Anthropocene, humans and animals are united in danger: they are, equally, threatened by global climate shifts, and must find new ways of working together in order to deal with that threat.

The development of a new 'relational framework' linking humans and the rest of the living world presents the opportunity to reconfigure that relationship so that there is a better balance between the two (Timothy, 2012). From this point of view, one of the most promising strategies is to restore a genuinely political relationship between humans and animals, so as to organise the living space available to the *polis*. In *Zoopolis* (2013), Sue Donaldson and Will Kymlicka put forward a political theory of domesticity: the relationship which we must ineluctably have with animals necessitates that we recognise animal rights. Some are negative rights (such as the right not to suffer), while some are positive (such as the right to be represented within the political community). Beyond this, though, Donaldson and Kymlicka lay the foundations for *animal citizenship*. Animals must be endowed with an extended form of status, which implies a type of citizenship: the categories of sovereignty, nationality (residence) and political participation can now be extended to them. Naturally, this raises important questions about the relation between *Animalism* and *Humanism*. However, such questioning must be adapted to take account of the dire situation now facing all earthlings, both human and non-human (Jean-Baptiste, 2020).

References

Barnosky, A. D., Koch, P. L., Feranec, R. S., Wing, S. L., & Shabel, A. B. (2004, January 10). Assessing the causes of late pleistocene extinctions on the continents. *Science, 306*(5693), 70–75.

Brook, B. W., & Bowman David, M. J. S. (2004). The uncertain blitzkrieg of Pleistocene megafauna. *Journal of Biogeography, 31*(4), 517–523.

Charles, D. (1868). *The variation of plants and animals under domestication.*

Corine, P. (2019). *Nourishment: A philosophy of the political body.* Bloomsbury Academic.

de Waal, F. (2017). *Are we smart enough to know how smart animals are?* W. W. Norton & Company.

Elizabeth, K. (2015). *The sixth extinction: An unnatural history.* Picador.

Jean-Baptiste, D. A. (2020). *Animalia.* Grove Press.

Marc, G., & Marie-Christine, G. (Eds.). (2016). *Styles, techniques and graphic expression in rock art.* British Archaeological Reports Ltd.

Philippe, D. (2014). *Beyond nature and culture.* University of Chicago Press.

Sue, D., & Will, K. (2013). *Zoopolis: A political theory of animal rights.* Oxford University Press.

Timothy, M. (2012). *The ecological thought.* Harvard University Press.

Vinciane, D. (2016). *What would animals say if we asked the right questions?* University of Minnesota Press.

Will, S., Crutzen, P. J., & McNeill, J. R. (2007). The anthropocene: Are humans now overwhelming the great forces of nature? *Ambio: A Journal of the Human Environment, 36*(8), 614–621.

Bruno Villalba is Professor of Political Science, at AgroParisTech (Paris-Saclay) and Member of Printemps (CNRS UMR 8085). His areas of research are Political Ecology, Sustainable Development and collapse studies. His research focuses on environmental political theory, notably through analysis of the capacity of the democratic system to reformulate its goals based on environmental constraints. He is the author or co-author of over ten books on politics and ecology. Books (selection): Political ecology in France, (ed. La découverte "Repères" 2022 in French); Collapsologists and their enemies (ed. Le Pommier, 2021, in French).

Asceticism

Inga Wiedemann

Abstract For centuries, scientists and philosophers have demanded that their contemporaries turn things round. We have to leave aside aspirations for the legendary wealth of Lydian King Croesus (563–546). Permanent growth must come to an end, because resources are limited. If over time we had listened to the many recommendations for a better equilibrium between Earth and humans, a less radical change than the one now predicted would have been possible. Now renunciation of everything we have become accustomed to is required. The hands of the Doomsday Clock are nearing twelve. How can we achieve the necessary life change?

Looking back, we see that in the last decades the various crises which are presently threatening us have been predicted again and again. What has prevented us from heeding the warnings? In this article, I explain the philosophy of contentedness or asceticism and cite the wisdom of Stoic ascetics. After presenting arguments for an ascetic, self-controlled life in our times with reference to the attractions of consumerism, I examine the research of anthropologists and sociologists in order to find explanations for the phenomena of delay and passivity prevalent in human behaviour.

The Philosophy of Asceticism

Time and again, philosophers have referred to the existential unity of human beings with their natural and spiritual environment. However, their teachings have found little resonance in the political and social world. Asceticism has its roots in a common heritage of the Hellenistic world. If we study Greek and Roman philosophers, we see that they compensated for renunciation with the joy of being released from unnecessary thoughts and acts. It is evident that self-control, temperance, or

I. Wiedemann (✉)
Free University of Berlin, Berlin, Germany

© The Author(s), under exclusive license to Springer Nature Switzerland AG 2023
N. Wallenhorst, C. Wulf (eds.), *Handbook of the Anthropocene*,
https://doi.org/10.1007/978-3-031-25910-4_165

abstinence (Greek: enkrateia), although often understood in a negative manner, for instance as hatred of the body, as the destruction of our instinctive urges, can also be interpreted in a more positive way, as the reintegration of the body and the transformation of the passions in their true and natural condition.

The urge for asceticism probably began in India as early as 2000 B.C. The earliest ideas of an asceticism which is in favour of protecting the environment can be traced backed to Jainism, which was revived by Mahavira in the sixth century B.C. in ancient India. An important point was the protection of life by non-violence, i.e. a symbiosis between all living beings and the five elements of earth, water, air, fire and space. Non-violence signifies that animals should not be slaughtered. Through similar motives members of the Orphic movement (sixth century B.C.) were moved to reject animal sacrifices and to lead a vegetarian and ascetic life. Living in this way, they wished to regain original human purity. Their purification had a specific eschatological meaning, because it released the soul from a burden of personal inherited guilt. The Graeco-Roman world cultivated many different forms of asceticism, with different meanings for those who practised them. Most of the ascetics fought against bodily indulgence and spiritual idleness. Their aim was to strengthen the health of body and soul by liberating the spirit from everyday, unnecessary worries. The Greek physician Galen (131–201) confirmed: "The character of the soul is corrupted by bad habits in food, drink, exercise, sights, sounds and music." (Galenus 1965). The members of an ascetic group felt united in the struggle against evil in its diverse forms. Evil could approach humans with improper insinuations, but possibly it had already taken possession of body and soul. Origenes (185–255 A.D.) described in his "De principiis" forty-two steps of asceticism (Origenes, 1979, pp. 245–269). He urged his readers: "Let us strive to go forward and to ascend one by one each of the steps of faith and the virtues. If we persist in them until we come to perfection, we shall be said to have made a stage at each of the steps of the virtues." (ibid., Hom. 27.3). The way in which Origenes interprets each of the forty-two way-stops mingles ascetic struggle and the attainment of virtues with contemplative experience. In the second stage, the struggle to gain virtues through self-denial takes place. It develops to a fight against demons, the devil and the opposing powers. Freud identified those "opposing powers" as laziness and indolence within our soul. After a long way over mountains and through valleys, the ascetic arrives at stage twenty-three, where he feels the "contemplation of amazement" (Greek: exstasis), when his mind is struck with amazement by the knowledge of great and marvellous visions (ibid., Hom. 27.12). The ascetic wishes to lead a self-responsible, rationalist, "good" life. The sort of asceticism advocated by the Stoics and practised by Marcus Aurelius, with its emphasis on ethics and the internal dynamics of the individual psyche, was constituted precisely to allow individuals to better perform their traditional social roles and functions. Indeed, it elevated these conventions to the level of moral obligations (Francis, p. 182). In contrast, the school of the Cynics preferred an asocial form of asceticism. Cynic values were based on explicitly denying and overturning the values and expectations of society. Cynics condemned the pursuit of ever more refined and extravagant pleasure (Francis, p. 65). In manifestly rejecting such institutions as property and marriage, the Cynics and other radical ascetics placed themselves in opposition to accepted values and thereby posed a threat to the social order.

The Stoics compiled many pieces of advice for a self-controlled life. The Stoic Musonius Rufus (*before 30 CE, + before 101/2) warned of luxury, wealth and extravagance, and indeed, his words question our Western life-style: […] "Man, in the image of God, when living in accord with nature, should be thought of as being like Him and being like Him being enviable, and being enviable he would forthwith be happy, for we envy none but the happy". (Musonius Rufus, 2020, p. 84)

To live in accord with nature brings inner peace and happiness. Wealth produces restlessness and dissatisfaction: "But if anyone thinks that wealth is the greatest consolation of old age, and that to acquire it is to live without sorrow, he is quite mistaken; wealth is able to procure for man the pleasures of eating and drinking and other sensual pleasures, but it can never afford cheerfulness of spirit nor freedom from sorrow in one who possesses it. Witnesses to this truth are many rich men who are full of sadness and despair and think themselves wretched – evidence enough that wealth is not a good protection for age". (ibid., p. 86)

We should avoid sumptuous meals: "We have come to such a point of delicacy in eating and gourmanderie that as some people have written books on music and medicine, so some have even written books on cooking which aim to increase the pleasure of the palate, but ruin the health". (ibid., p. 88)

He admits that it is hard to control the desires of palate and stomach: "Pleasure in eating is probably the hardest of all to combat. For other pleasures we encounter less often, and we can refrain from them for months and whole years, but of necessity we are tempted by this one every day and usually twice a day" (ibid., p. 91).

Greek philosopher Epicure (341 BC – 270 BC) compared the unassuming behaviour of animals with the constant needs of humans: "The human child lies naked on the ground, speechless, in need of every sort of aid to stay alive, when first Nature brings it forth by labour from its mother's womb into the shores of light. And it fills its place with mournful crying, as is appropriate for one who has such troubles ahead in life. But the various sorts of cattle and wild beasts grow up and have not need of rattles, nor does anyone have to speak to them in the gentle broken speech of the fostering nurse, nor do they look for different clothing for the changes of seasons. Finally, they have no need of arms, or of tall walls, to protect their own, since the earth itself supplies everything to all of them (cit. after Nussbaum, p. 254)".

Appeals for an Ascetic, Self-Controlled Life in Our Time

To plead for asceticism in our time appears absolutely out of the way. But there is a significant link between ancient ascetics and us. They knew that their lifestyle, overshadowed by worries and anxieties, occupied them too much, although they were envisioning a rational, "good" life. They felt equally guilty as nowadays people like Greta Thunberg and other climate activists, who in addition suffer from the indolence of their contemporaries. Our conscience tells us what we should do and what we should avoid. Little and big sins like consumerism, eating meat, polluting the environment and similar activities make us feel guilty. A radical life change would relieve us from remorse and scruple and enhance our tranquillity.

Arthur Schopenhauer (1788–1860) was the first to embrace asceticism as a practical answer to the riddle of life. His student Friedrich Nietzsche regarded spiritual self-mastery as a form of sublimation, as crucial for anyone who wants to give style to his or her life, and for individuals who wish to gain genuinely new inspirations (Van Ness, 1998, pp. 589–593). From the 1960s on, scientists warned of the permanent industrial growth, which causes ever more polluting and poisoning. They asked for restraint and modesty as a new way of asceticism. The international discussion focused on limits to food supplies, limits to an ecosystem's ability to absorb pollution, limits to population levels, and limits to natural resources. The space ventures between 1957 and 1961 brought about impressing pictures showing our small blue planet circulating in the dark universe. Adlai Stevenson (Stevenson, 1979, p. 828) compared in a speech for the U.N. our planet with a spaceship which requests our solicitude on our orbit through the universe. The metaphor of mankind as a spaceship crew has remained vivid as a request for conviviality.

Rachel Carson prophesied in *Silent Spring* (Carson, 1962) that because of the ever growing use of pesticides and other toxins sooner or later insects and birds would be exterminated. Carson called for the ascetic virtue of a strong human self-control. She pointed out that the biggest challenge for mankind was to end the destruction of nature. Her prophecy is still coming true. Recently, the World Wildlife Fund classified the progressing death of species as catastrophic. After reading Carson's book, P. Ehrlich called not only for sex education, birth control and abortion rights, but expressed at the same time the fear that the real threat was "improved technology" which increases the potential of war as a "population-control device" (Ehrlich, 1968). Ivan Illich dreamed of a convivial society of humans dedicated to an enlightened asceticism (Illich, 1975). Theologist Emil Küng (Küng, 1972) expected people to do away with the signs of abundance. Persons taking pride in excessive consumption should be despised. He proposed that immaterial qualities, not opulence, should in future raise a person's social status. Economist E. F. Schumacher's international bestseller *Small is Beautiful* points in a similar direction (Schumacher, 1973). Strongly influenced by Mahatma Gandhi, he developed the concept of a decentralized, ecological "people's economy". All these appeals can be compared to the urgings of V. Shiva (Shiva, 2016) to live in agreement with nature. She argues that ecological destruction and the marginalization of women set in motion a process of exploitation, inequality, and injustice. Whereas for many generations Indian women protected nature as a condition for human survival, the "white old man" overtook Indian agriculture with his way using pesticides and all kinds of harmful techniques. In a similar way, D. Haraway (Haraway, 2016) conceptualized our current epoch as what she calls the Chtulucene, which requires sym-poiesis, or making everything in harmony with nature, rather than auto-poiesis, or egoistic self-making.

In his book *An Inconvenient Truth* (Gore, 2007), former US-Vice-President Al Gore expressed an additional preoccupation. He predicted an "epic" destruction of

Earth, involving extreme weather, floods, droughts, epidemics and heat-waves beyond anything we have ever experienced. To underline this, he referred to the data of world's scientists which foresee that we have less than 10 years to avert a major catastrophe which could send our entire planet's system into a tailspin. Greta Thunberg started in 2018, anguished by the prognoses of Al Gore and others, her actions against climate change. In her speeches, she implored her audience again and again to close the gap between what is known about the climate crisis and how we behave. "Stop talking, start acting" is her motto. She wants us to lead a careful, attentive life. The same point is made by N. Klein, who pleads untiringly for a Green New Deal as a solution to support agricultural techniques and society (Klein, 2019).

Consumerism as a New Thread

Consumerism has become an important factor for industrial growth. Fairfield Osborn with *Our Plundered Planet* (Osborn, 1948), William Vogt with *Our Road to Survival* (Vogt, 1948) and Paul Ehrlich with *The Population Bomb* (Ehrlich, 1968) showed that we are consumers as much as we are producers, and that our growing appetites for new commodities have a tremendous cumulative impact on the planet. The same concern expressed Pope Francis in his Encyclical Letter *Laudato Sí* (Franciscus, 2015, p. 20): "The culture of discarding affects both excludes human beings and things that quickly become garbage."

It seems that humans have always yearned for more and more luxury. Wealth is still an indicator for the success of individuals, clans, and nations. Asceticism should now be understood in the context of the oppressive and life-denying-forces exemplified by powerful, contemporary corporations, governments, and professions. They advance their interests by encouraging and manipulating habitual, if not always addictive, forms of consumption (see Van Ness, pp. 589–593). Industrial growth must come to an end, and societies/consumers must be relieved from the pressure exercised by steady economic growth and the "duty" of consumerism. We should have in mind the words of Musonius and Epicure. Only ascetic regimens can undermine the pernicious habits of avid consumers and show a path to personal and communal health.

The Quest for Reversal

Meadows' *Limits to growth* (Meadows et al., 1972) was the first at5tempt to evaluate a huge body of data with the help of big MIT Computers. The Calculations proved that to continue as before would in 2100 lead to undesirable consequences, a prognosis affirmed by later verifications and actualizations. The report sold over 10 million copies in 25 languages, but most people would not believe in the gravity of the situation and saw its conclusions as doomsday prophecies. Why did it take so much time for reactions to the coming crises, although the predictions were increasingly threatening? Reacting to day-to-day challenges, we just don't have the time for thinking at future dangers (Francis, 1995, p. XIII). N. Luhmann wondered at the "exceptional determination of certain cultures to continue to function within the daunting prospects of environments proving increasingly perilous"(Luhmann, 1988, p. 40). Fighting against environmental perils can only be successful, if people take an interest in nature, and open their eyes for the problems next to them. A helpful motto might be German philosopher H. Jonas' ecological imperative: "Act in such a manner that the effects of your actions harmonize with the permanence of true human life on earth." (Jonas, 1979, p. 36) Crisis (Greek: krisis) originally means "decisive turning point". Having this in mind, it is obvious that it is time for a decisive reversal.

References

Carson, R. (1962). *Silent spring*. Mifflin.

Ehrlich, P. (1968). *The population bomb*. Ballantine Books.

Francis, J. A. (1995). *Subversive virtue: Asceticism and authority in the second-century Pagan World*. Pennsylvania State University Press.

Franciscus, P. (2015). *Encyclical letter Laudato Sí: On the care for our common home*. University of Santo Thomas Ecclesiastical Publications Office.

Galenus, C. (1965). De sanitude tuenda 1.8. In C. G. Kühn: Galeni Opera Omnia. Leipzig 1823–33, 6:40 repr. Olms.

Gore, A. (2007). *An inconvenient truth*. Viking Hardcover.

Haraway, D. (2016). *Staying with the trouble: Making Kin in the Chtulucene*. Duke University Press.

Illich, I. (1975). *Tools for conviviality*. Fontana Collins.

Jonas, H. (1979). *Das Prinzip Verantwortung*. Suhrkamp.

Klein, N. (2019). *On fire: The burning case for a green new deal*. Penguin Books.

Küng, E. (1972). *Wohlstand und Wohlfahrt: Von der Konsumgesellschaft zur Kulturgesellschaft* (St. Galler Wirtschaftswissenschaftliche Forschungen 28). Mohr.

Luhmann, N. (1988). *Die Wirtschaft der Gesellschaft*. Suhrkamp.

Meadows, D. H. / Meadows, D. L. / Randers, J. / Behrens III, W. W. (1972): The limits to growth. A report for the Club of Rome's project on the predicament of mankind. : Universe Books.

Musonius Rufus, G. (2020). *That one should disdain hardships. The Teachings of a Roman Stoic*. Trans. by Cora E. Lutz. Yale University Press.

Origenes. (1979). *An exhortation to Martyrdom. Book 4*. Trans. and introd. Rowan A. Greer. Paulist Press.

Osborn, F. (1948). *Our plundered planet*. Little, Brown.

Schumacher, E. F. (1973). *Small is beautiful*. Sphere Books.

Shiva, V. (2016). *Staying alive: Women, ecology, and development*. North Atlantic Books.

Stevenson, A. E. (1979). *The papers of Adlai E. Stevenson* (Vol. 8). Little, Brown.

Van Ness, P. H. (1998). Asceticism in philosophical and cultural-critical perspective. In V. L. Wimbush & R. Valantasis (Eds.), *Asceticism*. Oxford Press.

Vogt, W. (1948). *Our road to survival*. Yale University Press.

Inga Wiedemann, MfA, PhD, librarian. Currently research on asceticism at the Free Univ. Berlin.

Citizenship

Alain Pache

Abstract This text first offers a definition of citizenship and a brief historical overview of its evolution. Secondly, the text shows how the school is the institution of citizenship *par excellence*, but also a field of tension between the visions intended by the legislator and the interpretations made by the teaching staff on a daily basis. Finally, the text presents some perspectives for the school in the context of the Anthropocene. These are based on the models of the Whole-School Approach, transformative and transgressive social learning and resonance. Citizenship can thus contribute to the construction of a different relationship with the world.

Citizenship can be defined as a status that confers on its holder a set of rights and obligations because he or she is a member of a political community. In addition, there is an affective dimension that is more difficult to grasp, expressed in the idea of a feeling of belonging to a community of destiny that links the present, the past and the future (Audigier, 2007). For a long time, there was a dissociation between citizenship and nationality insofar as these political rights were only granted to a part of the nationals. This was the case for women in France until 1944, or in Switzerland until 1971 for federal elections (*ibid.*).

While citizenship necessarily refers to the political community to which one belongs, human rights stand out as the most universal reference with which the particular rights of each of these communities should not be in contradiction. However, this reference to human rights is subject to various interpretations (*ibid.*). Indeed, for some time now, there has been a strong tendency to emphasise only civil rights, seen as freedoms linked to the individual, to relegate political rights to the background and to try to cut up economic and social rights. This tendency is accentuated by the crisis of politics and its recomposition, caused in part by the extension of the market logic to ever wider areas of personal and social life, an extension that is also accompanied by an increase in inequalities and complex processes of

A. Pache (✉)
University of Teacher Education State of Vaud, Lausanne, Switzerland
e-mail: alain.pache@hepl.ch

© The Author(s), under exclusive license to Springer Nature
Switzerland AG 2023
N. Wallenhorst, C. Wulf (eds.), *Handbook of the Anthropocene*,
https://doi.org/10.1007/978-3-031-25910-4_166

disaffiliation and the disintegration of the social bond. Thus, the historical movement that has built up these different categories of rights as one, is now being countered by a movement that seeks to separate them (*ibid.*).

According to other authors, citizenship is above all a historical construction defined on the basis of two intellectual traditions: the separation of the balance of powers, invented by the English and Montesquieu; and the fusion between the individual and society through Rousseau's conception of the general will. These two currents of thought have nourished the development of two types of 'citizen', one that can be described as 'English-style' and the other as 'French-style' (Schnapper & Bachelier, 2000). At the end of the nineteenth century, however, the criticism against the idea of human equality and against a political project based on reason crystallised with the birth of the social sciences and the discovery of the individual and collective unconscious (*ibid.*).

Education is at the heart of the democratic project, because citizens must have the necessary means to exercise their rights in practice. Consequently, the school is the institution of citizenship *par excellence*. According to Schnapper and Bachelier (2000), it even has a double function: on the one hand, it provides a common language, culture, national ideology and historical memory through the content of the teaching. On the other hand, the school forms a fictitious space in the image of the political society itself. Indeed, students are treated in the same way, regardless of their historical and social background and beliefs. It is a place that is built against the real inequalities of social life, to resist the movements of civil society (*ibid.*).

But the school is not a society in miniature. In France, it is derived from the dominant model of the republican school. In England, on the other hand, the school inherited three distinct models: that of the aristocratic public schools; that of the schools set up for the people by the large rival religious foundations at the beginning of the nineteenth century; and finally, that of the pedagogical revival, which aimed to develop the 'whole child', with the objective of ensuring children's physical health, intellectual development, emotional and moral health, aesthetic sense, insight, practical skills, social competences and self-fulfilment (Raveaud, 2006).

However, a double movement is visible today in English and French schools: on the one hand, the perpetuation of a common fund of values, behaviours and representations taken for granted, which largely correspond to the citizen ideals of both countries; on the other hand, tensions and disagreements between the teaching profession and the legislator. In other words, while they convey a social model, teachers are not always aware of it, nor do they necessarily adhere to it (*ibid.*). In England, researchers denounce the effects of market mechanisms, which increase inequalities, and citizenship education is criticised sometimes for not being sufficiently anti-racist or multicultural, sometimes for insisting too much on the national dimension, when the critics do not underline the paradox of the very notion of 'citizen' in a country whose inhabitants are legally subjects of Her Majesty. In France, a first glance at the early grades might lead one to believe that the primary school conforms to the official instructions, that it still corresponds very strongly to the republican school model. But cracks are visible at several levels of the building. First of all, in its foundations. The ideal of citizenship that presides over the official instructions is very clear. But is the school able to ensure that it is put into practice? The sociology of Bourdieu and Passeron has revealed the republican model of equality of opportunity as a myth, not a description of reality (*ibid.*).

However, other tensions are perceptible in the school environment. Audigier (2007) mentions, for example, the priority given to 'living together', and the fact of leaving aside a large part of what makes up citizenship, such as, for example, a more systematic introduction to the legal universe both as a reference for everyday life, for example on the rules of life or on conflict resolution, and as work to open up to a universe that structures and organises our social life and therefore relations with others. Another tension concerns the place of citizenship education in education systems. While in England it is a subject in the National Curriculum for cycles 3 and 4 (pupils aged around eleven to sixteen), in other cases, citizenship education is included in a broader package such as Social Studies in Sweden (*ibid.*). Furthermore, it is worth recalling the fundamental difference between primary and secondary school. In the first case, interesting experiments have attempted to articulate experiences and knowledge, notably in the very organisation of the classroom and teaching. On the other hand, in the secondary levels, the combination of the reluctance of adults to allow pupils some freedom in the school setting, with the organisation of teaching and the plurality of teachers, as well as the importance of a conception of the school which attributes the main role to the transmission-construction of knowledge, all of this very quickly rendered obsolete the orientations of participation present in some official texts. This conception is based on the affirmation of a positive relationship between knowledge and social attitudes, as if the fact of mastering a lot of knowledge would lead to reasonable social attitudes. Finally, the focus on practices (setting up places for expression, project-based teaching, etc.) suggests that experience is sufficient in itself. On the contrary, it would be advisable to provide for moments of analysis, putting things into perspective or identifying the values at stake (*ibid.*).

In the context of the Anthropocene, humanity as a whole is now facing profound changes in its environment. This is an experience on a scale not seen since the emergence of agriculture, sedentarisation and the construction of the first cities some eleven thousand years ago. Human beings must therefore take into account the possibility of a collapse of civilisation and perhaps even the extinction of the human species (Wallenhorst & Pierron, 2019; Curnier, 2021). Therefore, the entire education system must adapt and evolve towards a Whole-School Approach (Henderson & Tilbury, 2004) and a transformative and transgressive model of social learning (Lotz-Sisitka et al., 2015). The Whole-School Approach model implies integrating sustainability issues in all areas of the educational institution, i.e. in the teaching, of course, but also in the mode of governance, infrastructure and finance or campus management. Student participation must therefore become more important, regardless of the level of study, and must be a solid basis for decision-making processes. The transformative and transgressive social learning model implies strengthening the competency approach, including competencies to think about complexity, to think critically, to identify values for sustainability and to prepare for action.

Thus, citizenship can be redefined both as a process of acculturation and as a process of changing the dominant model of society centred on the neo-liberal economy, which has resulted in the current environmental impasse and the widening of the gaps - in wages and treatment - between the richest and poorest.

But becoming a citizen in the twenty-first century also means building another relationship to the world, a relationship to the world based on resonance, which would thus become the 'essence' not only of human existence, but of all possible relationships to the world; it irrevocably precedes the ability to distance the world and make it available (Rosa, 2020). But, beware, resonance does not allow itself to be made available: therein lies the great source of irritation constitutive of this social formation, its fundamental contradiction, which produces, in ever new variants, "angry citizens" (*ibid.*, p. 51). The role of the school would thus be to accompany students in managing this frustration while proposing solutions to current problems from a perspective of social justice and respect for the environment.

References

Audigier, F. (2007). L'éducation à la citoyenneté dans ses contradictions. *Revue internationale de Sèvres, 44*, 25–34.

Curnier, D. (2021). *Vers une école éco-logique*. Le Bord de l'Eau.

Henderson, K., & Tilbury, D. (2004). *Whole-school approaches to sustainability: An international review of sustainable school programs*. Report prepared by the Australian research institution in education for sustainability (ARIES) for the Department of the Environment and heritage, Australian Government.

Lotz-Sisitka, H., Wals, A. E., Kronlid, D., & McGarry, D. (2015). Transformative, transgressive social learning: Rethinking higher education pedagogy in times of systemic global dysfunction. *Current Opinion in Environmental Sustainability, 16*, 73–80.

Raveaud, M. (2006). *De l'enfant au citoyen. La construction de la citoyenneté à l'école en France et en Angleterre*. Presses universitaires de France.

Rosa, H. (2018/ Trad. 2020). *Rendre le monde indisponible*. La Découverte.

Schnapper, D., & Bachelier, C. (collaboration) (2000). *Qu'est-ce que la citoyenneté?* Gallimard, Folio.

Wallenhorst, N., & Pierron, J.-P. (Eds.). (2019). *Eduquer en Anthropocène*. Le Bord de l'Eau.

Alain Pache is professor at the University of Teacher Education, State of Vaud (Switzerland) and head of the teaching and research Unit Didactics of Human and Social Sciences. His main fields of research are geographical education, education for sustainability, citizenship education, language practices, teacher training. He is member of the boards of the Swiss Society for Research in Education and of U Change, Swiss academies of arts and sciences.

Commons

Nina Gmeiner and Stefanie Sievers-Glotzbach

Abstract Commons' cannot only be understood as alternatives to private property regimes, but also as a practice of relating towards contemporaries, future generations and the environment. As such, commons have the potential to challenge the paradigms which characterize the Anthropocene and perpetuate unsustainable system dynamics. The three major paradigms of 'materialistic culture and growth', 'control and autonomy of humans over nature', and 'expert knowledge and specialization' are countered by practices of commoning, as these create scope for establishing social relationships within the present and with future generations, as well as human-nature relationships. This supports the cultivation of individual and collective competencies, virtues and values that strengthen agency for overcoming incumbent system paradigms. 'Commoning' promotes transformation processes by demonstrating the viability and desirability of alternatives to established, unsustainable paradigms in societal niches.

Humans around the world create communities to find their own local solutions to the most pressing ecological and social matters they face, and thereby to satisfy their needs. These practices and institutions are called 'commoning' and 'commons', respectively (Bollier & Helfrich, 2019). Over time, researchers' views on commons have shifted, taking into consideration more precisely and fundamentally in which ways practices of commoning question problematic logics of the Anthropocene. In the twentieth century, commons have been understood as a resource category for subtractable goods with limited excludability (Ostrom, 1990), such as village pastures or fishing grounds. It was assumed that they would best be managed by privatization or state control. Ostrom (1990) changed that view with her work on collective management of common-pool resources, where she redefined commons from being goods to being common-property institutions. From there, other scholars broadened the view to include other goods, ranging from

N. Gmeiner (✉) · S. Sievers-Glotzbach
Department of Business Administration, Economics and Law,
University of Oldenburg, Oldenburg, Germany
e-mail: nina.gmeiner1@uol.de; stefanie.sievers-glotzbach@uol.de

N. Wallenhorst, C. Wulf (eds.), *Handbook of the Anthropocene*,
https://doi.org/10.1007/978-3-031-25910-4_167

knowledge to health (Hess, 2008). Finally, the social processes of (re)production were brought to focus, culminating in a view of commoning as the collective act of need satisfaction (Bollier & Helfrich, 2015). In the following, commoning is discussed as a counter weight to increasing commodification and a living alternative to current socio-economic regimes and reductionist conceptions of the human being.

Why anthropological commons? While an economic understanding of commons first comes to mind with the seminal work 'Governing the commons' by Ostrom (1990), anthropologists have likewise been studying social systems of resource allocation and maintenance for decades (Olson, 1965). We argue that it is especially the insights of anthropologists, which are crucial to make commons a transformative space in the Anthropocene. In the following, we first explore economic and anthropological commons research. Then we outline how the latter's critiques towards the former led to a novel view of commons: commoning.

Commons is the collective (production and) governance of a resource by a group of people (commoners) through collectively designed rules, norms and practices. This definition shows that commons are not only a type of property regime, but a way of relating towards a resource and each other. The theoretical approaches in institutional-economic and anthropological commons research place different emphasis on this fact. Institutional-economic theory applies a more holistic view to humans than classical economic theory, but still views humans largely as rational decision makers (Nightingale, 2011). The theoretical leap here is that people will not only act competitively and pursue individual short-term economic wins, but will collectively and cooperatively plan for a long-term sustainable resource use if group size is moderate and communication possible (Ostrom, 1990). Contrary to Hardin's prediction of inevitable overexploitation of openly accessible resources (Hardin, 1968), commoners will organize for resource maintenance, even if that means non-optimal individual economic outcomes. To understand this repeated observation, institutional economists strive to generalize success standards for all commons by aggregating cases and distilling their design principles (Ostrom, 1990).

In turn, anthropological commons theory focuses on singular case studies and emphasizes the uniqueness of each case (Bardhan & Ray, 2006). Generalizations are only made for highly similar cases in the same local context, for example communal agricultural irrigation systems in northern India (Meinzen-Dick et al., 2002). Anthropologist's view of commoners' motives is complex and multi-layered, as they understand actors as being socio-culturally embedded. Besides (economic) utility, cultural norms, local environment, social relations, power, and historic path-dependencies are considered (Bardhan & Ray, 2006). More than that, a commons is not seen as a fixed institution, "a resource or place, but rather a set of more-than-human, contingent relations-in-the-making that result in collective practices of production, exchange and living with the world" (Nightingale, 2019, p. 18). Commons hence have to be understood as an ever-evolving process, where not the institution and not even the resource is determined, but rather the needs of the commoners are focused on. Both schools of thought have critiqued the other view repeatedly, until the discourse has finally culminated in a novel view of 'commoning', in which both parties seem to have found common ground in the past years.

Commoning is a social practice and is thus described with a verb. Central to commoning are practices of (re)producing (social) relations, as they are understood as "relational social frameworks" (Bollier & Helfrich, 2015). Commoning fulfils certain social functions, such as democratic participation, sovereignty and community building (Bollier & Helfrich, 2019; Vivero-Pol et al., 2019). By practicing new relations which fulfil these functions, commoners jointly create the living environments they desire (Tummers & MacGregor, 2019). As commoning does not describe a property regime, but rather a way of relating to others and the environment, they can form from within traditional property regimes (Sato & Alarcón, 2019) and change their logic.

Practices of commoning have potential to promote a social-ecological transformation by challenging incumbent paradigms, such as growth orientation, which manifest themselves at the level of system structures in economy, science and technology, and at the level of the individual through a reductionist understanding of the human being as self-regarding and independent being. The research on "leverage points" for transformation (Meadows, 1999) suggests that the most effective sustainability interventions are paradigms, which describe the major intents and mindsets in the current global social-ecological system, and the power to transcend paradigms. Sievers-Glotzbach and Tschersich (2019) identified three central, interlinked paradigms, which have to be overcome for a transformation towards sustainability: 'materialistic culture and growth', 'control and autonomy of humans over nature', and 'expert knowledge and specialization'.

Practices of commoning challenge these paradigms by initiating political debate as anti-enclosure movements, and by living alternatives to these paradigms in societal niches (Euler, 2018; Gmeiner et al., 2020). As commoning is oriented towards needs satisfaction and good life, and creates new scopes for action outside the market, it opposes the 'materialistic culture and growth' paradigm. Commoning includes activities of "(re)produsage" (Euler, 2018), such as reproduction of the resource, including its environment and care for other commoners (unless it is a digital commons), which contradict the 'control and autonomy of humans over nature' paradigm. Contrarily, commoning re-connects humans to nature and others (ibid.). Practices of knowledge commoning, such as in the development of open source software or open educational resources, governed by principles of peer-production, challenge the 'expert knowledge and specialization' paradigm: They transcend the division between scientific experts and societal actors, and acknowledge the importance of different knowledge systems (Benkler & Nissenbaum, 2006).

Further, the dominating ideal of the human as being independent and autonomous indicates how the incumbent paradigms shape economic theory and practice (Bina & Vaz, 2011). Therefore, Becker (2012) proposes establishing an understanding of the human being as relational, interdependent, and virtuous person related to contemporaries, future generations and to nature, and calls for a redesign of societal structures that support the realization of these three 'sustainability relations'. Commoning can be viewed as an important way of enacting these sustainable relations, especially when it is based on a change of perspective at an onto-epistemological level, assuming humans as being cooperative, social, and fundamentally relational (Bollier & Helfrich, 2019). Commons have been found to be practices that (re)produce relationships between humans and between humans and nature, which build on relational virtues of responsibility and care (Sato & Alarcón, 2019, p. 37), and also promote the cultivation of political virtues (Benkler & Nissenbaum, 2006). These individual virtues are reflected in the shared values and value-producing practices of commoners, including solidarity, sovereignty, sharing, re-democratization, and social-ecological sustainability (Gmeiner et al., 2020; Hardt & Negri, 2011; Wall, 2017).

An example in which all of these elements become visible in practice around the world are Seed Commons. They describe institutions based on collective management practices in plant breeding and seed production, where the community conducts the handling, growing, breeding, and sharing of seeds and varieties in a needs-oriented and self-organized way (Sievers-Glotzbach et al., 2020). They span from traditional seed sharing networks to recent anti-enclosure movements that resist intellectual property rights on varieties. The social practices in Seed Commons support or re-establish three core relationships in the context of sustainability: First, the building of community values among the commoners, such as support of rural livelihoods or organic agriculture, is promoted. Second, the relation to past and future generations (of farmers/breeders) is strengthened by honouring varieties as cultural goods, in which the knowledge and practice of generations of farmers and breeders is accumulated, and by taking the responsibility for maintaining crop genetic diversity for future humans. A close relationship between farmers/breeders and nature is nurtured by practicing the breeding and maintenance of varieties in the field based on a holistic approach to agroecosystems, and by recognizing cultural and intrinsic values of seeds beyond its economic value. Overall, Seed Commons promote relational competencies and virtues and challenge fundamental paradigms of the current economic system, such as commodification of nature – therewith combining two essential ingredients for a social-ecological transformation.

To conclude, institutional-economic commons research studies commons mainly as an alternative property regime and governance arrangement, while anthropological commons research puts emphasis on the fundamental socio-cultural and environmental embeddedness of collective action and perceives commons as an alternative form of societal organization. Bringing together the commoning approach with the discussions about leverage points for sustainability transformation, we showed that commoning creates scope for establishing new understandings of social relationships within the present and with future generations as well as human-nature relationships. These form the basis for cultivating individual and collective competencies, virtues and values that strengthen agency for overcoming the Anthropocene's incumbent system paradigms and pave ways to more sustainable futures.

References

Bardhan, P., & Ray, I. (2006). Methodological approaches to the question of the commons. *Economic Development and Cultural Change, 54*(3), 655–676. https://doi.org/10.1086/500032

Becker, C. U. (2012). *Sustainability ethics and sustainability research*. Springer. https://doi.org/10.1007/978-94-007-2285-9

Benkler, Y., & Nissenbaum, H. (2006). Commons-based peer production and virtue. *Journal of Political Philosophy, 14*(4), 394–419. https://doi.org/10.1111/j.1467-9760.2006.00235.x

Bina, O., & Vaz, S. G. (2011). Humans, environment and economies: From vicious relationships to virtuous responsibility. *Ecological Economics, 72*, 170–178. https://doi.org/10.1016/j.ecolecon.2011.09.029

Bollier, D., & Helfrich, S. (2015). *Patterns of Commoning. Common Strategies Group.* Levellers Press.

Bollier, D., & Helfrich, S. (2019). *Free, fair, and alive: The insurgent power of the commons.* New Society Publishers.

Euler, J. (2018). Conceptualizing the commons: Moving beyond the goods-based definition by introducing the social practices of commoning as vital determinant. *Ecological Economics, 143*, 10–16. https://doi.org/10.1016/j.ecolecon.2017.06.020

Gmeiner, N., Sievers-Glotzbach, S., & Becker, C. U. (2020). New values for new challenges: The emergence of progressive commons as a property regime for the 21st century. *Ethics, Policy & Environment*, 1–21. https://doi.org/10.1080/21550085.2020.1848194

Hardin, G. (1968). The tragedy of the commons. *Science, 162*(3859), 1243–1248. https://doi.org/10.1126/science.162.3859.1243

Hardt, M., & Negri, A. (2011). *Commonwealth.* Harvard University Press.

Hess, C. (2008). Mapping the new commons. *SSRN Electronic Journal.* https://doi.org/10.2139/ssrn.1356835

Meadows, D. (1999). *Leverage points: Places to intervene in a system.* The Sustainabiliy Institute.

Meinzen-Dick, R., Raju, K. V., & Gulati, A. (2002). What affects organization and collective action for managing resources? Evidence from canal irrigation systems in India. *World Development, 30*(4), 649–666. https://doi.org/10.1016/S0305-750X(01)00130-9

Nightingale, A. J. (2011). Beyond design principles: Subjectivity, emotion, and the (Ir)rational commons. *Society & Natural Resources, 24*(2), 119–132. https://doi.org/10.1080/08941920903278160

Nightingale, A. J. (2019). Commoning for inclusion? Commons, exclusion, property and socio-natural becomings. *International Journal of the Commons, 13*(1), 16. https://doi.org/10.18352/ijc.927

Olson, M. (1965). *The logic of collective action: Public goods and the theory of groups.* Harvard University Press.

Ostrom, E. (1990). *Governing the commons: The evolution of institutions for collective action.* Cambridge University Press. https://doi.org/10.1017/CBO9780511807763

Sato, C., & Alarcón, J. M. S. (2019). Toward a postcapitalist feminist political ecology' approach to the commons and commoning. *International journal of the commons, 13*(1), 36. https://doi.org/10.18352/ijc.933

Sievers-Glotzbach, S., & Tschersich, J. (2019). Overcoming the process-structure divide in conceptions of social-ecological transformation. *Ecological Economics, 164*, 106361. https://doi.org/10.1016/j.ecolecon.2019.106361

Sievers-Glotzbach, S., Tschersich, J., Gmeiner, N., Kliem, L., & Ficiciyan, A. (2020). Diverse seeds – Shared practices: Conceptualizing seed commons. *International Journal of the Commons, 14*(1), 418–438. https://doi.org/10.5334/ijc.1043

Tummers, L., & MacGregor, S. (2019). Beyond wishful thinking: A FPE perspective on commoning, care, and the promise of co-housing. *International Journal of the Commons, 13*(1), 62.
https://doi.org/10.18352/ijc.918

Vivero-Pol, Jose Luis, Tomaso Ferrando, Olivier de Schutter, and Ugo Mattei. 2019. The food
commons are coming. In *Routledge handbook of food as a commons,* eds. Jose Lius Vivero-Pol,
Tomaso Ferrando, Olivier de Schutter, and Ugo Mattei, 1–21. Routledge.

Wall, Derek. 2017. The commons in history: Culture, conflict, and ecology. : The MIT Press.

Nina Gmeiner is a researcher in the RightSeeds group, also located at the University of Oldenburg.
Her background is in Sustainability Economics and she is currently writing her PhD-thesis about
the connection of well-being and ownership regimes of seeds. She focus her present research on
commons as a governance model and its potential for social-ecological transformation.

Stefanie Sievers-Glotzbach is Junior Professor for the "Economy of the Commons" at the
University of Oldenburg, Germany, and leads the research group RightSeeds. She is a Sustainability
and Environmental Scientist. Nina Gmeiner is a researcher in the RightSeeds group, also located
at the University of Oldenburg. Her background is in Sustainability Economics and she is currently
writing her PhD-thesis about the connection of well-being and ownership regimes of seeds. Both
authors focus their present research on commons as a governance model and its potential for
social-ecological transformation.

Ecological Citizenship

Michel Bourban

Abstract This article explores the main features of ecological citizenship and explains why this form of post-national citizenship is better adapted to facing current environmental issues than traditional forms of bounded citizenship. It draws on Andrew Dobson's *Citizenship and the Environment* (2003), one of the most sustained attempts to examine citizenship from an ecological perspective, but also suggests modifying and complementing this influential account using three approaches: cosmopolitanism, limitarianism, and the planetary boundary framework. These three elements could contribute to giving a fresh start to ecological citizenship, a notion that was much debated from the mid-1990s to the mid-2000s, but that has been gradually marginalized in discussions within citizenship theory.

Ecological citizenship appeared in the mid-1990s as a renewed and expanded notion of citizenship that would help humanity to face global environmental problems, such as anthropogenic mass extinction, climate change, and ozone depletion (e.g., van Steenbergen, 1994; Christoff, 1996; Smith, 1998; Barry, 1999, 2002; Dobson, 2003; Valencia Sáiz, 2005; Dobson & Valencia Sáiz, 2005; Hayward, 2006; Dobson & Bell, 2006). In contrast with traditional liberal and civic republican approaches, which operated under the assumption that humans organize themselves into bounded political communities, ecological citizenship questions the long-standing association between citizenship and nationality. Due to global environmental changes, the relevant political communities today are not only delimited by national borders, but also by the influence of our economic activities and political choices on distant strangers.

Andrew Dobson's *Citizenship and the Environment* is, to date, the most influential book-length exploration of the citizenship–environment connection (Dobson, 2003). Dobson argues that the ecological model of citizenship is:

The original version of the chapter has been revised. A correction to this chapter can be found at https://doi.org/10.1007/978-3-031-25910-4_279

M. Bourban (✉)
University of Twente, Enschede, Netherlands
e-mail: m.bourban@utwente.nl

Anthropocentric: obligations of ecological citizenship are owed only to other human beings.

Relationalist: obligations of ecological citizenship are based on the particular relationships shared by ecological citizens.

Post-national: the community of ecological citizens expands beyond national borders.

Intergenerational: the community of ecological citizens includes members of future generations.

Asymmetric: obligations of ecological citizenship are owed only by those who make unfair use of their ecological space.

To begin with, ecological citizenship is anthropocentric (1). The main reason for this is that duties of citizenship and citizenship rights are a matter of justice, and justice is about relationships between human beings. The community of ecological citizens is a community of justice, and can therefore only be composed of humans. This implies that our relationships with non-human beings cannot be citizenly, even if they can be humanitarian. Although some green political theorists, in early contributions, proposed expanding citizenship rights and duties of citizenship to non-human beings (Twine, 1994; Smith, 1998), the anthropocentric view has become the most widespread in the literature (see e.g. contributions in Dobson & Bell, 2006).

More precisely, ecological citizenship is about the relationships shared by people occupying the same ecological space, with each person's share of that space being defined by their ecological footprint (2). The political community of ecological citizens is not given, but created by their ecological footprint, that is, the quantity of natural resources and services they appropriate to sustain their consumption and production patterns. Since our ecological footprint contributes to global environmental changes, the political community of ecological citizens extends beyond national borders (3). Like all forms of citizenship, ecological citizenship is a community of strangers. Given that ecological problems such as climate change are not only global, but also intergenerational, we can be fellow citizens with strangers distant in both space and time (4). Obligations of ecological citizenship "extend through time as well as space, towards generations yet to be born" (Dobson, 2003).

But what is the content of these obligations? The principal obligation of the ecological citizen is to have a sustainable ecological footprint. Drawing on the Brundtland report (WCED, 1987), Dobson (2003) highlights that "the ecological citizen will want to ensure that his or her ecological footprint does not compromise or foreclose the ability of others in present and future generations to pursue options important to them." Those who occupy more than their fair share of the available ecological space have an obligation (a) to reduce their ecological footprint and (b) to compensate those impacted by their unsustainable ecological footprint. These obligations are only owed by those who have an excessive ecological footprint; those who already make fair use of their ecological space have fulfilled their ecological obligations (5).

In the context of ecological citizenship, changes in behaviours and in underlying attitudes are therefore as important as political participation in the decision-making process that determines the terms of social cooperation. Since both public (e.g. political choices) and private (e.g. consumption choices) contribute to individuals' ecological footprint, the private realm becomes a site of citizenship activity. One can be a good ecological citizen not only as a voter, an elector, or an activist, but also as a consumer, a producer, a parent, or a worker (Barry, 2006; Bourban, 2020).

To sum up, ecological citizenship represents an expansion of traditional accounts of citizenship on three counts: from the public to the private realm; from the national to the post-national community; and from present to future generations. The public realm, the national community and present generations remain relevant, but to become properly ecological, the notion of citizenship needs to be extended beyond these features.

Despite these promising conceptual innovations, relatively few political theorists are working on ecological citizenship today. After a decade of sustained development from the mid-1990s to the mid-2000s, ecological citizenship has been progressively marginalized from normative debates on citizenship. The notion therefore needs a fresh start to make it the centre of attention again. Three approaches could prove helpful with this in mind.

The first approach is cosmopolitanism, which also supports a distinctive model of citizenship. Just like ecological citizenship, cosmopolitan citizenship is (a) unbounded and (b) less about vertical relations between citizens and the state than about the horizontal relationships between citizens themselves. Surprisingly, Dobson went to great length to distinguish his model of post-national citizenship from cosmopolitan citizenship, claiming that ecological citizenship is "a specific instantiation and interpretation of post-cosmopolitan citizenship" (Dobson, 2003). He criticizes in particular the "thin community" of common humanity advocated by cosmopolitans, which he contrasts with the "thick community" of historical obligation he supports, which he believes is the only one that can lead to a truly political community. It is nevertheless possible to see these two forms of citizenship – ecological and cosmopolitan – as mutually reinforcing, making it possible to go beyond national citizenship, which remains dominant both in theory and in practice. In particular, a cosmopolitan account would ground the principal obligation of ecological citizenship in the universal right to a fair (whether sufficient or equal) ecological space. A recognition of this right would reinforce the idea that the obligation to adopt a sustainable ecological footprint is a duty of global and intergenerational justice (Hayward, 2006).

The second approach is limitarianism, a view of distributive justice according to which "it is not morally permissible to have more resources than are needed to fully flourish in life" (Robeyns, 2017). While sufficientarianism sets a lower threshold on the quantity of goods people should own to live a decent or flourishing life, limitarianism sets an upper threshold on the quantity of goods people should be allowed to possess. So far, limitarianism has focused on economic resources such as income and wealth. But wealth limitarianism can be complemented by ecological limitarianism, which sets an upper threshold on the amount of ecological resources individuals ought to appropriate. Limitarianism is directly related to the core commitment of ecological citizenship: both are based on the idea that individuals should adapt their lifestyles to the natural limits to economic and population growth. By setting an upper threshold on the legitimate use of natural resources and services, ecological citizenship makes it possible to develop the content of ecological limitarianism. The use of the natural capital by a moral agent should not be so great that, if it were to be universalized, it would foreclose the ability of others in present and future generations to pursue options important to them or to live a fulfilling life.

Where would the upper threshold be set by such an ecological limitarian account? This is where the third approach proves useful: the planetary boundary framework, which highlights nine biophysical thresholds that should not be crossed (Steffen et al., 2015). The disruption of overarching, planetary-level systems causes environmental changes taking us out of the stable environmental conditions of the Holocene. Four planetary systems have already been pushed beyond their critical limits, marking our shattering entry into the Anthropocene: the climate system, biodiversity, the nitrogen and phosphorus cycles, and land use. The planetary boundary framework also makes it possible to separate the ecological footprint into its different components, such as greenhouse gas emissions contributing to climate change and ocean acidification, fertilizer use contributing to the disturbance of the nitrogen and phosphorus cycles, pesticide and herbicide use together with fishing and hunting practices contributing to biodiversity loss, and so on. Each time, a corresponding planetary boundary represents an upper limit not to cross, an absolute line below which humans can produce and consume sustainably.

A restatement of the notion of ecological citizenship would represent two major advantages. On the one hand, a reconceptualized and updated notion of ecological citizenship based on cosmopolitanism, limitarianism, and the planetary boundary framework could contribute to putting this original model of post-national citizenship at the centre of debates in political theory once again. On the other hand, drawing on the conceptual innovations of ecological citizenship theorists, especially their extension of citizenship activity beyond the public sphere, together with their inclusion in the political community of strangers distant in both space and time, could allow scholars in climate ethics, environmental ethics, and global justice to find new normative resources to move the debates in their fields forward. Since the most pressing ecological issues, such as climate change and biodiversity loss are genuinely global and intergenerational, we are in urgent need of theories that expand key normative notions such as justice, responsibility, and citizenship beyond national borders and present generations.

References

Barry, J. (1999). *Rethinking green politics: Nature, virtue and Progress.* Sage.
Barry, J. (2002). Vulnerability and virtue: Democracy, dependency, and ecological stewardship. In B. A. Minteer & B. P. Taylor (Eds.), *Democracy and the claims of nature* (pp. 133–152). Rowman & Littlefield.
Barry, J. (2006). Resistance is fertile: From environmental to sustainability citizenship. In A. Dobson & D. Bell (Eds.), *Environmental citizenship* (pp. 21–48). The MIT Press.

Bourban, M. (2020). Ethics, energy transition, and ecological citizenship. In *Reference module in earth systems and environmental sciences* (pp. 1–17). Elsevier.

Christoff, P. (1996). Ecological citizens and ecologically guided democracy. In B. Doherty & M. de Geus (Eds.), *Democracy and green political thought: Sustainability, rights and citizenship* (pp. 151–169). Routledge.

Dobson, A. (2003). *Citizenship and the environment*. Oxford University Press.

Dobson, A., & Bell, D. (2006). *Environmental citizenship*. MIT Press.

Dobson, A., & Valencia Sáiz, A. (2005). *Citizenship, environment, economy*. Routledge.

Hayward, T. (2006). Ecological citizenship: Justice, rights and the virtue of resourcefulness. *Environmental Politics, 15*(3), 435–446. https://doi.org/10.1080/09644010600627741

Robeyns, I. (2017). Having too much. In J. Knight & M. Schwartzberg (Eds.), *NOMOS LVI: Wealth. Yearbook of the American Society for Political and Legal Philosophy*. New York University Press.

Smith, M. J. (1998). *Ecologism: Towards ecological citizenship*. Open University Press.

Steffen, W., Richardson, K., Rockström, J., Cornell, S. E., Fetzer, I., Bennett, E. M., Biggs, R., et al. (2015). Planetary boundaries: Guiding human development on a changing planet. *Science, 347*(6223), 1259855. https://doi.org/10.1126/science.1259855

Twine, F. (1994). *Citizenship and social rights: The interdependence of self and society*. Sage.

Valencia Sáiz, A. (2005). Globalisation, cosmopolitanism and ecological citizenship. *Environmental Politics, 14*(2), 163–178. https://doi.org/10.1080/09644010500054848

van Steenbergen, B. (1994). Towards a global ecological citizen. In B. van Steenbergen (Ed.), *The condition of citizenship* (pp. 141–152). Sage.

WCED. (1987). *Our Common Future*. Oxford University Press.

Michel Bourban is a Postdoctoral Researcher at the University of Twente. He works on the project "Cosmopolitan Citizenship and Ecological Citizenship", funded by a SNSF research grant. He holds a PhD in Philosophy from the University of Lausanne and Paris-Sorbonne University (Paris IV). His work on climate justice, climate ethics, ecological citizenship, effective altruism, and innovation ethics has been published in journals such as *Ethics, Policy & Environment*, *Philosophy & Technology*, and *De Ethica*, and in edited volumes published by Springer, Routledge, Elsevier, and Göttingen University Press. He has also published two books on climate justice and climate ethics with the French publishers PUF and Vrin.

Environmental Ethics

Daphnée Valbrun

Abstract This article is a reflection on the serious environmental challenges (global warming, rampant pollution, etc.) that humanity faces today in order to contribute to an improvement of our own well-being and that of others. In other words, it questions the role of environmental ethics in protecting nature in order to improve the conditions of human existence. Indeed, today, our greatest challenge seems to be how to adjust our actions towards human dignity so that a kind treatment nature that can be beneficial to human can remain the goal of all human actions. Hence the importance of environmental ethics in directing human action towards a greater social responsibility.

Humanity is facing an urgent and serious environmental problem to which viable and effective solutions must be found to diminish the damage caused to the environment. For several decades, we have been witnessing a strong industrial development and an extreme growth of the world population which are not without consequences for the planet. This is accompanied by the exploitation and degradation of our natural resources and a massification of toxic waste that leads to the increase of CO_2 in the atmosphere. Add to this, several million people who face, on a daily basis, precarious environmental conditions: droughts, small living space for each human being, soil erosion and infertility, rising sea levels, floods, tropical storms, etc. This situation is a great concern especially for the poorest groups due to their limited access to resources.

Wallenhorst affirms: 'The warmer climate, the denser the atmosphere in CO_2 and methane, the more acidic oceans and the surface area they cover will leave lesser land for human. The smaller the ecumene, the sharply reduced biodiversity, the more frequent and higher the heat waves in which regions become uninhabitable for humans' (Wallenhorst, 2020, p. 45). This quote shows that the consequences of most of our actions on the planet are serious. However there is a need for our contemporaries to be aware of this in order to protect and keep the environment safe.

D. Valbrun (✉)
Catholic University of the West, Angers, France

© The Author(s), under exclusive license to Springer Nature
Switzerland AG 2023
N. Wallenhorst, C. Wulf (eds.), *Handbook of the Anthropocene*,
https://doi.org/10.1007/978-3-031-25910-4_169

This situation creates in many people great anxiety about the future of the planet. Therefore, it is time for humans to assume their responsibilities and to be more involved in environmental protection projects because some of the damage caused by us is sometimes comparable to that caused by nuclear weapons. It is from this perspective that Furio argues: 'Only nuclear weapons approach such definitional criteria, even if we take into account the differences between these two threats, which together represent the end of modern political rationality and the transition to a new and unknown era' (Cerutti, 2008, pp.107–108). The author shows through this quotation, that the situation is serious and requires an awareness of everyone before the planet is completely destroyed by technical and technological progress.

In order to make us aware of this situation, there has been much interesting researchlike that of the biochemist Crutzen (1995), Grivenald (2007), Wallenhorst, (2016), Hétier (2021). The wish of these authors is that people become aware of the harmful impacts of their actions on the whole planet and that they act in a healthier way for the environment. Unfortunately, in spite of the numerous illuminations brought by the scientific literature that fights for environmental ethics, some cold, stubborn researchers make speeches to show the importance of Homo Oeconomicus to the detriment of utilitarianism without underlining the deadly consequences of the latter. Could we say that these authors are acting out of ignorance? At least, would they have particular interests that push them to have this representation of these destructive forces? According to Barrère and Martuccelli, it is an 'unheard-of possibility of exploration of new dimensions for the individual and an unsuspected risk of misguidance and splintering of the self in an unrelated series of experiences' (Barrere, 2005P. 60). Probably, these splinters of self and experiences that modernity produces could have made sense if it was not for an economic aim to the detriment of our own self and of the other.

This will to promote the benefits of Homo Oeconomicus is not only found among academics, it is also evident among certain politicians who try to lull us into a false sleep with honeyed but dangerous speeches. Indeed, some so-called great power and capitalist countries show that the protection of the environment is a major preoccupation, because it is a serious political and moral problem that can destroy the planet and its components. Indeed, their speech makes feel a real will to change the order of things. However, the practice is quite different, because they are the biggest exploiters, degraders and polluters of the environment. They implement public policies that do not solve the real problems of the environment. If we apply the legal formula that the author of the crime is the one who benefits from the crime, we can believe that it is the will of a state not to change the order of things because they make big money at the expense of the most vulnerable.

This raises the question of what the purpose of politics is. Should politics have a destructive purpose or personal interests at the expense of the community? What they may not understand is that the hour is grave and 'it is no longer a matter of political opinion, but of pure necessity' (Zin, 2017, p.27). This necessity requires swift action and must impose ethical action in order to provide viable solutions to the environmental crisis that threatens the planet. It is also a state of emergency that we are facing that deserves to have a global awareness of the threats that currently face humans and natural species. Thereby the interest of an ethical action towards the environment. What then is environmental ethics?

According to Larrère, 'Environmental ethics exists as a philosophical reflection that has been able to combine classical moral questions (what is value? how do we distinguish between good and evil? is pluralism necessary?) and contemporary problems that make nature the object of philosophical debate' (Larrère, 2006, p.75). Developed in America from the twentieth century onwards, the question of environmental ethics was the subject of reflection by many researchers such as Peter Singer, Tom Rigan, Leopold Aldo, etc. Indeed, these authors emphasize the importance of moral dignity and the place of the human being in nature. Through his text entitled Ethical Practice published in 1979, Peter Singer postulates that contrary to the Kantian perception, human dignity is only a question of sensibility and not of reason. He even qualifies as speciesism any behaviour that prioritizes human interests over those of animals. Indeed, the human being would not be superior to any other natural species and must adopt a respectful behaviour towards them.

Tom Rigan brings out another dimension in the relationship between humans and nature: that of obligatory cohabitation. He places humans and animals in the same biotic community with a common vision. This vision is that of wanting to build its future, with the only difference, the animal would not be a being *rationae* and the human as animal endowed with reason should act more with dignity towards the natural species. This dignity is advocated today through the environmental ethics that advocate a better treatment of nature because it should not be a means of destruction but we must confer moral rights.

Today, environmental issues and their consequences call for important social transformations (acting and thinking differently). Therefore, thinking an environmental ethics should be a fundamental reflection that allows us to establish norms, rights and human duties towards nature. It refers to a triple relation: relation of oneself to the other, relation of oneself to the cosmos, relation of oneself to life. This triple relation results both directly or indirectly from the modes of relations that we establish between ideologies, cultures, human actions and environment. It must necessarily analyze the relations between nature and humans in a reflexive way by considering the proper needs of both while taking into account their impacts in space and in time with regard to moral values. In this perspective, we can ask ourselves, how can we think of an environmental ethics by emphasizing the citizen's responsibility?

Protecting the environment is the duty of all citizens in view of the well-being of all. This is why we all have a responsibility in the current environmental crisis. Today, the underdeveloped countries have heavy debts that keep them in a situation of underdevelopment due to the fact that their wealth is very often plundered by the so-called great powers, or at least poorly managed at the internal level. Moreover, the debts are accompanied by many threats to the planet, such as the environmental crisis, which is creating many ecological inequalities in the world. These situations should make us aware of a debt towards ecology. That of a protection and humility towards the environment. This environmental humility is recognition of one's own limits' (Goffi, 2009, p.165). For the freedom, that most actors feel towards nature is destructive. We must recognize our limitations despite all the great human achievements. Citot Vincent points out that 'freedom is a power of denaturation' (Citot, 2005, p. 37). So he adds, 'man has to do something with what nature has made of him' (Ibid, p. 37). This thing that humanity must do in my opinion is respect the environment more. This makes it important to think about environmental issues from the ethical perspective of human rights in order to consider the future of future generations. Since rights are universal and inherent, all individuals must have the same rights from birth. This imposes on the present generations to preserve the conditions to allow the future generation to exercise these rights in good condition.

Thus, environmental ethics is of paramount importance. It finds its meaning and relevance in a certain awareness of the damage that can be caused to the environment. To avoid that this damage has repercussions on future generations, a more respectful behaviour towards nature is necessary. Thinking of an environmental ethic with citizen responsibilities is an effort to create and justify a new relationship between humans and nature.

In conclusion, a universal ethic of environmental protection is vital to save the planet and the self of the future. This ethic must be built on a civic duty with both personal and collective responsibilities. It must not be built around an anthropocentric vision of humans because human life is threatened, but must also be built with a view to justice and respect for other beings in nature who also have a life like that of humans. An ethical vision of the environment invites us to go beyond our egoism to think about the future of the planet. Education appears today as the best way to promote this new orientation of human life based on the respect of the human being and their environment.

References

Barrere, A. e. D. M. (2005). La modernité et l'imaginaire de la mobilité: 'inflexion contemporaine. *Cahiers internationaux de sociologie, 118*(1).

Cerutti, F. (2008). Le réchauffement de la planète et les générations futures. *Pouvoirs, 127*(4).

Citot, V. (2005). Le processus historique de la Modernité et la possibilité de la liberté (universalisme et individualisme). *Le Philosophoire, 25*(2), 35–37.

Crutzen, P. J., & Graedel, T. E. (1995). *Athmosphère, climate, and change*. Scientific American library.

Goffi, J. (2009). *L'éthique des vertus et l'environnement*. Multitude N°36.

Grivenald, J. (2007). *Le contexte immédiat de la création de l'écologie intégrale*. Paris Etablie.

Hétier, R. (2021). *L'humanité contre l'anthropocène*. Paris PUF.

Larrère, C. (2006). *Ethique de l'environnement'*. Mulitude N°24.

Rist, G. (2010). *L'Homo oeconomicus: un fantôme dangereux, L'économie ordinaire entre songes et mensonges, sous la direction de Rist Gilbert*. Presses de Sciences Po.

Wallenhorst, N. (2016). *Politique et éducation en anthropocène*. Raison politique N°62.

Wallenhorst, N. (2020). Quel type de citoyenneté en Anthropocène? *Le Télémaque, 58*(2).

Zin, J. (2017). L'Anthropocène nous rend responsables du monde. *EcoRev, 44*(1).

Daphnée Valbrun is in her first-year of PhD in Educational Sciences at the Catholic University of the West and Strasbourg University. She is a lecturer in the department of Education Sciences at Catholic University of West. Her main research topics are the anthropocene, citizenship, educational pedagogies, autonomy and experience in education.

Environmental Protection

Augustine Pamplany and Bert Gordijn

Abstract The beginning of environmental consciousness and policy discussions on environmental protection can be traced back to the nineteenth century. Since the second half of the twentieth century, most policies and action plans towards environmental protection evolved under the leadership of the UN. While the early policies focused on sustainable development, recent policy discussions have moved beyond this, drawing on the idea of ecological resilience. Environmental protection initiatives face several ethical challenges.

The Rise of Environmental Consciousness

The framing of policies to preserve a healthy environment can be traced back to the nineteenth century, particularly in the US. The Yellow Stone National Park Protection Act – passed by the US Congress in 1872 – is considered as one of the first laws to protect the environment (US Statutes, 1872:22–23). It was meant to reserve a tract of land near the Yellowstone River, known for its wildlife and geothermal features, as a public park. Subsequently, the Sierra Club, an environmental organization founded in California in 1892, and the National Adudbon Society, a large environmental group, founded in 1905 in the US, were further seminal moves towards environmental protection.

The first half of the twentieth century saw at least two major developments reflecting the rising environmental consciousness. First, in 1933, the Convention Relative to the Preservation of Fauna and Flora in their Natural State, popularly known as the London Convention, was drafted. This agreement for the conservation

A. Pamplany (✉)
Institute of Advanced Interdisciplinary Studies, Aluva, India
e-mail: admin@iaiss.com

B. Gordijn
Institute of Ethics, Dublin City University, Dublin, Ireland
e-mail: bert.gordijn@dcu.ie

of nature, signed by Belgium, Italy, the UK, British India, Egypt, Anglo-Egyptian Sudan, Union of South Africa, Tanganyika and Portugal, was described as the *magna carta* of wildlife conservation (Boardman, 1981:34). Second, in 1948, the International Union for Conservation of Nature and Natural Resources was founded.

The second half of the twentieth century saw a series of disasters and diseases related to environment, which gave extra impetus to the rising public awareness of environmental issues. DDT, whose insecticidal traits had been discovered in 1939, increasingly showed its nefarious impact on the environment. Massive bird kills occurred from DDT spraying at Cape Cod in 1958 (Whitney, 2012), which alerted about the adverse impacts of pesticides on environment. Furthermore, the 1950s saw the Minmata disease, a neurological disease caused by mercury poisoning in Minmata in Japan. Rachel Carson's *Silent Spring* (Carson, 1962) was a trend setting publication that described both the environmental dangers from the indiscriminate use of pesticides as well as the Minmata disease. Subsequently, more research was conducted focused on the environment. Acid rain was first detected at Hubbard Brook in North America in the mid-1960s. It resulted from the emission of sulphur dioxide and nitrogen oxides into atmosphere (National Science Foundation, 2012). In 1985, Farman and colleagues published their paper about the loss of ozone in the atmosphere over the Antarctic (Farman et al., 1985). The rise of global media enabled better featuring of disasters such as famines in sub-Saharan Africa, the leak of poisonous gas at Bhopal in India in 1984, and the nuclear disaster at Chernobyl in 1986. Further studies showed a sharp increase in the disasters directly related to environment, such as droughts and floods attributed to development mismanagement (Brundtland, 1987:16).

Rising awareness of environmental issues gave birth to a series of new regulations concerning clean water and air. The US Clean Water Act in 1948 and 1972, and the Clean Air Act of 1963 and 1970 are examples. Sustainable development, conceptualized as "development that meets the needs of the present without compromising the ability of future generations to meet their own needs" (Brundtland, 1987:16), became a *mantra* in environmentalism.

UN Programmes for Environmental Protection

With the global surge in environmental consciousness a series of international conferences, some of them popularly known as the earth summits, were organised by the UN. Famous examples are the 1972 Conference on the Human Environment in Stockholm, the 1982 Earth Summit in Nairobi, the 1992 Conference on Environment and Development in Rio, the 2002 World Summit on Sustainable Development in Johannesburg, the 2009 Climate Change Conference in Copenhagen, the 2012 Conference on Sustainable Development in Rio, and the 21st Conference of the Parties in Paris (2015).

These and other UN conferences produced a series of policy documents and treaties such as the UNEP 1972, the CBD 1972, the Brundtland Report 1987, the Rio

Declaration 1992, UNFCC 1994, the Kyoto Protocol 1995, and the Paris Agreement 2015, thus providing a practical framework for sustainable development by advocating policies and regulations, particularly for emission control, culminating in the **Paris agreement with the** major goal to limit global warming to 2 degree Celsius and ideally to **1.5 degrees Celsius,** compared to pre-industrial levels.

Beyond Sustainable Development

Academic discussions on environmental protection have moved beyond the model of sustainable development towards new ideas. Consider the example of 'ecosystem resilience,' which has been gaining prominence in shaping environmental policies (Gunderson, 2000; Gibbs, 2009). The notion of ecological resilience introduced by Holling (1973) refers to the capacity of an ecosystem to return to equilibrium after disturbances, examples of which are loss of biodiversity, exploitation of natural resources, and climate change.

The resilience approach calls for a paradigm-shift from maximum sustainable yield to environmental resource management – the management of the interaction and impact of human societies on the environment. Sustainable development was understood quite abstractly in the Brundtland report. The post-Brundtland environmental thinkers became more interested in the analysis of the particularities of conceptualizing and practicing sustainable development in real world contexts (see, for example, Costanza & Patten, 1995; Costanza & Daly, 1992).

One of the emphases in post-Brundtland thinking on development was to distinguish between weak and strong sustainability. Weak sustainability, a concept in environmental economics introduced by Robert Solow (1993), holds that human capital can replace natural capital. Human capital covers skills, knowledge and infrastructure. Natural capital refers to natural resources such as fossil fuels, biodiversity, and water. Weak sustainability permits the decline of natural resources if it leads to sufficient increase in human capital. For example, a natural resource like coal may be used to produce electricity, which in turn increases the quality of life. In contrast, strong sustainability holds that one form of capital cannot be substituted for another form (Ang & Van Passel, 2012). In strong sustainability, natural capital should be maintained independent of human capital (Costanza & Daly, 1992).

The resilience approach criticizes weak sustainability as producing extreme sensitivity to natural disturbances (Van den Bergh, 2007). Weak sustainability is inadequate for development as it does not take account of the "limits to earth's resilience" (Ross, 2008:32). Thus, the resilience approach requires strong sustainability, which emphasizes the integrity of all ecological systems. Resilience becomes a major tool of ecological integrity along with stability and diversity of the ecosystems. The resilience approach also pushes forward the resilience of the vulnerable human populations as a critical factor in environmental policy making. Ross calls for a "shift towards ecological sustainability" (Ross, 2008:32) to replace the sustainable development approach.

Ethical Issues

Leading issues in contemporary environmental ethics are challenges to anthropo-centrism, disenchantment of nature, rights of non-human species, intergenerational equity, ethics of development, and distributive and procedural justice (Brennan & Lo, 2021). The huge divide between the lower and higher income countries deserves special mention. While environmental disaster is largely caused by the developmental patterns of the rich nations, its brunt is significantly pushed upon the poor countries that are less responsible for the environmental problems. This recognition led the Kyoto protocol (1992) to underscore the principle of "common but differentiated responsibility and respective capabilities" (Kyoto Protocol, 1992:4). The Brundtland report had already observed that "a world in which poverty and inequity are endemic will always be prone to ecological …crises" (Brundtland, 1987:30). Unfortunately, many prevalent policies of environmental protection are vulnerable to compounding the environmental injustice already meted out to the global poor (Islam & Winkel, 2017:149; Preston, 2016). Concepts like environmental racism and discrimination are invoked regarding the selection of sites for hazardous industries and waste disposal (see, for example, Hamilton, 1995). Just policies of environmental protection require complying with the principles of fair treatment, involvement of all peoples, and fair and just distribution of environmental benefits and harms (World Social Report, 2020).

References

Ang, F., & Van Passel, S. (2012). Beyond the environmentalist's paradox and the debate on weak versus strong sustainability. *Bioscience, 62*(3), 251–259.

Boardman, R. (1981). *International organization and the conservation of nature*. Macmillan.

Brennan, A., & Lo, Y.-S.. (2021). Environmental ethics, *the Stanford encyclopedia of philosophy* (Fall 2021 edition), Edward N. Zalta (ed.). https://plato.stanford.edu/archives/fall2021/entries/ethics-environmental/

Brundtland, G. H. (1987). Our common future—Call for action. *Environmental Conservation, 14*(4), 291–294.

Carson, R. (1962). *Silent Spring*. Houghton Mifflin Harcourt.

Costanza, R., & Daly, H. E. (1992). Natural capital and sustainable development. *Conservation Biology, 6*(1), 37–46.

Costanza, R., & Patten, B. C. (1995). Defining and predicting sustainability. *Ecological Economics, 15*, 193–196.

Farman, J. B., Gardiner, B., & Shanklin, J. (1985). Large losses of total ozone in Antarctica reveal seasonal ClO_x/NO_x interaction. *Nature, 315*, 207–210.

Gibbs, M. T. (2009). Resilience: What is it and what does it mean for marine policymakers? *Marine Policy, 33*(2), 322–331.

Gunderson, L. H. (2000). Ecological resilience – In theory and application. *Annual Review of Ecology and Systematics, 31*, 425–439.

Hamilton, J. T. (1995). Testing for environmental racism: Prejudice, profits, political power? *Journal of Policy Analysis and Management, 14*(1), 107–132.

Holling, C. S. (1973). Resilience and stability of ecological systems. *Annual Review of Ecology and Systematics, 4*, 1–23.

Islam, S. N., & Winkel, J.. (2017). *Climate change and social inequality.* UN Department of Economic & Social Affairs, 149.

Kyoto Protocol. (1992). https://unfccc.int/resource/docs/publications/08_unfccc_kp_ref_manual.pdf

National Science Foundation. (2012). *Acid rain: Scourge of the past or trend of the present?* https://www.nsf.gov/discoveries/disc_summ.jsp?cntn_id=124955

Preston, C. (2016). *Climate justice and geoengineering: Ethics and policy in the atmospheric Anthropocene.* Rowman and Littlefield.

Ross, A. (2008). Modern interpretations of sustainable development. *Journal of Law and Society, 36*(1), 32.

Solow, R. M. (1993). An almost practical step towards sustainability. *Resources Policy, 16*(3), 162–172.

U.S. Statutes at Large. (1872). Vol. 17, Chap. 24, pp. 32–33.

Van den Bergh, J. (2007). In G. Atkinson, S. Dietz, & E. Neumayer (Eds.), *Handbook of sustainable development.* Edward Elgar.

Whitney, C. (2012). The silent decade: Why it took ten years to ban DDT in the United States. *The Virginia Tech Undergraduate Historical Review, 1.* https://doi.org/10.21061/vtuhr.v1i0.5

World Social Report. (2020). *Inequality in a rapidly changing world*, by UN Department of Economic and Social Affairs.

Augustine Pamplany obtained a PhD in Ethics from Dublin City University with a thesis on the Justice in Geoengineering. He teaches Ethics, Science and Religion, Philosophy of Science, Indian Philosophy, and Scientific Cosmology at various institutions in India. Augustine has published seven books and several peer-reviewed articles covering the same areas.

Bert Gordijn is Full Professor of Ethics and Director of the Institute of Ethics at Dublin City University in Ireland. He has an extensive record of books, edited volumes, peer-reviewed publications and international lectures on a broad range of topics such as Bioethics, Disaster Ethics, Technology Ethics and Research Ethics.

Equity and Equality

Benjamin Lewis Robinson

Abstract The Anthropocene stresses existing inequalities among humans and human communities, broadens the scope of these questions across time, space, species, and the rest of nature, and emerges as an epoch in which the concept of 'equivalence' is in crisis. This article first considers intra-human inequality, sketching the genesis of environmental and climate justice movements before indicating certain conceptual problems with equity in international climate diplomacy. Focus then shifts to measures of inequality and criticism of the dominant economic model that assumes that all values can be monetized. Insofar as questions of inequality remain concentrated on humans, such concerns, however important, do not measure up to the rift in orders of scale that the Anthropocene brings about. A politics of the Anthropocene oriented to climate justice will have to negotiate the incommensurable scales that define the epoch.

In lectures titled *The Human Condition in the Anthropocene* (2015), Dipesh Chakrabarty addressed the phrase 'common but differentiated responsibilities' that since the Rio Earth Summit in 1992 has been a touchstone in international climate diplomacy. He argues that the terms 'common' and 'differentiated' refer to distinct concepts of humanity in the Anthropocene: what is 'common' is humankind as a species (*anthropos*), part of the history of life on the planet and recently a geophysical force; what is 'differentiated' are the humans (*homo*) of the human sciences, history, and politics. These two aspects are irrevocably entangled but not commensurable. However urgent and important, demands for 'climate justice', insofar as they are concerned with social, economic, and demographic inequalities among humans, remain confined to the perspective of *homo*. They have not yet come to terms with the reality of the Anthropocene in which *anthropos*, humanity as a species, has become an ambivalent agency in the earth system. I will return to Chakrabarty's claims at the end of this article to situate them in a general crisis of

B. L. Robinson (✉)
NYU, New York, NY, USA
e-mail: blr4141@nyu.edu

equivalence provoked by the Anthropocene. But first, one can accept his distinction and nonetheless argue that the advent of *anthropos* on the natural-political scene has exacerbated the inequalities of *homo*.

That the Anthropocene has stressed and extended inequalities among humans has long been acknowledged. In a vicious cycle, poorer countries and disadvantaged populations in all countries are disproportionately affected by climate related hazards because they are more exposed, more vulnerable, and least able to adapt. Owing to this multiplier effect of climate change on existing inequalities, the politics of the Anthropocene is inextricably entangled with questions of social and global justice.

With the rising awareness of anthropogenic changes to earth systems, the prejudice that saw an antagonism between socio-economic and environmental concerns is no longer tenable. On the contrary, one can trace a convergence of these concerns going back to the coining of the phrase 'environmental justice' at the first National People of Color Environmental Leadership Summit in Washington, DC (1991). The environmental justice movement in the United States emerged in response to racialized inequalities in hazardous exposure to toxic environmental harm. It can be related to global grassroots movements addressing the social, economic, and demographic disparities exacerbated by the 'slow violence' (Nixon, 2011) of environmental degradation. These intersectional movements including ecofeminism and indigenous resistance can be collected under the rubric 'environmentalism of the poor' (Guha & Martínez-Alier, 1997).

In the same years, the movement for 'climate justice' emerged primarily out of countries of the so-called developing world. This postcolonial critique argued that climate policy that did not acknowledge historical disparities in emissions and distinguish between 'survival' and 'luxury' emissions was in effect 'environmental colonialism' (Agarwal & Narain, 1991). The poor were not only being denied the 'freedoms' afforded by economic development that the rich, especially in rich nations, enjoyed, but made to pay for them as well. In response, demands were made for the acknowledgement of 'climate debt' in the name of restorative justice (World People's Conference, 2010). A further concern relates to 'just transitions' when enacting environmental measures, to ensure that particular communities historically dependent on, for example, carbon-intensive industries like coal mining do not suffer disproportionately. Climate justice was thus from the start related to calls for global social justice on the basis that ecological and social issues must be addressed in tandem. Similar sentiment propelled the rise of 'sustainable development' in international discourse.

In recent years, 'climate justice' has been adopted by youth activists to refer to intergenerational justice. While 'school strikes' have problematically shifted the focus back to Europe and North America, their use of the postcolonial phrase 'climate justice' can be legitimized on the basis that environmental exploitation in the present is a form of imperialism of the future. It has potentially devastating consequences for generations of people, distant in time, whom those responsible will never encounter in person. Inequalities in the Anthropocene have to be thought on a scale previously unheard of across space and time in ways that expose the limits of conventional politics. Additionally, the Anthropocene has radicalized critiques of anthropocentrism, prompting a shift in political sensibilities. 'Multispecies justice' attends to inequities affecting other beings and species.

As a planetary problem, the politics of the Anthropocene has focused on international diplomacy, which in turn has focused on climate change. While a 'principle of equity' has been invoked as a cornerstone of climate negotiations, they have been disappointing in practice. Here the technical-scientific question of how to measure differences in historical and ongoing emissions and their offsets meets the ethical-political question concentrated in the phrase 'common but differentiated responsibility'. The conceptual questions posed by these negotiations can be schematized in various ways (Boran, 2018; Jamieson, 2014; Gardiner, 2011). Steve Vanderheiden distinguishes between equity- and responsibility-based approaches (2008). The former includes 'resource sharing', where carbon is treated as a common resource, and 'cost-sharing', where the burdens of abatement are equally shared. While equity approaches are future-oriented, responsibility addresses past inequalities on the 'polluter pays' principle. Further adjustments involve 'survival emissions', 'ability to pay' and a fair 'baseline' date.

North-South inequities have been further exacerbated by a tendency to emphasize mitigation over adaptation. Mitigation involves reducing or removing greenhouse gases, while adaptation is about protecting populations from the harmful effects of climate change. The focus on mitigation is seen as freeze-framing current inequalities. Any equitable international policy would involve aiding poor communities to adapt to ecological changes that they are not responsible for so as to improve their welfare rather than compound their misery. A solution may lie in 'decarbonizing development' (Roberts & Parks, 2006), although affected communities increasingly advocate more radical alternatives to development drawing on gendered, indigenous, and decolonial forms of life (Kothari et al., 2014; Shiva, 1989).

And then there is the problem of the level of analysis: climate negotiations take place between states, so how should the responsibility of corporations, municipalities, and individuals be accounted for? To politicize these questions different measures have been proposed, most influentially the 'ecological footprint' which gauges the 'ecological space' appropriated by different actors. Efforts have been made to translate ecological costs, historically discounted as 'externalities', into the economy by monetizing 'ecosystem services'. The Stern Review by the former Chief Economist of the World Bank was key to generating acknowledgement in corporate and policy circles of the significance of climate change by providing a calculation of its economic costs. It also brought into focus a central contention of intergenerational policy, the 'social discount rate' by which the present value of future outcomes is calculated.

One can get the impression that climate justice is ultimately a monumental cost-benefit analysis distributing risks and resources. But this overlooks significant inequalities that are even harder to measure. Complementing the demand for distributive justice are concerns about inequalities in recognition, capabilities, and participation (Schlosberg, 2007). Nonetheless, climate justice has arguably become caught up in an economic reckoning that may be at cross-purposes with its project. The movement gathered steam in a period of financialization of the global economy and the 'neoliberal' extension of the logic of the market to all areas of human concern. It is striking therefore that already the earliest formulations of environmental justice sought explicitly to avoid the translation of just principles into monetary value (Harvey, 1996). While monetary calculations remain on the agenda, the climate movement has also contributed to a radical interrogation of the economic assumptions that undergird the logic of unlimited growth in the name of a finite conception of care for and belonging to the planet.

On the fringes of the discipline, ecological economists are advocating 'degrowth'. Contesting the assumption that all values can in principle be converted into monetary ones, they propose participatory 'multi-criteria evaluation' methods that seek to achieve equitable outcomes while acknowledging a plurality of distinct value systems (Martínez-Alier, 2012). Marxist critics take this suspicion of economic value further. Jason Moore argues that value-production in capitalism depends on a cheapening of nature, the devastating effects of which go structurally unaccounted for (Moore, 2015). As capitalism, for such critics, is the root of the problem, there can be no solutions within capitalism. Politics needs to be recalibrated beyond the economic calculus that is arguably a contributing factor to the Anthropocene.

To return to Chakrabarty's distinction, it can now be reformulated as a problem of scale: the politics of *homo* do not measure up to the dynamics of the earth systems in which *anthropos* is implicated. Chakrabarty describes this split condition as the 'epochal consciousness' of the Anthropocene, a consciousness that precedes any particular form of identification or belonging. This is also a consciousness of radically incommensurate scales. Some have argued that faced with the Anthropocene conventional ethics and politics are no longer possible because they continue to rely on values scaled to humans (*homo*) and so risk banalizing the 'Earth-shattering' realities at stake (Hamilton et al., 2015, 8).

Rather than abandoning inherited scales, it may be more promising to multiply them to include an array of forms of knowledge and experience and draw on the insights afforded by indigenous cosmologies, subjugated knowledges, and a deep archive of historical cultural forms (Horn & Bergthaller, 2019). Recent approaches to the epistemological problem of scale in the Anthropocene have abandoned the 'Russian doll model', according to which each order of scale is nested inside the next, for a messier model of 'sprawling scales' sensitive to 'friction between experiences, perspectives and values' (Thomas et al., 2020, 11). The project for equitable politics in the Anthropocene will be about negotiating multiple incommensurable scales for the sake of a varied conception of climate justice.

References

Agarwal, A., & Narain, S. (1991), Reprinted 2003. Global warming in an unequal world: A case of environmental colonialism. Centre for Science and Environment.

Boran, I. (2018). On inquiry into climate justice. In T. Jafry, M. Mikulewicz, & K. Helwig (Eds.), *Routledge handbook of climate justice* (pp. 26–41). Routledge.

Chakrabarty, D. (2015, February 18–19). *The human condition in the Anthropocene*. Tanner lectures in human values, Yale University.

First National People of Color Environmental Leadership Summit. (1991, October 24–27). *Principles of Environmental Justice*.

Gardiner, S. (2011). *A perfect moral storm: The ethical tragedy of climate change*. Oxford University Press.

Guha, R., & Martínez-Alier, J. (1997). *Varieties of environmentalism: Essays north and south*. Routledge.

Hamilton, C., Bonneuil, C., & Gemenne, F. (Eds.). (2015). *The Anthropocene and the global environmental crisis: Rethinking modernity in a new epoch*. Routledge.

Harvey, D. (1996). *Justice, nature and the geography of difference*. Blackwell.

Horn, E., & Bergthaller, J. (2019). *The Anthropocene: Key issues for the humanities*. Routledge.

Jamieson, D. (2014). *Reason in a dark time: Why the struggle against climate change failed – And what it means for our future*. Oxford University Press.

Kothari, A., Demaria, F., & Acosta, A. (2014). Buen Vivir, degrowth and ecological Swaraj: Alternatives to sustainable development and the green economy. *Development, 57*(3–4), 362–375.

Martínez-Alier, J. (2012). Environmental justice and economic degrowth: An alliance between two movements. *Capitalism Nature Socialism, 23*(1), 51–73.

Moore, J. (2015). *Capitalism and the web of life: Ecology and the accumulation of capital*. Verso.

Narain, S. (2019). Equity: The final frontier for an effective climate change agreement. In K.-K. Bhavnani, J. Foran, & P. A. Kurian (Eds.), *Climate futures: Re-imagining global climate justice* (pp. xxiv–xxx). Zed Books.

Nixon, R. (2011). *Slow violence and the environmentalism of the poor*. Harvard University Press.

Roberts, J. T., & Parks, B. C. (2006). *A climate of injustice: Global inequality, north-south politics, and climate policy*. Cambridge University Press.

Schlosberg, D. (2007). *Defining environmental justice: Theories, movements, and nature*. Oxford University Press.

Shiva, V. (1989). *Staying alive: Women, ecology, and development*. Zed Books.

Thomas, J. A., Williams, M., & Zalasiewicz, J. (2020). *The Anthropocene: A multidisciplinary approach*. Polity Press.

Vanderheiden, S. (2008). *Atmospheric justice: A political theory of climate change*. Oxford University Press.

World People's Conference on Climate Change and the Rights of Mother Earth. (2010, April 22). *Peoples Agreement*.

Benjamin Lewis Robinson is Assistant Professor of German at NYU. He is the author of *Bureaucratic Fanatics: Modern Literature and the Passions of Rationalization* (De Gruyter, 2019) and is currently engaged in a project at the intersections of biopolitics and ecopolitics titled *States of Need / States of Emergency*. He co-edited a recent special issue of *The Germanic Review* on "*Schuld* (guilt/debt) in the Anthropocene" and the volume *The Work of World Literature* (2021).

Ethics of Care

Miriam Tola

Abstract This essay explores the ambivalences of care while developing conceptual links between social and ecological care. Drawing on contemporary ecological
activism as well as feminist and anti-racist perspectives, it shows how theories and
practices of care focusing on interdependence as well as inequality problematize the
Anthropocene narrative that foregrounds the impact of an undifferentiated human
species on planetary processes. The essay contributes to rethinking care as a way of
engaging in the world to make it habitable again.

In September 2021 Extinction Rebellion staged a demonstration outside the Bank of
England in the City of London. The environmental activists demanded that the
financial institution and the British government declare an end to all new fossil fuel
funding and take concrete steps towards tackling climate breakdown. Dozens of
demonstrators engaged in civil disobedience by breaking bail conditions banning
them from the financial district. Images of the protest circulating in social networks
show young and not-so-young people holding placards that read: "Arrested for caring". This slogan raises a range of ethical and political questions. Is protest a form
of care? Who is providing care here and who is benefiting from it? What does it
mean to care in the so-called Anthropocene, the contested name for a geological
epoch marked by the irreversible impact of the human species on earth systems?

This essay places Extinction Rebellion's reference to care within the larger turn
to this concept in contemporary scholarship and activism. Amidst a global pandemic and the breakdown of social and ecological life support systems, care has
emerged as powerful counterpoint to the obsession with novelty, progress and
growth typical of modernity (Jackson, 2014). In recent years, it has inspired a flurry
of scholarly publications, conferences, political manifestoes, mutual-aid groups and
workshops (The Care Collective, 2020; Woodly et al., 2021). This chapter focuses
on theories and practices that help to develop conceptual links between social and

M. Tola (✉)
University of Lausanne, Lausanne, Switzerland
e-mail: miriam.tola@unil.ch

© The Author(s), under exclusive license to Springer Nature
Switzerland AG 2023
N. Wallenhorst, C. Wulf (eds.), *Handbook of the Anthropocene*,
https://doi.org/10.1007/978-3-031-25910-4_172

ecological care and, in so doing, open up spaces for repairing the damages of self-devouring growth, the pervasive economic model that consumes and reduces to waste the material substance of the planet (Livingston, 2019). It draws from diverse feminist and antiracist genealogies to interrogate the possibilities and the ambivalences of caring within and against the Anthropocene, a concept that illuminates the scope of the ecological crisis but largely obscures differential responsibilities and vulnerabilities in relation to it (Vergès, 2017; Pulido, 2018; Ferdinand, 2019). Ultimately, this essay contributes to reorienting care as practice for transforming modes of living together on earth.

From the Latin word, *cūra,* care means attention, concern but also uncertainty and apprehension. At its broadest, this term signals the efforts to keep something in good conditions. Extinction Rebellion's slogan suggests that the seemingly intractable environmental crisis and its attendant injustices require collective attention. The protest's immediate goal at the City of London was to expose "fossil finance" which pours millions each year into the fossil fuel industry with the support of governments. But on a deeper level, it expressed the refusal of carelessness in the face of disaster and an active concern toward the increasing precariousness of life on a planet damaged by socio-ecological relations organized around the paradigm of economic growth and profit-making. This points to two aspects of care that are worth exploring. First, caring in public, as a collective, is a response to feelings of anxiety and grief stemming from climate disruption, governmental inaction, and the increasing awareness of environmental injustices. It is an invitation to getting involved in concrete ways, investing time and energy into manifold struggles for reclaiming a future that is slipping out of hands. Second, caring includes both social and ecological dimensions. Ecological disobedience, with its insistence on the ethical-political obligation to respond to the devastating effects of rising seas, heat waves and floods, points to care as a process that acknowledges human dependence on a planetary milieu, the very condition of existence for humans and other living beings. In foregrounding the attachment to the material basis of human existence, it calls into question the Anthropocene as "a tragic story with only one real actor" (Haraway, 2016, 39), an undifferentiated human species, to shift focus on the specific responsibilities of the actors behind financial flows supporting fossil fuel infrastructures.

In a sense, Extinction Rebellion's protest suggests that everyone can take care of others and assume responsibility for repairing a planet devoured by unchecked growth. Yet, as valuable as this idea is, it is also crucial to recall that care itself is inseparable from systemic inequalities (Hobart & Kneese, 2020). As feminist and anti-racist perspectives have demonstrated, care is never innocent. A matter of relationality, often of intimacy and embodied practices, it is also a field of power and privilege. While some subjects, often white men, have been constructed as deserving care, others, usually gendered and racialized subjects, have been burdened with the task of providing it through a range of everyday activities, including cleaning, picking up trash, tending for children and the elderly, providing food and comfort. The labour of caring as well as the possibility of accessing care have been unevenly distributed creating acute asymmetries and exclusions along lines of gender, race, class and species.

The genealogies of care are complex, traversed by distinct concerns and political projects. In the 1980s, the feminist ethics of care challenged theories of justice anchored in the modern liberal model of the self-sufficient subject by focusing on the relational dimension of moral life. If early formulations proposed that women's moral choices are oriented towards preserving attachments rather than autonomy (Gilligan, 1982), others troubled the notion of care as feminine disposition. Building on insights from the 1970s campaign Wages for Housework, materialist feminists have problematized care as unpaid work and analyzed its central role in the social reproduction of capitalism (Costa & James, 1975; Federici, 1975; Fraser, 2016). Black feminist scholars and activists have directed attention to racialized women and their historical role as enslaved servants working in white households (Davis, 1981). This move has complicated approaches that, albeit unwittingly, centre white women's experience of oppression but fail to account for power dynamics that continue to shape care work in transnational migration (Boris & Parreñas, 2010; Raghuram, 2019). At the same time that diverse critiques of care work became relevant, the interest towards practices of self-care emerged to illuminate the survival strategies of racialized communities in contexts of state neglect and unequal access to health and social services. Describing her struggle with cancer and the institutional racism of the healthcare system in the United States, the queer Black feminist Audre Lorde evoked self-care as an act of political warfare and collective preservation (Lorde, 1988). Considering the ambiguities of care, Pascale Molinier aptly defined it as a "hot bed of conflict, tension, tugging, and of ambivalence" (2013, 4). Such a claim conveys a challenge to the conflation of care with positive affects. It captures the tensions *within* the concept while insisting on its vital role as life-making practice.

While most feminist and anti-racist approaches have focused on interhuman inequalities in the distribution of care, an emergent body of work provides insights into its socio-ecological dimension. In an often-quoted passage, the political theorist Joan Tronto provides a capacious definition of care "as a species activity that includes everything we do to maintain, continue, and repair our 'world' so that we can live in it as well as possible. That world includes our bodies, our selves, and our environment, all of which we seek to interweave in a complex, life-sustaining web. Caring thus consists of the sum total of practices by which we take care of ourselves, others and the natural world" (Tronto, 1993, 103). Tronto's expansive notion of care is helpful to reimagine it for the current time of planetary impasse. It envisions bodies and ecosystems as part of a web of life, emphasizing the interdependencies between human and nonhuman beings. Still, in this framework human beings are prioritized as ethical subjects of care, capable of repairing a world, holding together all aspects of it.

Maria Puig de la Bellacasa has revised this formulation proposing to decentre human agency and consider the invisible but indispensable labours and agencies of earth's beings (Puig de la Bellacasa, 2017). Ecological care, she observes, is not a species-specific activity but it takes place through encounters involving human and other-than-humans in a process of reciprocal transformation. Permaculture, a set of design method and ethics to organize human activities often applied to agriculture, offers an example of working *with* other living beings rather than against them (Puig de la Bellacasa, 2017). Permaculture, explains Laura Centemeri, is meant to repair rather than consume socio-ecological systems and to support forms of reinhabitation (Centemeri, 2019). From this perspective, care is not primarily conceived in terms of intersubjective relation but as a form of response-ability (Haraway, 2016), an openness to the influence that other-than-human beings exert in context-specific situations. As Thom van Dooren notes, caring in a time of extinction also comprises mourning. It refers to the concrete acts that sustain the life of a multitude of beings, including snails, butterflies and birds, but also to noticing what has been and will be lost and refusing to be silent in the face of processes of destruction and loss (Van Dooren, 2021).

Recalling the variegated genealogies of care, the ways in which care is distributed or denied, the distinctions between bodies that provide care and those who are used to receive it, allows to question historical hierarchies within the human and between the human and the rest of the living. Connecting social and ecological care requires us to pay attention to different modes of maintaining and regenerating and find ways to keep them in the same frame. It allows to acknowledge how longstanding inequalities inform the exposure to the "slow violence" (Nixon, 2011) of (un) natural disasters and hinder the work of reparation. Reorienting care within and against the Anthropocene means thinking with ecological activists to turn away from consumer-driven growth toward revaluing everyday acts of repairing, while at the same time questioning the relations of power and privilege that underpin prevalent modes of organizing care across lines of gender, race, class and species. In the present conjuncture, care is a way of engaging in the world to make it habitable again.

References

Boris, E., & Parreñas, R. S. (2010). *Intimate labors: Cultures, technologies, and the politics of care*. Stanford University Press.

Centemeri, L. (2019). *La permaculture ou l'art de réhabiter*. Editions Quae.

Costa, D., & James, M. e. S. (1975). *The power of women and the subversion of the community*. Falling Water Press.

Davis, A. (1981). *Women, race and class*. Random House.

Federici, S. (1975). *Wages against housework*. Falling Wall Press.

Ferdinand, M. (2019). *Une écologie decolonial. Penser l'écologie depuis le monde caribéen*. Seuil. 52, N°375.

Fraser, N. (2016). Contradictions of capital and care. *New Left Review, 100*, 99–117.

Gilligan, C. (1982). *In a different voice: Psychological theory and Women's development*. Harvard University Press.

Haraway, D. (2016). *Staying with the trouble: Making kin in the Chthulucene*. Duke University Press.

Hobart, H.'i. J. K., & Kneese, T. (2020). Radical care: Survival strategies for uncertain times. *Social Text, 38*(1), 1–16.

Jackson, S. (2014). Rethinking repair. In T. Gillespie, P. Boczkowski, & K. Foot (Eds.), *Media technologies: Essays on communication, materiality and society* (pp. 221–239). MIT Press.

Livingston, J. (2019). *Self-devouring growth: A planetary parable as told from southern Africa.* Duke University Press.

Lorde, A. (1988). *A burst of light.* Firebrand Books.

Molinier, P. (2013). *Le travail du care.* La Dispute.

Nixon, R. (2011). *Slow violence and the environmentalism of the poor.* Harvard University Press.

Puig de la Bellacasa, M. (2017). *Matters of care speculative ethics in more than human worlds.* University of Minnesota.

Pulido, L. (2018). Racism and the Anthropocene. In G. Mitman, R. Emmett, & M. Armiero (Eds.), *The remains of the Anthropocene* (pp. 116–128). University of Chicago Press.

Raghuram, P. (2019). Race and feminist care ethics: Intersectionality as method. *Gender, Place & Culture, 26*(5), 613–637.

The Care Collective. (2020). *The care manifesto: The politics of interdependence.* Verso.

Tronto, J. (1993). *Moral boundaries: A political argument for an ethic of care.* Routledge.

Van Dooren, T. (2021). *Mourning as Care in the Snail Ark.* https://culanth.org/fieldsights/mourning-as-care-in-the-snail-ark, Accessed 14 Nov 2021.

Vergès, F. (2017). Racial capitalocene: Is the Anthropocene racial? In G. T. Johnson & A. Lubin (Eds.), *Futures of black radicalism* (pp. 72–82). Verso.

Woodly, D., Brown, R., Marin, M., Threadcraft, S., Harris, C. P., Syedullah, J., & Ticktin, M. (2021). The politics of care. *Contemporary Political Theory., 20*, 890–925. https://doi.org/10.1057/s41296-021-00515-8

Miriam Tola is Assistant Professor at the University of Lausanne, Switzerland, where she is a member of the Institute of Geography and Sustainability. An interdisciplinary scholar of gender, race and the environmental humanities, she is the coeditor of the books *The Routledge Handbook of Ecomedia* and *Ecologie della cura.* Her articles have appeared in journals such as *Theory & Event, South Atlantic Quarterly, Feminist Review, Environmental Humanities* and *Feminist Studies.* Prior to her academic work, she was a journalist working in Italy and the United States.

Habitat

Sandra Wooltorton and Anne Poelina

Abstract How should we inhabit or 'live into' our bioregional places in our multi-species worlds? We are two Western Australian women concerned to promote an awareness of placeusing the wisdom of a culture practised for many millennia. We wonder how we, as a Western society, might be able to understand 'habitat' anew, within restrictions and contexts formed of the Anthropocene? We begin with the reminder that everyone has ancestors who are Indigenous, perhaps many thousand years ago, therefore everyone has the innate capacity to reconnect with living places in a relational way, as our ancestors knew. Within this worldview, we propose three Australian Indigenous notions to support inhabitation or *living into* our storied place-meanings. This is very different from occupation. Seasonality, *liyan* and place/time express Indigenous ideas of habitat characterized by simplicity, cooperation, localization, kinship and relationship.

In this chapter, we reflectively and creatively consider re-inhabiting our bioregional home in this multi-species world. What is the meaning of 'habitat' within this Anthropocene place/time, we ask? We begin with the reminder that everyone has ancestors who are Indigenous, perhaps many thousand years ago (Kimmerer, 2013), therefore everyone has the innate capacity to reconnect with living places in a relational way, as our ancestors knew.

By way of introduction, we are two Western Australian women concerned to promote a place-consciousness using the wisdom of a culture practiced for many millennia. Anne is a Nyikina and Warrwa Indigenous woman who belongs to the Martuwarra Fitzroy River in our state's north (Kimberley region). Anne's ancestors have inhabited our Country (when capitalized, denotes an Indigenous English language understanding of one's habitat, which is relational, spiritual, animate and inclusive of kinship) since the dawn of time. Sandra is a settler woman from the

S. Wooltorton (✉) · A. Poelina
Nulungu Research Institute, University of Notre Dame Australia, Broome Campus,
Fremantle, WA, Australia
e-mail: sandra.wooltorton@nd.edu.au; Anne@majala.com.au

N. Wallenhorst, C. Wulf (eds.), *Handbook of the Anthropocene*,
https://doi.org/10.1007/978-3-031-25910-4_173

southwest, now living in the Kimberley. Some of Sandra's ancestors have occupied Western Australia for at least six generations, while others are Indigenous to the United Kingdom, particularly the English northeast. We deliberately use different terms – inhabit or occupy – to illustrate place-relationships. As we write, we think about our own place as a case study perhaps socio-politically typical of other nations of the Global North – USA, for example.

The Anthropocene is a geological era characterized by human activities functioning as a force of greater power and universality than nature (Steffen et al., 2007). Like all researchers in this book, we are anxious about futures, considering floods, fires, famines and pandemics now rage across the planet. Similarly, we are concerned about colonial and other state-sponsored varieties of aggression (Norman et al., 2021; Paradies, 2020). Symptomatic of an aggressive lifeway characteristic of the Anthropocene, Australian corporations and governments often disregard Indigenous cultural heritage and wisdom. As an example of corporate indifference, Jukuun Gorge, a 46,000 year old rock shelter recording northern Australian human history, was detonated in a mining operation (Allam & Wahlquist, 2020). Similarly, our governments discourage remote Indigenous communities (Kagi, 2014) and undermine Indigenous interests on native title land (Poelina et al., 2020). Furthermore, Australians are major per capita carbon emitters at a time of accelerating climate change (Infrastructure Australia, 2019; Norman et al., 2021).

Regional and planetary assault is disproportionate among cultures however (Escobar, 2018), and sustainability crises are crises of relationship (Milgin et al., 2020). There is significant wisdom for changing the world within Indigenous onto-epistemological systems (Yunkaporta, 2019).

Transition discourses such as Escobar (2018) envision post carbon socio-ecological civilizations, and we ground our thinking within the paradigm shift from survival to relationality (Graham, 2008). Within this worldview, we propose three Australian Indigenous notions to support inhabitation or to *live into* our storied place-meanings. This is very different to occupation. Seasonality, *liyan* and place/time express Indigenous ideas of habitat characterized by simplicity, cooperation, localization, kinship and relationality. Seasonality refers to more-than-human beings (after Abram, 1996) inhabiting the seasons through observing, eating, hearing and caring-with deep relationships between multi-species co-inhabitants. The humid rains arrive, floods rise and fertility characterizes ecosystems, and then when the clouds pass, the ground dries and cool winds guide the interactions, all the time feeding and being food. Inhabiting seasons is to cultivate practical consciousness of the lives of plants and their interdependent ecosystem companions, and to eat within local limits. It is to feel, understand and know the changing energies, the vibrancies and interests of one's living rivers, trees, parks and wild places – and to support their rewilding.

Liyan is a Nyikina word referring to a feeling sometimes referred to as intuition or our moral compass. *Liyan* responds to living habitat which also holds *liyan,* and which responds reciprocally (Nyamba Buru Yawuru, 2018). Indigenous people know that caring for Country means it will return the care responsively. This action includes response-ability, meaning we all have the ability to respond (Bawaka Country et al., 2019). Place/time is a 'long now', a non-linear sense of time where past, present and future are always present in this place. It means that the ancestral characters, shadows and stories of the past continue to enliven the present in a seasonal, located, creative, relational way, meaning that the future is emergent, imagined and actively co-becoming (Bawaka Country et al., 2016). Connecting these notions are relationships within nature rather than seeing ourselves separate to our places and our multi-species peers. We relate to our places, through a kinship system that includes kin species with response-abilities. That is, our human nature is place-based; we are rooted in nature, which we constantly nurture and receive nurturing from in return. This is a human habitat, which we *in*habit. We need to re-inhabit our places if we have grown apart.

Being human does not imply we occupy place on its own. To emerge from the Anthropocene, we will know being human in new ways. We will know humans as embedded within multi-species communities, which include our places to which we need to relate as kin. Our habitats are capable of returning our care. We ourselves are assemblages of cooperative processes, rather than 'individuals' narrowly defined. We are holobiont by birth, "an organism plus its persistent communities of symbionts" (Gilbert, 2017, p. M73). Gilbert says symbiosis, a relationship between two different species, is the earth's life support strategy, and the way of life on earth. Humans *in*habit this way of life, rather than exist separately. This is a different human story to the one producing the Anthropocene. It is a multi-millennial way of understanding humans in our places. These stories still inhabit the Kimberley.

Like elsewhere on our damaged planet, Western Australia's Kimberley is a political battlefield of arguments across the socio-environmental sciences and humanities fields, mainly about the economics and dynamics of development and its sustainability or otherwise (Poelina et al., 2020). Neoliberal economists in Australia can trace their lineage to colonization, which remains the driving mindset (Strakosch, 2015). On the other hand, with Indigenous people, lawyers argue for the rights of living places and rivers, to serve the interests of ecosystems and their mutually interdependent Indigenous cultures and the native title holders themselves (RiverOfLife et al., 2020). There is much more to consider, such as first law, which directs cooperation and peace among Kimberley Indigenous nations (Redvers et al., 2020). Here, an Indigenous cultural approach to collaborative water governance, with a first law multi-millennial history, is being re-established (Poelina et al., 2019).

Within an argumentative, politically charged context, education and living Indigenous knowledge systems are keys to *re*-inhabiting our places as a human civilization on earth. This requires an enmeshed way of life wholly within our living habitats. This position necessitates much learning and action to ensure our human adventure will be able to continue in our environments in the Anthropocene. This is underway in the Kimberley (Allam & Earl, 2021) and around the world.

References

Abram, D. (1996). *The spell of the sensuous: Perception and language in a more-than-human world*. Pantheon Books.

Allam, L., & Earl, C. (2021, June 6th). A journey down WA's mighty Martuwarra, raging river and sacred ancestor. *The Guardian*. Retrieved from https://www.theguardian.com/australia-news/2021/jun/05/a-journey-down-was-mighty-martuwarra-raging-river-and-sacred-ancestor

Allam, L., & Wahlquist, C. (2020, December 13). Gobsmacked: How to stop a disaster like Juukan Gorge happening again. *The Guardian*. Retrieved from https://www.theguardian.com/australia-news/2020/dec/13/gobsmacked-how-to-stop-a-disaster-like-juukan-gorge-happening-again

Bawaka Country, Wright, S., Suchet-Pearson, S., Lloyd, K., Burarrwanga, L., Ganambarr, R., et al. (2016). Co-becoming Bawaka: Towards a relational understanding of place/space. *Progress in Human Geography, 40*(4), 455–475. https://doi.org/10.1177/0309132515589437

Bawaka Country, Suchet-Pearson, S., Wright, S., Lloyd, K., Tofa, M., Sweeney, J., et al. (2019). Goŋ Gurtha: Enacting response-abilities as situated co-becoming. *Environment and Planning D: Society and Space, 37*(4), 682–702. https://doi.org/10.1177/0263775818799749

Escobar, A. (2018). *Designs for the Pluriverse: Radical interdependence, autonomy, and the making of worlds*. Duke University Press Books.

Gilbert, S. F. (2017). Holobiont by birth: Multilineage individuals as the concretion of cooperative processes. In A. Tsing, H. Swanson, E. Gan, & N. Bubandt (Eds.), *Arts of living on a damaged planet: Ghosts and monsters of the Anthropocene* (pp. M73–M90). University of Minnesota Press.

Graham, M. (2008). Some thoughts about the philosophical underpinnings of aboriginal worldviews. *Australian Humanities Review, 45*, 181–194. Retrieved from http://press-files.anu.edu.au/downloads/press/p38881/pdf/eco04.pdf

Infrastructure Australia. (2019). *An assessment of Australia's future infrastructure needs: The Australian infrastructure audit 2019*. Australian Government.

Kagi, J. (2014, November 12). Plan to close more than 100 remote communities would have severe consequences, says WA Premier *ABC News*. Retrieved from http://www.parliament.wa.gov.au/publications/tabledpapers.nsf/displaypaper/3912881c1313c21e87d6392248257e46000fdf56/$file/tp-2881.pdf

Kimmerer, R. W. (2013). *Braiding sweetgrass: Indigenous wisdom, scientific knowledge and the teachings of plants* (1st ed.). Milkweed Editions.

Milgin, A., Nardea, L., Grey, H., Laborde, S., & Jackson, S. (2020). Sustainability crises are crises of relationship: Learning from Nyikina ecology and ethics. *People and Nature, Early View*. https://doi.org/10.1002/pan3.10149

Norman, B., Newman, P., & Steffen, W. (2021). Apocalypse now: Australian bushfires and the future of urban settlements. *npj Urban Sustainability, 1*(1), 2. https://doi.org/10.1038/s42949-020-00013-7

Nyamba Buru Yawuru. (2018). Mabu liyan, mabu buru, mabu ngarrungunil: strong spirit, healthy country, healthy community. Retrieved from http://www.yawuru.com/

Paradies, Y. (2020). Unsettling truths: Modernity, (de-)coloniality and indigenous futures. *Postcolonial Studies*, 1–19. https://doi.org/10.1080/13688790.2020.1809069

Poelina, A., Taylor, K. S., & Perdrisat, I. (2019). Martuwarra Fitzroy River council: An indigenous cultural approach to collaborative water governance. *Australasian Journal of Environmental Management, 26*(3), 236–254. https://doi.org/10.1080/14486563.2019.1651226

Poelina, A., Brueckner, M., & McDuffie, M. (2020). For the greater good? Questioning the social licence of extractive-led development in Western Australia's Martuwarra Fitzroy River region. *The Extractive Industries and Society*. doi:https://doi.org/10.1016/j.exis.2020.10.010.

Redvers, N., Poelina, A., Schultz, C., Kobei, D. M., Githaiga, C., Perdrisat, M., et al. (2020). Indigenous natural and first law in planetary health. *Challenges, 11*(29). https://doi.org/10.3390/challe11020029

RiverOfLife, M., Poelina, A., Bagnall, D., & Lim, M. (2020). Recognizing the martuwarra's first law right to life as a living ancestral being. *Transnational Environmental Law, 9*(3), 541–568. https://doi.org/10.1017/S2047102520000163

Steffen, W., Crutzen, P. J., & McNeill, J. R. (2007). The Anthropocene: Are humans now overwhelming the great forces of nature. *Ambio, 36*(8), 614–621. https://doi.org/10.1579/0044-744 7(2007)36[614:TAAHNO]2.0.CO;2

Strakosch, E. (2015). *Neoliberal indigenous policy: Settler colonialism and the 'post-welfare' state*. Palgrave Macmillan.

Yunkaporta, T. (2019). *Sand talk: How indigenous thinking can save the world*. Text Publishing.

Sandra Wooltorton is a Professor with the Nulungu Research Institute, University of Notre Dame Australia, and Adjunct Professor with the Centre for People, Place and Planet, Edith Cowan University. She holds a Doctor of Philosophy in Sustainability and Technology Policy, and qualifications in ecological philosophy, cultural geography and education. Her current work involves transformative learning and practices for anticolonial worldviews and ways of understanding place-time, multi-species relationship and healing for people and planet. Her focus is the collaborative development of just, healthy futures.

Anne Poelina is a Professor and Senior Research Fellow with Notre Dame University, Research Fellow with Northern Australia Institute, Charles Darwin University, and Visiting Fellow Australian National University and Member of the Water Justice Hub. Anne holds a Doctor of Philosophy (First Law), Doctor of Philosophy (Indigenous Wellbeing), Master Public Health and Tropical Medicine, Master Education, and Master of Arts (Indigenous Social Policy). Her current work explores First Law and the emergence of ancestral personhood, property rights, and equity through legal pluralism. Her focus is Bio-Regional Framework, regional governance, unity, and co-design planning and decision-making.

Home, Homeland

Werner Wintersteiner

Abstract Although the term homeland ǀ foyer ǀ domovina ǀ szülőföld does not have the same meaning or trigger the same emotional connotations in all languages (the German word *Heimat* is particularly rich in this respect) and although it has no equivalent in many languages, the constellation it describes is typical of modernity and even more relevant for the era of accelerated globalisation or the Anthropocene (Related concepts around the terms *fatherland*, *motherland*, or *mother country*, *land of ancestors* can be found in numerous languages around the world (https://en.wikipedia.org/wiki/Homeland [9/8, 2021])). Due precisely to the fact that acceleration, constant change and mobility are fundamental characteristics of modernity, the concepts of *home or homeland* are gaining in importance as an antipole. The term represents (lost) security, anchoring and stability. It is used in both a reactionary and a progressive sense. This article is also a plea for the recovery of a progressive, decolonial and cosmopolitan concept of homeland.

In his fascinatingly clear-sighted and elegant text *L'exil* (The Exile), the writer and artist John Berger (1985) takes a sobering look back at the last 200 years: "Never in history have so many people been uprooted as in our time. Emigration, imposed or chosen, across national borders or from the village to the metropolis, is the essential experience of our time." This uprooting, as Berger calls it, implies a break with traditions, self-evident cultural truths, customary practices and old certainties. It is a loss of what the term home expresses in multiple languages.

W. Wintersteiner (✉)
Klagenfurt University, Klagenfurt am Wörthersee, Austria
e-mail: werner.wintersteiner@aau.at

1059

Home as the Centre of the World[1]

The spread of capitalism, with all its upheavals, has already been described in detail by Karl Marx and Friedrich Engels in the *Communist Manifesto*. A certain ambivalence of evaluation is evident here. Although the great (societal) changes are booked as losses, at the same time they are also seen as an opportunity to reach a distanced and sober realisation of one's own situation through alienation: "Constant revolutionising of production, uninterrupted disturbance of all social conditions, everlasting uncertainty and agitation distinguish the bourgeois epoch from all earlier ones. All fixed, fast-frozen relations, with their train of ancient and venerable prejudices and opinions, are swept away, all new-formed ones become antiquated before they can ossify. All that is solid melts into air, all that is holy is profaned, and man is at last compelled to face with sober senses his real conditions of life, and his relations with his kind." (Marx & Engels, 2012 [1848], sect. 1, paragraph 18).

However, according to John Berger, one must examine the concept of *home* I *foyer* in its much broader, original meaning in order to appreciate the significance of the changes brought about by the spread of capitalism, the two world wars of the twentieth century, and lastly, the new impetus of globalisation in recent decades. To understand the concept of *home* is to understand what the age of migration, "uprooting" and displacement actually means: "Emigration is not only leaving a country, crossing the water, living among strangers, it is also undoing the meaning of the world – and at the extreme limit – abandoning oneself to the unreal which is the absurd." (Berger, 1985).

The *foyer/home* has always represented the centre of the world, not in a geographical sense, but in an existential sense. It is the cosmos, the ordered, real world, as opposed to the chaos, the "unreal" outside. Without the *foyer* there is no order and no sense. As Berger explains, it is, so to speak, the intersection between a vertical and a horizontal line. The vertical line connects people with heaven above and simultaneously with the realm of the dead, below the Earth. The horizontal line connects people with one another.

These remarks of Berger's already hint at the impact of colonialism, although he himself does not address this further. At the time of his analysis, Berger primarily had European migration in the 19th and 20th centuries in mind. However, his findings apply all the more to the millions of people who have literally been driven out of their world by European or Western imperialism and colonialism, and not just over the last 200 years, but since the conquest of the Americas. Western policy has not been the only policy, but it has been the decisive policy for these migratory changes on a global scale.

---

[1] https://www.uni-saarland.de/fileadmin/user_upload/Professoren/fr38_ProfBarboza/Aktuelles/CFP_TagungHeimat_2016_engl..pdf

Home as Resistance

The above-mentioned changes have led to many counter-reactions and counter-movements that vary greatly amongst themselves, but mostly counter the unsatisfactory present with an allegedly better past and often rely on the local factor and on the concept of *homeland* in a narrowly local or regional, but also in a national and nationalist sense. *Home* is conceived of as a return to the past and as a form of resistance to all the evils of modernity. Both the American and the European extreme right have seized the concept of *homeland* for themselves and are trying to gain legitimacy in this way. But it would be a mistake to abandon the concept altogether as a result. This would mean ignoring its ambivalence, thereby also overlooking positive elements that exist even in reactionary movements. In practice, this also entails discrediting local resistance against the negative effects of globalisation as patriotic, egoistic and nationalistic. The consequence is that these movements can no longer be effectively countered (Cf. Geiselberger, 2017).

For *homeland* is more than "an explosive topic between kitsch and utopia", as some observers believe. *Home | Homeland*, as a concrete utopia, in fact has the potential to regain a progressive political significance. However, in the age of the Anthropocene, characterized by fundamental incursions into nature and thus into the framework of conditions of human life caused by humans, there can be no direct return to lost models of *home*, even if these experiences should certainly be taken up and included in a dialogue together with the best achievements of modern-day knowledge.

One of the first and most effective authors who aimed to redeem the discredited concept of *homeland* and use it for a social vision was Ernst Bloch. Building upon Karl Marx's tenth Feuerbach thesis, which speaks of "socialised humanity", Bloch formulates *homeland* as a political programme that takes into account the harmonious connection between humanity and nature that has become so important today: "A socialised humanity bound to a sharing nature is the transformation of the world into homeland." (Bloch, 1995 [1954]). He continues: "True genesis is not at the beginning but at the end, and it begins to begin only when society and existence become radical, i.e, grasp their roots. But the root of history is the working, creating human being who reshapes and overhauls the given facts. Once he has grasped himself and established what is his, without expropriating and alienation, in real democracy, there arises in the world something which shines into the childhood of all and in which no one has yet been: homeland." (Bloch, 1995, 1375/76).

Homeland Earth (Terre-Patrie)

Today, however, the situation has shifted once more, or rather many crisis phenomena have dramatically intensified since the half-century in which Bloch's work was published. The discourse surrounding the Anthropocene is itself a symptom of this.

A principle of hope that fails to take into account all the destruction caused by the modern utopia of the infinite exploitation of nature is no longer sustainable. Edgar Morin identifies a threefold crisis that leads to great anxiety, and thus tends to lead to politically regressive behaviour: "The three factors of anxiety are the loss of the future, the loss of the past and the planetary crisis." (Morin, 2003, 191) For a long time, the author explains, faith in progress (of science, democracy and prosperity) was unbroken. Today we are confronted with all the fatal consequences of the ideology of progress and unbridled growth. Moreover, Morin notes, we have "lost the past" (ibid.), the security of traditions. And thirdly, it is precisely this multiple planetary crisis that we are now facing as a consequence of this development. The bottom line: "We have transformed the Earth, domesticated its vegetal surfaces, and gained mastery over its animals. We are not for all that masters of the cosmos, not even of the Earth." (Morin & Kern, 1998, 145).

Due to this crisis, in turn, according to Morin, an earthly community of destiny has emerged. In this planetary era, the Earth itself has become our common home – *Homeland Earth*, as the title of his book (1998) reads. It is important to note that the original French title, Terre-Patrie, carries other (more political) connotations than the English *Homeland Earth*. It becomes the common task of humanity – which, however, is far from being a community – to preserve the Earth in order to maintain our own living conditions: "We belong to the Earth which belongs to us. [...] We must preserve, we must save our Homeland." (Morin & Kern, 1998, 143 and 146). We need both solidarity among humans and the "solidarity" of humans with nature. John Berger also made the case for global solidarity as the only way out of the loss of homeland: "It is [...] impossible to return to the historical experience when each village was at the heart of reality. The only hope of making a centre again is to make a centre of the whole world. Only one thing can transcend the lack of a modern home; global solidarity. Brotherhood is too easy a term. Regardless of Cain and Abel, brotherhood gives hope that all problems will be solved. In reality, many are insoluble. Hence the eternal need for solidarity." (Berger, 1985).

This is a clear statement against any backward-looking, romanticising concept of *homeland*. Solidarity does not come from natural bonds, but is an act of political will. Morin's work adopts this idea, but substantially deepens and expands on it using the reference to the biosphere. Declaring the Earth a homeland means extending the emotional ties that people usually have to their closest surroundings, and often even to their nation, to the entire Earth – to other people and to all of nature. But in so doing, these ties are transformed into a political alliance. It is a common world, yet a "common world of strangers", as Étienne Tassin puts it (2003, 22). The local thus connects with the global; they are no longer thought of as opposites but as belonging together. This creates a link between *Homeland Earth* and corresponding postcolonial discourses, which will be discussed in the next section.

Cosmopolitan Localism

Cosmopolitan Localism, an expression originating from Wolfgang Sachs (1999), is a programme designed to connect the local (home in the traditional sense) with the global: "This symbiotic connection between different levels of scale of everyday life, from the local to the planet as a whole, would integrate two longstanding and distinct traditions – cosmopolitanism and localism – and would be the basis for a new kind of social, cultural, political and economic settlement, Cosmopolitan Localism." (Kossoff, 2019, 51) It is a vision in which power is transferred, more and more, to the hands of civil society and in which as much as possible is produced, shaped and decided at the local level, without renouncing global networks in the process. The aim is to thus enable a sustainable life with little domination.

A more radical version of cosmopolitan localism can be found in the work of postcolonial theorists such as Walter Mignolo. He views cosmopolitanism as an initially Western project, as the friendly side of colonial and imperial expansion, so to speak (Mignolo, 2011, esp. chap. 7). Mignolo supports and encourages all indigenous efforts to renew traditions that represent an alternative to Western colonialism, capitalism and the exploitation of nature. He indicates that a new, non-domineering cosmopolitanism could emerge from this. Central to this is "the communal as a connector", which allows for many different communal varieties:

"Cosmopolitanism in a decolonial vein shall aim at the communal not as a universal model but as a universal connector among different noncapitalist socioeconomic organisations around the world. Thus, communalism is not a model of society, but a principle of organisation. Many models will emerge, based on local histories, memories, embodiments, practices, languages, religions, categories of thought. The communal as a connector, rather than as a universal model, means, in the first place, to delink from both capitalism and communism (or socialism in its softer version), brothers of the same parents, the European Enlightenment." (Mignolo, 2011, 275) In Mignolo's vision, the local, the homeland, becomes the site of resistance and the new beginning of a world capable of establishing global justice among people and with all living things: "De-colonial localism is global or, if you wish, cosmopolitical. Thus, we arrive at the paradoxical conclusion that, if cosmopolitanism shall be preserved in the humanities goal towards the future, it should be 'cosmopolitan localism', an oxymoron no doubt, but the Kantian project of one localism being the universal is untenable today." (Mignolo, 2010, 127).

All translations from the French by the author.

References

Berger, J. (1985). L'exil. In: *Lettre internationale* n°1. https://www.imagespensees.org/memoires/exil-humanite-identite/article/l-exil [9/7, 2021].

Bloch, E. 1995 [1954]. *The principle of hope* (N. Plaice, S. Plaice, & Paul Knight, Trans.). MIT Press.

Geiselberger, H. (Ed.). (2017). *The great regression*. Polity Press.

Kossoff, G. (2019). Cosmopolitan localism: The planetary networking of everyday life in place. *Cuadernos del Centro de Estudios en Diseño y Comunicación, 73*, 51–66.

Marx, K., & Engels, F. (2012 [1848]). *The communist manifesto*. Verso.

Mignolo, W. (2010). Cosmopolitanism and the de-colonial option. *Studies in Philosophy and Education, 29*(2), 111–127.

Mignolo, W. (2011). *The darker side of Western modernity. Global futures, decolonial options*. Duke University Press.

Morin, E. (2003). Une anthroposociogenèse de la violence politique. In: M. Pagès (s.l.d.), La violence politique. *Erès | Sociologie clinique* (pp. 183–209).

Morin, E., & Kern. Br. (1998). *Homeland Earthe. A manifesto for the new Millenium*. Hampton Press.

Sachs, W. (1999). *Planet dialectics: Explorations in environment and development*. Fernwood Publications.

Tassin, É. (2003). *Un monde commun. Pour une cosmo-politique des conflits*. Éditions du Seuil.

Werner Wintersteiner, former founding director of the "Centre for Peace Research and Peace Education" at Klagenfurt University, Austria. Research fields: peace and global citizenship education, culture and peace, the Alps-Adriatic region, literature and peace, literature education. Recent publications: Die Welt neu denken lernen – Plädoyer für eine planetare Politik (2021). [Learning to Rethink the World – a Plea for a Planetary Politics]; (With Cristina Beretta and Mira Miladinović Zalaznik): Manifestlo Alpe Adria (2020); Herbert C. Kelman: Resolving deep-rooted conflicts. Essays on the Theory and Practice of Interactive Problem-Solving. Eds. Werner Wintersteiner and Wilfried Graf (2017).

Responsibility

Jean-Louis Genard

Abstract This article examines the philosophical and sociological meaning of 'responsibility'. It proposes to understand it as an interpretant – in the sense of Peirce (Collected papers of Charles Sanders Peirce. Edited by C. Hartshorne and P. Weiss. Harvard University Press, 1931) – of "what happens". From all ages, societies have sought to give meaning to this question, providing multiple answers: destiny, astral determinism, grace, Providence, impurity... Emerging progressively in the Middle Ages, the responsibilising interpretant came to impose itself in the seventeenth and eighteenth centuries, against its various competitors, in particular theological ones, at the same time as what was to become its main competitor, the deterministic interpretant, reinforced by the success of the natural sciences. Based on this presupposition, the article analyses the main tensions that are currently at work in the responsibilising interpretant, both its internal tensions, analysed on the basis of the grammar of personal pronouns, and its external tensions, analysed on the basis of the new configurations of the human and the life sciences.

The word "responsibility" is thoroughly familiar and many institutions such as the law, insurance or social policy are founded on its premise. Yet, if we are to understand responsibility from an anthropological point of view we need to situate it in its historical context and show its deep links with Western modernity.

Primo, Responsibility as a 'Cosmo-Anthropological' Interpretant Finding a sense in 'what happens': this is the question that each culture has faced; each time bringing is own replies, each time different: fate, an impurity (Douglas, 1966), grace, mana, fortuity, fortune, collapse, the unconscious, habitus, genetic or neuronal determinism, intention, the stars, Providence, responsibility, luck, law of series...

Translation by Gail Ann Fagen.

J.-L. Genard (✉)
Université libre de Bruxelles, Brussels, Belgique
e-mail: jean.louis.genard@ulb.be

which could be called 'cosmo-anthropological interpretants'. The ways to give sense to an order, or to a disorder, in the world, by attributing, or not attributing, human participation. It is also a way to organise what modernity has disjoined by opposing the category of nature to culture (Descola, 2005).

In relation to this, the specificity of Western modernity is to have proposed, certainly in contrast to, but also alongside, other competing interpretants – especially medieval notions of theology, the stars or Fortune – what could be called a "responsibility-assuming" or "accountable" interpretant, thereby making free will, intention, capacity, willingness, etc. the cause, or at least a main cause, of 'what happens' (Genard, 1999, 2011). Instead of 'cause', however, we should speak of 'motive'. Indeed, the 'words for it' came into use quite gradually. Only gradually did the French verb '*vouloir*' (to want/wish for) become a noun as well, to signify that individual capacity for 'will'. Only gradually, from the twelfth century, did the meaning of the word intention, *intentio*, evolve and expand to assume the meaning we ascribe in relation to free will (Cassin, 2004, 608s). These notions became stable quite gradually, whether in French or in English. For example it was not until the eighteenth and nineteenth centuries that the words 'responsible' and 'irresponsible' entered into current and shared vocabulary (Henriot, 1977).

The change, however, definitely went beyond semantics. The gradual incursion of the accountable interpretant coincided with deep-seated changes in institutional dispositifs that at the same time adapted to this interpretant and played a role in its emergence. Likewise, 'events' or 'facts' obviously do not have the same social 'effects' when they are interpreted from the register of impurity, fate… or responsibility. One example comes from law. It originally concerned itself primarily with the 'object stolen' in the aim to restore order or the state of things; it then became 'subjectivized', tending to sanction 'theft', thus punishing an intention (Villey, 1964, 1977). Or, as Max Weber underlined, social groups (*Verbände*) or, for example, urban citizenry, came to be based not so much on mere membership, but rather through commitment or a pledge (Weber, 1986). Thus a series of institutions, such as marriage, were reshaped by integrating the consent, the *consensus*, of the persons involved. This is how Hobbes, Rousseau and many others conceived of political order, along the model of a contract, a voluntary commitment. This is how individual liberties were guaranteed and protected, through declarations. This is how morals became meritocratic. These are only a few examples.

This deep-seated 'cosmo-anthropological' transformation, obviously, also plays a role in the irreducible specificity of Western modernity, among all other cultures. This is especially apparent in its separation between nature and culture, the latter's precise specificity being its grounding in the responsibility of those who form this culture (Descola, 2005).

This progression of the accountable interpretant, however, did meet with some resistance. Furthermore, at the very time it appeared to gain the upper hand over its former competitors, mainly theological, new interpretants, for the most part related to the emerging human sciences, began to contest its privileges. In his *Critique of*

Pure Reason (2003 (1781)), Kant was undoubtedly the first theoretician, making the opposition between freedom and determinism an antinomy that can be seen as a keystone of our modernity in general (he was a highly versed admirer of Laplacian determinism), but also, more particularly, of our anthropological modernity. Michel Foucault's discussion of this is certainly the clearest. Indeed, when he considered that man – meaning man in Western modernity – was 'born' in the seventeenth and eighteenth centuries, Foucault made direct reference to the antinomy, or rather to Kantian dualism. According to Foucault, modern man was born as an 'empirical-transcendental double' (1966). We could thus rewrite the history of the anthropology of Western modernity from the angle of tensions between responsibility, irresponsibility, determinism, probability, and so on. However, before discussing these 'external' tensions of responsibility, we should first look at some of its 'internal' tensions.

Secundo, the Internal Tensions of Responsibility The essential anthropological mutation linked to the emergence of the accountable interpretant is clearly the recognition of individual and autonomous will. In this view, responsibility is above all understood as the 'faculty to begin', to use Kant's own words. This means starting a chain of actions henceforth imputable to a 'subject'. It is a responsibility in the first person (I), but evidently reversible. This responsibility that I grant to myself, I also presume towards another person, whom – in principle, barring reasons to the contrary – I consider equally as a subject, endowed with will.

This is nonetheless not the only accentuation of responsibility. Just as responsibility incites me to answer for my actions, it also potentially obliges me, not to answer to another for my actions, but to answer *for* the other. This would be the responsibility in the second person (Thou), which Lévinas theorised, dramatized through reference to the Face. For me, in the first person, the Other is also a potential call or summons (*fordern auf*), as shown, for example, when we analyse people's reactions when they see a homeless person (Bidet et al., 2015).

This articulation of the different accentuations of responsibility can be enriched by pursuing this 'grammar of personal pronouns' which has guided us. Responsibility can also be conjugated in plural. It can become collective (We), as illustrated by expressions of collective solidarity or social protection schemes.

It can also be abstract, impersonal (He/She, Them) as in insurance programmes where the subjective facet of responsibility dissolves into a large collectivity. This obviously entails the risk of giving the impression that someone in fault too easily clear themselves of responsibility for any wrong done. We can also add the responsibility projected on to a scapegoat (Them, You).

Sociologically and historically, the accentuation of these different modes fluctuates over time. In our times, for example, accountability policies (I) are developing at the cost of State social solidarity instruments based on collective responsibility (We).

Tertio, Responsibility at the Heart of External Tensions As regards the tensions that we have called 'external' – even if spaces are still found where people believe in horoscopes or the stars and others hope for a streak of luck or fate to smile on them – the most significant tensions come into play when the accountable interpretant is called into question through interpretants linked to development of various scientific knowledge about human life. As Kant suspected, this tension is a keystone. It obviously has a history of its own. Human sciences have always questioned and posed the problem of responsibility. There are so many examples. Starting from the nineteenth century, we have Lombroso's works on the 'born criminal' and Freud's exploration of the unconscious. Sociology also contributed with the Gabriel de Tarde's model of hypnotic influence and Bourdieu's habitus.

Without a doubt, the strongest problems are posed by biological sciences. From the development of genetic knowledge that tends to find biological determinisms for a number of behaviours hitherto seen as arising from personal commitments. We also see the rapid development of neurosciences which tend to link physical-chemical, cerebral, processes to behaviours that apparently reflected the accountable interpretant. Furthermore, these new findings certainly deploy their performative effects, finding their way into unexpected realms, radically shifting concepts and evaluations of responsibility, to either attenuate or maximise it. This can be the case when courts invoke cerebral imagery so as to more correctly evaluate an accused person's responsibility or irresponsibility. Or even more, this can be the case when knowledge of genetic heritage serving to predict health risks has the effect of maximising the responsibility of those who do not adapt their lifestyles accordingly, resulting in partial loss of opportunities for access to social benefits.

References

Bidet, A., Boutet, M., Chave, F., Gayet-Viaud, C., & Le Méner, E. (2015). Publicité, sollicitation, intervention. *SociologieS*, Dossiers, Pragmatisme et sciences sociales: explorations, enquêtes, expérimentations. http://sociologies.revues.org/4941. Accessed 3 Feb 2020.

Cassin, B. (Ed.). (2004). *Vocabulaire européen des philosophes*. Seuil-Le Robert.

Descola, P. (2013). *Beyond nature and culture* (J. Lloyd, Trans.). University of Chicago Press. [Original French edition 2005].

Douglas, M. (1966). *Purity and danger: An analysis of concepts of pollution and Taboo*. Routledge and Keegan Paul.

Foucault, M. (1970). *The order of things: An archaeology of the human sciences*. Pantheon Books. [Original French edition 1966].

Genard, J.-L. (1999). *La grammaire de la responsabilité*. Cerf, coll. « Humanités ».

Genard, J.-L. (2011). Expliquer, comprendre, critiquer. *SociologieS*, La recherche en actes, Régimes d'explication en sociologie. http://journals.openedition.org/sociologies/3555. Accessed 3 April 2020.

Henriot, J. (1977). Note sur la date et le sens de l'apparition du mot « responsabilité ». *Archives de philosophie du droit*, n°22, 59–62.

Peirce, C. S.. (1931). *Collected papers of Charles Sanders Peirce*. Edited by C. Hartshorne and P. Weiss. Harvard University Press.

Villey, M. (1964). La genèse du droit subjectif chez Guillaume d'Occam. *Archives de philosophie du droit*, n°9, 97–127.

Villey, M. (1977). Esquisse historique sur le mot responsable, *Archives de philosophie du droit*, n° 22, 49–51.

Weber, M. (1986). *The City*. Free Press. [Original German edition 1921].

Jean-Louis Genard is a philosopher, doctor in sociology, professor at the Université libre de Bruxelles, co-editor-in-chief of the journal Sociologies. He has published numerous books, including *La Grammaire de la responsabilité* (Cerf, 2000), *Action publique et subjectivité* (with F. Cantelli, LGDJ, 2007), *L'éthique de la recherche en sociologie* (with M. Roca i Escoda, 2019), as well as numerous articles on anthropological transformations and developments, responsibility, capacitation and empowerment, ethics, public policies and epistemological and ethical issues specific to the human sciences.

Virtue Ethics

Corine Pelluchon

Abstract Virtue ethics, which is concerned with the set of representations, emotions and ways of being that drive us to act individually and collectively, is an important resource in the Anthropocene. Indeed, it helps to bridge the gap between theory and practice, which is striking in the environmental field, since individuals and States recognize the seriousness of global warming and the erosion of biodiversity, their human causes and their global consequences, but changes in lifestyles and public policies are non-existent or too slow. What self-transformation is required for the Anthropocene and what would be a virtue ethics adapted to our times?

Why is it so difficult to change our lifestyles and achieve environmental sustainability when no one can deny that our development model has a destructive ecological, sanitary and social impact and that it inflicts considerable violence on animals? What could contribute to a democratic answer to the ecological crisis we are confronted with?

Virtue ethics provides a relevant answer to these challenges because it helps reducing the gap between theory and practice by focusing on our concrete motivations, that is to say on the representations and affects that drive us to act both at the individual and collective level. Instead of insisting on principles or on the consequences of our actions, it underlies the moral traits that could lead us to be temperate and cooperate with others in order to promote a more sustainable model of development, including other ways of producing food, energy, farming the land, etc.

This does not mean that we have to renounce to the deontological approach with is based on principles and norms such as obligations, prohibitions, regulations and taxes. Nor do we have to stop paying attention to the consequences of our actions. However, it is necessary to supplement the mainstream approach of environmental ethics which does not take into account the emotional components of moral judgments and the set of representations, values and ways of being that cause

C. Pelluchon (✉)
Gustave Eiffel University, Paris, France

profound individual transformations and social changes. We actually have to explain how human beings can have pleasure in adopting ecologically virtuous lifestyles and participate at the local, national and international levels in the major structural transformations that are necessary for the reorientation of the economy, the sharing of resources and international cooperation (Pelluchon, 2018).

To be sure, if people are torn between duty and happiness, self-interest and general interest, they will try to evade ecological norms because the latter will be felt as mere constraints, as burdens. A government by fear and a war of each against each, of nations against nations, will exacerbate identity-based reflexes, generating disastrous political and geopolitical crises. However, virtue ethics does not signify that the solution to global warming and the erosion of biodiversity lies solely with individuals, as if everyone had to cultivate their own garden and adapt to the collapse in order to survive. Resorting to a virtue-based-approach of environmental ethics that articulates the individual level associated with lifestyles and the political level linked to our participation as economic actors and as citizens in the necessary transformations means that we need to be enlightened. Individuals must be capable of changing their consumption habits and of reworking their representations in order to live with others, both human and non-human, in a more sustainable and fairer manner. What self-transformation is expected for the Anthropocene and what resources does virtue ethics, which goes back to Plato and Aristotle, offer us? What would be a virtue ethics adapted to our times?

Respect for nature and animals does not derive from duties or norms. This assertion is common to both ecofeminism and virtue ethics: Responsibility and care for nature and other beings are not a matter of applying abstract principles, but they are rooted in ways of beings. (Plumwood, 1991). However, there is a difference between ecofeminism and virtue ethics. Whereas ecofeminists think that our capacity to care for others is linked to the particular bonds that tie us with them, virtue ethics advocate a more universalistic approach to morality and politics. The activists and mothers who, in America, in the 1980s, were outraged by the State's indifference to their health being endangered by nuclear facilities and who formed grassroots groups that gave rise to the ecofeminist movement have shown that becoming aware of what endangers nature and one's children health is an opportunity to turn one's situation of vulnerability into a strength. (Hache, 2016). Fear and cooperation actually drove them to organize themselves and put pressure on governments, obliging the latter to take into account the social, environmental, and health challenges they encountered in their daily lives. (Macy, 1983) However, particular relationships and emotions are not enough to give birth to moral traits or virtues that change our way of being and behaviour. Without a process of self-transformation, we will not succeed in changing our way of inhabiting the Earth and making room for others, including other species. Thus, the key to environmental virtues and to the alleviation of animal suffering is our relationship with ourselves.

Virtue ethics takes its source from the morals of antiquity which consider that the choice of some goods (money, glory or justice) determines the character and the behaviour, as seen in Book VIII of Plato's *Republic* and in Book V of *Laws* (1992, 2016). More precisely, Aristotle (2009), in *The Nicomachean Ethics,* explains that virtue is an acquired disposition (*hexis*), which explains why the honest person is always honest, the temperate person always temperate. There is an alliance between virtue and happiness, since acting well brings a sense of fulfilment, as in sobriety. For Aristotle, all virtues, be they moral or intellectual, refer to a global way of being that is essentially linked to a good deliberation, to the ability to aim for and maintain the golden mean in all circumstances. But if Aristotle highlights the centrality of prudence, which is an intellectual virtue, and the solidarity between prudence, wisdom, courage and justice, which designate the four cardinal virtues of the Classics, he does not explain how one becomes prudent, other than by watching prudent people act and by getting used to doing prudent actions. Nor does he say how to avoid the reversal of virtues into vices, of courage into rashness. How is it possible to maintain the art of moderation on all occasions?

To answer these questions, we have developed a virtues ethics by describing the process of individuation that can lead a person to integrate in his own good the good of other living beings, present and future, human and non-human. (Pelluchon, 2018, 9–13; 95–103). Not only does the ethic of consideration reject the essentialism of the ancient philosophers, for whom there is a fixed human nature and teleology providing a single model for all persons, but it is also based on an enlargement of the self and on an experience of the incommensurable we do not find in Aristotle.

The ethic of consideration borrows from Spinoza's *Ethics* the idea that only an enlargement of our self can change our relationship with others and create new affects, such as joy and gratitude. (Spinoza, 2008; Pelluchon, 2018, 83–86). As in Arne Næss, who always refers to Spinoza when he speaks of ecosophy, this enlargement of the self, which implies the knowledge of the links between life forms and also has an impact upon our affects, explains that we stop considering us as an empire within an empire. We also have pleasure by consuming less and enjoy contemplating nature instead of trying to objectify it or considering the other beings as mere ends for our means. (Næss, 2008, 171–178; Pelluchon, 2018, 86–90). However, the ethic of consideration differs from Spinoza's and Næss's thinking in that it places value on negative emotions such as fear and anxiety, which are often associated with global warming. Furthermore, and most importantly, it is conditioned on humility, which is the awareness of our limitations, and is also based on the recognition of our vulnerability and our engendered condition that reflects our dependence on others. The contribution of Christians to virtue ethics lies in this role of humility, which also means that we cannot possess virtue once and for all, contrary to what Aristotle and Spinoza thought.

I borrowed the notion of consideration from Bernard of Clairvaux (1091–1153). Consideration is not a virtue, but the global attitude upon which all the virtues are grounded. Encompassing prudence and all the other virtues, it is based on an experience of transcendence that changes the way one perceives oneself and behave. In *De consideratione*, which is a political treatise in which Bernard of Clairvaux addressed Pope Eugene III, he says that only faith can foster real concern for others and extirpate dominion. (Clairvaux, 2004) Consideration, in Bernard of Clairvaux, rests upon an experience of the incommensurable that presupposes faith understood as a particular relationship to God. By contrast, in our work, *considération,* which comes from *cum* (with) *sidus*, *sideris* (constellation of stars), and which implies looking carefully at a being while acknowledging his or her own value, is defined by transdescendence. (Pelluchon, 2018, 99–103). This word, which was coined by Jean Wahl and is also found in Levinas, does not describe a movement from bottom to top, as faith in God or contemplation (transascendence). Nonetheless, it is a way of speaking of something that overcomes us: It is the experience of the common world that entails the past, present, and future generations, as well as institutions and other species. When we are mindful of our engendered condition and experience our vulnerability, we better feel the bonds that tie us with other engendered and vulnerable beings. We understand that we belong to this common world that is older than us and welcomes us when we come to life. Trandescendence actually outlines a movement of deepening of oneself and of one's carnal condition that transforms the consciousness of one's belonging to the common world into a lived knowledge and changes from within one's way of being and interacting with others.

When I become truly aware, through my body, of my belonging to the common world and feel the connection to other life forms, I realize the community of destiny uniting me with them as well as my specific responsibilities. Instead of being a burden, the desire to pass on a habitable world by making room for all living beings becomes the horizon of my thoughts and actions. I do not live only for myself, but what I live for, by eating, by living somewhere, is a way of living with others, of saying whether or not I make room for them. To live is to live from, to live with and even to live for, to have as a horizon a world to transmit.

Such an approach is quite different from the deontological approach we find in Kant's philosophy of respect, which does not include animals and is opposed to affection. *Considération* is an attitude that binds our mind and our body, our knowledge and our emotions, and that makes us experience the flourishment of others as part of our own flourishment. There are different degrees of *considération*, from the concern for the future of one's children and the commitment to ones' political community to the concern for other living beings and the love of the world.

There is not one ethics of virtues for humans, another for the environment, and a third for animals. There is only one attitude that is rooted in a complete reworking of the mental maps that structure our relationship to ourselves, others and nature. Stressing the relevance of virtue ethics to the Anthropocene means we can only respond peacefully and effectively to the ecological, sanitary, economic and political crises of our time by effecting an anthropological revolution and thereby fostering consideration, not domination. In order to achieve such a transformation of one's representations, values, ways of being and behaviour, an ethics of virtues based on the deepening of the knowledge of oneself as a carnal being, begotten and linked to other living beings, restores one's capacity to act (one's power *to*), instead of exerting one's power *over* the others, be they human or non-human.

In a nutschell, the change of development model that is necessary to avoid environmental disasters, wars and a loss of meaning, supposes to decolonise our imaginary, to move from the Scheme of domination to the Scheme of consideration (Pelluchon, 2021, 98–100, 174–180). The Scheme designates the matrix or the set of conscious and unconscious representations that govern our society, orient our economic choices, determine our way of conducting ourselves in private and public life. To remove the Scheme of domination that transforms everything - agriculture, politics, technology, economics - into war, implies our making this inner change that is the key to initiating structural changes both at the local and global level, fighting the lobbies and seeking democratic and peaceful solutions to conflicts. Only then can ecology become a project for individual and collective emancipation in the Anthropocene.

References

Aristoteles. (2009). *The nicomachean ethics* (David Ross, Trans.) and revised by Lesley Brown. Oxford University Press.

Clairvaux, B. (of). (2004). On consideration (John D. Anderson and Elisabeth T. Kennan, Trans.). Cistertian Publications.

Hache, É. (2016). *Reclaim. Recueil de textes éco-féministes, choisis et présentés par É. Hache.* Cambourakis.

Macy, J. (1983). *Despair and personal power in the nuclear age.* New Society Publishers.

Næss, A. (2008). *Life's Philosophy: Reason and Feeling in a Deeper World* (Roland Huntford, Trans.), with Per Ingvar Haukeland. University of Georgia Press.

Pelluchon, C. (2018). *Éthique de la considération.* Seuil. German translation by H. Jatho, Ethik der Wertschätzung. Tugenden für eine ungewisse Welt. WBG.

Pelluchon C. (2021). *Les Lumières à l'âge du vivant.* Seuil. German translation by U. Bischoff, Das Zeitalter des Lebendigen. Eine neue Philosophie der Aufklärung. WBG.

Plato. (1992). *Republic* (G. M. A Grube, Trans.), revised by C.D. C Reeve. Hackett Publishing Company.

Plato. (2016). *Laws.* Translated by Tom Griffith. Cambridge University Press.

Plumwood, V. (1991). Nature, and gender, feminism, environmental philosophy, and the critique of rationalism. *Hyptia, 6*(1), 3–27. https://doi.org/10.1111/j.1527-2001.1991.tb00206.x

Spinoza. (2008). *Ethics* (David Eugene Smith, Trans.). Forgotten Books.

Corine Pelluchon A specialist in political philosophy and ethics, Corine Pelluchon is Professor at Gustave Eiffel University. She is the author of a dozen books, most of which have been translated into foreign languages. She develops a philosophy of corporeality that has two aspects, one centered on vulnerability, the other on our habitation of the Earth which is always a cohabitation with other living beings. This philosophy of corporeality underlines the relational dimension of the subject and leads to install ecology and the animal cause at the heart of ethics and politics. In this work, which is part of the heritage of the Enlightenment while overcoming the dualisms and anthropocentrism of past philosophies, ecology and the animal cause are never dissociable from the promotion of a new humanism. In 2020, she was awarded the Günther Anders Prize for Critical Thinking in Germany for her whole work. Books (selection): *Leo Strauss and the Crisis of Rationalism. Another Reason, Another Enlightenment,* translated by R. Howse, Suny, 2015; *Les Nourritures. Philosophie du corps politique,* Seuil, 2015, 2020 (*Nourishment. A Philosophy of the Political Body,* translated by J. E. H. Smith, Bloomsbury, 2019); *Manifeste animaliste. Politiser la cause animale,* Alma, 2017, Rivages, 2021; *Éthique de la considération,* Seuil, 2018 (*Ethik der Wertschätzung. Tugenden für eine ungewisse Welt,* translated by H. Jatho, WBG Academic, 2019); *Réparons le monde. Humains, animaux, nature,* Rivages, 2020; *Les Lumières à l'âge du vivant,* Seuil, 2021 (*Das Zeitalter des Lebendigen. Eine neue Philosophie der Aufklärung,* translated by U. Bischoff, WBG Academic, 2021).